D1573943

COMPREHENSIVE POLYMER SCIENCE

IN 7 VOLUMES

EDITORIAL BOARD

Sir Geoffrey Allen, FRS (Chairman)
Unilever Research and Engineering, London, UK

J. C. Bevington (Deputy Chairman)
University of Lancaster, UK

C. Booth
University of Manchester, UK

A. Ledwith
Pilkington plc Group Research, Ormskirk, UK

S. Russo
Università di Sassari, Italy

S. L. Aggarwal
GenCorp Inc., Akron, OH, USA

G. C. Eastmond
University of Liverpool, UK

C. Price
University of Manchester, UK

P. Sigwalt
Université Pierre et Marie Curie, Paris, France

INTERNATIONAL ADVISORY BOARD

H. C. Benoit
Université de Strasbourg, France

M. Hirooka
Sumitomo Chemical Co. Ltd., Tokyo, Japan

W. J. MacKnight
University of Massachusetts, Amherst, MA, USA

C. G. Overberger
University of Michigan, Ann Arbor, MI, USA

M. Tasumi
University of Tokyo, Japan

F. Danusso
Politecnico di Milano, Italy

Y. Imanishi
Kyoto University, Japan

J. H. O'Donnell
University of Queensland, St. Lucia, Australia

N. A. Platé
Academy of Sciences, Moscow, USSR

G. Wegner
Max-Planck-Institut für Polymerforschung, Mainz, FRG

COMPREHENSIVE POLYMER SCIENCE

The Synthesis, Characterization, Reactions & Applications of Polymers

CHAIRMAN OF THE EDITORIAL BOARD

SIR GEOFFREY ALLEN, FRS

Unilever Research and Engineering, London, UK

DEPUTY CHAIRMAN OF THE EDITORIAL BOARD

JOHN C. BEVINGTON

University of Lancaster, UK

Volume 5

Step Polymerization

VOLUME EDITORS

GEOFFREY C. EASTMOND

University of Liverpool, UK

ANTHONY LEDWITH

Pilkington plc Group Research, UK

SAVERIO RUSSO

Università di Sassari, Italy

PIERRE SIGWALT

Université Pierre et Marie Curie, Paris, France

PERGAMON PRESS

OXFORD · NEW YORK · BEIJING · FRANKFURT
SÃO PAULO · SYDNEY · TOKYO · TORONTO

U.K.	Pergamon Press plc, Headington Hill Hall, Oxford OX3 0BW, England
U.S.A.	Pergamon Press, Inc., Maxwell House, Fairview Park, Elmsford, New York 10523, U.S.A.
PEOPLE'S REPUBLIC OF CHINA	Pergamon Press, Room 4037, Qianmen Hotel, Beijing, People's Republic of China
FEDERAL REPUBLIC OF GERMANY	Pergamon Press GmbH, Hammerweg 6, D-6242 Kronberg, Federal Republic of Germany
BRAZIL	Pergamon Editora Ltda, Rua Eça de Queiros, 346, CEP 04011, Paraiso, São Paulo, Brazil
AUSTRALIA	Pergamon Press (Australia) Pty Ltd., P.O. Box 544, Potts Point, NSW 2011, Australia
JAPAN	Pergamon Press, 5th Floor, Matsuoka Central Building, 1-7-1 Nishishinjuku, Shinjuku-ku, Tokyo 160, Japan
CANADA	Pergamon Press Canada Ltd, Suite No. 271, 253 College Street, Toronto, Ontario, Canada M5T 1R5

Copyright © 1989 Pergamon Press plc

All Rights Reserved. No part of this publication may be reproduced, stored in a retrieval system or transmitted in any form or by any means: electronic, electrostatic, magnetic tape, mechanical, photocopying, recording or otherwise, without permission in writing from the publishers.

First edition 1989

Library of Congress Cataloging-in-Publication Data

Comprehensive polymer science.

Includes index.
Contents: v. 1. Polymer characterization/volume editors, Colin Booth & Colin Price—v. 2. Polymer properties/volume editors, Colin Booth & Colin Price—[etc.]—v. 7. Specialty polymers & polymer processing/volume editor, Sundar L. Aggarwal.
1. Polymers and polymerization.   I. Allen, G. (Geoffrey), 1928–   II. Bevington, J. C.
QD381.C66 1988   547.7   88-25548

British Library Cataloguing in Publication Data

Comprehensive polymer science.
Vol. 5: Step polymerization
1. Polymers
I. Allen, Geoffrey   II. Bevington, John C.
III. Eastmond, Geoffrey C.
547.7

ISBN 0-08-036209-5 (vol. 5)
ISBN 0-08-032515-7 (set)

Printed in Great Britain by A. Wheaton & Co. Ltd., Exeter

Contents

	Preface	ix
	Contributors to Volume 5	xi
	Contents of All Volumes	xv

General Characteristics of Step Polymerization

1	Polycondensation and Related Reactions F. Parodi, *EniChem SpA, Milano, Italy* and S. Russo, *Università di Sassari, Italy*	1
2	General Aspects P. Manaresi and A. Munari, *Università di Bologna, Italy*	11
3	Factors Affecting Rate of Polymerization A. Munari and P. Manaresi, *Università di Bologna, Italy*	35
4	Molecular Weight Distributions A. Munari and P. Manaresi, *Università di Bologna, Italy*	47
5	Formation of Cyclic Oligomers G. Montaudo and P. Maravigna, *Università di Catania, Italy*	63
6	Ring–Chain Equilibria U. W. Suter, *Massachusetts Institute of Technology, Cambridge, MA, USA* and *Eidgenössische Technische Hochschule, Zürich, Switzerland*	91
7	Constitutional Regularity in Linear Condensation Polymers F. T. Gentile and U. W. Suter, *Massachusetts Institute of Technology, Cambridge, MA, USA* and *Eidgenössische Technische Hochschule, Zürich, Switzerland*	97
8	Multifunctional Step Polymerization T. A. Vilgis, *Max-Planck-Institut für Polymerforschung, Mainz, Federal Republic of Germany*	117
9	Experimental Methods of Bulk Polymerization V. V. Korshak and V. A. Vasnev, *USSR Academy of Sciences, Moscow, USSR*	131
10	Experimental Methods of Solution Polymerization V. V. Korshak and V. A. Vasnev, *USSR Academy of Sciences, Moscow, USSR*	143
11	Interfacial and Dispersion Polymerization V. V. Korshak and V. A. Vasnev, *USSR Academy of Sciences, Moscow, USSR*	167
12	Condensation Polymerization in Non-aqueous Dispersions D. J. Walbridge, *ICI PLC, Paints Division, Slough, UK*	197
13	Solid-state Polymerization F. Pilati, *Università di Bologna, Italy*	201
14	Cyclobutane Polymers by Topochemical Photopolymerization M. Hasegawa, *University of Tokyo, Japan*	217
15	Topochemical Polymerization: Diynes D. Bloor, *Queen Mary College, London, UK*	233

16	Step Copolymerization E. Maréchal and A. Fradet, *Université Pierre et Marie Curie, Paris, France*	251

Synthesis by Step Polymerization

17	Polyesters F. Pilati, *Università di Bologna, Italy*	275
18	Polyarylates B. D. Dean, M. Matzner and J. M. Tibbitt, *Amoco Performance Products Inc., Bound Brook, NJ, USA*	317
19	Unsaturated Polyesters A. Fradet, *Université Pierre et Marie Curie, Paris, France* and P. Arlaud, *Norsolor, Groupe C.d.F. Chimie, Drocourt, France*	331
20	Polycarbonates D. C. Clagett and S. J. Shafer, *General Electric Company, Pittsfield, MA, USA*	345
21	Aliphatic Polyamides R. J. Gaymans, *Technische Hogeschool Twente, Enschede, Netherlands* and D. J. Sikkema, *Akzo Fibres and Polymers Research, Arnhem, Netherlands*	357
22	Aromatic Polyamides L. Vollbracht, *Akzo Fibres and Polymers Research, Arnhem, Netherlands*	375
23	Isocyanate-derived Polymers F. Parodi, *EniChem SpA, Milano, Italy*	387
24	Polyurethanes K. C. Frisch and D. Klempner, *University of Detroit, MI, USA*	413
25	Polyureas A. J. Ryan and J. L. Stanford, *University of Manchester Institute of Science and Technology, UK*	427
26	Polybenzyls C. P. Tsonis, *University of Petroleum and Minerals, Dhahran, Saudi Arabia*	455
27	Polyphenylenes M. B. Jones, *University of North Dakota, Grand Forks, ND, USA* and P. Kovacic, *University of Wisconsin-Milwaukee, Milwaukee, WI, USA*	465
28	Poly(phenylene oxide)s D. M. White, *General Electric Co., Schenectady, NY, USA*	473
29	Poly(ether ketone)s P. A. Staniland, *ICI PLC, Wilton, Cleveland, UK*	483
30	Polyimides and Other Heteroaromatic Polymers B. Sillion, *CEMOTA, Vernaison, France*	499
31	Polysulfides D. E. Vietti, *Morton Thiokol Inc., Woodstock, IL, USA*	533
32	Poly(phenylene sulfide)s J. F. Geibel and R. W. Campbell, *Phillips Petroleum Co., Bartlesville, OK, USA*	543
33	Polysulfones F. Parodi, *EniChem SpA, Milano, Italy*	561
34	Polysiloxanes J. J. Lebrun and H. Porte, *Rhône Poulenc, Saint Fons, France*	593

35	Phenol–Formaldehyde Polymers A. KNOP, *Rütgerswerke AG, Frankfurt, Federal Republic of Germany,* V. BÖHMER, *Johannes-Gutenberg-Universität, Mainz, Federal Republic of Germany* and L. A. PILATO, *Temecon Group International Inc., Edison, NJ, USA*	611
36	Urea–Formaldehyde and Melamine–Formaldehyde Polymers D. BRAUN and H.-J. RITZERT, *Deutsches Kunststoff-Institut, Darmstadt, Federal Republic of Germany*	649
37	Epoxy Resins K. A. HODD, *Brunel University, Uxbridge, UK*	667
38	Polymers with Main-chain Mesogenic Units E. CHIELLINI, *University of Pisa, Italy* and R. W. LENZ, *University of Massachusetts, Amherst, MA, USA*	701
39	Polymers with Side-chain Mesogenic Units R. ZENTEL, *Universität Mainz, Federal Republic of Germany*	723
Subject Index		733

Preface

It is only 60 years since Staudinger's model of the molecular nature of a polymer was becoming universally accepted and the physical states of rubbers, plastics and fibres understood. Unfortunately, for some time many academic chemists continued not to appreciate the full significance of polymerization reactions and physicists tended to regard polymeric materials as inevitably being of indeterminate composition and unamenable to study by conventional physical methods.

Nevertheless, in the 1930s the foundations were laid for the understanding of the main polymerization mechanisms. An industry based on synthetic rubbers, plastics and fibres was soon established. In World War II it played a major strategic role and afterwards grew to be one of the main elements of the heavy chemicals industry. It became recognized that synthetics may be superior to natural materials in their properties and that they may be used for completely new purposes.

Alongside the production of well-defined materials there grew the ability to characterize the structure of polymer molecules and to understand the relationships between methods of preparation and subsequent treatment, structure and properties, both chemical and physical. As a result, a vast literature of polymer science and technology has been generated and four Nobel prizes awarded specifically for contributions to polymer science. Add to this the fact that many biological molecules, including polypeptides, enzymes, antibodies, carbohydrates and so on, are polymers of varying degrees of complexity, then the universality of polymers in the physical and biological sciences and technologies forms a dominant modern theme.

Comprehensive Polymer Science is a series of volumes designed to set down the structure of this vast subject in such a way that researchers and teachers of polymer science and workers in associated fields can find an authoritative and comprehensive account of the topic of immediate interest. That topic is set out in a framework of related subjects. The text is focused on synthetic polymers with little reference to biological macromolecules *per se* but the science underpins both physical and biological systems.

To ensure that the wide coverage is maintained at an authoritative level, more than 250 authors from 20 countries have been enlisted. Their contributions have been organized into a series of major themes:

Volume 1	Polymer Characterization
Volume 2	Polymer Properties
Volumes 3–5	Polymerization Mechanisms
Volume 6	Polymer Reactions
Volume 7	Specialty Polymers & Polymer Processing

Because of the wide coverage the editors were presented with a particularly difficult decision with regard to symbols and nomenclature. The latter does not follow strictly the recommendations of IUPAC nor are symbols consistent throughout the whole work. However, usage in a particular chapter is consistent with the practice in the current literature. Thus a reader will be able to frame new publications in the context of the information presented in this series of volumes.

We should like to acknowledge the way in which the staff at the publisher, particularly Dr Colin Drayton (who initially proposed the project), Dr Helen McPherson and their editorial team, have supported the editors and authors in their endeavour to produce a text that is both complete and up-to-date and that will appeal to industrial and academic researchers alike. *Comprehensive Polymer Science* is a milestone in the literature of the subject in terms of coverage, clarity and a sustained high level of presentation.

GEOFFREY ALLEN
London

JOHN C. BEVINGTON
Lancaster

Contributors to Volume 5

Dr P. Arlaud
Norsolor, Groupe C.d.F. Chimie, Usine de Drocourt, Route d'Arras, F-62320 Rouvroy, France

Professor D. Bloor
Department of Chemistry, Queen Mary College, Mile End Road, E1 4NS, London, UK

Dr V. Böhmer
Johannes-Gutenberg-Universität, J. J. Becher-Weg 34/5B1, D-6500 Mainz, Federal Republic of Germany

Professor D. Braun
Deutsches Kunststoff-Institut, Schlossgartenstrasse 6R, D-6100 Darmstadt, Federal Republic of Germany

Dr R. W. Campbell
Research and Development, Phillips Petroleum Co., Bartlesville, OK 74004, USA

Professor E. Chiellini
Institute of Industrial Organic Chemistry, Università di Pisa, Via Risorgimento, I-56100 Pisa, Italy

Dr D. C. Clagett
General Electric Co., Plastics Business Group, 1 Plastics Avenue, Pittsfield, MA 02101, USA

Dr B. D. Dean
Amoco Performance Products Inc., PO Box 409, Bound Brook, NJ 08805, USA

Dr A. Fradet
Laboratoire de Synthèse Macromoléculaire, Université Pierre et Marie Curie, 12 rue Cuvier, F-75005 Paris Cedex 05, France

Professor K. C. Frisch
Polymer Technologies Inc., University of Detroit, 4001 West McNichols Road, Detroit, MI 48221-9987, USA

Dr R. J. Gaymans
Vakgroep Macromoleculaire Chemie en Materiaalkunde, Technische Hogeschool Twente, PO Box 127, 7500 AE Enschede, The Netherlands

Dr J. F. Geibel
Research and Development, Phillips Petroleum Co., Bartlesville, OK 74004, USA

Dr F. A. Gentile
Department of Chemical Engineering, Massachusetts Institute of Technology, Cambridge, MA 02139, USA

Professor M. Hasegawa
Department of Synthetic Chemistry, Faculty of Engineering, University of Tokyo, Hongo, Bunkyo-ku, Tokyo 113, Japan

Dr K. A. Hodd
Department of Materials Technology, Brunel University of West London, Uxbridge, Middlesex UB8 3PH, UK

Professor M. B. Jones
Department of Chemistry, University of North Dakota, Box 7185 University Station, Grand Forks, ND 58202, USA

Professor D. Klempner
Polymer Technologies Inc., University of Detroit, 4001 West McNichols Road, Detroit, MI 48221-9987, USA

Dr A. Knop
Rutgerswerke AG, Mainzer Landstrasse 217, 600 Frankfurt am Main 1, Federal Republic of Germany

Professor V. V. Korshak[†]
Institute of Organo-element Compounds, USSR Academy of Sciences, Vavilov Str. 28, 117813 Moscow V-334, USSR

Professor P. Kovacic
Department of Chemistry, University of Wisconsin-Milwaukee, Milwaukee, WI 53201, USA

Dr J. J. Lebrun
Rhône Poulenc Recherches, Centre de Recherches de Saint Fons, BP 62, F-69192 Saint Fons Cedex, France

Professor R. W. Lenz
Polymer Science and Engineering Department, University of Massachusetts, Amherst, MA 01003, USA

Professor P. Manaresi
Dipartimento di Chimica Applicata e Scienza dei Materiali, Facoltà di Ingegneria, Università di Bologna, viale Risorgimento 2, I-40136 Bologna, Italy

Dr P. Maravigna
Dipartimento di Scienze Chimiche, Dell Università di Catania, viale A Doria 6, I-95125 Catania, Italy

Professor E. Maréchal
Laboratoire de Synthèse Macromoléculaire, Université Pierre et Marie Curie, 12 rue Cuvier, F-75252 Paris Cedex 05, France

Dr M. Matzner
Amoco Performance Products Inc., PO Box 409, Bound Brook, NJ 08805, USA

Professor G. Montaudo
Dipartimento di Scienze Chimiche, Dell Università di Catania, viale A Doria 6, I-95125 Catania, Italy

Dr A. Munari
Dipartimento di Chimica Applicata e Scienza dei Materiali, Facoltà di Ingegneria, Università di Bologna, viale Risorgimento 2, I-40136 Bologna, Italy

Dr F. Parodi
EniChem SpA, R&D Division, via Medici del Vascello 26, I-20138, Milano, Italy

Professor F. Pilati
Dipartimento di Chimica Applicata e Scienza dei Materiali, Facoltà di Ingegneria, Università di Bologna, viale Risorgimento 2, I-40136 Bologna, Italy

Dr L. A. Pilato
598 Watchung Road, Bound Brook, NJ 08805, USA

Dr H. Porte
Rhône Poulenc Recherches, Centre de Recherches de Saint Fons, BP 62, F-69192 Saint Fons Cedex, France

Dr H.-J. Ritzert
Deutsches Kunststoff-Institut, Schlossgartenstrasse 6R, D-6100 Darmstadt, Federal Republic of Germany

Professor S. Russo
Dipartimento di Chimica, Università di Sassari, via Vienna 2, I-79100 Sassari, Italy and Centro CNR Macromolecole, corso Europa 30, I-16132 Genova, Italy

Dr A. J. Ryan
Department of Chemical Engineering and Materials Science, University of Minnesota, 151 Amundsen Hall, Minneapolis, MN 55455, USA

Dr S. J. Shafer
General Electric Co., Plastics Business Group, 1 Plastics Avenue, Pittsfield, MA 02101, USA

Dr D. J. Sikkema
Akzo Fibres and Polymers Research, Velperweg 76, 6824 BM Arnhem, The Netherlands

Dr B. Sillion
CEMOTA, Echangeur de Solaize-Vernaison, 69 Solaize-France, BP No. 3, F-69390 Vernaison, France

Dr J. L. Stanford
Department of Polymer Science and Technology, University of Manchester Institute of Science and Technology, PO Box 88, Sackville Street, Manchester M60 1QD, UK

Dr P. A. Staniland
ICI Chemicals and Polymers Group, Room N 219, PO Box No. 90, Wilton, Middlesbrough, Cleveland TS6 8JE, UK

Professor U. W. Suter
Department of Chemical Engineering, Massachusetts Institute of Technology, Cambridge, MA 02139, USA

Dr J. M. Tibbitt
Amoco Performance Products Inc., PO Box 409, Bound Brook, NJ 08805, USA

Dr C. P. Tsonis
Department of Chemistry, University of Petroleum and Minerals, Dhahran 31261, Saudi Arabia

Dr V. A. Vasnev
Institute of Organo-element Compounds, USSR Academy of Sciences, Vavilov Str. 28, 117813 Moscow V-334, USSR

Dr D. E. Vietti
New Products Research, Morton Chemical Division, Morton Thiokol Inc., 1275 Lake Avenue, Woodstock, IL 60098-7499, USA

Dr T. A. Vilgis
Max-Planck-Institut für Polymerforschung, PO Box 3148, D-6500 Mainz, Federal Republic of Germany

Dr L. Vollbracht
Akzo Fibres and Polymers Research, Velperweg 76, 6824 BM Arnhem, The Netherlands

Mr D. J. Walbridge
Paints Division, ICI Plc, Wexham Road, Slough SL2 5DS, UK

Dr D. M. White
General Electric Co., Corporate Research and Development, PO Box 8, Schenectady, NY 12301, USA

Dr R. Zentel
Institut für Organische Chemie, Universität Mainz, J. J. Becher-Weg 18-20, D-6500 Mainz, Federal Republic of Germany

Contents of All Volumes

Volume 1 Polymer Characterization

Introduction
1   Perspectives
2   Nomenclature
3   Averages and Distributions

Solution Methods
4   Colligative Properties
5   Scattering Properties: Light and X-Rays
6   Scattering Properties: Neutrons
7   Photon Correlation Spectroscopy: Technique and Scope
8   Photon Correlation Spectroscopy: Application to Polymer Solutions
9   Dilute Solution Viscometry
10 Sedimentation and Diffusion
11 Polyelectrolytes

Separation Methods
12 Size Exclusion Chromatography
13 Hydrodynamic Chromatography
14 Field-flow Fractionation
15 Fractionation
16 Chromatographic Cross-fractionation

Spectroscopic and Related Methods
17 Structure of Chains by Solution NMR Spectroscopy
18 Dynamics of Chains in Solutions by NMR Spectroscopy
19 NMR Spectroscopy of Polymers in the Solid State
20 IR Spectroscopy
21 Raman Spectroscopy
22 Emission Spectroscopy
23 ESR Spectroscopy
24 Characterization of Surfaces
25 Optical Activity
26 Mass Spectrometry
27 Pyrolysis GLC

Diffraction and Scattering Methods
28 Crystal Structure by X-Ray Diffraction
29 Electron Diffraction from Crystalline Polymers
30 Small-angle X-Ray Scattering from Crystalline Polymers
31 X-Ray Scattering from Non-crystalline and Liquid Crystalline Polymers
32 Neutron Scattering from Solid Polymers

Microscopy
33 Optical Microscopy
34 Electron Microscopy
35 Etching and Microstructure of Crystalline Polymers

Thermal Methods
36 Thermal Analysis
37 Thermogravimetric Analysis

Subject Index

Volume 2 Polymer Properties

1   Chain Statistics and Scaling Concepts
2   Chain Conformations
3   Polymer Solutions
4   Phase Separation and Pulse-induced Critical Scattering
5   Polymer Blends
6   Chain Segregation in Block Copolymers
7   Hydrodynamic Properties
8   Nonlinear Viscoelastic Behavior
9   Rubber Elasticity and Characterization of Networks
10 Glass Formation and Glassy Behavior
11 Crystallization and Melting
12 Crystallization and Morphology
13 Oriented Polymers
14 Textile Fibres
15 Strength and Toughness
16 Dynamic-mechanical Properties
17 Acoustical Properties
18 Dielectric Properties
19 Rheo-optical Properties
20 Permeation Properties
21 Ionic Conductivity
22 Electrical Conductivity
23 Surface Properties
24 Adsorption
25 Ionomers
Subject Index

Volume 3 Chain Polymerization I

General Characteristics of Chain Polymerization
1   General Concepts
2   Copolymerization
3   Block and Graft Copolymers
4   Molecular Weight Distributions
5   General Experimental Methods in Polymerization
Free-radical Polymerization
6   Overall Mechanisms
7   Experimental Techniques
8   Azo and Peroxy Initiators
9   Redox Initiators
10 Other Initiating Systems
11 Chemistry of Bimolecular Termination
12 Kinetics of Bimolecular Termination
13 Chain Transfer
14 Telomerization
15 Copolymer Composition
16 Rates of Copolymerization
17 Reactivities in Free-radical Polymerization
18 Molecular Weight Distributions
19 Template Polymerization
20 Polymerization in Aqueous Solution
21 Polymerization at High Pressure
22 Ring-opening Polymerization

23 Polymerization of Fluoro Monomers
24 Polymer Reaction Engineering
Anionic Polymerization
25 Carbanionic Polymerization: General Aspects and Initiation
26 Carbanionic Polymerization: Kinetics and Thermodynamics
27 Carbanionic Polymerization: Termination and Functionalization
28 Carbanionic Polymerization: Polymer Configuration and the Stereoregulation Process
29 Carbanionic Polymerization: Copolymerization
30 Carbanionic Polymerization: Hydrogen Migration Polymerization
31 Anionic Ring-opening Polymerization: General Aspects and Initiation
32 Anionic Ring-opening Polymerization: Epoxides and Episulfides
33 Anionic Ring-opening Polymerization: Stereospecificity for Epoxides, Episulfides and Lactones
34 Anionic Ring-opening Polymerization: Lactones
35 Anionic Ring-opening Polymerization: Lactams
36 Anionic Ring-opening Polymerization: *N*-carboxyanhydrides
37 Anionic Ring-opening Polymerization: Copolymerization
38 Experimental Aspects
Cationic Polymerization
39 Carbocationic Polymerization: General Aspects and Initiation
40 Carbocationic Polymerization: Alkenes and Dienes
41 Carbocationic Polymerization: Styrene and Substituted Styrenes
42 Carbocationic Polymerization: Vinyl Ethers
43 Carbocationic Polymerization: *N*-Vinylcarbazole
44 Carbocationic Polymerization: Copolymerization
45 Cationic Ring-opening Polymerization: Introduction and General Aspects
46 Cationic Ring-opening Polymerization: Thermodynamics
47 Cationic Ring-opening Polymerization: Formation of Cyclic Oligomers
48 Cationic Ring-opening Polymerization: Ethers
49 Cationic Ring-opening Polymerization: Acetals
50 Cationic Ring-opening Polymerization: Cyclic Esters
51 Cationic Ring-opening Polymerization: Sulfides
52 Cationic Ring-opening Polymerization: Amines and N-containing Heterocycles
53 Cationic Ring-opening Polymerization: Copolymerization
Subject Index

Volume 4 Chain Polymerization II

Transition Metal Initiated and Related Polymerizations
1 Monoalkene Polymerization: Ziegler–Natta and Transition Metal Catalysts
2 Monoalkene Polymerization: Mechanisms
3 Monoalkene Polymerization: Stereospecificity
4 Monoalkene Polymerization: Copolymerization
5 Conjugated Diene Polymerization
6 Metathesis Polymerization: Chemistry
7 Metathesis Polymerization: Applications
8 Polymerization of Acetylene
9 Polymerization of Mono- and Di-substituted Acetylenes
10 Group Transfer and Aldol Group Transfer Polymerization
Heterogeneous Free-radical Polymerization
11 Polymerization in Emulsions
12 Polymerization in Micelles and Microemulsions
13 Polymerization in Inverse Microemulsions
14 Aqueous Suspension Polymerizations
15 Polymerization in Non-aqueous Dispersions

16 Precipitation Polymerization

Solid-state and Radiation-induced Polymerizations
17 Solid-state Polymerization
18 Polymerization in Clathrates
19 Polymerization by High-energy Radiation
20 Photoinitiated Polymerization
21 Plasma Polymerization

Special Aspects of Chain Polymerization
22 Alternating Copolymerization
23 Cyclopolymerization
24 Electroinitiated Polymerization

Polymerization of Inorganic Ring Systems
25 Polymerization of Cyclosiloxanes
26 Phosphazene High Polymers

Subject Index

Volume 5 Step Polymerization

General Characteristics of Step Polymerization
1   Polycondensation and Related Reactions
2   General Aspects
3   Factors Affecting Rate of Polymerization
4   Molecular Weight Distributions
5   Formation of Cyclic Oligomers
6   Ring–Chain Equilibria
7   Constitutional Regularity in Linear Condensation Polymers
8   Multifunctional Step Polymerization
9   Experimental Methods of Bulk Polymerization
10 Experimental Methods of Solution Polymerization
11 Interfacial and Dispersion Polymerization
12 Condensation Polymerization in Non-aqueous Dispersions
13 Solid-state Polymerization
14 Cyclobutane Polymers by Topochemical Photopolymerization
15 Topochemical Polymerization: Diynes
16 Step Copolymerization

Synthesis by Step Polymerization
17 Polyesters
18 Polyarylates
19 Unsaturated Polyesters
20 Polycarbonates
21 Aliphatic Polyamides
22 Aromatic Polyamides
23 Isocyanate-derived Polymers
24 Polyurethanes
25 Polyureas
26 Polybenzyls
27 Polyphenylenes
28 Poly(phenylene oxide)s
29 Poly(ether ketone)s
30 Polyimides and Other Heteroaromatic Polymers
31 Polysulfides
32 Poly(phenylene sulfide)s
33 Polysulfones
34 Polysiloxanes

35 Phenol–Formaldehyde Polymers
36 Urea–Formaldehyde and Melamine–Formaldehyde Polymers
37 Epoxy Resins
38 Polymers with Main-chain Mesogenic Units
39 Polymers with Side-chain Mesogenic Units
Subject Index

Volume 6 Polymer Reactions

1 Chemical Modification of Synthetic Polymers
2 Chemical Modification of Cellulose and its Derivatives
3 Polymeric Reagents
4 Crosslinking of Rubbers
5 Photocrosslinking
6 Crosslinking: Chemistry of Surface Coatings
7 Photochemical and Radiation Sensitive Resists
8 Polymer Networks
9 Macromonomers, Oligomers and Telechelic Polymers
10 Synthesis of Block Copolymers by Transformation Reactions
11 Heterochain Block Copolymers
12 Synthesis of Graft Copolymers
13 Interpenetrating Polymer Networks
14 Rubber Modification of Plastics
15 Thermal Degradation
16 Pyrolytic Formation of High-performance Carbon Fibres
17 Ceramic Fibres
18 Photodegradation
19 Effects of Antioxidants and Stabilizers
20 Effects of Dyes and Pigments
21 Biodegradation
22 Polymer Degradation in Biological Environments
23 Mechanochemical Degradation
Subject Index

Volume 7 Specialty Polymers & Polymer Processing

Generic Polymer Systems and Applications
1 Block Copolymers
2 Rubber-modified Plastics
3 Glass-reinforced Thermosetting Polyester Molding: Materials and Processing
4 Polymer Blends
5 Synthetic Polymer Adhesives
6 Polymers for Electronic Applications
7 Biomedical Applications of Synthetic Polymers
8 Polymers for Fibers
Unit Operations of Polymer Processing
9 Mixing of Polymers
10 Extrusion of Rubber and Plastics
11 Injection Molding of Rubbers
12 Reaction Injection Molding
13 Computer Control of Extrusion
14 Computer-aided Engineering in Injection Molding
15 Blow Molding of Polymers
Cumulative Subject Index

1
Polycondensation and Related Reactions

FABRIZIO PARODI
EniChem SpA, Milano, Italy
and
SAVERIO RUSSO
Università di Sassari, Italy

1.1 GENERAL CONSIDERATIONS	1
1.2 HISTORICAL BACKGROUND OF POLYCONDENSATION	2
1.3 STEP-GROWTH POLYMERIZATION	5
1.3.1 Step Polycondensation	6
1.3.2 Step Polyaddition	6
1.4 REFERENCES	8

1.1 GENERAL CONSIDERATIONS

The class of macromolecular products commonly defined as condensation polymers includes most of the synthetic materials manufactured world-wide and used as high-strength and/or high-toughness plastics and fibers (*e.g.* polyamides, polyesters, polycarbonates, *etc.*), as well as almost all the hard resins (unsaturated polyesters, epoxy resins, urea–, melamine– and phenol–formaldehyde resins, *etc.*), covering a very wide range of applications.

Those polymeric materials used in relatively small but important industrial sectors as sealants, elastomers, foams, adhesives coatings (silicones, alkyd resins, polyimides, *etc.*), as well as several heavy-duty polymers (*e.g.* poly(arylene oxide)s, poly(arylene sulfide)s, aromatic polysulfones, *etc.*) and emerging engineering thermoplastics (such as aromatic poly(ether ketone)s, liquid-crystalline aromatic polyesters and polyamides, *etc.*) are also members of the same family. Moreover, important polymeric products of natural origin (cellulose, starch, proteins, *etc.*) can be considered as condensation polymers, by virtue of the processes by which they are most probably synthesized by living organisms.

The definition of such a family of condensation polymers does not imply that there is an intrinsic characteristic of all the polymers mentioned above, but is intended only as a reference to the condensation methods of polymerization by which they are *usually* synthesized, either in the laboratory or on an industrial scale. Indeed, a number of different criteria can be considered suitable for classifying either the polymers[1-5] or the synthetic procedures used to obtain them.[3,6-8] Perhaps the best way, still advisable for a comprehensive classification of polymers, is based on the chemistry and mechanism of the polymerization process. In such a context, the term *process* should be used to mean the overall system of individual reactions linked to a specific polymerization procedure (such as activation, initiation, propagation, termination, side reactions, *etc.*) and the term *polymerization* should be used only to identify the growth of a macromolecule, irrespective of the mechanism and chemistry involved.

From the standpoint of the polymer growth mechanism, two entirely different processes, step and chain polymerizations, are distinguishable. As is well known, step-growth polymerizations proceed *via* a step-by-step succession of elementary reactions between reactive sites, which are usually functional groups, but may also be ions, complexes or even free radicals. Each independent step causes the disappearance of two coreacting centres and creates a new linking unit between a pair of molecules. In order to obtain high polymers, the reactants must be at least difunctional; monofunctional ones act as chain-stoppers.

In chain polymerizations (see also Volume 3, Chapter 1), propagation is caused by direct reaction of a species, bearing a suitably generated active center, with a monomer molecule. The active center (free radical, ionic, *etc.*) is regenerated chainwise by each act of growth, and the monomer itself constitutes the feed, progressively converted into the polymer.

With both the step and chain mechanism, the polymerization processes can be based on either one of two types of propagation reaction: condensation or addition (*i.e.* with or without elimination of various by-products, respectively); hereafter, condensation and addition polymerizations are called simply 'polycondensations' and 'polyadditions'.

Most of the polycondensation processes, which occur *via* a suitable choice of condensation reactions (such as esterification, amidation, acetalization, nucleophilic or electrophilic substitutions, *etc.*), follow a step-growth mechanism. In contrast, although a chain mechanism is the usual feature of polyaddition processes, several of them proceed by a stepwise addition of pairs of functional groups brought by the reactants (isocyanate/hydroxyl, isocyanate/amine or Diels–Alder reaction, Michael-type adducts formation, *etc.*). Such polyadditions display the same functionality criteria and the same features as any step polymerization.

As detailed in the following chapters, both step polycondensations and step polyadditions are characterized, in comparison to chain polymerization, by the presence of intermediate reaction products, which have the same structure as the final macromolecules but a lower degree of polymerization; these are generally stable and may be readily isolated. Furthermore, a very important dependence of the average molecular weight value on monomer conversion is found (high polymers are achieved only as the reaction approaches completion).

As already mentioned, polymers of the most important families, to which this volume devotes specific sections in Part 2, are synthesized by step-growth processes. With the notable exception of polyurethanes and polyureas (typically obtained by step polyaddition from diisocyanates and diols or diamines, respectively), their preparation is almost always accomplished through condensation reactions and, on these grounds, they are currently known as 'condensation polymers'.

For many years, the term polycondensation has been improperly used as a synonym for step polymerization, or even extended to cover certain chain-growth processes. The resultant confusion has prevented the use of proper classifications in the field of polymerization processes and related polymers. Therefore, some 'historical' considerations may be appropriate and helpful in order to survey the general features of step-growth polymerization.

1.2 HISTORICAL BACKGROUND OF POLYCONDENSATION

The earliest classification into condensation and addition polymers, proposed by Carothers[9,10] and used for a long time, was based on the classical definitions given in organic chemistry of condensation and addition reactions, through which polymers are synthesized from suitable monomers or reactants.

Condensation and addition are the terms used to define inter- or intra-molecular reactions which can give covalent bonds between unjoined atoms or groups, and occur *with* or *without* elimination of simple molecules, respectively. The by-products can be water, alcohols, halohydric acids, alkali metal halides, nitrogen, carbon dioxide, hydrogen, ammonia, *etc.*

Based on such a neat stoichiometric definition (loss or not of side-products), polymerization processes consisting of a succession of condensation or addition reactions have been termed condensation or addition polymerizations. This classification, accepted regardless of the chemistry, mechanism and kinetics of the process, is still currently adopted.[11-14]

Carothers' subdivision of the polymers thus synthesized into condensation and addition polymers implies a cross correlation between the polymer and the process by which it is obtained. He was the first to introduce the classical definition of monomer functionality and its requirements for polymerization,[9,10] functionalities higher than two customarily leading to branched or crosslinked products.

Carothers described as condensation polymers those arising from a succession of condensation reactions between reactants which are at least difunctional, *i.e.* those bearing two (or more) functional groups A and B, able to react with each other giving a linking group E plus by-products of various kinds. A simple difunctional scheme is depicted by the AABB-type polycondensation of equation (1) and by the AB-type self-polycondensation of equation (2).

$$nA-Y-A \;+\; nB-Z-B \longrightarrow -(Y-E-Z-E)_n- \;+\; \text{by-products} \qquad (1)$$

$$nA-Y-B \longrightarrow -(Y-E)_n- \;+\; \text{by-products} \qquad (2)$$

Typical examples are the (1)-type polycondensation of dicarboxylic acids with diols or diamines, which leads to polyesters or polyamides, and the (2)-type self-polycondensation of hydroxy- or amino-carboxylic acids, similarly affording polyesters or polyamides; in these cases, water is the reaction by-product evolved. It is suggested that the term *monomers* be restricted to AB-type molecules, whereas AA- and BB-type ones should preferably be called *reactants*, the true monomer being the product of their first reaction.

An important feature of many condensation polymers, pointed out by Carothers and regarded as an additional identification criterion, is that the monomers or reactants can be regenerated from the polymers by chemical reactions, typically by hydrolysis (but ammono- or amino-lysis, alcoholysis, *etc.*, may also be suitable); the commonest example, in fact, is given by polyesters, which are rather easily hydrolyzed back to the corresponding diols and dicarboxylic acids.

Since polycondensations occur with evolution of by-products, the interlinking unit E formed has fewer atoms than the sum of the groups A and B; moreover, A, B and E display very different features (consider, for example, $-OH$, $-CO_2H$ and $-CO_2-$). This implies that the molecular formula of the constitutional repeating unit is quite different from that of the starting monomer or reactants.

A reversed situation is found for polyadditions, which involve no by-product elimination and yield polymers with a structural unit identical to that of the monomer (as in the polymerization of vinyl compounds), or having at least the same atoms even though connected with a different structural linkage.

Carothers' classification, based on stoichiometric criteria and implying polymer-structure/polymerization-process identity, proved to be rather inconsistent when the many existing exceptions of polymers attainable *via* condensation as well as by addition processes were considered; such products should thereby belong to both classes at the same time. For example, polyurethanes can be synthesized either from a diamine and a bis(haloformate), as in Scheme 1, path (a) (a Schotten–Baumann-type polycondensation reaction, evolving halohydric acid), or by reacting a diisocyanate with a diol as in Scheme 1, path (b) (polyaddition).

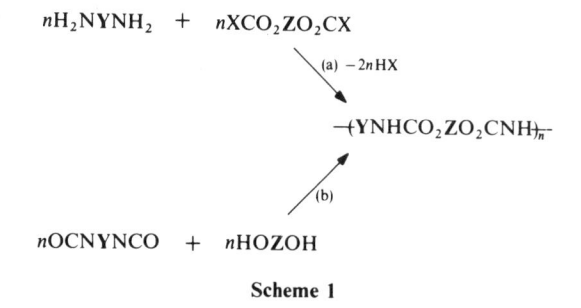

$$nH_2NYNH_2 \;+\; nXCO_2ZO_2CX$$
$$\searrow (a)\; -2nHX$$
$$-(YNHCO_2ZO_2CNH)_n-$$
$$\nearrow (b)$$
$$nOCNYNCO \;+\; nHOZOH$$

Scheme 1

Analogously, two synthetic routes are available for polyureas, either by polycondensation of diamines with phosgene (hydrochloric acid is eliminated) or by polyaddition of diamines and diisocyanates.

Even linear polyethylene can be obtained, though with very different molecular weights, by addition polymerization of ethylene (equation 3), by polycondensation of diazomethane (today known as proceeding *via* a chain mechanism, equation 4) or by Wurtz step-polycondensation of α,ω-dihalo-*n*-alkanes with sodium metal (equation 5).

$$nCH_2=CH_2 \xrightarrow{\text{Ziegler–Natta catalyst}} -(CH_2CH_2)_n- \qquad (3)$$

$$nCH_2N_2 \xrightarrow{BF_3} -(CH_2CH_2)_{n/2}- \;+\; nN_2\uparrow \qquad (4)$$

$$nX-(CH_2)_m-X \;+\; 2nNa \longrightarrow -(CH_2CH_2)_{(mn)/2}- \;+\; 2nNaX \qquad (5)$$

As a further example, polyaziridine, which is commonly prepared by chain ring-opening of aziridine (equation 6), can also be synthesized by polycondensation of 1,2-dibromoethane with ammonia (equation 7).

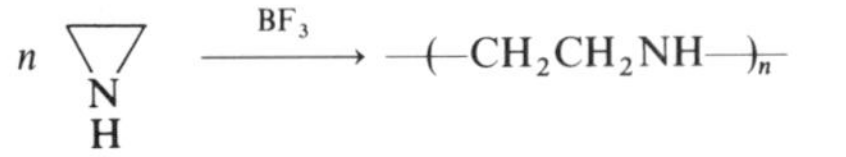

$$n \underset{\underset{H}{N}}{\triangle} \xrightarrow{BF_3} -(-CH_2CH_2NH-)_n- \quad (6)$$

$$n\,BrCH_2CH_2Br + 3n\,NH_3 \longrightarrow -(-CH_2CH_2NH-)_n- + 2n\,NH_4Br \quad (7)$$

Exceptions such as those mentioned above and a deeper knowledge of the polymerization processes made it imperative to define more precise criteria of differentiation.

Flory was the first to criticize Carothers' classification and indicated the need for a distinction between polymer structure and polymerization process,[15,16] a criterion also stressed by Mark[17] and soon adopted.[18]

Flory pointed out the fact that polycondensations occur as a succession of independent condensation steps between functional groups, whereas polyadditions usually proceed *via* a chain mechanism which involves the presence of active centers of various types. By virtue of such a marked difference, condensation and addition polymerizations had therefore to be separately considered as synonyms of two types of growth mechanisms, and not only in a simple stoichiometric sense.

Flory recognized, however, that certain polyadditions proceed *via* processes formally resembling typical polycondensations (*i.e.* with a step-growth mechanism), but without liberation of any by-product. Examples of this kind are given, in fact, by the synthesis of polyurethanes and polyureas from diisocyanates and diols or diamines, respectively. Thus, in order to rationalize all the above concepts and statements, Flory decided to consider as a polycondensation any polymerization process proceeding 'by a reaction between pairs of functional groups with the formation of a type of inter-unit functional group not present in the monomer(s)'.[16] Hence, condensation polymers should be all those macromolecular products synthesized by polycondensation reactions in the sense described above, regardless of whether by-products are evolved or not.

On the other hand, it is well known that many cyclic compounds, such as lactams, lactones, cyclic urethanes, cyclic ureas, *etc.*, can be converted by polyaddition (in terms of stoichiometry) into polymers whose structures are identical to those attainable by condensation (in terms of the broad Flory definition) of suitable difunctional reactants. For instance, ω-amino acids and their ω-lactams can afford the same polyamide by condensation and addition polymerization, respectively, as in Scheme 2.

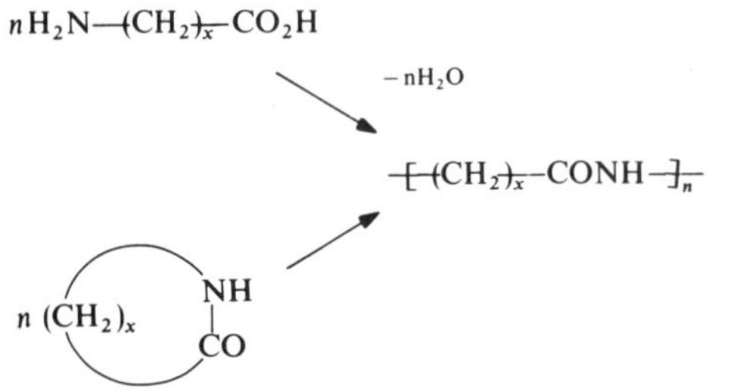

Scheme 2

Similar considerations are applicable to ω-hydroxy acids and their ω-lactones; cyclic (*x*-methylene)ureas and (*x*-methylene)diamines/(*x*-methylene) diisocyanates (Schemes 3); (*x*-methylene)diols/phosgene and cyclic (*x*-methylene) carbonates (Scheme 4), and so on.

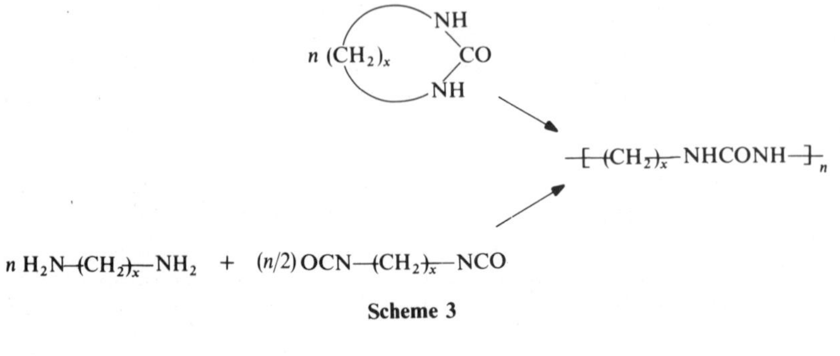

Scheme 3

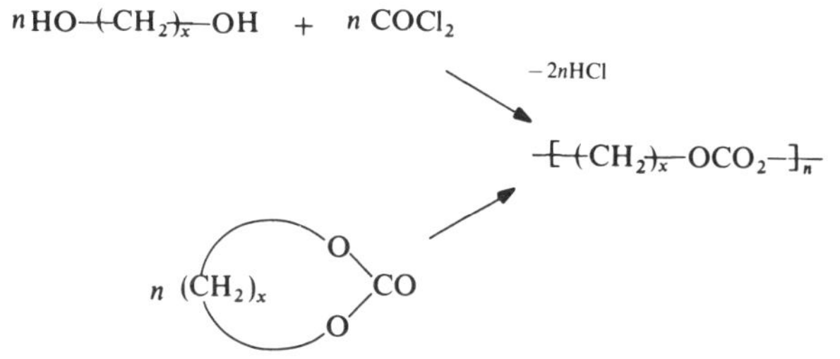

$$n\,HO\text{-}(CH_2)_x\text{-}OH \quad + \quad n\,COCl_2$$

$$\downarrow -2n\,HCl$$

$$n\,(CH_2)_x\underset{O}{\overset{O}{\diagup}}CO \longrightarrow \text{-}[\text{-}(CH_2)_x\text{-}OCO_2\text{-}]_n\text{-}$$

Scheme 4

The above polymers have the general feature, typical of Carothers' condensation polymers, of being degraded to monomeric fragments by suitable chemical means (not restoring, however, the cyclic monomers from which they could have been synthesized). It should be noticed that cyclic compounds such as those mentioned above polymerize without generation of a new interlinking chemical group; on the contrary, the latter is already a characteristic element of their ring structure. Thus, their polymerization cannot be considered as a polycondensation reaction in the sense first given by Flory to the term.

On these grounds, facing the above contradictions and in order to match as far as possible the separate concepts of polymerization process and polymer structure, Flory proposed a broader definition for condensation polymers. This class should therefore include not only products arising from polycondensations according to the original definition, but also polymers which, by chemical degradation (*e.g.* hydrolysis), give end products 'with chemical structure different from that of the structural units',[16] due to addition or loss of atoms or to isomeric changes in structure. This re-definition would therefore enable all traditional condensation polymers (poly-amides, -esters, -urethanes, -ureas, -carbonates, *etc.*) to be dealt with, regardless of whether they are prepared by true condensation processes or by ring-opening polyaddition.

In this sense, moreover, a number of naturally occurring polymers which cannot be synthesized in a laboratory by a condensation process (cellulose, starch, lignin, *etc.*) can be considered as condensation polymers from 'hypothetical' monomers (*i.e.* from the products to which they can be degraded) by 'hypothetical' processes of polymerization.[3]

All the above considerations represent a review of the 'roots' and origins of some terms currently in use in polymer chemistry, but are still imperfect in terms of a proper classification of polymers: the distinction between condensation and addition processes is based on intrinsic characteristics (such as degradability of products) of rather ambiguous meaning.

1.3 STEP-GROWTH POLYMERIZATION

The first detailed classification of polymerization processes based on growth mechanism, as given in Section 1.1, was reported by Lenz[19] and, almost simultaneously, by Sokolov.[20]

Lenz and, in more detail, Sokolov (who adopts the term 'polycondensation' in place of step polymerization) have adequately described the features of step-growth processes from a general point of view. A difunctional scheme for step polycondensation has already been given by equations (1) and (2). However, difunctional step-polymerizations should more correctly be illustrated by equations such as (8) and (9), where end groups are indicated.

$$n\,A\text{-}Y\text{-}A \; + \; n\,B\text{-}Z\text{-}B \longrightarrow A\text{-}(\text{-}Y\text{-}E\text{-}Z\text{-}E\text{-})_n Y\text{-}E\text{-}Z\text{-}B \; + \; \text{by-products} \qquad (8)$$

$$n\,A\text{-}Y\text{-}B \longrightarrow A\text{-}(\text{-}Y\text{-}E\text{-})_{n-1} Y\text{-}B \; + \; \text{by-products} \qquad (9)$$

Each reaction step causes disappearance of two reactive sites A and B (converted into a linking unit E) and frees two reactive sites at the ends of any growing molecule, irrespective of its size. Therefore, equations such as (8) and (9) are the correct way to write a step-polymerization reaction, though reactive chain-ends are often omitted for practical reasons.

The reaction conversion at any stage can be expressed by the fraction of total reactive sites consumed and titration of them on a polymer can give a number-average value of the achieved molecular weight.

Reactive sites usually display comparable reactivity regardless of the size of the molecule to which they are linked, and the polymerization process has, throughout its course, the characteristics of a statistical combination of fragments. Dimer formation will be largely predominant in the earliest stages of the process; trimers, tetramers and so on, will be preferentially produced in the subsequent stages. Molecular weight distributions will simply follow this random scheme, unless marked changes in group reactivity from monomers to dimers, trimers, *etc.* (as observed in many cases) takes place.

The above features imply four important characteristics of step polymerization: (i) the important tendency to produce cyclic oligomers *via* intramolecular reaction of reactive sites at the ends of the same oligomeric molecule; (ii) the slow increase of molecular weights with reaction conversion, due to the requisite of combining the highest number of molecular fragments; (iii) the need for equimolecularity of reactive sites as close as possible to the theoretical value, in order to get the highest molecular weights; and (iv) the opportunity to deactivate the polymer end-groups, in order to stabilize molecular weight values and polymer properties.

Step polymerization can proceed reversibly (*e.g.* polyesterification from diacids or diesters and glycols, polyamidation from diacids and diamines) or irreversibly (*e.g.* polyesterification from diacyl chlorides and diols, formation of polyureas from diisocyanates and diamines). Reversibility can be evaluated in kinetic terms of direct and reverse reactions (thermodynamic equilibria are seldom reached).

Polymerization rates of free-radical chain polymerizations are often enhanced by increasing viscosities (the well-known 'gel effect'). By contrast, step polymerizations are invariably slowed down by higher viscosities of the reaction medium, which limit the free motion of the active centers until, above certain critical values, the overall kinetics are governed and described by rheological parameters. This is almost always the case for step polymerizations performed in bulk or in concentrated solutions.

Multifunctional step-polymerizations, based on monomers or monomer mixtures with functionality greater than two, usually lead to branching, gelation or even complete crosslinking. Functionalites greater than two (accidentally present as impurities or generated by side reactions) are most undesirable in the synthesis of thermoplastic high polymers because of the loss of processability (*e.g.* melt spinning of fibers, *etc.*) caused by very small amounts of the gelled product. In contrast, controlled multifunctional polymerization is the basis for the synthesis of many resins (urea–, melamine–, phenol–formaldehyde, alkyd or epoxy resins) or polymers (poly-urethanes, -ureas, -siloxanes), which are fully crosslinked *via* further reactions during injection molding, casting, coating, foaming and other processing steps.

1.3.1 Step Polycondensation

Table 1 gives a survey of typical step polycondensations; they follow the criteria described above in general and are described in detail in the specific sections of this volume.

It should also be mentioned that, by means of step-polycondensation processes, some products can be synthesized which belong to the class of inorganic polymers; macromolecular compounds whose backbone does not contain sequences of carbon atoms.

Polysiloxanes, with a skeletal repeat unit —SiO—, are the main members of the family and Volume 5, Chapter 34 deals with them. Other less important polymers, typically synthesized *via* step-condensation methods, are at least worth mentioning in this context. They include, for example, polysilazanes (repeat unit —SiN—), polysilthianes (—SiS—), polyphosphoxanes (—PO—), polyphosphazanes (—PN—), *etc.*

A number of books and reviews[21-25] have been devoted to inorganic polymers and the reader is referred to them for a deeper insight.

1.3.2 Step Polyaddition

Examples of this subclass of step-growth processes are shown in Table 2; some of them (such as the synthesis of polyurethanes and polyureas) are particularly important and are extensively treated in the appropriate chapters of this volume.

A brief mention should be made here of the hydrogen-transfer polymerization of acrylamide and C-substituted acrylamides by anionic initiators, which leads to poly(β-alanine) and substituted nylon-3 polymers by what is now considered a step-growth process.[26,27] Indeed, their overall mechanism of polymerization is still open to controversial interpretations. This subject has been included in Volume 4, but the reader should be aware that, depending on experimental conditions, either inter- or intra-molecular hydrogen-transfer may occur.

Table 1 Typical Step-growth Condensation Polymerizations

Reactants	Reaction by-product	Characteristic linkage formed	Type of polymer
Dicarboxylic acids + diols	H_2O	$-CO_2-$	Polyesters
Dialkyl esters of dicarboxylic acids + diols	ROH	$-CO_2-$	Polyesters
Diacyl chlorides + diols	HCl	$-CO_2-$	Polyesters
Diacyl chlorides + diamines	HCl	$-CONH-$	Polyamides
Dicarboxylic acids + diamines	H_2O	$-CONH-$	Polyamides
Bis(chloroformate)s + diamines	HCl	$-O_2CNH-$	Polyurethanes
Bis(chloroformate)s + diols	HCl	$-OCO_2-$	Polycarbonates
Diols (or bisphenols) + phosgene	HCl	$-OCO_2-$	Polycarbonates
Bisphenols + diphenyl carbonate	PhOH	$-OCO_2-$	Poly(arylene carbonate)s
Dichlorosilanes + H_2O	HCl	$-SiO-$	Polysiloxanes
Organic dichlorides + Na_2S_x	NaCl	$-S_x-$	Polysulfides
Phenols + O_2	H_2O	$-ArO-$	Poly(arylene ether)s
Urea (or melamine) + formaldehyde	H_2O	$-NHCH_2-$	'Amino-resins'
Phenols + formaldehyde	H_2O	$-ArCH_2-$	'Phenolic resins'
Diols + aldehydes	H_2O	$-OCH(R)O-$	Polyacetals
Diketones + diamines	H_2O	$>C=N-$	Poly(azomethine)s
Aromatic sulfonyl chlorides	HCl	$-ArSO_2-$	Poly(arylene sulfone)s
Aromatic acyl chlorides	HCl	$-ArCO-$	Poly(arylene ketone)s
Benzyl chlorides	HCl	$-C_6H_4CH_2-$	Polybenzyls
Diisocyanates	CO_2	$-N=C=N-$	Poly(carbodiimide)s
Tetracarboxylic acid + diamines	H_2O	$-CO\diagdown N- \atop -CO\diagup$	Polyimides
Dinitriles + hydrazine	H_2	imidazole ring with $N-N$, NH_2	Poly(aminotriazole)s
Dicarboxylic acid diphenyl esters + dihydrazines	PhOH	$-CONHNH-$	Polyhydrazides

Table 2 Typical Step-growth Addition Polymerizations (AABB-type)

Reactants	Characteristic linkage formed	Type of polymer
Diisocyanates + diols	$-O_2CNH-$	Polyurethanes
Diisocyanates + diamines	$-NHCONH-$	Polyureas
Diisothiocyanates + diamines	$-NHCSNH-$	Poly(thiourea)s
Diepoxides + diamines	$-CH_2CHN< \atop OH$	
Diepoxides + diisocyanates	oxazolidone ring	Poly(2-oxazolidone)s
Dithiols + unconjugated dienes	$-CH_2CH_2S-$	Polysulfides
Conjugated bis(diene)s + bis(dienophile)s	Various	
Divinyl sulfones + diols	$-OCH_2CH_2SO_2-$	
Organic dihalides + diamines	$-\overset{|}{\underset{|}{N}}{}^{\pm}\ X^-$	Poly(ammonium halide)s
Dinitriles + diols	$-CONH-$	Polyamides

For two other interesting examples of step polyaddition, the following systems can be mentioned:
(a) Compounds containing two (or more) electron-deficient alkenic bonds (such as divinyl sulfones, acrylic diesters, *etc.*,) can add diprotic nucleophiles of general formula H—Z—H (*e.g.* primary or secondary diamines, diols, dithiols, *etc.*) following the general AABB-polymerization scheme of equation (10), each step of which implies formation of a Michael-type adduct; **Y** is an electron-withdrawing group.

$$n \,{>}C{=}C{-}Y{-}C{=}C{<} \;+\; nH{-}Z{-}H \longrightarrow -(\!CCH{-}Y{-}CHC{-}Z\!)_n \qquad (10)$$

(b) Diels–Alder adduct formation from a conjugated diene and a dienophile, as in equation (11), can be considered as the elementary step of the so-called Diels–Alder polymerization processes, on which several reviews have been published.[28–31]

$$\text{(diene)} \;+\; \text{(dienophile)} \longrightarrow \text{(adduct)} \qquad (11)$$

Either AABB-type polyaddition of bis(diene)s with bis(dienophile)s, or AB-type self-polyaddition of diene–dienophile compounds can be suitable, as shown, for example, in equations (12) and (13), respectively.

$$(12)$$

$$(13)$$

If both the diene and the dienophile are part of a ring system, polycyclic products result and perfect ladder polymers (as in equation 13) can be synthesized.

More detailed approaches and theories, as well as specific examples of step-growth polymerizations will be described in the following chapters.

1.4 REFERENCES

1. O. Leuchs, 'The Classifying of High Polymers', Butterworths, London, 1968.
2. R. B. Fox, in 'Encyclopedia of Polymer Science and Technology', ed. N. M. Bikales, H. F. Mark and N. G. Gaylord, Wiley, New York, 1968, vol. 9, p. 336.
3. G. Odian, 'Principles of Polymerization', 2nd edn., Wiley, New York, 1981, chap. 1.
4. L. G. Donaruma, in 'Encyclopedia of Materials Science and Engineering', ed. M. B. Bever, Pergamon Press, Oxford, 1986, vol. 5, p. 3732.
5. H. S. Kaufman, in 'Encyclopedia of Materials Science and Engineering', ed. M. B. Bever, Pergamon Press, Oxford, 1986, vol. 5, p. 3739.
6. N. M. Bikales, in 'Encyclopedia of Polymer Science and Technology', ed. N. M. Bikales, H. F. Mark and N. G. Gaylord, Wiley, New York, 1965, vol. 3, p. 762.
7. 'Encyclopedia of Polymer Science and Engineering', 2nd edn., ed. H. F. Mark, N. M. Bikales, C. G. Overberger and G. Menges, Wiley, New York, 1985, vol. 3, p. 549.
8. R. W. Lenz, in 'Kirk-Othmer Encyclopedia of Chemical Technology', 3rd edn., ed. M. Grayson and D. Eckroth, Wiley, New York, 1982, vol. 18, p. 720.
9. W. H. Carothers, *J. Am. Chem. Soc.*, 1929, **51**, 2548; *Chem. Rev.*, 1931, **8**, 353.
10. H. Mark and G. S. Whitby (eds.), 'Collected Papers of W. H. Carothers on High Polymeric Substances', Interscience, New York, 1946, p. 4.
11. G. F. D'Alelio, 'Fundamental Principles of Polymerization', Wiley, New York, 1952, chap. 2.
12. 'IUPAC Basic Definitions of Terms Relating to Polymers', *Pure Appl. Chem.*, 1974, **40**, 477.
13. 'Encyclopedia of Polymer Science and Engineering', 2nd edn., ed. H. F. Mark, N. M. Bikales, C. G. Overberger and G. Menges, Wiley, New York, 1986, vol. 4, p. 118.
14. 'Encyclopedia of Polymer Science and Engineering', 2nd edn., ed. H. F. Mark, N. M. Bikales, C. G. Overberger and G. Menges, Wiley, New York, 1985, vol. 1, p. 470.
15. P. J. Flory, *Chem. Rev.*, 1946, **39**, 137.
16. P. J. Flory, 'Principles of Polymer Chemistry', Cornell University Press, Ithaca, NY, 1953, chap. 2.
17. H. Mark and A. V. Tobolsky, 'Physical Chemistry of High Polymeric Systems', Interscience, New York, 1950.
18. B. Golding, 'Polymers and Resins', Van Nostrand, Princeton, NJ, 1959, chap. 1.

19. R. W. Lenz, 'Organic Chemistry of Synthetic High Polymers', Interscience, New York, 1967, chap. 1.
20. L. B. Sokolov, 'Synthesis of Polymers by Polycondensation', Israel Program for Scientific Translations, Jerusalem, 1968, chap. 1.
21. F. G. A. Stone and W. A. G. Graham (eds.), 'Inorganic Polymers', Academic Press, New York, 1962.
22. D. L. Venezky, in 'Encyclopedia of Polymer Science and Technology', ed. N. M. Bikales, H. F. Mark and N. G. Gaylord, Wiley, New York, 1967, vol. 7, p. 664.
23. J. R. MacCallum, in 'Step-Growth Polymerizations', ed. D. H. Solomon, Dekker, New York, 1972, chap. 7.
24. N. H. Ray, 'Inorganic Polymers', Academic Press, London, 1978.
25. A. Rheingold, in 'Encyclopedia of Polymer Science and Engineering', 2nd edn., ed. H. F. Mark, N. M. Bikales, C. G. Overberger and G. Menges, Wiley, New York, 1987, vol. 8, p. 138.
26. J. P. Kennedy and T. Otsu, *J. Macromol. Sci., Rev. Macromol. Chem.*, 1972, **C6**, 237.
27. G. Camino, L. Costa and L. Trossarelli, *J. Polym. Sci., Polym. Chem. Ed.*, 1980, **18**, 377.
28. E. A. Kraiman, in 'Encyclopedia of Polymer Science and Technology', ed. N. M. Bikales, H. F. Mark and N. G. Gaylord, Wiley, New York, 1966, vol. 5, p. 23.
29. W. J. Bailey, in 'Encyclopedia of Polymer Science and Technology', ed. N. M. Bikales, H. F. Mark and N. G. Gaylord, Wiley, New York, 1967, vol. 8, p. 97.
30. A. D. Delman, *J. Macromol. Sci., Rev. Macromol. Chem.*, 1968, **C2**, 153.
31. W. J. Bailey, in 'Step-Growth Polymerizations', ed. D. H. Solomon, Dekker, New York, 1972, chap. 6.

2
General Aspects

PIERO MANARESI and ANDREA MUNARI
University of Bologna, Italy

2.1 INTRODUCTION	11
2.2 CHEMICAL CLASSIFICATION OF REACTION MECHANISMS	13
2.2.1 Reactions Involving Carbonyls	13
2.2.2 Nucleophilic Substitution Reactions	16
2.2.3 Miscellaneous Reactions	17
2.3 STEP POLYMERIZATION OF BIFUNCTIONAL REACTANTS	18
2.3.1 Extent of Reaction and Number-average Degree of Polymerization	19
2.3.2 Degree of Polymerization in Conditions of Chemical Equilibrium	24
2.4 KINETICS OF STEP POLYMERIZATION	26
2.4.1 Equal Reactivity of Functional Groups	26
2.4.2 Rate of Step Polymerization	27
2.4.3 Kinetics of Reversible Polymerization	32
2.5 REFERENCES	33

2.1 INTRODUCTION

As considered in Chapter 1 of this volume, step-growth polymerization is a process involving at the beginning one or more reactants (monomers) carrying at least two reactive functional groups. These monomers, on reacting between themselves, with or without the elimination of small molecules, give rise to oligomeric products which still contain functional groups and which are able to react again, either between themselves or with the monomers. In this way products of ever growing molecular weight are formed, from monomer to dimer, trimer, tetramer, pentamer and so on, as illustrated in Scheme 1.

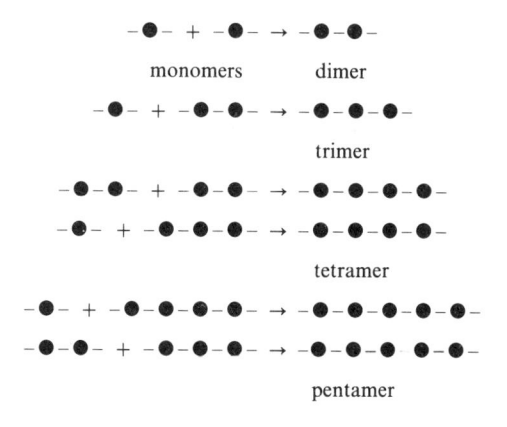

$$-\bullet- \; + \; -\bullet- \; \rightarrow \; -\bullet-\bullet-$$
$$\text{monomers} \quad \text{dimer}$$

$$-\bullet- \; + \; -\bullet-\bullet- \; \rightarrow \; -\bullet-\bullet-\bullet-$$
$$\text{trimer}$$

$$-\bullet-\bullet- \; + \; -\bullet-\bullet- \; \rightarrow \; -\bullet-\bullet-\bullet-\bullet-$$
$$-\bullet- \; + \; -\bullet-\bullet-\bullet- \; \rightarrow \; -\bullet-\bullet-\bullet-\bullet-$$
$$\text{tetramer}$$

$$-\bullet- \; + \; -\bullet-\bullet-\bullet-\bullet- \; \rightarrow \; -\bullet-\bullet-\bullet-\bullet-\bullet-$$
$$-\bullet-\bullet- \; + \; -\bullet-\bullet-\bullet- \; \rightarrow \; -\bullet-\bullet-\bullet-\bullet-\bullet-$$
$$\text{pentamer}$$

Scheme 1

If the initial polymerization system contains a single monomer carrying two mutually reacting A and B functional groups, the constitutional repeating unit (CRU)[1] of the polymer will contain only one monomeric unit; the structure of the CRU will generally be derived from that of the monomer through the elimination of a small molecule. For example, in the case of the polyesterification of lactic acid, a polymeric chain with an ester repeating unit will be formed (equation 1).

$$n\text{HOCHCO}_2\text{H} \longrightarrow \text{H}\mathord{-}\!\left[\text{OCHCO}\right]_{n-1}\!\!\mathord{-}\text{OCHCO}_2\text{H} + (n-1)\text{H}_2\text{O} \qquad (1)$$
$$\quad\;\;\text{Me} \qquad\qquad\qquad\;\;\text{Me} \qquad\;\;\,\text{Me}$$

lactic acid   poly(lactic acid)

However, if the initial polymerization system contains two different monomers, each carrying two equal functional groups, of type A for the first monomer and of type B for the second one (A reacting only with B and *vice versa*), the CRU will contain two monomeric units, of composition equal to those of the two monomers or differing from these by only a few atoms. For example, this is the case for the polyurethane obtained from 1,4-butanediol and 4,4'-diphenylmethane diisocyanate. and the polyamide (Nylon 6,6) synthesized from hexamethylenediamine and adipic acid, shown in equations (2) and (3) respectively.

$$n\text{OCN}\text{—}\langle\text{C}_6\text{H}_4\rangle\text{—CH}_2\text{—}\langle\text{C}_6\text{H}_4\rangle\text{—NCO} + n\text{HO(CH}_2)_4\text{OH} \longrightarrow$$

4,4'-diphenylmethane diisocyanate   1,4-butanediol

$$\text{OCN}\text{—}\langle\text{C}_6\text{H}_4\rangle\text{—CH}_2\text{—}\langle\text{C}_6\text{H}_4\rangle\text{—NHCO}\mathord{-}\!\left[\text{O(CH}_2)_4\text{OCONH}\text{—}\langle\text{C}_6\text{H}_4\rangle\text{—CH}_2\text{—}\langle\text{C}_6\text{H}_4\rangle\text{—NHCO}\right]_{n-1}\!\!\mathord{-}\text{O(CH}_2)_4\text{OH} \qquad (2)$$

polyurethane

$$n\text{H}_2\text{N(CH}_2)_6\text{NH}_2 + n\text{HO}_2\text{C(CH}_2)_4\text{CO}_2\text{H} \longrightarrow \text{H}\mathord{-}\!\left[\text{NH(CH}_2)_6\text{NHCO(CH}_2)_4\text{CO}\right]_n\!\!\mathord{-}\text{OH} + (2n-1)\text{H}_2\text{O} \qquad (3)$$

hexamethylenediamine   adipic acid   poly(iminohexamethyleneiminoadipoyl) or poly(hexamethylene adipamide)

A special situation arises when one at least of the monomers bears more than two reactive functional groups. This is the case for multifunctional step polymerization (see Volume 5, Chapter 8), where the resulting macromolecular chains are not generally linear, and, according to the initial composition of the system and to the extent of reaction, the formation of a branched or even of a crosslinked polymer is possible. Moreover, in some specific cases there is also the possibility of obtaining linear polymers starting from monomers having functionality greater than two. This is the case for the synthesis of a polyimide[2] by cyclopolycondensation of pyromellitic dianhydride (a tetrafunctional reactant) and 4,4'-diaminodiphenyl ether (see Volume 5, Chapter 30).

In summary, we can define step-growth polymerization as a process involving in each step a bimolecular reaction between two functional groups belonging to two molecules with functionality greater than or equal to two and of arbitrary length, with the disappearance of an active center for each reacting molecule. The initial system can be formed by a single monomer carrying both A and B mutually reactive functional groups (case a of Scheme 2), as is the case for the polyesterification of lactic acid.

Alternatively, the initial system may contain two distinct kinds of monomers each carrying two equal functional groups, of the A type for the first kind of monomer and of the B type for the second one (case b of Scheme 2). This is the case for polyamides obtained from diamines and dicarboxylic acids, and for polyurethanes. There is also a third possibility, when the initial system is formed by monomers carrying functional groups of a unique type able to react between themselves (case c of Scheme 2). An example is the ester-interchange process used commercially to prepare poly(ethylene terephthalate). The reaction between dimethyl terephthalate and ethylene glycol affords a monomer that contains 2-hydroxyethyl end groups and polymerization occurs subsequently by reaction of these groups between themselves (see Scheme 3).

General Aspects

$$nA-R-B \rightarrow A-(R)_n-B \quad \text{(a)}$$

$$nA-R_1-A + mB-R_2-B \rightarrow A-(R_1R_2)_x-B,\ A-(R_1R_2)_yR_1-A,\ B-(R_2R_1)_zR_2-B \quad \text{(b)}$$

$$nA-R-A \rightarrow A-(R)_n-A \quad \text{(c)}$$

Scheme 2

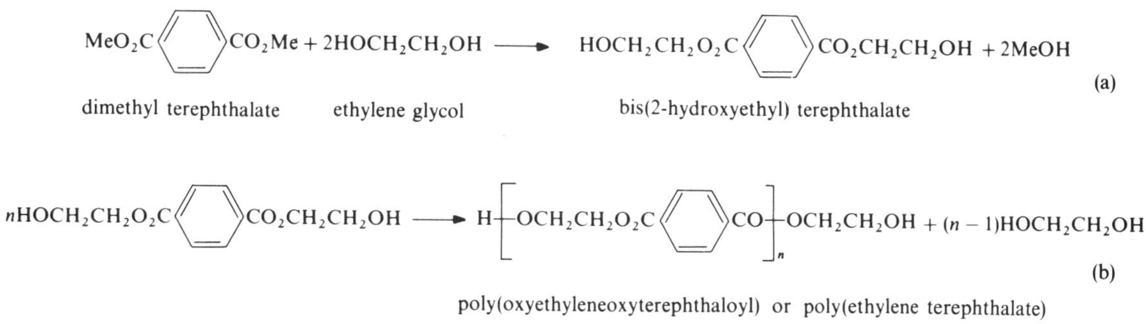

$$\text{MeO}_2\text{C}-\text{C}_6\text{H}_4-\text{CO}_2\text{Me} + 2\text{HOCH}_2\text{CH}_2\text{OH} \longrightarrow \text{HOCH}_2\text{CH}_2\text{O}_2\text{C}-\text{C}_6\text{H}_4-\text{CO}_2\text{CH}_2\text{CH}_2\text{OH} + 2\text{MeOH} \quad \text{(a)}$$

dimethyl terephthalate   ethylene glycol   bis(2-hydroxyethyl) terephthalate

$$n\text{HOCH}_2\text{CH}_2\text{O}_2\text{C}-\text{C}_6\text{H}_4-\text{CO}_2\text{CH}_2\text{CH}_2\text{OH} \longrightarrow \text{H}\text{–}[\text{OCH}_2\text{CH}_2\text{O}_2\text{C}-\text{C}_6\text{H}_4-\text{CO}]_n\text{–}\text{OCH}_2\text{CH}_2\text{OH} + (n-1)\text{HOCH}_2\text{CH}_2\text{OH} \quad \text{(b)}$$

poly(oxyethyleneoxyterephthaloyl) or poly(ethylene terephthalate)

Scheme 3

In the situation depicted in (a) and (c) of Scheme 2, the molecules present in the system during the polymerization will all be of the same type with regard to the functional groups, namely A–B type in case (a) and A–A type in case (c). In case (b), after some time the polymerization system will contain three types of molecules, namely A–A, B–B and A–B, the relative amounts of which depend on the initial stoichiometry of the system and on the extent of reaction.

The step polymerization reaction is perhaps the most versatile means for the synthesis of various categories of polymers.[3-5] In fact many reactions familiar to organic chemists may be utilized to carry out step polymerizations, and several different conditions may be chosen: sometimes it is possible to work in the melt, in solution or in the solid state at high temperature; in other cases the polymerization can be performed at low temperatures in solution or in inhomogeneous liquid systems[5-7] (see Volume 5, Chapters 10 and 11). In Tables 1 and 2 of Volume 5, Chapter 1, some typical examples of step polymerization reactions are shown.

2.2 CHEMICAL CLASSIFICATION OF REACTION MECHANISMS

For reactants having functionalities of two or higher, many reactions may be suitable for step polymerization; from a practical point of view these may be grouped into three categories, namely those involving carbonyls, nucleophilic substitution and miscellaneous reactions.[3,6] Since in these polymerizations polyfunctional reactants behave like their monofunctional analogs, most of the data available in the literature on the reaction mechanisms have been obtained from studies on monofunctional models. In this section some basic chemical features of these groups of reactions will be considered, referring to Volume 5, Chapters 17–39, for specific treatments of the most important processes of step polymerization.

2.2.1 Reactions Involving Carbonyls

Reactions of this type are employed for the synthesis of a large number of commercial polymers, such as polyesters, polyamides, polyurethanes, polycarbonates, aldehyde-based polymers and others.[4,8,9] The commonly accepted general mechanism for these reactions[10-14] consists of the initial nucleophilic addition of an active hydrogen compound to the electron poor carbonyl carbon atom of the substrate, with the formation of a metastable intermediate that can undergo a subsequent elimination reaction (reaction a of Scheme 4). In some cases, such as in the synthesis of polyurethane, the initial addition to the carbonyl carbon atom is not followed by elimination (reaction b of Scheme 4); in other cases the attacking agent carries no hydrogen atom on Y, an example being the reaction of a phenoxide ion with an acid chloride in the synthesis of polycarbonates (reaction c of Scheme 4).

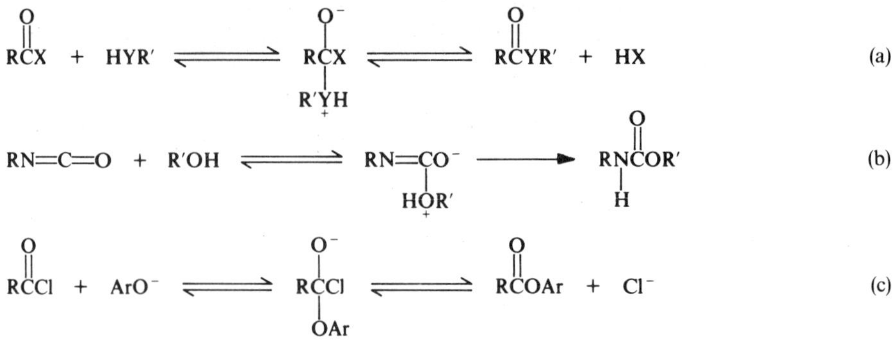

$$\underset{\substack{\|\\O}}{RCX} + HYR' \rightleftharpoons \underset{\substack{|\\R'YH\\+}}{\overset{O^-}{RCX}} \rightleftharpoons \underset{\substack{\|\\O}}{RCYR'} + HX \qquad (a)$$

$$RN{=}C{=}O + R'OH \rightleftharpoons \underset{\substack{|\\HOR'\\+}}{RN{=}CO^-} \longrightarrow \underset{\substack{|\\H}}{\overset{O}{\underset{\|}{RNCOR'}}} \qquad (b)$$

$$\underset{\substack{\|\\O}}{RCCl} + ArO^- \rightleftharpoons \underset{\substack{|\\OAr}}{\overset{O^-}{RCCl}} \rightleftharpoons \underset{\substack{\|\\O}}{RCOAr} + Cl^- \qquad (c)$$

Scheme 4

Typical active hydrogen compounds are alcohols, thiols, amines and phenols, while among carbonyl compounds carboxylic acids, esters, amides, acid halides, isocyanates, urethanes and ureas are the most commonly used. The attacking agent (not necessarily an active hydrogen compound) is generally a Lewis base, which carries an unshared pair of electrons on the Y atom. Hence an increase in its nucleophilic character should facilitate the addition to the electron poor carbonyl carbon atom of the substrate, and should make easier the elimination of X as a negative ion. Also, it is expected that a decrease in electron density at the carbonyl carbon atom, due, for instance, to the presence of electron-withdrawing substituents on R, makes the attack by the Lewis base easier.

In this way some experimental facts can be explained, such as the higher reactivity of aromatic acyl chlorides or isocyanates than that of the corresponding aliphatic compounds, and some effects of substituents in either R or X on the reaction rate,[15-17] in agreement with the Hammett equation relating reactivity to structure in substituted aromatic compounds.[10,11] Steric factors and the polarity of the solvent are also important in controlling the rate of these polymerizations.[14,18-21] Furthermore, the use of catalysts is frequently necessary in order to obtain high molecular weight products, due to their selective activity on the chain growth reactions.

Catalysts may be organic or inorganic acids or bases, or different types of metal compounds. Strong protonic acids (mainly sulfuric acid and aromatic sulfonic acids) or carboxylic organic acids are frequently employed as catalysts. Though the detailed reaction mechanisms may vary, it is suggested that the acids can promote electron withdrawal from the carbonyl carbon atom through coordination to the oxygen atom (Scheme 5).[3]

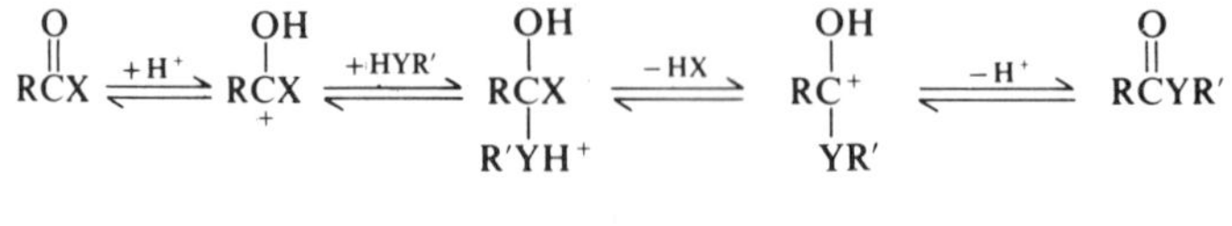

$$\underset{\substack{\|\\O}}{RCX} \underset{}{\overset{+H^+}{\rightleftharpoons}} \underset{+}{\overset{OH}{\underset{|}{RCX}}} \overset{+HYR'}{\rightleftharpoons} \underset{\substack{|\\R'YH^+}}{\overset{OH}{\underset{|}{RCX}}} \overset{-HX}{\rightleftharpoons} \underset{\substack{|\\YR'}}{\overset{OH}{\underset{|}{RC^+}}} \overset{-H^+}{\rightleftharpoons} \underset{\substack{\|\\O}}{RCYR'}$$

Scheme 5

A typical acid-catalyzed step polymerization reaction is the direct esterification of a dicarboxylic acid with a diol: kinetic studies on polyesterification and polyester hydrolysis have shown that in most cases, in the presence of strong protonic acids, the reaction follows an Ingold $A_{AC}2$ mechanism (Scheme 6).[22]

If the catalyst is a carboxylic acid (for instance the monomer itself), such a mechanism may be complicated by the protonation equilibrium of the acid, which is in turn strongly affected by the reaction medium and temperature.

Among the basic catalysts, tertiary amines are frequently employed. It is generally thought that amines act as Lewis bases, forming an initial coordination complex with the carbonyl carbon atom, with subsequent displacement by the active-hydrogen compound;[6] this happens in the synthesis of polyurethanes (Scheme 7).

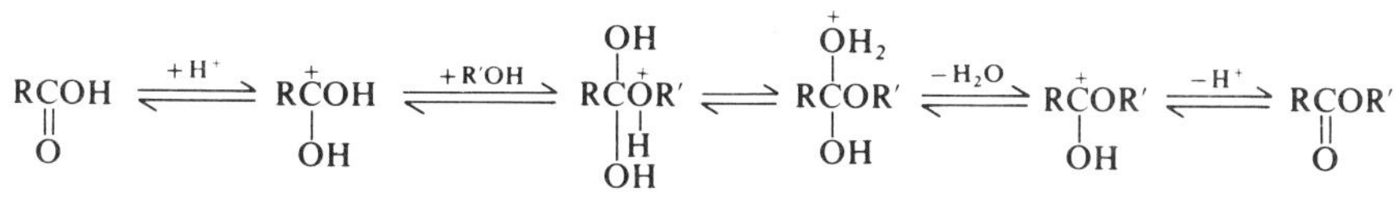

$$RCOH \underset{}{\overset{+H^+}{\rightleftharpoons}} R\overset{+}{C}OH \underset{}{\overset{+R'OH}{\rightleftharpoons}} RCOR' \underset{}{\overset{}{\rightleftharpoons}} R\overset{+}{C}OR' \underset{}{\overset{-H_2O}{\rightleftharpoons}} R\overset{+}{C}OR' \underset{}{\overset{-H^+}{\rightleftharpoons}} RCOR'$$

Scheme 6

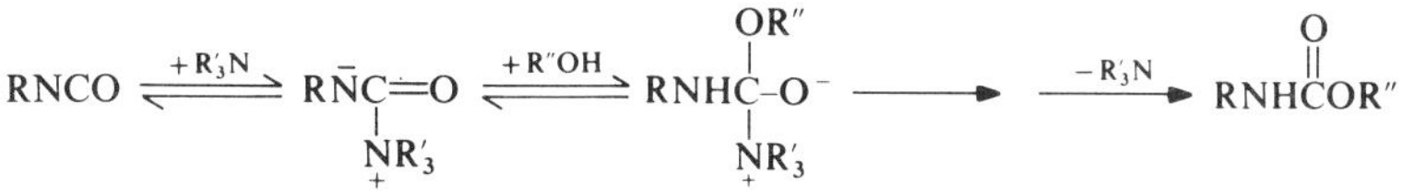

$$RNCO \underset{}{\overset{+R_3'N}{\rightleftharpoons}} R\bar{N}\overset{}{C}=O \underset{}{\overset{+R''OH}{\rightleftharpoons}} RNH\overset{OR''}{\underset{\overset{+}{N}R_3'}{C}}-O^- \longrightarrow \xrightarrow{-R_3'N} RNHCOR''$$

Scheme 7

In the case of the reaction of phenyl isocyanate with methanol, it has been shown that for a series of amine catalysts, in the absence of steric effects, the activity of the amine increases with its base strength.[23,24] The catalytic effect of tertiary amines on the reactions of phenols with acid chlorides or with phosgene and chloroformates, utilized for the synthesis of polyarylates and polycarbonates, has been analogously explained, supposing the initial formation of a complex between the amine and the carbonyl compound.

Metal compounds frequently employed as catalysts are alkali metal and alkaline earth metal acetates, zinc and manganese acetates, antimony, germanium, lead and tin oxides or organometallic compounds, titanium compounds and others. These catalysts, which in many cases show a very powerful effect, may act as Lewis acids. For instance, titanium alkoxides are frequently used to prepare polyesters by the process of direct esterification or of ester interchange.[22] In these cases, according to the literature data on model compounds,[25] the catalytic mechanism may involve the initial formation of an adduct between a carbonyl group and the titanium atom (Scheme 8), and the consequent decrease of the electron density at the carbonyl carbon atom should facilitate the subsequent attack.

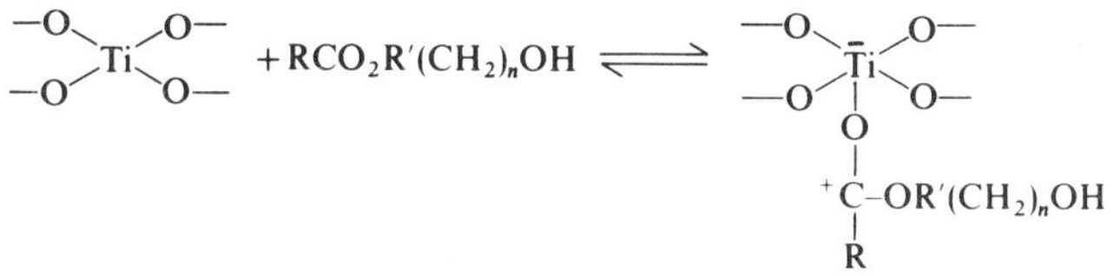

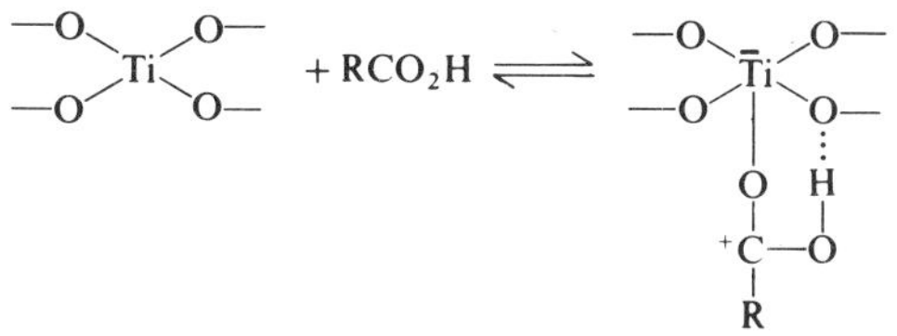

Scheme 8

Catalysts based on organometallic compounds of tin, lead, bismuth and cobalt are frequently employed in the synthesis of polyurethanes; it is generally supposed that the catalytic mechanism requires a coordination of the metal with both the oxygen atoms of the isocyanate and of the alcohol.[6]

There are some step reactions, more precisely polymerizations involving carbonyls, called activated polycondensations, which occur easily in the absence of catalysts. For instance, polyamides can be obtained at room temperature in polar solvents when diesters activated by the presence of heteroatom groups (such as ether or hydroxyl groups) in the α or β position are employed in the synthesis. The acceleration effect has been ascribed to both intra- and intermolecular interactions between ester and amino groups, that may facilitate the approach of an amine to an ester carbonyl group (Scheme 9).[26,27]

Another way to perform an activated synthesis of polyamides and polyesters involves the use of precursors or coreactants such as phosphorylating agents, which are able to produce activated intermediates *in situ* under mild reaction conditions.[28] These intermediates contain P–N and

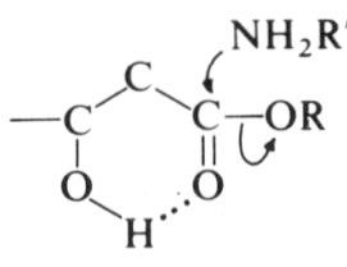

intramolecular interaction

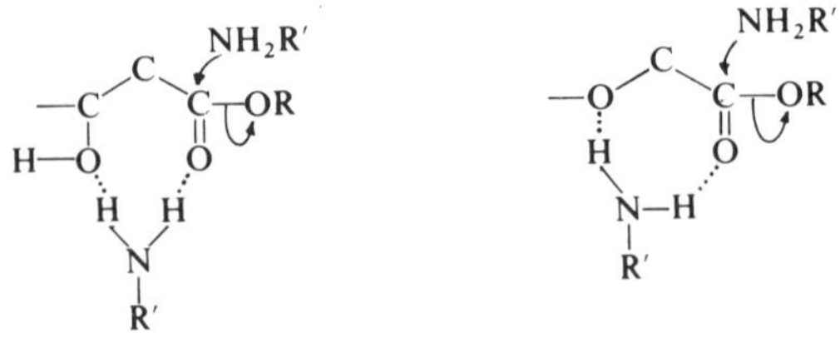

intermolecular interactions

Scheme 9

P–OCO bonds, which are easily susceptible to acidolysis, aminolysis or alcoholysis reactions. For instance, using aryl phosphites or phosphonites and pyridine as coreactants, it is possible to obtain the self-polycondensation of aminocarboxylic acids;[29-32] analogously it has been reported that a complex of diphenylchlorophosphate with lithium bromide in pyridine may promote direct polycondensation between aromatic dicarboxylic acids and diphenols.[33]

2.2.2 Nucleophilic Substitution Reactions

Reactions of this type are employed for the synthesis of polyesters, polysulfides, polysulfones, polysiloxanes, polyamines and of many other macromolecular substances.[8] For instance, polyethers or polysulfones are easily prepared in polar aprotic solvents employing a diphenol such as bisphenol A (in the form of its phenoxide ion) as the nucleophilic agent, and an aryl or aralkyl halide as the substrate of the leaving group (equations 4 and 5).[3,28,34,35]

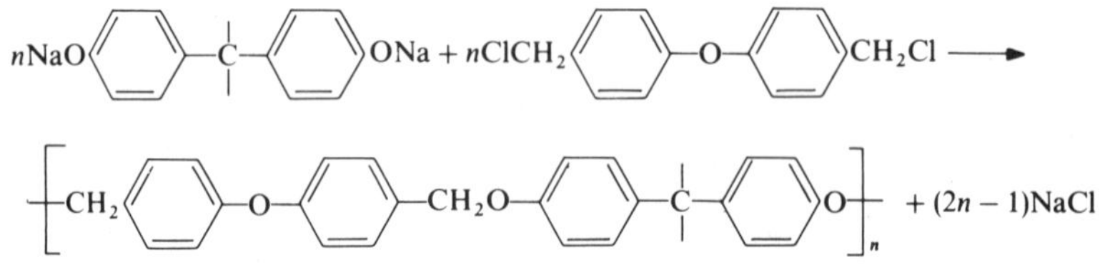

$$n\text{NaO}\langle\text{Ar}\rangle\text{—C—}\langle\text{Ar}\rangle\text{ONa} + n\text{ClCH}_2\langle\text{Ar}\rangle\text{—O—}\langle\text{Ar}\rangle\text{CH}_2\text{Cl} \longrightarrow$$

$$\left[\text{CH}_2\langle\text{Ar}\rangle\text{—O—}\langle\text{Ar}\rangle\text{—CH}_2\text{O—}\langle\text{Ar}\rangle\text{—C—}\langle\text{Ar}\rangle\text{—O}\right]_n + (2n-1)\text{NaCl} \qquad (4)$$

$$n\text{NaO}\langle\text{Ar}\rangle\text{—C—}\langle\text{Ar}\rangle\text{ONa} + n\text{Cl}\langle\text{Ar}\rangle\text{—SO}_2\text{—}\langle\text{Ar}\rangle\text{Cl} \longrightarrow$$

$$\left[\langle\text{Ar}\rangle\text{—SO}_2\text{—}\langle\text{Ar}\rangle\text{—O—}\langle\text{Ar}\rangle\text{—C—}\langle\text{Ar}\rangle\text{—O}\right]_n + (2n-1)\text{NaCl} \qquad (5)$$

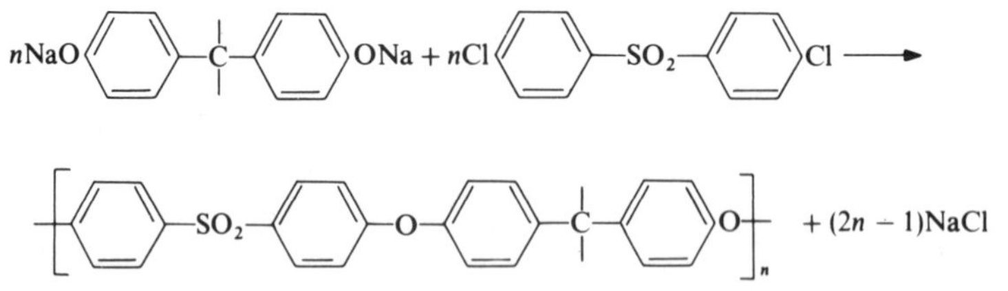

It has been verified that in many cases these reactions follow one or both of the two Ingold mechanisms designated by S_N1 (first-order kinetics) or S_N2 (second-order kinetics).[6,10] The slow rate-controlling step in the reactions of S_N1 type is the halide ionization (reaction a of Scheme 10), so that the rate is proportional to the halide concentration. In the S_N2 type reactions the rate-controlling step is the formation of an intermediate (reaction c of Scheme 10), so that the kinetics are first order with respect to both the reactants. Therefore, the reaction will be facilitated by a relative weakness of the bond between the substrate and the leaving group (generally a halide ion); further, if the mechanism followed is at least partly of the S_N2 type, the nucleophilic character of the attacking

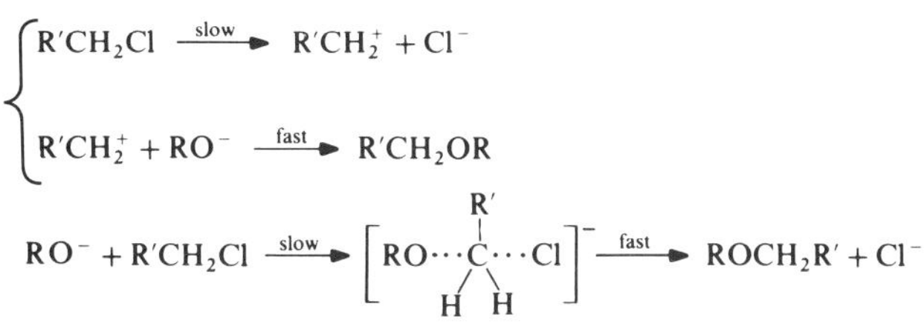

$$\begin{cases} R'CH_2Cl \xrightarrow{slow} R'CH_2^+ + Cl^- & \text{(a)} \\ R'CH_2^+ + RO^- \xrightarrow{fast} R'CH_2OR & \text{(b)} \end{cases}$$

$$RO^- + R'CH_2Cl \xrightarrow{slow} \left[RO \cdots \underset{\underset{H}{|}}{\overset{\overset{R'}{|}}{C}} \cdots Cl \right]^- \xrightarrow{fast} ROCH_2R' + Cl^- \quad \text{(c)}$$

Scheme 10

agent will also be very important. Moreover, other factors such as steric hindrances and the nature of the medium frequently affect the kinetics.

Another example of a nucleophilic substitution reaction of industrial importance is the synthesis of poly(phenylene sulfide)s from *p*-halobenzenethiols (equation 6). In this case the reaction generally follows approximately second-order kinetics; nevertheless, reactive groups in the monomers strongly affect each other, unlike the case for polymer chains.[36] Thus the monomers have a different reactivity if compared to dimers, trimers and higher oligomers.

The polysiloxanes (commonly known as silicones) are another class of important commercial polymers which are prepared by a nucleophilic substitution, the reaction being that between an organohalosilane and water (equation 7).

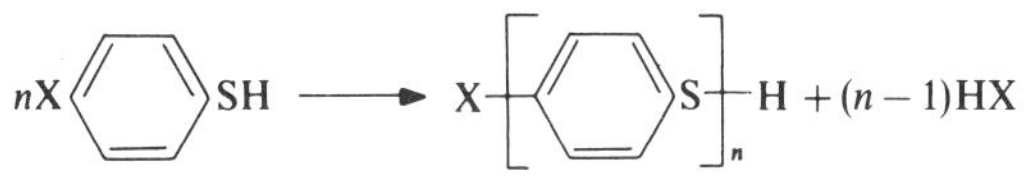

$$nX\text{-}C_6H_4\text{-}SH \longrightarrow X\text{-}[C_6H_4\text{-}S]_n\text{-}H + (n-1)HX \quad (6)$$

$X = F, Cl, Br, I, \ldots$

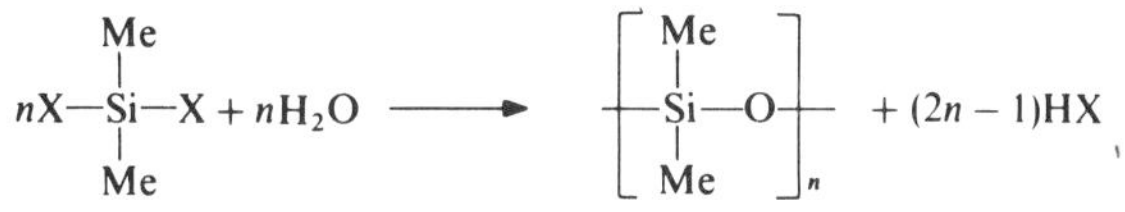

$$nX\text{-}\underset{\underset{Me}{|}}{\overset{\overset{Me}{|}}{Si}}\text{-}X + nH_2O \longrightarrow \left[\underset{\underset{Me}{|}}{\overset{\overset{Me}{|}}{Si}}\text{-}O\right]_n + (2n-1)HX \quad (7)$$

2.2.3 Miscellaneous Reactions

Many electrophilic substitution reactions fall into this class, such as the Friedel–Crafts reactions.[28] An instance is the synthesis of polybenzyls by the self-condensation of benzyl chloride in the presence of Lewis acids as catalysts.[37–41] This reaction is usually complicated by the multi-functional nature of the monomer, because the first and subsequent substitution steps enhance the reactivity of the benzene ring in both *ortho* and *para* positions; consequently the polymer formed is in general highly branched (equation 8).

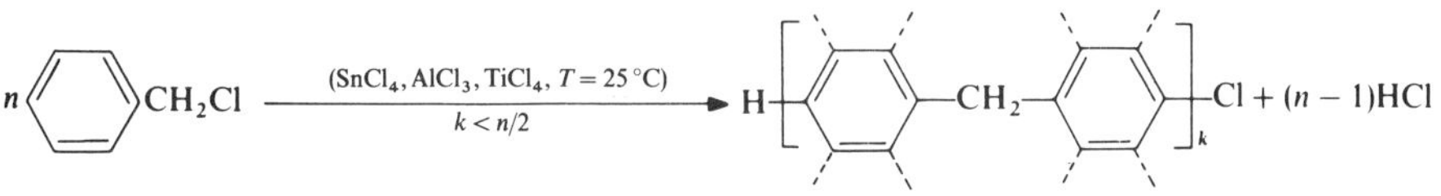

$$n\,C_6H_5CH_2Cl \xrightarrow[k < n/2]{(SnCl_4, AlCl_3, TiCl_4, T=25\,°C)} H\text{-}[C_6H_4\text{-}CH_2]_k\text{-}Cl + (n-1)HCl \quad (8)$$

If the reaction is conducted starting from α-methylbenzene chloride as the monomer and at low temperature, the electrophilic attack occurs only at the *para* position: the polymer obtained is a linear poly(*p*-phenyleneethylidene) (equation 9).[42]

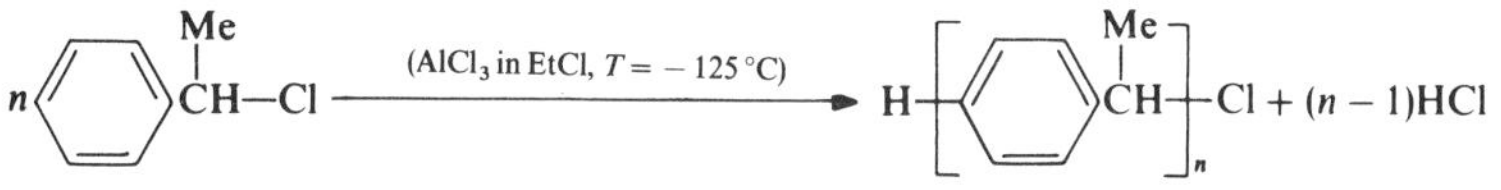

$$n\,C_6H_5\overset{\overset{Me}{|}}{C}H\text{-}Cl \xrightarrow{(AlCl_3 \text{ in EtCl}, T = -125\,°C)} H\text{-}[C_6H_4\text{-}\overset{\overset{Me}{|}}{C}H]_n\text{-}Cl + (n-1)HCl \quad (9)$$

Aromatic polyketones and poly(ether ketone)s also can be prepared by Friedel–Crafts type acylation reactions between activated substrates, such as diphenyl ether or biphenyl, and dicarboxylic acid chlorides, in the presence of $AlCl_3$ or $AlCl_3$–polyphosphoric acid mixtures (equations 10 and 11).[43–45] These syntheses may also be performed starting from carboxylic acids, if mixtures of

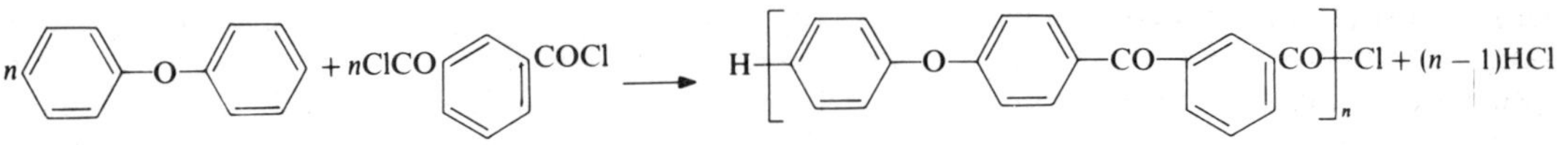

$$\text{(10)} \quad n\,\text{Ph-O-Ph} + n\text{ClCO-Ar-COCl} \longrightarrow \text{H}[\text{-Ph-O-Ph-CO-Ar-CO-}]_n\text{Cl} + (n-1)\text{HCl}$$

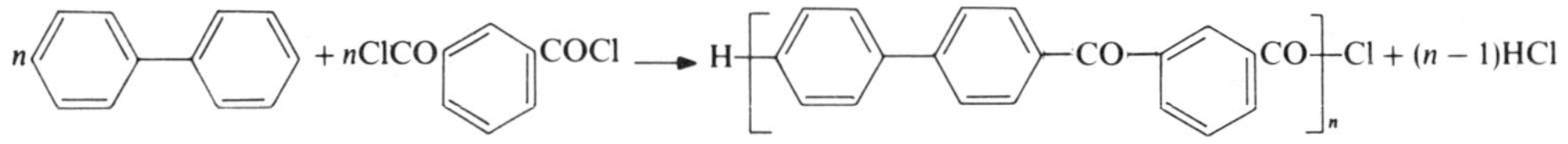

$$\text{(11)} \quad n\,\text{Ph-Ph} + n\text{ClCO-Ar-COCl} \longrightarrow \text{H}[\text{-Ph-Ph-CO-Ar-CO-}]_n\text{Cl} + (n-1)\text{HCl}$$

HF and BF_3 or P_4O_{10} and methanesulfonic acid are used as catalysts (equation 12).[46–48] Similarly, linear poly(aryl sulfone)s may be prepared by Friedel–Crafts type polycondensation of aromatic sulfonyl chlorides with aromatic hydrocarbons in the presence of a strong Lewis acid as catalyst.[49]

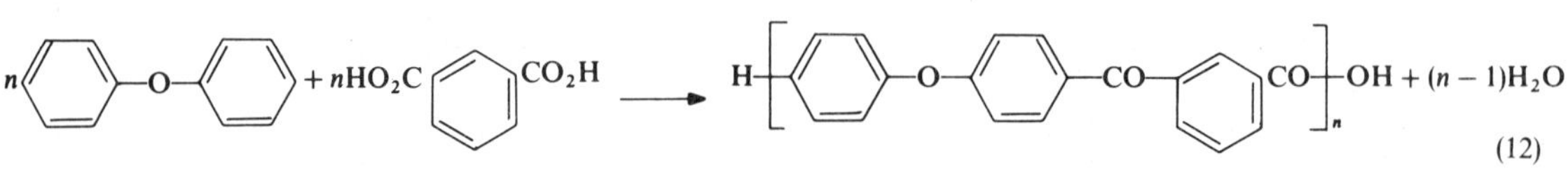

$$\text{(12)} \quad n\,\text{Ph-O-Ph} + n\text{HO}_2\text{C-Ar-CO}_2\text{H} \longrightarrow \text{H}[\text{-Ph-O-Ph-CO-Ar-CO-}]_n\text{OH} + (n-1)\text{H}_2\text{O}$$

A different type of reaction, now a well-established industrial process, is the synthesis of poly(phenylene ether)s by oxidative coupling of phenols. This reaction, which involves a free radical step-growth mechanism, is generally conducted at room temperature in the presence of a complex amine–Cu-based catalyst (equations 13 and 14).[4]

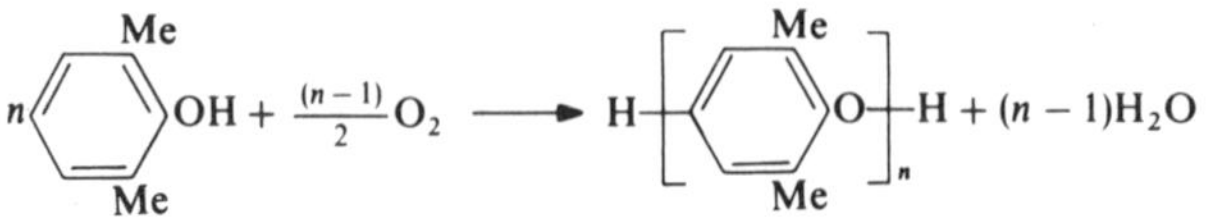

$$\text{(13)} \quad n\,\text{(2,6-Me}_2\text{-C}_6\text{H}_3\text{)OH} + \tfrac{(n-1)}{2}\text{O}_2 \longrightarrow \text{H}[\text{-(2,6-Me}_2\text{-C}_6\text{H}_2\text{)-O-}]_n\text{H} + (n-1)\text{H}_2\text{O}$$

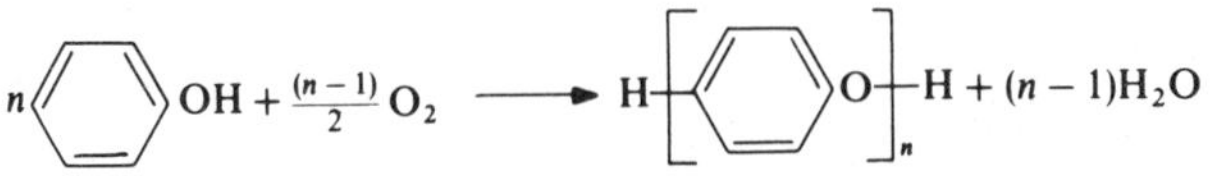

$$\text{(14)} \quad n\,\text{C}_6\text{H}_5\text{OH} + \tfrac{(n-1)}{2}\text{O}_2 \longrightarrow \text{H}[\text{-C}_6\text{H}_4\text{-O-}]_n\text{H} + (n-1)\text{H}_2\text{O}$$

Several other step polymerizations based on unusual radical coupling reactions have been reported in the literature;[3] for instance, the polyrecombination (at high temperature and in the presence of peroxides) of p-diisopropylbenzene (equation 15)[50,51] and the oxidation of benzene-1,4-dithiol by Fe^{3+} compounds to poly(p-phenylene disulfide) (equation 16).[52]

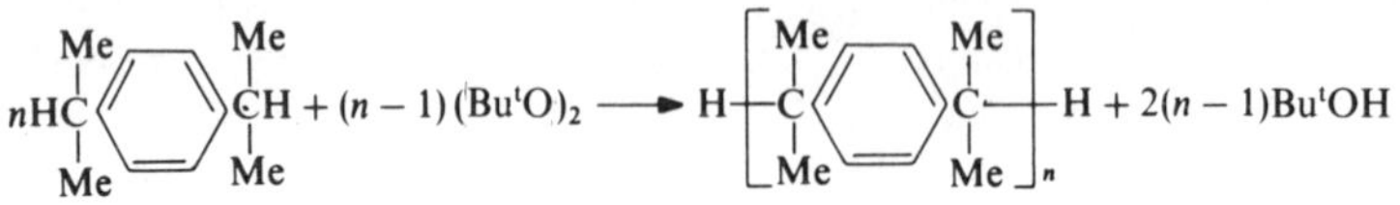

$$\text{(15)} \quad n\,\text{Me}_2\text{HC-C}_6\text{H}_4\text{-CHMe}_2 + (n-1)(\text{Bu}^t\text{O})_2 \longrightarrow \text{H}[\text{-CMe}_2\text{-C}_6\text{H}_4\text{-CMe}_2\text{-}]_n\text{H} + 2(n-1)\text{Bu}^t\text{OH}$$

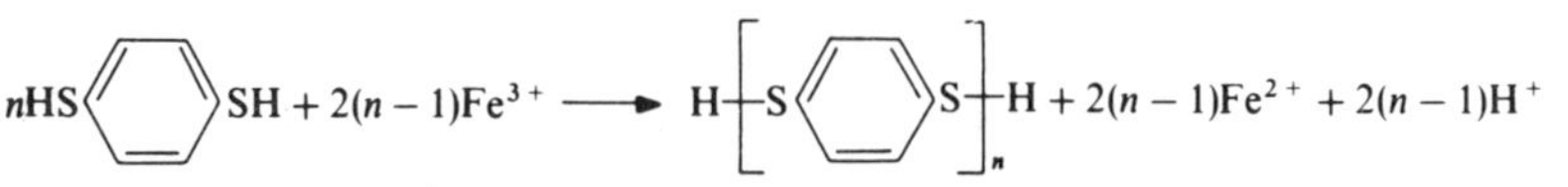

$$\text{(16)} \quad n\,\text{HS-C}_6\text{H}_4\text{-SH} + 2(n-1)\text{Fe}^{3+} \longrightarrow \text{H}[\text{-S-C}_6\text{H}_4\text{-S-}]_n\text{H} + 2(n-1)\text{Fe}^{2+} + 2(n-1)\text{H}^+$$

2.3 STEP POLYMERIZATION OF BIFUNCTIONAL REACTANTS

Although, as shown above, a very large number of chemical reactions may in principle be suitable for step polymerization, in practice not all the reactions known allow the synthesis of high molecular weight linear polymers, owing to several possible hindrances which limit or prevent the growth of the chains.

A first limitation may be the presence of side reactions which give rise to undesirable products with a possible change of the functionality of the system. For instance, as will be seen below, the polymerization may stop at low conversions (and consequently at low molecular weights) if side reactions produce monofunctional substances. Moreover, in many cases monomers or oligomers can undergo cyclization reactions which produce ring compounds, generally formed of a small number (from one to four) of monomeric units lacking in functional groups. The cyclization becomes very important and competitive with linear polymerization, if cyclic structures of high thermodynamic stability with five-, six- and seven-membered rings are formed.[3,5,53,54] This point will be specifically treated in Chapters 5 and 6 of this volume. Another practical feature of most polymerizations is a reaction rate which is usually low at room temperature. For this reason, important processes (such as many industrial syntheses of polyesters and polyamides) are carried out at high temperatures (from 150 °C to about 300 °C). In these conditions it may be necessary to take into consideration not only the unavoidable and irreversible side reactions of polymer thermal degradation, but also the reverse reaction between the polymer and the eventual small molecular weight by-product (for instance the hydrolysis reaction of polyesters), as well as the redistribution reactions between the chains (for instance the ester interchange reaction in polyesters).[53]

Thus the polymerization may become an equilibrium process (the so-called equilibrium step polymerization), and the equilibrium can be displaced to the product side by removing the by-product, if present, so that high conversions and high molecular weights are achieved. Furthermore, as will be seen, the stoichiometry of the system must be carefully assured, maintaining during the process a balance between the concentrations of mutually reacting functional groups.

Conversely, if the rate of polymerization is also high at room temperature, generally the macromolecular chains do not undergo degradation and reverse reactions are absent. Thus the polymerization process has a typical character of irreversibility (the so-called nonequilibrium step polymerization). An example is the interfacial synthesis of polycarbonates from phosgene and bisphenol A (see Volume 5, Chapter 20): the polymerization system is usually biphasic and the mechanism is controlled by the slow diffusion of the reactants and not by the very fast chemical reactions. In these circumstances an exact stoichiometric balance between the concentrations of the functional groups is generally not required in order to achieve high molecular weights.[5,7]

2.3.1 Extent of Reaction and Number-average Degree of Polymerization

In designing and controlling a step-growth polymerization, it is very useful to estimate the number-average degree of polymerization, since this quantity is simply correlated with the extent of reaction at any point of the process, as well as to the initial composition of the system.[5,9,53,54]

As an example, the rather general case of a step polymerization that initially involves a mixture of molecules, each bearing two functional groups of type A (A–A monomer), and of other molecules each bearing functional groups of type B (B–B monomer), A reacting only with B and *vice versa* (case b of Scheme 2) is considered.

If $N_{0,AA}$ and $N_{0,BB}$ are the initial numbers of molecules of A–A and B–B monomer respectively, the ratio $r = N_{0,AA}/N_{0,BB}$ can be taken as the parameter of the stoichiometry of the initial system.

The extent of reaction (or conversion) with respect to functional groups of type A can be defined as

$$p_A = \frac{\text{number of reacted A groups}}{\text{initial total number of A groups}} = \frac{2N_{0,AA} - F_A}{2N_{0,AA}} \tag{17}$$

where F_A is the number of functional groups of type A actually present. Similarly, the extent of reaction with respect to the functional groups of type B can be defined as

$$p_B = \frac{\text{number of reacted B groups}}{\text{initial total number of B groups}} = \frac{2N_{0,BB} - F_B}{2N_{0,BB}} \tag{18}$$

where F_B is the number of functional groups of type B actually present.

As can be easily demonstrated, p_A and p_B are simply related; since at any point during the process the number of A groups reacted must be equal to that of reacted B groups, it follows that

$$2N_{0,AA} - F_A = 2N_{0,BB} - F_B \tag{19}$$

and, combining with equations (17) and (18)

$$2N_{0,AA}p_A = 2N_{0,BB}p_B \tag{20}$$

Introducing the parameter r gives

$$rp_A = p_B \tag{21}$$

Thus, the extent of reaction can be taken as either p_A or p_B, taking into consideration that in general $p_A \neq p_B$ and that $p_A = p_B$ only in the so-called stoichiometric case ($r = 1$). It must also be noted that the extent of reaction is usually defined with respect to functional groups not in excess (A groups if $r < 1$).

The number-average degree of polymerization $\bar{X}_n$ is defined as the number of monomeric units which on the average constitutes a macromolecule; it is therefore defined as

$$\bar{X}_n = \frac{\text{total number of monomeric units}}{\text{actual number of molecules}} \qquad (22)$$

In order to express the denominator of equation (22) in a simple way, the following assumption must be made, which is justified in many circumstances: any secondary reactions are neglected, in particular the intramolecular ones which lead to cyclic molecules without terminal functional groups. With this assumption equation (22) can be written as

$$\bar{X}_n = \frac{N_{0,AA} + N_{0,BB}}{N_{0,AA} + N_{0,BB} - (2N_{0,AA} - F_A)} \qquad (23)$$

where the term in parentheses represents the number of type A functional groups that have reacted and also the decrease in the number of molecules due to the chain growth.

Using equation (17) and introducing the parameter r, equation (23) may be written as

$$\bar{X}_n = \frac{1+r}{1+r-2rp_A} \qquad (24)$$

Similarly $\bar{X}_n$ can be expressed as a function of p_B which gives

$$\bar{X}_n = \frac{1+r}{1+r-2p_B} \qquad (25)$$

It is necessary to point out that the definition of $\bar{X}_n$ refers to the monomeric units, not to the CRU of the polymer, and does not include the by-product molecules eventually eliminated from the reaction mixture. For instance, in the case of polyesterification of dimethyl terephthalate and ethylene glycol (in two steps, see Scheme 3), each repeating unit (CRU) of the polymer has the formula: $-OCH_2CH_2O_2C-C_6H_4-CO-$ and corresponds to the sum of the two monomeric units of the acid and of the glycol.

Moreover, equation (24) holds only when the functional groups disappear from the system through step polymerization, and not through cyclization or other side reactions, or through physical processes such as evaporation or distribution between two liquid phases, as happens in inhomogeneous systems. Conversely, the redistribution reactions between two chains, present in many high temperature step polymerization processes, do not affect $\bar{X}_n$ because they do not change the number of polymer molecules.

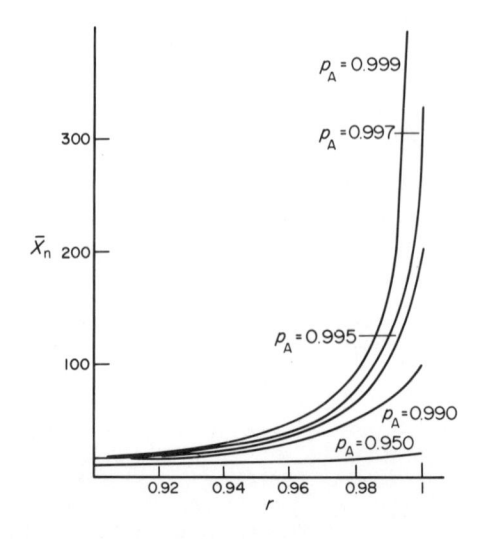

Figure 1 $\bar{X}_n$ vs. r for different values of the conversion p_A

In Figure 1 $\bar{X}_n$ is plotted *vs.* r for different values of p_A. It can be seen that high polymers (for instance with $\bar{X}_n > 100$) can be obtained only if the initial concentrations of A and B type groups are close to stoichiometric ones and the conversion is sufficiently high ($>99\%$).

From the number-average degree of polymerization $\bar{X}_n$, the number-average molecular weight $\bar{M}_n$ can be easily calculated as

$$\bar{M}_n = \bar{M}_{n,0} \bar{X}_n + M_{SMWP} \quad (26)$$

where M_{SMWP} is the molecular weight of that part of the terminal functional group which is not part of the CRU and which through reaction may give rise to a small molecular weight by-product (SMWP); its value is generally negligible for $\bar{X}_n > 10$. $\bar{M}_{n,0}$ represents the number-average molecular weight of the two monomeric units, and can easily be expressed in terms of the initial composition of the system as equation (27), where M_A and M_B are the molecular weights of the monomeric units of type A and of type B respectively. It must be noted that as r usually assumes values rather close to 1 $\bar{M}_{n,0}$ generally has a value not very different from the arithmetic mean of M_A and M_B.

$$\bar{M}_{n,0} = M_A \frac{N_{0,AA}}{N_{0,AA} + N_{0,BB}} + M_B \frac{N_{0,BB}}{N_{0,AA} + N_{0,BB}} \quad (27)$$

Two cases illustrate the application of equation (24). The first is the stoichiometric one; in this case $r = 1$ and equation (24) becomes

$$\bar{X}_n = \frac{1}{1-p} \quad (28)$$

where, of course, $p = p_A = p_B$. This simple relationship, sometimes called Carothers' equation,[55] allows verification that in order to obtain polymers with values of $\bar{X}_n$ (for instance) of 10, 100, 1000 and 10 000, starting from a stoichiometric monomeric mixture, it is necessary to reach values of the extent of reaction of 0.9, 0.99, 0.999 and 0.9999 respectively.

It must be pointed out that equation (28) gives an infinite value of $\bar{X}_n$ at $p = 1$; this result has no physical meaning: in fact, p never succeeds in assuming the value of unity, not only for well-known practical reasons (presence of side and reverse reactions, *etc.*), but also for a theoretical one. Indeed if, overcoming all above difficulties, all functional groups of A and B type had reacted (avoiding cyclizations, as in the above assumption), then one large molecule constituted of all the monomeric units present and terminated with one functional group of type A and one of type B should remain. In this limiting condition the value of p should be very close to unity, but not exactly one, as

$$p = 1 - \frac{1}{2N_{0,AA}} = 1 - \frac{1}{2N_{0,BB}} \quad (29)$$

The corresponding value of $\bar{X}_n$ is very large (of the order of Avogadro's number, as is easily calculated by substituting equation (29) into equation (28), but not infinite.

Equation (28) is very useful because it holds not only if the initial system consists of molecules of a single monomer terminated with A and B type functional groups (A–B monomer, A reacting only with B and *vice versa*), but also for a monomer terminated with two type A functional groups (A–A monomer, A reacting with A), *i.e.* in both cases (a) and (c) of Scheme 2.

The second illustrative case is that of complete reaction of the functional groups not in excess (*i.e* if $r < 1$, the A groups), obviously starting from a system with nonstoichiometric composition. In this case p_A is exactly equal to unity, whilst $p_B = r$ and equation (24) may be written as equation (30), which permits calculation of the maximum attainable degree of polymerization for a fixed non-stoichiometric system. For example, employing an initial 5% molar excess of B–B type monomer ($r = 0.952$), the maximum degree of polymerization is about 41. This clearly shows the effect of limitation of molecular weight due to the nonstoichiometry of the initial mixture.

$$\bar{X}_n = \frac{1+r}{1-r} \quad (30)$$

For step-growth polymerizations many difficulties are frequently met in practice either when reaching very high conversions, or, as in the case with two different reactants, in order to maintain a stoichiometric balance during the reaction. Moreover, in a few cases (for instance, for unsaturated polyesters) it is sufficient to obtain $\bar{X}_n$ values of 10–20, and thereafter the conduction of the reaction is easy: if two different monomers are employed, the stoichiometry may be controlled starting with an excess of the more volatile one, this excess being subsequently removed by distillation during the polymerization.

In other cases (polyesters and polyamides to be used as fibers or films, *etc.*) $\bar{X}_n$ values higher than 200 are frequently required. Thus the polymerization is generally carried out employing a single monomer at gradually increasing temperatures and under vacuum to facilitate the removal of the low molecular weight by-products (if present) and the displacement of the related equilibrium, as will be seen in the following. Another feature that must be considered is the effect on the number-average degree of polymerization of monofunctional compounds, incidentally present as impurities or deliberately added. These, if able to react with terminal groups of the monomers, act as molecular weight regulators.

Consider the rather general system outlined above (initial system composed of A–A and B–B monomers, A reacting only with B and *vice versa*), with $N_{0,AA} = N_{0,BB}$ to which a certain number $N_{0,RB}$ of molecules, each bearing only one functional group of type B, is added; in this case $\bar{X}_n$ can be expressed as

$$\bar{X}_n = \frac{N_{0,AA} + N_{0,BB} + N_{0,RB}}{N_{0,AA} + N_{0,BB} + N_{0,RB} - F_{A,r}} \tag{31}$$

where $F_{A,r}$ is the number of reacted A-functional groups. Introducing the composition parameter $r' = N_{0,AA}/(N_{0,AA} + N_{0,RB})$, equation (31) becomes

$$\bar{X}_n = \frac{1 + r'}{1 + r' - 2p_A r'} \tag{32}$$

It must be noted that equation (32) is formally identical to equation (24) with the substitution of r' for r; therefore the effect of a monofunctional reactant in limiting the number-average degree of polymerization is analogous to that due to an excess of one of the bifunctional reagents in the absence of the monofunctional one. Furthermore, the presence of a monofunctional reactant gives rise to a certain number of polymer chains terminated with no reactive functional groups: this fact may increase the chemical stability of the polymer.

Useful deductions can be made by substituting $p_A = 1$ in equation (32); that is analyzing the dependence of the maximum attainable number-average degree of polymerization on the initial percentage of monofunctional reactant. With this condition equation (32) becomes

$$\bar{X}_n = \frac{2N_{0,AA} + N_{0,RB}}{N_{0,RB}} \tag{33}$$

which, for $N_{0,RB} \ll 2N_{0,AA}$, as happens in practice, gives

$$\bar{X}_n \simeq \frac{2N_{0,AA}}{N_{0,RB}} \tag{34}$$

Equation (34) shows that the maximum attainable value of $\bar{X}_n$ is inversely proportional to the molar fraction of monofunctional reactant, explaining the effect of this latter as a limiting factor on the degree of polymerization.

If the initial system is composed of a monomer of type A–B (A reacting only with B and *vice versa*) plus a monofunctional one of type R–B, $\bar{X}_n$ can be calculated as

$$\bar{X}_n = \frac{N_{0,AB} + N_{0,RB}}{N_{0,AB} + N_{0,RB} - F_{A,r}} \tag{35}$$

where $N_{0,AB}$ is the number of monomeric units contained in the A–B monomers, while $N_{0,RB}$ is the number contained in R–B molecules; they are also the initial number of A–B and of R–B monomer molecules respectively. In this case it is easy to obtain the following expression for $\bar{X}_n$

$$\bar{X}_n = \frac{1 + r''}{1 + r'' - 2p_A r''} \tag{36}$$

where the composition parameter r'' is defined as

$$r'' = \frac{(N_{0,AB}/2)}{(N_{0,AB}/2) + N_{0,RB}} \tag{37}$$

It must be noted that equation (36) is formally identical to equation (32), with the substitution of r'' for r'.

In the simple case of an initial mixture of A–A type bifunctional monomer molecules and of R–A type monofunctional monomer molecules (A reacting with A) $\bar{X}_n$ can be calculated as

$$\bar{X}_n = \frac{N_{0,AA} + N_{0,RA}}{N_{0,AA} + N_{0,RA} - (F_{A,r}/2)} \tag{38}$$

If $N_{0,RA} \ll N_{0,AA}$ and consequently $p_A \simeq F_{A,r}/2N_{0,AA}$, then

$$\bar{X}_n = \frac{1 + r'''}{1 + r''' - 2p_A r'''} \tag{39}$$

where

$$r''' = \frac{(N_{0,AA}/2)}{(N_{0,AA}/2) + N_{0,RA}} \tag{40}$$

Equation (39) is also formally identical to equations (32) and (36), but has been derived in an approximate way.

The validity of the equations relating the degree of polymerization to the composition parameter and the extent of reaction has been experimentally verified in many cases of step polymerizations.[56]

All the equations given above for the calculation of $\bar{X}_n$ in the different cases can be summarized using the method of Carothers,[55] i.e. introducing the 'functionality of the system', defined as the average number of functional groups which can react per molecule of reagent, and defining only one kind of extent of reaction p

$$p = \frac{\text{number of reacted functional groups}}{\text{total initial number of functional groups able to react}} = \frac{2(N_0 - N)}{fN_0} \tag{41}$$

where N_0 and N are the initial and actual total number of molecules respectively and f is the functionality of the system as defined above. From the definition of $\bar{X}_n$ equations (42) and (43) result.

$$p = \frac{2}{f}\left(1 - \frac{1}{\bar{X}_n}\right) \tag{42}$$

$$\bar{X}_n = \frac{2}{2 - pf} \tag{43}$$

Great care must be made in calculating f according to the definition above, especially when nonstoichiometric initial systems are considered. In fact in the case of A–A type monomer reacting with B–B type monomer, with $r < 1$ (B groups in excess), f is expressed by equation (44) as not all the B groups can react.

$$f = \frac{2(N_{0,AA} + N_{0,BB}) - 2(N_{0,BB} - N_{0,AA})}{N_{0,AA} + N_{0,BB}} = \frac{4N_{0,AA}}{N_{0,AA} + N_{0,BB}} = \frac{4r}{r+1} \tag{44}$$

Substituting equation (44) in equation (43), the independently calculated equation (24) for $\bar{X}_n$ is obtained taking p_A as p.

If $N_{0,AA} = N_{0,BB}$, but a monofunctional monomer R–B is initially added, f is

$$f = \frac{2(N_{0,AA} + N_{0,BB}) + N_{0,RB} - N_{0,RB}}{N_{0,AA} + N_{0,BB} + N_{0,RB}} = \frac{4N_{0,AA}}{2N_{0,AA} + N_{0,RB}} = \frac{4r'}{r'+1} \tag{45}$$

which, substituted in equation (43), gives the already demonstrated equation (32). These two examples show the general applicability of Carothers' method based on the functionality of the system.

All the equations developed above have been obtained under the assumption of the absence of side reactions, particularly the intramolecular ones, which latter lead to the disappearance of functional groups without diminishing the number of molecules (thus making equation (23) invalid).

In practice a certain content of cyclic molecules, particularly cyclic oligomers, ranging from one to about ten wt. %, are formed during the polymerization; this content depends on the type of polymer, the reaction conditions and the degree of polymerization. Consequently, as the most common method for the determination of $\bar{X}_n$ is that of calculating it from the concentration of terminal functional groups of the polymer (through titration or spectroscopic methods), it is clear that, if an appreciable content of cyclic molecules is present, the $\bar{X}_n$ value calculated on the basis of the above expressions contains a certain error. Appropriately corrected expressions for $\bar{X}_n$ as well as for the other kinds of average degree of polymerization should then be employed.

2.3.2 Degree of Polymerization in Conditions of Chemical Equilibrium

In many step polymerizations, if the temperature is sufficiently high and/or the time of reaction is rather long, chemical equilibrium can be actually reached. In these cases, in which monomer molecules, polymer chains and other incidental products are in equilibrium, the process is called 'equilibrium' or 'reversible polymerization'.[5,54]

The polymerization reaction, in the case of the presence of two types (A, B) of mutually reacting functional groups, can be schematically represented as

$$-A + B - \overset{K}{\rightleftharpoons} -AB- + \text{SMWP (small molecular weight product, if any)} \qquad (46)$$

If K is the equilibrium constant of the reaction, a mathematical relationship, considering the system as ideal, involving the concentrations of reagents and products may be written as

$$K = \frac{[-AB-][\text{SMWP}]}{[A][B]} \qquad (47)$$

If the initial system is composed of A–A and B–B monomers, $[A]_0 = 2N_{0,AA}/V$ and $[B]_0 = 2N_{0,BB}/V$ are the initial concentrations of A and B groups respectively; in the absence of other kinds of reactions, supposing that the total volume of the system V remains constant during the reaction, equations (48) and (49) may be written, where $[A]_r$ and $[B]_r$ are respectively the number of reacted A and B groups per unit volume.

$$[A] = [A]_0 - [A]_r = (1 - p_A)[A]_0 \qquad (48)$$

$$[B] = [B]_0 - [B]_r = (1 - p_B)[B]_0 \qquad (49)$$

Since, in a closed system

$$[\text{SMWP}] = [-AB-] = p_A[A]_0 = p_B[B]_0 \qquad (50)$$

equation (47) can be written as

$$K = \frac{(p_A[A]_0)^2}{(1 - p_A)(1 - p_B)[A]_0[B]_0} \qquad (51)$$

and finally

$$K = \frac{rp_A^2}{(1 - p_A)(1 - rp_A)} \qquad (52)$$

In the stoichiometric case ($r = 1$, and consequently $p_A = p_B = p$) equation (52) becomes

$$K = \frac{p^2}{(1 - p)^2} \qquad (53)$$

which, solved for p and considering only the solution greater than zero, gives

$$p = \frac{K - K^{0.5}}{K - 1} = \frac{K^{0.5}}{K^{0.5} + 1} \qquad (54)$$

By substitution of equation (54) in equation (28), equation (55) results. This expression is very useful for evaluating the number-average degree of polymerization from the value of the equilibrium constant for a specific reaction.

$$\bar{X}_n = K^{0.5} + 1 \qquad (55)$$

In many cases equilibrium constant values are rather low (in the cases of esterification and transesterification reactions, they are not larger than ten)[6,8,57-60] and the maximum degree of polymerization is consequently very low, as can be seen from Table 1, where as an example K values ranging from 0.1 to 1000 are reported along with the corresponding $\bar{X}_n$ calculated with equation (55). The low values generally attainable for $\bar{X}_n$ consequently result in low values for $\bar{M}_n$; for instance if $\bar{M}_{n,0} = 100$ (see equation 26) and neglecting M_{SMWP}, with $K = 1000$, the corresponding equilibrium value of $\bar{M}_n$ is 3260.

Table 1 Number-average Degree of Polymerization for Different Values of the Equilibrium Constant

K	$\bar{X}_n$
0.1	1.3
1	2.0
10	4.2
50	8.1
100	11.0
500	22.4
1000	32.6

All the above calculations were carried out for a closed system, *i.e.* a system in which no displacement of the equilibrium is promoted by external factors; the results obtained demonstrate that for such a system a high molecular weight can not be reached. Thus the polymerization must be carried out in an open system so that by removing continuously one or more products of the forward reaction, the equilibrium, which is supposed to be always maintained, is driven to higher molecular weights.

In practice, the small molecular weight by-products (water, methanol, *etc.*) are usually removed; in order to obtain good results in this sense, temperature must be taken above the boiling point of the SMWP and/or pressure must be reduced or purging with an inert gas must be carried out. Beyond a certain limit, the removal of SMWP is controlled by its diffusion in the reaction mixture: this is particularly true at high molecular weights, when the viscosity of the system is very high. In these circumstances, the polymerization is diffusion controlled and, as will be seen in a detailed manner in the following discussion, the maximum molecular weight attainable is strictly correlated to the SMWP content in the system.[61]

In the case of an open system, in which equilibrium is maintained, it is useful to write equation (47) in the form

$$K = \frac{p_A [A]_0 [SMWP]}{(1-p_A)(1-p_B)[A]_0 [B]_0} \tag{56}$$

As $[SMWP] = N_{SMWP}/V$ (where N_{SMWP} is the number of molecules of the product maintained in the system) and defining $Y_{SMWP} = N_{SMWP}/(N_{0,AA} + N_{0,BB})$ gives

$$K = \frac{Y_{SMWP} p_A (1+r)}{2(1-p_A)(1-rp_A)} \tag{57}$$

which, in the stoichiometric case, reduces to

$$K = Y_{SMWP} \frac{p}{(1-p)^2} \tag{58}$$

Rearranging equation (58) with equation (28) gives

$$\bar{X}_n = \frac{1}{2} + \left(\frac{1}{4} + \frac{K}{Y_{SMWP}}\right)^{0.5} \tag{59}$$

which for $K/Y_{SMWP} \gg 1/4$ reduces to

$$\bar{X}_n \simeq \left(\frac{K}{Y_{SMWP}}\right)^{0.5} \qquad (60)$$

From this equation, obtained in the stoichiometric case, it can be seen that when $K = 10$, for instance, Y_{SMWP} must be maintained at less than 10^{-3} to reach values of $\bar{X}_n > 100$, and lower than 2.5×10^{-4} if an $\bar{X}_n$ value of 200 must be reached.

2.4 KINETICS OF STEP POLYMERIZATION

2.4.1 Equal Reactivity of Functional Groups

The kinetics of step polymerization proceed with innumerable separate reactions (see Scheme 1) and would be very difficult to analyze in a rigorous way. In the rate equations a large number of different kinetic constants (one for each reaction step) should be introduced, since the frequency of collision between molecules at a given temperature is normally inversely proportional to the square of their mass. In the case of polymerization reactions this would involve considering the reactivity of functional groups depending on the mass of the reactants involved in the reaction step and then introducing innumerable kinetic constants in the rate equations. Fortunately, in many cases the simplifying assumption of the equal reactivity of functional groups of the same kind can be made.[53] This means that a functional group belonging to a bifunctional reactant reacts independently both of the length of the chain to which it is attached and of whether or not the other group on the same molecule has reacted.

This latter assumption needs no commenting on, since it is rather obvious, and experimentally supported, that when the polymer chain is constituted of at least some monomeric units, the reactivity of a functional group is not influenced by the eventual reaction of the group at the other extremity of the polymer chain.

The assumption of the independence of the reactivity of functional groups from the length of the chain is experimentally observed in many cases, mainly in the melt, in concentrated solution and in emulsion polymerization, where it appears that the reaction rate constants are independent from the reaction time and the polymer molecular weight.

As an example taken from the data available in the literature,[62,63] the rate constant values for the esterification reaction of a series of homologous carboxylic acids, which differ from each other only in molecular weight (equation 61), are reported in Table 2. These data, even if referred to nonpolymeric molecules, are very useful in understanding the limits of the above assmption in the case of step polymerization reactions; in fact, the reactivity of carboxylic functional groups in this specific case strongly decreases with molecular weight, but this effect is vanishing for $x > 3$.

$$H(CH_2)_x CO_2H + EtOH \rightarrow H(CH_2)_x CO_2Et + H_2O \qquad (61)$$

Table 2 Rate Constant for Esterification in Homologous Carboxylic Acids

Molecular size (x)	k ($\times 10^4$ l mol^{-1})	Molecular size (x)	k ($\times 10^4$ l mol^{-1})
1	22.1	9	7.4
2	15.3	11	7.6
3	7.5	13	7.5
4	7.5	15	7.7
5	7.4	17	7.7
8	7.5		

Such a result has the following theoretical explanation. The reactivity depends on the collision frequency of the groups and not of the whole molecules, and a terminal functional group has a much greater mobility than the macromolecule to which it is attached, due to the rearrangements that occur in nearby segments of the polymer chain. Furthermore, though the increase of molecular weight may reduce the ease with which reactive ends come together, the large molecule reduces the ease with which reactive ends diffuse away from each other, so that they may collide repeatedly. Therefore, except for very high molecular weights, in the melt polymerization the overall collision rate of functional groups with neighboring groups will be about the same as for small molecules.

In some specific cases, when physical processes control the growth of polymer chains, the above assumption may not be valid and the kinetic treatment must take this into account: as a consequence the overall rate may depend on many parameters, for instance on chain length, diffusion coefficients, *etc.* This is particularly true for a melt polymerization in which the system has a very high viscosity (in general, because of the high molecular weight of the polymer); in this case,[64] as well as when macromolecules have a rod-like structure,[65] the diffusion processes may become the slowest steps and tend to control the polymer growth kinetics.

The same happens in dilute solution polymerization (particularly in a θ solvent[66-68]), when the diffusion of the whole polymer in the solvent is necessary in order to react, or in interfacial polymerization, where the mass transfer processes generally become the rate-determining steps.[7]

2.4.2 Rate of Step Polymerization

Referring to Scheme 1, the rate of any step of polymerization, that is the overall rate of formation of the *i*-product (polymer species constituted by *i* monomeric units), is $R_i = d[i\text{-product}]/dt$. This rate is given by a simple kinetic equation in terms of the kinetic rate constant k and as a function of the concentrations of the two reacting species. Under the assumption of equal reactivity of functional groups of the same kind, the form of the kinetic equation and the value of k will be the same for all reaction steps; thus the overall polymerization rate will be the sum of the rates of each step. The concentrations of the individual species of various size, which are generated and disappear during the polymerization, usually are not measurable; therefore it is necessary to follow the progress of the reaction by evaluating the concentrations of functional groups not yet reacted at each moment. As a consequence, a step polymerization, involving for instance mutually reactive functional groups of A and B type, will be treated kinetically as a process occurring between independent A and B groups, with an overall rate $R = \Sigma_i R_i = -d[A]/dt = -d[B]/dt$, where [A] and [B] represent the overall concentrations of A and B type terminal groups respectively, independent of the length of the chain to which they are attached.

Experimental data show that frequently, but not necessarily always, for a large range of conversions step polymerization reactions in homogeneous systems follow an overall kinetic order of two, which becomes three if a foreign catalyst is present or if one of the reactants itself behaves as a catalyst. In this latter case the rate of polymerization will be

$$R = -\frac{d[A]}{dt} = k[\text{catalyst}][A][B] \tag{62}$$

Kinetic equations of this general form are frequently encountered in step polymerization: the case of the polyesterification can be taken as a very explanatory example.

It is well known that simple esterifications are catalyzed by acids, metal derivatives, or are self-catalyzed reactions;[69] the same behavior is followed by polyesterification reactions. Kinetics and mechanisms of the polyesterification of diacids with diols have been studied extensively in experimental conditions specifically chosen to continuously shift the equilibrium in the direction of polymer formation, either in the presence or in the absence of foreign acids as catalysts.[22]

In Flory's fundamental work[70,71] it was found that the uncatalyzed irreversible reaction, at least in the last stages, shows an overall third-order, with second-order in carboxyl and first-order in hydroxyl group concentration. A simplified mechanism consistent with these results is the common Ingold $A_{AC}2$ mechanism, which involves the autoprotolysis of the carboxylic acid with the formation, in this case, of an ion pair, owing to the low dielectric constant of the medium[22]

$$2\ \text{-}\underset{\text{O}}{\overset{\text{O}}{\|}}\text{C-OH} \overset{K_1}{\rightleftarrows}\ \text{-}\underset{\text{O}}{\overset{\text{O}}{\|}}\text{C-O}^-,\ ^+\text{C} \overset{\text{OH}}{\underset{\text{OH}}{\diagdown}} \tag{63}$$

$$[\text{RCO}_2^-,\ \text{R'C(OH)}_2^+]$$

ion pair

followed by the attack of the alcohol (which represents the rate-determining step), and the elimination of water (Scheme 11).

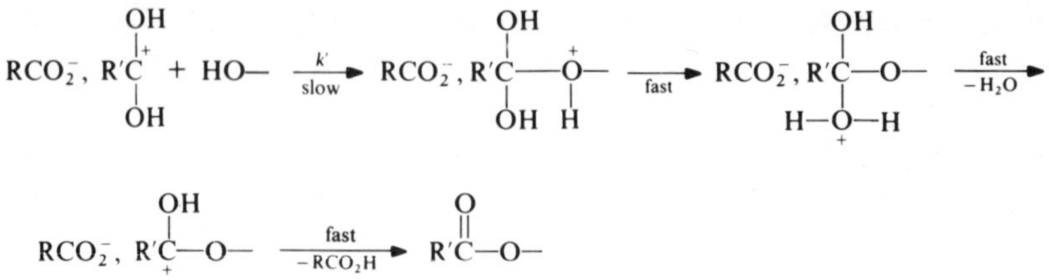

$$\text{RCO}_2^-, \text{R'C}(\text{OH})_2^+ + \text{HO}- \xrightarrow[\text{slow}]{k'} \text{RCO}_2^-, \text{R'C}\underset{\underset{\text{OH H}}{|}}{\overset{\overset{\text{OH}}{|}}{-}}\text{O}- \xrightarrow{\text{fast}} \text{RCO}_2^-, \text{R'C}\underset{\underset{\text{H-O-H}}{|}}{\overset{\overset{\text{OH}}{|}}{-}}\text{O}- \xrightarrow[-\text{H}_2\text{O}]{\text{fast}}$$

$$\text{RCO}_2^-, \text{R'C}\overset{\text{OH}}{\underset{+}{|}}-\text{O}- \xrightarrow[-\text{RCO}_2\text{H}]{\text{fast}} \text{R'C}\overset{\text{O}}{\overset{\|}{-}}\text{O}-$$

Scheme 11

The rate of polymerization will be

$$-\frac{d[\text{CO}_2\text{H}]}{dt} = -\frac{d[\text{OH}]}{dt} = k'[\text{OH}][\text{RCO}_2^-, \text{R'C(OH)}_2^+] \tag{64}$$

and, as it is also

$$K_1 = \frac{[\text{RCO}_2^-, \text{R'C(OH)}_2^+]}{[\text{CO}_2\text{H}]^2} \tag{65}$$

it may be seen that

$$-\frac{d[\text{CO}_2\text{H}]}{dt} = k'K_1[\text{OH}][\text{CO}_2\text{H}]^2 \tag{66}$$

and also

$$-\frac{d[\text{CO}_2\text{H}]}{dt} = k[\text{CO}_2\text{H}]^2[\text{OH}] \tag{67}$$

with $k = k'K_1$. From the integration of equation (67), using the mass balance

$$[\text{CO}_2\text{H}]_0 - [\text{CO}_2\text{H}] = [\text{OH}]_0 - [\text{OH}] \tag{68}$$

there results

$$\frac{1}{([\text{OH}]_0-[\text{CO}_2\text{H}]_0)^2}\left\{\log\left(\frac{[\text{CO}_2\text{H}][\text{OH}]_0}{[\text{CO}_2\text{H}]_0[\text{OH}]}\right) + ([\text{OH}]_0-[\text{CO}_2\text{H}]_0)\left(\frac{1}{[\text{CO}_2\text{H}]} - \frac{1}{[\text{CO}_2\text{H}]_0}\right)\right\} = kt \tag{69}$$

which expressed in terms of p_A (conversion with respect to $-\text{CO}_2\text{H}$ groups) through equations (17) and (21) gives

$$\frac{1}{(1-r)^2}\left\{\log\left(\frac{1-p_A}{r(1-p_A)+1-r}\right) + \left(\frac{1}{r}-1\right)\frac{p_A}{(1-p_A)}\right\} = k[\text{OH}]_0^2 t \tag{70}$$

where $r = [\text{CO}_2\text{H}]_0/[\text{OH}]_0$, or in terms of $\bar{X}_n$ using equation (24)

$$\frac{1}{(1-r)^2}\left\{\log\left(\frac{\bar{X}_n^\infty - \bar{X}_n}{r(\bar{X}_n^\infty + \bar{X}_n)}\right) + \frac{(1+r)(\bar{X}_n - 1)}{r(\bar{X}_n^\infty - \bar{X}_n)}\right\} = k[\text{OH}]_0^2 t \tag{71}$$

where $\bar{X}_n^\infty$ is given by $(1+r)/(1-r)$ and is the limit value of the number-average degree of polymerization.

When $r = 1$ (stoichiometric case), starting from the kinetic equation (67) with $[\text{CO}_2\text{H}] = [\text{OH}]$, it is easy to obtain

$$\frac{1}{2}\left(\frac{1}{[\text{OH}]^2} - \frac{1}{[\text{OH}]_0^2}\right) = kt \tag{72}$$

or in terms of p

$$\frac{1}{2(1-p)^2} - 1 = k[\text{OH}]_0^2 t \tag{73}$$

or in terms of $\bar{X}_n$ (see equation 28)

$$\bar{X}_n^2 = 1 + 2[OH]_0^2 kt \tag{74}$$

The results for $\bar{X}_n$ obtained with equation (74) ($r=1$) and equation (71) for r ranging from 0.95 to 0.99 are represented in Figure 2. It can be seen that beyond a certain time dependent on the specific r value appreciable differences appear in the results obtained with the two equations; therefore, beyond this time, the general equation (71) must be employed. As an example, if $r=0.99$ (1% excess of OH groups only), an $\bar{X}_n$ value of 100 is reached, and the use of equation (74) results in a calculated $\bar{X}_n$ of about 200.

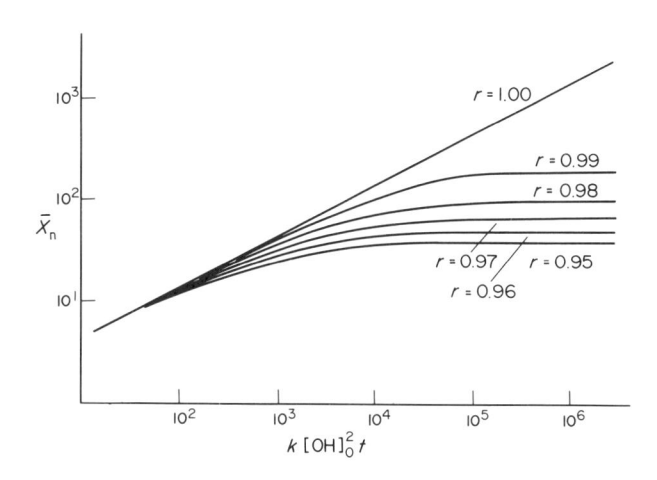

Figure 2 Self-catalyzed polyesterification: $\bar{X}_n$ vs. $k[OH]_0^2 t$ for different values of the stoichiometry parameter r

The data reported by Flory[70,71] show that irreversible stoichiometric polyesterification of adipic acid and diethylene glycol follows a linear plot of $1/(1-p)^2$ vs. time, with a small deviation in the initial region at low extent of reaction ($p<0.80$) and in the later stages ($p>0.93$). Many other investigations into the kinetics of self-catalyzed polyesterifications have been carried out and several different conclusions have been reached regarding the overall reaction order. Even reaction orders which are not constant throughout the course of the reaction have been proposed, and also dependences have been suggested to be, for instance, of a second, a third, or of an intermediate order. Consequently, different reaction mechanisms have been proposed which, for example, take into account the formation of free ions instead of ion pairs by autoprotolysis of the acid. Other proposed mechanisms involve association equilibria between either carboxyl or hydroxyl groups, or the protonation of the alcohol, or the inclusion of a reverse reaction due to the presence of unremoved water; a review by Fradet and Marechal is devoted to a careful examination of the literature concerning this subject.[22]

Many of the above reported discrepancies arise from a failure to appreciate the complexity of the kinetics under conditions usually encountered in polyesterifications. Hamman et al.[72] carried out experiments in order to obtain more accurate kinetic data for the last stages of esterification and polyesterification reactions. Conditions which avoid the loss of reactant by side reactions or volatilization and media based on dilute solutions of a preformed ester or polyester were chosen, in order to minimize changes in polarity of the medium during the reaction. The experiments confirmed that these reactions are second order in carboxyl and first order in hydroxyl group concentration also in the region of $p=0.80$–0.99. Similar conclusions have been reached by Fradet and Marechal, studying esterifications and polyesterifications in bulk or in benzophenone solutions.[73,74]

The deviations from linearity in the plot of $1/(1-p)^2$ vs. time, generally observed in the low conversion region ($p<0.80$), have been ascribed either to a possible failure of the equal reactivity assumption for monomers and oligomers or, prevalently, to the large polarity change that takes place in the reaction medium. Indeed, if solvent is absent, the system is initially a concentrated solution of alcohol and acid, whereas at high conversions it becomes a dilute solution of carboxyl and hydroxyl groups in an ester or polyester. Therefore in the low conversion region the polarity of the medium changes very rapidly and consequently the reaction mechanism or the value of the kinetic constants may change.

Another important effect, which is generally neglected in the kinetic approach, is that linked to the volume variations during the course of polyesterification due to the continuous removal of SMWP, water in this case. In order to take it into account, several authors[75-77] have proposed the use of corrected values of the concentrations and consequently of the extent of reaction p. In a recent paper Helias,[78] for the case of irreversible stoichiometric third-order polycondensations of A–B and A–A plus B–B type and assuming volume additivity, derived equations which predict a nonlinearity of the $1/(1-p)^2$ plot vs. time for low extents of reaction, and a linearity for higher conversions. In contrast to the traditional assumptions, the slope of the linear part does not give the true rate constant; the extent of nonlinearity at small conversions and the deviation of the apparent rate constant value from the true one depend on the initial number of A groups and on the density of the leaving molecule. Furthermore, the range of the nonlinear region increases with decreasing rate constant values, for instance by decreasing temperature, in agreement with experimental observations.

Moreover, a kinetic analysis based on thermodynamic activities instead of concentrations would require the knowledge of the activity coefficients, which in general vary with the composition of the system: only in the later stages of the reaction may the solution of carboxyl and hydroxyl groups be considered ideal or at least with constant values of the activity coefficients.

All these remarks take into consideration only the kinetic data obtained at relatively high conversions: this is also of practical significance, because only at high conversions is high molecular weight polymer formed.

Many authors have studied the kinetics of polyesterification in the presence of an externally added strong acid as catalyst, i.e. in reaction conditions feasible to obtain high polymers in relatively short times.[70-74] In many cases it has been found that at a fixed catalyst concentration the reaction is first order in both carboxyl and hydroxyl concentrations. A mechanism consistent with these results involves a proton transfer from the strong acid with the formation of an ion pair (and not free ions).[22]

$$AH + \underset{\substack{\| \\ O}}{-C-OH} \overset{K_2}{\rightleftarrows} A^-, \underset{\substack{| \\ OH}}{-\overset{OH}{\overset{|}{C}^+}} \quad (75)$$

$$[R'(OH)_2^+]$$

The subsequent steps are the same as in the preceding case; therefore the reaction rate will be given by

$$\frac{-d[CO_2H]}{dt} = \frac{-d[OH]}{dt} = k'[OH][A^-, R'(OH)_2^+] \quad (76)$$

and, as it is

$$K_2 = \frac{[A^-, R'(OH)_2^+]}{[AH][CO_2H]} \quad (77)$$

it results that

$$\frac{-d[CO_2H]}{dt} = k'K_2[AH][OH][CO_2H] \quad (78)$$

and finally, with $k = k'K_2[AH]$

$$\frac{-d[CO_2H]}{dt} = k[CO_2H][OH] \quad (79)$$

Equation (79), using the stoichiometric balance given by equation (68), can be integrated to give

$$\frac{1}{[OH]_0 - [CO_2H]_0} \log\left(\frac{[CO_2H]_0[OH]}{[CO_2H][OH]_0}\right) = kt \quad (80)$$

which can be written in terms of p_A as

$$\frac{1}{1-r} \log\left(\frac{r(1-p_A)+1-r}{1-p_A}\right) = k[OH]_0 t \quad (81)$$

General Aspects

Introducing the number-average degree of polymerization $\bar{X}_n$ it may be shown that

$$\bar{X}_n = \bar{X}_n^\infty \frac{1 - r\exp\{-(1-r)[OH]_0 kt\}}{1 + r\exp\{-(1-r)[OH]_0 kt\}} \tag{82}$$

In many cases the stoichiometry of the system is assured and equations (80), (81) and (82) become simpler; in fact putting $[A] = [B]$ in equation (79) and integrating results in

$$\frac{1}{[OH]} - \frac{1}{[OH]_0} = kt \tag{83}$$

or in terms of p

$$\frac{1}{(1-p)} - 1 = [OH]_0 kt \tag{84}$$

or finally, in terms of $\bar{X}_n$

$$\bar{X}_n = 1 + [OH]_0 kt \tag{85}$$

In Figure 3 the values of $\bar{X}_n$ obtained with equation (85) ($r = 1$) and equation (82) for r ranging from 0.95 to 0.99 are reported. Analogously to the uncatalyzed (or self-catalyzed) reaction, it must be noted that after a certain time appreciable differences appear in the results obtained with the two equations.

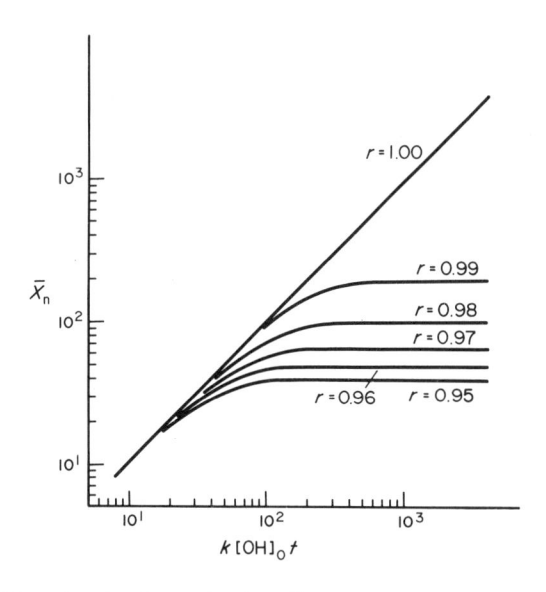

Figure 3 Strong protonic acid-catalyzed polyesterification: $\bar{X}_n$ vs. $k[OH]_0 t$ for different values of the stoichiometry parameter r

It must also be pointed out that rigorously equation (79) is incomplete, because the uncatalyzed reaction, always present, has been neglected: a second-order term of the kind in equation (67) should be incorporated (with a different kinetic constant, of course). However the contribution of the uncatalyzed term to the reaction is often negligible, due to the very high overall rate constant of the catalyzed reaction.

Data from Flory[70,71] show that in the case of the irreversible stoichiometric polymerization of diethylene glycol with adipic acid, the presence of the catalyst (p-toluenesulfonic acid) strongly increases the reaction rate; moreover, the plot of $\bar{X}_n$ or of $1/(1-p)$ vs. time is linear, as predicted by equations (84) and (85), with a small deviation only in the region of low conversions. Similar results have been obtained by many authors for simple esterifications and polyesterification, in the presence of strong protonic catalysts.[22] With few exceptions[76,79,80] the reaction order, when determined, is one in acid, alcohol and catalyst: the mechanisms suggested may involve either ion pairs or free ions.

The small deviations from the pseudo-second-order kinetics in the initial region, which are common either for the cases of simple esterifications or polyesterifications, may be attributed, as in the preceeding case, mainly to changes in the polarity of the medium and to volume variations during the reaction.[22,53,72,78]

All the above results substantially confirm the validity of the assumption of the equal reactivity of functional groups of the same kind, irrespective of the molecular size.

2.4.3 Kinetics of Reversible Polymerization

In the above section the kinetics of polyesterification were considered in a detailed manner, but a complete treatment must include a study of the polymerization which also takes into account possible reverse reactions (equation 86). Very often the polymerization conditions are such that they make the effect of the reverse reaction negligible (for instance when the eventual equilibrium is driven toward the polymer), but the investigation of this latter case is nevertheless useful.

$$-CO_2H \; + \; HO- \; \underset{k'}{\overset{k}{\rightleftarrows}} \; -CO_2- \; + \; H_2O \tag{86}$$

As above, attention will be focused on the special case of polyesterification, taken as an example of a polycondensation reaction, but the same approach could be used for other types of polymerizations, such as ester interchange or polyamidation reactions.

In the case of external acid-catalyzed polyesterification (see Section 2.3), if k and k' are the overall kinetic constants for the forward and the reverse reactions respectively, the kinetic equation can be written as

$$-\frac{d[CO_2H]}{dt} = k[CO_2H][OH] - k'[H_2O][-CO_2-] \tag{87}$$

where, for the reverse reaction, a first order in both H_2O and $-CO_2-$ concentrations is also assumed.

In the simple stoichiometric case ($r=1$) $[CO_2H]=[OH]$, and as $[-CO_2-]=[H_2O]=[CO_2H]_0-[CO_2H]$, equation (87) can be simply integrated to give, in terms of $\bar{X}_n$, using equations (17) and (28)

$$\frac{K^{0.5}}{2} \log\left(\frac{\bar{X}_n^E - 2 + \bar{X}_n}{\bar{X}_n^E - \bar{X}_n}\right) = k[OH]_0 t \tag{88}$$

where $\bar{X}_n^E$ is the value of the number-average degree of polymerization when chemical equilibrium is

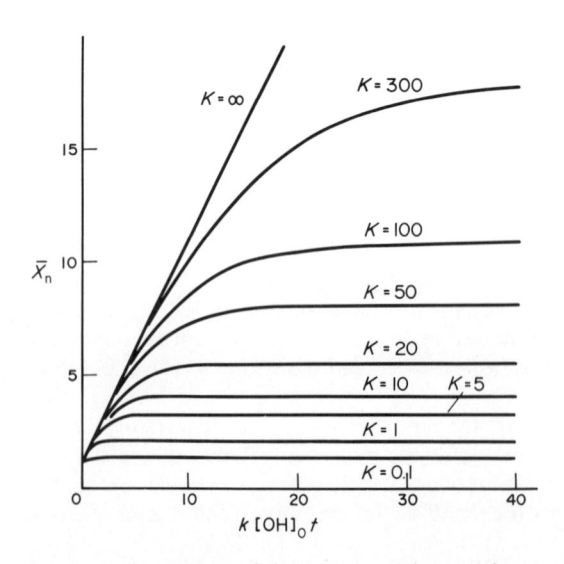

Figure 4 Kinetics of polyesterification in the presence of the reverse reaction: number-average degree of polymerization vs. $k[OH]_0 t$ for various values of the equilibrium constant

reached (equation 55), obviously independent of the reaction kinetics, and $K = k/k'$ is the equilibrium constant of the reaction.

In Figure 4 $\bar{X}_n$ is plotted vs. $k[OH]_0 t$ for K ranging from 0.1 to infinity; even if K for polyesterification reactions is usually no greater than ten, this plot is very useful to visualize the influence of the reverse reaction on the increase of $\bar{X}_n$. Similar behavior can be found for ester interchange and polyamidation reactions.

2.5 REFERENCES

1. IUPAC Commission on Macromolecular Nomenclature, 'Nomenclature of Regular Single-Strand Organic Polymers', IUPAC Information Bulletin 29, November 1972; *Macromolecules*, 1973, **6**, 149.
2. C. E. Sroog, in 'Encyclopedia of Polymer Science and Technology', ed. H. F. Mark, N. G. Gaylord and N. M. Bikales, Interscience, New York, 1969, vol. 11, p. 247.
3. R. W. Lenz, 'Organic Chemistry of Synthetic High Polymers', Interscience, New York, 1967.
4. S. R. Sandler and W. Karo, in 'Organic Chemistry', ed. A. T. Blomquist and H. Wasserman, Academic Press, New York, 1974, vol. 1.
5. L. B. Sokolov, 'Synthesis of Polymers by Polycondensation', translated by J. Schmorak, Israel Program for Scientific Translation, Jerusalem, 1968.
6. J. H. Saunders and F. Dobinson, in 'Comprehensive Chemical Kinetics', ed. C. H. Bamford and C. F. H. Tipper, Elsevier, New York, 1965, vol. 15, chap. 7.
7. P. W. Morgan, 'Condensation Polymers by Interfacial and Solution Methods', Interscience, New York, 1965.
8. H. F. Mark, N. G. Gaylord and N. M. Bikales (eds.), 'Encyclopedia of Polymer Science and Technology', Interscience, New York, 1969, vol. 10.
9. H. G. Elias, 'Macromolecules', Plenum Press, New York, 1977, vol. 2.
10. E. S. Gould, 'Mechanism and Structure in Organic Chemistry', Holt, New York, 1959.
11. L. P. Hammett, 'Physical Organic Chemistry', McGraw Hill, New York, 1940.
12. J. D. Roberts and M. C. Caserio, 'Basic Principles of Organic Chemistry', Benjamin, New York, 1965, p. 530.
13. S. Patai, 'The Chemistry of the Carbonyl Group', Interscience, London, 1966.
14. J. W. Baker and J. B. Holdsworth, *J. Chem. Soc.*, 1947, 713.
15. S. Ephraim, A. E. Woodward and R. B. Mesrobian, *J. Am. Chem. Soc.*, 1958, **80**, 1326.
16. J. H. Saunders and K. C. Frisch, 'Polyurethanes: Chemistry and Technology', Interscience, New York, 1962, Part 1.
17. J. Burkus and C. F. Eckert, *J. Am. Chem. Soc.*, 1958, **80**, 5948.
18. L. C. Case, *J. Chem. Eng. Data*, 1960, **5**, 347.
19. L. C. Case, *J. Appl. Polym. Sci.*, 1964, **8**, 533.
20. L. C. Case and K. W. Li, *J. Appl. Polym. Sci.*, 1964, **8**, 935.
21. M. S. Newman, *J. Am. Chem. Soc.*, 1950, **72**, 4783.
22. A. Fradet and E. Marechal, *Adv. Polym. Sci.*, 1982, **43**, 51.
23. J. W. Baker, M. M. Davies and J. Gaunt, *J. Chem. Soc.*, 1949, 24.
24. J. W. Baker and J. Gaunt, *J. Chem. Soc.*, 1949, 9, 19, 27.
25. F. Pilati, P. Manaresi, B. Fortunato, A. Munari and P. Monari, *Polymer*, 1983, **24**, 1479.
26. N. Ogata, K. Sanui, T. Ohtake and H. Nakamura, *Polym. J.*, 1979, **11**, 827.
27. E. Marechal, *Bull. Soc. Chim. Fr.*, 1987, **4**, 713.
28. I. Goodman, in 'Developments in Polymerization', ed. R. N. Haward, Applied Science, London, 1979, vol. 2, chap. 4.
29. N. Yamazaki, F. Higashi and J. Kawabata, *Makromol. Chem.*, 1974, **175**, 1825.
30. N. Yamazaki, M. Matsumoto and F. Higashi, *J. Polym. Sci., Polym. Chem. Ed.*, 1975, **13**, 1373.
31. N. Yamazaki and F. Higashi, *J. Macromol. Sci., Chem.*, 1975, **9**, 761.
32. C. I. Chiriac and J. K. Stille, *Macromolecules*, 1977, **10**, 710.
33. F. Higashi, A. Hoshio, Y. Yamada and M. Ozawa, *J. Polym. Sci., Polym. Chem. Ed.*, 1985, **23**, 69.
34. R. N. Johnson, A. G. Farnham, R. A. Clendinning, W. F. Hale and C. N. Merriam, *J. Polym. Sci., Part A-1*, 1967, **5**, 2375.
35. J. B. Rose, *Polymer*, 1974, **15**, 456.
36. R. W. Lenz, C. E. Handlovits and H. A. Smith, *J. Polym. Sci.*, 1962, **58**, 351.
37. H. C. Haas, D. I. Livingston and M. Saunders, *J. Polym. Sci.*, 1955, **15**, 503.
38. N. Grassie and G. I. Meldrum, *Eur. Polym. J.*, 1971, **7**, 629.
39. H. Lee, D. Stoffey and K. Neville, 'New Linear Polymers', McGraw Hill, New York, 1967.
40. G. Montaudo, P. Finocchiaro, S. Caccamese and F. Bottino, *J. Polym. Sci., Part A-1*, 1970, **8**, 2475.
41. J. Kuo and R. W. Lenz, *J. Polym. Sci., Polym. Chem. Ed.*, 1976, **14**, 2749.
42. J. Kuo and R. W. Lenz, *J. Polym. Sci., Polym. Chem. Ed.*, 1977, **15**, 119.
43. I. Goodman, J. E. McIntyre and W. Russel (ICI Inc.), *Br. Pat.* 971 227 (1964) (*Chem. Abstr.*, 1964, **61**, 14 805a).
44. J. E. McIntyre (ICI Inc.), *Br. Pat.* 994 583 (1965) (*Chem. Abstr.*, 1965, **63**, 5 815h).
45. K. Niume, F. Toda, K. Uno and Y. Iwakura, *J. Polym. Sci., Polym. Lett.*, 1977, **15**, 283.
46. Du Pont de Nemours, *Br. Pat.* 1 102 679 (1968) (*Chem. Abstr.*, 1968, **68**, 87 757c).
47. K. J. Dahl (Raychem Corp.), *Germ. Pat.* 2 206 836 (1972) (*Chem. Abstr.*, 1973, **78**, 98 766m).
48. M. Veda and T. Kano, *Makromol. Chem., Rapid Commun.*, 1985, **5**, 833.
49. H. A. Vogel, *J. Polym. Sci., Part A-1*, 1970, **8**, 2035.
50. A. S. Hay, *J. Org. Chem.*, 1960, **25**, 1276; 1962, **27**, 5320.
51. M. M. Koton, *J. Polym. Sci.*, 1961, **52**, 97.
52. V. C. Parekh and P. C. Gulia, *J. Ind. Chem. Soc.*, 1934, **11**, 95.
53. P. J. Flory, 'Principles of Polymer Chemistry', Cornell University Press, Ithaca, NY, 1953.
54. G. Odian, 'Principles of Polymerization', 2nd eds., Wiley–Interscience, New York, 1981.
55. W. H. Carothers, *Trans. Faraday Soc.*, 1936, **32**, 39.

56. V. V. Korshak, *Pure Appl. Chem.*, 1966, **12**, 101.
57. D. H. Solomon, in 'Step-Growth Polymerizations', ed. D. H. Solomon, Dekker, New York, 1972, chap. 1.
58. D. B. Jacobs and J. Zimmerman, in 'Polymerization Processes', ed. C. E. Schildknecht and I. Skeitst, Wiley–Interscience, New York, 1977, chap. 12.
59. M. Katz, in 'Polymerization Processes', ed. C. E. Schildknecht and I. Skeist, Wiley–Interscience, New York, 1977, chap. 13.
60. H. Sawada, 'Thermodynamics of Polymerization', Dekker, New York, 1976, chap. 6.
61. H. G. Elias, *J. Macromol. Sci., Chem.*, 1978, **12**, 183.
62. B. V. Bhide and J. J. Sudborough, *J. Indian Inst. Sci., Sect. A*, 1925, **8**, 89.
63. L. Rand, B. Thir, S. L. Reegen and K. C. Frisch, *J. Appl. Polym. Sci.*, 1965, **9**, 1787.
64. G. Rafler, A. D. Sparing, B. Otto, K. Stein, and C. Muehlhaus, *Acta Polym.*, 1987, **38**, 161.
65. D. B. Cotts and G. C. Berry, *Macromolecules*, 1981, **14**, 930.
66. S. I. Kuchanov, M. L. Keshtov, P. G. Khalatur, V. A. Vasnev, S. V. Vinogradova and V. V. Korshak, *Makromol. Chem.*, 1983, **184**, 105.
67. E. Turska, A. Dems and B. Borthowska–Barela, *J. Polym. Sci., Polym. Symp.*, 1973, **42**, 419.
68. A. Dems, *Faserforsch. Textiltech.*, 1977, **28**, 595.
69. S. Patai, 'The Chemistry of Carboxylic Acids and Esters', Interscience, London, 1969.
70. P. J. Flory, *J. Am. Chem. Soc.*, 1939, **61**, 3334.
71. P. J. Flory, *J. Am. Chem. Soc.*, 1940, **62**, 2261.
72. S. D. Hamann, D. H. Solomon and J. D. Swift, *J. Macromol. Sci., Chem.*, 1968, **2**, 153.
73. A. Fradet and E. Marechal, *Eur. Polym. J.*, 1978, **14**, 761.
74. A. Fradet and E. Marechal, *J. Polym. Sci., Polym. Chem. Ed.*, 1981, **19**, 2905.
75. E. Szabó-Réthy, *Eur. Polym. J.*, 1971, **7**, 1485.
76. C. C. Lin and K. H. Hsieh, *J. Appl. Polym. Sci.*, 1977, **21**, 2711.
77. A. Fradet and E. Marechal, *Fortschr. Hochpolym. Forsch.*, 1981, **43**, 51.
78. H. G. Elias, *Makromol. Chem.*, 1985, **186**, 847.
79. S. A. Chen and K. C. Wu, *J. Polym. Sci., Polym. Chem. Ed.*, 1982, **20**, 1819.
80. A. F. Shaaban, M. A. Salem and N. N. Messiha, *J. Appl. Polym. Sci.*, 1987, **33**, 2203.

3
Factors Affecting Rate of Polymerization

PIERO MANARESI and ANDREA MUNARI
University of Bologna, Italy

3.1 INTRODUCTION	35
3.2 DIFFERENT INITIAL REACTIVITIES AND SUBSTITUTION EFFECT	37
3.2.1 *Presence of Initial Asymmetry Only*	38
3.2.2 *Presence of Induced Asymmetry Only*	39
3.3 DEPENDENCE ON MOLECULAR SIZE	41
3.3.1 *First Model*	42
3.3.2 *Second Model*	43
3.4 REFERENCES	45

3.1 INTRODUCTION

In Volume 5, Chapter 2, Section 2.4, the simplifying principle of equal reactivity of functional groups of the same kind has been applied in describing the kinetics of step polymerization of bifunctional monomers. The principle requires that, in the case of reactants carrying two functional groups of the same type, the reactivity of each group is equal and independent of whether the other group of the same molecule has reacted or not and independent of the length of the chain to which it is attached. While the assumption of independence of reactivity from the chain size is generally valid (at least for polycondensations carried out in the melt or in concentrated solution, at not too low conversion values and in the absence of physical processes that control the growth of polymer chains), there are many cases of bifunctional reactants (monomers and, less frequently, oligomers) in which the reactivities of the two groups of the same kind are different or interdependent.

Different reactivities may occur in principle with asymmetrical monomers, such as, for instance, 2,4-tolylene diisocyanate (**1**), a reactant frequently employed in the industrial synthesis of polyurethanes,[1] or glycols, such as propylene glycol (**2**).

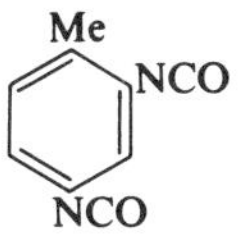

        Me
        ⌬-NCO   MeCHCH$_2$OH
           |
     NCO   OH

      (**1**) 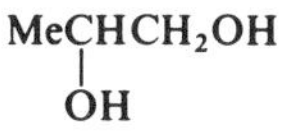   (**2**)

While for aliphatic bifunctional monomers the asymmetry brings about small differences (generally caused by steric factors) in the reactivities of the two groups of the same kind, in the case of aromatic monomers, owing to either electronic or steric effects, these reactivities may be very different.[2-4] This fact is supported, for instance, by comparison between the reactivities with *n*-butanol of the two isocyanate groups of various phenylene and tolylene diisocyanates and with those of phenyl and tolyl isocyanates.[5,6] Data from the literature, reported in Table 1, show that the presence of methyl groups decreases, by electron donation, the reactivity of the isocyanate group; the decrease is greater, perhaps owing to an additional steric effect, when the methyl group is in the *ortho* position. On the other hand, in diisocyanates, the reactivity of the more reactive isocyanate group is greatly increased by electron withdrawal by the other isocyanate group.

The change in reactivity of one functional group upon reaction of the other one in the same molecule is called the 'substitution effect' and, being well known in organic chemistry, is not specific for the case of polymers. This interdependence of reactivities, which usually requires a close proximity of the two groups, almost invariably occurs in aromatic monomers and less frequently in aliphatic ones. For instance, in the above example of diisocyanates (see Table 1), whether symmetrical or not, the reactivity of the second group strongly decreases (by about 10 times) once the first one has reacted, since the urethane group has weaker electron-withdrawing properties than the isocyanate one.[7]

Table 1 Reactivities of Various Mono- and Di-isocyanates in Reaction with *n*-Butanol[5]

Isocyanate reactants	Rate constant[a]	
	First isocyanate group	Second isocyanate group, after reaction of the the first one
Phenyl isocyanate	1	—
o-Tolyl isocyanate	0.15	—
p-Tolyl isocyanate	0.50	—
m-Phenylene diisocyanate	11.5	1.2
p-Phenylene diisocyanate	7.5	0.76
2,4-Tolylene diisocyanate	4.9	0.39
2,6-Tolylene diisocyanate	2.3	0.35

[a] Relative to phenyl isocyanate; at 24.4 °C in toluene solution, with triethylamine as catalyst.

Analogously, for the synthesis of polyether–polysulfones from bis(4-chlorophenyl) sulfone and alkali metal salts of bisphenols (equation 1), an interdependence of the reactivities of the two groups of the same kind in both monomers, of A–A and B–B type, respectively, has been found.[8–10]

$$n \; Cl\text{–}\phi\text{–}SO_2\text{–}\phi\text{–}Cl \; + \; n \; NaO\text{–}\phi\text{–}R\text{–}\phi\text{–}ONa \longrightarrow (2n-1) \; NaCl \; + \; Cl\text{–}(\phi\text{–}SO_2\text{–}\phi\text{–}O\text{–}\phi\text{–}R\text{–}\phi\text{–}O)_n\text{–}Na \quad (1)$$

Substitution effects may also arise for bifunctional reactants with dissimilar functional groups (monomers of A–B type). A well-known example is the synthesis of poly(phenylene sulfide)s from *p*-halobenzenethiols (equation 2).

$$n \; Cl\text{–}\phi\text{–}SH \longrightarrow Cl\text{–}(\phi\text{–}S)_n\text{–}H \; + \; (n-1)HCl \quad (2)$$

It has been demonstrated[11] that the reactivity of the halophenyl groups in this case is greater in the growing polymer chain than in the monomer, owing to the strong electron-donating effect of the –S– group in the monomer, with a consequent decrease in the electron deficiency of the *para* carbon atom. A similar situation occurs during the self-condensation of benzyl chloride to polybenzyl, where the oligomers are more activated than the monomer toward substitution.[12–16]

The substitution effect may also occur in many cases of multifunctional step-polymerization, such as the condensation of formaldehyde with phenol, cresol, urea and melamine.[17–22]

All these effects, being typical 'near order', are generally restricted to monomers, so that there is a change in reactivity only at the beginning of the polymerization. A dependence of reactivity on molecular size may also occur at relatively high values of the degree of polymerization, but in the absence of physical processes controlling the growth of the chains, for instance when step polymerizations are carried out in dilute solutions. It has been proposed by Kuchanov et al.[23] that, in this case, the reactivity of a functional group may be influenced by molecular fragments located relatively far from it along the polymer chain ('far order' effect) but at a small distance in space owing to favourably coiled molecular conformations. Since the effective local concentration of these fragments of the same polymer molecule depends on the degree of polymerization in addition to the flexibility of the chain, the reactivity of a functional group depends, in this case, on the size of polymer chain to which it is attached. Depending on temperature, solvent and nature of the polymeric chains, both a gradual increase or decrease of reactivity with polymer chain size is possible: the dependence law may be complex, as found for the esterification of phenolphthalein oligoterephthalates of various degrees of polymerization with benzoyl chloride under Θ conditions.[23] In this case, the reactivity of the hydroxyl end-groups, due to the solvation effects of the ester groups of the same macromolecule, increases weakly at the beginning of the polymerization, then increases strongly and then increases weakly again with increasing degree of polymerization.

3.2 DIFFERENT INITIAL REACTIVITIES AND SUBSTITUTION EFFECT

Theoretical models of the polycondensation of bifunctional monomers, where the principle of equal reactivity of functional groups of the same kind fails, have received attention, especially in recent years. Several workers[24-27] have studied the reactions where the rate of reaction at a site depends on whether the adjacent site has reacted or not. A mathematical model of the polymerization has also been made,[28] assuming a linear dependence of the rate constant on the degree of polymerization, but such a model cannot be applied directly to real systems.

In order to understand the effect of non-equivalence of functional groups of the same kind on the kinetics of polymerization, it is interesting to take a relatively simple example into consideration:[29-32] the case of a bifunctional monomer of A_1–A_2 type, showing initial and induced asymmetry simultaneously, which can react with a monomer of B–B type, that has neither initial nor induced asymmetry.

The reactions which take place between the various pairs of functional groups can be represented as shown in Scheme 1. The stars on the A_1 and A_2 groups indicate that A_1 and A_2 are at the end of a polymer chain and, for induced asymmetry, have reactivities different from those of A_1 and A_2 groups, respectively, belonging to monomer molecules.

$$A_2A_1 + BB \xrightarrow{k_1} *A_2A_1BB—$$

$$A_1A_2 + BB \xrightarrow{k_2} *A_1A_2BB—$$

$$—A_2A_1^* + BB \xrightarrow{k_1^*} —A_2A_1BB—$$

$$—A_1A_2^* + BB \xrightarrow{k_2^*} —A_1A_2BB—$$

Scheme 1

If $[A_1]$, $[A_2]$, $[A_1^*]$, $[A_2^*]$, $[A]$, $[B]$ and $[M]$ represent the molar concentrations of A_1, A_2, A_1^*, A_2^* functional groups, of total A and B groups and of A_2–A_1 monomer molecules, respectively, under the assumption of overall second-order kinetics for all the reactions reported in Scheme 1, the kinetics equations can be written as

$$-\frac{d[A_1^*]}{dt} = k_1^*[A_1^*][B] - k_2[B][M] \qquad (3)$$

$$-\frac{d[A_2^*]}{dt} = k_2^*[A_2^*][B] - k_1[B][M] \qquad (4)$$

$$-\frac{d[M]}{dt} = (k_1 + k_2)[M][B] \qquad (5)$$

As the material balance is

$$[A]_0 - [A] = [B]_0 - [B] \qquad (6)$$

it results that

$$-\frac{d[B]}{dt} = -\frac{d[A]}{dt} \tag{7}$$

and, as

$$[A] = [A_1^*] + [A_2^*] + 2[M] \tag{8}$$

it results finally that

$$-\frac{d[B]}{dt} = (k_1 + k_2)[M][B] + (k_1^*[A_1^*] + k_2^*[A_2^*])[B] \tag{9}$$

The set of differential equations (3)–(5) and (9) can be solved numerically if values of the four kinetic constants k_1, k_2, k_1^*, k_2^* are available. It is of interest to investigate in detail two special cases in which only one effect is present, initial asymmetry or induced asymmetry.

3.2.1 Presence of Initial Asymmetry Only

In this case

$$k_1 = k_1^* \neq k_2 = k_2^* \tag{10}$$

and equations (3), (4) and (9) reduce to

$$-\frac{d[A_1]}{dt} = k_1[A_1][B] \tag{11}$$

$$-\frac{d[A_2]}{dt} = k_2[A_2][B] \tag{12}$$

$$-\frac{d[A]}{dt} = -\frac{d[B]}{dt} = k_1[A_1][B] + k_2[A_2][B] \tag{13}$$

In the simple stoichiometric case ($[A]_0 = [B]_0$ and consequently $[A_1]_0 = [A_2]_0 = [B]_0/2$); introducing the dimensionless variables

$$\alpha_1 = \frac{[A_1]}{[A_1]_0} \tag{14}$$

$$\alpha_2 = \frac{[A_2]}{[A_2]_0} \tag{15}$$

$$\beta = \frac{[B]}{[B]_0} \tag{16}$$

$$\tau = [A_1]_0 k_2 t \tag{17}$$

and rearranging equations (11)–(13) gives the following differential equation:

$$\frac{d\alpha_1}{d\tau} = -\frac{1}{s}\alpha_1^2(1 + \alpha_1^{(s-1)}) \tag{18}$$

where $s = k_2/k_1$. It can be easily demonstrated that the following equations also hold:

$$\alpha_2 = (\alpha_1)^s \tag{19}$$

$$\beta = (\alpha_1 + \alpha_2)/2 \tag{20}$$

Equation (18) has been numerically evaluated, with the initial condition given in equation (21) and giving appropriate values to the parameter s, using a routine based on the fifth- and sixth-order Runge–Kutta–Verner method (IMSL routine DVERK[39]).

$$\alpha_1(0) = 1 \tag{21}$$

With the aid of equations (19) and (20), the functions, α_1, α_2 and β have been determined. In terms of the number-average degree of polymerization ($\bar{X}_n$), with

$$p = 1 - \frac{[B]}{[B]_0} \tag{22}$$

it results that

$$\bar{X}_n = 1/(1 - p) = [B]_0/[B] = 1/\beta \tag{23}$$

(see also Volume 5, Chapter 2, Section 2.3.1).

The results obtained by integrating equation (18) for s ranging from one to infinity are plotted as $\bar{X}_n$ vs. τ in Figure 1. The results for $0 < s < 1$ have not been reported, since A_1 and A_2 groups have an interchangeable role, due to the symmetry of the reaction system.

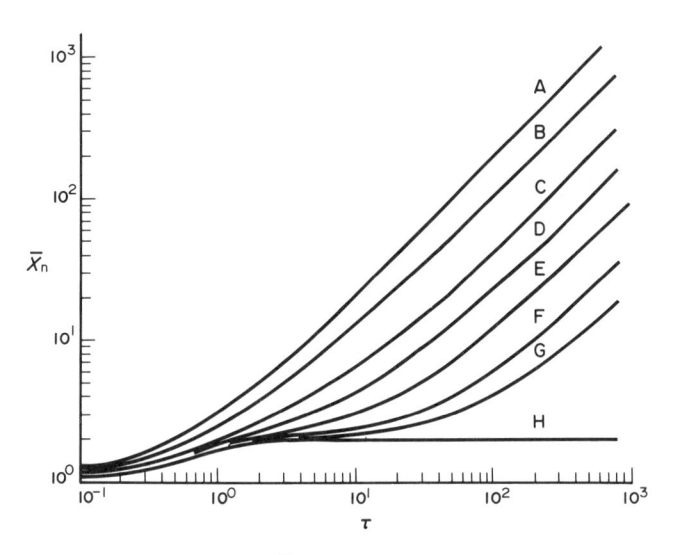

Figure 1 Number-average degree of polymerization $\bar{X}_n$ vs. adimensional time τ for different values of the ratio $s = k_2/k_1$ (case of initial asymmetry): A, 1; B, 2; C, 5; D, 10; E, 20; F, 50; G, 100; H, ∞

As τ contains the kinetic constant k_2, all interpretation of the data in Figure 1 must be for a fixed value of this kinetic constant. Firstly, it can be noted that the curves with $s > 1$ lie under that corresponding to $s = 1$ ($k_2 = k_1$). This is obvious, since $s > 1$ means $k_2 > k_1$ and, for a fixed value of k_2, it means that k_1 is smaller than in the case of $s = 1$ and, consequently, the decrease of [B] (and the increase in $\bar{X}_n$) is slower than in the case of equal reactivities. It must also be noted that, for s equal to infinity, $\bar{X}_n$ tends to the value of two, since $s = \infty$ means $k_1 = 0$. This is the case of monofunctional A-type monomer reacting with B–B monomer. The final situation in the stoichiometric case corresponds to the presence of 'A–AB–B molecules plus an equal number of 'A–AB–BA–A' and B–B molecules, where A' indicates an A group which cannot react since the corresponding kinetic constant (k_1) is zero; in this case obviously $\bar{X}_n = 2$. In Figure 1, an appreciable effect of the different reactivity of A_1 and A_2 groups on the kinetics of polymerization is shown even for $s = 2$: for instance, a period of time twice that for $s = 1$ is required to reach an $\bar{X}_n$ of 100. When $s = 10$ (still a reasonable value), the time taken to reach the same $\bar{X}_n$ is 10 times greater than that for the case of equal reactivity.

3.2.2 Presence of Induced Asymmetry Only

In this case

$$k_1 = k_2 = k \tag{24}$$

$$k_1^* = k_2^* = k^* \tag{25}$$

The general Scheme 1 reduces to Scheme 2 and the kinetic equations (3), (4) and (9) become

$$-\frac{d[A_1^*]}{dt} = k^*[A_1^*][B] - k[A_1][B] \qquad (26)$$

$$-\frac{d[A_1]}{dt} = 2k[A_1][B] \qquad (27)$$

$$-\frac{d[B]}{dt} = k^*[A_1^*][B] + k[A_1][B] \qquad (28)$$

$$A_1A_1 + BB \xrightarrow{k} {}^*A_1A_1BB{-}$$

$$-A_1A_1^* + BB \xrightarrow{k^*} -A_1A_1BB{-}$$

Scheme 2

Then, introducing the dimensionless variables

$$\alpha_1^* = \frac{[A_1^*]}{[A]_0} \qquad (29)$$

$$\alpha_1 = \frac{[A_1]}{[A]_0} \qquad (30)$$

$$\beta = \frac{[B]}{[B]_0} \qquad (31)$$

$$\tau' = k[B]_0 t \qquad (32)$$

equations (26)–(28) become, in the stoichiometric case,

$$\frac{d\alpha_1^*}{d\tau'} = -(s'\alpha_1^*\beta - \alpha_1\beta) \qquad (33)$$

$$\frac{d\alpha_1}{d\tau'} = -2\alpha_1\beta \qquad (34)$$

$$\frac{d\beta}{d\tau'} = -(s'\alpha_1^*\beta + \alpha_1\beta) \qquad (35)$$

where $s' = k^*/k$, with the initial conditions

$$\alpha_1^*(0) = 0 \qquad (36)$$

$$\alpha_1(0) = 1 \qquad (37)$$

$$\beta(0) = 1 \qquad (38)$$

It must be noted that in equation (27) a factor of two appears on the right-hand side; it arises from the fact that when an A_1 group reacts with a B group the other A_1 belonging to the same monomer molecule transforms into an A_1^*, since the monomer molecule becomes part of a polymer chain.

The set of differential equations (33)–(35) has been solved with the same numerical method used for integrating equation (18) (routine DVERK) and the results are plotted in Figure 2 as $\bar{X}_n$ (see equation 23) vs. the dimensionless variable τ' for s' ranging from 0 to 100. It must be noted that, for $s' = 0$, $\bar{X}_n$ tends to the limiting value of two. This corresponds to $k^* = 0$ (unreactive A groups belonging to polymer molecules) and then to the formation of B–BA–A* molecules, together with an equal number of trimers *A–AB–BA–A* and of B–B unreacted monomer molecules in the stoichiometric case. This situation is similar to that of the preceding case at high τ values when s is infinite. As can be seen in Figure 2, a strong effect on the kinetics of polymerization arises at values of s' not very different from one, that is when the reactivities of A groups belonging to a polymer molecule are not too different from those of A groups belonging to monomer molecules. This effect is very evident for $s' < 1$, whilst for $s' > 5$ (due to the fact that, when k^* is sufficiently high, the

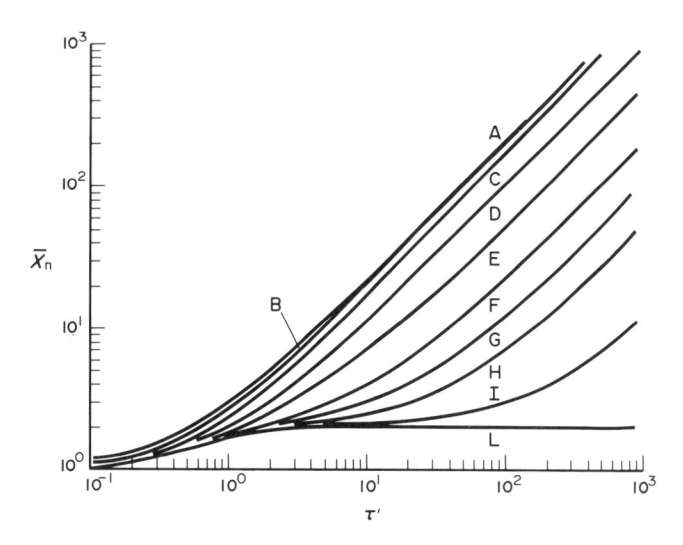

Figure 2 Number-average degree of polymerization $\bar{X}_n$ vs. adimensional time τ' for different values of the ratio $s' = k^*/k$ (case of induced asymmetry): A, 100; B, 5; C, 2; D, 1; E, 0.5; F, 0.2; G, 0.1; H, 0.05; I, 0.01; L, 0.00

polymerization is controlled by the reaction of A functional groups belonging to A–A monomer molecules), no appreciable effect of s' on the increase of $\bar{X}_n$ appears. Other interesting considerations involve the molecular weight distribution, but they will be discussed in Volume 5, Chapter 4.

All of the above discussion involved only rather specific cases; a more general one could be considered (system initially composed of A_1–A_2 and B_1–B_2 monomers, with initial and induced asymmetry in both A and B groups), but a rather complex kinetic scheme requiring several rate constants would need to be introduced. A considerable number of parameters (ratios between kinetic constants) should appear in the kinetic equations in their dimensionless form and so the various contributions of initial and induced asymmetry make a simple analysis of the results difficult. Moreover, in writing the kinetic equations, a well-defined order (overall second-order in cases treated here) has been assumed. As reported in Volume 5, Chapter 2, Section 2.4, this is not true in all cases for all conversions. The treatment of complex cases has therefore been avoided because of the assumptions, not always experimentally confirmed, which must be made in writing the kinetic equations.

3.3 DEPENDENCE ON MOLECULAR SIZE

In general, polymerizations in which the reactivity of the functional groups depends on the length of the chain to which they are attached are difficult to treat from a mathematical point of view: in fact, one kinetic constant, a function, for instance, of the polymer chain length, or several constants, each relating to a well-defined range of sizes of polymer molecules, should be introduced. In a very simple way, the situation can be modelled introducing only two rate constants, one associated with the reaction between monomer and all the species, and one characterizing all the other reactions.[33,34] Another scheme implies one rate constant for the reaction between monomer and monomer, and one for all the other reactions.[35–37] These models work well in some cases, while for others they represent a good first-order approximation.

In the following sections, these two models will be developed, referring to the self-polymerization of an A–R–B monomer (A reacting only with B and *vice versa*), since this case is simple and more illustrative of the problem than that of the polymerization of A–A plus B–B type monomers (this last case has been treated in describing the substitution effect in Section 2.2).

For the self-polycondensation of an A–R–B monomer, the reactions may be represented by

$$P_m + P_n \xrightarrow{k_{p,mn}} P_{m+n} + \text{condensation product}$$

$$m, n = 1, 2, \ldots \ldots \quad (39)$$

where P_m represents a polymer molecule having m repeating units and $k_{p,mn}$ is a rate constant that in general depends on m and on n.

3.3.1 First Model

Let k_{11} be the rate constant associated with the reaction between monomer and all the other species (monomer included), and k_p that associated with the reaction between two molecules different from the monomer. In this case the reactions can be written as Scheme 3*.[33]

$$P_1 + P_1 \xrightarrow{k_{11}/2} P_2 \qquad (a)$$

$$P_1 + P_n \xrightarrow{k_{11}} P_{n+1}, \quad n > 1 \qquad (b)$$

$$P_n + P_m \xrightarrow{k_p} P_{m+n}, \quad m, n > 1, \; m \neq n \qquad (c)$$

$$\xrightarrow{k_p/2} P_{2m}, \quad m > 1, \; m = n \qquad (d)$$

Scheme 3

The kinetic equations, after rearrangement, may be written as[33,34]

$$\frac{d[P_1]}{dt} = -2k_{11}[P_1][P] \qquad (40)$$

$$\frac{d[P_2]}{dt} = k_{11}[P_1]^2 - 2(k_{11} - k_p)[P_1][P_2] - 2k_p[P_2][P] \qquad (41)$$

$$\frac{d[P_n]}{dt} = 2(k_{11} - k_p)[P_1][P_{n-1}] - 2(k_{11} - k_p)[P_1][P_n]$$

$$+ k_p \sum_{1}^{n-1} {}_m[P_m][P_{n-m}] - 2k_p[P_n][P]; \quad n = 3, 4, \ldots \qquad (42)$$

where $[P_j]$, $j = 1, 2, \ldots, n, \ldots$, represents the molar concentration of the individual j-mer and $[P] = \sum_j [P_j]$.

Summing the individual equations over all n values one obtains

$$\frac{d[P]}{dt} = (k_{11} - k_p)[P_1]^2 - k_p[P]^2 - 2(k_{11} - k_p)[P_1][P] \qquad (43)$$

---

* In reactions (a) and (d) of Scheme 3 (as in the subsequent Scheme 5), the kinetic constants k_{11} and k_p (for $m = n$) appear divided by two in order to avoid counting twice the molecular collisions when writing the terms relative to the reactions between molecules of the same size in the kinetic equations. To clarify this point, it is useful to consider two kinds of elementary reactions (Scheme 4). In case (a) of Scheme 4, the initial system is composed of two types of molecules; each of these can be thought of as labelled: $X_1, X_2, \ldots, X_i, \ldots, X_m$ and $Y_1, Y_2, \ldots, Y_j, \ldots, Y_n$. Each X_i molecule can collide with a Y_j molecule, so the possible collisions are $X_1 \to Y_1, X_1 \to Y_2, X_1 \to Y_3, \ldots; X_2 \to Y_1, X_2 \to Y_2, X_2 \to Y_3, \ldots$, and so on. It is clear that the number of collisions $X \to Y$ per unit time and unit volume is proportional to both X and Y concentrations.

$$X + Y \longrightarrow \text{products} \qquad (a)$$

$$X + X \longrightarrow \text{products} \qquad (b)$$

Scheme 4

Reaction (b) of Scheme 4 is substantially different; in fact, the possible collisions of X_1 are $X_1 \to X_2$, $X_1 \to X_3$, $X_1 \to X_4, \ldots$; those of X_2 are $X_2 \to X_1, X_2 \to X_3, X_2 \to X_4, \ldots$; those of X_3 are $X_3 \to X_1, X_3 \to X_2, X_3 \to X_4, \ldots$, and so on. It must be noted that, in this case, each collision between two molecules is counted twice (that between X_1 and X_2 is counted as $X_1 \to X_2$ and $X_2 \to X_1$, that between X_1 and X_3 is counted as $X_1 \to X_3$ and $X_3 \to X_1$, and so on). Therefore, in writing the number of collisions per unit time and unit volume (N) in this case, a factor of $1/2$ needs to be introduced in the constant of proportionality between N and $[X]^2$. This can be seen quantitatively by comparing, for instance, the expressions of N in the two cases for a simple system constituted by molecules of ideal gases.[38]

With regard to Schemes 3 and 5, it must also be emphasized that the constants k_{11}, k_p, $k_{11}/2$ and $k_p/2$ refer to one of the functional groups of a monomer or polymer molecule. Since each molecule brings one A group and one B group, which can react independently, a factor of two multiplying each contribution to the rate of formation of each j-mer must be introduced when writing the kinetic equations referred to the molar concentrations of the various j-mers (which are obviously molecules).[34,37]

Equations (40) and (43), put in adimensional form, have been solved numerically[33] with the initial conditions

$$[P_1] = [P] = [P_1]_0 \quad (44)$$

$$[P_2] = [P_3] = \ldots = 0 \quad (45)$$

The results in terms of $\bar{X}_n$ vs. the adimensional parameter τ'', where

$$\tau'' = 2k_p[P_1]_0 t \quad (46)$$

are shown in Figures 3 and 4 for different values of the ratio $R = k_{11}/k_p$.

Both $\bar{X}_n$ and $\bar{X}'_n$ (this latter calculated excluding the monomer contribution) for the same τ'' value increase with increasing R, assuming an asymptotic behaviour for $R \geq 25$. The spread of the curves is larger for $\bar{X}_n$ than for $\bar{X}'_n$, especially at low R values, due to the monomer content contribution to the number-average degree of polymerization.

3.3.2 Second Model

Let k_{11} be the rate constant associated with the reaction of the monomer with itself, and k_p that associated with the reaction of the monomer with all other species and of all other species with each other. In this case, the reactions can be written as Scheme 5,[35,36] and the kinetic equations are[35,36]

$$\frac{d[P_1]}{dt} = -2(k_{11} - k_p)[P_1]^2 - 2k_p[P_1][P] \quad (47)$$

$$\frac{d[P_2]}{dt} = k_{11}[P_1]^2 - 2k_p[P_2][P] \quad (48)$$

$$\frac{d[P_n]}{dt} = k_p \sum_1^{n-1} {}_m[P_m][P_{n-m}] - 2k_p[P_n][P]; \quad n = 3, 4, \ldots \quad (49)$$

$$\frac{d[P]}{dt} = -(k_{11} - k_p)[P_1]^2 - k_p[P]^2 \quad (50)$$

$$P_1 + P_1 \xrightarrow{k_{11}/2} P_2 \quad \text{(a)}$$

$$P_1 + P_n \xrightarrow{k_p} P_{n+1}, \quad n > 1 \quad \text{(b)}$$

$$P_n + P_m \xrightarrow{k_p} P_{m+n}, \quad m, n > 1, \ m \neq n \quad \text{(c)}$$

$$\xrightarrow{k_p/2} P_{2m}, \quad m > 1, \ m = n \quad \text{(d)}$$

Scheme 5

The set of differential equations, (47) and (50), put in adimensional form, has been solved numerically[35,36] with the initial conditions given in equations (44) and (45); the results in terms of $\bar{X}_n$ vs. the adimensional parameter τ'' (see equation 46) are shown in Figures 5 and 6. Figure 5 shows a smaller spread of curves when compared with the first model, due to the presence of only one special rate constant associated with the monomer, whilst in Figure 6 it can be seen that the curve of $\bar{X}'_n$ is the same for all R values.

For both models it is clear that, beyond a certain value, the influence of R on the rate of polymerization is no longer appreciable; this is due to the fact that high R values mean high k_{11} values (for a fixed k_p) and beyond a certain k_{11} value the polymerization is controlled by the reaction of polymer molecules, and consequently is not influenced by the value of k_{11}. Moreover, a strong influence of R on the increase in $\bar{X}_n$ is shown for R values less than one; with diminishing R the polymerization rate becomes slower, and, especially for the first model, tends to become almost negligible when R is very low.

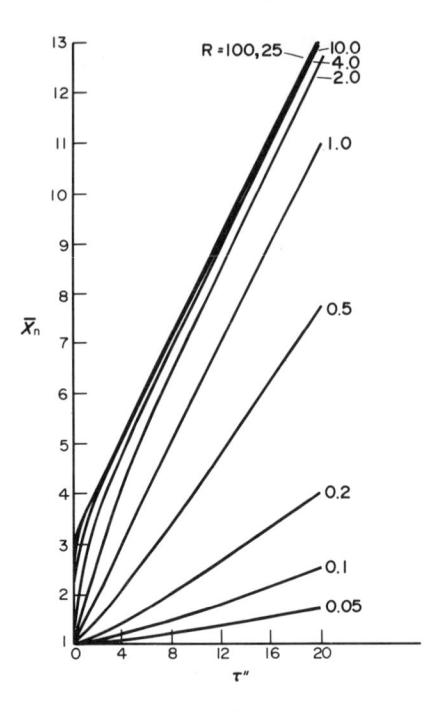

Figure 3 Number-average degree of polymerization $\bar{X}_n$ vs. adimensional time τ'' for different values of the ratio $R = k_{11}/k_p$ (dependence of reactivity on molecular size, first model)[33]

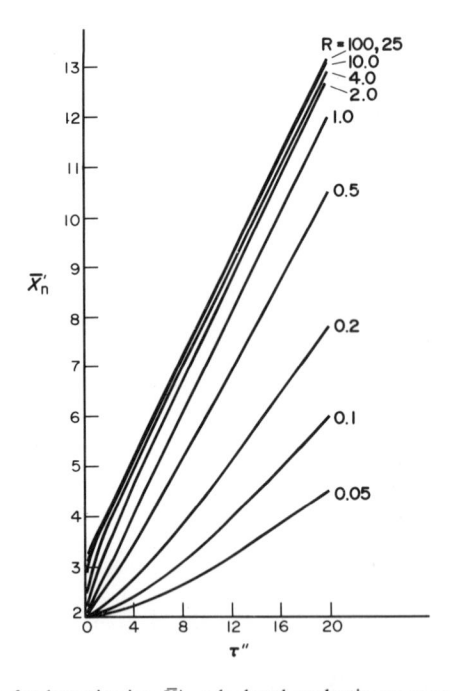

Figure 4 Number-average degree of polymerization $\bar{X}'_n$, calculated neglecting monomer content contribution, vs. adimensional time τ'' for different values of R (dependence of reactivity on molecular size, first model)[33]

The extension of these two models to include the presence of the reverse reaction was carried out by the same authors.[34,37] However, the results obtained will not be discussed here, because the number of parameters required in the kinetic equations hinders a simple discussion. Also, in practice, equilibrium polycondensations are usually carried out in an open driven system (see Volume 5, Chapter 2, Section 2.3.2), in which the equilibrium is reached in a short time; in this case, the kinetics of the process, that should also take into account the removal of the low molecular weight by-product, usually has little importance.

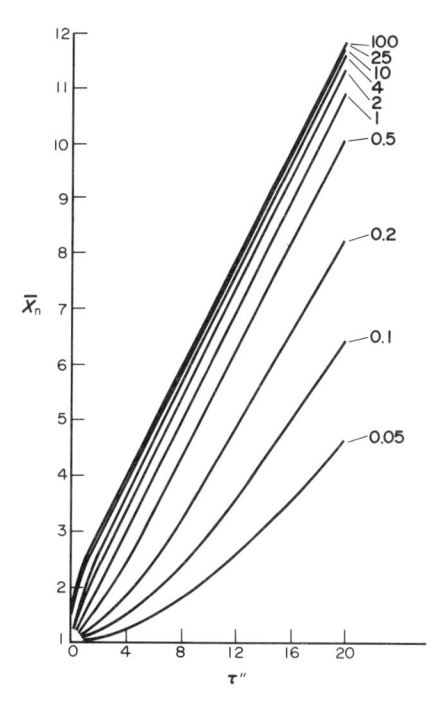

Figure 5 Number-average degree of polymerization $\bar{X}_n$ vs. adimensional time τ'' for different values of R (dependence of reactivity on molecular size, second model, reproduced with permission of Wiley-Interscience from ref. 30)

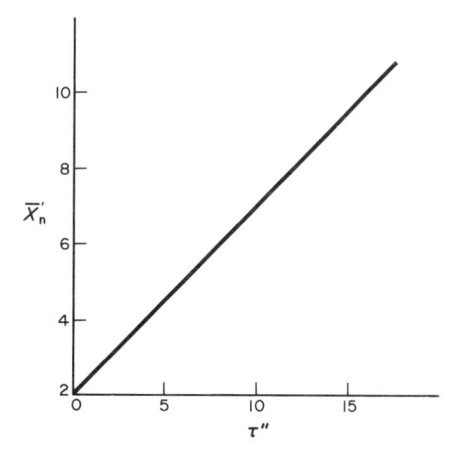

Figure 6 Number-average degree of polymerization $\bar{X}'_n$, calculated neglecting monomer content contribution, vs. adimensional time τ''. The curve is a straight line for all R values (dependence of reactivity on molecular size, second model, from the results of Gupta et al.[36])

3.4 REFERENCES

1. J. H. Saunders and K. C. Frisch, 'Polyurethanes: Chemistry and Technology', Interscience, New York, 1962, part 1.
2. L. B. Sokolov, 'Synthesis of Polymers by Polycondensation', translated by J. Schmorak, Israel Program for Scientific Translation, Jerusalem, 1968.
3. R. W. Lenz, 'Organic Chemistry of Synthetic High Polymers', Interscience, New York, 1967.
4. J. H. Saunders and F. Dobinson, in 'Comprehensive Chemical Kinetics', ed. C. H. Bamford and C. F. H. Tipper, Elsevier, New York, 1965, vol. 15, chap. 7.
5. J. Burkus and C. F. Eckert, *J. Am. Chem. Soc.*, 1958, **80**, 5948.
6. F. Brock, *J. Org. Chem.*, 1959, **24**, 1802.
7. L. H. Peebles, Jr., *Macromolecules*, 1974, **7**, 872.
8. S. R. Schulze and A. L. Baron, *Adv. Chem. Ser.*, 1969, **91**, 692.
9. T. -Y. Yu, S. -K. Fu, S. -J. Li, C. -G. Ji and W. Cheng, *Polymer*, 1984, **25**, 1363.
10. A. Kh. Bulai, V. N. Klyuchnikov, Ya. G. Urman, I. Ya. Slonim, L. M. Bolotina, V. A. Kozhina, M. M. Gol'der, S. G. Kulichikhin, V. P. Beghishev and A. Ya. Malkin, *Polymer*, 1987, **28**, 1349.

11. R. W. Lenz, C. E. Handlovits and H. A. Smith, *J. Polym. Sci.*, 1962, **58**, 351.
12. H. C. Haas, D. I. Livingston and M. Saunders, *J. Polym. Sci.*, 1955, **15**, 503.
13. L. Valentine and R. W. Winter, *J. Chem. Soc.*, 1956, 4768.
14. G. Montaudo, R. Passerini, F. Bottino, S. Caccamese and P. Finocchiaro, *Ann. Chim. (Rome)*, 1967, **57**, 879, 905.
15. D. B. V. Parker, W. G. Davies and K. D. South, *J. Chem. Soc. B*, 1967, 471.
16. G. Montaudo, P. Finocchiaro, S. Caccamese and F. Bottino, *J. Polym. Sci., Part A-1*, 1970, **8**, 2475.
17. A. G. Ryabukhin, *Vysokomol. Soedin., Ser. A*, 1969, **11**, 2562.
18. G. Schiemann and E. Hartmann, *Makromol. Chem.*, 1963, **63**, 174.
19. K. C. Eapen and L. M. Yeddanapalli, *Makromol. Chem.*, 1968, **119**, 4.
20. D. J. Francis and L. M. Yeddanapalli, *Makromol. Chem.*, 1969, **125**, 119.
21. K. Sato, *Bull. Chem. Soc. Jpn.*, 1968, **41**, 7.
22. J. W. Aldersley and M. Gordon, *J. Polym. Sci., Part C*, 1967, **16**, 4567.
23. S. I. Kuchanov, M. L. Keshtov, P. G. Halatur, V. A. Vasnev, S. V. Vinogradova and V. V. Korshak, *Makromol. Chem.*, 1983, **184**, 105.
24. J. B. Keller, *J. Chem. Phys.*, 1962, **37**, 2584.
25. T. Alfrey, Jr. and W. G. Lloyd, *J. Chem. Phys.*, 1963, **38**, 318.
26. C. B. Arends, *J. Chem. Phys.*, 1963, **38**, 322.
27. J. B. Keller, *J. Chem. Phys.*, 1963, **38**, 325.
28. V. S. Nanda and S. C. Jain, *J. Chem. Phys.*, 1968, **49**, 1318.
29. E. Ozizmir and G. Odian, *J. Polym. Sci., Polym. Chem. Ed.*, 1980, **18**, 1089.
30. G. Odian, 'Principles of Polymerization', 2nd edn., Wiley-Interscience, New York, 1981.
31. K. S. Gandhi and S. V. Babu, *AIChE J.*, 1979, **25**, 266.
32. K. S. Gandhi and S. V. Babu, *Macromolecules*, 1980, **13**, 791.
33. S. K. Gupta, A. Kumar and A. Bhargawa, *Eur. Polym. J.*, 1979, **15**, 557.
34. S. K. Gupta, N. L. Agarwalla, P. Rajora and A. Kumar, *J. Polym. Sci., Polym. Phys. Ed.*, 1982, **20**, 933.
35. R. Goel, S. K. Gupta and A. Kumar, *Polymer*, 1977, **18**, 851.
36. S. K. Gupta, A. Kumar and A. Bhargawa, *Polymer*, 1979, **20**, 305.
37. A. Kumar, P. Rajora, N. L. Agarwalla and S. K. Gupta, *Polymer*, 1982, **23**, 222.
38. K. J. Laidler, 'Chemical Kinetics', McGraw-Hill, New York, 1965.
39. T. E. Hull, W. H. Enright and K. R. Jackson, Technical Report No. 100, Department of Computer Science, University of Toronto, October 1976.

4
Molecular Weight Distributions

ANDREA MUNARI and PIERO MANARESI
University of Bologna, Italy

4.1 INTRODUCTION	47
4.2 MOLECULAR WEIGHT DISTRIBUTION IN AN A–B TYPE SELF-POLYMERIZATION	47
4.3 A–A PLUS B–B TYPE POLYMERIZATION	52
4.3.1 Classical Theories	52
4.3.2 A Recent Approach	55
4.3.3 Presence of a Monofunctional Reactant	57
4.4 EXPERIMENTAL VERIFICATIONS	58
4.5 EFFECT OF NON-EQUAL REACTIVITY OF FUNCTIONAL GROUPS OF THE SAME KIND	59
4.6 REFERENCES	62

4.1 INTRODUCTION

In Volume 5, Chapter 2, the expressions relating the number-average degree of polymerization $\bar{X}_n$ to the extent of reaction were derived for different situations. These relations hold, whatever the kinetic mechanism of the process may be, provided that the reactive functional groups disappear from the system only through step polymerization and not through cyclization or other side reactions or through physical processes.

Since during a step polymerization a mixture of polymer molecules of various degrees of polymerization is formed, it is also important to estimate the molecular weight distribution (which, unlike $\bar{X}_n$, may depend on the kinetic mechanism) in order to obtain a more complete picture of the process. The importance of the molecular weight distribution is due to the fact that many polymer properties (melt viscosity, dilute solution viscosity, mechanical characteristics, *etc.*) are deeply influenced by the shape and width of the distribution.

Using a logical procedure, the molecular weight distribution function can be derived from a complete analysis of the kinetic scheme; the solution of the consequent set of differential equations is possible, provided that some approximations may be made.

Since the step-polymerization process includes a very large number of reactions, statistical treatments may also be employed, at least when the probability factors, related to the reactivities of the different functional groups, may be considered constant in the course of the process.

In the following sections, the most important cases of linear step-polymerization will be examined and the molecular weight distribution functions, or some of their parameters, will be derived using either statistical or kinetic methods.

4.2 MOLECULAR WEIGHT DISTRIBUTION IN AN A–B TYPE SELF-POLYMERIZATION

The derivation of the molecular weight distribution in the case of a step polymerization starting from an A–B type monomer (A reacting only with B and *vice versa*), was developed by Flory using a statistical method.[1,2] The calculations performed were based on the assumption of equal reactivity of functional groups of the same kind and the absence of intramolecular reactions or other side reactions or physical processes leading to the disappearance of functional groups.

In this type of step polymerization, molecules of the kind shown in equation (1) are formed.

$$A-B(A-B)_{x-2}A-B \tag{1}$$

The probability of finding a molecule consisting of x monomeric units is that of finding $(x-1)$ reacted A groups and 1 unreacted A group (the same reasoning could equally be applied to B groups). If p is the probability that an A group has reacted (and, therefore, $p = (F_{A,0} - F_A)/F_{A,0}$, which is the ratio between the number of reacted A groups and the initial number of A groups, where F_A and $F_{A,0}$ are the number of A groups actually and initially present, respectively), the probability that $(x-1)$ A groups have reacted to give an x-mer is the product of $(x-1)$ independent probabilities p, and is expressed by $p^{(x-1)}$. It must be noted that the probability p corresponds to the conversion or extent of reaction defined elsewhere (see Volume 5, Chapter 2, Section 2.3).*

In order to obtain the probability of finding a molecule consisting of x monomeric units, it is necessary to multiply $p^{(x-1)}$ by the probability that the remaining A groups have not reacted, i.e. by $(1-p)$. Therefore the probability N'_x of finding an x-mer is[2]

$$N'_x = p^{(x-1)}(1-p) \tag{2}$$

If N is the total number of polymer molecules, the number of molecules N_x having x monomeric units is

$$N_x = Np^{(x-1)}(1-p) \tag{3}$$

N can be then expressed in terms of p and the total number of monomeric units N_0 (which, in this case, is also the initial number of monomer molecules).

$$N = N_0(1-p) \tag{4}$$

Substituting this expression into equation (3), one obtains

$$N_x = N_0 p^{(x-1)}(1-p)^2 \tag{5}$$

Then, neglecting the weight of the two end-groups if a low molecular weight product is eliminated in the polymerization, the weight fraction of x-mer is[2]

$$w_x = xN_x/N_0 \tag{6}$$

and, substituting equation (3) into equation (6) gives

$$w_x = x(1-p)^2 p^{(x-1)} \tag{7}$$

Equations (5) and (7) represent the number- and weight-distribution functions, respectively, for the step polymerization of an A–B monomer at the extent of reaction p. In Figures 1 and 2 plots of the two distribution functions for various values of p are shown. As can be seen, the mole fraction of monomer molecules is greater than that of any polymer species, regardless of the extent of reaction, whereas, on a weight basis, the content of low molecular weight molecules is small and decreases as the extent of reaction increases. Maxima in the weight distribution occur[1] for $x = -(1/\ln p)$ which is near to the value of the number-average degree of polymerization, $\bar{X}_n = 1/(1-p)$ (see Volume 5, Chapter 2, Section 2.3), provided that the conversion p has reached a sufficiently high value.

The distribution function reported above was first obtained by Flory and is usually known as the 'most probable distribution' or the 'Shultz–Flory distribution'. This latter denomination is due to the fact that equation (5) can be considered as a special case of a more general distribution function, called the Shultz distribution, that holds for some chain polymerizations (see Volume 1, Chapter 3 and Volume 3, Chapter 4).

Another way to derive the 'most probable' distribution is through the kinetic method,[3,4] which consists of writing the kinetic equation concerning the formation of each x-mer and then, after solving the set of differential equations so obtained, expressing the number of molecules of each x-mer as a function of the number of monomeric units x.

---

* From a rigorous point of view, p, as defined above, should be named p_A, but in the case of the self-polymerization of an A–B type monomer, since the stoichiometry of the system is assured, $p_A = p_B$ and, therefore, each of the two conversions can equally be used with the common symbol p.

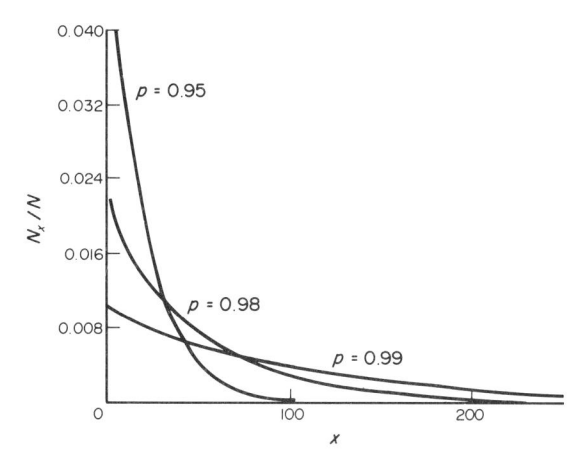

Figure 1 Mole fraction distribution of chain molecules for various extents of reaction p (reproduced from ref. 2 with permission of Cornell University Press)

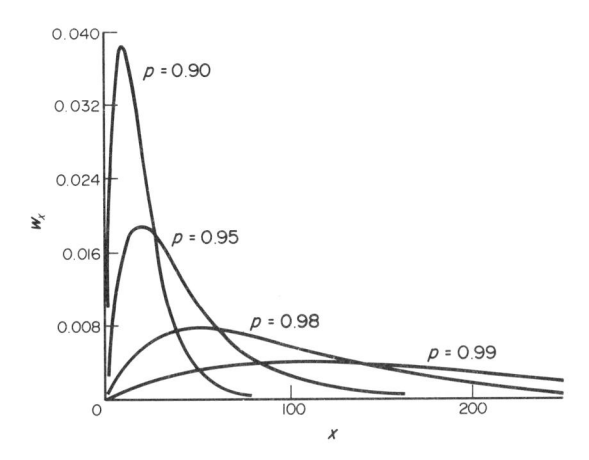

Figure 2 Weight fraction distribution of chain molecules for various extents of reaction p (reproduced from ref. 2 with permission of Cornell University Press)

If P_1 represents a monomer molecule, P_2 a dimer molecule, *etc.*, and $[P_1]$, $[P_2]$, *etc.*, their respective molar concentrations, *i.e.* for the generic x-mer

$$[P_x] = N_x/VL \tag{8}$$

where V is the total system volume and L is Avogadro's number, the total polymer concentration (monomer included) is

$$[P] = \sum_{i}^{\infty} x[P_x] \tag{9}$$

Assuming an overall second-order for all the reactions that each x-mer undergoes, the kinetic equations can be written

$$\frac{d[P_1]}{dt} = -2k[P_1][P] \tag{10}$$

$$\frac{d[P_x]}{dt} = k\sum_{1}^{x-1} j[P_j][P_{x-j}] - 2k[P_x][P] \tag{11}$$

with $x \geq 2$.

It must be noted that the factor 2 that appears in equations (10) and (11) is due to the fact that each x-mer has two functional groups (one of type A and one of type B) which can react independently. This factor is not present in the first term of equation (11), since the summation counts each reaction, except for $x = j$; a factor 0.5 should be included when a species reacts with itself (more details on this subject are given in Volume 5, Chapter 3).

Since it is easy to demonstrate that

$$[P] = [P]_0/\bar{X}_n \tag{12}$$

where

$$[P]_0 = N_0/VL \tag{13}$$

is the initial overall concentration of monomer and polymer molecules ($[P]_0$ and, as at the beginning only monomer molecules are present, coincides with the initial concentration of monomer $[P_1]_0$) and as

$$dt = d\bar{X}_n/k[P]_0 \tag{14}$$

(this result can be easily obtained by rearranging equation (43) or equation (50) in Volume 5, Chapter 3 under the equal reactivity assumption $k_p = k_{11} = k$), then equation (10) becomes

$$-\frac{d[P_1]}{[P_1]} = 2\frac{d\bar{X}_n}{\bar{X}_n} \tag{15}$$

The solution, with the condition that, at $t = 0$, $\bar{X}_n = 1$ and $[P_1] = [P]_0$, is[4]

$$[P_1] = [P]_0/\bar{X}_n^2 \tag{16}$$

Equation (11), with $x = 2$ and introducing equations (12) and (14) into it, becomes

$$\frac{d[P_2]}{d\bar{X}_n} = \frac{[P_1]^2}{[P]_0} - 2\frac{[P_2]}{\bar{X}_n} \tag{17}$$

and, after substituting the expression for $[P_1]$ (equation 16)

$$\frac{d[P_2]}{d\bar{X}_n} + 2\frac{[P_2]}{\bar{X}_n} = \frac{[P]_0}{(\bar{X}_n)^4} \tag{18}$$

Equation (18) is a linear differential equation and can be integrated with the initial condition

$$[P_2]_{\bar{X}_n=1} = 0 \tag{19}$$

giving[4]

$$[P_2] = \frac{[P]_0}{(\bar{X}_n)^2}(1 - 1/\bar{X}_n) \tag{20}$$

Analogously, one finds that the differential equation for each $[P_x]$ is linear, differing from equation (18) only in the term on second member, and integration gives[4]

$$[P_x] = \frac{[P]_0}{(\bar{X}_n)^2}(1 - 1/\bar{X}_n)^{(x-1)} \tag{21}$$

In terms of the extent of reaction (or conversion) p (see Volume 5, Chapter 2)

$$[P_x] = [P]_0(1 - p)^2 p^{(x-1)} \tag{22}$$

Assuming that the total volume V remains constant for the entire course of the reaction, substituting equations (8), (13) and (4) into equation (22), one obtains

$$N_x = N(1 - p)p^{(x-1)} \tag{23}$$

which is identical to equation (3) obtained by Flory with a statistical approach and from which equation (7) can be easily obtained.

Experimental determinations of molecular weight distributions may be obtained in an integral form in which a 'cumulative weight fraction' I_x of all polymer molecules having degrees of polymerization up to x (x included) are expressed as functions of x. This can be done by integration of equation (7) from 1 to x (equation 24). Sometimes, one can find in the literature another, but less correct, expression for I_x that can be derived by integrating equation (7) from 0 to x (equation 25).

$$I_x = 2 - p - [1 + (1 - p)x]p^{(x-1)} \tag{24}$$

$$I_x = \frac{1}{p} - [1 + (1 - p)x]p^{(x-1)} \tag{25}$$

Equations (24) and (25) give practically the same results when x is not too low. Both equations (24) and (25) are valid for values of p that are sufficiently high, since they have been derived using the approximation

$$\ln p \simeq p - 1 \tag{26}$$

Once the distribution functions have been determined, the nth-order moment can be calculated (equation 27).

$$Q_n = \sum_{1}^{\infty} x^n N_x \quad (n = 0, 1, 2, \ldots) \tag{27}$$

By using the ratio between the appropriate pair of moments, all the different kinds of average degree of polymerization can be obtained.

$$\bar{X}_n = Q_1/Q_0 = \sum_{1}^{\infty} x N_x \bigg/ \sum_{1}^{\infty} N_x \tag{28}$$

$$\bar{X}_w = Q_2/Q_1 = \sum_{1}^{\infty} x^2 N_x \bigg/ \sum_{1}^{\infty} x N_x \tag{29}$$

$$\bar{X}_z = Q_3/Q_2 = \sum_{1}^{\infty} x^3 N_x \bigg/ \sum_{1}^{\infty} x^2 N_x \tag{30}$$

$$\bar{X}_{z+1} = Q_4/Q_3 = \sum_{1}^{\infty} x^4 N_x \bigg/ \sum_{1}^{\infty} x^3 N_x \tag{31}$$

Substituting into equations (28), (29), (30) and (31) the expression for N_x obtained in equation (5), one can derive for the 'most probable distribution' the following expressions for the number-, weight-, z- and ($z+1$)-average degrees of polymerization, respectively*

$$\bar{X}_n = 1/(1 - p) \tag{32}$$
$$\bar{X}_w = (1 + p)/(1 - p) \tag{33}$$
$$\bar{X}_z = (1 + 4p + p^2)/(1 - p^2) \tag{34}$$
$$\bar{X}_{z+1} = (1 + 11p + 11p^2 + p^3)/(1 + 4p + p^2)(1 - p) \tag{35}$$

which have been calculated utilizing the following mathematical relationships[5]

$$\sum_{1}^{\infty} p^{(x-1)} = 1/(1 - p) \tag{36}$$

$$\sum_{1}^{\infty} x p^{(x-1)} = 1/(1 - p)^2 \tag{37}$$

$$\sum_{1}^{\infty} x^2 p^{(x-1)} = (1 + p)/(1 - p)^3 \tag{38}$$

---

* It must be noted that the expression (32) of the number-average degree of polymerization can be obtained independently, with only the assumption of the absence of intramolecular reactions. In fact, the definition

$$\bar{X}_n = N_0/N = F_{A,0}/F_A = F_{B,0}/F_B = 1/(1 - p) \tag{42}$$

where F_A and F_B are the number of A- and B-functional groups actually present and $F_{A,0}$ and $F_{B,0}$ are those in the initial system, holds whatever the distribution function may be and even if the principle of equal reactivity of functional groups of the same kind is not valid (see also Volume 5, Chapter 3, Section 3.2)

$$\sum_{x=1}^{\infty} x^3 p^{(x-1)} = (1 + 4p + p^2)/(1 - p)^4 \tag{39}$$

$$\sum_{x=1}^{\infty} x^4 p^{(x-1)} = (1 + 11p + 11p^2 + p^3)/(1 - p)^5 \tag{40}$$

and, in general, if a higher-order moment must be calculated[5]

$$\sum_{x=1}^{\infty} x^n p^{(x-1)} = \frac{d}{dp}\left[p \sum_{x=1}^{\infty} x^{(n-1)} p^{(x-1)} \right] \tag{41}$$

In equations (28)–(31), the expressions for the various kinds of average degree of polymerization have been given as functions of the extent of reaction p when the 'most probable distribution' holds. The importance of these kinds of average lies in the fact that the corresponding average molecular weights, which can be easily derived in the case of the self-polymerization of an A–B type monomer by multiplying the average degree of polymerization by the mass of the monomeric unit M_0, are experimentally measurable and have an important influence on some polymer properties. In fact $\bar{M}_n$, $\bar{M}_w$ and $\bar{M}_z$ can be experimentally determined from measurements of colligative properties, light-scattering analysis and from sedimentation equilibrium measurements in the ultracentrifuge, respectively, whilst $\bar{M}_{(z+1)}$ has been proposed,[6] in addition to $\bar{M}_w$ and $\bar{M}_z$, to be one of the parameters of the molecular weight distribution influencing the viscoelastic properties of polymers.

To visualize the polydispersity of the polymer, i.e. the width of the molecular weight distribution, some ratios between averages can be used. The most common is the polydispersity index (PDI) that is expressed by

$$\text{PDI} = \bar{X}_w/\bar{X}_n = 1 + p \tag{43}$$

whilst another significant ratio is

$$\bar{X}_z/\bar{X}_w = (1 + 4p + p^2)/(1 + p)^2 \tag{44}$$

It must be noted that all kinds of average molecular weight are initially equal and become greater and greater with increasing p; the ratios (43) and (44), which are initially equal to 1, tend to diverge and when p is close to 1 they assume approximately the values*

$$\bar{X}_w/\bar{X}_n \simeq 2 \tag{45}$$

$$\bar{X}_z/\bar{X}_w \simeq 1.5 \tag{46}$$

4.3 A–A PLUS B–B TYPE POLYMERIZATION

4.3.1 Classical Theories

In the A–A plus B–B polymerization (A reacting only with B and *vice versa*), three types of molecular species are present (structures **1**, **2** and **3**) depending on whether the degree of polymerization x is even or odd: all molecules consisting of an even number of monomeric units will

---

* The assertion that $\bar{X}_w/\bar{X}_n \simeq 2$ and $\bar{X}_z/\bar{X}_w \simeq 1.5$ for p close to 1, while being experimentally confirmed when the 'most probable distribution' holds and the molecular weight of the polymer is sufficiently high, has some limits from a theoretical point of view, which will be underlined in the following discussion.

Under the assumption of the absence of intramolecular reactions (and, therefore, of cyclic molecules), in the case of the self polymerization of an A–B monomer, the limiting value of the conversion is $p_{\lim} = (N_0 - 1)/N_0 = 1 - 1/N_0$, which corresponds to the limiting case of the presence of only one very large linear molecule (rings are avoided) terminated with one A- and one B-group. In such a situation, it is clear that all the kinds of average degree of polymerization are identical and both the ratios $\bar{X}_w/\bar{X}_n$ and $\bar{X}_z/\bar{X}_w$ are equal to 1. It is evident that when p is extremely close to 1 the expression of the 'most probable distribution' does not hold; this happens because Flory neglected the fact that the reaction system has a finite size in deriving the expression for his distribution function, i.e. the number of monomeric units is not infinite. So, when a very large molecule is formed (with x near N_0), the possibility of forming other very large molecules must be excluded; moreover, the number fraction of molecules with $x = N_0$ when $p = p_{\lim}$ must necessarily be 1.

A more correct approach, based on combinatorial methods,[7,8] shows that Flory's distribution holds up to very high conversions, but for p extremely close to $p_{\lim}$ the rigorous expression of the distribution function diverges from Flory's one and, for $p = p_{\lim}$, reduces to the case of monodisperse polymer. Moreover, it must again be emphasized that, in practical cases, the values of p attainable are never high enough to require the application of the rigorous approach.

necessarily be of type (1), whilst those consisting of an odd number of units will be type (2) or (3).

$$(\text{A—AB—B})_i \qquad (\text{A—AB—B})_j\text{A—A} \qquad \text{B—B}(\text{A—AB—B})_j$$

(1) $i = x/2$ where x is an even integer

(2) $j = (x-1)/2$ where x is an odd integer

(3) $j = (x-1)/2$ where x is an odd integer

If the starting monomeric mixture is stoichiometric (*i.e.* the number of initial A- and B- groups is the same) and the equal reactivity principle holds, both even and odd molecules will be present, and odd molecules will be equally divided between types (2) and (3). If the initial mixture is not stoichiometric, the relative numbers of the three types of molecules may be various, depending on the initial composition and on the value of the conversion; for instance, if the initial number of B–B monomer molecules is greater than that of A–A monomer molecules, *i.e.* in the presence of an excess of B groups, there will be more odd molecules (including both types) than even molecules. In particular, at completion of the reaction ($p_A = 1$) only type (3) molecules will remain and the corresponding weight fraction can be calculated following the approach used in the case of A–B type step-polymerization, which gives[1,2]

$$w_x = xr^{(x-1)/2}(1 - r)^2/(1 + r) \tag{47}$$

where x is restricted to odd integral values and r is the usual ratio between the initial number of A and B groups.

In the general case of non-stoichiometry of the initial mixture, for whatever value of the extent of reaction (non-complete reaction), the distribution function for the three types (**1, 2** and **3**) of molecules must be separately derived in the same way as has been shown for A–B self-polymerization. It can be demonstrated[1] that the expressions of the distribution functions in terms of weight fraction, under the usual assumptions of equal reactivity of functional groups of the same kind and of the absence of intramolecular reactions, are

$$w_x(\text{even}) = xp_A^{(x-1)}r^{x/2}2(1 - p_A)(1 - p_B)/(1 + r) \tag{48}$$

$$w_x(\text{odd-A}) = xp_A^{(x-1)}r^{x/2}(1 - p_A)^2 r^{-1/2}/(1 + r) \tag{49}$$

$$w_x(\text{odd-B}) = xp_A^{(x-1)}r^{x/2}(1 - p_B)^2 r^{-1/2}/(1 + r) \tag{50}$$

Starting from these equations, or from those using the mole fraction distribution of three types of molecules,[1] one can derive the expressions for the various average molecular weights (or of the average degrees of polymerization); moreover, it is easier to derive $\bar{M}_w$, using the formula obtained by Stockmayer[9] in a more general case. In fact Stockmayer, considering a system initially composed by $A_1, A_2, \ldots, A_i, \ldots$ molecules of reactants each bearing $f_1, f_2, \ldots, f_i, \ldots$ functional groups of type A and by $B_1, B_2, \ldots, B_j, \ldots$ molecules of other reactants each bearing $g_1, g_2, \ldots, g_j, \ldots$ functional groups of type B (A reacting only with B and *vice versa*), respectively, and under the usual assumptions of equal reactivity of functional groups of the same kind and the absence of intramolecular reactions, obtained (through statistical methods) the following expression of the distribution function[9]

$$N(m_i, n_j) = \frac{K\left(\sum_i f_i m_i - \sum_i m_i\right)!\left(\sum_j g_j n_j - \sum_j n_j\right)! \prod_i \left(y_i^{m_i}/m_i!\right)\prod_j (z_j^{n_j}/n_j)!}{\left(\sum_i f_i m_i - \sum_i m_i - \sum_j n_j + 1\right)!\left(\sum_j g_j n_j - \sum_j n_j - \sum_i m_i + 1\right)!} \tag{51}$$

where

$$z_j = \frac{g_j B_j}{\sum_j g_j B_j} \frac{p_A(1 - p_B)^{g_j - 1}}{(1 - p_A)} \tag{52}$$

$$y_i = \frac{f_i A_i}{\sum_i f_i A_i} \frac{p_B(1 - p_A)^{f_i - 1}}{(1 - p_B)} \tag{53}$$

$$K = \left(\sum_i f_i A_i\right)(1 - p_A)(1 - p_B)/p_B = \left(\sum_j g_j B_j\right)(1 - p_A)(1 - p_B)/p_A \tag{54}$$

and where $N(m_i, n_j)$ is the number of molecules of a species consisting of $m_1, m_2, \ldots, m_i, \ldots$

monomeric units of type A and of $n_1, n_2, \ldots, n_j, \ldots$ monomeric units of type B. From equation (51) the following simple expressions for $\bar{M}_w$ can be derived[10]

$$\bar{M}_w = \bar{M}_{w0} + \frac{\bar{f}_{n0}p_A[p_A(\bar{f}_{w0} - 1)\bar{M}_{B0}^2 + p_B(\bar{g}_{w0} - 1)\bar{M}_{A0}^2 + 2\bar{M}_{A0}\bar{M}_{B0}]}{\bar{M}_{n0}[1 - p_A^2 r(\bar{f}_{w0} - 1)(\bar{g}_{w0} - 1)]} \tag{55}$$

where

$$\bar{M}_{w0} = \frac{\sum_i M_i^2 x_{Ai} + \sum_j M_j^2 x_{Bj}}{\sum_i M_i x_{Ai} + \sum_j M_j x_{Bj}} \tag{56}$$

$$\bar{M}_{n0} = \sum_i M_i x_{Ai} + \sum_j M_j x_{Bj} \tag{57}$$

$$\bar{M}_{A0} = \frac{\sum_i M_i f_i A_i}{\sum_i f_i A_i} \tag{58}$$

$$\bar{M}_{B0} = \frac{\sum_j M_j g_j B_j}{\sum_j g_j B_j} \tag{59}$$

$$\bar{f}_{n0} = \sum_i f_i x_{Ai} \tag{60}$$

$$\bar{f}_{w0} = \frac{\sum_i f_i^2 A_i}{\sum_i f_i A_i} \tag{61}$$

$$\bar{g}_{w0} = \frac{\sum_j g_j^2 B_j}{\sum_j g_j B_j} \tag{62}$$

$$x_{Ai} = \frac{A_i}{\sum_i A_i + \sum_j B_j} \tag{63}$$

$$x_{Bj} = \frac{B_j}{\sum_i A_i + \sum_j B_j} \tag{64}$$

It must be noted that in equations (51)–(64) all the summations $\sum_i$ and $\sum_j$ are extended to all the kinds of A-type monomeric units and of B-type monomeric units, respectively.*

Equation (55) has great importance in calculating the weight-average molecular weight in the case of polyfunctional step-polymerization (see Volume 5, Chapter 8), but is useful also for the case of bifunctional A–A plus B–B type polymerization; in fact, in this case equation (55) reduces to

$$\bar{M}_w = \bar{M}_0 \frac{\left(1 + r p_A^2 + \frac{4r}{r+1} p_A\right)}{(1 - r p_A^2)} \tag{65}$$

with the simplifying assumptions that both the masses of the A–A and B–B monomeric units are equal to M_0; this assumption, although not realistic, has been made in order to elucidate in a simple way the effect of the non-stoichiometry of the polymerization system, without introducing other parameters, such as the ratio between the masses of the monomeric units of the two kinds, which could obscure this effect.

From this equation it is easy to obtain the expressions for the weight-average degree of polymerization, since in this case it is $\bar{X}_w = \bar{M}_w/M_0$, and to evaluate the strong effect of the non-stoichiometry of the initial mixture of $\bar{X}_w$. In Figure 3, $\bar{X}_w$ vs. r is shown for some values of the extent

---

* In equations (56)–(59) some molecular masses appear (M_i, M_j). If no product is eliminated in the polycondensation reaction, the masses of the monomers and of the corresponding monomeric units are identical and no error is possible. When a small molecular weight product (SMWP) forms in the polycondensation, then some caution must be used in applying the above equations. In fact, Stockmayer's equation (55) was derived without taking into account the formation of a SMWP; in order to take it into account two approximate routes can be followed: (i) to consider as M_i and M_j the corresponding monomer mass (the results obtained are rigorously valid only at zero conversion, i.e. when only monomer molecules are present); and (ii) to consider as M_i and M_j the mass of the corresponding monomeric units (the results obtained should be rigorously valid if all functional groups have reacted). In practice, as the values of the conversion are usually rather high, little error is committed if the latter approximation is used, as shown in ref. 10 for some specific cases.

A rigorous correction to take into account the elimination of the SMWP, whatever the value of the conversion may be, is simple for $\bar{M}_n$ (it can be easily demonstrated[10] that in this case $\bar{M}_n = (\bar{M}_{n,0} - p_A \bar{f}_{n,0} M_{SMWP})/(1 - p_A \bar{f}_{n,0})$, where in calculating $\bar{M}_{n,0}$ (see equation 57) the masses of monomers must be used and obviously M_{SMWP} is the molecular mass of the SMWP), but is rather complex for $\bar{M}_w$.[11,12]

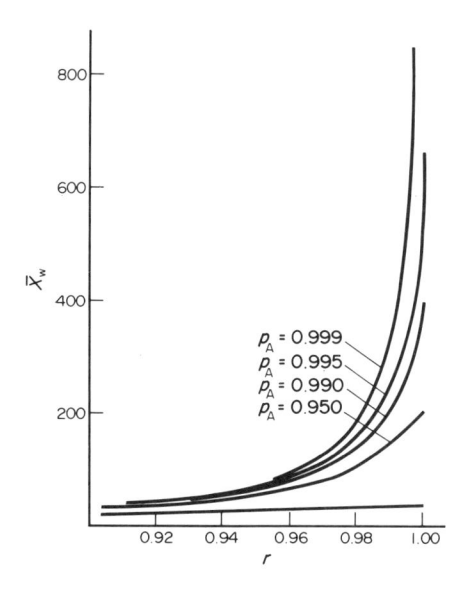

Figure 3 Weight-average degree of polymerization $\bar{X}_w$ vs. composition parameter r for various extents of reaction p_A.

of reaction p_A; from the figure it can be seen that an excess of only 10% in B groups has a strong limiting effect on $\bar{X}_w$, especially at high values of p_A.

The effect of an excess of functional groups of one kind on the distribution function can be simply evaluated by calculating the PDI. Since in the non-stoichiometric case

$$\bar{X}_n = \frac{r + 1}{r + 1 - 2rp_A} \tag{66}$$

(see also Volume 5, Chapter 2, Section 2.3), then the PDI is given by

$$\text{PDI} = \bar{X}_w/\bar{X}_n = \frac{\left(1 + rp_A^2 + \frac{4r}{r+1}p_A\right)(r + 1 - 2rp_A)}{(1 - rp_A^2)(r + 1)} \tag{67}$$

Calculations carried out for p_A ranging from 0.7 to 1.0 and r ranging from 0.9 to 1.0 show only a slight influence of this latter parameter on the PDI and, therefore, on the distribution function; for instance, with $r = 0.9$ and $p_A = 0.95$, equation (67) gives a PDI value of 1.92 instead of that of 1.95 reached with $r = 1$ at the same value of p_A (this effect becomes less and less with increasing conversion).

4.3.2 A Recent Approach

A large number of treatments other than that of Flory have been carried out in the past on the molecular weight distribution in linear step-polymerization. Most of them[13-35] were developed in order to obtain general expressions valid in the case of the formation of copolymers of various types, also taking into account the effect of a possible difference in reactivity among functional groups of the same kind; other efforts were made in the field of polyfunctional stepwise polymerization.[11,12,36-39] These two classes of model are cited here because the expressions obtained can be reduced to the simpler case of bifunctional polymerization in which the equal reactivity principle holds, which is the subject of the present chapter. Since the knowledge of only some average molecular weights (usually at least $\bar{M}_w$ and $\bar{M}_n$) is often sufficient to correlate the effect of the shape of the distribution to many polymer properties, the treatment developed by Macosko and Miller, which is simple and allows calculation of the average properties without deriving the molecular weight distribution function, will be developed in the following discussion.

The treatment of Macosko and Miller,[12,24,37-39] which will be applied in the case of A–A plus B–B polymerization with the usual assumptions of equal reactivity of functional groups of the same type and the absence of intramolecular reaction, is based on the recursive nature of a step polymerization and on an elementary law of conditional expectation.

If A is an event and $\bar{A}$ its complement, and Y is a random variable, $E(Y)$ its expectation (or average value) and $E(Y|A)$ its conditional expectation given the event A has occurred, then the law of total probability for the expectations[40] is

$$E(Y) = E(Y|A)P(A) + E(Y|\bar{A})P(\bar{A}) \tag{68}$$

where $P(A)$ and $P(\bar{A})$ are the probabilities that an event A and its complement $\bar{A}$, respectively, occur.

$$\text{A—A} + \text{B—B} \longrightarrow \text{—BA—}\overset{1}{\overrightarrow{A'}}\overset{2}{\overrightarrow{B'}}\text{—}\overset{3}{\overrightarrow{B''}}\overset{4}{\overrightarrow{A}}\text{—}\overset{1}{\overrightarrow{A}}\text{B—}$$

Scheme 1

The case of A–A plus B–B polymerization is represented in Scheme 1; let $N_{0,AA}$ and $N_{0,BB}$ be the initial number of A–A and B–B monomer molecules, respectively, and suppose that the system polymerizes until some fraction p_A of the A groups and some fraction p_B of the B groups have reacted. Taking an A at random and picked up, A' in Scheme 1, we want to calculate the weight $W_{A'}^{out}$ of the chain attached to A', looking out from its parent molecule, in the direction $\overset{1}{\rightarrow}$. It will be equal to zero if A' has not reacted, while if A' has reacted it is equal to $W_{B'}^{in}$; that is, the weight attached to B' looking in direction $\overset{2}{\rightarrow}$ (i.e. looking into the part of the molecule that begins with B')

$$W_{A'}^{out} = \begin{cases} 0 & \text{if A' does not react} \\ W_{B'}^{in} & \text{if A' does react (with B')} \end{cases} \tag{69}$$

The law of total expectation (equation 68) implies

$$E(W_A^{out}) = p_A E(W_B^{in}) + (1 - p_A)\cdot 0 \tag{70}$$

In a similar way we proceed along the structure of Scheme 1 until the recursive nature of the structure itself bring us back to equation (70).*

$$E(W_B^{in}) = M_{BB} + E(W_B^{out}) \tag{71}$$

$$E(W_B^{out}) = p_B E(W_A^{in}) + (1 - p_B)\cdot 0 \tag{72}$$

$$E(W_A^{in}) = M_{AA} + E(W_A^{out}) \tag{73}$$

Let W_{AA} be the total molecular weight of the molecule of which a randomly chosen A–A unit is a part, and analogously for W_{BB}, and then

$$E(W_{AA}) = E(W_A^{in}) + E(W_A^{out}) \tag{74}$$

$$E(W_{BB}) = E(W_B^{in}) + E(W_B^{out}) \tag{75}$$

In fact equation (74) means that, taking an A group at random, the weight of the polymer molecule of which it is a part is the sum of the weights of the two parts of the molecule, one on the left- and one on the right-hand side of the group A; similarly for equation (75). Solving the set of equations (70)–(75), one obtains

$$E(W_{AA}) = M_{AA} + 2p_A \frac{M_{BB} + p_B M_{AA}}{1 - p_A p_B} \tag{76}$$

$$E(W_{BB}) = M_{BB} + 2p_B \frac{M_{AA} + p_A M_{BB}}{1 - p_B p_B} \tag{77}$$

---

* Equations (70) and (72) have been reported extensively, even if the second term in both cases is obviously equal to zero, in order to show that the two equations derive from a correct applications of the law of total expectation (equation 68). In equation (70) p_A corresponds to $P(A)$ (the event is the reaction of the A group), $(1 - p_A)$ corresponds to $P(\bar{A})$ (the complement to the event is that the A group has not reacted). Similarly in equation (72) it is p_B that corresponds to $P(A)$ and $(1 - p_B)$ to $P(\bar{A})$.

Equation (71) can be obtained in the following way: if in Scheme 1 we take the B group picked-up as B' and look in the direction $\overset{2}{\rightarrow}$, the expected weight of the molecule, of which B' is a part, in this direction $[E(W_B)]$ is the sum of that part attached to B'' looking in the direction $\overset{3}{\rightarrow}$ plus the weight of that part of the molecule included between B' and B'', i.e. M_{BB}. A similar procedure can be used to obtain equation (73).

Molecular Weight Distributions

To find the weight-average molecular weight, we pick a unit of mass at random and compute the expected weight of the molecule of which it is a part

$$\bar{M}_w = w_{AA} E(W_{AA}) + w_{BB} E(W_{BB}) \tag{78}$$

where

$$w_{AA} = \frac{M_{AA} N_{0,AA}}{M_{AA} N_{0,AA} + M_{BB} N_{0,BB}} \tag{79}$$

$$w_{BB} = \frac{M_{BB} N_{0,BB}}{M_{AA} N_{0,AA} + M_{BB} N_{0,BB}} \tag{80}$$

Substituting the expressions for $E(W_{AA})$ and $E(W_{BB})$ (equations 76 and 77) into equation (78), one obtains

$$\bar{M}_w = \frac{(1 + rp_A^2)(rM_{AA}^2 + M_{BB}^2) + 4rp_A M_{AA} M_{BB}}{(rM_{AA} + M_{BB})(1 - rp_A^2)} \tag{81}$$

which coincides with the results attainable starting from the Stockmayer distribution; in particular, when the masses of the A–A and B–B type monomeric units, M_{AA} and M_{BB} respectively, are identical (and equal to M_0), equation (81) reduces to equation (65).

4.3.3 Presence of a Monofunctional Reactant

The effect of the presence in the polymerization system of a monofunctional reactant is discussed in Volume 5, Chapter 2, Section 2.3, where it is shown that it limits greatly the number-average degree of polymerization. The question that subsequently arises is how does the presence of a monofunctional reactant influence the molecular weight distribution in a linear step-polymerization? In order to answer this, the case of a step polymerization starting from a bifunctional monomer of the A type (A–A) and one of B type (B–B) plus a monofunctional reactant carrying a B functional group (R–B) (A reacting only with B and *vice versa*) will be examined, with the usual assumptions of equal reactivity of functional groups of the same kind and the absence of intramolecular reactions. Analogous procedures could be developed in the case of A–B self polymerization in the presence of an R–B (or R–A) monofunctional reactant.

As some mathematical difficulties arise in deriving the exact distribution function in this case, one can be satisfied by determining only the effect of the monofunctional reactant on $\bar{X}_w$, or, better, on the PDI. In this regard, the value of the weight-average degree of polymerization can be calculated starting from equation (55), that in the present case reduces to

$$\bar{X}_w = 1 + \frac{2 p_A r (p_A (1 + r - r \rho_{RB}) + 2)}{(1 + r + \rho_{RB})(1 - r p_A^2 (1 - \rho_{RB}))} \tag{82}$$

which is valid when the masses of the three kinds of monomeric unit are equal to M_0 (and, therefore, $\bar{X}_w = \bar{M}_w / M_0$), and where

$$\rho_{RB} = \frac{N_{0,RB}}{2 N_{0,BB} + N_{0,RB}} \tag{83}$$

and $N_{0,RB}$ is the initial number of molecules of the monofunctional reactant (ρ_{RB} is the fraction of initial B functional groups belonging to monofunctional reactant molecules).

On the other hand, in the same case, one obtains the following expression from the definition of $\bar{M}_n$.

$$\bar{X}_n = \frac{r + 1 + \rho_{RB}}{1 + \rho_{RB} + r(1 - 2 p_A)} \tag{84}$$

As PDI $= \bar{X}_w / \bar{X}_n$, using equations (2) and (84) it is easy to calculate the expression for the PDI as a function of p_A and ρ_{RB}.

In order to fully elucidate the effect of a monofunctional reactant on the PDI, a specific case will be examined, which is when the initial number of bifunctional monomer molecules of A–A type is equal to that of bifunctional monomer molecules of B–B type, with in addition a certain number of monofunctional R–B type molecules. In this case, as

$$r = 1 - \rho_{RB} \tag{85}$$

equations (82) and (84) become respectively

$$\bar{X}_w = 1 + \frac{p_A(1 - \rho_{RB})[p_A(1 + (1 - \rho_{RB})^2) + 2]}{1 - p_A^2(1 - \rho_{RB})^2} \tag{86}$$

$$\bar{X}_n = \frac{1}{1 - p_A(1 - \rho_{RB})} \tag{87}$$

(equation 87 is equivalent to equation 31 in Volume 5, Chapter 2, in which a different composition parameter had been introduced in order to obtain an expression for $\bar{X}_n$ formally identical to others previously obtained in different cases).

From equations (86) and (87) the effect of the monofunctional reactant on the PDI can be evaluated; calculations carried out for ρ_{RB} values up to 0.1 and for p_A ranging from 0.9 to 1.0 showed little influence of ρ_{RB} on the PDI and, therefore, on the width of the molecular weight distribution. Some of the results obtained are collected in Figure 4, in terms of PDI vs. ρ_{RB} for different values of the conversion p_A.

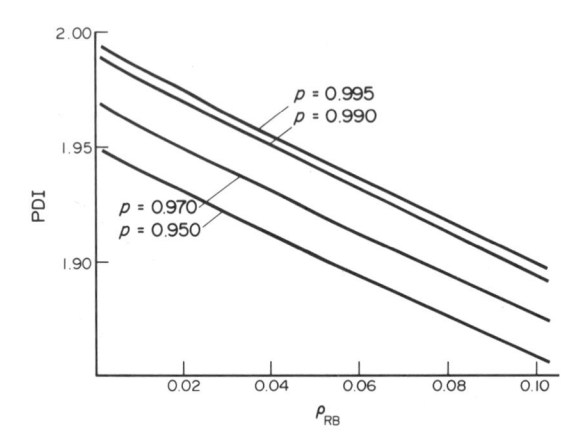

Figure 4 Polydispersity index PDI vs. monofunctional content ρ_{RB} for various extents of reaction p in the stoichiometric case (r = 1).

Irrespective of the conversion, the presence of the monofunctional reactant produces a slight narrowing of the distribution so that the PDI diminishes by about 5% when ρ_{RB} is 0.1 (*i.e.* the mole fraction of monofunctional reactant, calculated with respect to B type monomers, is more than 0.18, as can be seen from an easy calculation).

Another specific case has been investigated, that of an exact overall stoichiometry of the system (r = 1). In this situation the effect of the monofunctional reactant on PDI is analogous to that of the previous case, only slightly stronger; with $\rho_{RB} = 0.1$, the PDI diminishes by about 7% for a fixed (high) value of the conversion.

4.4 EXPERIMENTAL VERIFICATIONS

In deriving the molecular weight distribution function or in calculating directly its most important parameters, using either statistical or kinetic methods, it has been assumed that once a given pair of functional groups have reacted with each other, they remain bonded permanently. However, in many cases of step polymerization *via* elimination of small molecules (*i.e.* polycondensations) this assumption may not be valid, because rupture of the formed bonds can take place under appropriate conditions; it is well known, for instance, that polyesters, polyamides, polysulfides, polysiloxanes, *etc.*, can undergo intermolecular reactions between a terminal functional group (or an interunit repeating linkage) of a polymer molecule and an interunit repeating linkage of another polymer molecule during polymerization or processing (in general at high temperature).[41] These reactions do not lead to a variation in the number of terminal functional groups and polymer molecules, and thus do not affect either the value of the conversion p or the number-average degree of polymerization $\bar{X}_n$, therefore involve an interchange between the two polymer molecules. In the important example of polyesters, the interchange may take place between an ester and a terminal hydroxyl group (alcoholysis reaction), or between an ester and a terminal carboxyl group (acidolysis reaction) or between two ester groups belonging to two distinct polymer molecules (esterolysis reaction).

In general, these reactions are reversible and, in many cases, all the interunit linkages have the same reactivity, independent of the size of the molecule of which they are part and their position in the chain. Under these conditions the process of interchange of skeletal bonds occurs in a completely random way; as a consequence, it favours the 'most probable distribution', if for some reason, e.g. an eventual difference in reactivity between functional groups of the same kind, the distribution function is of a different shape.[42,43] In this regard, it is possible to demonstrate that a polymerization system, whatever its initial molecular weight distribution, in which a random scission of linkages in polymer chains takes place, goes toward the 'most probable distribution'. This may happen, for instance, with polyesters and polyamides, due to the hydrolysis and alcoholysis reactions.

The assumption of the absence of intramolecular reactions that was also made in deriving the 'most probable distribution' is not rigorously valid in general, since cyclic molecules always occur to some extent, both *via* the possible reaction between two terminal functional groups of the same molecule, and *via* a possible interchange between a terminal groups and an interunit repeating linkage of the same molecule. It can be shown that cyclization generally leads to slight broadening of the distribution[2] and that it can be neglected except in the very early stages of a step polymerization or when the polymerization is carried out in dilute solution. When the intramolecular reactions cannot be neglected, it can be demonstrated that Flory's distribution applies to linear molecules, considered as a separate subset.[2]

The 'most probable distribution' has been experimentally verified as valid in many cases within the limits of experimental error, particularly for polyamides and polyesters obtained by step polymerization in homogeneous phase at high temperature. We can take, as an example, the fractionation data reported in ref. 44, which confirms this in the cases of poly(hexamethyleneadipamide) and of other polycondensates. Also, data obtained with more modern techniques, such as gel permeation chromatography, show that for poly(butylene terephthalate) and poly(ethylene terephthalate) the 'most probable distribution' is valid; in particular, the value of the PDI is close to 2, as expected from the theory, within an error of about 10%.[45–47] It must be emphasized that the shape of the distribution function usually depends on the polymerization conditions as well; as an example, it has been reported that in the case of poly(hexamethylenesebacamide) the PDI is close to 2 only in the cases of homogeneous equilibrium processes in the melt, or in solid phase or in high temperature solution processes.[48]

4.5 EFFECT OF NON-EQUAL REACTIVITY OF FUNCTIONAL GROUPS OF THE SAME KIND

In Volume 5, Chapter 3 the effect of non-equal reactivity of functional groups of the same kind on the kinetics of polymerization has been analyzed. As pointed out there, the kinetic aspect is not very important; from a practical point of view, the effect on the molecular weight distribution is more relevant, as can be visualized in a simple manner by studying the influence of non-equal reactivity on the PDI.

First, the case of A–A plus B–B polycondensation in the presence of both initial asymmetry and a substitution effect on the two A groups belonging to the A–A type monomer molecule will be treated. For simplicity it will be assumed that the B–B type monomer has neither initial asymmetry nor undergoes a substitution effect.

In this case the reaction scheme becomes as in Scheme 2. The kinetic approach allows, in addition to some statistical considerations, expression of the PDI as a function of conversion, e.g. of the B groups.[25,26] In Figure 5, PDI is plotted in the stoichiometric case vs. p for two different specific cases of unequal reactivity (different values of the kinetic constants of Scheme 2) along with the curve for the equal reactivity case. In the same figure the modified polydispersity index (MPDI), obtained by not taking into account the contributions of the monomer molecules on $\bar{M}_n$ and on $\bar{M}_w$, is reported for the same examples. As can be seen, the effect of both initial and induced asymmetry on the PDI is rather strong, especially at conversions in the range 0.9–1.0, which is usually the range of practical interest, resulting in a broadening of the molecular weight distribution and consequently an increase in the PDI when k_1^* and k_2^* are greater than k_1 and k_2 respectively; narrowing of the distribution can be observed if k_1^* and k_2^* are smaller than k_1 and k_2, respectively. The effect on the MPDI is almost inappreciable, since it does not take into account the contribution due to the two monomers, whose concentrations are strongly affected by the presence of the substitution effect.

$$A_2A_1 + BB{-} \xrightarrow{k_1} *A_2A_1BB{-}$$

$$A_1A_2 + BB{-} \xrightarrow{k_2} *A_1A_2BB{-}$$

$$-A_2A_1^* + BB{-} \xrightarrow{k_1^*} -A_2A_1BB{-}$$

$$-A_1A_2^* + BB{-} \xrightarrow{k_2^*} -A_1A_2BB{-}$$

Scheme 2

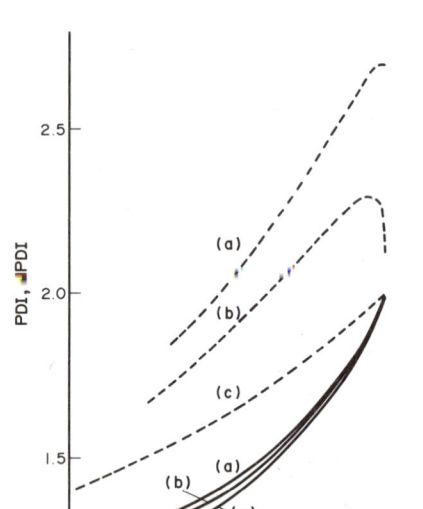

Figure 5 Polydispersity index PDI (----) and modified polydispersity index MPDI (——) vs. p in the stoichiometric case, in different situations of initial and induced asymmetry. (a) $k_2/k_1 = 1.25$, $k_1^*/k_1 = 3$, $k_2^*/k_2 = 4$; (b) $k_2/k_1 = 1.25$, $k_1^*/k_1 = 3$, $k_2^*/k_2 = 1.5$; (c) case of equal reactivity ($k_2/k_1 = k_1^*/k_1 = k_2^*/k_2 = 1$) (from the results of ref. 26 with permission of the American Chemical Society)

As previously demonstrated in Volume 5, Chapter 3 for the kinetic aspect, the effect on the molecular weight distribution of the dependence of the reactivity of functional groups on the length of the chain to which they are attached will now be examined. In particular, one of the two models proposed there will be used, *i.e.* the case of the self-polymerization of an A–B type monomer, in which the reactivity of the functional groups of the monomer in the reaction with all species present (monomer included) is different from that of all other species (see Scheme 3 and also Volume 5, Chapter 3). The kinetic treatment[31,34] leads to expressions of the PDI and of MPDI as functions of adimensional time $\tau'' = 2k_p[P_1]_0$, where k_p is the kinetic constant related to the functional groups of all species, monomer excluded, reacting with each other, and $[P_1]_0$ is the initial concentration of monomer molecules. In Figures 6 and 7, the PDI and MPDI, respectively, are plotted vs. τ'' for different values of the ratio $R = k_{11}/k_p$. From these figures a strong effect can be seen on the PDI, whilst a lesser effect on the MPDI is shown. The reason for this difference is the same as for the substitution effect on A–A plus B–B polymerization; monomer concentration is more affected by the difference between k_{11} and k_p values (that is between the reactivity of monomer and polymer molecules) and, as the MPDI does not take it into account, it is scarcely influenced by this difference in reactivity.

$$P_1 + P_1 \xrightarrow{k_{11}/2} P_2$$

$$P_1 + P_n \xrightarrow{k_{11}} P_{n+1} \quad (n > 1)$$

$$P_n + P_m \xrightarrow{k_p} P_{m+n} \quad (m, n > 1, m \neq n)$$

$$\xrightarrow{k_p/2} P_{2m} \quad (m > 1, m = n)$$

Scheme 3

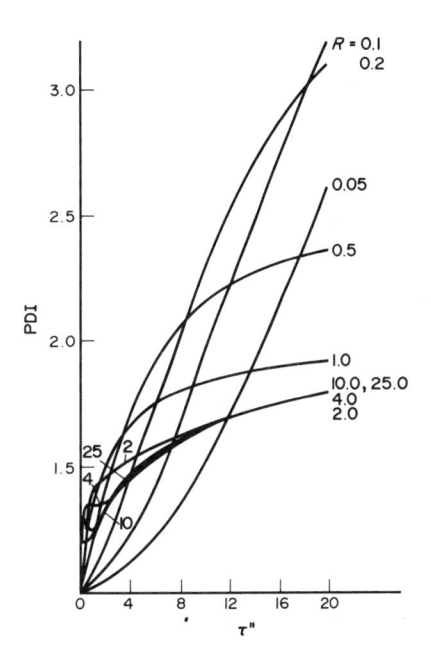

Figure 6 Polydispersity index PDI vs. adimensional time $\tau'' = 2k_p[P_1]_0 t$ for several values of the ratio $R = k_{11}/k_p$ (reproduced from ref. 32)

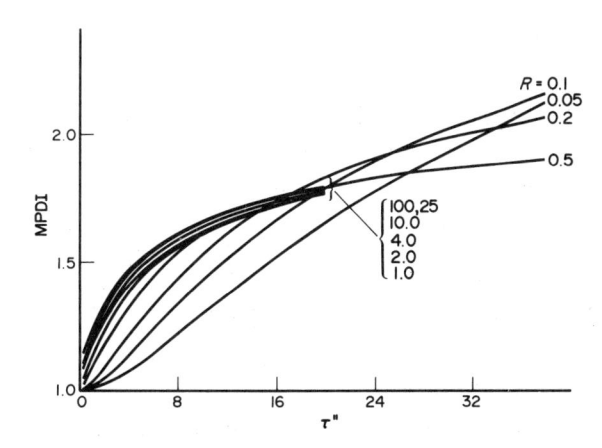

Figure 7 Modified polydispersity index MPDI vs. adimensional time $\tau'' = 2k_p[P_1]_0 t$ for several values of the ratio $R = k_{11}/k_p$ (reproduced from ref. 32)

Analogous conclusions can be drawn from the results obtained with other models such as that considering a different reactivity of the monomer only when reacting with another monomer molecule,[32,33,35] or that considering the reactivity of the polymer chain depending in a linear way on its length.[27]

4.6 REFERENCES

1. P. J. Flory, *J. Am. Chem. Soc.*, 1936, **58**, 1877.
2. P. J. Flory, 'Principles of Polymer Chemistry', Cornell University Press, Ithaca, New York, 1953.
3. H. Dostal and R. Raff, *Monatsch. Chem.*, 1936, **68**, 188.
4. C. Tanford, 'Physical Chemistry of Macromolecules', Wiley, New York, 1963.
5. J. Brandrup and E. H. Immergut (eds.), 'Polymer Handbook', Interscience, New York, 1966.
6. J. D. Ferry, 'Viscoelastic Properties of Polymers', Wiley, New York, 1970.
7. V. S. Nanda, *J. Polym. Sci., Part A*, 1964, **2**, 2275.
8. V. S. Nanda, *J. Macromol. Sci., Phys.*, 1980, **18**, 685.
9. W. H. Stockmayer, *J. Polym. Sci.*, 1952, **9**, 69; 1953, **11**, 424.
10. A. Munari and P. Manaresi, *Chim. Ind. (Milan)*, 1984, **66**, 755.
11. D. Durand and C. M. Bruneau, *Br. Polym. J.*, 1981, **13**, 33..
12. C. W. Macosko and D. R. Miller, *Macromolecules*, 1976, **9**, 199.
13. J. G. Watterson and J. W. Stafford, *J. Macromol. Sci.., Chem.*, 1971, **5**, 679.
14. D. Durand and C. M. Bruneau, *Makromol. Chem.*, 1979, **180**, 2947.
15. D. Durand and C. M. Bruneau, *Makromol. Chem.*, 1980, **181**, 421.
16. D. Durand and C. M. Bruneau, *Makromol. Chem.*, 1980, **181**, 1673.
17. D. Durand and C. M. Bruneau, *Eur. Polym. J.*, 1981, **17**, 707.
18. D. Durand and C. M. Bruneau, *Eur. Polym. J.*, 1981, **17**, 715.
19. L. C. Case, *J. Polym. Sci.*, 1958, **29**, 455.
20. L. C. Case, *J. Polym. Sci.*, 1959, **37**, 147.
21. L. C. Case, *J. Polym. Sci.*, 1959, **39**, 175.
22. L. C. Case, *J. Polym. Sci.*, 1959, **39**, 183.
23. L. C. Case, *J. Polym. Sci.*, 1960, **48**, 27.
24. L. Lopez-Serrano, J. M. Castro, C. W. Macosko and M. Tirrel, *Polymer*, 1980, **21**, 263.
25. K. S. Gandhi and S. V. Babu, *AIChE J.*, 1979, **25**, 266.
26. K. S. Gandhi and S. V. Babu, *Macromolecules*, 1980, **13**, 791.
27. V. S. Nanda and S. C. Jain, *J. Chem. Phys.*, 1968, **49**, 1318.
28. J. L. Stanford and R. F. T. Stepto, *J. Chem. Soc., Faraday Trans. 1*, 1975, **71**, 1292.
29. T. M. Orlova, S. S. A. Pavlova and L. V. Dubrovna, *J. Polym. Sci., Polym. Chem. Ed.*, 1979, **17**, 2209.
30. H. G. Elias, *J. Macromol. Sci., Chem.*, 1978, **12**, 183.
31. R. Goel, S. K. Gupta and A. Kumar, *Polymer*, 1977, **18**, 851.
32. S. K. Gupta, A. Kumar and A. Bhargawa, *Eur. Polym. J.*, 1979, **15**, 557.
33. S. K. Gupta, A. Kumar and A. Bhargawa, *Polymer*, 1979, **20**, 305.
34. S. K. Gupta, N. L. Agarwalla, P. Rajora and A. Kumar, *J. Polym. Sci., Polym. Phys. Ed.*, 1982, **20**, 933.
35. A. Kumar, P. Rajora, N. L. Agarwalla and S. K. Gupta, *Polymer*, 1982, **23**, 222.
36. M. Gordon, *Proc. R. Soc. London, Ser. A*, 1962, **268**, 240.
37. D. R. Miller and C. W. Macosko, *Macromolecules*, 1978, **11**, 656.
38. D. R. Miller, E. M. Valles and C. W. Macosko, *Polym. Eng. Sci.*, 1979, **19**, 272.
39. D. R. Miller and C. W. Macosko, *Macromolecules*, 1980, **13**, 1063.
40. S. M. Ross, 'Introduction to Probability Models', Academic Press, New York, 1972.
41. T. Davies, in 'Chemical Reactions of Polymers', ed. E. M. Fettes, Interscience, New York, 1964, chap. 7.
42. P. J. Flory, *J. Am. Chem. Soc.*, 1942, **64**, 2205.
43. J. B. Carmichael, in 'Encyclopedia of Polymer Science and Technology', ed. H. F. Mark, N. G. Gaylord and N. M. Bikales, Interscience, New York, 1970, vol. 13.
44. G. J. Howard, in 'Progress in High Polymers', ed. J. C. Robb and F. W. Peaker, Iliffe Books, London, 1961, vol. 1.
45. F. J. Muller, U. Altenhofen, F. J. Wortmann and H. Zahn, *Angew. Makromol. Chem.*, 1979, **83**, 171.
46. M. Naoki, I. Park, S. L. Wunder and B. Chu, *J. Polym. Sci., Polym. Phys. Ed.*, 1985, **23**, 2567.
47. H. Inata and S. Matsamura, *J. Appl. Polym. Sci.*, 1987, **33**, 3069.
48. Y. Arai, M. Watanabe, K. Sanui and N. Ogata, *J. Polym. Sci., Polym. Chem. Ed.*, 1985, **23**, 3081.

5
Formation of Cyclic Oligomers

PIETRO MARAVIGNA and GIORGIO MONTAUDO
University of Catania, Italy

5.1 INTRODUCTION	63
5.2 GENERAL AND MECHANISTIC ASPECTS	64
5.3 SEPARATION AND IDENTIFICATION METHODS	65
5.4 CYCLIC OLIGOMERS	69
5.4.1 *Esters and Lactones*	70
5.4.2 *Carbonates*	70
5.4.3 *Amides and Lactams*	74
5.4.4 *Urethanes and Ureas*	75
5.4.5 *Formals*	76
5.4.6 *Sulfides*	78
5.4.7 *Siloxanes*	84
5.4.8 *Miscellaneous*	85
5.5 FORMATION OF LINEAR OLIGOMERS	85
5.6 REFERENCES	89

5.1 INTRODUCTION

The behavior and properties of cyclic molecules are attracting great attention these days, with major emphasis on their conformational characteristics, phase-transfer catalysis, molecular inclusion and ion complexation capability, among others. Several reviews and books covering various aspects of this subject have been published.[1-27]

The formation of cyclic oligomers in polymerization reactions is a more widespread phenomenon than was once believed. Cyclic oligomers are formed under a variety of experimental conditions, which are often deliberately chosen to avoid polymer formation. This is the case for the well-known 'high dilution method' of Ruggli and Ziegler,[1,4] which was devised to minimize the amount of polymer formed when monomers carrying very reactive end-groups react together.

It is now recognized that the production of the high molecular weight polymers is often accompanied by the formation of sizeable amounts of low molecular weight cyclic oligomers.[8,10]

Cyclic oligomers have been frequently obtained in the preparation of condensation polymers,[5,6,8,10,13,17,24-26] or in the ring-opening polymerization of heterocyclic monomers.[11,12,21,22]

In the following, we shall discuss some general aspects concerning the mechanism of oligomer formation in polycondensation reactions, emphasizing the modern separation and identification methods which have been of key importance in the advancement of studies in this area of polymer science. Finally, we shall present in some detail the specific oligomers produced in the most relevant classes of step-growth polymers.

A certain degree of uncertainty exists, even among specialists, when discussing the details of the mechanisms of cyclic oligomer formation during polymerization reactions. This can be related to the historical development of research in this field, which originated, not from the study of intramolecular reactions, but from the theory of macrocyclization equilibria.

Until recently, intramolecular cyclization processes have been mainly studied by organic chemists, whereas macrocyclization equilibria were studied by polymer scientists, using a statistical thermodynamics approach. Now that the results obtained by the two groups have been brought together, a more thorough understanding of the matter has been reached.

Macrocyclization equilibria are able to describe the behavior of linear polymer chains when they are allowed to reach thermodynamic equilibrium and generate cyclic oligomers. Depolymerization of a high molecular weight linear chain is involved in the formation of cyclic oligomers by ring–chain equilibration. Therefore the bimodal distributions of molecular weights, which are characteristic of these equilibrating systems, are perfectly acceptable to a polymer chemist.

The situation is different when a polymerization reaction takes place. Since the molecular weights obtained in step-growth reactions depend solely on the extent of reaction p, and the molecular weight distributions are Gaussian at any p, no sizeable amounts of oligomers should be present beyond a p value of 0.97–0.98. If they are present, they might originate from a reaction (intramolecular cyclization) competing with the polymerization (intermolecular reaction), or from a concurrent depolymerization of the linear polymer formed earlier in the reaction (ring–chain equilibrium). The latter process is thermodynamically controlled, and therefore the actual oligomer concentrations should be different from those produced by the intramolecular cyclizaton reaction, which is kinetically controlled.

Although it is sometimes possible to clearly distinguish between the two mechanisms, this is not generally the case, as discussed below.

5.2 GENERAL AND MECHANISTIC ASPECTS

The cyclization of a chain molecule is a process which is in competition with the polymerization reaction. The formation of cyclic oligomers is the result of an intramolecular reaction (first-order kinetics), whereas polymerization is a second-order reaction. This fact constitutes the basis of the dilution principle,[1,4] which has been exploited to obtain higher yields of cyclic oligomers by partially reducing the polymerization rate.

However, at high monomer concentrations and high conversions, *i.e.* under conditions when only polymers should be obtained, the formation of cyclic oligomers is still observed. The mechanism of formation of these cyclic species has long been debated, and attention has been focused on the content and distribution of macrocyclic compounds produced during depolymerization reactions.[10]

Historically, a satisfactory theory of macrocyclization equilibria was formulated by Jacobson and Stockmayer,[3] long before the availability of experimental data. Experiments have been mainly concerned with the (catalyzed) depolymerization of macromolecules to generate cyclic oligomers (ring–chain equilibrium).[10,24]

The Jacobson–Stockmayer theory includes the following assumptions: chains in solution obey Gaussian statistics; all rings formed are strainless; reactivity of all reactive sites along the chain is the same; a thermodynamic distribution is obtained; and end-group effects are negligible.

Therefore, the cyclization probability is related to the mean separation of the reacting sites, and the equilibrium concentration of each cyclic oligomer is predicted to decrease proportionally to $n^{-2.5}$, where n is the ring size expressed as the number of repeating units present.

Systems which depend entirely on ring–chain equilibria have been found.[24] The Jacobson–Stockmayer theory has been verified in detail, and it is generally accepted that it also provides a convenient testing ground for theories on chain conformation and dynamics in solution.[21,24–26]

In principle, the macrocyclization equilibria might also be attained during polymerization reactions. However, the distribution of cyclic compounds resulting in the latter case may deviate from thermodynamically controlled distributions because of kinetic factors. Indeed, the Jacobson–Stockmayer theory is based on the assumption of equal reactivity of reaction sites. This condition is usually fulfilled for the functional groups attached to the inner portion of a flexible chain molecule, so that the mechanism of cyclization resulting in a thermodynamic distribution of cyclic oligomers is an *intramolecular back-biting* process (equation 1, where A are the functional groups along the chain and E are the end-groups, unreactive towards the functional groups A). Cyclic polymers are ideal examples of systems capable of undergoing only back-biting processes and of generating a thermodynamically controlled distribution of cyclic oligomers.

$$E\sim\sim A-A-A-A\sim\sim E \tag{1}$$

If a cyclic polymer does not obey ring–chain equilibration statistics, this is due to conformational preferences or restrictions present along its chain (see below).

When the end-groups are capable of reacting between themselves or with the functional groups A along the chain at a higher rate compared to the reaction rates in equation (1), an *end-biting process* takes place (equation 2), which violates the equal reactivity principle,[3] and a kinetically controlled distribution results. In this case, the concentration of each cyclic oligomer is predicted to decrease proportionally to $n^{-1.5}$, where n is the ring size expressed in number of repeat units present.[21,23]

$$E\text{---}A\text{---}A\text{---}A\text{---}A\text{---}A\text{---}E \qquad (2)$$

Another situation where the thermodynamic distribution may not be achieved occurs in systems where preferred or restricted conformations arise along the chain molecule, since the Jacobson–Stockmayer theory requires strainless rings.[3] This condition is never fulfilled for real systems in which the formation of small rings is concerned.

In several polymerizing systems, the concentration of certain cyclic oligomers is found to exceed considerably the values theoretically predicted.[11,12,28]

In the majority of real polymerization systems, the influence of preferred conformations and/or kinetic factors is so important that the observed distributions cannot be described in terms of the Jacobson–Stockmayer theory. A further complication, encountered when dealing with polymerization reactions, is related to the fact that it is not usually known to what extent cyclization was advanced when the polymerization was stopped.[21]

Both polymerization and catalyzed depolymerization reactions have been exploited to synthesize cyclic oligomers corresponding to numerous step-growth polymers (see Section 5.4). Furthermore, it has been observed with increasing frequency that cyclic oligomers are formed in the thermal decomposition processes of several polymers.[15,27] Although the latter phenomenon has always been associated with the practical problems of the thermal stability of polymers, disregarding any particular interest in the synthesis of rings, thermal depolymerization has great potential as a synthetic method, as described in Section 5.4.

Thermal decomposition of polymers is a bulk process (in the solid or melt state), inherently different from solution depolymerization reactions. It has been recently observed that the thermal cleavage of condensation polymers is often selective, and that intramolecular cyclization reactions occur quite frequently, producing cyclic oligomers.[27]

5.3 SEPARATION AND IDENTIFICATION METHODS

Current methods of detecting oligomers contained in polymer samples are based on gas, liquid and size-exclusion chromatography [GC, high performance liquid chromatography (HPLC) and gel-permeation chromatography (GPC)], combined with various structural identification techniques.

Cyclic oligomers contained in polysiloxanes are typical examples of compounds that can be separated by GC.[10,24] However, in recent times, the versatility of the liquid chromatography techniques (HPLC and GPC) has reduced the use of GC, which can only be applied to volatile and thermally stable molecules.

The GPC traces given in Figure 1 show that stable materials extracted from crude polysulfide samples contain a series of low molecular weight compounds, which have been identified as cyclic oligomers (Table 1).[29] Of course, chromatography only separates the oligomers, and their identification remains a problem that must be dealt with using suitable instrumental techniques (nuclear magnetic resonance (NMR), mass spectrometry (MS), infrared spectroscopy (IR), *etc.*)

MS is particularly suitable for the detection of small amounts of oligomers contained in polymeric samples, and has recently been applied to the separation and characterization of oligomers formed in polymerization reactions.[27,30,31]

The conventional MS method for the detection of oligomers contained in polymer samples is based on the direct introduction of the crude polymerization products into the ion source of a mass spectrometer by the insertion probe for solid samples.[30,31] The probe temperature is then gradually increased and the evolving products are analyzed by repeated mass scans. Low molecular weight compounds, which evaporate undecomposed in the high vacuum of the ion source, are usually detected as separate ion current peaks.

In Figure 2 are shown the distillation curves (single ion currents) for a sample of poly(*m*-phenylene sulfide) [*m*-PPS, poly(thio-1,3-phenylene)] containing a mixture of cyclic oligomers, which were fractionated by a heating rate of $10\,°C\,\text{min}^{-1}$ and detected and identified by their molecular ions.[32]

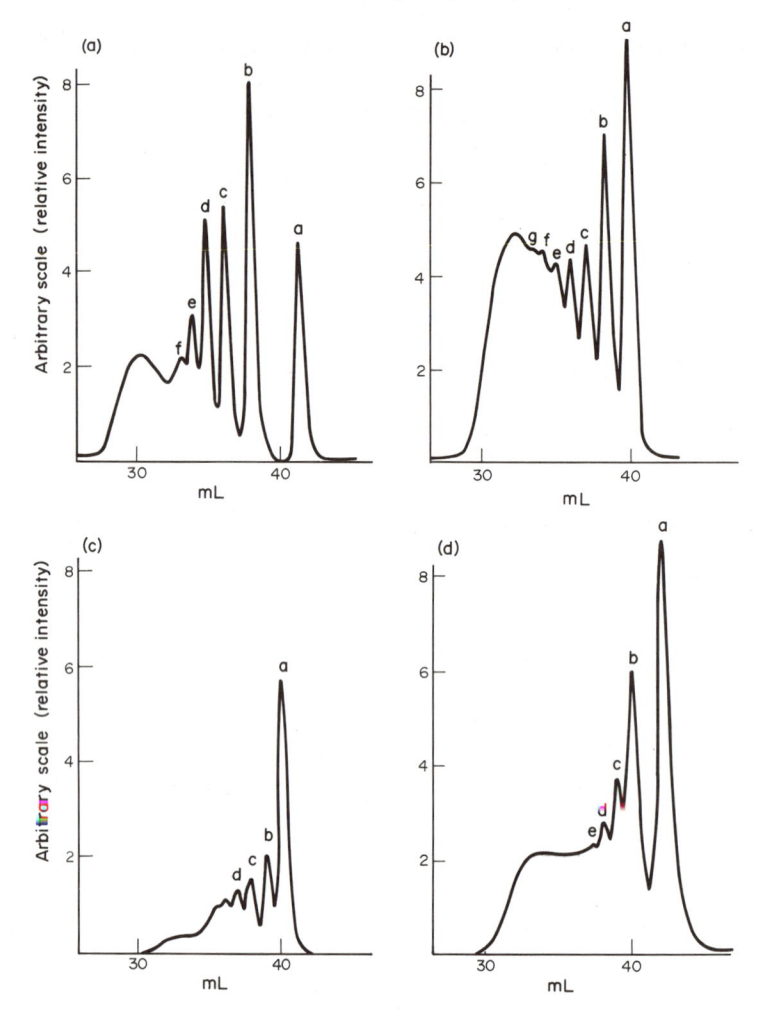

Figure 1 GPC trace (toluene) of cyclic oligomers extracted from polysulfides: (a) polymer I; (b) polymer II; (c) polymer V; and (d) polymer VI. Assignments and structural formulas for GPC peaks refer to oligomers with increasing molecular weight as specified in Table 1[29]

However, if the oligomers are thermally labile or their molecular weights are too high, they are subject to degradation and cannot be detected. Furthermore, if the oligomers are not stable under electron impact (EI-MS) or chemical ionization (CI-MS) modes, their molecular ions are absent in the mass spectra, making identification difficult. Fast atom bombardment mass spectrometry (FAB-MS), a recently introduced technique, appears to be a suitable method for the rapid analysis of complex mixtures of oligomers contained in synthetic polymers.[33-35]

The FAB-MS spectrum of a mixture of oligomers extracted from a crude sample of poly(butylene-isophthaloyl) (Figure 3), shows several peaks (Table 2) corresponding to the protonated molecular ions of cyclic oligomers, as well as to those of linear oligomers with hydroxy and carboxyl end-groups, up to a mass range of about 1450 daltons.[34] The FAB-MS spectrum allows the detection and identification of about twenty compounds present in the polyester extract, whilst only six peaks are detectable in the HPLC trace.[34] The FAB-MS spectrum in Figure 3 allows the detection of five cyclic oligomers and also the three series of linear oligomers expected to be produced by hydrolytic cleavage of the cyclic esters originally produced during the polymerization process (Table 2). In fact, cyclic esters are subject to hydrolysis during extraction and work-up procedures, so that open-chain oligomers are often produced.[34-36] Of course, the presence of linear oligomers complicates the HPLC or GPC tracings, leading to broad peaks and low resolution.

Manolova et al.[36] have recently reported an interesting method which allows an improved HPLC resolution. The mixture of cyclic and open-chain oligomers obtained by extraction of a caprolactone sample was reacted with phenyl isocyanate, therefore transforming the open-chain oligomers into carbamates. The change in molecular weight and polarity considerably improved the HPLC

Table 1 Cyclic Oligomers Formed in the Synthesis of Aromatic–Aliphatic Polysulfides[29]

Number	Cyclic compound	Method of detection	1	2	3	4	5	6	7	8
I	$[-C_6H_4-S-(CH_2)_6-S-]_n$	MS	X	X	X					
		GPC	X	X	X	X	X	X		
II	$[-C_6H_4-S-(CH_2)_3-S-]_n$	MS		X	X					
		GPC		X	X	X	X	X	X	X
III	$[-C_6H_4-S-(CH_2)_3-S-]_n$	MS		X	X					
		GPC		X	X	X	X	X	X	
IV	$[-C_6H_4-S-(CH_2)_3-S-]_n$	MS	X	X	X	X				
		GPC	X	X	X	X	X	X	X	
V	$[-C_6H_4-S-(CH_2)_2-S-]_n$	MS		X	X					
		GPC		X	X	X	X	X		
VI	$[-C_6H_4-SCH_2S-]_n$	MS	X	X	X					
		GPC	X	X	X	X	X	X		

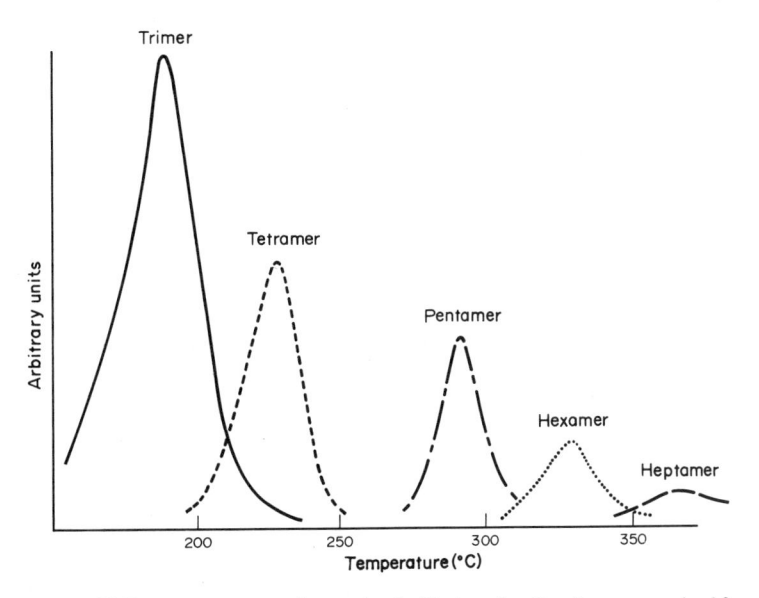

Figure 2 Single ion current (SIC) curves corresponding to the distillation of cyclic oligomers evolved from the MS of a crude sample of m-PPS[32]

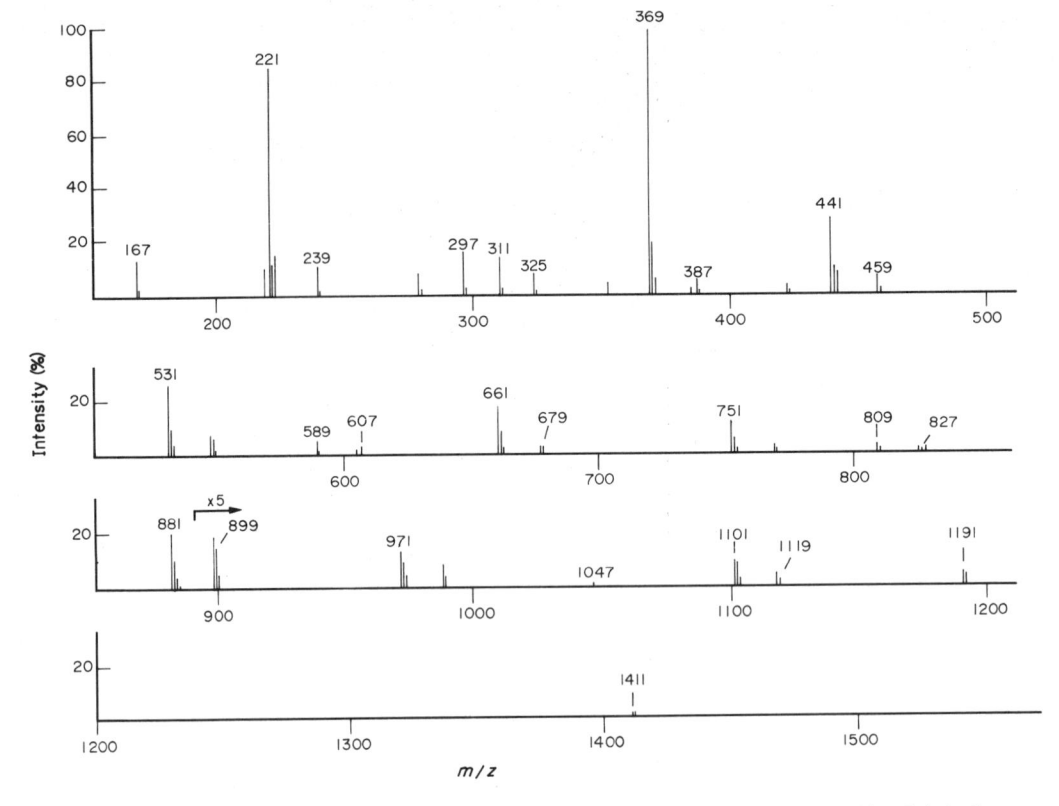

Figure 3 Positive FAB mass spectrum of the oligomer mixture extracted from crude poly(butylene isophthaloyl); structural assignments are given in Table 2[34]

Table 2 Cyclic and Open-chain Oligoesters Present in Poly(butylene-isophthaloyl)

Compound[a]		MH^+ (m/z)
⌐COPhCOO—(CH$_2$)$_4$—O⌐$_x$	$x=1$	221
	$x=2$	441
	$x=3$	661
	$x=4$	881
	$x=5$	1101
HO⌐COPhCOO—(CH$_2$)$_4$—O⌐$_x$H	$x=1$	239
	$x=2$	459
	$x=3$	679
	$x=4$	899
	$x=5$	1119
HO—(CH$_2$)$_4$—O⌐COPhCOO—(CH$_2$)$_4$—O⌐$_x$H	$x=1$	311
	$x=2$	531
	$x=3$	751
	$x=4$	971
	$x=5$	1191
	$x=6$	1411
HO⌐COPhCOO—(CH$_2$)$_4$—O⌐$_x$COPhCO$_2$H	$x=0$	167
	$x=1$	387
	$x=2$	607
	$x=3$	827
	$x=4$	1047

[a] Ph = (meta-disubstituted benzene ring)

resolution, allowing distinction between cyclic and linear oligomers.[36] Therefore, selective derivatization of linear oligomers is a useful way of increasing the resolution in HPLC analysis.

As mentioned above, the thermal decomposition of several polymers leads to the formation of cyclic oligomers. A unique application of MS analysis is the detection of oligomers generated by the direct pyrolysis of polymers in the mass spectrometer.[27,31] The latter is characterized by the simultaneous evaporation of all products formed in each decomposition stage.[37]

It has been seen (Figure 2) that the distillation temperature of preformed oligomers contained in a polymer sample is dependent on the molecular weight.[33] In contrast, the pyrolysis products show isochronous volatilization curves despite the fact that the molecular weight of the oligomers formed may be subject to a several-fold change.

A typical example is given by the pyrolysis of poly(m-hydroxybenzoic acid) [poly(oxy-1,3-phenylenecarbonyl)], which produces a mixture of cyclic oligomers from dimer to heptamer (M_2–M_7).[37] The single ion currents corresponding to the cyclic oligomers (Figure 4) indicate that they are all evolved simultaneously (with a maximum rate at about 500 °C), as would be expected for a series of compounds generated at once.

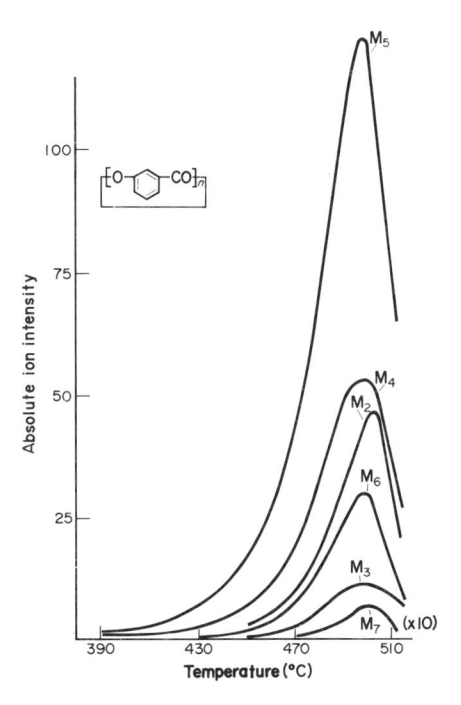

Figure 4 Single ion current (SIC) curves corresponding to the simultaneous evolution of cyclic oligomers formed in the thermal decomposition of poly(m-hydroxybenzoic acid) in MS[37]

Flash pyrolysis of polymers followed by GC–MS is a versatile and reproducible technique, and has been widely used for the analysis of thermal decomposition products of polymers.[38] However, only non-reactive thermally stable low molecular weight oligomers are usually detected by this method.[27,31]

5.4 CYCLIC OLIGOMERS

As mentioned above, the cyclic oligomers of interest in this chapter are those formed in polymerization reactions and in catalyzed or thermal depolymerizations.

This section has been devised to illustrate a series of tables that, for each class of condensation polymers, list the cyclic oligomers that have been reported in the literature in the period 1970–1986. A few references to 1987 articles have also been included. Earlier literature is not comprehensively covered here, but is occasionally quoted. Various aspects of cyclic oligomer formation have already been discussed in previous articles,[2-27] and a compilation of cyclic oligomers synthesized up to 1970 is available.[9]

5.4.1 Esters and Lactones

Cyclic oligomers are often formed in the course of polymerization reactions, or when heating polyesters above their melting temperatures. Indeed, extracts of crude aliphatic polyester samples often reveal the presence of cyclic oligomers. Also depolymerization reactions can be easily induced in several polyesters, leading to the formation of cyclic oligomers.[10,15,39-43]

The structures of cyclic esters and lactones formed under these conditions have been extensively investigated in the recent literature[36,37,43-66] with the help of different techniques (described in Section 5.3), and they are listed in Table 3.

The thermal decomposition of poly(glycolic acid) was reported[44-46] to yield mainly a cyclic dimer (Table 3; compound **1**), but recent studies[43] have shown that a series of higher oligomers are simultaneously generated in the pyrolysis (Table 3; compounds **2–10**). Analogously, the thermal decompositions of poly(lactic acid) and of poly(δ-valerolactone) yield a series of cyclic oligomers (Table 3; compounds **24–33** and **11–16**, respectively).[43,45,48] The mass spectrum of the pyrolysis products of poly(lactic acid) is shown in Figure 5. The pyrolysis produces a series of cyclic oligomers (labeled from M_2 to M_{10} in Figure 5) whose molecular ions appear with decreasing intensity in the spectrum.[48]

An investigation of the thermal decomposition products of poly(ε-caprolactone) showed[43] that, in this case, a series of cyclic oligomers are also formed (Table 3; compounds **17–23**); their distribution is different from that of the cyclic oligomers formed in the polymerization of ε-caprolactone.[36,47]

Cyclic oligomers have been obtained from poly(pivalolactone) by pyrolysis MS (Table 3; compounds **34–44**).[15,43,49] Furthermore, these oligomers have been separated and characterized by GPC.[49]

The cyclic dimer of poly(thioglycolic acid) (Table 3; compound **45**) has been isolated by HPLC from a pyrolysate of this polymer.[50]

Cyclic ethylene glycol and poly(ethylene glycol) succinates, previously prepared by Carothers et al[40] by thermal decomposition of the corresponding polymers, have been recently obtained (Table 3; compounds **46, 47, 50–57**) in the attempted polymerization of ethylene succinates.[51] Cyclic oligomers have also been detected in the pyrolysis MS of poly(ethylene succinate) (Table 3; compounds **46–49**).[43]

Numerous cyclic oligomers have been detected in the ring–chain equilibration of poly(ethylene adipate) (Table 3; compounds **68–71**),[43] poly(butylene adipate) (Table 3; compounds **72–76**)[53] and poly(decamethylene adipate) (Table 3; compounds **77–81**).[52] Cyclic oligoesters (Table 3; compounds **72–76**) have been detected in the toluene extract of a polyurethane synthesized from methylene/-di-p-phenyl diisocyanate (MDI) and a polyester–polyol.[53]

Aromatic oligoesters from succinic acid and hydroquinone or resorcinol have been detected by pyrolysis MS (Table 3; compounds **82–85** and **88–90**, respectively);[54,55] similar oligomers are obtained when succinic acid is replaced by adipic acid (Table 3; compounds **86, 87** and **91**).[54,55] Poly(ethyleneterephthaloyl) (PETP) oligomers (Table 3; compounds **92–99**) have been characterized by several research workers.[55-60] Cyclic PETP has also been recently reported.[56] In commercial PETP, a cyclic dimer containing three units of ethylene glycol and two units of terephthalic acid (Table 3; compound **114**) is also present.[59,64] Cyclic oligomers of poly(butyleneterephthaloyl) (PBTP) have been isolated by HPLC (Table 3; compounds **101–106**).[61-63]

In the high temperature synthesis of poly(butyleneisophthaloyl) (PBIP) a series of oligomers have been detected (Table 3; compounds **107–111**).[34] In the polymerization of 5-hydroxymethyl-2-furancarboxylic acid, several cyclic lactones were formed (Table 3; compounds **115–119**) which were separated by HPLC.[65,66]

Totally aromatic cyclic oligoesters have been obtained by thermal decomposition of aromatic polyesters (Table 3; compounds **120–133**).[37]

5.4.2 Carbonates

Several cyclic carbonates, both aliphatic–aromatic and totally aromatic, have been reported in the literature (Table 4).[67-73] The majority of cyclic carbonates discussed in this section have been obtained by pyrolysis of the corresponding polymers. However, in the case of poly(2,2-bis-(4-hydroxyphenyl)propane carbonate) [poly(oxy-1,4-phenylenedimethylmethylene-1,4-phenyleneoxycarbonyl)], the cyclic oligomers have also been obtained by extraction of the crude polymer (Table 4; compounds **146–148**). The formation of very large macrocyclic carbonates has been

Table 3 Cyclic Esters and Lactones

Compound number	Oligomer structure		n	Origin[a]	Detection[b]	Ref.
(1–10)	$\text{-[-(CH}_2\text{)}_m\text{-CO}_2\text{-]}_n$	$m = 1$	2–11	D	A, E	43–46
(11–16)	$\text{-[-CHMeCO}_2\text{-]}_n$	$m = 4$	1–6	D	E	43
(17–23)		$m = 5$	1–7	D	E	36, 43, 47
(24–33)			2–11	D	E	48
(34–44)	$\text{-[-CH}_2\text{CHMe}_2\text{CO}_2\text{-]}_n$		2–12	D	A, E	49
(45)	$\text{-[-CH}_2\text{COS-]}_n$		2	D	A	50
(46–49)	$\text{-[-OCCH}_2\text{CH}_2\text{CO}_2\text{-(CH}_2\text{)}_m\text{-O-]}_n$	$m = 2$	1–4	S, D	A, E	43, 51
(50), (51)		$m = 3$	1, 2	S	A	51
(52), (53)		$m = 4$	1, 2	S	A	51
(54), (55)		$m = 5$	1, 2	S	A	51
(56), (57)		$m = 10$	1, 2	S	A	51
(58–64)	$\text{-[-OCCH}_2\text{CH}_2\text{CO}_2\text{-(CH}_2\text{CH}_2\text{O)}_m\text{-]}_n$	$m = 3$	1–7	CD	C	52
(65–67)		$m = 4$–6	1	S	A	51
(68–71)	$\text{-[-OC(CH}_2\text{)}_4\text{CO}_2\text{-(CH}_2\text{)}_m\text{-O-]}_n$	$m = 2$	1–4	D	E	43
(72–76)		$m = 4$	2–6	S	E	53
(77–81)		$m = 10$	1–5	CD	C	52
(82–85)	$\text{-[-OC-(CH}_2\text{)}_m\text{-CO}_2\text{-C}_6\text{H}_4\text{-O-]}_n$	$m = 2$	2–4	D	E	55, 54
(86), (87)		$m = 4$	2, 3	D	E	55, 54

Table 3 (continued)

Compound number	Oligomer structure		n	Origin[a]	Detection[b]	Ref.
(88–90)	$\text{\textemdash OC-}\langle\text{C}_6\text{H}_4\rangle\text{-CO}_2\text{-(CH}_2)_m\text{-O-}]_n$	m = 2	2–4	S, D	A, E	54
(91)		m = 4	2	D	E	54
(92–99)	$\text{\textemdash OC-}\langle\text{C}_6\text{H}_4\rangle\text{-CO}_2\text{-(CH}_2)_m\text{-O-}]_n$	m = 2	2–9	S, CD	A	56, 59, 60
(100)			very large	CD	C	52
(101–106)		m = 4	2–7	S	A, B	61, 62, 63
(107–111)	$\text{\textemdash OC-}\langle\text{C}_6\text{H}_4\rangle\text{-CO}_2\text{-(CH}_2)_m\text{-O-}]_n$		1–5	S	E	48
(112)	$\text{\textemdash OC-}\langle\text{C}_6\text{H}_4\rangle\text{-CO}_2\text{CH}_2\text{CH}_2\text{O-}]_n$	m = 2	2	S	A	57
(113)		m = 4	2	S	A	57
(114)	—OCH$_2$CH$_2$		2	S	A	59, 64
(115–119)	$\text{\textemdash OC-}\langle\text{furan}\rangle\text{-CH}_2\text{O-}]_n$		2–6	S	A, C	65, 66

Formation of Cyclic Oligomers

Compound	Structure	n	Method	Ref.
(120–125)	$\{OC\text{-}C_6H_4\text{-}O\}_n$	–7	D, E	37
(126)	$\{OC\text{-}C_6H_4\text{-}CO_2\text{-}C_6H_4\text{-}O\}_n$	2	D, E	37
(127), (128)	$\{OC\text{-}C_6H_4\text{-}CO_2\text{-}C_6H_4\text{-}O\}_n$	2, 3	D, E	37
(129), (130)	$\{OC\text{-}C_6H_4\text{-}CO_2\text{-}C_6H_4\text{-}O\}_n$	2, 3	D, E	37
(131–133)	$\{OC\text{-}C_6H_4\text{-}CO_2\text{-}C_6H_4\text{-}O\}_n$	1–3	D, E	37

[a] S = oligomers obtained during polymerization conditions; CD = oligomers obtained in catalyzed solution depolymerization; D = oligomers obtained by thermal decomposition (pyrolysis conditions).
[b] A = identified and characterized; B = separated by HPLC; C = separated by GPC; E = analyzed by MS.

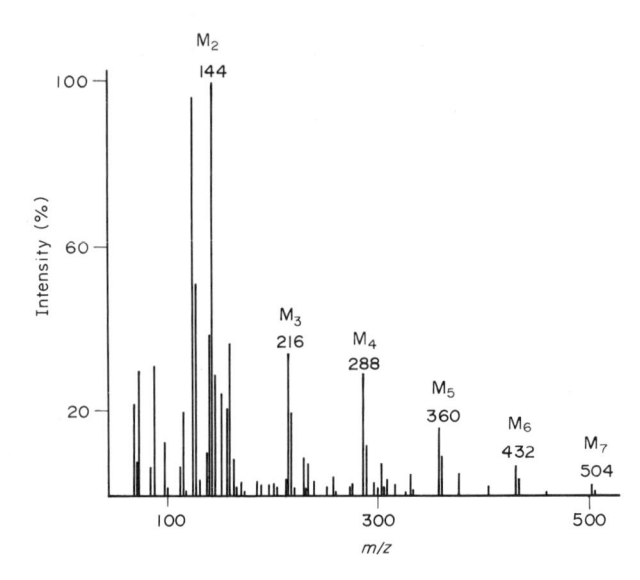

Figure 5 Ammonia negative chemical ionization spectrum of poly(lactic acid) taken at 360 °C. Cyclic oligomers (M_2–M_{10}) appear with decreasing intensity in the spectrum[48]

observed in the interfacial polymerization of bisphenol *A* (2,2-bis(4-hydroxyphenyl)propane) chloroformate with bisphenol *A*.[72] Catalyzed depolymerization of the corresponding chlorinated polymer afforded the cyclic trimer and tetramer (Table 4; compounds **150, 151**). The thermal decomposition of several polycarbonates yielded[67] numerous cyclic oligomers (Table 4; compounds **134–137**).

5.4.3 Amides and Lactams

Cyclic amides and lactams are among the most important molecules in this class of compounds, and have been widely studied.[10, 27, 74–84] As already mentioned, cyclic amides obtained in ring-opening polymerization reactions[21] will not be considered here.

The formation of lactams (Table 5; compounds **157–167**) has been reported in the thermal reorganization of some copolyamides obtained from aliphatic and aromatic amino acids.[35] Many studies on cyclic oligomers present in aliphatic polyamides have been concerned with Nylon 6 (Table 5; compounds **161–167**). Recently, HPLC and MS have been used to separate and identify the cyclic oligomers of Nylon 6.[33, 34] In Figure 6, an HPLC trace shows the separation of cyclic oligoamides of a Nylon 6 methanolic extract, up to the hexamer.[34]

The FAB-MS in Figure 7 shows the protonated molecular ions of cyclic oligomers contained in the above mixture (from the monomer to the decamer), indicating that FAB is useful for the detection of high molecular weight oligomers (above 1100 daltons) present in traces in the mixture.[34] Gradual heating of the above mixture into the ion source of the mass spectrometer does not lead to comparable results. In fact, low molecular weight oligomers (from monomer to pentamer) are detected by EI-MS, whereas oligomers from hexamer to decamer do not volatilize undecomposed and therefore remain undetected.[34]

The hydrogen-transfer polymerization of acrylamide is believed to be a step-growth process, and a cyclic trimer has been isolated from the polymerization batch (Table 5; compound **158**).[78, 79]

Several cyclic oligomers (Table 5; compounds **168–214**) have been detected in the thermal decomposition of a series of poly(α-amino acids) and copoly(α-amino acids).[81, 82]

A cyclic ester amide (Table 5; compound **215**) has been isolated and characterized during the polymerization reaction leading to the corresponding polymer.[83]

Cyclic oligomers are formed during polymerization leading to cyclic aliphatic polyamides (Table 5; compounds **216–218**), and their concentration is a function of temperature and water content.[84]

Formation of Cyclic Oligomers

Table 4 Cyclic Carbonates

Compound number	Oligomer structure	n	Origin[a]	Detection[b]	Ref.
(134), (135)	[−O(CH₂)₄OCO₂−C₆H₄−O−CO−]ₙ / −CO₂(CH₂)₄O−	1, 2	D	C	67
(136)	[−O−C₆H₄−CMe₂−C₆H₄−O−CO−]ₙ / −CO₂(CH₂)₄O−	2	D	C	67
(137)	[−O−C₆H₄−OCO₂(CH₂)₄O−CO−]ₙ	2	D	C	67
(138), (139)	[−O−C₆H₄−O₂C(CH₂)₄O−CO−]ₙ / −CO₂−C₆H₄−O−	1, 2	D	C	67
(140), (141)	[−O−C₆H₄−CMe₂−C₆H₄−(OCO₂(CH₂)₄O−CO)ₘ−]ₙ, $m=1$	1, 2	D	C	67
(142)	$m=2$	1	D	C	67
(143–145)	[−O−C₆H₄−O−CO−]ₙ	3–5	D	C	67
(146–148)	[−O−C₆H₄−CMe₂−C₆H₄−O−CO−]ₙ	2–4	S, D	A, C	67–71
(149)		very large	S	B	72
(150), (151)	[−O−C₆H₄−C(CCl₂)−C₆H₄−O−CO−]ₙ	3, 4	CD	A	73

[a] S = oligomers obtained during polymerization conditions; CD = oligomers obtained in catalyzed solution depolymerization; D = oligomers obtained by thermal decomposition (pyrolysis conditions).
[b] A = identified and characterized; B = separated by GPC; C = analyzed by MS.

5.4.4 Urethanes and Ureas

Some cyclic urethane esters (Table 6; compounds **219–226**) have been detected in a polyurethane based on MDI and a polyol obtained from butanediol and adipic acid.[53]

In the thermal decomposition of several polyurethanes and poly(urethane carbonates)[85–87] numerous cyclic oligomers have been detected (Table 6; compounds **228–239**).

The separation and characterization of cyclic oligomers formed in step-growth reactions leading to polyurethanes and polyureas has been reported.[86] In some cases, simultaneous detection of the macrocycles directly from the polymerization mixtures has been achieved by MS.[86] Cyclic com-

Table 5 Cyclic Amides and Lactams

Compound number	Oligomer structure	m	n	Origin[a]	Detection[b]	Ref.
	$\mathrm{{-}[CO{-}(CH_2)_m{-}NH]_n{-}}$					
(152–156)		1	2–6	D	G	81
(157), (158)		2	1, 3	D, S	A	77–79
(159)		3	1	D	G	77
(160)		4	1	D	G	77
(161–166)		5	1–6	D, CD	C, G	13, 77
(167)			> 6	S	G	34, 80
(168–174)	$\mathrm{{-}[COCH_2NMe]_n{-}}$	—	2, 4–9	D	G	81
(175–181)	$\mathrm{{-}[CO{-}N{\frown}]_n{-}}$	—	2–8	D	G	81
(182–209)	$\mathrm{{-}(CO{-}N{\frown})_m(COCH_2NMe)_n{-}}$	1–8	1–9	D	G	82
(210–214)	$\mathrm{{-}(COCHMeNH)_m(CO{-}N{\frown})_n{-}}$	1–3	1, 2	D	G	82
(215)	$\mathrm{{-}[CO(CH_2)_5OCO(CH_2)_5NH]_n{-}}$	—	1	S	A	83
(216–218)	—HN⟨⟩CH$_2$⟨⟩NH— / —(COCH$_2$)$_{10}$—CO— *cis-cis; cis-trans; trans-trans*	—	—	S	A	84

[a] S = oligomers obtained during polymerization conditions; CD = oligomers obtained in catalyzed solution depolymerization; D = oligomers obtained by thermal decomposition (pyrolysis conditions).
[b] A = identified and characterized; B = separated by HPLC; C = separated by GPC; E = separated by GC; F = separated by TLC; G = analyzed by MS; H = analyzed by GC-MS.

pounds (**227**) and (**245**)–(**247**) (Table 6), subsequent to detection by MS, were isolated and characterized. In Figure 8 are shown the total ion current (TIC) *vs.* temperature curves for the crude polymers analyzed. These curves somewhat resemble chromatographic tracings and, actually, each peak corresponds to the evolution of a cyclic oligomer.[86] In contrast to polyurethanes, polyureas do not yield cyclic oligomers by thermal decomposition, except in one case (Table 6; compounds **240–244**). However, although the latter polymer can be formally a polyurea, it is really a poly(aminoformal).[88]

5.4.5 Formals

The majority of cyclic ethers and formals which are reported in the literature are formed during ring-opening polymerization.[89–101] However, some cyclic formals are also formed by thermal decomposition (Table 7; compounds **281–290**)[102] and by catalyzed depolymerization (Table 7; compounds **248–280**)[103–107] of pure polymers.

In the case of poly(oxy-1,4-phenylenedimethylmethylene-1,4-phenyleneoxymethylene), two preformed oligomers (Table 7; compounds **289, 290**) were detected by MS analysis.[102] Figure 9(a) shows the TIC *vs.* temperature curve of a crude polymer sample. The two maxima at 220 °C and 320 °C correspond to the evolution of the cyclic oligomers (**289**) and (**290**), respectively.[102] Once detected and identified by MS, the two cyclic oligomers were separated from the polymer by solvent extraction. The TIC curve in Figure 9(b), corresponding to a sample of the polymer after oligomer extraction, shows a negligible ion current below 400 °C.[102]

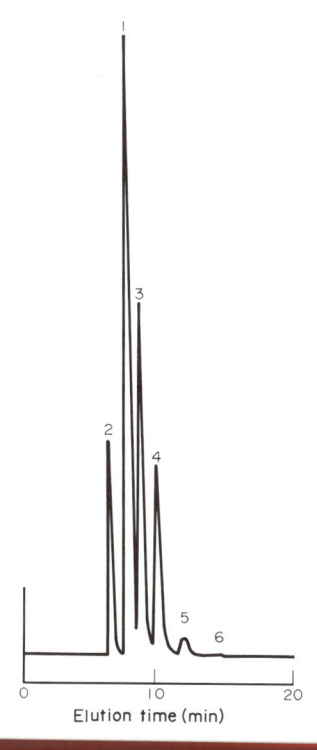

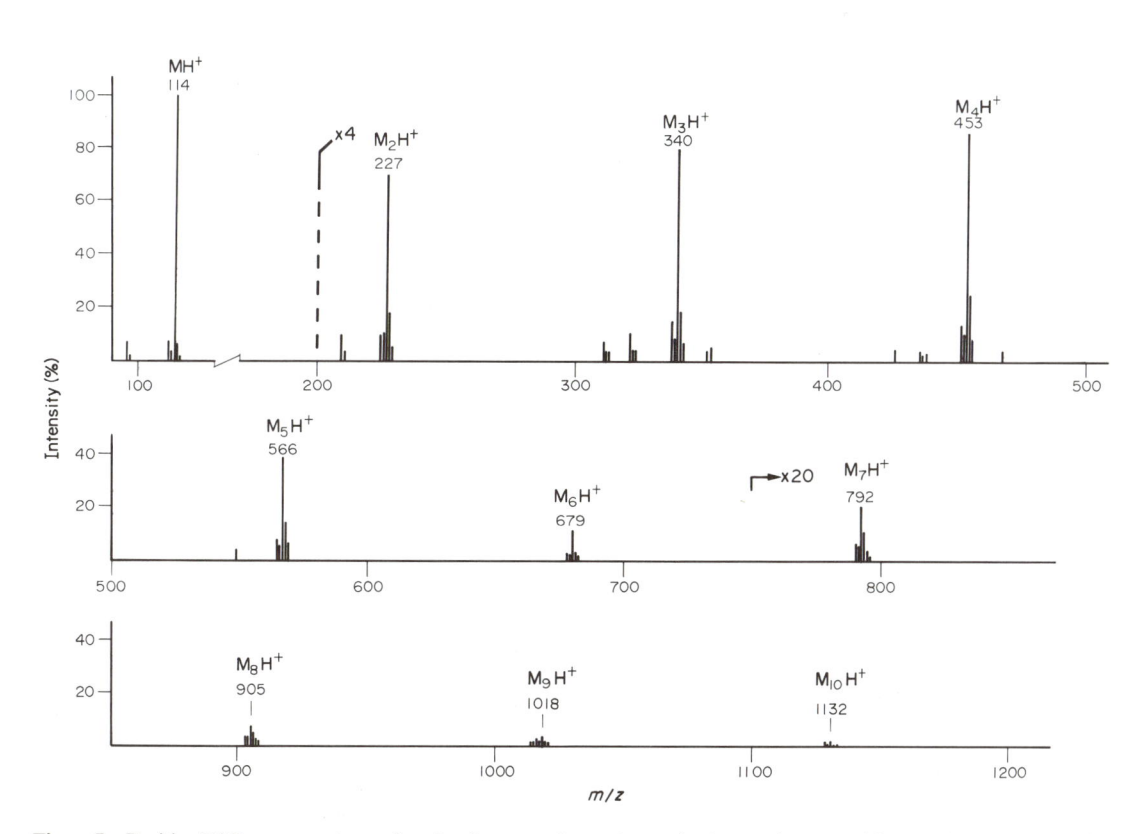

Figure 7 Positive FAB mass spectrum of cyclic oligomer mixture (up to the decamer) extracted from Nylon 6 (same sample as in Figure 6)[34]

Table 6 Cyclic Urethanes and Ureas

Compound number	Oligomer structure[a]	n	Origin[b]	Detection[c]	Ref.
	$\left[\begin{array}{l}\text{-[O(CH}_2)_4\text{OCONHArNHCO]}_m\text{-} \\ \text{-[CO(CH}_2)_4\text{CO}_2\text{(CH}_2)_4\text{O]}_n\text{-}\end{array}\right]$				
(219–223)	m = 1	1–5	S	B	53
(224–226)	m = 2	1–3	S	B	53
(227)	[-OCH$_2$CMe$_2$CH$_2$OCONMeArNMeCO-]$_n$	2	S	A	86
(228–230)	[-O-C$_6$H$_4$-OCONMe(CH$_2$)$_6$NMeCO-]$_n$	1–3	D	B	87
(231), (232)	[-O-C$_6$H$_4$-OCON(piperazine)CO-]$_n$	2, 3	D	B	87
(233)	[(-O-C$_6$H$_4$-OCO-)$_2$-N(piperazine)NCO-]$_n$	1	D	B	68
(234)	[-O-C$_6$H$_4$-OCONMeAr'NMeCO-]$_n$	1	D	B	87
(235), (236)	[-OAr'OCONMe(CH$_2$)$_6$NMeCO-]$_n$	1, 2	D	B	87
(237), (238)	[-CON(piperazine)N-(CO$_2$Ar'O)-]$_n$	1, 2	D	B	68, 87
(239)	[-OAr'OCONMeArNMeCO-]$_n$	1	D	A	85
(240–244)	[-N(ring with CO)NCH$_2$-]$_n$	2–6	D	B	88
(245)	[-NMeArNMeCO-]$_n$	2	S	A	86
(246), (247)	[-NMeArNMeCONMe(CH$_2$)$_6$NMeCO-]$_n$	1, 2	S	A	86

[a] Ar = –C$_6$H$_4$–CH$_2$–C$_6$H$_4$–, Ar' = –C$_6$H$_4$–CMe$_2$–C$_6$H$_4$–

[b] S = oligomers obtained during polymerization conditions; D = oligomers obtained by thermal decomposition (pyrolysis conditions).

[c] A = identified and characterized; B = analyzed by MS.

5.4.6 Sulfides

Cyclic oligomers with sulfide (or disulfide) bridges have been reported in the literature, originating from both step-growth polymerization reactions and thermal or catalyzed depolymerizations.[28,29,32,108–115]

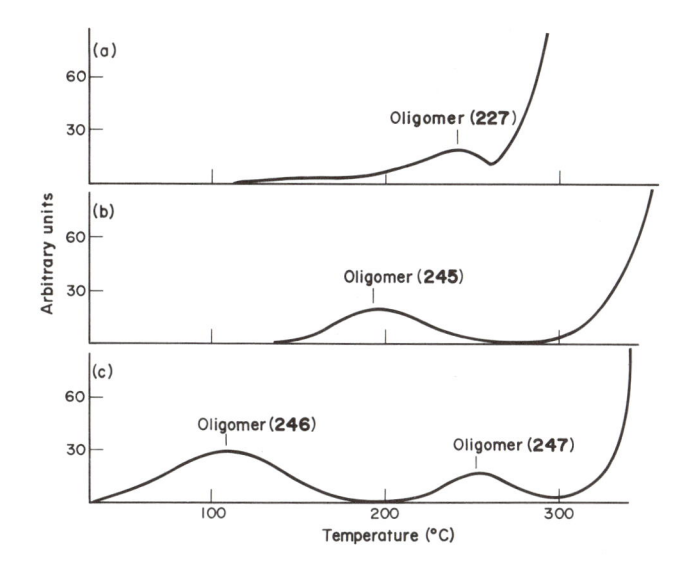

Figure 8 Total ion current (TIC) curves corresponding to the distillation of the cyclic urethanes and ureas evolved from the MS. Oligomer structures are (227), (245) and (247), respectively, in Table 6[86]

Table 7 Cyclic Formals

number	structure	m	n	Original[a]	Detection[b]	Ref
(248), (249)	$\{CH_2O-(CH_2CH_2O)_m\}_n$	1	2, 3	CD	B, C	106
(250–254)		2	2–6	CD	A	107
(255–274)		3	1–20	CD	A, B	103–105, 107
(275–277)		4	2–4	CD	A	107
(278–280)		5	2–4	CD	A	107
(281–284)	$\{CH_2O-C_6H_4-O\}_n$ (para)		2–5	D	C	102
(285–288)	$\{CH_2O-C_6H_4-O\}_n$ (meta)		2–5	D	C	102
(289), (290)	$\{CH_2O-C_6H_4-CMe_2-C_6H_4-O\}_n$		2, 3	S, D	A, C	95, 102

[a] S = Oligomers obtained during polymerization conditions; CD = Oligomers obtained in catalyzed solution depolymerization; D = Oligomers obtained by thermal decomposition (Pyrolysis conditions).
[b] A = identified and characterized; B = separated by GPC; C = analyzed by MS.

Aliphatic sulfides (Table 8; compounds **291–328**) have been extensively investigated.[28,108–111] A series of aliphatic polysulfides has been obtained from the polycondensation of dibromoalkanes with dithiols, and GPC and MS have been used to detect and identify the cyclic oligomers formed in the polymerization reactions.[28,29,32] The GPC trace given in Figure 10 shows that the soluble fraction extracted from the crude sample of poly(thiotrimethylene) contains a series of low molecular weight compounds. The identification of these compounds was achieved by fractional distillation of the oligomers in the MS.[28] Only cyclic oligomers with an even number of repeat units were detected (Table 9), and a linear dependence of elution volume on $\log M$ could be established for these compounds (Figure 11). The fact that only an *even* number of repeating units are contained in the

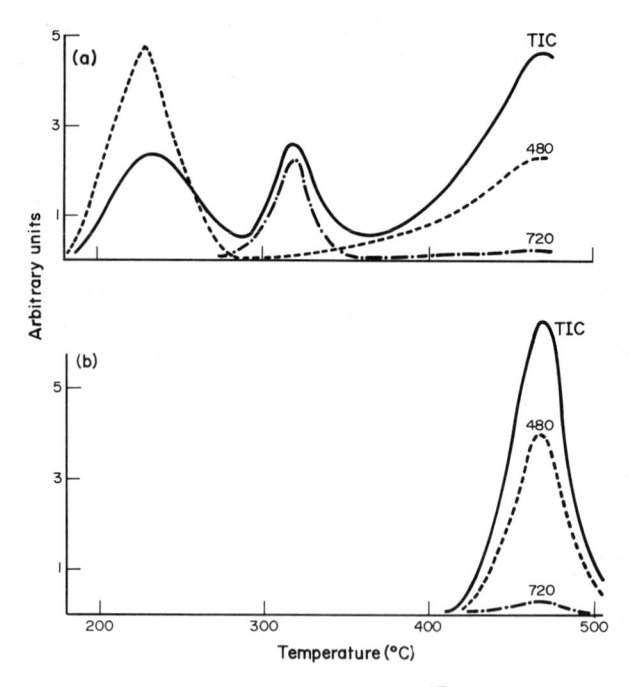

Figure 9 (a) Total ion current (TIC) curves of a crude sample of $\text{-}(\text{O-}\langle\bigcirc\rangle\text{-CMe}_2\text{OCH}_2)_{\overline{n}}$ showing two maxima which correspond to the evolution of preformed cyclic dimer (m/z 480) and cyclic trimer (m/z 720); single ion curves (SIC) for these molecular ions are also shown; (b) TIC and SIC curves as above, corresponding to a sample of the polymer after oligomer extraction; in this case, the cyclic oligomers originate from the polymer pyrolysis[102]

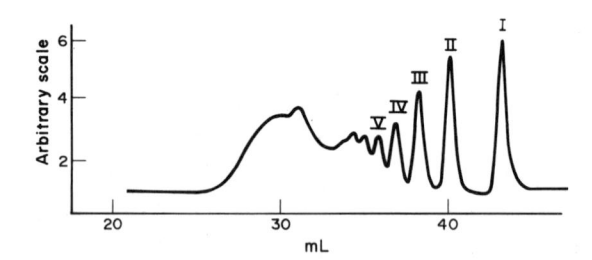

Figure 10 GPC trace (toluene) of the cyclic oligomers extracted from a crude sample of poly(trimethylene sulfide) $\text{-}\!\!\left[\text{-}(\text{CH}_2)_3\text{-S}\right]_{\overline{n}}$; structural assignments of GPC peaks refer to oligomers with increasing molecular weight as specified in Table 9[28]

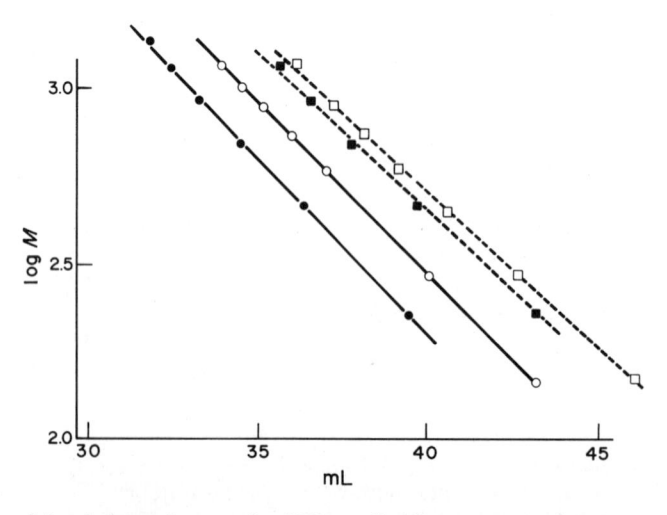

Figure 11 Dependence of the elution volume on log M for cyclic oligomers: —, toluene; ------, THF; ● ■, poly(hexamethylene sulfide); ○ □, poly(trimethylene sulfide)[28]

Table 8 Cyclic Sulfides

Compound number	Oligomer structure	m	n	Origin[a]	Detection[b]	Ref.
(291–296)	$\mathrm{+(CH_2)_m\!-\!S\!\!+_n}$	1	2–7	S, D	C, E, F	28, 108
(297–303)		2	1–7	S, D	E	108, 109
(304–314)		3	1–8, 10, 12, 14	S, D	C, E	28, 108, 109
(315–320)		6	2, 4, 6, 8, 10, 12	S	C, E	28
(321–325)	$\mathrm{+CHMeCH_2S_m+_n}$	1	2; 4–7	CD	C, F	110
(326)		2	1	CD	C, F	110
(327)	$\mathrm{+CMeEtCH_2S_m+_n}$	1	2	CD	C, F	111
(328)		2	1	CD	C, F	101
(329–334)	[phenyl-S–(CH$_2$)$_m$–S]$_n$	1	2–7	S, D	C, E	29, 112
(335–339)		2	2–6	S	C, E	29
(340–346)		3	2–8	S	C, E	29
(347–352)		6	1–6	S	C, E	29
(353–355)	[phenyl with S bridge–(CH$_2$)$_m$–S]$_n$	3	1–3	D	E	112
(356), (357)	[phenyl-SS–(CH$_2$)$_m$–S]$_n$	1	2, 3	D	E	112
(358), (359)		6	1, 2	D	E	112

Table 8 (continued)

Compound number	Oligomer structure	m	n	Origin[a]	Detection[b]	Ref.
(360–362)	H–[S–(CH$_2$)$_m$–S–C$_6$H$_4$–S–C$_6$H$_4$–S]$_n$ (ring with two phenylene-S units)	1	1–3	D	E	112
(363), (364)		2	1, 2	D	E	112
(365), (366)		3	1, 2	D	E	112
(367–369)	H–[S–(CH$_2$)$_m$–SS–C$_6$H$_4$–S–C$_6$H$_4$]$_n$	3	1–3	D	E	112
(370), (371)	H–[C$_6$H$_4$–S]$_n$	1	1, 2	D	E	112
(372), (373)		2	0, 2	D	E	112
(374–378)		—	2–6	S, D	E	32
(379–383)	H–[C$_6$H$_4$–S]$_n$	—	3–7	S, D	C, E	32
(384–391)	H–[C$_6$H$_4$–S]$_n$	—	3–10	S, D	A, E, B+E	32, 113, 115
(392)	H–[C$_6$H$_4$–C$_6$H$_4$–S]$_n$	—	4	S	A	115

[a] S = oligomers obtained during polymerization conditions; CD = oligomers obtained in catalyzed solution depolymerization; D = oligomers obtained by thermal decomposition (pyrolysis conditions).
[b] A = identified and characterized; B = separated by HPLC; C = separated by GPC; E = analyzed by MS; F = analyzed by GC–MS.

Table 9 Cyclic Oligomers Formed in the Synthesis of Aliphatic Polysulfides[28]

Cyclic compound	Method of detection	Number of repeating units in the cyclic oligomers												
		2	3	4	5	6	7	8	9	10	11	12	13	14
$\text{-}(\text{CH}_2)_6\text{S-}_n$	MS	X		X		X								
	GPC	X	X	X		X				X		X		
$\text{-}(\text{CH}_2)_3\text{S-}_n$	MS	X		X		X		X						
	GPC	X		X		X		X		X		X		X
$\text{-}(\text{CH}_2)_2\text{S-}_n$	MS	X												
	GPC	X	X											
$\text{-CH}_2\text{S-}_n$	MS	X	X	X	X									
	GPC													

cyclic oligomers (Table 9; compounds **304–320**) suggests that these cyclic oligomers are formed through an end-biting process of the type shown in equation (3). Oligomers containing up to 14 units were separated by GPC, and the higher molecular weight products are most likely also macrocycles.[28]

$$Br\text{\textasciitilde}R\text{\textasciitilde}Br + HS\text{\textasciitilde}R\text{\textasciitilde}SH \xrightarrow{Na} \begin{array}{c}Br\\R\\S\text{\textasciitilde}R\text{\textasciitilde}S^-Na^+\end{array} \xrightarrow{-NaBr} R\bigcirc R \quad (3)$$

This hypothesis is supported by the fact that thermal degradation of the above polysulfides causes the formation of cyclic oligomers containing both even and odd numbers of repeating units.[108] In fact, the theory predicts that the absence of reactive end-groups would result in a *back-biting* process with the production of all possible ring species.[3]

Aliphatic–aromatic cyclic sulfides (Table 8; compounds **329–352**) have been obtained in step-growth polymerization reactions[29] and also by pyrolysis MS (Table 8; compounds **353–373**).[112] Remarkably, the thermal decomposition of the above sulfides[112] afforded oligomers with disulfide bridges (Table 8; compounds **353–355, 360–362, 370–373**).

GPC and MS have been used to detect and identify the oligomers (Table 8; compounds **374–391**) formed in the polymerization reactions leading to poly(thio-1,2,3 or 4)phenylene).[32] Furthermore, the thermal degradation of these polymers was investigated by MS and the results indicate that they decompose producing the same cyclic oligomers (Table 8; compounds **374–391**). The distribution of the latter compounds is dependent on the structure of each isomeric polymer, as shown in Figure 12.[32]

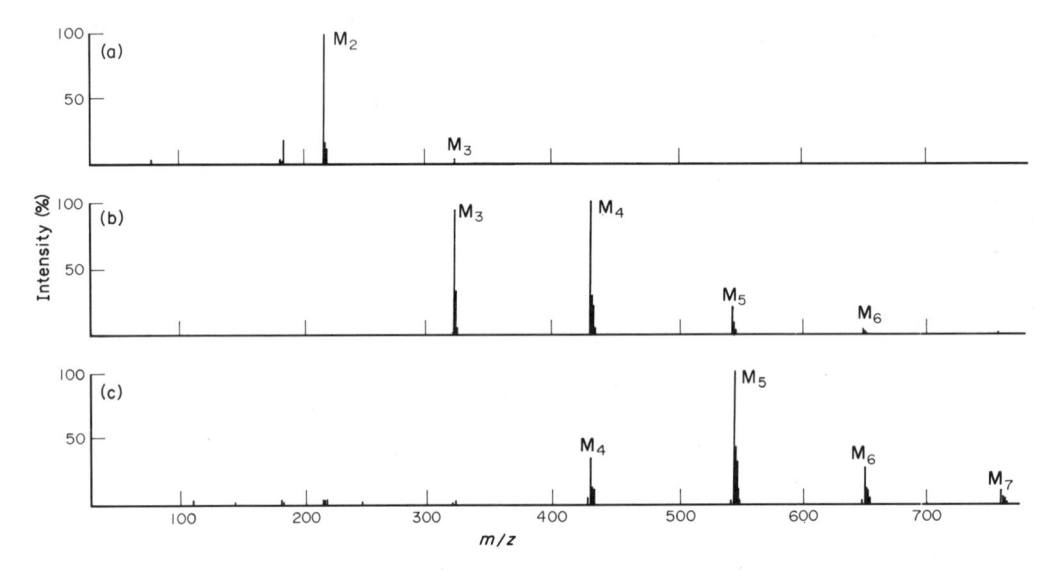

Figure 12 Mass spectra of the cyclic oligomers produced by thermal degradation of: (a) *o*-PPS; (b) *m*-PPS; (c) *p*-PPS. The relative abundance of the cyclic compounds formed is dependent on the structure of the polymers[32]

The cyclic sulfide (**392**; Table 8) has been obtained in the synthesis of poly(thiobiphenylene).[115]

5.4.7 Siloxanes

The formation of cyclic oligomers in the polymerization reactions leading to polysiloxanes is well documented.[10,25,116–119] Also, it is well known that polysiloxanes and many of their copolymers thermally decompose *via* an intramolecular exchange process leading to the production of cyclic oligomers.[118–122] Polysiloxanes equilibrate under the influence of heat and catalysts to yield a thermodynamically controlled distribution of ring and chain molecules.[10] Conversion to cyclic oligomers also occurs under slow and flash pyrolysis conditions, but a kinetically controlled distribution is obtained in these cases (Table 10).[118]

Table 10 Distribution of Cyclic Products Generated in the Thermal Degradation of Poly(dimethylsiloxanes)[a]

n	Equilibrium conditions[b]	TG Exp.[c]	Flash pyrolysis[d]	Slow pyrolysis[e]
3	1.7	185.9	277.8	243.9
4	100	100	100	100
5	61	41.1	36	35.6
6	18.5	46.3	22.2	37.6
7	3.8	30.7	16.6	13.1
8	2.5	—	—	1.1
9	1.7	—	—	0.2

[a] Octamethylcyclotetrasiloxane ($n = 4$) arbitrarily assigned the value 100.
[b] Data taken from ref. 126.
[c] Data taken from ref. 127.
[d] Data taken from ref. 128.
[e] Data taken from ref. 118.

Cyclic siloxanes (Table 11; compounds **393–423**) have been obtained by thermal or catalytic decomposition of the corresponding polymers.[10,116–122] The formation of large macrocyclic siloxanes has been reported for poly(dimethylsiloxane) (Table 11; compound **402**)[10,18] and for poly(phenylmethylsiloxane) (Table 11; compound **420**).[123]

Cyclic silarylene–siloxane oligomers (Table 11; compounds **424–464**) have been reported to originate from thermal decomposition of several copolymers.[119,121,124] An Si—C/Si—O and/or an Si—O/Si—O exchange process occurs in these cases and the corresponding oligomers have different structures from those of the original polymer repeat units.[27,119,121,124]

Various substituted cyclic siloxanes have been recently reviewed, emphasizing the preparative aspects and also the separation of the cyclic siloxanes by preparative GPC.[25]

Cyclic siloxanes up to about 10 000 daltons were recently detected by time-of-flight secondary-ion mass spectrometry (TOF-SIMS).[125]

5.4.8 Miscellaneous

Cyclic phosphazenes have been obtained by thermal or catalyzed depolymerization of various polyphosphazenes (Table 12; compounds **465–472**).[129–131]

A cyclic dimer (Table 12; compound **473**) has been found in the polymerization reaction leading to the corresponding poly(azomethine).[132]

Several cyclic oligomers (Table 12; compounds **474–532**) have been identified by MS in a Wittig polycondensation reaction leading to polymers containing trifluoromethylvinylene, 1,4-phenylene and/or 2,5-thiophenediyl units.[133,134]

5.5 FORMATION OF LINEAR OLIGOMERS

The formation of linear oligomers in polymerization reactions has been occasionally mentioned in the foregoing sections. Apart from the trivial case of low degrees of monomer conversion, which necessarily results in low molecular weight linear oligomers, their formation might be difficult to explain in step-growth polymerization reactions at high monomer conversion, where the molecular weight distribution is shifted to relatively high weights.

A further complication ought be taken into account, *i.e.* the simultaneous presence of linear and cyclic oligomers in many of the systems investigated. However, the latter observation is the key to the explanation of the problem.

In fact, it has been found that cyclic oligomers are first produced in the polymerization processes, and then hydrolyzed to linear oligomers both in the polymerization reaction or, more often, during the work-up process (extraction, precipitation, washing, drying, *etc.*).

The most common cases are with polyesters[34,36] and polycarbonates[69,70] (very sensitive to acids and humidity), and with Nylons.[135]

Table 11 Cyclic Siloxanes

Compound number	Oligomer structure	m	n	Origin[a]	Detection[b]	Ref.
(393–401) (402)	$\left[\begin{array}{c}\text{Me}\\-\text{SiO}-\\\text{Me}\end{array}\right]_n$		3–11 very large	D S, CD	C, E, B A	116–118, 120 16, 25
(403–411)	$\left[\begin{array}{c}\text{Me}\\-\text{SiO}-\\\text{Et}\end{array}\right]_n$		11–19	CD	B	16
(412–419) (420)	$\left[\begin{array}{c}\text{Ph}\\-\text{SiO}-\\\text{Me}\end{array}\right]_n$		3–10 very large	D, CD CD	A, C A	118, 122 113, 123
(421–423)	![structure with Me–Si–Me groups on phenylene]		2–4	D	C, E	119, 124
(424), (425)	![structure with Me–Si–Me and SiO–SiO linkages on phenylene]		2, 3	D	C	119

Formation of Cyclic Oligomers

Structure	n	m	Method 1	Method 2	Ref
(426–429)	1	2–5	D	C	119
(430), (431)	2	1, 2	D	C	119
(432)	3	1	D	C	119
(433–435)	1	1–3	D	C	119
(436), (437)	2	1, 2	D	C	119
(438–444)	1	3–9	D	E	121
(445–454)	2	1–10	D	E	121
(455–457)	3	1–3	D	E	121
(458), (459)	1	1, 2	D	C	119
(460)	2	1	D	C	119
(461–463)	1	1–3	D	C	119
(464)	2	1	D	C	118

[a] S = oligomers obtained during polymerization conditions; CD = oligomers obtained in catalyzed solution depolymerization; D = oligomers obtained by thermal decomposition (pyrolysis conditions).

[b] A = separated by GPC; B = separated by GC; C = analyzed by GC; MS; E = analyzed by MS; GC-MS.

Table 12 Miscellaneous

Compound number	Oligomer structure	n	Origin[a]	Detection[b]	Ref.
(465–467)	$\{-N=P(OCH_2CF_3)_2-\}_n$	3–5	D	B+C	129
(468–470)	$\{-N=P(OPh)_2-\}_n$	3–5	D	B+C	130
(471), (472)	$\{-N=P(O-2\text{-naphthyl})_2-\}_n$	3, 4	S, D	C	131
(473)	$\{-C_6H_4-CH=N(CH_2)_6N=CH-\}_n$	2	S	C	132
(474–489)	$\{-C_6H_4-C(CF_3)=CH-\}_n$	3–18	S	A, C	133
(490–502)	$\{-\text{thienyl}-C(CF_3)=CH-\}_n$	3–15	S	A, C	133
(503–510)	$\{-C_6H_4-C(CF_3)=CH-C_6H_4-CH=C(CF_3)-\}_n$	2–9	S	A, C	134
(511–515)	$\{-C_6H_4-C(CF_3)=CH-\text{thienyl}-CH=C(CF_3)-\}_n$	2–6	S	A, C	134
(516–520)	$\{-\text{thienyl}-C(CF_3)=CH-C_6H_4-CH=C(CF_3)-\}_n$	2–6	S	A, C	134
(521–532)	$\{-\text{thienyl}-C(CF_3)=CH-\text{thienyl}-CH=C(CF_3)-\}_n$	2–13	S	A, C	134

[a] S = oligomers obtained during polymerization conditions; D = oligomers obtained by thermal decomposition (pyrolysis conditions).
[b] A = separated by HPLC; B = separated by GPC; C = analyzed by MS.

Unfortunately, literature data on linear oligomers are still rather limited since it was difficult to differentiate between linear and cyclic oligomers before the advent of modern analytical tools, as described in Section 5.3.

5.6 REFERENCES

1. P. Ruggli, *Justus Liebigs Ann. Chem.*, 1912, **392**, 92.
2. W. Kuhn, *Kolloid-Z.*, 1934, **68**, 2.
3. H. Jacobson and W. H. Stockmayer, *J. Chem. Phys.*, 1950, **18**, 1600.
4. K. Ziegler, in 'Methoden der Organischen Chemie (Houben-Weil)', ed. E. Muller, Thieme, Stuttgart, 1955, vol. 412.
5. H. Zahn and G. B. Gleitsman, *Angew. Chem., Int. Ed. Engl.*, 1963, **2**, 410.
6. J. B. Carmichael, *J. Macromol. Chem.*, 1966, **1**, 207.
7. C. J. Pedersen, *J. Am. Chem. Soc.*, 1967, **89**, 7017.
8. H. R. Allcock, *J. Macromol. Sci., Rev. Macromol. Chem.*, 1970, **C4**, 149.
9. M. Rothe, in 'Polymer Handbook', 2nd edn., ed. J. Brandrup and E. H. Immergut, Wiley, New York, 1975.
10. J. A. Semlyen, *Adv. Polym. Sci.*, 1976, **21**, 41.
11. J. Chojnowsky, M. Ścibiorek and J. Kowalski, *Makromol. Chem.*, 1977, **178**, 1351.
12. E. J. Goethals, *Adv. Polym. Sci.*, 1977, **23**, 103.
13. J. S. Bradshaw, G. E. Maas, R. M. Izatt and J. J. Christensen, *Chem. Rev.*, 1979, **79**, 37.
14. Y. Imanishi, *Macromol. Rev.*, 1979, **14**, 1.
15. R. H. Wiley, *Macromol. Rev.*, 1979, **14**, 379.
16. A. J. Kirby, *Adv. Phys. Org. Chem.*, 1980, **17**, 183.
17. V. Rossbach, *Angew. Chem., Int. Ed. Engl.*, 1981, **20**, 831.
18. J. A. Semlyen, *Pure Appl. Chem.*, 1981, **53**, 1797.
19. G. W. Gokel and S. H. Korzeniowski, 'Macrocyclic Polyether Syntheses', Springer, Berlin, 1982.
20. L. Rossa and F. Vogtle, *Top. Curr. Chem.*, 1983, **113**, 1.
21. S. Penczek, P. Kubisa and K. Matyjaszewski, *Adv. Polym. Sci.*, 1985, **68/69**, 35.
22. S. Slomkowski, *Makromol. Chem.*, 1985, **186**, 2581.
23. L. Mandolini, *Adv. Phys. Org. Chem.*, 1986, **22**, 1.
24. J. A. Semlyen, in 'Cyclic Polymers', ed. J. A. Semlyen, Elsevier, London, 1986, p. 1.
25. P. V. Wright and M. S. Bevers, in 'Cyclic Polymers', ed. J. A. Semlyen, Elsevier, London, 1986, p. 85.
26. H. Hocker, in 'Cyclic Polymers', ed. J. A. Semlyen, Elsevier, London, 1986, p. 197.
27. G. Montaudo and C. Puglisi, in 'Development in Polymer Degradation', ed. N. Grassie, Applied Science, London, 1987, vol. 7, p. 35.
28. G. Montaudo, C. Puglisi, E. Scamporrino and D. Vitalini, *Macromolecules*, 1986, **19**, 2689.
29. G. Montaudo, C. Puglisi, E. Scamporrino and D. Vitalini, *J. Polym. Sci., Polym. Chem. Ed.*, 1987, **25**, 1653.
30. S. Foti and G. Montaudo, in 'Analysis of Polymer Systems', ed. L. S. Bark and N. S. Allen, Applied Science, London, 1982, p. 103.
31. G. Montaudo, *Br. Polym. J.*, 1986, **18**, 231.
32. G. Montaudo, C. Puglisi, E. Scamporrino and D. Vitalini, *Macromolecules*, 1986, **19**, 2157.
33. A. Ballistreri, D. Garozzo, M. Guiffrida, G. Montaudo, *Anal. Chem.*, 1987, **59**, 2024.
34. A. Ballistreri, D. Garozzo, M. Giuffrida, G. Montaudo, A. Filippi, C. Guaita, P. Manaresi and F. Pilati, *Macromolecules*, 1987, **20**, 1029.
35. G. Montaudo, C. Puglisi, E. Scamporrino and D. Vitalini, *Macromolecules*, 1988, **21**, in press.
36. N. E. Manolova, I. Gitsov, R. S. Velichkova and I. B. Rashkov, *Polym. Bull. (Berlin)*, 1985, **13**, 285.
37. S. Foti, M. Giuffrida, P. Maravigna and G. Montaudo, *J. Polym. Sci., Polym. Chem. Ed.*, 1984, **22**, 1201.
38. W. J. Irwin, 'Analytical Pyrolysis: A Comprehensive Guide', Dekker, New York, 1982, and references therein.
39. J. W. Hill and W. H. Carothers, *J. Am. Chem. Soc.*, 1933, **55**, 5031.
40. E. W. Spanagel and W. H. Carothers, *J. Am. Chem. Soc.*, 1935, **57**, 929.
41. F. W. Billmeyer, Jr. and A. D. Eckard, *Macromolecules*, 1969, **2**, 103.
42. F. W. Billmeyer, Jr. and I. Katz, *Macromolecules*, 1969, **2**, 105.
43. D. Garozzo, M. Giuffrida and G. Montaudo, *Macromolecules*, 1986, **19**, 1643.
44. D. R. Cooper, G. J. Sutton and B. J. Tighe, *J. Polym. Sci., Polym. Chem. Ed.*, 1973, **11**, 2045.
45. E. Jacobi, I. Lüderwald and R. C. Schulz, *Makromol. Chem.*, 1978, **179**, 429.
46. I. C. McNeill and H. A. Leiper, *Polym. Degradation Stab.*, 1985, **11**, 309; *Polym. Degradation Stab.*, 1985, **12**, 373.
47. I. B. Rashkov, I. Gitsov, I. M. Panayotov and J. P. Pascault, *J. Polym. Sci., Polym. Chem. Ed.*, 1983, **21**, 923.
48. D. Garozzo, M. Giuffrida and G. Montaudo, *Polym. Bull. (Berlin)*, 1986, **15**, 353.
49. I. Lüderwald and W. Sauer, *Makromol. Chem.*, 1981, **182**, 861.
50. H. R. Kricheldorf, K. Bösinger and G. Schwarz, *Makromol. Chem.*, 1973, **173**, 43.
51. A. Bachrach and A. Zilkha, *Eur. Polym. J.*, 1982, **18**, 421.
52. F. R. Jones, L. E. Scales and J. A. Semlyen, *Polymer*, 1974, **15**, 738.
53. R. P. Lattimer and K. R. Welch, *CHEMTECH*, 1981, **11**, 366.
54. S. Foti, M. Giuffrida, P. Maravigna and G. Montaudo, *J. Polym. Sci., Polym. Chem. Ed.*, 1984, **22**, 1217.
55. C. Aguilera and I. Lüderwald, *Makromol. Chem.*, 1978, **179**, 2817.
56. D. R. Cooper and J. A. Semlyen, *Polymer*, 1973, **14**, 185.
57. L. M. R. Crawford and D. A. Sutton, *Chem. Ind.(London)*, 1970, 1232.
58. W. S. Ha and Y. K. Choun, *J. Polym. Sci., Polym. Chem. Ed.*, 1979, **17**, 2103.
59. S. Shiono, *J. Polym. Sci., Polym. Chem. Ed.*, 1979, **17**, 4123.
60. H. Zahn and J. F. Repin, *Chem. Ber.*, 1970, **103**, 3041.
61. G. C. East and A. M. Girshab, *Polymer*, 1982, **23**, 323.
62. F. J. Müller, P. Kusch, J. Windeln and H. Zahn, *Makromol. Chem.*, 1983, **184**, 2487.
63. E. Meraskentis and H. Zahn, *J. Polym. Sci., Part A-1*, 1966, **4**, 1890; *Chem. Ber.*, 1970, **103**, 3034.

64. J. Goodman and B. F. Nesbitt, *Polymer*, 1960, **1**, 384; *J. Polym. Sci.*, 1960, **48**, 423.
65. H. Hirai, K. Naito, T. Hamasaki, M. Goto and H. Koinuma, *Makromol. Chem.*, 1984, **185**, 2347.
66. H. Hirai, *J. Macromol. Sci., Chem.*, 1984, **A21**, 1165.
67. S. Foti, M. Giuffrida, P. Maravigna and G. Montaudo, *J. Polym. Sci., Polym. Chem. Ed.*, 1983, **21**, 1567.
68. S. Foti, M. Giuffrida, P. Maravigna and G. Montaudo, *J. Polym. Sci., Polym. Chem. Ed.*, 1983, **21**, 1599.
69. H. Schnell, L. Bottenbruch, *Makromol. Chem.*, 1962, **57**, 1.
70. R. H. Wiley, *Macromolecules*, 1980, **13**, 1081.
71. H. Sotobayashi, S. L. Lie, J. Springer and K. Ueberreiter, *Makromol. Chem.*, 1968, **111**, 172.
72. A. Horbach, H. Vernaleken and K. Weirauch, *Makromol. Chem.*, 1980, **181**, 111.
73. J. E. Hallgren and R. O. Matthews, *J. Polym. Sci., Polym. Chem. Ed.*, 1979, **17**, 3781.
74. J. Mařik, J. Mitera, J. Králiček and J. Stehlíček, *Eur. Polym. J.*, 1977, **13**, 961.
75. M. Rothe, *Angew. Chem.*, 1962, **74**, 725.
76. M. Rothe, R. Hossbach, *Makromol. Chem.*, 1964, **70**, 150.
77. H. R. Kricheldorf and E. Leppert, *Makromol. Chem.*, 1974, **175**, 1731.
78. G. Camino, M. Guaita and L. Trossarelli, *J. Polym. Sci., Polym. Lett. Ed.*, 1977, **15**, 417.
79. G. Camino, L. Costa and L. Trossarelli, *J. Polym. Sci., Polym. Chem. Ed.*, 1980, **18**, 377.
80. J. M. Andrews, F. R. Jones and J. A. Semlyen, *Polymer*, 1974, **15**, 420.
81. A. Ballistreri, M. Giuffrida, P. Maravigna and G. Montaudo, *J. Polym. Sci., Polym. Chem. Ed.*, 1985, **23**, 1145.
82. A. Ballistreri, M. Giuffrida, P. Maravigna and G. Montaudo, *J. Polym. Sci., Polym. Chem. Ed.*, 1985, **23**, 1731.
83. I. Goodman and R. N. Vachon, *Eur. Polym. J.*, 1984, **20**, 529.
84. G. E. Hahn, Dissertation, Technische Hochschule Aachen, 1978.
85. S. Foti, P. Maravigna and G. Montaudo, *J. Polym. Sci., Polym. Chem. Ed.*, 1981, **19**, 1679.
86. S. Foti, P. Maravigna and G. Montaudo, *Macromolecules*, 1982, **15**, 883.
87. S. Foti, M. Giuffrida, P. Maravigna and G. Montaudo, *J. Polym. Sci., Polym. Chem. Ed.*, 1983, **21**, 1583.
88. G. Montaudo, E. Scamporrino and D. Vitalini, *J. Polym. Sci., Polym. Chem. Ed.*, 1983, **21**, 3321.
89. J. M. McKenna, T. K. Wu and G. Pruckmayr, *Macromolecules*, 1977, **10**, 877.
90. I. M. Robinson and G. Pruckmayr, *Macromolecules*, 1979, **12**, 1043.
91. D. R. Roberts and A. L. Gatzke, *J. Polym. Sci., Polym. Chem. Ed.*, 1978, **16**, 1211.
92. Y. Kawakami and Y. Yamashita, *Macromolecules*, 1977, **10**, 837.
93. Y. Yamashita, K. Iwao and K. Ito, *J. Polym. Sci., Polym. Lett. Ed.*, 1979, **17**, 1.
94. Y. Kawakami, A. Ogawa and Y. Yamashita, *J. Polym. Sci., Polym. Chem. Ed.*, 1979, **17**, 3785.
95. J. M. Hammond, J. F. Hooper and W. G. P. Robertson, *J. Polym. Sci., Part A-1*, 1971, **9**, 265.
96. S. G. Entelis and G. V. Korovina, *Makromol. Chem.*, 1974, **175**, 1253.
97. S. L. Malhotra, A. Leborgne and L. P. Blanchard, *J. Polym. Sci., Polym. Chem. Ed.*, 1978, **16**, 561.
98. K. Hashimoto, H. Sumitomo and S. Kawase, *J. Polym. Sci., Polym. Chem. Ed.*, 1978, **16**, 2327.
99. J. Starr and O. Vogl, *J. Macromol. Sci., Chem.*, 1978, **A12**, 1017.
100. A. S. Hay, F. J. Williams, H. M. Relles and B. M. Boulette, *J. Macromol. Sci., Chem.*, 1984, **A21**, 1065.
101. M. Bucquoye and E. J. Goethals, *Makromol. Chem.*, 1978, **179**, 1681.
102. G. Montaudo, C. Puglisi, E. Scamporrino and D. Vitalini, *Macromolecules*, 1986, **19**, 870.
103. K. Albrecht, D. Fleischer, A. Kane, C. Rentsch, Q. V. Tran Thi, H. Yamaguchi and R. C. Schulz, *Makromol. Chem.*, 1977, **178**, 881.
104. C. Rentsch and R. C. Schulz, *Makromol. Chem.*, 1977, **178**, 2335; *Makromol. Chem.*, 1978, **179**, 1403.
105. K. Albrecht, D. Fleischer, C. Rentsch, H. Yamaguchi and R. C. Schulz, *Makromol. Chem.*, 1977, **178**, 3191.
106. J. M. Andrews and J. A. Semlyen, *Polymer*, 1972, **13**, 141.
107. Y. Yamashita, J. Majumi, Y. Kawakami and K. Ito, *Macromolecules*, 1980, **13**, 1075.
108. G. Montaudo, C. Puglisi, E. Scamporrino and D. Vitalini, *J. Polym. Sci., Polym. Chem. Ed.*, 1987, **25**, 475.
109. J. P. Machon and A. Nicco, *Eur. Polym. J.*, 1972, **8**, 547.
110. R. P. Simonds and E. J. Goethals, *Makromol. Chem.*, 1978, **179**, 1689.
111. R. Simonds, W. Van Craeynest, E. J. Goethals and S. Boileau, *Eur. Polym. J.*, 1978, **14**, 589.
112. G. Montaudo, E. Scamporrino, C. Puglisi and D. Vitalini, *Polymer*, 1987, **28**, 477.
113. W. D. Reents, Jr. and M. L. Kaplan, *Polymer*, 1982, **23**, 310.
114. M. L. Kaplan and W. D. Reents, Jr., *Tetrahedron Lett.*, 1982, **23**, 373.
115. J. Franke and F. Vogtle, *Tetrahedron Lett.*, 1984, **25**, 3445.
116. J. Chojnowski and L. Wilczek, *Makromol. Chem.*, 1979, **180**, 117.
117. B. G. Zavin, A. A. Zhdanov, M. Ścibiorek and J. Chojnowski, *Eur. Polym. J.*, 1985, **21**, 135.
118. A. Ballistreri, D. Garozzo and G. Montaudo, *Macromolecules*, 1984, **17**, 1312.
119. A. Ballistreri, G. Montaudo and R. W. Lenz, *Macromolecules*, 1984, **17**, 1848.
120. N. Grassie and I. G. MacFarlane, *Eur. Polym. J.*, 1978, **14**, 875.
121. N. Grassie and S. R. Beattie, *Polym. Degradation Stab.*, 1984, **7**, 109; 231; *Polym. Degradation Stab.*, 1984, **8**, 177.
122. M. S. Beevers and J. A. Semlyen, *Polymer*, 1971, **12**, 373.
123. S. J. Clarson and J. A. Semlyen, *Polymer*, 1986, **27**, 1633.
124. B. Zelei, M. Blazsó, S. Dobos, *Eur. Polym. J.*, 1981, **17**, 503.
125. I. V. Blestos, D. M. Hercules, D. VanLeyen and A. Benninghoven, *Macromolecules*, 1987, **20**, 407.
126. J. B. Carmichael and J. B. Kinsinger, *Can. J. Chem.*, 1964, **42**, 1996.
127. T. H. Thomas and T. C. Kendrick, *J. Polym. Sci., Part A-2*, 1969, **7**, 537; *J. Polym. Soc., Part A-2*, 1970, **8**, 1823.
128. J. C. Kleinert and C. J. Weschler, *Anal. Chem.*, 1980, **52**, 1245.
129. H. R. Allcock and W. J. Cook, *Macromolecules*, 1974, **7**, 284.
130. H. R. Allcock, G. Y. Moore and W. J. Cook, *Macromolecules*, 1974, **7**, 571.
131. A. Ballistreri, S. Foti and G. Montaudo, *Makromol. Chem.*, 1981, **182**, 1319.
132. A. Ballistreri, D. Garozzo, M. Giuffrida, P. Maravigna and G. Montaudo, *J. Polym. Sci., Polym. Chem. Ed.*, 1986, **24**, 331.
133. G. Holzmann, G. Kossmehl and R. Nuck, *Makromol. Chem.*, 1982, **183**, 1711.
134. G. Holzmann, T. Kobilike and G. Kossmehl, *Colloid Polym. Sci.*, 1986, **264**, 1.
135. S. Russo, *Chim. Ind. (Milan)*, 1981, **63**, 412.

6
Ring–Chain Equilibria

ULRICH W. SUTER

*Massachusetts Institute of Technology, Cambridge, MA, USA
and Eidgenössische Technische Hochschule, Zürich, Switzerland*

6.1 INTRODUCTION	91
6.2 SYSTEMS CONSIDERED	91
6.3 THE ROLE OF THE SOLVENT	93
6.4 UNDER θ CONDITIONS	93
6.4.1 *Model Chains*	93
6.4.2 *Rotational Isomeric State Chains*	93
6.4.3 *An Example*	94
6.5 REFERENCES	95

6.1 INTRODUCTION

Cyclic species are amazingly common amongst the products from polymerization reactions and occur in many different circumstances. Step reaction polymerization almost invariably leads to a sizeable fraction of cyclics of various sizes besides linear chains, and a considerable portion of a high molecular weight material may consist of rings. Of interest here are the concentrations of cyclics under conditions where chain–ring equilibria are established.

Ring–chain equilibria baffled researchers until, in 1950, Jacobson and Stockmayer[1] presented a simple and elegant Ansatz which yielded a concise and general set of results. While comparison of the estimates with experimental values indicates that there are quantitative weaknesses of the Jacobson–Stockmayer theory, it is nevertheless qualitatively correct and describes virtually all known experimental findings in a melt or θ solvent (or not too far from θ conditions); it also makes interpretation of results far from θ conditions possible.

Macrocyclization equilibria have been accurately determined in many cases, and excellent reviews of the subject have been written; we would like to point the reader's attention to those by Semlyen.[2,3] Here we will give an introductory overview and present comparison with experimental data for one example, the equilibrium ring concentration distribution of metathesis products of a medium size cycloalkene, as determined by Höcker and co-workers.[4-7]

6.2 SYSTEMS CONSIDERED

Jacobson and Stockmayer[1] developed an elegant theory of macrocyclization equilibria in 1950, based on consideration of the following set of reaction equilibria (x and j can assume any positive value). The equilibrium of immediate interest

$$\text{X—M}_x\text{—Y} \rightleftharpoons \text{cyclic-M}_x + \text{XY} \qquad (1)$$

(X—M$_x$—Y is an x-meric linear chain molecule, cyclic-M$_x$ the x-meric cyclic compound, and XY a low molecular product that does not, however, need to be present) is compared to the corresponding acyclic equilibrium

$$\text{X—M}_x\text{—Y} + \text{X—M}_j\text{—Y} \rightleftharpoons \text{X—M}_{x+j}\text{—Y} + \text{XY} \qquad (2)$$

Combination of both reactions yields

$$\text{X—M}_{x+j}\text{—Y} \xrightleftharpoons{K_x} \text{cyclic-M}_x + \text{X—M}_j\text{—Y} \tag{3}$$

The concentration-based equilibrium constant K_x for reaction (3) is

$$K_x = [\text{cyclic-M}_x]/p^x \tag{4}$$

where p is the extent of reaction, since $[\text{X—M}_{x+j}\text{—Y}]/[\text{X—M}_j\text{—Y}] \approx p^x$ (see Chapter 2 of this volume). Experimental values of K_x are thus available once the concentration of rings and the extent of reaction at equilibrium is known.[2,3]

This concept can be universally applied to step polymerization products. Although there are extreme cases, such as rigid rod polymers with very high persistence lengths that will not form perceptible amounts of cyclics at equilibrium, these rings are relevant in most structures of importance. The practical impact comes from the fact that cyclics, especially the more frequently occurring small rings, have an (often deleterious) effect on the properties of the materials that contain them. Knowledge of the ring–chain equilibria is, therefore, necessary.

For many experimental circumstances the extent of reaction is very close to unity and $p^x \approx 1$ for x that are not too large. Then $K_x \approx [\text{cyclic-M}_x]$. Since K_x is a thermodynamic equilibrium *constant*, an immediate consequence is that for rings that are not too large, *i.e.* for the overwhelming number of rings and the vast majority of the mass occurring in cyclics (see below), the absolute concentration of each cyclic species is *constant*, and, in particular, not dependent on the total concentration of polymeric material in the system.[1] This implies that a certain concentration exists for each macrocyclization equilibrium system below which *essentially only cyclics* are present, and this is born out by experiment.[2,8] This 'critical' concentration may be surprisingly high, *e.g.* a value of approximately 85 vol.% poly(methyltrifluoropropylsiloxane) in toluene at 383 K is the concentration below which no linear species can be found.[8,9]

The distribution of cyclic species will be given further attention below. In all cases, however, it will be assumed that the equireactivity principle of step polymerization holds (see also Volume 5, Chapters 2 and 3); as a consequence the distribution of the linear species that (might) coexist with the rings at equilibrium may always be expected to follow the 'most probable' distribution.[10]

Jacobson and Stockmayer started from the concept that the density of a chain end of the linear chain at the locus of the other end of the same chain, $W(0)$, is the sole factor in determining the propensity for cyclization, and obtained the simple expression[1]

$$K_x = W(0)/(s_x N_A) \tag{5}$$

where N_A is Avogadro's number and s_x the symmetry number of the ring of degree of polymerization x (usually equal to $2x$). $W(0)\delta r$ (where r is the end-to-end vector of the linear chain) is the probability for the two chain ends to meet.

One can imagine that the requirement for meeting of the chain ends is not the only one necessary for the successful formation of a cyclic species, rather, two further qualifications must be met:[11] (i) the direction of the first and last bond of the linear chain must be such that the newly formed bond in cyclization will yield acceptable bond angles; and (ii) the conformation of the newly formed bond, specified by the positions of the two terminal bonds of the linear chain, must be acceptable. The first of these two criteria is quantified by the conditional probability for the newly formed bond angle to be free of angle strain θ, termed $\Gamma_0(1)\delta \cos \theta$; $\Gamma_0(1)$ is a special value of the conditional probability density function $\Gamma_r(\cos \theta)$ that describes the probability per unit range of $\cos \theta$ to assume the specified value of angle strain θ, provided the end-to-end vector has the indicated value r; if all values of θ are equally probably, $\Gamma_0(1) = 1/2$. The second criterion gives rise to a conditional probability for the conformation around the newly formed bond to be acceptable, termed $\Phi_{0,1}(\phi)\delta\phi$; $\Phi_{0,1}(\phi)$ is a special value of the probability density function $\Phi_{r,\cos\theta}(\phi)$ per unit range of the torsion angle of the newly formed bond, ϕ, provided the end-to-end vector has the indicated value r and the angle strain the indicated value θ; if all values of ϕ are equally probable, $\Phi_{0,1}(\phi) = 1/2\pi$. Since usually a bond can assume more than one conformation, a value averaged over the range of $0 < \phi < 2\pi$, $\bar{\Phi}_{0,1}$, must be used.

Combination of these factors yields a somewhat refined expression for the ring–chain equilibrium constants[11]

$$K_x = W(0) \cdot 2\Gamma_0(1) \cdot \bar{\Phi}_{0,1}/(s_x N_A) \tag{6}$$

One can argue[11] that torsion angles usually can assume values in several ranges, and that even in one of those ranges the requirements for the torsional angle should be less stringent than those for the bond angles, and that therefore $\bar{\Phi}_{0,1} \approx 1$. Hence, for the vast majority of cases the appropriate formula for K_x is

$$K_x = W(0) \cdot 2\Gamma_0(1)/(s_z N_A) \tag{7}$$

6.3 THE ROLE OF THE SOLVENT

One would expect the solvent to exert its influence on the ring–chain equilibria primarily through $W(0)$ which reflects the propensity for the chain ends to meet. Good solvents that expand the coil (by a factor $\alpha = (\langle r^2 \rangle / \langle r^2 \rangle_0)^{0.5}$, where the subscript 0 indicates unperturbed or θ conditions) should reduce the density of one chain end in the vicinity of the other and hence reduce K_x. For the smaller rings, i.e. cyclics corresponding to linear chains of dimensions too small to be significantly affected by the effects of excluded volume, one would expect the influence of the solvent quality to be small, but increasingly important for the larger rings.

$W(0)$ of a sufficiently long, unperturbed coil in a θ state depends on the degree of polymerization with a power of $-3/2$, i.e. $K_x \sim x^{-2.5}$.[1,12,13] In a good solvent, the chain length dependent effects of the excluded volume change this power law, since it is expected that $W(0)$ is quite sensitive to the quality of the solvent.[14,15] Scaling arguments lead to the dependence $K_x \sim x^{-2.8}$,[14] or even $K_x \sim x^{-2.9}$.[16] To date, published experimental data does not warrant deduction of scaling exponents for the ring–chain equilibria, but the available data show clearly that the chain length dependence of K_x is steeper for large degrees of polymerization, x, in good solvents than in the vicinity of the θ state.[8,17] From these data it is also possible to show that $K_x \approx \alpha^3 K_x(\theta)$.[8]

The most prominent effect of changing solvent, however, is the changes in the *overall* concentration of cyclics that it induces, i.e. especially in the smallest members of a distribution.[2,8,17] The smallest cyclics of poly(dimethylsiloxane), for instance, exhibit significantly higher values for K_x in a good solvent than in bulk or a θ solvent; at a certain ring size the equilibrium concentrations are the same for the different solvents, and larger rings are less frequent in the good solvent, in accordance with the above mentioned chain length dependence. These solvent effects on the concentration of the smallest rings must be explained with differences in the thermodynamic stability of the small rings from larger (more flexible) ones as well as in interactions of these cyclics with the solvent.[8,17]

6.4 UNDER θ CONDITIONS

6.4.1 Model Chains

The earliest model for ring–chain equilibria by Jacobson and Stockmayer[1] employed an unperturbed chain with Gaussian segments. In other words, they implied that the factor accounting for the conditional probability for the newly formed bond angle to be free of angle strain, $2\Gamma_0(1)$, was unity, i.e. that equation (5) could be used. Furthermore, since the density distribution of the chain vector for an unperturbed Gaussian chain is well known, they were able to calculate equilibrium concentrations for cyclics from knowledge only of the infinite chain characteristic ratio, $C = \langle r^2 \rangle_0 / nl^2$ ($n \to \infty$, where n is the number of skeletal bonds in the chain), of linear chains under θ conditions, by

$$K_x = (3/\pi C)^{3/2} (1/16 l^3 N_A) x^{-5/2} \tag{8}$$

Values calculated by this elegant theory are in general higher than the experimental values for rings exceeding a certain minimum size. Slight improvements can be obtained by using not the infinite chain characteristic ratio, but one that is more realistic, e.g. the one calculated for a freely jointed chain. The computed values are, however, not in significantly better agreement with experiment than those computed with Jacobson and Stockmayer's approach. Improvements require, therefore, more realistic chain models.

6.4.2 Rotational Isomeric State Chains

The consistent deviations of values computed with the Jacobson–Stockmayer theory from experimental data, especially in the regime of 'medium-sized' rings, have been attributed to two causes: (i) the failure of the Gaussian density distribution for infinite chains to properly account for the density of a chain end at the locus of the other end of the same chain; and (ii) the neglect of the geometrical constraints posed upon a ring by the bond angle at the joining atom (see above). The shortcomings of the density distribution have been addressed in different ways. Flory and Semlyen[13] computed the Gaussian density with a realistic rotational isomeric state (RIS) model for a chain of the same degree of polymerization as the corresponding ring, while Semlyen and co-workers[2,3,18,19] evaluated the exact RIS density by a total enumeration method which proved very successful for those ring sizes for which it can be applied (computational requirements become prohibitive for rings with numbers of bonds exceeding *ca.* 30). Flory, Suter and Mutter[11,20-23] took the position that the density distribution could be well approximated by a simple correction to the Gaussian distribution (by the first Hermitian correction term),[12] *i.e.* by

$$W_x(0) \approx (3/2\pi \langle r_x^2 \rangle_0)^{3/2} \cdot H(0)$$
$$H(0) \approx (1/8)(9\langle r_x^4 \rangle_0 / \langle r_x^2 \rangle_0^2 - 7) \quad (9)$$

or be estimated in a Monte Carlo procedure by the number of chain termini falling within a small sphere, *e.g.* of radius $0.3\langle r_x^2 \rangle_0^{1/2}$, around the other chain end, but that the deviation from randomness of the mutual orientation of the chain ends about to form a cyclic may be strong, and that, therefore, the probability for the chain ends to be in the correct arrangement for ring stability may deviate considerably from its random value. Consequently, the appropriate expression for K_x was found to be equation (6). They found that the term $2\Gamma_0(1)$, implicitly assumed to be unity by Jacobson and Stockmayer[1] and Semlyen and co-workers[18,19] can be estimated, in good approximation, by[11,23]

$$2\Gamma_0(1) \approx 1 + \{3/8H(0)\}\{35\langle P_1 \rangle_0 - 42\langle P_1 r^2 \rangle_0 / \langle r^2 \rangle_0 + 9\langle P_1 r^4 \rangle_0 / \langle r^2 \rangle_0^2\} \quad (10)$$

where P_1 is the cosine of the deviation of the bond angle formed from its exactly unconstrained value, $P_1 = \cos\theta$, and $H(0)$ is used in the same approximation as given above (see equation 9). All quantities needed can be computed by exact matrix generation methods,[20] or very conveniently by Monte Carlo techniques.[20-25] For an overview of cases treated by these methods, we refer the reader to Semlyen's review.[2] We present a recent example in the next section.

6.4.3 An Example

The approach using a low order Hermite correction to $W(0)$ and an estimate for the potential bond angle strain, as discussed above, has proven reasonably successful for rings that are not too small.[11,20-25] As an example, we address metathesis products from cyclododecene (C_{12}). Refs. 4–6 provide values for [cyclic-M_x]; they are listed in Table 1, together with the experimentally determined fraction of double bonds in *trans* configuration.

Table 1 Ring–Chain Equilibrium Data for Polymers Made by Metathesis Reaction from Cyclododecene (from refs. 4–6)

x	No. of bonds	Concentrations of cyclics (mmol l^{-1})	trans double bonds (%)
2	24	6.6	75
3	36	3.6	80
4	48	2.1	82
5	60	1.3	83
6	72	0.73	84
7	84	0.51	85

Theoretical values were computed from a recently developed[7] RIS model for polyalkenes with at least three single bonds between double bonds. The symmetry number for rings with regularly spaced double bonds is $s_x = 2x$. The statistical mechanical averages required for computation of $2\Gamma_0(1)$ according to equations (10) and (9) were obtained from Monte Carlo calculations. $W(0)$ was estimated directly from the Monte Carlo runs by the number of chain termini falling within a sphere of radius $0.3\langle r_x^2\rangle_0^{1/2}$ of the other chain end. The frequency of *trans* configurations were kept constant at 80%. The results of these computations, together with the experimental values,[4-6] are displayed in Figure 1.

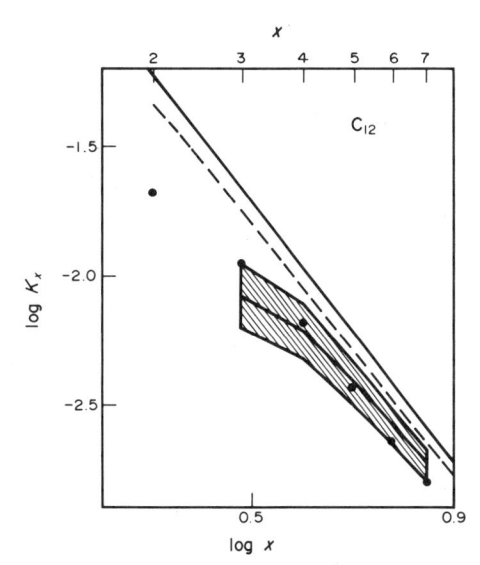

Figure 1 Macrocyclization equilibrium constant K_x for the metathesis products of cylododecene. Points correspond to data from Table 1 (refs. 4–6): the full line was calculated with the Jacobson–Stockmayer scheme; the dashed line contains, in addition, a correction for non-Gaussian density distribution of the chain vector at $r = 0$, obtained from Monte Carlo calculations; the shaded area represents values to which further corrections for the directional correlations of the chain ends [$2\Gamma_0(1)$ from equations (9) and (10)] have been applied, and the associated probable error

The Jacobson–Stockmayer equilibrium line is clearly too high, but gives asymptotically roughly the correct slope. The correction to the simple Gaussian density approximation, implicit in the Jacobson–Stockmayer scheme, by 'true' Monte Carlo densities W(0) is small, in all cases. The directional correlation between first and last bond of the linear chains, *i.e.* the probability for the bond angle formed by cyclization to assume exactly its unconstrained value provided the end-to-end vector has assumed the value zero, causes a much greater correction to the theoretical values. These corrections are, however, fraught with error, and a relatively broad band containing the most likely value and its probable error is used to indicate the results of these computations. Higher moments than those used in equations (9) and (10) were computed and used to assess the reliability of the prediction; below *ca.* 25 bonds per ring no reliable estimates were possible, and the shaded band in Figure 1 is truncated before this lower limit. In the range where a prediction can be made, however, the deviations of theory from experimental data is hardly more than the combined uncertainties. It is important to realize that the theoretical treatment involved is completely free of adjustable parameters. In light of this, the agreement between theory and experiment is quite satisfactory.

6.5 REFERENCES

1. H. Jacobson and W. H. Stockmayer, *J. Chem. Phys.*, 1950, **18**, 1600.
2. J. A. Semlyen, *Adv. Polym. Sci.*, 1976, **21**, 41.
3. J. A. Semlyen, in 'Cyclic Polymers', ed. J. A. Semlyen, Elsevier, London, 1986, chap. 1, as well as other chapters in this book.
4. H. Höcker, L. Reif, W. Reimann and K. Riebel, *Recl. Trav. Chim. Pays-Bas*, 1977, **96**, 279.
5. H. Höcker, W. Reimann, K. Riebel and Z. Szentivanji, *Makromol. Chem.*, 1976, **177**, 1707.
6. L. Reif and H. Höcker, *Macromolecules*, 1984, **17**, 952.
7. U. W. Suter and H. Höcker, *Makromol. Chem.*, in press.
8. P. V. Wright and M. S. Beevers, in 'Cyclic Polymers', ed. J. A. Semlyen, Elsevier, London, 1986, chap. 3.
9. P. V. Wright and J. A. Semlyen, *Polymer*, 1970, **11**, 462.
10. P. J. Flory, 'Principles of Polymer Chemistry', Cornell University Press, Ithaca, New York, 1953, p. 622.
11. P. J. Flory, U. W. Suter and M. Mutter, *J. Am. Chem. Soc.*, 1976, **98**, 5733.
12. P. J. Flory, 'Statistical Mechanics of Chain Molecules', Interscience, New York, 1969.
13. P. J. Flory and J. A. Semlyen, *J. Am. Chem. Soc.*, 1966, **88**, 3209.
14. P. -G. de Gennes, 'Scaling Concepts in Polymer Physics', Cornell University Press, Ithaca, New York, 1979, chap. 1.

15. K. F. Freed, 'Renormalization Group Theory of Macromolecules', Wiley Interscience, New York, 1987, chap. 13.
16. J. des Cloiseaux, *J. Phys., Colloq., (Orsay, Fr.)*, 1978, **C2**, 135.
17. P. V. Wright, *J. Polym. Sci., Polym. Phys. Ed.*, 1973, **11**, 51.
18. M. S. Beevers and J. A. Semlyen, *Polymer*, 1972, **13**, 385.
19. L. E. Scales and J. A. Semlyen, *Polymer*, 1976, **17**, 601.
20. U. W. Suter, M. Mutter and P. J. Flory, *J. Am. Chem. Soc.*, 1976, **98**, 5733.
21. M. Mutter, U. W. Suter and P. J. Flory, *J. Am. Chem. Soc.*, 1976, **98**, 5733.
22. M. Mutter, *J. Am. Chem. Soc.*, 1977, **99**, 8307.
23. U. W. Suter and M. Mutter, *Makromol. Chem.*, 1979, **180**, 1761.
24. W. L. Mattice, *J. Am. Chem. Soc.*, 1980, **102**, 2242.
25. W. L. Mattice, *J. Am. Chem. Soc.*, 1982, **104**, 5942.

7
Constitutional Regularity in Linear Condensation Polymers

FRANK T. GENTILE and ULRICH W. SUTER

Massachusetts Institute of Technology, Cambridge, MA, USA and Eidgenössische Technische Hochschule, Zurich, Switzerland

7.1	INTRODUCTION	97
7.2	THEORETICAL ASPECTS	99
	7.2.1 *Principal Considerations*	99
	7.2.2 *Quantitative Description*	99
	7.2.3 *Thermodynamics vs. Kinetics*	100
	7.2.4 *Kinetically Controlled Constitutional Isomerism*	101
	7.2.4.1 Kinetic parameters	101
	7.2.4.2 Modes of addition of monomer to the reaction mixture	103
	7.2.4.3 Cases where feed rate need not be considered	103
	7.2.4.4 Finite feed rates	105
	7.2.5 *Sequence Statistics*	107
	7.2.6 *Summary of Theoretical Development*	108
7.3	EXPERIMENTAL FACTS	109
	7.3.1 *Systems Investigated*	109
	7.3.2 *Random and Maximally Ordered Polymers*	109
	7.3.3 *Continuous Control Over Constitutional Order*	112
	7.3.4 *Constitutional Regularity and Properties*	113
7.4	CONCLUSIONS	114
7.5	REFERENCES	114

7.1 INTRODUCTION

Constitutional isomerism in chemical compounds arises from differences in the nature or sequence of bonding (*e.g.* Me$_2$O is a constitutional isomer of EtOH). Linear polymer chains give rise to a constitutional isomerism if they contain 'nonsymmetric' monomeric units such that arrangements with different mutual orientations can be distinguished (a sufficient condition for this is that the 'nonsymmetric' units have no two-fold rotation axes unless they are parallel to the polymer backbone). This type of structural isomerism has been known to exist in vinyl polymers for almost 50 years[1] and has been well studied; the terms 'head-to-tail' and 'head-to-head/tail-to-tail' are associated with it. For vinyl polymers it has been learned that constitutional isomerism is generally not controllable, since vinyl polymerization is almost always characterized by a high regioselectivity,[2-4] which forces nearly perfect head-to-tail arrangements of monomeric units. It is possible to obtain regular head-to-head/tail-to-tail alignment by special synthetic routes (see, for instance, work by Vogl and collaborators[5,6]). The properties of head-to-head/tail-to-tail vinyl polymers were observed to be remarkably similar to those of the usual head-to-tail versions, but for a vinylidene polymer stark differences in properties (such as melting point and glass transition temperature) were observed,[6] thus the question of the potential influence of this type of isomerism on properties cannot be answered definitely.

Constitutional isomerism can also exist in the products of step-growth polymerization. This possibility has been used in discussing possible explanations of unusual experimental results since the mid 1950s.[7–14] In 1952 Izard explained the differences in the properties of polycondensates from an asymmetric dicarboxylic acid with ethylene glycol by assuming they were of different constitution.[7] Shortly thereafter Doak and Campell noted the existence and possible effects of head-to-head/tail-to-tail and head-to-tail isomerism while studying the effect of alkyl substituents on the melting points of polyesters.[8] Later, Preston and Smith remarked that constitutional isomerism was possible in copolymers synthesized from diaminobenzanilides and diacid chlorides; they were the first to realize that the order of introduction of monomers into the reaction mixture is a crucial factor.[9] In 1967 Saotome and Schulz synthesized polymers from adipoyl dichloride and racemic or optically active lysin by different methods,[10] and attributed the observed variations in melting points to constitutional isomerism. Similar arguments were used that same year by Yesipova and Topchieva for optically active polyamides from *trans*-cyclopropanedicarboxylic acid and 1,2-propylenediamine.[11]

The potential usefulness of constitutional isomerism for the products of step polymerization lies in the observed dependence of polymer properties on constitution. A number of occurrences of this phenomenon have recently been observed. Morgan *et al.*, for instance, synthesized a variety of aromatic poly(azomethines) with ring substituents;[15] the placement of these moieties on the ring provides a useful range of melting and softening points, but the effects of some substituents (single ring substitution on both the diamine and aldehyde) is not additive as expected, presumably because of the possibility of 'ordered placement' of the moieties. Deschenaux, Neuenschwander and Pino[16] recently extended the range of properties for which constitutional isomerism is relevant when they reported on the synthesis and characterization of liquid crystalline poly(azomethines) with controlled constitution.

Quantitative experimental work relating polymer structure with properties began in 1972 with Murano's work on polyesters from malonic dimethyl ester and propylene glycol.[17] Three bands in the ^{1}H NMR spectrum of the resulting polymer were assigned to isomeric arrangements around the methylene protons. Morrison *et al.* synthesized polyamides from terephthaloyl dichloride and *p*-aminobenzhydrazides.[12] The reactions causing different polymer structures were monitored by NMR and the final constitution was measured. Later Lenz and Go investigated 'random' and 'blocky' condensation copolymers while studying the ester-exchange reaction in poly(ethylene 2-methylsuccinate).[14]

Detailed control of this isomerism has been attempted only recently.[18–26] Some of these attempts have been quite successful and several systems have been theoretically investigated. This work will be discussed in greater detail below. The most important questions addressed in these endeavors are (i) whether constitutional isomerism obtained in step-growth polymerization has a strong effect on the properties of the polymer, and (ii) whether the isomerism of a polymer can be controlled in an easy and preselectable manner. If the answers to both of these questions were in the affirmative, the control of structure would imply control over important properties.

Structure–property relationships arising from constitutional isomerism are not well known for condensation polymers. One might conjecture, however, that the variation of crystallinity with regularity and intermolecular cohesion are important factors. A simple visualization has been attempted in Figure 1; the head-to-tail and head-to-head/tail-to-tail regular structures are expected to pack more efficiently than the disordered structures, and should therefore be of higher crystal-

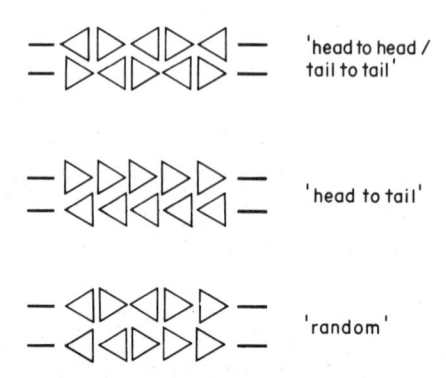

Figure 1 Schematic representation of structures from constitutionally ordered and random polymers

linity and of higher cohesiveness in general. Experimental evidence indicates that this may be true at least in part: solubility and softening temperature (indicative of crystallinity and intermolecular forces) are strongly influenced by structural regularity.[21,23,24]

It is, of course, possible to prepare polymers with highly ordered constitution in a series of synthetic steps; Takatsuka et al.[27] have, for instance, prepared polymers with completely alternating orientation by sequential synthesis. Here, however, we focus on the whole range of constitutional order, not only the regular extremes, which leads to considerations of the influence of the details of the mode of preparation on the resulting constitution, and only polymerizations in one synthetic step are considered.

7.2 THEORETICAL ASPECTS

7.2.1 Principal Considerations

Many different types of condensation polymer systems exhibiting constitutional isomerism can be imagined, since the only requirement is that at least one of the monomeric units in the chain can be distinguished in different orientations. An arbitrary but expedient classification into (1) homopolymers, (2) bipolymers and (3) polymers made from more than two different monomers is made. For the considerations of the influence of the mode of preparation on the resulting constitution, however, only polymerizations in a homogeneous reaction phase and in one synthetic step are considered.

The homopolymer system involves a single homobifunctional monomer represented by XabX. Here —ab— represents the nonsymmetric monomeric unit in the chain and X the leaving group in the polymerization reaction. This type of system has been theoretically investigated;[26] it was found that the relative reactivities of the functional groups completely determine the (average) constitution of the resulting chains. In the context of this section, therefore, they are not of interest.

The bipolymers either contain one monomer of distinguishable orientation in the chain, for instance the system with the monomers XabX + YccY, or two, e.g. the system XabX + YcdY. Here YccY represents a homobifunctional symmetric monomer and YcdY a similar but nonsymmetric monomer. Y is again a leaving group. Both systems have been analyzed for the case of functional groups whose reactivities are independent of each other;[26] it was found that very effective control over the constitutional regularity of the product polymer exists, and this has been exploited in subsequent work. The system of XabX + YccY, where the monomers can only condense with each other, is one where extensive experimental data exists and it will be considered theoretically here in some detail.[28]

The multipolymer systems have at least one nonsymmetrical monomer and three or more monomers totally. They provide a host of different possibilities, but exceed the scope of this text considerably. However, the methodology used below can be applied to these systems also.

For all systems one must consider the molecular weight distribution of the product polymer. Case has elaborated on the system XabX + YccY in detail.[29] From that work it may be concluded that all polymers studied here will conform to an average 'most probable' distribution. This and a set of other simplifying assumptions are listed in Table 1.

7.2.2 Quantitative Description

When XabX is reacted with YccY such that the only reactions allowed are —cY + —aX → —ac— + XY and —cY + —bX → —bc— + XY (XY is a low molecular weight product that does not

Table 1 Simplifying Assumptions

(A) One-step synthesis (no side reactions)
(B) Irreversible (kinetically controlled) reactions
(C) No more than two different monomers
(D) First order reactions with respect to the concentrations of the functional groups involved in the polycondensation
(E) Chain length independent reactivity
(F) Homogeneous reaction phases
(G) Most probable distribution ($\bar{M}_w/\bar{M}_n \approx 2$) at high extent of reaction
(H) Equimolarity of reacted functional groups at the end of the polycondensation
(I) Rings can be ignored

---

need to be produced) the polymer chains obtained contain alternately —ab— and —cc— groups. The shortest structured elements affected by the mutual orientation of —ab— units in the chains are —acca—, —accb—, —bcca— and —bccb—. Since —accb— arrangements are usually indistinguishable from —bcca— structures, the probability of two adjacent units pointing in the same direction s is given by:

$$s = [accb]/([acca] + [accb] + [bccb]) \qquad (1)$$

where the square brackets denote molar concentrations and [accb] includes both —accb— and —bcca— units.

A simple schematic is shown in Figure 2 (a triangle denotes an —ab— and a circle a —cc— unit); obviously, three general cases may be distinguished. Note that polymers which have a 'random' constitution (polymers which are 'disordered') will have an s value of $\frac{1}{2}$, but that the converse is not true. For example, polymers which have a very 'blocky' structure can have a great deal of local order along the chain but still have a value of $s = 0.5$. Whether or not s is sufficient to describe the polymers synthesized in single step reactions will be discussed below.

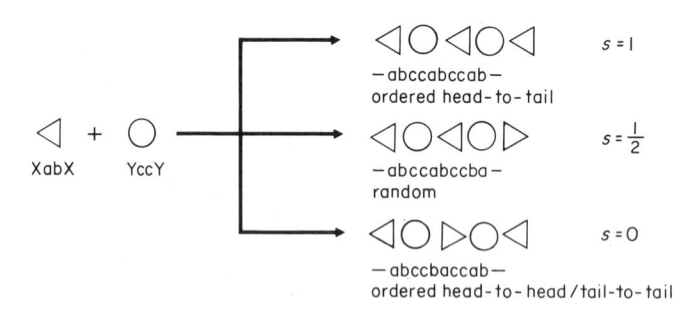

Figure 2 Schematic representation of polymers with different values of s

7.2.3 Thermodynamics vs. Kinetics

Kinetic factors can determine structural regularity to a very large extent, but constitutional isomerism could, in principal, also be determined by thermodynamic factors only. In fact, two polymers consisting of macromolecules of equal chain length and having regular head-to-tail and heat-to-head/tail-to-tail arrangements respectively can be expected to have somewhat different thermodynamic stability. The difference might be small, however, and if a polymer were equilibrated with respect to the orientation of the nonsymmetric units one would expect the structural regularity also to be small.

To assess the importance of thermodynamic regulation, a simple estimation of the differences in stability in terms of pair interactions is useful. One estimates what differences in free energy of interaction would be necessary to generate a polymer of a given s. This calculation can be simplified by assuming that entropy and kinetic energy both need not be considered because differences in them arising from changes in constitution will probably be small. Thus the difference in free energy will be approximated by the difference in total potential energy, and one can calculate the equilibrium constant for the two reactions

$$\left. \begin{array}{c} \text{—accb—} \rightleftharpoons \text{—acca—} \quad \text{and} \quad \text{—bccb—} \\ K_{eq} = \{[acca] + [bccb]\}/[accb] = (1 - s)/s \end{array} \right\} \qquad (2)$$

The free energy difference is given by $\Delta G = -RT\{\ln(1-s) - \ln(s)\}$. A plot of ΔG vs. s is shown in Figure 3.

One question is how large the interaction energies actually are. Preliminary calculations, not reported here, show that interactions between units with strong permanent dipole moments are likely to be the most dominant. An estimate can be obtained by calculating the difference in the Coulombic interactions between the dipoles for —accb— and —acca— or —bccb— units.[30] If the

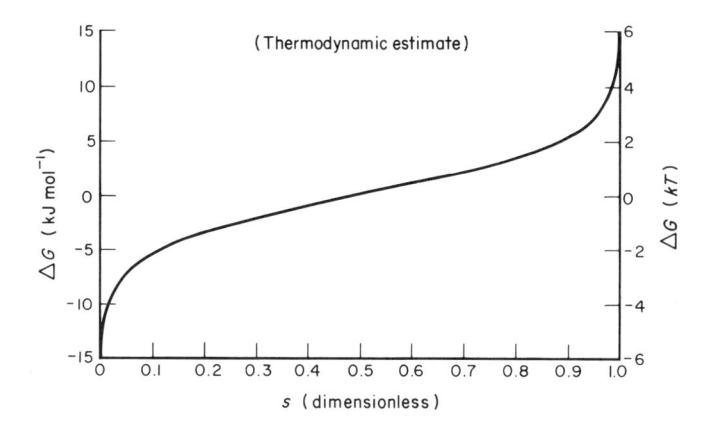

Figure 3 Rough estimate of the free energy necessary to achieve a given order through thermodynamic control at ambient temperature

—ab— units have dipole moments μ and are a distance r apart, these values are

$$\text{for —accb—} \qquad -2\mu^2/4\pi\varepsilon_0 r^3 \qquad (3a)$$

$$\text{for —acca— and —bccb—} \qquad +2\mu^2/4\pi\varepsilon_0 r^3 \qquad (3b)$$

$$\text{and hence the difference is} \qquad \mu^2/\pi\varepsilon_0 r^3 \qquad (3c)$$

Typical values yield an 'upper-bound' of about 1 kT at ambient temperature, approximately the order of magnitude of a Van der Waals interaction.[31]

By comparing this number with Figure 3 it is noted that values for s in the range 0.3–0.7 are not unlikely to occur through an equilibrium mechanism, but that considerably smaller or larger values are hardly reasonable. Thus a kinetic mechanism is a prerequisite for high degrees of order.

7.2.4 Kinetically Controlled Constitutional Isomerism

7.2.4.1 Kinetic parameters

The relevant kinetic parameters affecting constitutional isomerism are the relative reactivities of the functional groups in the nonsymmetric monomer and the mutual dependence of the reactivity of the functional groups in the symmetric monomer. These reactivities are represented by r, the ratio of reaction rate constants[26] for the reactions

$$\text{—aX} + \text{Yc—} \xrightarrow{k_{aX}} \text{—ac—} + \text{XY}$$

$$\text{—bX} + \text{Yc—} \xrightarrow{k_{bX}} \text{—bc—} + \text{XY}$$

so that

$$r = k_{bX}/k_{aX} \qquad 0 \leq r \leq 1 \qquad (4)$$

and by the reactivity ratios g_a and g_b that characterize the kinetic tendencies of the symmetric monomer (the —aX group of the nonsymmetric monomer has been arbitrarily chosen to be more reactive). g_a is the ratio of the reaction rate constants of the second reacting functional group on the YccY monomer to that of the first reacting functional group, if the first one has reacted with an —aX group. The definition of g_b is analogous. Thus

$$\left.\begin{array}{l} g_a = k_{cY,\,\text{second}}/k_{cY,\,\text{first}(-aX)} \quad (0 \leq g_a) \\ g_b = k_{cY,\,\text{second}}/k_{cY,\,\text{first}(-bX)} \quad (0 \leq g_b) \end{array}\right\} \qquad (5)$$

These differences in the reactivities of —cY groups arise from electronic or steric effects (the phenomena is known as 'induction'). If there were no induction, then $g_a = g_b = 1$. Deactivating induction means that g_a (or g_b) < 1, while activating induction means g_a (or g_b) > 1.

It is conceivable that induction also exists in the nonsymmetric monomer. However, since the cases of interest involve $r \ll 1$ (and often less than 0.01), the actual values for g_a and g_b are moderate ($0.1 < g < 10$) and the likely chemical structure of the nonsymmetric monomer is probably not conducive to strong chemical induction, this case does not need to be considered here.

Obviously, there cannot be any constitutional regularity if the chain-forming reaction does not distinguish between the two functional groups —aX and —bX of the nonsymmetric monomer. A value of unity for r will always give rise to $s = 1/2$. A value of $r < 1$ is therefore a prerequisite in all further considerations.

The relevant elementary reactions and their relative rate constants are (the product XY is not shown)

$$\text{—aX} + \text{YccY} \xrightarrow{1} \text{—accY}$$

$$\text{—bX} + \text{YccY} \xrightarrow{r} \text{—bccY}$$

$$\text{—accY} + \text{Xa—} \xrightarrow{g_a} \text{—acca—}$$

$$\text{—accY} + \text{Xb—} \xrightarrow{rg_a} \text{—accb—}$$

$$\text{—bccY} + \text{Xa—} \xrightarrow{g_b} \text{—bcca—}$$

$$\text{—bccY} + \text{Xb—} \xrightarrow{rg_b} \text{—bccb—}$$

The rate equations for this reaction scheme can be integrated and the value of s determined. One finds that, depending on the values of r, g_a and g_b, constitutional order can be varied over the range of $0 \leq s \leq 1$.

It is convenient to express the rate equations in terms of the dimensionless variables

$$\left. \begin{array}{rcl} (A) &=& [\text{—aX}]/[a]_{sto} \\ (B) &=& [\text{—bX}]/[b]_{sto} \\ (C) &=& [\text{—cX}]/[c]_{sto} \end{array} \right\} \quad (6)$$

where the subscript stands for 'stoichiometric.' In cases where all the a groups are present at the beginning of the reaction, $[a]_{sto} = [\text{—aX}]_0$. If XabX is added during the course of the reaction, $[a]_{sto}$ is the concentration corresponding to the total —a— present at the end of the addition (at equimolarity). $[b]_{sto}$ and $[c]_{sto}$ are defined analogously. Note that $[c]_{sto} = 2[a]_{sto} = 2[b]_{sto}$ due to equimolarity of functional groups. The relative concentrations of the other components are defined by

$$\left. \begin{array}{rclcrcl} (CC) &=& [\text{YccY}]/[a]_{sto} & \quad (ACC) &=& [\text{—accY}]/[a]_{sto} \\ (BCC) &=& [\text{—bccY}]/[a]_{sto} & \quad (ACCA) &=& [\text{—acca—}]/[a]_{sto} \\ (ACCB) &=& [\text{—accb—}]/[a]_{sto} & \quad (BCCB) &=& [\text{—bccb—}]/[a]_{sto} \end{array} \right\} \quad (7)$$

Further simplification can be achieved by the introduction of a dimensionless time θ such that

$$\theta = [a]_{sto} k_{aX} t \quad (8)$$

For the specification of the mode and rate of addition of reactants to the polymerization mixture, two monomer feed functions need to be defined

$$\left. \begin{array}{rcl} dY/d\theta &=& d[\text{XabX}]_{added}/([a]_{sto}^2 k_{aX} dt) \\ dZ/d\theta &=& d[\text{YccY}]_{added}/([a]_{sto}^2 k_{aX} dt) \end{array} \right\} \quad (9)$$

such that Y and Z are the fractions of XabX and YccY respectively that have already been added to the reaction system. The system can then be represented by the following set of ordinary differential equations

$$d(A)/d\theta = -2(A)(CC) - g_a(A)(ACC) - g_b(A)(BCC) + dY/d\theta \quad (10a)$$

$$d(B)/d\theta = -2r(B)(CC) - 2rg_a(ACC)(B) - rg_b(B)(BCC) + dY/d\theta \quad (10b)$$

$$d(CC)/d\theta = -2(A)(CC) - 2r(B)(CC) + dZ/d\theta \quad (10c)$$

$$d(ACC)/d\theta = 2(A)(CC) - g_b(A)(BCC) - rg_a(B)(ACC) \quad (10d)$$

$$d(BCC)/d\theta = 2r(B)(CC) - g_b(A)(BCC) - rg_b(B)(BCC) \quad (10e)$$

$$d(ACCA)/d\theta = g_a(ACC)(A) \quad (10f)$$

$$d(BCCB)/d\theta = rg_b(BCC)(B) \quad (10g)$$

$$d(ACCB)/d\theta = rg_a(ACC)(B) + g_b(BCC)(A) \quad (10h)$$

This system of equations becomes homogeneous when $dY/d\theta = dZ/d\theta = 0$, e.g. at $\theta > 0$ if the two monomers are mixed together 'infinitely fast' at the beginning of the reaction ($\theta = 0$). It becomes stiff as $r \to 0$, but can be integrated numerically and the results checked using a Monte-Carlo technique. Finally the number average degree of polymerization X_n is given by

$$X_n = 2(Y + Z)/[(A) + (B) + 2(C)] \quad (11)$$

7.2.4.2 Modes of addition of monomer to the reaction mixture

Since high constitutional regularity can only be obtained through kinetic control, the most relevant parameters are the kinetic rate constants and the feed rates of the two monomers. Therefore, an examination of the influence of the mode of addition of the reactants to the reaction mixture is necessary. For simplicity, very high degrees of polymerization ($X_n \to \infty$) are assumed here.

One can distinguish three principal modes of mixing of the monomers: (1) XabX and YccY mixed all at once (or 'infinitely fast mixing'), (2) XabX added to YccY, and (3) YccY added to XabX. Analysis of cases (2) and (3) can be further refined by considering the rate of addition [one can, of course, consider case (1) to be the limit of fast addition for both cases (2) and (3)], but the extremes of 'infinitely slow' mixing will be investigated first for simplicity.

7.2.4.3 Cases where feed rate need not be considered

(i) When functional groups are independent, i.e. $g_a = g_b = 1$

In this case, since no induction in the symmetric monomer exists, the constitution can only be influenced through r and the mode of addition of reactants to the monomer mixture.[26] Below we evaluate the effects of the mixing modes discussed above. It will be seen that $s = s(r)$ and $0 \leq s \leq \frac{1}{2}$.

(a) 'All at once'. If XabX monomer is mixed infinitely fast with YccY monomer, only disordered chains can be obtained. The faster reacting group (—aX) will react first, but the intermediates are of random structure and only random polymers can be obtained.[26]

(b) The nonsymmetric monomer is added to the symmetric monomer. If XabX monomers are added infinitely slowly to the mixture containing all of the YccY monomer, then both —aX and —bX groups will react at the same (concentration dependent) rate. This again yields a completely disordered polymer.

(c) The symmetric monomer is added to the nonsymmetric monomer. When YccY monomer is added infinitely slowly to XabX monomer, constitutional order is possible. This happens because YccY monomers, present only in very small concentration, react preferentially with —aX groups. If r is small, only XbaccabX will be formed, and when all —aX groups have been exhausted the reaction of the remaining (—bX) functional groups forces head-to-head/tail-to-tail order ($s \to 0$).

Results of computations are shown in Figures 4 and 5. Figure 4 shows that, no matter what the value of r is, if XabX and YccY are mixed all at once at the beginning of the reaction, only polymers of s approaching 1/2 can be obtained. Figure 5 illustrates the strong dependence of s on r for the case where YccY is being slowly added to XabX. A good approximation to the value of s at $1/X_n \to 0$

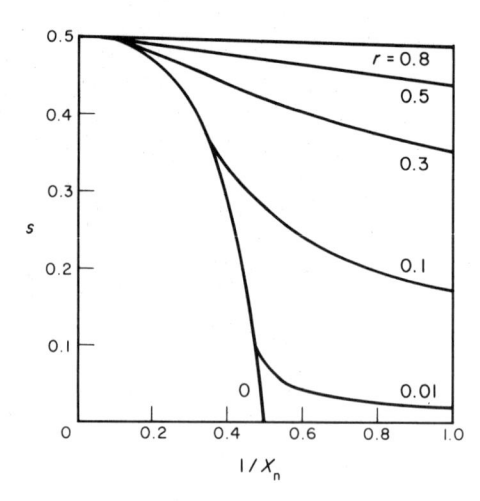

Figure 4 Degree of constitutional regularity s as a function of degree of polymerization (measuring extent of reaction) $1/X_n$ when all reactants are mixed at once, for various values of r. All functional groups react independently

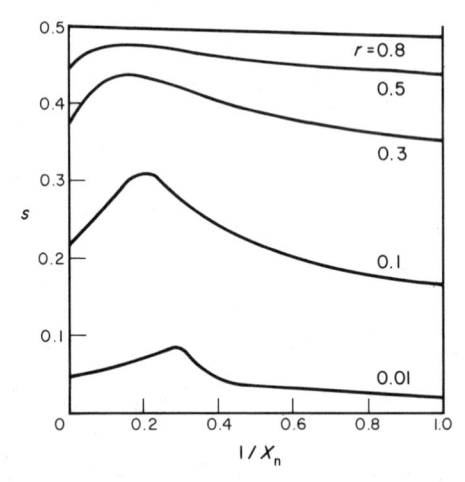

Figure 5 s as a function of degree of polymerization (measuring extent of reaction) $1/X_n$ when the symmetric monomer YccY is added very slowly to the reaction mixture containing all XabX, for various values of r. All functional groups react independently

which holds within 2% for $r < 0.1$ for these cases is[26]

$$s \approx 1 - r^{r/(1-r)} \qquad (12)$$

(ii) When functional groups in the symmetric monomer have interdependent reaction rates but equal induction, i.e. $g_a = g_b \neq 1$

(a) 'All at once'. If the monomers are mixed infinitely fast and there is induction, constitutional order is achievable through either of the following two mechanisms. If r is sufficiently small and the second reacting group on the YccY monomer is strongly deactivated by reaction of the first (i.e. $g_a = g_b \to 0$), then s will approach unity. This happens because the dominant early intermediate is XbaccY, i.e. a pseudo-monomer of the AB-type which later forms head-to-tail arrangements. If, on the other hand, the induction is activating (i.e. $g_a = g_b \to \infty$) then s will approach zero, since the dominant early intermediate is XbaccabX. Sample cases of s as a function of $g = g_a = g_b$ are shown in Figure 6. Obviously, $s = s(r,g)$, and $0 \leq s \leq 1$.

(b) One monomer is added to the other. If one of the monomers is added 'infinitely slowly' to the reaction mixture, induction is not operative on the YccY monomer. Thus the results of these cases are identical to the ones discussed above.

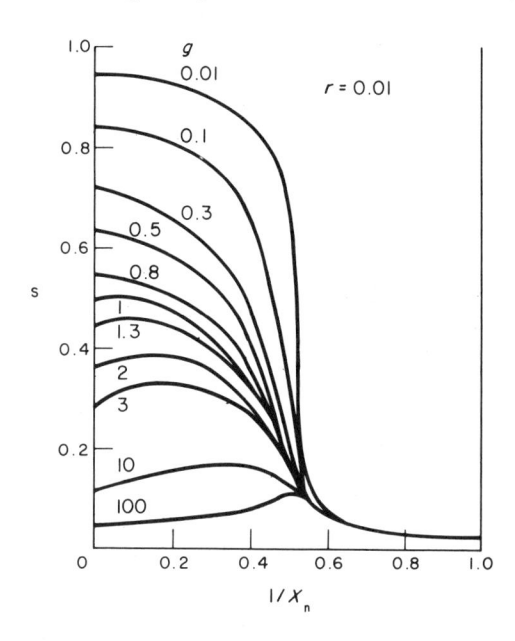

Figure 6 s as a function of degree of polymerization (measuring extent of reaction) $1/X_n$ when all reactants are mixed at once, for various values of r. The functional groups on the symmetric monomer YccY are dependent on each other, but show equal degrees of induction g ($g = g_a = g_b$)

(iii) When functional groups in the symmetric monomer have interdependent reaction rates and unequal induction, i.e. $g_a \neq 1$, $g_a \neq g_b$

For the extreme addition rates considered in this section ('infinitely slowly' or 'all at once') values of s are similar to those of the preceding case, except that for values of $r \ll 1$ only the value of g_a affects the constitution. This is true since constitution is controlled by the —aX groups reacting faster than —bX and is therefore only greatly affected by —accY groups.

7.2.4.4 Finite feed rates

The model described in Section 7.2.4.1 can be used with any specified type of monomer addition (*e.g.* constant feed rate, stepwise addition, *etc.*). Here we examine[28] the case of constant rates of addition of YccY to a mixture containing all XabX, and of XabX to a mixture containing all YccY. A constant rate of addition is attractive because it is easily employed both in the laboratory and in larger scale applications.

The rate of addition of one monomer to the other is quantified by the dimensionless parameters $dY/d\theta$ (XabX added to a mixture containing all YccY) and $dZ/d\theta$ (YccY added to a mixture containing all XabX). The discussion is structured as in the previous section.

(i) When functional groups are independent, i.e. $g_a = g_b = 1$

Figure 7 depicts a series of computations for s as a function of the constant feed rates $dY/d\theta$ and $dZ/d\theta$ for various values of r. In all cases the constitution described is the one of a polymer of degree of polymerization 100. Monomer feed rates vary continuously from 'very slow' to 'very fast' addition. The interesting range is the one where s varies quite steeply when YccY is added to XabX. Here polymer order can be easily controlled from the maximum to the minimum possible, within the limits of the given kinetic parameters; $s = s(r, \text{feed rate})$ and $0 \leq s \leq 1/2$. Note that the cases where XabX is added to YccY always yield a disordered polymer.

(ii) When functional groups in the symmetric monomer have interdependent reaction rates but equal induction, i.e. $g_a = g_b \neq 1$

Figure 8 depicts the effect of changes in g_a for both types of addition at a constant value of $r = 0.01$. For $g_a = g_b = 0.067$, varying the addition rate of the YccY monomer to XabX monomer

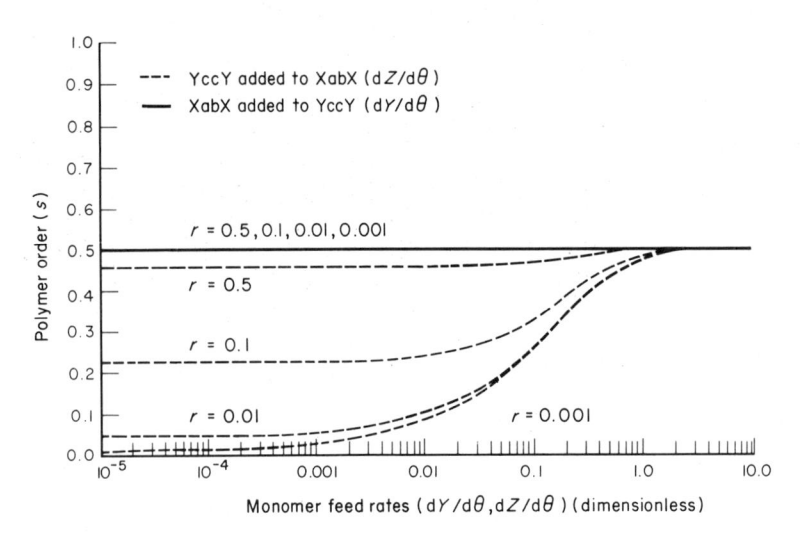

Figure 7 s as a function of monomer feed rates $dY/d\theta$ (when XabX is fed continuously) and $dZ/d\theta$ (when YccY is fed continuously) for various values of r; $ga = gb = 1$. All functional groups react independently

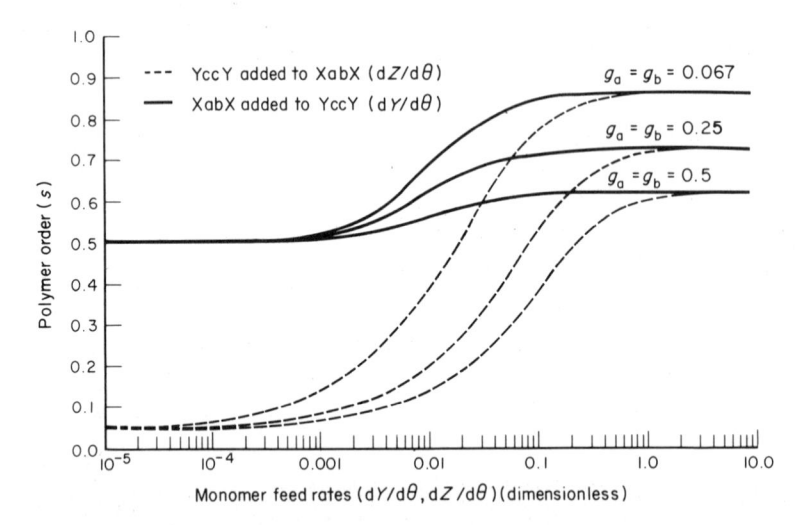

Figure 8 s as a function of monomer feed rates $dY/d\theta$ (when XabX is fed continuously) and $dZ/d\theta$ (when YccY is fed continuously) for $r = 0.01$. The functional groups on the symmetric monomer YccY are dependent on each other, but show equal degrees of induction g ($g = g_a = g_b$)

controls constitutional isomerism from $s = 0.049$ to $s = 0.864$. This represents a change from a structure that is of high head-to-head/tail-to-tail regularity to one that is of mostly head-to-tail structure. Note that mixing all at once yields the maximum value of s.

(iii) When functional groups in the symmetric monomer have interdependent reaction rates and unequal induction, i.e. $g_a \neq 1$, $g_a \neq g_b$

In Section 7.2.4.3 it was noted that in all cases where induction was present only the value of g_a caused significant changes in polymer constitution. For finite feed rates this is no longer always true. As shown in Figure 9, when YccY is added to XabX, the results are identical to those obtained for equal induction with respect to changes in g_a, as expected from reasons already discussed. However, when XabX is added to YccY, values of s much lower than the previous minimum of 0.5 are obtained at low (but not 'infinitely slow') feed rates. The cause for this is the formation of more —bccY units than in either of the two extremes. These units have a deactivating induction lower than that of —accY units and this causes more head-to-head/tail-to-tail arrangements due to the YccabccY intermediates reacting preferentially on the Ycca— side. Note that only a particular combination of r, g_a, g_b and feed rate of XabX will cause this phenomena. Table 2 summarizes the influence of kinetic parameters and mixing modes on constitutional order.

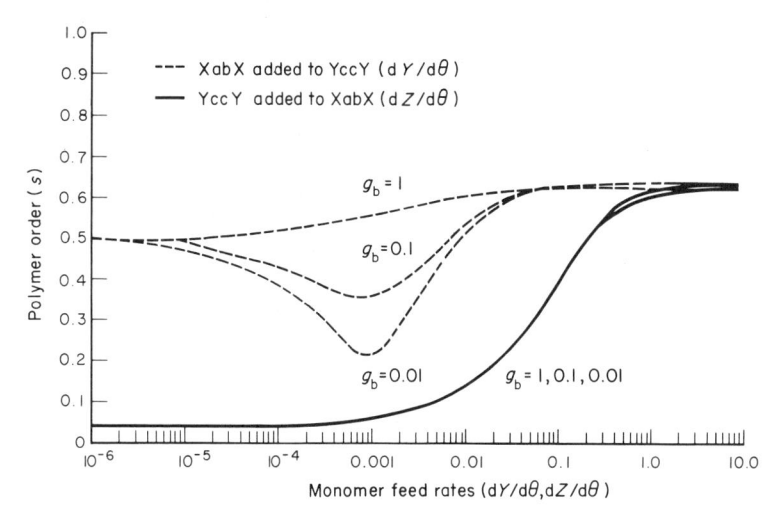

Figure 9 s as a function of monomer feed rates $dY/d\theta$ (when XabX is fed continuously) and $dZ/d\theta$ (when YccY is fed continuously) for $r = 0.01$. The functional groups on the symmetric monomer YccY are dependent on each other and have different degrees of induction: $g_a = 0.5$, g_b varies

Table 2 Kinetic Control of Constitutional Isomerism

Cases where feed rate need not be considered (very slow or very fast)	
Independent functional groups: $g_a = g_b = 1$	$(0 \leq s \leq 0.5)$
YccY slowly added to XabX	$s = s(r)$
XabX slowly added to YccY or immediate mixing of reactants	$s = 0.5$
Functional groups in symmetric monomer are interdependent: $g_a, g_b \neq 1$	$(0 \leq s \leq 1)$
YccY slowly added to XabX	$s = s(r)$ (as above)
XabX slowly added to YccY	$s = 0.5$ (as above)
Immediate mixing of reactants	$s = s(g_a, r)$
	$g_a \to 0, s \to 1$
	$g_a \to \infty, s \to 0$
Finite feed rates	
Independent functional groups: $g_a = g_b = 1$	$(0 \leq s \leq 0.5)$
YccY added to XabX	$s = s(r, \text{feed rate})$
XabX added to YccY	$s = 0.5$
Functional groups in symmetric monomer are interdependent: $g_a, g_b \neq 1$	$(0 \leq s \leq 1)$
YccY added to XabX	$s = s(r, g_a, \text{feed rate})$
XabX added to YccY	$s = s(r, g_a, g_b, \text{feed rate})$

7.2.5 Sequence Statistics

Additional information on the polymer structure (as well as a check of the numerical integration methods) can be obtained by Monte-Carlo simulations. The value of s is determined by the frequency of —acca—, —accb— and —bccb— diad units in the polymer, the smallest arrangements with distinguishable orientation. The next level of refinement would involve triads, *i.e.* arrangements like —accabcca—, —accabccb—, etc. In order to determine if knowledge of s is sufficient to describe the polymer structure, information is required on the fractions of triads in the system. If polymers of equivalent s also have comparable triad fractions, then the value of s could be exclusively used to describe the polymer structure. This is particularly important when comparing polymers of equal s, synthesized from the same monomers, but *via* different synthetic routes. For example, it is seen in Figure 9 that for a given set of kinetic parameters polymers of equal s can be obtained by very different addition modes.

In the Monte-Carlo method one simulates the reaction between a large number (*e.g.* 100 000) of monomer molecules of each type (*e.g.* 200 000 monomer molecules total). During the course of the simulation of the polymerization reaction the number of chain ends, diads and triads formed are counted. Once the reaction is completed, the resulting numbers are analyzed and one is able to generate the structure of a 'typical chain' or calculate the 'blockiness' of the chains.[28]

Calculations of this type show that s is sometimes not sufficient to describe the polymer chain. An example is shown in Table 3.[28] Two polymers, PI and PII, were made from monomers that react with the relative rate constants $r = 0.01$, $g_a = 0.5$ and $g_b = 0.1$ to a final degree of polymerization of 100. Polymer PI was made by the very slow addition of XabX to YccY, while polymer PII was generated by a much faster addition of YccY to XabX. Polymer PI is very disordered, having nearly equal amounts of all triads possible and blocks of nearly equal lengths, while polymer PII is much less disordered (it shows a tendency towards high amounts of —accb— blocks). Diad statistics as expressed by s may not always be sufficient to describe a polymer with nonsymmetric units.

Table 3 Comparing Polymers with Equal s, Prepared Differently from the Same Monomers[a]

Polymer PI: XabX added to YccY ($dY/d\theta = 1 \times 10^{-5}$)	
Triad	Fraction (± 0.02)
—accabccb— —bccbacca—	0.27
—accabcca— —accbacca—	0.27
—accbaccb— —bccabcca—	0.23
—bccabccb— —bccbaccb—	0.23
Average head-to-tail block length:	3.3 ± 0.2
Average head-to-head/tail-to-tail block length:	3.5 ± 0.2
Polymer PII: YccY added to XabX ($dZ/d\theta = 2 \times 10^{-1}$)	
Triad	Fraction (± 0.02)
—accabccb— —bccbacca—	0.32
—accabcca— —accbacca—	0.19
—accbaccb— —bccabcca—	0.32
—bccabccb— —bccbaccb—	0.17
Average head-to-tail block length:	2.0 ± 0.2
Average head-to-head/tail-to-tail block length:	4.0 ± 0.2

[a] $r = 1 \times 10^{-2}$, $g_a = 0.5$, $g_b = 0.1$.

7.2.6 Summary of Theoretical Development

This section suggests that constitutional isomerism in step polymerization products can be controlled in an easy and preselectable manner. It has been shown that high degrees of order can only be obtained *via* kinetically controlled reactions and that the same single reaction step can be used to generate polymers of continuously differing order, within the limits of the relative reaction rate constants, *i.e.* r, g_a and g_b.

In general, if the symmetric monomer (YccY) is fed very slowly to the reaction mixture containing all of the nonsymmetric monomer (XabX), the highest possible head-to-head/tail-to-tail regularity is achieved. If the nonsymmetric monomer (XabX) is added very slowly to the reaction mixture containing all of the symmetric monomer (YccY), a random polymer is invariably obtained. Finally, if the reactants are mixed all at once ('infinitely fast'), the highest possible degree of head-to-tail constitution is achieved; this might, however, only be as high as the random case, *i.e.* if no interdependence of functional groups exists, no significant head-to-tail regularity can be obtained.

The constitutional regularity of macromolecules from step-growth polymerizations is succinctly expressed by s, the probability of two adjacent nonsymmetric monomeric units pointing in the same direction along the chain. It is, however, possible to obtain polymers from different one-step syntheses that are characterized by the same s but have clearly different constitutional statistics. In that case, more detailed statistics are needed.

7.3 EXPERIMENTAL FACTS

7.3.1 Systems Investigated

Attempts to prepare polymers with controlled constitution have been appearing in the literature for some time, but only the advent of modern analytical methods has made it possible to draw definite conclusions and to evaluate theoretical considerations. This era started with Pino et al.[18] who succeeded in combining synthesis with analysis (by mild, quantitative degradation and subsequent analysis of the fragments) of polyamides from the nonsymmetric p-carboxybenzyl-carboxymethylene thioether dichloride (similar to **1** and corresponding to XabX) and ethylenediamine (**2**, corresponding to YccY). Almost simultaneously, Korshak and collaborators[19] published the preparation and NMR analysis of polyesters from terephthalic dichloride and 2,2-(p-hydroxyethoxyphenyl)(p-hydroxyphenyl)propane.

$$O_2NC_6H_4O_2CCH_2SCH_2-C_6H_4-CO_2C_6H_4NO_2 \qquad H_2NCH_2CH_2NH_2$$

$$(1) \qquad\qquad (2)$$

Since then, several polymer types have been prepared in a single synthetic step, in some instances with very high degrees of constitutional order. They comprise polyesters,[24] polyureas[24,32] and polyamides,[18,21,23,24,28,33,34] among them fully aromatic rigid-rod polyamides.[28,33,34] The results from these efforts are summarized below.

7.3.2 Random and Maximally Ordered Polymers

In Section 7.2 it was pointed out that polymers with the highest achievable constitutional regularities can be made by either adding the symmetric monomer very slowly to the reaction mixture containing all of the nonsymmetric monomer (the highest possible head-to-head/tail-to-tail regularity is achieved), or by mixing the reactants all at once (the highest possible degree of head-to-tail constitution is reached, although this is only the one corresponding to a random polymer if the functional groups in the symmetric monomer react independently of each other). Slowly feeding the nonsymmetric monomer to the reaction mixture containing all of the symmetric monomer yields a random polymer.

Pino et al. first demonstrated[21,23] these principals in the preparation of polyamides prepared by aminolysis of the bis(p-nitrophenyl) ester (**1**) with ethylenediamine (**2**). The value of r was measured to be 1.65×10^{-2} in DMSO/DMF (2:1) as solvent at $0\,°C$, and induction was estimated to be absent ($g_a = g_b = 1$). From this a minimum possible value of $s = 0.067$ was calculated for quantitative conversion and for an 'infinitely slow' addition of (**2**) to the reaction mixture; the maximum value of s was estimated at 0.5. The syntheses were carried out under conditions of very slow addition of (**2**) to (**1**) as well as by adding (**1**) to (**2**). Table 4 shows the constitutional regularity and some properties of the resulting polyamides; the lowest observed value of s, 0.079, corresponds to 92% head-to-head/tail-to-tail regularity. Theoretical values agreed well with experiment. Polyamides of much greater degrees of order than those in Table 4 have been obtained in the polycondensations of ethylenediamine with monomers such as (**3**) or (**4**) by Steinmann et al.[23] Values of r are very low (1.15×10^{-5} for **3** and 2.8×10^{-4} for **4**, in DMSO at $25\,°C$). The lowest value of s observed in these cases (i.e. <0.005 for polyamides from **2** and **3**) corresponds to the highest regularity ever reported for condensation polymers of this type; indeed, since the estimated degree of polymerization for these macromolecules is of the order of 100, the very low value of s implies that the majority of chains in the sample are perfectly regular. Results for a regular and a random polyamide from (**2**) and (**4**) are shown in Table 5.[23]

Meyer, Gentile and Suter[33,34] prepared rigid-rod fully aromatic polyamides (aramids) from monosubstituted and 2,6-disubstituted p-phenylenediamines (**5**) and terephthaloyl dichloride (**6**).

$$O_2NC_6H_4O_2C(CH_2)_nO-\underset{OMe}{\overset{OMe}{C_6H_3}}-CO_2C_6H_4NO_2$$

(**3**) $n = 1$
(**4**) $n = 10$

Table 4 Polyamides Obtained from (1) and (2)[21,23]

Process[a]	Expected	s Found Cleavage	s Found NMR	$[\eta]$[b]	M.P. (°C)
Very slow addition of (2) to (1)	0.067	0.079	0.095	0.24	258
Addition of (1) to (2)	0.5	0.493	0.48	0.41	224

[a] A solution of one monomer in DMSO/DMF (2:1) was added to an equimolar solution of the second monomer in the same solvent mixture. The reaction was carried out under vigorous stirring at 0 °C.
[b] In units of dl g^{-1}, in m-cresol at 30 °C.

Table 5 Properties of a Regular and an Irregular Polyamide from (2) and (4)[23]

s	$[\eta]$[a]	M.P. (°C)	Solubility[b] DMSO	Solubility[b] DMF	Moisture content[c] (%)
0.03	1.6	142.5	140	10	0.45
0.46	1.1	105	200	180	0.7

[a] In units of dl g^{-1}, in m-cresol at 30 °C.
[b] In units of mg ml^{-1}, maximum amount of polymer giving an apparently homogeneous solution, at room temperature.
[c] In air of 55% relative humidity at 25 °C, at equilibrium.

The monomers were chosen to yield low values of r (≤ 0.01) and values of g_a and g_b below unity to allow for the widest possible range of s. Low temperature amidation reactions in homogeneous phase were employed for full control of the polymerization without side reactions in order to obtain high molecular weight samples. Typically, mixtures of THF and a traditional amidation solvent, such as NMP, were employed such that the aromatic acid chloride would come in contact with the amidation solvent as little as possible.

The fact that two regular polymers with different constitution can be prepared by simply changing the mode of monomer feed was shown first[32] in polyureas from bis(p-nitrophenyl) carbonate (7) and 2-(4-aminophenyl)ethylamine (8). Here the induction was shown to be very large or absent, depending on the first reactant; the relative reaction rate constants for (7) are $g_a \approx 3.6 \times 10^{-3}$ and $g_b \approx 1$. Since r is also small (6.3×10^{-6}), a polymer with substantially head-to-tail order could be produced (the maximum value of s is 0.89). Table 6 shows the results.

$$O_2N-\langle\!\!\!\bigcirc\!\!\!\rangle-OCO_2-\langle\!\!\!\bigcirc\!\!\!\rangle \qquad H_2NCH_2CH_2-\langle\!\!\!\bigcirc\!\!\!\rangle-NH_2$$

(7)   (8)

Parenthetically it shall be noted here that attempts to control the constitution of polyesters by the same methods have also been made. The difficulty in polyesters lies in the rapid and hard-to-avoid transesterification reactions. The highest regularity achieved so far corresponds to a value of s of ca. 0.3 and was obtained[24] in the transesterification polymerization between ethylene glycol and an

Table 6 Constitutionally Regular and Irregular Polyureas from (7) and (8)[32]

Method of preparation	s	$[\eta]$[a]	$\bar{M}_n$
(7) added slowly to (8)	0.05	0.38	3700
(7) and (8) mixed rapidly	0.89	0.33	1700
(8) added slowly to (7)	0.47	0.57	4300

[a] In units of dl g^{-1}, in DMSO/LiCl (2%), at 30 °C.

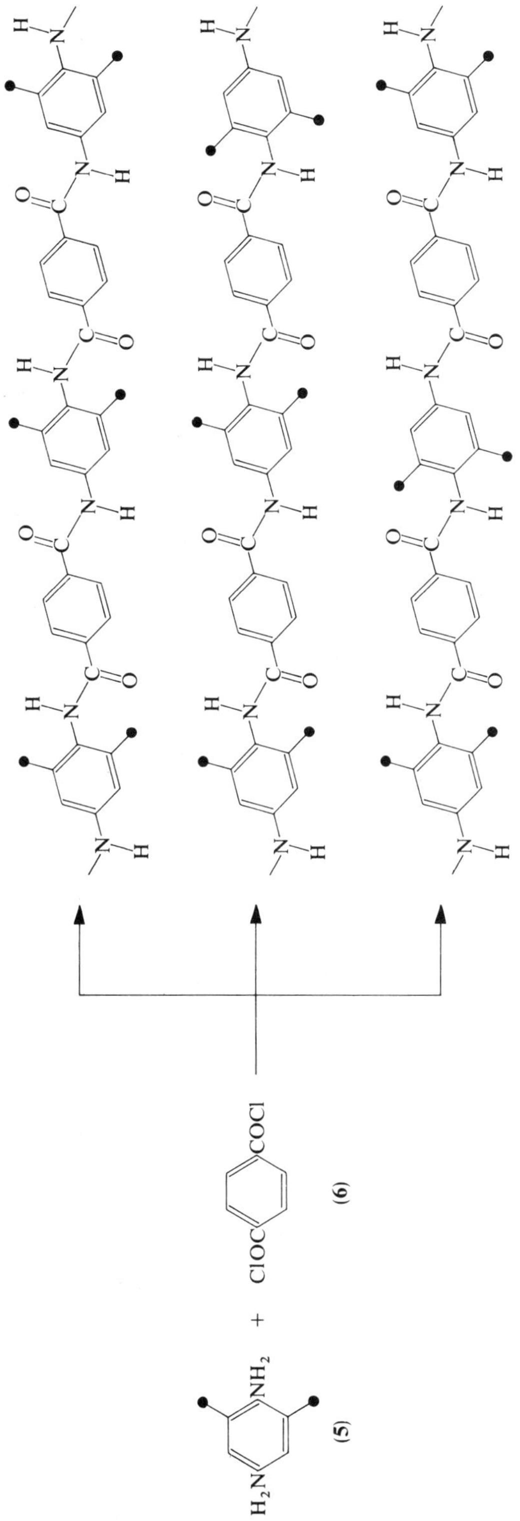

Constitutional Regularity in Linear Condensation Polymers

Scheme 1

aliphatic nonsymmetric divinyl ester in the presence of a mixture of carefully balanced basicity and nucleophilicity (containing N,N,N',N'-tetramethyl-1,8-diaminonaphthalene and 1,8-diazabicyclo-[5.4]undec-7-ene).

7.3.3 Continuous Control over Constitutional Order

Polyamides from monomers (1) and (2) were used first[23] to attempt the continuous control over constitutional order following the principles outlined in Section 7.2.4.4. The addition of the diamine monomer was carried out such that the free amino group concentration was constant (until the equimolar amount had been added) at a preset fraction q of the total stoichiometric value. Table 7 lists the measured values of s vs. q. Note that the addition method is not exactly the same as the one presented above in Section 7.2.4.4 where the case of a constant rate of addition is discussed, but is very similar; $q = 0$ represents infinitely slow addition and $q = 1$ represents instantaneous mixing at the start of the reaction. In Figure 10 the experimental values of s are plotted together with two theoretically determined curves for values of r bracketing the experimental value. Indeed, constitutional order is controllable in real systems.

Table 7 Detailed Control over Constitution in Polyamides from (1) and (2)[23]

Process[a]	s	$[\eta]$[b]	M.P. (°C)
Very slow addition of			
(2) to (1) ($q = 0$)	0.079	0.24	258
(2) to (1) ($q = 0.05$)	0.100	—	249
(2) to (1) ($q = 0.10$)	0.137	—	246
(2) to (1) ($q = 0.15$)	0.178	—	241
(2) to (1) ($q = 0.20$)	0.226	—	238
(2) to (1) ($q = 0.24$)	0.260	—	236
Addition of			
(1) to (2) ($q = 1$)	0.493	0.41	224

[a] The rate of addition is controlled so that the free amino group concentration is equal to the fraction q of the concentration required for equimolarity, until the necessary amount of diamine is added.
[b] In units of dl g^{-1} in m-cresol at 30 °C.

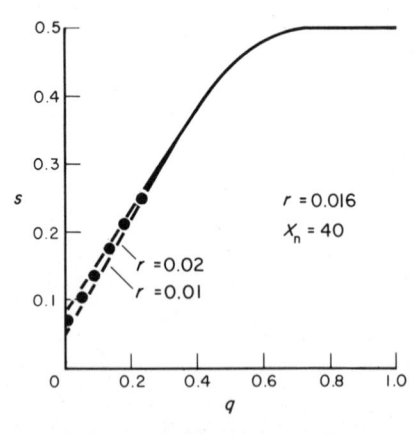

Figure 10 Comparison of experimental data (dots) and theoretical prediction (lines) for the constitutional regularity of polyamides prepared from (1) and (2) at different feed rates q (see text for details). Data from ref. 23

Knowledge of the absolute rate constant k_{aX} (in addition to r, g_a and g_b) allows prediction of s as a function of absolute monomer feed rate (*e.g.* in units of ml h^{-1} at known concentration and volume). For the system of terephthaloyl dichloride and 2,6-dichloro-p-phenylenediamine, Figure 11 shows a comparison between theoretically predicted order and values actually measured.[33,34] Deviations

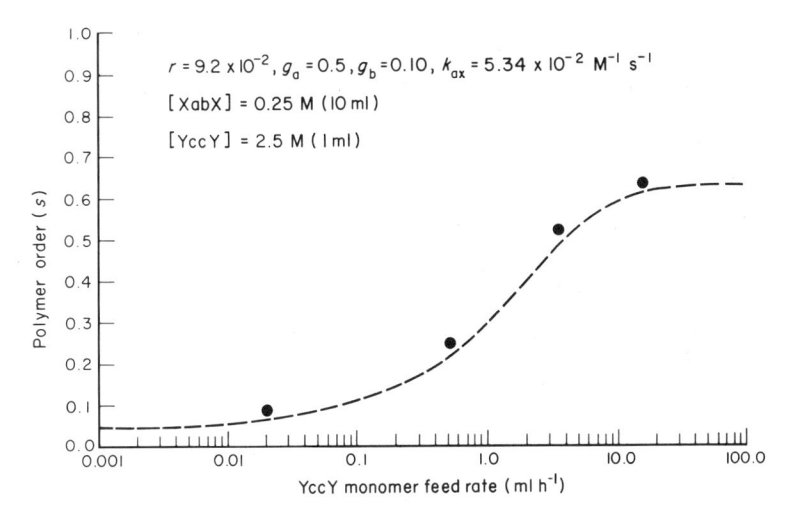

Figure 11 Comparison of experimental data (dots) and theoretical prediction (lines) for poly(2,6- dichloro-p-phenyleneterephthalamide), the terephthalic dichloride being fed continuously at various rates to the reaction mixture containing the nonsymmetric diamine

might be due to errors in molecular weight (they might be slightly lower than those used for the theory, $X_n = 100$) and to uncertainties in the determination of s (constitutional order was measured by ^{1}H NMR[34]). Even so, agreement between calculation and experiment is quite good.

7.3.4 Constitutional Regularity and Properties

Besides information on the mode of preparation of polymers and the degree of constitutional regularity achieved, Tables 4 to 7 also contain data on properties of the products. Tables 4, 5 and 7 show values for the melting point; inspection reveals that in all cases the more regular samples melt at higher temperatures and that this fact is not at all connected to molecular weight. The increase is particularly impressive in the values in Table 7 where it is monotonic and it can be seen that a polymer with the desired melting point (within the values spanned) can be prepared easily.

Table 5 also contains information on 'solubility' (*i.e.* the maximum polymer concentration giving an apparently homogeneous solution at room temperature) and the equilibrium moisture content of a sample in contact with moist air. Note that there is a distinct change in properties when s changes. Differences are not always very large, but it is clear that many polymer characteristics are affected by constitutional order.

Often it is important that some properties can be changed while others are invariant to the structural change. As an example of a system where such a behavior is observed (and advantageous for the application characteristics of the polymer) we consider aramids capable of exhibiting constitutional isomerism as introduced in the preceding sections. Figure 12 shows the tendency of poly(2,6-dichloro-p-phenyleneterephthalamide) to form a liquid crystal phase in concentrated (96%) sulfuric acid.[35] Conformational analysis of these chains indicates that the constitutional regularity should not influence the persistence length of the macromolecules, *i.e.* the rigid-rod character, and the tendency for solution anisotropy should therefore be equally unaffected. The propensity for formation of an anisotropic phase (a 'liquid crystalline phase') is measured by the critical volume fraction of polymer for phase heterogeneity v_p^*.[36] The theoretical curve based on this concept and calculated using Flory's 1956 theory for rigid-rod liquid crystalline polymers[36] is displayed in Figure 12. The corresponding experimental values are shown for highly regular and completely random polymers, as well as for intermediate cases.[35] It does not seem that the orientational tendencies of the moieties on the ring are important for the ability to form an anisotropic phase.

When the same set of aramids and two other sets, poly(nitro-p-phenyleneterephthalamide) and poly(methoxy-p-phenyleneterephthalamide), are used to determine the apparent 'solubility' in a solvent (NMP), *i.e.* the maximum concentration of polymer giving an apparently homogeneous solution, all polymers exhibit a substantial change with order in the maximum amount of material that can be dissolved in the solution. The differences must be due to constitution and not to changes

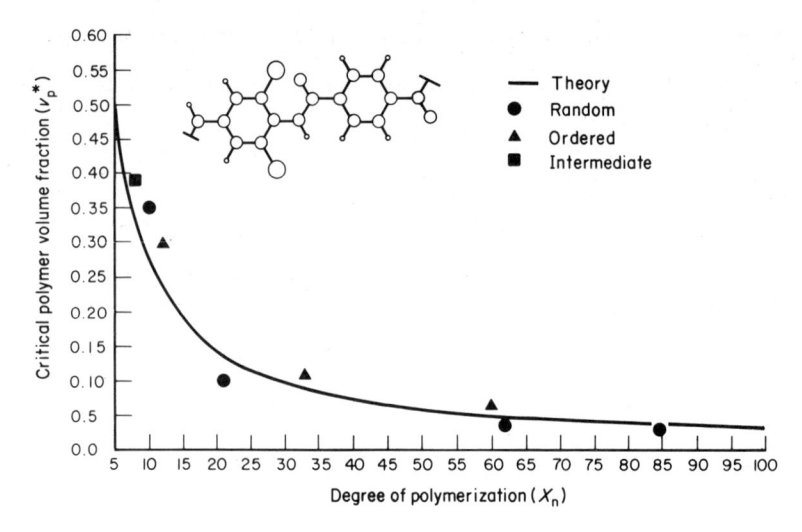

Figure 12 Comparison of experimental data (●, ▲, ■ symbols) and theoretical prediction (line following Flory's 1956 theory) for the volume fraction of polymers at the threshold between homogeneous isotropic and biphasic isotropic/nematic solution v_p^* as a function of the estimated degree of polymerization X_n for several poly(2,6-dichloro-p-phenyleneterephthalamide)s

in molecular weight: high molecular weight random polymers are equally as soluble as their much lower molecular weight ordered counterparts.[33,34]

7.4 CONCLUSIONS

Constitutional isomerism strongly affects many properties of macromolecules obtained from step polymerization and it can be controlled in an easy and preselectable manner.

7.5 REFERENCES

1. H. Staudinger and A. Steinhofer, *Justus Liebigs Ann. Chem.*, 1935, **517**, 35; C. S. Marvel and C. L. Levesque, *J. Am. Chem. Soc.*, 1938, **60**, 280; C. S. Marvel and C. E. Denoon, Jr., *J. Am. Chem. Soc.*, 1938, **60**, 1045; C. S. Marvel, J. H. Sample and M. F. Roy, *J. Am. Chem. Soc.*, 1939, **61**, 3241.
2. P. J. Flory, 'Principles of Polymer Chemistry', Cornell University Press, Ithaca, NY, 1953, chap. 6.
3. F. A. Bovey, 'Chain Structure and Conformation of Macromolecules', Academic Press, New York, 1982, chap. 6.
4. P. J. Flory and F. S. Leutner, *J. Polym. Sci.*, 1948, **3**, 880; 1950, **5**, 267.
5. O. Vogl, *Polym. Prepr., Am. Chem. Soc., Div. Polym. Chem.*, 1979, **20(1)**, 154, and references cited therein; S. Grossman, A. Yamada and O. Vogl, *J. Macromol. Sci., Chem.*, 1981, **16**, 897; S. Grossman, A. Stolarczyk and O. Vogl, *Monatsh. Chem.*, 1981, **112**, 1279; T. Kondo, M. Kitayama and O. Vogl, *Polym. Bull. (Berlin)*, 1982, **8**, 9; H. Kawaguchi, Y. Sumida, J. Muggee and O. Vogl, *Polymer*, 1982, **23**, 1805; T. Tanaka and O. Vogl, *Macromol. Synth.*, 1982, **8**, 21.
6. M. T. Malanga and O. Vogl, *Polym. Bull. (Berlin)*, 1983, **8**, 236.
7. E. F. Izard, *J. Polym. Sci.*, 1952, **8**, 503.
8. K. W. Doak and H. N. Campbell, *J. Polym. Sci.*, 1955, **18**, 215.
9. J. Preston and R. Smith, *J. Polym. Sci., Part B*, 1966, **4**, 1033.
10. K. Saotome and R. C. Schulz, *Makromol. Chem.*, 1967, **109**, 239.
11. N. G. Yesipova and I. N. Topchieva, *Polym. Sci. USSR (Engl. Transl.)*, 1967, **8**, 190.
12. R. W. Morrison, J. Preston, J. C. Randall and W. B. Black, *J. Macromol. Sci., Chem.*, 1973, **7(1)**, 99.
13. S. G. Cottis, J. Economy and R. S. Storm, *Ger. Pat.*2 248 127, (1973) (*Chem. Abstr.*, 1973, **79**, 5887).
14. R. Lenz and S. Go, *J. Polym. Sci., Polym. Chem. Ed.*, 1974, **12**, 1.
15. P. W. Morgan, S. L. Kwolek and T. C. Pletcher, *Macromolecules*, 1987, **20**, 729.
16. R. Deschenaux, P. Neuenschwander and P. Pino, *Helv. Chim. Acta*, 1986, **69**, 1349.
17. M. Murano, *Polym. J.*, 1972, **3**, 663.
18. P. Pino, P. G. Casartelli, J. A. Quiroga, and G. P. Lorenzi, *Int. Symp. Macromol. Chem., Madrid, Prepr.*, 1974, **1**, 255.
19. V. V. Korshak, S. V. Vinogradova, S. I. Kuchanov, V. A. Vasnev, G. D. Markova and A. I. Tarasov, *Vysokomol. Soedin., Ser. A*, 1974, **16**, 1992.
20. V. V. Korshak, S. V. Vinogradova, V. A. Vasnev, G. D. Markova and T. V. Lecae, *J. Polym. Sci., Polym. Chem. Ed.*, 1975, **13**, 2741.
21. P. Pino, G. P. Lorenzi, U. W. Suter, P. G. Casartelli, A. Steinmann, F. J. Bonner and J. Quiroga, *Macromolecules*, 1978, **11**, 624.
22. S. V. Vinogradova, *Vysokomol. Soedin., Ser. A*, 1977, **19**, 667.
23. A. Steinmann, U. W. Suter and P. Pino, *Int. Symp. Macromol. Chem., Florence, Prepr.*, 1980, **2**, 228.

24. A. Steinmann, A. Arber, M. Schmucki, U. W. Suter, G. P. Lorenzi and P. Pino, *Int. Symp. Macromol. Chem., Strasbourg, Prepr.*, 1981, **1**, 103.
25. U. W. Suter, in 'Giornate di Studio Sulla Policondensazione', Italian Association of the Science and Technology of Macromolecules, 1981, p. 7.
26. U. W. Suter and P. Pino, *Macromolecules*, 1984, **17**, 2248.
27. R. Takatsuka, K. Uno, F. Toda and Y. Iwakura, *J. Polym. Sci., Polym. Chem. Ed.*, 1977, **15**, 1905.
28. F. T. Gentile, W. R. Meyer and U. W. Suter, in preparation.
29. L. C. Case, *J. Polym. Sci.*, 1958, **29**, 455.
30. G. C. Maitland, R. Rigby, E. B. Smith and W. A. Wakeham, 'Intermolecular Forces, Their Origin and Determination', Clarendon Press, Oxford, 1981, chap. 1.
31. From the following table, derived from J. N. Israelachvili, 'Intermolecular and Surface Forces, Applications to Colloidal and Biological Systems', Academic Press, New York, 1985.

Interaction	*Energy* (kT)
C—H bond strength	172
Ion/dipole	39
Hydrogen bond	4–16
Van der Waals	1

32. M. A. Schmucki, P. Pino and U. W. Suter, *Macromolecules*, 1985, **18**, 823.
33. W. R. Meyer, F. T. Gentile, J. A. Gado and U. W. Suter, *Polym. Prepr., Am. Chem. Soc., Div. Polym. Chem.*, 1986, **27** (2), 186.
34. W. R. Meyer, F. T. Gentile and U. W. Suter, *Polym. Prepr., Am. Chem. Soc., Div. Polym. Chem.*, 1987, **28** (2), 290.
35. F. T. Gentile, W. R. Meyer and U. W. Suter, *Polym. Prepr., Am. Chem. Soc., Div. Polym. Chem.*, 1987, **28** (2), 288.
36. P. J. Flory, *Adv. Polym. Sci.*, 1984, **59**, 1.

8
Multifunctional Step Polymerization

THOMAS A. VILGIS

Max-Planck-Institut für Polymerforschung, Mainz, FRG

8.1	INTRODUCTION	117
8.2	THE CLASSICAL THEORY OF GELATION	118
	8.2.1 *The Flory–Stockmayer Model*	118
	8.2.2 *The Gel Point of Various Systems*	121
	8.2.2.1 Step reaction polymerization	121
	8.2.2.2 Classical Flory–Stockmayer theory and equal reactivity	121
	8.2.2.3 Classical theory and different reactivities	121
	8.2.2.4 Vulcanization of preformed polymers	121
	8.2.2.5 Star-shaped polymers	122
	8.2.2.6 Cyclization	122
	8.2.3 *The Size of the Clusters*	122
	8.2.4 *The Kinetic Approach*	124
	8.2.5 *The Rate Theory of Random Polymerization*	125
8.3	REFINEMENT OF THE CLASSICAL THEORY	125
	8.3.1 *Percolation*	125
	8.3.2 *Excluded Volume Forces and the Size of the System*	127
	8.3.3 *Recent Advances in Kinetics*	127
8.4	REFERENCES	128

8.1 INTRODUCTION

Multifunctional step polymerization is the fundamental process in network formation. The detailed knowledge of the reaction process, the resulting structure, as well as the kinetics are most essential questions. Their answers — even if only crude and approximate — may roughly predict the macroscopic physical behaviour of the resulting polymeric material, *e.g.* the mechanical properties of a gel. The prediction of macroscopic behaviour, such as mechanical, thermal or dynamic properties from microscopic structure, *i.e.* the connectivity of the chains, molecular weight distributions, *etc.* from microscopic processes is not solved in general. Theorists have developed simplified models, which enable detailed calculations for the polymerization process and the resulting macroscopic behaviour.

The probably simplest model is to start from a liquid of φ-functional molecules and initiate a polymerization of these units and let them polymerize through. This is one realization of the classical Flory–Stockmayer model,[1-4] where the question of gelation theory has been raised for the first time.

Another realization of this transition is crosslinking from preformed polymer.[1] In this case one starts with a polymer melt or a concentrated polymer solution, where φ-functional crosslinks have been added. This process is called vulcanization.

Most important is to realize that gelation and vulcanization processes describe a sort of liquid–solid transition, since the final material exhibits a finite shear modulus, rather than a viscosity. During the multifunctional polymerization reaction larger and larger molecules are formed. The viscosity η increases and close to the gel point it can be written as a power law[5,6] $\eta \sim (p_c - p)^{-k}$, where p is a measure for the extent of reaction. Right above the gel point, indicated by p_c, at the critical extent of reaction one observes a shear modulus[6] $G_\infty \sim (p - p_c)^t$ and the reaction bath responses with properties typical for solids. The boundary which separates liquid and solid behaviour is the gel point. Its determination is another very important question and generally unsolved.

It is intuitively clear that both gelation and vulcanization belong to the same class of this type of liquid–solid transitions,[1,7,8] and there are only small differences, which will be mentioned at appropriate places later.

In the next section the classical theory of gelation will be described briefly. But it must be noticed that by having a closer look to the critical properties of this type of transition, which can be formulated as a usual second order phase transition,[9,10–13] strong deviations from theory and experiment will be present. Presumably this problem can be solved by percolation theory.[5,6,15] One important feature of the percolation theory is that the object ('cluster') right at the critical point — the gel point — turns out to be scale invariant or self-similar,[5,12–16] i.e. at any scale of observation it has, on average, the same structure. A rapidly developing area of modern statistical physics is devoted to these random fractals.[15–18] In this chapter we will not spend much space on the fractal aspect of percolation and step polymerization. We refer the reader to Chapter 8, Volume 6 in this work, by the same author, where this aspect is discussed in more detail.

Later kinetic aspects are considered here,[18,19] and recent developments will be reported. Finally we will report on recent developments, which face the problems risen from classical theory. These considerations will discuss percolation theory, excluded volume effects, which have strong effects on the radius of gyration, and later developments in kinetics. We will see that these latter points will give rise to new open questions, to be answered in future.

8.2 THE CLASSICAL THEORY OF GELATION

8.2.1 The Flory–Stockmayer Model

The essence of the classical theory of gelation and vulcanization is summarized in Flory's book.[1] It has been developed by Flory[2] and Stockmayer.[3,4] In order to have non-linear polymerization one should start with reactants based on φ-functional units as in (**1**; where $\varphi = 4$) which reacts with a branched molecule, i.e. schematically shown by (**2**) or alternatively, consider a φ-functional monomer and a bifunctional unit B—B which connects to a part of the final macromolecule like as in (**3**).

$$
\begin{array}{ccc}
\text{A} & \text{A A A} & \text{A A} \\
\text{A—A} & \text{A—AA—AA—A} & \text{A—AB—BA—A} \\
\text{A} & \text{A A A} & \text{A B A} \\
& \text{A} & \text{BA—A} \\
& \text{A—A} & \text{A} \\
& \text{A} & \\
(1) & (2) & (3)
\end{array}
$$

The critical properties of the polymerization are of central importance.[5,6,14,18,19] If it is assumed that there is a gel point p_c, i.e. the extent of reaction where the macromolecule is so large that it extends throughout the entire volume for the first time, most of the macroscopic quantities can be expanded in terms of the distance from the gel point $|p - p_c|$. This expansion only makes sense not too far from the gel point. To reach this state a model has to be developed, which relates the nature of the reaction process to the macroscopic properties, like the degree of polymerization, the molecular weight distribution or the radius of gyration. A general model is very complicated, and the first step is an idealized and simplified model. The most simple picture is the Flory–Stockmayer theory. In Chapter 8, Volume 6 of this work, the graph theory approach, using generating functions, is presented. This is equivalent to the more elementary approach used by Flory, but it should be remembered that the classical theory is not only the basis of the graph theory approach, but is identical and produces results within the same limitations.

The original theory[2–4] starts from estimating the probability that a B—B group has reacted with a φ-functional A group, like in the reaction scheme above. Moreover if A—A and B—B groups are able to react with an A_φ group the probability that the A group of a branch unit is connected to a sequence $A_\varphi(B—BA—A)_m B—BA_\varphi$ is given by $w_A w_B \rho \{w_B w_B (1 - \rho)\}^m$; where ρ is the density of all end units of the φ-functional monomers and w_A, w_B are the elementary probabilities of finding an A—A or B—B group. By summing over all possibilities $0 \le m \le \infty$, which is a geometric series,

Flory's fundamental expression is derived:

$$p = w_A w_B \rho / \{1 - [w_A w_B (1 - \rho)]\} \qquad (1)$$

Notice that the formation of loops has been neglected.

From this expression most of the cases of interest can be derived. Consider for example the situation where a number of A—A units reacts with φ-functional R—A_φ units this general equation reduces to

$$p = w_A \rho / \{1 - [w_A (1 - \rho)]\} \qquad (2)$$

Notice that this equation is independent of the functionality φ.

One question is whether the system is able to form an infinite network of a huge, branched molecule, or if the reaction stops and only finite branched molecules are present. This decision can be made by analyzing the values of p. Gelation will occur if p exceeds a critical value p_c

$$p_c = 1/(\varphi - 1) \qquad (3)$$

Flory's argumentation was that if the probability p is larger than p_c enough new end groups at the clusters are produced so that the reaction is likely to continue indefinitely to form one large cluster. By using graph theory it can be demonstrated that equation (3) for the gel point is rigorous within the Flory–Stockmayer theory (see also Chapter 8 in Volume 6).

In order to connect measurable quantities, such as the degree of polymerization, with these considerations we will introduce the graph theory approach briefly.

Let us assume that each of the φ-functional groups has reacted with another one with some probability p. p is generally a function of temperature, concentration, etc. As qualitatively noted above, the value of p determines whether one gets an infinite network or only finite clusters. Therefore we define, in the regime below the gel point, a probability $P_n(p)$ to find a molecule with n monomers. $P_n(p)$ defines the molecular weight distribution. This probability can be written as $P_n(p) = c_n/c$, where c_n is the concentration of the n-mers and c the total concentration of monomers present. The correct normalization is[5]

$$\sum_{n=1} n P_n(p) = 1 \quad \text{for} \quad p < p_c \qquad (4)$$

Hence the gel fraction can be defined as

$$P_\infty(p) = 1 - \sum_{n=1} n P_n(p) \quad \text{for} \quad p > p_c \qquad (5)$$

This is the probability that a monomer is part of the gel fraction, i.e. the infinite cluster.

Further define a new distribution $\Omega_n(p) = n P_n(p)$ which is, according to equation (4) immediately normalized. $\Omega_n(p)$ is the probability that any monomer has reacted and is part of a cluster of $n + 1$ monomers. We define a generating function $F_0(\theta)$[20-22]

$$F_0(\theta) = \sum_n \Omega_n(p) \theta^n \qquad (6)$$

θ is an auxiliary variable required for any generating function. It can be seen immediately that the generating function contains all information required. For example the weight average molecular weight, which is measured usually by light scattering[23] is given by

$$\bar{P}_w = \frac{\partial}{\partial \theta} \log F_0(\theta) \Big|_{\theta=1} \qquad (7)$$

Higher moments can be calculated by taking higher derivatives of the generating function.

If loop formation is neglected F_0 can be rewritten in terms of another function F_1, which is the probability $(1 - p)$ that a unit has not reacted at one branch, but is connected to n further units, i.e.

$$F_0(\theta) = [1 - p + p \theta F_1(p, \theta)]^\varphi \qquad (8)$$

This can be written out further if one assumes that the network has only a tree-like structure and loops are forbidden. It is essential to realize that this is a mean field assumption. A disadvantage of the graph theory approach is that loops can only be taken into account in lowest approximation,[21,27] *i.e.* small loops.

Using the tree approximation F_1 can be rewritten as

$$F_1(p, \theta) = [1 - p + p\theta F_1(p, \theta)]^{\varphi - 1} \qquad (9)$$

and by iteration one finds

$$F_0(\theta) = \theta(1 + \theta(1 + \theta(1 + \theta(\cdots)^{\varphi-1})^{\varphi-1})^{\varphi-1})^{\varphi} \qquad (10)$$

which describes a tree with branching index φ.

The solution of equation (6) and equations (8, 9) is given by

$$F_0 = (1 - \varphi)^{\varphi} p \sum_{n=1}^{\infty} \frac{[(\varphi - 1)n]!}{[(\varphi - 2)n + 2]!(n - 1)!} X^{n-1} \qquad (11)$$

with $X = p^{\theta}(1 - p)^{\varphi - 2}$. From this equation one can easily find the probabilities Ω_n by taking the nth coefficient in θ.

Let us first show how one confirms the gel point (equation 3). Therefore we rewrite equation (10) to $F_1 \equiv w^{\varphi - 1}$ and find

$$\frac{w - 1}{p\theta} = w^{\varphi - 1} - \frac{1}{\theta} \qquad (12)$$

which has to be solved. The results are obvious: If $p < 1/(\varphi - 1)$, all clusters are finite and a sol phase is obtained, whereas if $p > 1/(\varphi - 1)$, the infinite network has been formed. This result agrees with the qualitative argumentation of Flory.

Thus the mean field theory predicts a gel point at $p_c = 1/(\varphi - 1)$.[2-4]

Consider now the weight average of the degree of polymerization (see equation 7). It turns out immediately by using the generating function

$$X_w = X_{w0} \frac{1 + p}{1 - p(\varphi - 1)} \sim \frac{1}{p_c - p} \sim (p_c - p)^{-\gamma} \qquad (13)$$

X_{w0} is the degree of polymerization of the initial molecules, *e.g.* $X_{w0} = 1$ if one starts with polyfunctional monomers. Such a behaviour is typical for a phase transition[28] and indeed this defines the critical exponent γ, which is 1 within the classical theory (see Chapter 8, Volume 6 of this work and the references therein).

Finally consider the probability of having an n-mer $P_n(n)$. This can be worked out within the same frame work, but it can be done very simply by elementary methods,[5] so that only the result is quoted

$$P_n(p) = n^{-5/2} \exp[-n(p - p_c)^2] \qquad (14)$$

Thus $P_n(p)$ obeys a typical scaling form, which led to a scaling theory of percolation.[5,6,29]

Similarly all quantities of interest can be expanded in powers of $|p - p_c|$ and the classical critical exponents can be determined.[5-7] It is known from experiments that this classical theory is not valid in three dimensions.[5,28-30] This will be considered later in this chapter.

Recent comprehensive summaries for the extended use of the graph theoretical formalism in the context of gelation are given by Burchard[23] and Dusek.[31,32] In these references more complicated cases have been treated, *i.e.* components with unlike functional groups of different reactivity, copolymerization and copolymerization of unlike functional groups. Moreover the mean field theory of multistage processes during network formation has been reported.[33] The first stage is the reaction of the monomers of type 1 giving a product 1, the second stage is the reaction of the monomers of type 2 with the product 1, and so on. This procedure can be applied to the reaction of preformed polymers as well.

Further it has been worked out in refs. 21 and 23 how other physical quantities like the scattering function or diffusion constants can be very easily found from the generating function. But one should keep in mind that a non-mean field version of graph theory is extremely difficult.[27,34,35]

8.2.2 The Gel Point for Various Systems

The gel point is not a universal quantity and depends strongly on the system considered, *i.e.* all details of chemistry, thermodynamic conditions, *etc.* will more or less define the actual value of p_c. Theoretically, this problem is only solved within some limits. It depends on the types of reaction, on the environment of the molecules, on the lattice type, if one works with lattice models (computer simulations), *etc.* Only approximate formulae from very idealized models can be derived and only a brief summary is given here.

8.2.2.1 Step reaction polymerization

Here higher functional units are added to themselves to form clusters. In an early approach[36] Carothers predicted

$$p_c = \frac{2}{\varphi}(1 - 1/X_n) \tag{15}$$

where X_n is the number average of the degree of polymerization. It is given by

$$X_n = \frac{2}{\varphi(2/\varphi - p)} \tag{16}$$

for $p < p_c$. X_n diverges at p_c, and the divergence is (apart from prefactors) of the same form as in equation (13). The exponent γ is characteristically 1, as in all mean field theories for the gelation problem. Since X_n becomes larger and larger p_c can almost be approximated by $p_c = 2/\varphi$ in Carother's approach.

8.2.2.2 Classical Flory–Stockmayer theory and equal reactivity

The gel point is predicted to $p_c = 1/(\varphi - 1)$ as already mentioned. Note that one has to assume an equal reactivity of each functional group, no matter where it is placed in the polymer.

It is important to realize that the value for the gel point predicted by Carothers is always larger than that of Flory and Stockmayer, *i.e.* $2/\varphi \geq 1/(\varphi - 1)$, $\forall \varphi \geq 2$. Another consequence from this inequality is given by the comparison of X_w from equation (13) and X_n from equation (16). In general one finds $X_w/X_n > 2$, *i.e.* broad molecular weight distributions for $p < p_c$.

8.2.2.3 Classical theory and different reactivities

Macosko and Miller[37] derived equations for the gel point in various limits. For the step polymerization of homopolymers the same equation as in the Flory–Stockmayer theory was derived. The extent of reaction at the gel point for systems containing two types of mutually reactive groups can be estimated by[37,38,39]

$$p_c^2 = \frac{1}{(\varphi_1 - 1)(\varphi_2 - 1)} \tag{17}$$

where φ_1 and φ_2 are the average functionalities of the two elements.

8.2.2.4 Vulcanization of preformed polymers

In the case of gelation with preformed linear polymers the gel point is given by

$$p_c = 1/X_{ow} \tag{18}$$

where X_{ow} is the weight average degree of polymerization of the preformed polymer.[1,7,40,41]

8.2.2.5 Star-shaped polymers

The equations derived for the reactions defined above can be applied to reactions involving star-shaped polymers. An example is (4) with $m_1 \ll m_2$.

$$m_1 \; A\!\!-\!\!\underset{A}{\overset{A}{\underset{|}{\overset{|}{A}}}}\!\!-\!\!A \;+\; m_2 \; A\!-\!B \;\longrightarrow\; \cdots A\!-\!BA\!-\!BA\!\underset{\underset{\vdots}{\underset{A}{\underset{|}{\overset{A}{\underset{|}{\overset{B}{\underset{|}{\overset{A}{\underset{|}{\overset{B}{\underset{|}{}}}}}}}}}}}}{\overset{\overset{\vdots}{\overset{B}{\overset{|}{\overset{A}{\overset{|}{\overset{B}{\overset{|}{\overset{A}{\overset{|}{}}}}}}}}}}{-}}\!AB\!-\!AB\!-\!A \qquad (4)$$

Similar equations for the gelation process as discussed above can be derived. The number average of the degree of polymerization is in this case

$$X_n = \frac{1+r}{1+r-p} \qquad (19)$$

where r is the ratio of reaction constants. Mostly one has $r \ll 1$, and $p_c \approx 1$. One advantage of such reactions is that molecular weight distributions are narrow. The ratio X_w/X_n is less than 2, i.e.

$$X_w/X_n = 1 + \frac{p\varphi}{\{p(\varphi-1)+1\}^2} \underset{p \to 1}{\approx} 1 + 1/\varphi \qquad (20)$$

It should be remembered that all values given here are based on the assumption that no cyclization takes place and so they are not very reliable and derivations from experiments have been reported.[31–33,42]

8.2.2.6 Cyclization

Within the limit of small loops, i.e. small fluctuations in structure one can take into account cyclization within graph theory[27,34,35] and the gel point can be estimated[43,44,45]

$$p_c = \frac{1}{(\varphi-1)(1-\Lambda)} \qquad (21)$$

$0 \leq \Lambda < 1$ is a ring-forming parameter, which is extremely difficult to calculate. It has been reported that it depends on the functionality as well.[46,47]

Computer simulations provide a more accurate basis or the gel point determination. Therefore one has to take into account that the gel point itself depends on the lattice on which the simulations are carried out. For simple lattices the gel points can be derived mathematically[48,49] and most of the values are collected in Stauffer's book.[5] An empirical rule for example is $p_c z =$ constant, where z is the coordination number of the lattice.[5]

In the case of urethane formation Shy and Eichinger[50] have shown how computer simulation can be used to characterize the gelation process of a particular example. They studied the influence of molecular weight, molecular weight distributions and the effect of side reactions, as well as the cyclization on the gel point. Computer simulations to compare step reaction and vulcanization have been made by Boots et al.[51] The shift of the gel point is studied directly under various 'experimental' conditions.

8.2.3 The Size of the Clusters

In this section the size of the branched polymer within the mean field limit is derived. This was first derived by Stockmayer and Zimm.[52] A more mathematically elegant derivation has been provided by Kramers.[53] The radius of gyration of a general object containing N monomers is defined as

$$\langle R^2(N) \rangle = 1/(2N^2) \sum_{i,j}^{N} \langle (r_i - r_j)^2 \rangle \tag{22}$$

where the $\{r_i\}$, $1 \leq i \leq N$ are the set of coordinates for each monomer on the polymer. The average $\langle \ \rangle$ is the weighted average over all configurations the polymer is able to take. To do this in the tree approximation select two arbitrary points i and j on the tree. Both points are connected by one path only. Now label this path by an index, say μ. The distance between two neighbouring monomers is then given by $b \equiv |r^{\mu+1} - r^{\mu}|$. This is the analogue of the Kuhn length in linear polymers. Define then vectors a_{ij}^{μ}, which label each monomer in the corresponding path μ. Thus the radius of gyration is given by

$$\langle R^2(N) \rangle = 1/(2N^2) \sum_{i,j}^{N} \sum_{\mu} (a_{ij}^{\mu})^2 \tag{23}$$

Note that $a_{ij}^{\mu\,2} \equiv b^2$. According to Kramers, one may cut the polymer at monomer μ, and count the probabilities of getting two parts, *i.e.* with n and $N - n$ monomers. These probabilities are related to a partition function Z_n and Z_{N-n}. Thus equation (22) can be rewritten as

$$\langle R^2 \rangle = (b^2/n) \left\{ \sum_{n=1}^{N-1} n(N - n) Z_n Z_{N-n} \right\} \Big/ \sum_{n=1}^{N-1} Z_n Z_{N-n} \tag{24}$$

The partition function is generally determined by the 'propagator' $G(r - r', n)$ connecting two points r and r' with n steps on the tree. Usually the relation between Z_n and G is $Z_n = \int dr \int dr' G(r - r', n)$. If no branching is present, *i.e.* if we would have a linear chain, G would be the random walk propagator[54] but changes are expected due to branching. de Gennes[55] pointed out that this problem can be solved self-consistently by deriving a Dyson equation for the propagator. The basic property can be expressed in a Feynman diagram analogously to quantum field theory or many body statistical mechanics.[56] Let us define therefore the exact propagator as $G(r, n)$ or its Fourier/Laplace transform $G(k, E)$, where the k vector is the Fourier variable of r and E is the Laplace variable of n. Define in the same way $G_0(r, n)$ or $G_0(k, E)$ as the propagator for the random walk, *i.e.* a polymer without branching. Look at a representative diagram (equation 25)[55,57]

$$\underset{G}{\text{———}} = \underset{G_0}{\text{———}} + \underset{G_0 \quad G}{\diagup^{G}} \tag{25}$$

Each thick line corresponds to the exact propagator, whereas each thin line corresponds to the ideal random walk propagator. In Fourier Laplace space this diagram is written as

$$G(k, E) = G_0(k, E) + G_0(k, E) G(k, E) G(0, E) \tag{26}$$

where the last term is due to branching, which can occur on an arbitrary place between i and j. Note that the exact G appears on both sides of the equation and with different arguments. Therefore a general solution becomes very difficult. Let us restrict ourselves on the discussion of the partition function required for the calculation of the radius of gyration. By noting that the Laplace transform partition function Z_n with respect to n is related to the Fourier Laplace transformation of the propagator by $Z(E) = G(0, E)$ we find the Laplace transform of the partition function

$$Z(E) = Z_0(E) + Z_0(E) Z^2(E) \tag{27}$$

where $Z_0(E)$ is the Laplace transform of the partition function of a random walk,[7,55] *i.e.*

$$Z_0(E) = E^{-1} \tag{28}$$

Equation (27) is now a simple quadratic equation for the partition function, which can be solved easily. For large molecules (large N limit or small E limit) $Z(N)$ becomes

$$Z(N) \cong N^{-3/2} e^{2N} \tag{29}$$

The fast growth with N is typical for trees. Using this expression in equation (24) we find for the radius of gyration

$$\langle R^2 \rangle \sim N^{1/2} b^2 \tag{30}$$

in contrast to the random walk where we have $\langle R^2 \rangle \sim Nb^2$. This law, which was derived by Zimm and Stockmayer,[52] can only be valid for phantom polymers, since if we look at the density $\rho = N/R^d$ in d dimensional space, we find $\rho \sim N^{1-d/4}$. This means that in $d=3$ the density diverges if N tends to infinity, which is unphysical. This problem has to be solved later in this chapter by taking excluded volume effects into account (see also Chapter 8 in Volume 6 of this work).

8.2.4 The Kinetic Approach

Here kinetic models of gelation are discussed briefly. The aim is to derive all the properties given by the Flory–Stockmayer, or graph theoretical method, from a rate equation, corresponding to a reaction of general type $A_n + A_{n'} \to A_{n+n'}$, where the A_n can be single elements, like particles or polymer chains, or already clusters or branched macromolecules consisting of n monomers. If $n' = 1$ such a rate equation corresponds to a step polymerization model, where a monomer is added to an already formed macromolecule, or cluster. Notice the generality of such models. For reasons of simplicity we almost restrict ourselves to non-reversible reactions, where the reaction is only possible in one direction.

Basically there are two limits in growth models. One is the diffusion-limited case, where the particles react immediately if they are close together (within a certain range), the other one is the reaction-controlled case, where the sticking probability is very low and the particles have to bounce together several times before they react. Further generalizations are consideration of different trajectories, *i.e.* the particles are moving on ballistic or Levy flight trajectories.[58] It is clear that the structure and density (*i.e.* the fractal dimension) of the resulting cluster depends on such details and we refer to ref. 58 where detailed discussions can be found.

Smoluchowski[59] suggested his basic equation where, within some limitations, all cases can be discussed. The time dependence of the cluster size distribution $P_n(t)$ is given by

$$\frac{\partial P_n}{\partial t} = \frac{1}{2} \sum_{i+j=n} K_{ij} P_i P_j - P_n \sum_{j=1}^{k} K_{nj} P_j \tag{31}$$

The kernels K_{ij} are generalized reaction constants where details of the model which is considered enter. It is possible to work out the results of the Stockmayer–Flory theory from this equation if the proper kernels K_{ij} are used.

Let us discuss this and try to find the K_{ij} for simple models.[60] For the RA_φ model, *i.e.* the A monomers are φ-functional, we have in the tree approximation $K_{ij} = \sigma_i(\varphi)\sigma_j(\varphi)$ with $\sigma_i(\varphi) = (\varphi - 2)i + 2$. Thus the reaction kernel becomes

$$K_{ij} = \{(\varphi - 2)i + 2\}\{(\varphi - 2)j + 2\} \tag{32}$$

which is in the limit of large functionalities $\varphi \to \infty$, $K_{ij} \sim ij$. Consider the kinetics of this large functionality (RA_∞) model with the corresponding kernel $K_{ij} \sim ij$ and the initial condition $P_n(0) = \delta_{1k}$. The corresponding Smoluchowski equation is

$$\frac{\partial P_n}{\partial t} = \frac{1}{2} \sum_{i+j=n} ij\, P_i P_j - nP_n \sum_{j=1}^{k} j P_j \tag{33}$$

Let us consider the molecular weight distribution Ω_n introduced above as $\Omega_n = nP_n$ first. We can introduce the generating function $F_0(E) = \sum \Omega_n e^{-nE}$ which is essentially the same as before if we put $\theta = e^{-E}$. The usual trick to solve such equations is to multiply them on both sides with ne^{-nE} and transform it to an equation for the generating function, which is hopefully simpler. Proceeding this way, we find for the generating function

$$\frac{\partial F_0(E, t)}{\partial t} = \frac{\partial F_0(E, t)}{\partial E} \{F_0(0, t) - F_0(E, t)\} \tag{34}$$

because of the normalization, (see equation 4) we notice that $F_0(0, t) \equiv 1$. Let us introduce a function

$$f = \sum_n P_n e^{-nE} \tag{35}$$

which satisfies

$$F_0(E, t) = -\partial f(E, t)/\partial E \tag{36}$$

Therefore the differential equation for F_0 can be transformed to a differential equation in $f(E, t)$

$$\partial f/\partial t = \frac{1}{2}(\partial f/\partial E)^2 + \partial f/\partial E \tag{37}$$

which is Euler's equation in hydrodynamics.[61] Its solution is known. It can be shown that there exists a critical time t_c, which is the gel time. Above t_c a infinite network is present, whereas below t_c a sol phase with finite polymers exists. A scaling analysis of this equation and using the definition of f shows that the solution for $P_n(t)$ is of the form

$$P_n(t) \sim n^{-5/2} \exp[-\alpha n\{(t - t_c)/t_c\}^2] \tag{38}$$

which is indeed identical to equation (14), since $t \sim p$ and $t_c \sim p_c$. The validity of this model is on the same level as the Flory–Stockmayer theory, since it starts from the same assumptions and has the same limitations.

8.2.5 The Rate Theory of Random Polymerization

The rate theory was derived independently from the cascade or graph theory, but is based on the same physical assumptions. There are extensive recent reviews in the literature,[45–47] so that we can be very brief here. This approach starts from estimating the existence probability for any configuration, i.e. $[A_\varphi]_n$, and its rate of change. The rate theory includes small rings and cycles by estimating their probability of occurence and their rate of change. Reaction kinetic equations, together with the probabilities give information on ring formation, gel point and structure in a similar way as graph theory. The limitations of this approach are again the loop formation. From a physical point of view large loops can be taken into account formally, but the mathematical complexity is too high to derive proper results for such large systems.

8.3 REFINEMENT OF THE CLASSICAL THEORY

Most of the modern theories are summarized more extensively in Chapter 8, Volume 6 of this work, but for completeness we will mention some general aspects in this chapter here.

8.3.1 Percolation

The classical theory of gelation assumes that there are no loops in the structure, i.e. the branched molecule has the structure of a tree. This is not realistic since the formation of loops becomes more probable the higher the generation of the tree is. In the classical theory quantities like the average molecular weight diverge if the percolation threshold is approached. This is characteristic for phase transitions. Thus critical exponents can be defined in close analogy to second order phase transitions.[30,62] The most convenient model for such a connectivity transition is the bond percolation model.[5] In order to visualize the nature of the model let us imagine a lattice in two, three, etc. dimensions and colour the bonds of the lattice at random. These coloured bonds symbolize monomers or preformed polymer chains. This procedure will produce connected finite clusters if the concentration of the coloured bonds is finite. At a threshold concentration p_c finally one big cluster will be obtained, which connects two opposite planes of the lattice. There is no restriction on the formation of such bonds, and the model will produce loops, dangling ends, single connecting bonds between parts of the large cluster, i.e. large fluctuations in connectivity will be produced. It is also important to realize that the gel point in the percolation model, i.e. the percolation threshold, is larger compared to that from the classical theory. This is because loop formation is allowed, as already seen in equation (21), i.e. $\Lambda > 0$.

The analogy to phase transitions suggests the definition of the following critical exponents.[5,29,63,64] The role the order parameter plays is the gel fraction, or the probability that a monomer belongs to the infinite cluster

$$P_\infty(p) \sim (|p - p_c|)^\beta \qquad (39)$$

The analogue of the susceptibility is the weight average degree of polymerization which one can define as

$$S \equiv X_w \sim |p - p_c|^{-\gamma} \qquad (40)$$

The total number of finite clusters can be expressed as

$$K(p) \sim |p - p_c|^{2-\alpha} \qquad (41)$$

This is an extensive quantity and can be related to the free energy in thermal phase transitions[30] and the specific heat exponent α appears. The correlation length, which is the linear size of a typical cluster, or in other words the distance over which two monomers are connected, diverges as

$$\xi \sim |p - p_c|^{-\nu} \qquad (42)$$

Similar divergencies can be given for the viscosity in the sol phase and the elastic modulus in the gel phase.

Motivated from the result for the classical theory Stauffer[29] proposed a scaling form for the cluster size distribution $P_n(p)$

$$P_n(p) \sim n^{-\tau} g\{n(p - p_c)^{1/\sigma}\} \qquad (43)$$

where g is an unknown scaling function. In the classical theory the exponents have been $\tau = 5/2$ and $\sigma = 1/2$, and the function g was an exponential. Scaling laws can be recovered by using the definition of the quantities in terms of the cluster size distribution function, for example one finds

$$\alpha + 2\beta + \gamma = 2 \qquad (44)$$

which is well known in every second order phase transition.[28,30] An important scaling relation is the hyperscaling law[62,65]

$$\nu d = 2 - \alpha = 2\beta + \gamma \qquad (45)$$

which connects the correlation length exponent and the space dimension d with the other exponents. To derive the hyperscaling relation one has to assume that the clusters do not interpenetrate.[5,64] Then the free energy can be written generally as $K \sim \xi^{-d} \sim (p - p_c)^{\nu d}$. On the other hand K has to scale in the form given by equation (40). Thus relation between the exponents is given by the hyperscaling law.

Consider now the exponents calculated in the classical limit, i.e. in Cayley tree approximation. One finds then the exponents which are valid for the Flory–Stockmayer theory. These are given by[2-6]

$$-\alpha = \beta = \gamma = 1$$
$$\nu = 1/2 \qquad (46)$$
$$\tau = 5/2, \quad \sigma = 1/2$$

If we insert these exponents in the scaling relations equation (44) and equation (45) we see that the relation (44) is satisfied, while the hyperscaling is only valid for $d = 6$. This is general and we have to conclude that the hyperscaling relation breaks down at the upper critical dimension $d_c = 6$.[64,65] In dimensions equal and higher than the upper critical dimension (UCD) the mean field solutions are valid. Generally we have to conclude that the Flory–Stockmayer, if applied to percolation theory works only in dimensions equal or larger than six.[7,65]

Moreover it has been shown numerically[66] that in $d \geq 6$ more than one infinite cluster exists, so that the conditions that the clusters do not overlap is no longer satisfied, i.e. hyperscaling breaks down.

In three dimensions the non-classical exponents (predicted by computer simulations)[5] are given by

$$\gamma = 1.8, \quad \nu = 0.88, \quad \beta = 0.43, \quad \alpha = 0.64 \tag{47}$$

which are very different from the predictions of the Flory–Stockmayer theory. In the case of vulcanization, *i.e.* gelation with preformed polymers, the non-classical nature is not as severe as in the RA_φ reaction since the width of the critical region is proportional to the inverse length of the preformed polymer to the power $1/3$.[7,67]

8.3.2 Excluded Volume Forces and the Size of the System

The calculation of the radius of gyration of the branched molecule in the tree approximation predicts the $R^4 \sim N$ relationship. As already mentioned this would predict an anomalous density mass relation. This is due to the phantom nature of the polymer, since the chains and clusters are allowed to pass through each other. In addition we have to consider excluded volume forces. These forces may alter the size–mass dependence, and will lead to a swelling effect.[7,68,69] We expect the chains or clusters to become larger. The dominant contribution for the excluded volume interaction comes from the two body collisions,[7] *i.e.* if two parts of the molecule come close together, they will be repelled. The total energy of the repulsion is proportional to the probability of finding these two parts close together. This probability of finding the two body interaction is given by the square of the local concentration ρ. Thus the repulsive energy can be estimated by $U \sim v \int \rho^2 d^d R$.[7] v is called the excluded volume parameter. The crudest mean field estimate of U will be sufficient. It is given by the replacement of ρ by the overall monomer density within the volume of the cluster N/R^d in d dimensions. Therefore U becomes

$$U = v N^2/R^d \tag{48}$$

The free energy of the 'animal', or branched polymer, has then two contributions, *i.e.* the entropy $S = -R^2/N^{1/2}$ and U, *i.e.* $F = U - TS$. Minimization of the free energy leads to the swollen size of the animal in a good solvent[73]

$$R \sim N^{\nu_a} \quad \text{with} \quad \nu_a = \frac{5}{2(d + 2)} \tag{49}$$

Thus the three dimensional size exponent in dilute solvent is given by $\nu_a = 1/2$. The inverse relation $R^{1/\nu_a} \sim N$ defines the fractal dimension of the object, here $d_f = 1/\nu_a$.

Despite the crudeness of this 'Flory–de Gennes' theory it provides a powerful tool to estimate the size of the object under various conditions. Screening of excluded volume forces can be treated as well. In dense systems excluded volume forces become screened, as first shown rigorously by Edwards.[54] In the Flory approach this can be done by replacing v by v/N^y where $y = 1$ in melt clusters and $y = 1/2$ near the percolation threshold.[70,71,72]

Finally we note that all these result can be put in a general context[74] using the idea of polymeric fractals[75] (see Chapter 8, Volume 6 of this work).

Further it is important to realize that for all these qualitative arguments discussed so far, a renormalizable field theory can be developed. This has been been done recently by Lubensky *et al.*[76-78] Because of limitations in space we cannot discuss these results here.

8.3.3 Recent Advances in Kinetics

Here we summarize briefly some recent developments in kinetics. The most critical assumption in the classical approach described above is the equal reactivity in the step polymerization process, *i.e.* still unreacted functional units inside the polymer have the same reactivity as reactive groups at the perimeter. Intuitively one may argue that reactions at the perimeter are more probable than reactions inside the molecule. In order to take this effect into account a new reaction kernel has been proposed.[79-82] As an example let us look at

$$K_{ij} = (ij)^w \tag{50}$$

which replaces equation (32). w is related to the fraction of most active bonds in a i-mer. The corresponding Smoluchowski equation becomes

$$\frac{\partial P_n}{\partial t} = \frac{1}{2} \sum_{i+j=n} (ij)^w P_i P_j - n^w P_n \sum_{j=1}^{k} j^w P_j \qquad (51)$$

which can be solved along the same lines as those described above. Therefore one uses $F_0(E)$ and defines a new generating function $F_w = \Sigma\, n^{w-1} \Omega_n(t)\, e^{-nE}$. Hence $\Omega_n = nP_n$.

Using all definitions introduced above in Section 8.2.4, the transformed equation is

$$\frac{\partial F_{w=0}(E, t)}{\partial t} = 1/2\, F_w^2(E, t) - F_w(E, t)\, F_w(0, t) \qquad (52)$$

with the assumption that the solution has the scaling form proposed by Stauffer, *i.e.* in equation (14) the exponents τ and σ can be related to w

$$\sigma + \tau = 2w + 1 \qquad (53)$$

and the other exponents can be expressed in terms of w.

This is a rather strange result; firstly it suggests that the exponents τ and σ are related to each other. This is a contradiction to the scaling hypothesis. Secondly since w depends on the detailed chemistry of the monomers and on other details as well, all other exponents will depend on w and so on details of the chemistry, too. This is a contradiction to the universality hypothesis. There is a further the restriction. It can be shown that $w > 1/2$ to get a gel phase. If $w \leq 1/2$ one is in the flocculation regime and no gel phase exists (see Chapter 8 in Volume 6 of this work). These problems remain to be solved in the future.

ACKNOWLEDGEMENT

The author is very grateful to Professor S. Russo for various suggestions and critical reading of the manuscript.

8.4 REFERENCES

1. P. J. Flory, 'Principles of Polymer Chemistry', Cornell University Press, 1959.
2. P. J. Flory, *J. Am. Chem. Soc.*, **63**, 1941, 3083, 3091, 3096.
3. W. H. Stockmayer, *J. Chem. Phys.*, 1943, **11**, 45.
4. W. H. Stockmayer, *J. Chem. Phys.*, 1944, **12**, 125.
5. D. Stauffer, 'Introduction to Percolation Theory', Taylor and Francis, London, 1985.
6. D. Stauffer, A. Coniglio and M. Adam, *Adv. Polym. Sci.*, 1982, **44**, 74.
7. P. G. de Gennes, 'Scaling Concepts in Polymer Physics', Cornell University Press, Ithaca, 1979.
8. T. C. Lubensky, in 'Ill Condensed Matter', eds. R. Balian, R. Maynard and G. Toulouse, North-Holland, Amsterdam, 1979.
9. J. W. Essam, in 'Phase Transitions and Critical Phenomena', eds. C. Domb and M. B. Green, Academic Press, New York, 1972, vol. 2.
10. G. Deutscher, R. Zallen and J. Adler (eds.), 'Percolation Structures and Processes', Adam Hilger, Bristol, 1983.
11. J. W. Essam, *Rep. Prog. Phys.*, 1983, **43**, 843.
12. R. Zallen, 'The Physics of Amorphous Solids', Wiley, New York, 1983.
13. S. Kirkpatrick, *Rev. Mod. Phys.*, 1973, **45**, 574.
14. H. E. Stanley, in 'Disordered Systems and Localization', ed. C. Castellani, C. Di Castro and L. Peliti, Springer-Verlag, Heidelberg, 1981.
15. H. E. Stanley and N. Ostrovsky (eds.), 'On Growth and Form', Elsevier, Amsterdam, 1985.
16. B. B. Mandelbrot, 'The Fractal Geometry of Nature', Freeman, San Fransisco, 1982.
17. L. Pietronero and E. Tosatti (eds.), 'Fractals in Physics', North-Holland, Amsterdam, 1985.
18. F. Family and D. P. Landau (eds.), 'Kinetics of Aggregation and Gelation', North-Holland, Amsterdam, 1984.
19. H. J. Herrmann, *Phys. Rep.*, 1986, **136**, 153.
20. A. T. Balaban (ed.), 'Chemical Applications of Graph Theory', Academic Press, London, New York, 1976.
21. M. Gordon, in 'Chemical Applications of Graph Theory', ed. A. T. Balaban, Academic Press, London, New York, 1976.
22. F. Harary, 'Graph Theory', Addision Wesley, Reading, USA, 1971.
23. W. Burchard, *Adv. Polym. Sci.*, 1983, **48**, 1.
24. M. Gordon, *Proc. R. Soc. London, Ser. A*, 1962, **268**, 240.
25. G. R. Dobson and M. Gordon, *J. Chem. Phys.*, 1964, **41**, 2389.
26. G. R. Dobson and M. Gordon, *J. Chem. Phys.*, 1965, **43**, 705.
27. M. Gordon and G. R. Scantlebury, *J. Polym. Sci., Part C*, 1968, **16**, 3933.

28. H. E. Stanley, 'Introduction to Phase Transitions and Critical Phenomena', Oxford University Press, Oxford, 1982.
29. D. Stauffer, *Phys. Rep.*, 1979, **54**, 1.
30. C. Domb and M. B. Green (eds.), 'Phase Transitions and Critical Phenomena', Academic Press, New York, 1976, vol. 6.
31. K. Dusek, *Adv. Polym. Sci.*, 1986, **78**, 1.
32. K. Dusek, in 'Physics of Finely Divided Matter', eds. N. Boccara and M. Daoud, Springer Verlag, Heidelberg, 1985.
33. K. Dusek, *Polym. Bull.*, 1987, **17**, 239.
34. M. Gordon and W. B. Temple, *Makromol. Chem.*, 1972, **152**, 277.
35. K. Dusek and V. Vojta, *Br. Polym. J.*, 1977, **9**, 164.
36. W. H. Carothers, *Trans. Faraday Soc.*, 1936, **32**, 39.
37. C. W. Macosko and D. R. Miller, *Macromolecules*, 1976, **9**, 199.
38. D. R. Miller and C. W. Macosko, *Macromolecules*, 1976, **9**, 206.
39. D. R. Miller, E. M. Valles and C. W. Macosko, *Polym. Eng. Sci.*, 1979, **19**, 272.
40. J. E. Mark, R. R. Rahalkar and J. L. Sullivan, *J. Chem. Phys.*, 1979, **70**, 1794.
41. M. Gottlieb, C. W. Macosko, G. S. Benjamin and E. W. Merril, *Macromolecules*, 1981, **14**, 1039.
42. K. Dusek, in 'Developments in Polymerization—3', ed. R. N. Haward, Applied Science, London, 1982.
43. R. W. Kilb, *J. Phys. Chem.*, 1958, **62**, 969.
44. R. T. F. Stepto, *Faraday Discuss. Chem. Soc.*, 1974, **57**, 69.
45. Z. Ahmad and R. T. F. Stepto, *Colloid Polym. Sci.*, 1980, **258**, 663.
46. R. T. F. Stepto, in 'Developments in Polymerization—3', ed. R. N. Haward, Applied Science, London, 1982.
47. A. C. Lloyd and R. T. F. Stepto, *Br. Polym. J.*, 1985, **17**, 190.
48. J. W. Essam, D. S. Gaunt and A. J. Guttmann, *J. Phys. A: Math. Gen.*, 1978, **11**, 1983.
49. H. Kesten, 'Percolation Theory for Mathematicians', Birkhauser, Boston, 1982.
50. L. Y. Shy and B. E. Eichinger, *Br. Polym. J.*, 1985, **17**, 200.
51. H. M. J. Boots, J. G. Kloosterboer, G. M. M. van de Hei and R. B. Pandey, *Br. Polym. J.*, 1985, **17**, 219.
52. B. H. Zimm and W. H. Stockmayer, *J. Chem. Phys.*, 1949, **17**, 1301.
53. H. A. Kramers, *J. Chem. Phys.*, 1946, **14**, 415.
54. M. Doi and G. F. Edwards, 'The Theory of Polymer Dynamics', Oxford University Press, Oxford, 1986.
55. P. D. de Gennes, *Biopolymers*, 1968, **6**, 715.
56. A. L. Fetter and J. D. Walecka, 'Quantum Theory of Many-Particle Systems', McGraw-Hill, New York, 1971.
57. M. Daoud and J. F. Joanny, *J. Phys.*, 1981, **42**, 1309.
58. R. Jullien and R. Botet, 'Aggregation and Fractal Aggregates', World Scientific, Singapore, 1987.
59. M. von Smoluchowski, *Phys. Z.*, 1916, **17**, 565.
60. M. H. Ernst, in 'Fundamental Problems in Statistical Mechanics', ed. E. G. D. Cohen, North-Holland, Amsterdam, 1985.
61. R. M. Ziff and G. Stell, *J. Chem. Phys.*, 1980, **73**, 3496.
62. P. Pfeuty and G. Toulouse, 'Introduction to Renormalization Group and Critical Phenomena', Wiley, New York, 1977.
63. S. Kirkpatrick, in 'Ill Condensed Matter', eds. R. Balian, R. Maynard and G. Toulouse, North-Holland, Amsterdam, 1979.
64. A. Coniglio, in 'Physics of Finely Divided Matter', eds. N. Boccara and M. Daoud, Springer Verlag, Heidelberg, 1985.
65. G. Toulouse, *Nuovo Cimento Soc. Ital. Fis. B*, 1974, **23**, 234.
66. L. de Arcangelis, *J. Phys. A: Math. Gen.*, 1987, **20**, 3057.
67. P. G. de Gennes, *J. Phys. Lett. (Orsay, Fr.)*, 1976, **37**, L1.
68. J. des Cloizeaux and G. Jannink, 'Les Polymeres en Solution: Modelisation et leur Structures', Edition de Physique, Les Ulis, 1987.
69. S. Alexander, G. S. Grest, H. Nakanishi and T. A. Witten, *J. Phys. A: Math. Gen.*, **17**, 1984, L158.
70. M. Daoud and F. Family, *J. Phys. (Orsay, Fr.)*, 1984, **45**, 151.
71. T. A. Vilgis, *Phys. Rev. A.*, 1987, **36**, 1506.
72. M. Daoud, E. Bouchaud and G. Jannink, *Macromolecules*, 1986, **19**, 1955.
73. J. Isaacson and T. C. Lubensky, *J. Phys. Lett. (Orsay, Fr.)*, 1981, **42**, 175.
74. T. A. Vilgis, *Makromol. Chem., Rapid Commun.*, 1988, in press.
75. M. E. Cates, *J. Phys. (Orsay, Fr.)*, 1985, **46**, 1059.
76. T. C. Lubensky and J. Isaacson, *Phys. Rev. B: Condens. Matter*, 1979, **20**, 2130.
77. T. C. Lubensky and J. Wang, *Phys. Rev. B: Condens. Matter*, 1986, **33**, 4998.
78. T. C. Lubensky and A. M. S. Tremblay, *Phys. Rev. B: Condens. Matter*, 1986, **34**, 568.
79. P. Leyvraz, in 'On Growth and Form', eds. H. E. Stanley and N. Ostrovsky, Elsevier, Amsterdam, 1985.
80. M. H. Ernst, in 'Fractals in Physics', eds. L. Pietronero and E. Tosatti, North-Holland, Amsterdam, 1985.
81. P. G. J. van Dongen and M. H. Ernst, in 'Fractals in Physics', eds. L. Pietronero and E. Tosatti, North-Holland, Amsterdam, 1985.
82. R. M. Ziff, in 'Kinetics of Aggregation and Gelation', ed. F. Family and D. P. Landau, North-Holland, Amsterdam, 1984.
83. E. M. Hendriks, in 'Kinetics of Aggregation and Gelation', eds. F. Family and D. P. Landau, North-Holland, Amsterdam, 1984.

9
Experimental Methods of Bulk Polymerization

VASILI V. KORSHAK and VALERI A. VASNEV
USSR Academy of Sciences, Moscow, USSR

9.1	INTRODUCTION	131
9.2	GENERAL PRINCIPLES	132
9.3	POLYCONDENSATION WITH PARTICIPATION OF POLYMERS	133
	9.3.1 Reactions of Monomers with Polymers	133
	9.3.1.1 Synthesis of homopolymers via chain extension	133
	9.3.1.2 Synthesis of copolymers	134
	9.3.2 Reactions of Polymers with Polymers	135
	9.3.2.1 Synthesis of homopolymers	135
	9.3.2.2 Synthesis of copolymers	136
9.4	POLYCONDENSATION AFFECTED BY PRESSURE	137
9.5	POLYCONDENSATION IN THE PRESENCE OF A FILLER	138
9.6	POLYCYCLOTRIMERIZATION	138
9.7	OTHER METHODS	140
9.8	REFERENCES	141

9.1 INTRODUCTION

Since in the process of polymerization in bulk the initial compounds and the polymer are in the molten state, the reaction can be carried out in the absence of solvent. This is one of the most thoroughly studied and available methods. It has now been applied in the laboratory for more than a hundred years.

Polycondensation in bulk is mainly performed to implement the process of equilibrium polycondensation when the values of the equilibrium constants are in the range of 1 to 10.[1,2] These processes include polyesterification of dicarboxylic acids or their alkyl esters with diols, polyamidation of dicarboxylic acids with diamines, polycoordination of metal compounds with polydentate ligands and other reactions.[1]

Polycondensation in bulk is also used successfully in the case of irreversible processes.[3] Polyurethanes from diisocyanates and diols,[4] polyheteroarylenes from dicarboxylic acids or their derivatives and polynucleophilic monomers such as tetraamines[3,5] and other polymers are also produced by means of this method.

Prospects for further development of polycondensation in bulk include the application of polymers as initial compounds. Such polymers may enter the reaction either at the end groups or due to exchange interaction of chains, for example following the reactions of alcoholysis, aminolysis, acidolysis and others.[1]

Thermodynamic compatibility of the polymers is an important factor affecting the rate of interchain exchange,[6] although partial or even full incompatibility of polymers retards these reactions, but does not exclude their occurrence. The compatibility of polymers improves during the course of the reaction.

Polycondensation in bulk has a number of advantages compared to other methods, *viz.* relative simplicity of the technological scheme, possibilities for synthesizing polymers of high purity, direct utilization of the polymer melts obtained for production of films and fibres, an increasing yield per unit of the equipment capacity due to the absence of solvents, substantial reduction of the production costs and technological processes which are ecologically more sound.

Disadvantages of the method include high consumption of energy, longer duration of the process and requirements for a high thermostability of the initial monomers and the polymers obtained.

At present polycondensation in bulk is applied in industry to produce polyesters [poly(ethylene terephthalate), poly(butylene terephthalate), alkyd oligomers, unsaturated polyesters, polycarbonates and others], polyamides [poly(hexamethyleneadipamide), poly(hexamethylenesebacamide) and others], polyurethanes, polybenzimidazoles, polyimides and some other polymers.

9.2 GENERAL PRINCIPLES

One of the major peculiarities of polycondensation in bulk is the application of high temperatures which exceed the melting point of the polymer thus produced by 10–20 °C and are found in practice to be in the temperature range 150–350 °C. A temperature rise increases the reaction rate and may result in higher molecular weights of the polymer. However, the temperature should not be raised over a certain limit, above which side reactions, including the reactions of thermal and thermooxidative degradation of monomers and polymers, become noticeable. To reduce the effect of thermooxidation, polycondensation is often performed in an inert gas flow, which, in addition, promotes the removal of low molecular weight products (water, alcohol, acids, *etc.*) from the reaction area.

The necessity for efficient removal of the low molecular weight product is often an important factor in bulk polycondensation. It mostly refers to the reversible processes, and the equilibrium of the reaction must be shifted towards the formation of high molecular weight compounds. During the later stages of the reaction, due to the high viscosity of the melt, the rate of removal of the low molecular weight product may become a limiting stage determining the rate of the entire process.

Thus the poly(ethylene terephthalate) polycondensation rate depends on the melt layer thickness.[7] Because in this case the removal of low molecular weight product from the melt layer is a limiting process stage, the reaction in a thin layer proceeds considerably more rapidly. Diffusive retardation can also be connected with lower mobility of macromolecules at high conversion stages of polycondensation in bulk.[8]

The removal of the low molecular weight product from the reaction medium can be improved by several existing procedures: (a) application of vacuum at the final stages of the process; (b) stirring of the melt, the molecular weight of the polymer being affected by the shape and the type of the stirring rod, which determine the higher or lower intensity of stirring and the thickness of the layer formed; and (c) in some cases, addition of solvents able to interact with the low molecular weight reaction product to yield an azeotrope, the boiling point of which is lower than that of the solvent used (azeotropic polycondensation). Thus, in order to remove the water formed in the process of producing polyesters, some aromatic hydrocarbons (benzene, toluene) in an amount up to 10% of the initial compound's weight are introduced into the reaction system. In the case of toluene (boiling point 110.8 °C) water is removed in the form of azeotrope (water content 20%) at 84.1 °C.

Often polycondensation in bulk proceeds very slowly and can take ten or more hours, and this slowness may be related to the low reactivity of the monomer. Catalysts are often used to accelerate the reaction. With the help of catalysts, not only the rate of the process can be increased, but also the quality of the polymer can be improved owing to the decrease of the reaction temperature and to the minimization of side reactions. However, this is not always the case because successfully solving the first part of the problem connected with acceleration of the condensation process may give polymers contaminated by catalyst degradation products and other impurities.

The problem of catalysis of polycondensation in bulk has received much attention.[2,9-11] According to the papers published in this field, priority has been given to polyesterification[9,12-24] and to the synthesis of poly(alkylene terephthalate)s.[12-15,18,19,21-24] Of great importance in this respect is the development of a highly efficient process for producing polymers directly from terephthalic acid and glycols. Compounds of titanium and tin were proposed as catalysts of the reaction in the case of poly(butylene terephthalate).[23]

Specific features of the equilibrium process require special apparatus for polycondensation in bulk. It is necessary to perform the reaction in an inert atmosphere or *in vacuo* as well as to set up the liquid film regime at the final stages of the process since the removal of the low molecular weight product from the high viscosity melt is facilitated in thin films. Polyesterification of adipic acid with pentaerythrite exemplifies that the rate constant for the film set-up more than twice exceeds that of the process in the open reactor with mixing.[25] Development of a reactor for both batch and continuous processes of polycondensation in bulk is a very important problem technologically,[26] requiring further experimental and theoretical studies.

For this purpose a mathematical model of the process can be used that allows for choice of optimum reactor configuration and optimum process conditions, predicting its basic parameters (temperature, vacuum, vapour phase composition and so on) and determining the characteristics of the resulting polymer (molecular weight, acid and hydroxyl numbers and others).[8,27-29]

At present, the most relevant aspects of various reactions of polymer synthesis in bulk (including polyesterification, polyamidation, the reactions for producing polyurethanes and others) have been established.[1-5,30] Further development of this type of polycondensation on the one hand depends both on modification and improvement of already known processes (e.g. development of more efficient catalysts), and on the application of new approaches based on the effects of physical factors (pressure, sound, light, etc.), and on the other hand, amelioration of the reaction in the presence of fillers and polymers actively participating in the synthesis of macromolecules, etc. is required.

9.3 POLYCONDENSATION WITH PARTICIPATION OF POLYMERS

The idea of using polymers as initial compounds for polycondensation originated from several considerations. First, up to now a great number of polymers with various properties have been synthesized from quite a lot of monomers, including easily available ones. In this connection, the problem of creating polymers with required properties can be approached differently. The alternatives are either to produce a completely new polymer or to make an attempt to 'build' such a polymer using already known macromolecules, and, primarily, polymers manufactured in industry. Second, it is known that the synthesis of copolymers enables fine regulation of the properties of the resultant high molecular weight products. Third, the application of commercial polymers for the synthesis of new copolymers makes it possible to produce copolymers directly in the course of processing, for example in the extruder. All these considerations together promoted the idea of using manufactured polymers for the production of new structures, very attractive from technological, economic and ecological points of view.

9.3.1 Reactions of Monomers with Polymers

Reactions of monomers with polymers may yield homopolymers or copolymers. In the first case, the reaction with polymers involves monomers (extenders of the chain) which join the end groups of the macromolecule. In the second case, the polymer participates in polycondensation with monomers or reacts with monomers in exchange reactions which finally yield copolymers.

9.3.1.1 Synthesis of homopolymers via chain extension

Various compounds are used as chain extenders of already formed macromolecules: diesters (diphenyl carbonate, diphenyl terephthalate, phenyl orthocarbonate and others), bisepoxides, diisocyanates, dianhydrides of tetracarboxylic acids, ethylene carbonate, bisheterocyclic compounds (bis-2-oxazolines) and others. The most effective chain extenders in the reaction with polyesters were found to be ethylene carbonate[31] and heterocyclic compounds.[32-34] Diisocyanates are widely used as chain extenders.[35-37] In this case polyesters, polyethers, polysiloxanes and others can be applied as oligomers.

Introduction of a small amount of ethylene carbonate (1%) into the melt of poly(ethylene terephthalate) at 285 °C for 70 min results in an appreciable increase of the polymer molecular weight (intrinsic viscosity increases from 0.46 to 0.75 dl g^{-1}) and in a decrease of the amount of its carboxyl groups[31] (Scheme 1).

Bisheterocyclic compounds were found to be more effective chain extenders, viz. bisoxazolines, bisoxazines, bisoxazolones and bisbenzoxazinones.[32-34] After a 5 min exposure of the polyethylene terephthalate) melt with a small addition (0.5 mol %) of 2,2-bis(2-oxazoline) (1) at 280 °C the intrinsic viscosity of the polymer increases from 0.66 to 1.06 dl g^{-1} (Scheme 2).[32] In the case of poly(butylene terephthalate) similar results are obtained. The increasing amount of the chain elongator from 0.5

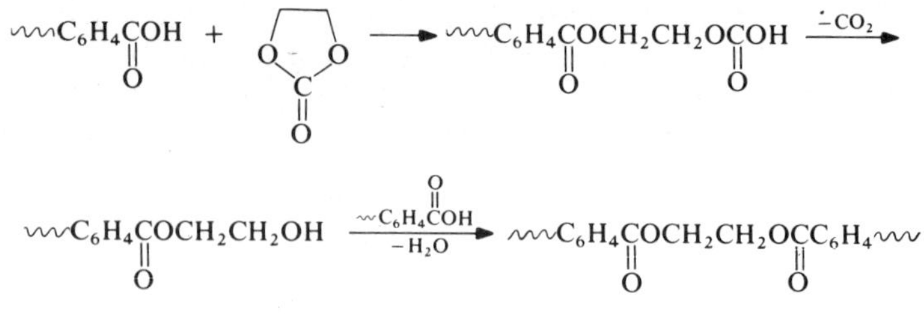

$$\sim\sim C_6H_4\underset{\underset{O}{\|}}{C}OH + \underset{\underset{\underset{O}{\|}}{C}}{\overset{O\quad O}{\diagup\diagdown}} \longrightarrow \sim\sim C_6H_4\underset{\underset{O}{\|}}{C}OCH_2CH_2O\underset{\underset{O}{\|}}{C}OH \xrightarrow{-CO_2}$$

$$\sim\sim C_6H_4\underset{\underset{O}{\|}}{C}OCH_2CH_2OH \xrightarrow[-H_2O]{\sim C_6H_4\overset{\overset{O}{\|}}{C}OH} \sim\sim C_6H_4\underset{\underset{O}{\|}}{C}OCH_2CH_2O\underset{\underset{O}{\|}}{C}C_6H_4\sim\sim$$

Scheme 1

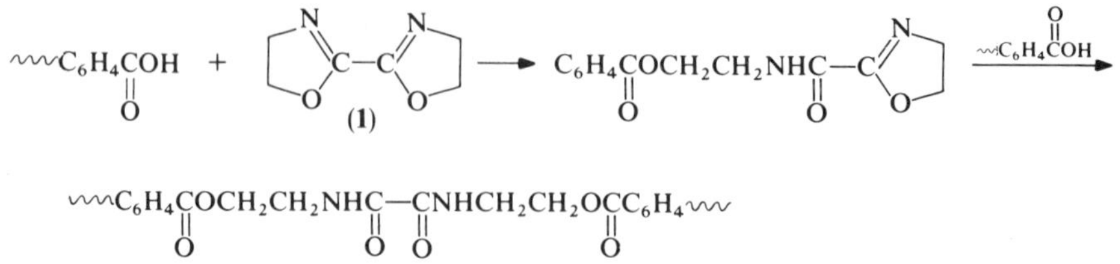

$$\sim\sim C_6H_4\underset{\underset{O}{\|}}{C}OH + \underset{(1)}{\overset{\overset{N}{\diagup}\quad\overset{N}{\diagdown}}{\underset{O}{\diagdown}C-C\underset{O}{\diagup}}} \longrightarrow C_6H_4\underset{\underset{O}{\|}}{C}OCH_2CH_2NH\underset{\underset{O}{\|}}{C}-C\underset{O}{\overset{N}{\diagdown\diagup}} \xrightarrow{\sim C_6H_4\overset{\overset{O}{\|}}{C}OH}$$

$$\sim\sim C_6H_4\underset{\underset{O}{\|}}{C}OCH_2CH_2NH\underset{\underset{O}{\|}}{C}-\underset{\underset{O}{\|}}{C}NHCH_2CH_2O\underset{\underset{O}{\|}}{C}C_6H_4\sim\sim$$

Scheme 2

to 1% has practically no effect on the polymer molecular weight, which depends only on the initial characteristics of the polymer.

Polycondensation of monomers with polymers includes a great number of reactions related to the cure of oligomers and is considered in the appropriate chapters of Volume 5.

9.3.1.2 *Synthesis of copolymers*

As in the case of the synthesis of homopolymers (Section 9.3.1.1.), one of the traditional methods for producing block copolymers is based on the application of chain extenders, for example diisocyanates.[38] The interaction of poly(hexamethylenesebacate) and poly(2-methyl-2-ethyl-1,3-propylenesebacate) with hexamethylenediisocyanate to produce block copolyesters is a typical example of such a reaction.[39]

Another route for the synthesis of copolymers, including block copolymers, is the application of polymers as comonomers.[38] Such polymers contain terminal groups capable of participating in polycondensation and their chains do not enter exchange reactions. Poly(alkylene glycols) serve as an example of such polymers.[40,41] Polycondensation of dimethyl terephthalate with ethylene glycol in the presence of a small amount (2.6 mol %) of poly(ethylene glycol) with molecular weight of 400 has been performed in order to produce modified poly(ethylene terephthalate), the fibres of which are readily dyed. Thermoelastoplastics are based on oligourethanes and oligomers of other types, primarily polyethers and polyesters.[35,37] During operation these block copolymers are similar to elastomers; they are capable of reversible deformation and flow as thermoplastics at elevated temperatures, for instance during processing.

Another example of polycondensation in bulk of the same type is the utilization of poly(arylene ether sulfones) with terminal acetylated OH groups in the reaction with *n*-acetoxybenzoic acid or with bisphenol diacetate and terephthalic acid.[42] This interaction yields segmentated heat resistant copolymers (see Volume 5, Chapter 33).

A very promising method for the production of new copolymers is the procedure based on exchange reactions of polymers with monomers.[1,38] The synthesis of copolymers in this process is due both to exchange interactions and to the reaction of chain growth which proceed in a consecutive–parallel manner. In the first stage, due to exchange reactions, the monomers break the macromolecules and enter into their structure. Then the fragments of polymers and monomers thus formed can undergo the condensation reactions causing the chain growth. Being repeated many times, these reactions finally result in the formation of copolymers.

In 1972, Hamb[43] was one of the first to show that the synthesis of copolymers in such conditions was possible. In the process he proposed poly(ethylene terephthalate) reacts with 2,2-bis(4-hydroxyphenyl)propane diacetate and terephthalic acid in the presence of the catalyst (dibutyl tin oxide) at 280 °C to yield random copolyesters (Scheme 3).

Conditioning the poly(ethylene terephthalate) melt in a mixture with terephthalic acid,[43] 2,2-bis-(4-hydroxyphenyl)propane diacetate,[43] 1,4-butanediol[44] or 1,3-propanediol[45] also results in copoly-

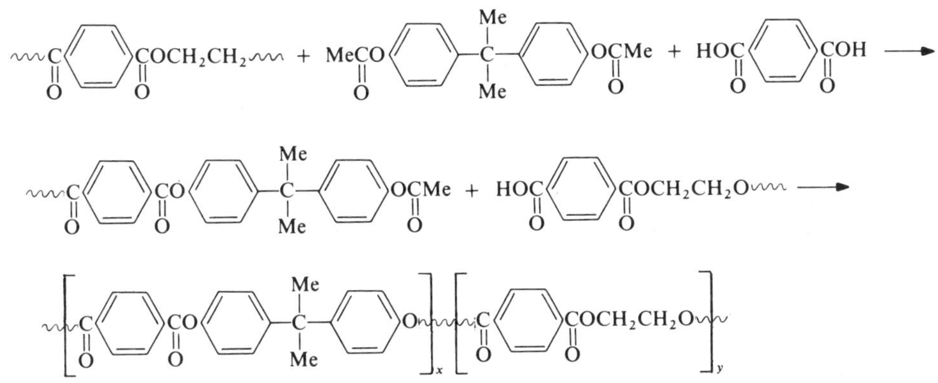

Experimental Methods of Bulk Polymerization

[Scheme 3 structures]

Scheme 3

mers, though the latter have small molecular weights. This is due to violation of the equivalence rule of interacting groups, because in exchange reactions the monomers with the same type of functionality take part, *i.e.* either dicarboxylic acid or diol.

The research work pertaining to the interaction of poly(ethylene terephthalate) with hydroxybenzoic acids has been considerably developed.[46]

The interest in the subject has been determined by the fact that the reaction yields liquid crystalline copolymers which can be processed into high strength materials.[47] Thus the interaction between poly(ethylene terephthalate) and *p*-acetoxybenzoic acid resulted in statistical copolymers capable of yielding mesomorphic melts in the case where the former mixture contains a critical amount of oxybenzoic component[46,48–51] (equation 1). In addition to *p*-acetoxybenzoic acid, 3-methoxybenzoic acid[52] and 3-bromo- and 3,5-dibromo-*p*-acetoxybenzoic acids[51] have been used to synthesize copolymers.

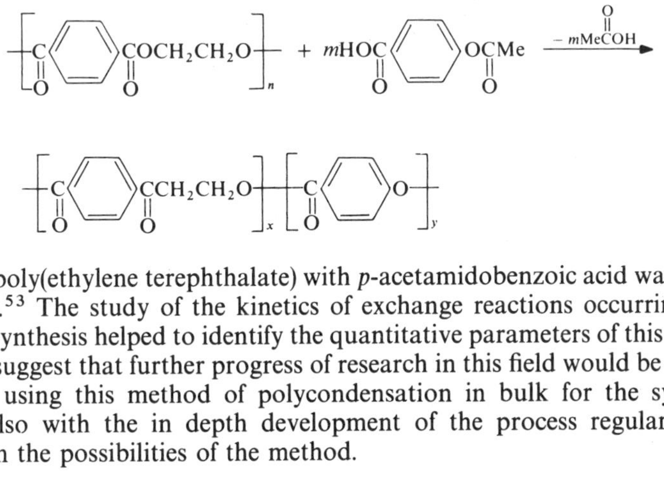

[equation 1 structures] (1)

The reaction of poly(ethylene terephthalate) with *p*-acetamidobenzoic acid was used to synthesize poly(amido esters).[53] The study of the kinetics of exchange reactions occurring in the course of poly(amido ester) synthesis helped to identify the quantitative parameters of this complex process.[56]

It is possible to suggest that further progress of research in this field would be connected not only with the ways of using this method of polycondensation in bulk for the synthesis of various copolymers but also with the in depth development of the process regularities, which would substantially widen the possibilities of the method.

9.3.2 Reactions of Polymers with Polymers

Reactions of polymers with polymers, similar to reactions of monomers with polymers, are performed with two purposes in view: first, to increase the molecular weight of the performed polymer and second, to obtain new valuable copolymers and derived materials.

9.3.2.1 *Synthesis of homopolymers*

In the absence of an extender, keeping the polymer melt at an elevated temperature may also result in an increase in its molecular weight. Thus after heating poly(ethylene terephthalate) for 75 min at 285 °C the intrinsic viscosity of the polymer increases from 0.46 to 0.58 dl g^{-1}.[41] However this method is inefficient and time consuming.

Successful development of the reactions of direct polycondensation (see this volume, Chapter 10, Section 10.4) has made it possible to employ the condensing reagents found there for raising the molecular weight of the polymers, such as polyamides, polyesters and others.[55-57] Heating the polymer melt in the presence of a small amount of triphenyl phosphite (0.5–2.5%) for a short time results in a noticeable increase in the molecular weight. This enables this method to be used in the short time of processing (extrusion) of polyamides-6 and -66,[55,56] poly(ethylene terephthalate)[57] and other polymers of a similar type. Thus, after heating poly(ethylene terephthalate) in an extruder (diameter 25 mm, length:diameter = 24 mm) for 10 min in the presence of 1% triphenyl phosphite at 285 °C the intrinsic viscosity increases from 0.68 to 1.07 dl g^{-1}.[57]

The application of other, more developed condensing reagents or systems may result in much greater efficiency of such reactions.

9.3.2.2 Synthesis of copolymers

One of the trends in the synthesis of copolymers from homopolymers is based on the mutual interaction of their terminal groups.[58,59] Such a synthesis can be exemplified by the reaction between oligo(alkylene glycol)s and oligo(dimethylsiloxanes) containing terminal isocyanate groups.[59]

Another approach is based on the exchange reactions between the polymers resulting first in block copolymers and then yielding statistical copolymers (Figure 1).

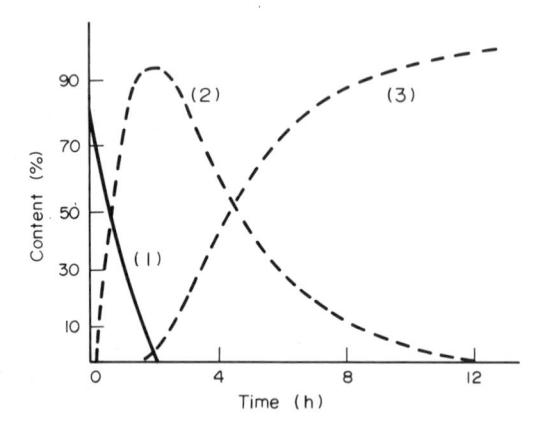

Figure 1 Scheme of the change in the composition of the reaction mixture of two homopolymers under heating:[1] (1) homopolymers; (2) block copolymer; and (3) statistical copolymer

The results given in Figure 1 were obtained for the interaction of poly(hexamethyleneadipamide) with poly(hexamethyleneisophthalamide),[60] that is, for a specific system, and therefore cannot be automatically used for other cases. However Figure 1 clearly shows the changes in the composition of the reaction medium with the course of time, *viz.* the mixture of homopolymers becomes a block copolymer, which changes into a statistical copolymer.

If during exchange reactions between polymers no problems that increase the molecular weight of the resulting copolymer, the process can be carried out at a standard pressure without removal of low molecular weight products. The use of vacuum enables removal of both low molecular weight products and volatile impurities and the products of thermal degradation.

Interchain exchange was first successfully employed as a method for the synthesis of copolymers by Brubaker, Cofman and McGrew[61] in 1944. They produced copolyamides by mixing the melts of various polyamides at 285 °C. In later years, using the exchange interaction of polymers, various copolyesters, copolyamides, poly(amido ester)s and other types of copolymers were synthesized.[1,62]

At present the method of synthesis of copolymers by means of interchain interaction of polymers is undergoing a so-called second birth. In the 1980s the number of publications on this subject has substantially increased. The main reason for this is the need for new materials, for which the demand has considerably increased in recent years. One of the most promising directions is based on the utilization of the commercial polymers for production of mixtures and polymers with the given properties. Though at first sight this looks like a practical problem, it has given rise to a series of in-depth scientific studies.

As a result of a great number of investigations, a variety of copolyesters in the melt were obtained by means of exchange reactions, namely from mixtures of aliphatic polyesters,[63–66] poly(ethyleneterephthalate) with poly(alkylene terephthalate),[67–69] poly(alkyleneterephthalate) with polycarbonate,[70–73] poly(alkylene terephthalate)s with polyarylate (copolymers of tere- and iso-phthalic acids and 2,2-bis(4-hydroxyphenyl)propane),[74–77] polycarbonate with polyarylate,[78,79] and polyarylate with poly(arylene oxide),[80,81] as well as copolyamides were synthesized from mixtures of aliphatic polyamides.[56]

Many authors have confirmed that as a result of exchange reactions, the mixture of homopolymers passes into block copolymers which, in their turn, change into statistical copolymers. The structure and properties of the products obtained have been examined by means of modern physical methods by which it was possible to define the most important characteristics of copolymers and associate them with the microstructure of the macromolecules.[63–81] It has been found, among other things, that heating the mixtures of poly(ethylene terephthalate) with polyarylate for 16 hours at 280 °C results in amorphous statistical copolymers (equation 2) with a glass transition temperature (glass point) which ranges between the glass points of the initial polymers.[75]

$$\left[\text{—C}\bigcirc\text{COCH}_2\text{CH}_2\text{O—}\right]_n + \left[\text{—C—Ar—CO}\bigcirc\text{C(Me)}_2\bigcirc\text{O—}\right]_m \longrightarrow$$

$$\left[\text{—C}\bigcirc\text{COCH}_2\text{CH}_2\text{O—}\right]_x \left[\text{—C—Ar—CO}\bigcirc\text{C(Me)}_2\bigcirc\text{O—}\right]_y \quad (2)$$

$$\text{Ar} = \text{—}\bigcirc\text{—} \quad \text{and} \quad \text{—}\bigcirc$$

Important results were obtained in studies of transesterification of poly(ethylene terephthalate) with its deuterated analogue.[82,83] The investigation performed has led to an interesting assumption according to which the course of the exchange reactions may benefit the polymer crystallization.[83]

A number of authors came to an opposite conclusion concerning the reactions of copolyesters in a heterophase solid–liquid system induced by crystallization.[84–89] They noticed an unusual phenomenon whereby the block structures were found to form upon heating the statistical copolymers. This process, followed by a decrease of the system entropy, proceeds due to crystallization of the blocks formed, which as a result leave the reaction area. The rate of the exchange reactions induced by crystallization increases with a decrease of the value of the molecular weight of the initial polymers, while the reactions themselves may proceed both in the solid and liquid state.[86,89]

There were also marked phenomena related to the effect of the chemical factor on the physical structure formation on the one hand, and to the effect of the physical factor on the route of the chemical reaction on the other. These phenomena are opening an attractive prospect of a fine regulation of the properties of polymeric materials in the required direction. Thus, for example, it may become of great importance for the creation of high strength products based on thermotropic liquid crystalline copolymers.

9.4 POLYCONDENSATION AFFECTED BY PRESSURE

In the 1970s Enikolopyan *et al.* showed by numerous examples that the effect of high pressure and shear deformations can promote various chemical reactions including those which do not proceed under the usual conditions. Thus, under extreme conditions of high pressure in combination with shear deformation, maleic acid enters into reaction due to the addition of its carboxyl groups to the available double bonds.[90]

The above method was successfully applied to the synthesis of various poly(heteroarylene)s, such as polyimides, poly(oxadiazole)s and poly(phenylquinoxaline)s.[91] For example, a pressure of 2 GPa affects an equimolar mixture of pyromellitic acid dianhydride and 4,4′-diaminodiphenyl oxide at an angle of the anvil slope equal to 1000° to yield polyimide with a reduced viscosity of 0.4 dl g^{-1}. The value of the molecular weight of the poly(heteroarylene)s formed increases with an increase in the pressure and shear deformations.

The reactions of epoxides are greatly affected by plastic flow under pressure. The effect of high pressure (to 5 GPa) and big shear deformations (the angle of the anvil slope to 2500°) results in polymerization of the diglycidyl ether of hydroquinone.[92] Under these conditions, the pressure has the effect of a catalyst of the Lewis acid type, redistributing the charge of the C—O bond of the epoxide group.

In the interaction between the diglycidyl ether of hydroquinone and diamines under a pressure of 2 GPa, the activity of the latter forms a series corresponding to that found for the corresponding reaction under the usual conditions.[93]

Further development in research and studies of the effect of high pressure and shear deformations would allow identification of the possibilities of the method more precisely. However it is obvious even now that the above-described approach could be successfully employed to produce polymers from very varied monomers, including ones of low reactivity.

9.5 POLYCONDENSATION IN THE PRESENCE OF A FILLER

Polycondensation in the presence of a filler is a new and very promising procedure of polycondensation in bulk. In this case, the filled polymers which are formed may be processed into variously shaped articles immediately after the synthesis is completed. As a result, a very complicated and labour-consuming stage of producing the composite material from the ready polymer and filler becomes unnecessary. The problem of the effect of a filler on the synthesis of the polymer is of interest, that is the effect on the reaction rate, extent of reaction and macrostructure of the macromolecules formed. It may be suggested that the nature of the filler, its amount, specific surface and modifying additives would markedly affect these factors. Available (though few) literature data prove the validity of such an assumption.[94,95]

The rate of polycondensation in bulk is also greatly affected by the amount of filler in the system. Thus the kinetics of the reaction of N,N'-m-phenylenebismaleimide with 4,4′-diaminodiphenylmethane preformed in the melt in the presence of graphite, tetrafluoroethylene and molybdenum disulfide considerably depends on the volume fraction of the filler in the system.[94]

The state of the filler surface is an important factor influencing the rate of the processes. In fact, the interaction of macromolecules with the filler surface (glass microspheres) decreases the solidification rate of N,N'-tetraglycidyl-4,4′-diaminodiphenylmethane with 4,4′-diaminodiphenylmethane at 190–210 °C.[95] The assumption made is indirectly supported by the results of studying the effect of the surface on the properties of liquid crystalline polymers.[96] A considerable acceleration of the orientation of thermotropic polyester is observed after the glass surface is treated with the micron-sized diamond powder.

The great practical importance of polycondensation in the presence of a filler determines the necessity of most serious studies in this field which, in the long term, will demonstrate the actual limits of the method.

9.6 POLYCYCLOTRIMERIZATION

Polycyclotrimerization appears to be a very interesting reaction of polymer synthesis. The process is characterized by the stepwise mechanism of the polymer chain growth and the absence of the yield of low molecular weight product. Polycyclotrimerization in bulk is used to produce polymers based on monomers containing nitrile, cyanate, cyanamide, isocyanate and some other groups with C—N multiple bonds (Scheme 4). The major regularities of this method are described in a number of reviews.[97–101]

In this process, the chain grows during cyclotrimerization of functional groups. As a result crosslinked systems with the same knot-to-knot fragments and six-membered cycles are formed as a knot network. Strictly speaking, this process is not polycondensation, because the cyclotrimerization proceeds by polymerization of unsaturated C—N bonds with a very short chain length. Formal resemblance of polycyclotrimerization to polycondensation consists of the stepwise character of the macromolecular network frame formation. The polycyclotrimerization of dicyanate has made it possible to find satisfactory agreement of experimental with theoretical values of the gel formation point, which confirms that the process occurs in a stepwise fashion.[101,102]

Synthesis of polymers by means of polycyclotrimerization in the melt is usually performed in the presence of a catalyst, such as a Lewis acid at 100–260 °C (Table 1).

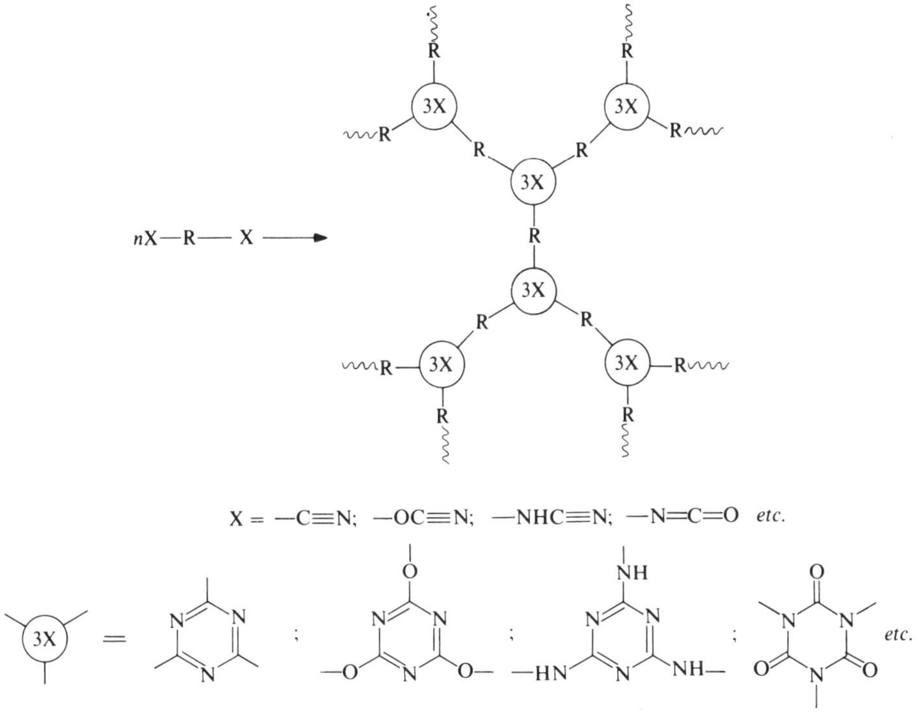

$X = -C{\equiv}N;\ -OC{\equiv}N;\ -NHC{\equiv}N;\ -N{=}C{=}O$ etc.

Scheme 4

Table 1 Conditions of Cyclotrimerization in Bulk of Various Monomers

Monomers	Temperature (°C)	Catalysts	Refs.
Oxibis(perfluoropropionitrile)	90–100	$Bu_3Sb(OH)_2$	a
2,2-Bis(4-cyanophenyl)propane	150–200	$ZnCl_2$	b
Mixture of 4,4′-diisocyanodiphenylmethane with hexamethylene diisocyanate	120–260	Mixture of triethylenediamine with phenylglycidyl ether	c
4,4′-Dicyanoamidodiphenylmethane	200	Without catalyst	d

[a] J. L. Zollinger, J. R. Throckmorton, S. T. Ting, S. A. Mitsch and D. E. Elrick, *J. Macromol. Sci., Chem.*, 1969, **3**, 1443.
[b] V. A. Pankratov, A. A. Ladovskaya, S.-S. A. Pavlova, G. I. Timofeeva, V. V. Korshak and S. V. Vinogradova, *Vysokomol. Soedin, Ser. A*, 1978, **20**, 1074.
[c] V. V. Korshak, S. V. Vinogradova, G. L. Slonimskii, V. A. Pankratov, A. A. Askadskii, Ts. M. Frenkel, L. F. Larina and K. A. Bychko, *Vysokomol. Soedin., Ser. A*, 1981, **23**, 1244.
[d] V. V. Korshak, V. A. Pankratov, S. V. Vinogradova, N. P. Antsiferova and D. F. Kutepov, *Dokl. Akad. Nauk SSSR*, 1975, **220**, 1081.

Polycyclotrimerization of monomers with the C—N multiple bonds is followed by side reactions, *viz.* polymerization to chains, dimerization and others. The only exception concerns dicyanates because their cyclotrimerization reactions proceed with high selectivity and with practically quantitative conversion, which results in the formation of network polymers of regular chemical structure, capable of crystallization.[97]

Both monomers and oligomers can be used as initial compounds for polycyclotrimerization. In the latter case there is the possibility of regulating the length and chemical structure of the knot-to-knot fragment, thus changing the properties of the polymer in the direction required.[97–101,103]

Therefore, with an increase in the number of silicon atoms in the chains between the network knots from 2 to 102 the glass transition temperature of the polymer based on dicyanate esters of oligo(dimethylsiloxane)s decreases from 108 °C to −120 °C, while the initial temperature of weight decrease in the air increases from 250 °C to 360 °C.[97]

Introduction of cyanate and other similar groups as side groups in macromolecules allows crosslinked polymers with new valuable properties to be obtained, due to cyclotrimerization of these groups. For example, thermal treatment of poly(phenyltriazine)s and poly(phenylenequinoxaline)s with nitrile and cyanate side groups results in crosslinked polymers of high heat resistance.[104]

A wide range of modifications to the properties of the prepared materials is possible by using several monomers for polycyclotrimerization. Changing the amount, structure and combination of various comonomers, it is possible to change the parameters and characteristics of copolymers, viz. their heat resistance and thermal stability, mechanical properties and others. Moreover, monomer mixtures, as compared to the initial monomers, melt at lower temperatures and also, as a rule, are prone to overcooling. All these features together provide the substantial technological advantages of copolycyclotrimerization.[97-101]

Also very promising is polycyclotrimerization carried out in the presence of a filler, since in this case the composite materials can be obtained directly in the process of polymer synthesis.

In conclusion, it should be stressed that the availability of the initial monomers, great technological advantages of polycyclotrimerization and the unique properties of the polymers formed make the method widely used for solving applied problems.[101]

9.7 OTHER METHODS

A search for new types of reactive monomers has led to the development of 'the method of silylated monomers', which is now successfully used in polycondensation in bulk. It was first used in the synthesis of polyesters. Thus polyesters were obtained from silylated 2,2-bis(4-hydroxyphenyl)-propane and diacid dichlorides for 2–5 hours at 150–250 °C (equation 3).[105]

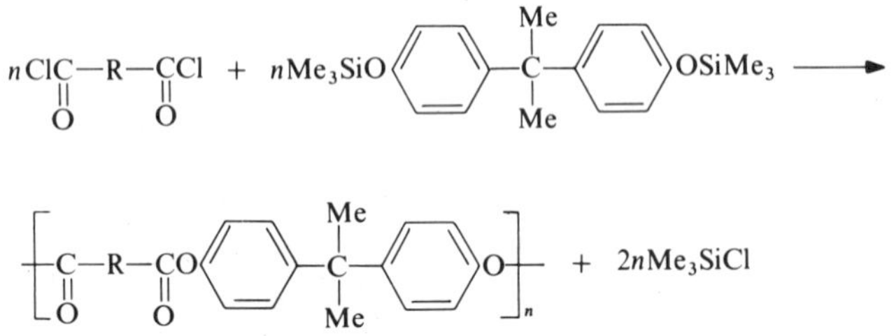

$$n\text{ClC}-\text{R}-\text{CCl} + n\text{Me}_3\text{SiO}\langle\text{C}_6\text{H}_4\rangle\text{C}(\text{Me})_2\langle\text{C}_6\text{H}_4\rangle\text{OSiMe}_3 \longrightarrow$$

$$\left[-\text{C}(\text{O})-\text{R}-\text{CO}\langle\text{C}_6\text{H}_4\rangle\text{C}(\text{Me})_2\langle\text{C}_6\text{H}_4\rangle\text{O}-\right]_n + 2n\text{Me}_3\text{SiCl} \tag{3}$$

The application of silylated hydroxybenzoic acids has turned out to be particularly successful for the synthesis of polyesters and copolyesters[105,106] (Scheme 5).

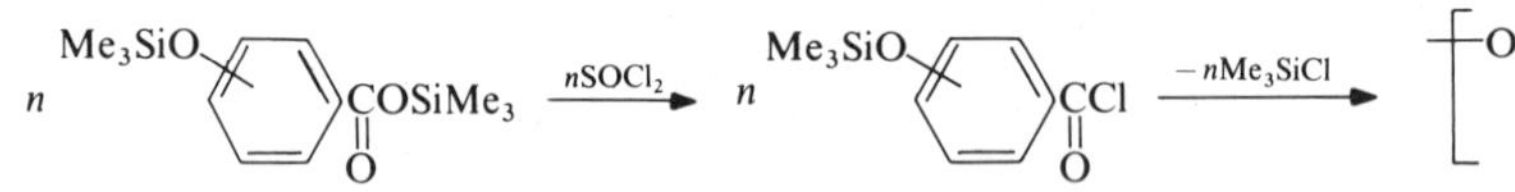

Scheme 5

In contrast to ordinary analogues, chlorides of silylated hydroxybenzoic acids (2) are stable at room temperature and their utilization in the reaction of polycondensation in bulk yields polymers with molecular weights two to three times that of polymers based on p-acetoxybenzoic acids. For example, polycondensation of m-trimethylsiloxybenzoyl chloride at 250 °C yielded polyester with $\bar{M}_n = 11\,800$, while heating the melt of m-acetoxybenzoic acid in the presence of catalytic amounts of magnesium at 220–300 °C enabled a polymer with $\bar{M}_n = 3850$ to be obtained.[105]

'The method of isomerizational polycyclization' worked out by Pravednikov et al.[107] is another interesting variant of polycondensation in bulk.

During synthesis of poly(heteroarylene)s by means of isomerization polycyclization the formation and cyclization of the prepolymer does not result in liberation of low molecular weight volatile products, which allows this reaction to be used for the production of monolithic articles.

The wide range of possibilities offered by this method were demonstrated by the synthesis of poly(iminoquinazoline)s and poly(iminobenzoxazinone)s by the interaction of bisnitrile-containing diamines and diols with diisocyanates[107] (Scheme 6). The polymers obtained exhibit high thermal stability and can be used as glues, plastics, laminated materials and for other purposes.

In conclusion it should be noted that the great advantages of polycondensation in bulk do determine further developments of the method. These developments are being carried out in various directions, in connection with applications of new initial compounds, including oligomers and polymers, new active functional groups, new catalysts, etc. The approach based on the effect of physical factors seems to be very promising. This approach may lead to both a rate increase in the

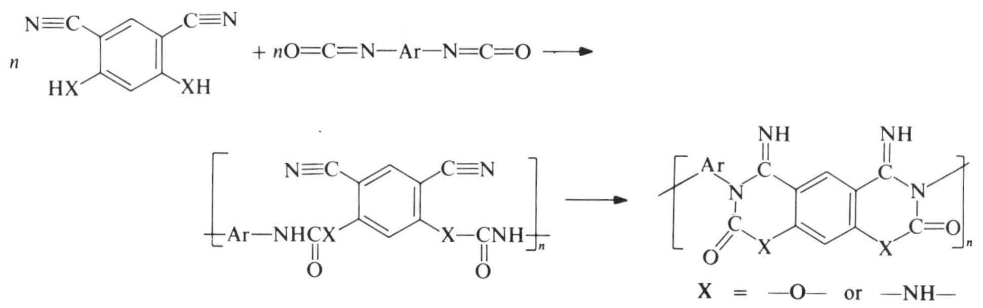

Scheme 6

synthesis of polymers and to improvements in their properties. Polycondensation in bulk carried out in the presence of fillers, both inorganic and organic, including polymeric ones, also seems to be very promising for the future. In the long run, the pace of the development of one or another direction will be to a great extent determined by the urgency of possible practical applications.

9.8 REFERENCES

1. V. V. Korshak and S. V. Vinogradova, 'Equilibrium Polycondensation', Nauka, Moscow, 1968 (in Russian).
2. V. V. Korshak, *Usp. Khim.*, 1984, **53**, 3.
3. V. V. Korshak and S. V. Vinogradova, 'Non-equilibrium Polycondensation', Nauka, Moscow, 1972 (in Russian).
4. J. H. Sounders and K. C. Frisch, 'Polyurethanes: Chemistry and Technology. Part 1. Chemistry', Interscience, New York, 1962.
5. V. V. Korshak, A. L. Rusanov and L. Ch. Plieva, *Faserforsch. Textiltech.*, 1977, **18**, 371.
6. D. R. Paul and S. Newman (eds.), 'Polymer Blends', Academic Press, New York, 1978.
7. T. M. Pell and I. J. Davis, *J. Polym. Sci., Part A-2*, 1973, **11**, 1671.
8. K. Ravindranath and R. A. Mashelkar, *Chem. Eng. Sci.*, 1986, **41**, 2969.
9. A. Fradet and E. Marechal, *Adv. Polym. Sci.*, 1982, **43**, 51.
10. S. Kotanti, *Sen'i Gakkaishi*, 1984, **40**, 141.
11. Yu. N. Nizel'skii, in 'Physical Chemistry of Polyurethane', ed. S. I. Omel'chenko, Nauka, Kiev, 1981, p. 3 (in Russian).
12. K. Tomita, *Polymer*, 1976, **17**, 221.
13. K. H. Wolf and H. Herlinger, *Angew. Makromol. Chem.*, 1977, **65**, 187.
14. T. H. Shah, J. I. Bhatty, G. A. Gamlen and D. Dollimore, *J. Macromol. Sci., Chem.*, 1984, **21**, 431.
15. T. H. Shah, J. I. Bhatty and G. A. Gamlen, *Polymer*, 1984, **25**, 1333.
16. K. A. Joshi and I. Kattaura, *Man-Made Text. India*, 1984, **27**, 516.
17. B. Lu, H. He, Y. Men, G. Liu, D. Li and X. Zhang, *Chin. Appl. Chem.*, 1985, **2**, 53.
18. F. Pilati, A. Munari, P. Manaresi and V. Bonora, *Polymer*, 1985, **26**, 1745.
19. J. Hsu and K. Y. Choi, *J. Appl. Polym. Sci.*, 1986, **32**, 3117.
20. T. Sasakawa, H. Takasaki, T. Ikeda and S. Tazuke, *Makromol. Chem.*, 1986, **187**, 547.
21. Ch.-Ch. Lin and S. Baliga, *J. Appl. Polym. Sci.*, 1986, **31**, 2483.
22. K. Ravindranath and R. A. Mashelkar, *Chem. Eng. Sci.*, 1986, **41**, 2197.
23. M. Tanaka, H. Iida, H. Ikeuchi and H. Komatsu, *Sen'i Gakkaishi*, 1987, **43**, 35.
24. F. Leverd, A. Fradet and E. Marechal, *Eur. Polym. J.*, 1987, **23**, 695, 699, 705.
25. A. S. Fomin, V. D. Korzhov, A. V. Mogunov, L. N. Gureeva and V. A. Ignatov, *Lakokras. Mater. Ikh Primen.*, 1984, **5**, 9.
26. K. Ravindranath and R. A. Mashelkar, *Chem. Eng. Sci.*, 1986, **41**, 2969.
27. K. Ravindranath and R. A. Mashelkar, *Polym. Eng. Sci.*, 1982, **22**, 610.
28. S. K. Gupta, *Proc. Indian Acad. Sci., Chem. Ser.*, 1983, **92**, 1983.
29. T. Yamada, Y. Imamura, O. Makimura and H. Kamatani, *Proc. World Congr. III Chem. Eng., Tokyo*, 1986, **4**, 104.
30. E. Roerdink, P. J. De Jong and J. Warnier, *Polym. Commun.*, 1984, **25**, 194.
31. V. V. Shevchenko, L. V. Polyakov and A. A. Afinogenov, *Khim. Volokna*, 1983, **2**, 42.
32. H. Inata and S. Matsumura, *J. Appl. Polym. Sci.*, 1985, **30**, 3325.
33. H. Inata and S. Matsumura, *J. Appl. Polym. Sci.*, 1986, **32**, 5193.
34. H. Inata and S. Matsumura, *J. Appl. Polym. Sci.*, 1986, **32**, 4581.
35. J. M. Buist (ed.), 'Developments in Polyurethane', Applied Science, London, 1978.
36. V. P. Kuznetsova, N. N. Zaskovenko and K. V. Zalunnya, 'Organo-Silicon Polyurethanes', Naukova Dumka, Kiev, 1984 (in Russian).
37. I. M. Raigorodsky and E. Sh. Goldberg, *Usp. Khim.*, 1987, **56**, 1891.
38. P. M. Valetskii and I. P. Storozhuk, *Usp. Khim.*, 1979, **48**, 75.
39. J. J. O'Malley, *J. Polym. Sci., Polym. Lett. Ed.*, 1974, **12**, 381.
40. D. S. Varma, A. Maheswari, V. Gupta and I. K. Varma, *Angew. Makromol. Chem.*, 1980, **90**, 23.
41. V. V. Shevchenko, A. S. Chegolya and E. M. Aizenshtein, *Vysokomol. Soedin., Ser. A*, 1985, **27**, 2333.
42. J. M. Lambert, E. Yilgor, I. Yilgor and G. L. Wilkes, *Polym. Prepr., Am. Chem. Soc., Div. Polym. Chem.*, 1985, **26**, 275.
43. F. L. Hamb, *J. Polym. Sci., Polym. Chem. Ed.*, 1972, **10**, 3217.

44. T. Balakrishnan, E. Ponnusamy and H. K. Kothandaraman, *Polym. Bull.*, 1981, **5**, 187.
45. T. Balakrishnan, E. Ponnusamy, C. T. Vijayakumar and H. Kothandaraman, *Polym. Bull.*, 1981, **6**, 195.
46. J. Economy and W. Volksen, in 'The Strength and Stiffness of Polymers', ed. A. Z. Zachariades and R. S. Poeter, Dekker, New York/Basel, 1983, p. 293.
47. R. W. Lenz and J. L. Jin, *Polym. News*, 1986, **11**, 200.
48. W. J. Jackson, Jr. and H. F. Kuhfuss, *J. Polym. Chem., Polym. Chem. Ed.*, 1976, **14**, 2043.
49. M. R. Mackley, F. Pinaud and C. Siekmann, *Polymer*, 1981, **22**, 437.
50. L. Quach, W. Volksen, R. Herbold and J. Economy, *Polym. Prepr., Am. Chem. Soc., Div. Polym. Chem.*, 1986, **27**, 307.
51. M. Balachandar, T. Balakrishnan and H. Kothandaraman, *J. Polym. Sci., Polym. Chem. Ed.*, 1979, **17**, 3713.
52. C. Vasiliu–Oprea, *Mater. Plast.*, 1985, **22**, 215.
53. W. J. Jackson, Jr. and H. F. Kuhfuss, *J. Appl. Polym. Sci.*, 1980, **25**, 1685.
54. M. J. Han, H. C. Kang and K. E. Choi, *Macromolecules*, 1986, **19**, 1649.
55. S. M. Aharoni, *Polym. Bull.*, 1983, **10**, 210.
56. Y. P. Khanna and E. A. Turi, *Polym. Prepr., Am. Chem. Soc., Div. Polym. Chem.*, 1984, **25**, 98.
57. S. M. Aharoni, C. E. Forbes, W. B. Hammond, D. M. Hindenlang, F. Mares, K. O. Brien and R. D. Sedgwick, *J. Polym. Sci., Polym. Chem. Ed.*, 1986, **24**, 1281.
58. A. Noshay and J. E. McGrath, 'Block Copolymers: Overview and Critical Survey', Academic Press, New York, 1977.
59. D. C. Allport and W. H. Janes (eds.), 'Block Copolymers', Applied Science, London, 1973.
60. V. V. Korshak, T. M. Frunze and I.-N. Lu, *Vysokomol. Soedin.*, 1960, **2**, 984.
61. M. M. Brubaker, D. D. Cofman and F. C. McGrew (Du Pont), US Pat. 2 339 237 (1944) (*Chem. Abstr.*, 1944, **38**, 3853).
62. A. Noshay and J. E. McGrath, 'Block Copolymers', Academic Press, New York, 1977.
63. H. G. Ramjit and R. D. Sedgwick, *J. Macromol. Sci., Chem.*, 1976, **10**, 815.
64. L. Lüderwald, *J. Macromol. Sci., Chem.*, 1979, **13**, 869.
65. H. G. Ramjit, *J. Macromol. Sci., Chem.*, 1983, **19**, 41.
66. H. G. Ramjit, *J. Macromol. Sci., Chem.*, 1983, **20**, 659.
67. D. S. Varma, R. A. Agarwal and I. K. Varma, *Br. Polym. J.*, 1985, **17**, 83.
68. D. S. Varma, R. Agarwal and I. K. Varma, *Polym. Commun.*, 1985, **26**, 346.
69. A. Misra and S. N. Gard, *J. Polym. Sci., Polym. Phys. Ed.*, 1986, **24**, 983.
70. A. W. Birley and X. Y. Chen, *Br. Polym. J.*, 1985, **17**, 297.
71. Z. H. Huang and L. H. Wang, *Makromol. Chem., Rapid Commun.*, 1986, **7**, 255.
72. P. Godard, J. M. DeKoninck, V. Devlesaver and J. Devaux, *J. Polym. Sci., Polym. Chem. Ed.*, 1986, **24**, 3301.
73. P. Godard, J. M. DeKoninck, V. Devlesaver and J. Devaux, *J. Polym. Sci., Polym. Chem. Ed.*, 1986, **24**, 3315.
74. M. Kimura and R. S. Porter, *J. Polym. Sci., Polym. Phys. Ed.*, 1983, **21**, 367.
75. M. Kimura, G. Salee and R. S. Porter, *J. Appl. Polym. Sci.*, 1984, **29**, 1629.
76. J. I. Eguiazábal, G. Ucar, M. Cortázar and J. J. Iruin, *Polymer*, 1986, **27**, 2013.
77. A. Ausin, J. I. Eguiazábal, M. E. Munoz, J. J. Pena and A. Sanbamaria, *Polym. Eng. Sci.*, 1987, **27**, 529.
78. J. Devaux, P. Devaux and P. Godard, *Makromol. Chem.*, 1985, **186**, 1227.
79. I. Mondragon and J. Nazabal, *J. Appl. Polym. Sci.*, 1986, **32**, 6191.
80. J. I. Eguiazábal, J. J. Iruin and M. Cortazar, *J. Appl. Polym. Sci.*, 1986, **32**, 5945.
81. I. Mondragon, P. M. Remiro and J. Nazabal, *Eur. Polym. J.*, 1987, **23**, 125.
82. K. P. McAlea, J. M. Schultz, K. H. Gardner and G. D. Wignall, *Polymer*, 1986, **27**, 1581.
83. J. Kugler, J. W. Gilmer, D. Wiswe, H.-G. Zachmann, K. Hahn and E. W. Fischer, *Macromolecules*, 1987, **20**, 1116.
84. R. W. Lenz, E. Martin and A. N. Schuler, *J. Polym. Sci., Polym. Chem. Ed.*, 1973, **11**, 2265.
85. R. W. Lenz, K. Ohata and J. Funt, *J. Polym. Sci., Polym. Chem. Ed.*, 1973, **11**, 2273.
86. R. W. Lenz and S. Go, *J. Polym. Sci., Polym. Chem. Ed.*, 1973, **11**, 2927.
87. R. W. Lenz and S. Go, *J. Polym. Sci., Polym. Chem. Ed.*, 1974, **12**, 1.
88. R. W. Lenz and A. N. Shuler, *J. Polym. Sci., Polym. Symp.*, 1978, **63**, 343.
89. G. Chen and R. W. Lenz, *Polymer*, 1985, **26**, 1307.
90. A. B. Solovyeva, V. A. Zhorin and N. S. Enikolopyan, *Dokl. Akad. Nauk SSSR*, 1978, **240**, 125.
91. V. A. Zhorin, A. E. Chesnokova, G. L. Berestneva, V. V. Korshak and N. S. Enikolopyan, *Vysokomol. Soedin., Ser. B*, 1984, **26**, 140.
92. E. L. Indoleva, V. A. Zhorin, A. N. Zelenetskii and N. S. Enikolopyan, *Dokl. Akad. Nauk SSSR*, 1986, **287**, 373.
93. E. L. Indoleva, V. A. Zhorin, A. N. Zelenetskii and E. N. Enikolopyan, *Vysokomol. Soedin., Ser. B*, 1986, **28**, 338.
94. A. V. Khabenko, V. E. Kalinchikov, L. I. Marinyuk, T. V. Rakhmanova and S. A. Dolmatov, *Zh. Prikl. Khim. (Leningrad)*, 1984, **57**, 1827.
95. J. Mijovic, *J. Appl. Polym. Sci.*, 1986, **31**, 1177.
96. P. G. Martin, J. S. Moore and S. I. Stupp, *Macromolecules*, 1986, **19**, 2459.
97. V. A. Pankratov, S. V. Vinogradova and V. V. Korshak, *Usp. Khim.*, 1977, **46**, 530.
98. V. A. Sergeev, V. K. Shitikov and V. A. Pankratov, *Usp. Khim.*, 1979, **48**, 148.
99. V. A. Pankratov, Ts. M. Frenkel and A. M. Fainleib, *Usp. Khim.*, 1983, **52**, 1018.
100. V. A. Pankratov, D. F. Kutepov and G. I. Shitikov, *Plastmassy*, 1983, **2**, 12.
101. G. M. Pogosyan, V. A. Pankratov, V. N. Zaplishnyi, G. I. Matsoyan, 'Polytriazines, Armenian SSR Academy of Sciences', Erevan, 1987 (in Russian).
102. V. V. Korshak, V. A. Pankratov, A. A. Ladovskaya and S. V. Vinogradova, *J. Polym. Sci., Polym. Chem. Ed.*, 1978, **16**, 1697.
103. V. V. Korshak, G. I. Shukurov, V. A. Pankratov and D. F. Kutepov, *Plastmassy*, 1983, **2**, 10.
104. P. M. Hergenrother, *Polym. Prepr., Am. Chem. Soc., Div. Polym. Chem.*, 1974, **15**, 781.
105. H. R. Kricheldorf and G. Schwarz, *Polym. Bull.*, 1979, **1**, 383.
106. H. R. Kricheldorf and G. Schwarz, *Macromol. Chem.*, 1983, **184**, 475.
107. I. V. Vasilyeva, L. N. Kurkovskaya, A. N. Flerova, R. M. Gitina, Ye. L. Zaitseva, E. N. Teleshov and A. N. Pravednikov, *Vysokomol. Soedin., Ser. A*, 1979, **21**, 1114.

10
Experimental Methods of Solution Polymerization

VASILI V. KORSHAK and VALERI A. VASNEV
USSR Academy of Sciences, Moscow, USSR

10.1	INTRODUCTION	143
10.2	ACCEPTOR-CATALYTIC POLYCONDENSATION	144
	10.2.1 Polyesterification	145
	10.2.2 Polyamidation	146
10.3	ACTIVATED POLYCONDENSATION	147
	10.3.1 Activated Esters	147
	10.3.1.1 Polyamides	148
	10.3.1.2 Polyesters	148
	10.3.1.3 Polyurethanes and polyureas	149
	10.3.2 Activated Amides	152
	10.3.3 Silylated Monomers	152
10.4	DIRECT POLYCONDENSATION	153
	10.4.1 Polyamides	154
	10.4.2 Polypeptides and Polyureas	155
	10.4.3 Polyesters	156
	10.4.4 Polyheteroarylenes	159
10.5	MATRIX POLYCONDENSATION	160
10.6	OTHER METHODS	161
	10.6.1 Polycondensation in an Ordered Medium	161
	10.6.2 Precipitation Polycondensation	162
	10.6.3 Electrochemical Polycondensation	162
10.7	REFERENCES	162

10.1 INTRODUCTION

Solution polycondensation is a widely used polycondensation procedure, in which the monomers and the resultant polymer are placed in a single-phase solution. Different variations of this technique are known, for when a monomer or a polymer is not completely soluble in the reaction mixture. As a rule, however, it is for the case when the monomers and the polymer are completely soluble in the medium that the most favourable conditions are created for high molecular weight polymers to be obtained. This can be accomplished by using a mixture of two or more solvents and/or by raising the reaction temperature. Usually, the temperature of polycondensation in solution is in the range 20–250 °C. In the course of solution polycondensation the polymer can form thermodynamically unstable (metastable) solutions. After isolating the polymer from such a solution it may be impossible to dissolve it again in a given solvent. Precipitation of a highly crystalline polymer, poorly swollen by the reaction medium, from the polycondensation solution results in the termination of the growth of macromolecules (physical chain termination). In a precipitated amorphous or slightly crystalline polymer, capable of swelling, the macromolecule growth reactions continue.

Solution polycondensations are distinguishable on the basis of the solvent influence on molecular weight and structure of the polymer formed. The solvent itself, as well as the impurities in it (*e.g.* moisture), can give rise to undesirable side effects: exchange reactions, reactions of terminal groups, deactivation and blocking. Of special significance among the side reactions which reduce the polymer molecular weight is cyclization, whose contribution increases with a decrease in reaction solution concentration. Polymer precipitation from solution in the course of the synthesis can lead to its crystallization.

Solution polycondensation has a number of characteristic features that distinguish it favourably from other polycondensation techniques: (1) it allows the process to be carried out in relatively mild conditions, which is especially important in the synthesis of high-melting polymers, when a high reaction temperature in the melt can cause the destruction of monomers and polymers; (2) the solvent can act as reaction catalyst; (3) the removal of the low molecular product (alcohol, acid, *etc.*) from the reaction vessel is simplified; (4) it assures good heat transfer, which is especially important for exothermic reactions; and (5) the obtained solutions of polymers can be directly used to manufacture films and fibres.

In its turn, the use of a solvent in the synthesis of polymers gives rise to a number of economic, technological and ecological problems, associated with the relatively low concentrations of polymers in solution (0.2–1 mol l^{-1}): the use of large quantities of polymer precipitating medium; the recovery and reutilization of solvent and precipitant after separation and washing of the polymer; and clearing the polymer from solvent and low molecular weight reaction products, such as salts of metals and hydrochlorides of tertiary amines. All these factors complicate the potential application of solution polycondensation on an industrial scale, especially in the large scale production of polymers. Solution polycondensation is used in industry to produce aromatic polyamides, polyarylates, polycarbonates, poly(arylene sulfones)s, poly(phenylene sulfide)s, poly(ether ketone)s, polyurethanes, polyureas, polyheteroarylenes and a number of other polymers. Indeed, the relevance of solution polycondensation is much wider, if one takes into account that the first stages of poly(alkylene terephthalate) production are conducted in an excess of glycol (in solution), and that the first stages of the production of aliphatic polyamides (of the nylon-6,6 type) as well as of phenol–formaldehyde and urea–formaldehyde resins are performed in aqueous solution.

In laboratory practice, solution polycondensation has no limitation and finds very wide applications in the synthesis of a large number of polymers.

The rules governing solution polycondensation, the different techniques used to perform it and the opportunities it provides for the synthesis of polymers have been examined and summarized in a number of monographs[1-4] and reviews.[5-23]

10.2 ACCEPTOR-CATALYTIC POLYCONDENSATION

In the 1960s a large group of reactions relating to solution polycondensation began to be subdivided, depending on temperature, into low and high temperature ones.[1-3] The polymer synthesis processes conducted at temperatures below 100 °C were designated as low temperature polycondensations. These reactions include the widely used low temperature polyesterification (equations 1 and 2), polyamidation (equation 3), formation of polyurethanes (equation 4) and polyureas (equation 5), as well as a number of others.

$$n\text{ClCOCl} + n\text{HOR''OH} \xrightarrow[-2n\text{R}_3\text{N}\cdot\text{HCl}]{2n\text{R}_3\text{N}} \left[-\text{COR''O}- \right]_n \tag{1}$$

$$n\text{ClCOR'COCl} + n\text{HOR''OH} \xrightarrow[-2n\text{R}_3\text{N}\cdot\text{HCl}]{2n\text{R}_3\text{N}} \left[-\text{CR'COR''O}- \right]_n \tag{2}$$

$$n\text{ClCOR'COCl} + n\text{NH}_2\text{R''NH}_2 \xrightarrow[-2n\text{B}\cdot\text{HCl}]{2n\text{B}} \left[-\text{CR'CNHR''NH}- \right]_n \tag{3}$$

B = tertiary amine or amide

$$n\text{OCNR'NCO} + n\text{HOR''OH} \longrightarrow \left[-\text{R'NHCOR''OCNH}- \right]_n \tag{4}$$

$$n\text{OCNR'NCO} + n\text{NH}_2\text{R''NH}_2 \longrightarrow \left[-\text{R'NHC(=O)NHR''NHC(=O)NH}- \right]_n \quad (5)$$

In the case of reactions (1)–(3) a necessary condition for them to proceed in solution at low temperatures is the use of organic bases.

Only in the case of polyamides are organic bases aimed at binding HCl, as free HCl interacts with NH_2 groups, converting them into poorly active salts, and ultimately results in the formation of low molecular weight polymer. Acting as acceptors of the acid, organic bases (tertiary amines, amides) also perform the function of reaction catalysts in polyamidation. However, the activity of monomers in these processes is sufficiently high for them to interact at quite low temperatures even without catalysts.[1] Polyamides can actually be synthesized in the presence of a 'pure' acceptor of an acid. For instance, polyamides have been obtained from 4,4′-bis(chloroformyl)bibenzoyl and diamines at $-20\,°C$ in the presence of propylene oxide as the acceptor.[24]

In the synthesis of polycarbonates and polyesters the organic bases (tertiary amines) primarily perform the function of reaction catalysts, assuring the reactions occur at reduced temperatures. In this case HCl binding is not a necessary condition for a high molecular polymer to be obtained, since the reaction of phosgene and dicarboxylic acid dichlorides with diols is practically irreversible.[3]

Because of the dual function of organic bases in polyesterification and polyamidation, it was suggested that these and similar reactions (equations 1–3) should be called acceptor-catalytic polycondensation.[3] This designation makes possible a more accurate attribution of a given solution polycondensation technique to a specific chemical interaction. The necessity of this classification became especially obvious after the development of new techniques of conducting polycondensation in solution: activated and direct polycondensation, which can proceed both below and above $100\,°C$.

In the 1970s and 1980s the interest in acceptor-catalytic polycondensation noticeably increased because this technique was used as the basis for the synthesis of a large number of highly thermostable polyheteroarylenes,[25] liquid crystalline polyesters[26] and other relevant polymers.

10.2.1 Polyesterification

Acceptor-catalytic polycondensation was first used as a method of synthesizing polymers, polycarbonates in the given case, by Einhorn in 1898 (equation 1). It was, however, more than 50 years later before this technique was used again to obtain polycarbonates and polyesters.[27,28]

Starting from the 1960s, as a result of intensive research on kinetics, mechanism and the rules governing of reactions (1) and (2), acceptor-catalytic polyesterification began to be widely used for the synthesis of a large variety of polymers.[29] This research effort made it possible to formulate a number of specific conditions whose observance is necessary for high molecular polyesters to be obtained.

Thus a study of the rules governing acceptor-catalytic polyesterification showed polyesters with a maximum molecular weight to be formed when there is an equimolar ratio of diacid dichlorides and diols.[30] The optimum amount of tertiary amine is determined by the structure of the initial reagents. In going from bisphenols to glycols and from triethylamine to pyridine the molar ratio of tertiary amine to diol rises from 2:1 to 8:1.[31] An increase in the acidity (activity) of bisphenols[32] and in the basicity of tertiary amines[33] facilitates the formation of polymers with a higher molecular weight. Increasing activity of diacid dichlorides, as a rule, also results in an increase of polyester molecular weight.[34]

When choosing the optimum temperature conditions for acceptor-catalytic polyesterification, one has to take into account the fact that the polymer molecular weight vs. the reaction temperature dependence is expressed by a curve with two maxima.[31,35] This allows the synthesis of high molecular weight polymers in different temperature regions, e.g. at -20 or $+50\,°C$ (Figure 1). The shape of curve 2 in Figure 1, with two maxima, is associated with a complex temperature dependence of the reaction rate constant (curve 1), which in turn is caused by the multistage nature of acceptor-catalytic polyesterification.[35]

The molecular weight value of polyesters is noticeably dependent on the concentration of monomers.[31] In the case of highly active initial compounds, polymers with the highest molecular weight are formed when the concentrations of reactants are equal to 0.15–$0.20\,\text{mol}\,l^{-1}$. A low activity of at least one of the initial compounds raises the optimum concentration values of monomers to 1–$2\,\text{mol}\,l^{-1}$.

A great impact on the results of acceptor-catalytic polyesterification is exerted by the nature of the solvent. A trend is observed for the molecular weight of the formed polyester to decrease with

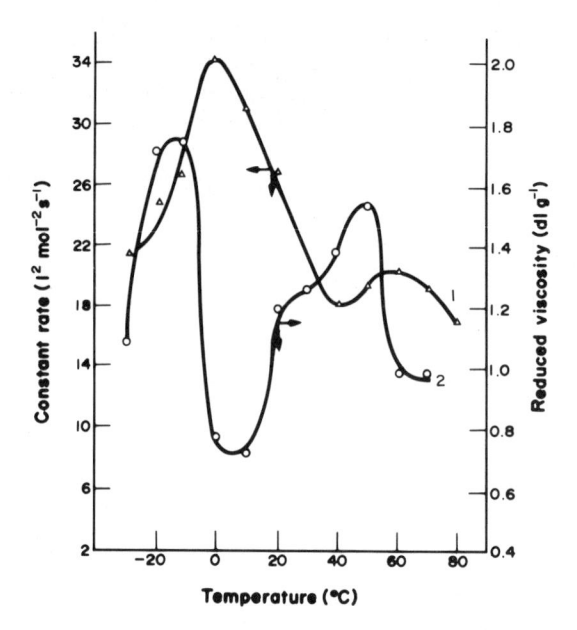

Figure 1 Rate constant of the reaction of o-chlorophenol with p-chlorobenzoyl chloride (curve 1) and the reduced viscosity of polyarylate from 2,2-bis(4-hydroxy-3-chlorophenyl)propane and terephthaloyl dichlorides (curve 2) vs. the reaction temperature (toluene, triethylamine)[35]

increasing dielectric constant of the medium.[36] When polycondensation is performed in a metastable solution, the aggregation of macromolecules creates more favourable conditions for the formation of high molecular compounds than when the reaction is conducted in a thermodynamically stable solvent with a similar value of the dielectric constant.[36]

Precipitation of an amorphous or slightly crystalline polymer from solution under the conditions of acceptor-catalytic polyesterification does not hinder the subsequent increase of its molecular weight.[36,37] In this case the optimum conditions for the synthesis consist of good solubility of the initial compounds in the reaction medium, considerable swelling capacity of the polymer in a low polarity medium, or high polarity of the medium when the polymer-swelling capacity of the solvent is low.[37]

An interesting peculiarity of acceptor-catalytic polyesterification is the conformational specificity of this process, making it possible to obtain conformationally regular polyarylates (polyesters based on dicarboxylic acids and bisphenols), whose conformational tacticity is formed during the course of polycondensation, and the formed macromolecules contain regular sequences of one rotational isomer.[29]

Conformationally regular polymers of the above type were first synthesized from o,o'-substituted derivatives of 2,2-bis(4-hydroxyphenyl)propane and terephthalic acid dichlorides.[38] The macromolecules of this polymer were shown to contain, depending on the reaction conditions, regular sequences of *cis* (**1**) or *trans* (**2**) rotational isomers of the bisphenol fragment.[39,40]

In recent years, various types of polyesters have been synthesized by acceptor-catalytic polyesterification: liquid crystalline ones,[26] functionalized ones[41,42] and others.

Along with polyesters, acceptor-catalytic polycondensation produces polyphosphonates based on diacid dichlorides of phosphonic acids and bisphenols.[43,44]

10.2.2 Polyamidation

Research into acceptor-catalytic polyamidation in solution (equation 3) began to develop in the 1960s, largely due to Morgan's fundamental work.[1,45] The results of his and other studies facilitated the widespread application of acceptor-catalytic polyamidation as one of the principal methods of synthesizing aromatic polyamides and their derivatives.[46]

An important feature of acceptor-catalytic polyamidation, distinguishing it from polyesterification, is that in the synthesis of polyamides the role of the acceptor for the evolving acid (HCl) is often played by the amide solvent, *e.g.* dimethylacetamide or *N*-methylpyrrolidone. If the dissolving capacity of such single component solvents is not sufficient, multicomponent dissolving systems are used: mixtures of amide solvents[47] or amide–salt mixtures.[48] The use of such systems not only makes it possible to improve the solubility of the formed polymers but also raises the reaction rate; in particular, it produces a significant catalytic effect in acceptor-catalytic polyamidation.[49] The introduction of a tertiary amine into amide–salt systems increases the reaction rate even more.[50]

An interesting peculiarity of acceptor-catalytic polyamidation is the different role of water in this process. For the reaction of diacid dichlorides with diamines, the presence of even slight concentrations of moisture (more than 0.01%) in solution leads to a decrease in the polymer molecular weight.[51] Substitution of diacid difluoride for dichloride makes it possible to obtain high molecular polyamides at a considerable water content (up to 50%), which not only raises the degree of swelling in the precipitated polymer but also accelerates the reaction.[52] The efficiency of polyamidation catalysis with carboxylic acids also depends on the structure of the monomers: in contrast to diacid dichlorides, the reaction of difluorides with diamines is noticeably accelerated by benzoic acid, which results in the formation of polyamides with an even higher molecular weight.[53]

The optimum reaction temperature and concentration of monomers in acceptor-catalytic polyamidation reactions are determined by numerous factors: the activity of initial compounds, the properties of the reaction medium, the solubility of the polyamide formed, *etc.* Aromatic polyamides with a maximum molecular weight are formed when the concentration of monomers lies within $0.25-1.0 \, \text{mol} \, \text{l}^{-1}$.[54] The optimum temperatures of polyamidation usually lie in the range 0 to 40 °C. As exemplified by the acceptor-catalytic polyamidation of 2,7-diaminofluorene with isophthaloyl dichloride in dimethylacetamide, the temperature dependence of polymer molecular weight was found to have two maxima, at 0 and 30 °C.[55]

The use of active (amide, *etc.*) solvents in acceptor-catalytic polyamidation results in this process acquiring a number of specific peculiarities. In order to reduce the contribution of side reactions and raise the polymer molecular weight, the diacid dichlorides are charged in the solid form with a certain intensity of mixing and feed rate.[56] For the synthesis of polyamides the diamines can not only be used in the form of free bases, known to be easily oxidizable, but also as dichlorohydrates.[57]

A high rate of acceptor-catalytic polyamidation makes it possible to perform the polymer synthesis in media that are not solvents for it.[58] The faster the reaction proceeds, the higher the degree of oversaturation of the formed metastable solutions will be. This fact considerably expands the number of possible solvents for polymer synthesis.

Successful development of acceptor-catalytic polyamidation in solution, on the whole completed in the 1970s, allows at present a wide use of this technique for the synthesis of various polyamides, including substituted polyamides.[59-61] A great role is assigned to acceptor-catalytic polyamidation in the synthesis of thermostable polyheteroarylenes by means of two-stage polycondensation,[62] as well as from monomers containing heterocycles in their structure.[63-65]

10.3 ACTIVATED POLYCONDENSATION

The use of highly active diacid dichlorides in acceptor-catalytic polycondensation for the synthesis of polymers has a number of significant limitations. These limitations are associated with the easy hydrolyzability of diacid dichlorides, which makes it necessary to observe certain precautions in their synthesis and storage. Moreover, a high reactivity of these monomers gives rise in a number of cases to a considerable contribution from side reactions, which facilitate the formation of low molecular polymers.

The search for new solution polycondensation methods resulted in the 1970s in the development of activated polycondensation, based on the activation of electrophilic (activated diesters or diamines) or nucleophilic (silylated diamines or diols) monomers. In the latter case diacid dichlorides, activated diesters, *etc.* can be used as the electrophilic monomer.

10.3.1 Activated Esters

The activated ester technique is based on the well known concept that insertion of an electron-acceptor substituent into the ester alkoxy residue casues a considerable increase in its activity with respect to nucleophiles (amines, alcohols, *etc.*).

In 1969 Overberger and Sebenda[66] were the first to use activated dicarboxylic acid diesters (equation 6) for the synthesis of polyamides based on piperidine.

$$nXOCR'COX + nH_2NR''NH_2 \longrightarrow \left[-CR'CNHR''NH- \right]_n + 2nXOH \qquad (6)$$
$$\underset{O}{\|}\underset{O}{\|} \qquad\qquad\qquad \underset{O}{\|}\underset{O}{\|}$$

$$X = -\!\!\left\langle\!\!\bigcirc\!\!\right\rangle\!\!-\!NO_2, \quad -\!\!\left\langle\!\!\bigcirc\!\!\right\rangle\!\!-\!NO_2, \quad -ON\!\!\!\bigcirc\!\!\!, \quad -ON\!\!\!\bigcirc$$

During the 1970s and early 1980s activated polycondensation has been actively developed: new types of activated diesters, catalysts and reaction conditions were investigated. All these improvements allow synthesis of various polyamides, polyesters, polyurethanes and polyureas with the help of the activated ester technique.

10.3.1.1 Polyamides

In the 1970s research on the synthesis of polyamides by activated polycondensation was oriented to a considerable extent towards searching for new activated diesters whose use would make it possible to expand considerably the synthetic potential of this technique and obtain high molecular polymers. Altogether more than 30 activated diesters were studied: chlorine-, fluorine- and nitro-substituted aryl esters,[66-73] methyl-, nitro- and un-substituted phenyl thioesters,[67,69,74] heterocyclic esters,[67,74-79] hydroxylamine derivatives[66,67,80] and a number of others. Some of these esters are shown in equation (6).

Research into activated polycondensation showed pentafluorophenyl and 4-nitrophenyl diesters (Table 1) to be highly effective from the point of view of the molecular weight value of the polyamide formed. From the data in Table 1 it is clearly seen that the use of fluorine- and nitro-substituted phenyl esters, as compared with unsubstituted ones, results in polymers with a much higher molecular weight.

No direct connection was found to exist between the activated diester reactivity and the polyamide molecular weight, *viz.* highly active diesters can produce a polymer with not very high molecular weight. For instance, the polycondensation of bis-2,4-dinitrophenyl adipate with 4,4'-diaminodiphenylmethane proceeds much faster than in the case of bis-4-nitrophenyl adipate; the latter, however, produces a polymer with a higher molecular weight (Table 1).

Activated diesters, in distinction to diacid dichlorides, act as milder, selective acylating agents, which makes it possible to obtain polyamides based on cyclization prone dicarboxylic acids, *e.g.* succinic acid and natural amino acids. For instance, as a result of the reaction of bis-4-nitrophenyl succinate with diamines in hexamethylphosphorotriamide at 30 °C, after 8 h a high molecular polyamide was obtained, with a reduced viscosity of $1.24\,\text{dl}\,\text{g}^{-1}$.[81]

With the help of the activated polycondensation technique the authors succeeded in obtaining polyamides with rather high molecular weights from the activated esters of aspartic and glutamic acids and hexamethylenediamine.[82]

In conclusion, it should be emphasized that the investigation of polymers synthesized by activated polycondensation in solution showed them to be not inferior but, in a number of cases, superior in their properties to similar polymers obtained by other methods. This opens up interesting perspectives for successful applications of activated diesters in the synthesis of polyamides.

10.3.1.2 Polyesters

In the early 1980s it was shown that the technique of activated polycondensation in solution can also be successfully applied to the synthesis of polyesters.[83] The necessary reaction conditions were studied for the polycondensation of activated diesters of adipic and isophthalic acids with phenolphthalein in 1,2-dichloroethane solution in the presence of triethylamine (equation 7).

$$n\text{XOCR'COX} + n\text{HO}\text{-Ar-C(phenolphthalein)-Ar-OH} \xrightarrow{2n\text{Et}_3\text{N}} \left[-\underset{\underset{O}{\|}}{C}R'\underset{\underset{O}{\|}}{C}O\text{-Ar-C-Ar-O-}\right]_n + 2n\text{XOH} \quad (7)$$

R' = —(CH$_2$)$_4$—, —C$_6$H$_4$—;

X = 2,4-dinitrophenyl, 4-nitrophenyl, 2,4,5-trichlorophenyl, N-succinimidyl, 2-nitrophenyl

A polyarylate in quantitative yield with the highest molecular weight (reduced viscosity 0.94 dl g^{-1}) was obtained in the case of bis-2,4-dinitrophenyl adipate, when the reaction was conducted at 25 °C for 3 h. In equation (7) activated diesters are arranged according to their capacity to form high molecular polymers.

In going from adipates to isophthalates the molecular weight of polyarylates decreases. Glycol, when interacting under these conditions with the activated diester, produces only oligomeric products. Low molecular polyarylates are formed when the synthesis is accomplished in amide solvents and dimethyl sulfoxide. Apart from dichloroethane, high molecular polymers have been of diols obtained in methyl ethyl ketone and benzene.[83]

10.3.1.3 Polyurethanes and polyureas

In 1981, high molecular weight polyurethanes were synthesized by polycondensation of diol-activated biscarbonates with diamines.[84] Various nitro- and chloro-phenyl esters (equation 8) were used as activated carbonates.

$$n\text{XOCOR'OCOX} + n\text{H}_2\text{NR''NH}_2 \longrightarrow \left[-\underset{\underset{O}{\|}}{C}OR'O\underset{\underset{O}{\|}}{C}NHR''NH-\right]_n + 2n\text{XOH} \quad (8)$$

X = 4-nitrophenyl, 2-nitrophenyl, 2,4,5-trichlorophenyl, 2,3,5-trichlorophenyl, 4-nitrophenyl, phenyl

In the reaction with hexamethylenediamine in the solvent dimethylformamide at 110 °C, substituted bisphenyl carbonates of 1,3-propanediol form the series shown in equation (8). In accordance with this series, polyurethanes with the highest molecular weight (reduced viscosity of 1.40 dl g^{-1}) are formed using bis-4-nitrophenyl carbonate. A change from aliphatic to less basic aromatic diamines (e.g. 4,4'-diaminodiphenylmethane) results in the formation of polyurethanes with not very high molecular weights. Nevertheless, the method of polyurethane synthesis by means of activated polycondensation in solution is sufficiently versatile and simple and does not require the use of catalysts. This method will be indispensable in the cases when bischloroformates of diols are not readily available, and the reaction of the diols themselves with diisocyanates is complicated by side reactions.

In 1982, the polycondensation of carbonic-acid-activated diesters with aliphatic and aromatic diamines was used to accomplish the synthesis of high molecular weight polyurea (equation 9).[85]

$$n\text{NO}_2\text{-C}_6\text{H}_3\text{X-OCO-C}_6\text{H}_3\text{X-NO}_2 + n\text{NH}_2\text{R''NH}_2 \longrightarrow [-\underset{\underset{O}{\|}}{C}\text{NHR''NH}-]_n + 2n\text{NO}_2\text{-C}_6\text{H}_3\text{X-OH} \quad (9)$$

X = H, NO$_2$

Table 1 Results of Activated Polycondensation in Hexamethylphosphorustriamide Solution[a,b]

XOCR'COX		$NH_2R''NH_2$	T (°C)	Reaction duration (h)	Polymer yield (%)	Reduced viscosity (dl g^{-1})
R'	X	R''				
—(CH$_2$)$_4$—	tetrafluorophenyl	—C$_6$H$_4$—CH$_2$—C$_6$H$_4$—	80	4	100	2.20
—(CH$_2$)$_4$—	p-NO$_2$-C$_6$H$_4$	—C$_6$H$_4$—CH$_2$—C$_6$H$_4$—	80	18	100	1.67
—(CH$_2$)$_4$—	2,4-dinitrophenyl	—C$_6$H$_4$—CH$_2$—C$_6$H$_4$—	80	4	99	1.21
—(CH$_2$)$_4$—	phenyl	—C$_6$H$_4$—CH$_2$—C$_6$H$_4$—	80	40	68	0.10
m-C$_6$H$_4$	tetrafluorophenyl	—C$_6$H$_4$—CH$_2$—C$_6$H$_4$—	80	4	98	1.05

![m-tolyl]	![NO₂-C₆H₄]	![-CH₂-biphenyl-]	80	40	93	0.76
![m-tolyl]	![NO₂-C₆H₄]	–(CH₂)₆–	30	6	100	1.42
![m-tolyl]	![F₄-C₆]	–(CH₂)₆–	30	4	97	0.98
–(CH₂)₄–	![NO₂-C₆H₄]	–(CH₂)₆–	30	6	99	1.02
–(CH₂)₄–	![F₄-C₆]	–(CH₂)₆–	30	6	97	0.87

[a] Concentration of monomers = 0.6 mol l^{-1}. [b] R. D. Katsaeava, D. P. Kharadze and L. M. Avalishvili, *Polish J. Appl. Chem.*, 1986, **30**, 187.

The reaction was conducted in dimethylacetamide solution, in a number of cases in the presence of LiCl, at 90 °C for 2–8 h. Polycondensation of bis-4-nitrophenyl carbonate, both with hexamethylenediamine and 4,4'-diaminophenyl oxide, produced polymers with much higher molecular weights than in the case of bis-2,4-dinitrophenyl carbonate. Along with diamines, their silylated analogues can also be used in this reaction.[85]

Thus the aforementioned results of synthesizing polyamides, polyesters, polyurethanes and polyureas show the activated ester technique to be highly promising for the synthesis of a large variety of heterochain polymers. In most cases polymers with the highest molecular weight are formed with the use of readily available bis-4-nitrophenyl esters. Mild reaction conditions, high selectivity of synthesis and correspondingly high yields and molecular weights of the polymers formed constitute a good basis for the further successful development of this technique.

From a practical point of view application is somewhat limited by the necessity to withdraw a low molecular reaction product, *e.g.* 4-nitrophenol, from the polymer, as well as by the problem of reutilizing the reaction solution after isolating and purifying the polymer. All these difficulties are less relevant if the problem arises of performing a highly selective synthesis process and/or synthesizing polymers based on complex polyfunctional monomers.

10.3.2 Activated Amides

The activated amide technique is based on raising the electrophilicity of monomers (dicarboxylic acids) by using their corresponding *N*-acyl derivatives (equation 10).

$$n \bigcirc NCR'CN \bigcirc \underset{O\ O}{\parallel\ \parallel} + nH_2NR''NH_2 \longrightarrow \left[-CR'CNHR''NH- \right]_n + 2n \bigcirc NH \qquad (10)$$

In 1951, Staab[86] was the first to use dicarboxylic acid diimidazole derivatives (3) to synthesize polyamides, but obtained only low molecular polymers. In the late 1960s it was shown that for the synthesis of high molecular polyamides with the help of activated amides it is necessary to use different catalysts: imidazole and pyridine hydrochlorides, 1-hydroxybenzotriazole, HCl, $AlCl_3$, $BF_3 \cdot Et_2O$ and others.[87]

In the 1970s polyamides and polyhydrazides based on various diacylbislactams (4) (including diacylbiscaprolactam),[88,89] diacylbisthiolactams (5),[89,90] diacylbissuccinimide (6)[91] and other analogous compounds[92,93] began to be successfully produced. In the 1980s a number of diacyl derivatives of benzazoles were proposed for use as activated amides,[94–98] among them the derivatives of benzoxazole-2-thione (7).[95]

Based on the above, as well as on a number of other, activated diamides, a large variety of high molecular aliphatic and aromatic polyamides[87–98] and polyhydrazides[91] have been synthesized. In some cases there has been success in raising the polymer molecular weight by using catalysts, among which 1-hydroxybenzotriazole proved to be especially effective. Examples are known, however, where the same compound accelerates one reaction with the participation of activated diamines and decelerates another one. Thus aluminum chloride appreciably speeds up the polyamidation with diacylbisimidazoles taking part[87] and inhibits markedly the reaction of diacylbislactams with diamines.[88]

It can be assumed that the further development of this technique will be associated with a search for more effective activated diamides and polycondensation catalysts.

10.3.3 Silylated Monomers

An interesting possibility for performing solution polycondensation under mild conditions lies in using the method, widely known in organic chemistry,[99] of silylating the initial compounds, primarily nucleophilic monomers: diamines, diols and a number of others (equation 11). Dicarboxylic acid derivatives (diacid dichlorides, dianhydrides, activated esters) and other reagents, *e.g.* p-xylylene dichloride, can be used as electrophilic monomers in the reaction with silylated monomers. The reaction of p-xylylene dichloride with N,N'-bistrimethylsilylated diamines was used to produce high molecular polyamines.[100]

$$n\text{XCR'CX} + n\text{Me}_3\text{SiYR''YSiMe}_3 \longrightarrow \left[-\underset{\underset{\text{O}}{\|}}{\text{CR'}}\underset{\underset{\text{O}}{\|}}{\text{CYR''Y}}- \right]_n + 2n\text{Me}_3\text{SiCl} \qquad (11)$$

$$X = \text{Cl, OCR}, -\text{O}\!-\!\!\!\langle\!\!\!\bigcirc\!\!\!\rangle\!\!-\!\text{NO}_2 \text{ and others; } Y = \text{NH, O}$$

Diacid dichlorides are most often used in the synthesis of polymers with the help of silylated monomers, in particular the bischloroformate of 1,3-butanediol[101] and trimethylsiloxybenzoyl chloride.[102] Along with diacid dichlorides, activated diesters are used for polymer synthesis based on silylated diamines.[81,85]

To date, various high molecular polymers have been obtained by means of the silylated monomer technique: aromatic polyamides,[103,104] aromatic polyesters,[102,105] polyurethanes,[101] polyureas,[85] polyamines,[100] linear polyimides,[106] heterochain polymers based on natural amino acids,[81] poly(ether sulfone)s[107] and poly(ether ketone)s.[108]

It should be emphasized that the use of silylated monomers makes it possible to accomplish the reactions that proceed very poorly with the participation of unsilylated analogues. For instance, with the help of N,N'-bistrimethylsilylated N,N'-diphenyl-1,4-phenylenediamine in the reaction with diacid dichlorides, the authors[103] succeeded in obtaining N-phenylated aromatic polyamides. This fact, as well as the mild conditions under which the polycondensation is conducted, the high molecular weight of the polymers obtained and a wide combination of the end polymer structures, makes it possible to speak of the good prospects of this polycondensation technique for the synthesis of polymers.

10.4 DIRECT POLYCONDENSATION

Considerable progress in the field of acceptor-catalytic and activated polycondensation cannot in the final analysis circumvent the problem associated with the necessity of using in these processes the specifically synthesized and often unstable derivatives of dicarboxylic acids — their diacid dichlorides and activated diesters. One of the ways out of this situation could lie in creating a method of polycondensation in which it would be possible to use the dicarboxylic acids themselves, or some other specially unactivated monomers, and the polymer formation process would then take place at moderate temperatures with a high rate. The solution was found in the direct polycondensation technique, whose intensive development began in the 1970s. Within the framework of direct polycondensation, dicarboxylic acids in the presence of condensing reagents, *e.g.* phosphorus compounds, react directly, 'instantaneously', with various nucleophiles (diamines, *etc.*) in solution under sufficiently mild reaction conditions.

In subsequent years the development of research in the field of direct polycondensation was primarily associated with the work of Japanese scientists (Yamazaki, Ogata, Higashi and others), which resulted in a wide application of this technique for the synthesis of a large variety of polymers: polyamides, polypeptides, polyureas, polyesters, polyheteroarylenes, polyketones, polysulfones, *etc.*[109,110] Used as condensing reagents in direct polycondensation are the compounds of tri- and penta-valent phosphorus, as well as those containing no phosphorus (tosyl chlorides, thionyl chlorides, *etc.*).

The general principles of performing direct polycondensation have been established. Thus, in the synthesis of polyamides in the presence of phosphorus-containing condensing reagents it is necessary to take into account that: (i) the reaction proceeds better in aprotic amide solvents in the presence of organic bases (pyridine) and metal chlorides (LiCl); and (ii) polymers with the highest molecular weight are formed, as a rule, at temperatures of 80–120 °C; one succeeds in obtaining fully aliphatic polyamides with a relatively low molecular weight. It should not be forgotten, however, that in each particular case these conditions require verification, and the future development of research can introduce substantial changes to them.

10.4.1 Polyamides

In the first publications that appeared in the early 1970s the condensing system used in polyamide synthesis was a *system with trivalent phosphorus*: triphenyl phosphite/imidazole[111,112] (equation 12), previously used for the synthesis of peptides.[113] However, all the attempts at obtaining high molecular polymers with its help ended in failure. Neither did the use, instead of triphenyl phosphite, of its substituted analogues,[114] and polyphosphates[115] in combination with imidazole result in a noticeable increase in the molecular weight of polyamides.

$$n\text{HOCR'COH} + n\text{NH}_2\text{R''NH}_2 + 2n\text{P(OPh)}_3 \longrightarrow \left[-\underset{\underset{O}{\|}}{C}\text{R'}\underset{\underset{O}{\|}}{C}\text{NHR''N} - \right]_n + 2n\text{PhOH} + 2n\text{OPH(OPh)}_2 \quad (12)$$

It became possible to synthesize polyamides with higher molecular weights by using such condensing reagents as pyridine N-phosphonium salts, formed upon the oxidation of phosphites with mercury salts in pyridine,[116–118] or by diphenyl phosphite diphenoxylation in the absence of an oxidizer.[119] By direct polycondensation in the presence of the pyridine N-phosphonium salt (**8**) it is possible to obtain polyamides with sufficiently high molecular weights from aliphatic dicarboxylic acids and aromatic diamines in N-methylpyrrolidone solution at 100 °C.[120] Based on aliphatic diamines, only low molecular polyamides are formed. In going to aromatic dicarboxylic acids, high molecular polymers were not obtained in the case of either aromatic or aliphatic diamines.[120]

(**8**)

The development of these studies has made it possible to develop a new promising method for the synthesis of polyamides (Yamazaki's method), based on using pyridine instead of imidazole in a system with triphenyl phosphite (equation 12).[121] As well as with phosphites, a similar reaction can be accomplished with other phosphorus compounds, such as phosphinites (R_2POAr), phosphonites (RP(OR)$_2$), phosphonates (RPO(OAr)$_2$) and others, with aryl esters being the most effective in attaining a high molecular weight.[122] In all the systems, the established regularities of polymer formation proved to be similar to those discussed above: high molecular polyamides are formed only in the case of aliphatic dicarboxylic acids, and the changeover from aliphatic to aromatic diamines decreases the polymer molecular weight. One of the causes of this was associated with the poor solubility of aromatic compounds and their precipitation from the reaction solution.

To solve this problem it was proposed that direct polycondensation should be conducted in the presence of metal salts, improving the solubility of polyamides, which has ultimately resulted in creating the most advantageous and universal condensing system: triphenyl phosphite/pyridine/LiCl.[123] As exemplified by the polycondensation of aminobenzoic acids, the optimum reaction conditions were found to be: LiCl concentration = 4%, pyridine content in the mixture with N-methylpyrrolidone = 40 vol. %, temperature = 100 °C, monomer concentration = 0.6 mol l^{-1}, molar ratio of monomer:triphenyl phosphite = 1:1, reaction duration 6 h. A moderate temperature, acceptable reaction duration and a high molecular weight of the polymer formed are indicative of a sufficiently high effectiveness of direct polyamidation in the presence of the above condensing reagents.

A further improvement of the condensing system, the conditions under which the reaction is conducted (the sequence of feeding the reagents, the temperature variations, *etc.*) and the selection of initial monomers have made it possible for different workers to obtain polyamides with an even higher molecular weight from both aminobenzoic acids and dicarboxylic acids and diamines.[124–129] In particular, poly(p-phenyleneterephthalamide) with an intrinsic viscosity of 6.2 dl g^{-1} has been synthesized.[129]

Development of the chemistry of direct polycondensation showed that it was possible to synthesize high molecular polyamides in the presence of the triphenyl phosphite–LiCl system, with the reaction conducted in amide solvents without pyridine (Higashi's method).[130]

An interesting trend in raising the molecular weight of polymers lies in carrying out direct polycondensation in the presence of the additions of functional polymers: poly(4-vinylpyridine),[131] poly(ethylene oxide)[132] and poly(vinylpyrrolidone).[133] These systems, exemplifying matrix polycondensation, make it possible to obtain polymers with a higher molecular weight, which is especially important in the cases when without a matrix such polymers are not formed at all. In each particular case the mechanism of polymeric matrix action can be different. For instance, poly(4-vinylpyridine), acting within the framework of Yamazaki's method, raises the local concentration of dicarboxylic acid as a result of its adsorption and raises its activity via the N-phosphonium salt.[131] In contrast to poly(vinylpyridine), poly(ethylene oxide) does not activate dicarboxylic acids, but only increases their local concentration via the hydrogen bonds.[132]

In 1981 a new condensing system was proposed for the synthesis of polyamides and other polymers by direct polycondensation. The system consists of triphenylphosphine, a polyhalo compound and pyridine.[134-137] In the case of hexachloroethane the reaction proceeds according to equation (13). When the reaction is performed in pyridine at 20 °C the polycondensation is completed in an hour.

$$n\text{HOCR'COH} + n\text{NH}_2\text{R''NH}_2 + 2n\text{Ph}_3\text{P} + 2n\text{C}_2\text{Cl}_6 \xrightarrow[-2n\text{py}\cdot\text{HCl}]{2n\text{py}} [-\text{CR'CNHR''NH}-]_n + 2n\text{Ph}_3\text{PO} + 2n\text{C}_2\text{Cl}_4 \quad (13)$$

It is noteworthy that in the case of such a halogen-containing compound as tetrabromomethane it was for the first time possible to obtain fully aliphatic polyamides with a good molecular weight (for nylon-6,10 the reduced viscosity is equal to 0.26 dl g^{-1}).[137] Unfortunately, however, CBr$_4$ leads to poor results in synthesizing aromatic polyamides.[134]

As well as trivalent phosphorus compounds, *compounds of pentavalent phosphorus* are also used in the synthesis of polyamides by direct polycondensation. Among these condensing reagents, the first to take note of is the readily available and inexpensive P$_2$O$_5$.[138] Using P$_2$O$_5$ in N-methylpyrrolidone solution one can obtain polyamides with a good reduced viscosity (up to 0.8 dl g^{-1}).

An opposite case to P$_2$O$_5$ is quite an exotic compound, N,N'-phenylphosphonobis-2(3H)-benzothiazolone (**9**), which has proved to be a very effective condensing reagent. In the presence of this reagent the reaction proceeds very fast; at 20 °C it reaches completion within 10 min and results in the formation of high molecular aromatic polyamides.[139]

(9)

The *condensing reagents containing no phosphorus* used in the synthesis of polyamides are thionyl chloride,[140-142] arenesulfonyl chlorides,[110] picryl chloride,[143,144] SiCl$_4$[145] and activated carbonic acid derivatives based on benzazoles.[146,147]

An interesting condensing agent is SiCl$_4$, in whose presence high molecular polyamides were obtained from aminobenzoic acids and from phenylenedicarboxylic acids and phenylenediamines (equation 14). The reaction was conducted in pyridine at 0 °C for 24 h. An important aspect of direct polycondensation with the participation of SiCl$_4$ is the fact that in the course of the reaction, polymers filled with SiO$_2$ are formed. This seems to be the first example of obtaining filled polymers via a low molecular reaction product.

$$n\text{HOCR'COH} + n\text{NH}_2\text{R''NH}_2 + n\text{SiCl}_4 \xrightarrow[-4n\text{py}\cdot\text{HCl}]{4n\text{py}} [-\text{CR'CNHR''NH}-]_n + n\text{SiO}_2 \quad (14)$$

10.4.2 Polypeptides and Polyureas

The synthesis of polypeptides from amino acids with the use of traditional techniques is complicated by side reactions and therefore results in the formation of low molecular polymers.[148]

The first successful step in this direction was made as a result of the synthesis of polypeptides from amino acids by means of direct polycondensation in the presence of the iodine/phosphite/pyridine or imidazole system.[149] Under these conditions polypeptides were isolated with a degree of polymerization reaching 50.

The further development of polypeptide synthesis is associated with the use of diphenyl phosphite/pyridine,[150] triphenyl phosphite/pyridine,[121] phosphinite, phosphonite or phosphonate/pyridine[122] systems. In all cases the polypeptides were obtained with good yields, but not in a high molecular weight. This is caused by side reactions, including cyclization, and steric hindrance from the amino acid residues. It proved necessary to use polymeric matrices to overcome these obstacles. High molecular polypeptides were thus obtained by matrix polycondensation in the presence of the triphenyl phosphite/LiCl system and poly(vinylpyrrolidone).[151,152] An increase in the polymeric matrix molecular weight from 10^4 to 3.6×10^5 raises the molecular weight of poly(α-leucine) from 10^3 to 3.5×10^4.[152] Based on DL-methionine, matrix polycondensation in N-methylpyrrolidone produced a polypeptide with the highest molecular weight (intrinsic viscosity $1.09\,dl\,g^{-1}$).[152]

Systems based on diphenyl phosphite (triphenyl phosphite) and pyridine act as sufficiently effective condensing systems in the synthesis of polyureas.[153] Mild conditions of synthesis and a high selectivity of phosphorus-containing reagents make it possible to obtain aromatic polyureas and polythioureas directly from CO_2 and CS_2 and diamines (equation 15).

$$n CO_2 + n NH_2R''NH_2 + (n/2)P(OPh)_3 \xrightarrow{py} \left[-\underset{\underset{O}{\|}}{C}NHR''NH- \right]_n + (n/2)HOPH(OPh)_2 + n PhOH \quad (15)$$

For the synthesis of polyurea and polythiourea based on CO_2 and CS_2 one can use, besides the above-mentioned condensing reagents, various phosphorus halogen compounds: PCl_3, dichlorophenyl- and ethyl-phosphines, chlorodiphenyl and diethyl phosphites, etc.[154] For instance, direct polycondensation of aromatic diamines with CO_2 and CS_2 in the presence of ethylene chlorophosphite in pyridine at 50–60 °C and 15–20 atm pressure for 6–7 h produced polyureas and polythioureas with a quantitative yield and intrinsic viscosity of up to $1.1\,dl\,g^{-1}$.[155,156]

10.4.3 Polyesters

The synthesis of polyesters by means of direct polycondensation began to be performed in the late 1970s and early 1980s, i.e. much later than the synthesis of polyamides. This was caused by the relatively low activity of diols, which affected the specificity of the developed condensing systems.

Among the condensing *trivalent phosphorus compounds* the first to be used for direct polyesterification were the systems of direct polyamidation: triphenyl phosphite and chlorides of trivalent phosphorus.[157] These systems were tested in the polycondensation of dicarboxylic acids and diols, as well as hydroxybenzoic acids, e.g. 4-hydroxy-3,5-dimethoxybenzoic acid (**10**; equation 16). The reaction was conducted at 140 °C for 4 h in the presence of pyridine. In all cases polymers with a relatively low molecular weight and a reduced viscosity of up to $0.36\,dl\,g^{-1}$ were isolated.

$$n HO-\underset{MeO}{\overset{MeO}{C_6H_2}}-CO_2H + n P(OPh)_3 \xrightarrow{py} \left[-\underset{MeO}{\overset{MeO}{C_6H_2}}-\underset{\underset{O}{\|}}{C}- \right] + n OPH(OPh)_2 + n PhOH \quad (16)$$

(**10**)

Similar results were obtained with the use of another well known condensing system: triphenylphosphine/polyhalo compound/pyridine.[158] Especially noteworthy here is the fact that the use of matrix polymers [poly(ethylene oxide), poly(4-vinylpyridine) and poly(2-vinylpyridine)] made it possible to raise markedly the molecular weight of the formed polyesters.[159]

The relative failure of the attempts to synthesize high molecular polyesters in the presence of condensing systems based on trivalent phosphorus compounds made it necessary to search for new systems, including those containing *pentavalent phosphorus compounds*. Among the first systems to be studied were poly(ethylene phosphate)/imidazole[160] and hexachlorocyclotriphosphazene/pyridine/LiCl.[160–162] However, it was only in the case of the latter system that it was possible to obtain polymers with a sufficiently high molecular weight (for compound (**10**) with a reduced viscosity of $0.8\,dl\,g^{-1}$).

An important step in the development of direct polyesterification was the suggestion of using triphenylphosphine dichloride to perform the reaction.[163] With the help of the triphenylphosphine dichloride/tertiary amine (triethylamine, pyridine, *etc.*) system in chlorobenzene at 130 °C, polyarylates were obtained with very high molecular weights (with $\bar{M}_w$ up to 103 200). It is of interest that by replacement of pyridine with triethylamine the ratio between $\bar{M}_w$ and $\bar{M}_n$ of the polymers decreases from 5–6 to 2–3.[163]

Successful application of triphenylphosphine dichloride for the synthesis of polyesters opens up good prospects for the practical implementation of direct polycondensation, since the triphenylphosphine oxide formed in the course of the reaction can be easily regenerated into the initial triphenylphosphine dichloride under the action of various compounds, *e.g.* oxalyl chloride (Scheme 1).

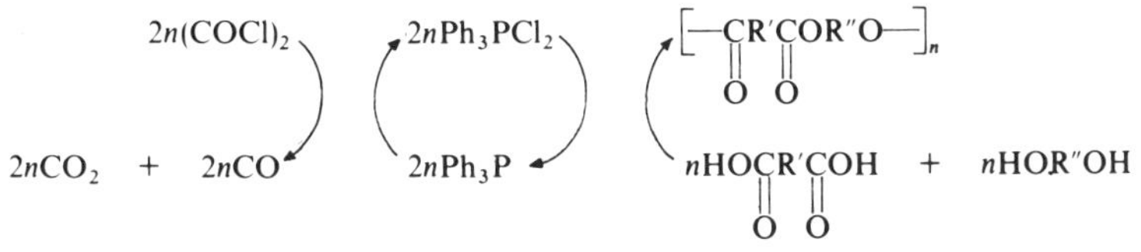

Scheme 1

Also of great interest in this respect are the condensing systems: diphenyl chlorophosphate/pyridine/LiCl,[164] diphenyl chlorophosphate/pyridine/LiBr,[165] diphenyl chlorophosphate/pyridine/ether[166] and diphenyl chlorophosphate/dimethylformamide.[167,168] In all these systems diphenyl phosphate, formed in the conditions of direct polycondensation under the action of thionyl chloride, can be easily converted again into diphenyl chlorophosphate (Scheme 2). No less important is the fact that, based on these systems, high molecular polymers can be obtained. Thus, when the reaction is conducted in the presence of diphenyl chlorophosphate at 120 °C in pyridine with LiCl, high molecular polyarylates with $\bar{M}_w$ up to 86 000 have been obtained.[165] The use of poly(ethylene oxide) as the ether also allows the synthesis of polymers with a very high molecular weight, which increases with an increase in the polymeric matrix molecular weight.[166]

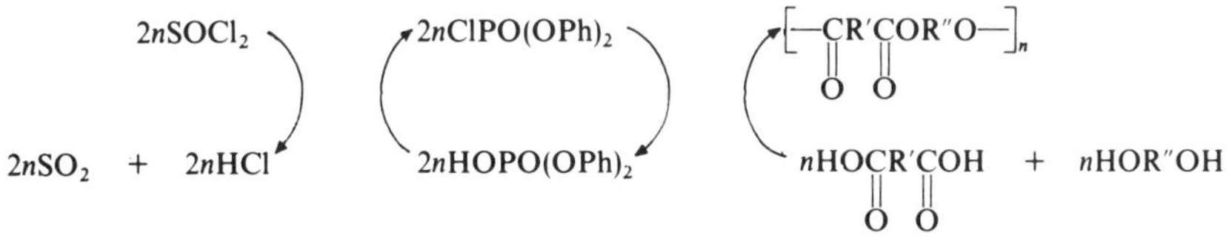

Scheme 2

It is noteworthy that copolymers obtained in the presence of dimethylformamide have a better solubility than similar copolymers synthesized under the action of hexachlorocyclotriphosphazene.[167] If the reason for this difference is associated with the different structure of the copolymers, then by varying the nature of the condensing reagent one could change the microstructure and therefore the properties of polymers. This result opens up a new direction in solving the major problem of synthesizing polymers with preassigned properties.

Along with phosphorus-containing condensing reagents, also used for direct polyesterification, are *compounds containing no phosphorus in the molecule*: thionyl chloride, picryl chloride and arenesulfonyl chlorides.

One of the known ways of producing diacid dichlorides is the action of thionyl chloride on dicarboxylic acids. Therefore, it appears quite logical to use thionyl chloride in the presence of organic bases (pyridine, dimethylformamide, *etc.*) as the condensing reagent in direct polyesterification.[169–172] Data have appeared recently, indicating that the mechanism of this reaction is much more complex.[173] However, the complexity of the mechanism is well compensated for by the simplicity of the process and its remarkable efficiency, making it possible to obtain high molecular polymers both from aromatic dicarboxylic acids and bisphenols and from aromatic hydroxycarboxylic acids, with 4-hydroxy-3-methoxycinnamic acid among them.[171] When the reaction is conducted in the previously prepared pyridine and thionyl chloride mixture, polyesters are formed with a very high molecular weight, higher than the molecular weight of similar polymers synthesized in the presence of phosphorus-containing condensing systems (Table 2).

Table 2 Results of Direct Polycondensation for a Mixture of 4-Hydroxy- and 4-Hydroxy-3,5-dimethoxy-benzoic Acids (50:50 mol)

Condensing system	Solvent	Reaction conditions Temperature (°C)	Duration (h)	Intrinsic viscosity (dl g^{-1})	Ref.
Diphenyl chlorophosphate/pyridine	Pyridine	120	3	0.86	a
Diphenyl chlorophosphate/dimethylformamide	Pyridine	120	3	1.42	b
Thionyl chloride/pyridine	Pyridine	80	3	3.78	c
Tosyl chloride/dimethylformamide	Pyridine	120	2	2.99	d

[a] F. Higashi, N. Kekubo and M. Gato, *J. Polym. Sci., Polym. Chem. Ed.*, 1983, **21**, 3241.
[b] F. Higashi, Y. Yamada and A. Hoshio, *J. Polym. Sci., Polym. Chem. Ed.*, 1984, **22**, 2181.
[c] F. Higashi and T. Mashimo, *J. Polym. Sci., Polym. Chem. Ed.*, 1986, **24**, 1697.
[d] F. Higashi, T. Mashimo, I. Takahashi and N. Akiyama, *J. Polym. Sci., Polym. Chem. Ed.*, 1985, **23**, 3095.

An interesting condensing system for the synthesis of polyesters is the picryl chloride/pyridine system (equation 17).[174] With the help of this system it becomes possible to obtain high molecular polyesters from aromatic dicarboxylic acids and glycols. Poly(hexamethylene terephthalate) with a reduced viscosity of 0.43 dl g^{-1} has been synthesized in this way. However, on the basis of molecular weight values of polyarylates, picryl chloride is inferior to thionyl chloride and arenesulfonyl chlorides.

$$n\text{HOCR'COH} + n\text{HOR''OH} + 2n\text{Cl-C}_6\text{H}_2(\text{NO}_2)_3 \xrightarrow[-2n\text{py} \cdot \text{HCl}]{2n\text{py}} [-\text{CR'COR''O}-]_n + 2n\text{HO-C}_6\text{H}_2(\text{NO}_2)_3 \quad (17)$$

Condensing systems based on arenesulfonyl chlorides very promising systems for direct polyesterification. As exemplified by the arenesulfonyl chlorides/pyridine/LiCl system, it was shown that polyarylates with the highest molecular weights can be obtained in the presence of benzene- and p-toluenesulfonyl chlorides.[175] The use of p-toluenesulfonyl chloride (tosyl chloride) in the condensing system with dimethylformamide and pyridine has considerably expanded the possibilities of direct polycondensation and raised its efficiency.[176-179] In the literature this fact is associated with the reaction mechanism, according to which tosyl chloride reacting with dimethylformamide forms an adduct (Vilsmeier's adduct **11**). At the next stage the adduct (**11**) interacts with carboxylic acid, forming a mixed anhydride (**12**) that selectively acylates the hydroxyl-containing compound (equation 18).

$$\text{ArSO}_2\text{Cl} + \text{HC(O)NMe}_2 \xrightarrow{\text{py}} [\text{ArSO}_2\text{OCH}=\overset{+}{\text{N}}\text{Me}_2]\text{Cl}^- \xrightarrow[-\text{py} \cdot \text{HCl}]{+\text{R'COOH}} \text{R'COSO}_2\text{Ar} \xrightarrow{+\text{R''OH}}$$

$$(11) \qquad\qquad (12)$$

$$\text{R'COR''O} + \text{ArSO}_2\text{OH} \quad (18)$$

The final result of polycondensation is greatly affected by the structures of arenesulfonyl chloride and formamide, the ratio of reagents in the condensing system, the time during which Vilsmeier's adduct stays in the tosyl chloride solution, the rate of diol addition and other factors.[176-179] Under optimum conditions one can synthesize polyesters with a very high molecular weight (Table 2). Unfortunately, the condensing systems based on arenesulfonyl chlorides do not make it possible to obtain high molecular polyesters from aromatic dicarboxylic acids and glycols.[180]

As a result of extensive research, promising condensing systems have thus been created, with the help of which high molecular polyarylates are synthesized from aromatic dicarboxylic acids and bisphenols and from aromatic oxycarboxylic acids. The most promising in this respect appear to be the systems containing thionyl chloride and arenesulfonyl chlorides. Unfortunately, however, all the above-mentioned systems are poorly effective for direct polyesterification of aromatic dicarboxylic acids and glycols and are ineffective in the reactions with the participation of aliphatic dicarboxylic acids. These problems are still to be solved.

10.4.4 Polyheteroarylenes

The synthesis of polyheteroarylenes by means of direct polycondensation is a new trend in direct polycondensation. Several possible methods for effecting such syntheses are known. On the one hand, polyheteroarylenes can be obtained by using monomers containing heterocycles in their structure. In this case the polymer synthesis reaction comprises a further version of reactions already considered, *e.g.* direct polyamidation. In this way poly(amidophenylquinoxaline)s have been obtained.[181]

Varieties of direct polyamidation are also represented by the formation of poly(amidohydrazide)s,[124,182,183] polyhydrazides[124,182,183] and poly(amido acid)s,[184] *i.e.* polymers formed at the first stage of a two-stage technique for the synthesis of polyheteroarylenes. Cyclization of the obtained polymers at the second stage results in poly(amidooxadiazole)s, poly(oxadiazole)s and polyimides respectively.

In the synthesis of poly(amidohydrazide)s and polyhydrazides condensing systems are used which are known as efficient direct polyamidation systems: triphenyl phosphite/pyridine/LiCl,[124] triphenyl phosphite/LiCl[182] and diphenyl phosphite/pyridine/LiCl.[183] In all cases the reaction is performed in N-methylpyrrolidone.

Poly(amidohydrazide)s with the highest molecular weights have been obtained with the use of diphenyl phosphite, pyridine and LiCl[183] (equation 19). Thus, as a result of direct polycondensation of terephthalic acid with p-aminobenzoic acid hydrazide at 120 °C for 4 h, poly(amidohydrazide) with an intrinsic viscosity of 4.68 dl g^{-1} was obtained.

$$n\text{HOCR'COH} + n\text{NH}_2\text{-C}_6\text{H}_4\text{-CNH}_2\text{NH}_2 + 2n\text{HOP(OPh)}_2 \xrightarrow{\text{py: LiCl}}$$

$$\left[-\text{CR'CNH-C}_6\text{H}_4\text{-CNH}_2\text{NH}-\right]_n + 2n(\text{HO})_2\text{P(OPh)} + 2n\text{PhOH} \quad (19)$$

$$R' = -\text{C}_6\text{H}_4-, -\text{C}_6\text{H}_4-, (-\text{CH}_2-)_4$$

All attempts to synthesize polyhydrazides by means of direct polycondensation have resulted in polymers with a relatively low molecular weight.[124,182,183] Thus, with the help of the system considered to be the best for the synthesis of these polymers (based on triphenyl phosphite and LiCl), a polyhydrazide with an intrinsic viscosity of 0.82 dl g^{-1} was obtained at 140 °C after 2 h from terephthalic acid and isophthalic acid hydrazide.

Direct polycondensation of aromatic tetracarboxylic acids with aromatic diamines in the presence of triphenyl phosphite or diphenyl sulfite and tertiary amine produced high molecular poly(amido acid)s.[184]

On the other hand, the synthesis of polyheteroarylenes by direct polycondensation in one stage, based on monomers containing functional groups capable of cyclization, appears to be very promising. A mixture of phosphorus pentoxide with methanesulfonic acid has proved to be a very effective condensing agent for such reactions. With the help of this system the reaction of 3,3'-diaminobenzidine tetrachlorohydrate with aliphatic dicarboxylic acids at 120–140 °C produced high molecular poly(benzimidazole)s.[185] Under similar conditions the reaction of 2,5-diamino-1,4-benzenedithiol dihydrochloride with aliphatic and aromatic dicarboxylic acids made it possible to synthesize high molecular poly(benzothiazole)s.[186]

The data presented in Section 10.4 thus show that noticeable progress has been made in the development of direct polycondensation techniques for the synthesis of polymers. Very effective condensing systems have been created, making it possible to obtain high molecular polymers in sufficiently mild conditions with high yield. However, a high consumption of activating reagents, their relatively high cost, the necessity of purifying the polymer after synthesis and a number of other factors, including ecological ones, hinder extensive practical application of the developed systems. A serious limitation for a wider use of the method is the low molecular weight of the resultant fully aliphatic polyamides and polyesters, as well as polyarylates based on aliphatic dicarboxylic acids.

It can be assumed that the solution of these problems will be associated with developing new, highly selective reaction systems distinguished by being readily available, highly efficient and regeneratable. A promising approach lies in developing fundamentally new activating reagents that occupy an intermediate place between the true catalysts and the conventional systems of direct polycondensation.

10.5 MATRIX POLYCONDENSATION

Matrix polycondensation is a new, as yet not very well studied, technique of polymer synthesis by means of solution polycondensation. In matrix polycondensation the polymer synthesis is achieved in the presence of previously formed macromolecules (polymeric matrices), which raises the reaction rate and the molecular weight of the polymers formed. The introduced matrix exerts its effect by interacting with the monomer or the formed macromolecule, which can lead to the activation of reacting groups and an increase in their local concentration, as well as to the spatial orientation of reagents, convenient for the interaction.

Extensive potentialities of matrix polycondensation were convincingly demonstrated in direct polycondensation. By conducting the synthesis in the presence of poly(ethylene oxide),[132,159,166] poly(4-vinylpyridine),[131,159] poly(2-vinylpyridine)[159] and poly(vinylpyrrolidone)[133,151,152] it was possible to obtain polymers with a higher molecular weight. In particular, the use of the above polymers as matrices has resulted in the formation of polyamides with a markedly higher molecular weight (Table 3).

Table 3 Results of the Synthesis of Polyamides by Direct Polyamidation of Terephthalic Acid with 4,4'-Diaminodiphenylmethane in the Presence of Polymeric Matrices

Polymeric matrix		Polyamide	Ref.
Structure	Molecular weight	intrinsic viscosity	
Poly(ethylene oxide)	3×10^5	1.10	a
Poly(4-vinylpyridine)	1×10^5	1.17	b
Poly(vinylpyrrolidone)	4×10^5	1.55	c
Without polymeric matrix		0.33	d

[a] F. Higashi, Y. Nakano, M. Goto and H. Kakinoki, *J. Polym. Sci., Polym. Chem. Ed.*, 1980, **18**, 1099.
[b] F. Higashi, M. Goto, Y. Nakano and H. Kakinoki, *J. Polym. Sci., Polym. Chem. Ed.*, 1980, **18**, 851.
[c] F. Higashi and Y. Taguchi, *J. Polym. Sci., Polym. Chem. Ed.*, 1980, **18**, 2875.
[d] N. Yamazaki, M. Matsumoto and F. Higashi, *J. Polym. Sci., Polym. Chem. Ed.*, 1975, **13**, 1373.

As exemplified by direct polycondensation of amino acids, the polymeric matrix, poly(vinylpyrrolidone), was shown to interact with condensing reagents (triphenyl phosphite and LiCl) and the monomer.[152] The activation and the increased local concentration of amino acid molecules influence not only the reaction rate and the molecular weight of the formed polypeptide, but also the reaction direction, leading to the formation of a linear polymer. The matrix effect increases noticeably with increasing molecular weight of poly(vinylpyrrolidone).[152]

Interesting possibilities of matrix polycondensation have been demonstrated in the reactions with the participation of monomers activated by the presence of heteroatoms or groups in their structure.[187-194] Among such compounds are diethyl tartrate (13), diethyl muconate (14), diethyl chelidonate (15) and others. The compounds that have been investigated as polymeric matrices include poly(2-vinylpyridine)s and poly(4-vinylpyridine)s, poly(vinylpyrrolidone), poly(vinyl alcohol), polyacrylonitrile, poly(nitrostyrene), sugars, 4-vinylpyridine copolymers with styrene and others. In the presence of a polymeric matrix the rate of the reaction of dimethyl tartrate and dimethyl muconate with hexamethylenediamine in dimethyl sulfoxide at 60 °C proved to increase with increasing molecular weight of the introduced polymer.[187-189,191,192,194]

(15)

When the polymeric matrix is sensitive to external effects, *e.g.* to UV light, this leads to an even greater matrix effect. This kind of dependence was established to exist for the polycondensation of diethyl chelidate with diamines in dioxane at 30 °C in the presence of poly(vinylcarbazole).[190]

A necessary condition for the matrix to influence the polycondensation process is its interaction with the monomers, reacting groups or fragments of the formed macromolecules. The bonds arising here can be of a highly varied nature: coordination, donor–acceptor, hydrogen and other bonds. Insertion of different groups, capable of such interactions, into the polymeric matrix raises the matrix effect on polycondensation. Thus, the polycondensation of 3-hydroxybutyric acid derivatives containing nucleic bases with diamines is noticeably accelerated when a polystyrene matrix carrying a complementary base is added to the solution.[195]

A multifunctional polymer, polyacrylic acid, has recently been shown to exert great influence on the results of urea polycondensation with formaldehyde.[196,197]

All these results show that matrix polycondensation has good prospects as a method of polymer synthesis, though numerous problems are still unresolved. It is, for instance, absolutely unclear what the possibilities of polymeric matrices are in the synthesis of polymers with a preassigned structure, in stereospecific processes, *etc.* Another problem to be taken care of is that of clearing the obtained polymer from the matrix. The answer to all these questions will make it possible to define more clearly the role and the place of matrix polycondensation in the overall list of methods used in the chemistry of high molecular weight compounds.

10.6 OTHER METHODS

Along with such all-embracing methods as the acceptor-catalytic, activated, direct and matrix polycondensation there exist a number of other solution polycondensation techniques that are either of a more specific nature or have not yet been studied well enough to be regarded as being of importance. Among such methods are the following: polycondensation in an ordered medium, precipitation polycondensation and electrochemical polycondensation.

10.6.1 Polycondensation in an Ordered Medium

In 1973 Oakenfull[198] found that an increase in the length of aliphatic radicals in amines and esters brings about a very significant growth of the aminolysis reaction rate, which can be associated with the formation of an organized, ordered medium, owing to the aggregation of the aliphatic portion of the molecules.

This effect was successfully utilized to obtain high molecular polypeptides, *e.g.* as in refs. 199–203. Thus the polycondensation of hexadecyl and octadecyl esters of alanine and glycine in monolayers obtained from benzene solution on an aqueous substrate was found to proceed at room temperature in the absence of a catalyst.[201] The polyamidation rate in the monolayer is considerably higher than the polycondensation rate in the melt, which is caused by the formation of organized assemblies with an arrangement of functional groups that is conducive to polyamidation (equation 20).

$$\text{wwww} = C_{18}H_{37}$$

Although this way of conducting the reaction does not fully conform to the concepts of solution polycondensation, it can nevertheless be assumed that in the conditions of a lyotropic, organized polycondensation system a similar situation can arise.

Indeed, it has been recently found that when tricyclodecenetetracarboxylic acid dianhydride is subjected to polycondensation with diamines in dioxane at 50–70 °C, in the conditions of mesophase formation, the poly(amido acid) macromolecules are intensively growing.[204] This fact makes it possible to hope that in the not too distant future polycondensation in an ordered medium will disclose its secrets to the investigators and occupy the relevant place it deserves among the other polycondensation techniques.

10.6.2 Precipitation Polycondensation

In the 1960s the idea was advanced of the possibility of synthesizing high molecular weight compounds under the conditions of precipitation polycondensation.[205] This idea is founded on the concept that in the polycondensation process, firstly, all the polymer formation reactions are reversible, and secondly, in the precipitated polymer the growth of macromolecules stops. Then, in order to shift the equilibrium towards the formation of high molecular polymers, it is necessary either to remove effectively the low molecular products (acid, water, *etc.*) from the reaction medium or to select a precipitant–solvent system in which macromolecules with a high molecular weight would not dissolve and would therefore precipitate. When this happens in the unprecipitated polymer, as a result of chain propagation reactions, the equilibrium MWD will be restored, and the formed macromolecules with a high molecular weight will again precipitate from the reaction mixture. At the end of polycondensation all of the polymer formed will then have an average degree of polycondensation considerably higher than in the equilibrium process.

It was later established, however, that the growth of macromolecules in the precipitated polymer does not always stop. This was shown to be true for acceptor-catalytic polyesterification[29] and high temperature polyheterocyclization.[22] At present the term 'precipitation polycondensation' is applied to variants of solution polycondensation when the polymer precipitates from the solution at the very beginning of the reaction and continues to increase its molecular weight.[22] This process has unquestionable technological advantages since the stage of precipitating the polymer from the solution is not needed here.

10.6.3 Electrochemical Polycondensation

In contrast to electrochemical polymerization, electrochemical polycondensation is at present only at an early stage in its development. In the electrochemical process of polymer production the active particles are being generated by the participation of the monomer, the solvent and a specially introduced electrochemical reagent in the electrode process.[206,207]

An example of successful electrochemical polycondensation is the electro-oxidative reaction of 2,6-disubstituted phenols on a platinum electrode, as a result of which poly(2,6-disubstituted 1,4-phenylene oxide)s have been obtained.[207] In the presence of a solvent capable of being electrolyzed (dichloroethane), and with an increased current density, the polymer yield increases. It is important that, in contrast to oxidative polycondensation, electro-oxidative polycondensation conducted in the presence of the CuCl/pyridine catalytic system can produce polymers based on 2,6-dichlorophenol and unsubstituted phenol.[207] This means that the electrochemical methods of synthesis do not simply duplicate the conventional methods of polymer production, but expand their scope.

It can be assumed that, despite all the complexity of electrochemical reactions, in the near future the electrochemical polycondensation technique will find a wider application in the synthesis of various polymers.

The method of polymer synthesis founded on the use of enzymes also has considerable potential. It is not difficult to visualize enzymatic polycondensation as providing wide opportunities for the synthesis of polymers, having very high reaction rates and extremely high selectivity. At present, however, only a few individual examples of polymer synthesis with the help of enzymes are known.[208]

10.7 REFERENCES

1. P. W. Morgan, 'Condensation Polymers: by Interfacial and Solution Methods', Interscience, New York, 1965.
2. V. V. Korshak and S. V. Vinogradova, 'Equilibrium Polycondensation', Nauka, Moscow, 1968.

3. V. V. Korshak and S. V. Vinogradova, 'Nonequilibrium Polycondensation', Nauka, Moscow, 1972.
4. L. B. Sokolov, 'The Fundamentals of Polymer Synthesis by the Polycondensation Method', Khimiya, Moscow, 1979.
5. J. Ferguson, in 'Macromolecule Chemistry', 1984, vol. 3, p. 76.
6. V. V. Korshak, *Usp. Khim.*, 1984, **53**, 3.
7. V. A. Sergeev and L. I. Vdovina, *Vysokomol. Soedin., Ser. A*, 1984, **26**, 2019.
8. I. K. Miller and J. Zimmerman, *ACS Symp. Ser.*, 1985, **285**, 159.
9. W. G. Dorner, *Kunst. Plast.*, 1985, **32**, 26.
10. S. V. Vinogradova, *Vysokomol. Soedin., Ser. A*, 1985, **27**, 2243.
11. V. V. Korshak and N. M. Kozyreva, *Usp. Khim.*, 1985, **54**, 1841.
12. Y. Imai and M. Ueda, *Yuki Gosei Kagaku Kyokaishi*, 1982, **40**, 58.
13. V. V. Korshak, *Vysokomol. Soedin., Ser. A*, 1982, **24**, 1571.
14. V. V. Korshak, *Acta Polym.*, 1983, **34**, 603.
15. L. B. Sokolov, *Plast. Massy*, 1983, no. 10, 3.
16. S. V. Vinogradova, in 'Advances in Polymer Chemistry', ed. V. V. Korshak, MIR Publishers, Moscow, 1986, p. 75.
17. A. L. Rusanov, in 'Advances in Polymer Chemistry', ed. V. V. Korshak, MIR Publishers, Moscow, 1986, p. 159.
18. V. N. Ignatov, V. A. Vasnev and S. V. Vinogradova, *Vysokomol. Soedin., Ser. A*, 1987, **29**, 899.
19. F. J. Serna, J. G. de la Campa and J. de Abajo, *Rev. Plast. Mod.*, 1986, **37**, no. 360, 735.
20. J. de Abajo, F. J. Serna and J. G. de la Campa, *Rev. Plast. Mod.*, 1986, **37**, no. 361, 71.
21. S. Nozawa, *Kobunshi*, 1986, **35**, 1074.
22. A. L. Rusanov, *Vysokomol. Soedin., Ser. A*, 1986, **28**, 1571.
23. P. M. Hergenrother, *Polym. J.*, 1987, **19**, 73.
24. F. Akutsu, H. Takeyama, M. Miura and K. Nogakubo, *Makromol. Chem.*, 1985, **186**, 483.
25. V. V. Korshak and A. L. Rusanov, in 'Polymer Yearbook', ed. R. Pethrick and G. Zaikov, Academic Press, New York, 1986, vol. 3, p. 115.
26. V. P. Shibayev and N. A. Platé, *Zh. Vses. Khim. Ova.*, 1983, **28**, 165.
27. C. S. Marvell and A. Kotch, *J. Am. Chem. Soc.*, 1951, **73**, 1100.
28. H. Schell, *Angew. Chem.*, 1956, **68**, 633.
29. V. A. Vasnev and S. V. Vinogradova, *Usp. Khim.*, 1979, **48**, 30.
30. S. V. Vinogradova, V. A. Vasnev and V. V. Korshak, *Vysokomol. Soedin., Ser. B*, 1967, **9**, 522.
31. V. V. Korshak, V. A. Vasnev, S. V. Vinogradova and A. V. Vasil'ev, *Vysokomol. Soedin:, Ser. A*, 1974, **16**, 502.
32. S. V. Vinogradova, V. A. Vasnev, T. I. Mitaishvili and A. V. Vasilev, *J. Polym. Sci., Ser. A-1*, 1971, **9**, 3321.
33. S. V. Vinogradova, T. I. Mitaishvili, V. A. Vasnev, V. V. Korshak and M. B. Melamud, *Vysokomol. Soedin., Ser. A*, 1971, **13**, 912.
34. V. V. Korshak, V. A. Vasnev, M. G. Keshelava, S. V. Vinogradova and L. N. Gvozdeva, *Vysokomol. Soedin., Ser. A*, 1978, **20**, 139.
35. V. A. Vasnev, A. I. Tarasov, S. V. Vinogradova and V. V. Korshak, *Vysokomol. Soedin., Ser. A*, 1975, **17**, 1212.
36. S. V. Vinogradova, V. V. Korshak, A. V. Vasil'ev and V. A. Vasnev, *Vysokomol. Soedin., Sér. A*, 1973, **15**, 2015.
37. V. V. Korshak, S. V. Vinogradova, V. A. Vasnev and T. I. Mitaishvili, *Vysokomol. Soedin., Ser. A*, 1969, **11**, 81.
38. V. V. Korshak, S. V. Vinogradova and V. A. Vasnev, *Faserforsch. Textiltech.*, 1977, **28**, 491.
39. V. V. Korshak, S. V. Vinogradova, V. A. Vasnev, E. B. Musayeva, A. P. Gorshkov, G. K. Semin and L. N. Gvozdeva, *Dokl. Akad. Nauk. SSSR*, 1976, **226**, 350.
40. V. V. Korshak, S. V. Vinogradova, V. A. Vasnev and A. V. Vasil'ev, *J. Polym. Sci., Polym. Lett. Ed.*, 1972, **10**, 429.
41. W. Deits and O. Vogl, *J. Macromol. Sci., Chem.*, 1981, **16**, 1145.
42. P. M. Gomez and O. Vogl, *Polym. J.*, 1986, **18**, 429.
43. K.-S. Kim, *J. Appl. Polym. Sci.*, 1983, **28**, 1119.
44. A. Natansohn, *J. Appl. Polym. Sci.*, 1986, **32**, 2961.
45. J. Preston and W. B. Black, *Polym. Prepr., Am. Chem. Soc., Div. Polym. Chem.*, 1976, **17**, 40.
46. L. B. Sokolov, V. D. Gerasimov, V. M. Savinov and V. K. Belyakov, 'Thermostable Aromatic Polyamides', Khimiya, Moscow, 1975.
47. A. A. Fedorov, V. M. Savinov, L. B. Sokolov, M. L. Zlatogorsky and V. S. Grechishkin, *Vysokomol. Soedin., Ser. B*, 1973, **15**, 74.
48. A. A. Fedorov, V. M. Savinov, L. B. Sokolov and I. G. Lukyanenko, *Vysokomol. Soedin., Ser. B*, 1969, **11**, 129.
49. V. S. Naumov and L. B. Sokolov, *Kinet. Katal.*, 1971, **12**, 762.
50. L. B. Sokolov, Yu. B. Rotenberg and V. M. Savinov, *Dokl. Akad. Nauk SSSR*, 1980, **250**, 1402.
51. B. Jingsheng, Y. Anji, Z. Shengqing, Z. Shufan and H. Chang, *J. Appl. Polym. Sci.*, 1981, **26**, 1211.
52. V. I. Logunova, L. B. Sokolov and V. M. Savinov, *Vysokomol. Soedin., Ser. A*, 1976, **18**, 450.
53. V. S. Naumov, L. B. Sokolov, D. F. Sokolova and N. V. Novozhilova, *Vysokomol. Soedin., Ser. A*, 1969, **11**, 2141.
54. V. M. Savinov, L. B. Sokolov and V. M. Ivanov, *Plast. Massy*, 1967, no. 6, 25.
55. O. Ya. Fedotova, V. V. Korshak and E. I. Nesterova, *Vysokomol. Soedin., Ser. A*, 1973, **15**, 80.
56. V. M. Ivanov, V. M. Savinov and L. B. Sokolov, *Vysokomol. Soedin., Ser. A*, 1978, **20**, 1722.
57. V. M. Savinov, Yu. B. Rotenberg and L. B. Sokolov, *Vysokomol. Soedin., Ser. B*, 1981, **23**, 151.
58. L. B. Sokolov, *Vysokomol. Soedin., Ser. A*, 1971, **13**, 1425.
59. A. Kehayoglou, G. Karayannidis and I. Sideridou-Karayannidou, *Makromol. Chem.*, 1982, **183**, 293.
60. F. R. Diaz, L. H. Tagle and M. E. Padilla, *J. Polym. Sci., Polym. Chem. Ed.*, 1985, **23**, 2043.
61. F. R. Diaz, L. H. Tagle, A. Godoy, C. Hodgson and J. P. Olivares, *J. Polym. Sci., Polym. Chem. Ed.*, 1985, **23**, 2757.
62. J. Gopal and M. Srinivasan, *J. Polym. Sci., Polym. Chem. Ed.*, 1986, **24**, 1577.
63. M.-A. Kakimoto, Y. S. Negi and Y. Imai, *J. Polym. Sci., Polym. Chem. Ed.*, 1985, **23**, 1797.
64. Y. Imai, N. N. Maldar and M.-A. Kakimoto, *J. Polym. Sci., Polym. Chem. Ed.*, 1985, **23**, 2077.
65. M.-A. Kakimoto, Y. S. Negi and Y. Imai, *J. Polym. Sci., Polym. Chem. Ed.*, 1986, **24**, 1511.
66. C. G. Overberger and J. Sebenda, *J. Polym. Sci., Part A-1*, 1969, **7**, 2875.
67. N. Ogata, K. Sanui and K. Jijima, *J. Polym. Sci., Polym. Chem. Ed.*, 1973, **11**, 1095.
68. M. Ueda, M. Okada and Y. Imai, *J. Polym. Sci., Polym. Chem. Ed.*, 1976, **14**, 2665.
69. M. Ueda, A. Sato and Y. Imai, *J. Polym. Sci., Polym. Chem. Ed.*, 1979, **17**, 783.

70. R. D. Katsarava, D. P. Kunchulia, L. M. Avalishvili, G. G. Andronikashvili and M. M. Zaalishvili, *Vysokomol. Soedin., Ser. A*, 1979, **21**, 2696.
71. M. S. Jacovic, J. Djonlagic and R. W. Lenz, *Polym. Bull. (Berlin)*, 1982, **8**, 295.
72. B. A. Zhubanov and N. P. Lyubchenko, *Vysokomol. Soedin., Ser. A*, 1982, **24**, 1474.
73. R. D. Katsarava, D. P. Kharadze, L. M. Avalishvili and M. M. Zaalishvili, *Makromol. Chem., Rapid Commun.*, 1984, **5**, 585.
74. M. Ueda, A. Sato and Y. Imai, *J. Polym. Sci., Polym. Chem. Ed.*, 1978, **16**, 475.
75. K. Sanui, S. Tanaka and N. Ogata, *J. Polym. Sci., Polym. Chem. Ed.*, 1977, **15**, 1107.
76. M. Ueda, Y. Miyazawa, A. Sato and Y. Imai, *Polym. J.*, 1976, **8**, 609.
77. M. Ueda, A. Sato and Y. Imai, *J. Polym. Sci., Polym. Chem. Ed.*, 1979, **17**, 2013.
78. M. Ueda, T. Harada, S. Aoyama and Y. Imai, *J. Polym. Sci., Polym. Chem. Ed.*, 1981, **19**, 1061.
79. M. Ueda, K. Seki and Y. Imai, *Macromolecules*, 1982, **15**, 17.
80. C. Lu, P. Liu and C. Hu, *J. Polym. Sci., Polym. Chem. Ed.*, 1981, **19**, 2091.
81. R. D. Katsarava, D. P. Kharadze and L. M. Avalishvili, *Makromol. Chem.*, 1986, **187**, 2053.
82. R. D. Katsarava, D. P. Kharadze, L. M. Avalishvili, T. N. Omiadze and M. M. Zaalishvili, *Vysokomol. Soedin., Ser. B*, 1986, **28**, 518.
83. R. D. Katsarava, D. P. Kharadze, L. M. Avalishvili and M. M. Zaalishvili, *Vysokomol. Soedin., Ser. B*, 1982, **24**, 198.
84. R. D. Katsarava, T. M. Kartvelishvili and M. M. Zaalishvili, *Vysokomol. Soedin., Ser. B*, 1981, **23**, 460.
85. R. D. Katsarava, T. M. Kartvelishvili, Yu. A. Davidovich, M. M. Zaalishvili and S. V. Rogozhin, *Dokl. Akad. Nauk SSSR*, 1982, **266**, 363.
86. H. A. Staab, *Chem. Ber.*, 1957, **90**, 1326.
87. K. Kuze, *Kogyo Kagaku Zasshi*, 1969, **72**, 1603 (*Chem. Abstr.*, 1969, **71**, 12 504).
88. N. Ogata, K. Sanui and K. Konishi, *Kobunshi Kagaku*, 1973, **30**, 202.
89. V. V. Korshak, T. M. Frunze, V. V. Kurashev, L. B. Danilevskaya and T. V. Volkova, *Vysokomol. Soedin., Ser A*, 1975, **17**, 1409.
90. M. Ueda, S. Aoyama and Y. Imai, *Makromol. Chem.*, 1979, **180**, 2807.
91. N. Ogata, K. Sanui and T. Nohmi, *J. Polym. Sci., Polym. Chem. Ed.*, 1974, **12**, 1327.
92. Y. Imai, H. Okunoyama and M. Ohkoushi, *Polym. J.*, 1975, **7**, 130.
93. Y. Imai, *Kobunshi*, 1978, **27**, 723.
94. M. Ueda, T. Harada, S. Aoyama and Y. Imai, *J. Polym. Sci., Polym. Chem. Ed.*, 1981, **19**, 1061.
95. M. Ueda, K. Seki and Y. Imai, *Macromolecules*, 1982, **15**, 17.
96. P. Ferruti, M. C. Tanzi, L. Rusconi and R. Cecchi, *Makromol. Chem.*, 1981, **182**, 2183.
97. L. Rusconi, M. C. Tanzi, C. Zambelli and P. Ferruti, *Polymer*, 1982, **23**, 1689.
98. Y. Saegusa, S. Nakamura, N. Chau and Y. Iwakura, *J. Polym. Sci., Polym. Chem. Ed.*, 1983, **21**, 637.
99. J. F. Klebe, *Adv. Org. Chem.*, Interscience, New York–London, 1972, **8**, 98.
100. J. F. Klebe, *J. Polym. Sci., Part A*, 1964, **2**, 2673.
101. R. D. Katsarava and T. M. Kartvelishvili, *Vysokomol. Soedin., Ser. B*, 1986, **28**, 377.
102. H. R. Kricheldorf and G. Schwarz, *Makromol. Chem.*, 1983, **184**, 475.
103. M.-A. Kakimoto, Y. Oishi and Y. Imai, *Makromol. Chem., Rapid Commun.*, 1985, **6**, 557.
104. Y. Oishi, M.-A. Kakimoto and Y. Imai, *Macromolecules*, 1987, **20**, 703.
105. H. R. Kricheldorf and G. Schwarz, *Polym. Bull. (Berlin)*, 1979, **1**, 383.
106. J. R. Bouser, P. J. Williams and H. Kurz, *J. Org. Chem.*, 1983, **48**, 4111.
107. H. R. Kricheldorf and G. Bier, *J. Polym. Sci., Polym. Chem. Ed.*, 1983, **21**, 2283.
108. H. R. Kricheldorf and G. Bier, *Polymer*, 1984, **25**, 1151.
109. N. Yamazaki and F. Higashi, *Adv. Polym. Sci.*, 1981, **38**, 1.
110. M. Ueda, *Kobunshi*, 1986, **35**, 131.
111. N. Ogata and H. Tanaka, *Polym. J.*, 1971, **2**, 672.
112. N. Ogata and H. Tanaka, *Polym. J.*, 1972, **3**, 365.
113. Yu. V. Mitin and O. V. Glinskaya, *Tetrahedron Lett.*, 1969, **60**, 5267.
114. N. Ogata and H. Tanaka, *Polym. J.*, 1974, **6**, 461.
115. N. Ogata, K. Sanui and M. Harada, *J. Polym. Sci., Polym. Chem. Ed.*, 1979, **17**, 2401.
116. N. Yamazaki and F. Higashi, *Tetrahedron Lett.*, 1975, no. 5, 415.
117. N. Yamazaki and F. Higashi, *Bull. Chem. Soc. Jpn.*, 1973, **46**, 1235.
118. N. Yamazaki and F. Higashi, *Bull. Chem. Soc. Jpn.*, 1973, **46**, 1239.
119. N. Yamazaki, F. Higashi and J. Kawabata, *Makromol. Chem.*, 1974, **175**, 1825.
120. N. Yamazaki and F. Higashi, *J. Polym. Sci., Polym. Lett. Ed.*, 1974, **12**, 185.
121. N. Yamazaki, F. Higashi and J. Kawabata, *J. Polym. Sci., Polym. Chem. Ed.*, 1974, **12**, 2149.
122. N. Yamazaki, M. Niwano, J. Kawabata and F. Higashi, *Tetrahedron*, 1975, **31**, 665.
123. N. Yamazaki, M. Matsumoto and F. Higashi, *J. Polym. Sci., Polym. Chem. Ed.*, 1975, **13**, 1373.
124. J. Preston and W. Hofferbert, *J. Polym. Sci., Polym. Symp.*, 1978, **65**, 13.
125. F. Higashi, Y. Aoki and Y. Taguchi, *Makromol. Chem., Rapid Commun.*, 1981, **2**, 329.
126. F. Higashi, S. Ogata and Y. Aoki, *J. Polym. Sci., Polym. Chem. Ed.*, 1982, **20**, 2081.
127. J. Asrar, J. Preston and W. R. Krigbaum, *J. Polym. Sci., Polym. Chem. Ed.*, 1982, **20**, 79.
128. J. Preston, W. R. Krigbaum and R. Kotek, *J. Polym. Sci., Polym. Chem. Ed.*, 1982, **20**, 3241.
129. W. R. Krigbaum, R. Kotek, Y. Mihara and J. Preston, *J. Polym. Sci., Polym. Chem. Ed.*, 1984, **22**, 4044.
130. F. Higashi, M. Goto and H. Kakinoki, *J. Polym. Sci., Polym. Chem. Ed.*, 1980, **18**, 1711.
131. F. Higashi, M. Goto and H. Kakinoki, *J. Polym. Sci., Polym. Chem. Ed.*, 1980, **18**, 851.
132. F. Higashi, Y. Nakano, M. Goto and H. Kakinoki, *J. Polym. Sci., Polym. Chem. Ed.*, 1980, **18**, 1099.
133. F. Higashi and Y. Toguchi, *J. Polym. Sci., Polym. Chem. Ed.*, 1980, **18**, 2875.
134. G.-C. Wu, H. Tanaka, K. Sanui and N. Ogata, *J. Polym. Sci., Polym. Lett. Ed.*, 1981, **19**, 343.
135. G.-C. Wu, H. Tanaka, K. Sanui and N. Ogata, *Polym. J.*, 1982, **14**, 571.
136. G.-C. Wu, H. Tanaka, K. Sanui and N. Ogata, *Polym. J.*, 1982, **14**, 797.
137. N. Ogata, K. Sanui and Sh. Tan, *Polym. J.*, 1984, **16**, 569.

138. M. Ueda and N. Kawahazasaki, *Makromol. Chem., Rapid Commun.*, 1983, **4**, 801.
139. M. Ueda and A. Mochizuki, *Macromolecules*, 1985, **18**, 2353.
140. M. Ueda, S. Aoyama, M. Konno and Y. Imai, *Makromol. Chem.*, 1978, **179**, 2089.
141. Y. Imai, S. Aoyama, T.-Q. Nguyen and M. Ueda, *Makromol. Chem., Rapid Commun.*, 1980, **1**, 655.
142. F. Higashi, *Kobunshi*, 1986, **35**, 1098.
143. H. Tanaka, G.-C. Wu, Y. Iwanaga, K. Sanui and N. Ogata, *Polym. J.*, 1982, **14**, 331.
144. H. Tanaka, G.-C. Wu, Y. Iwanaga, K. Sanui and N. Ogata, *Polym. J.*, 1982, **14**, 635.
145. P. Strohriegl, W. Heitz and G. Weber, *Makromol. Chem., Rapid Commun.*, 1985, **6**, 111.
146. M. Ueda, H. Oikawa, N. Kawaharasaki and Y. Imai, *Bull. Chem. Soc. Jpn.*, 1983, **56**, 2483.
147. M. Ueda, N. Kawaharasaki and Y. Imai, *Bull. Chem. Soc. Jpn.*, 1984, **57**, 85.
148. C. H. Bamford, A. Elliot and T. Homby, 'Synthetic Polypeptides', Academic Press, New York, 1956.
149. L. Le Cuilly, A. Brack and G. Spach, *Makromol. Chem.*, 1978, **179**, 2829.
150. N. Yamazaki, J. Kawabata and F. Higashi, *J. Polym. Sci., Polym. Chem. Ed.*, 1977, **15**, 1511.
151. F. Higashi, K. Sano and H. Kakinoki, *Polym. Prepr. Jpn.*, 1979, **28**, 52.
152. F. Higashi, K. Sano and H. Kakinoki, *J. Polym. Sci., Polym. Chem. Ed.*, 1980, **18**, 1841.
153. N. Yamazaki, F. Higashi and T. Iguchi, *Tetrahedron Lett.*, 1974, no. 13, 1191.
154. N. Yamazaki, T. Tomioka and F. Higashi, *Bull. Chem. Soc. Jpn.*, 1976, **49**, 3104.
155. C. I. Chiriac, *Polym. Bull. (Berlin)*, 1986, **15**, 65.
156. C. I. Chiriac, *Polym. Bull. (Berlin)*, 1986, **15**, 143.
157. F. Higashi, N. Kokubo and M. Goto, *J. Polym. Sci., Polym. Chem. Ed.*, 1980, **18**, 2879.
158. N. Ogata, K. Sanui, H. Tanaka and S. Yasuda, *Polym. J.*, 1981, **13**, 989.
159. S. Yasuda, G.-C. Wu, H. Tanaka, K. Sanui and N. Ogata, *J. Polym. Sci., Polym. Chem. Ed.*, 1983, **21**, 2609.
160. F. Higashi, K. Kubota, M. Sekizuka and M. Goto, *J. Polym. Sci., Polym. Lett. Ed.*, 1980, **18**, 385.
161. F. Higashi, Y. Ito and K. Kubota, *Makromol. Chem., Rapid Commun.*, 1981, **2**, 29.
162. F. Higashi, K. Kubota, M. Sekizuka and M. Higashi, *J. Polym. Sci., Polym. Chem. Ed.*, 1981, **19**, 2681.
163. Sh. Kitayama, K. Sanui and N. Ogata, *J. Polym. Sci., Polym. Chem. Ed.*, 1984, **22**, 2705.
164. F. Higashi, A. Hoshio and J. Kiyoshige, *J. Polym. Sci., Polym. Chem. Ed.*, 1983, **21**, 3241.
165. F. Higashi, A. Hoshio, Y. Yamada and M. Ozawa, *J. Polym. Sci., Polym. Chem. Ed.*, 1985, **23**, 69.
166. F. Higashi and Y. Yamada, *J. Polym. Sci., Polym. Chem. Ed.*, 1985, **23**, 2709.
167. F. Higashi, Y. Yamada and A. Hoshio, *J. Polym. Sci., Polym. Chem. Ed.*, 1984, **22**, 2181.
168. F. Higashi, A. Hoshio and H. Ohtani, *J. Polym. Sci., Polym. Chem. Ed.*, 1984, **22**, 3983.
169. Y. Imai, S. Aoyama, T.-Q. Nguyen and M. Ueda, *Makromol. Chem., Rapid Commun.*, 1980, **1**, 655.
170. H.-G. Elias and R. J. Warner, *Makromol. Chem.*, 1981, **182**, 681.
171. H.-G. Elias and J. A. Palacios, *Makromol. Chem.*, 1985, **186**, 1027.
172. F. Higashi and T. Mashimo, *J. Polym. Sci., Polym. Chem. Ed.*, 1986, **24**, 1697.
173. F. Higashi, *Kobunshi*, 1986, **35**, 1098.
174. H. Tanaka, Y. Iwanaga, G. Wu, K. Sanui and N. Ogata, *Polym. J.*, 1982, **14**, 643.
175. F. Higashi, N. Akiyama and T. Koyama, *J. Polym. Chem., Polym. Chem. Ed.*, 1983, **21**, 3233.
176. F. Higashi, N. Akiyama, I. Takahashi and T. Kayama, *J. Polym. Sci., Polym. Chem. Ed.*, 1984, **22**, 1653.
177. F. Higashi, I. Takahashi, N. Akiyama and T. Chang, *J. Polym. Sci., Polym. Chem. Ed.*, 1984, **22**, 3607.
178. F. Higashi and T. Mashimo, *J. Polym. Sci., Polym. Chem. Ed.*, 1985, **23**, 2999.
179. F. Higashi, T. Mashimo, I. Takahashi and N. Akiyama, *J. Polym. Sci., Polym. Chem. Ed.*, 1985, **23**, 3095.
180. F. Higashi and T. Mashimo, *J. Polym. Sci., Polym. Chem. Ed.*, 1985, **23**, 2715.
181. V. V. Korshak, E. S. Krongauz and N. M. Belomoina, *Vysokomol. Soedin., Ser. B*, 1984, **26**, 793.
182. F. Higashi and N. Kokubo, *J. Polym. Sci., Polym. Chem. Ed.*, 1980, **18**, 1639.
183. F. Higashi and M. Ishikawa, *J. Polym. Sci., Polym. Chem. Ed.*, 1980, **18**, 2905.
184. G. I. Nosova, M. M. Koton, L. A. Laius, Yu. N. Sazanov, V. M. Denisov, P. P. Nefedov and M. A. Lazareva, *Vysokomol. Soedin., Ser. A*, 1985, **27**, 812.
185. M. Ueda, M. Sato and M. Mochizuki, *Macromolecules*, 1985, **18**, 2723.
186. M. Ueda, S. Yokote and M. Sato, *Polym. J.*, 1986, **18**, 117.
187. N. Ogata, K. Sanui, H. Nakamura and M. Kuwahara, *Polym. Prepr., Am. Chem. Soc., Div. Polym. Chem.*, 1979, **20**, 463.
188. N. Ogata, K. Sanui, H. Nakamura and H. Kishi, *J. Polym. Sci., Polym. Chem. Ed.*, 1980, **18**, 933.
189. N. Ogata, K. Sanui, H. Nakamura and M. Kuwahara, *J. Polym. Sci., Polym. Chem. Ed.*, 1980, **18**, 939.
190. N. Ogata, K. Sanui and M. Abe, *J. Polym. Sci., Polym. Chem. Ed.*, 1981, **19**, 1361.
191. N. Ogata, K. Sanui, H. Tanaka, H. Matsuo and F. Iwaki, *J. Polym. Sci., Polym. Chem. Ed.*, 1981, **19**, 2609.
192. N. Ogata, K. Sanui, F. Iwaki and A. Nomiyama, *J. Polym. Sci., Polym. Chem. Ed.*, 1984, **22**, 793.
193. N. Ogata, K. Sanui, M. Yoshikawa, H. Baba and K. Goto, *Polym. J.*, 1985, **17**, 821.
194. N. Ogata, K. Sanui, M. Yoshikawa and Y. Saigou, *Polym. J.*, 1985, **17**, 1221.
195. M. Hattori and M. Kinoshita, *Mem. Fac. Eng., Osaka City Univ.*, 1979, **19**, 211.
196. A. A. Litmanovich, S. V. Markov and I. M. Papisov, *Dokl. Akad. Nauk SSSR*, 1984, **278**, 676.
197. A. A. Litmanovich, S. V. Markov and I. M. Papisov, *Vysokomol. Soedin., Ser. A*, 1986, **28**, 1271.
198. D. Oakenfull, *J. Chem. Soc., Perkin Trans. 2*, 1973, **7**, 1006.
199. K. Hanabusa, Y. Miwa, K. Kondo and K. Takemoto, *Makromol. Chem., Rapid Commun.*, 1980, **1**, 433.
200. K. Hanabusa, K. Kondo and K. Takemoto, *Makromol. Chem.*, 1981, **182**, 9.
201. K. Fukuda, Y. Shibasaki and H. Nakahara, *J. Macromol. Sci., Chem.*, 1981, **15**, 999.
202. K. Hanabusa, H. Shirai, K. Hojo, K. Kondo and K. Takemoto, *Makromol. Chem.*, 1982, **183**, 1101.
203. K. Kondo, Y. Miwa and K. Takemoto, *Makromol. Chem.*, 1983, **184**, 1171.
204. B. A. Zhubanov, V. A. Solomin, E. N. Lyakh and A. Sh. Cherdabayev, *Vysokomol. Soedin., Ser. B*, 1987, **29**, 24.
205. E. Turska and A. Dems, *J. Polym. Sci., Part C*, 1968, no. 22, 407.
206. D. C. Phillips, S. Spewock and W. M. Alvino, *J. Polym. Sci., Polym. Chem. Ed.*, 1976, **14**, 1137.
207. E. Tsuchida, H. Nishide and T. Maekawa, *J. Macromol. Sci., Chem.*, 1984, **21**, 1081.
208. S. Fukui, *Kobunshi*, 1986, **35**, 535.

11
Interfacial and Dispersion Polymerization

VASILI V. KORSHAK and VALERI A. VASNEV

USSR Academy of Sciences, Moscow, USSR

11.1 INTRODUCTION	167
11.2 CLASSIFICATION	168
11.3 INTERFACIAL POLYCONDENSATION	170
11.3.1 Liquid–Liquid Systems	170
11.3.1.1 Governing factors	171
11.3.1.2 Non-aqueous systems	174
11.3.1.3 Interfacial catalysis	175
11.3.1.4 Synthesis of polymers	178
11.3.2 Gas–Liquid Systems	180
11.3.2.1 Governing factors	181
11.3.2.2 Possibilities of synthesis	182
11.3.3 Solid–Liquid Systems	183
11.4 DISPERSION POLYCONDENSATION	184
11.4.1 Emulsion Polycondensation	184
11.4.1.1 Governing factors	184
11.4.1.2 Synthesis of polymers	187
11.4.2 Suspension Polycondensation	187
11.4.2.1 Governing factors	188
11.4.2.2 Possibilities of synthesis	191
11.5 CONCLUSIONS	193
11.6 REFERENCES	193

11.1 INTRODUCTION

The method of interfacial polycondensation was first applied by Einhorn[1] in 1898 to obtain a polycarbonate by means of the interaction of phosgene dissolved in toluene with a hydroquinone water–alkali solution. Subsequently, however, this work was forgotten, and only in the 1950s did it become clear that interfacial polycondensation opens up extensive possibilities for the synthesis of a great variety of polymers.[2,3]

The essence of interfacial polycondensation lies in the monomers being dissolved in immiscible liquids and the reaction proceeding at the interface or near it. If one of the monomers is gaseous, the polymer is formed at the gas–liquid interface.

In the 1960s, dispersion polycondensation gained wide acceptance; the growth of macromolecules proceeds in a liquid or solid disperse phase distributed in a liquid dispersion medium. These processes are called emulsion and suspension polycondensation, respectively.[4,5]

A distinguishing feature of polycondensation in the above heterogeneous systems is the high rate of the reaction, which is conducted, as a rule, at temperatures that are close to room temperature. This allows the use of thermally unstable monomers, preserves unsaturated bonds and other reactive groups in the macromolecules and avoids the thermal destruction of polymers. Moreover, interfacial polycondensation offers the possibility of manufacturing finished products (films, membranes, fibres, capsules) directly during the course of polymer formation.

Among the disadvantages of interfacial and dispersion polycondensation, the following are of the greatest relevance: the use of large quantities of organic solvents in conducting the process and in the subsequent purification of the polymer, toxicity of these solvents, the high energy costs associated with intensive stirring of the system, the necessity of removing completely the low molecular weight reaction products (inorganic salts) from the polymers and corrosion of the equipment.

Despite these shortcomings, the positive aspects of polycondensation in heterophase systems, its simplicity and high efficiency have served as a sound basis for intensive development of this method, enabling the synthesis of widely different types of polymers, including polyamides, polyesters and polyethers, poly(thioester)s and polysulfides, polysulfonamides, polyphosphonamides, polyurethanes, polyureas, poly(heteroarylene)s, organoelement polymers, *etc.*

The extensive possibilities of interfacial and dispersion polycondensation, and the promising future of these methods, have attracted the attention of researchers. In recent years, numerous publications on this problem have appeared, including a number of reviews[3, 5–11] and monographs.[2, 4, 12, 13]

11.2 CLASSIFICATION

Depending on both the structure of the monomers and the reaction conditions, the growth of macromolecules in interfacial and dispersion polycondensation takes place in different reaction zones: at the interface, or in a thin layer of either disperse phase or the dispersion medium, or in the bulk of the disperse phase or dispersion medium. This zone is comparatively easily identified for gas–liquid interfacial and suspension polycondensation (Scheme 1).

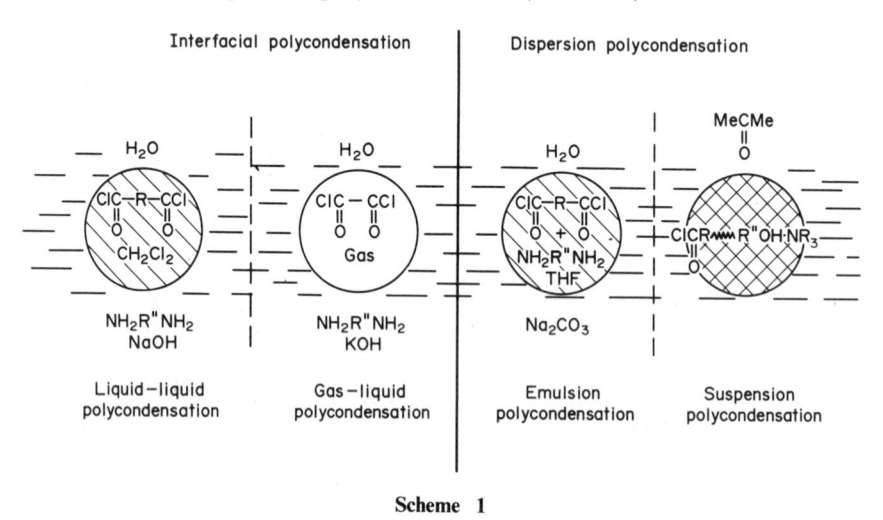

Scheme 1

In gas–liquid interfacial polycondensation, because of the high rate of the process and practically complete insolubility of the gaseous monomer in the dispersion medium (water), the reaction proceeds at the interface. However, the situation becomes more complicated if one uses less active monomers or if the dispersion medium is not water but an organic solvent in which the gaseous monomer is soluble.

In the case of suspension polycondensation, the macromolecules grow in the precipitated polymer, *i.e.* in a disperse phase. It should be noted that precipitation of the polymer occurs at high conversion of monomer, and the initial polycondensation stages proceed in solution. If a sufficiently large quantity of the formed polymer remains in solution, a process that can be regarded as intermediate between solution and suspension polycondensation takes place in the subsequent stages of the reaction.

It is more difficult to differentiate between interfacial polycondensation in a liquid–liquid system and emulsion polycondensation (Scheme 1).

In interfacial polycondensation, the interface or a part (layer) of a phase constitutes the reaction zone. In emulsion polycondensation, the whole volume of one of the phases constitutes the reaction zone. However, a clear-cut classification is complicated by the fact that intermediate variants are possible, when the reaction is taking place neither at the interface nor in the whole phase volume, but in a layer.

The depth of reaction zone penetration δ into the reacting phase depends on the effective diffusion coefficient D^*, as well as on the rate constant of the second-order reaction k and the monomer concentration $[M]$: $\delta = (D^* k^{-1} [M]^{-1})^{1/2}$.[14,15]

Depending on the relationship between the radius of drops r_d in the disperse phase and δ, different variants of the process are possible. Only two of them can be defined with sufficient clarity: $\delta > r_d$, emulsion polycondensation; and $\delta \ll r_d$, interfacial polycondensation. Acceleration of the mass transfer of monomer to the reaction zone (an increase in D^*), a decrease in polycondensation rate (a decrease in $k[M]$) and a decrease in the size of drops promote the emulsion nature of the process.

One of the experimentally determined criteria of the heterogeneity of a system, making possible its classification, is based on the proportionality between D^* and the distribution coefficient K_D of monomers in the disperse phase (concentration = C_1) and the dispersion medium (concentration = C_2): $K_D = C_1 C_2^{-1}$. With increasing K_D, the conditions are created for the transition from interfacial to emulsion polycondensation.

A changeover from the interfacial (surface) to the emulsion (volume) polycondensation process eventually depends on the type of solvent used, which influences the general course of the reaction and the K_D value. The use of solvents that are characterized by a high distribution coefficient value decreases the optimum ratio of monomers to the equimolar one and increases the molecular weight of the polymers formed (Figure 1). With increasing K_D, the process changes from interfacial (point 1) to emulsion (point 5).

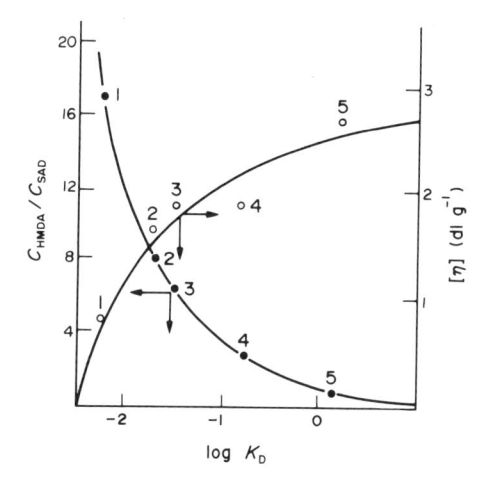

Figure 1 ($\bullet$) Optimum ratio between the concentrations of hexamethylenediamine and sebacoyl dichloride (C_{HMDA}/C_{SAD}) and ($\circ$) the intrinsic viscosity $[\eta]$ of polyamide vs. the coefficient of diamine distribution in the organic solvent–water system K_D.[2,4] Organic solvent: 1, cyclohexanone; 2, p-xylene; 3, CCl$_4$; 4, CCl$_4$–chloroform (30:70); 5, chloroform

The reaction zone can change its location not only in going from one solvent to another, but also with a change in the concentration of reagents. This is clearly illustrated by the results of the copolycondensation of tetra- and deca-methylenediamines with isophthaloyl dichloride in a chloroform–water system.[16] In this case, the copolyamide composition changes markedly with an increase in dichloride concentration from 0.01 to 0.2 mol l^{-1}, which is the consequence of a change in the microkinetic regime of the process.[17] With dichloride concentrations lower than 0.02 and higher than 0.15 mol l^{-1}, the copolymer contains a considerably larger amount of decamethylenediamine. The diphilic long-chain diamine being well adsorbed on the interface, such a composition of the copolymer indicates that the process is taking place on the surface. In the 0.02–0.15 mol l^{-1} range of dichloride concentration, the copolymer is enriched with tetramethylenediamine, which is indicative of the reaction taking place in a layer, or even in the bulk of the phase. This result demonstrates that the analysis of copolymer composition is one way of establishing the location of the reaction zone in interfacial polycondensation.

In addition to this, heterogeneous systems are classified in terms of the mutual dependence between the molecular weight of the polymer and the conversion yield, the ratio of monomers, the quantity of monofunctional impurity, etc.[4] All these methods use the deviation of the general rules of heterophase processes from those of the processes taking place in a homophase system. Thus, it is only in a heterogeneous system that one can obtain a high molecular weight polymer at low conversions of the reacting groups, i.e. at a low polymer yield (Figure 2). A high molecular weight at a low polymer yield indicates the diffusive nature of the process, since in this case the concentration

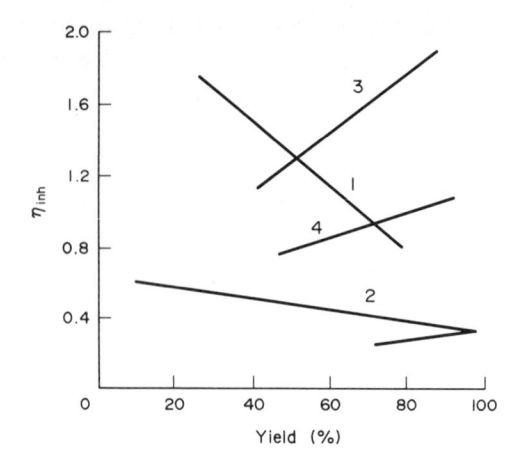

Figure 2 Inherent viscosity η_{inh} vs. the yield of polyamide for polycondensation in a liquid–liquid system.[18] Inherent viscosity and the yield of polymer are determined as their function of a polycondensation parameter (P). Systems: 1, hexamethylenediamine (water)–adipoyl dichloride (CCl_4), P = ratio of comonomers; 2, ethylenediamine hydrochloride (water)–terephthaloyl dichloride (benzene), P = alkali concentration in aqueous phase; 3, hexamethylenediamine (water)–terephthaloyl dichloride (CCl_4), P = ratio of monomer concentrations; 4, hexamethylenediamine (water)–sebacoyl dichloride (CCl_4), P = diamine concentration in aqueous phase

of monomer in the reaction zone will differ from the concentration in the whole volume, which is what actually leads to the above dependence.

For static conditions, i.e. in the absence of stirring, a clear idea of the zone where the reaction is taking place can be obtained by using dyed powders as markers.[2] Inserting such a powder at the interface between the two phases makes it possible to establish from what side, the aqueous or the organic, the polymer film is growing, i.e. in which zone the polycondensation is taking place.

The above methods do not exhaust all the possibilities of establishing the location of the reaction zone and the stage determining the reaction rate in a heterophase system. Using model compounds, conducting detailed spectral and kinetic studies and other approaches offer unlimited possibilities in this respect, which makes finding the optimum conditions for the process ultimately possible. Unfortunately, in recent years, studies in the field of interfacial and dispersion polycondensation have been mainly performed with practical purposes in mind, and have been concentrating on the synthesis of new polymers; in particular, liquid-crystalline polymers. It can be assumed, however, that the current fundamental research into interfacial catalysis, as well as the studies on kinetics and reaction mechanism, will provide a sound foundation for further development of both the theory and the practice of polycondensation in heterophase systems.

11.3 INTERFACIAL POLYCONDENSATION

The general concept of interfacial polycondensation in liquid–liquid and gas–liquid systems is now sufficiently comprehensive to allow formulation of the principal laws governing the process and utilize actively its extensive synthetic possibilities. However, a changeover to new reaction systems and the use of modern interfacial transfer catalysts raises highly diverse, and frequently quite complicated, problems for the researcher. This necessitates not only conducting new experiments but also performing a detailed comprehensive analysis of previously obtained results.

As is known, interfacial polycondensation can be conducted in heterophase systems both with and without stirring.[2] Systems without stirring find a limited application; they are used, for example, to obtain certain types of membranes.[3,19] The principal emphasis at present is placed on the interfacial processes conducted under intensive stirring.

11.3.1 Liquid–Liquid Systems

Interfacial polycondensation in a liquid–liquid system is one of the most widely used polycondensation techniques. As a rule, one of the solvents in the system is water. However, in a number of cases, interfacial polycondensation has been successfully performed in non-aqueous two-phase systems consisting only of organic solvents, which introduces certain peculiarities into the polymer synthesis process.

11.3.1.1 *Governing factors*

The principal laws governing the functioning of liquid–liquid systems without stirring were considered by Morgan.[2] At present, such processes are used much more rarely than those with stirring. That is why in this and subsequent sections most attention will be given to the systems with stirring.

Fundamental research into the general rules of interfacial polycondensation was carried out in the 1960s and 1970s.[2,4,12,13] In recent years, research in this direction has been primarily associated with studying the problem of interfacial catalysis and the synthesis of new polymers.

The kinetics of interfacial polycondensation are determined to a considerable extent by the structure of the initial compounds and the reaction conditions. The fastest processes have very high rate constants, equal to 10^2–10^6 $l\,mol^{-1}\,s^{-1}$.[2] Among such reactions is the interaction of aliphatic diamines with dichlorides of dicarboxylic acids. Common features of high-rate reactions in a liquid–liquid system are low activation energy values (from 4 to 40 $kJ\,mol^{-1}$) and strong exothermicity (up to 300 $kJ\,mol^{-1}$).

The rate of interfacial polycondensation is substantially dependent on the nature of the organic solvent. For instance, the rate constants of the reaction of benzoyl chloride with aniline rise substantially in the following series of solvents: benzene (1) < dichloromethane (2.8) < nitrobenzene (15.5) < tetrahydrofuran (67.6) < acetonitrile (138.1).[5,20]

Interfacial polycondensations of acyl dichlorides with diamines and bisphenols, in such low polarity solvents as CCl_4 and *n*-heptane, are characterized by fairly low rate constant values.[2,21,22] Thus, the rate constants of the polyamidation and polyesterification of terephthaloyl dichloride with piperazine and 2,2-bis(4-hydroxyphenyl)propane (bisphenol *A*) in the CCl_4–water system are equal to less than 10^{-2} $l\,mol^{-1}\,s^{-1}$.[21] Substitution of dihalogenated alkanes for acyl dichlorides decreases the interfacial polycondensation rate even further. This is indirectly confirmed by the considerable time (up to 12 h) and high temperature (85 °C) required for the reaction between 1,9-dibromononane and bisphenols in the nitrobenzene–water system.[23]

As already noted above, the rate of polycondenstion in a heterophase system can be influenced not only by kinetic but also by diffusion factors. A low rate of mass transfer to the reaction zone can be associated with the formation of a poorly permeable polymer film hindering the diffusion of monomers. Moreover, the mass transfer rate decreases at small values of the coefficient of monomer distribution between the disperse phase and the dispersion medium ($K_D \ll 1$). The relationship between reaction and mass transfer rates, in the final analysis, determines the reaction zone location: at the interface or in part of the volume of the phases. The problem is quite complicated, since the mechanism of interfacial polycondensation depends on numerous factors, including the sturcture of monomers, the interfacial transfer catalysts, *etc*.[2,4,12] This, in turn, influences the dependence of the yield and the molecular weight of the formed polymer on the reaction duration. Thus, in the interfacial polycondensation of adipoyl dichloride with hexamethylenediamine, a high molecular weight polyamide is formed at a comparatively low conversion of monomer.[2] The polymer yield rises with increasing reaction time. This result testifies to the fact that the process of polyamide formation is of a clearly defined diffusive nature.

In contrast to this, in the polycondensation of bisphenol *A* with the dichlorides of tere- and isophthalic acids in a dichloroethane–water system in the presence of benzyltriethylammonium chloride, the conversion of bisphenoxide in the initial stages of the reaction is very high (90%), while the polymer molecular weight is not high and increases gradually.[24] It can be assumed that, in this case, at least in the initial stages of the process, the polycondensation proceeds similarly to solution polycondensation. A sharp increase in the copolyester molecular weight in the final stages of the reaction, with conversion exceeding 98%, is associated by the authors with the local concentration effect, arising as a result of an increase in the organic phase viscosity caused by the increase in the polymer molecular weight. The diffusive nature of the process is indirectly confirmed by the dependence of the polymer molecular weight on the bisphenol:dichloride molar ratio. Polymers with a high molecular weight were obtained with the ratio of the above monomers equal to 1.6, *i.e.* much higher than equimolarity.[24]

The heterogeneous nature of interfacial polycondensation is the cause of an appreciable dependence of both the molecular weight and the yield of polymer on the surface tension at the interface. Addition of emulsifiers, *i.e.* surface-active substances reducing the surface tension, as a rule leads to an increase in the polymer yield.[4,12] The effect of surface tension on the polymer molecular weight is more complex and can result in both a direct and an inverse dependence.[2,4,12] The molecular weight dependence on surface tension has been most clearly established in the case of the polycondensation of aromatic diamines with dicarboxylic acid dichlorides. Aromatic diamines being surface-active substances, an increase in surface tension raises the concentration of these monomers at the interface, which facilitates the course of interfacial polycondensation. The data in Figure 3 corroborate this conclusion, according to which an increase in the surface tension between the phases leads to an increase in the molecular weight of polyamides.[25] The problems of interfacial catalysis in heterophase polycondensation are considered in Section 11.3.1.3.

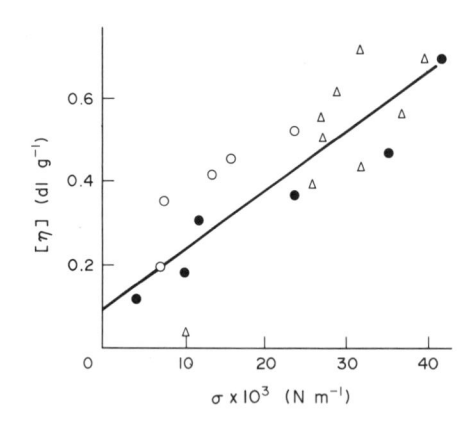

Figure 3 Intrinsic viscosity $[\eta]$ of poly(p-phenylene terephthalamide) $vs.$ the surface tension σ at the interface; σ was varied by adding emulsifiers to octane–water (○) and CCl_4–water (●) systems and by changing the nature of the organic phase (△)[25]

The size of the interfacial surface can be regulated by changing the intensity of stirring. As a rule, interfacial polycondensation is accomplished with fairly low rotational speeds of the stirrer (10^3–10^4 r.p.m.). With an increase in the stirring rate, the polymer yield and molecular weight usually increase.[2,4,12] However, this dependence can be of a more complex nature; it is therefore very important to study the effect exerted by the size of the interfacial surface on the polycondensation results. Of interest in this respect is the technique developed for determining the size of drops in the disperse phase and, correspondingly, the size of the interfacial surface.[26]

In interfacial polycondensation, the properties of the formed polymeric film can play a major role not only under static but also under dynamic conditions. In the latter case, the preservation of the film at the interface during the whole polycondensation process can be the reason for the diffusion of monomers through the polymer film becoming the rate-limiting stage.[27] It is interesting that a polymer obtained in thin layers of reagents (up to 10^3 Å) has a higher molecular weight than a polymer obtained with dropwise polycondensation.[28] Investigation of the process of acyl dichloride copolycondensation with diamine in a thin layer led to the conclusion that, at the beginning of the reaction, the limiting stage is the dichloride mass transfer from the organic disperse phase to the reaction zone. The formed diphilic macromolecules of polyamide lie between the aqueous and the organic phase, and the subsequent stages proceed according to the mechanism of diffusional deceleration through the swollen polymeric film.[28] Unfortunately, the data available at present do not allow the clear definition of the influence of both the structure of the polymers and the properties of the formed polymeric films on the mechanism of interfacial polycondensation.

The kinetics and mechanism of polycondensation in a two-phase system depend to a considerable extent on the nature of the solvents. One of the solvents is usually water, which dissolves both the reagents (diamines, bisphenoxides, acceptors) and the low molecular weight reaction products (salts) well. The availability and harmless ecological effects of water are also important factors. The organic solvent plays an important part in interfacial polycondensation: it determines the distribution of the reagents between the phases (Figure 1), the surface tension between the phases (Figure 3), the diffusion rate of reagents, the swelling capacity and permeability of the formed polymer, the extent of side reactions, $etc.$[2,4,12] Depending on the properties of the solvent used, the polymer can be present in the reaction system in the form of a swollen or an unswollen precipitate or in solution. There is a school of thought according to which the most favourable conditions are when the polymer swells, or is even soluble, in the organic phase. Sometimes a search for the optimum interfacial polycondensation conditions results in using a mixture of organic solvents as the organic phase. For instance, in the synthesis of poly(thioester)s from dithiols and acyl dichlorides, a mixture of benzene with n-hexane is used.[29,30]

Interfacial polycondensation is mostly conducted at temperatures that are close to room temperature. As a rule, a rise in temperature increases the contribution of side reactions (*e.g.* hydrolysis) and decreases the polymer yield and molecular weight. However, in each particular case the optimum temperature of the process can differ noticeably from room temperature, *e.g.* in the synthesis of various poly(thioester)s, the temperature used has been equal to 6 and 25 °C.[29,30] The use of poorly active monomers can markedly raise the reaction temperature.[23]

The heterogeneity of polycondensation in a liquid-liquid system is very clearly shown by the nature of the dependence between the molecular weight of the polymers and the ratio of the monomers. When the process is taking place in the diffusion region, a different ratio of the volume of the phases at a constant concentration of monomers in these phases does not affect the molecular weight of the formed polymer (Figure 4). As the diffusion rate depending only on the concentration of reagents, and these are the same, an excess of one of the monomers (*i.e.* excess volume of the phase) does not affect the molecular weight of the polymer.[31]

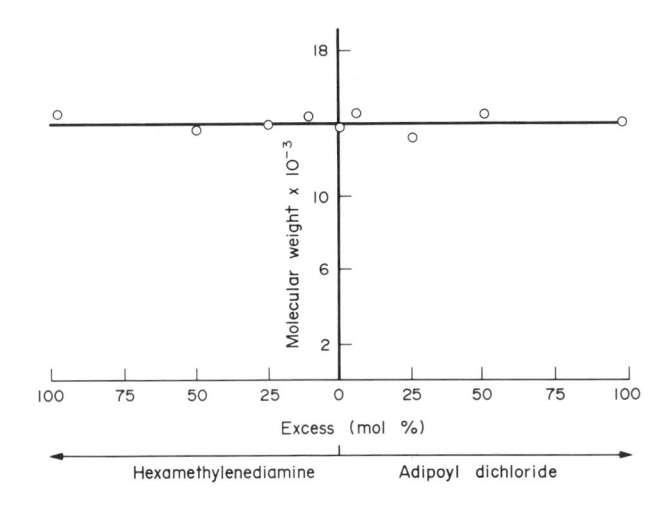

Figure 4 Influence of the ratio of monomers on the molecular weight of poly(hexamethyleneadipamide)[31]

A change in the concentration of monomer in the phases can noticeably affect the polymer weight. However, in contrast to polycondensation in a homogeneous system, the optimum polymer molecular weight is reached with a non-equimolar ratio of the reagents, and the effect of an excess of one of the monomers over its optimum quantity is not so pronounced.[2,4,12,24,29,30]

One of the conditions of successful interfacial polycondensation is the use of acceptors of the acid (HCl) evolved in the course of the reaction. The function of the acid acceptor in interfacial polycondensation can be varied. Strongly basic monomers, *e.g.* aliphatic diamines, are converted into poorly active salts under the action of acids. Acceptors, binding the acid, facilitate the existence of the monomer in its active form. Compared with diamines, bisphenols have a much lower nucleophilic reactivity. Under the action of the acid acceptor (a strong base) bisphenol is converted into a very active bisphenoxide anion that interacts with the second monomer, *e.g.* with acyl dichlorides.

The quantity of acceptor should not be lower than that required to neutralize the evolving acid, and may even exceed this quantity. Thus, in the synthesis of poly(thioester)s, a 100% excess of alkali is used.[29,30] However, as the acid can perform the functions of a catalyst, addition of a certain amount of acid to the aqueous phase at the beginning of the reaction raises the polymer molecular weight.[32] The use of organic bases (tertiary amines) as acceptors, when compared with inorganic bases, promotes, as a rule, the formation of polymers with lower yield and molecular weight.

As in any polycondensation process, in interfacial polycondensation the influence of different factors on the polymer molecular weight is eventually determined by the relationship between the rates of chain propagation and termination, *i.e.* by the contribution of side reactions. In the case of such monomers as acyl dichlorides, bis(chloroformate)s and diisocyanates, the hydrolysis of their functional groups represents the principal side-reaction. The ratio between the rates of chain propagation and hydrolysis is, as a rule, very high. In the case of polyamidation, they differ by more than two orders of magnitude.[33] It is, nevertheless, hydrolysis that can be the main cause of the polymers having a low molecular weight. The contribution of hydrolysis depends on both the reactivity of the initial monomers and the conditions under which the polycondensation is conducted: the nature of the solvent and the acid acceptor, the temperature, the concentration of reagents and other factors.[2,12,13]

Thus, the hydrolysis rate rises with an increase in temperature and increasing alkali content in the aqueous phase,[34] as well as with increasing organic solvent polarity.[35] In the latter case, an important role is played by the degree of acyl dichloride distribution between the phases. An increase in dichloride content in the aqueous phase promotes its hydrolysis. Various impurities, including monofunctional ones, as well as the oligomer cyclization processes, constitute an important factor influencing both the molecular weight and the molecular weight distribution of polymers.[2,4,12,13]

The synthesis of high molecular weight polymers being a problem of major importance, a researcher always faces the task of searching for the optimum polycondensation conditions. An attempt at solving this problem has resulted in the development of interfacial polycondensation in non-aqueous media.

11.3.1.2 Non-aqueous systems

The aim of conducting interfacial polycondensation in non-aqueous binary systems is to obtain high molecular weight polymers by reducing the contribution of hydrolysis reactions, as well as by raising the concentration of the monomer that is poorly soluble in water. Ethylene glycol, glycerol, acetonitrile and a number of other solvents are usually employed instead of water in two-phase systems. At present, polyamides and polyesters have been synthesized by this method (Table 1).

Table 1 Results of the Synthesis of Polymers by Interfacial Polycondensation in Non-aqueous Systems

Monomers		$[\eta]$ (dl g^{-1})	Ref.
Hexamethylenediamine + sebacoyl dichloride	Ethylene glycol–tetrachloroethylene	1.05	a
Hexamethylenediamine + sebacoyl dichloride	Glycerol–cyclohexane	1.11	a
Bisphenol A + terephthaloyl dichloride	Acetonitrile–CCl$_4$	0.9	b
Resorcinol + terephthaloyl dichloride	Acetonitrile–CCl$_4$	0.23	c

^a P. W. Morgan and S. L. Kwolek, *J. Polym. Sci.*, 1962, **62**, 33.
^b N. Ogata, K. Sanui, T. Onozaki and S. Imanishi, *J. Macromol. Sci., Chem.*, 1981, **A15**, 1059.
^c H. Nakamura, S. Imanishi, K. Sanui and N. Ogata, *Polym. J.*, 1979, **11**, 661.

In the case of polyamidation, polymers with quite high molecular weights have been obtained in the glycerol–cyclohexane and ethylene glycol–tetrachloroethylene systems.[36] For polyesterification, the acetonitrile–CCl$_4$ system[36,37] proved to be the most effective. In this system, polyesters whose molecular weight is sufficiently high are formed in a number of cases in quantitative yield within 5 min in the presence of triethylamine at room temperature.[37] By means of interfacial polycondensation in non-aqueous media, copolyterephthalates of bisphenol A and resorcinol have been synthesized.[36,37] The copolymers with the highest molecular weight ($[\eta] = 0.8$ dl g^{-1}) have been obtained at an equimolar ratio of bisphenols, since it is in this case that the formed copolyester swells best of all in the reaction system.[37] A different distribution of monomers, in particular bisphenols,[36] between the organic phases provides new opportunities for the synthesis of copolymers with a different macromolecular structure.

Interfacial polycondensation in non-aqueous media has the disadvantage of involving costly (as compared with water) and toxic organic solvents. Moreover, since low molecular weight salt products of the reaction have a poorer solubility in organic solvents, clearing the polymers of impurities that worsen their quality becomes more complicated. The future of interfacial polycondensation in non-aqueous media depends on the the results of further research that can reveal new specific features of this process.

11.3.1.3 Interfacial catalysis

In the 1950s, Schnell[39] used tertiary amines and quaternary ammonium salts to speed up interfacial polycondensation. This seems to have been the first example of using interfacial transfer catalysts in the synthesis of polymers (polycarbonates). In recent years, along with ammonium salts, phosphonium, arsonium, selenonium and sulfonium compounds and tertiary amines have been used as catalysts.[2]

In the 1960s, interfacial catalysis began to be intensively used to solve various synthetic problems, including problems of polymer synthesis. In recent years, systematic research has been carried out on the mechanism of this process, and the possibilities for its further development have been elaborated.[48-53] The use of interfacial catalysis in heterophase processes for the synthesis of polymers has considerably expanded their application.[48-53]

In interfacial polycondensation, as a result of this catalysis, macromolecule formation is accelerated, thus increasing polymer yield and molecular weight. The general principle of interfacial catalysis is presented in Scheme 2(a). The most frequently used onium compounds are those of ammonium and phosphonium. The onium cation (Q^+) interacts in the aqueous phase with the anion (Y^-), forming a Q^+Y^- complex, very soluble in the organic phase and easily passing into it. In the organic phase, RX reacts with Q^+Y^-, forming the reaction product and the initial Q^+X^- complex, which is distributed between the phases. It has been established that, along with onium salts, such compounds as crown ethers (cyclic ethers), cryptands (cyclic amino ethers) and podands [linear polyethers, e.g. poly(ethylene glycol)s] can be used as interfacial transfer catalysts.

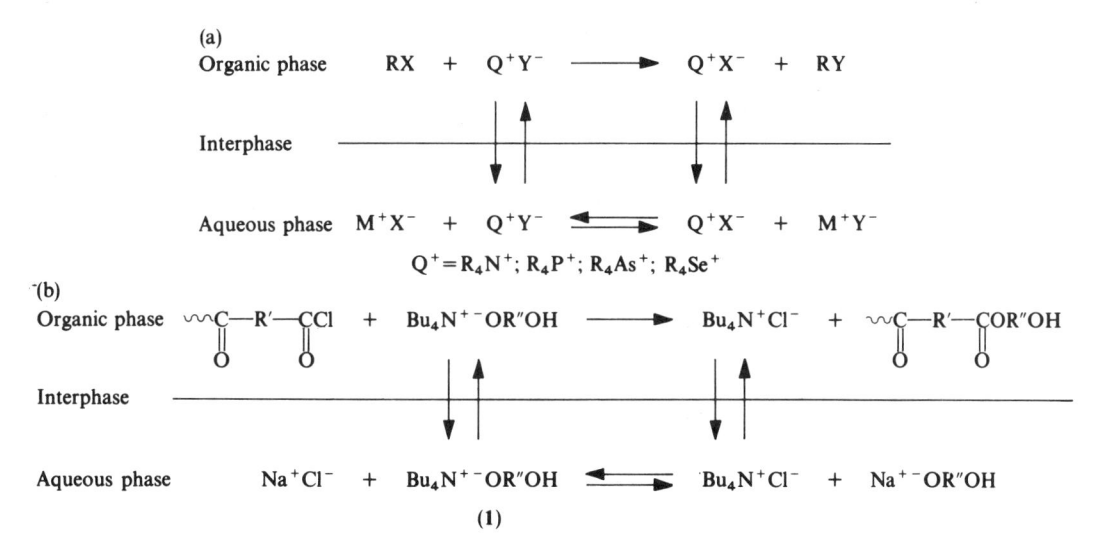

Scheme 2 Generalized phase-transfer reaction scheme (a) and biphasic reaction scheme of polyester synthesis (b) with onium salt

In interfacial polycondensation, e.g. interfacial polyesterification (Scheme 2b), conducted in the presence of a tetrabutylammonium salt, the bisphenoxide anion interacts in the aqueous phase with the onium compound, yielding a complex (1) that is readily soluble in the organic phase. This very fact causes the fast transfer of complex (1) into the organic phase where it reacts with the acyl chloride. Tetrabutylammonium chloride, formed as a result of this reaction, takes part again in the interfacial catalysis. This kind of mechanism was suggested for the interfacial polycondensation of isophthaloyl dichloride with bisphenols in the dichloromethane–water–NaOH system, conducted in the presence of benzyltriethylammonium chloride.[54,55]

The mechanism of the action of ammonium salts and other onium compounds seems to be more complicated. Along with an increase in the solubility of monomers and the interfacial transfer catalysis, the action of onium compounds can result in an increase in polymer solubility, emulsification of liquid phases, micellar catalysis, etc. The similarity of the structure of interfacial transfer catalysts and surface-active substances makes the idea of micellar catalysis in interfacial polycondensation quite feasible. In this catalysis, the reactions are being accelerated owing to an increase in both the concentration and the dissociation constants of the reagent molecules in the micelles. At present, however, no convincing proof has been found of micellar catalysis in interfacial polycondensation.

Moreover, the data available suggest that, in the presence of interfacial transfer catalysts, the reactions that follow the micellar catalysis mechanism are not taking place.[42] Indicative of this is a linear reaction rate vs. catalyst concentration dependence in the organic phase, as well as a high catalytic efficiency of salts that act poorly, or not at all (tetrabutylammonium chloride) as surface-active substances.[46]

Of considerable interest in this respect are the results obtained from studying the interfacial polycondensation of bisphenol A with terephthaloyl dichloride in the chloroform–water–NaOH system in the presence of different ammonium compounds and sodium dodecanesulfonate (SDS).[62] When small catalyst concentrations (5×10^{-4} mol l^{-1}) are used, an appreciable effect is only observed for decyltrimethylammonium bromide ($[\eta] = 1.58$ dl g^{-1}, whereas in the absence of catalyst $[\eta] = 0.50$ dl g^{-1}). Ammonium salts with shorter and longer chains (tetramethyl- and hexadecyltrimethyl-ammonium bromides), as well as SDS, do not raise the polymer molecular weight ($[\eta] = 0.48$ dl g^{-1}). An increase in the concentration of catalyst to 100×10^{-4} mol l^{-1}, when a strong emulsifying effect is observed, and above 240×10^{-4} mol l^{-1}, when micelles are formed, although promoting an increase in polymer molecular weight, does not raise it as much as one would expect in the case of micellar catalysis. Thus, a rise in dodecyltrimethylammonium bromide concentration from 5×10^{-4} to 100×10^{-4} and 248×10^{-4} mol l^{-1}, i.e. twenty- and fifty-fold, raises the polymer intrinsic viscosity from 0.58 to 1.05 and 1.27 dl g^{-1}. These results indicate the absence of the micellar effect in the reaction in question. It can nevertheless be assumed that under certain conditions the interfacial polycondensation processes will actually be following the micellar catalysis mechanism.

An important criterion, determining the choice of a certain salt as the catalyst, is the molecular weight of the formed polymer. In many respects, it depends on the type of chemical reaction and the particular conditions under which it is accomplished. Examples are known when the highest effectiveness is shown by one certain type of onium salt, e.g. arsonium, phosphonium or ammonium.[3, 7, 56, 57]

The reason for such differences in the effectiveness of salts could arise from a greater or lesser stability of the onium salt with respect to temperature or the presence of alkali. For instance, in the presence of hydroxide ions, phosphonium cations begin to decompose irreversibly at 50 °C, and in a strongly alkaline medium this decomposition takes place even at room temperature (equation 1).[58–67]

$$[R_4P^+]OH^- \longrightarrow R_3PO + RH \quad (1)$$

Ammonium salts are stable in a weakly alkaline medium at 80 °C. Under the action of alkali, ion exchange, e.g. OH$^-$ for Cl$^-$,[63] takes place in quaternary ammonium salts, which does not affect the ammonium cation catalytic activity, but can lead to further transformations. Under more severe conditions, the intensive decomposition of salts by the Hofmann reaction (equation 2) or the reverse Menschutkin reaction (equation 3) occur. Thus, benzyltriethylammonium chloride, when heated at 110 °C in concentrated alkali, is converted in high yield (72%) into N,N-diethylbenzylamine with 20 h.[64] At 40 °C, benzyltriethylammonium chloride is decomposed in a water–acetone–NaOH solution into triethylamine and benzyl chloride.[65] At room temperature in concentrated alkali solutions, ammonium salts remain stable for several days. It is necessary to bear in mind that under certain conditions onium salts can be formed in the course of the reaction. In particular, an ammonium salt is formed as a result of the interaction of triethylamine with chloroform.[66]

$$[C_6H_4CH_2\overset{+}{N}Et_3]OH^- \longrightarrow C_6H_4CH_2NEt_2 + CH_2=CH_2 + H_2O \quad (2)$$

$$[C_6H_4CH_2\overset{+}{N}Et_3Cl^-] \rightleftharpoons C_6H_4CH_2Cl + NEt_3 \quad (3)$$

An important problem, from the point of view of obtaining high molecular weight polymers, is that of the optimum quantity of onium catalyst. Usually this quantity is from 2 to 40 mol %, in terms of the monomer. In the interfacial polycondensation of bisphenol A with terephthaloyl dichloride, the optimum quantity of tetraethylammonium chloride in the reaction has been shown to equal 20% of the polymer weight.[3] In the absence of a catalyst, the polymer formed has the intrinsic viscosity of 0.36 dl g^{-1}. When this quantity rises to 50 wt %, the polymer intrinsic viscosity drops from 1.2 to 0.4 dl g^{-1}.

In interfacial polycondensation, compounds that form complexes with metal ions compete successfully with onium compounds. Such complex-forming compounds are crown ethers (e.g. **2**), cryptands (e.g. **3**) and podands (e.g. **4**). As seen from Scheme 3, both the cation and the anion pass from the aqueous into the organic phase by the action of a crown ether. In the case of an onium salt, only the anion passes into the organic phase (Scheme 2). Therefore, when the counterion plays an important part in the reaction the effectiveness of onium salts and crown ethers will differ noticeably.

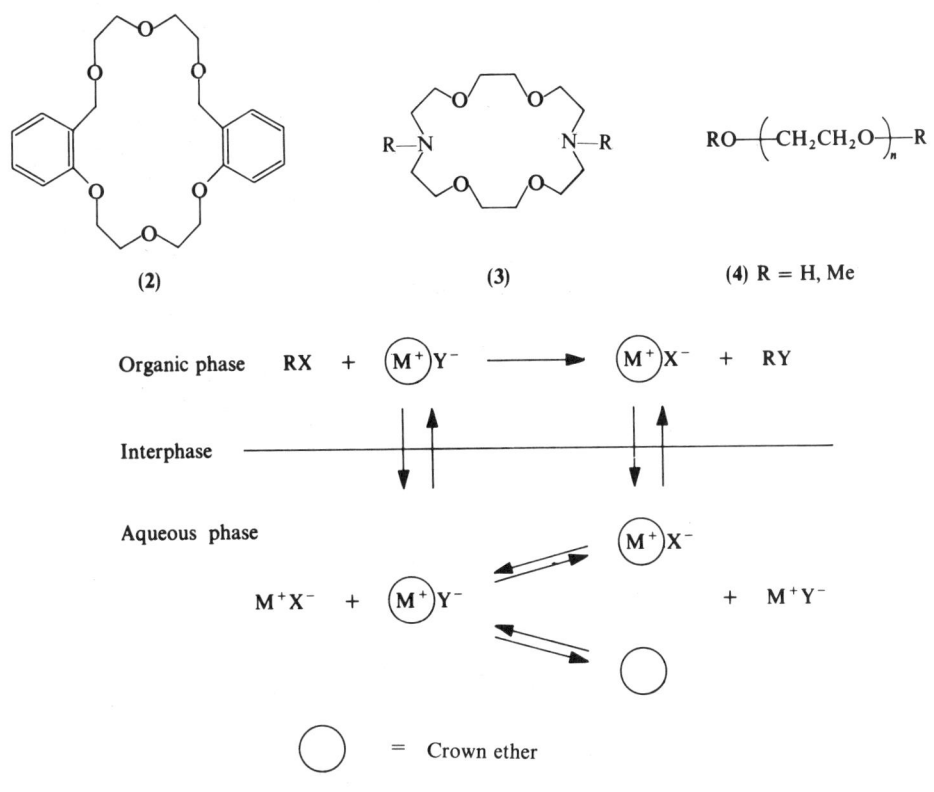

(2)

(3)

(4) R = H, Me

Organic phase   RX + (M$^+$)Y$^-$ ⟶ (M$^+$)X$^-$ + RY

Interphase

Aqueous phase

M$^+$X$^-$ + (M$^+$)Y$^-$ ⇌ (M$^+$)X$^-$ + M$^+$Y$^-$

◯ = Crown ether

Scheme 3 Generalized phase-transfer reaction scheme with crown ether

Interfacial transfer catalysts are, at present, chosen empirically. Table 2 shows the results of synthesizing various polymers by means of interfacial polycondensation in the presence of some ammonium salts and crown ethers. From the data in Table 2, it follows that, on the one hand, the use of interfacial transfer catalysts, in comparison with a non-catalytic process, results in the formation of polymers with considerably higher molecular weights. On the other hand, the catalyst efficiency depends in many respects on the type and conditions of the reaction. Thus, in the synthesis of aromatic polysulfonates, the most efficient catalyst was tetrabutylammonium chloride,[56] for aromatic polyphosphonates it was hexadecyltrimethylammonium chloride,[56] for aromatic polyethers it was dicyclohexyl-[18]-crown-6[67] and for aliphatic polysulfides, it was dibenzo-[18]-crown-6.[68] The reasons for such a difference have not been properly analyzed as yet, but without such information a judicious choice of the optimum catalyst is impossible.

Table 2 Results of the Synthesis of Polymers under the Conditions of Interfacial Catalysis

Catalysts	Aromatic polysulfonates [η] (dl g^{-1})	Ref.	Aromatic polyphosphonates [η] (dl g^{-1})	Ref.	Aromatic polyethers [η] (dl g^{-1})	Ref.	Aliphatic polysulfides [η] (dl g^{-1})	Ref.
Without catalyst	0.20	a	0.05	a	—	—	0.51	c
Tetrabutylammonium chloride	1.44	a	0.36	a	0.42	b	0.83	c
Hexadecyltrimethyl-ammonium chloride	1.19	a	0.70	a	0.26	b	0.58	c
[15]-Crown-5	1.31	a	0.21	a	0.36	b	0.73	c
Dibenzo-[18]-crown-6	0.95	a	0.59	a	0.37	b	0.93	c
Dicyclohexyl-[18]-crown-6	1.28	a	0.64	a	0.84	b	0.84	c

[a] Y. Imai, *J. Macromol. Sci., Chem.*, 1981, **A15**, 833.
[b] Y. Imai, M. Ueda and M. Ii, *J. Polym. Sci., Polym. Lett. Ed.*, 1979, **17**, 85.
[c] Y. Imai, A. Kato, M. Ii and M. Ueda, *J. Polym. Sci., Polym. Lett. Ed.*, 1979, **17**, 579.

It was established for the first time in 1979 that poly(ethylene glycol)s (PEG) can be used to catalyze interfacial transfer in interfacial polycondensation.[67] This was shown for the synthesis of aromatic polyethers in the CH_2Cl_2–water–KOH system in the presence of PEG with $\bar{M}_n$ from 600 to 20 000. Polymers with the highest molecular weight ($[\eta] = 0.47$ dl g^{-1}) were obtained with PEG having $\bar{M}_n = 2000$.[67]

As a result of the progress made in the field of interfacial catalysis, this method has been widely used in recent years to obtain a great variety of polymers. The potential of interfacial catalysis has not been exhausted, however, and in forthcoming years catalytic interfacial polycondensation will certainly develop further.

11.3.1.4 Synthesis of polymers

Interfacial polycondensation is a widely applied method of synthesizing numerous hetero- and carbo-chain polymers. The development of interfacial transfer catalysts has expanded substantially the possibilities of the method, which has begun to be used for the reaction between monomers with moderately active groups, e.g. dihaloalkanes and bisphenols. Many of the related problems of polymer synthesis have been analyzed in monographs.[2,12] In this section we shall therefore examine only the recent developments.

Polyamides are obtained by means of interfacial polycondensation from acyl dichlorides and diamines or their salts (equation 4). This method has recently been used to synthesize aliphatic and aromatic polyamides and copolyamides,[69-72] including N-methyl-substituted polyamides.[73,74] The nature of the organic solvent was found to influence the microstructure of N-methyl-substituted copolyamides. In going from n-hexane to chloroform, the structure of the copolymers changes from statistical to a block structure.[73] Filling aliphatic polyamides, e.g. nylon 6,6, with metal silicates under conditions of interfacial polycondensation makes it possible to control the thermal characteristics of polymers.[75] For instance, an increase in the degree of filling from 1 to 10% is accompanied, respectively, by an increase and a decrease in the temperature of composite melting and softening, as compared with the unfilled sample. A promising variant of the practical use of interfacial polycondensation under static conditions is the production of polyamide membranes on a polysulfonic substrate, which proved to be highly efficient in water purification.[19] The polyamide layer is obtained by impregnating the substrate successively, first with an aqueous solution of a diamine or poly(ethyleneimine) and an inorganic acceptor, and then with the solution of sebacoyl dichloride in n-hexane.

$$n\text{ClC(O)}-R'-\text{C(O)Cl} + n\text{NHR}-R''-\text{NHR} \xrightarrow[-2n\text{NaCl}]{2n\text{NaOH}} \text{\{C(O)}-R'-\text{C(O)NR}-R''-\text{NR\}}_n \quad (4)$$

R = H, Me

By means of interfacial polycondensation in a liquid–liquid system, polyurethanes, polyureas and polyhydrazides can be obtained.[2,14] The synthesis of N-methyl-substituted and unsubstituted polyurethanes (equation 5)[76] and polyhydrazides (equation 6)[77] made it possible to carry out a systematic study of the thermal properties of these polymers. Interesting possibilities of interfacial polycondensation were shown in the synthesis of polyurethanes by the interaction of diisocynates with diols, performed in the absence of a solvent under intentive stirring of the reaction system.[78]

$$n\text{ClC(O)O}-R'-\text{OC(O)Cl} + n\text{NHR}-R''-\text{NHR} \xrightarrow[-2n\text{NaCl}]{2n\text{NaOH}} \text{\{C(O)O}-R'-\text{OC(O)NR}-R''-\text{NR\}}_n \quad (5)$$

R = H, Me

$$n\text{ClC(O)}-R'-\text{C(O)Cl} + n\text{NHR}-\text{NMR}\cdot 2\text{HCl} \xrightarrow[-4n\text{NaCl}]{4n\text{NaOH}} \text{\{C(O)}-R'-\text{C(O)NR}-\text{NR\}}_n \quad (6)$$

R = H, Me

Interfacial polycondensation is widely used for the synthesis of a great variety of polyesters and poly(thioester)s: polycarbonates[76,79,80] and poly(thiocarbonate)s (equations 7 and 8),[81] polyarylates (polyesters based on bisphenols) (equation 9)[3,62,79,82,83] and others. To raise the polyesterification efficiency, the reaction is usually performed in the presence of interfacial transfer catalysts.

$$n\text{ClCCl} + n\text{HOR''OH} \xrightarrow[-2n\text{NaCl}]{2n\text{NaOH}} \text{\{COR''O\}}_n \qquad (7)$$
$$\underset{\text{X}}{\|} \hspace{4.5cm} \underset{\text{X}}{\|}$$

$$\text{X = O, S}$$

$$n\text{ClCO—R'—OCCl} + n\text{HOR''OH} \xrightarrow[-2n\text{NaCl}]{2n\text{NaOH}} \text{\{CO—R'—OCOR''O\}}_n \qquad (8)$$
$$\underset{\text{O}}{\|}\hspace{1cm}\underset{\text{O}}{\|} \hspace{4cm} \underset{\text{O}}{\|}\hspace{2cm}\underset{\text{O}}{\|}$$

$$n\text{ClC—R'—CCl} + n\text{HXR''XH} \xrightarrow[-2n\text{NaCl}]{2n\text{NaOH}} \text{\{C—R'—CXR''X\}}_n \qquad (9)$$
$$\underset{\text{O}}{\|}\hspace{1cm}\underset{\text{O}}{\|} \hspace{4cm} \underset{\text{O}}{\|}\hspace{2cm}\underset{\text{O}}{\|}$$

$$\text{X = O, S}$$

An interesting example of the use of interfacial catalysis is the synthesis of polycarbonate–polysiloxane block copolymers in a CH_2Cl_2–water–KOH system in the presence of tetraethylammonium chloride.[80] To produce poly(thiocarbonate)s from thiophosgene and bisphenol A in a CH_2Cl_2–water–NaOH system, tetrabutylammonium bromide[81] proved to be the most efficient catalyst out of the group of onium salts and crown ethers.

Interfacial transfer catalysts have been used to synthesize poly(thioester)s (equation 9)[29,30,84] and poly(amidoester)s.[85] In the case of poly(thioester)s, the use of the catalyst (benzyltriethylammonium bromide) produced both positive[30] and negative results. In the latter case, the polymer yield and molecular weight decreased in the presence of the catalyst.[29]

Interfacial catalysis has been used to obtain aromatic polysulfonates (equation 10)[7,56,82] and polyphosphonates (equation 11; Table 2).[7,86,87] It was shown for these polymers that the efficiency of the interfacial transfer catalyst is influenced by the nature of the alkali metal cation in the acid acceptor.[7] In systems with a crown ether, polymers with a higher molecular weight have been obtained with KOH, whereas in the case of quaternary ammonium salts the polymer molecular weight is only slightly affected by the type of metal hydroxide. This can be associated with the size of the cation and, therefore, its ability to form complexes with the crown ether, as well as with influence of the cation on the reaction course in the organic phase, which does not happen in the case of an ammonium catalyst (Schemes 2 and 3).

$$n\text{ClSO}_2\text{R'SO}_2\text{Cl} + n\text{HOR''OH} \xrightarrow[-2n\text{KCl}]{2n\text{KOH}} \text{\{SO}_2\text{R'SO}_2\text{OR''O\}}_n \qquad (10)$$

$$n\text{ClPCl}(\text{C}_6\text{H}_5) + n\text{HOR''OH} \xrightarrow[-2n\text{KCl}]{2n\text{KOH}} \text{\{POR''O(C}_6\text{H}_5)\}_n \qquad (11)$$

$$n\text{HalR'Hal} + n\text{HXR''XH} \xrightarrow[-2n\text{KCl}]{2n\text{KOH}} \text{\{R'XR''X\}}_n \qquad (12)$$

$$\text{Hal = Cl, Br;} \quad \text{X = O, S}$$

With the help of interfacial transfer catalysts, various polyethers and poly(thioether)s (equation 12) have been synthesized. The interaction of bis(4-chloro-3-nitrophenyl) sulfone with bisphenol A in the presence of different catalysts has produced aromatic poly(ether sulfone)s.[7,67] The polymer with the highest molecular weight is formed in a CH_2Cl_2–water–KOH system in the presence of dicyclohexyl-[18]-crown-6. Polyethers with a high molecular weight have also been obtained from the reaction of p-xylene dihalides with bisphenol A in the presence of different catalysts.[88,89] Under these conditions, benzyltriethylammonium chloride showed itself to be the most efficient catalyst.

A promising trend in the development of interfacial polycondensation lies in the synthesis of thermotropic liquid-crystalline polymers.[23,79,90] In particular, the interaction of bisphenol A and 4,4'-dihydroxybiphenyl with 1,9-dibromononane in a nitrobenzene (o-dichlorobenzene)–

water–NaOH system in the presence of a tetrabutylammonium salt has produced thermotropic copolyethers with a smectic mesophase.[23]

A method has been developed, based on interfacial catalysis, for the synthesis of poly(hydroxy ether)s from bisphenol A and α-epichlorohydrin (2-chloromethyloxirane) in a dioxane–water–KOH system in the presence of a catalyst (equation 13).[57] The polymer with the highest molecular weight ($[\eta] = 0.34$ dl g^{-1}) was formed in the presence of benzyltriethylammonium chloride; the authors, however, proposed the use of cheap and less toxic PEGs as catalysts. An increase in PEG $\bar{M}_n$ from 300 to 14 000 only influences the yield slightly (56–64%) and molecular weight ($[\eta] =$ 0.18–0.23 dl g^{-1}) of the poly(hydroxy ether). The developed process, as compared with the industrial one, takes place at a lower temperature (86 °C instead of 100 °C) and is much faster (6 h instead of 72 h).

$$n\text{HOR''OH} + n\,\underset{O}{\triangle}\!\!-\!\text{CH}_2\text{Cl} \xrightarrow[-n\text{KCl}]{n\text{KOH}} \left[\text{ROCH}_2\underset{\underset{\text{OH}}{|}}{\text{CH}}\text{CH}_2\text{O}\right]_n \quad (13)$$

The use of more active dithiols, instead of bisphenols, in interfacial polycondensation allows the synthesis of high molecular weight aliphatic and aromatic polysulfides (equation 12).[67,68,91–93] Conducting the reaction in the presence of interfacial transfer catalysts, primarily in the presence of crown ethers and ammonium salts, raises the rate and the molecular weight of the polymer formed.

Thus, the use of interfacial transfer catalysts raises considerably the efficiency of the reaction between dibromoalkanes and Na$_2$S dissolved in water.[92] Performing the polycondensation in the presence of hexadecyltrimethylammonium chloride raises the polysulfide intrinsic viscosity from 0.08 to 0.71 dl g^{-1}.

Various metal-containing polymers have been obtained by means of interfacial polycondensation,[94] including those containing titanium.[95–97] For instance, high molecular weight titanium-containing polyoximes with $\bar{M}_w$ up to 3×10^5 (equation 14) have been synthesized by the interaction of bis(cyclopentadienyl)titanium dichloride with dioximes in a chloroform–water–NaOH system.[96] The use of various dyes, with hydroxy and sulfonate groups, as nucleophilic monomers resulted in the formation of titanium-containing polymeric dyes.[97]

$$n\,\text{Cl}\underset{\underset{\text{Cp}}{|}}{\overset{\overset{\text{Cp}}{|}}{\text{Ti}}}\text{Cl} + n\text{HOR''OH} \xrightarrow[-2n\text{NaCl}]{2n\text{NaOH}} \left[\underset{\underset{\text{Cp}}{|}}{\overset{\overset{\text{Cp}}{|}}{\text{Ti}}}\text{OR''O}\right]_n \quad (14)$$

Cp = cyclopentadienyl group

With the help of interfacial polycondensation, one can obtain not only heterochain but also carbochain polymers. For the synthesis of such polymers it is necessary to use compounds with an activated CH$_2$ group (5). As a result of the reaction of compound (5) with p-xylene dichloride in a benzene–water–NaOH system in the presence of benzyltriethylammonium chloride, high molecular weight polymers (equation 15) are formed.[98–100]

$$n\text{ClCH}_2\!\!-\!\!\langle\!\!\bigcirc\!\!\rangle\!\!-\!\text{CH}_2\text{Cl} + n\underset{\underset{R}{|}}{\overset{\overset{\text{CN}}{|}}{\text{CH}}}_2 \xrightarrow[-2n\text{NaCl}]{2n\text{NaOH}} \left[\text{CH}_2\!\!-\!\!\langle\!\!\bigcirc\!\!\rangle\!\!-\!\text{CH}_2\underset{\underset{R}{|}}{\overset{\overset{\text{CN}}{|}}{\text{C}}}\right]_n \quad (15)$$

(5) R=Ph, CO$_2$CMe$_3$

The extensive synthetic possibilities of interfacial polycondensation are indicative of a promising future for a heterophase process in a liquid–liquid system. Further development of this method is to a great extent associated with progress in interfacial catalysis. It can be anticipated that the development of micellar catalysis will also facilitate the further progress of interfacial polycondensation.

11.3.2 Gas–Liquid Systems

When interfacial polycondensation is conducted in a gas–liquid system, one of the monomers is in the gaseous state and the other is in the liquid phase, usually as an aqueous solution (Scheme 1). In liquid-phase interfacial polycondensation, cases are also known when one of the monomers, *e.g.* phosgene, is in the gaseous state, but after its dissolution in one of the liquid phases the polymerization takes place in a liquid–liquid system at the interface or in a layer in one of the phases. In gas-phase polycondensation, polymer formation cannot take place in the gaseous phase. If an aqeuous solution of the monomer serves as the second phase, these reactions take place mainly at the interface.

In 1960, Sokolov and co-workers[17,101] were the first to use the method of gas-phase polycondensation for the synthesis of poly(oxamide)s, using gaseous oxalyl chloride with aqueous solutions of diamines.

11.3.2.1 *Governing factors*

In polycondensation at the liquid–gas interface, highly reactive pairs of monomers are used. Different acyl halides (oxalyl chloride, oxalyl fluoride, phosgene, thiophosgene, *etc.*), carbon suboxide, formaldehyde, *etc.* are used in the gaseous phase. In the liquid phase the aqueous solutions of diamines, aliphatic dithiols, and some other compounds can be used. The main governing factors of gas-phase polycondensation will be discussed using the oxalyl chloride–diamine system (hexamethylenediamine, phenylenediamine)[4,17] as an example.

The contact of monomers during the polycondensation in a gas–liquid system is achieved by sparging a gaseous monomer (in a mixture with nitrogen, air or some other inert diluent) through the solution of a second monomer. The formed polymeric film envelops the gas bubble passing through the liquid phase, where this film eventually remains.

High activity of the initial monomers determines a high polycondensation rate and, therefore, a short process duration, which is proportional to the height of the bubbling layer.[103,104] In the case of oxalyl chloride and hexamethylenediamine this height must be no less than 10 mm (contact time ~ 0.05 s).

An aqueous solution of one of the monomers most often used in gas-phase polycondensation. It is, however, possible to use organic solvents as well (Figure 5). The use of organic solvents makes it possible to reduce the contribution of the principal side-reaction, *i.e.* hydrolysis of the acyl halide, and also to use monomers that are insoluble in water. As can be seen from Figure 5, the nature of the solvent significantly affects the polycondensation results. Polyamides with the highest molecular weight ($\eta_{red} = 1.08$ dl g^{-1}) are formed in water. This result is due to a number of reasons: (1) Polyamidation in water proceeds at a very high rate.[105] (2) The gaseous monomer is practically insoluble in water, and it is the passage of this monomer into the organic phase that disturbs the equivalence of the reacting groups. Hydrolysis of the gaseous monomer in water reduces to a minimum its content in the liquid phase, which facilitates the process taking place on the surface. (3) Low molecular weight reaction products (salts) are readily soluble in water, which speeds up their withdrawal from the interface. (4) The formed polyamide readily swells in water, which assures a sufficient mobility of its macromolecules and promotes the chain propagation reactions.

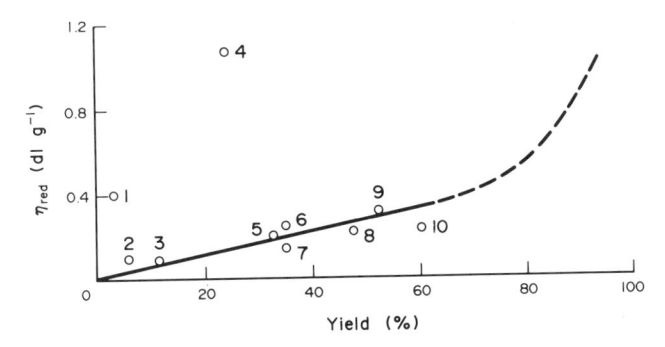

Figure 5 Reduced viscosity η_{red} *vs.* the yield of poly(hexamethyleneoxamide) obtained in different gas–liquid systems.[102] Liquid phase: 1, dimethylformamide; 2, ethanol; 3, *n*-butanol; 4, water; 5, dibutyl ether; 6, *p*-xylene; 7, dioxane; 8, nitrobenzene; 9, *n*-octane; 10, chlorobenzene.

Raising the temperature of gas-phase polycondensation increases the polymer yield and molecular weight.[106] This distinguishes it from polycondensation in a liquid–liquid system. The reason for this difference is, on the one hand, a decrease in the solubility of the gaseous monomer in the liquid phase, which reduces the contribution made by the side reaction of acyl halide hydrolysis, and, on the other hand, an increase in the mobility of the macromolecules.

Continuous sparging of the gaseous monomer through the aqueous solution of the other monomer excludes, in principle, the preservation of an equilvalent ratio between the interacting groups. However, the fact that polycondensation is taking place at the interface, *i.e.* the diffusive nature of the process, promotes the formation of a high molecular weight polymer, since there is a constant inflow of fresh portions of monomer from the gaseous and liquid phases. This mechanism of gas-phase polycondensation allows the synthesis of polymers with high molecular weights but in small yields (Figure 5).

The dependence of polymer yield and molecular weight on both the monomer concentration in the liquid phase and the partial pressure of the gaseous monomer has the shape of a curve with a maximum. The maximum of polyamide molecular weight corresponds to the diamine concentration that is 30–80 times higher than the monomer concentration in the gaseous phase.[102] The optimum concentration values depend on the activity of the reagents and the conditions of the process. For instance, for poly(*p*-phenyleneoxamide) the optimum concentrations lie within the interval of 0.02–0.05 mol l^{-1}.[70]

Both the nature of the acid acceptor and the pH of the aqueous phase produce an appreciable effect on the polymer molecular weight.[106,107] In the case of aliphatic poly(oxamide)s, the best results have been obtained in a weakly alkaline medium.[106] The aromatic polyamides with the maximum yield and molecular weight are formed in an acid medium at pH 4–5.

A high rate of polymer formation and the absence of an organic solvent in the system undoubtedly constitute the advantages of gas-phase polycondensation. Nevertheless, by no means all the reaction systems can be used within the framework of this method. In particular, the considerable solubility of phosgene in water prevents its gas-phase polycondensation with bisphenols or aromatic diamines, whose activity is inferior to that of aliphatic diamines. A low pressure of the gaseous monomer vapour also prevents a successful reaction. Nevertheless, the synthetic possibilities of polycondensation in a gas–liquid system are sufficiently high, and further studies in this field could considerably expand its scope.

11.3.2.2 Possibilities of synthesis

Gas–liquid polycondensation was first applied to obtain polyamides based on oxalyl chloride and diamines (equation 16).[101] Later on, the development of this method made it possible to obtain a great variety of aliphatic and aromatic poly(oxamide)s.[70,102] The sparging of oxalyl chloride through a chloroform–water system raised the poly(oxamide) yield from 60 to 99%.[3] In the synthesis of polyamides, instead of oxalyl chloride, oxalyl fluoride, perfluoroadipoyl dichloride and carbon suboxide have been used.[108]

$$n\text{ClC—CCl} + n\text{NH}_2\text{R}''\text{NH}_2 \xrightarrow[-2n\text{NaCl}]{2n\text{Na}_2\text{CO}_3} \text{—[C—CNHR}''\text{NH]}_n \quad (16)$$
$$\;\;\;\;\|\;\;\;\;\|\;\|\;\;\;\;\|$$
$$\;\;\;\;\text{O}\;\;\;\text{O}\;\text{O}\;\;\;\text{O}$$

Along with polyamides, gas-phase polycondensation has also produced polyureas, by the reaction of phosgene or thiophosgene with diamines, and poly(thioester)s, by the reaction of oxalyl chloride with dithiols.[108]

An interesting possibility for the use of gas–liquid polycondensation was demonstrated in the manufacturing of semipermeable membranes. Polyamine or polyphenol membranes were obtained by treating with formaldehyde vapour a polysulfonic substrate impregnated with a solution of 1,3-diaminobenzene or resorcinol, respectively.[19]

A new trend in the use of polycondensation in gas–liquid systems lies in the synthesis of polycarbonates and poly(thiocarbonate)s by the reaction of CO_2 or CS_2 with *p*-xylene dibromide and potassium glycolates (equation 17).[109-111] The reaction can be conducted both in aprotic dipolar solvents and in dioxane, benzene and other non-polar media in the presence of a crown ether ([18]-crown-6). The authors believe that, in the first stage, a fast reaction is taking place between CO_2 and potassium glycolate, after which the formed salt ($KOCO_2RO_2COK$) slowly interacts with *p*-xylene dibromide.[110] It should be noted that the above example represents the most complicated case of interfacial polycondensation, since the reaction is taking place in a system containing, along with a gas and a liquid, a solid phase (potassium glycolate). Obviously, in such a system the role of interfacial transfer catalysts is especially great. Polymer synthesis based on gaseous CO_2 is not only a very interesting example of a new approach to the problem of gas–liquid polycondensation and interfacial catalysis, but it also allows the synthesis of polycarbonates without using poisonous phosgene.

$$nCX_2 + nBrCH_2\text{-}\langle\text{C}_6H_4\rangle\text{-}CH_2Br + nKOR''OK \xrightarrow[-2nKBr]{\text{crown ether}} {+\!\!\!\!\!\!\!\!\!\!-}CXCH_2\text{-}\langle\text{C}_6H_4\rangle\text{-}CH_2XCOR''O{-\!\!\!\!\!\!\!\!\!\!+}_n \quad (17)$$

$$X = O, S; R'' = \text{-}\langle\text{C}_6H_{10}\rangle\text{-}, \text{-}CH_2\text{-}\langle\text{C}_6H_4\rangle\text{-}CH_2\text{-}$$

The extensive synthetic possibilities of gas–liquid polycondensation have been shown in the synthesis of polysiloxanes. With gaseous methyltrichlorosilane sparged through a water layer, soluble oligomers (equation 18) were obtained in quantitative yield.[112] If this process is conducted in a liquid–liquid system the formed products are insoluble and non-melting. This and other examples are indicative of good prospects for future development of interfacial polycondensation in gas–liquid systems.

$$(n + m)\,MeSiCl_3 + xH_2O \xrightarrow{-yHCl} [MeSiO_{1.5}]_n[MeSiO]_m\underset{OH}{|} \quad (18)$$

11.3.3 Solid–Liquid Systems

A very interesting interfacial polycondensation technique, conducted in a solid–liquid system, has recently found extensive application. One of the monomers in this case does not dissolve in the organic medium and is situated in the solid phase. The use of interfacial transfer catalysts makes it possible to intensify the polycondensation between this monomer and the one situated in solution. The resulting polymers have a sufficiently high molecular weight. The mechanism of this process has not yet been fully studied. It can be assumed, however, that in this case, as in the case of polycondensation in a gas–liquid system, the reaction can take place both at the interface and in the volume of the liquid phase.

An interesting example of polycondensation in a solid–liquid system is the production of polycarbonate from K_2CO_3 and p-xylylene dibromide (equation 19).[113,114] The use of [18]-crown-6 as the interfacial transfer catalyst has made it possible to obtain in diglyme [bis(2-methoxyethyl) ether] a polymer with a 50% yield and a molecular weight of 12 900. In this reaction lithium and sodium carbonates proved to be poorly effective, and silver, zinc, barium, strontium and calcium carbonates were absolutely ineffective. The replacement of p-xylylene dibromide with 1,6-dibromohexane produced polycarbonates with a higher molecular weight ($\bar{M}_n = 32\,000$).

$$nK_2CO_3 + nBrCH_2\text{-}\langle\text{C}_6H_4\rangle\text{-}CH_2Br \xrightarrow{\text{crown ether}} {+\!\!\!\!\!\!\!\!\!\!-}COCH_2\text{-}\langle\text{C}_6H_4\rangle\text{-}CH_2O{-\!\!\!\!\!\!\!\!\!\!+}_n \quad (19)$$

Aromatic polyethers (equation 20) have been obtained by means of nucleophilic substitution of hexafluorobenzene with bisphenols under the conditions of interfacial catalysis.[114,115] The solid–liquid system consisted in this case of solid K_2CO_3 and a solution of the reagents: hexafluorobenzene and bisphenol. Heating the acetone solution of hexafluorobenzene and bisphenol A in the presence of K_2CO_3 and [18]-crown-6 at 55 °C for 48 h led to the formation of a polymer in quantitative yield and $[\eta] = 0.58$ dl g^{-1}.[115]

$$nC_6F_6 + nHOR''OH \xrightarrow[-2nKF]{nK_2CO_3} {+\!\!\!\!-}C_6F_4OR''O{-\!\!\!\!+}_n \quad (20)$$

As already noted in Section 11.3.2.2, one example of polycondensation in a solid–liquid system is the synthesis of polycarbonates directly from CO_2 in the presence of glycolates of metals.[109-111]

Interfacial polycondensation in a solid–liquid system is, in principle, not a new method of polymer synthesis, since a large number of cases are known where one of the monomers is poorly soluble in the reaction solution. The fundamental difference in the above examples lies in, firstly, a very low solubility of reagents, in particular inorganic salts, and, secondly, in a considerable intensification of polycondensation under the action of interfacial transfer catalysts. Further progress in this field will make it possible to expand the scope of the problems that can be successfully solved with the help of interfacial polycondensation.

11.4 DISPERSION POLYCONDENSATION

In dispersion polycondensation, conducted in a heterophase system, the reaction proceeds in the full volume of one of the phases: a liquid phase in emulsion polycondensation or a solid phase in suspension polycondensation (Scheme 1). In the former case, the system consists of two liquid phases: organic solvent and water. However, despite the seeming heterogeneity of the system, polycondensation is actually taking place in the organic solvent droplets, and in essence is a variety of solution polycondensation. In the latter case, the system consists of a solid phase (polymer precipitated from solution at the very beginning of the reaction) and a liquid phase (organic solvent). In this process, growth of the polymer chain occurs mostly in the solid phase. Intermediate variants are also possible, e.g. when the polymer precipitates in the course of emulsion polycondensation but the growth of its macromolecules continues, or when only a part of the polymer precipitates at the very beginning of the reaction conducted in the organic solvent medium. Interpretation of these cases is even more difficult than for the 'pure' variants of dispersion polycondensation.

11.4.1 Emulsion Polycondensation

Emulsion polycondensation began its development in the late 1950s in the works of Morgan et al.,[2,3] who used organic solvents that were miscible with water. Systems of this kind are still mostly used in the process. However, emulsion polycondensation can also proceed within the classical version of interfacial polycondensation, when two-phase systems are used, consisting of water and an organic solvent which is immiscible with water.

11.4.1.1 Governing factors

A characteristic feature of emulsion polycondensation, associated with the reaction proceeding in organic phase droplets, poses certain requirements for the choice of the two-phase system. One of the principal conditions here is a degree of reagent distribution between the phases, ensuring that the chain propagation reaction proceeds in the organic phase and that binding of the evolving acid takes place in the aqueous phase. It means that the value of the distribution coefficient for the concentrations of reagents in the organic and aqueous phases ($K_D = C_{\text{org phase}}/C_{\text{aq phase}}$) has to be $\gg 1$ for the monomers and $\ll 1$ for the acceptor. In one of the most widely used reaction systems, acyl dichloride–diamine, the first monomer, as a rule, is wholly in the organic phase, whilst the K_D value for the second monomer depends on its structure, the organic phase properties and other factors. In particular, water-soluble bicyclic diamines pass almost entirely into the organic phase.[4] However, for many of the diamines widely used for polymer synthesis, the K_D values amount to only 0.005–1.0.[116] To reduce the diamine solubility in water, inorganic salts (salting-out agents) are introduced into the system. Under the action of the salting-out agent (Na_2CO_3) the K_D value of m-phenylenediamine in a tetrahydrofuran (THF)–water system becomes larger than unity (8.7).[117] In emulsions, the water content in the organic solvent can be quite high (up to 50%), which also facilitates the passage of diphilic diamine molecules into the organic phase. The K_D values for acid acceptors usually lie within 0.005–0.02.[4,117]

As polymer formation in emulsion polycondensation proceeds in the bulk of the organic phase, the principal governing factors of this process are similar to those of solution polycondensation. This similarity is, in turn, a necessary criterion for assigning a heterophase process to emulsion polycondensation.

The laws governing emulsion polycondensation have been studied in the most detail for polyamidation and polyesterification using dichlorides of aromatic dicarboxylic acids with aromatic diamines and bisphenols, respectively.[4,5,69]

Characteristic of emulsion polycondensation is the high rate of polymer formation.[5,118,119] The rate constant of polyamidation of terephthaloyl dichloride with m-phenylenediamine at 25 °C in a THF–water–Na_2CO_3 system is equal to 52.9 $l\,mol^{-1}\,s^{-1}$.[118] An increase in water content in the organic solvent raises the reaction rate, which can be associated both with an increase in the polarity of the medium and with an increase in its ability to undergo specific solvation, speeding up the decomposition of the intermediate complex.[69] For instance, in going from anhydrous THF to THF containing 4.5 and 35.2% of water, the rate constants of the reaction of terephthaloyl dichloride with m-phenylenediamine form the following series: 1:3:25.[119] Hydrolysis of the acyl dichlorides is also accelerated in such media, but to a considerably smaller extent, which promotes the formation of high molecular weight polyamides.[5]

High molecular weight polymers can be obtained under the conditions of an emulsion process only with intensive stirring of the heterophase system, when the rotational speed of the stirrer amounts to 2000–5000 r.p.m.[4,5] This is caused by the necessity of accelerating the monomer (e.g. diamine) mass transfer, which is a necessary condition for the reaction to proceed in the emulsion droplets.

The fact that polycondensation is taking place in the bulk of the organic phase means that the nature of the organic solvent has a considerable effect on the polymer yield and molecular weight. The best results have been obtained when the process is conducted in such water-miscible solvents as THF, cyclohexane, acetone, 2,4-dimethyltetramethylene sulfone (2,4-dimethylthiolane 1,1-dioxide), dioxane, 1,2-propylene oxide and a number of others.[3–5] In the above solvents not only are sufficiently high K_D values obtained for the monomers, but the necessary conditions are also created for high molecular weight polymers, dissolving or swelling in the organic phase, to be formed. An increase in the water content of THF reduces the solubility of polyarylates in the organic phase, and, therefore, reduces their molecular weight.[120] The ratio between the volumes of the aqueous and the organic phases is also an important factor, exerting a noticeable influence on the polycondensation results. This ratio is usually 1:1–1:3 by volume.[3,121]

An increase in the reaction temperature, as a rule, lowers the molecular weight of the formed polymer. The main reason for this is associated with an increased contribution of side reactions, primarily the hydrolysis of acyl chlorides and the interaction between chlorides and solvent.[2,4] Emulsion polycondensation is usually conducted at room temperature.

Under the conditions of emulsion polycondensation, high molecular weight polymers can only be obtained with an equimolar ratio of monomers.[69,120] As in the case of solution polycondensation, even a slight deviation from equimolarity of reagents leads to a noticeable decrease in the molecular weight of the polymers (Figure 6). This dependence is preserved both upon a change in the concentration of monomers in the phases, with the volume of phases remaining constant, and upon a change in the volume of phases, with the monomer concentration in them remaining constant. This confirms the mechanism of emulsion polycondensation, according to which polymer formation takes place in the entire volume of the organic phase. A similar conclusion is supported by the dependence of the molecular weight on the polymer yield, which has the shape characteristic of polycondensation in a homogeneous system: the polymer molecular weight rises with yield.[4]

The use of salting-out agents and acceptors of the evolving acid is one of the necessary conditions for emulsion polycondensation to be successfully achieved in water-miscible solvents. With the help

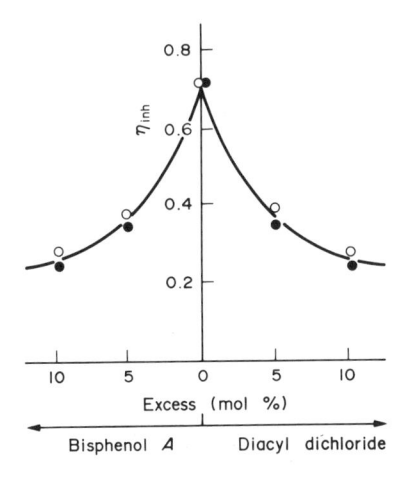

Figure 6 Inherent viscosity η_{inh} of bisphenol A copolytereisophthalate vs. the ratio of initial monomers with a change in their concentration (●) and with a change in the volume of the phases (○)[120]

of salting-out agents (inorganic salts), the content of monomers is increased and the quantity of water decreased in the organic phase. With increasing concentration of salting-out agent in the aqueous phase, the polymer molecular weight increases (Figure 7). It should be noted that, as the polycondensation proceeds, the binding of the evolving acid (HCl) by the acceptor may result in an increase in the salting-out capacity of organic salts in the aqueous phase. In some cases, *e.g.* in polyamidation, the same compound (Na_2CO_3) acts as the salting-out agent and the acceptor. The low salting-out capacity of such an acceptor as NaOH makes it necessary to add a salting-out agent (NaCl) to the system (Figure 8). A too high acceptor content in a system (more than 3 mol mol^{-1} of HCl) results in a noticeable hydrolysis of the acyl chlorides and, correspondingly, to a decrease in the molecular weight of the polymers.

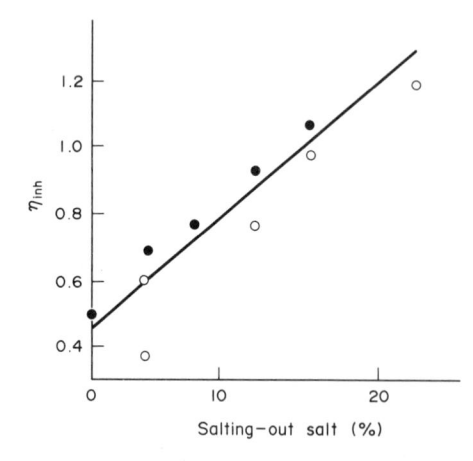

Figure 7 Inherent viscosity η_{inh} of bisphenol *A* copolytereisophthalate *vs.* the initial concentration of salting-out agent in the THF–water–NaOH–salting-out agent system.[120] Salting-out agent: (○) KCl, (●) NaCl

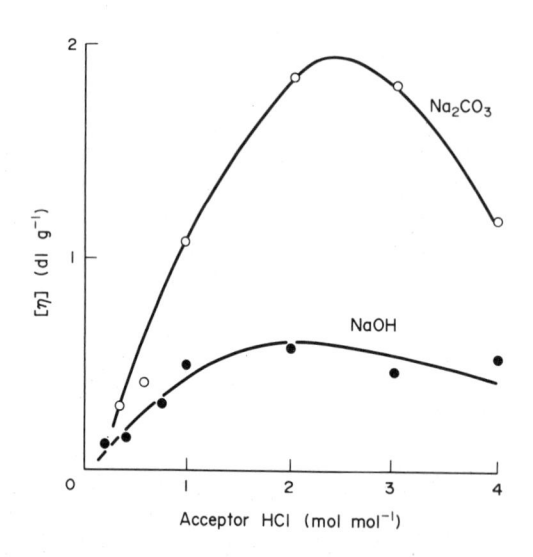

Figure 8 Intrinsic viscosity [η] of poly(*m*-phenylene isophthalamide) *vs.* the quantity of HCl acceptor in the THF–water–acceptor system[69]

Emulsion polycondensation is a very promising method for the synthesis of polymers both on a laboratory and an industrial scale.[3] Unfortunately, the number of publications on research into the governing factors of this process and the possibilities for its improvement and development has decreased considerably in recent years. The future of emulsion polycondensation depends in many respects on the progress achieved in this research.

11.4.1.2 Synthesis of polymers

With the help of emulsion polycondensation, a large number of various polymers have been synthesized. Some of the polymers obtained are listed in Table 3. This method has been very successfully used to obtain aromatic and aromatic–aliphatic polyamides (equation 4).[3,4,69,72,74,122] Aromatic copolyamides have recently been obtained in a THF–water–Na_2CO_3 system, with the main emphasis placed on those based on tere- and iso-phthaloyl dichlorides and m-phenylenediamine.[118,123,124] It has been found that, where acyl dichlorides are successively introduced into the reaction system, by varying their concentration and ratio one can synthesize copolymers with a different structure, from the statistical to the block type.[123]

Table 3 Results of the Synthesis of Polymers by Emulsion Polycondensation in the THF–Water–Na_2CO_3 System

Polymers	Monomers	Yield (%)	$[\eta]$ (dl g^{-1})	Ref.
Polyamide	Terephthaloyl dichloride + trans-2,5-dimethylpiperazine	94	3.4	a
Polyarylate	Terephthaloyl dichloride + phenolphthalein	99	1.20	b
Polyurethane	Bisphenol A + hexamethylenediamine	98	0.91	c
Polyurea	p,p'-Diaminodiphenylmethane bis(carbamyl chloride) + N,N'-dimethyl-p,p'-diaminodiphenylmethane	80	0.14	c
Polyhydrazide	Isophthaloyl dichloride + hydrazine hydrate	100	0.22	d

a P. W. Morgan, J. Macromol. Sci., Chem., 1981, **A15**, 683.
b T. V. Kudim and L. B. Sokolov, Vysokomol. Soedin., Ser. A, 1978, **20**, 1802.
c S. Foti, P. Maravigna and G. Montaudo, Chim. Ind. (Milan), 1983, **65**, 337.
d A. Ballisteri, D. Garozzo, G. Montaudo, A. Polliano and M. Giuffrida, Polymer, 1987, **28**, 139.

Another important class of polymers produced by emulsion polycondensation is represented by polyesters, primarily polyarylates.[4,120,125,126] Along with polyarylates, various copolyarylates have been obtained based on bisphenols and mixtures of dichlorides of aromatic dicarboxylic acids.[120]

Besides polyamides and polyesters, emulsion polycondensation has produced polysulfonamides,[4] polyurethanes,[76] polyureas,[76] polyhydrazides[77] and polyamidrazones.[3] The synthetic possibilities of this process are by no means confined to the above set of polymers. However, as already emphasized, the further development of emulsion polycondensation for polymer synthesis is directly related to the progress in studying the fundamental principles of the method, as well as in solving numerous technological problems (intensive stirring, solvent regeneration, clearing the polymer from inorganic salt, etc.).

11.4.2 Suspension Polycondensation

In the 1960s, reports appeared of polycondensation being accompanied under special conditions by the separation of a polymer precipitate. As exemplified by the synthesis of poly(terephthaloyl-trans-2,5-dimethylpiperazine) in mixtures of a good and a poor solvent, it was established that the polymeric precipitate is reactive and that macromolecules can grow in it.[2] According to other data, however, polymers (polycarbonates) in the precipitate are inactive and do not take part in chain propagation reactions. It has been proposed that polycondensation in the course of which polymer is separated as a precipitate should be called precipitative polycondensation (see Volume 5, Chapter 10, Section 10.6.2).[124] Later, this term began to be used to designate all the polycondensation processes accompanied by polymer precipitation, although it had been established that macromolecule growth can actually proceed in such precipitates.

Starting in the late 1960s, a substantial contribution to precipitative polycondensation research has been made by Korshak and co-workers.[12,128–130] As a result of these studies, the principal laws governing this type of polycondensation have been established. It was concluded that, with regard to the reaction zone location, the process consists of two stages. In the first stage, the macromolecules grow in the bulk of the solvent, i.e. in homogeneous conditions. In the second stage, practically all the low molecular weight polymer formed (at conversions higher than 90%) precipitates, and the reaction system consists of two phases: the solid (polymer suspension) and the liquid (solvent) phase. In such a heterophase system, the polymer chain propagation takes place mainly in a solid disperse phase. The solvent (disperse medium) performs important functions: it provides for the separation of the system into phases; it causes the polymer swelling, assuring in this way the necessary mobility of

11.4.2.1 Governing factors

The governing factors in suspension polycondensation have been mainly studied as exemplified by the synthesis of polyarylates from dichlorides of aromatic dicarboxylic acids and bisphenols, conducted in the presence of tertiary amines.[12,129,131-135] The initial reagents most often used were terephthaloyl dichlorides, phenolphthalein and triethylamine (equation 21). The chosen process, that was called acceptor-catalytic polycondensation (see Volume 5, Chapter 10, Section 10.2), proceeds at a high rate, noticeably dependent on the type of the solvent used. Thus, for the reaction of benzoyl chloride with p-nitrophenol, conducted at 30 °C in the presence of triethylamine, the rate constants (k, $l^2 \, mol^{-2} \, s^{-1}$) rise in the following series of solvents: dioxane (6.5) < benzene (43) < acetone (1.7×10^3) < dichloroethane (3.6×10^4).[136]

$$\text{(21)}$$

To accomplish suspension polycondensation, it is necessary to use solvents in which the initial monomers are readily soluble, and in which the polymer formed is insoluble. In the case of the synthesis of polyarylates, one such solvent is acetone. When polycondensation is conducted in acetone, phenolphthalein polyterephthalate precipitates from solution in quantitative yield (95%) within the first minute of the reaction, but the growth of its molecular weight continues (Figure 9, curve 1).[131] The polyarylate molecular weight continues to increase to practically the same extent if the precipitated polymer is filtered off, washed and dispersed in a new portion of acetone (Figure 9, curve 3). Since the initial compounds that have not entered into the reaction are completely removed after such treatment, the result obtained is indicative of the polymer molecular weight increasing because of the interaction of the terminal groups of the macromolecules in the precipitate. This is also indicated by the large increase of the molecular weight (the reduced viscosity rises from 0.80 to 1.57 dl g^{-1}) after the precipitated polymer has been washed, dried and treated with an acetone solution of terephthaloyl dichloride and triethylamine.[131]

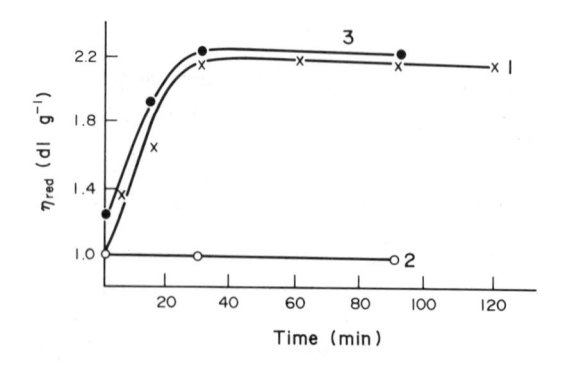

Figure 9 Reduced viscosity η_{red} of phenolphthalein polyterephthalate vs. the reaction time at 50 °C:[131] 1, synthesis in acetone; 2, synthesis in acetone with the removal of solvent after the first minute of the reaction; 3, synthesis in acetone with the removal of acetone after the first minute of the reaction, washing the precipitate with acetone and introducing pure acetone

An important question is the reason for the formation of high molecular weight polymers in the disperse phase. Investigations into polycondensation in binary mixtures of solvents, consisting of a good (dichloroethane) and a poor (acetone) solvent, enable one to make a fairly well-grounded assumption (Figure 10).[134] In this process, an increase in acetone (precipitant) quantity lowers the dissolving capability of the mixture, and, with a certain composition of the mixture, the polymer precipitates from the reaction solution, which, as shown in Figure 10, is accompanied by a noticeable increase in its molecular weight. One can assume the observed effect to be associated with the aggregation of macromolecules, preceding the polymer precipitation from solution. This aggregation raises the local concentration of terminal groups, which increases the polycondensation rate and decreases the relative content of impurities and, correspondingly, the contribution of side reactions. Both these factors are conducive to the formation of polymers with a higher molecular weight. Similar effects seem to make possible the production of high molecular weight polymers in the precipitate. On the strength of this reasoning, the great influence exerted on the molecular weight by the capability of the solvent to cause polymer swelling becomes quite explicable: the higher the degree of swelling, the greater the mobility of the macromolecules and, therefore, the higher the molecular weight of the polymer formed.[134]

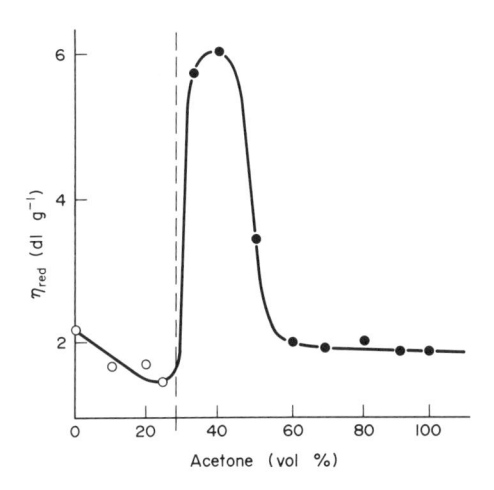

Figure 10 Reduced viscosity η_{red} of phenolphthalein polyterephthalate *vs.* the composition of the mixture of dichloroethane with acetone. Polymer yield 92–98%,[134] (○) solution polycondensation, (●) suspension polycondensation

Along with the ability of the solvent to cause polymer swelling, the results of suspension polycondensation are strongly affected by the influence of the solvent on the reaction kinetics. Under the conditions of a dispersion process, high kinetic constants of the chain propagation reaction promote the formation of high molecular weight polymers. This is well illustrated by the synthesis of a polyarylate in a mixture of two poor solvents (benzene and acetone), differing from each other both in their ability to cause polymer swelling and in their influence on the reaction rate constant (in acetone this constant is almost two orders of magnitude higher than in benzene[136]). As seen from Figure 11, with increasing acetone content in the mixture the polyarylate swelling capacity decreases, and the dielectric constant (ε) of the medium increases (the growth of ε is parallel to the growth of the reaction rate constant). The opposing actions of these factors result in the marked dependence of the polymer molecular weight on the mixture composition, with a maximum at 30–40 vol% acetone content.

Research into the dynamics of polyarylate accumulation in the precipitate and the increase of its molecular weight has made possible the estimation of the kinetics of the propagation of polymeric chains in both phases. The reaction rate constant in solution and in the polymer precipitate were found to be respectively equal, in a mixture of dichloroethane with n-heptane (80:20 vol%), to 0.014 and 0.034 l mol^{-1} s^{-1}, and in a mixture of THF with acetone (64:36 vol%), to 0.070 and 0.080 l mol^{-1} s^{-1}.[135] This result shows that polymer precipitation from solution does not lower, but may even raise the kinetic constant of polycondensation.

In distinction to solution polycondensation, in suspension polycondensation there is no clear-cut effect of reaction temperature on the polymer molecular weight (see Volume 5, Chapter 10, Section 10.2). Since in acceptor-catalytic polyesterification in acetone there is an interval of temperatures (20–50 °C) where the polyarylates obtained have approximately the same weight,[131] the

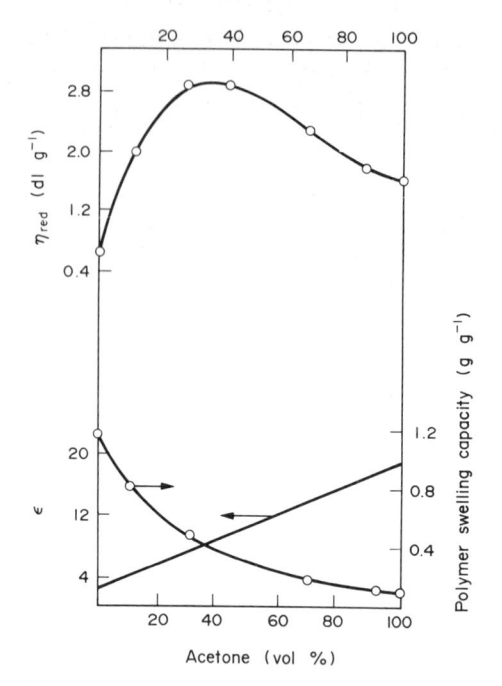

Figure 11 Reduced viscosity η_{red} of phenolphthalein polyterephthalate, polymer swelling capacity and dielectric constant ε of the medium vs. the composition of the mixture of benzene with acetone; polymer yield 96–98%[132]

molecular weight of the formed polymer is much more sensitive to the concentration of monomers and the sequence of their introduction into the reaction medium. However, with optimum concentrations of the monomers (0.1–0.2 mol l^{-1}), irrespective of the sequence of their introduction into the reaction mixture, the polyarylates formed have similar molecular weights.[131]

The course of suspension polycondensation is greatly affected by the physical structure of the precipitated polymer. When the polymer has an amorphous or a slightly crystalline structure, its molecular weight can continue to grow in the precipitate. A characteristic example of this is the above-mentioned phenolphthalein polyterephthalate whose amorphous structure allows study of the factors governing the suspension polycondensation process. Crystallization of the precipitated polymer is a serious obstacle for the polycondensation to proceed successfully, since this precipitate will be poorly reactive. That is why polymer precipitation from solution is sometimes called 'physical' chain termination.[69] The first to be mentioned among the polymers forming highly crystalline precipitates are aromatic polyamides.[4,69]

There exist several methods of raising the activity of polymeric precipitates. One of them consists of adding triethylamine hydrochloride to the reaction system containing the precipitated polyamide [poly(m-phenyleneisophthalamide)]. The hydrochloride, interacting with the amide groups, raises the mobility and, correspondingly, the activity of the macromolecules.[137] Another approach is based on using catalysts (tertiary amines) that raise the reactivity of the terminal groups of aromatic polyamides.[138]

More complicated, in comparison with suspension polycondensation, is the process where the molecular weight of the polymers increases both in the solution and in the precipitate, and, with the reaction approaching completion, the high molecular weight compound passes from the solution into the disperse phase. This process is called two-phase polycondensation.[139,140]

Under the conditions of two-phase polycondensation, good opportunities are created for the molecular weights and molecular weight distributions (MWD) of polymers to vary within broad limits.[135,139,140] An example of two-phase polycondensation is the synthesis of poly(phenylquinoxaline) from 1,4-bis(phenylglyoxalyl)benzene and 3,3',4,4'-tetraaminodiphenyl oxide in a mixture of N-methylpyrrolidone with methanol.[140] At the end of the process, the solid phase contains 85% of poly(phenylquinoxaline), whose molecular weight is higher and MWD narrower, as compared with the molecular weight and MWD of the polymer in solution. For this system the molecular weights and MWD of the polymers were shown to depend on the degree of polycondensation completion at which the phase separation starts and on the ratio of the rate constants of macromolecule growth in both phases. This result opens extensive opportunities for controlling the polymer molecular weight and MWD by selecting the appropriate organic medium and

11.4.2.2 Possibilities of synthesis

Suspension polycondensation, as a method of polymer synthesis, is at present used on a limited scale. This appears to be strange, because the formation of a polymer in the form of a suspension has a number of unquestionable advantages: relatively low reaction mixture viscosity and stirring rate, elimination of the polymer precipitation stage, the polymer being formed as a finely dispersed powder, which simplifies its clearing from impurities and subsequent processing, *etc*. Especially important, in our opinion, is the absence of the stage of polymer precipitation from solution in the method, since it is this stage that usually causes a number of serious technological, economic and ecological problems.

Some examples of polymer synthesis with the help of suspension polycondensation are presented in Table 4.

Table 4 Results of the Synthesis of Polymers by Suspension Polycondensation

Polymer	Monomers	Reaction medium, temperature	η_{red} (dl g^{-1})	Ref.
Polyarylate	Terephthaloyl dichloride + phenolphthalein	Acetone, 40 °C	1.8	a
Polyamide	Isophthaloyl dichloride + *m*-phenylenediamine	Chloroform + Et$_3$N·HCl, 30 °C	1.9	b
Poly(naphthoylene-benzimidazole)	Napthalene-1,4,5,8-tetracarboxylic acid + 3,3′,4,4′-tetraaminodiphenyl oxide	*m*-Cresol, 160 °C	2.5	c
Poly(phenylquinoxaline)	1,4-Bis(phenylglyoxalyl)benzene + 3,3′,4,4′-tetraaminodiphenyl oxide	N-methylpyrrolidone + acetic acid, 20 °C	0.8	c

[a] V. V. Korshak, S. V. Vinogradova, V. A. Vasnev and T. I. Minaishvili, *Vysokomol. Soedin., Ser. A*, 1969, **11**, 81.
[b] P. W. Morgan, *Macromolecules*, 1977, **10**, 1381.
[c] V. V. Korshak, A. L. Rusanov, A. Ya. Chernikhov, G. V. Kazakova, A. M. Berlin, S. Kh. Fidler and T. V. Lekae, *Dokl. Akad. Nauk SSSR*, 1985, **282**, 375.

Suspension polycondensation has been successfully applied to various polyarylates (equation 21) and copolyarylates.[12,128,129,141] An interesting feature of suspension acceptor-catalytic polyesterification is its high conformational specificity (see Volume 5, Chapter 10, Section 10.2.1).[129] The use of di- and poly-substituted binuclear monomers with central bridge groups in this process, decelerating the rotation of individual fragments of molecules, makes it possible to obtain macromolecules containing regular sequences of rotational isomers in the chains. A fundamentally new feature of such syntheses consists of conformationally regular polymers being formed directly in the course of polymer chain propagation.[129,142] Regularity of the spatial microstructure of macromolecules rises noticeably in a heterophase system, when the formed conformationally regular polymer is insoluble in the reaction solution and precipitates in the course of the synthesis. The properties of conformationally regular polymers, *e.g.* their thermal stability, differ from those of polymers with a statistical content of conformational isomers.[142] This opens up new opportunities for controlled variation of polymer characteristics directly in the course of the synthesis.

Extensive opportunities are offered by suspension copolycondensation for varying the structure and properties of copolymers over wide ranges. The principal laws governing this process were studied using copolyesters obtained by acceptor-catalytic copolyesterification of aromatic acyl dichlorides (intermonomers) with bisphenol and glycol or two different bisphenols (comonomers) (equation 22).[143-146] At the very beginning of the reaction, the formed copolymers precipitate in quantitative yield from the reaction mixture, and the process is therefore of a suspension nature.

$$n\text{ClC}-\text{R}'-\text{CCl} + \begin{array}{c}(n-m)\text{HOR}''\text{OH}\\ m\text{HOR}'''\text{OH}\end{array} \xrightarrow[-2n\text{R}_3\text{N·HCl}]{2n\text{R}_3\text{N}} [\text{C}-\text{R}'-\text{COR}''\text{O}]_x \sim \sim [\text{C}-\text{R}'-\text{COR}'''\text{O}]_y \quad (22)$$

$$\text{R}'' = \text{Ar}; \quad \text{R}''' = \text{Ar}', \text{alkyl}$$

For a quantitative estimation of the microstructure of copolymers, the K_M parameter (microheterogeneity coefficient) was used,[141,147] which is similar in its content to the parameter B frequently used in the literature.[148] For statistical copolymers $K_M = 1$, for a regularly alternating copolymer $K_M = 2$, while for a mixture of homopolymers $K_M = 0$; the region of block copolymers lies within the $0 < K \ll 1$ range.

The studies performed showed that the microstructure of the polymers, obtained after the addition of a tertiary amine to the solution of intermonomer and comonomers, depended on the type of organic medium. Varying the quantity of n-heptane (a strong precipitant) in mixtures with acetone, dichloroethane and toluene makes it possible to change the K_M of copolymers over a wide range (from 0.8 to 1.3).[144] For the same monomers in a homophase system, the use of solvents with radically different properties, $e.g.$ dichloroethane and dioxane, exerts practically no effect on the K_M value of the copolymers ($K_M = 1.12$ and 1.09, respectively).[129]

Research into the properties of solvents and their mixtures has shown the cause of this phenomenon to lie in the different abilities of organic media to precipitate polymers from solution and give rise to their swelling, which eventually influences macromolecule growth in suspension copolycondensation. In particular, an increase in the precipitating capability of a mixture of acetone with n-heptane lowers the K_M of the copolymers (Figure 12).[144]

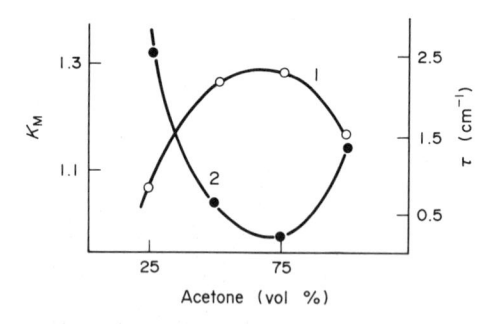

Figure 12 K_M of 2,2-bis(3-chloro-4-hydroxyphenyl)propane and hexamethylene glycol copolyterephthalate (curve 1) and precipitating capacity of the medium (curve 2) $vs.$ the composition of the mixture of acetone with n-heptane (τ = turbidity with the precipitant volume fraction equal to 0.45)[144]

In contrast to solution copolycondensation, in suspension copolycondensation an increase in the concentration of monomers decreases the K_M of the copolymers. Here, an increase in the concentration of monomers and an increase in the precipitating capability of the organic phase create the conditions for the formation of block copolymers.[144]

In single-stage copolycondensation, in both heterophase and homophase systems, a gradual introduction of intermonomer into the reaction mixture promotes the formation of copolymers with a block microstructure.[146] However, in contrast to a homophase system, in suspension copolycondensation the relative activity of comonomers scarcely affects the K_M value of copolymers and, therefore, the length of the formed blocks.

Analysis of the results obtained allows the conclusion that the laws governing the formation of copolymer microstructure in heterophase copolycondensation depend on a larger number of factors when compared with homophase copolycondensation. The reason for this lies in the copolymer formation process in a heterophase medium being complicated by the action of diffusive, absorptive and other factors inherent in such systems. It is the action of these factors, however, that gives rise to the wider opportunities provided by suspension copolycondensation, permitting the variation of copolymer structure over wider ranges. Establishing the general principles of forming the microstructure of copolymers in heterophase processes constitutes a complex and important problem that requires further theoretical and experimental study.

The process of synthesizing polyamides, $e.g.$ poly(m-phenyleneisophthalamide) (equation 23), by means of polycondensation in solution is often regarded as a suspension process, since the polymers formed at the very beginning of the reaction can precipitate from solution.[4,69,137] The possibility of further growth of macromolecules in the precipitated polymer depends in many respects on its physical structure. The rate of molecular weight growth is sharply decelerated in a crystalline polymeric precipitate.[149] In the case of polymers with an amorphous or a slightly crystalline structure, the polycondensation can proceed in the precipitate.[4,137]

$$nClCOC_6H_4COCl + nNH_2C_6H_4NH_2 \xrightarrow[-2nR_3N \cdot HCl]{2nR_3N} [-COC_6H_4CONH-C_6H_4-NH-]_n \quad (23)$$

A new trend in the use of suspension polycondensation is the synthesis of poly(heteroarylene)s, e.g. poly(naphthoylene benzimidazole)s (equation 24).[130,150] In the 1980s it was found that, under the conditions of a catalytic process, poly(heteroarylene)s that have precipitated from the reaction solution in quantitative yield continue to increase their molecular weight.[150] An interesting feature of suspension polyheterocyclization is the high temperature of some of the reactions. All the previously known examples of interfacial and dispersive polycondensation represented low temperature processes, conducted at temperatures, as a rule, much lower than 100 °C.[2,4,12] The high temperature suspension method of synthesizing poly(heteroarylene)s has clearly demonstrated the extensive potential of heterophase processes.

(24)

11.5 CONCLUSIONS

Interfacial and dispersion processes represent major laboratory methods for synthesizing numerous polycondensation polymers. Widespread application of the method is facilitated by its unquestionable advantages: sufficiently mild reaction conditions, quantitative yield and high molecular weight of the polymers formed. The future of heterophase polycondensation is associated with expanding the range of the polymers obtained, with ascertaining the specific possibilities of interfacial reactions, as well as with solving very important technological, economic and ecological problems (high stirring intensities, solvent regeneration, polymer purification, etc.). By analogy with other polymerization processes, one can assume that research into the specificity of polycondensation reactions at the interface and in micellar systems, with the participation of lyotropic and other organized media, will give a new impetus to the further development of interfacial and dispersion polycondensation. In particular, it is probable that one of the possible variants of stereospecific polycondensation will be accomplished in a heterophase process.

The progress of interfacial and dispersion polycondensation is inseparably connected with the use of these methods in industry to produce polycarbonates, polyarylates, aromatic polyamides and other polymers. The future will show the true possibilities of heterophase processes in solving particular practical problems, including those of producing new polymers.

11.6 REFERENCES

1. A. Einhorn, *Justus Liebigs Ann. Chem.*, 1898, **300**, 135.
2. P. W. Morgan, 'Condensation Polymers: by Interfacial and Solution Methods', Interscience, New York, 1965.
3. P. W. Morgan, *J. Macromol. Sci., Chem.*, 1981, **A15**, 683.
4. L. B. Sokolov, 'Fundamentals of Synthesis by the Polycondensation Method', Khimiya, Moscow, 1979 (in Russian).

5. G. Reinisch, U. Gohlike and H. -H. Ulrich, *Makromol. Chem., Suppl.*, 1979, **3**, 177.
6. E. Z. Casassa, *J. Macromol. Sci., Chem.*, 1981, **A15**, 787.
7. Y. Imai, *J. Macromol. Sci., Chem.*, 1981, **A15**, 833.
8. L. J. Mathias, *J. Macromol. Sci., Chem.*, 1981, **A15**, 853.
9. O. Ya. Fedotova, M. L. Kerber and N. M. Kozyreva, *Plast. Massy*, 1983, **2**, 15.
10. Y. Imai, *Yuki Gosei Kagaku Kyokaishi*, 1984, **42**, 1095.
11. C. M. Paleos, *Chem. Soc. Rev.*, 1985, **14**, 45.
12. V. V. Korshak and S. V. Vinogradova, 'Nonequilibrium Polycondensation', Nauka, Moscow, 1972 (in Russian).
13. F. Millich and C. E. Carraher, Jr. (eds.), 'Interfacial Synthesis', Dekker, New York, 1977, vols. 1 and 2.
14. S. L. Kiperman, 'Introduction to the Kinetics of Heterogeneous Chemical Reactions', Nauka, Moscow, 1964 (in Russian).
15. D. A. Frank-Kamenetsky, 'Diffusion and Heat Transfer in Chemical Kinetics', Nauka, Moscow, 1967 (in Russian).
16. L. B. Sokolov, V. Z. Nikonov and G. N. Shilyakova, *Vysokomol. Soedin., Ser. A*, 1969, **11**, 616.
17. L. B. Sokolov and V. Z. Nikonov, in 'Interfacial Synthesis', ed. F. Millich and C. E. Carraher, Jr., Dekker, New York, 1977, vol. 1, p. 141.
18. L. B. Sokolov, *Vysokomol. Soedin., Ser. A*, 1970, **12**, 971.
19. J. E. Cadotte, R. S. King, R. J. Majerle and R. J. Petersen, *J. Macromol. Sci., Chem.*, 1981, **A15**, 727.
20. L. V. Kuritsyn, L. B. Sokolov and S. S. Gitis, *Vysokomol. Soedin., Ser. A*, 1978, **20**, 1093.
21. E. M. Hodnett and D. A. Holmer, *J. Polym. Sci.*, 1962, **58**, 1415.
22. J. H. Bradburg, P. J. Crawford and A. N. Hambly, *Trans. Faraday Soc.*, 1968, **64**, 1337.
23. T. D. Shaffer, M. Jamaludin and V. Percec, *J. Polym. Sci., Polym. Chem. Ed.*, 1985, **23**, 2913.
24. H. -B. Tsai and Y. -D. Lee, *J. Polym. Sci., Polym. Chem. Ed.*, 1987, **25**, 1505.
25. L. B. Turetsky and L. B. Sokolov, *Vysokomol. Soedin.*, 1961, **3**, 1449.
26. A. E. Golland, O. S. Matyukhina, M. I. Siling, N. I. Gelperin, L. A. Karpenko and G. I. Faidel, *Vysokomol. Soedin., Ser. A*, 1985, **27**, 884.
27. V. Z. Nikonov and L. B. Sokolov, *Vysokomol. Soedin., Ser. B*, 1968, **10**, 337.
28. L. B. Sokolov and V. B. Igonin, *Dokl. Akad. Nauk SSSR*, 1979, **245**, 860.
29. W. Podkoscielny and W. Kowalewska, *J. Polym. Sci., Polym. Chem. Ed.*, 1984, **22**, 1579.
30. W. Podkoscielny and A. Kultys, *J. Polym. Sci., Polym. Chem. Ed.*, 1984, **22**, 2265.
31. T. M. Frunze, V. V. Korshak, V. V. Kurashev and P. A. Aliyevsky, *Vysokomol. Soedin.*, 1959, **1**, 1795.
32. L. B. Sokolov and T. V. Kudim, *Vysokomol. Soedin.*, 1960, **2**, 699.
33. P. J. Crawford and J. P. Bradbury, *Trans. Faraday Soc.*, 1968, **64**, 185.
34. V. V. Korshak, S. V. Vinogradova, T. M. Frunze, A. S. Lebedev and V. V. Kurashev, *Vysokomol. Soedin.*, 1961, **3**, 984.
35. T. I. Shein, G. I. Kudryavtsev and L. N. Vlasova, *Khim. Volokna*, 1963, **1**, 17.
36. P W. Morgan and S. L. Kwolek, *J. Polym. Sci.*, 1962, **62**, 33.
37. H. Nakamura, S. Imanishi, K. Sanui and N. Ogata, *Polym. J.*, 1979, **11**, 661.
38. N. Ogata, K. Sanui, T. Onozaki and S. Imanishi, *J. Macromol. Sci., Chem.*, 1981, **A15**, 1059.
39. H. Schnell, 'Chemistry and Physics of Polycarbonates', Interscience, New York, 1962.
40. W. P. Weber and G. W. Gokel, 'Phase Transfer Catalysis in Organic Synthesis', Springer, New York, 1977.
41. E. V. Dehmlow, *Angew. Chem.*, 1977, **89**, 521.
42. C. M. Starks and C. Liotta, 'Phase Transfer Catalysis: Principles and Techniques', Academic Press, New York, 1978.
43. W. E. Keller (ed.), 'Compendium of Phase Transfer Reactions and Related Synthetic Methods', Fluka, Switzerland, 1979.
44. M. Hiraoka, 'Crown Compounds. Their Characteristics and Applications', Elsevier, New York, 1982.
45. S. S. Yufit, 'The Mechanism of Interfacial Catalysis', Nauka, Moscow, 1984 (in Russian).
46. L. J. Mathias and C. E. Carraher, Jr. (eds.), 'Crown Ethers and Phase Transfer Catalysis in Polymer Science', Plenum Press, New York, 1984.
47. M. M. Sharma, *Sadhana*, 1985, **8**, 387.
48. E. Z. Casassa, *J. Macromol. Sci., Chem.*, 1981, **A15**, 787.
49. L. J. Mathias, *J. Macromol. Sci., Chem.*, 1981, **A15**, 853.
50. Y. Imai, *Yuki Gosei Kogaku Kyokaishi*, 1984, **42**, 1095.
51. Y. Imai, *Kobunshi Kako*, 1984, **33**, 581.
52. D. C. Sherrington, *Macromol. Chem.*, 1984, **3**, 303.
53. T. Nishikubo, *Kobunshi*, 1986, **35**, 132.
54. Z. K. Brzozynski, J. Dubczynski and J. Petrus, *J. Macromol. Sci., Chem.*, 1979, **A13**, 875.
55. Z. K. Brzozowski, J. Petrus and J. Dubczynski, *J. Macromol. Sci., Chem.*, 1979, **A13**, 887.
56. Y. Imai, M. Ueda and M. Ii, *Makromol. Chem.*, 1978, **179**, 2085.
57. A. K. Banthia, D. Lunsford, D. C. Webster and J. E. McGrath, *J. Macromol. Sci., Chem.*, 1981, **A15**, 943.
58. W. E. McEwen, K. F. Kumli, A. Blade-Font, M. Zanger and C. A. Vanderwerf, *J. Am. Chem. Soc.*, 1964, **86**, 2378.
59. W. E. McEwen, A. Axelrad, M. Zanger and C. A. Vanderwerf, *J. Am. Chem. Soc.*, 1965, **87**, 3948.
60. R. V. Pagilagan and W. E. McEwen, *J. Chem. Soc., Chem. Commun.*, 1966, 652.
61. D. Landini, A. Maia and G. Padda, *J. Org. Chem.*, 1982, **47**, 2264.
62. E. Z. Casassa, D. -Y. Chao and M. Hehson, *J. Macromol. Sci., Chem.*, 1981, **A15**, 799.
63. I. A. Yesikova and S. S. Yufit, *Izv. Akad. Nauk SSSR, Ser. Khim.*, 1981, **11**, 2693.
64. E. V. Dehmlow, *Angew. Chem., Int. Ed. Engl.*, 1977, **16**, 493.
65. S. Yufit and I. A. Yesikova, *Izv. Akad. Nauk SSSR, Ser. Khim.*, 1981, **9**, 1996.
66. K. Isagawa, Y. Kimura and S. Kwon, *J. Org. Chem.*, 1974, **39**, 3171.
67. Y. Imai, M. Ueda and M. Ii, *J. Polym. Sci., Polym. Lett. Ed.*, 1979, **17**, 85.
68. Y. Imai, A. Kato, M. Ii and M. Ueda, *J. Polym. Sci., Polym. Lett. Ed.*, 1979, **17**, 579.
69. L. B. Sokolov, V. D. Gerasimov, V. M. Savinov and V. K. Belyakov, 'Thermostable Aromatic Polyamides', Khimiya, Moscow, 1975 (in Russian).
70. W. Deits, S. Grossman and O. Vogl, *J. Macromol. Sci., Chem.*, 1981, **A15**, 1027.
71. G. Cum, R. Gallo, F. Severini and A. Spadaro, *Angew. Makromol. Chem.*, 1986, **138**, 111.

72. A. Ballistreri, D. Garozzo, M. Giuffrida, P. Maravigna and G. Montaudo, *Macromolecules*, 1986, **19**, 2693.
73. S. J. Huang and J. Kozakiewicz, *J. Macromol. Sci., Chem.*, 1981, **A15**, 821.
74. A. Ballistreri, D. Garozzo, P. Maravigna, G. Montaudo and M. Giuffrida, *J. Polym. Sci., Polym. Chem. Ed.*, 1987, **25**, 1049.
75. O. V. Syrkova, D. A. Kabanov and I. P. Gavrilina, *Zh. Prikl. Khim.*, 1986, **59**, 1641.
76. S. Foti, P. Maravigna and G. Montaudo, *Chim. Ind. (Milan)*, 1983, **65**, 337.
77. A. Ballistreri, D. Garozzo, G. Montaudo and A. Pollicino, *Polymer*, 1987, **28**, 139.
78. R. G. Pearson, *J. Polym. Sci., Polym. Chem. Ed.*, 1985, **23**, 9.
79. R. Ohta, Y. Kanai, F. Funaki, Y. Amano, N. Koide and K. Iimura, *Kobunshi Ronbunshu*, 1986, **43**, 301.
80. J. S. Riffle, R. G. Freelin, A. K. Banthia and J. E. McGrath, *J. Macromol. Sci., Chem.*, 1981, **A15**, 967.
81. L. H. Tagle, F. R. Diaz, J. C. Vega and P. F. Alquinta, *Makromol. Chem.*, 1985, **186**, 915.
82. Z. Brzozowski, *Polimery (Warsaw)*, 1984, **29**, 415.
83. Y. Imai and S. Tassavori, *J. Polym. Sci., Polym. Chem. Ed.*, 1984, **22**, 1319.
84. W. Podkoscielny and S. Szubinska, *J. Appl. Polym. Sci.*, 1986, **32**, 3277.
85. Y. Imai, S. Abe and M. Ueda, *J. Polym. Sci., Polym. Chem. Ed.*, 1981, **19**, 3285.
86. Y. Imai, N. Sato and M. Ueda, *Makromol. Chem., Rapid Commun.*, 1980, **1**, 419.
87. Y. Imai, H. Kamata and M. Kakimoto, *J. Polym. Sci., Polym. Chem. Ed.*, 1984, **22**, 1259.
88. N. Yamazaki and Y. Imai, *Polym. J.*, 1983, **15**, 603.
89. N. Yamazaki and Y. Imai, *Polym. J.*, 1985, **17**, 377.
90. T. D. Shaffer and V. Percec, *Makromol. Chem.*, 1986, **187**, 111.
91. M. Ueda, N. Sakai, M. Komatsu and Y. Imai, *Makromol. Chem.*, 1982, **183**, 65.
92. M. Ueda, Y. Oishi, N. Sakai and Y. Imai, *Macromolecules*, 1982, **15**, 248.
93. M. Ueda, R. Takasawa and Y. Imai, *Makromol. Chem., Rapid Commun.*, 1983, **3**, 905.
94. C. E. Carraher, Jr., J. E. Sheats and C. U. Pittman, Jr. (eds.), 'Organometallic Polymers', Academic Press, New York, 1978.
95. C. E. Carraher, Jr., in 'Interfacial Synthesis', ed. F. Millich and C. E. Carraher, Jr., Dekker, New York, 1977, vol. 2, chap. 21.
96. C. E. Carraher, Jr., L. P. Torre and H. M. Molloy, *J. Macromol. Sci., Chem.*, 1981, **A15**, 757.
97. C. E. Carraher, Jr., R. A. Schwarz, J. A. Schroeder and M. Schwarz, *J. Macromol. Sci., Chem.*, 1981, **A15**, 773.
98. Y. Imai, T. Q. Nguen and M. Ueda, *J. Polym. Sci., Polym. Lett. Ed.*, 1981, **19**, 205.
99. Y. Imai, A. Komeyama, T. Q. Nguyen and M. Ueda, *J. Polym. Sci., Polym. Chem. Ed.*, 1981, **19**, 2997.
100. Y. Imai and M. Ueda, *Polym. Prepr., Am. Chem. Soc., Div. Polym. Chem.*, 1982, **23** (1), 164.
101. L. B. Sokolov, L. V. Turetsky and T. V. Kudim, *Vysokomol. Soedin.*, 1960, **2**, 1744.
102. L. B. Sokolov, T. V. Kudim and L. V. Turetsky, *Vysokomol. Soedin.*, 1961, **3**, 1370.
103. L. B. Sokolov, L. V. Turetsky and L. I. Tugova, *Vysokomol. Soedin.*, 1962, **4**, 1817.
104. I. V. Parfenov, L. B. Sokolov and S. S. Novokreshchenov, *Zh. Prikl. Khim.*, 1966, **39**, 208.
105. L. B. Sokolov, V. I. Logunova and D. F. Sokolova, *Dokl. Akad. Nauk SSSR*, 1969, **189**, 347.
106. L. B. Sokolov, *J. Polym. Sci.*, 1962, **58**, 1253.
107. H. J. Chang and O. Vogl, *J. Polym. Sci., Polym. Chem. Ed.*, 1977, **15**, 1043.
108. L. B. Sokolov, *Vysokomol. Soedin.*, 1964, **6**, 2117.
109. K. Soga, Y. Toshida, S. Hosoda and S. Ikeda, *Makromol. Chem.*, 1977, **178**, 2747.
110. K. Soga, Y. Toshida, S. Hosoda and S. Ikeda, *Makromol. Chem.*, 1978, **179**, 2379.
111. K. Soga, Y. Toshida, I. Hattori, K. Nagata and S. Ikeda, *Makromol. Chem.*, 1980, **181**, 979.
112. V. V. Kireev, T. V. Vasilyeva, I. P. Stakhanov, V. A. Ignatov, A. S. Kuprin, L. A. Prikhod'ko and L. A. Petrovina, *Dokl. Akad. Nauk SSSR*, 1986, **290**, 859.
113. K. Soga, S. Hosoda and S. Ikeda, *J. Polym. Sci., Polym. Chem. Ed.*, 1979, **17**, 517.
114. D. J. Gerbi, R. F. Williams, R. Kellman and J. C. McPheeters, *Polym. Prepr., Am. Chem. Soc., Div. Polym. Chem.*, 1981, **22** (2), 387.
115. R. Kellman, D. J. Gerbi, J. C. Williams, R. F. Williams and R. B. Bates, *Polym. Prepr., Am. Chem. Soc., Div. Polym. Chem.*, 1982, **23** (1), 174.
116. P. W. Morgan and S. L. Kwolek, *J. Polym. Sci.*, 1959, **40**, 300.
117. L. B. Sokolov, *Vysokomol. Soedin.*, 1965, **7**, 601.
118. H.-H. Ulrich, J. Moskalenko and G. Reinisch, *Acta Polym.*, 1980, **31**, 734.
119. L. B. Sokolov, L. V. Kuritsyn, Yu. A. Fedotov and L. A. Bobko, *Vysokomol. Soedin., Ser. A*, 1982, **24**, 606.
120. T. V. Kudim and L. B. Sokolov, *Vysokomol. Soedin., Ser. A*, 1978, **20**, 1802.
121. L. B. Sokolov and T. V. Kudim, *Vysokomol. Soedin., Ser. A*, 1965, **7**, 1899.
122. L. B. Sokolov and V. Z. Nikonov, in 'Interfacial Synthesis', ed. F. Millich and C. E. Carraher, Jr., Dekker, New York, 1977, vol. 1, p. 167.
123. Yu. A. Fedotov, V. A. Subbotin, L. B. Sokolov, S. S. Gitis, N. I. Zotova and G. N. Troshin, *Vysokomol. Soedin., Ser. A*, 1985, **27**, 2137.
124. Yu. A. Fedotov, V. A. Subbotin, S. S. Gitis, V. D. Gerasimov, N. I. Zotova and G. Ye. Troshin, *Khim. Volokna*, 1986, **5**, 43.
125. S. Foti, M. Giuffrida, P. Maravigna and G. Montaudo, *J. Polym. Sci., Polym. Chem. Ed.*, 1984, **22**, 1201.
126. S. Foti, M. Giuffrida, P. Maravigna and G. Montaudo, *J. Polym. Sci., Polym. Chem. Ed.*, 1984, **22**, 1217.
127. E. Turska and A. Dems, *J. Polym. Sci., Part C*, 1968, **22**, 407.
128. V. A. Vasnev and S. I. Kuchanov, *Usp. Khim.*, 1973, **42**, 2194.
129. V. A. Vasnev and S. V. Vinogradova, *Usp. Khim.*, 1979, **48**, 30.
130. A. L. Rusanov, *Vysokomol. Soedin., Ser. A*, 1986, **28**, 1571.
131. V. V. Korshak, S. V. Vinogradova and V. A. Vasnev, *Vysokomol. Soedin., Ser. A*, 1968, **10**, 1329.
132. V. V. Korshak, S. V. Vinogradova, V. A. Vasnev and T. I. Mitaishvili, *Vysokomol. Soedin., Ser. A*, 1969, **11**, 81.
133. L. V. Dubrovina, S. A. Pavlova, V. A. Vasnev, S. V. Vinogradova and V. V. Korshak, *Vysokomol. Soedin., Ser. A*, 1970, **12**, 1308.
134. S. V. Vinogradova, V. V. Korshak, A. V. Vasilyev and V. A. Vasnev, *Vysokomol. Soedin., Ser. A*, 1973, **15**, 2015.

135. I. V. Blagodatskikh, T. P. Bragina, L. V. Dubrovina, S. A. Pavlova and V. V. Korshak, *Acta Polym.*, 1986, **37**, 156.
136. V. A. Vasnev, A. I. Tarasov, S. V. Vinogradova and V. V. Korshak, *Vysokomol. Soedin., Ser. A*, 1975, **17**, 721.
137. P. W. Morgan, *Macromolecules*, 1977, **10**, 1381.
138. Yu. B. Rotenberg, T. I. Tarasova and V. M. Savinov, *Khim. Volokna*, 1986, **2**, 26.
139. V. V. Korshak, S. A. Pavlova, G. I. Timofeyeva, S. A. Kroyan, E. S. Krongaus and A. P. Travnikova, *Acta Polym.*, 1984, **35**, 662.
140. V. V. Korshak, S. A. Pavlova, G. I. Timofeyeva, S. A. Kroyan, E. S. Krongaus and A. P. Travnikova, *Vysokomol. Soedin., Ser. A*, 1985, **27**, 763.
141. V. V. Korshak, S. V. Vinogradova, S. I. Kuchanov and V. A. Vasnev, *J. Macromol. Sci., Rev. Macromol. Chem.*, 1976, **14**, 27.
142. V. V. Korshak, S. V. Vinogradova, V. A. Vasnev, A. V. Vasilyev, A. A. Askadsky, T. A. Babushkina, G. L. Slonimsky, G. K. Semin, Yu. K. Godovsky and E. S. Obolonkova, *Vysokomol. Soedin., Ser. A*, 1974, **16**, 291.
143. V. V. Korshak, I. Ya. Slonim, S. A. Vinogradova, Ya. G. Urman, V. A. Vasnev, A. Kh. Bulai and T. M. Gogiashvili, *Vysokomol. Soedin., Ser. B*, 1980, **22**, 4.
144. V. A. Vasnev, T. M. Gogiashvili, B. D. Lavrukhin, S. V. Vinogradova and V. V. Korshak, *Vysokomol. Soedin., Ser. A*, 1981, **23**, 2537.
145. V. V. Korshak, V. A. Vasnev, S. V. Vinogradova, T. M. Gogiashvili, I. Ya. Slonim, Ya. G. Urman, A. Kh. Bulai and R. G. Koshelava, *Vysokomol. Soedin., Ser. A*, 1981, **23**, 2567.
146. V. V. Korshak, S. V. Vinogradova, V. A. Vasnev, T. M. Gogiashvili, I. Ya. Slonim, Ya. G. Urman and A. Kh. Bulai, *Vysokomol. Soedin., Ser. A*, 1982, **24**, 325.
147. S. I. Kuchanov, 'Methods of Kinetic Calculations in Polymer Chemistry', Khimiya, Moscow, 1978 (in Russian).
148. R. Yamadera and M. Murano, *J. Polym. Sci., Part A-1*, 1967, **5**, 2259.
149. L. B. Sokolov, Yu. B. Rottenberg and V. M. Savinov, *Dokl. Akad. Nauk SSSR*, 1980, **250**, 1402.
150. V. V. Korshak, A. L. Rusanov, A. Ya. Chernikhov, G. V. Kazakova, A. M. Berlin, S. Kh. Fidler and T. V. Lekae, *Dokl. Akad. Nauk SSSR*, 1985, **282**, 375.

12
Condensation Polymerization in Non-aqueous Dispersions

DEREK J. WALBRIDGE
ICI PLC, Paints Division, Slough, UK

12.1	INTRODUCTION	197
12.2	DISPERSION POLYMERS FROM SOLUBLE REACTANTS	197
12.3	DISPERSION POLYMERS FROM EMULSIONS OF LIQUID REACTANTS	198
12.4	DISPERSION POLYMERS FROM SOLID REACTANTS IN AN INERT LIQUID	199
12.5	PARTICLE STABILITY	199
12.6	KINETICS OF CONDENSATION DISPERSION POLYMERIZATION	200
12.7	REFERENCES	200

12.1 INTRODUCTION

No detailed studies of condensation dispersion polymerizations in non-aqueous media have been published, although processes describing the preparation of such dispersions have appeared in the patent literature. Two reviews by Thompson,[1,2] which include summaries of the extensive research carried out by his own group on polyester and polyamide dispersions, provide the only source of readily accessible information on the topic.

Dispersions of condensation polymers of colloidal dimensions may be protected against flocculation by steric stabilization mechanisms identical in nature to those described for addition polymer dispersions in Volume 4, Chapter 15. Although the same types of block or graft copolymer dispersants may be employed, the criteria governing the adsorption of the dispersant on the polymer particle are more complex than in the case of addition polymerization (see Section 12.3).

Dispersions of condensation polymers may be formed (a) by the reaction of the monomeric species dissolved in an initially homogeneous solution in the dispersion medium, (b) by preemulsification of the reactants in the medium before the condensation reaction, or (c) by milling solid reactants to produce a fine dispersion in the medium before reaction. In each case the continuous phase of the dispersion should be such that it does not interfere with the progress of the polymerization. Much of the reported work has been conducted in inert hydrocarbon diluents of appropriate boiling points, to facilitate fast reactions and allow azeotropic removal of the condensation reaction by-product.

The dispersion route to condensation polymers offers the advantages of faster reactions, lower processing temperatures and shorter process times, in comparison with melt condensation processes. The product is a low viscosity fluid, which may contain 60–70% of polymer. It is thus suitable for thin film applications. A potential disadvantage of the dispersion process is the presence of significant quantities of dispersant, which may adversely affect the properties of fibres, films or bulk materials fabricated from the polymer.

12.2 DISPERSION POLYMERS FORMED FROM SOLUBLE REACTANTS

Stable dispersions can be produced by dissolving the reactants, together with a polymeric dispersant or its precursor, in an organic liquid from which the polymer precipitates as it is formed.

Dispersions of polyesters, polyamides and polyethers were prepared by the homogeneous solution route early in the development of non-aqueous polymer dispersions, but no subsequent developments have been reported.[1,3,4] The process differed from an addition dispersion polymerization in that the polymerization proceeded in a slow stepwise fashion until the solubility threshold was reached, whereupon virtually all the polymer precipitated. The particle size was determined by the amount of stabilizer present at this stage in relation to the amount of insoluble polymer precipitated.

A typical example was the condensation of bis(4-hydroxyphenyl-2,2-propane) with a mixture of the acid chlorides of terephthalic and isophthalic acid in a mixed chlorinated hydrocarbon/ester solvent. The dispersant was a copolymer of methyl methacrylate, lauryl methacrylate and methacrylic acid.

2,6-Lutidine was found to be an effective acid scavenger, which separated as a coarse crystalline precipitate which could be removed by filtration. Problems of particle stabilization could almost certainly be reduced by utilizing the graft copolymer dispersant stabilizers which were developed subsequently.[5]

12.3 DISPERSION POLYMERS FROM EMULSIONS OF LIQUID REACTANTS

Coarse suspensions of condensation polymers have been prepared in which the reactants were dispersed in aliphatic hydrocarbon in the presence of a swelling agent such as tetraethylene glycol dimethyl ether or tetramethylene sulfone.[6] It was claimed that the swellant conferred a degree of stability on the dispersion. Possibly local solubilization of surface layers of polymer provided stability in a manner analogous to the stabilization of poly(vinyl chloride) organosols by plasticizer absorption. Coarse suspensions of poly(propylene terephthalates), with particle diameters of *ca.* 200 μm, have also been prepared by heating a 20% w/w suspension of bis(hydroxypropyl) terephthalate in tetralin in the presence of a small percentage of polystyrene and an antimony acetate catalyst.[7] The process was carried out by refluxing under pressure for 40 minutes at 270 °C.

In order to prepare fine particle dispersions with diameters down to the sub-micron level, it is necessary to have present a dispersing agent capable of providing steric stabilization by forming a solvated polymeric monolayer at the particle–liquid interface. This solvated layer prevents particle flocculation and coagulation. The problem of achieving stabilization is two-fold. First, it is necessary to make a fine particle emulsion of the reactants. The particle size of this emulsion should be close to that desired for the final polymer dispersion, because the only change in particle size during the process should be that caused by loss of the reaction by-products. Block or graft copolymers similar to those developed for addition dispersion polymerization are suitable in principle. However, it was demonstrated by Nicks and Osmond[8,9] that an additional requirement for the stability of liquid-in-liquid emulsions was that the anchor (non-solvated) component of the dispersant should not only be insoluble in the dispersion medium, but should be fully soluble in the disperse phase. If the anchor group is not soluble, it is easily rejected from the mobile droplet surface, which is then free to coalesce with another droplet. However if the anchor group dissolves in the droplet, coalescence is less probable because of the extra energy required to desolvate the anchor group. This energy may be largely entropic in nature but may be partly enthalpic when there are specific polar, *e.g.* acid/base, interactions, between the anchor and the droplet phase. The second problem of stabilization is encountered once the condensation polymerization has commenced. The dispersed droplets undergo a rapid change in character as the molecular weight increases, which can lead to the rejection of the anchor group of the stabilizer if it is no longer compatible. The problem can be overcome by the inclusion of reactive groups in the anchor group of the stabilizer, which participate in the condensation reaction. The stabilizer is then covalently linked into the particle surface.[2]

Graft copolymer stabilizers comprising a methacrylate or styrene copolymer backbone to which side chains of poly(12-hydroxystearic acid) are attached have proved to be effective stabilizers for polyester and polyamide dispersion polymerizations in aliphatic hydrocarbon diluents.[10] Reaction between the anchor component and the particle was brought about by incorporating a small proportion of glycidyl methacrylate or methacrylic acid in the copolymer backbone. The use of a condensation polymer as a stabilizing soluble chain in a condensation dispersion polymerization is questionable because of the possibility of ester or amide interchange reactions causing a breakdown of the stabilizer. There is some evidence that the use of soluble esterification catalysts such as tetraalkyl titanates can lead to a loss of dispersion stability, but in the main it has proved possible to prepare stable concentrated dispersions. Ideally, however, an unreactive soluble component would be desirable.

A typical example of the preparation of a polyester dispersion is the condensation of bis(hydroxyethyl) terephthalate (BHET) to form poly(ethylene terephthalate).[1] 37.5 parts BHET, 56 parts of aliphatic hydrocarbon (b.p. 250 °C), 6.5 parts of a 40% solution of dispersion stabilizer in aliphatic hydrocarbon (b.p. 150–170 °C) and 0.02 parts calcium acetate were charged in a stirred flask fitted with a Dean and Stark apparatus and condenser. The dispersion stabilizer was a poly[(12-hydroxystearate)-co-(methyl methacrylate/glycidyl methacrylate)] polymer in which the components were in the weight ratio 100:90:10. The charge was heated with gentle stirring to 140 °C, the melting point of BHET, at which point the stirring rate was increased to produce a fine particle emulsion. The temperature was raised to distil off ethylene glycol until a final reflux temperature of 250 °C was reached after 2.5 hours. The product was a 31% dispersion of polymer of particle sizes in the range 2–20 μm. In an alternative procedure, the dispersant was dissolved first in the molten monomer, followed by the slow addition of the hydrocarbon diluent, under conditions of rapid stirring, until the inversion occurred of the diluent-in-monomer emulsion formed initially. The polymerization was then carried out as before giving a particle size range of 1–10 μm.

The process is not limited to hydrocarbon media. A variation has been described in which BHET was polymerized in silicone oil using a dispersant of poly(dimethylsiloxane) grafted to a polyacrylate.[11] The graft was either of a comb type, prepared by copolymerizing a siloxane, carrying a terminal methacryloyloxy or other reactive group, with acrylic monomers, or was of the more random structure produced by replacing the terminal reactive group by reactive groups distributed along the siloxane chain.

Polymer dispersions may also be prepared from mixtures of two or more reactants, e.g. a glycol/dicarboxylic acid condensation polymer. When the starting components are solids or are not compatible they may be dissolved in a common solvent which is itself insoluble in the dispersion medium. Thus nylon salts or ω-amino acids may be dissolved in ethylene glycol before emulsifying in aliphatic hydrocarbon with a suitable graft copolymer. The ethylene glycol and the water of reaction may then both be removed by azeotropic distillation to yield a polyamide or copolyamide.

The reactants need not necessarily be in the same phase initially. In one example an emulsion of an aqueous solution of hexamethylenediamine in aliphatic hydrocarbon was reacted with dissolved dimethyl terephthalate to yield poly(hexamethylene terephthalamide).[1] Polyurethanes also have been produced by emulsifying a polyol in hydrocarbon with a graft copolymer dispersant and then reacting with a dissolved diisocyanate.[12]

12.4 DISPERSION POLYMERS FROM SOLID REACTANTS IN AN INERT LIQUID

Solid reactants, such as nylon 6,6 salt or 11-aminoundecanoic acid, may be milled in hydrocarbon with a graft copolymer dispersant to form fine particle dispersions and then polymerized in dispersion.[13] In the case of aminoundecanoic acid, the preferred method when the initial dispersion was prepared in a ball mill, was to charge the mill with a 1:1 v/v mixture of reactant and diluent together with about 5% of dispersant. The mill was inspected at 24 hour intervals and if thickening was apparent a further 2% dispersant was added. The process was continued until the particle size was less than 2 μm, by which time 10–14% of dispersant had been added. A further 1–2% dispersant was added and then the solids content was reduced to 30% before transferring to a reactor for the polymerization stage.[2]

12.5 PARTICLE STABILITY

A general feature of condensation dispersion polymerizations, when compared with addition polymer processes, is the much higher level of dispersant, usually two to three times, required to give colloidal stability at equivalent particle size. There are several factors which lead to this higher dispersant level. The initial disperse phase volume of the reactants in a condensation process is normally higher than the final polymer volume, particularly when glycols and carrier solvents have to be removed. Secondly, an excess of dispersant is necessary in the emulsification stage to maintain monolayer coverage of the deformable droplets and to stabilize freshly generated surface before coalescence can occur. Finally it has been observed that in condensation dispersion processes the particles frequently are cenospheres.[1] This may be the result of emulsification of the diluent into the disperse phase which would lead to a higher stabilizer usage.

An additional problem of dispersion stability has been observed in those preparations in which the polymer can crystallize. While the dispersion remained above the crystalline melting point, colloidal stability was maintained. On cooling the dispersion slowly through the melting point, the latex flocculated. Rapid quenching to well below the melting point prevented flocculation. Studies of particle crystallization by IR, DSC and X-ray diffraction techniques confirmed that latex flocculation was associated with polymer crystallization, whilst the crash-cooled dispersions remained amorphous.[2]

12.6 KINETICS OF CONDENSATION DISPERSION POLYMERIZATION

The elimination of by-products from a highly viscous reacting mass is much more efficient and rapid when it is in the form of a thin layer or a finely divided solid. It would be expected that a polycondensation reaction would be faster in a dispersion than in a conventional melt process, that the rate should not be diffusion controlled[14] and that it should be possible to produce high molecular weight polymer.[15] Qualitatively, dispersion reactions do proceed faster than melt reactions and it is possible to operate processes at 170–200 °C rather than *ca.* 220–250 °C. However, the only quantitative results which have been published were for nylon-11.[16] After approximately five hours reaction the dispersion had reached 99% conversion against 75% for the melt process.

12.7 REFERENCES

1. K. E. J. Barrett and M. W. Thompson, in 'Dispersion Polymerization in Organic Media', ed. K. E. J. Barrett, Wiley, London, 1975, p. 220.
2. M. W. Thompson, in 'Polymer Colloids', ed. R. Buscall, T. Corner and J. F. Stageman, Elsevier Applied Science, London, 1985, p. 24.
3. ICI Ltd., *Br. Pat.* 1 095 931 (1967) (*Chem. Abstr.*, 1965, **62**, 12 022).
4. ICI Ltd., *Br. Pat.* 1 095 932 (1967) (*Chem. Abstr.*, 1965, **62**, 12 021).
5. D. J. Walbridge, in ref. 1, p. 45.
6. Rohm and Haas Ltd., *US Pat.* 3 329 653 (1967) (*Chem. Abstr.*, 1967, **67**, 65 323).
7. Mitsubishi Rayon Co. Ltd., *Jpn. Pat.* 85 139 717 (1985) (*Chem. Abstr.*, 1986, **104**, 69 330).
8. ICI Ltd., *Br. Pat.* 1 211 532 (1970) (*Chem. Abstr.*, 1969, **71**, 13 840).
9. D. J. Walbridge, in ref. 1, p. 67.
10. ICI Ltd., *Br. Pat.* 1 373 531 (1971) (*Chem. Abstr.*, 1973, **78**, 125 352).
11. Mitsubishi Rayon Co. Ltd., *Eur. Pat.* 212 862 (1987) (*Chem. Abstr.*, 1987, **107**, 78 482).
12. Union Carbide Corp., *US Pat.* 4 000 218 (1976) (*Chem. Abstr.*, 1977, **86** 91 443).
13. ICI Ltd., *Br. Pat.* 1 403 794 (1975) (*Chem. Abstr.*, 1975, **83** 165 125).
14. P. Cefelin, *Vysokomol. Soedin., Ser. A*, 1973, **15**, 423.
15. V. S. Nanda, *J. Polym. Sci., Part A*, 1964, **2**, 2275.
16. K. E. J. Barrett and H. R. Thomas, in ref. 1 p. 197.

13

Solid-state Polymerization

FRANCESCO PILATI
Università di Bologna, Italy

13.1 INTRODUCTION	201
13.2 SOLID-STATE POLYCONDENSATION OF MONOMERS	202
13.2.1 General Aspects	202
13.2.2 Polymers Obtained by Solid-state Polymerization of Monomers	204
13.2.3 Crystalline Order and Reaction Kinetics	206
13.3 SOLID-STATE POLYCONDENSATION OF PREPOLYMERS	207
13.3.1 General Aspects	207
13.3.2 Factors Affecting Postcondensation	208
13.3.3 Postcondensation of Polyesters	209
13.3.4 Crystallization-induced Reactions	213
13.3.5 Postcondensation of Polyamides	214
13.4 REFERENCES	215

13.1 INTRODUCTION

Experimental observations of solid-state organic reactions are somewhat dispersed throughout the literature. This subject seems particularly attractive for reactions leading to polymers because it may give polymeric materials of relevant scientific and industrial interest, with properties which are not obtainable by the usual processes or which cannot be prepared by any other method. The only comprehensive review on this topic is more than 20 years old.[1]

Although the first evidence of polymers obtained by solid-state polymerization (SSP) was reported in the literature in the 1930s,[2-4] it was only in the 1960s that a more systematic exploration of this field was carried out. Among the many papers which have been published, a number considered the synthesis of polyalkynes[5] and polyacrylamides,[6] but these subjects, as well as the other polymers that are usually obtained by chain reactions, are not within the scope of this chapter (see Volume 4, Part 3). Also the polymer-supported synthesis of polypeptides, often referred to as solid-phase polymerization, will not be considered here.

In this chapter we will limit the discussion to polymers which are usually obtained by polycondensation; however, the number of possible reactions in this category seems unlimited and no attempt can be made to be exhaustive. Thus the large group of resins (phenol formaldehyde, urea formaldehyde, melamine formaldehyde, epoxy resins, *etc.*), the crosslinking reactions of which occur in the solid state, at least in the final stage, will not be considered here. Polyesters and polyamides will be discussed in detail, as their SSPs have received major attention, and only a few examples of other polymers will be reported.

Various aspects of solid-state polycondensation were reviewed about 20 years ago.[7] The papers which have been published so far in this area suggest that the subject can be better discussed in two distinct parts according to whether the starting materials are crystalline monomers or semi-crystalline prepolymers. Nevertheless, one usually refers to both cases as solid-state polymerization even though the terms postcondensation or postpolymerization are sometime used for the latter case.

In the first case the monomer is transformed into a polymer at a temperature lower than the melting point of both monomer and polymer by a reaction which rarely takes place in a real solid state and, as will be seen below, an overlapping of chain and step growth mechanisms may occur during the polymerization.

In the second case the polymerization is carried out on low or medium molecular weight semicrystalline prepolymers at a temperature below the melting point; it is generally accepted that the polymerization proceeds by step reactions in the amorphous regions of the semicrystalline polymer.

The main interest in SSP of monomers has been confined up to now to laboratories, where it is chiefly used to obtain polymers which cannot be prepared by any other polymerization method or which are in an otherwise not achievable highly oriented form.

SSP of prepolymers, on the other hand, has received attention particularly because of the commercial interest in the high quality polymers obtained by this process. It is usually used to prepare polymers with extremely high molecular weight, generally not attainable by polycondensation in the molten state.

Polymers produced by SSP often have improved properties because side reactions leading to defects in the chemical structure or to undesirable by-products are limited or avoided.

Polycondensation in the solid state is a rather complex process involving chemical reactions, the rate of which may be affected by the restricted molecular mobility of the reactive end groups, and diffusion of volatile by-products. A complete description of SSP should therefore take into account the appropriate kinetics of the chemical reactions as well as the diffusion of by-products out of the particles and their removal from the particle surface; the whole process is complicated by the high temperature at which SSP is usually performed and by the many variables involved. As it is difficult or impossible to consider all these features at once, most papers have usually treated only some of these; most frequently, only the factors affecting the overall process rate have been investigated and more or less simplified models used when considering the kinetic aspects. All of this, along with the frequent absence of a complete set of data for the characterization of the polymerizing system, makes a quantitative comparison among different papers difficult or impossible.

In the first part of this chapter, devoted to SSP of monomers, the general aspects, the various polymers prepared and the kinetics of the process are reported and discussed. The second part is concerned with the SSP of prepolymers; both the factors affecting postcondensation and the kinetics of the process will be discussed in detail for aliphatic polyamides and for polyesters of terephthalic acid.

13.2 SOLID-STATE POLYCONDENSATION OF MONOMERS

13.2.1 General Aspects

It has been known for a long time that polymers may be obtained by polymerization of crystalline monomers below their melting point.[2,3] Most of the work in the literature refers to unsaturated monomers polymerized by radiation;[8] however, a number of papers have been concerned with polycondensation reactions.

There are many reasons why researchers are attracted to SSP of crystalline monomers. One of these is that the process is potentially of great scientific and commercial interest, as it may lead to the direct production of highly oriented polymer fibers from needle-like crystals, thus eliminating the expensive spinning process.

Polymers having a higher constitutional order than that achievable by melt or solution polymerization might be obtained from SSP of suitable monomers, provided that the reaction mechanism is controlled by both the restricted mobility and the crystalline order of the monomer. However, evidence of the feasibility of this hypothesis has not been reported so far.

This technique can also be used in order to obtain polymers with crystalline structures that would be difficult or impossible to achieve by other means. Moreover, it may be possible to prepare a polymer in the solid state when the analogous polycondensation in the melt or in solution is limited or precluded by the high temperature required, by the low solubility of the polymer, by steric hindrance or the excessive cyclization of the monomer. In other cases, for polymers that can easily be prepared in the melt, macromolecules with fewer defects in the chemical structure, such as spurious moieties in the chains or chain branching (generally originating from undesirable side reactions), can be obtained by SSP due to the lower temperature or to the mechanism involved. For example, linear poly(1,4-phenylene sulfide) (2) can be prepared (equation 1) by SSP from the *para*-halogen-substituted benzenethiol salts (1)[9] whilst the polymerization is accompanied by chain branching and crosslinking when the same reaction is carried out in the melt.[10]

$$n\text{MS}\langle\text{C}_6\text{H}_4\rangle\text{X} \longrightarrow [\langle\text{C}_6\text{H}_4\rangle\text{S}]_n + n\text{MX} \quad (1)$$

(1)   (2)

M = Li, Na, K; X = F, Cl, Br, I

Finally, polymers prepared by SSP usually contain a lower amount of cyclic oligomers than the analogous ones prepared in other ways.[11]

Copolymers from an intimate mixture of crystals[12,13] or mixed crystals[14,15] of different monomers can be obtained by SSP. In principle, it is conceivable that the ordering of different monomers in a mixed crystalline phase favors a copolymer chain growth with a well defined regular structure. The results of some authors seem to support this idea; indeed, an alternate copolymer (5) was obtained from mixed crystals of N-carboxy anhydrides of L-leucine (3) and L-valine (4),[15] probably according to equation (2). On the other hand, the synthesis of an alternating copolypeptide was attempted, unsuccessfully, from a racemic mixture of the N-carboxy anhydride of γ-benzyl glutamate (6);[14] in this latter case the chemical structure of the resulting copolymer seems to consist of fairly long blocks of D and L isomers.

$$n\left[\begin{array}{c}\text{Me}_2\text{CHCH}_2\text{-NCA} \\ (3)\end{array} + \begin{array}{c}\text{Me}_2\text{CH-NCA} \\ (4)\end{array}\right] \longrightarrow \begin{array}{c}-[\text{NHCHCONHCHCO}]_n- \\ |\quad\quad\quad| \\ \text{CH}_2\quad\text{CHMe}_2 \\ | \\ \text{CHMe}_2 \\ (5)\end{array} + 2n\text{CO}_2 \quad (2)$$

mixed crystals

(6) PhCH$_2$O$_2$CCH$_2$CH$_2$-NCA

For a crystalline monomer to be polymerized in the solid state, the reaction has to take place at sufficiently high rate at a temperature below the melting point of both monomer and polymer. For this reason the choice of the appropriate temperature is a critical point; too high temperatures may lead to partial melting with concomitant sticking and a change in the nature of the process, while too low temperatures may prevent chain growth. Usually the SSP is carried out at a temperature 5–40 °C below the melting point of the crystalline monomer. As a consequence of the very high activation energy of the overall SSP process,[16,17] it is possible that an abrupt increase in reaction rate happens close to the melting point of the monomer by raising the temperature by only a few degrees.[17–19] In some cases, in order to have higher reaction rates, it may be convenient to increase the temperature during SSP, following the increase of the melting point of the polymer with chain growth. Depending on the monomer and the temperature, the SSP time may range from a few hours to several days or weeks, and the monomers submitted to SSP can be in the form either of single crystals[20–22] or of polycrystalline aggregates.

As a rule, during SSP low molecular weight by-products have to be removed to shift the equilibrium towards the polymer. This can be accomplished in different ways; most frequently by static or dynamic vacuum or by a stream of inert gas, or sometimes by distilling off the by-products along with an inert liquid in which both the crystalline monomer and the polymer are suspended.[15,18,23–27] A general rule for the choice of method does not exist; for small amounts of monomers SSP is often performed in sealed ampoules under vacuum or in a controlled atmosphere. When the SSP is carried out in a static atmosphere, it has been found that an increase in the pressure of the inert gas depresses or prevents the polycondensation.[21] On the other hand, for other monomers the presence of water is needed in order to perform SSP.[28,29] The presence of water in a closed system does not limit the molecular weight attainable in a significant way.[21]

In the case of an inert liquid, used as the medium in which the insoluble crystalline monomer is dispersed, some other variables such as the nature of the non-solvent, the monomer to solvent ratio and the distillation rate have been considered.[23] None of these factors was found to affect the process to a significant extent.

The dimensions of the crystal are another factor which may be thought to be important. As a rule, this aspect has been disregarded for SSP of monomers; when it was considered,[23] it was found that it did not have any significant effect when the grain size was kept below 20–25 mesh.

Another important factor that seems to be able to improve, limit or prevent SSP is the effect of surfaces on the nucleation of the new crystalline polymeric phase.[21,30,31]

No data are available concerning the molecular weight distribution of polymers obtained by SSP of monomers.

13.2.2 Polymers Obtained by Solid-state Polymerization of Monomers

In theory, each reaction which takes place in the molten state or in solution can also occur in the solid state. In practice most of these reactions are not suitable for SSP, because the low melting point of the monomers restrains the useful temperature range, making the reaction rate too slow.

Most of the papers that have appeared in the literature on the SSP of monomers deal with the synthesis of polyamides, with several on polypeptides and only a few examples reported for other polycondensates. Polyamides have been obtained mostly from α,ω-amino acids or from salts of dicarboxylic acids and diamines.

Regarding the SSP of α,ω-amino acids, single crystals of 6-aminohexanoic acid (AHA) (m.p. 204–205°C) heated *in vacuo* for 16 h at 168°C were found to give highly oriented poly(1,6-hexanamide) (nylon 6).[20,21] Single crystals of the linear dimer and trimer of AHA behaved in a manner qualitatively similar even though the reaction rates were distinctly lower. An increase of the reaction rate and polymerization degree was obtained by performing the SSP of the dimer and trimer in an atmosphere of AHA vapor. The same authors tried unsuccessfully to polymerize 4-aminobutyric acid (m.p. 206°C) in the solid state at 170°C and 5-aminovaleric acid (m.p. 157°C) at 120°C.[21]

11-Aminoundecanoic acid (m.p. 188°C) has been transformed into poly(α,ω-undecanamide) (nylon 11) by SSP either from single crystals at 160°C under vacuum,[22] or from the polycrystalline monomer at a temperature closer to its melting point.[16,32,33]

Many other polyamides have been obtained by SSP of α,ω-amino acids[12,16,17,33,34] and the overall reaction rate seems to be a complex function of the chemical structure; accordingly, the rate of SSP is found to decrease in the order nylon 11 > nylon 7 > nylon 9.[33]

Aliphatic polyamides have also been obtained by SSP of cyclic diamides[28,29] or from salts of dicarboxylic acids and diamines;[12,17–19,23–26,33,35] in this latter case distillation of diamine out of the system may limit the maximum molecular weight attainable.[17,18,33]

Fully aromatic polyamides have been obtained by SSP: poly(p-benzamide) (**8**) of fairly high molecular weight has been prepared from (p-phenoxycarbonylamino)benzoic acid (**7**) and p-(p-phenoxycarbonylaminobenzamido)benzoic acid (**9**)[36] with evolution of phenol and carbon dioxide, according to equations (3) and (4) respectively; poly[4,4″-(p-terphenylene)amide] (**11**) has been obtained from 4-amino-4″-carboxy-p-terphenyl (**10**)[37] according to equation (5); and a carboxylated polyamide (**14**) from SSP under high pressure of pyromellitic anyhydride (**12**) and 1,2- or 1,3-diaminobenzene (**13**)[38] according to equation (6).

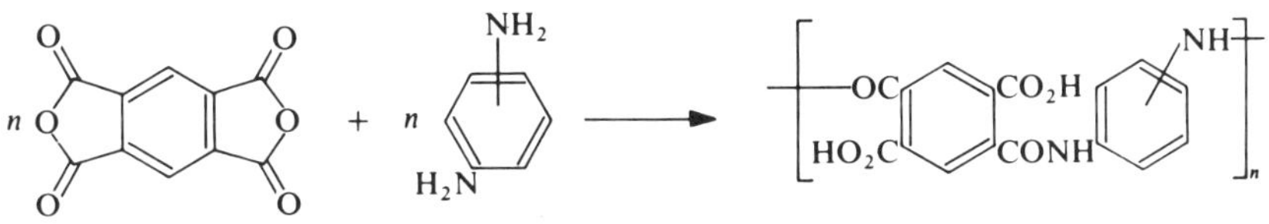

$$\text{(12)} + n \text{(13) } o\text{- or } m\text{-} \longrightarrow \text{(14)} \quad (6)$$

When the low reaction rate is a critical factor, catalysts can help to speed up the process. Many substances have been reported to be effective in this respect for polyamidation[17,25,26,39,40] and among these H_3BO_3 has been recognized to be the most powerful.[17,25,39,40]

Polypeptides have been obtained by SSP from the N-carboxy anhydride (NCA) derivatives of α-amino acids with evolution of carbon dioxide,[14,15] or from the methyl esters of α-amino acids with evolution of methanol.[30,31]

Linear aliphatic polyesters are normally low melting materials and are not suitable to be polymerized by SSP; conversely, polyalkylene arylates have sufficiently high melting points and can be prepared, in theory, by SSP of their monomers.

Bis(2-hydroxyethyl terephthalate) (15) (m.p. 110°C) has been shown to undergo SSP,[41] but owing to its low melting point, the reaction is very slow indeed and it is impossible to produce a high molecular weight poly(ethylene terephthalate) (PET). In practice this process is prevented for all the bis(ω-hydroxyalkyl terephthalate) monomers owing to their low melting points.

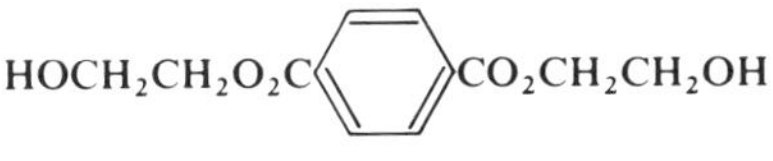

$$HOCH_2CH_2O_2C\text{—}C_6H_4\text{—}CO_2CH_2CH_2OH$$

(15)

The lower carboxyl-terminated oligomers of PET, on the other hand, melt at higher temperatures than the polymer[42,43] and therefore offer the possibility of a process for converting terephthalic acid into PET in the solid state.[43,44] Several poly(ethylene arylene dicarboxylate)s have indeed been prepared by polymerizing in the solid state the corresponding aromatic dicarboxylic acids, with oxirane contained as a vapor in a controlled amount in the inert gas stream passing through the reactor (Scheme 1).[43]

$$P_xCO_2H + \triangle_O \rightarrow P_xCO_2CH_2CH_2OH \quad (7)$$

$$P_xCO_2CH_2CH_2OH + P_yCO_2H \rightleftharpoons P_xCO_2CH_2CH_2O_2CP_y + H_2O \quad (8)$$

$$2P_xCO_2CH_2CH_2OH \rightleftharpoons P_xCO_2CH_2CH_2O_2CP_x + HOCH_2CH_2OH \quad (9)$$

Scheme 1

If the hydroxyethyl-terminated products of equation (7) are rapidly removed by reactions (8) or (9), then the hydroxyethyl-ended intermediates would never attain a high concentration and the polycondensation would proceed through the much higher-melting carboxyl-ended oligomers.

Little or no reaction takes place between the acids and oxirane in the absence of catalyst at temperatures up to 230°C; effective catalysts for this reaction include triphenylphosphine, in combination with a tetraalkyl titanate or antimony trioxide, and phosphorous acid. Polymers of high molecular weight have been produced by this process at 180–200°C without loss of particulate form.[43] It is noteworthy that this heterogeneous process, unlike homogeneous polycondensation reactions, produces polyesters of high molecular weight, while unreacted aromatic dicarboxylic acid is still present, a feature which is characteristic of polyaddition reactions.

The preparation of linear poly(1,4-phenylene sulfide), free of branching and crosslinking, has been performed by SSP of para-halogen-substituted benzenethiol salts[9] according to equation (1).

Random copolyamides have been obtained by SSP of several mixtures of two different monomers[12,13] and it is interesting to note that the copolycondensation rate in the solid state is higher than the homopolymerization rate of the corresponding monomers.

Finally, as described in the previous section, copolymers with a more ordered sequence distribution have been prepared by SSP of mixed crystals.[14,15]

13.2.3 Crystalline Order and Reaction Kinetics

In order for a crystalline monomer to react in the solid state, the molecules have to move to meet each other. It seems reasonable that both the kind of crystalline order in the growing polymer and the reactivity depend to a considerable degree on the characteristics of the crystal structure of the parent monomer.[45]

The nature of these reactions may eliminate side reactions and lead to highly ordered crystalline structures, but may also hinder crystallization if no single path for nucleation and growth exists from the monomer crystal to the polymer crystal. From the very few examples available in the literature it is not possible to predict in general whether (and what kind of) crystalline order will result for a certain polymer. Polymers with crystalline order higher than that usually achieved by melt or solution polymerization have been obtained in some cases;[1,20-22,46] X-ray diffraction patterns like those obtained from melt polymerization have been observed after SSP of some other polymers[28,29] and crystallinity is completely prevented, probably by the restraint connected with the solid-state reaction, in another case.[14]

It is remarkable that ordered growth of polymers can take place in spite of the need to provide for the diffusion of the low molecular weight molecules eliminated in the polycondensation reaction out of the crystal. For instance, single crystals of AHA (m.p. 204–205°C) have been converted at 170°C to highly biaxially oriented poly(1,6-hexanamide).[20,21] This is particularly remarkable in view of the fact that the polycondensation was accompanied by an 18% reduction of volume and a 17% contraction in the direction of the chain axis.

As another example, the polycondensation of a single crystal of 11-aminoundecanoic acid (m.p. 187–188°C), sealed in evacuated ampoules and heated at 160°C, gives a biaxially oriented polycrystalline aggregate of poly(α,ω-undecanamide).[22] It is again surprising that this polymerization reaction can proceed with a remarkable retention of orientation considering that the water eliminated during the process has to diffuse through the crystal and that there is an 11% contraction along the chain direction and a 16% overall reduction in volume.

Crystalline structures similar to those obtained by melt polycondensation have been reported for polymers prepared by SSP of cyclic diamides.[28,29]

The polypeptides prepared by SSP of α-amino acid NCA[14] show only diffuse halos in the X-ray diffraction photographs and the authors have concluded that the restraints inherent in solid-state reactions prevent the growth of crystallites for these intrinsically crystallizable polymers.

As for the usual crystallizations, from the melt or solution, a slow nucleation step followed by growth on the crystal is expected for the polymer crystallization during SSP. This generally results in an initial induction period that can be reduced by powdering the monomer crystals, probably by providing an increase in the number of surface defects on which the reaction is assumed to start preferentially.[30,31,46]

If molecules are to react, they have to collide with each other. It is thus natural to expect that restrictions to molecular mobility in the crystalline state would be reflected in a corresponding reduction of reactivity. This is in agreement with the vast majority of cases of solid-state reactions of monomers.[11-13,19]

In cases where the polymer chains are restricted to a lattice position from the beginning of the SSP, step-growth is impossible and a reaction which follows a step-growth pathway in solution or in the melt must change to a chain-growth pathway in which one monomer unit is added after another to the growing chains.

Reaction mechanisms in the crystalline state are in general quite complex,[1,45] but the polycondensation mechanism of crystalline monomers is expected to be even more complicated since it has to account for the small molecules which are formed and have to escape from the crystal.

As well as for the relationship between crystalline order in polymers and in their parent monomers, even for the kinetics of SSP it is impossible to deduce general conclusions from the limited amount of data available at present. In particular, no information is available for the rate of such processes in single crystals due to the crystal imperfections which make the kinetic measurements barely reproducible. On the contrary, some results have been reported for polycrystalline samples for which a reasonably good reproducibility of reaction rates is achievable due to statistical averaging of the behaviour of the individual crystals.

From a detailed study of the kinetic mechanism for SSP of AHA to nylon 6[21] it has been observed that the process is characterized by three stages: an induction period, a stage in which the monomer disappears at a constant rate while polymer of relatively low molecular weight is formed, and finally a slow polycondensation of the polyamide chains. These results suggest that in the first period the polymeric phase has to be nucleated and both conversion and polymer formation are negligible. In the second stage, the monomer is transported through the vapor phase to the propagating polyamide and mainly gives rise to new nucleii. Thus, after 5 h and 9 h at 170 °C, 6% and 32% of monomer is converted to nylon 6 with a number-average degree of polymerization of 16.5 and 22 respectively. Finally, after exhaustion of the monomer, the molecular weight is increased by polycondensation involving terminal groups of the polymer chains. Obviously the last part of the process takes place in a liquid-like phase in order to enable diffusion of chain ends. It is found[21] that in the latter stage the reaction does not follow the conventional second- or third-order kinetics expected for polyamidation in the melt.[47,48] A double logarithmic plot of degree of polymerization against reaction time is found to be linear for this last stage.[21] This three-stage model can account also for the results previously reported for cyclic diamides[29] and for the copolycondensation of mixtures of crystalline monomers.[12,13]

Other authors have studied and proposed a mechanism for the SSP of dodecamethylene diammonium adipate (m.p. 151–152 °C).[18,24,25] The polymerization was carried out on the crystalline monomer dispersed in an inert liquid (or in a liquid mixture), where both the monomer and the polymer are insoluble, at an appropriate temperature determined by the boiling point of the liquid medium. On the basis of their experimental results a three step mechanism was suggested. In the first step (the induction period), the experimental data fit a model of nucleation and growth. In the second, the reaction proceeds according to second-order kinetics in liquid phase areas resulting from the accumulation of water produced from the amidation reaction, which gives rise to local low melting domains. Finally, in the third step, in which the formation of products of higher molecular weight occurs, the reaction shows a change in the reaction mechanism and the data fit a diffusion-controlled model best, even though other kinetic models are fitted within experimental error.[18]

13.3 SOLID-STATE POLYCONDENSATION OF PREPOLYMERS

13.3.1 General Aspects

Whilst solid-state polycondensation of monomers has so far received only scientific interest, the further polymerization of semicrystalline prepolymers at temperatures below their melting point has been developed chiefly for commercial purposes. This process is today used in industry in order to obtain PET, poly(butylene terephthalate) (PBT), nylon 6 and nylon 6,6 of a particular high molecular weight and quality grade. In fact, when polyesters and polyamides have to be used as engineering plastic materials, the molecular weight plays an important role in order to reach the desired properties. Certain processing techniques, blow-molding and extrusion in particular, also require that the molten polymer has a suitably high melt viscosity to prevent collapse in the soft preformed state. However, such high melt viscosity requires high molecular weights which can be obtained only with great difficulty by conventional melt polycondensation processes in bulk. This is because polymer melts with such a high melt viscosity become difficult to handle and to remove from the commonly used bulk polymerization equipment and because at the high temperatures needed in order to lower the viscosity side reactions occur which counteract the molecular weight increase and lead to by-products. These disadvantages can be limited or avoided in solid-state polycondensation which is, in principle, a straightforward process. Since simple reactors are generally used, the production cost of this additional step is generally low and is largely recovered as the finished product has more useful characteristics.

Postcondensation processes are carried out by heating low or medium molecular weight polycondensates (prepolymers) in the form of chips or powders in a stream of gas or under vacuum at a temperature above the polymer glass transition but below the melting point. Under these conditions the functional groups are sufficiently mobile to give reactions in which by-products such as water or glycol are released. As these are generally equilibrium reactions characterized by low equilibrium constants, the removal of the by-products is necessary to build up high molecular weight chains; this involves diffusion through the particles and desorption of by-products from the particle surfaces. The SSP rate therefore may depend on chemical or physical processes or both, being determined by the slower process.

The main factors that can affect SSP include particle size, initial molecular weight, number and type of chain end groups, catalyst employed, crystallinity, reaction temperature and time, and the technique used to remove by-products. These factors are also interrelated to each other and may depend on the apparatus used. Because of the differences in reactor design and production capacity, reaction conditions are usually optimized on line.

The types of apparatus that can be used to perform postcondensation are simple and depend on the method employed to remove volatile by-products. Test tubes, stationary bed reactors or rotatory vacuum reactors are commonly used in laboratories; tumble dryers, fixed or moving bed reactors are usually employed in industry.

Since the capability of polyamides to further polymerize in the solid state was recognized more than 40 years ago[4] a number of papers and many patents have been published on SSP of prepolymers. The major attention has been devoted to polyesters of terephthalic acid (PET and PBT in particular) and to aliphatic polyamides (namely, nyon 6 and nylon 6,6) for their commercial interest and because of the stringent requirement of high melting point to perform SSP effectively.

In the class of SSP processes one can include also the exchange reactions involving reorganization of random copolymers to multiple block copolymers.[49-55] These have recently received special attention for the transformation of copolymers in the liquid crystal state[54,55] and are commonly referred to as crystallization-induced reactions (CIR).

In the followng paragraphs a general overview of the most important factors involved in the postcondensation processes and the peculiar features of the kinetic aspects of SSP for polyesters and polyamides will be discussed. A brief examination of the main aspects of crystallization-induced reactions will be also given.

13.3.2 Factors Affecting Postcondensation

The temperature is probably the most important factor in SSP due to its interrelation with almost all other aspects of the process. An increase of the SSP temperature usually results in an increase of the overall rate of the process as a consequence of the increment in both reaction and diffusion rates. This may affect SSP so much as to change the controlling step of the process.[56,57] Too low temperatures require too long reaction times; the lowest temperature at which SSP takes place in a detectable manner depends on the polymer considered and is usually around 150–170°C for polyesters and 90–120°C for polyamides. Higher temperatures, on the other hand, favor undesired side reactions and may cause problems connected with particle sticking. Commonly, SSP is carried out at a temperature close to the melting point of the polymer (10–40°C lower).

In some cases it may be convenient to preheat the polymer in order to dry it and avoid sticking; this procedure is particularly important when the starting polymer has a high moisture content (polyamides) or when it has a low initial crystallinity (PET). Usually the prepolymer is charged as chips of about 0.03 cm^3 volume; the different geometry of the chips (cubes, cylinders, small plates etc.) may affect their packing tendency. Addition of glass beads has been claimed to avoid adhesion of PET particles to the reactor walls.[58]

The degree of crystallinity may also affect the diffusion coefficients[56] of by-products and the concentration of end groups in the amorphous region.[59]

Another important aspect of SSP is the purging of by-products produced during polycondensation. In this respect two factors have to be considered; the diffusion of the volatile products through the particles and their removal from the particle surface.

As expected, it is observed that the SSP rate rises on decreasing the particle diameter,[41,57,60-62] *i.e.* by increasing the surface/volume (S/V) ratio; for this purpose the employment of a porous starting prepolymer to speed up the process has also been proposed.[63] The rate increment, following the increase of the S/V ratio, is obviously dependent on some other variables of the process and on the polymer considered.

In order to remove the by-products, either vacuum or a stream of an inert gas can be used. A detailed comparison of the two methods has not been reported so far; however, nonsystematic evidence of their nearly equivalent effectiveness is scattered through the literature; for instance, it has been found that there is little or no difference in the global rate of molecular weight increase of PET using vacuum or dry nitrogen.[57,64,65] The choice between the two methods is usually made on the basis of general convenience rather than for the final properties of the polymer or the reaction rate. Industrially the choice is chiefly governed by the overall cost.

When an inert gas is used, two aspects have to be considered; the nature of the gas and its flow rate. Nitrogen is the most frequently used gas, but also H_2, He, air, CO_2 and exhausted combustion gases are employed effectively.[57,60] In some cases a better quality product was claimed when SSP of polyesters was carried out in a stream of inert gas containing the corresponding glycol.[66] A comparison of the effectiveness of different gases on the molecular weight increase of PET shows that at 250°C the effectiveness decreases from He to CO_2 to N_2.[60]

Studies on the effects of gas flow rate have shown that the molecular weight increases with gas flow rate up to a limit which is probably dependent on factors such as the geometry of the apparatus, temperature, particle diameter etc.[60,64]

In general, during SSP the molecular weight rises to a maximum after which it may decrease, usually for long reaction times, if the scission reactions become predominant.[60,67] As an example, the change of the intrinsic viscosity with reaction time is shown in Figure 1 for samples removed from the surface, the center and an intermediate depth from 3 mm thick planar sheets of PBT submitted to SSP.[67] As shown in Figure 1, and for SSP of large particles in general, a molecular weight gradient between the surface and the center may occur,[57,67-69] depending of course on the process parameters and the starting polymer.

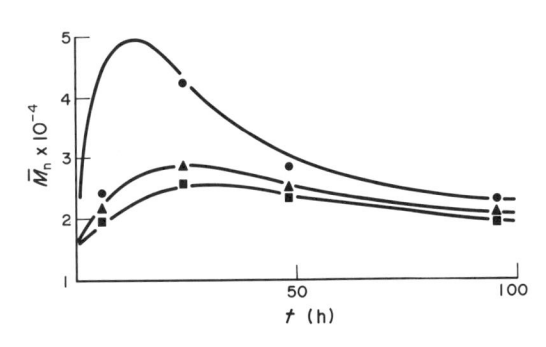

Figure 1 Number-average molecular weight $vs.$ SSP reaction time for slices removed from the surface (●), the center (■) and at intermediate depth (▲) of planar sheets of 3 mm thick PBT. Continuous lines are predicted by calculation[69] (reproduced by permission of Marcel Dekker Inc. from *Polym. Proc. Eng.*, 1986, **4**, 303)

It may be expected that, owing to the molecular weight gradient or to the kinetics of the process in general, the overall molecular weight distribution (MWD) deviates from the most probable MWD usually obtained in melt polycondensation.[47]

The scarce experimental data on MWD of polymers submitted to SSP[70-72] suggest that this process may lead to a slightly broader MWD, although a systematic dependence of molecular weight on various process parameters has not been observed.[71]

Calculations based on the assumption that the most probable MWD holds locally for each different depth in planar sheets of PBT (3 mm thick) give values of the overall polydispersity index ($\bar{M}_w/\bar{M}_n$) ranging from 2 to 2.2, depending on the molecular weight gradient.[73] On the other hand, Meyer has calculated that the MWD of postcondensate polyamides is not very different from the most probable one expected for equilibrium melt polycondensation.[74]

Finally it has to be mentioned that the polymers obtained by SSP generally have a lower content of cyclic oligomers than the analogous ones from solution or melt polycondensation.[75,76]

13.3.3 Postcondensation of Polyesters

Polyesters are commonly synthesized in melt condensation processes either by transesterification of dimethyl esters, or by direct esterification of carboxylic acids, with glycols (equations 10–12). These reactions proceed at a significant rate only at high temperature even in the presence of catalysts. However, during SSP the temperature has to be lower than the melting point of the polymer. For this reason the polycondensation in the solid state is limited to semicrystalline polyesters with high melting points. Among these, poly(alkylene terephthalate)s are the most interesting, and, due to their commercial importance, the SSP of PET and PBT are by far the most studied. With regard to the effect of the alkylene moiety, it was found that at 215 °C the SSP rate decreased in the order PBT > poly(1,3-propylene terephthalate) > PET[57] probably as a result of a complex combination of many effects, such as the different reactivity of glycols, different volatility,

$$MeO_2CR'CO_2Me + 2HOROH \rightleftharpoons HORO_2CR'CO_2ROH + 2MeOH \quad (10)$$

$$2\sim R'CO_2ROH \rightleftharpoons \sim R'CO_2RO_2CR'\sim + HOROH \quad (11)$$

$$\sim R'CO_2H + HOR\sim \rightleftharpoons \sim R'CO_2R\sim + H_2O \quad (12)$$

Scheme 2

chain end mobility, percent of crystallinity *etc*. Patents concerning the SSP of copolyesters have also been reported in the literature.[77]

All the reactions that take place during the melt polymerization of polyesters can in theory also occur during SSP. Ester interchange with elimination of glycol (equation 11) and direct esterification of hydroxyl and carboxyl end groups (equation 12) are the main reactions responsible for the rise of molecular weight. However, alcoholysis (equation 13), esterolysis (equation 14) and acidolysis (equation 15) may take part in the process. In addition to these, ester scission reactions[78–82] and side reactions concerning the end groups[79,83,84] should be considered in a complete kinetic scheme. Reaction mechanisms and kinetic parameters for these reactions are discussed in more detail in the chapter on the synthesis of polyesters.

$$P_xCO_2ROH + P_yCO_2RO_2CP_z \rightleftharpoons P_xCO_2RO_2CP_y + HORO_2CP_z \tag{13}$$

$$P_xCO_2RO_2CP_y + P_zCO_2RO_2CP_w \rightleftharpoons P_xCO_2RO_2CP_z + P_yCO_2RO_2CP_w \tag{14}$$

$$P_xCO_2H + P_yCO_2RO_2CP_z \rightleftharpoons P_xCO_2RO_2CP_y + P_zCO_2H \tag{15}$$

Except for ester scission and side reactions, all the above interchange reactions are equilibrium ones, and their equilibrium constants are close to unity, so they may be responsible for an increase of molecular weight as long as they give linear low molecular weight by-products which can be removed from the polymer particles. Due to the higher diffusion coefficient of water and glycols, equations (11) and (12) are by far the most important. Regarding the phase where the reactions occur, it is generally accepted that these take place in the amorphous regions where all the reactive end groups are assumed to be located.

The reaction kinetics, the chain end mobility or the diffusion of the low molecular weight by-products out of the polymer particle may become the rate-controlling step, depending on the SSP conditions.

A general treatment for SSP has to consider all the above reactions and the diffusion of by-products, as well as the two-phase nature of semicrystalline polymers and the chain end mobility. The resultant kinetic model has to account for the experimentally observed molecular weight modification, as well as for the change of end group concentration with time.

Although the reactions that take place during the melt polycondensation of PET have been widely studied (see Volume 5, Chapters 9 and 17), a complete kinetic model has not yet been developed. Most of the papers dealing with SSP of PET have only considered the distinct effects of various factors, or have interpreted the experimental results on the basis of simplified kinetic models.

In the first article dealing with SSP kinetics of PET, only the reaction shown in equation (11) leading to the formation of ethylene glycol was considered even though evidence of direct esterification and of an effect of particle size on the rate of glycol evolution were reported.[41] The authors found that the rate of glycol evolution follows a second-order reaction for both low and medium molecular weight of the starting PET. From the rate constants, evaluated in the range 170–250°C, activation energies of 74 and 81 kJ mol^{-1} were derived for low and medium molecular weight samples respectively. It was also noted that the rate constants are higher than those expected from extrapolation of the melt polymerization values, probably as a consequence of the higher concentration of end groups in the amorphous region. It has to be emphasized that, in the absence of antimony trioxide as catalyst, SSP does not proceed at all. In the same paper it is reported that chain orientation in drawn samples does not lead to any considerable change in the rate of reaction.

Some other authors have developed a kinetic model based on diffusion and on a general power law kinetic equation. Their calculations, for different geometries of the PET particles (sphere[85] plane sheet[85] and cube[56]), were compared with the experimental results of intrinsic viscosity. They concluded that SSP is controlled by diffusion when the PET particle size is greater than 100 mesh and the temperature is higher than 210°C[56] whilst at a lower temperature (160°C) the reaction becomes the rate-controlling step.[56,85] They also suggested that in the temperature range 160–210°C SSP is controlled by both diffusion and chemical reaction. The diffusivity was found to be linearly proportional to the mass fraction of the amorphous phase in PET[56] and an activation energy for the diffusion of about 126 kJ mol^{-1} was derived.

The above models did not take into account all different reactions which can occur and consequently cannot explain the effect of different hydroxyl to carboxyl end group ratios[64] and the molecular weight decrease at long SSP times.[60]

In order to explain the effect of the end group ratio, Zimmermann and co-workers proposed a kinetic model based on two chemical reactions, disregarding diffusion.[64] They considered the ester interchange and esterification-growing reactions (equations 11 and 12 respectively) for the molecular weight increase and a generic parameter for 'degradation' reactions. At 223°C they found that the kinetic constant of the esterification in equation (12) was more than twice that of the ester interchange reaction of equation (11); this explained the effect of end group ratio and on this basis they concluded that the optimum value is for $(OH/CO_2H) = 2$. They also found a global activation energy for the molecular weight increase of 125 kJ mol^{-1}, very close to that derived by Chang[56] for diffusion.

Other authors fitted their experimental data by plotting number-average molecular weight against the square root of the reaction time for PET, both catalyst free[86-88] and with different catalysts;[65] in the latter case different activation energies were obtained for different catalysts. In their work the authors also investigated the effect of some SSP conditions on the final content of acetaldehyde. They found that the latter decreases with increasing gas flow rate, being higher under vacuum than under nitrogen; in any case, the acetaldehyde content after SSP was always lower than after polycondensation in the molten state, resulting in only 2 ppm, in the best case, against 24 ppm of the starting PET.

Recently a model based on reactions (11) and (12) and taking into account end group diffusion, with or without by-product diffusion, has been proposed.[89]

This model can describe the decrease of the hydroxyl and carboxyl end group concentrations at various temperatures, reaction times and particle dimensions. In the absence of any by-product diffusion control (particle dimensions lower than 80–100 mesh), the decrease of the kinetic constant during SSP has been ascribed, analogously to what Gaymans has proposed for polyamides,[90] to the change in the distribution of end group distances and was accounted for by equation (16), where $k_{i,eff}$ is the effective kinetic constant, $k_{i,0}$ the initial kinetic constant, when there is a homogeneous distribution of the end groups, and $k_{i,1}$ a dimensionless kinetic constant related to diffusion rates of end groups; y refers either to hydroxyl or carboxyl end group concentration (for ester interchange and esterification reactions respectively), and Δ and 0 indicate already reacted and initial concentration respectively.

$$k_{i,eff} = k_{i,0}\{1 - (1/k_{i,1})(\Delta[y]/[y]_0 - \Delta[y])\} \qquad (16)$$

For larger particles (14–18 mesh) and higher temperatures (210–240 °C) the diffusion of by-products may become a limiting factor. The analysis of the experimental results lead the authors to conclude that the diffusion coefficient of water is much larger than that of ethylene glycol and the rate limitation due to water diffusion for reaction (12) is negligible. As a consequence, the esterification reaction, which for small particles at lower temperatures gives a contribution to the molecular weight increase close to that of ester interchange reaction (11), becomes predominant for larger particles and at higher temperatures.

These results seem to be able to account for the effect of the hydroxyl to carboxyl end group ratio,[64] as well as for the higher rate of SSP after remelting[41] previously reported for SSP of PET.

In general, from all the results reported for SSP of PET it can be concluded that both reactions (11) and (12) have to be considered for the molecular weight increase, and that their relative importance depends on the particle dimensions and temperature.

The SSP process for PBT is rather similar to that for PET even though several specific features distinguish PBT from PET. Thus because of the lower melting point of PBT, its SSP is commonly carried out at a lower temperature (200–225 °C), but the SSP rate is nevertheless higher. Unlike PET, precrystallization is not required for PBT which melt-crystallizes faster than PET. In addition, PBT being less sensitive to oxidation than PET, industrial exhaust gases can be used instead of nitrogen as the inert gas for the removal of volatile by products.[57] However, the most important difference probably lies in the reactions involving the two polymers. The main chain-growing reactions are similar in nature, although different catalysts are usually employed, but side reactions are rather different. Whilst for PET these are quite complex and lead mainly to acetaldehyde, 2-methyloxolane and diethylene glycol moieties in the chains,[65,78,79] for PBT the most important ones are the intramolecular cyclization of terminal 4-hydroxybutyl esters[83,84] giving THF and carboxyl end groups, and ester chain scission,[80-82] the latter occurring with a similar mechanism but at higher rates than for PET, giving two carboxyl end groups for each scission. As a consequence, the relative importance of reactions (11) and (12) may change during SSP for these side reactions also and it is not surprising that the initial hydroxyl to carboxyl end group ratio (r) is a very important parameter for SSP of PBT.[57,61,66-69,91]

Many patents and several papers have been published on SSP of PBT. Some of the latter only deal with the role of various parameters[57,61,91] whereas in a few other cases[67-69] a kinetic model, based on five reactions and the diffusion of three volatile species (1,4-butanediol, water and terephthalic acid), has been developed and used in order to interpret the experimental results. In the latter case,

SSP was performed at 214 °C on 3 mm thick planar sheets of PBT and the molecular weight at different depths in the sheets was measured over a wide range of reaction times. The curves for surface, center and intermediate depth are shown in Figure 1.

The model proposed to interpret these results is based on various assumptions, the most important of which are that all chain end groups are located in the amorphous regions, to which the reactions are confined, and that the most probable molecular weight distribution holds locally. The latter hypothesis is based on the assumption that an uninterrupted stochastic process of molecular weight redistribution occurs as a consequence of the fast ester-exchange reactions in the presence of a titanium catalyst. This process, which occurs for the chains in the amorphous regions as well as the chain-folded part of the macromolecules in the crystalline regions, is supposed to retain locally the most probable molecular weight distribution and also a homogeneous distribution of chain end groups, i.e. it is assumed that the end groups can move easily via exchange reactions. Consequently, the end group diffusion is neglected as a limiting factor for the reaction rate. Calculations are developed by taking kinetic and equilibrium constants from the literature and using only the diffusion coefficients of water, 1,4-butanediol and terephthalic acid as adjustable parameters. This model, developed for PBT in the form of both planar sheets[67,68] and spheres[92] fits the experimental data over a wide range of reaction times. It has to be emphasized that it also fits the experimental results previously reported by other authors.[91] It can also account for the molecular weight gradient previously observed[57,66] in the PBT particles and for the effect of the hydroxyl to carboxyl end group ratio r. The curves calculated from this model fit, in a fairly good manner, even the measured change of end group concentration with time.[92] From this model one can also infer information concerning the relative importance of the several reactions considered during SSP.

The relative role played by chemical kinetics and diffusion can be easily estimated for each reaction by the Thiele modulus (i.e. the ratio between reaction and diffusion characteristic rates). At 214 °C, for 3 mm thick planar sheets, it has been found that both reactions (11) and (12), leading to the formation of 1,4-butanediol and water respectively, are diffusion controlled, whilst the acidolysis reaction (15), which becomes important at long reaction times only, is controlled by both diffusion and chemical kinetics.

The model has also been used for predictive purposes in order to inspect the effect of different operating conditions;[69] some of the more interesting results are summarized below.

For very fine particles, where the effect of diffusion can be ruled out, the roles of end group ratio r, of initial molecular weight and reaction time are shown in Figures 2–4.

The maximum achievable value of the number-average molecular weight ($\bar{M}_{n,max}$) and the time needed to reach it are shown in Figures 2 and 3 against r for different initial number-average molecular weights $\bar{M}_n^0$. It appears that a significant increase of $\bar{M}_n$ can be obtained only when r is higher than 0.5, and the higher r is, the less the effect of $\bar{M}_n^0$ on $\bar{M}_{n,max}$. However the time needed to obtain $\bar{M}_{n,max}$ becomes larger as $\bar{M}_n^0$ decreases.

The effect of sample thickness is shown in Figure 4; as expected, the presence of diffusive resistances both decreases the maximum overall $\bar{M}_n$ and increases the time taken to reach it.

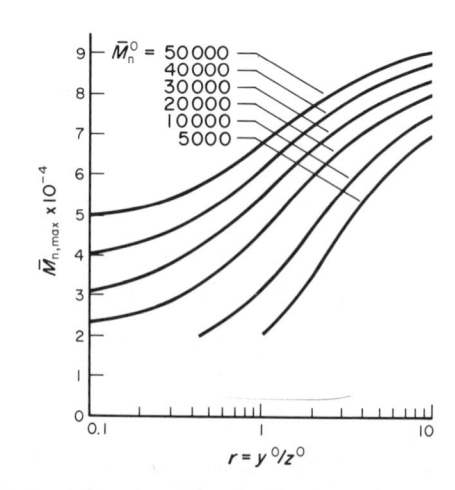

Figure 2 Maximum number-average molecular weight ($\bar{M}_{n,max}$) achievable in PBT postpolymerization vs. r (the hydroxyl to carboxyl end group ratio), with the initial molecular weight ($\bar{M}_n^0$) as a parameter (reproduced by permission of Marcel Dekker Inc. from Polym. Proc. Eng., 1986, **4**, 303)

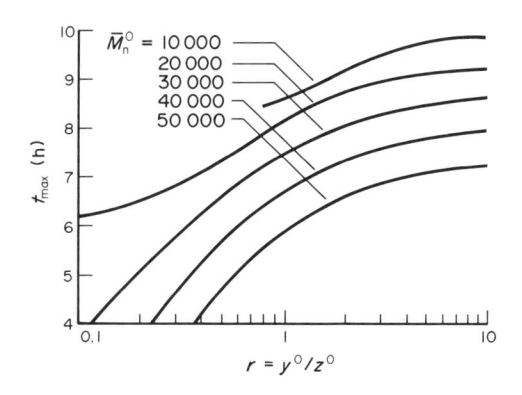

Figure 3 Time needed to reach $\bar{M}_{n,\,max}$ vs. r with the initial molecular weight $\bar{M}_n^0$ as a parameter (reproduced by permission of Marcel Dekker Inc. from *Polym. Proc. Eng.*, 1986, **4**, 303)

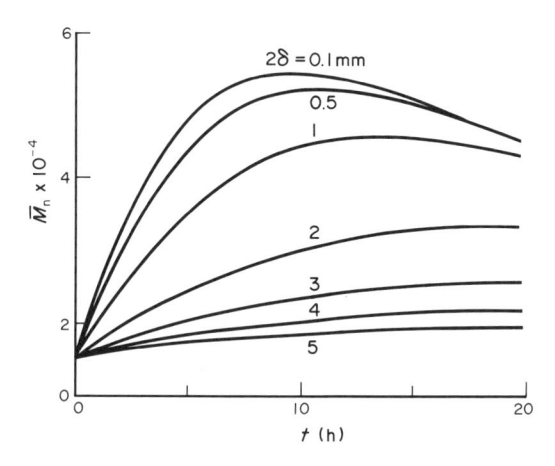

Figure 4 Overall number-average molecular weight vs. time for PBT sheets of different thickness (2δ) ($\bar{M}_n^0 = 15\,800$, $r = 2.94$; reproduced by permission of Marcel Dekker Inc. from *Polym. Proc. Eng.*, 1986, **4**, 303)

13.3.4 Crystallization-induced Reactions

Evidence of reorganization from random to multiple block copolymers was reported more than 20 years ago,[49] but it was only 10 years after that Lenz and co-workers undertook a deeper investigation of this process called crystallization-induced reaction (CIR).[50-54]

The publications on this subject have been concerned with copoly(ester/acetal)s, *cis-/trans-*copoly(butadiene) and, mainly, with copolyesters; in the latter case, more pertinent to the subject of this chapter, the process involves one or more of the exchange reactions (13)–(15).

The reorganization occurs in partially crystalline copolymers at temperatures below the melting point of the crystalline component, so it can be included in the class of solid-state processes and is discussed as a particular case of postcondensation of polyesters. As the exchange reactions take place only at high temperatures, the CIR process is restricted to copolymers having adequate type and content of crystallizable units.[52] CIR reorganization can also occur above the melting point in liquid crystal melts,[54,55] but it has never been found to take place in isotropic melts.[54]

The exchange reactions preferentially occur in the presence of a suitable catalyst, whose choice becomes an important variable as different catalysts may activate the exchange reactions (13)–(15) and/or degradation in distinct ways depending on copolymer type and temperature. Among the various catalysts tested, Lenz and co-workers found that Sb_2O_3 and *p*-toluenesulfonic acid were the most effective for the systems investigated.[53,54]

During CIR reorganization, crystallizable sequences formed through stochastic exchange reactions are taken into the crystalline regions and become inaccessible to further reaction. This obviously requires that the two (or more) monomeric units of the copolymer are not isomorphous. The process leads to an increase of crystalline domain sizes and of the overall degree of crystallinity as well as of the sequence length of the crystallizable units. These effects, in turn, are reflected in a higher melting point and lower solubility. Thus, for instance, when copoly(ethylene terephthalate/2-methylsuccinate) having about 80% of terephthalate units is heated at 220 °C for 30 h, the number-average sequence length of the terephthalate units increases from 5.3 to 6.8 and the melting point from 228 to 237 °C.[52]

The rate of the process is probably a complex result of crystallization and reaction kinetics: at temperatures well below the melting point the reaction seems to be the rate-controlling step, whilst the relative insensitivity to the amount of catalyst and the effect of temperature suggest that crystallization probably becomes the rate-controlling step when the melting point (or the clearing transition temperature for the liquid crystal state) is approached.[53,54]

13.3.5 Postcondensation of Polyamides

As with polyesters, polyamide prepolymers can be further polymerized in the solid state in an oxygen free atmosphere at temperatures 10–160 °C below their melting point. This postcondensation process is widely used in industry to produce high molecular weight materials for tire cords and engineering plastics; however, despite its widespread application, very little information, confined to aliphatic polyamides, has been published so far about mechanism and kinetics.[85,90,93-95]

Only the amidation reaction (17) is usually considered for SSP of polyamides. Kinetic studies performed on this reaction in an anhydrous melt are somewhat contradictory; however, most of them support a second-order kinetic equation ($[CO_2H][NH_2]$) for an extent of reaction lower than about 0.9 and a third-order dependence ($[CO_2H]^2[NH_2]$) for higher conversions.[48]

$$P_xCO_2H + H_2NP_y \rightleftharpoons P_xCONHP_y + H_2O \tag{17}$$

In the kinetic treatment of the data, it is generally assumed that all end groups are located in the amorphous regions. This hypothesis is supported by the results of a study on the effect of the degree of crystallinity on the molecular weight of polyamides in equilibrium with a constant pressure of water.[59] The increase of molecular weight with crystallinity has been explained by assuming that the end groups are located in the amorphous phase; as a consequence, their concentration in these regions would be higher than if they were distributed over the total volume and the reaction (17) is shifted to the right in order to reestablish equilibrium within this region. A higher degree of crystallinity, i.e. a lower volume fraction of amorphous phase, will result in a greater increase of molecular weight.

The authors who first dealt with the kinetics of SSP of polyamides studied nylon 6,6[85,93] and nylon 6,10[85] in the temperature range 90–180 °C. They concluded that the chemical reaction and the diffusion of volatile products are the controlling steps for nylon 6,6 and nylon 6,10 respectively.[85] However, other mechanisms could probably also explain their experimental results.

The SSP of nylon 6 was carefully examined by Gaymans and co-workers.[90] They found not only that the reaction rate was affected by the starting molecular weight (as it had been found previously[96]), but also that different heat treatments, performed before submitting the sample to SSP, did not affect the rate of the process and that SSP proceeded faster, under the same conditions, when 0.1% H_3PO_4 was added. They also found that the rate of the process was enhanced when SSP started again after an intermediate quick melting.

These observations lead the authors to conclude that the diffusion of reactive end groups is the limiting step of the process. Indeed, their experimental data can be explained by assuming that there is a change in the distribution of the reactive end groups in the amorphous region during SSP.

The starting polymer has an initial stochastic distribution of end groups, but as SSP proceeds, the end groups with the smallest end-to-end distances react whilst the others remain virtually frozen in. The change of the chain end distribution accounts for the reduction of SSP rate with time, being a consequence both of the decrease of the nearest groups and of the slow diffusion rate of carboxyl end groups, which catalyze the chain growth reaction. When H_3PO_4 is used as the catalyst, the reaction proceeds faster because H_3PO_4, which is not bonded to macromolecules, can diffuse freely. However, when a sample previously submitted to SSP is melted, the stochastic distribution of chain ends is restored accounting for the faster subsequent SSP.

On the basis of this model the authors proposed a kinetic equation which is first order in end group concentration and inversely proportional to the reaction time (equation 18).[90]

$$-(dc/dt) = k(c/t) \tag{18}$$

A similar kinetic model seems to account for the results of experiments accomplished on nylon 4,6.[94] Even in that case the authors concluded that the diffusion of the reactive end groups is the controlling step even though both the chemical reaction and the diffusion of the by-products can have an effect on the overall rate.

Postcondensation has also been successfully performed for poly(1,4-tetramethylene terephthalamide) (nylon T4) in order to obtain a high molecular weight not otherwise achievable.[95]

Very high molecular weight poly(*p*-benzamide) was prepared by SSP of oligomers probably bearing carboxyl and isothiocyanate end groups at temperatures increasing from 180 to 400 °C.[97] Poly(*m*-benzamide) was prepared in the same way.[97]

Finally it has to be recalled that most of the SSP of monomers are in effect postcondensation processes in the last stage of the polymerization.

13.4 REFERENCES

1. H. Morawetz, in 'Physics and Chemistry of the Organic Solid-State', ed. D. Fox, M. M. Labes and A. Weissberger, Interscience, New York, 1965, vol. II, p. 853.
2. H. W. Kohlschutter, *Justus Liebigs Ann. Chem.*, 1930, **482**, 75.
3. H. W. Kohlschutter and L. Sprenger, *Z. Phys. Chem., Abt. B*, 1932, **16**, 284.
4. P. J. Flory (Du Pont), *US Pat.* 2 172 374 (1939) (*Chem. Abstr.*, 1940, **34**, 198g).
5. R. H. Baughman and K. C. Yee, *J. Polym. Sci., Macromol. Rev.*, 1978, **13**, 219.
6. W. M. Thomas, in 'Encyclopedia of Polymer Science and Technology', ed. H. F. Mark, N. Gaylord and N. M. Bikales, Interscience, New York, 1964, vol. 1, p. 177.
7. L. B. Sokolov, 'Synthesis of Polymers by Polycondensation', Jerusalem Program for Scientific Translation, Jerusalem, 1968, chap. 8, p. 192.
8. A. Chapiro, in 'Encyclopedia of Polymer Science and Technology', ed. H. F. Mark, N. G. Gaylord and N. M. Bikales, Interscience, New York, 1969, vol. 11, p. 725.
9. R. W. Lenz, C. E. Handlovits and H. A. Smith, *J. Polym. Sci.*, 1962, **58**, 351.
10. R. W. Lenz and C. E. Handlovits, *J. Polym. Sci.*, 1960, **43**, 167.
11. H. Morawetz, *J. Polym. Sci., Part C*, 1966, **12**, 79.
12. E. I. Levites, A. V. Volokhina and G. I. Kudryavtsev, *Polym. Sci. USSR (Engl. Transl.)*, 1963, **4**, 1594.
13. B. A. Bagramyants, A. V. Volokhina, G. I. Kudryavtsev and N. S. Yenikolopian, *Polym. Sci. USSR (Engl. Transl.)*, 1967, **9**, 200.
14. G. Kovacs, E. Kovacs and H. Morawetz, *J. Polym. Sci., Part A-1*, 1966, **4**, 1553.
15. M. Oya, K. Uno and Y. Iwakura, *Makromol. Chem.*, 1972, **154**, 309.
16. A. V. Volokhina and G. I. Kudryavtsev, *Dokl. Akad. Nauk SSSR*, 1959, **127**, 1221 (*Chem. Abstr.*, 1960, **54**, 282e).
17. A. V. Volokhina, G. I. Kudryavtsev, S. M. Skuratov and A. K. Bonetskaya, *J. Polym. Sci.*, 1961, **53**, 289.
18. C. D. Papaspyrides and E. M. Kampouris, *Polymer*, 1984, **25**, 791.
19. F. Chambret, *Bull. Soc. Chim. Fr.*, 1947, 283.
20. N. Morosoff, D. Lim and H. Morawetz, *J. Am. Chem. Soc.*, 1964, **86**, 3167.
21. E. M. Macchi, N. Morosoff and H. Morawetz, *J. Polym. Sci., Part A-1*, 1968, **6**, 2033.
22. E. M. Macchi and A. A. Giorgi, *Makromol. Chem.*, 1979, **180**, 1603.
23. E. M. Kampouris, *Polymer*, 1976, **17**, 409.
24. E. M. Kampouris and C. D. Papaspyrides, *Polymer*, 1985, **26**, 413.
25. C. D. Papaspyrides and E. M. Kampouris, *Polymer*, 1986, **27**, 1433.
26. C. D. Papaspyrides and E. M. Kampouris, *Polymer*, 1986, **27**, 1437.
27. California Research Corp., *Br. Pat.* 852 672 (1960). (*Chem. Abstr.*, 1961, **55**, 10 968a).
28. Y. Iwakura, K. Uno, M. Akiyama and K. Haga, *J. Polym. Sci., Part A-1*, 1969, **7**, 657.
29. Y. Iwakura, K. Uno, K. Haga and K. Nakamura, *J. Polym. Sci., Polym. Chem. Ed.*, 1973, **11**, 367.
30. L. A. Ae. Sluyterman and H. J. Veenendaal, *Recl. Trav. Chim. Pays-Bas*, 1952, **71**, 137.
31. L. A. Ae. Sluyterman and M. Kooistra, *Recl. Trav. Chim. Pays-Bas*, 1952, **71**, 277.
32. P. Guyot, *C.R. Hebd. Seances Acad. Sci.*, 1951, **233**, 1604.
33. A. V. Volokhina and G. I. Kudryavtsev, *Khim. Volokna*, 1959, **5**, 13 (*Chem. Abstr.*, 1960, **54**, 9754c).
34. A. V. Volokhina, M. N. Bogdanov and G. I. Kudryavtsev, *Vysokomol. Soedin.*, 1960, **2**, 92 (*Chem. Abstr.*, 1960, **54**, 22 479a).
35. A. V. Volokhina, G. I. Kudryavtsev, M. V. Raeva, M. N. Bogdanov, V. D. Kalmykova, F. M. Mandrosova and N. P. Okromchedlidze, *Khim. Volokna*, 1964, **6**, 30 (*Chem. Abstr.*, 1965, **62**, 10 361d).
36. Y. Iwakura, K. Uno and N. Chau, *J. Polym. Sci., Polym. Chem. Ed.*, 1973, **11**, 2391.
37. P. Costa Bizzarri, C. Della Casa and A. Monaco, *Polymer*, 1980, **21**, 1065.
38. M. Prince and J. Hornyak, *J. Polym. Sci., Polym. Lett. Ed.*, 1966, **4**, 601.
39. A. V. Volokhina and G. I. Kudryavtsev, *Vysokomol. Soedin.*, 1959, **1**, 1724 (*Chem. Abstr.*, 1960, **54**, 23 411e).
40. E. G. Khripkov, V. M. Kharitonov and G. I. Kudryavtsev, *Proizvod. Sin. Volokon*, 1971, 63 (*Chem. Abstr.*, 1972, **77**, 35 035g).
41. C. H. Bamford and R. P. Wayne, *Polymer*, 1969, **10**, 661.
42. H. Zahn and G. B. Gleitsman, *Angew. Chem., Int. Ed. Engl.*, 1963, **2**, 410.
43. A. A. B. Browne and J. E. McIntyre, *Polymer*, 1983, **24**, 1615.
44. A. A. B. Browne and J. E. McIntyre, *Ger. Pat.* 2 337 288 (1974) (*Chem. Abstr.*, 1974, **81**, 13 977).
45. B. Wunderlich, *Adv. Polym. Sci.*, 1968, **5**, 568.
46. H. Zeng and L. Feng, *Gaofenzi Tongxun*, 1983, **5**, 321 (*Chem. Abstr.*, 1984, **100**, 139 663).

47. P. J. Flory, 'Principles of Polymer Chemistry', Cornell University Press, Ithaca, NY, 1953, p. 83.
48. W. Sweeny and J. Zimmerman, in 'Encyclopedia of Polymer Science and Technology', ed. H. F. Mark, N. G. Gaylord and N. M. Bikales, Interscience, New York, 1969, vol. 10, p. 498.
49. R. M. Schulken, Jr., R. E. Boy, Jr. and R. H. Cox, *J. Polym. Sci., Part C*, 1964, **6**, 17.
50. R. W. Lenz, E. Martin and A. N. Schuler, *J. Polym. Sci., Polym. Chem. Ed.*, 1973, **11**, 2265.
51. R. W. Lenz and S. Go, *J. Polym. Sci., Polym. Chem. Ed.*, 1973, **11**, 2927.
52. R. W. Lenz and S. Go, *J. Polym. Sci., Polym. Chem. Ed.*, 1974, **12**, 1.
53. R. W. Lenz and A. N. Schuler, *J. Polym. Sci., Polym. Symp.*, 1978, **63**, 343.
54. R. W. Lenz, J-Il Jin and K. A. Feichtinger, *Polymer*, 1983, **24**, 327.
55. E. R. George and R. S. Porter, *Macromolecules*, 1986, **19**, 97.
56. T. M. Chang, *Polym. Eng. Sci.*, 1970, **10**, 364.
57. L. H. Buxbaum, *J. Appl. Polym. Sci., Appl. Polym. Symp.*, 1979, **35**, 59.
58. W. Gey, W. Langhauser, H. Heinze, H. J. Rothe and P. Freund, *Ger Pat.* 2 453 577 (1976) (*Chem. Abstr.*, 1976, **85**, 63 647).
59. J. Zimmerman, *Polym. Lett.*, 1964, **2**, 955.
60. Li-C. Hsu, *J. Macromol. Sci., Phys.*, 1967, **1**, 801.
61. B. Fortunato, F. Pilati and P. Manaresi, *Polymer*, 1981, **22**, 655.
62. S. Chang, M. F. Sheu and S. M. Chen, *J. Appl. Polym. Sci.*, 1983, **28**, 3289.
63. K. H. Berger, H. Toll, W. Fichtner, U. Froboess and E. Schoebel, *Ger. (East) Pat.* 146 610, 148 781 (1981) (*Chem. Abstr.*, 1981, **95**, 44 146, 187 900).
64. E. Schaaf, H. Zimmermann, W. Dietzel and P. Lohmann, *Acta Polym.*, 1981, **32**, 250.
65. S. A. Jabarin and E. A. Lofgren, *J. Appl. Polym. Sci.*, 1986, **32**, 5315.
66. W. F. H. Borman (General Electric Co.), *US Pat.* 3 953 404 (1976); *Ger. Pat.* 2 503 000 (1975) (*Chem. Abstr.*, 1975, **83**, 194 045).
67. C. Gostoli, F. Pilati, G. C. Sarti and B. Di Giacomo, *J. Appl. Polym. Sci.*, 1984, **29**, 2873.
68. C. Gostoli, F. Pilati and G. C. Sarti, in 'Preprints of the Third International Conference on Reactive Processing of Polymers', Strasbourg, 1984, p. 49.
69. F. Pilati, C. Gostoli and G. C. Sarti, *Polym. Proc. Eng.*, 1986, **4**, 303.
70. C.-Y. Cha, *Polym. Prepr., Am. Chem. Soc., Div. Polym. Chem.*, 1965, **6**, 84.
71. S. A. Jabarin and D. C. Balduff, *J. Liq. Chromatogr.*, 1982, **5**, 1825.
72. B. Langlà and C. Strazielle, *Makromol. Chem.*, 1986, **187**, 591.
73. F. Pilati and C. Gostoli, unpublished results.
74. K. Meyer, *Angew. Makromol. Chem.*, 1973, **34**, 165.
75. D. R. Cooper and J. A. Semlyen, *Polymer*, 1973, **14**, 185.
76. G. Wick and H. Zeitler, *Angew. Makromol. Chem.*, 1983, **112**, 59.
77. W. F. H. Borman, D. I. Craft and M. Kramer (General Electric Co.), *Ger. Pat.* 2 613 649 (1976) (*Chem. Abstr.*, 1977, **86**, 5976); S. W. Go (Owens-Illinois Inc.), *US Pat.* 4 330 661 (1982) (*Chem. Abstr.*, 1982, **97**, 93 035); Toray Industries Inc. *Jpn. Pat.* 82 32 926 (1982) (*Chem. Abstr.*, 1982, **97**, 73 030); R. R. Smith and J. R. Wilson (Goodyear Tire and Rubber Co.) *Eur. Pat. Appl.* 174 265 (1986) (*Chem. Abstr.*, 1986, **104**, 225 454).
78. H. Zimmermann, in 'Developments in Polymer Degradation', ed. N. Grassie, Applied Science, London, 1984, vol. 5, p. 79.
79. E. P. Goodings, *Soc. Chem. Ind. (London) Monogr.*, 1961, **13**, 211.
80. L. H. Buxbaum, *Angew. Chem., Int. Ed. Engl.*, 1968, **7**, 182.
81. V. Passalacqua, F. Pilati, V. Zamboni, B. Fortunato and P. Manaresi, *Polymer*, 1976, **17**, 1044.
82. J. Devaux, P. Godard and J. P. Mercier, *Makromol. Chem.*, 1978, **179**, 2201.
83. R. M. Lum, *J. Polym. Sci., Polym. Chem. Ed.*, 1979, **17**, 203.
84. F. Pilati, P. Manaresi, B. Fortunato, A. Munari and V. Passalacqua, *Polymer*, 1981, **22**, 1566.
85. F. C. Chen, R. G. Griskey and G. H. Beyer, *AIChE J.*, 1969, **15**, 680.
86. M. Droscher and G. Wegner, *Polymer*, 1978, **19**, 43.
87. M. Droscher, *J. Appl. Polym. Sci., Appl. Polym. Symp.*, 1981, **36**, 217.
88. M. Droscher, *Ind. Eng. Chem., Prod. Res. Dev.*, 1982, **21**, 126.
89. S. A. Chen and F. L. Chen, *J. Polym. Sci., Polym. Chem. Ed.*, 1987, **25**, 533.
90. R. J. Gaymans, J. Amirtharay and H. Kamp, *J. Appl. Polym. Sci.*, 1982, **27**, 2513.
91. H. D. Dinse and E. Tucek, *Acta Polym.*, 1980, **31**, 108.
92. M. Chiappani, Ph.D. Thesis, University of Bologna, 1985.
93. R. G. Griskey and B. I. Lee, *J. Appl. Polym. Sci.*, 1966, **10**, 105.
94. R. J. Gaymans and J. Schuijer, *ACS Symp. Ser.*, 1979, **104**, 137.
95. R. J. Gaymans, *J. Polym. Sci., Polym. Chem. Ed.*, 1985, **23**, 1599.
96. G. C. Monroe, *US Pat.*, 3 031 433 (1962) (*Chem. Abstr.*, 1963, **58**, 588g).
97. W. Memeger, Jr., *Polym. Prepr., Am. Chem. Soc., Div. Polym. Chem.*, 1976, **17**, 163.

14
Cyclobutane Polymers by Topochemical Photopolymerization

MASAKI HASEGAWA
University of Tokyo, Japan

14.1 OVERVIEW OF FOUR-CENTER TYPE PHOTOPOLYMERIZATION	217
14.2 PREPARATION OF POLYMERS	218
14.3 REACTION MECHANISM OF FOUR-CENTER TYPE PHOTOPOLYMERIZATIONS	224
14.3.1 Step-growth Mechanisms	224
14.3.2 Crystallographic Interpretation	226
14.4 CHARACTERISTIC PROPERTIES OF THE POLYMERS FORMED	229
14.5 REFERENCES	231

14.1 OVERVIEW OF FOUR-CENTER TYPE PHOTOPOLYMERIZATION

As thermal motion of molecules is restricted in a crystal, the crystal structures of the starting compound and the resulting product are expected to afford the most pertinent information for characterizing the majority of the features of a solid state reaction. The relative orientation of two reacting molecules is most important in determining their reactivity in a crystalline state bimolecular reaction. In fact, as successive reactions are required throughout a large number of monomer molecules, the molecular arrangement must be restricted to that favoring the crystalline state polymerization in order to give a high polymer.

In spite of early reports on solid state photoreactions, it was in a rather recent article that the correlation between the reaction behavior and the molecular arrangement in the crystal was clearly described. The reason is, of course, that before the development of computer science, crystallographic analysis had been very difficult even for a simple organic molecule.

For example, the solid state photodimerization of cinnamic acid crystals was reported by Liebermam in 1889, and the photochemical reaction of two types of cinnamic acid crystals was interpreted in terms of a crystal-lattice-controlled reaction by Bernstein and Quimby in 1943.[1] In 1964, Schmidt and co-workers demonstrated the correlation between the crystal structures of two types of cinnamic acid crystals and the configurations of the corresponding photoproducts, and they proposed a topochemical rule for the crystal structure of photoreactive alkenic compounds.[2] There are three types of crystal structures of photodimeric alkene molecules: the α type crystal in which the double bonds of neighboring molecules are arranged in a parallel fashion and make contact at a distance of approximately 4 Å across a center of symmetry; the β type, characterized by a lattice with one axial length of 4 Å between translationally related molecules; and the γ type, in which no double bonds of neighboring molecules are within 4.7 Å of each other. On photoirradiation an α type crystal of cinnamic derivatives gives a centrosymmetric dimer ($\bar{1}$-dimer) which is an α truxillic derivatives, and a β type crystal gives a dimer of mirror symmetry (m-dimer) which is a β truxinic derivative. However, no crystal correlations were demonstrated between the starting alkene and the cyclobutane product in the work of Schmidt and co-workers.

Solid state polymerization has been studied extensively, mostly of the vinyl and cyclic ether derivatives during the 1950s. In the ring-opening polymerization of crystals of cyclic oligoethers, *e.g.* trioxane, a certain correlation was confirmed between the molecular arrangement in the monomer crystal and the growth direction of the polymer chain.[3] In the polymerization of the solid solution composed of 3,3-bis(bromomethyl)- and 3-bromomethyl-3-ethyl-oxetanes, the composition of the two monomers in the resultant copolymer is determined by the monomer ratio in the solid solution of the monomers, and not by the monomer reactivity ratio.[4] Although the term 'topochemical polymerization' was used for the first time for the polymerization of the trioxane crystal, no correlation had then been demonstrated between the monomer and the polymer in the dimension of the crystal unit cell.

The first study of the topochemical [2+2] photopolymerization was carried out on a crystal of 2,5-distyrylpyrazine (DSP) in 1967.[5]

DSP was first prepared in 1905,[6] and a photochemical change of the DSP crystal was briefly reported in 1958.[7]

The photopolymerization that the DSP crystal undergoes in forming the cyclobutane in the main chain occurs by a step-growth mechanism, and produces a highly crystalline linear polymer in quantitative yield.

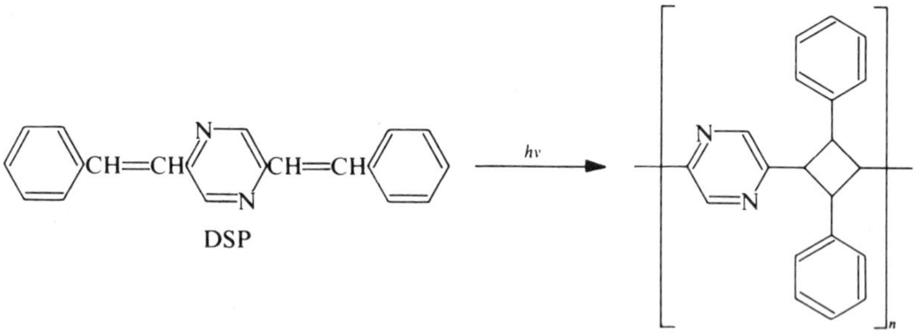

Scheme 1

The reaction behavior of DSP crystal and the configuration of poly-DSP (PDSP) were thoroughly interpreted in terms of topochemical concepts by Schmidt. In the DSP crystal the double bonds are related to the center of symmetry and are separated by 3.939 Å (see Table 3).[8] The polymerization proceeds by retention of the space group of the crystal (*Pbca*), accompanied by a slight approach of the centers of gravity of the monomer units (approximately 1.8% shrinkage in the direction of the c axis).[9]

The correlation between the crystal unit cells of the starting compound and the final product was originally demonstrated for DSP and PDSP crystals in organic reactions. In addition, the reaction of DSP crystal was the first example of photopolymerization by a step-growth mechanism.

The polymerization of DSP crystal was named 'four-center type photopolymerization'. The same type of polymerization was reported for a series of α,α'-bis(4-acetoxy-3-methoxybenzylidene)-1,4-phenylenediacetonitrile.[10]

The reactivities of dialkene crystals and their product configurations are almost all interpreted satisfactorily from the crystal structure of the starting monomers by Schmidt's rule. However, several recent results deviate from the above rule. These deviations are related to the distance and parallelism of reactive double bonds.[11]

Another example of topochemical polymerization in which the crystal correlation has been confirmed between the starting compound and the product is that of diacetylene derivative crystals. The reactions were reported by Dunitz and by Seher,[12] and the lattice-controlled mechanism of these reactions, which was suggested by Hirshfeld and Schmidt,[13] has been confirmed by Wegner and his collaborators.[14] In contrast to the polymerization of dialkene crystals, the polymerization of diacetylene crystals proceeds by a chain mechanism.

There are quite a few other examples of polymerization reactions which apparently proceed in the solid state above the glass transition temperature, but not in the crystalline state. In such solid state polymerizations, no appreciable correlations are observed between the crystal structure and the reactivity. A ring-closure into the aromatic polybenzimidazole from the polyaminoamide at the final stage of polymerization[15] or a further growth of molecular weight of aromatic polyester by annealing[16] may be included in this category.

14.2 PREPARATION OF POLYMERS

Synthetic study of four-center type photopolymerization established the empirical rule that the crystalline state [2+2] photodimerization of alkenes can be broadly extended to the crystalline state photopolymerization of symmetric and unsymmetric dialkenes, *e.g.* stilbazole extended to the DSP series, cinnamic derivatives to the dialkyl 1,4-phenylenediacrylates, and to the pyridylethenyl cinnamates,[17,18] as is illustrated in Scheme 2.

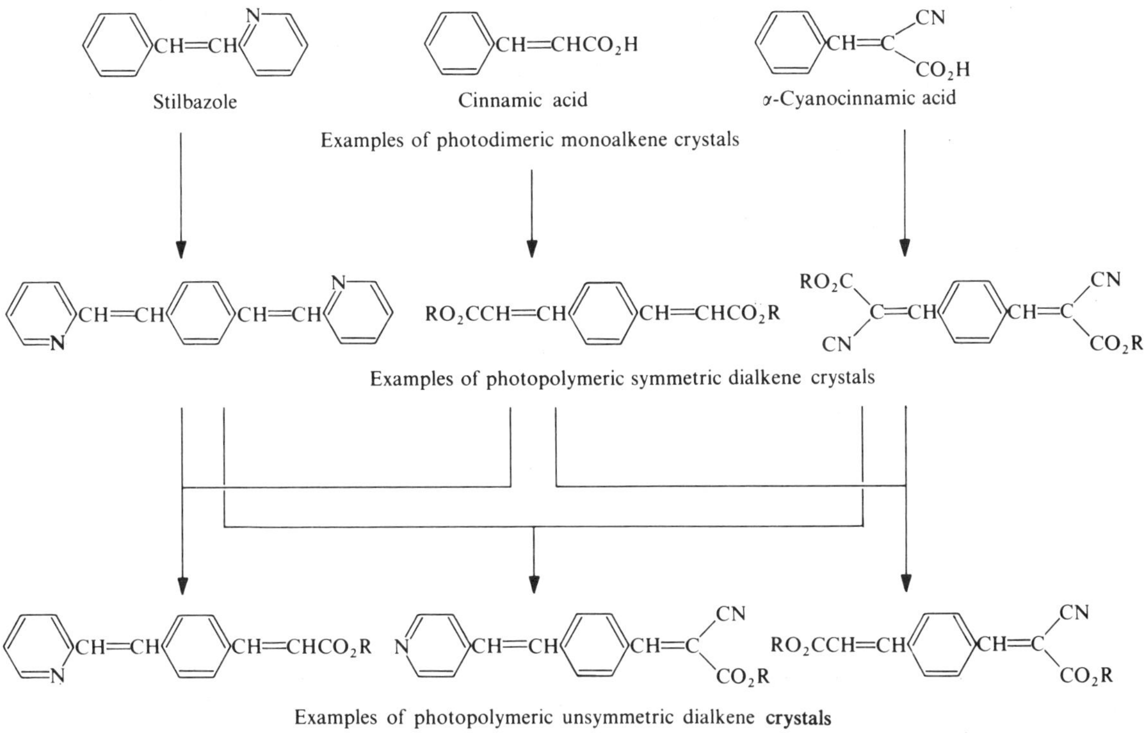

Scheme 2 The relation between the chemical structures of photoreactive monoalkene and dialkene crystals

This rule suggested the similarity of the molecular arrangement between the photoreactive crystals of monoalkenic and dialkenic compounds. Since then, use of the empirical rule has allowed a great number of symmetric and unsymmetric photoreactive dialkene crystals to be found.

A few unsymmetric dialkene monomers, which were prepared according to this rule, have been converted into the optically active oligomeric photoproducts by chiral crystal formation.[19]

With a few exceptions, the photoreactive dialkenes contain two double bonds which are conjugated to each other. Most of these monomers are prepared by aldol condensation between appropriate aldehyde and methyl groups, for example between benzaldehyde and 2,5-dimethyl pyrazine to give the DSP, or between terephthalaldehyde and malonic acid followed by decarboxylation to give 1,4-phenylenediacrylic acid (1,4-PDA). Instead of an aromatic methyl derivative, Wittig's reagent is more convenient to use in some cases. For preparation of unsymmetric dialkenes, successive two-step condensations are carried out.

These conjugated dialkene monomers have a rigid linear molecular shape which is presumably favored for potentially reactive orientation between intermolecular double bonds. The exceptional examples of topochemically photoreactive monomers include dialkenic dimers which are isolated from the photoproducts at the intermediate stage of polymerization. These monomers are not conjugated, but are still rigid rod-like structures.

A similar series of the conjugated bisbutadiene type derivatives, such as 1,4-phenylenebutadienoic ester has been prepared and subjected to four-center type photopolymerization.[20]

The molecular arrangement of nearly all the symmetric dialkene compounds is that of the α type crystal, whereas the arrangement of several unsymmetric compounds is the β type crystal. In the α type crystal, molecules are stacked along an approximately 7 Å axis, displaced by half a molecule in the direction of the long molecular axis, to form a characteristic plane-to-plane stack. In many of the photopolymerizable β type crystals, the molecules are piled up infinitely along the shortest crystal axis to form a parallel plane-to-plane stack.

In assuming the topochemical process, the configuration of resulting photoproducts is precisely predicted from these crystal structures of the monomer; the linear polymers are derived from the α type crystals and the zigzag type polymers or tricyclic cyclophanes from the β type crystals. Sometimes dialkenic dimers are obtained both from the α and the β type crystals, as illustrated in Scheme 3.

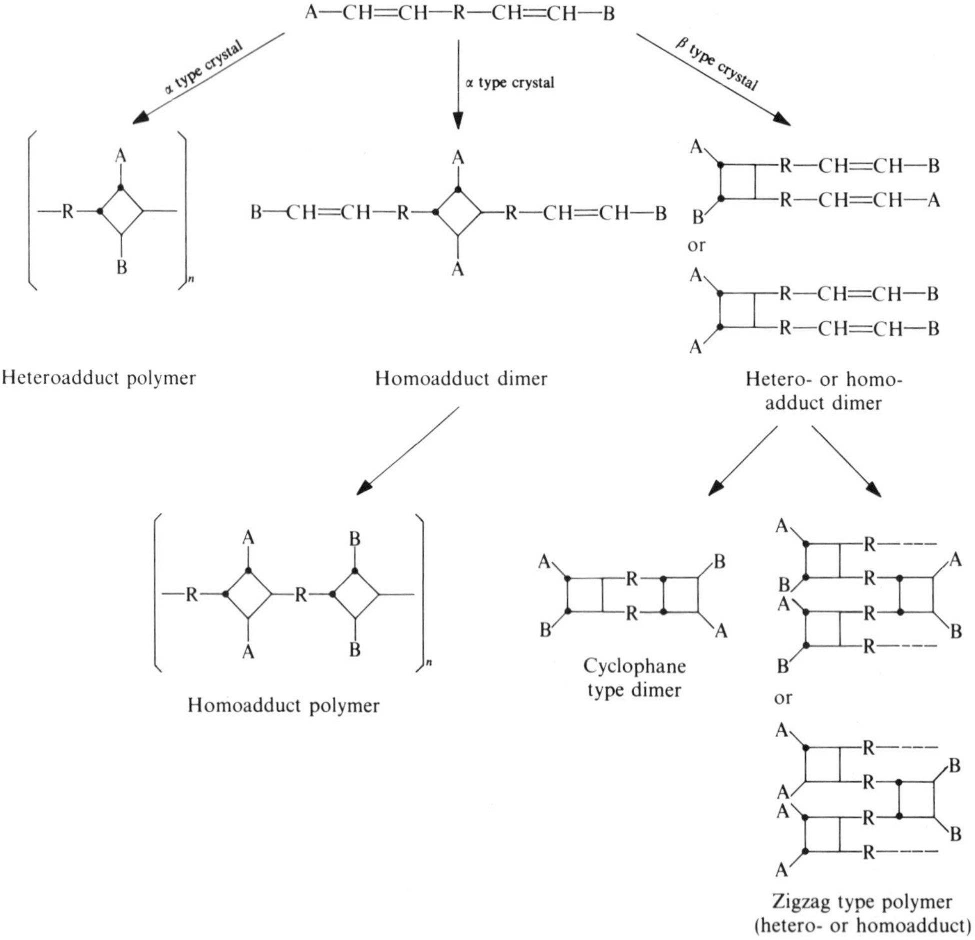

Scheme 3 Structures of the photoproduct from the topochemical reaction of symmetric (A = B) and unsymmetric (A ≠ B) dialkene crystals. The terms hetero- and homo-adduct denote photoproducts derived from the unsymmetric dialkene crystals

For large scale polymerization, each monomer is dispersed as fine crystals in a suitable dispersant that does not dissolve any trace amounts of monomer, such as water or water/ethanol. The dispersed monomer crystals are irradiated in a quartz flask with a 100 W or 500 W high-pressure mercury lamp (or with a xenon lamp) for an appropriate period while being vigorously stirred.

Many of the monomer crystals absorb in the visible region of wavelength and are colored orange to yellow, but these crystals fade to opaque white during the reaction.

Almost all of the photoproducts from the α type crystal are of a highly crystalline linear polymeric nature. However, many of the products from the β type crystal are amorphous and oligomeric. Several typical examples of topochemical behavior and product properties are shown in Table 1.

Polymerization proceeds smoothly if the temperature is sufficiently less than the crystal melting point of the monomer, as is illustrated in the case of the 1,4-PDA Et crystal (m.p. 96 °C) in Table 1.

The cyclobutane formation is easily monitored by IR and ^{1}H NMR spectroscopy. The strong IR absorption peak at around 1650 cm^{-1} (C=C, aliphatic) and the medium peak at 970 cm^{-1} (*trans* HC=CH) are gradually diminished during the reaction. In addition, a clear shift of the conjugated carbonyl peak into the nonconjugated carbonyl peak region (approximately 1650 cm^{-1} → 1720 cm^{-1}) is seen for the reaction of carbonyl derivatives. The steric configuration of the cyclobutane ring is determined from the NMR peaks at around $\tau = 4.5$–5.0, and it accords with that predicted from the crystal structure of the monomer. MS analysis is very useful for the assignment of hetero- and homo-adduct type products.

Table 1 Examples of Topochemical Behavior of Dialkene Crystals[a]

Structure of dialkenes	Irradiation temperature	Reactivity[b] (Yield)[c]	Type of adduct[d]	Morphology	η inh[e]	Ref.

Ar'—CH=C—Ar—C=CH—Ar''
          | |
          X X						

$Ar = 1,4$-pyrazylene, $X = H$						
Ar', Ar'' = phenyl (DSP)[f]	r.t.	++	Linear	Crystalline	~10	23
Ar'=4-ethoxycarbonylphenyl, Ar''= phenyl	r.t.	++	Linear, hetero	Crystalline	0.96	24(a)
$Ar = 1,4$-phenylene						
Ar', Ar'' = 2-pyridyl, $X = H$ (P2VB)	r.t.	+−	Linear	Crystalline	0.3∼2.0	23
Ar', Ar''=4'-MeCO$_2^-$, 3'-methoxy-phenyl, $X = CN$[f]	r.t.	+(−)	−	−	0.44(η)	10
$\begin{array}{c} X \\ \diagdown \\ RCO_2 \end{array} C{=}CH{-}Ar{-}CH{=}C \begin{array}{c} CO_2R' \\ \diagup \\ X \end{array}$						
$Ar = 1,4$-phenylene						
$X = H, R, R' = Et$ (1,4-PDA Et)	r.t.[g]	+−(50%)	Linear	Amorphous	0.16	25
	−25 °C	+	Linear	Crystalline	1.4	25
$X = H, R = Et, R' = Me$ (1,4-PDA EtMe)	r.t.	+	Linear, homo	Amorphous	3.12[h]	26
$X = CN, R, R' = Pr^n$	0–5 °C	++	Linear	Crystalline	3.0	17
$Ar = 2,6$-naphthalene						
$X = H, R, R' = Et$	−	−(−)	Linear	−	−	27
$RCO_2{-}CH{=}CH{-}1,4{-}C_6H_4{-}CH{=}C \begin{array}{c} CN \\ \diagdown \\ CO_2R' \end{array}$						
$R = 3$-pentyl, $R' = Me$,	−	−(−)	Linear, hetero	−	Oligomer[i]	19
$R = 2$-butyl, $R' = Et$	−	−(−)	Linear, hetero	−	Oligomer[i]	19
$R, R' = Me$	r.t.	+(−)	Zigzag, homo	Amorphous	Oligomer	28
$Ar{-}CH{=}CH{-}1,4{-}C_6H_4{-}CH{=}C \begin{array}{c} X \\ \diagdown \\ CO_2R \end{array}$						

Table 1 (continued)

Structure of dialkenes	Irradation temperature	Reactivity[b] (Yield)[c]	Type of adduct[d]	Morphology	η inh[e]	Ref.
Ar = 2-pyridyl						
X = CN, R = Me	r.t.	++	Zigzag, hetero	Amorphous	Dimer	29
Ar = 4-pyridyl						
X = H, R = Me	r.t.	++	Linear, hetero	Amorphous	0.47	30
X = CN, R = Et	r.t.	+	Linear, homo	Amorphous	0.36	30
X = CN, R = Pr^n	r.t.	++	Linear, homo	Crystalline	Dimer[j]	31
Ar = pyrazyl						
X = H, R = Et	r.t.	++	Linear, hetero	Crystalline	8.19	24(b)
X = CN, R = Me	r.t.	+	Zigzag, hetero	Amorphous	Oligomer	24(b)

[a] Results are given for a typical experimental run.
[b] ++, + and +− represent 'highly reactive', 'reactive' and 'fairly reactive' respectively.
[c] Final yield is quantitative unless otherwise stated.
[d] Refer to Scheme 2.
[e] 0.2 ~ 0.3 g dL^{-1} solvent, 30 °C.
[f] A photostable crystal is obtained by recrystallization under specified conditions.
[g] Room temperature is not satisfactorily low for the topochemical photopolymerization of the monomer (m.p. 96 °C).
[h] Acetone-insoluble part.
[i] The photoproduct from a single crystal is optically active.
[j] The dimer crystal recrystallized from a 1-propanol solution is photoreactive.

Two types of unusual topochemical reaction have been reported. In the crystal of 1,4-dicinnamoylbenzene (1,4-DCB) there are two topochemically allowed reaction routes, to give a cyclophane type dimer or to give a zigzag type polymer. However, the relative orientation of the two pairs of intermolecular reactive double bonds is not parallel and in addition, within each pair, the double bonds do not have equivalent orientations. On photoirradiation of the 1,4-DCB crystal the cyclophane is obtained as a major product in which four substituents on the cyclobutane ring are all *trans* to each other (δ type configuration) with a small amount of zigzag type oligomers.[21]

Another example is the topochemical reaction of a mixed crystal. The crystals of ethyl and propyl α-cyano-4-[2-(4-pyridyl)ethenyl]cinnamates (C4-PyEC Et and Pr^n) have nearly the same crystal structure, and they give a homoadduct type linear polymer and the same type of dimer respectively. However, the mixed crystal of these two ester compounds over a wide range of molar ratios has a quite different crystal structure and, on photoirradiation, the crystal gives the [2.2] paracyclophane crystal quantitatively.[22]

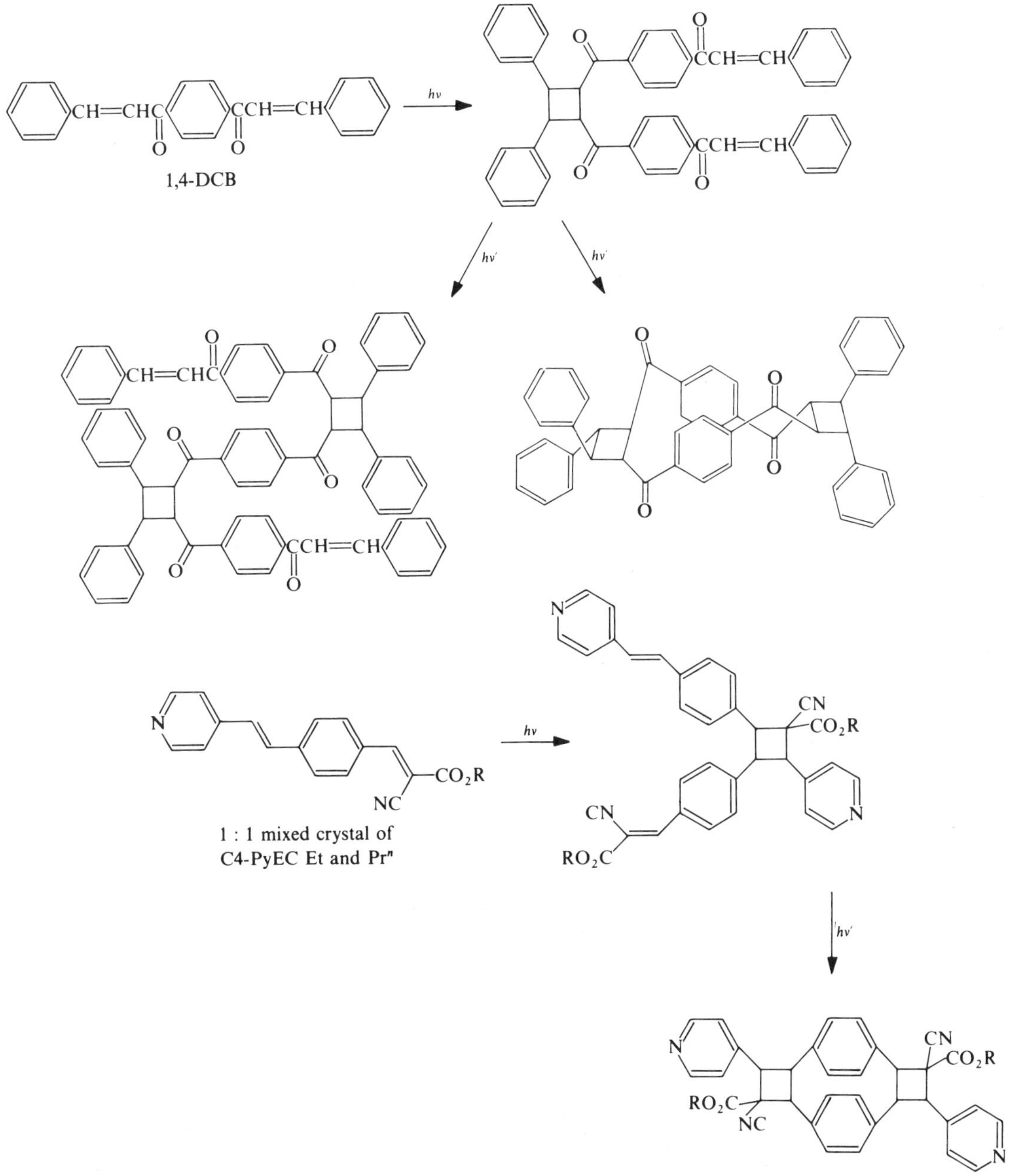

Scheme 4

14.3 REACTION MECHANISM OF FOUR-CENTER TYPE PHOTOPOLYMERIZATIONS

14.3.1 Step-growth Mechanism

The four-center type photopolymerization proceeds with irradiation by visible to UV light. As all the polymerizable monomers contain two conjugated double bonds, the π–π* electronic transition of the growing terminal alkene (hv') is shifted to a higher energy level than that of the alkene in the monomer (hv). Therefore, two excited species are required to obtain a high polymer in high conversion, and if the alkene in the monomer is excited exclusively leaving the terminal alkene in the ground state then the reaction is generally saturated at the oligomer stage. Such behavior in the two-step mechanism is observed for all the conjugated monomers.[32]

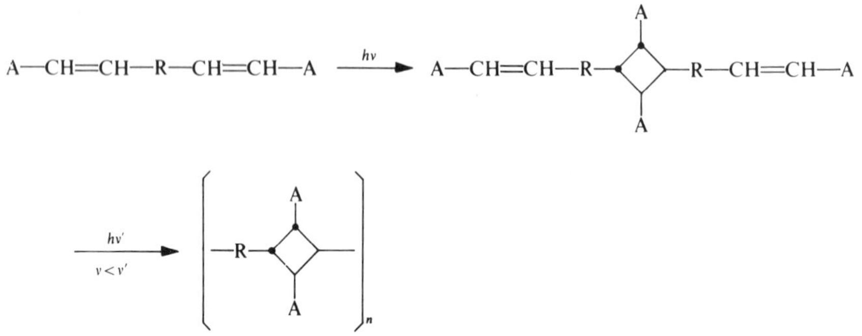

Scheme 5

For example, upon the irradiation of DSP crystals with 430 nm monochromatic light, the UV absorption spectrum changes and the maximum peaks are shifted to 350 nm and 290 nm for the oligomer crystals from 420 nm and 370 nm for the monomer crystals. The average degree of polymerization of the oligomer thus obtained approximately corresponds to a pentamer. Further details of the oligomerization process have been studied from the viewpoint of molecular weight distribution and features of the crystal transition.[33]

On further irradiation of the DSP oligomer crystals at wavelengths shorter than 400 nm, PDSP crystals with an absorption maximum at 280 nm are obtained.

The quantum yields were separately measured for the oligomerization of the monomer and subsequent polymerization of the resulting oligomer by excitation with monochromatic light for DSP, P2VB and 1,4-PDA Me (Table 2).[32]

Quantum yields (<1.7), which are based on the number of double bonds that disappear, are consistent with polymerization by a step-growth mechanism.

On irradiation by a xenon lamp or a high-pressure mercury lamp, the alkene bonds are converted to cyclobutane rings and the molecular weight of the photoproduct increases continuously with irradiation time. This type of chain growth is typical of a step-growth mechanism.

Under constant irradiation conditions the polymerization rate increases if the temperature is raised in a sufficiently low temperature range. However, it begins to decrease if the melting point of the reacting crystal is approached.

Table 2 Quantum Yields (Φ) of Oligomerization and Subsequent Polymerization of DSP, P2VB and 1,4-PDA Me[a]

	Wavelenth used in irradiation (oligomerization) (nm)	Φ^c	Wavelenth used in irradiation (polymerization) (nm)	Φ^c
DSP	436	1.2	365	1.6
P2VB[b]	405	0.04	—	—
1,4-PDA Me	365	1.2	313	0.7

[a] T. Tamaki, Y. Suzuki and M. Hasegawa, *Bull. Chem. Soc. Jpn.*, 1972, **45**, 1988.
[b] 2,2'-(1,4-divinylbenzene-β,β'-ylene) dipyridine.
[c] On the basis of the number of double bonds that disappear.

The highest reaction rate is observed superficially at the specified temperature for individual crystals (T_{opt}), whilst a higher average molecular weight is always observed, at the same conversion, for the product from the α type crystal irradiated at a lower temperature. The T_{opt} for the reaction of a 1,4-PDA Et crystal is seen around $-20\,°C$.[34] The result reflects the fact that four-center type photopolymerization is a topochemical reaction where thermal diffusion of the molecules is not essential for chain growth.

Such unique kinetic behavior of the temperature-dependent reaction rate is interpreted by assuming that the rate is proportional to the reaction probability, which is greatly influenced by the extent of thermal deviation of two reactive alkene bonds from their optimal positions in the crystal.[35]

As is obvious from a two-step mechanism by exclusive excitation of the monomer and from the temperature-dependent behavior, the molecular weight of the photoproduct can be controlled over a wide range by controlling the wavelength of the incident light and/or the reaction temperature.

The temperature-dependent behavior has been throughly investigated in the polymerization of 1,4-PDA Et in the temperature range from 4.2 K (liquid helium)[36] to temperatures above the crystal melting point (96 °C).[34]

In contrast to the symmetric dialkene monomers, in the unsymmetric dialkene crystals which give a homoadduct type product the topochemical environment is different between two pairs of intermolecular reactive double bonds and therefore a single type of dimer is very often accumulated spontaneously at the intermediate stage of polymerization without any control by the wavelength of the incident light or by the reaction temperature. One clear cut example is the 'even-numbered polymerization mechanism' of the photopolymerization of C4-PyEC Et crystals, where the monomer reacts only with itself but not with any other species.[18] Therefore, in the reaction of C4-PyEC Et all the growing species are of an even-numbered degree of polymerization, as is seen from GPC curve at the intermediate stage of polymerization in Figure 1.

$$M + M \xrightarrow{h\nu} Dimer \xrightarrow{h\nu} Polymer\ (M-M)_n$$

$$M + Dimer \xrightarrow[\times]{h\nu} Trimer$$

$$M + Trimer \xrightarrow[\times]{h\nu} Tetramer$$

Scheme 6

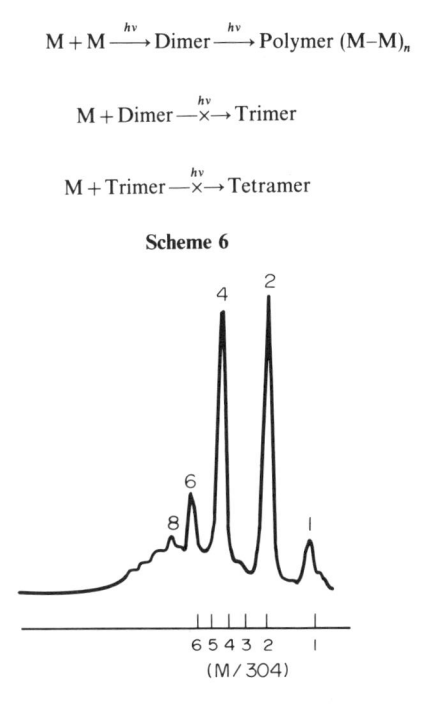

Figure 1 GPC curve at the intermediate stage of polymerization of a C4-PyEC Et crystal (M. Hasegawa, *Pure Appl. Chem.*, 1986, **58**, 1179)

In addition, in the even-numbered polymerization mechanism, all the reactive species except the monomer should have one type of alkenic group at the end.

In contrast to the C4-PyEC Et crystal, the C4-PyEC Pr^n crystal is quantitatively converted into the Ī-dimer crystal, which is photostable. The striking difference of behavior between the two crystals is satisfactorily explained by the crystallographic analysis (Figure 2 and Table 3).[18,31]

The schematic illustration in Figure 2 explains the exclusive formation of the dimer from C4-PyEC Et. In the dimer, the intermolecular double bonds on the pyridyl side come closer to each

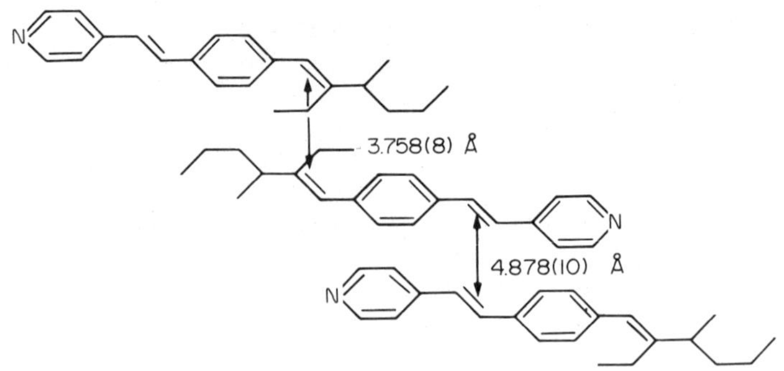

Figure 2 Schematic drawing of the exclusive formation of the homoadduct Ī-dimer from C4-PyEC Et (M. Hasegawa, H. Harashina, S. Kato and K. Saigo, *Macromolecules*, 1986, **19**, 1276)

other and within the necessary reactive distance (4.878 Å → 4 Å).[37] On the other hand, in the dimer of n-propyl ester, the distance is still of a photostable length although it does become slightly shorter (5.088 Å → 5.047 Å).[31]

14.3.2 Crystallographic Interpretation

In the polymerization of α type monomer crystals, independent of whether they are symmetric or unsymmetric, linear high polymers are usually obtained in quantitative yields. This is because there is only one reactive spot for each double bond in the α type crystals and in addition there is essentially no topochemical termination except crystal defects. However in many of the β type crystals each double bond has two possibilities for giving [2+2] cyclodimerization, which inevitably produces a topochemically terminating spot. No examples have been reported of the formation of high polymers having a zigzag type of main chain.

In many of the photoreactive crystals, the electron-rich nitrogen or carbonyl group approaches the electron-deficient benzene ring. In addition, the molecular shape and packing of photoreactive monoalkene and dialkene crystals are very similar both in the α and β type crystals, as is illustrated in schematic drawings of cinnamic acid (α), DSP, P2VB and 1,4-PDA Me for the α type crystals in Figure 3.[38]

These two common features may govern the formation of photopolymerizable crystals and thus allow an empirical rule to be established for similar photoreactivity between monoalkene and dialkene crystals, as is illustrated in Scheme 2.

Among the crystallographic data of photoreactive dialkene crystals several examples are tabulated in Table 3.

The space group of the crystals is retained in all the photoreactions to give crystalline linear polymers. The unit cell of the monomer crystal shrinks in some cases and expands in others during the photoreaction. All the crystals in Table 3 have the intermolecular reactive carbon-to-carbon distances within the reactive distance ($\simeq 4$ Å). The double bonds are not arranged in parallel in 1,4-DCB and 4-PyEC Me crystals although these are within the reactive distance. Intermolecularly, the two pairs of double bonds are in different arrangements to each other in 1,4-PDA EtMe, 1,4-DCB and C4-PyEC Pr^n crystals and these pairs inevitably show different reactivities.

From the oscillation and Weissenberg photographs, PDSP, poly-P2VB, and poly-1,4-PDA Me and Et have been confirmed to be orientated three-dimensionally in the crystals as prepared. Furthermore, the 1,3-*trans* configuration of the cyclobutane can be determined from the space group of these polymers since these space groups absolutely require the center of symmetry to be in the center of the planar cyclobutane ring.

The relative orientation of monomer and polymer crystals was determined by rotating single crystals of several photopolymerizable α type dialkenes on photoirradiation (Table 4 and Figure 4).[38]

In the DSP(α) and P2VB crystals, the directions of three crystal axes of the polymer coincide with those of the monomer. By comparison of the crystal data of monomer and dimer[31] the relative orientation of C4PyEC Pr^n and its dimer crystals must belong to Group 1 in Table 4.

After the first reports on the crystal structure correlations between DSP and PDSP crystals,[9] a different crystallographic result was reported by Wegner and co-workers claiming that the space group of DSP (*Pbca*) is not maintained, but transformed to $P2_1ca$ in the PDSP crystal.[27] However, the first structural analysis was confirmed to be correct, as reported in 1979.[44]

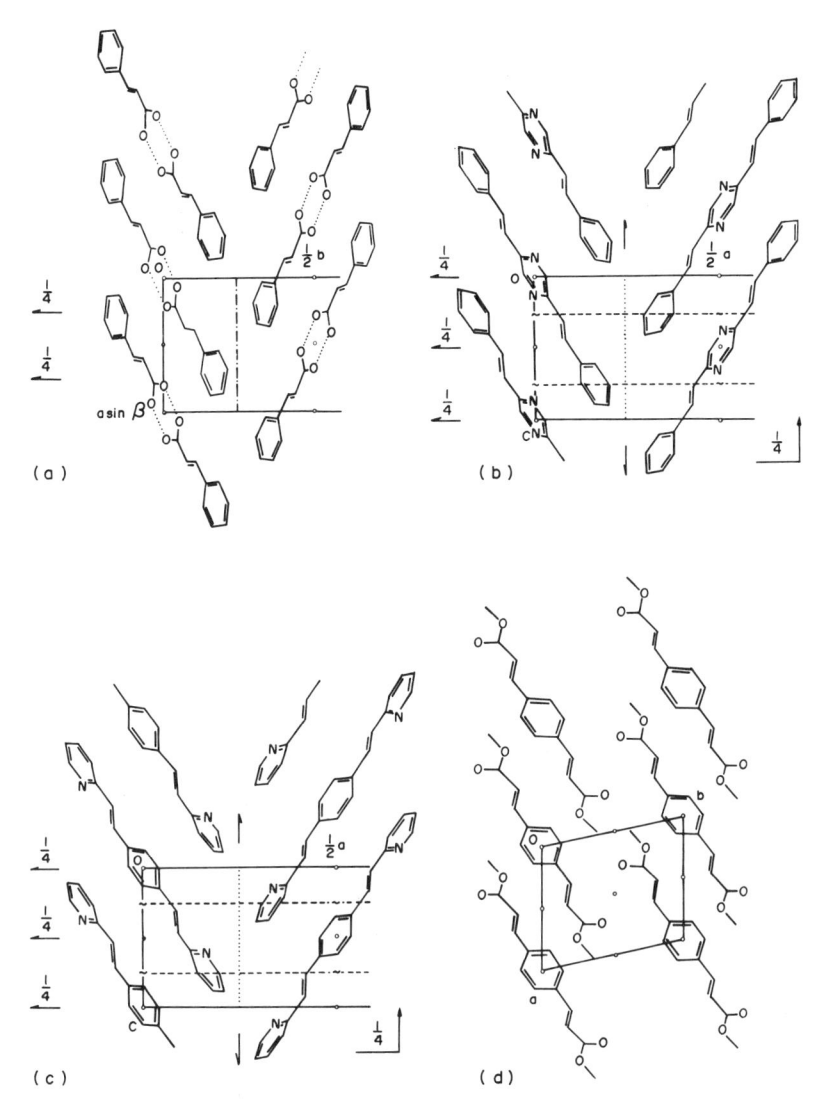

Figure 3 Crystal structures of (a) cinnamic acid (α), (b) DSP (α), (c) P2VB, and (d) 1,4-PDA Me (H. Nakanishi, M. Hasegawa and Y. Sasada, *J. Polym. Sci., Polym. Ed.*, 1977, **15**, 173)

A different type of relative orientation is seen in 1,4-PDA Ph, 1,4-PDA Et and 1,4-CPAPr^n where the crystal system and space group are retained but the direction of only one of the unit axes is maintained during the polymerization. However, the growth direction of the polymer chain deviates from the expected direction of chain growth in the monomer crystal, as is seen for 1,4-PDA Et in Figure 4. In the case of 1,4-PDA Me crystals the directions of all axes of the polymer do not coincide with those of the monomer. The movements of the monomeric units in the compounds in Groups 2 and 3 are comparably large, *e.g.* elongation of 8.4% for 1,4-PDA Ph, 11.1% for 1,4-CPAPr^n and 8.8% for 1,4-PDA Me. On the other hand, no significant density changes are observed in the polymerization of all the monomers investigated (less than 4%). In addition the temperature dependencies of the reaction behaviors and the continuous change of the X-ray diffraction diagrams for all the monomers support the idea that the polymerization proceeds by a typical crystal-lattice-controlled mechanism. In the photostable DSP(γ) crystal, molecules form a characteristic layer type packing without any overlap of adjacent molecules. The γ form of DSP is thermally transformed into the α form without any appreciable differential scanning calorimetry (DSC) peak.[45]

In conclusion, topotaxies found in four-center type photopolymerization are classified into three groups according to the topotactic control, relating to the concidence of the direction of the three axis of the monomer and polymer crystals, as shown in Table 4 and illustrated in Figure 4.

Almost the same results and discussion as contained in ref. 38, which are related to crystallographic results and to the relative orientations, have been described in another article by Thomas and co-workers.[46]

Table 3 Crystallographic Data for Dialkenic Monomers and their Products

Compound		Space group	a, Å (α, °)	b, Å (β, °)	c, Å (γ, °)	Z	Dx (g cm^{-3})	C–C Distance[a] (Å)	Ref.
DSP(α)	Monomer	$Pbca$	20.638	9.599	7.655[f]	4	1.244	3.939	8
	Polymer	$Pbca$	18.36	10.88	7.52[g]	4	1.257	—	9
DSP(γ)[b]	Monomer	$P2_1/a$	13.833	18.615	5.823	4	1.261	4.187, 4.369	39
P2VB	Monomer	$Pbca$	21.060	9.567	7.311[f]	4	1.281	3.910	40
	Polymer	$Pbca$	18.9	10.5	7.53[g]	4	1.26	—	38
1,4-PDA Me	Monomer	$P\bar{1}$	7.148[f] (94.97)	8.382 (116.85)	5.844 (78.06)	1	1.339	3.957	41
	Polymer	$P\bar{1}$	7.82[g] (107.8)	7.42 (106.0)	6.04 (78.8)	1	1.29	—	38
1,4-PDA Et	Monomer	$P2_1/a$	7.399[f]	9.894 (99.74)	10.167	2	1.242	3.970	42
1,4-PDA EtMe	Polymer	$P2_1/a$	8.16[g]	9.98	8.62	2	1.30	3.891, 4.917	42
	Monomer	$P\bar{1}$	12.460 (77.43)	9.713 (87.09)	5.958 (80.75)	2	1.24	—	18b, c
1,4-CPAPr[n]	Monomer	$P2_1/n$	5.341	26.112 (103.81)	6.882	2	1.265	3.931	43
1,4-DCB[c]	Monomer	$P\bar{1}$	5.798 (89.12)	7.923 (82.12)	19.307 (88.67)	2	1.280	3.903, 3.955[h] 3.973, 4.086[h]	21
4-PyEC Me[d]		$P2_1/a$	7.107	35.869 (114.62)	5.948	4	1.278	4.023, 4.097[h]	30
C4-PyEC Et[e]	Monomer	$P\bar{1}$	11.664 (85.84)	9.151 (104.44)	7.814 (80.05)	2	1.279	3.759, 4.878[i]	30
C4-PyEC Pr[n]	Monomer	$P\bar{1}$	8.919 (90.24)	12.429 (93.18)	7.781 (98.11)	2	1.241	3.729, 5.088[i]	31
	Dimer	$P\bar{1}$	8.908 (87.38)	11.926 (91.43)	8.060 (102.55)	1	1.266	5.047	31

[a] Intermolecular distance between reacting carbons.
[b] Photostable crystal (γ) is obtained by sublimation under a reduced pressure while α type crystal is obtained by recrystallization from benzene solution.
[c] 1,4-Dicinnamoylbenzene.
[d] Methyl 4-[2-(4-pyridyl)ethenyl]cinnamate.
[e] Ethyl α-cyano-4-[2-(2-pyridyl)ethenyl]cinnamate.
[f] Direction of chain growth.
[g] Direction of polymer chain.
[h] Intermolecular reactive double bonds are not in parallel.

Table 4 Topotaxies in the Four-center Type Photopolymerization of Dialkenic Monomer Crystals[a]

Group	Coincidence of crystal symmetry between monomer and polymer	Monomer	Crystal system	Space group
1	Crystal system, space group and directions of three axes	DSP	Orthorhombic	$Pbca$
		P2VB	Orthorhombic	$Pbca$
2	Crystal system, space group and direction of unit axis	1,4-PDA Et	Monoclinic	$P2_1/a$
		1,4-PDA Ph	Monoclinic	$P2_1/c$
		1,4-CPA Pr^{n} [b]	Monoclinic	$P2_1/n$ and $P2_1$
3	Crystal system and space group	1,4-PDA Me	Triclinic	$P\bar{1}$

[a] H. Nakanishi, H. Hasegawa and Y. Sasada, *J. Polym. Sci., Polym. Phys. Ed.*, 1977, **15**, 173.
[b] Dipropyl 3,3'-(1,4-phenylene)bis(2-cyanoacrylate).

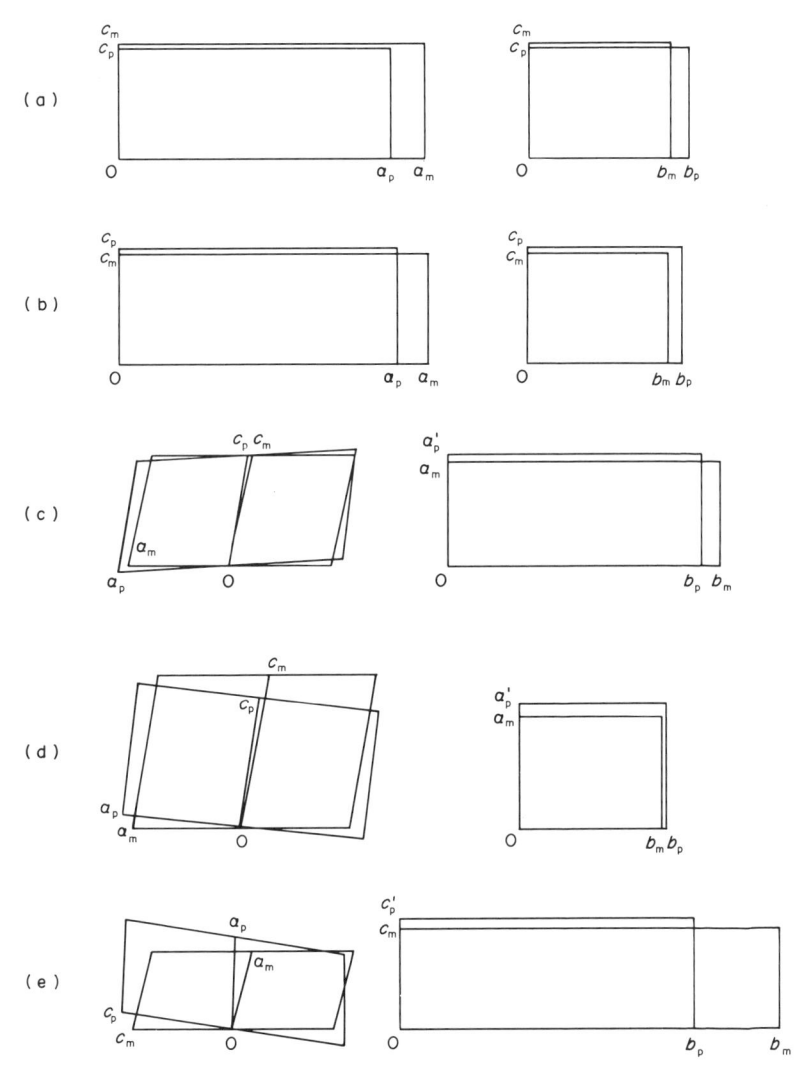

Figure 4 Relative crystallographic orientations of the monomers and the polymers: (a) DSP (α); (b) P2VB; (c) 1,4-PDA Ph; (d) 1,4-PDA Et; and (e) 1,4-CPA Pr^{n} (H. Nakanishi, M. Hasegawa and Y. Sasada, *J. Polym. Sci., Polym. Phys. Ed.*, 1977, **15**, 173)

14.4 CHARACTERISTIC PROPERTIES OF THE POLYMERS FORMED

The polymers prepared by four-center type photopolymerization depolymerize photochemically and thermally in solution to the corresponding monomers,[47,48] as is expected from a ring cleavage reaction yielding two alkenes from a number of cyclobutane derivatives to which the chromophore groups are directly attached. Highly crystalline linear polymers show no crystalline melting point but thermally depolymerize in the crystalline state. The X-ray diffraction pattern, spectrometric and DSC analyses indicate that thermal depolymerization occurs in the crystalline state throughout the reaction steps from the high polymer to the oligomers. At the oligomer stage, the crystalline melting point becomes lower than the cleavage temperature of the cyclobutane ring. In conclusion, a reversible topochemical process, which is a monomer crystal-lattice-controlled photopolymerization and a polymer crystal-lattice-controlled thermal depolymerization, has been established, as is shown in the DSP and PDSP schematic in Scheme 7.[49]

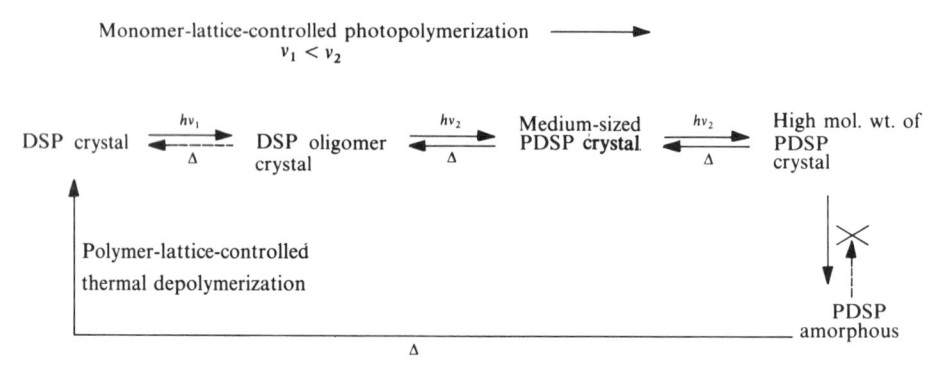

Monomer-lattice-controlled photopolymerization ⟶
$v_1 < v_2$

DSP crystal $\underset{\Delta}{\overset{hv_1}{\rightleftharpoons}}$ DSP oligomer crystal $\underset{\Delta}{\overset{hv_2}{\rightleftharpoons}}$ Medium-sized PDSP crystal $\underset{\Delta}{\overset{hv_2}{\rightleftharpoons}}$ High mol. wt. of PDSP crystal

Polymer-lattice-controlled thermal depolymerization

PDSP amorphous

Scheme 7 Reversible topochemical process between DSP and PDSP

All the linear polymers derived from α type crystals contain alternating 1,3-*trans* cyclobutane and 1,4-arylene units in the main chain. It follows that these polymers are stiff rods as is illustrated for PDSP in Figure 5.[50]

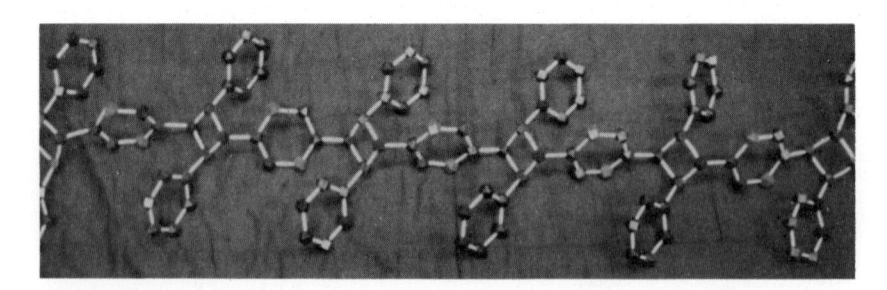

Figure 5 Molecular model of PDSP (M. Hasegawa, H. Nakanishi and T. Yurugi, *Chem. Lett.*, 1975, 497)

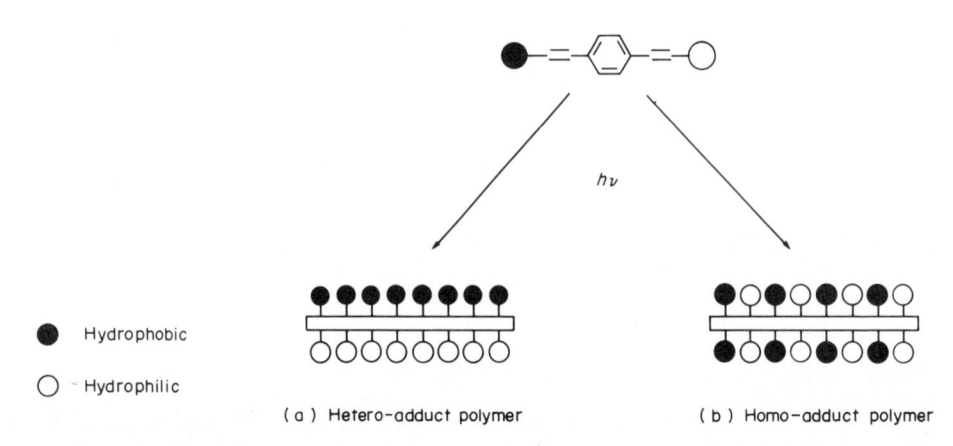

● Hydrophobic
○ Hydrophilic

(a) Hetero-adduct polymer   (b) Homo-adduct polymer

Figure 6 Alternating polymer structures of hydrophobic and hydrophilic groups (M. Hasegawa, *Pure Appl. Chem.*, 1986, **58**, 1179)

As was observed for several linear polymers, the higher molecular weight polymers are thermally less stable both in the crystalline state and in solution.[48–51] Recently, such an unusual, thermal stability dependency on molecular weight was characterized theoretically to be a feature of polymers having a rigid rod-like molecular shape.[52]

In addition to the rigid rod-like chain structures, linear polymers from the unsymmetric dialkenes have either 'heteroadduct' or 'homoadduct' structures. In these polymers two different types of branching units, such as hydrophobic and hydrophilic groups (*e.g.* alkoxycarbonyl and pyridyl groups) are arranged on opposite sides (a), or in alternating order (b) in each branch of the polymer chain (Figure 6).

Several recent articles have dealt with the application of unsymmetric dialkenic monomers to the copolymerization of mixed crystals[22,53] or to LB-film techniques.[54,55]

14.5 REFERENCES

1. H. I. Bernstein and W. C. Quimby, *J. Am. Chem. Soc.*, 1943, **65**, 1845.
2. M. D. Cohen and G. M. J. Schmidt, *J. Chem. Soc.*, 1964, 1996; G. M. J. Schmidt, *J. Pure Appl. Chem.*, 1971, **27**, 647.
3. S. Okamura, K. Hayashi and M. Nishii, *J. Polym. Sci., Part C*, 1963, **4**, 839.
4. H. Watanabe, K. Hayashi and S. Okamura, *J. Polym. Sci., Part B*, 1963, **1**, 397.
5. M. Hasegawa and Y. Suzuki, *J. Polym. Sci., Part B*, 1967, **5**, 813.
6. R. Franke, *Chem. Ber.*, 1905, **38**, 3727.
7. C. F. Koelsch and W. H. Gumprecht, *J. Org. Chem.*, 1958, **23**, 1603.
8. Y. Sasada, H. Shimanouchi, H. Nakanishi and M. Hasegawa, *Bull. Chem. Soc. Jpn.*, 1971, **44**, 1262.
9. H. Nakanishi, M. Hasegawa and Y. Sasada, *J. Polym. Sci., Part A-2*, 1972, **10**, 1537.
10. M. J. Holm and F. Zienty (Monsanto Chem. Co.), US Pat. 3 312 688 (*Chem. Abstr.*, 1967, **67**, 12 151); *J. Polym. Sci., Part A-1*, 1972, **10**, 1311.
11. For reviews see (a) M. Hasegawa, K. Saigo and T. Mori, *ACS Symp. Ser.*, 1984, **266**, 255; (b) V. Ramamurthy and K. Venkatesan, *Chem. Rev.*, 1987, **87**, 433.
12. (a) J. D. Dunitz and J. M. Robertson, *J. Chem. Soc.*, 1947, 1145; (b) A. Seher, *Justus Liebigs AP. Chem.*, 1954, **589**, 222.
13. F. L. Hirshfeld and G. M. J. Schmidt, *J. Polym. Sci., Part A.*, 1964, **2**, 2181.
14. For a review see H. Bassler, H. Sixil and V. Enkelmann, *Adv. Polym. Sci.*, 1984, **63**, and refs. therein.
15. H. Vogel and C. S. Marvel, *J. Polym. Sci.*, 1961, **50**, 511.
16. M. Hasegawa and F. Suzuki, *Kogyo Kagaku Zasshi*, 1963, **66**, 1230.
17. F. Nakanishi and M. Hasegawa, *J. Polym. Sci., Part A-1*, 1970, **8**, 2151.
18. For reviews see (a) M. Hasegawa, *Chem. Rev.*, 1983, **83**, 507; (b) M. Hasegawa, *Pure Appl. Chem.*, 1986, **58**, 1179; (c) M. Hasegawa, K. Saigo, S. Kato and H. Harashina, *ACS Symp. Ser.*, 1987, **337**, 44; (d) M. Hasegawa, in 'Organic Solid State Chemistry', ed. G. R. Desiraju, Elsevier, Amsterdam, 1987, chap. 5, p. 153.
19. (a) L. Addadi and M. Lahav, *J. Am. Chem. Soc.*, 1978, **100**, 2838; *Pure Appl. Chem.*, 1979, **51**, 1269; (b) L. Addadi, J. van Mil and M. Lahav, *J. Am. Chem. Soc.*, 1982, **104**, 3422.
20. H. Nakanishi, H. Suzuki and F. Nakanishi, *J. Polym. Sci., Polym. Lett. Ed.*, 1982, **20**, 653.
21. (a) M. Hasegawa, K. Saigo, T. Mori, H. Uno, M. Nohara and H. Nakanishi, *J. Am. Chem. Soc.*, 1985, **107**, 2788; (b) H. Nakanishi, M. Hasegawa and T. Mori, *Acta Crystallogr., Sect. C*, 1985, **41**, 70.
22. M. Hasegawa, Y. Maekawa, S. Kato and K. Saigo, *Chem. Lett.*, 1987, 907.
23. M. Hasegawa, Y. Suzuki, F. Suzuki and H. Nakanishi, *J. Polym. Sci., Part A-1*, 1969, **7**, 743.
24. (a) M. Hasegawa, T. Katsumata, Y. Ito and Y. Iitaka, *Macromolecules*, 1988, in press; *cf.* ref.18(a); (b) M. Hasegawa, M. Aoyama, H. Ohhashi and Y. Maekawa, paper submitted for publication; *cf.* ref. 18(a).
25. F. Suzuki, Y. Suzuki, H. Nakanishi and M. Hasegawa, *J. Polym. Sci., Part A-1*, 1969, **7**, 2319.
26. M. Hasegawa, S. Kato, N. Yonezawa and K. Saigo, *J. Polym. Sci., Polym. Lett. Ed.*, 1986, **24**, 153; *cf.* refs. 18(b) and 18(c).
27. (a) W. Meyer, G. Liser and G. Wegner, *J. Polym. Sci., Polym. Phys. Ed.*, 1978, **16**, 1365; (b) V. Enkelmann, H. Kapp and W. Meyer, *Acta Crystallogr., Sect. B*, 1978, **34**, 2350.
28. (a) H. Nakanishi and Y. Sasada, *Acta Crystallogr., Sect. B*, 1978, **34**, 332; (b) F. Nakanishi, T. Tanaka, F. Tsunoda and H. Nakanishi, *Nippon Kagaku Kaishi*, 1981, 412.
29. S. Kato, M. Nakatani, H. Harashina, K. Saigo, M. Hasegawa and S. Sato, *Chem. Lett.*, 1986, 847.
30. M. Hasegawa, H. Harashina, S. Kato and K. Saigo, *Macromolecules*, 1986, **19**, 1276.
31. M. Hasegawa, S. Kato, K. Saigo, S. R. Wilson, C. L. Stern and I. C. Paul, *J. Photochem. Photobiol., Sect. A, Chem.*, 1988, **41**, 385.
32. T. Tamaki, Y. Suzuki and M. Hasegawa, *Bull. Chem. Soc. Jpn.*, 1972, **45**, 1988.
33. H.-G. Brown and G. Wegner, *Makromol. Chem.*, 1983, **184**, 1103; *Mol. Cryst. Liq. Cryst.*, 1983, **96**, 121.
34. H. Nakanishi, F. Nakanishi, Y. Suzuki and M. Hasagawa, *J. Polym. Sci., Polym. Chem. Ed.*, 1973, **11**, 2501.
35. M. Hasegawa and S. Shiba, *J. Phys. Chem.*, 1982, **86**, 1490.
36. G. N. Gerasimov, O. B. Mikova, E. B. Kotin, N. S. Nekhoroshev and A. D. Abkin, *Dokl. Akad. Nauk SSSR*, 1974, **216**, 1051.
37. K. Kamiya, private communication, to be published.
38. H. Nakanishi, M. Hasegawa and Y. Sasada, *J. Polym. Sci., Polym. Phys. Ed.*, 1977, **15**, 173.
39. H. Nakanishi, K. Ueno and Y. Sasada, *Acta Crystallogr., Sect. B*, 1976, **32**, 3352.
40. H. Nakanishi, K. Ueno, M. Hasegawa and Y. Sasada, *Chem. Lett.*, 1972, 301.
41. K. Ueno, H. Nakanishi, M. Hasegawa and Y. Sasada, *Acta Crystallogr., Sect. B*, 1978, **34**, 2034.
42. (a) H. Nakanishi, K. Ueno and Y. Sasada, *Acta Crystallogr., Sect. B*, 1978, **34**, 2209; (b) H. Nakanishi, K. Ueno and Y. Sasada, *J. Polym. Sci., Polym. Phys. Ed.*, 1978, **16**, 767.
43. H. Nakanishi, K. Ueno and Y. Sasada, *Acta Crystallogr., Sect. B*, 1976, **32**, 1616.

44. H. Nakanishi, Y. Sasada and M. Hasegawa, *Polym. Lett.*, 1979, **17**, 459.
45. H. Nakanishi, G. M. Parkinson, W. Jones, J. M. Thomas and M. Hasegawa, *Isr. J. Chem.*, 1979, **18**, 261.
46. H. Nakanishi, W. Jones, J. M. Thomas, M. Hasegawa and W. L. Rees, *Proc. R. Soc. London, Ser. A*, 1980, **369**, 307.
47. Y. Suzuki, T. Tamaki and M. Hasegawa, *Bull. Chem. Soc. Jpn.*, 1974, **47**, 210.
48. M. Hasegawa, H. Nakanishi and T. Yurugi, *Polym. Lett.*, 1976, **14**, 47.
49. M. Hasegawa, H. Nakanishi, T. Yurugi and K. Ishida, *Polym. Lett.*, 1974, **12**, 57.
50. M. Hasegawa, H. Nakanishi and T. Yurugi, *Chem. Lett.*, 1975, 497.
51. M. Hasegawa, H. Nakanishi and T. Yurugi, *J. Polym. Sci., Polym. Chem. Ed.*, 1978, **16**, 2113.
52. E. Hanamura, *Solid State Commun.*, 1987, **63**, 1097.
53. M. Hasegawa, Y. Maekawa, Y. Endo and K. Saigo, *RAC-ACS Copolymerization Symposium*, Sydney, Australia, 1987, preprint, p. 20.
54. F. Nakanishi, S. Okada, K. Ichimura and M. Suda, *Annual Symposium of Res. Inst. for Polym. and Textiles*, Tokyo, Japan, 1985, preprint, p. 159.
55. H. Matsuda, M. Haruta, H. Munakata, Y. Tomita and T. Hamamoto (Canon Co.), *Jpn. Pat.* (1985) 83 237, 83 238 and 85 448.

15
Topochemical Polymerization: Diynes

DAVID BLOOR
Queen Mary College, London, UK

15.1 INTRODUCTION	233
15.2 ROUTES TO MACROSCOPIC POLYMER CRYSTALS	234
15.3 STRUCTURAL ASPECTS OF DIYNE POLYMERIZATION	235
15.3.1 *Polymerization of Single Crystals*	235
15.3.2 *Polymerization in Less Perfect Environments*	240
15.4 POLYMERIZATION KINETICS AND INTERMEDIATES	241
15.5 CONCLUSIONS	246
15.6 REFERENCES	247

15.1 INTRODUCTION

The solid state polymerization of disubstituted diynes has been extensively studied because it yields, for certain substituent groups, macroscopic single crystals containing extended, conjugated polymer chains. The physical properties of these quasi-one-dimensional, semiconducting polymers are striking and have been the subject of considerable interest in their own right.[1-3] The commensurate interest in the reaction leading to these unique materials has been reflected in the numerous reviews that have appeared over the last decade.[1-23]

In common with many areas of current interest, observations of the effects of the solid state polymerization of diynes were made in the last century but not properly understood. The photoreactivity of synthetic and natural diynes was originally reported nearly 100 years ago.[24,25] Further reports appeared in the 1930s[26-28] and 1940s.[29-31] Of these, the work of Strauss *et al.*[26,27] and Dunitz and Robinson[30] were of particular significance. The former authors attributed the colouration of dihalogen diynes to a reaction leading to bond formation between individual molecules in the crystal lattice. They attributed the intense colour to the formation of bonds with 'weakly bound electrons' and suggested that X-ray studies of structure would provide further insights. Unfortunately the compounds they reported were explosive and apparently no further experiments were carried out. Dunitz and Robinson determined the structure of diyne dicarboxylic acid; this was almost certainly the first observation of diyne polymerization produced by ionizing radiation. They observed rapid blackening of the crystals, which was attributed to decomposition. The absence of powder diffraction rings, the then accepted signature of a polymer, was taken to mean that the level of decomposition was low. In retrospect, it seems more likely that this was due to the perfect register of the monomer and polymer in the lattice of the partially polymerized crystal. Since the first microscopic polymer single crystals were not obtained until the late 1950s[32,33] it is not surprising that this probable outcome of Dunitz and Robertson's early experiment was not recognized at the time.

In the late 1940s, Jones and co-workers initiated an extensive and fruitful programme of synthesis of acetylenic compounds. The colouration of diyne compounds was noted in one of the first reports[31] but the cause of this was not discussed until later papers. Armitage *et al.*[34] suggested that the insoluble reaction products were the result of the photoinduced crosslinking between the monomers, which was favoured by their packing in the crystal lattice. Polymer formation was also suggested by Bohlmann[35] and was identified as a common cause of colouration in natural and synthetic diynes by Black and Weedon.[36] Seher[37] proposed a polyacene structure for the product of the polymerization of diyne dicarboxylic acids. This incorrect model was adopted by Bohlmann,[38] though he correctly deduced that the polymer chains should be aligned in the monomer lattice.

In 1964 Hirshfeld and Schmidt, in the course of a review of topochemical polymerization, put forward a model in which polymerization could occur by the rotation of monomers on fixed lattice sites.[39] They suggested that such polymerization had occurred in the diyne dicarboxylic acid studied by Dunitz and Robertson, however they failed to make the connection with the synthetic work described above.[31,34–38] It was another five years before the work of Wegner[40] brought these separate strands together. Wegner's studies of a range of diyne crystals showed unambiguously the occurrence of a topochemical reaction leading to the formation of extended conjugated chains. Furthermore, he demonstrated that the product could be a macroscopic crystal and that a fixed lattice dimension is not an essential requirement for the production of these polymer single crystals. The conclusion that many diynes, not just those with an exact lattice match between monomer and polymer, could react in the solid state to give a single crystal product, stimulated interest in the preparation and properties of macroscopic, single crystal, diyne polymers. This brief review focusses on the solid state reaction of diynes. Discussion of physical properties of the products can be found elsewhere.[1–3,15]

15.2 ROUTES TO MACROSCOPIC POLYMER CRYSTALS

Following the discovery of microscopic polymer crystals,[32,33] the challenge of the production of larger, chain-extended polymer crystals was tackled by many workers. The crystallization of preformed polymers leads to chain folds with a fold length which is dependent on the crystallization conditions.[41] Crystallization under high pressure and temperature results in chain-extended crystals several micrometres thick. The crystals are, however, small with high defect density.[42] A variety of uniaxial materials have been produced by shear-induced orientation in solution, in the melt, and in solid samples. None of these techniques is capable of producing macroscopic single crystals with low defect densities.

Routes, by which either the polymer chain and the crystal lattice are formed simultaneously, or the lattice is preformed, were suggested.[43] The former involves the simultaneous polymerization and crystallization of monomer molecules at a gas–solid or liquid–solid interface. A number of examples of this process have been considered.[44] The only clear example is the chain extension of poly(oxymethylene) microcrystals by insertion of monomers at the fold surface of lamellar crystals.[45,46] This lack of success is not surprising since the conditions necessary for this method to succeed are severe. First, the free energy change on crystallization should be small, i.e. crystallization close to the ceiling temperature for the polymer. Secondly, it is necessary to maintain active chain ends on the crystal surface and, finally, the reaction must be carried out near equilibrium.

The use of preformed lattices offers two alternatives: (a) monomer single crystals; and (b) inclusion compounds. The latter can lead to chain-extended polymers with specific chemical structures.[47,48] These polymer chains are separated in the host matrix and are not a continuous lattice. On dissolution of the matrix, the polymer chains can aggregate, but the result is not a single crystal. Monomer single crystals have been investigated, as part of the wider field of solid state chemistry, for a number of years.[49–54] Synthesis in the solid state can lead to products not readily accessible by solution or melt synthesis.[55] This follows since the crystal lattice selects one, or a very small number, of the possible molecular conformations and orients the molecules in well-defined close packing. The problems encountered in predicting crystal structures, and which particular polymorphs will be found in practice, have limited the use of solid state reactions. Modern crystallographic and computational methods[56] now offer better prospects for the design of reactive crystals with specified reaction products, extending the pioneering work of Lahav et al.[55,57] and Thomas and co-workers.[49,50]

Dimerization is a common reaction which occurs between molecules in close contact in a crystal lattice.[49–51] Polymerization requires that a chain reaction propagates along an array of monomers situated in a reactive arrangement. Many examples of solid state polymerization are known with products with morphologies from amorphous to highly crystalline. Amorphous products result when the lattice is destroyed after the polymerization is initiated, e.g. if the reaction is strongly exothermic. If an ordered polymer is to be produced the lattice must enforce either a topotactic or a topochemical reaction. For a topotactic reaction the product phase has a small number of crystallographically equivalent orientations, ideally just one orientation, with respect to the initial crystal lattice. A topochemical reaction is, in contrast, controlled by the geometry and spatial symmetry of adjacent molecules and the reaction pathway is uniquely defined by the original crystal structure. A topochemical reaction will often lead to a topotactic product, but topotaxy can result from a non-topochemical reaction. Topochemical reactions can occur either within a perfect lattice or at lattice defects, *e.g.* dislocations, stacking faults, *etc*. The special packing at defects can provide a reactive arrangement even though the normal lattice cannot sustain a reaction. Thus defects can act as initiation sites and, by deforming the surrounding lattice, nucleate reaction throughout the crystal. Though this can result in polymerization, the formation of a single crystal product is more likely to result when reaction is possible in the undeformed crystal.

Several solid state reactions resulting in highly crystalline polymers are known. The ring-opening polymerization of trioxane and tetraoxane has been identified as a topotactic rather than a topochemical process.[44,58] The four centre solid state polymerization of dialkenes, *e.g.* distyrylpyrazine, has been extensively investigated and shown to be a topochemical reaction.[59,60] The solid state polymerization of S_2N_2 to give poly(sulfur nitride) is also a topochemical reaction.[61] In all these cases the end product is not a perfect crystal as the monomer crystals fibrillate during the course of the reaction. Though this also happens in some instances for diynes, there are numerous cases where the crystal structure remains intact. The range of substituents which have been employed is large so that the diynes provide a unique class of monomers with the ability to form polymer single crystals of high perfection and with macroscopic dimensions. Such single crystals enable the intrinsic properties of the conjugated polymer chains to be studied directly, avoiding the problems of interpretation which result from the complex morphologies found in amorphous and semicrystalline polymers. The interest in the process by which the single crystals are obtained has resulted in the solid state polymerization of diynes becoming one of the most thoroughly investigated topics in solid state chemistry.

The criteria necessary for the polymerization of diyne crystals have been established through critical assessment of the relationship between reactivity and the lattice packing of the monomers. The kinetics of the reaction have been extensively studied and interpreted using theoretical models. In several cases the course of the reaction has been probed by spectroscopy and an extremely detailed knowledge of the process has been obtained. Polymerization has also been observed in less well-ordered situations, *e.g.* Langmuir–Blodgett films, liquid crystals and copolymers incorporating diyne units. These topics will be surveyed briefly in the following sections.

15.3 STRUCTURAL ASPECTS OF DIYNE POLYMERIZATION

15.3.1 Polymerization of Single Crystals

Several criteria must be satisfied if the solid state polymerization of a diyne crystal is to occur and if the product is to retain the perfection of the initial monomer crystal. First, it must be possible for a reaction to occur between the diyne units in the monomer crystal. Secondly, this reaction must proceed along a unique direction so that only one topotactic product is obtained. Finally, the reaction must not result in either segregation of the product or build-up of internal strain or both, since these are factors likely to result in the disintegration of the polymerizing crystal.

The topochemical polymerization of a linear array of diyne monomers is shown schematically in Figure 1. The arrangement of the monomer units can be characterized by the separation of the monomers in the direction of closest contact, d, and the angle between the axis of the diyne units and the polymerization direction, γ. Crystal structures of diyne polymers show that the polymer chain structure is close to that shown in Figure 1 and that the reaction must be a 1,4-polymerization. The distance which determines whether this reaction is possible is the separation of the 1 and 4' carbons, D. This is essentially the model proposed by Hirschfeld and Schmidt[39] but without the restriction of the exact lattice match of monomer and polymer.

The reaction illustrated in Figure 1 can be initiated thermally, by UV or ionizing radiation and chemically at the crystal surface. Mechanical deformation of the crystal lattice can affect the reaction, *e.g.* shear of the monomer crystal lattice can initiate the reaction and hydrostatic pressure can influence, or induce, polymerization. Thus polymerization can be made to occur in crystals with negligible reactivity in their normal state. Polymerization can occur both in the bulk of the crystal and at lattice defects, though the latter is likely to lead to a less perfect crystal.

For diynes which adopt an extended conformation, the crystal lattice packing favours linear arrays which may be converted into a single polymer chain by a concerted reaction. The criterion of a unique reaction direction allowing only a single topotactic product is frequently met. Thus, the criterion of a reactive lattice packing is generally more important. Two approaches to the definition

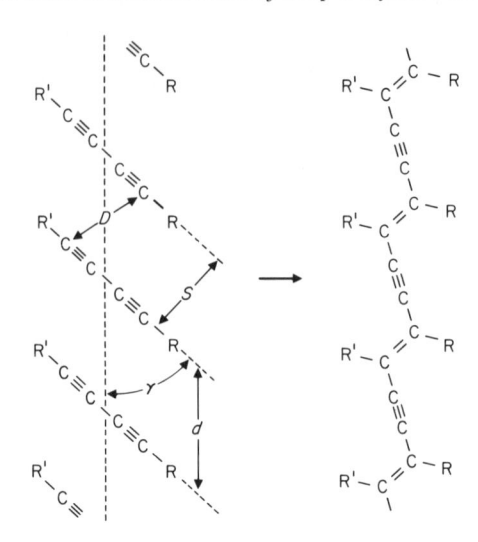

Figure 1 Schematic representation of the solid state polymerization of disubstituted diynes: left, an array of monomers with a reactive packing and a unique polymerization direction; right, the yne–ene polymer chain product (for definitions of the symbols see the text)

of this criterion have been considered. The simpler of the two is that of Schmidt.[62] This is based on observation of the separation of reacting atoms in a number of different reactive organic crystals. Schmidt noted that the carbon atoms between which bonds are formed must be separated by less than 0.4 nm in the crystal lattice. For larger separations a reaction is unlikely so that this can be taken as a critical value. An alternative criterion was proposed by Baughman,[8] who considered the case in which both rotation and translation of the monomer units occurs. The relative reactivity of different diynes was assessed by the calculation of least motion reaction pathways. The diyne units were treated as rigid rods with their motion confined to the plane containing these rigids rods, *i.e.* the plane of the paper in Figure 1. Alternative motions were considered for: (1) the centres of the monomer units remaining fixed on the linear axis of the monomer array; and (2) the individual diyne units maintaining a fixed orientation and reacting by translational motion. The former accords with the experimental observation that the polymer chains have the same orientation as the monomer arrays. The latter gives a larger least motion path length and can be discarded on these two counts.

A critical comparison between the predictions of these two models and X-ray structural data for reactive and unreactive diynes was first made by Bloor[14,15,63] and has subsequently been extended by other authors.[12,23] In Figure 2, the values of d and γ deduced from crystal structure data are plotted, with × denoting a reactive monomer and ○ an inactive one. S, the perpendicular distance between the diyne units (see Figure 1), has a minimum value of 3.4 Å, twice the van der Waals radius for triply bonded carbon atoms. This is shown by the full curve. Also shown are the limits set by Schmidt's criterion (dashed curve), and a root-mean-square displacement of 0.1 nm, calculated for the reaction path (1) described above (chain line). The distribution of the points for reactive monomers suggests that the criterion that D must be less than 0.4 nm is a more adequate limiting condition for solid state reactivity than that of a least motion reaction path as calculated by Baughman. Subsequent compilations have reinforced this view though a small number of apparent exceptions have been reported. These arise in cases where polymerization occurs in a lattice deformed by either external forces or the occurrence of a phase transition during the polymerization process. Thus, in order to assess reactivity, *e.g.* by the use of molecular models, the criterion of Schmidt is a reasonable guide. Note that, though some early papers suggested notionally fixed values for γ, in fact for each value of d the application of the limiting curves in Figure 2 sets a range of values γ for which reaction is possible.[64] It should also be noted that compilations of d and γ that appear in the literature contain discrepancies. The origin of these is unclear but probably result from incorrect calculation of d and γ from the original crystallographic data.

Baughman[52] also considered the conditions on the site symmetry of the monomer units for a unique polymerization direction to be guaranteed. These conditions are critical only if the site symmetry is high. Most organic molecules occur in tri- and mono-clinic lattices and the monomer site symmetry is usually low so that this is not an important factor, as noted above. In fact, the only symmetry requirement for the formation of the structure shown in Figure 1 is that the diyne moieties

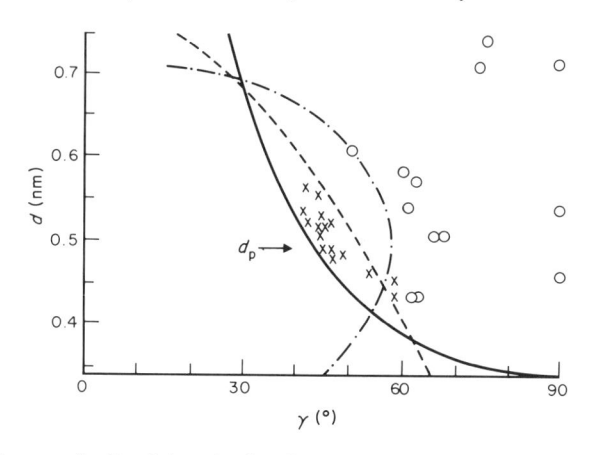

Figure 2 Plot of d vs. γ with curves for $D = 0.4$ nm (broken line) and a least motion reaction path of 0.1 nm (chain line). The typical polymer repeat distance is indicated by the arrow (d_p). Points are the experimental values for reactive ($\times$) and unreactive ($\bigcirc$) diynes (reproduced from ref. 15 with permission of Kluwer Academic Publishers)

should be parallel to one another. If the reaction can proceed in several different directions the product can contain an appreciable amorphous component.[65] Only isolated examples of reactivity have been reported for non-parallel diyne monomers.[66,67] Similarly a few examples exist of polymerization in diyne arrays with unequal spacing.[68] For details of other crystal structure studies compilations can be found in refs. 12 and 23.

The criteria for the occurrence of solid state reactivity and a single reaction direction are more readily satisfied by diynes with large end groups with either planar or hydrogen bonding components. This is illustrated in Table 1 which lists examples of compounds whose polymerization behaviour has been studied. The substituent groups are important since it is the interactions between them that principally determine the crystal lattice structure of the monomer and, hence, reactivity and its consequences. Additionally, the substituents form the interface between the growing polymer chains and the monomer lattice in which they are embedded. Since diyne polymers have repeat distances close to 0.49 nm in most instances the changes in lattice dimensions on polymerization are considerable, cf. Figure 2. Furthermore, monomers with dimensions similar to those of the polymer are not necessarily the most reactive monomers nor do they provide the most perfect polymer crystals.

Two extreme cases have been considered.[69] Either the strain produced by lattice mismatch is concentrated at the interface between polymer and monomer or it is more evenly distributed through the crystal lattice. In the former case the polymer formed first acts as a nucleus for further polymerization resulting in heterogeneous polymerization as shown in Figure 3(a). Interfacial strains

Table 1 Substituent Groups of Some Reactive Disubstituted Diynes in Relation to the Quality of the Polymer Crystals Obtained by Solid State Polymerization (the General Chemical Formula is $R_1-C\equiv C-C\equiv C-R_2$.)[14]

R_1	R_2	Abbreviation	Polymer crystal quality
$-CH_2OH$	$-Me$	1OH	Poor
$-CH_2OH$	$=R_1$	HD	Very poor
$-(CH_2)_2OH$	$=R_1$	OD	Poor
$-(CH_2)_9Me$	$-(CH_2)_8CO_2H$	TCDA	Poor
$-C_6H_4OCO(CH_2)_3OCOC_6H_4-$	(Cyclic)	BPG	Good
$-CH_2OCONHPh$	$=R_1$	PU	Good
$-CH_2OCONHEt$	$=R_1$	ETU	Good
$-(CH_2)_4OCONHPh$	$=R_1$	TCDU	Good
$-(CH_2)_nOCONHCH_2CO_2Bu$	$=R_1$	nBCMU	Good
$-CH_2OSO_2C_6H_4OMe$	$=R_1$	MBS	Good
$-(CH_2)_4OSO_2C_6H_4Me$	$=R_1$	TS-12	Good
$-CH_2OSO_2C_6H_4F$	$=R_1$	FBS	Excellent
$-CH_2OSO_2C_6H_4Me$	$=R_1$	TS	Excellent
$-CH_2NC_{12}H_8$	$=R_1$	DCH	Excellent

are built up leading to disruption of the crystal during polymerization and a fibrillar product, cf. the examples given in Section 15.2 above. In the latter case, the interfacial energy is low and single polymer chains can form supercritical nuclei[70] and propagate by a chain reaction leading to a homogeneous polymerization. In this extreme the polymerizing crystal remains in a single phase, *i.e.* it behaves as a solid solution of polymer in the monomer matrix (Figure 3b). The gradual change in crystal lattice parameters during polymerization allows quite large changes in lattice dimensions between monomer and polymer to be tolerated without disruption of the crystal. In practice real crystals will show intermediate behaviour. An example which is close to the ideal solid solution limit is provided by the bis(*p*-toluenesulfonate) diyne (TS). The smooth evolution of lattice parameters during thermal polymerization of this monomer is illustrated in Figure 4.

These processes can be described in terms of the changes in energy during polymerization. Homogeneous polymerization corresponds to a continuous decrease in free energy, *i.e.* the reduction in the free energy of the monomer molecules by polymerization is larger than the free energy contribution from intermolecular interactions. Heterogeneous polymerization results when the intermolecular contribution creates an energy barrier to further reaction in a single phase system but which can be circumvented by the appearance of a second phase. Intermediate cases can occur when an energy minimum is reached at intermediate conversion and, in the absence of a phase transition, the partially polymerized crystal is stable.[71] If there is a small difference between the free energies of the alternative one and two phase systems, then the exact polymerization behaviour will be determined by the relative kinetics. Examples are known where polymerization with ionizing radiation results in polymer crystals while thermal polymerization leads to phase separation.[65,72-75] While such generalizations are easy to make, precise calculations of the energetics of polymerization are much more difficult.[17,76] To date only one detailed calculation, for the monomer 2,4-hexadiynediol (HD), has been reported.[77] Unfortunately, though the molecules are simple enough to allow reasonable fundamental calculations to be made, this is a case where several polymerization directions occur.[78]

The models described above allow the polymerization behaviour of diynes to be explained once the crystal structure of the diyne is known. Consider the three monomers with simple end groups,

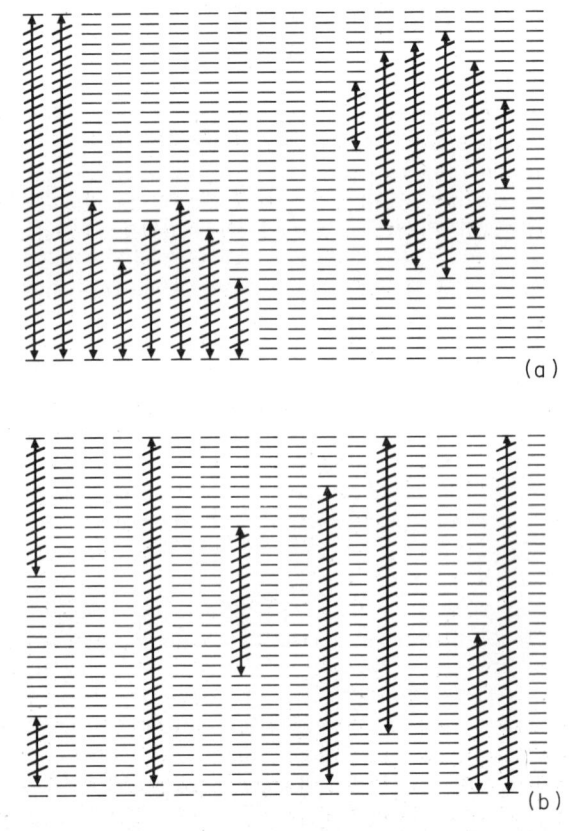

Figure 3 Models of solid state polymerization by (a) heterogeneous and (b) homogeneous chain nucleation (reproduced from ref. 7 by permission of the Royal Society of Chemistry)

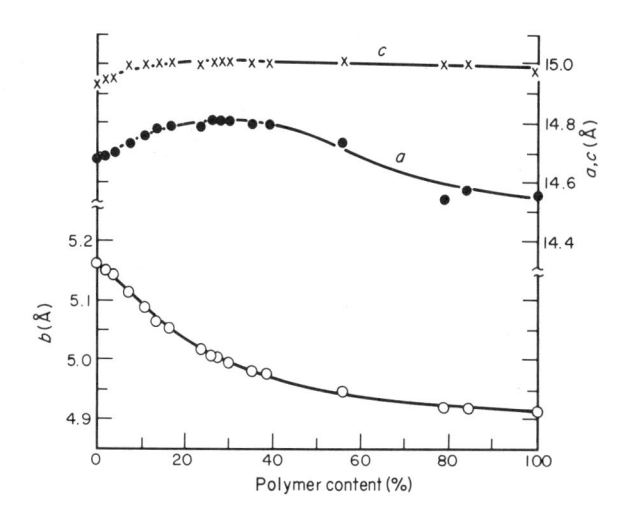

Figure 4 Lattice parameters of bis(toluenesulfonate) diyne (TS) crystals during thermal polymerization; the lattice parameters were determined at 300 K (reproduced from ref. 15 with permission of Kluwer Academic Publishers)

labelled 1OH, HD and OD in Table 1. Strong hydrogen bonding between the alcohol groups in 1OH brings the monomers into a reactive configuration with a single reaction direction.[79] The interaction between the hydrogen-bonded molecules is strong while that through the methyl groups is weak. It is not surprising, therefore, that the crystals fibrillate during polymerization. In HD there is hydrogen bonding and more than one crystallographic direction along which reaction can occur.[78] This is likely to lead to high internal stress and indeed disruption of X-ray polymerized crystals is observed at 40% polymer content after an initial single phase polymerization.[74] Thermal polymerization leads to an amorphous product as a result of competition between the different possible polymerization modes, phase separation and crosslinking reactions.[65] OD has a complex hydrogen-bonded network but only one possible direction for polymerization.[80] Even though the polymerization produces less than 2% expansion of the crystal a-axis this results in a considerable strain in the hydrogen bond network. Complete single phase polymerization is possible but as a result of the internal strain the crystals disintegrate when subjected to an external stress.

The prediction of the polymerization behaviour of diynes simply on the basis of their chemical composition remains a difficult problem since, as noted earlier, the calculation of crystal lattice structures for large molecules is not routine. Figure 2 shows that the configuration space in which reaction is possible is quite small which compounds the problem. These difficulties are highlighted by consideration of the crystal structures of closely related monomers. For example, the diynes with end groups of the general form $-CH_2OSO_2C_6H_4X$, where X = —H (BS), —Me (TS), —OMe (MBS), —Cl (CBS) and —Br (BBS), and with multiple methyl substitution of the terminal benzenoid ring, e.g. the bis(2-mesitylenesulfonate) diyne (MSHD) and the bis(pentamethylbenzenesulfonate) diyne (PMHD).[81-87] The crystal structures reveal that these monomers can crystallize in one, or more, of three different conformations which result from rotation of the substituent group about the C—O bond, as shown in Figure 5. Of these conformations only that adopted by BS and TS gives rise to a packing suitable for solid state polymerization. The more linear structure adopted by CBS, BBS and MSHD produces either a much larger molecular separation between adjacent diynes or a much larger tilt angle. In all cases the distance between the 1 and 4' carbon atoms is greater than 0.4 nm. The folded structure of MBS and PMHD results in a large separation of the diyne units due to the steric interference of the substituents.

The related bis(naphthalenesulfonate) diyne (NS) adopts the same conformation as TS but the larger substituents rotate the diyne units to a large tilt angle and reaction is not possible.[88] Finally the additional degrees of freedom provided by additional CH_2 groups in a series of analogous compounds results either in a face-to-face packing of benzenoid rings between molecules in alternant stacks or a folded unreactive conformation.[64] The calculated lattice energies of the different structures are equal, within the uncertainty that the choice of interatomic potentials places on the calculation.[89] This is highlighted by the fact that the crystal growth of CBS and MBS at high supersaturation produces reactive crystal modifications.[90,91] That these have similar conformations to those of TS is shown by the ability of CBS and MBS to cocrystallize with TS[91,92] and by the essentially identical structures of TS and MBS polymers.[93,94]

15.3.2 Polymerization in Less Perfect Environments

Though the requirements for solid state reactivity of diynes discussed above are quite strict, there is some leeway which indicates that polymerization is possible in less ordered environments. Examples that have appeared in the literature are Langmuir–Blodgett (LB) films, copolymers containing diynes and liquid crystalline phases. Of these, the first two are situations in which the packing of the diyne units can be close to that encountered in single crystals.

LB films made from linear amphiphilic molecules can adopt near-crystalline packing with the molecular axes tilted at considerable angles to either the subphase or substrate surface. Thus, polymerization of diyne-containing amphiphiles is likely. This was observed first for diynoic acid films,[95] which were subsequently shown to consist of crystal-like domains with sizes in the range 10 nm to 100 μm.[96,97] Both phospholipids and lipids containing diynes have been shown to polymerize[98–100] as have amphiphilic pyridinium and bipyridinium salts.[101] Polymerization of LB films and liposomes is achieved by UV irradiation and the nature of the photopolymerization process has

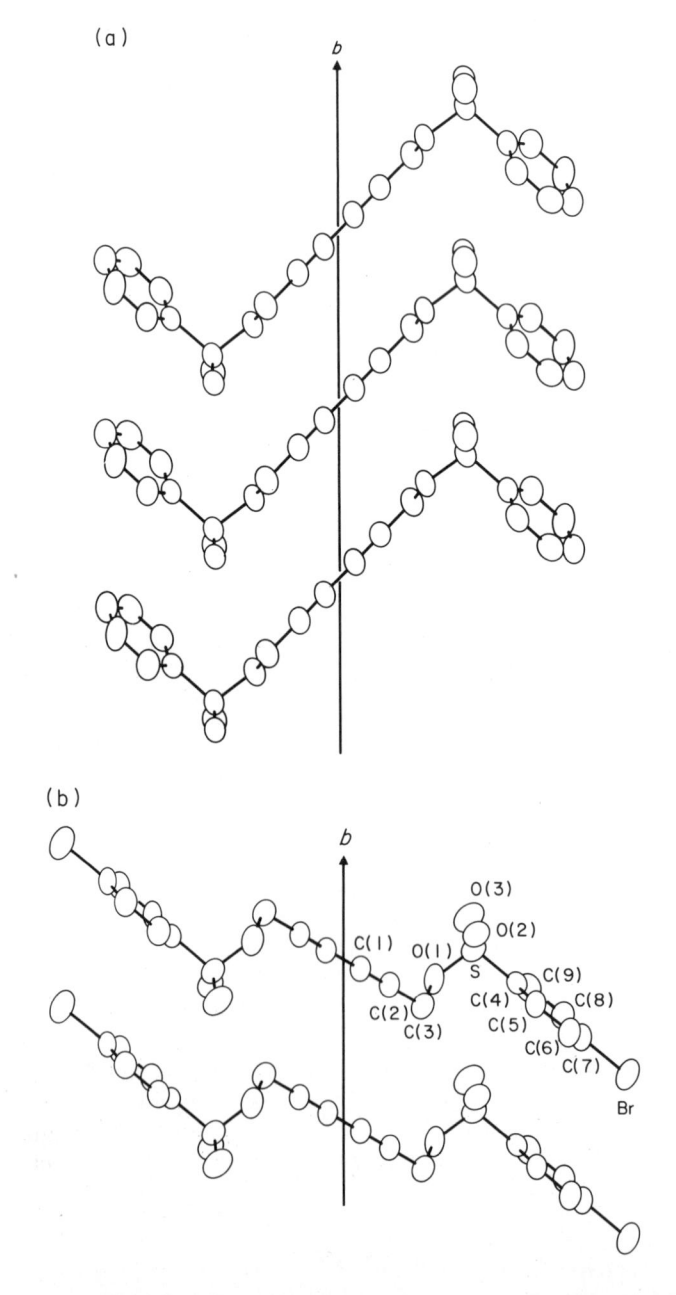

Figure 5 Crystal structures of (a) BS and (b) BBS; see text for meaning of abbreviations (reproduced from refs. 81, 85 and 87 by permission of the International Union of Crystallography)

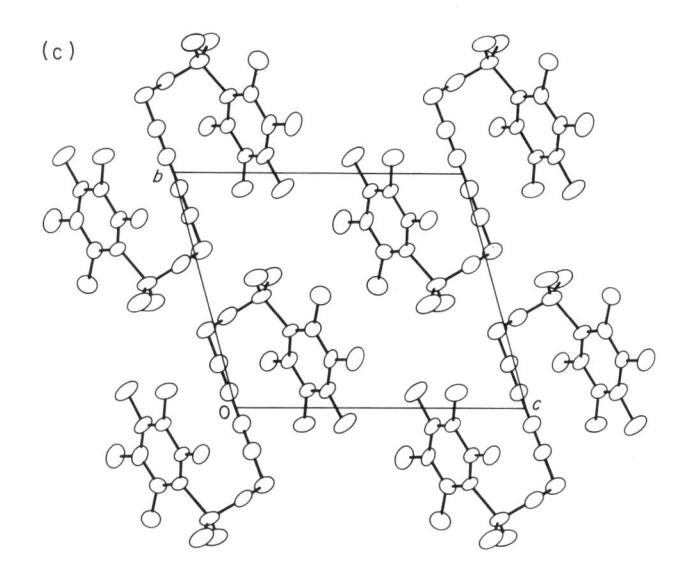

Figure 5 (c) PMHD; see text for meaning of abbreviations (reproduced from refs. 81, 85 and 87 by permission of the International Union of Crystallography)

been extensively studied.[102-106] The use of dye sensitizers has been demonstrated. There is also a self-sensitization process in which the initially formed polymer absorbs radiation and initiates further polymerization. Excessive irradiation leads to degradation and ultimately bleaching of the polymer films. The structure of the films has been studied by microscopy,[96,97] IR and ESR spectroscopy[107,108] and resonant Raman spectroscopy.[109,110] All these studies indicate that the polymerization reaction in LB films and liposomes is identical with that observed in single crystals. This is also the case in thin crystalline films of the bis(toluenesulfonate) diyne prepared by spreading solutions on a water surface in a LB trough.[111]

The crosslinking of copolymers containing diyne units was noted shortly after interest was created by Wegner's reports of polymerization in single crystals.[112,113] These early findings went largely unnoticed but were extended by Patel et al.[114] These studies of polyurethane-diyne copolymers have been considerably extended by the work of Rubner and co-workers.[115-117] He reported the synthesis of a range of segmented copolymers and their characterization by differential scanning calorimetry, and IR and visible spectroscopy. The diynes form rigid blocks which segregate in the more flexible urethane matrix. Polymerization occurs in these crystal-like regions. The polydiyne crosslinks are sensitive to the reordering of the urethane matrix by external influences, *e.g.* heat and mechanical deformation. Diyne polymerization has also been observed in the simpler system poly(1,8-nonadiyne).[118] In this instance elongation by 100% prior to polymerization resulted in a higher crosslinking density which was attributed to higher crystallinity in the stretched film. The closely related macromonomer, poly(1,11-dodecadiyne), has been studied.[119] The X-ray structures of macromonomer and the cross-polymerized material show that the diyne reaction is identical to that in diyne monomer crystals.

Related studies have probed the liquid crystal forming properties of diyne-containing monomer and polymers. A series of diynic diesters were observed to have varying reactivity in their solid phases, while those that had nematic liquid crystal phases were unreactive in that phase.[120] Of the polymeric derivatives of 10,12-docosadiyne-1,22-dioic acid studied, two displayed smectic liquid crystal phases. The polymers were highly reactive in the solid state but not in the smectic phase.[121] Diphenyldiyne polymers exhibit liquid crystal phases at high temperatures. These are unstable due to reaction of the triple bonds.[122] It is not clear from the literature whether this is analogous to the solid state reaction or is a manifestation of polymerization that is often observed in diyne melts. The products of the latter reaction are black, brittle glasses that bear little resemblance to the product obtained from reactive diyne crystals. Thermal polymerization in a number of liquid crystalline diynes has been reported recently.

15.4 POLYMERIZATION KINETICS AND INTERMEDIATES

There have been extensive studies of both the kinetics and the intermediate steps of diyne polymerization.[12,14,17,18,23,76] A model of the polymerization kinetics in the perfect crystal lattice has been developed by Baughman[124] and Baughman and Chance.[125] This considers that since, in general, the polymer does not match the monomer lattice, the kinetics are strongly influenced by the lattice strain produced during polymerization. This model is effective in fitting the kinetics of a number of diynes. There has, however, been considerable controversy about details of the kinetics, e.g. isotope effects,[126] polymerization under pressure,[127] etc.[17,18,76] In fact, the experimental kinetics curves can be generated by both single and multiphase models.[63,76] Thus, other experimental studies are necessary to determine which model is applicable before attempts are made to discuss the kinetics. In this brief review it is not possible to adequately cover all the studies of kinetics and the reader should consult the many reviews of this topic referenced above.

Thermal polymerization of diynes is characterized by an activation energy of the order 20 kcal mol^{-1} (84 kJ mol^{-1}).[128–135] In the case of the bis(toluenesulfonate) diyne (TS) the heat of polymerization is -32 ± 1 kcal mol^{-1} (-134 kJ mol^{-1}).[129–135] Photopolymerization studies of the urethane substituted diyne 4BCMU (see Table 1) give a much lower activation energy of 7 kcal mol^{-1} (29 kJ mol^{-1}) and a slightly smaller heat of reaction of -22 ± 1 kcal mol^{-1} (-92 kJ mol^{-1}).[136,137] These results can be understood in terms of the energy diagram shown in Figure 6 which displays schematic curves for the TS and 4BCMU derivatives. For TS the activation barrier for thermal polymerization is lower than that of 4BCMU, which displays negligible thermal polymerization. Conversely, decay from the monomer photoexcited state to the dimer and higher oligomers is favoured for 4BCMU which is more photoactive than TS. The weak activation for photopolymerization is attributed to a small barrier between higher oligomers.

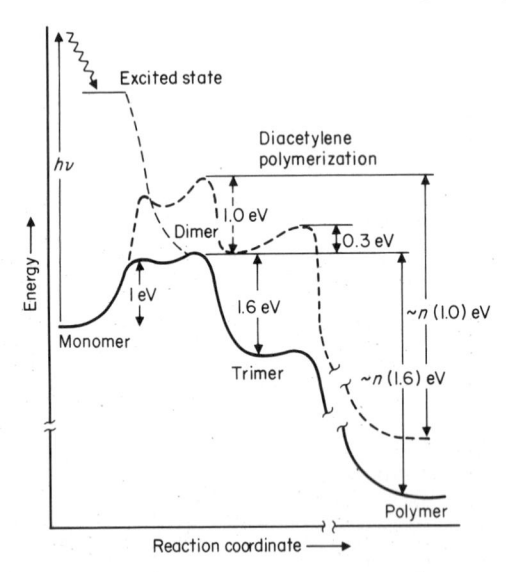

Figure 6 Reaction diagram for thermal and photopolymerization of diynes. The full curve models the case of TS and the dash curve that of 4BCMU (reproduced from ref. 138 by permission of the American Chemical Society)

Figure 6 implies that an active chain end exists during the propagation of the polymer chain. Two possibilities were suggested by Wegner,[128] as shown in Figure 7. The first is a two step process with a single radical electron at the chain end and a second unpaired electron at the bond alternation defect between butatriene and yne–ene structures. Such radicals give rise to a doublet state ($S = 1/2$). The alternative is a carbene, a two electron radical at the end of an yne–ene chain. In the latter case the electrons occupy either a triplet state ($S = 1$) or a singlet state ($S = 0$).

Though the singlet carbene radicals ($S = 0$) cannot be observed by electron spin resonance spectroscopy (ESR), both the doublet radicals and the triplet carbenes give characteristic ESR spectra. Doublet radicals give a single line, typically with a g-value close to two, while triplet carbenes have a three line ESR spectrum. The splitting of the lines in the spectrum is a measure of the dipolar interaction between the triplet electrons.[139]

Unlike many other conjugated polymers, single crystal diyne polymers have very weak ESR signals at $g \sim 2$ for both γ-ray[140] and thermally polymerized samples.[141,142] This is attributed to a much smaller number of paramagnetic defects in crystalline diyne polymer than in other less perfect conjugated polymers. ESR spectra recorded during the polymerization show a smooth increase in the number of unpaired spins. If the spins are assumed to reside on the polymer there is actually a decrease in the spin concentration per gramme of polymer as polymerization proceeds.[141] If the

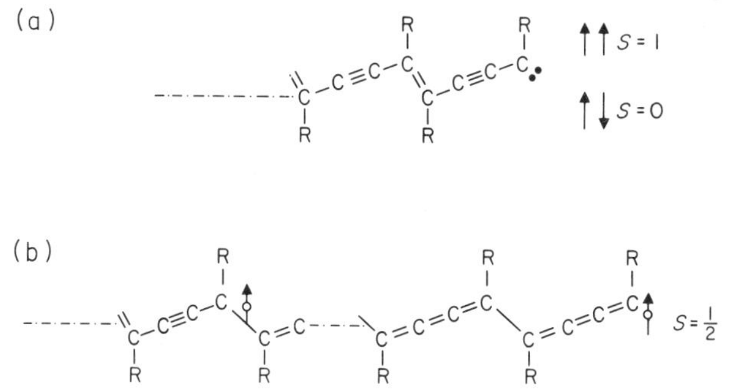

Figure 7 Models for the active chain end species on an yne–ene polymer chain: (a) a carbene radical; and (b) localized singlet radicals (reproduced from ref. 14 by permission of Applied Science Publishers, Ltd.)

predominant active chain ends were doublet radicals, a maximum spin concentration coincident with the maximum polymerization rate would be expected. No such maximum was observed and it was shown that the residual unpaired electrons in the diyne polymer crystals were due to surface and bulk defects.[143-145] Thermally polymerizing bis(toluenesulfonate) diyne (TS) crystals were shown to have the widely separated ESR absorption characteristic for a carbene triplet.[146] The angular dependence of these lines was used, together with observations at low temperatures, to obtain values for the spin Hamiltonian parameters, D, E and g of: $D/hc = +(0.2731 \pm 0.0005)\,\text{cm}^{-1}$; $E/hc = -(0.0048 \pm 0.0005)\,\text{cm}^{-1}$; $g_z = 2.004 \pm 0.002$, $g_y = 2.003 \pm 0.002$ and $g_x = 2.004 \pm 0.002$.[142,147] The principal magnetic axes z and y are oriented in the plane of the polymer chain with the z axis at an angle of 24° to the chain axis. These values indicate that the triplet electrons adopt a disk shape orthogonal to the axis of the adjacent triple bond.

The carbene ESR signals have maximum strength at peak polymerization rate. The spectra become too weak to detect when polymerization is complete, indicating that either few chain ends are left in the polymer crystal or that an end-capping reaction occurs. Oxygen scavenging can be ruled out since the same effects are observed for samples prepared under nitrogen and even under normal atmospheric conditions oxygen cannot diffuse far into the single crystals.[148]

These results show that carbene radicals are the predominant active chain ends present on growing polymer chains but do not provide information about either the initiation step or the short, oligomeric intermediates. On energetic and mechanistic grounds a dimer diradical with a butatriene structure is the most likely first reaction product.[134,149] Initial observations of diradical species during thermal and X-ray polymerization of TS were not conclusive.[144,150] Further observations of diradicals in X-ray polymerized urethane substituted diynes[151] and triplets in samples of the same monomer UV polymerized at low temperature[152] did not clarify the situation. Triplets were also observed in similar experiments for other diyne monomers.[153,154]

These conflicting results have been resolved by a series of detailed measurements of the intermediates produced by UV irradiation of bis(toluenesulfonate) diyne (TS) monomer crystals at low temperature. At first, measurements made at 4.2 K showed that irradiation produced a series of carbene radicals at the ends of oligomers containing from two to five repeat units.[155] The hyperfine structure observed in the ESR spectra show that the radical electrons are delocalized on to the side group. Optical spectra recorded on similarly irradiated samples showed that non-paramagnetic species, singlet carbenes, were also produced.[156,157] The growth and decay of these species as a function of irradiation time was studied.[157,158] Irradiation into the singlet absorption bands leads to conversion of the singlets into triplets. UV irradiation at 77 K was found to give rise, in addition to the carbenes, to a series of diradical species.[159] Both species were observed to disappear at about 110 K when irradiated crystals were warmed. This was coincident with the rapid local polymerization of the irradiated portion of the crystal. Thermal conversion of the radicals at 100 K produces other carbene species and also quintet dicarbene radicals.[160] The dicarbene radicals also appear in deuterated TS UV irradiated at 4.5 K.[161]

These early studies have been consolidated by detailed spectroscopic investigations and the development of theoretical models with which to interpret the experimental findings. Optical spectra have been reported by Gross and Sixl.[162,163] The ESR spectra have been analyzed in detail.[164-166] This work has been reviewed by Sixl.[18] Configuration, continuum and Huckel self-consistent field methods have been used to calculate the electronic energies of the intermediates.[167,168] Complementary studies have been reviewed by Schwoerer and Niederwald.[20] These include detection of quintet ($S = 2$) states due to dicarbenes,[161] real time spectroscopic studies of UV-induced polymerization of TS at room temperature[169] and detailed studies of the structure of triplet state dicarbene radicals by electron nuclear double resonance (ENDOR).[170,171]

The various radicals that have been identified are shown schematically in Figure 8. Photopolymerization at 4.2 K leads to: (a) diradicals (DR) occurring on oligomers with from two to six repeat units, the ground state is the $S = 0$ singlet; (b) asymmetric carbene radicals (AC), with lengths of from three to ten repeat units, in the $S = 1$ ground state; and (c) stable oligomers (SO) with from four to eleven repeat units.

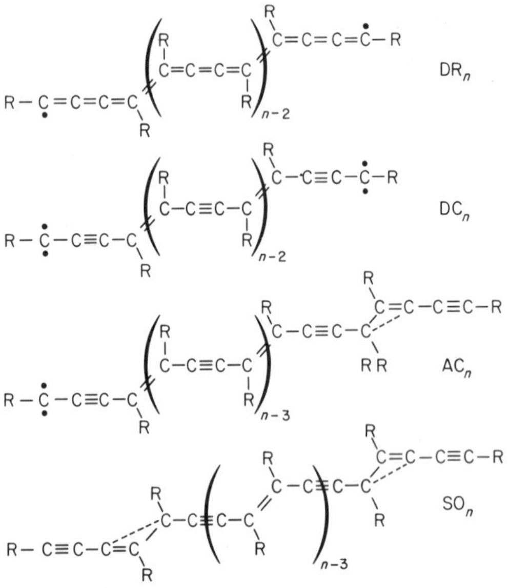

Figure 8 Radical intermediates identified in partially polymerized diyne crystals at low temperatures: DR = diradicals, DC = dicarbenes, AC = asymmetric carbenes and SO = stable oligomers (reproduced from ref. 162 by permission of Elsevier Science Publishers)

The pseudo-cyclopropene rings illustrated for the stable oligomers in Figure 8 are a conjecture,[18] there being no direct experimental evidence to identify the terminating group. The evolution of the optical spectra is shown in Figure 9. The initial stage of the polymerization is the production of the diradical dimer (Figure 10). Subsequent irradiation leads to the addition of one further unit but can either lead to a similar radical, e.g. diradical dimer to diradical trimer, or a modified species, e.g. diradical dimer to asymmetric carbene trimer to stable oligomer tetramer. Irradiation at room temperature produces a series of diradicals up to five repeat units long.[169] Thermal excitation leads to conversion from short to long oligomers on a microsecond time-scale. After a few milliseconds an intermediate with a polymer-like spectrum appears but with a longer wavelength absorption. Finally after many seconds, the normal polymer spectrum appears. By comparison with the low temperature spectra, the intermediate is identified as a polymer with reactive carbene end groups.

The dicarbene illustrated in Figure 8 is one of the indirect products which are generated thermally above 80–100 K. These are: (d) dicarbene radicals (DC), with lengths from seven to fourteen repeat units, a singlet ($S = 0$) ground state and a triplet ($S = 1$) excited state; and (e) carbene radicals on long chains, where the interaction of the carbenes is insufficient to form a dicarbene. These radicals have an $S = 1$ ground state similar to the low temperature asymmetric carbenes. The detailed structure of the dicarbenes have been determined by ENDOR of samples annealed above 80 K and recooled to 4.2 K.[161,171,172] Thermal propagation of all these radicals is observed above 100 K until long polymer chains are produced.

It is worth noting that much of the discussion concerning thermal polymerization kinetics of TS has centred on the length of chains produced at low conversion.[17,64] The spectroscopic studies of thermal polymerization after UV initiation[169] show unambiguously that these cannot be very short, as suggested by some indirect methods, but are probably a few tens of repeat units long at least.

Studies of several other diynes have revealed the presence of the same intermediates (see Figure 11).[18,172] Thus, the reaction scheme described above has general applicability to diyne solid state polymerization.

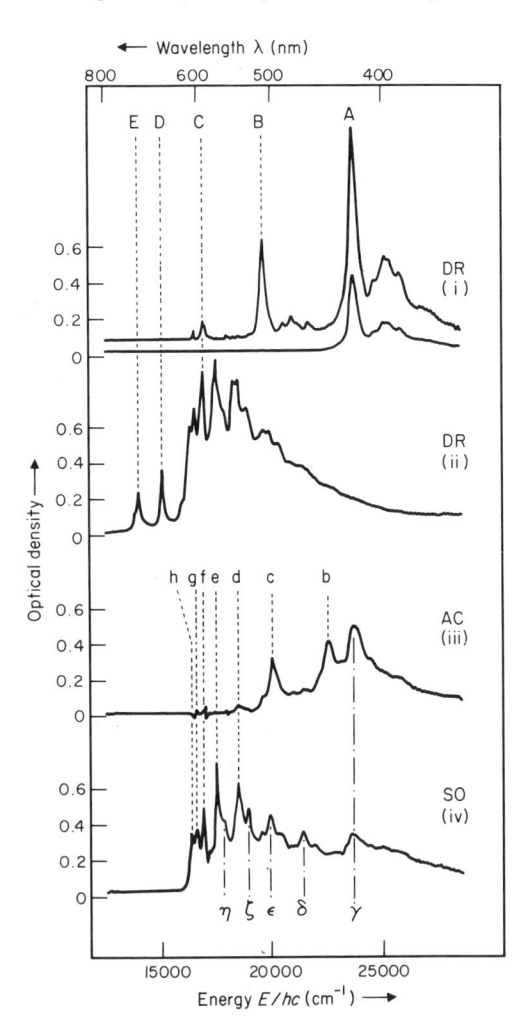

Figure 9 Optical absorption spectra of a bis(toluenesulfonate) diyne crystal at 4.2 K for: (i) one eximer laser pulse and 60 s of 367 nm irradiation; (ii) as (i) plus 1 h of 367 nm irradiation; (iii) three eximer laser pulses plus 2 min of 423 nm irradiation; and (iv) as (iii) plus 75 min of 367 nm irradiation (reproduced from ref. 162 by permission of Elsevier Science Publishers)

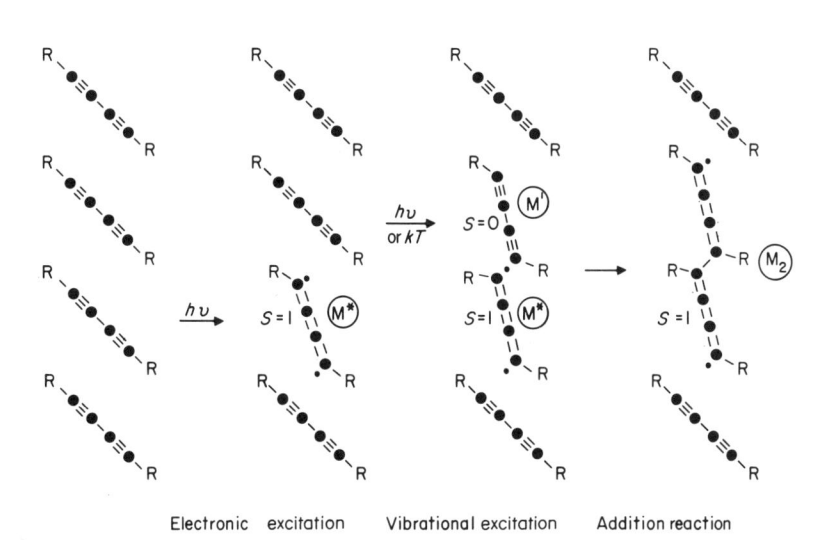

Figure 10 Diyne dimer initiation reaction: M*, metastable monomer produced by irradiation; M', distorted adjacent monomer; and M_2, diradical dimer product of 1,4-addition reaction (reproduced from ref. 18 by permission of Springer-Verlag)

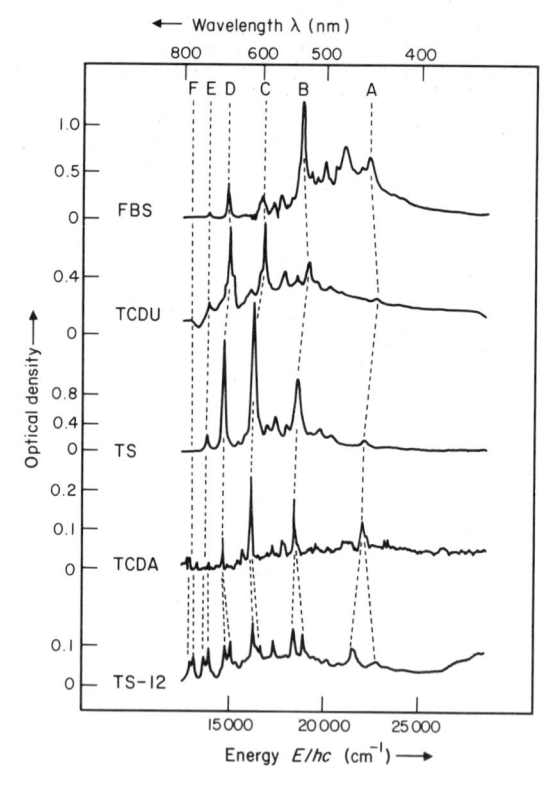

Figure 11 Diradical oligomers produced by UV irradiation of different diynes at 4.2 K; see Table 1 for identification (reproduced from ref. 162 by permission of Elsevier Science Publishers)

15.5 CONCLUSIONS

Our knowledge of the solid state polymerization of diynes is more detailed than that for any other solid state reaction. In summary:

(a) The reactivity of disubstituted diynes depends on the approach of the reacting carbons to within less than 0.4 nm.

(b) If the monomers are aligned along a unique reaction direction and homogeneous polymerization occurs, the product is a macroscopic polymer crystal.

(c) The kinetics of homogeneously polymerizing crystals is controlled principally by the strain within the lattice due to mismatch of the dimensions of the monomer and polymer lattices.

(d) The reaction mechanism for thermal and photo-polymerization proceeds *via* diradicals with a butatriene structure, and dicarbenes and asymmetric carbenes with an yne–ene structure.

Despite the wealth of knowledge concerning diyne polymerization, we cannot routinely and accurately predict whether a single crystal polymer can be obtained from a monomer of known chemical composition. To do so demands advances in our ability to predict the crystal structures for large organic molecules which can adopt a number of different conformations.

It is still unclear how thermal polymerization can be initiated in a perfect lattice. Though defects can play a role, the defect density in good crystals is too low for this to be significant.[173] The kinetics in more complex inhomogeneous reactions has not been studied in detail. There is a continuing interest in the physical properties of polydiynes which stimulates studies of the polymerization of known materials[174] and the synthesis of new substituted diynes with a view to producing new polymer materials.[175–181] The polymerization of diynes and the properties of the product polymers have produced a few surprises in the past and are likely to do so in the future.

Pressure of space means that some topics closely related to the theme of this review have been omitted and others discussed very briefly. It may be that some important work has been overlooked. The author apologizes to anyone whose work has not been reported.

15.6 REFERENCES

1. D. Bloor and R. R. Chance (eds.), 'Polydiacetylenes', Martinus Nijhoff Publishers, Dordrecht, 1985.
2. D. Bloor, *Philos. Trans. R. Soc. London, Ser. A.*, 1985, **314**, 51.
3. D. J. Sandman, *ACS Symp. Ser.*, 1987, **337**.
4. G. Wegner, in 'Chemistry and Physics of One-dimensional Metals', ed. H. J. Keller, Plenum Press, New York, 1977, p. 297.
5. G. Wegner, *Pure Appl. Chem.*, 1977, **49**, 443.
6. G. Wegner, in 'Molecular Metals', ed. W. E. Hatfield, Plenum Press, New York, 1979, p. 209.
7. G. Wegner, *Faraday Discuss., Chem. Soc.*, 1980, **68**, 494.
8. R. H. Baughman, *J. Polym. Sci., Polym. Phys. Ed.*, 1974, **12**, 1511.
9. R. H. Baughman and K. C. Yee, *Macromol. Rev.*, 1978, **13**, 219.
10. R. H. Baughman and R. R. Chance, *Ann. N. Y. Acad. Sci.*, 1978, **313**, 705.
11. V. Enkelmann, *Springer Lecture Notes in Phys.*, 1980, **113**, 1.
12. V. Enkelmann, *Adv. Polym. Sci.*, 1984, **63**, 91.
13. D. Bloor, *Springer Lecture Notes in Phys.*, 1980, **113**, 14.
14. D. Bloor, in 'Developments in Crystalline Polymers I', ed. D. C. Bassett, Applied Science, London, 1982, p. 151.
15. D. Bloor, in 'Quantum Chemistry of Polymers, Solid State Aspects', ed. J. Ladik, J. M. Andre and M. Seel, D. Reidel, Dordrecht, 1984, p. 191.
16. D. Huntsman, in 'The Chemistry of Functional Groups', Supplement C, ed. S. Patai and Z. Rappoport, Wiley, Chichester, 1983, p. 917.
17. H. Baessler, *Adv. Polym. Sci.*, 1984, **63**, 1.
18. H. Sixl, *Adv. Polym. Sci.*, 1984, **63**, 49.
19. H. Sixl and W. Neumann, *Mol. Cryst. Liq. Cryst.*, 1984, **105**, 41.
20. M. Schwoerer and H. Niederwald, *Makromol. Chem., Suppl.*, 1985, **12**, 61.
21. D. J. Sandman, B. S. Elman, G. P. Hamill, C. S. Velazquez and L. A. Samuelson, *Mol. Cryst. Liq. Cryst.*, 1986, **134**, 89.
22. G. A. Vinogradov, *Russ. Chem. Rev. (Engl. Transl.)*, 1984, **53**, 77.
23. V. M. Misin and M. I. Cherkashin, *Usp. Khim.*, 1985, **54**, 956; *Russ. Chem. Rev. (Engl. Transl.)*, 1985, **54**, 562.
24. A. Baeyer and L. Landsberg, *Ber. Dtsch. Chem. Ges.*, 1882, **15**, 57.
25. A. Herbert, *C. R. Hebd. Seances Acad. Sci.*, 1896, **122**, 1550.
26. F. Straus, L. Kollek and W. Heyn, *Ber. Dtsch. Chem. Ges.*, 1930, **63**, 1868.
27. F. Straus, L. Kollek and H. Hauptmann, *Ber. Dtsch. Chem. Ges.*, 1930, **63**, 1886.
28. A. Steger and J. Van Loon, *Recl. Trav. Chim. Pay-Bas*, 1933, **52**, 593.
29. A Castille, *Bull. Acad. R. Méd. Belg.*, 1941, **6**, 152.
30. J. D. Dunitz and J. M. Robertson, *J. Chem. Soc.*, 1947, 1145.
31. K. Bowden, S. I. Heilbron, E. R. H. Jones and K. H. Sargent, *J. Chem. Soc.*, 1947, 1579.
32. A. Keller, *Philos. Mag.*, 1957, **2**, 1171.
33. P. H. Geil, 'Polymer Single Crystals', Interscience, New York, 1963.
34. J. B. Armitage, C. L. Cook, N. Entwistle, E. R. H. Jones and M. C. Whiting, *J. Chem. Soc.*, 1952, 1998.
35. F. Bohlmann, *Chem. Ber.*, 1951, **84**, 785.
36. H. K. Black and B. C. L. Weedon, *J. Chem. Soc.*, 1953, 1785.
37. A. Seher, *Justus Liebigs Ann. Chem.*, 1954, **589**, 222.
38. F. Bohlmann, *Angew. Chem.*, 1957, **69**, 82.
39. F. L. Hirshfeld and G. M. J. Schmidt, *J. Polym. Sci., Part A*, 1964, **2**, 2181.
40. G. Wegner, *Z. Naturforsch., Teil B*, 1969, **24**, 824.
41. A. Keller, *Rep. Prog. Phys.*, 1968, **31**, 623.
42. D. C. Bassett, in 'Developments in Crystalline Polymers 1', ed. D. C. Bassett, Applied Science, London, 1982, p. 115.
43. B. Wunderlich, *Fortschr. Hochpolym.-Forsch.*, 1968, **5**, 568.
44. M. Iguchi, H. Kanetsuna and T. Kawai, *Br. Polym. J.*, 1971, **3**, 177.
45. G. Wegner, E. W. Fischer and A. Munoz-Escalona, *Makromol. Chem., Suppl.*, 1975, **1**, 521.
46. M. Droescher and G. Wegner, *Ind. Eng. Chem., Process. Res. Dev.*, 1979, **18**, 259.
47. B. Tieke and G. Chapuis, in ref. 3, p. 61.
48. M. Farina, G. di Silvestro and P. Sozzani, in ref. 3, p. 79.
49. J. M. Thomas, *Pure Appl. Chem.*, 1979, **51**, 1065.
50. H. Nakanishi, W. Jones, J. M. Thomas, M. Hasegawa and W. L. Rees, *Proc. R. Soc. London, Ser. A*, 1980, **369**, 307.
51. A. Gavezzotti and M. Simonetta, *Chem. Rev.*, 1982, **82**, 1.
52. P. N. Prasad and J. Swiatkiewicz, *Mol. Cryst. Liq. Cryst.*, 1983, **93**, 25.
53. C. Braeuchle, *Mol. Cryst. Liq. Cryst.*, 1985, **93**, 83.
54. D. Y. Curtin and I. C. Paul, *Chem. Rev.*, 1981, **81**, 525.
55. L. Addadi and M. Lahav, *Pure Appl. Chem.*, 1979, **51**, 1269.
56. S. Ramdas and J. M. Thomas, *Chem. Br.*, 1985, **21**, 49.
57. B. S. Green, M. Lahav and D. Rabinovich, *Acc. Chem. Res.*, 1979, **12**, 191.
58. E. H. Andrews and G. E. Martin, *J. Mater. Sci.*, 1973, **8**, 1315.
59. M. Hasegawa, *Adv. Polym. Sci.*, 1982, **42**, 1.
60. M. Hasegawa, K. Saigo, S. Kato and H. Hasegawa, in ref. 3, p. 44.
61. M. M. Labes, P. Love and L. F. Nichols, *Chem. Rev.*, 1979, **79**, 1.
62. G. M. J. Schmidt, in 'Reactivity of the Photoexcited Organic Molecule', Wiley, New York, 1967, p. 227.
63. D. Bloor, *Mol. Cryst. Liq. Cryst.*, 1983, **93**, 183.
64. D. Bloor, in ref. 3, p. 128.
65. D. Bloor and G. C. Stevens, *J. Polym. Sci., Polym. Phys. Ed.*, 1977, **15**, 703.
66. G. N. Patel, E. N. Duesler, D. Y. Curtin and I. C. Paul, *J. Am. Chem. Soc.*, 1980, **102**, 461.
67. D. J. Ando and S. Mann, personal communication.
68. V. Enkelmann, G. Wenz, M. A. Mueller, M. Schmidt and G. Wegner, *Mol. Cryst. Liq. Cryst.*, 1984, **105**, 11.

69. J. Kaiser, G. Wegner and E. W. Fischer, *Isr. J. Chem.*, 1972, **10**, 157.
70. I. M. Papissov and V. A. Kabanov, *J. Polym. Sci., Part C*, 1967, **16**, 911.
71. G. Wegner, *Chimia*, 1974, **28**, 475.
72. V. Enkelmann, R. J. Leyrer, G. Schleier and G. Wegner, *J. Mater. Sci.*, 1980, **15**, 168.
73. K. C. Yee and R. R. Chance, *J. Polym. Sci., Polym. Phys. Ed.*, 1978, **16**, 431.
74. R. H. Baughman, *J. Appl. Phys.*, 1972, **43**, 4362.
75. C. Galiotis, R. J. Young, D. J. Ando and D. Bloor, *Makromol. Chem.*, 1983, **184**, 1083.
76. D. Bloor, in ref. 1, p. 1.
77. M. V. Basilevskii, G. N. Gerasimov and S. I. Petrochenko, *Chem. Phys.*, 1985, **97**, 331.
78. E. Haedicke, K. Penzien and H. W. Schnell, *Angew. Chem., Int. Ed. Engl.*, 1971, **10**, 940.
79. D. A. Fisher, D. N. Batchelder and M. B. Hursthouse, *Acta Crystallogr., Sect. B*, 1978, **34**, 2365.
80. D. A. Fisher, D. J. Ando, D. N. Batchelder and M. B. Hursthouse, *Acta Crystallogr., Sect. B*, 1978, **34**, 3799.
81. D. J. Ando, D. Bloor, M. B. Hursthouse and M. Motevalli, *Acta Crystallogr., Sect. C*, 1985, **41**, 224.
82. V. Enkelmann and G. Wegner, *Angew. Chem.*, 1977, **89**, 432.
83. D. A. Fisher, D. J. Ando, D. Bloor and M. B. Hursthouse, *Acta Crystallogr., Sect. B*, 1979, **35**, 2075.
84. J. J. Mayerle and T. C. Clarke, *Acta Crystallogr., Sect. B*, 1978, **34**, 143.
85. R. L. Williams, D. J. Ando, D. Bloor, M. Motevalli and M. B. Hursthouse, *Acta Crystallogr., Sect. B*, 1982, **38**, 2078.
86. A. R. Werninck, E. Blair, H. W. Milburn, D. J. Ando, D. Bloor, M. Motevalli and M. B. Hursthouse, *Acta Crystallogr., Sect. C*, 1985, **41**, 227.
87. M. Motevalli, P. A. Norman, M. B. Hursthouse, A. R. Werninck, H. W. Milburn, E. Blair, D. Bloor and D. J. Ando, *Acta Crystallogr., Sect. C*, 1986, **42**, 1049.
88. R. L. Williams, D. J. Ando, D. Bloor and M. B. Hursthouse, *Acta Crystallogr., Sect. B*, 1979, **35**, 2072.
89. R. L. Williams, Ph. D. Thesis, University of London, 1982.
90. D. J. Ando, D. Bloor, C. L. Hubble and R. L. Williams, *Makromol. Chem.*, 1980, **181**, 453.
91. V. Enkelmann, *J. Mater. Sci.*, 1980, **15**, 951.
92. D. J. Ando, D. Bloor and B. Tieke, *Makromol. Chem., Rapid Commun.*, 1980, **1**, 385.
93. D. Kobelt and E. F. Paulus, *Acta Crystallogr., Sect. B*, 1974, **30**, 231.
94. R. L. Williams, D. J. Ando, D. Bloor and M. B. Hursthouse, *Polymer*, 1980, **21**, 1269.
95. B. Tieke, G. Wegner, D. Naegele and H. Ringsdorf, *Angew. Chem., Int. Ed. Engl.*, 1976, **15**, 764.
96. B. Tieke, G. Lieser and K. Weiss, *Thin Solid Films*, 1983, **99**, 95.
97. B. Tieke and K. Weiss, *J. Colloid Interface Sci.*, 1984, **101**, 129.
98. H. H. Hub, B. Hupfer, H. Koch and H. Ringsdorf, *Angew. Chem.*, 1980, **92**, 962.
99. D. S. Johnston, S. Sanghera, M. Pons and D. Chapman, *Biochim. Biophys. Acta*, 1980, **602**, 57.
100. E. Lopez, D. F. O'Brien and T. H. Whitesides, *J. Am. Chem. Soc.*, 1982, **104**, 305.
101. B. Tieke and G. Lieser, *Macromolecules*, 1985, **18**, 327.
102. J. P. Fouassier, B. Tieke and G. Wegner, *Isr. J. Chem.*, 1979, **18**, 227.
103. C. Bubeck, B. Tieke and G. Wegner, *Ber. Bunsenges. Phys. Chem.*, 1982, **86**, 495.
104. C. Bubeck, B. Tieke and G. Wegner, *Ber. Bunsenges. Phys. Chem.*, 1982, **86**, 499.
105. C. Bubeck, B. Tieke and G. Wegner., *Mol. Cryst. Liq. Cryst.*, 1983, **96**, 109.
106. H. Nakahara, K. Fukuda, K. Seki, S. Asada and H. Inokuchi, *Chem. Phys.*, 1987, **118**, 123.
107. F. Kajzar and J. Messier, *Chem. Phys.*, 1981, **63**, 123.
108. F. Kajzar and J. Messier, *Thin Solid Films*, 1983, **99**, 109.
109. B. Tieke and D. Bloor, *Makromol. Chem.*, 1979, **180**, 2275.
110. I. R. J. Lyall and D. N. Batchelder, *Br. Polym. J.*, 1985, **17**, 372.
111. R. R. McCaffrey, P. N. Prasad, M. Fornalik and R. Baier, *J. Polym. Sci., Polym. Phys. Ed.*, 1985, **23**, 1523.
112. A. S. Hay, D. A. Bolon, K. R. Leimer and R. F. Clark, *J. Polym. Sci., Polym. Lett. Ed.*, 1970, **8**, 97.
113. G. Wegner, *Makromol. Chem.*, 1970, **134**, 219.
114. A. O. Patil, D. D. Deshpande, S. S. Talwar and A. B. Biswas, *J. Polym. Sci., Polym. Chem. Ed.*, 1981, **19**, 1155.
115. M. F. Rubner, *Macromolecules*, 1986, **19**, 2114.
116. M. F. Rubner, *Macromolecules*, 1986, **19**, 2129.
117. M. F. Rubner, D. J. Sandman and C. Velazquez, *Macromolecules*, 1987, **20**, 1296.
118. K. E. Knol, L. W. van Horssen, G. Challa and E. E. Havinga, *Polym. Commun.*, 1985, **26**, 71.
119. M. Thakur and J. B. Lando, *Macromolecules*, 1983, **16**, 143.
120. Y. Ozcayir and A. Blumstein, *Mol. Cryst. Liq. Cryst.*, 1986, **135**, 237.
121. Y. Ozcayir, J. Asrar, S. B. Clough and A. Blumstein, *Mol. Cryst. Liq. Cryst.*, 1986, **138**, 167.
122. Y. Ozcayir and A. Blumstein, *J. Polym. Sci., Polym. Chem. Ed.*, 1986, **24**, 1217.
123. J. Tsibouklis, Ph. D. Thesis, Napier College, Edinburgh, 1988.
124. R. H. Baughman, *J. Chem. Phys.*, 1978, **68**, 3110.
125. R. H. Baughman and R. R. Chance, *J. Chem. Phys.*, 1980, **73**, 4113.
126. C. Kröhnke, V. Enkelman and G. Wegner, *Chem. Phys. Lett.*, 1980, **71**, 38.
127. J. K. Lochner, H. Baessler and T. Hinrichsen, *Ber. Bunsenges. Phys. Chem.*, 1979, **83**, 899.
128. G. Wegner, *Makromol. Chem.*, 1972, **154**, 35.
129. A. R. McGhie, P. S. Kalyanaraman and A. F. Garito, *J. Polym. Sci., Polym. Lett. Ed.*, 1978, **16**, 335.
130. A. F. Garito, A. R. McGhie ad P. S. Kalyanaraman, in 'Molecular Metals', ed. W. E. Hatfield, Plenum Press, New York, 1979, p. 255.
131. A. R. McGhie, P. S. Kalyanaraman and A. F. Garito, *Mol. Cryst. Liq. Cryst.*, 1979, **50**, 287.
132. E. M. Barrall, T. C. Clarke and A. R. Gregges, *J. Polym. Sci., Polym. Phys. Ed.*, 1978, **16**, 1355.
133. R. R. Chance, G. N. Patel, E. A. Turi and Y. P. Khanna, *J. Am. Chem. Soc.*, 1978, **100**, 1307.
134. G. N. Patel, R. R. Chance, E. A. Turi and Y. P. Khanna, *J. Am. Chem. Soc.*, 1978, **100**, 6644.
135. G. N. Patel, *J. Polym. Sci., Polym. Phys. Ed.*, 1979, **17**, 1591.
136. R. R. Chance and M. L. Shand, *J. Chem. Phys.*, 1980, **72**, 948.
137. A. Prock, M. L. Shand and R. R. Chance, *Macromolecules*, 1982, **15**, 238.
138. H. Eckhardt, T. Prusik and R. R. Chance, *Macromolecules*, 1983, **16**, 732.

139. S. P. McGlynn, T. Azumi and M. Kinoshita, 'Molecular Spectroscopy of the Triplet State', Prentice-Hall, Englewood Cliffs, NJ, 1969.
140. R. H. Baughman, G. J. Exarhos and W. M. Risen, *J. Polym. Sci., Polym. Phys. Ed.*, 1974, **12**, 2189.
141. G. C. Stevens and D. Bloor, *J. Polym. Sci., Polym. Phys. Ed.*, 1975, **13**, 2411.
142. H. Eichele, M. Schwoerer, R. Huber and D. Bloor, *Chem. Phys. Lett.*, 1976, **42**, 342.
143. G. C. Stevens and D. Bloor, *Phys. Status Solidi A*, 1978, **45**, 483.
144. G. C. Stevens and D. Bloor, *Phys. Status Solidi A*, 1978, **46**, 141.
145. G. C. Stevens and D. Bloor, *Phys. Status Solidi A*, 1978, **46**, 619.
146. G. C. Stevens and D. Bloor, *Chem. Phys. Lett.*, 1976, **40**, 37.
147. R. Huber, M. Schwoerer, C. Bubeck and H. Sixl, *Chem. Phys. Lett.*, 1978, **53**, 35.
148. D. N. Batchelder, N. J. Poole and D. Bloor, *Chem. Phys. Lett.*, 1981, **81**, 560.
149. R. R. Chance and G. N. Patel, *J. Polym. Sci., Polym. Phys. Ed.*, 1978, **16**, 859.
150. Y. Hori and L. D. Kispert, *J. Chem. Phys.*, 1978, **69**, 3826.
151. Y. Hori and L. D. Kispert, *J. Am. Chem. Soc.*, 1979, **101**, 3173.
152. H. Gross, H. Sixl, C. Kroehnke and V. Enkelmann, *Chem. Phys.*, 1980, **45**, 15.
153. C. Bubeck, H. Sixl, D. Bloor and G. Wegner, *Chem. Phys. Lett.*, 1979, **63**, 574.
154. C. Bubeck, T. H. Nguyen Xuan, H. Sixl, B. Tieke and D. Bloor, *Ber. Bunsenges. Phys. Chem.*, 1983, **87**, 1149.
155. C. Bubeck, H. Sixl and H. C. Wolf, *Chem. Phys.*, 1978, **32**, 231.
156. H. Sixl, W. Hersel and H. C. Wolf, *Chem. Phys. Lett.*, 1978, **53**, 39.
157. W. Hersel, H. Sixl and G. Wegner, *Chem. Phys. Lett.*, 1980, **73**, 288.
158. C. Bubeck, H. Sixl and W. Neumann, *Chem. Phys.*, 1980, **48**, 269.
159. W. Neumann and H. Sixl, *Chem. Phys.*, 1980, **50**, 273.
160. C. Bubeck, W. Hersel, W. Neumann, H. Sixl and J. Waldmann, *Chem. Phys.*, 1980, **51**, 1.
161. R. Huber and M. Schwoerer, *Chem. Phys. Lett.*, 1980, **72**, 10.
162. H. Gross and H. Sixl, *Chem. Phys. Lett.*, 1982, **91**, 262.
163. H. Gross, H. Sixl, S. F. Fichert and E. W. Knapp, *Chem. Phys.*, 1984, **84**, 321.
164. H. Benk and H. Sixl, *Mol. Phys.*, 1981, **42**, 779.
165. H. Sixl and W. Neumann, *Mol. Cryst. Liq. Cryst.*, 1984, **105**, 41.
166. H. Sixl, W. Neumann, R. Huber, V. Denner and E. Sigmund, *Phys. Rev. B*, 1985, **31**, 142.
167. C. Kollmar and H. Sixl, *J. Chem. Phys.*, 1987, **87**, 1396.
168. C. Kollmar and H. Sixl, *J. Chem. Phys.*, 1988, **88**, 1343.
169. H. Niederwald and M. Schwoerer, *Z. Naturforsch., Teil A*, 1983, **38**, 749.
170. W. Hartl and M. Schwoerer, *Chem. Phys.*, 1982, **69**, 443.
171. R. Mueller-Nawrath, R. Angstl and M. Schwoerer, *Chem. Phys.*, 1986, **108**, 121.
172. H. Gross, H. Sixl, C. Krohnke and V. Enkelman, *Chem. Phys.*, 1980, **45**, 15.
173. M. Dudley, J. N. Sherwood, D. J. Ando and D. Bloor, *Mol. Cryst. Liq. Cryst.*, 1983, **93**, 223.
174. Z. Iqbal, N. S. Murthy, Y. P. Khanna, J. S. Szobota, R. A. Dalterio and F. J. Owens, *J. Phys. C*, 1987, **20**, 4283.
175. G. N. Patel, *Macromolecules*, 1981, **14**, 1170.
176. D. Bloor, D. J. Ando, J. S. Obhi, S. Mann and M. R. Worboys, *Makromol. Chem., Rapid Commun.*, 1986, **7**, 665.
177. P. Strohriegl, *Makromol. Chem. Rapid Commun.*, 1987, **8**, 437.
178. M. Bertault, L. Toupet, J. Canceill and A. Collet, *Makromol. Chem., Rapid Commun.*, 1987, **8**, 443.
179. P. Strohriegl, H. Schultes, D. Heindl, P. Gruner-Bauer, V. Enkelmann and E. Dorman, *Ber. Bunsenges. Phys. Chem.*, 1987, **91**, 918.
180. C. Krohnke, *Ber. Bunsenges. Phys. Chem.*, 1987, **91**, 982.
181. J. Veciana, A. Galan, J. Riaza, O. Armet, E. Molins and C. Miravitlles, *Mol. Cryst. Liq. Cryst.*, 1988, **156**, 289.

16
Step Copolymerization

ERNEST MARÉCHAL and ALAIN FRADET
Université Pierre et Marie Curie, Paris, France

16.1 INTRODUCTION	251
16.2 DISTRIBUTIONS IN COPOLYCONDENSATES	252
16.2.1 Critical analysis of Several Copolycondensation Theoretical Studies	252
16.2.2 Explicit calculation of $\bar{M}_n$, $\bar{M}_w$ and Sequence Length Distribution	254
16.2.2.1 Average molecular weights	254
16.2.2.2 Comonomer sequence lengths	256
16.2.3 Monte Carlo Simulations	257
16.2.4 Kinetic Study	258
16.3 APPLICATIONS OF COPOLYCONDENSATION	263
16.3.1 Introduction	263
16.3.2 Copolycondensations Involving Two Reactions of the Same Type	264
16.3.2.1 Ester bond formation	264
16.3.2.2 Amide bond formation	265
16.3.2.3 Carbonate bond formation	267
16.3.2.4 Aromatic ether bond formation	268
16.3.2.5 Urethane bond formation	268
16.3.2.6 Aromatic ketone bond formation	269
16.3.3 Copolycondensations Involving Reactions of a Different Nature	269
16.3.3.1 Simultaneous formation of ester and carbonate bonds	269
16.3.3.2 Simultaneous formation of ester and amide bonds	270
16.3.3.3 Simultaneous formation of amide and urea bonds	270
16.3.3.4 Simultaneous formation of amide and urethane bonds	271
16.3.3.5 Simultaneous formation of urea and urethane bonds	271
16.3.3.6 Simultaneous formation of amide and hydrazide bonds	272
16.4 CONCLUSION	272
16.5 REFERENCES	272

16.1 INTRODUCTION

In step copolymerization, copolymer chains are formed from at least three AA type monomers (for example two diacids and one diol) or at least two AB type monomers (for example two amino acids). Step copolymerization has been the subject of very few fundamental studies although it is used in the production of many materials (polyurethanes, polyesters, *etc.*) which are considered in Chapters 17 to 25 of this volume. Moreover, many structures classified as '*copolycondensates*' are in fact prepared by *homopolycondensation*. Thus, most block copolycondensates result from the homopolycondensation of two telechelic oligomers. In the same way, alternate copolymers are very often obtained by homopolycondensation of monomers containing two different monomer units; the preparation of poly(ester-co-imide) may thus involve the synthesis of dianhydride with aromatic ester bonds followed by the homopolycondensation of A with an aromatic diamine.

On the other hand, the preparation of block copolycondensates by step polymerization of a telechelic oligomer with the precursors of the other block is really a step copolymerization (Scheme 1) where B and C are able to react with A and with each other.

$$A-(block\ 1)-A + BB + CC \text{ or } A-(block\ 1)-A + BC$$

Scheme 1

Although fundamental studies of step copolymerization are far less numerous than those on chain copolymerization, the first part of this chapter is devoted to sequence distributions in copolycondensates; this includes a critical analysis of copolycondensation theoretical studies, the explicit calculation of $\bar{M}_n$, $\bar{M}_w$ and of sequence lengths for AA, BB, CC and A_1B_1, A_2B_2 systems, Monte–Carlo theoretical simulations and kinetic studies.

Several examples of structures obtained by step copolymerization are described in the second part of this chapter.

16.2 DISTRIBUTIONS IN COPOLYCONDENSATES

Step copolymerization theoretical studies are much less numerous than those relative to chain copolymerization for two reasons: (i) as products with high molecular weights can be obtained only when the conversion is close to 100%, most copolymers have the composition of the feed; and (ii) many step copolymerizations are equilibrated processes which means that the sequence distribution resulting from the relative reactivities is rapidly changed.

However, irreversible step copolymerizations are an important development [e.g. polyurethane prepared by reaction injection molding (RIM)] and the knowledge of the sequence distribution is important as the distribution of the soft and hard segments in the chain has a determining influence on the material properties. Several treatments have been proposed[1–4] which we will analyze briefly before developing the treatment of Macosko et al.[5]

16.2.1 Critical Analysis of Several Copolycondensation Theoretical Studies

All approaches in refs. 1–4 model the polymerization as a Markov chain process. Flory[1] discovered the Markov chain nature of most polymerization processes and proposed relations to describe the distribution of chain lengths and molecular weights.

For the linear polycondensation of an AB type monomer the probability P_x to obtain an xmer by adding exactly $(x-1)$ AB units to an original AB, without further addition, is

$$P_x = p^{x-1}(1-p); \quad \sum_{x=1}^{\infty} P_x = 1 \tag{2}$$

where p (fractional conversion of A or B groups) is the probability that any A or B group has reacted, that is to say, the reaction (or transition) probability.

The number N_x of xmers is

$$N_x = N_0 p^{x-1}(1-p)^2 \tag{3}$$

where N_0 is the initial number of AB units. The number average and weight average chain lengths ($\bar{X}_n$ and $\bar{X}_w$ respectively) are

$$\bar{X}_n = \frac{1}{1-p} \tag{4}$$

$$\bar{X}_w = \frac{1+p}{1-p} \tag{5}$$

Case[2] has applied these principles to the *step copolymerization* of the system AA, BB, CC where BB does not react with CC; in such a system six types of polymeric chains exist classified with respect to their end groups: A∼A, B∼B, C∼C, A∼B, A∼C, B∼C. Each has a distribution of chain lengths which can be obtained by the reasoning which leads to equation (3). The overall molecular weight distribution (MWD) and chain length distribution (CLD) are the sums of the individual distributions weighted by their respective mole fraction in the reaction mixture. Unfortunately this results in relations which are very difficult to handle.

Lowry[3] obtained CLD by matrix manipulation. Let us consider the reaction medium at some extent of reaction; it contains already formed chains. A particular transient state is characterized by the presence of a particular comonomer unit at some position along the chain; the transition probability between two transient states is the probability that a particular transient state, characterized by a specific distribution of comonomer units, is followed by another particular transient state characterized by another specific distribution of monomer units. When the reaction between two monomers is forbidden, the transition probability is zero. If the total number of comonomer molecules in the reaction mixture is N, there are N transient states. Let $\mathbf{Q}$ be the $N \times N$ matrix of the probabilities of transition between the various transient states.

Since the chains have finite lengths, it is necessary to define a matrix $\mathbf{R}$ of probabilities of absorption (or of termination) which corresponds to the state where no comonomer unit addition follows a particular state. Matrix $\mathbf{R}$ is a $N \times 1$ matrix of probabilities of absorption. Matrix $\mathbf{O}$ is a $1 \times N$ matrix containing only zeros. It corresponds to the impossibility of an absorption state generating a transient state. $I = 1$, which assures that once the system has entered the absorbed state (termination) it no longer evolves.

Lowry[3] showed that the $(N+1) \times (N+1)$ transition matrix $\mathbf{P}$ is

$$\mathbf{P} = \begin{pmatrix} I & O \\ R & Q \end{pmatrix} \tag{6}$$

In addition to $\mathbf{P}$ it is necessary to know the vector $\overset{\circ}{q}$ of the probabilities of each of the transient states initiating a chain.

The CLD is given by

$$P_x = \overset{\circ}{q} \mathbf{Q}^{x-1} (I' - \mathbf{Q}) I \tag{7}$$

where I is a N-dimensional column vector of ones and I' is the identity matrix of rank $N \times N$.

Lowry[3] determined $\bar{X}_n$ and $\bar{X}_w$ from equation (6) as

$$\bar{X}_n = \overset{\circ}{q} (I' - \mathbf{Q})^{-1} I \tag{8}$$

$$\bar{X}_w = \overset{\circ}{q} [2(I' - \mathbf{Q})^{-1} - I'](I' - \mathbf{Q})^{-1} I \tag{9}$$

In fact equations (7) to (9) are a matrix generalization of equations (2), (4) and (5); thus for the case of an AB polymerization $\mathbf{Q} = p$ and $\overset{\circ}{q} = 1$ so that $(I' - \mathbf{Q})^{-1} = 1/(1-p)$ and equations (7) to (9) become respectively

$$P_x = p^{x-1}(1-p); \quad \bar{X}_n = \frac{1}{1-p}; \quad \bar{X}_w = \frac{1+p}{1-p} \tag{10}$$

Unfortunately there is no obvious way to introduce the molecular weight of each comonomer in Lowry's treatment which makes the determination of MWD impossible. This is a problem because in most cases the experimental determinations relative to *copolymers* give the molecular weight and not the total number of units incorporated into a chain.

Peller[4] used a matrix $\mathbf{M}$ of sequential probabilities which are different from but related to the Markov chain transition probabilities. Unfortunately this results in relations of high unwieldiness and here too there is no clear way to calculate MWD.

Kuchanov and Brun[6,7] proposed a theory of equilibrium copolycondensation considering (or not considering) substitution effects, which relates to cases when the formation of intramolecular bonds between monomer units affects the activity of end or internal adjacent functional groups. In the first order substitution effect, it is assumed that the free energy of any functional group depends only on the adjacent group, while in a second order substitution effect it depends on the two nearest groups on each side.

Kuchanov and Brun showed that the microstructure of products of any process may be described using a Markov chain. They established relations between the statistical characteristics of copolymers (molecular weight, composition, structure and molecular distributions) the equilibrium constants and the composition of the monomer mixture.

16.2.2 Explicit Calculation of $\bar{M}_n$, $\bar{M}_w$ and Sequence Length Distribution

Macosko et al.[5] developed a method leading to explicit and easy to handle relations for both CLD and MWD. They considered the system of comonomers AA, BB, CC where functional groups B and C can react with A but not with each other. They determined average molecular weights and comonomer sequence lengths. This work was summarized by Tirrell.[8]

16.2.2.1 Average molecular weights

Let us consider a randomly chosen functional group A* in the copolymer chain.
The two chain arms attached to A* are labelled 'looking in' and 'looking out'

$$\xrightarrow{\text{out}}$$
$$\sim\!\text{A--A}^*\!\sim$$
$$\xleftarrow{\text{in}}$$

Scheme 2

Let $E(W_{AA})$ be the expected (or average) value of the weight of polymer attached to A

$$E(W_{AA}) = E(W_A^{\text{out}}) + E(W_A^{\text{in}}) \tag{11}$$

$E(W_A^{\text{out}})$ and $E(W_A^{\text{in}})$ are the weights of the chain arms 'looking out' and 'looking in'

$$E(W_A^{\text{out}}) = \begin{cases} 0 \text{ if A is unreacted} \\ E(W_B^{\text{in}}) \text{ if A has reacted with B} \\ E(W_C^{\text{in}}) \text{ if A has reacted with C} \end{cases}$$

Application of the so-called law of total probability of expectation[9,10] leads to

$$E(W_A^{\text{out}}) = E(W_B^{\text{in}}) p_{AB} + E(W_C^{\text{in}}) p_{AC} \tag{12}$$

where p_{AB} and p_{AC} are the probabilities that A has reacted with B and C respectively.

If A_0, B_0, C_0 and A, B, C are the concentrations of species A, B and C at time 0 and time t, reaction conversions p, q_1, q_2 can be defined as

$$p = \frac{A_0 - A}{A_0}; \quad q_1 = \frac{B_0 - B}{B_0}; \quad q_2 = \frac{C_0 - C}{C_0} \tag{13}$$

These quantities can be determined by kinetic analysis.
Obviously

$$p = p_{AB} + p_{AC} \tag{14}$$

and

$$p_{AB} = \frac{B_0 - B}{B_0 - B + C_0 - C}; \quad p = \frac{q_1 B_0}{q_1 B_0 + q_2 C_0} p \tag{15}$$

$$p_{AC} = \frac{C_0 - C}{B_0 - B + C_0 - C}; \quad p = \frac{q_2 C_0}{q_1 B_0 + q_2 C_0} p \tag{16}$$

For stoichiometric reasons

$$A_0 p = B_0 q_1 + C_0 q_2 \tag{17}$$

and therefore

$$p_{AB} = r_1 q_1; \quad p_{AC} = r_2 q_2 \tag{18}$$

where

$$r_1 = B_0/A_0; \quad r_2 = C_0/A_0 \tag{19}$$

The expected weight $E(W_B^{in})$ on any group B looking into its parent molecule is the molecular weight M_{BB} of BB plus the expected weight $E(W_B^{out})$ of the other arm

$$E(W_B^{in}) = M_{BB} + E(W_B^{out}) \qquad (20)$$

In the same way

$$E(W_C^{in}) = M_{CC} + E(W_C^{out}) \qquad (21)$$

From the law of total probability of expectation

$$E(W_B^{out}) = E(W_A^{in}) q_1 \qquad (22)$$

$$E(W_C^{out}) = E(W_B^{in}) q_2 \qquad (23)$$

and

$$E(W_A^{in}) = M_{AA} + E(W_A^{out}) \qquad (24)$$

Combination of equation (11) and equation (24) gives

$$E(W_{AA}) = M_{AA} + 2E(W_A^{out}) \qquad (25)$$

In the same way

$$E(W_{BB}) = M_{BB} + 2E(W_B^{out}) \qquad (26)$$

$$E(W_{CC}) = M_{CC} + 2E(W_C^{out}) \qquad (27)$$

Combination of equation (12), equation (20), equation (21) and equations (22)–(26) gives

$$E(W_{AA}) = M_{AA} + 2\left(\frac{M + QM_{AA}}{1 - Q}\right) \qquad (28)$$

$$E(W_{BB}) = M_{BB} + 2q_1\left(M_{AA} + \frac{M + QM_{AA}}{1 - Q}\right) \qquad (29)$$

$$E(W_{CC}) = M_{CC} + 2q_2\left(M_{AA} + \frac{M + QM_{AA}}{1 - Q}\right) \qquad (30)$$

where

$$M = M_{BB} p_{AB} + M_{CC} p_{AC} \qquad (31)$$

and

$$Q = q_1 p_{AB} + q_2 p_{AC} \qquad (32)$$

The weight average molecular weight is given by

$$\bar{M}_w = \sum_{i=A,B,C} E(W_{ii}) \omega_{ii} \qquad (33)$$

where ω_{ii} is the weight fraction of monomer ii in the initial reaction mixture. Substitution of equations (28)–(30) into equaton (33) results in

$$\bar{M}_w = W + 2R + 2G(\omega_{AA} + \omega_{BB} q_1 + \omega_{CC} q_2) \qquad (34)$$

where

$$W = \sum_{i=A,B,C} \omega_{ii} M_i; \quad R = \sum_{i=B,C} q_i M_i \omega_{ii} \qquad (35)$$

$$G = \frac{M + QM_{AA}}{1 - Q} \qquad (36)$$

Equation (34) was obtained by selecting units at random which favours the long chains. Calculation of $\bar{M}_n$ requires that each chain in the mixture is sampled with equal probability whatever its size, which can be accomplished by selecting chain ends. Since $E(W_i^{out})=0$ for a chain end

$$\bar{M}_n = \sum_{i=A,B,C}^{i} E(W_i^{in}) x_{ii} \tag{37}$$

where x_{ii} is the number fraction of end groups derived from monomer ii.

$$x_{AA} = A_0(1-p)/\Sigma \tag{38}$$
$$x_{BB} = B_0(1-q_1)/\Sigma \tag{39}$$
$$x_{CC} = C_0(1-q_2)/\Sigma \tag{40}$$

where

$$\Sigma = A_0(1-p) + B_0(1-q_1) + C_0(1-q_2) \tag{41}$$

and

$$\bar{M}_n = N + R'G(x_{AA} + x_{BB}q_1) + x_{CC}q_2) \tag{42}$$

where

$$N = \sum_{i=A,B,C} x_{ii} M_i; \quad R' = \sum_{i=A,B,C} q_i M_{ii} x_{ii} \tag{43}$$

16.2.2.2 Comonomer sequence lengths

Macosko et al.[5] defined the sequence length as follows

$$\sim\!\!\text{AA–BB–AA–BB–AA–BB}\!\sim \quad \sim\!\!\text{AA–CC–AA–CC}\!\sim$$

BB sequence length 3   CC sequence length 2

Scheme 3

Let us consider a randomly chosen A; the expected number $E(N_{A,BB}^{out})$ of BB units attached to a randomly chosen A depends on three possibilities:

$$E(N_{A,BB}^{out}) = \begin{cases} 0 \text{ if A is unreacted} \\ E(N_{B,BB}^{in}) \text{ if A has reacted with B} \\ 0 \text{ if A has reacted with C} \end{cases}$$

Then

$$E(N_{A,BB}^{out}) = E(N_{B,BB}^{in}) p_{AB} \tag{44}$$

By analogy with equations (20), (22) and (24)

$$E(N_{B,BB}^{in}) = 1 + E(N_{B,BB}^{out}) \tag{45}$$
$$E(N_{B,BB}^{out}) = E(N_{A,BB}^{in}) q_1 \tag{46}$$
$$E(N_{A,BB}^{in}) = E(N_{A,BB}^{out}) \tag{47}$$

The weight average sequence length for BB, $\bar{N}_{w,BB}$, is one plus the sum of the lengths of BB sequences extending out from a BB unit selected at random

$$\bar{N}_{w,BB} = 1 + 2E(N_{B,BB}^{out}) \tag{48}$$

$E(N_{B,BB}^{out})$ can be calculated from equations (44)–(47) which gives

$$\bar{N}_{w,BB} = \frac{1 + r_1 q_1^2}{1 - r_1 q_1^2} \tag{49}$$

The number average sequence length is calculated in the same way but with the BB selected at the end of a sequence to weigh all the sequences evenly

$$\bar{N}_{n,BB} = 1 + E(N_{B,BB}^{out}) \tag{50}$$

$$\bar{N}_{n,BB} = \frac{1}{1 - r_1 q_1^2} \tag{51}$$

Similar relations can be established for $\bar{N}_{w,CC}$ and $\bar{N}_{n,CC}$.

Workers using similar calculations to those above[5] determined $\bar{M}_w$, $\bar{M}_n$ and the weight average and number average sequence lengths of A_1B_1 and of A_2B_2 in a $A_1B_1 + A_2B_2$ step copolymerization.

Until now only linear copolycondensations have been considered. However Macosko and Miller[10] analyzed systems involving monomers with a functionality higher than two. Following Stockmayer's notations[11] this can be represented schematically as in Scheme 4. They established relations giving the value of $\bar{M}_w$ and of conversion at gel point.

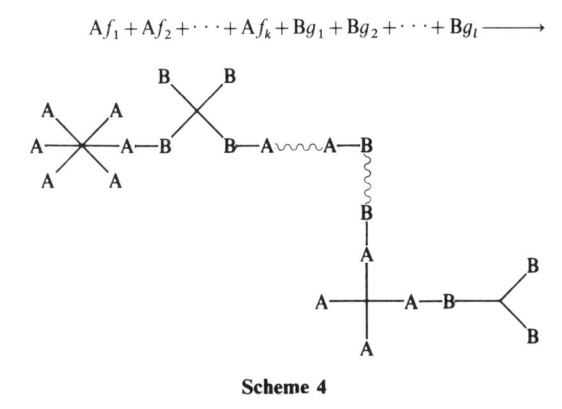

$$Af_1 + Af_2 + \cdots + Af_k + Bg_1 + Bg_2 + \cdots + Bg_l \longrightarrow$$

Scheme 4

16.2.3 Monte Carlo Simulations

Chaumont *et al.*[12] studied the simulation of polycondensation by a Monte Carlo type method. Their process was tested only on two rather simple cases: the self-condensation of AB type monomers and the chain extension involving AA type macromolecules and BB type micromolecular coupling agents. However, it could be applied to more sophisticated systems including copolymerization of several monomers.

In fact, we found only a few Monte Carlo simulations of sequence distributions in step-growth copolymerization.[13,14,15] They are relevant to copolycondensations without interchange reactions

$$AA + BB + XY \rightarrow Copolymer \tag{52}$$

A and B react only with X or Y as, for example, in the synthesis of copolyurethanes

$$HO\text{\textasciitilde}OH + HO-R-OH + O=C=N-R'-N=C=O \rightarrow Copolyurethane \tag{53}$$

Johnson and O'Driscoll[13] developed a model for the synthesis of polyurethane block copolymers under various ideal reaction conditions. The authors considered a set of AA, BB and XY molecules where the possible reactions are as given in Scheme 5. They studied three systems by using the rate constants given in Table 1. Rate constants in I are chosen to test a system with random behaviour. The rate constants of systems II and III approximately correspond to those expected[16] for the copolycondensation of a polyol (AA), butanediol (BB) and 4,4'-diphenylmethane diisocyanate (MDI) as XY in system II or toluene diisocyanate (TDI) as XY in system III.

$$\sim A + \sim X \rightarrow \sim A-X\sim \quad \text{rate} = k_1[A][X]$$
$$\sim A + \sim Y \rightarrow \sim A-Y\sim \quad \text{rate} = k_2[A][Y]$$
$$\sim B + \sim X \rightarrow \sim B-X\sim \quad \text{rate} = k_3[B][X]$$
$$\sim B + \sim Y \rightarrow \sim B-Y\sim \quad \text{rate} = k_4[B][Y]$$

Scheme 5

Table 1 Kinetic Constants Used in Simulated Systems[13]

System	k_1	k_2	k_3	k_4
I	1	1	1	1
II	1	1	2	2
III	1	7	2	14

This simulation[13] provides information not only on copolymer molecular weight distribution but also on the size distribution of the sequences within the copolymer. It shows that the kinetic factors are unimportant in determining intrachain sequences compared with the relative concentrations (A/B) of the coreactants. Moreover it appears that the sequence length is below four units.

Simulated results obtained for system I at 99% conversion and corrected for chain ends were compared to those obtained from the deterministic equation (50) showing satisfactory agreement.

Sorta and Melis[17] discussed block length distribution in copolycondensates but using a system which, according to Johnson and O'Driscoll[13], was (implicitly) an equilibrium system rather than one with structure determined by the kinetics of polymerization.

16.2.4 Kinetic Study

There are few quantitative results relative to the kinetics of copolycondensation. Some of the more recent results were published by M. J. Han;[18-20] they relate to interchange reactions studied at the beginning of the copolycondensation. He studied[18] the copolycondensation of two different diol esters of dibasic diacids in conditions where hydroxy chain ends attack the terminal ester groups of the chain. No catalyst was added and diols HOR^1OH and HOR^2OH were continuously removed (see Scheme 6).

$$\sim COOR^1OH + HOR^1OOC\sim \xrightarrow{k_{11}} \sim COOR^1OOC\sim + HOR^1OH$$

$$\sim COOR^1OH + HOR^2OOC\sim \xrightarrow{k_{12}} \sim COOR^2OOC\sim + HOR^1OH$$

$$\sim COOR^2OH + HOR^1OOC\sim \xrightarrow{k_{21}} \sim COOR^1OOC\sim + HOR^2OH$$

$$\sim COOR^2OH + HOR^2OOC\sim \xrightarrow{k_{22}} \sim COOR^2OOC\sim + HOR^2OH$$

Scheme 6

Let $[OH]_1$, $[OH]_2$, $[COO]_1$ and $[COO]_2$ be the concentrations of hydroxyl and terminal ester groups in chain ends 1 and chain ends 2 respectively. From homopolycondensation studies, Han assumed that these reactions are second order, which leads to equation (54) which can be rearranged into equation (55).

$$\frac{d[HOR^1OH]}{d[HOR^2OH]} = \frac{k_{11}[COO]_1[OH]_1 + k_{12}[COO]_1[OH]_2}{k_{21}[COO]_2[OH]_1 + k_{22}[COO]_2[OH]_2} \tag{54}$$

$$\frac{d[HOR^1OH]}{d[HOR^2OH]} = \frac{[COO]_1}{[COO]_2} \frac{k_{11}[OH]_1/[OH]_2 + k_{12}}{k_{21}[OH]_1/[OH]_2 + k_{22}} \tag{55}$$

In general the following conditions are satisfied at the initial stage

$$\frac{[OH]_1}{[OH]_2} = \frac{[COO]_1}{[COO]_2} = a \tag{56}$$

Equation (55) becomes

$$b = a\frac{ak_{11} + k_{12}}{ak_{21} + k_{22}} \tag{57}$$

$$ak_{11} - \frac{b}{a}k_{22} = bk_{21} - k_{12} \tag{58}$$

where $b = d[HOR^1OH]/d[HOR^2OH]$ is the ratio of the diols formed during time dt. Values of k_{11} and k_{22} are obtained from homopolycondensation studies. If the left term of equation (58) is plotted against b the rate constants k_{21} and k_{12} of the cross reactions can be obtained from the slope and intercept.

In a similar manner, the concentrations of diol units incorporated in the copolyester during time dt are given by

$$-\frac{d[-OR^1O-]}{dt} = k_{21}[COO]_2[OH]_1 + k_{11}[COO]_1[OH]_1 \tag{59}$$

$$-\frac{d[-OR^2O-]}{dt} = k_{12}[COO]_1[OH]_2 + k_{22}[COO]_2[OH]_2 \tag{60}$$

which leads to

$$c = \frac{d[-OR^1O-]}{d[-OR^2O-]} = a\frac{k_2 + ak_{11}}{ak_{12} + k_{22}} \tag{61}$$

The azeotropic monomer ratio (the composition of the chain is the same as that of the feed) corresponds to

$$c = a = (k_{21} - k_{22})/(k_{12} - k_{11}) \tag{62}$$

These relations were applied to the copolycondensation of bis(2-hydroxyethyl) terephthalate (HET, monomer 1, index 1) with bis(2-hydroxy-1-propyl) terephthalate (HPT, **1**, monomer 2, index 2). The rate constants of the cross reactions were obtained by copolymerizing monomer mixtures with determined values of $a = [HET]_0/[HPT]_0$ and analyzing the mixture of diols (1,2-dihydroxyethane and 1,2-dihydroxypropane) recovered before the conversion reaches 10%. From the resulting values of b it was possible to calculate $10^3[ak_{11} - (b/a)k_{22}]$ (left hand side of equation 58) which, when plotted against b, gave a straight line allowing the calculation of the reactivity ratios $r_1 = k_{11}/k_{12}$ and $r_2 = k_{22}/k_{21}$. The results are given in Table 2.

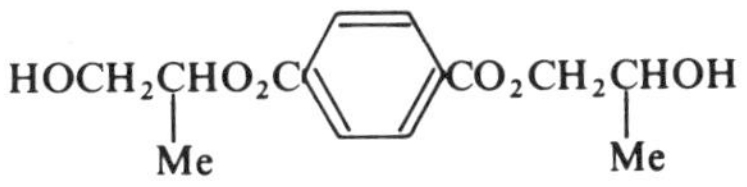

$$HOCH_2\underset{Me}{CH}O_2C\text{—}\underset{}{\bigcirc}\text{—}CO_2CH_2\underset{Me}{CH}OH$$

(**1**) HPT

Table 2 Rate Constants and Reactivity Ratios Relative to the Copolycondensation of HET and HPT in the Bulk at 160 °C[18]

Rate constant $\times 10^3$ (mol^{-1} kg h^{-1})		Reactivity ratios
k_{11}	1.69	$r_1 = 4.23$
k_{22}	1.31	$r_2 = 1.70$
k_{21}	0.77	
k_{12}	0.40	

The curve (Figure 1) obtained by plotting the molar percentage of 1,2-ethanediyl units (ED) in the copolymer against the molar percentage of HET in the feed is quite similar to the chain copolymerization diagram in the case $r_1 > 1$ and $r_2 > 1$. The experimental azeotropic composition corresponds to 29.5% of HET in the monomer mixture.

Similarly[19] low molecular weight amino-terminated polyamides were synthesized via amide interchange reactions of diamine amides of diacids

$$\sim CONHR^1NH_2 + H_2NR^1NHCO\sim \xrightarrow{k_{11}} \sim CONHR'NHCO\sim + H_2NR^1NH_2 \text{ etc...} \tag{63}$$

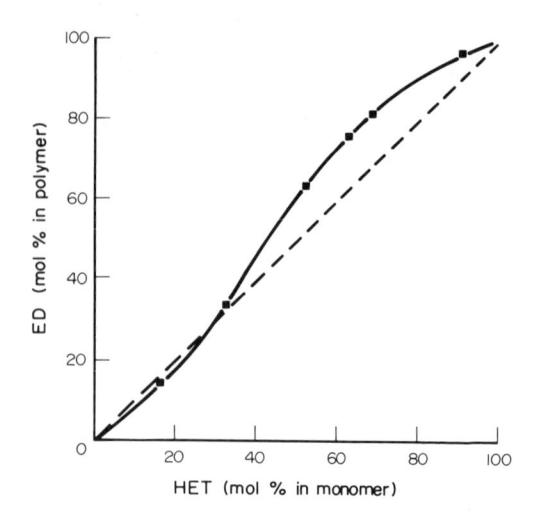

Figure 1 Composition of the copolymer as a function of feed composition[18]

Experimental determinations were carried out on the reaction between N,N'-bis(2-aminoethyl)decandiamide (AES, index 1) and N,N,N',N'-bisdiethyleneiminodecandiamide (DEIS, **2**, index 2). It was found that the polycondensation followed second order kinetics with respect to amino and amide groups in the monomers. The rate constants and reactivity ratios are given in Table 3.

(2) DEIS

Table 3 Rate Constants and Reactivity Ratios Relative to the Copolycondensation of AES and DEIS in the Bulk at 200 °C[19]

Rate constant × 10^3 (mol^{-1} kg h^{-1})		Reactivity ratios
k_{11}	3.01	$r_1 = 1.30$
k_{22}	1.70	$r_2 = 0.59$
k_{12}	2.31	
k_{21}	2.89	

The curve (Figure 2) obtained by plotting the molar percentage of 1,2-diaminoethane units (DAE) in the copolymer against mol % AES in the feed is quite similar to the chain copolymerization diagram when $r_1 > 1$ and $r_2 < 1$. No azeotrope can be found in the diagram. Another interesting example[20] is the synthesis of polyester amides by copolycondensation of a diamine amide of a diacid (see Scheme 7). The kinetics of these copolycondensations was investigated by using N,N'-bis-

$\sim$CONHR^{1}NH$_2$ + H$_2$NR^{1}NHOC$\sim$ $\xrightarrow{k_{11}}$ $\sim$CONHR^{1}NHOC$\sim$ + H$_2$NR^{1}NH$_2$

$\sim$CONHR^{1}NH$_2$ + H$_2$NR^{2}NHOC$\sim$ $\xrightarrow{k_{12}}$ $\sim$CONHR^{2}NHOC$\sim$ + H$_2$NR^{1}NH$_2$

$\sim$CONHR^{2}NH$_2$ + H$_2$NR^{1}NHOC$\sim$ $\xrightarrow{k_{21}}$ $\sim$CONHR^{1}NHOC$\sim$ + H$_2$NR^{2}NH$_2$

$\sim$CONHR^{2}NH$_2$ + H$_2$NR^{2}NHOC$\sim$ $\xrightarrow{k_{22}}$ $\sim$CONHR^{2}NHOC$\sim$ + H$_2$NR^{2}NH$_2$

Scheme 7

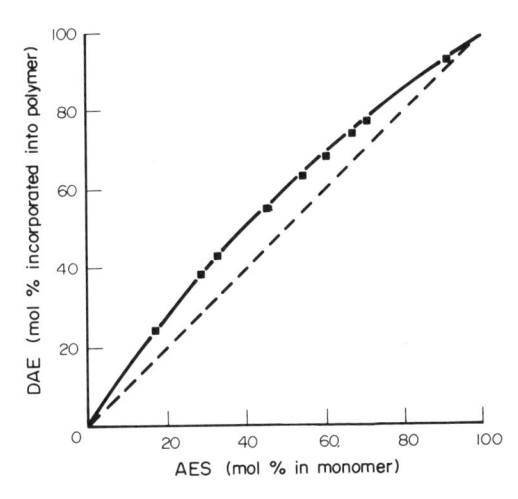

Figure 2 Composition of the copolymer as a function of feed composition[19]

(2-aminoethyl)decandiamide (AES, index 1) and bis(2-hydroxyethyl)decandioate (HES, index 2) as monomers. The values of rate constants and reactivity ratios are given in Table 4.

Table 4 Rate Constants and Reactivity Ratios Relative to the Copolycondensation of AES and HES in the Bulk at 200 °C[20]

Rate constant × 10^3 (mol^{-1} kg h^{-1})		Reactivity ratios
k_{11}	3.01	$r_1 = 10.03$
k_{22}	34.70	$r_2 = 0.19$
k_{12}	0.30	
k_{21}	186.0	

Figure 3 shows the composition diagram in which the molar percentage of DAE in the copolymer is plotted against mol. % AES in the feed. It is similar to the diagram for a chain copolymerization where $r_1 > 1$ and $r_2 < 1$.

In refs. 18–20 it is assumed that the concept of equireactivity applies, *i.e.* that all functional end groups attached to the same residue have the same reactivity. However, if one of the species shows a

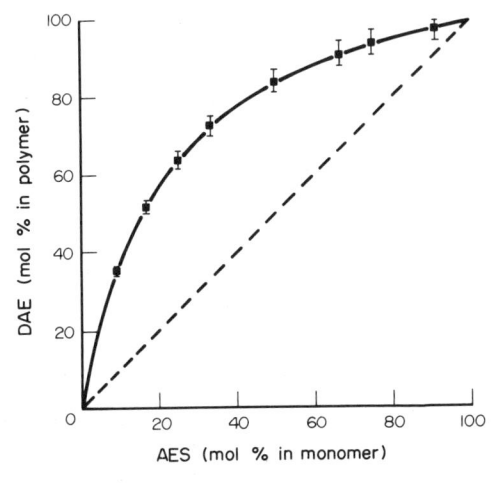

Figure 3 Composition of the copolymer as a function of feed composition[20]

high penultimate effect, as in phosgene, chloroformates and phthalates, this assumption is no longer valid. Mackey et al.[21] developed a treatment of copolycondensation taking into account the non-equireactivity of functional groups; it was applied to the irreversible polyesterification of terephthaloyl chloride with a mixture of bisphenol A and neopentyl glycol. Their main purpose was to calculate the fraction of alternating triads and to compare this value with that obtained from NMR determinations.

Let T, D_1 and D_2 be terephthaloyl chloride or terephthaloyl residue and two diols respectively. The copolyester formed in the reaction

$$T + xD_1 + (1-x)D_2 \rightarrow \sim D_1TD_1TD_2TD_1TD_2TD_2TD_1T\sim \tag{64}$$

can be described as a sequence of triads shown in Scheme 8.

$$\sim D_1TD_1\sim \text{ (fraction } f_{11}\text{)}; \quad \sim D_2TD_2\sim \quad (f_{22}); \quad \sim D_1TD_2\sim \text{ or } \sim D_2TD_1\sim \quad (f_{12})$$

Scheme 8

A kinetic model is used to calculate f_{12} assuming a homogeneous and irreversible copolyesterification where ester interchange can be neglected. Equireactivity is supposed to apply for the reactions of hydroxy functions but not for the reactions of acid chloride whose rate constants depend on nearest neighbour residue. The kinetic equations involve six rate constants:

$$\sim T + D_1 \sim \xrightarrow{k_{TD_1}} \sim TD_1\sim$$

$$\sim D_1T + D_1 \sim \xrightarrow{k_{D_1TD_1}} \sim D_1TD_1\sim$$

$$\sim T + D_2 \sim \xrightarrow{k_{TD_2}} \sim TD_2\sim$$

$$\sim D_2T + D_2 \sim \xrightarrow{k_{D_2TD_2}} \sim D_2TD_2\sim$$

$$\sim D_2T + D_1 \sim \xrightarrow{k_{D_2TD_1}} \sim D_2TD_1\sim$$

$$\sim D_1T + D_2 \sim \xrightarrow{k_{D_1TD_2}} \sim D_1TD_2\sim$$

Scheme 9

Let r be the relative reactivity ratio of T and DT with alcohol groups and δ the relative reactivity of D_1, and D_2 with T functions. Since the reactivity ratio of the two hydroxy groups does not depend on the reaction site, there is only one coefficient of type δ. As found for the system studied it is supposed that $k_{TD_1}/k_{D_1TD_1} = k_{TD_2}/k_{D_2TD_2}$ which means that there is only one ratio of type r.

In the case of random copolycondensation ($r \neq 1$ or $\delta \neq 1$) Flory's treatment of polycondensation can be applied and the diol sequencing can be calculated from simple probability chains. When the fraction r of all reacted functions approaches unity, the diol distribution does not depend on the relative reactivity of the diols and the triad frequencies are given by

$$f_{11} = x^2$$
$$f_{12} = 2x(1-x)$$
$$f_{22} = (1-x)^2$$

Scheme 10

In non-random copolycondensations ($r \neq 1$ and $\delta \neq 1$) two cases can be distinguished: one step reaction and sequential addition.

The calculation of the diols sequences is difficult in the case of the one step reaction, except when $\delta \rightarrow \infty$, which corresponds to a sequential addition of diols with very different reactivity. In the first step of sequential addition, one mole of T reacts with x moles of D_1 to give a product which then reacts with $(1-x)$ moles of D_2 to form a high molecular weight copolymer ($\delta \rightarrow \infty$). After the first step, which is described by $r = k_{TD_1}/k_{D_1TD_1}$, the oligomer contains $(1-x)$ moles of species with all T end groups corresponding to $2(1-x)$ unreacted T functions.

If α is the number of residual T monomers, then the number of oligomer T-end groups will be

$2(1-x-\alpha)$, and after condensation of oligomers with D_2 the frequencies will be:

$$f_{22} = \alpha$$
$$f_{12} = (1-x-\alpha)$$
$$f_{11} = \alpha + 2x - 1$$

Scheme 11

The value of α can be obtained from kinetic considerations leading to

$$1 - x = \frac{r-1}{2r-1}\alpha + \frac{r}{2r-1}\alpha^{1/2r} \tag{65}$$

which allows calculation of α in terms of x and r. The pure alternating copolycondensate results from the reaction of TD_1T with a stoichiometric amount of D_2; the same result could be obtained for $r \to \infty$ ($x = 1/2$) in the sequential addition procedure ($f_{11} = f_{22} = 0; f_{12} = 1$).

The pure block copolymer is obtained by reacting the product formed by condensation of x moles of T and x moles of D_1 with a mixture containing y moles of D_2 and y moles of T. It contains D_1T and D_2T polyester blocks in the ratio x/y with a negligible alternating content. The same results could be obtained with sequential addition if $r \to 0$ and $y = 1 - x$ ($f_{22} = 1 - x; f_{11} = x; f_{12} = 0$).

Mackey et al.[21] compared these theoretical relations with the experimental values obtained from NMR study of the reaction between isophthaloyl or terephthaloyl chloride with a mixture of bisphenol A and neopentyl glycol. Experiment and theory are in good agreement in the case of pure sequential addition. However, in the case of one step processes the experimental f_{12} values are significantly lower than expected; this discrepancy was assigned to the non-instantaneous mixing of the reactants.

Turska et al.[22-26] studying copolycondensation in solution showed that the composition of the copolymer is not constant during the course of the reaction as it depends on the relative reactivity of the monomers and on the extent p of the reaction. Theoretical considerations lead these authors[23] to the conclusion that from three monomers A, B and C, it is possible to obtain copolymers of the same composition — but of different structure of macromolecule chains — if the relative reactivity of the monomers can be controlled by altering the reaction conditions. In particular, if homopolycondensations of A with B and C satisfy the Arrhenius relationship and have different activation energies, then there should be a temperature at which the rate constants are equal to each other. Straight lines corresponding to $\ln k = f(1/T)$ intersect at the reciprocal value of this temperature T_i, which is called the *isokinetic temperature*. At the isokinetic temperature there should be formed a product of invariant composition independent of the extent of reaction and of random distribution of B and C units, with

$$T_i = \frac{E_{AB} - E_{AC}}{R(B_{AB} - B_{AC})} \tag{66}$$

where E_{AB}, E_{AC}, and B_{AB}, B_{AC} are respectively the activation energies and preexponential factors in Arrhenius equations relative to homopolycondensations $A+B$ and $A+C$. R is the ideal gas constant.

Turska et al.[26] carried out experimental determinations on two systems: (I) phenolphthalein; dichloro-3,3'-bisphenol A; terephthaloyl dichloride ($T_i = 242$ °C); and (II) phenolphthalein; dimethyl-3,3'-bisphenol A; terephthaloyl dichloride ($T_i = 200$ °C). Copolycondensations were carried out in α-chloronaphthalene. Experimental results show that the composition of the copolycondensation product depends on the reactivity ratios of the monomers. Thus at 180 °C (i.e. below T_i) phenolphthalein ($k = 0.56 \times 10^{-3}$ l mol^{-1} s^{-1}) is more reactive than dichloro-3,3'-bisphenol A ($k = 0.56 \times 10^{-3}$ l mol^{-1} s^{-1}). However at isokinetic temperature (242 °C) the reactivity of both monomers should be equal; analytical determinations have shown that at this temperature the copolyester composition does not depend on p and equals the composition of the initial mixture at the start of the reaction.

16.3 APPLICATIONS OF COPOLYCONDENSATION

16.3.1 Introduction

At least two competitive reactions occur in copolycondensation; these are: (i) analogous reactions between at least three different reactants (Scheme 12); and (ii) reactions of different types between at least two reactants (Schemes 13–14).

$$AR^1A + AR^2A + BR^3B$$

Scheme 12

$$AR^1B + CR^2C$$

Scheme 13

$$AR^1A + BR^2B + CR^3C$$

Scheme 14

Most applications relate to Scheme 12, which is, in fact, relatively close to the corresponding homopolycondensations. Thus most unsaturated polyesters are obtained by copolyesterification between several acids and diols resulting in copolymers exhibiting a wide range of physical and chemical properties; this is also the case for copolyamides, copolycarbonates and, above all, for polyurethanes, which are obtained by reacting a diisocyanate with a mixture of diols and triols. As most of these compounds are described in specific chapters we will study only examples where copolycondensation brings very specific characteristics to the product. This is the case for copolycondensation involving ARA type monomers where R is an oligomer or a group with a specific property, such as colour in polymeric dyes,[27] or a group which is not formed during the polycondensation, such as a bifunctional aromatic ether.

Copolycondensation exhibits shortcomings which are difficult to avoid: (i) most of the resulting copolymers are statistical and their properties can be very different from those of their alternative homologues, particularly when a microphase segregation is required as in thermoplastic elastomers; (ii) poorly controlled side-reactions can take place particularly when copolycondensation involves two different competitive reactions (Schemes 13 and 14). In spite of these shortcomings, copolycondensation is a powerful method of synthesis with many advantages. The most important is probably that it is a 'one pot one step' method which can be carried out using simple techniques.

16.3.2 Copolycondensations involving two reactions of the same type

16.3.2.1 Ester bond formation

Copolyesterifications are widely used to prepare polyesters of very different types: unsaturated polyesters (Vol. 5, Chapter 19); aromatic thermotropic polyesters;[28] and block copolymers whose polyester block can be prepared in the presence of the α,ω-dihydroxy (or α,ω-dicarboxy) oligomer corresponding to the other block.[29,30] This is the case for Du Pont de Nemours Hytrel[31] which is a poly(tetramethylene terephthalate)-*block*-poly(oxytetramethylenoxy) (see equation 67). Such a copolymer is characterized by a statistical distribution of the blocks; however the same statistical distribution can be obtained by heating the copolymer resulting from the polycondensation of the corresponding telechelic oligomers (3) and (4). The statistical quality of the copolymer is due to rearrangements resulting from transesterification reactions.

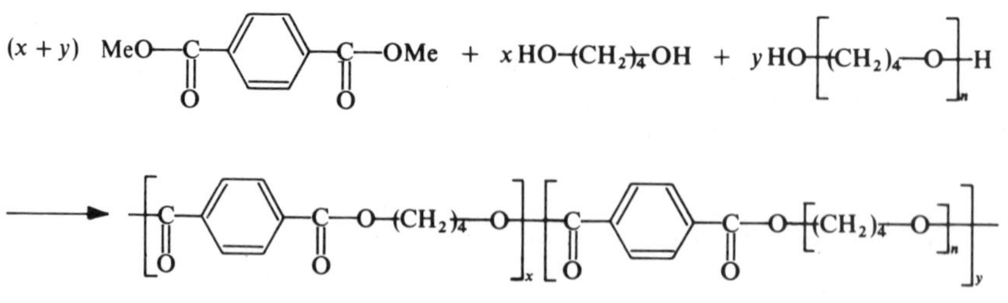

$$(x+y)\ \text{MeO-CO-C}_6\text{H}_4\text{-CO-OMe} + x\,\text{HO-(CH}_2)_4\text{OH} + y\,\text{HO-[(CH}_2)_4\text{-O]}_n\text{H}$$

$$\longrightarrow \left[\text{[CO-C}_6\text{H}_4\text{-CO-O-(CH}_2)_4\text{-O]}_x\text{[CO-C}_6\text{H}_4\text{-CO-O-[(CH}_2)_4\text{O]}_n\text{]}_y\right] \quad (67)$$

$$\text{HO-[OC-C}_6\text{H}_4\text{-CO}_2(\text{CH}_2)_4\text{O]}_x\text{-CO-C}_6\text{H}_4\text{-CO}_2\text{H} \qquad \text{HO-[(CH}_2)_4\text{O]}_n\text{-H}$$

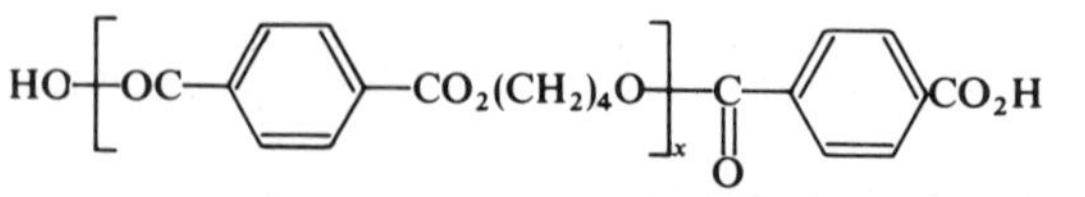

(3) 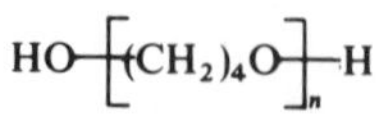   (4)

When transesterification reactions involve the formation of a volatile low molecular weight compound, this can be eliminated, thus increasing the molecular weight of the final product as observed in the case of aliphatic polyesters.[32]

Polyesteramide-*block*-polyether can be prepared in the same way[33] using an amide group containing dimethylic ester (**5**). Solution copolycondensation of diols and diacid chlorides can also give statistical copolymers. Thus copoly(ester sulfone)s were prepared by copolycondensation of dihydroxydiphenyl sulfone, terephthaloyl chloride and aliphatic and aromatic diols[34] (equation 68), the resulting copolymer being characterized by a statistical distribution of poly(ester sulfone) and aromatic and aliphatic–aromatic polyester blocks.

Copolymers of the type described above can be obtained by reacting an α,ω-diacid chloride sulfone[35] (**6**) with a bisphenol or a diol. Polycondensation of diacid chloride mixtures with chiral diols leads to chiral copolyesters (equation 69).[36] Copolythioesters[37] were obtained by interfacial polycondensation of an aromatic diacid chloride with 1,4- and 1,5-dimercaptonaphthalenes.

16.3.2.2 Amide bond formation

Classical copolyamides were obtained by reacting diacids and diamines in the bulk; in most cases one of the diacids is aromatic, such as isophthalic or terephthalic acid.[38] Polyamide-*block*-polyethers[39] were prepared by reacting an α,ω-diaminopolyether and a diamine with a diacid chloride (equation 70). The polyether blocks improve the flexibility of the polyamide.

Maréchal et al.[27,40–42] used statistical copolycondensation to prepare polymeric dyes. Thus adipic acid, 1,6-diaminohexane and α,ω-diamino dyes such as (**7**) or (**8**) were copolycondensed at 270 °C in the bulk. Plots of [η] against the molar percentage of dye in the feed very often exhibit a

$$(x+y)\ \text{Cl}-\underset{O}{\overset{\|}{C}}-(CH_2)_8-\underset{O}{\overset{\|}{C}}-\text{Cl} + x\text{HN}\underset{}{\diagdown\!\!\!\diagup}\text{NH} + y\text{H}_2\text{NCHCH}_2\text{(OCH}_2\text{CH)}_n\text{-NH}_2$$
$$\qquad\qquad\qquad\qquad\qquad\qquad\qquad\qquad\qquad\qquad\qquad\quad \text{Me} \qquad\qquad \text{Me}$$

$$\longrightarrow \left[\left[\underset{O}{\overset{\|}{C}}-(CH_2)_8-\underset{O}{\overset{\|}{C}}-N\diagdown\!\!\!\diagup N\right]_x\left[\underset{O}{\overset{\|}{C}}-(CH_2)_8-\underset{O}{\overset{\|}{C}}\text{NHCHCH}_2\text{(OCH}_2\text{CH)}_n\text{NH}\right]_y\right] \quad (70)$$

(7)

(8)

maximum whose position depends on the nature of the dye. The same phenomenon was observed for other aromatic polycondensates.[41]

Solution copolycondensation of diacid chlorides with diamines permits the preparation of copolycondensates containing functional groups such as sulfones.[43] However they are mainly used to prepare block copolyamides such as polyamide-*block*-poly(propyleneoxide).[44]

$$\text{ClCO(CH}_2)_8\text{CO}_2\text{[CHCH}_2\text{O]}_n\text{CO(CH}_2)_8\text{COCl} \xrightarrow[\text{in CHCl}_3/\text{H}_2\text{O}]{\text{ClCO(CH}_2)_8\text{COCl} \atop \text{NH}_2\text{(CH}_2)_6\text{NH}_2} \left[\text{[O(CHCH}_2\text{O)}_n\text{CO(CH}_2)_8\text{CO}-\right.$$
$$\qquad\qquad\text{Me} \qquad\qquad\qquad\qquad\qquad\qquad\qquad\qquad\qquad\qquad\qquad\text{Me}$$
$$\qquad\qquad\qquad\qquad\qquad\qquad\qquad\qquad\qquad\qquad\qquad\qquad\qquad\qquad\text{[NH(CH}_2)_6\text{NHCO(CH}_2)_8\text{CO]}_m\text{]}_l \quad (71)$$

Copyamides containing short polythiophenylene blocks[45] have also been prepared by emulsion polymerization (equation 72). Polyamide-*block*-polysiloxanes have been prepared by reacting an aromatic dianhydride with a mixture of an aromatic diamine and a diaminopolysiloxane[46] (equation 73) and heating the resulting polyamidic acid gives the polyimide-*block*-polysiloxane (9).

Activation processes such as the use of triphenyl phosphite in solution copolycondensation permitted the synthesis of polyamide-*block*-polybutadiene (equation 74).[47]

The microstructure of this statistical copolymer is close to those of the block copolymers resulting from the polycondensation of prepared telechelic oligoamides and telechelic polybutadiene; mechanical properties of the films are roughly the same although the initial modulus of the statistical copolyamide is slightly below those of the block copolymers. On the other hand the properties of the statistical polyaramide-*block*-polyoxyethylene prepared in the same way from α,ω-dicarboxypoly-oxyethylene are very different from those of the corresponding block copolymers.[48] Thus an increase of the glass transition temperature T_g of polyoxyethylene block and a decrease of the tensile strength

$$\text{Cl}-\underset{O}{\overset{\|}{C}}-\!\!\diagup\!\!\!\!\diagdown\!\!-\underset{O}{\overset{\|}{C}}-\text{Cl} + \text{Cl}-\underset{O}{\overset{\|}{C}}-\!\!\diagup\!\!\!\!\diagdown\!\!-\underset{O}{\overset{\|}{C}}-\text{Cl} + \text{H}_2\text{N}-\!\!\diagup\!\!\!\!\diagdown\!\!-\text{S}\!\!-\!\!\diagup\!\!\!\!\diagdown\!\!\text{NH}_2$$

$$\xrightarrow[\text{H}_2\text{O/NaOH}]{\text{Cyclohexanone}} \text{Copyamide-}block\text{-PPS} \qquad\qquad\qquad (72)$$

[Scheme/equation (73) showing reaction of a bis-anhydride (benzophenone tetracarboxylic dianhydride) with a diaminobenzophenone and an α,ω-diaminopropyl polydimethylsiloxane in DMAC/THF, giving a poly(amic acid) intermediate, which converts to polyimide–polysiloxane block copolymer (9).]

[Equation (74): reaction of isophthalic acid with (n+1) H$_2$N–C$_6$H$_4$–O–C$_6$H$_4$–NH$_2$ and HO$_2$C(CH$_2$CH=CHCH$_2$)$_x$CO$_2$H in presence of (PhO)$_3$P, Pyridine/NMP at 100 °C to give a polyamide block copolymer.]

and elongation at break are observed in the case of statistical copolymers. This suggests that they contain irregular structures due to the random coupling of the blocks, in contrast to the highly structural regularity of the copolymers prepared by polycondensation of telechelic oligomers.

16.3.2.3 Carbonate bond formation

Copolycarbonates can be easily prepared by phosgenation of phenol-group-containing molecules such as bisphenols A and E[49] or of telechelic oligomers which give block copolycarbonates; thus polycarbonate-*block*-polyesters have been prepared from dihydroxypolyesters, phenolphthalein and phosgene (equation 75)[50] as well as polycarbonate-*block*-polyoxyethylene (equation 76).[51]

[Equation (75): HO–(CH$_2$CH$_2$O–C(O)(CH$_2$)$_4$C(O)–O)–CH$_2$CH$_2$OH + phenolphthalein + COCl$_2$ → Polyester-*block*-polycarbonate]

16.3.2.4 Aromatic ether bond formation

Copolyetherification has been mainly used to prepare sulfone-group-containing polymers (**10**)[52] or polymers containing both aromatic ketones and sulfones,[53] as shown in equation (77) for the preparation of a copoly(ether ketone)-*block*-copoly(ether ketone sulfone).

(**10**)

$$\text{(77)}$$

Copolyether ketone-*block*-copolyether ketone sulfone

16.3.2.5 Urethane bond formation

Reaction of diisocyanates with mixtures of hydroxylated compounds is widely used in the syntheses of polyurethanes (see Volume 5, Chapters 23 and 24). Specific applications worth mentioning include synthesis of multiblock copolymers (equation 78)[54] and synthesis of self-coloured polyurethanes.[55] These latter were prepared by reaction of a α,ω-dichloroformate polyether with a mixture of an α,ω-diamino dye and 1,6-diaminocyclohexane.

$$\text{(78)}$$

16.3.2.6 Aromatic ketone bond formation

Friedel–Crafts copolycondensation of diacid chlorides with aromatic compounds is widely used in the synthesis of thermostable materials,[56-58] such as poly(ether sulfone)-*block*-polyketone-*block*-polyether

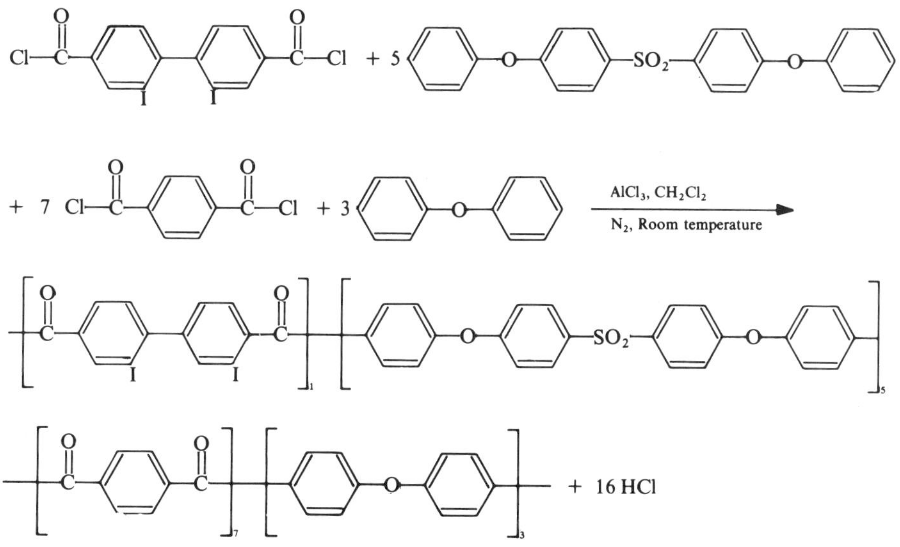

$$\text{[diacid chloride with iodine substituents]} + 5 \text{[bis(phenoxyphenyl)sulfone]}$$

$$+ 7 \text{Cl-[diacid chloride]-Cl} + 3 \text{[diphenyl ether]} \xrightarrow[\text{N}_2, \text{Room temperature}]{\text{AlCl}_3, \text{CH}_2\text{Cl}_2}$$

$$\left[\text{ketone-biphenyl-iodo unit}\right]_1 \left[\text{ether sulfone ether unit}\right]_5$$

$$\left[\text{ketone unit}\right]_7 \left[\text{ether unit}\right]_3 + 16 \text{ HCl} \quad (79)$$

16.3.3 Copolycondensations Involving Reactions of a Different Nature

Examples of this type of copolycondensation are far less numerous than those described in Section 16.3.2. This is partly due to the fact that the rates of the different reactions must be of the same order of magnitude. Thus, if two reactions with very different rates take place in a three-reactant-containing system, two of the compounds polycondense preferentially to form a telechelic oligomer which reacts with the third compound to form a BAB type block copolymer. Such a reaction is not really a copolycondensation as the two reactions are not simultaneous during the whole of the process.

Several examples of copolycondensations involving reactions of a different nature are shown in Scheme 15.

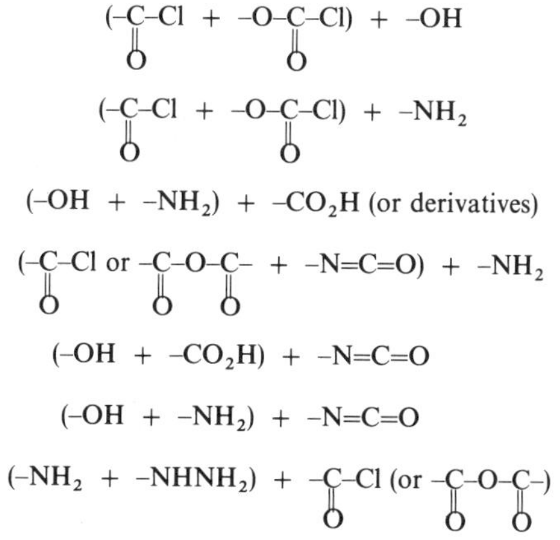

$$(-\underset{\underset{O}{\|}}{C}-Cl \ + \ -O-\underset{\underset{O}{\|}}{C}-Cl) \ + \ -OH$$

$$(-\underset{\underset{O}{\|}}{C}-Cl \ + \ -O-\underset{\underset{O}{\|}}{C}-Cl) \ + \ -NH_2$$

$$(-OH \ + \ -NH_2) \ + \ -CO_2H \text{ (or derivatives)}$$

$$(-\underset{\underset{O}{\|}}{C}-Cl \text{ or } -\underset{\underset{O}{\|}}{C}-O-\underset{\underset{O}{\|}}{C}- \ + \ -N=C=O) \ + \ -NH_2$$

$$(-OH \ + \ -CO_2H) \ + \ -N=C=O$$

$$(-OH \ + \ -NH_2) \ + \ -N=C=O$$

$$(-NH_2 \ + \ -NHNH_2) \ + \ -\underset{\underset{O}{\|}}{C}-Cl \text{ (or } -\underset{\underset{O}{\|}}{C}-O-\underset{\underset{O}{\|}}{C}-)$$

Scheme 15

16.3.3.1 Simultaneous formation of ester and carbonate bonds

Polyester-*block*-polycarbonates have been prepared according to equation (80)[59]

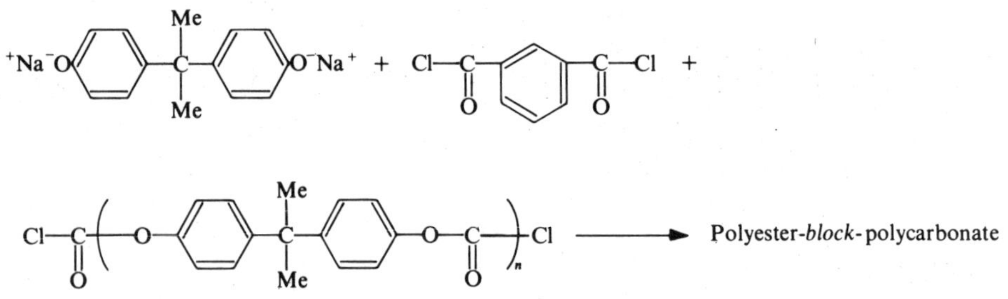

16.3.3.2 Simultaneous formation of ester and amide bonds

Aromatic carboxylic acids react with aromatic NH_2 and OH groups at the same rate in the presence of compounds such as diphenyl chlorophosphate (DCP)[60] or tosyl chloride.[61,62] This results in the formation of high molecular weight copoly(ester amides). Thus the copolycondensation of p-aminobenzoic and p-hydroxybenzoic acids gives soluble poly(ester amide)s[60] with a statistical distribution of ester and amide functions.

In the same way the copolycondensation of diacids with diphenols and diamines gives statistical copolymers. However in this case amide and ester linkages are head to head and tail to tail (equation 82) and not head to tail as in the product of reaction (81)[61,62]:

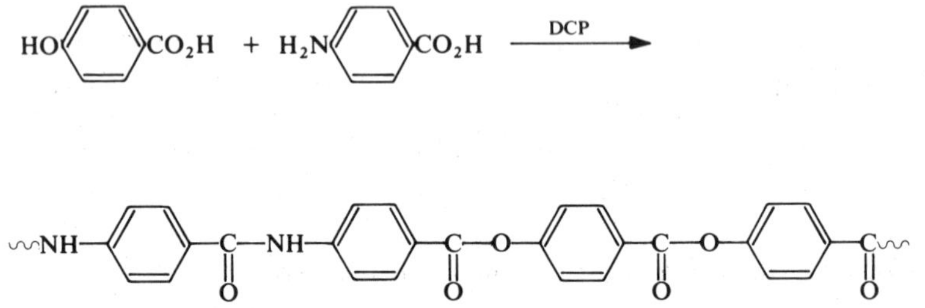

16.3.3.3 Simultaneous formation of amide and urea bonds

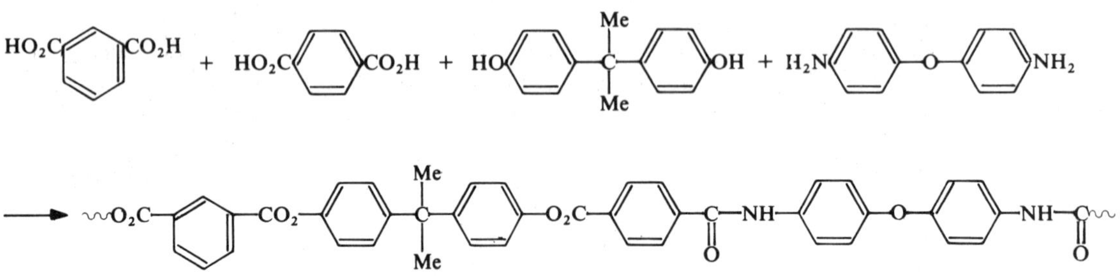

When a dianhydride and a diisocyanate are reacted simultaneously with a diamine, a copolymer containing urea, acid and amide bonds is formed; on heating it gives a poly(imide-co-urea) (Scheme 16).[63]

The relative reactivities of anhydrides and isocyanates when reacted with 4,4′-diaminodiphenyl ether follow the sequence: pyromellitic dianhydride > benzophenone tetracarboxylic dianhydride > 3,3′-dimethyl-4,4′-diphenylmethane diisocyanate > 4,4′-diphenylmethane diisocyanate > 9-chlorotricyclo[4.2.2.0^{2,5}]dec-9-ene-3,4,7,8-tetracarboxylic dianhydride.

Copolymers containing the same bonds as those described in Scheme 16 can be obtained by simultaneous reaction of diisocyanates and aromatic diacid chlorides with aromatic diamines.[64]

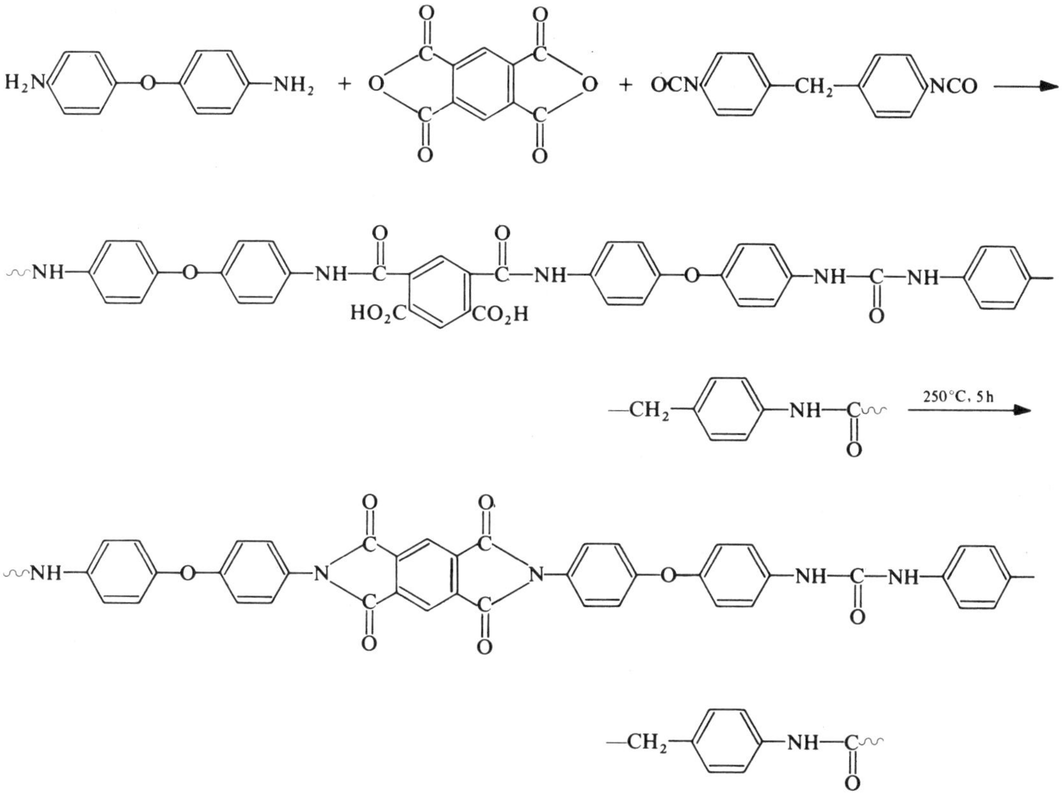

Scheme 16

16.3.3.4 Simultaneous formation of amide and urethane bonds

Diisocyanates can react with an acid or with an alcohol in the same experimental conditions[65]

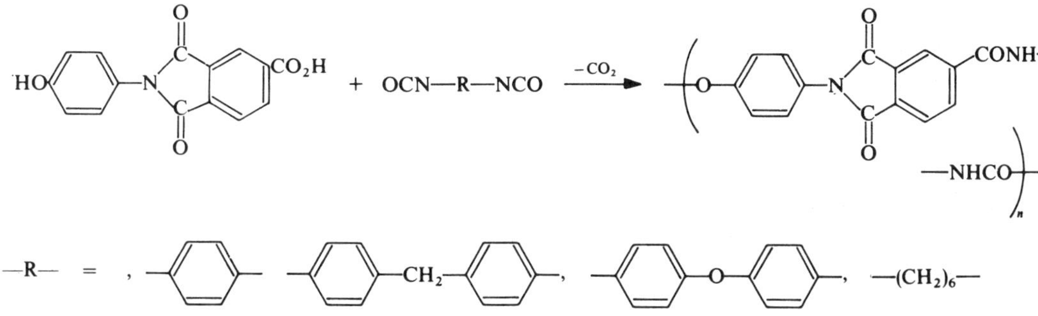

Copolymers containing bonds of the same type as those described in equation (83) are formed when a diacid chloride and a dichloroformate are reacted with a diamine[66]

$$H_2N(CH_2)_6NH_2 + Cl-\underset{O}{C}-(CH_2)_4-\underset{O}{C}-Cl + Cl-\underset{O}{C}-O-(poly-\varepsilon\text{-caprolactone})-O-\underset{O}{C}-Cl \rightarrow \text{Polyamide-}block\text{-polyether} \quad (84)$$

16.3.3.5 Simultaneous formation of urea and urethane bonds

A diisocyanate can react at the same time with alcohol and amine functions to give a poly(urethane-urea). This has been applied to the synthesis of block copolymers[67]

$$(n+1) \underset{\text{NCO}}{\overset{\text{Me}}{\underset{}{\bigcirc}}}\text{NCO} + n\text{H}_2\text{NCH}_2\text{CH}_2\text{NH}_2 + \text{HO-[(CH}_2)_4\text{O]}_n\text{H} \longrightarrow$$

$$\left[\left[\underset{\text{O}}{\text{C}}-\text{NH}-\underset{\text{Me}}{\bigcirc}-\text{NH}-\underset{\text{O}}{\text{C}}-\text{NHCH}_2\text{CH}_2\text{NH}\right]_m\left[\underset{\text{O}}{\text{C}}-\text{NH}-\underset{\text{Me}}{\bigcirc}-\text{NH}-\underset{\text{O}}{\text{C}}-\text{O-[(CH}_2)_4\text{O]}_n\right]_p\right] \quad (85)$$

16.3.3.6 Simultaneous formation of amide and hydrazide bonds

Both functions of *p*-aminobenzhydrazide can react with one (or a mixture of more than one) acid chloride[68] to form a polyamide-*block*-polyhydrazide

$$\text{H}_2\text{N}-\bigcirc-\underset{\text{O}}{\text{C}}-\text{NHNH}_2 + \text{Cl}-\underset{\text{O}}{\text{C}}-\bigcirc-\underset{\text{O}}{\text{C}}-\text{Cl} \longrightarrow$$

$$\sim\underset{\text{O}}{\text{C}}-\bigcirc-\underset{\text{O}}{\text{C}}\left[\text{NH}-\bigcirc-\underset{\text{O}}{\text{C}}-\text{NHNH}-\underset{\text{O}}{\text{C}}-\bigcirc-\underset{\text{O}}{\text{C}}\right]_n\sim \quad (86)$$

16.4 CONCLUSION

Many synthesis processes reported as copolycondensations are in fact homopolycondensations. This confusion arises from two facts: (i) many block copolymers are prepared by homopolycondensation of telechelic oligomers; and (ii) even when several monomers are involved, the reactions which take place can be successive homopolycondensations if their kinetics are very different (see Section 16.3.3).

To control successfully the distribution of monomer units in the chain it is essential to improve knowledge of the mechanisms and of the kinetics, to characterize all the side reactions and to determine their dependence on experimental conditions.

16.5 REFERENCES

1. P. J. Flory, 'Principles of Polymer Chemistry', Cornell University Press, Ithaca, NY, 1953, chap. 8.
2. L. C. Case, *J. Polym. Sci.*, 1958, **29**, 455.
3. G. G. Lowry, in 'Markov Chains in Monte-Carlo Calculations in Polymer Science', ed. G. G. Lowry, Dekker, New York, 1969.
4. L. Peller, *J. Chem. Phys.*, 1962, **36**, 2976.
5. F. Lopez–Serrano, J. M. Castro, C. W. Macosko and M. Tirrell, *Polymer*, 1980, **21**, 263.
6. S. I. Kuchanov and Ye. B. Brun, *Vysokomol. Soedin., Ser. A*, 1979, **21**, 700.
7. Ye. B. Brun and S. I. Kuchanov, *Vysokomol Soedin., Ser. A*, 1979, **21**, 691.
8. D. A. Tirrell, in 'Encyclopaedia of Polymer Science and Engineering', Wiley, New York, 1986, vol. 4, p. 226.
9. S. M. Ross, in 'Introduction to Probability Models', Academic Press, New York 1972, chap. 3.
10. C. W. Macosko and D. R. Miller, *Macromolecules*, 1976, **9**, 199.
11. W. H. Stockmayer, *J. Polym. Sci.*, 1952, **9**, 69; 1963, **11**, 424.
12. Ph. Chaumont, Y. Gnanou, G. Hild and P. Rempp, *Makromol. Chem.*, 1985, **186**, 2321.
13. A. F. Johnson and K. F. O'Driscoll, *Eur. Polym. J.*, 1984, **20**, 979.
14. T. A. Speckhard, J. A. Miller and S. L. Cooper, *Macromolecules*, 1986, **19**, 1558.
15. J. A. Miller, T. A. Speckhard and S. L. Cooper, *Macromolecules*, 1986, **19**, 1568.
16. D. R. Miller and C. W. Macosko, *Macromolecules*, 1978, **11**, 656.
17. E. Sorta and A. Melis, *Polymer*, 1979, **19**, 1153.
18. M. J. Han, *Macromolecules*, 1980, **13**, 1009.
19. M. J. Han, *Macromolecules*, 1982, **15**, 438.
20. M. J. Han, H. Ch. Kang and K. B. Choi, *Macromolecules*, 1986, **19**, 1649.

21. J. H. Mackey, V. A. Pattison and J. A. Pawlak, *J. Polym. Sci., Polym. Chem. Ed.*, 1978, **16**, 2849.
22. E. Turska, S. Boryniec and L. Pietrzak, *J. Appl. Polym. Sci.*, 1974, **18**, 667.
23. E. Turska, S. Boryniec and A. Dems, *J. Appl. Polym. Sci.*, 1974, **18**, 671.
24. E. Turska and L. Pietrzak, *23rd IUPAC Int. Symp. Macromol.*, 1974, **1**, 125.
25. E. Turska, S. Boryniec and R. Jantas, *J. Appl. Polym. Sci.*, 1976, **20**, 1849.
26. E. Turska, L. Pietrzak and R. Jantas, *J. Appl. Polym. Sci.*, 1979, **23**, 2409.
27. E. Maréchal, *Prog. Org. Coat.*, 1982, **10**, 25.
28. L. L. Chapoy, 'Recent Advances in Liquid Crystalline Polymers', Elsevier, London, 1985.
29. W. K. Witsiepe, *Polym. Prepr., Am. Chem. Soc., Div. Polym. Chem.*, 1972, **13**, 588.
30. C. M. Boussias, R. H. Peters and R. H. Stills, *J. Appl., Polym. Sci.*, 1980, **25**, 855.
31. W. K. Witsiepe (E. I. du Pont de Nemours), *Ger. Pat.* 2 213 128 (1972) (*Chem. Abstr.*, 1973, **78**, 17 337).
32. A. Fradet and E. Maréchal, *Eur. Polym. J.*, 1978, **14**, 755.
33. E. Sorta and G. della Fortuna, *Polymer*, 1980, **21**, 728.
34. Y. Ding and Q. Chen, *Kexue Tongbao*, 1986, **31**, 501.
35. E. S. M. E. Mansour, A. M. I. Khalifa and L. Rateb, *J. Macromol. Sci., Chem.*, 1981, **16**, 651.
36. E. Chiellini and G. Galli, *Macromolecules*, 1985, **18**, 1652.
37. W. Podkoscielny and W. Charmas, *J. Polym. Sci., Polym. Chem. Ed.*, 1979, **17**, 2429.
38. N. Kato and T. Susuki (Toray Industries Inc.) *Jpn. Pat.* 60 135 429 (1985) (*Chem. Abstr.*, 1986, **104**, 34 509).
39. J. K. Rasmussen and H. K. Smith, *J. Appl. Polym. Sci.*, 1983, **28**, 2473.
40. Ph. Gangneux and E. Maréchal, *Bull. Soc. Chim. Fr.*, 1973, 1466, 1483.
41. J. C. Bonnet and E. Maréchal, *Bull. Soc. Chim. Fr.*, 1972, 3561.
42. E. Bonnet, Ph. Gangneux and E. Maréchal, *Bull. Soc. Chim. Fr.*, 1976, 504, 507.
43. E. S. M. E. Mansour, A. M. I. Khalifa and L. Rateb, *J. Macromol. Sci., Chem.*, 1981, **16**, 669.
44. N. Ogata and N. Yui, *J. Macromol. Sci., Chem.*, 1984, **21**, 1097.
45. M. Aritomi and M. Terauchi (Mitsubishi Petrochemicals Co., Ltd) *Eur. Pat. Appl.*, 162 606 (1985) (*Chem. Abstr.*, 1986, **104**, 130 501).
46. J. D. Summers, C. A. Arnold, R. H. Bott, L. T. Taylor, T. C. Ward and J. E. Mac Grath, *Polym. Prepr., Am. Chem. Soc., Div. Polym. Chem.*, 1986, **27**, 403.
47. S. Ogata, M. Kakimoto and Y. Imai, *Macromolecules*, 1985, **18**, 851.
48. Y. Imai, M. Kajiyama, S. Ogata and M. Kakimoto, *Polym. J.*, 1985, **17**, 1173.
49. M. Takamatsu, K. Hashimoto, A. Manabe and H. Ichihana (Teijin Chemicals Ltd.), *Jpn. Pat.* 60 243 115 (1985) (*Chem. Abstr.*, 1986, **104**, 207 920).
50. N. A. Memon and H. L. Williams, *Macromol. Synth.*, 1982, **8**, 31.
51. T. Suzuki and T. Kotaka, *Macromolecules*, 1980, **13**, 1495.
52. T. E. Attwood, D. A. Barr, T. King, A. B. Newton and J. B. Rose, *Polymer*, 1977, **18**, 359.
53. J. L. Freeman and J. B. Rose (ICI Ltd) *Ger. Pat.* 2 705 587 (1977) (*Chem. Abstr.*, 1977, **87**, 136 624).
54. I. A. Volegova, Yu. K. Gdovskii, A. N. Aksenov, I. P. Storozhuk and V. V. Korskak, *Vysokomol. Soedin., Ser. B*, 1983, **25**, 792.
55. B. Petit and E. Maréchal, *Bull. Soc. Chim. Fr.*, 1974, 1597.
56. N. G. Gileva, M. G. Zolutukhin, S. N. Salazkin, S. R. Rafikov, H. H. Hoerhold and D. Raabe, *Acta Polym.*, 1984, **35**, 282.
57. R. L. Frentzel and C. S. Marvel, *J. Polym. Sci., Polym. Chem. Ed.*, 1979, **17**, 1073.
58. A. Somers and C. S. Marvel, *J. Polym. Sci., Polym. Chem. Ed.*, 1980, **18**, 1511.
59. Idemitsu Kosan Co, Ltd, *Jpn. Pat.* 58 147 423 (1983) (*Chem. Abstr.*, 1984, **100**, 52 225).
60. F. Higashi, M. Ozawa, A. Hoshio and A. Mochizuki, *J. Polym. Sci., Polym. Chem. Ed.*, 1985, **23**, 1699.
61. F. Higashi, Y. Mihara, I. Takahashi, W. H. Chen and T. C. Chang, *J. Polym. Sci., Polym. Chem. Ed.*, 1985, **23**, 2851.
62. F. Higashi, T. Nishi, W. H. Chen and T. C. Chang, *J. Polym. Sci., Polym. Chem. Ed.*, 1986, **24**, 187.
63. B. A. Zhubanov and L. B. Rukhina, *Izv. Akad. Nauk. Kaz. SSR., Ser. Khim.*, 1978, **28**, 40 (*Chem. Abstr.*, 1979, **90**, 72 503).
64. Y. M. Ahn and I. S. Hong, *Hanguk Sumyu Konghakhoe Chi.*, 1986, **23**, 130 (*Chem. Abstr.*, 1986, **105**, 153 623).
65. K. Kurita and H. Murakoshi, *Polym. Commun.*, 1985, **26**, 179.
66. J. M. Huet and E. Maréchal, *Eur. Polym. J.*, 1974, **10**, 771.
67. C. S. Paik Sung, T. W. Swith and N. H. Sung, *Macromolecules*, 1980, **13**, 117.
68. K. Karakulski and H. Wojcikiewicz, *Polimery (Warsaw)*, 1984, **29**, 226 (*Chem. Abstr.*, 1985, **103**, 6793).

17
Polyesters

FRANCESCO PILATI
Università di Bologna, Italy

17.1	INTRODUCTION	275
17.2	POLYESTERIFICATION REACTIONS	276
	17.2.1 General Aspects	276
	17.2.2 Polyesters from Direct Esterification	279
	17.2.2.1 Direct esterification at high temperature	279
	17.2.2.2 Direct esterification under mild conditions	285
	17.2.3 Polyesters from Alcoholysis	289
	17.2.4 Exchange Reactions between Ester and Carboxyl Groups	292
	17.2.5 Ester–Ester Exchange Reactions	293
	17.2.6 Polyesters from Acyl Chlorides	294
	17.2.7 Polyesters from Anhydrides	296
	17.2.8 Polyesters from Cyclic Esters	297
	17.2.9 Polyesters from Biosynthesis	299
	17.2.10 Other Reactions for the Preparation of Polyesters	299
	17.2.11 Degradation Reactions	300
17.3	CLASSES OF POLYESTERS	302
	17.3.1 Polyesters from Aliphatic Diols and Diacids	302
	17.3.2 Polyesters from Aliphatic Hydroxy Acids	303
	17.3.3 Polyesters from Aliphatic Diols and Aromatic Diacids	305
	17.3.3.1 Poly(ethylene terephthalate)	305
	17.3.3.2 Poly(1,4-butylene terephthalate)	307
	17.3.4 Polyesters from Aromatic Diols and Aliphatic Diacids	309
17.4	REFERENCES	309

17.1 INTRODUCTION

The first studies on the synthesis of polyesters date back to the last century;[1-5] however, it was only in the late 1920s that Carothers undertook an extensive study of this subject.[6] However, the linear aliphatic polyesters prepared by his pioneering work did not find great commercial interest due to their low melting points, and it was only about 15 years after, when Whinfield prepared poly(ethylene terephthalate) (PET),[7] that polyesters became widely used for fibers and films. In more recent years, technological improvements have made high molecular weight PET also suitable for blow-molded beverage bottles, and another polyester, poly(1,4-butylene terephthalate) (PBT), has found wide application as a plastic molding material.

Among the other polyesters, poly(ε-caprolactone) (PCL) is of commercial interest because of its remarkable ability to form truly compatible blends with a large number of other polymers such as poly(vinyl chloride) (PVC), poly(styrene–acrylonitrile) (SAN), poly(acrylonitrile–butadiene–styrene) (ABS), *etc.*[8] In addition, hydroxy-terminated PCLs of relatively low molecular weight have been used for the preparation of polyurethanes and in coating industries. Applications where environmental effects are of primary importance have stimulated a renewed interest in poly(β-hydroxyalkanoate)s, thermoplastic polyesters completely biodegradable and exhibiting properties similar to polypropylene,[9,10] which have been known for a long time and are now commercially available. Poly(β-hydroxybutyrate) (PHB) is produced as a carbon reserve by many different bacteria and can be regarded as the first commercial polymer produced using biotechnology.[11]

Polyarylates, which usually show high heat distortion temperatures and good mechanical properties, and copolyesters containing aromatic blocks interlinked with flexible spacer groups, which may display thermotropic liquid-crystalline character, have received particular scientific consideration in the last few decades (see also Volume 5, Chapters 18, 38 and 39).[12] Great attention has also been paid to the syntheses of copolyesters; these can generally be prepared through the usual reactions for preparation of homopolyesters, and different chemical structures may be obtained provided that suitable monomers and proper reaction strategies are chosen. Random copolyesters prepared from aliphatic diols and mixtures of aliphatic and aromatic dicarboxylic acids have found applications as hot-melt adhesives,[13] while segmented block copolymers, poly(ester–ether)s and poly(ester–urethane)s in particular, have received much attention as thermoplastic elastomers.[14]

Difunctional monomers bearing functional groups such as carboxyl or its derivatives (esters, chlorides and anhydrides) and hydroxy compounds, or their acetates, are mainly used as starting materials for the syntheses of linear polyesters; small amounts of comonomers (usually less than 0.01 mol %) with functionalities higher than two can be added to obtain branched polyesters. The methods of preparation and chemical characteristics of a large number of monomers used for the syntheses of polyesters have been reported in the literature.[15]

Few chemically novel reactions for preparing polyesters have been reported in recent years when compared to those first employed, but esterification reactions carried out in solution under mild conditions in the presence of suitable activating agents are an important exception (see Section 17.2.2.2). However, most of the studies of the last few decades have been chiefly devoted to a better understanding of the traditional processes, with particular regard to kinetics, the role of the catalyst and side reactions, in order to obtain better control of the overall process and to improve the quality of the resulting polyesters. A review of the research trends of the last decade in polycondensation, dealing also with areas of interest in polyester synthesis, has appeared recently.[16]

In this chapter we intend to review the main aspects of the more extensively studied reactions, describe the methods of syntheses usually employed to prepare the most popular polyesters (PET and PBT) and discuss the more convenient routes for some particular classes of polyesters. Several chapters in this volume are specifically devoted to certain classes of polyesters (unsaturated polyesters, polyarylates, polyacetones, polycarbonates) or deal extensively with polyesters and copolyesters (bulk, solution, interfacial and solid-state step-polymerization, step copolymerization, liquid-crystal polymers *etc.*); these subjects will not be considered or will only be marginally taken into account in this chapter.

17.2 POLYESTERIFICATION REACTIONS

17.2.1 General Aspects

As a general requisite, for a reaction to be useful for preparing polyesters, it must lead to chains of high molecular weight, regular chemical structure and, possibly, with well-defined types of terminal groups.

The high molecular weight needed for good end-properties can only be achieved for conversions of the functional groups of the monomers close to unity (see Volume 5, Chapter 2), which are only attainable by using starting monomers of high purity, controlling the stoichiometry (in the case of AA/BB-type polymerization) and reducing side reactions by an appropriate choice of reaction conditions. Monofunctional impurities with low volatility are particularly harmful because they strongly limit the maximum molecular weight attainable. An appropriate choice of reaction conditions is also a basic requisite for the control of chemical structure and type of terminal group. From the stringent requirements necessary to perform polymerization successfully, it follows that, among the many reactions that can be used for the preparation of esters in classic organic chemistry,[17] relatively few can be exploited to prepare polyesters. Many reactions are not used because the appropriate monomers are not readily available, while, in other cases, extension to the synthesis of polyesters is limited by too many side reactions which prevent the attainment of high molecular weights. Economic factors often make other reactions, successfully employed for the preparation of simple esters, worthless when the same polyester can be prepared through a more economical route or the high cost of the resulting polyester is not counterbalanced by improvement of properties.

Many good reviews in encyclopedias, books and papers have been concerned with various reactions for the synthesis of polyesters in general[18-25] or of some classes of polyesters in particular.[26,27]

By analogy with classic organic chemistry, most polyesterification reactions are believed to occur by substitution at carbonyl carbon atoms through addition–elimination mechanisms,[28] as represented schematically in equation (1), where X = —OH, —OR″, —OCOR‴ and —Cl, Y can be a neutral or negatively charged nucleophilic agent (R′OH and R′O⁻ or R′CO$_2^-$, respectively) and R, R′, R″ and R‴ are alkyl or aryl groups.

$$R-\underset{X}{\overset{O}{C}} + Y \underset{k_{i,1}}{\overset{k_1}{\rightleftarrows}} \left[R-\underset{Y^{\delta+}}{\overset{O^{\delta-}}{C}}-X\right] \text{ or } \left[R-\underset{Y^{\delta-}}{\overset{O^{\delta-}}{C}}-X\right] \underset{k'_{i,1}}{\overset{k'_1}{\rightleftarrows}} R-\underset{Y}{\overset{O}{C}} + X \quad (1)$$

(1a)   (1b)

The addition intermediates (**1a**) and (**1b**), for neutral and negatively charged nucleophilic agents respectively, cannot be isolated or detected in any direct manner, and are therefore postulated on the basis of isotopic oxygen-exchange reactions and by extrapolating evidence of stable addition compounds observed for anhydrides, amides, *etc.*

Experimental evidence also suggests that formation of the addition intermediate probably occurs through a perpendicular approach to the carbonyl carbon atom by the attacking nucleophile.[28]

The overall reaction (1) is stepwise in nature and the relative rates of formation and partition of the addition intermediate determine the overall rate and the equilibrium between reactants and products. In general, the formation of the addition intermediate (**1**) has been found to be the slow step, and a catalyst is often used to increase the rate of this stage. The chemical structure of R, R′, R″, R‴, X and Y may influence both the rate of formation of (**1**) and its partitioning and, consequently, the overall rate and equilibrium.

According to the mechanism proposed, an increase in the electron-withdrawing character of R will result in easier formation of (**1**). Resonance interaction of R with the carbonyl group tends, on the contrary, to stabilize the ground state with respect to the transition state, which must be similar to the tetrahedral intermediate, and hence reduces the rate of formation of (**1**). Bulky groups on R can hinder the nucleophilic attack and therefore reduce the rate of the first step.

Structural changes in X are related to both inductive and resonance effects and are more difficult to interpret. The electron-withdrawing power of X increases both the rate of formation of (**1**) and its partition towards products. Increased resonance of X with the carbonyl group increases the stability of the ground state with respect to the transition state and will result in a lower rate; resonance interaction increases in the order —Cl < —O$_2$CR′ < —OR′ < —NR$_2''$. The effectiveness of the nucleophile Y is, of course, related to the reactivity of the carboxylic acid derivatives; the more reactive the latter, the wider the range of nucleophiles which can be used effectively. The overall rate of the reaction will depend on the rate of formation of (**1**) and on its partition between reactants and products; when X and the attacking nucleophile Y are equally good leaving groups, the increase in nucleophilicity parallels the increase in the overall rate.

According to the above discussion, polymerizations in which X is a particularly good leaving group and Y a strong nucleophile should proceed under mild conditions, whilst it is expected that a high temperature is required for weak nucleophiles and poor leaving groups. Coreactants, capable of transforming *in situ* one type of functional group (usually the carboxylic group) into a highly activated structure which can readily undergo reaction with the other reactive function, have been successfully used to improve reactivity; examples of direct esterification carried out under mild conditions will be discussed in Section 17.2.2.2.

The most widely used reactions for the preparation of polyesters are direct esterification (X = OH, Y = R′OH) and alcoholysis (X = OR″, Y = R′OH), usually performed at high temperature in the melt, and reactions of acyl chlorides (X = Cl) with hydroxy compounds (R′OH) or phenolates (R′O⁻), generally carried out at medium/low temperature in solution or by interfacial synthesis.

Exchange reactions, such as acidolysis (X = OH, Y = R′CO$_2$R‴) and ester–ester exchange (X = OR″, Y = R′CO$_2$R‴), can also take place at high temperature in the presence of suitable catalysts; however, they are rarely used for the convenient preparation of polyesters. For this reason these reactions are little studied and frequently disregarded, even though they may sometimes play an important role in the control of the resulting chemical structures, in particular when copolymers with non-random chemical structures are prepared or processed at high temperature or when different polyesters are blended in the molten state. In such cases, these exchange reactions can contribute significantly to driving the system towards a final random copolymer, or, in a similar way, to restore the most probable molecular weight distribution (MWD) when polymers having a MWD narrower or wider than that predicted by the Flory–Schultz theory are melt mixed.

Other methods of synthesis have been suggested as particularly suitable for preparing particular classes of polyesters, such as, for example, poly(α-ester)s,[27] biodegradable polymers from biosyntheses[10] and oligoesters with a very narrow MWD.[29] These reactions, as well as other reactions seldom used for polyesterification, will be briefly discussed in the following sections. The preparation of copolymers with different chemical structures is mainly a problem of strategy and control of common reactions rather than of new reactions. This subject is covered in general for step copolymerization in Volume 5, Chapter 16, and will be only marginally discussed in this chapter.

The synthesis of a particular type of polyester can often be successfully performed by different methods: in bulk (usually at high temperature), in solution, by an interfacial method or in the solid state; the choice is primarily one of convenience and of availability of starting materials. The main features of the polycondensation methods have been widely discussed in Chapters 9–13 of this volume.

When performed at low temperature, either in solution or by an interfacial method, polyesterification may present problems related to solubility or removal of the residual catalyst, by-products or solvent that may induce harmful side-reactions either during processing or service. When polymerizations are carried out at high temperature, chain-scission reactions may compete with the growing-chain reactions and become the most important limiting factor for the molecular weight increase. Other side reactions, leading to spurious moieties in the chains or to unwanted by-products in the final polymer, may accompany esterification reactions; the formation of acetaldehyde during the preparation of PET is a well-known example. Some degradation reactions occur with a general mechanism valid for most polyesters and will be discussed in this section, whereas some side reactions specific for a class or for single polyesters will be considered in distinct sections dealing with that particular class of polyesters.

As a result of reaction (1), an equilibrium between reactants and products would be reached in a closed system. As discussed above, the equilibrium constant is dependent on the chemical structure of X and Y; most of the more commonly used reactions have equilibrium constants close to or less than unity, and are therefore considered as being 'of equilibrium', but for X = —Cl and Y = R′OH or R′O$^-$ the equilibrium is almost completely shifted to the right[30] and reactions involving acyl chlorides are usually considered as being 'of non-equilibrium'. In general, the low molecular weight products resulting from the polycondensation are removed from the equilibrium; for reactions involving acyl chlorides the removal of hydrogen chloride (as chloride salt) is mainly performed in order to avoid side reactions during polymerization or processing. In the case of equilibrium reactions, the removal of low molecular weight products is necessary in order to allow the reaction to proceed to high conversion. Either a vacuum or bubbled gas is usually used for the latter purpose, but sometimes the by-products can be removed from the reactor as an azeotropic mixture with the solvent.

Intramolecular condensation and exchange reactions yield cyclic oligomers which usually accompany linear polyesters. The extent of cyclization and the number of monomeric units in cyclic oligomers depends primarily on the chemical structure of the monomeric units and on the polycondensation method (solid-state polycondensation usually leads to a lower oligomer content[31]) and on reaction conditions. An extensive list of the type and amount of cyclic oligoesters that have been found has been assembled[32] for many polyesters. Theoretical correlations between chemical structure and the equilibrium concentration of oligomers, and recent analytical methods to determine cyclic oligomers, are discussed in Chapters 5 and 6 of this volume.

Since the knowledge of correct rate equations and of relative kinetic parameters is of primary importance for the elucidation of reaction mechanisms and for optimization of industrial processes, many papers in the scientific literature have been devoted to kinetic studies.

In these studies, it is generally accepted that, if the system is a true solution and the neighborhood of all the functional groups does not change during polymerization, the reactivity of a functional group does not depend on the chain length, at least after the first few steps of the reaction, as first stated by Flory[33,34] (see also Volume 5, Chapter 3). Accordingly, the kinetic constants are generally considered to have the same value in each step of the reaction, independent of the chain length.

Some problems related to the polymeric nature of the reactants, such as high viscosity of the medium, low solubility of reactants, separation and identification of reaction products, high temperature required to perform polycondensation in the molten state and the simultaneous occurrence of several reactions, can make kinetic studies difficult. When the reaction is carried out in bulk in a molten state, the high viscosity of the reaction medium may limit or hinder the removal of volatile by-products and, depending on the reactor geometry, the diffusion of volatile by-products out of the polymerization system can increasingly contribute to the overall rate until it becomes the rate-controlling step.

Therefore, for the kinetic description of polycondensation processes, the rate of chemical reactions (forward and reverse) as well as the rate of physical transport processes should be taken into account. Although polycondensations carried out in a thin layer[35,36] have shown that the latter aspect is not always negligible, it is usually disregarded by assuming that the reaction by-products are removed so fast that their concentration does not enter into the rate equations. This assumption, which greatly simplifies calculations, may lead in some cases to disputable kinetic results.

Some further complications derive from the high temperature needed to maintain polymers in the molten state or to perform polycondensation at a sufficiently high rate. At high temperatures, many reactions can occur simultaneously, leading to chain growth or to chain scission or changing the chemical nature of end groups; the relative role of these reactions is usually determined by the reaction conditions and by the catalysts employed. Kinetic constants of growing reactions are generally much higher than those of degradation; however, when conversion becomes close to unity, the low concentration of terminal groups and the high concentration of ester groups within the chains make the overall rate of molecular weight increase and of chain scission similar, and the equation referring to the latter should be included in the kinetic scheme. The several possible reactions make correct kinetic studies quite complex and most papers in the literature involve a number of simplifying assumptions that sometimes make the reported results disputable.

Some additional difficulty in kinetic studies may result from the low solubility of reactants in the reaction medium and from the separation and identification of reaction products. Suitable low molecular weight monofunctional 'model' compounds are often used to reduce the number of possible reactions and for easier identification of the reaction products. Using 'model' compounds, only a few components are present in the reacting system instead of the complex mixture of oligomers which is always present, even at a low extent of reaction, during polymerization, and the usual chromatographic techniques of separation can be usefully employed to overcome difficulties in the analysis of the various components, thus ambiguous interpretations of the results are reduced or do not occur. The choice of the 'model' requires, of course, particular care, in order that the reactivity of the functional groups in the chosen 'model' does not differ from that in the polymer.

Most polyesterification reactions only occur at a sufficiently high rate in the presence of suitable catalysts. Due to the academic and industrial relevance of catalysts, an enormous number of publications have appeared, particularly in the patent literature, on this subject. Most of these references seem, however, to be of uncertain value and are often of doubtful originality.

Although it can be assumed in general that the role of the catalyst is to increase the positive charge on the carbonyl carbon atom, many questions are still unanswered about the true catalytic mechanism of most metal catalysts, which are by far the most widely used. One reason is that the original catalyst often undergoes exchange reactions with reactants leading to products of unknown concentration and catalytic activity; the overall catalytic activity is therefore the result of a complex set of reactions and coordination equilibria involving catalyst and functional groups. The type and role of catalysts will be further considered in connection with the discussion of separate reactions.

In addition to their effect on growing reactions, catalysts may also affects side reactions, and compounds with good catalytic activity towards growing reactions can be unsuitable owing to their effect on side reactions. This is the case for titanium alkoxides, which are very good catalysts for PET polymerization but are not used owing to the yellowing of PET.

For the sake of clarity, the various reactions used to prepare polyesters will be considered separately in the following sections; however, the reader should bear in mind that other reactions can often take place simultaneously. This is particularly important when polymerizations are carried out at high temperature where most of the exchange reactions are operating and side reactions can also occur.

17.2.2 Polyesters from Direct Esterification

17.2.2.1 Direct esterification at high temperature

Most polyesters can be obtained by direct esterification at high temperature from dicarboxylic acids and diols (equation 2) and from hydroxy acids (equation 3).

$$n\text{HO}_2\text{CRCO}_2\text{H} + n\text{HOROH} \rightleftharpoons \text{H}(\text{ORO}_2\text{CRCO})_n\text{OH} + (2n-1)\text{H}_2\text{O} \qquad (2)$$

$$n\text{HO}_2\text{CROH} \rightleftharpoons \text{H}(\text{ORCO})_n\text{OH} + (n-1)\text{H}_2\text{O} \qquad (3)$$

Although reaction (2) usually proceeds smoothly with aliphatic diols bearing primary and secondary hydroxyl groups, it is not generally suitable for diols having aromatic and tertiary hydroxyl groups because of poor nucleophilicity and competing elimination reactions, respectively. Similarly, reaction (3) has been used to prepare many polyesters from hydroxy acids,[18] but it is often unsuitable in many other cases. In fact, except glycolic and lactic acids, α-hydroxy acids can rarely be used for direct esterification[27] and, likewise, because of the competing dehydration reaction leading to unsaturated carboxylic acids, the attainment of polyesters by this route is prevented for most of the 3-hydroxy acids.[37] The high temperature required for melt polymerization and the low nucleophilicity of the phenol group make this reaction generally unsuitable even for the preparation of polyesters from aromatic hydroxy acids, unless the reaction is properly activated. In general, the chemical structure of the radical to which the carboxylic group is bonded has a minor effect, and the reaction occurs easily for both aromatic and aliphatic carboxylic acids; however, bulky substituents may sterically hinder the nucleophilic addition of the hydroxy derivative to the carboxyl group.

When polymerizations have been carried out with different straight-chain diols, it has been found that both reaction rate and equilibrium constant increase from ethylene glycol (EG) to 1,4-butanediol (BD) to 1,6-hexanediol (HD); on the other hand, an increase in the number of methylene groups in the aliphatic carboxylic acid leads to minor changes.[38]

Both reactions (2) and (3) are 'of equilibrium', with equilibrium constants generally close to or less than unity, and, consequently, the equilibrium has to be shifted to the right by continuously removing the water throughout the course of the polymerization, in order to obtain a conversion very close to unity required to reach high molecular weights. The water elimination can be accomplished either by vacuum or by a stream of inert gas or by distilling it out of the reactor along with a part of the solvent, whose total amount is then restored in the reactor with dry solvent.

An excess of diols is often employed in order to increase the initial reaction rate; the excess diol is subsequently removed, in monomeric form, *via* an alcohol–ester exchange reaction in the last period of polymerization, so that the initial imbalance of the functional groups does not affect the molecular weight of the final polymer.

Direct esterification can proceed at high temperature (180 °C–230 °C) even in the absence of added catalysts; in this case, the carboxyl groups of the monomer provide protons which can catalyze the reaction. Small amounts (0.1–0.5 wt %) of external catalyst are nevertheless added in order to increase the reaction rate significantly. Hundreds of compounds have been claimed in the patent literature as effective catalysts; strong protonic acids (H_2SO_4, benzene-, naphthalene- and p-toluene-sulfonic acids are the most popular), oxides or salts of heavy metal ions (often acetates are preferred for their higher solubility) and organometallic compounds of titanium, tin, zirconium and lead are the most frequently reported catalysts. An extensive list of catalysts employed in direct esterification has been recently reported in a review by Fradet and Marechal.[39]

According to classic organic chemistry, direct esterification can be catalyzed by either acidic or basic compounds, and Ingold[40] has proposed eight different mechanisms, four for acid-catalyzed and four for base-catalyzed processes. Basic compounds are seldom used as catalysts for polyesterification and, among the acid-catalyzed mechanisms, $A_{AC}2$ is by far the most frequently observed. The mechanisms proposed for direct esterification of low molecular weight esters have been investigated in detail by many workers and have been discussed in detail in a review by Bender.[28] According to these investigations, Scheme 1 is generally accepted for proton-catalyzed reactions.

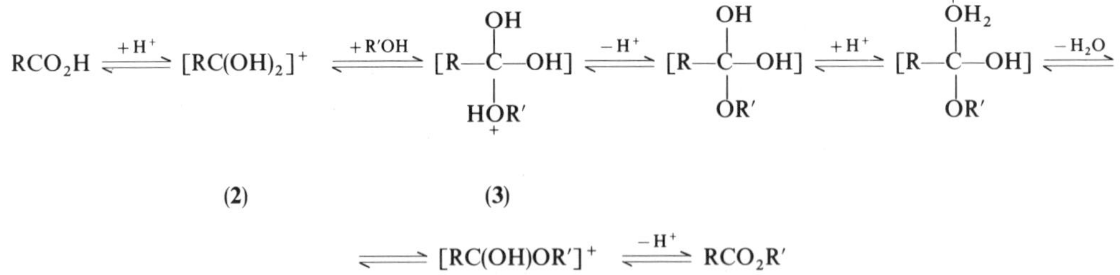

$$RCO_2H \underset{}{\overset{+H^+}{\rightleftharpoons}} [RC(OH)_2]^+ \underset{}{\overset{+R'OH}{\rightleftharpoons}} [R-\underset{H\overset{+}{O}R'}{\overset{OH}{\underset{|}{C}}}-OH] \underset{}{\overset{-H^+}{\rightleftharpoons}} [R-\underset{OR'}{\overset{OH}{\underset{|}{C}}}-OH] \underset{}{\overset{+H^+}{\rightleftharpoons}} [R-\underset{OR'}{\overset{\overset{+}{O}H_2}{\underset{|}{C}}}-OH] \underset{}{\overset{-H_2O}{\rightleftharpoons}}$$

(2) (3)

$$\rightleftharpoons [RC(OH)OR']^+ \underset{}{\overset{-H^+}{\rightleftharpoons}} RCO_2R'$$

Scheme 1

In this scheme, the reaction of the protonated form (2) of the carboxylic acid with the hydroxy compound to give the addition intermediate (3) is usually taken as the rate-controlling step. This mechanism is usually extrapolated to proton-catalyzed direct polyesterification.

Owing to the low basicity of substrates such as carboxylic acids, the concentration of protonated species (2) can be extremely low, and alternative mechanisms, involving a nucleophilic attack assisted by compounds able to form hydrogen bonds in cyclic transition states such as (4) or (5), have also been considered.[28,39,41]

                    (4)   (5)

The polarity of the reaction medium and the presence of basic substances which can be more easily protonated than the carboxyl group may strongly affect the concentration of the free ion species (2). Media of increasing polarity may shift the acid–base equilibria from indissociate forms towards ionization, *i.e.* leading to formation of indissociate ion-pairs, and, possibly, to dissociation, *i.e.* with formation of free ion species.[42] Such changes lead, of course, to variation in reaction rate constants and kinetic equations.

The strong change in polarity of the medium, expected when conversion increases during direct esterification in the absence of solvent, is probably the main reason for the contradictory results reported in the literature with regard to partial orders of reaction,[39,43] although other causes, discussed in Section 17.2.1, can explain the conflicting results reported by different authors. Inappropriate choice of reactions conditions, such as temperature range of measurements, molar ratio of reactants and poor analytical results, can be misleading in other cases.

For reactions performed in closed systems or whenever the water is not efficiently removed, the contribution of the reverse reaction must also be considered. It should also be remembered that the kinetics of the early stages of polyesterification are often complicated by the initial heterogeneity of the reacting system, caused by the very low solubility of the acids.[44–46] The main techniques that can be used to perform polyesterification and the methods used to evaluate kinetic parameters have recently been critically discussed.[39] Correction of kinetic equations for volume or weight changes during polycondensation should be properly introduced to obtain more precise kinetic data.[39,47–50]

In the absence of external catalysts, the usually accepted kinetic equation is overall third-order, being first and second order in hydroxyl and carboxyl groups, respectively, according to equation (4).

$$\text{rate} = k^\circ [RCO_2H]^2 [R'OH] \qquad (4)$$

The partial order of two for the carboxyl groups suggests that they behave as reactant and catalyst at the same time. Such a kinetic equation was first proposed by Flory,[33,34,51–53] using a stoichiometric ratio of reactants, for the last stage of the reaction when the concentration of reactive endgroups has been greatly decreased and the dielectric properties of the medium no longer change with conversion. Stoichiometric conditions make calculations easier for correlation of conversion with the overall rate, but it is impossible to achieve unambiguous information on partial orders from such results. Since it is impossible to perform polyesterification in an excess of one of the reactants, because stoichiometry is required to reach high molecular weights, the order in acid and alcohol were determined by several authors on monofunctional low molecular weight compounds under non-stoichiometric conditions. Hamann[43] and Fradet[54] confirmed partial orders of two and one for RCO_2H and $R'OH$, respectively, for reactions of 1-dodecanol with *n*-dodecanoic acid in *n*-dodecyl decanoate[43] and of 1-octadecanol with *n*-octadecanoic acid in benzophenone.[54] These partial orders were confirmed on polymeric monofunctional reactants for the reaction of ω-hydroxy- and ω-carboxy-poly(oxyethylene),[54] for the further polymerization of poly(1,10-decylene adipate) and poly(diethylene glycol adipate) of low molecular weights,[43] and for the reaction of adipic acid with pentaerythritol either in solution[55] or in the melt.[56]

Scheme 2 has been proposed in order to explain the results of uncatalyzed reactions in media of low dielectric constant (equivalent to those found at high conversion during polyesterification), where the presence of free ions can be reasonably excluded. According to this scheme, the reaction rate is given by equation (8) and, assuming that the concentration of ion pairs with respect to the overall concentration of CO_2H and OH is negligible, as expected for media of such low polarity,[42] the concentration of (6) is derived through the equilibrium constant K_5 (equation 9) and, substituted in equation (8) leads to equation (10) which is analogous to that originally proposed by Flory.[51–53]

$$2RCO_2H \xrightleftharpoons{K_5} (RCO_2^-, RC(OH)_2^+) \quad (5)$$
$$(6)$$

$$(6) + R'OH \xrightleftharpoons{k_6, \text{slow}} (RCO_2^-, RC(OH)(OR')^+) + H_2O \quad (6)$$
$$(7)$$

$$(7) + R'OH \xrightleftharpoons{\text{fast}} RCO_2R' + RCO_2H \quad (7)$$

Scheme 2

$$v = k_6[(6)][R'OH] \quad (8)$$
$$[(6)] = K_5[RCO_2H]^2 \quad (9)$$
$$v = k_6 K_5[RCO_2H]^2[R'OH] \quad (10)$$

An overall third-order mechanism could, however, also be interpreted by assuming hydrogen-bond adducts[39,41] according to Scheme 3.

$$2RCO_2H \xrightleftharpoons (RCO_2^-, RC(OH)_2^+) \quad (11)$$
$$(6)$$

$$RCO_2H + R'OH \xrightleftharpoons (RCO_2^-, R'OH_2^+) \quad (12)$$
$$(8)$$

(6) + R'OH
                               ⇌ (slow) [cyclic H-bonded structure (9)] ⇌ (fast) $RCO_2R' + RCO_2H + H_2O$ (13)
(8) + RCO_2H

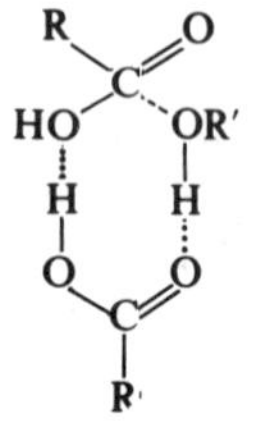

Scheme 3

Although overall third-order is mostly accepted, several studies of uncatalyzed polyesterification have been interpreted on the basis of different kinetics: overall second-order;[38,57] an overall order of 2.5;[58–61] second followed by third order in the last stage of reaction;[41,57,62] and even an overall order of six has been reported.[63]

As described above, Fradet and co-workers confirmed equation (10) for the reaction of 1-octadecanol with n-octadecanoic acid in benzophenone[58] at 160–210 °C, but their experimental results agree better with an overall order of 2.5 (1.5 and 1.0 for CO_2H and OH, respectively) for the same reaction in octadecyl octadecanoate.[60] In the latter case they also found that the reaction rate is higher than when performed in a more polar medium such as benzophenone. To explain these results, a significant dissociation of the ion pairs, with the formation of free ions, was assumed. As the reaction medium, octadecyl octadecanoate, is non-polar, this seemed rather unlikely even to the authors, who supposed that the polar groups, OH, CO_2H and water, concentrate in certain areas where they create a very polar medium which favors the dissociation of ion pairs.[60]

Scheme 4, based on dissociation into free ions, can account for partial orders of 1.5 and 1.0, for CO_2H and OH respectively. Let S be the solvent (or any other species which is more easily protonated than RCO_2H) whose protonated form (SH^+) is largely predominant over the protonated carboxyl species $RC(OH)_2^+$. From the rate equation (17) and, assuming $[SH^+] = [RCO_2^-] + [RC(OH)_2^+] \simeq [RCO_2^-]$, from equilibrium constants (18) and (19), equation (20) will result.

$$S + RCO_2H \xrightleftharpoons{K_{14}} SH^+ + RCO_2^- \quad (14)$$

$$SH^+ + RCO_2H \xrightleftharpoons{K_{15}} S + RC(OH)_2^+ \quad (15)$$

$$RC(OH)_2^+ + R'OH \xrightleftharpoons{k_{16}} (3) \xrightleftharpoons RCO_2R' + H_3O^+ \quad (16)$$

Scheme 4

$$v = k_{16}[\text{R'OH}][\text{RC(OH)}_2^+] \quad (17)$$

$$K_{14} = [\text{SH}^+][\text{RCO}_2^-]/[\text{S}][\text{RCO}_2\text{H}] \quad (18)$$

$$K_{15} = [\text{S}][\text{RC(OH)}_2^+]/[\text{SH}^+][\text{RCO}_2\text{H}] \quad (19)$$

$$v = k_{16}K_{15}[\text{SH}^+][\text{RCO}_2\text{H}][\text{R'OH}]$$

$$= k_{16}(K_{15}/[\text{S}])(K_{14}[\text{S}][\text{RCO}_2\text{H}])^{0.5}[\text{RCO}_2\text{H}][\text{R'OH}]$$

$$= k_{16}K_{15}(K_{14}/[\text{S}])^{0.5}[\text{RCO}_2\text{H}]^{1.5}[\text{R'OH}] \quad (20)$$

Based on similar considerations of free-ion catalysis, an overall order of 2.5 was assumed to hold for the whole course of polyesterification.[61] Although this is probably not true for most cases, nevertheless it is possible that dissociation occurs in the earlier stages of polyesterification when a highly polar medium, due to the excess of glycol, is present.

The higher the partial orders of the reactants, the stronger the reaction rate decrease in the last period of polycondensation, when the concentration of functional groups is greatly reduced. In order to avoid a reaction rate that is too low in the last stage, a catalyst is usually added.

When a strong protic acid has been used as the catalyst, most authors have found that the experimental data fit an overall third-order reaction quite well (first order in CO_2H, OH and catalyst),[43,52,53,60,61,64-72] according to equation (21). More correctly, two distinct independent contributions are generally considered in the rate equation, which assumes the form of equation (22), where the second term in the right-hand side represents the contribution from the added catalyst.

$$v = k_{\text{cat}}[\text{AH}][\text{RCO}_2\text{H}][\text{R'OH}] \quad (21)$$

$$v = k^\circ[\text{RCO}_2\text{H}]^2[\text{R'OH}] + k_{\text{cat}}[\text{AH}][\text{RCO}_2\text{H}][\text{R'OH}] \quad (22)$$

However, at high conversion the contribution of the uncatalyzed reaction usually becomes negligible and the reaction rate again assumes the form of equation (21). This equation can be interpreted on the basis of an ion-pair mechanism, as shown in Scheme 5.

$$\text{AH} + \text{RCO}_2\text{H} \xrightleftharpoons{K_{23}} (\text{A}^-, \text{RC(OH)}_2^+) \quad (23)$$
$$(\mathbf{10})$$

$$(\mathbf{10}) + \text{R'OH} \xrightleftharpoons{k_{24}} [\text{A}^-, \text{RC(OH)}_2(\text{OR'})^+] \quad (24)$$
$$(\mathbf{11})$$

$$(\mathbf{11}) \rightleftharpoons \text{RCO}_2\text{R'} + \text{H}_2\text{O} \quad (25)$$

Scheme 5

The protonated species RC(OH)_2^+ is formed by proton transfer from the strong protic acid AH and is independent of the ion-pair concentration (R'OH_2^+, RCO_2^-); by replacing the concentration of (**10**) in equation (26) with the equivalent expression derived from equilibrium (23), equation (21) is obtained with $k_{\text{cat}} = k_{24}K_{23}$.

$$v = k_{24}[(\mathbf{10})][\text{R'OH}] \quad (26)$$

In the hypothetical case that RC(OH)_2^+, R'OH_2^+ and A^- free ions would be formed after proton transfer, a kinetic equation with a partial order of 0.5 for both AH and R'OH and 1.0 for RCO_2H should result. Experimental data seem to exclude such a mechanism.

Although strong protic acids can be very effective catalysts of direct esterification, they usually also catalyze side reactions, and this may lead to polyesters of poor quality; metal catalysts are therefore generally preferred in industrial applications. Many metal derivatives have been found to be effective catalysts for direct esterification at high temperature, as has been reported in innumerable patents and in many papers (an extensive list is reported in the literature[39]), but the knowledge of their mechanism and kinetics is far from being definitive.

When various metal catalysts have been compared, titanium alkoxides and tin compounds have been found to be the most efficient.[46,73,74]

While proton-catalyzed direct esterification has been widely studied for both low molecular weight esters and polyesters, very little information is available for the catalytic mechanism of metal compounds, despite the large number of patents that have appeared on this subject.

Analogously to proton catalysis, which introduces a positive charge into a substrate, distorting the electronic distribution in the molecule and thus making reactions feasible, metal compounds, when they can form a complex with a substrate, may introduce a positive charge into a molecule. However, many experimental results suggest that the behavior of these catalysts in the reacting system is rather more complex. Examples of exchange reactions between the metal ligands with compounds bearing OH or CO_2H groups are well known in the literature;[75] additional evidence is the absence of catalytic activity for metal derivatives which cannot undergo exchange reactions with the substrate, implying that exchange of ligands is a prerequisite to catalytic activity.[73,76-79] Accordingly, $Sn(Bu^n)_4$, which does not exchange, does not display catalytic activity, whereas $Sn(CO_2)_2$, Bu_2SnO and other tin derivatives do.[73] Additionally, the role of exchange reactions has been clearly shown for titanium and zirconium alkoxides;[77,78] it has been found that for a series of titanium derivatives with different ligands the catalytic activity decreases in the order $Ti(OBu^n)_4 \gg Ti(OPr^i)_2(O_2C_6H_3NO_2) > Ti(O_2C_6H_3NO_2)_2$. The decrease in catalytic activity has been attributed to the limited exchange occurring with compounds containing p-nitro-o-phenylenedioxy ligands, since it is known that cateohol (1,2-benzenediol) derivatives of titanium do not give exchange reactions.[75,80]

On the other hand, when exchange is possible the catalytic effect does not depend on the nature of the ligands [$Ti(OBu^n)_4$ has the same catalytic activity as $Ti(OPr^i)_4$] unless the ligands resulting from the exchange are too bulky; in this latter case they can hinder coordination with the metal resulting in a decrease of catalytic effectiveness.[77] Association or condensate forms of the catalyst can be either already present in the initial catalyst[81-85] or can be formed during the reaction[39,77,78,86-88] and are probably in equilibrium with the species bearing OH and CO_2H groups and with water.[89]

According to the above discussion, the catalyst can be present in various forms which may change with the composition of the reaction system and therefore with conversion; the effective number of catalytic sites and the catalytic activity of the various forms are generally unknown and can differ from that corresponding to the amount of catalyst introduced. The presence of condensate forms of titanium and zirconium alkoxides have been suggested in order to explain kinetic results.[39,77,78,86,87,89,90]

On summarizing the above features, it appears that for a metal catalyst to be effective it has to be able to exchange ligands and to coordinate with reactants forming a complex that is not too strong. When the catalytic activities of many metal compounds have been compared, it has been found that alkoxide derivatives of titanium and zirconium are the most efficient.[75,77] Scheme 6 proposes some of the possible reactions that metal alkoxides can undergo in reacting systems such as those encountered in direct esterification. Equations (27) and (28) refer to exchange reactions and equations (29)–(31) refer schematically to the formation of condensate forms. Exchange reactions are usually assumed to occur faster than the formation of the ester which may proceed through different paths and transition states. In the absence of direct evidence, both intra- or inter-molecular reactions between ligands and free or coordinated reactants (R'OH or RCO_2H) may be suggested, all accounting for the kinetic data. Marechal and co-workers[76] compared four different mechanisms and arrived at the conclusion that a mechanism based on the intermediate (12) is slightly to be preferred as it gives less scattered values for both titanium and zirconium catalysts.

$$M(OR)_n + xR'OH \rightleftharpoons M(OR)_{(n-x)}(OR')_x + xROH \qquad (27)$$

$$M(OR)_n + yR''CO_2H \rightleftharpoons M(OR)_{(n-y)}(O_2CR'')_y + yROH \qquad (28)$$

$$M(OR)_{(n-y)}(O_2CR'')_y \rightleftharpoons MO(OR)_{(n-y-1)}(O_2CR'')_{(y-1)} + R''CO_2R \qquad (29)$$

$$2M(OR)_{(n-y)}(O_2CR'')_y \rightleftharpoons (OR)_{(n-y-1)}(O_2CR'')_y MOM(OR)_{(n-y)}(O_2R'')_{(y-1)} \qquad (30)$$

$$M(OR)_n + H_2O \rightleftharpoons (OR)_{(n-1)}MOM(OR)_{(n-1)} + 2ROH \qquad (31)$$

M = metal

Scheme 6

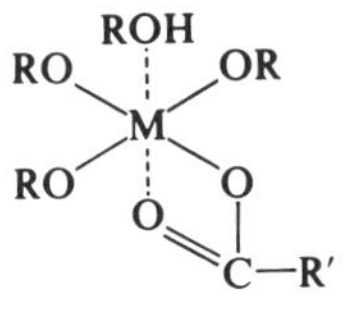

Polyesters 285

(12) M = metal

In many cases it is no longer possible to speak of kinetic orders, owing to the complexity of the kinetic equations; however, under particular conditions, the kinetic equations reduce to simpler forms. In these cases, a partial first-order seems more reliable[39,80,87,91] for the catalyst, even though partial orders of 0.5[73,91] and 0[74] have been reported.

Many kinetic data for uncatalyzed, proton- and metal-catalyzed direct esterification have been collected in a review by Fradet and Marechal;[39] values for the activation energy are usually 40 to 100 kJ mol^{-1} for reactions with or without catalyst and irrespective of the type of catalyst.

17.2.2.2 *Direct esterification under mild conditions*

Direct esterification of low molecular weight compounds can be performed at low temperature using both strong acids and bases as catalysts;[92] however, these conditions are not generally suitable for polyesterification. Many other catalysts have been effectively used for the preparation of esters under mild conditions, but only a few have been successfully extended to polyesters; trifluoroacetic anhydride (TFAAn), 1,1'-carbonyldiimidazole (CDI) and, more recently, phosphorus derivatives, picryl chloride and tosyl chloride (TsCl) have been used as effective promoters of direct polyesterification.

TFAAn is known as a useful rapid esterification promotor of acids with both alcohols and phenols for low molecular weight compounds;[92] however, very few examples of extension of this reaction to polymerizations have been reported. Preparation of poly(oxy-1,4-phenylenecarbonyl) has been reported from *p*-hydroxybenzoic acid in TFAAn at 75 °C for 15 min.[93,94] Similarly, a polyester with $\eta_{inh} = 0.86$ was obtained when 3-propyl-4-hydroxybenzoic acid was heated at 100 °C for 3 h in sealed tubes in the presence of TFAAn;[94] the same procedure was adopted for the preparation of polyesters from 3,5-diisopropyl-4-hydroxybenzoic acid.[95] For low molecular weight substances, the reaction seems to proceed through a mixed acid anhydride[92,93,96] and, by analogy, the same mechanism may also be assumed to operate with polymers.

Staab[97] prepared PET and poly(hexamethylene terephthalate) (PHMT) of low molecular weight by combining terephthalic acid and the corresponding diols with CDI in the presence of small amounts of alkoxides. The reaction occurs, in solutions of THF, DMF, chloroform and similar inert solvents, according to Scheme 7. The *N*-acylimidazole (13), initially formed from CDI and the acid, undergoes alcoholysis according to equation (33). This reaction is quite slow at room temperature and the mixture must be heated at 60–70 °C, or a catalyst, such as an alkoxide, needs to be added. In the presence of sodium ethoxide, for example, the reaction is complete in a few minutes. The alkoxide, a strong nucleophilic agent, reacts rapidly with the *N*-acylimidazole (13) forming the ester and imidazolylsodium; the latter, in equilibrium with alcohol, regenerates sodium alkoxide.[98]

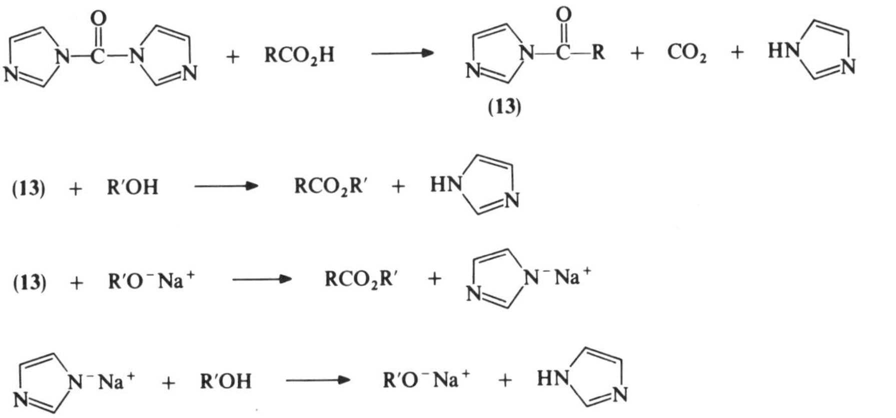

$$\text{Im-CO-Im} + RCO_2H \longrightarrow \text{Im-CO-R} + CO_2 + H\text{Im} \qquad (32)$$
$$(13)$$

$$(13) + R'OH \longrightarrow RCO_2R' + H\text{Im} \qquad (33)$$

$$(13) + R'O^-Na^+ \longrightarrow RCO_2R' + \text{Im}^-Na^+ \qquad (34)$$

$$\text{Im}^-Na^+ + R'OH \longrightarrow R'O^-Na^+ + H\text{Im} \qquad (35)$$

Scheme 7

Recently, several methods of synthesis, successfully employed for the preparation of aromatic polyamides under mild conditions, have been extended, after appropriate modifications, to the synthesis of polyesters by direct esterification at low temperature.[99-117] Although these reactions have sometimes been used to prepare polyesters and copolyesters from aliphatic dicarboxylic acids and diols,[100,102,103,105,110,116] they are particularly convenient for the synthesis of polyarylates. These methods are particularly relevant for the polymerization and copolymerization of hydroxybenzoic acids, for which the usual methods of solution and interfacial polymerization are unsuitable, owing to the difficulty of preparing the corresponding acyl chlorides. These polymers could otherwise be prepared at high temperature through exchange reactions from their acetates or phenyl esters.[118]

Substituted or unsubstituted hydroxybenzoic acids, as well as bisphenols with aromatic dicarboxylic acids, have been polymerized by these reactions in quantitative yields, at temperatures of 30–130 °C, to polyesters of high molecular weight. High molecular weight polyesters containing the cinnamoyl group in the polyester backbone have also been easily prepared without affecting the thermally labile —C=C— bond of the monomer.[101,111,112] Various copoly(ester–amide)s have also been prepared successfully by these solution methods.[119-123]

These reactions are based on the activation of the carboxyl group by condensing agents; phosphorus derivatives,[99-101,105,106,108,109,111,112,117] tosyl chloride[107,110,113-116] and picryl chloride[102,103] have been successfully employed with an excess of pyridine (or other tertiary nitrogen bases), and, often, with coactivating agents. Triphenylphosphine dichloride (TPPCl$_2$),[100,105,108] diphenyl chlorophosphate (DPCP),[106,109,111,112] hexachlorocyclotriphosphatriazene (HCTPT)[99,101] and phosphorus oxychloride (POCl$_3$)[117] were found to be the most effective phosphorus compounds, and it is preferable to use these rather than the highly reactive and readily hydrolyzed acid chlorides such as PCl$_5$, OSCl$_2$ and O$_2$SCl$_2$.

TPPCl$_2$ and DPCP seem particularly attractive for industrial applications of the method, because they can be easily and quantitatively regenerated from their unreactive final by-products. Triphenylphosphine oxide (TPPO) can be transformed into TPPCl$_2$ through a reaction with phosgene[124] or oxalyl chloride[125] according to Scheme 8.[108] Diphenyl phosphate (DPP) regenerates DPCP by reacting with thionyl chloride[106,126] according to Scheme 9.

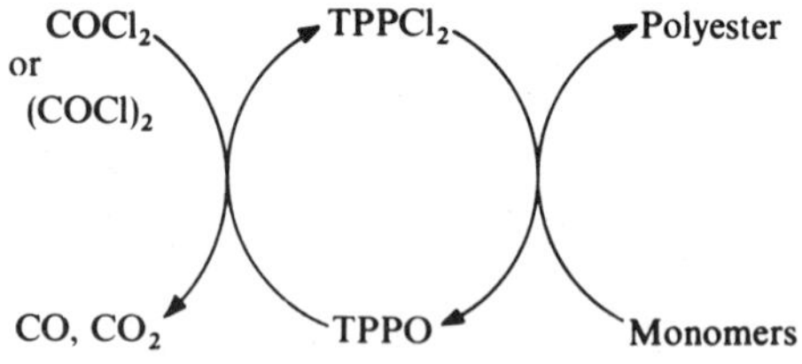

Scheme 8

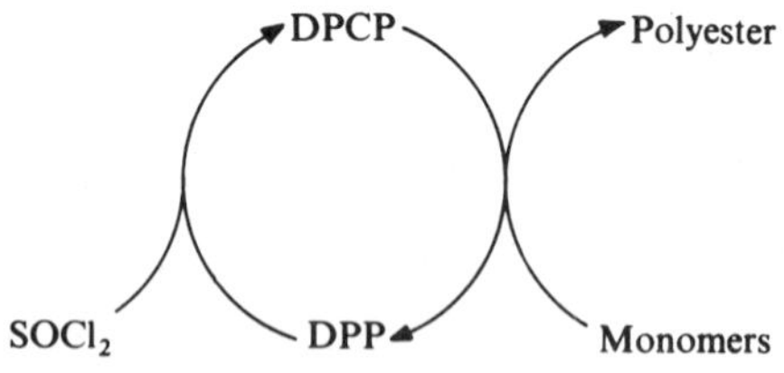

Scheme 9

TPPCl$_2$ can be prepared either from triphenylphosphine (TPP) with polyhalo compounds[100,104,105,127] or from TPPO with phosgene[108,124] or oxalyl chloride.[108,125] When TPPCl$_2$ was prepared *in situ* from TPP,[100,105] hexachloroethane was found to give the best results, for both yield and molecular weight of the resulting polyesters, among several polyhalo compounds tested.

It is well known[128,129] that TPPCl$_2$ exists as an ion pair and that its phosphorus atom is attacked by nucleophilic reagents to form phosphonium salts; accordingly, it is generally assumed that in the presence of pyridine (or other tertiary nitrogen bases) TPPCl$_2$ leads to N-phosphonium salts of pyridine (**14**)[104,105,108] according to equation (36). Salt (**14**) then reacts with carboxyl groups to give an acyloxy N-phosphonium salt (**15**) according to equation (37). The salt (**15**) produces the corresponding esters or anhydrides when attacked by hydroxy or carboxy compounds, respectively, in the presence of an acid acceptor[104,127] according to Scheme 10.

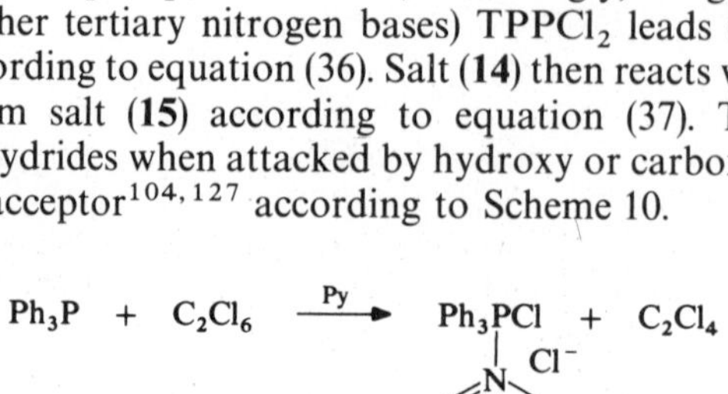

$$Ph_3P + C_2Cl_6 \xrightarrow{Py} Ph_3PCl + C_2Cl_4 \quad (36)$$

(**14**)

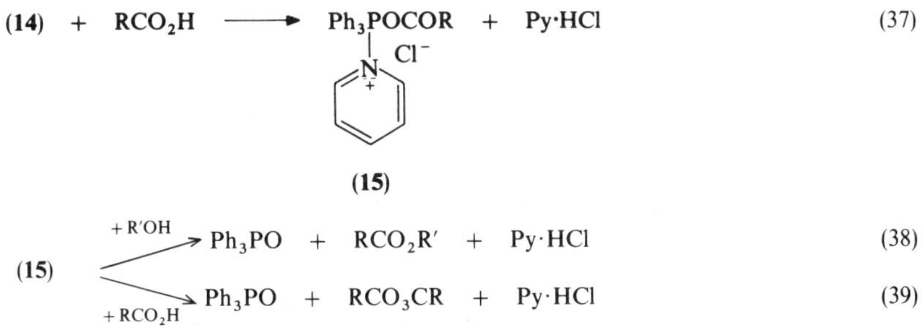

Scheme 10

$$(14) + RCO_2H \longrightarrow Ph_3\overset{+}{P}OCOR \cdot \underset{Cl^-}{\underset{|}{N}}\!\!\bigcirc + Py\cdot HCl \quad (37)$$

(15)

$$(15) \underset{+RCO_2H}{\overset{+R'OH}{\diagup\diagdown}} \begin{array}{l} Ph_3PO + RCO_2R' + Py\cdot HCl \quad (38) \\ Ph_3PO + RCO_3CR + Py\cdot HCl \quad (39) \end{array}$$

The reaction of the acyloxy N-phosphonium salt (15) with carboxyl groups with the formation of anhydride, according to equation (39), limits the maximum molecular weight attainable by changing the stoichiometric molar ratio between hydroxyl and carboxyl groups.[104,108] It is probably for this reason that aging times of more than 15 min, after TPPCl$_2$ and the acid are reacted and before the addition of bisphenol, lead to a decrease in the molecular weight of the final polyesters.[108]

DPCP in pyridine can activate a carboxylic acid through the formation of a mixed anhydride (formula 16) which gives the corresponding ester on alcoholysis with phenols, according to Scheme 11. Reaction (43), concurrent with reaction (42), may lead to the unreactive product (19) and, consequently, to lower yields and molecular weights, particularly when reaction (41) is slow and not selective.

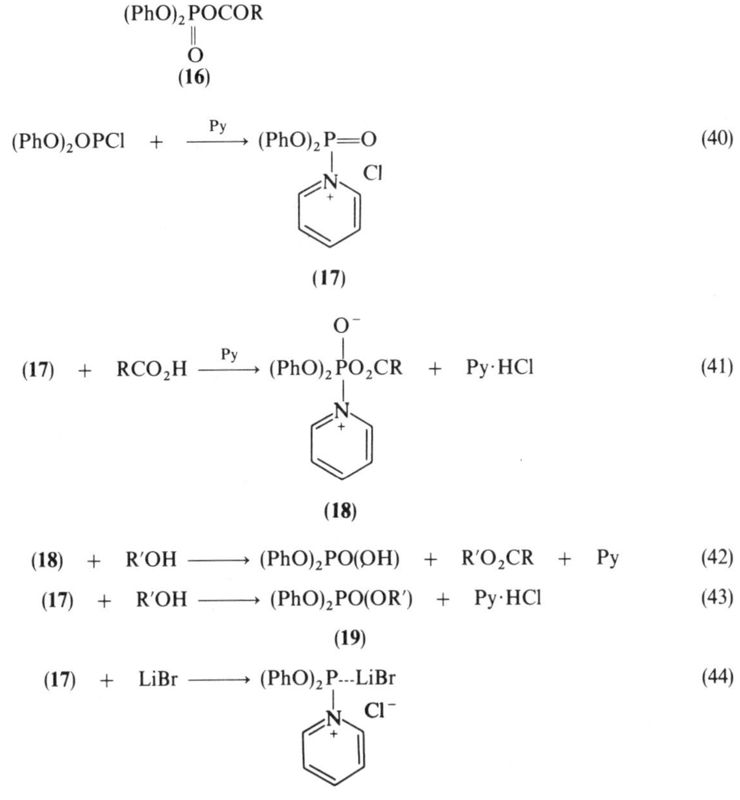

$$\underset{(16)}{(PhO)_2\overset{\overset{O}{\|}}{P}OCOR}$$

$$(PhO)_2OPCl + \overset{Py}{\longrightarrow} (PhO)_2P\!\!=\!\!O \quad (40)$$
$$\underset{(17)}{}$$

$$(17) + RCO_2H \overset{Py}{\longrightarrow} (PhO)_2\overset{O^-}{\underset{|}{P}}O_2CR + Py\cdot HCl \quad (41)$$
$$\underset{(18)}{}$$

$$(18) + R'OH \longrightarrow (PhO)_2PO(OH) + R'O_2CR + Py \quad (42)$$

$$(17) + R'OH \longrightarrow (PhO)_2PO(OR') + Py\cdot HCl \quad (43)$$
$$(19)$$

$$(17) + LiBr \longrightarrow (PhO)_2P\cdots LiBr \quad (44)$$

Scheme 11

It has been found that DPCP in pyridine is a very useful condensing agent for the synthesis of polyarylates directly from aromatic dicarboxylic acids and bisphenols, but not so effective for the polymerization of hydroxybenzoic acids.[106] This latter process was improved by carrying out the reaction with DPCP in the presence of other substances as coactivators: DMF,[109] ethers[111] and lithium halides[112] were found to be effective for this purpose.

Lithium bromide was found to give the best results among various halides tested[112] and it is assumed that it forms a complex with (17), decreasing the electron density on the phosphorus atom which becomes more susceptible to a nucleophilic attack of RCO_2H, according to equation (44).

DPCP can even be activated by *N,N*-disubstituted amides; *N,N*-dialkyl-substituted amides were found to be more effective than *N*-aryl-substituted ones, and, in particular, DMF was the most effective.[109] It has been suggested that DPCP reacts with DMF forming a Vilsmeier–Haack adduct (**20**);[126,130] the activated reaction of carboxylic acids with DPCP in the presence of DMF may be assumed to proceed through a similar adduct followed by hydrolysis by phenol. The reaction can be represented as shown in Scheme 12. The effect of ethers was attributed to a more favorable disposition of carboxyl groups to undergo a reaction with (**17**) in the presence of ethereal oxygen atoms, owing to oxonium complexes formed *via* hydrogen bonding.[111]

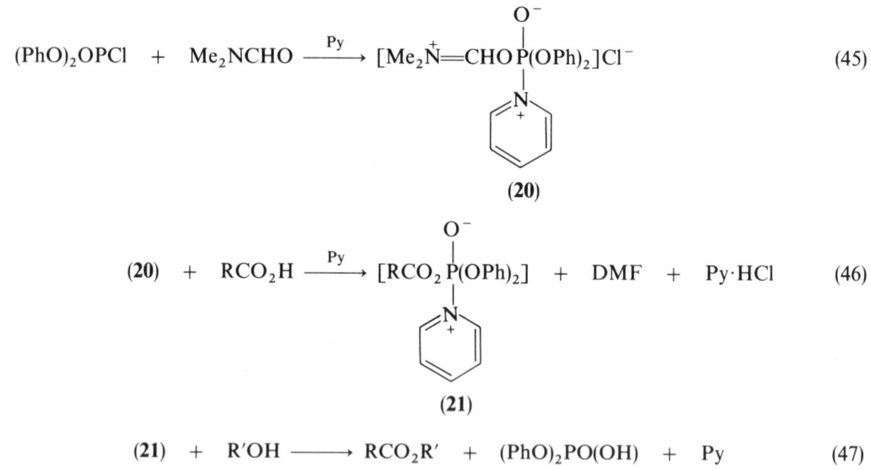

$$(PhO)_2OPCl + Me_2NCHO \xrightarrow{Py} [Me_2\overset{+}{N}=CHO\overset{O^-}{\underset{\underset{Py^+}{N}}{P}}(OPh)_2]Cl^- \quad (45)$$

(**20**)

$$(20) + RCO_2H \xrightarrow{Py} [RCO_2\overset{O^-}{\underset{\underset{Py^+}{N}}{P}}(OPh)_2] + DMF + Py\cdot HCl \quad (46)$$

(**21**)

$$(21) + R'OH \longrightarrow RCO_2R' + (PhO)_2PO(OH) + Py \quad (47)$$

Scheme 12

Alkyl ethers work better than aryl ones and in particular poly(ethylene oxide)s (PEO), among various ethers tested, gave polyesters of higher molecular weight, especially when the molecular weight of PEO was $1-2.5 \times 10^5$, suggesting a partial matrix effect.

For all coactivators, aging times of 20–30 min at room temperature of the solution of DPCP and coactivator in pyridine and its slow addition to the acid solution result in the highest molecular weights.[106,109,111,112]

Activating systems consisting of HCTPT in pyridine[106,109,111,112] or of phosphorus oxychloride in the presence of lithium chloride monohydrate[117] behave similarly to DPCP.

Picryl chloride has also been used, in the presence of tertiary nitrogen bases, as an activator for direct polyesterification.[102,103] It was suggested that the reaction proceeds *via* a trinitrophenyl ester which reacts with hydroxylated compounds to form esters and picric acid.[103] This system is less effective than those discussed above and, as a rule, leads to lower yields and molecular weights; however, as a characteristic feature, it also seems to be able to activate, in some cases, the reaction of aliphatic diols with dicarboxylic acids.

Finally, Higashi and co-workers have reported several examples of direct esterification under mild conditions by using tosyl chloride (TsCl) in pyridine, alone or in the presence of other coactivators.[107,110,113–116] The overall behavior of these systems resembles, to some extent, that of DPCP; coactivators such as DMF[110,115] and *N*-methylimidazole[113] have been found to improve both yield and molecular weight of the polyester.

As a general rule, for all the above-mentioned activating systems, many factors affect both the yield and molecular weight of the polyester, and quantitative yields and high molecular weights can only be obtained by an appropriate choice of reaction conditions. Side reactions, such as those reported in Scheme 10, equation (39) and Scheme 11, equation (43), are probably the main limiting factors.

These or other similar side reactions[127] may be responsible for the reduction of both yield and molecular weight, particularly when the order of addition of reactants is not the most appropriate.[103,105,107,113,116,117] Among the various factors affecting these types of polyesterification, some are quite general for all activated systems and can be summarized as follows. The best amount of activating agent is always that which is stoichiometrically equivalent (or in a slight excess) to the overall concentration of functional groups.

Tertiary nitrogen bases other than pyridine can be used for polymerizations; using TPPCl$_2$ as an activating agent, both triethylamine (TEA) and pyridine were found to be suitable for the polycondensation of bisphenol *A* (BPA) with an equimolar mixture of terephthalic acid (TPA) and isophthalic acid (IPA). However, while a mild reaction takes place and the MWD is rather broad ($\bar{M}_w/\bar{M}_n = 5$–6) when pyridine is used, a vigorous reaction and a polyester with the most probable MWD are obtained when TEA is used.[108] When the polymerization of 2,5-pyridinedicarboxylic acid (25PyA) with 1,10-decanediol was performed using picryl chloride as the condensing agent,[102,103] high molecular weights were obtained in the presence of pyridine whilst no polymer was obtained with TEA. Similarly, strong differences in both yield and molecular weight were observed, using 2-, 3- or 4-methylpyridine as tertiary bases,[101,103,105,113] for the polymerization of 3,5-dimethoxy-4-hydroxybenzoic acid and for the reaction of benzoic acid with p-chlorophenol,[113] with HCTPT and TsCl, respectively.

Poly(vinylpyridine)s, behaving like matrix polymers, give a strong molecular weight increase when added during the polymerization of 3-hydroxybenzoic acid (3HBA) and the polymerization of BPA with IPA.[105]

When polymerizations are performed over a range of temperature and monomer concentrations, yield and molecular weight usually pass through a maximum; according to the monomers considered, either room temperature, for the polymerization of BPA with 25PyA[105] and of 1,10-decanediol with 25PyA,[103] or a higher temperature (130 °C), for the polymerization of BPA with IPA/TPA (50/50),[108] were found to give the best results.

0.6–1.2 mol l^{-1} of monomer is, in general, the best range of concentration,[103,105,108] at least when TPPCl$_2$ is used as the activating agent. Reactions performed with DPCP are usually better carried out at 120 °C because a precipitate is formed on adding hydroxy acids to DPCP solutions in pyridine, which dissolves on heating at 120 °C while the solution becomes gradually viscous.

Finally, appropriate choice of the solvent may improve polymerization either by increasing solubility, and thus changing the nature of the polymerization from heterogeneous to homogeneous, or by affecting the reaction rate itself.

17.2.3 Polyesters from Alcoholysis

Most of the polyesters that can be made by direct esterification can also be prepared, and often advantageously, through an exchange reaction between ester and hydroxyl groups, usually called alcoholysis. This is a reaction of great commercial importance as most of the industrial processes of polyester manufacture proceed exclusively or in large part *via* such a reaction, which is schematically shown in equations (48) and (49) for AA/BB- and AB-type monomers, respectively.

$$nR''O_2CRCO_2R'' \;+\; (n + 1)HOR'OH \;\rightleftharpoons\; H(OR'O_2CRCO)_nOR'OH \;+\; 2nR''OH \tag{48}$$

$$nHORCO_2R'' \;\rightleftharpoons\; H(ORCO)_nOR'' \;+\; (n - 1)R''OH \tag{49}$$

Ester derivatives have lower melting points, higher solubility in diols and can usually be obtained at a higher purity grade than the corresponding acids and therefore often allow the production of better quality products and easier process control. However, higher costs, resulting from more expensive raw materials, higher energy consumption and more expensive plants, sometimes make direct esterification more convenient, as in the case of PET.[132]

According to the general equation (1), for addition–elimination reactions, alcoholysis is assumed to occur through a nucleophilic attack of a hydroxy compound on the carbonyl carbon atom. The reaction rate will therefore be determined by both k_1, the rate of nucleophilic attack, and $k_{i,1}/k'_1$, the partitioning of the addition intermediate, and the equilibrium constant will be given by $K = k_1 k'_1/k_{i,1} k'_{i,1}$. Reaction rate and equilibrium constants will therefore depend on the chemical structure of R' and R''. It is found that aliphatic diols can react with both alkyl and aryl esters, whilst phenolic compounds, which are poorer nucleophiles and better leaving groups than aliphatic hydroxy compounds, require R'' = aryl to yield polyesters. As found for direct esterification, compounds with tertiary hydroxyl groups are generally not suitable for polyesterification *via* alcoholysis.

When R''O is a better leaving group than R'O and R'OH is more nucleophilic than R''OH, the equilibria (48) and (49) will be shifted towards the products and it might be thought that polyesters could be obtained under mild conditions and without removal of by-products, analogously to the situation for polyamides.[133,134] Ogata and co-workers[135] have carried out trials to prepare polyesters starting from 'active diesters', *i.e.* from diesters where R'' is a good leaving group. They indeed observed an increase in the equilibrium constants from values lower than unity, usually observed for both R' and R'' aliphatic radicals,[44b,136–144] to values of 6–17 for phenoxy, thiophenyl and 3-oxypyridyl leaving groups. However, the reactions of 'active' adipates, terephthalates and isophthalates with EG and BD occur only in the presence of suitable catalysts and, when performed in a closed system, yield only low molecular weight polyesters. The authors concluded that, with these 'active' diesters, application of a vacuum to remove R"OH is also required in order to achieve high molecular weights.[135]

The removal of volatile products is, obviously, even more important when the equilibrium constants are low, as for aliphatic R' and R", and can be accomplished by the different methods described before, although evaporation under reduced pressure is generally the preferred method and R" = Me is usually chosen.

While the stoichiometric ratio of functional groups is implicit for AB monomers, different molar ratios of reactants are generally used for reaction (48), depending on the volatility of HOR'OH and R"OH. For less volatile diols, such as bisphenols or high molecular weight glycols, a stoichiometric ratio of diester and diol is generally employed, whilst, when both R"OH and HOR'OH are low molecular weight compounds easily removed from the reaction system, such as methanol and EG, a large excess of diol (e.g. $OH/CO_2R = 2.2$) is commonly employed. A lower excess of diol (20–50 mol%) is used for by-products of intermediate volatilities or to reduce the loss of glycol through side reactions, as happens in the case of BD which is partly transformed into THF during the synthesis of PBT. When an excess of diol is used, the reaction is commonly performed in two stages according to Scheme 13; in the first, which is often called the transesterification stage and is carried out at atmospheric pressure in the temperature range 150–210 °C, the low molecular weight alcohol is removed and low molecular weight hydroxyalkyl esters, mainly bis(hydroxyalkyl) esters, are formed. In the second stage, which is often called the polycondensation stage, the reaction is carried out under reduced pressure (10–30 Pa) at a sufficiently high temperature (220–280 °C) to keep the polyesters in the molten state and the excess of diol is removed to form polyesters.

$$R''O_2CRCO_2R'' + 2HOR'OH \rightleftharpoons HOR'O_2CRCO_2R'OH + 2R''OH \qquad (50)$$

$$nHOR'O_2CRCO_2R'OH \rightleftharpoons H(OR'O_2CRCO)_nOR'OH + (n-1)HOR'OH \qquad (51)$$

Scheme 13

According to equation (51), all the chains during polymerization should be hydroxy-terminated; however, due to the high temperature at which the polymerization is usually performed, chain-scission reactions and other side reactions can occur and lead to carboxyl end-groups. Control of the OH/CO_2H end-group ratio is of primary importance for good final properties, for a higher rate in a subsequent solid-state process (see Volume 5, Chapter 13) or when the polyester is intended for the preparation of polyurethanes.

In contrast to direct esterification, alcoholysis proceeds very slowly in the absence of catalysts,[74,89,145–147] even at high temperature. Strong protic acids, such as p-toluenesulfonic acid, sulfuric acid, etc., can catalyze alcoholysis, but they are not as effective as they are for direct esterification.[145,146] Furthermore, since they are also catalysts for side reactions, the resulting polyesters are generally of poorer quality and are more prone to hydrolysis than compared to those prepared using metal derivatives, which are therefore the preferred catalysts ($1–5 \times 10^{-4}$ mol of metal catalyst per mol of diester is commonly used). Due to the economic relevance of alcoholysis, a large number of compounds of almost all the elements have been tested as catalysts and an enormous number of patents has been published; however, despite the fact that many patents frequently claim effective new catalysts, most of these are of uncertain value or novelty. Comparison of the catalytic activity of various metal catalysts has been reported by several authors;[74,88,146,148,149] among these, acetates of lead(II), lead(IV), zinc, manganese, calcium, cobalt and cadmium, and oxides such as Sb_2O_3 and GeO_2, for the first and second stages of reaction, respectively, and titanium alkoxides for both stages have generally been found to be the most effective catalysts.

Their overall catalytic activity is probably the result of many different events; solubility in the reaction medium,[145,150] exchange-reaction capability of the original ligands with reactants and effects on concomitant reactions are probably the most relevant. All these features can obviously be affected by reaction conditions such as type and concentration of functional groups and reaction temperature.

The complex interrelations between these factors make investigation of the role of the catalyst in the alcoholysis reaction difficult and the catalytic mechanism is not yet fully understood. In the absence of direct evidence, more than one mechanism can be suggested to account for experimental results. Coordination of the ester group to the metal atom and subsequent reaction with one of the alkoxy ligands seems the more obvious mechanism. Two different reaction paths, shown in equations (52) and (53), have been suggested[75] as more plausible. On the basis of spectroscopic evidence for low molecular weight compounds,[149,151-153] coordination to the carbonyl oxygen atom seems to be more likely. Tomita and coworkers[149] found a correlation between the catalytic activity of various metal catalysts and the shift produced by these catalysts in the carbonyl stretching band of dibenzoylmethane. They found a 'volcano-shaped' activity order, which means that coordination with the metal has to be neither too weak nor too strong, and have suggested the measurement of this shift as a possible method for predicting catalytic activity. For this latter purpose, other authors[154] have proposed a differential microcalorimetric technique.

$$\text{(52)}$$

$$\text{(53)}$$

Experimental evidence has shown that different catalysts behave in different ways according to the composition of the medium. It has been found that carboxyl groups strongly inhibit the catalytic activity of $Ti(OBu)_4$[74,147] and of acetates of zinc, calcium and manganese.[74,144,155] On the other hand, using a hydroxyl-rich reaction medium, a decrease in the catalytic activity of Sb_2O_3 has been found.[74,144,156,157] This evidence, which shows that disputable results may be obtained from oversimplified interpretations of kinetic data, may be explained by assuming that carboxyl or hydroxyl groups have different coordination and/or exchange capabilities towards different metal catalysts. Strong coordination complexes, difficult reactant/ligand exchange reactions or association of the catalyst may result in a decrease in catalytic activity. The different behavior of various types and concentrations of acids,[88,155] effects of different ligands in the starting catalysts[14f,149] and inhibiting effects of acids[144,147,155] and water[74,89] can be interpreted on the basis of coordination and exchange reactions involving the catalyst. Additionally, effects of the catalyst on side reactions may result in an apparent inhibiting effect or in unwanted by-products. Zimmermann[158] has found that titanium, calcium, zinc and manganese catalysts increase the rate of chain scission for PET, which, in contrast, is little affected by Sb_2O_3. The first group of catalysts, on the other hand, does not affect the degradation rate of other poly(alkylene terephthalate)s[159-160] such as PBT, poly(hexamethylene terephthalate) (PHMT) and poly(decamethylene terephthalate) (PDMT).

In the light of the above discussion, it can be understood why metal catalysts are frequently classified as 'transesterification' and 'polycondensation' catalysts. Acetates of lead, zinc, calcium and manganese, whose catalytic activity is reduced in the presence of carboxyl end-groups, are some examples belonging to the first group, whereas Sb_2O_3 and GeO_2, which are active only in media with a low hydroxyl group concentration, belong to the second group. For these reasons, two different catalysts with different dependences on the nature of the reaction medium are frequently used. Titanium alkoxides are very good catalysts for both stages, but may sometimes lead to discoloration, as in the case of PET.

From the above discussion and from analogous considerations to those seen for direct esterification, it can be seen that kinetic studies are usually quite complex. Several concomitant reactions, uncertain concentration and catalytic activity of species originating from exchange reactions, inhibiting effects of functional groups and contribution of diffusion of glycol to the overall rate are some of the most relevant factors which may cause incorrect results. Appropriate choice of reaction conditions or the use of 'model' compounds can simplify kinetic studies.

Despite the complexity of the systems considered and the limited number of papers published, it is generally accepted that the kinetic equation of metal-catalyzed alcoholysis is first order in both ester and hydroxyl group concentrations according to equation (54), which becomes second order in hydroxyl groups when all the end groups of the chains are hydroxyalkyl ester groups.

$$v \;=\; k[\text{OH}][\text{Ester}] \;=\; k[\text{OH}]^2 \tag{54}$$

For low catalyst concentrations, a partial kinetic order of or close to unity has usually been found for the catalyst;[74,87,88,147] however, at higher catalyst concentrations, its kinetic order has been found to decrease and this has been ascribed to association of the catalyst.[87,88] When inhibiting effects of carboxyl[89,144,147] or hydroxyl groups[144] must be considered, or when independent contributions of uncatalyzed and carboxyl-catalyzed reactions are significant, more complex kinetic equations should be used, and the concept of kinetic order no longer applies.

A certain interest has been devoted to distinguishing between the reactivity of hydroxyl groups in half-esterified glycols and that in free glycols. Using a ratio of EG to dimethyl terephthalate (DMT) greater than or equal to two, and assuming that the reactivity of the methyl ester groups is the same in DMT and in its oligomers, some authors have calculated that oligomerization reactions do not proceed to any significant extent during transesterification, at least up to 90% conversion of methyl ester groups, because the hydroxyl groups in the free glycol are 100 times more reactive than those in the half-esterified glycols.[161,162] Other authors, however, conclude that the reactivity of hydroxyl groups in free glycols is only two times higher than in oligomers,[140] and that the first methyl ester group reacts about two times faster than the second.[163]

Challa[139] has studied the alcoholysis reaction, which involves reaction of a terminal hydroxyl group with an internal ester group. This redistribution reaction, which only changes the distribution of the chains and leaves the number of polymer molecules as well as the type of functional groups unchanged, has been found to occur at the same rate as the reaction involving two hydroxyl end-groups, but to have a higher activation energy (130 kJ mol^{-1} against 96.4 kJ mol^{-1}).

Activation energies for metal-catalyzed reactions are usually in the range of 40–110 kJ mol^{-1}; E_a values of 63–71 kJ mol^{-1} and 47 kJ mol^{-1} have been reported for zinc acetate[140,163,164] and Ti(OBu^n)$_4$,[74,147] respectively.

Equilibrium constants for aliphatic R' and R'' in equations (48) and (49) range, in general, from 0.2 to 0.8 for both polymers[136,137,140,142,144] and model compounds,[143,144] but, from the few data available, it is difficult to correlate the values with the chemical structure of reactants. Some authors found nearly the same value for different glycols,[142] whereas slightly lower values have been found when methyl ester groups are involved in the equilibria.[136,137] Almost all the authors found these reactions to be athermal or slightly exothermic.

17.2.4 Exchange Reactions between Ester and Carboxyl Groups

The exchange reaction between carboxyl and ester groups, commonly called acidolysis, is schematically represented in equation (55) for AA/BB-type monomers and, of course, is also valid for AB-type monomers, where R, R' and R'' can be aliphatic or aromatic. The limited number of applications of this reaction for the synthesis of polyesters is reflected in the relatively scarce literature, far less abundant than for direct esterification or alcoholysis.[165]

$$n\text{R''CO}_2\text{RO}_2\text{CR''} \;+\; n\text{HO}_2\text{CR'CO}_2\text{H} \;\xrightleftharpoons{K_{55}}\; \text{R''CO}_2(\text{RO}_2\text{CR'CO}_2)_n\text{H} \;+\; (n-1)\text{R''CO}_2\text{H} \tag{55}$$

As for other ester-exchange reactions, it is generally accepted that reaction (55) is 'of equilibrium'. Equilibrium constants are rarely reported in the literature; however, their values can be calculated from the equilibrium constants of hydrolysis of the esters in equation (55). In fact, it results that $K_{55} = K_h/K_{h'}$, where K_h and $K_{h'}$ are the equilibrium constants of the hydrolysis of R''CO$_2$R— and —R'CO$_2$R—, respectively. If the chemical structures of the esters are not too different, it can be assumed that K_h and $K_{h'}$ have similar values and consequently K_{55} can be reasonably approximated to unity.

In accordance with the expected low value of K_{55}, acidolysis can be successfully employed for the synthesis of polyesters, provided that R''CO$_2$H in equation (55) can be easily removed from the reacting system. The lower volatility of benzoic acid is probably one of the reasons why benzoates are less suitable than acetates.[166] For reasons connected with volatility and monomer cost, acidolysis has only been used practically when R″CO— is an acetyl group. The reaction can be carried out by heating a bisphenol diacetate with the dicarboxylic acid[167-169] at high temperature in the melt, or in solution in a high boiling solvent or even in the solid state, and acetic acid is removed under reduced pressure or distilled out with the solvent. It has been found that bulky aklyl groups, such as *t*-butyl groups, adjacent to the phenolic function can sterically hinder the polyesterification by acidolysis;[167] the same author also observed that, when diacetates cannot undergo polymerization, the corresponding bisphenols are insoluble in alkali. Analogously to AA/BB-type monomers, acetoxyarenecarboxylic acids have been converted to AB polyesters[118,170-172] or included, as comonomers, in the chains of previously prepared polyesters.[170-172] When poly(oxy-1,4-phenylenecarbonyl) has been obtained by heating *p*-acetoxybenzoic acid in the presence of magnesium turnings at 220–280 °C under reduced pressure and with a flow of argon, it has been found that etherification and decarboxylation side-reactions also occur.[118] For the preparation of copolymers, a polyester is melted and heated at high temperature with the acetoxyarenecarboxylic acid[171,172] (or with an equimolar mixture of a dicarboxylic acid and a bisphenyl diacetate); after an initial decrease in the molecular weight of the initial polyester, as a result of acidolysis and ester–ester exchange reactions, the molecular weight increases again in the subsequent polymerization step where acetic acid is removed under reduced pressure.

Acidolysis reactions can also be performed on acetoxy derivatives of aliphatic diols, but owing to the fact that (as expected) they react with more difficulty[166] than diacetates of aromatic compounds and that the corresponding diols are less expensive and can easily react with either dicarboxylic acids or their esters, acidolysis is not usually employed for aliphatic diols. It has also been found that, for the acetoxy derivatives of aliphatic hydroxyl groups, substituents may hinder polymerization by acidolysis, whilst the non-acetylated monomer can polymerize by direct esterification and alcoholysis.[173]

An aspect, often neglected, of acidolysis is its role in redistribution (of molecular weight and of comonomeric units) during melt blending at high temperature. Evidence of a sharp decrease in molecular weight when polyesters are heated with low molecular weight compounds bearing carboxyl groups[166,174] suggests that this reaction occurs to a significant extent under the conditions commonly encountered in high temperature polyesterification, and, in certain cases, may make a significant contribution to the overall rate of molecular weight increase.

Many different substances have been reported to be effective catalysts for this reaction: H_2SO_4, BF_3, magnesium and $SnOBu_2$ are some examples; however, this aspect has not been widely and systematically studied and a comparison of the catalytic activity of various compounds is therefore impossible at present.

The mechanism of acidolysis has been studied for low molecular weight compounds in a temperature range well below that commonly used for polymerization; evidence of both alkyl–oxygen and acyl–oxygen fission has been reported.[175] For polymerization by acidolysis, two mechanisms have been proposed: in the first,[167] a reaction occurring by acyl–oxygen fission and involving an intermediate anhydride, analogous to that reported for low molecular weight compounds, was proposed; in the second,[176] the first mechanism was rejected and it was instead postulated that the reaction proceeds *via* a four-membered-ring transition state. Both mechanisms are, however, based on uncertain evidence.

As mentioned above for equilibrium constants, very few data have been published about the kinetics; Korshak and co-workers[174] suggested that the reaction between ethyl stearate and acetic acid in trioxane solution at 164 °C was overall first-order. The reaction of *p*-butylbenzoic acid with bisphenol *A* polycarbonate was assumed to be second order, and an activation energy of 98 kJ mol^{-1} was calculated.[177]

The scarcity of kinetic data is perhaps one of the reasons why the contribution of this reaction is usually disregarded in polymerization when direct esterification and alcoholysis can also occur. However, the presence of dicarboxylic acids has been often observed among the volatile products,[178-180] suggesting that a contribution to the overall molecular weight increase derives from acidolysis. This contribution, which becomes particularly relevant when the hydroxyl to carboxyl end-group ratio becomes very low, had to be taken into account for a correct interpretation of experimental results of solid-state polymerization of PBT.[181]

17.2.5 Ester–Ester Exchange Reactions

Another type of ester-exchange reaction can occur between two ester groups according to equation (56). This reaction, also called ester interchange, double ester exchange or esterolysis, is perhaps even less studied than acidolysis,[24,165,182,183] and has not found any practical application in the preparation of polyesters. The main reason is probably because no advantages are expected with respect to the reactions discussed before, as against more expensive starting materials. Nevertheless, ester–ester exchange reactions may play an important part in determining the chemical structure of copolyesters prepared or processed at high temperature and in influencing the products obtained from melt blending of different polyesters. Redistribution of chain lengths and randomization of chemical units are the consequence of intra- and inter-molecular ester–ester exchange reactions. The control of these reactions may provide a new method for the preparation of copolymers with a wide variation in microstructure directly within processing equipment.[177,184-191]

$$RCO_2R'' + R'CO_2R''' \rightleftharpoons R'CO_2R'' + RCO_2R''' \tag{56}$$

Difficulties in separating the contribution of ester–ester exchange from those of other ester-exchange reactions make its study problematic and disputable results can be obtained unless end-capped chains are used and scission reactions avoided.[33,165,182] The failure to use end-capped reactants may invalidate or make ambiguous results attributed to ester–ester exchange, as the same reaction products can also be formed from consecutive alcoholysis or acidolysis reactions.

Ester–ester exchange has been reported for the reaction of poly(hexamethylene sebacate) with deuterated diethyl succinate,[183,192] for the reaction of end-capped oligoesters with PET,[193] and has been suggested as the most plausible reaction path for the formation of cyclic oligomers.[194] The same reaction has been found to be the main source of randomization during melt mixing of PBT/PC[177,184-186] and PET/PC[189-191] when titanium alkoxides are added as the catalyst. Ester–ester exchange may play an important role in analogous systems (PBT/copolyarylate[188] and PET/copolyarylate[187]) and is responsible for the change in deuterated block length observed by small-angle neutron scattering (SANS) after annealing at 250–280 °C of deuterated and non-deuterated PET.[195,196] Evidence for the occurrence of this reaction has also been observed by mass spectrometry on various poly(alkylene adipate)s.[197-199]

In the absence of suitable catalysts, the reaction is very slow and H_2SO_4 and alkali are used to catalyze the reaction;[183] however, titanium alkoxides are by far the most powerful catalysts.[177,184-191] Sb_2O_3, which is a good catalyst for the polycondensation of PET, shows only a limited catalytic activity towards ester–ester exchange reactions.[189-191] Arsenic[200] and phosphorus compounds[186,190,191] are able to reduce dramatically the catalytic activity of titanium alkoxides, providing a method for avoiding redistribution.

Few data have been reported for the kinetics and equilibrium of this reaction. A second order (first for both ester groups) is generally assumed[190,197,198] and activation energies of 130–150 kJ mol^{-1} have been reported.[190,196,198] A mechanism involving an association complex has been postulated.[197] No data are available for the equilibrium constant; however, it can be expressed as the product of the equilibrium constants of the alcoholysis reaction of the two esters in the left-hand side of equation (56). For similar chemical structures of the esters, the equilibrium constant is expected to have a value close to unity.[175]

17.2.6 Polyesters from Acyl Chlorides

Reactions of monomers bearing acyl chloride groups with compounds having alcoholic or phenolic groups provide a useful method for polyester synthesis according to equations (57) and (58). These reactions can be performed either at high or low temperature, though the low temperature method is by far the most widely used.[18-24]

$$nClOCRCOCl + nHOR'OH \rightarrow {\it (}OCRCO_2R'O{\it)}_n + 2nHCl \tag{57}$$

$$nHORCOCl \rightarrow {\it (}ORCO{\it)}_n + nHCl \tag{58}$$

Under mild conditions, due to the high value of the equilibrium constant, these reactions do not suffer from the reversibility found for the other reactions so far discussed, and are therefore generally called 'non-equilibrium' polyesterifications. Actually, equilibrium constants of 4.3×10^3 and 4.7×10^3 have been reported by Vinogradova[30] for reactions yielding arylates and polyarylates, respectively.

Temperatures ranging from 100 to >300 °C can be used,[201] depending on the reactivities of the components and the physical properties of the polymer. Under these conditions, the reaction also proceeds rapidly without catalysts, and can be carried out either in bulk, (when the reactants and the polymer are low melting and miscible), or in solution, (employing a high boiling solvent and a flow of an inert gas to remove hydrogen chloride) or even in the solid state (for high melting polymers).[202,203] In these cases, the progress of the reaction can be followed by titrating the hydrogen chloride removed.

At high temperature, however, side reactions can seriously limit the attainment of high molecular weights and lower the quality of the final product. Solvent, carrier gas and reactants must be carefully dried in order to avoid traces of moisture that can lead to hydrolysis of chloroformyl groups with the formation of less reactive carboxyl groups.

For all these reasons, the reaction is preferably carried out at low temperature by interfacial polycondensation or in solution. Differences between low and high temperature polycondensation[20] and between low temperature interfacial and solution methods[20,23] have been discussed in the literature as well as in Volume 5, Chapters 10 and 11 of this work.

Interfacial polymerization is usually performed in water-immiscible organic solvent systems,[20,204–206] but has also been carried out using two immiscible organic solvents.[208,209] In aqueous systems, the diacyl chlorides are dissolved in an organic solvent immiscible with water, and vigorously stirred with an aqueous solution of alkaline bisphenolates. This method is particularly suitable for the preparation of polyesters from bisphenols, but inappropriate for glycols which have higher pK_a values and therefore less tendency to form glycolate ions. The method is widely applicable to a large number of bisphenols and dibasic acyl chlorides, although aromatic acyl dichlorides give generally higher molecular weights than aliphatic acyl dichlorides.[204] Substituents, which reduce the basicity of the phenoxide ions[204,210] or sterically hinder the reaction of the phenol group,[95] may limit the attainment of high molecular weight.

It has been proposed that the reaction occurs by nucleophilic attack of a phenoxide ion on the acyl chloride according to an S_N2 substitution mechanism;[211] however, diffusion of reactants in the reaction locus may play an important part in the overall mechanism of interfacial polycondensation which is still rather uncertain. In contrast with the interfacial synthesis of polyamides, there is evidence that the reaction occurs near the interface in the aqueous phase;[23] the actual location depending on reaction conditions.

Experiments performed to ascertain whether the process is kinetic- or diffusion-controlled led Sokolov[212] to conclude that the overall rate is controlled by the chemical reaction rather than by the diffusion of monomer to the locus of reaction.

Several factors may affect the overall rate: type of solvent, rate of stirring, temperature, the addition of detergents, quaternary ammonium salts and accelerators are some of the most important; a detailed discussion of each factor can be found in the literature.[20,23] In particular, quaternary ammonium salts or tertiary alkylamines are assumed to increase the polymerization rate, probably by making the transfer of phenoxide ions into the organic phase easier. Thus the reaction with acyl chloride is promoted and, after the reaction has taken place, the salt is then extracted by the aqueous phase and the cation is free to start the cycle again. When interfacial polycondensations take place in a water-immiscible organic solvent system, hydrolysis of the acyl halide is possible as a side reaction and may prevent the attainment of high molecular weights. Therefore, interfacial polycondensation in non-aqueous systems has been proposed for monomers of low reactivity towards acyl chlorides and with low solubility in water.[208,209] It has been found that this latter method can be more convenient for preparing aromatic polyesters in terms of yield and molecular weight. Among various combinations of immiscible systems, the adiponitrile/CCl_4 system has been found to be superior to other solvent systems. When a mixture of bisphenols is used, the sequence distribution of the copolyester can be controlled from random to block types by appropriate selection of the binary solvents.[208,209]

In solution, at room temperature, bisphenols react very slowly with acyl chlorides and the reaction is usually performed in the presence of an excess of a basic substance (tertiary amines, pyridine, *etc.*) which activates the reaction in several possible ways[20,23,30] and combines with the liberated hydrogen chloride. The mechanism of the reaction in solution is probably a combination of a complex set of reactions and is not completely understood. An initial overall reaction order of two has been reported;[30,213] however, at higher temperatures (120–150 °C), an overall order of 1.5 (1.0 and 0.5 with respect to bisphenol and acyl chloride, respectively) was found for the reaction of terephthaloyl chloride with 9,9-bis(4-hydroxyphenyl)fluorene.[30] The reaction rate as well as the molecular weight of the polymer may depend on many factors. The nature of the solvent, the base used as acid acceptor and activator and the temperature are the most important factors for a given reaction.

The dielectric constant of the solvent seems to have a dramatic effect on the rate of reaction; it has been reported[214] that the rate increased by more than two orders of magnitude on passing from benzene to dichloroethane. On the other hand, it has been found that the reaction rate decreases when the solvent can form hydrogen bonds with the hydroxy compound used as the monomer.[214] In addition, solvents of low dielectric constant may sometimes favor the formation of high molecular weight polymers if they display a high swelling power with respect to the polymer.[215] Often a mixture of solvents with low and high swelling powers and dielectric constants may result in the optimal solution for a given polymerization, as the increased swelling power balances the decreased polarity of the medium.

As the molecular weight and the reaction rate are probably the result of complex equilibria and reactions, their profile can be quite complex on changing the temperature.[214] In the temperature range -30 to $+80\,°C$, similar complex profiles were obtained for both polyarylates and their corresponding low molecular weight model;[214] this result seems to exclude an effect of the molecular weight.

It has also been found that the type of base added to the reacting system can have an important influence on the reaction rate, even reversing the rank of reaction rates for glycols and phenols. At $40\,°C$ in dichloroethane, bisphenols react with terephthaloyl chloride, in the presence of triethylamine, faster than 1,6-hexanediol, whilst, when the weaker base pyridine is used, the order of reaction rate is reversed.[30] To account for this observation it has been assumed that in the presence of triethylamine and pyridine different catalytic mechanisms are prevailing.

Non-equilibrium polyesterifications are particularly suitable for the preparation of copolyesters; in fact, as the reaction is carried out at low temperature, redistribution reactions (occurring through exchange reactions) are limited or excluded and a copolymer with a non-random chemical structure can be obtained. In this respect, for difunctional monomers, the difference in reactivity of a functional group, before and after the other is reacted, becomes the determining factor. It was found that, after the first functional group has reacted, the reactivity of the second can increase, remain constant or decrease, sometimes quite considerably.[30,216] The effect of substituents on this aspect, as well as on the overall reaction rate, may be the determining factor. It has been reported, for example, that when terephthaloyl chloride reacts with *para*-substituted phenols, in dioxane at $30\,°C$ in the presence of triethylamine, the ratio between the first and second rate constant varies with the nature of the substituents on the phenol. Electron-withdrawing substituents decrease this ratio while the opposite effect is observed for electron-donors.[30] In the case of BPA and 4,4'-oxydiphenol reacting with substituted benzoyl chlorides, the reactivity of the second hydroxyl group is practically the same as the first, regardless of the nature of the substituents.[30] By a proper choice of reaction conditions (solvents, activating base, temperature and order of addition of reactants), microstructures ranging from random to alternating or block can be obtained. For reactions occurring in homogeneous systems, it is possible to summarize many experimental observations into some rules.[30,217,218]

When a monomer xAx can react with two comonomers yBy and zCz, the microstructure of the resulting copolymer depends on both the reactivity of the functional groups and on the order of addition of the comonomers. If the reactivity of the second functional group x in xAx does not change after the first has reacted, its addition at one time at the beginning of the copolymerization results in a random copolymer, regardless of differences in the reactivity of yBy and zCz. Under these conditions, the addition of the two comonomers yBy and zCz, even in subsequent stages, results in random copolymers. In contrast, when the reactivity of the second functional group in A changes, the microstructure may be shifted towards alternating or block copolymers. If the reactivity of the second group decreases, the fraction of alternate sequences increases, while the average length of the blocks increases when the reactivity of the second group increases. Block copolymers can also be prepared when the comonomers yBy and zCz differ significantly in reactivity and the comonomer xAx is supplied in the reaction zone at a rate lower than the reaction rate with the more reactive monomer.

Temperature[213,219,220] and reaction medium[221] can have a powerful effect on the reactivity and can be used to control the composition of copolymers.[220] Obviously, an excess of the monomer bearing chloroformyl groups may be a general method for the preparation of polyesters with terminal chloroformyl groups, suitable for the synthesis of block copolymers.

17.2.7 Polyesters from Anhydrides

Cyclic anhydrides can be useful acylating agents for the synthesis of polyesters; diols and diepoxides, or similar compounds of higher functionality, are mostly used as comonomers.

The reaction with diols is shown in Scheme 14. It proceeds in two steps because the free acid, formed in the first step, is much less reactive than the original anhydride. When the reaction is carried out without a catalyst, it usually lasts many hours even at the high temperature (180–240 °C) at which it is usually performed;[223] catalysts used for direct esterification are also generally suitable for this reaction. The method is of little importance for the preparation of linear polyesters and its application is practically limited to the use of phthalic or maleic anhydride in the synthesis of random copolyesters and of alkyd and glycerophthalic resins. The latter are commonly prepared from phthalic anhydride and monofunctional fatty acids reacted with mixtures of aliphatic di- and poly-ols. Maleic anhydride is reacted with aliphatic diols to form unsaturated polyesters which are subsequently transformed into crosslinked polymers through a free-radical polymerization with styrene.

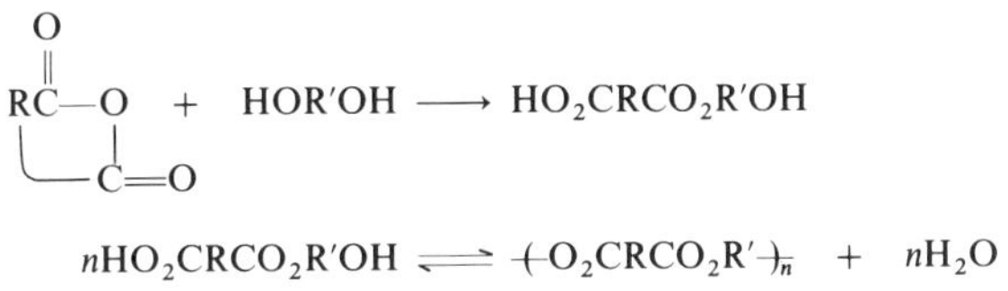

$$RC(=O)-O-C(=O)R + HOR'OH \longrightarrow HO_2CRCO_2R'OH \tag{59}$$

$$nHO_2CRCO_2R'OH \rightleftharpoons (O_2CRCO_2R')_n + nH_2O \tag{60}$$

Scheme 14

High molecular weight linear polyesters have been obtained from the reaction of 1,6-hexanediol and succinic anhydride.[224,225] Polyesters containing free carboxyl groups have been obtained by reacting pyromellitic anhydride ($1H,3H$-benzo[1,2-c: 4,5-c']difuran-1,3,5,7-tetrone) and ethylene glycol, first in acetone and, after evaporation of the solvent, for 4 h at 100 °C under reduced pressure.[226] The resulting polyester was soluble in aqueous sodium hydroxide.

Recently, a new method for the preparation of low molecular weight oligoesters, having hydroxyl end-groups at both ends and a polydispersity index $\bar{M}_w/\bar{M}_n < 1.1$, has been reported.[29] These oligomers, which may be interesting macromers for the preparation of model block copolymers, were prepared in high yield, under mild conditions, by a two-stage reaction from phthalic anhydride and aliphatic diols. In the first stage, phthalic anhydride is reacted with a symmetrical diol (molar ratio 2:1) in the presence of 4-(dimethylamino)pyridine (4DMAP) and triethylamine. In the second stage, the carboxyl-ended diester obtained in the first stage is reacted with an unsymmetrical diol and dicyclohexyldicarbodiimide (molar ratios 1:2:2.04) in the presence of a catalytic amount of p-toluenesulfonic acid. The hindered secondary hydroxyl groups in the unsymmetrical diol allow chain extension to be avoided and practically monodisperse oligoesters to be obtained, while, by the conventional method at high temperature and the same molar ratios of reactants, values of $\bar{M}_w/\bar{M}_n$ ranging from 1.3 to 1.7 would be obtained.

Another reaction which leads to polyesters is that between anhydrides and diepoxides in the presence of suitable catalysts.[227,228] However, as this reaction is not strictly considered as a method of synthesis of polyesters, the reader should refer to the chapter concerning epoxides for its discussion (see Volume 5, Chapter 37).

17.2.8 Polyesters from Cyclic Esters

Since the pioneering work of Carothers,[6] lactones, cyclic diesters and their alkyl- and aryl-substituted derivatives have been known to be suitable starting materials for the preparation of linear aliphatic polyesters.

Some problems associated with the polymerization of linear monomers at high temperature, such as by-product removal, exact stoichiometry and side reactions, may be avoided in the polymerization of cyclic esters.

Their polymerization can be initiated by various types of substances and, according to the initiator employed, it may proceed *via* cationic, anionic, coordination–insertion and 'alcoholysis' mechanisms. Many chapters in this work deal with various aspects of ring-opening polymerization; in particular, anionic and cationic mechanisms are covered in Chapters 34 and 49 of Volume 3, respectively; this section, therefore, will be limited to polymerizations proceeding by an alcoholysis mechanism which resembles a chain rather than a step mechanism and which usually leads to low molecular weight, low melting polyesters[229–231] and is therefore far less studied.

The tendency of these substances to undergo polymerization has been widely reported in many books and reviews[18–26,232–234] and is thoroughly discussed in Chapters 34 and 49 of Volume 3; therefore, we shall briefly summarize a few generalizations here.[26,232]

Polymerization of lactones and cyclic esters generally depends, to a considerable degree, on their chemical structure; that is, in particular, on their ring size and on the type and position of the substituents. High ring-strain, deriving from both angle distortion (three- and four-membered rings) or hydrogen-atom crowding within the ring (rings with more than seven or eight atoms), favors polymerization, while substituents diminish the polymerizability of these monomers by increasing the ring stability with respect to the open chain. Consequently, most of the four-, seven- and eight-membered ring cyclic esters and carbonates are polymerizable, although some substituted ones can resist polymerization.[232] Accordingly, γ-butyrolactone, a cyclic ester with a five-membered ring, does not polymerize under the usual reaction conditions,[6,232] even though it has been reported that it polymerizes in 20% yield to a low molecular weight polyester at 160 °C under a pressure of 2000 MPa.[235] Lactones with a greater number of atoms in the ring can polymerize readily (δ-valerolactone and ε-caprolactone),[6] or with more difficulty (3-n-propyl-δ-valerolactone **22** and 6,6-dimethyl-δ-valerolactone **23**[24–26,232]) or do not polymerize at all (pentadecanolide[6,18]). A similar behavior was found for cyclic diesters (polymerizability of diglycolide **24** > dilactide **25** ≫ tetraphenyldiglycolide **26** and tetramethyldiglycolide **27**).[18,24,26,232]

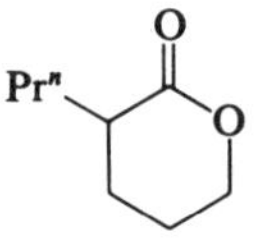

(22) 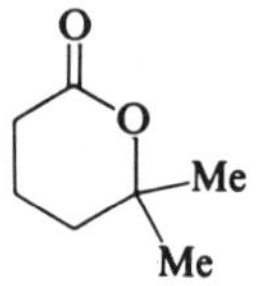   (23) 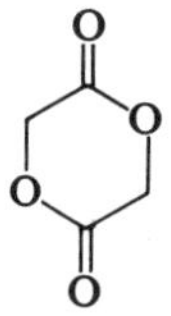   (24) 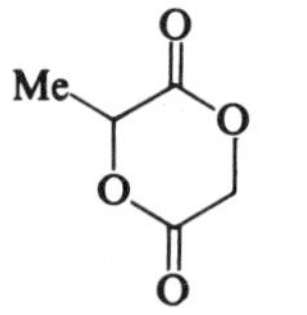   (25) 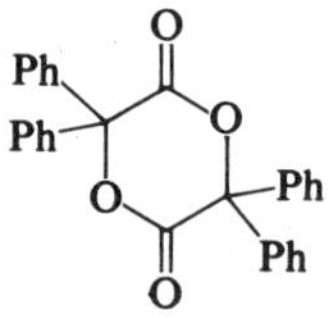   (26) 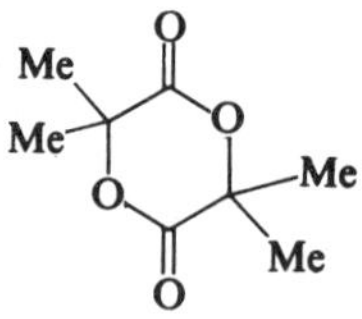   (27)

The size of the ring also has a very important influence on the extent to which the cyclic monomers can be converted into polyesters.[6] For example, a substantial amount of δ-valerolactone was reported to exist in equilibrium with the polymer at temperatures exceeding 150 °C.[27]

Favored conformations of the ester group in the ring[237] and skeletal ring-strain[234] have been considered as the main factors for the ease of polymerization in lactones with different ring-sizes, but no clear correlation has been found.

Equilibria occurring in polymerization proceeding *via* alcoholysis are shown in Scheme 15. Active hydrogen donors, such as water, alcohols, amines and similar substances, can be conveniently used to start the hydrolytic polymerization; a proper choice of initiator may provide a useful method of controlling the nature of the end groups. Alcohols, amines and similar monofunctional initiators lead to macromolecules with a functional hydroxyl group at one end only of the chain; water, aliphatic diols and other similar difunctional initiators give macromolecules with two functional groups and the chains grow at both ends. Initiators with more than two active functional groups may be usefully employed for the preparation of branched polyesters.[238] In principle, the molecular weight of the resulting polyesters may be controlled by the ratio of lactone concentration to that of initiator.

$$XH + OC\text{---}R \rightleftharpoons XCOROH \qquad (61)$$

$$XCOROH + nOC\text{---}R \rightleftharpoons XCO(RCO_2)_nROH \qquad (62)$$

Scheme 15

It has been reported that uncatalyzed polymerizations initiated with hydrogen donors occur at a relatively slow rate and give only low molecular weight polyesters;[26,230] for example, the polymerization of ε-caprolactone at 190 °C initiated with ethylene glycol requires 35 h for quantitative conversion of the monomer and produces a molecular weight of only approximately 5000.[239]

The polymerization is probably started by nucleophilic attack of the initiator on the carbonyl group of the monomer and proceeds by subsequent nucleophilic attacks of the resulting hydroxyl end-groups.[236]

Metal ion salts or titanium and tin alkoxides are effective catalysts for this reaction[230,240–242] and lead to an increase in both reaction rate and molecular weight.[230,231] These catalysts, however, usually lead to broader MWD values, because redistribution by exchange reactions occurs simultaneously with the stepwise chain-growth.[231]

When water is used as the initiator, direct esterification may occur during polymerization at high

17.2.9 Polyesters from Biosynthesis

It is now well established that poly(β-hydroxybutyrate) (PHB) (**28**) can be synthesized by many bacteria.[243-245] It exists, as a carbon reserve, in cytoplastic fluid in the form of crystalline granules about 0.5 μm in diameter,[244,246-248] from which PHB can be isolated, as native granules[246,247,249,250] or by solvent extraction,[247] in an optically active form usually having a molecular weight of 10^6 or more[245,250-252] and properties which resemble those of isotactic polypropylene.[250-255] In addition, it is noteworthy that PHB is completely biodegradable to innocuous compounds (D($-$)3-hydroxybutyric acid) and that it is a potential source of crotonic acid by pyrolysis.[255]

$$\text{\textendash}(\text{OCHMeCH}_2\overset{\text{O}}{\overset{\|}{\text{C}}})_n\text{\textendash}$$

(**28**)

PHB is just one of a family of bacterial poly(β-hydroxyalkanoate)s[256,257] which can be derived using different feedstocks. Acetic acid, derived from these feedstocks, is the building material for PHB monomeric units, whereas propionic acid results in poly(β-hydroxyvalerate) (PHV) units and, similarly, higher homologs of the acids in different poly(β-hydroxyalkanoate)s.[256]

Random copolymers can also be obtained *via* bacterial synthesis provided that a suitable feedstock is added to the growing culture. For example, poly(HB-*co*-HV) (**29**) has been prepared from a nitrogen-free medium containing both sodium acetate and propionate as carbon sources.[249,258,259] Because of their easy biodegradation and high thermal stability, these materials are particularly interesting for applications where environmental effects are of primary importance and are now available in a range of compositions from the Agricultural Division of ICI.

$$\text{\textendash}(\text{OCHMeCH}_2\text{CO})_n\text{\textendash}(\text{OCHEtCH}_2\text{CO})_m\text{\textendash}$$
$$\quad\quad\quad\text{HB}\quad\quad\quad\quad\quad\quad\text{HV}$$

(**29**)

17.2.10 Other Reactions for the Preparation of Polyesters

In this section, some other examples of reactions that have been successfully applied to polyester synthesis are described. Some of these are merely modifications of the reactions previously discussed, involving different products and reactions only for the initial stage of polymerization, after which the polymerization proceeds by one of the previous reactions. High costs of starting monomers, difficult control of the reaction and low molecular weight often characterize these reactions and make them of little practical importance. A number of routes have pure academic interest, allowing the production of only a limited range of polyester structures with low molecular weights. In some cases, the real mechanism has not been investigated and it is doubtful which kind of mechanism, chain or stepwise, is followed.

For all these reasons, these reactions have generally found few applications and, therefore, this section is limited to a brief listing of some examples; some more examples are reported in the literature.[24]

Diammonium and dialkylammonium salts of terephthalic acid can react with ethylene glycol at high temperature to give bis(2-hydroxyethyl) terephthalate which further polymerizes to PET by alcoholysis.[260]

Dialkyl halides have been reacted with dicarboxylic acid salts (silver or alkali metals are the more suitable metal ions) to give polyesters according to equation (63).[261,262]

$$n\text{MO}_2\text{CRCO}_2\text{M} + n\text{XR}'\text{X} \longrightarrow \text{\textendash}(\text{O}_2\text{CRCO}_2\text{R}')_n\text{\textendash} + 2n\text{MX} \quad\quad (63)$$
$$\text{X} = \text{Cl, Br, I} \quad \text{M} = \text{Ag, Na, K}$$

The methods discussed before often fail for the preparation of poly(α-ester)s and specific alternative methods have been developed.[27] A more detailed discussion of these methods is reported in Section 17.3.2.

Oxychloroformyl groups (—O—CO—Cl) can react with carboxyl groups[263,264] either by interfacial methods in the presence of sodium hydroxide or in solution in the presence of an excess of pyridine, according to equation (64).

$$nHO_2CRCO_2H\ +\ nClOCOR'OCOCl\ \longrightarrow\ {+\!O_2CRCO_2R'\!+}_n\ +\ 2nCO_2\ +\ 2nHCl \qquad (64)$$

Poly(tetramethylene adipate) and poly(2,2-bis(4-hydroxyphenyl)propane adipate) were prepared from tetramethylenebis(chloroformate) and adipic acid in the presence of sodium hydroxide and of catalytic amounts of amine[265] and from phosgene and equivalent amounts of BPA and adipic acid in pyridine.[263] However, it has been reported[266-269] that this reaction does not proceed quantitatively to ester groups, but anhydride and carbonate groups may also be formed and, consequently, contained in the final polymer.[270]

Polyesters can also be obtained by polymerization of ketenes. These can react with oxygen,[271] with aldehydes, ketones or carboxyl esters[272-274] or by themselves[275-278] and bisketenes with diols.[279] It is noteworthy that a number of highly substituted poly(α-ester)s and poly(β-ester)s, otherwise inaccessible or scarcely obtained, are made available by this route.

Cyclic carbonates or sulfites can give polyesters by reaction with aliphatic or aromatic dicarboxylic acids.[280,281]

A random copolyester consisting of 50/50 of units (**30**) and (**31**) was obtained in 90% yield from terephthalaldehyde by refluxing it in a cyclohexane solution for 16 h with triethylaluminum as the catalyst[282] (a Cannizzaro–Tischtschenko reaction).

$$-(\!OCH_2-\!\!\bigcirc\!\!-CO\!)_x\qquad -(\!OCH_2-\!\!\bigcirc\!\!-CH_2O_2C-\!\!\bigcirc\!\!-CO\!)_y$$

X   Y

(**30**)   (**31**)

Finally, polymers containing ester groups in the backbone chain can be obtained in a number of cases through reactions of difunctional compounds that already contain the ester group. For example, alternating copoly(ester–amide)s and copoly(ester–carbonate)s can be prepared by reacting ester-containing monomers, having chloroformyl groups at both ends, with diamines[283,284] or ester-containing bisphenols with phosgene.[285] As a further example,[286] addition of dithiols to ester-containing divinyl monomers leads to poly(ester–thioether)s.

Coupling reactions of previously prepared functionalized polymers is an obvious way for preparing a practically unlimited number of block copolymers.

17.2.11 Degradation Reactions

With the term 'degradation reactions', we consider here only those reactions which occur during the synthesis of polyesters leading to a decay in properties. In some cases, these reactions counterbalance the molecular weight increase either by chain scission or as a consequence of the transformation of reactive terminal groups into less reactive or unreactive ones; in other cases, they give unwanted low molecular weight by-products or spurious moieties in the chains.

Hydrolysis, glycolysis, ammonolysis, hydrazinolysis and similar reactions, which are well known and widely used to destroy polyester chains for recovering monomers or for analytical purposes, will be not considered here because their discussion is beyond the scope of this chapter.

Some degradation reactions, characteristic of a particular polyester or of a class of polyesters, will be treated in subsequent sections when that particular polymer or class of polyesters is considered. Accordingly, the formation of by-products such as acetaldehyde and insertion of diethylene glycol moieties into PET chains, which are some examples of degradation reactions occurring during the production of PET, will be discussed in the section concerning the preparation of PET. Analogously, the formation of THF in the polymerization of PBT and some other examples are mentioned in other sections.

On the other hand, a few reactions are quite general for polyesters and will be discussed here. They usually occur at high temperature and become increasingly more important as the temperature rises because of their high activation energy.

The rate of thermal chain-scission and the mechanisms by which it occurs depends on the chemical structure of the polyester.[158,160,287-289] In the case of poly(alkylene terephthalate)s, it is quite generally accepted that the main chain-scission reaction occurs randomly by abstraction of a β hydrogen from the glycol moiety *via* a cyclic transition state, according to equation (65), which leads to the formation of a carboxyl end-group and a vinyl end-group. A less important reaction involving thermal homolytic cleavage of the ester bond was deduced using mass spectroscopy.[288]

$$\sim\sim RC\overset{O\cdots H}{\underset{O-CH_2}{\diagdown CHR'}}\sim\sim \longrightarrow \sim\sim RCO_2H + CH_2=CHR'\sim\sim \qquad (65)$$

Reaction (65) occurs, at a significant rate, only above 220–250 °C. Experimental evidence for it has been obtained either directly on the polymer, from intrinsic viscosity measurements, carboxyl group titration and mass spectrometry, or on model compounds by separation and analysis of the degradation products. In all cases, a first-order kinetic equation was observed and very similar kinetic data were obtained for polymers and their corresponding models.[158,159,178,179,290,291] From the mechanism proposed, it is expected, and has indeed been found, that polyesters without β-hydroxy atoms in the β position on the diol moiety are more thermally stable.[287,292]

The kinetics of degradation of poly(alkylene terephthalate)s, and of PET in particular, are by far the most studied for both polymers themselves[158,160,178,179,288,290-300] and their model compounds.[291,294,298-303]

The catalysts used for polymerization usually remain in the polyester, and it has been found that they may sometimes affect the thermal degradation. For PET, it has been observed that the rate of scission mainly depends on the catalyst used for the transesterification,[158,294,295,299,303] but it has also been noted that the catalyst does not affect the thermal degradation of PBT and poly(hexamethylene terephthalate) (PHMT).[158-160,299]

The kinetic constant of thermal degradation of PET is more than one order of magnitude lower than those for poly(alkylene terephthalate)s having higher numbers of methylenic groups.

According to the nature of R' in equation (65), the reaction can further proceed through another β-hydrogen abstraction or *via* a different path. In the case of PBT and PHMT, a further β-hydrogen abstraction yields 1,3-butadiene[158-160,290,291,297] and 1,5-hexadiene,[158,160] respectively, as the main by-products, whilst the reaction paths followed in the case of PET are more complex; the vinyl ester end-groups produced in the first scission can further react with substances and functional groups present in the reaction system, leading mainly to acetaldehyde and polyene sequences.[158,178,179,304]

Polyesters from aliphatic dicarboxylic acids and diols can undergo two different mechanisms simultaneously; the most favored is the cleavage of the acyl–oxygen bond of the ester group following the abstraction of a hydrogen atom from the α position of the methylenic moiety of the acid; this reaction leads to ketene and hydroxyl end-groups[288,305] according to equation (66). A mechanism of β-hydrogen abstraction, like that of equation (65), seems to occur to a lesser extent[288,305] for these polyesters.

$$\sim\sim CH_2\overset{O}{\underset{H}{\overset{\|}{C}}}\!\!\diagup(OCH_2CH_2\sim\sim \longrightarrow \sim\sim CH_2CH=C=O + HOCH_2CH_2\sim\sim \qquad (66)$$

Thermal degradation of polyesters from hydroxycarboxylic acids or lactones depends strongly on the ring size and on the nature of the substituents, cyclic oligomers often being the major products of chain scission.[288]

Decarboxylation is another side reaction that may occur, according to equation (67), during the synthesis of polyesters. The rate of splitting out of CO_2, at a given temperature, depends strongly on the chemical nature of R, aliphatic acids usually being more prone to decarboxylation than aromatic acids. The disappearance of carboxyl groups with the formation of unreactive alkyl or aryl end-groups corresponds to a decrease in the overall functionality of the system with a consequent effect on the maximum molecular weight attainable.

$$\mathit{\sim\!\!\sim} RCO_2H \longrightarrow \mathit{\sim\!\!\sim} RH + CO_2 \qquad (67)$$

Decarboxylation reactions may be particularly important for the preparation of polyesters from aliphatic dicarboxylic acids and glycols. Korshak and co-workers have studied the effect of chemical structure on decarboxylation[306] and found, for pure aliphatic dicarboxylic acids, that the reaction follows first-order kinetics and is usually characterized by a high activation energy (for example, for adipic acid, $k_{280\,°C} = 1.7 \times 10^{-6}\,s^{-1}$ and $E_a = 239\,kJ\,mol^{-1}$). The temperature at which this reaction becomes significant decreases on decreasing the number of carbon atoms in the acid and is particularly low for oxalic and malonic acids. It was found, in addition, that the rate of decarboxylation of these acids further increases in the presence of glycols and salts, whilst it decreases for the corresponding esters.[306] For this reason, poly(alkylene oxalate) and poly(alkylene malonate) can barely be obtained by direct esterification at high temperature.

17.3 CLASSES OF POLYESTERS

In Section 17.2, the main general methods for preparing polyesters have been discussed, but a comparison of the convenience of the various possible reactions and some specific reactions of certain polyesters or classes of polyesters have not been considered. More information on the polymerization processes of the most popular polyesters and the more convenient reactions for a certain polyester or class of polyesters will be reported below in separate sections.

The reactivity of monomers and the properties of the resulting polyesters are both strongly dependent on the aliphatic or aromatic nature of the bivalent radicals bearing the functional groups: the introduction of aromatic units in the chains has, in fact, a profound influence on virtually all important properties of polyesters; thus, the replacement of terephthalic acid in linear polyesters for a C_8 aliphatic diacid leads to an increase of T_g and T_m of more than 100 degrees and to a dramatic improvement in solvent resistance. The more useful methods of synthesis are therefore more conveniently discussed by classifying polyesters on the basis of the aliphatic or aromatic nature of the hydroxy and carboxy derivatives used as starting monomers. Accordingly, we have divided this section into four subsections: (i) polyesters from aliphatic dicarboxylic acids and diols; (ii) polyesters from aliphatic hydroxy acids; (iii) polyesters from aliphatic diols and aromatic dicarboxylic acids; and (iv) polyesters from aliphatic dicarboxylic acids and aromatic diols. Polyarylates, which represent an important class of polyesters, are the subject of Volume 5, Chapter 18 and will not be considered here. The number of publications which have appeared in the literature for each class of polyesters is so large that only a very cursory treatment, limited to a few examples, is possible in the space available.

17.3.1 Polyesters from Aliphatic Diols and Diacids

Except for oxalic and malonic acids, which decarboxylate to an appreciable extent even at temperatures as low as 170 °C,[306,307] aliphatic dicarboxylic acids are suitable starting monomers for direct esterification with aliphatic diols. Because of their low melting temperature, polyesters from aliphatic diols and dicarboxylic acids are most conveniently prepared by direct esterification in the melt at high temperature (180–230 °C). In order to avoid decarboxylation, polyoxalates and polymalonates are more conveniently prepared from the corresponding diesters rather than from the diacids. Alcoholysis of diesters in the presence of suitable catalysts can also be used to prepare aliphatic polyesters from dicarboxylic acids with a higher number of carbon atoms, but the higher cost of diesters makes this method of synthesis less suitable unless extremely low values of carboxyl end-groups in the resulting polyester are desired.

Anhydrides can replace the parent acids in the reaction with glycol; both succinic and maleic anhydrides are readily available and are used alone or in combination with other comonomers to prepare saturated and unsaturated polyesters, respectively.

Aliphatic polyesters can also be prepared from diols and diacyl chlorides in solution, preferably at high temperature, whilst the interfacial procedure is not used because glycols do not usually give an effective concentration of alkoxide ions in the presence of water.[308]

This class of polyesters is characterized by low melting points and low hydrolytic stability and, consequently, these polymers have only limited industrial applications, mainly as copolymers.

Many aliphatic polyesters have been prepared from C_2–C_{19} aliphatic dicarboxylic acids and various diols, and have been characterized for thermal transitions[306,307] and crystalline structure.[309–311]

The chemical structure of both diol and diacid moieties obviously influences the properties of the resulting polyesters, and odd–even effects lead to a zigzag alternation of melting points on changing the length of diol and diacid.

More rigid glycols, such as 1,4-cyclohexanedimethanol, may be used to increase the melting point of polyesters. Neopentyl glycol, which does not have hydrogen atoms in the β position with respect to the ester groups, leads to polyesters with higher thermal and hydrolytic stabilities[292] and many formulations use it to replace all or part of other glycols. For example, hydroxy-terminated polyesters of adipic acid with neopentyl glycol are suitable for preparing polyurethane foams which are tough, have good resistance to discoloration and possess superior hydrolytic and thermal stabilities.[312] Highly hindered diols such as 2,2,4,4-tetramethyl-1,3-cyclobutanediol or 2,2,4-trimethyl-1,3-pentanediol were found to be exceptionally stable to hydrolysis.[313]

By contrast, polyesters from oxalic and malonic acid are more susceptible to hydrolytic and thermal degradation and have therefore found few applications in the polymer field; however, the dimethyl-substituted malonic acid leads to polymers of higher melting points and stabilities.[314]

Although homopolymers have not found industrial applications, copolymers obtained by an appropriate balance of different monomers are used as intermediates for polyurethanes or as alkyd resins for varnishes and coatings or as hot-melts.

17.3.2 Polyesters from Aliphatic Hydroxy Acids

Linear aliphatic polyesters of general formula $+(ORCO)+$, whose functional groups are separated by nine or more atoms, can be easily polymerized from the corresponding ω-hydroxy acids by direct esterification. In fact, when the number of carbon atoms, x, between the two functional groups is greater than seven, the hydroxy acids yield linear polymers almost exclusively, as the intramolecular dehydration leading to lactones is hindered because of the crowding of hydrogen atoms within the rings[6,33,232–234] when $7 < x > 14$, and because of the statistical probability that the end groups will meet becomes negligible[33,34] when x is greater than 15.

On heating C_5–C_7 aliphatic hydroxy acids, both linear polyesters and lactones are obtained, in a ratio which depends on the reaction conditions;[6] in view of the fact that the corresponding lactones polymerize readily via one of the several methods available for the addition polymerization of lactones, linear polyesters of high molecular weight are therefore more conveniently prepared from lactones. Stable lactones are the main products when one tries to polymerize γ-hydroxy acids under the usual reaction conditions.[24–26]

β-Hydroxy acids having hydrogen atoms in the α position with respect to the carboxyl group usually undergo dehydration to unsaturated acids upon heating, according to equation (68). However, 3-hydroxy-2,2-dimethylpropanoic acid ($HOCH_2CMe_2CO_2H$), which does not have α-hydrogen atoms, gives the corresponding polyester upon heating.[315]

$$HOCH_2CH_2CO_2H \longrightarrow CH_2{=}CHCO_2H + H_2O \qquad (68)$$

As described in a previous section, some poly- and copoly-(β-hydroxyalkanoate)s have been successfully synthesized by enzymatic action of microorganisms.[10]

The poly(α-ester)s, i.e. linear aliphatic polyesters having only one main-chain carbon atom separating repeating ester units (formula **32**), have received comparatively little attention within the wide class of linear polyesters, probably owing to the lack of general methods of synthesis and to their relative ease of hydrolysis which precludes their use in most applications. They may, in principle, be derived by a direct esterification reaction, but they are generally not easily obtained in this way because their cyclic dimers, diglycolide (**24**) and dilactide (**25**), are obtained instead of linear polyesters.[27] Several alternative methods of synthesis for the preparation of poly(α-ester)s are discussed below.

$$+(O\overset{R^1}{\underset{R^2}{C}}CO)+$$

(**32**)

Although polyesters from glycolic acid can be prepared by direct esterification,[6] this reaction seems to be effectively limited to the formation of polymers from glycolic and lactic acid.

A more satisfactory procedure of synthesis consists in converting the α-hydroxy acids in the corresponding dilactides (**33**) and then to polymerize the latter in the presence of a suitable initiator by an anionic or cationic mechanism.

(**33**)

However, as seen before in Section 17.2.8, the polymerization of substituted diglycolides and dilactides is not always feasible, being subjected to certain structural limitations. This fact, coupled with the failure of some α-hydroxy acids to cyclize smoothly to glycolides, again limits the method to the preparation of a few polyglycolides and polylactides. The polyester from mandelic acid (2-hydroxy-2-phenylacetic acid; $R^1 = H$, $R^2 = Ph$) was prepared by polymerization of the sodium salt of α-chlorobenzilic acid[261] in accordance with the method of synthesis using salts of carboxylic acids and dihalides. The reaction takes place by heating the salt at 100–135 °C for several hours; the method has also been used for the preparation of polyesters from α-chlorolactic acid and 2-chloro-2-methylpropionoic acid salts,[261] but it seems to be unreliable for general use.

An interesting alternative way to synthesize poly(α-ester)s is through the so-called anhydrocarboxylates (**34**) and anhydrosulfites (**35**), the equivalent for α-hydroxy acids of the N-carboxy anhydrides of α-amino acids.

(**34**)   (**35**)

The conversion of α-hydroxy carboxylic acids to the cyclic compounds (**34**) and (**35**) involves the use of phosgene[316,317] and thionyl chloride,[27,317–319] respectively, through a ring-closure reaction. Also, in this case, side reactions may hinder the preparation of (**35**), particularly when aromatic substituents are involved.[27,317–319] These monomers can be polymerized by three distinct mechanisms. The first of these can be used for both (**35**) and (**34**) when at least one of the substituents is hydrogen. The polymerization takes place by a bimolecular hydroxy-initiated propagation reaction according to equation (69). The rate-controlling step is the attack of terminal hydroxyl groups on the carbonyl carbon atom of the ring and is affected by the solvent polarity and, to a lesser extent, by temperature (as expected from the relatively low activation energy, 50 kJ mol^{-1}).[320–322] As the concentration of nucleophile affects the rate and molecular weight in opposite ways, high molecular weight polymers cannot be prepared at a reasonable rate by this route.

$$\sim\!\!\sim\!\!OH + (\mathbf{35}) \longrightarrow (\mathbf{36}) \longrightarrow OCOCOH + SO_2 \qquad (69)$$

(**36**)

When the monomers have two substituents R^1 and R^2 more bulky than hydrogen, this mechanism seems to be sterically hindered; in such a case, the reaction occurs by thermal decomposition of the ring with the formation of an intermediate polymerizable α-lactone (**37**). This decomposition seems to be favored by increasing bulkiness of substituents and retarded by their increasing polarity. The polymerization proceeds according to Scheme 16, the first reaction being the rate-controlling step. According to this mechanism, solvent polarity has little or no effect on the rate, while the temperature plays an important role, the activation energy for decomposition being about

125 kJ mol^{-1}.$^{323-325}$ Side reactions leading to ketones, carbon oxides and sulfur dioxides become increasingly competitive at high temperatures.^{326,327}

A third mechanism, involving the use of an aprotic base such as pyridine, is particularly useful for anhydrocarboxylates. The reaction probably proceeds according to Scheme 16 and the role of pyridine may be regarded as a catalytic means of generating the α-lactone.^{328}

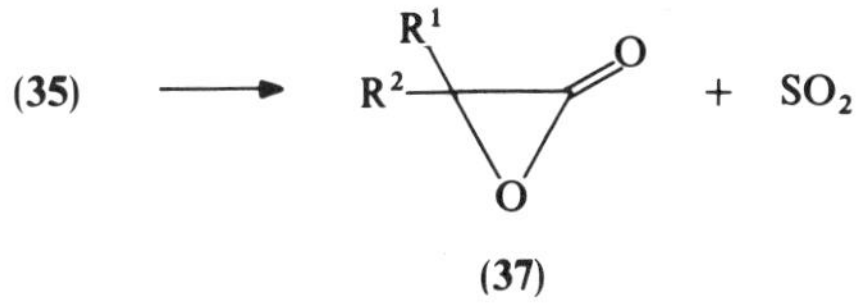

$$(35) \longrightarrow \underset{(37)}{R^2 \overset{R^1}{\underset{O}{\triangle}} O} + SO_2 \qquad (70)$$

$$(37) + HO\sim \longrightarrow \sim O_2C\underset{R^2}{\overset{R^1}{\underset{|}{C}}}OH \qquad (71)$$

Scheme 16

The thermal degradation of poly(α-ester)s has recently been reviewed and discussed.^{27}

Except for a few cases, polyesters obtained from aliphatic hydroxy acids have little commercial importance because of their relatively low melting points, their tendency to cyclize and their sensitivity to moisture. In general they have found limited use, alone or in combination with other materials, for hot-melts, varnishes or as intermediates for polyurethane synthesis. Nevertheless, some polyesters of this class, namely poly(ε-caprolactone) (PCL) and several poly- and copoly-esters of α- and β-hydroxy acids have found applications in some specific uses.

The thermodynamic compatibility of PCL with many other polymers^8 has made it particularly attractive for applications in polymeric blends, and bishydroxy-terminated PCLs have also found an application in the preparation of polyurethanes with low brittle points and high hydrolytic stability.

Poly(α-ester)s and poly(β-ester)s are characterized, in particular, by their ease of hydrolysis and biodegradation. Their intrinsically higher hydrolysis rate, which precludes many applications, is, however, the reason behind the use of poly(glycolic acid), poly(lactic acid) and their copolymers as biomedical materials. Their biodegradability, which has been discussed elsewhere,^{329} is the reason for their application as adsorbable sutures and seems to offer wider opportunities in the design of drug delivery systems.^{330} Some poly(β-hydroxyalkanoate)s exhibit properties resembling those of conventional polyalkenes, and are also completely biodegradable. PHB and some copolyesters are now commercially available from biotechnological processes.^{11}

17.3.3 Polyesters from Aliphatic Diols and Aromatic Diacids

Among the many polyesters belonging to this class, PET, most of all, and PBT have achieved primary importance because of their good mechanical and electrical properties, excellent solvent resistance and good hydrolytic stability, and have found applications as fibres, films and molding materials.$^{24,331-333}$ Main and side reactions occurring during their preparation will be discussed below in separate sections.

Synthesis of the many other polyesters and copolyesters of this class does not present particular differences, except for side reactions and cyclic oligomer content which strictly depend on the chemical structure of the reactants; free acids, or their derivatives, esters, chlorides and anhydrides, can be used as the raw materials reacted with glycols.

Polymerizations are preferably carried out at high temperature in the molten state. For polymers which melt below 200 °C, a temperature between 210–230 °C is often a good compromise for sufficiently high reaction rate and low extent of side reactions; temperatures of 15–30 °C above the melting point are commonly used for high melting polyesters.

17.3.3.1 Poly(ethylene terephthalate)

PET was first prepared in 1946^7 and was commercially introduced in 1953 as a textile fiber and soon thereafter as a film. Its low crystallization rate makes it unsuitable as a molding material unless appropriate chemical modification or nucleating agents provide a commercially acceptable rate of crystallization. Recently, a dramatic development in PET production has resulted from the application of blow-molding technology to the production of beverage bottles for carbonated soft drinks where the barrier properties of PET are exploited.

Although PET can be prepared by methods such as reactions of EG with terephthaloyl chloride and of terephthalic acid (TPA) with ethylene oxide, direct esterification and alcoholysis, from EG with TPA and DMT respectively, are by far the most important and widely studied reactions.

For many years PET has been prepared exclusively from DMT and EG because free terephthalic acid (TPA) was not available to a sufficiently high degree of purity,[15,132] and because of the poor reaction control due to the heterogeneous reaction system. Recently, the availability of TPA of higher purity and improvements in the reaction process have made direct esterification the preferred process because of higher reaction rate, reduced catalyst requirements and elimination of methanol as a by-product.[132,332] PET made from TPA contains, however, higher amounts of diethylene glycol (3-oxapentamethylene) units (DEG) which result in lower melting point, reduced fiber and film strength and poorer thermal, oxidative and UV-light stability.[24,332,334–339]

The manufacture of PET is performed in two steps for both DMT and TPA as starting materials. While the second stage (polycondensation) is nearly identical in both cases, there are substantial differences in the first step.

When DMT is reacted with EG, the two steps are schematically as shown in Scheme 13, where R = 1,4-phenylene, R' = $-(CH_2)_2-$ and R'' = Me. The first step of transesterification is carried out at atmospheric pressure at 170–210 °C using one or more acetates of lead, zinc, manganese, calcium, cadmium, etc. as catalyst.[24,88,132,146,158,332]

In principle, PET could also be obtained using an EG/DMT molar ratio of unity; however, a large excess of EG (EG/DMT = 2.2 mol mol^{-1}) is commonly used for increasing the reaction rate and to obtain complete evolution of methanol at the end of the first stage (equation 50) because the presence of methyl ester groups in the second stage reduces the rate of the process and limits the polymerization degree of the final polymer.[132,158] At the end of the first step, bis(hydroxyethyl) terephthalate (BHET) is present in the reactor as a major component along with small amounts of low molecular weight oligomers.

Upon completion of the first stage, a phosphorus compound is frequently added[132,158,340,341] and it has been suggested that it deactivates the catalyst used in the first stage which would otherwise contribute to thermal degradation[158,341] at the higher temperature at which polycondensation is carried out. Recently, however, this interpretation has been rejected[340] and a mechanism involving the presence of aromatic phosphites has been proposed.[342]

When the polymerization is performed using TPA and EG, the first stage of reaction is carried out under pressure (0.3–0.5 MPa) at high temperature (230–260 °C) with elimination of water. Since the carboxyl groups of TPA can catalyze the direct esterification, a metal catalyst is not required. In the absence of methyl ester groups, EG/TPA molar ratios lower than two can be used, resulting in a more economical process.[132] A molar ratio of EG/TPA = 1.5 is frequently used, which gives, at the end of the first stage, a mixture of oligomers with an average degree of polymerization equal to three.

Before starting the second stage, a second catalyst, usually Sb_2O_3, antimony acetate or GeO_2, is added and the reaction is carried out, at 275–290 °C, under reduced pressure (10–50 Pa). Titanium alkoxides, which would be very good catalysts for both stages, are not suitable because highly yellow PET would result.

The excess of EG is eliminated in this stage, according to equation (51) (with R = 1,4-phenylene and R' = $-(CH_2)_2-$). The high melt viscosity of the polymer in the final stages of polymerization may lead to a diffusion-controlled process and special reactors which can generate large interfacial areas by mechanical means are used.[132]

The high temperature which is required to maintain the reactants in a molten state makes the scission reaction relatively more important because of its high activation energy (175–220 kJ mol^{-1}).[158,179,291,302] Chain scission occurs according to equation (72), via the β-hydrogen abstraction mechanism previously discussed in Section 17.2.11. The reaction is first order in ester group concentration with a kinetic constant of about 5×10^{-7} s^{-1} at 280 °C. The residual metal catalyst may increase the rate of chain scission, and Zimmerman[158,294,295] has suggested that this effect derives from an electron-withdrawing interaction of the metal with the carbonyl oxygen atom adjacent to the ester bond which undergoes scission.

$$\sim\!\!\langle\!\!\bigcirc\!\!\rangle\!\!-CO_2CH_2CH_2O_2C-\!\!\langle\!\!\bigcirc\!\!\rangle\!\!\sim \longrightarrow \sim\!\!\langle\!\!\bigcirc\!\!\rangle\!\!CO_2H + CH_2\!\!=\!\!CHO_2C\!\!\langle\!\!\bigcirc\!\!\rangle\!\!\sim \quad (72)$$

As can be seen in equation (**72**), the chain-scission reaction leads to carboxyl groups and to vinyl ester end-groups, which can undergo further reactions[132,158,178,179] leading to acetaldehyde,[132,158,178,179,343,344] to polyene chains[158,304] and to other by-products.[158,178,179,304,341,345]

The polyene chains are assumed to be chiefly responsible for PET discoloration. Acetaldehyde, for which various formation paths have been proposed,[158,178,179,343,344] is undesirable for applications in contact with food and beverages. Contents of acetaldehyde as low as 2 p.p.m. have been obtained by solid-state postpolymerization, against about 24 p.p.m. from melt processes, and the amount increases in the presence of oxygen.[345]

Another side reaction which occurs during the preparation of PET, leading to the formation of DEG moieties in the chains, has been widely investigated. It has been suggested that the DEG units originated: (i) from hydroxyethyl ester terminal groups through an intramolecular ortho ester intermediate,[343] followed by alcoholysis with alkyl–oxygen fission; (ii) from a reaction of hydroxyethyl ester terminal groups with ethylene oxide,[343] the latter produced as a by-product from minor side-reactions; and (iii) from direct dehydration from two alcoholic groups.[346–348] Experimental evidence[349] seems to fit the first reaction path best, shown in Scheme 17.

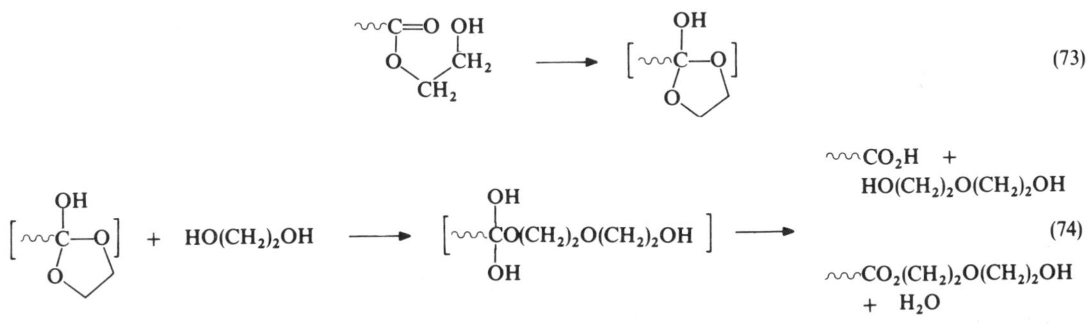

Scheme 17

The ester-scission reaction often limits the maximum molecular weight that can be achieved by melt processes and, when PET of higher molecular weight is required, a suitable prepolymer is subjected to a subsequent solid-state polymerization process (see Volume 5, Chapter 13).

The presence of a number of side reactions and diffusion effects accompanying the main chemical reactions complicates the kinetic analysis of polycondensation of PET. Reimschuessel[343,350,351] has clearly pointed out that various contributions deriving from uncatalyzed, carboxyl-catalyzed and metal-catalyzed reactions, as well as from side reactions and mass transfer, should be considered for the polymerization of TPA and EG. It is therefore clear that an adequate analytical treatment of the process is rather problematic, and often oversimplified schemes have been considered making the relative conclusions disputable.

Use of computer modeling and simulation techniques in the development of comprehensive kinetic schemes has allowed the consideration of main and side reactions and the achievement of significant technological improvements in the production of PET.[132] Calculations have been reported for the polymerization of PET, from both DMT and TPA, for semi-batch and continuous reactors,[132,352–359] and for model compounds.[343,350,351]

17.3.3.2 Poly(1,4-butylene terephthalate)

Poly(1,4-butylene terephthalate) (PBT), whose synthesis was first reported by Whinfield in 1949,[360] is a thermoplastic polyester closely resembling PET, from which it differs by only two methylene groups in the aliphatic moiety of the repeating unit. However, while PET underwent a rapid growth in applications as fibers and films, PBT remained relatively unknown until 1970 when it was commercially introduced because of its facile moldability, good electrical properties, toughness, dimension stability and solvent resistance.

As they have similar chemical structures, it is not surprising that PET and PBT are essentially prepared in the same way even though some differences in side reactions distinguish the two processes. Similarly to PET, PBT is commonly prepared by a two-stage process at high temperature in the molten state. Either DMT or TPA can be used with 1,4-butanediol (BD) as starting monomers, but, in contrast to PET, PBT is at present commercially produced almost exclusively from DMT; the ease with which BD undergoes an acid-catalyzed dehydration reaction to form THF is one of the reasons which make the process from TPA less attractive.

The spatial conformation of both BD and 4-hydroxybutyl ester end-groups favors reactions (75) and (76). Both reactions lead to an overall loss of BD, which results in an increased overall cost because THF cannot be conveniently rehydrated to BD.

$$HO(CH_2)_4OH \longrightarrow THF + H_2O \qquad (75)$$

$$\sim\!\!\langle\bigcirc\rangle\!\!-\!CO_2(CH_2)_4OH \longrightarrow THF + \sim\!\!\langle\bigcirc\rangle\!\!-\!CO_2H \qquad (76)$$

BD dehydration, equation (75), is acid catalyzed[361,362] and is characterized by an activation energy of 127 kJ mol^{-1}.[361] THF formation from reaction (76) is not affected by either the catalyst Ti(OBu)$_4$ or carboxyl groups[159] and probably occurs through an intramolecular reaction of alkyl–oxygen ester-bond scission, analogous to the mechanism proposed by Hovenkamp[349] for the formation of DEG units during the synthesis of PET. Reaction (76) was studied in a closed system on 4-hydroxybutyl benzoate, used as a model for the terminal hydroxybutyl groups of PBT, and it was found that it occurred at a significant rate even in a relatively low temperature range (150–190 °C; $k_{2(180)} = 9.4 \times 10^{-7}$ s^{-1}) following first-order kinetics with an activation energy of 121 kJ mol^{-1}, very close to the value found for THF formation from PBT.[297] Also, this reaction, which cannot be avoided, leads to the formation of carboxyl-end groups and, if not controlled by an appropriate choice of reaction conditions, may result in PBT of low molecular weight and poor properties.

At the high temperature (250–260 °C) at which the second stage of the polymerization must be carried out in order to maintain PBT in a molten state or because it is needed for processing, a random scission reaction of the internal ester bonds occurs through the β-hydrogen abstraction mechanism previously described in Section 17.2.11. The 3-butenyl ester end-groups formed in the first scission (equation 77) undergoes an analogous reaction leading to 1,3-butadiene according to equation (78). From experiments on both PBT and on 1,4-butylene dibenzoate, as a model of an internal ester group, reaction (77) has been found to follow first-order kinetics with respect to the ester groups[158,159,290,291,297,300] and reaction (78) has been found to proceed faster than (77), possibly due to the activation induced by the neighboring vinyl group.[159,290]

The overall result of reaction (77) and (78) is therefore the formation of 1 mol of 1,3-butadiene and of two carboxyl end-groups for each chain scission. Activation energies of 172–208 kJ mol^{-1} have been reported by different authors,[158,159,290,291,297,300] with very good agreement between PBT and its model.

$$\sim\!\!\langle\bigcirc\rangle\!\!-\!CO_2(CH_2)_4O_2C\!-\!\langle\bigcirc\rangle\!\!\sim \longrightarrow CH_2\!=\!CH(CH_2)_2O_2C\!\langle\bigcirc\rangle\!\sim + \sim\!\!\langle\bigcirc\rangle\!\!-\!CO_2H \qquad (77)$$
$$(38)$$

$$(38) \longrightarrow \sim\!\!\langle\bigcirc\rangle\!\!-\!CO_2H + CH_2\!=\!CHCH\!=\!CH_2 \qquad (78)$$

In contrast to PET, no evidence of ether formation has been reported in the literature during the polymerization of PBT, and reaction (77) is not affected by the catalyst employed during polymerization, although it occurs at a rate which is about 10 times higher than that found for PET.[158,159]

The main differences between the polymerization processes of PBT and PET are in connection with the different characteristics of the side reactions. Thus, titanium alkoxides [Ti(OBu^n)$_4$ and Ti(OPr^i)$_4$ mainly], which cannot be used for the polymerization of PET because of their effect on side reactions which induces yellowing, are the best catalysts for the polymerization of PBT, showing a very good catalytic activity for both stages of polymerization without discoloration of the final product. Additionally, in contrast to antimony catalysts, titanium alkoxides are also effective in the presence of methyl ester end-groups and therefore the complete elimination of methanol in the first stage is not required. This allows lower BD/DMT molar ratios which in turn results in a lower loss of BD as THF. The first stage is therefore carried out according to equation (50) (with R = 1,4-phenylene, R′ = $-(CH_2)_4-$ and R″ = Me) in the presence of titanium catalysts at atmospheric pressure, at 150–220 °C, until about 90–95% of the theoretical amount of methanol has been distilled out.

In the second stage, the excess of BD is removed from the alcoholysis equilibrium (corresponding to equation 51) under reduced pressure (10–50 Pa) at about 250 °C. Due to the formation of an increasing amount of carboxyl end-groups from reactions (76)–(78), the polymerization probably proceeds with a significant contribution of direct esterification,[147] particularly in the last stages of reaction. Optimal reaction conditions are required to control reactions (75)–(78) in order to obtain PBT with sufficiently high molecular weight, low CO_2H/OH end-group ratio and good general properties.

17.3.4 Polyesters from Aromatic Diols and Aliphatic Diacids

Polyesters having rigid and flexible segments in the main chains can be prepared from aliphatic dicarboxylic acids and aromatic diols or their derivatives. Most of these polymers are particularly attractive because they may display liquid-crystalline properties, and their phase-transition temperatures may be controlled through an appropriate choice of bisphenol moieties, which behave as mesogenic groups when incorporated into a polymer as a part of the main chain, and of the number of methylene units in the flexible diacid moiety.[363,364]

These types of polyester can be prepared at high temperature in the melt, either by acidolysis from bisphenol diacetates and dicarboxylic acids or by alcoholysis from bisphenols and the diphenyl esters of the dicarboxylic acids, as well as, at low temperature, in solution or by an interfacial method, from bisphenols and diacyl chlorides. The gross differences between high and low temperature polycondensation methods[20,23] and between solution and interfacial methods[23] have been discussed in terms of reaction rate and yield, product recovery, reactant purity, stoichiometric requirements *etc.* In the absence of systematic studies comparing the various methods with respect to molecular weight, a generalization cannot be attempted; however, it has been reported that when poly(4,4'-diphenyl sebacate) was prepared by different methods (in solution, interfacial and by acidolysis in the melt), the highest molecular weight was obtained from polycondensation by acidolysis of 4,4'-diacetoxybiphenyl and sebacic acid.[365]

Many examples of polyesters belonging to this class are listed in ref. 366. A number of polyesters exhibiting thermotropic liquid-crystalline behavior have been reported in recent years.[363,365,367–373] Substituents in the benzene ring of the mesogenic units usually lead to more soluble and relatively low-melting polyesters.[371,372] Polyesters prepared from the same bisphenol and from different diacids having odd and even numbers of methylene units may show different types of mesophases, indicating that changing the flexible portion of the repeating unit may affect the liquid-crystalline properties.[370,372]

17.4 REFERENCES

1. J. Berzelius, *Rapp. Ann.*, 1847, 26.
2. A. V. Laurenco, *Ann. Chim. Phys.*, 1863, **67**, 293.
3. K. Kraut, *Justus Liebigs Ann. Chem.*, 1869, **150**, 1.
4. Davidoff, *Ber.*, 1886, **19**, 406.
5. Vorlander, *Ann.*, 1894, **280**, 167.
6. (a) W. H. Carothers *et al.*, *J. Am. Chem. Soc.*, 1929, **51**, 2560; 1930, **52**, 3292; 1930, **52**, 711; 1932, **54**, 761; 1932, **54**, 1557; 1932, **54**, 1559; 1933, **55**, 4714; 1935, **57**, 935; 1936, **58**, 654; (b) H. Mark and G. S. Whitby, 'Collected papers of Wallace Hume Carothers', Interscience, New York, 1940.
7. (a) J. R. Whinfield, *Nature (London)*, 1946, **158**, 930; (b) J. R. Whinfield and J. T. Dickson, Br. Pat. 578 079 (1946) (*Chem. Abstr.*, 1947, **41**, 1495).
8. J. V. Koleske, in 'Polymer Blends', ed. D. R. Paul and S. Newman, Academic Press, New York, 1978, vol. 2, p. 369.
9. J. N. Baptist and F. X. Weber, *SPE Trans.*, 1964, **4**, 245.
10. R. H. Marchessault, *CHEMTECH*, 1984, 542.
11. E. R. Howells, *Chem. Ind. (London)*, 1982, 508.
12. S. L. Kwolek, P. W. Morgan and J. R. Schaefgen, in 'Encyclopedia of Polymer Science and Engineering', ed. J. I. Kroschwitz, Wiley, New York, 1987, vol. 9, p. 1.
13. S. C. Temin, in 'Encyclopedia of Polymer Science and Engineering', ed. J. I. Kroschwitz, Wiley, New York, 1985, vol. 1, p. 572.
14. (a) G. Holden, in 'Encyclopedia of Polymer Science and Engineering', ed. J. I. Kroschwitz, Wiley, New York, 1986, vol. 5, p. 426; (b) I. Goodman, in 'Development in Block Copolymers', ed. I. Goodman, Applied Science, London, 1982, vol. 1, p. 153; (c) R. J. Cella, in 'Encyclopedia of Polymer Science and Technology', ed. N. M. Bikales, Interscience, New York, 1977, suppl. vol. 2, p. 485.

15. J. K. Stille and T. W. Campbell, in 'High Polymers', ed. H. Mark, C. S. Marvel and H. W. Melville, Interscience, New York, 1972, vol. 27.
16. H. K. Reimschuessel, in 'Polymer Handbook', ed. H. G. Elias and R. A. Petrick, Harwood Academic, Chur, Switzerland, 1984, p. 183.
17. M. A. Ogliaruso and J. F. Wolfe, in 'The Chemistry of Acid Derivatives', ed. S. Patai, Wiley, New York, 1979, suppl. B, part 1, p. 411.
18. V. V. Korshak and S. V. Vinogradova, in 'Polyesters', ed. J. Burdon, Pergamon Press, Oxford, 1965.
19. I Goodman and J. A. Rhys, 'Polyesters', Iliffe Books, London, 1965, vol. 1.
20. P. W. Morgan, *Polym. Rev.*, 1965, **10**.
21. R. W. Lenz, 'Organic Chemistry of Synthetic High Polymer', Interscience, New York, 1967.
22. L. B. Sokolov, 'Synthesis of Polymers by Polycondensation', Israel Program for Scientific Translation, Jerusalem, 1968.
23. C. S. Temin, in 'Interfacial Synthesis', ed. F. Millich and C. E. Carraher, Jr., Dekker, 1977, New York, vol. 2, p. 27.
24. I. Goodman, in 'Encyclopedia of Polymer Science and Technology', ed. N. M. Bikales, Interscience, New York, 1969, vol. 11, p. 62.
25. I. Goodman, in 'Developments in Polymerization', ed. R. N. Haward, Applied Science, London, 1979, vol. 2, p. 149.
26. D. B. Johns, R. W. Lenz and A. Luecke, in 'Ring Opening Polymerization', ed. J. Ivin and T. Saegusa, Elsevier, London, 1984, vol. 1, chap. 7.
27. B. J. Tighe, in 'Developments in Polymer Degradation', ed. N. Grassie, Applied Science, London, 1984, vol. 5, p. 31.
28. M. L. Bender, *Chem. Rev.*, 1960, **60**, 53.
29. F. N. Jones and D. D.-L. Lu, 'Macromolecular Preprints of the International Conference on Functional Polymers and Biopolymers', Butwell, Oxford, 1986, p. 27.
30. S. V. Vinogradova, *Polym. Sci. USSR (Engl. Transl.)*, 1977, **19**, 769.
31. D. R. Cooper and J. A. Semlyen, *Polymer*, 1973, **14**, 185.
32. G. Wick and H. Zeitler, *Angew. Makromol. Chem.*, 1983, **112**, 59.
33. P. J. Flory, *Chem. Rev.*, 1946, **39**, 137.
34. P. J. Flory, 'Principles of Polymer Chemistry', Cornell University Press, Ithaca, NY, 1953.
35. T. M. Pell, Jr. and T. G. Davis, *J. Polym. Sci., Polym. Phys. Ed.*, 1973, **11**, 1671.
36. P. J. Hoftizer, *Appl. Polym. Symp.*, 1975, **26**, 349.
37. V. V. Korshak and S. V. Vinogradova, in 'Polyesters', ed. J. Burdon, Pergamon Press, Oxford, 1965, p. 23.
38. I. Vancso-Szmercsanyi and E. Makay-Bodi, *J. Polym. Sci., Part C*, 1968, **16**, 3709.
39. A. Fradet and E. Marechal, *Adv. Polym. Sci.*, 1982, **43**, 51.
40. C. K. Ingold, 'Structure and Mechanisms in Organic Chemistry', 2nd edn., Cornell University Press, Ithaca, NY, 1969, p. 1129.
41. Y. R. Fang, C.-G. Lai, J.-L. Lu and M.-K. Chen, *Sci. Sin. (Engl. Ed.)*, 1975, **18**, 72.
42. I. M. Kolthoff and S. Bruckenstein, *J. Am. Chem. Soc.*, 1956, **78**, 1.
43. S. D. Hamann, D. H. Solomon and J. D. Swift, *J. Macromol. Sci., Chem.*, 1968, **A2**, 153.
44. (a) M. Krumploc and J. Malek, *Makromol. Chem.*, 1973, **168**, 119; (b) M. Krumploc and J. Malek, *Makromol. Chem.*, 1973, **171**, 69.
45. M. Krumploc, V. Bazant and J. Malek, *Collect. Czech. Chem. Commun.*, 1974, **39**, 1158.
46. L. Nondek and J. Malek, *Makromol. Chem.*, 1977, **178**, 2211.
47. L. H. Peebles, Jr. and W. S. Wagner, *J. Am. Chem. Soc.*, 1959, **78**, 1206.
48. E. Szabo-Rethy, *Eur. Polym. J.*, 1971, **7**, 1485.
49. C. C. Lin and K. H. Hsien, *J. Appl. Polym. Sci.*, 1977, **21**, 2711.
50. A. Fradet and E. Marechal, *Polym. Bull. (Berlin)*, 1980, **3**, 441.
51. P. J. Flory, *J. Am. Chem. Soc.*, 1937, **59**, 466.
52. P. J. Flory, *J. Am. Chem. Soc.*, 1939, **61**, 3334.
53. P. J. Flory, *J. Am. Chem. Soc.*, 1940, **62**, 2261.
54. A. Fradet and E. Marechal, *J. Polym. Sci., Polym. Chem. Ed.*, 1981, **19**, 2905.
55. M. Gordon and G. R. Scantlebury, *J. Chem. Soc., Part B*, 1967, 1.
56. M. Gordon and C. G. Leonis, *J. Chem. Soc., Faraday Trans. 1*, 1975, **71**, 161.
57. M. Davies, *Research (London)*, 1949, **2**, 544.
58. C. Y. Huang, Y. Simono and T. Onizuka, *Kobunshi Kagaku*, 1966, **23** (254), 408 (*Chem. Abstr.*, 1967, **66** 86060).
59. M. Imoto, C. Y. Huang, J. Hashizume, S. Sano and K. Amamiya, *Kogyo Kagaku Zasshi*, 1960, **63**, 1807 (*Chem. Abstr.*, 1962, **57**, 9707h).
60. A. Fradet and E. Marechal, *J. Macromol. Sci., Chem.*, 1982, **A17**, 859.
61. A. C. Tang and K. S. Yao, *J. Polym. Sci.*, 1959, **35**, 219.
62. A. J. Amass, *Polymer*, 1979, **20**, 515.
63. K. Uberreiter and W. Hager, *Makromol. Chem.*, 1979, **189**, 1697.
64. A. Blaga, R. Vladea, M. Zver and M. Simon, *Rev. Chim. (Bucharest)*, 1978, **29**, 629 (*Chem. Abstr.*, 1979, **90**, 38234).
65. G. A. Gareev and V. P. Men'shutin, *Kinet. Katal.*, 1967, **8**, 1369 (*Chem. Abstr.*, 1968, **68**, 86486b).
66. M. Sumoto and Y. Hasegawa, *Kogyo Kagaku Zasshi*, 1965, **68**, 1900 (*Chem. Abstr.*, 1966, **64**, 12796d).
67. V. A. Ignatov *et al.*, *Izv. Vyssh. Uchebn. Zaved., Khim. Khim. Tekhnol.*, 1978, **21**, 419 (*Chem. Abstr.*, 1978, **89**, 110551).
68. M. T. Pope and R. J. Williams, *J. Chem. Soc.*, 1959, 3579.
69. F. G. Baddar *et al. Eur. Polym. J.*, 1971, **7**, 1621.
70. L. M. Gumenchuk, E. M. Krom, V. R. Kartashov and G. V. Moikin, *Izv. Vyssh. Uchebn. Zaved., Khim. Khim. Tekhnol.*, 1978, **21**, 844 (*Chem. Abstr.*, 1979, **90**, 54118).
71. N. Ivanoff, *Bull. Soc. Chim. Fr.*, 1950, 347.
72. C. E. H. Bawn and M. B. Huglin, *Polymer*, 1962, **3**, 257.
73. O. M. O. Habib and J. Malek, *Collect. Czech. Chem. Commun.*, 1976, **41**, 1158.
74. J. Otton and S. Ratton, 'Preprints of the European Symposium on Polymeric Materials, Lyon (F), September 14–18, 1987', volume on Chemical Aspects, p. CC01.
75. D. C. Bradley, R. C. Mehrotra and D. P. Gaur, 'Metal Alkoxides', Academic Press, London, 1978.
76. F. Leverd, A. Fradet and E. Marechal, *Eur. Polym. J.*, 1987, **23**, 695.

77. F. Leverd, A. Fradet and E. Marechal, *Eur. Polym. J.*, 1987, **23**, 699.
78. F. Leverd, A. Fradet and E. Marechal, *Eur. Polym. J.*, 1987, **23**, 705.
79. F. Pilati, P. Manaresi and A. Munari, *Polym. Commun.*, 1984, **25**, 187.
80. G. H. Dahl and B. P. Block, *Inorg. Chem.*, 1966, **5**, 1394.
81. C. N. Caughlan, H. S. Smith, W. Katz, W. Hodgson and R. W. Crowe, *J. Am. Chem. Soc.*, 1951, **73**, 5652.
82. R. L. Martin and G. Winter, *Nature (London)*, 1963, **197**, 687.
83. D. C. Bradley and C. E. Holloway, *Inorg. Chem.*, 1964, **3**, 1163.
84. V. T. Athavale, N. Mahadevan and R. M. Sathe, *Indian J. Chem.*, 1968, **6**, 462.
85. R. H. Stanley, in 'Kirk-Othmer Encyclopedia of Chemical Technology', Interscience, New York, 1969, vol. 20, p. 424.
86. F. Pilati, P. Manaresi, B. Fortunato, A. Munari and P. Monari, *Polymer*, 1983, **24**, 1479.
87. B. Fortunato, P. Manaresi, A. Munari and F. Pilati, *Polymer*, 1986, **27**, 29.
88. K. Tomita and H. Ida, *Polymer*, 1973, **14**, 55.
89. F. Pilati, A. Munari, P. Manaresi and V. Bonora, *Polymer*, 1985, **26**, 1745.
90. A. Fradet and E. Marechal, *J. Macromol. Sci., Chem.*, 1982, **A17**, 881.
91. A. Fradet and E. Marechal, *Eur. Polym. J.*, 1978, **14**, 761.
92. M. A. Ogliaruso and J. F. Wolfe, in 'The Chemistry of Acid Derivatives', ed. S. Patai, Wiley, New York, 1979, suppl. B, part 1, p. 411.
93. M. Stacey, E. J. Bourne, J. C. Tatlow and T. M. Tedder, *J. Chem. Soc.*, 1949, 2976.
94. P. W. Morgan, *Polym. Rev.*, 1965, **10**, 366.
95. D. R. Stevenson and J. E. Mulvaney, *J. Polym. Sci., Part A-1*, 1972, **10**, 2713.
96. P. W. Morgan, *J. Am. Chem. Soc.*, 1951, **73**, 860.
97. H. A. Staab, *Chem. Ber.*, 1957, **90**, 1326.
98. H. A. Staab, *Angew. Chem., Int. Ed. Engl.*, 1962, **1**, 351.
99. F. Higashi, K. Kubota, M. Sekizuka and M. Goto, *J. Polym. Sci., Polym. Chem. Ed.*, 1980, **18**, 385.
100. N. Ogata, K. Sanui and H. Tanaka, *Polym. J.*, 1981, **13**, 989.
101. F. Higashi, K. Kubota, M. Sekizuka and M. Higashi, *J. Polym. Sci., Polym. Chem. Ed.*, 1981, **19**, 2681.
102. H. Tanaka, G.-C. Wu, K. Sanui and N. Ogata, *Polym. J.*, 1982, **14**, 331.
103. H. Tanaka, Y. Iwanaga, G.-C. Wu, K. Sanui and N. Ogata, *Polym. J.*, 1982, **14**, 643.
104. G.-C. Wu, H. Tanaka, K. Sanui and N. Ogata, *Polym. J.*, 1982, **14**, 797.
105. S. Yasuda, G.-C. Wu, H. Tanaka, K. Sanui and N. Ogata, *J. Polym. Sci., Polym. Chem. Ed.*, 1982, **14**, 797.
106. F. Higashi, A. Hoshio and J. Kiyoshige, *J. Polym. Sci., Polym. Chem. Ed.*, 1983, **21**, 3241.
107. F. Higashi, N. Akiyama and T. Koyama, *J. Polym. Sci., Polym. Chem. Ed.*, 1983, **21**, 3233.
108. S. Kitayama, K. Sanui and N. Ogata, *J. Polym. Sci., Polym. Chem. Ed.*, 1984, **22**, 2705.
109. F. Higashi, Y. Yamada and A. Hoshio, *J. Polym. Sci., Polym. Chem. Ed.*, 1984, **22**, 2181.
110. F. Higashi, N. Akiyama, I. Takahashi and T. Koyama, *J. Polym. Sci., Polym. Chem. Ed.*, 1984, **22**, 1653.
111. F. Higashi and Y. Yamada, *J. Polym. Sci., Polym. Chem. Ed.*, 1985, **23**, 2709.
112. F. Higashi, A. Hoshio, Y. Yamada and M. Ozawa, *J. Polym. Sci., Polym. Chem. Ed.*, 1985, **23**, 69.
113. F. Higashi, M. Ozawa and Teh-Chon Chang, *J. Polym. Sci., Polym. Chem. Ed.*, 1985, **23**, 1361.
114. F. Higashi, T. Mashimo, I. Takahashi and N. Akiyama, *J. Polym. Sci., Polym. Chem. Ed.*, 1985, **23**, 3095.
115. F. Higashi and T. Mashimo, *J. Polym. Sci., Polym. Chem. Ed.*, 1985, **23**, 2999.
116. F. Higashi and T. Mashimo, *J. Polym. Sci., Polym. Chem. Ed.*, 1985, **23**, 2715.
117. F. Higashi, Y. Fujiwara and Y. Yamada, *J. Polym. Sci., Polym. Chem. Ed.*, 1986, **24**, 589.
118. J. Economy, R. S. Storm, V. I. Matkovitch, S. G. Cottis and B. E. Nowak, *J. Polym. Sci., Polym. Chem. Ed.*, 1976, **14**, 2207.
119. F. Higashi, A. Hoshio and H. Ohtani, *J. Polym. Sci., Polym. Chem. Ed.*, 1984, **22**, 3983.
120. F. Higashi and I. Takahashi, *J. Polym. Sci., Polym. Chem. Ed.*, 1985, **23**, 1369.
121. F. Higashi, M. Ozawa, A. Hoshio and A. Mochizuki, *J. Polym. Sci., Polym. Chem. Ed.*, 1985, **23**, 1699.
122. F. Higashi, Y. Mihara, I. Takahashi and W.-H. Chen, *J. Polym. Sci., Polym. Chem. Ed.*, 1985, **23**, 2851.
123. F. Higashi, M. Ozawa and A. Mochizuki, *J. Polym. Sci., Polym. Chem. Ed.*, 1986, **24**, 637.
124. R. Appel and W. Heinzelman, *Ger. Pat.* 1 192 205 (1965) (*Chem. Abstr.*, 1965, **63**, 8405).
125. M. Masaki and T. Fukui, *Chem. Lett.*, 1977, 151.
126. Z. Arnold and A. Holy, *Collect. Czech. Chem. Commun.*, 1962, **27**, 2886.
127. R. Appel, *Angew. Chem. Int. Ed. Engl.*, 1975, **14**, 801.
128. G. A. Wiley and W. R. Stine, *Tetrahedron Lett.*, 1967, **24**, 2321.
129. D. B. Denney, D. Z. Denney and B. C. Chang, *J. Am. Chem. Soc.*, 1968, **90**, 6332.
130. R. Ratz and O. J. Sweeting, *J. Org. Chem.*, 1963, **28**, 1608.
131. L. Caglioti, M. Poloni and G. Rosini, *J. Org. Chem.*, 1968, **33**, 2979.
132. K. Ravindranath and R. A. Mashelkar, in 'Developments in Plastic Technology', ed. A. Whelan and J. L. Craft, Elsevier, London, 1985, vol. 2, p. 1.
133. G. C. Overberger and J. Sebenda, *J. Polym. Sci., Part A-1*, 1969, **7**, 2875.
134. N. Ogata, K. Sanui and K. Iijima, *J. Polym. Sci., Polym. Chem. Ed.*, 1973, **11**, 1095.
135. N. Ogata and S. Okamoto, *J. Polym. Sci., Polym. Chem. Ed.*, 1973, **11**, 2537.
136. G. Challa, *Recl. Trav. Chim. Pays-Bas*, 1960, **79**, 90.
137. G. Challa, *Makromol. Chem.*, 1960, **38**, 105.
138. G. Challa, *Makromol. Chem.*, 1960, **38**, 123.
139. G. Challa, *Makromol. Chem.*, 1960, **38**, 138.
140. C. M. Fontana, *J. Polym. Sci., Part A-1*, 1968, **6**, 2343.
141. S. G. Hovenkamp, *J. Polym. Sci., Part A-1*, 1969, **7**, 3428.
142. E. Bonatz and G. Rafler, *Acta Polym.*, 1980, **31**, 402.
143. A. S. Chegolya, V. V. Shevchenko and G. D. Mikhailov, *J. Polym. Sci., Polym. Chem. Ed.*, 1979, **17**, 889.
144. S. G. Hovenkamp, *J. Polym. Sci., Part A-1*, 1971, **9**, 3617.
145. J. P. Flory, *J. Am. Chem. Soc.*, 1940, **62**, 2261.
146. W. Griehl and G. Schnock, *Faserforsch. Textiltech.*, 1957, **8**, 408.

147. F. Pilati, P. Manaresi, B. Fortunato, A. Munari and V. Passalacqua, *Polymer*, 1981, **22**, 799.
148. K. Yoda, K. Kimoto and T. Toda, *Kogyo Kagaku Zasshi*, 1964, **67**, 909.
149. K. Tomita and H. Ida, *Polymer*, 1975, **16**, 185.
150. G. Torraca and R. Turriziani, *Chim. Ind. (Milan)*, 1962, **44**, 483.
151. M. F. Lappert, *J. Chem. Soc.*, 1962, 542.
152. M. F. Lappert, *J. Chem. Soc.*, 1961, 817.
153. K.-H. Wolf and H. Herlinger, *Angew. Makromol. Chem.*, 1977, **65**, 133.
154. K.-H. Wolf, B. Kuster and H. Herlinger, *Angew. Makromol. Chem.*, 1978, **68**, 23.
155. C. C. Walker, *J. Polym. Sci., Polym. Chem. Ed.*, 1983, **21**, 623.
156. R. W. Stevenson and H. R. Nettleton, *J. Polym. Sci., Part A-1*, 1968, **6**, 889.
157. R. W. Stevenson, *J. Polym. Sci., Part A-1*, 1968, **6**, 889.
158. H. Zimmerman, in 'Developments in Polymer Degradation', ed. N. Grassie, Applied Science, London, 1984, vol. 5, p. 79.
159. F. Pilati, P. Manaresi, B. Fortunato, A. Munari and V. Passalacqua, *Polymer*, 1981, **22**, 1566.
160. G. Rafler, J. Blaesche, B. Moller and M. Stromeyer, *Acta Polym.*, 1981, **32**, 608.
161. J. Yamanis and M. Adelman, *J. Polym. Sci., Polym. Chem. Ed.*, 1976, **14**, 1945.
162. J. Yamanis and M. Adelman, *J. Polym. Sci., Polym. Chem. Ed.*, 1976, **14**, 1961.
163. K. V. Datye and H. M. Raje, *J. Appl. Polym. Sci.*, 1985, **30**, 205.
164. K. Ravindranath and R. A. Mashelkar, *J. Polym. Sci., Polym. Chem. Ed.*, 1982, **20**, 3447.
165. A. M. Kotliar, *Macromol. Rev.*, 1981, **16**, 367.
166. F. L. Hamb, *J. Polym. Sci., Polym. Chem. Ed.*, 1972, **10**, 3219.
167. M. Levine and S. C. Temin, *J. Polym. Sci.*, 1958, **28**, 179.
168. S. C. Temin, *J. Org. Chem.*, 1961, **26**, 2518.
169. E. R. Wallsgrove and F. Reeder, *Br. Pat.* 636 429 (1950) (*Chem. Abstr.*, 1950, **44**, 7878d).
170. R. Gilkey and J. R. Caldwell, *J. Appl. Polym. Sci.*, 1959, **2**, 198.
171. W. J. Jackson, Jr. and H. F. Kuhfuss, *J. Polym. Sci., Polym. Chem. Ed.*, 1976, **14**, 2049.
172. H. F. Kuhfuss and W. J. Jackson, Jr., *US Pat.* 3 778 410 (1973) (*Chem. Abstr.*, 1974, **80**, 14 6894t).
173. H. K. Reimschuessel and T. DeBona, *Macromolecules*, 1981, **13**, 1582.
174. V. V. Korshak and S. V. Vinogradova, in 'Polyesters', ed. J. Burdon, Pergamon Press, Oxford, 1965, p. 261.
175. J. Koskikallio, in 'The Chemistry of the Carboxylic Acids and Esters', ed. S. Patai, Interscience, New York, 1969, chap. 3, p. 103.
176. D. F. Loncrini, *J. Polym. Sci., Part A-1*, 1966, **4**, 1531.
177. J. Devaux, P. Goddard and J. P. Mercier, *J. Polym. Sci., Polym. Phys. Ed.*, 1982, **20**, 1901.
178. E. P. Goodings, *SCI Monogr.*, 1961, **13**, 211.
179. L. H. Buxbaum, *Angew. Chem. Int. Ed. Engl.*, 1968, **7**, 182.
180. B. Fortunato, F. Pilati and P. Manaresi, *Polymer*, 1981, **22**, 655.
181. F. Pilati, G. C. Gostoli and G. C. Sarti, *Polym. Process. Eng.*, 1986, **4**, 303.
182. T. Davies, in 'High Polymers', ed. E. M. Fettes, Interscience, New York, 1964, vol. XIX, p. 501.
183. V. V. Korshak and S. V. Vinogradova, in 'Polyesters', ed. J. Burdon, Pergamon Press, Oxford, 1965, p. 266.
184. J. Devaux, P. Goddard, J. P. Mercier, R. Touillaux and J. M. Dereppe, *J. Polym. Sci., Polym. Phys. Ed.*, 1982, **20**, 1881.
185. J. Devaux, P. Goddard and J. P. Mercier, *J. Polym. Sci., Polym. Phys. Ed.*, 1982, **20**, 1885.
186. J. Devaux, P. Goddard and J. P. Mercier, *Polym. Eng. Sci.*, 1982, **22**, 229.
187. M. Kimura, R. S. Porter and G. Salee, *J. Polym. Sci., Polym. Phys. Ed.*, 1983, **21**, 367.
188. M. Kimura, G. Salee and R. S. Porter, *J. Appl. Polym. Sci.*, 1984, **29**, 1629.
189. F. Pilati, E. Marianucci and C. Berti, *J. Appl. Polym. Sci.* 1985, **30**, 1267.
190. P. Goddard, J. M. Dekoninck, V. Devlesaver and J. Devaux, *J. Polym. Sci., Polym. Chem. Ed.*, 1986, **24**, 3301.
191. P. Goddard, J. M. Dekoninck, V. Devlesaver and J. Devaux, *J. Polym. Sci., Polym. Chem. Ed.*, 1986, **24**, 3315.
192. D. M. Kursanov, V. V. Korshak and S. V. Vinogradova, *Izv. Akad. Nauk SSSR, Ser. Khim.*, 1953, 140 (*Chem. Abstr.*, 1953, **48**, 3912g).
193. M. Droscher and F. G. Schmidt, *Polym. Bull. (Berlin)*, 1981, **4**, 261.
194. I. Goodman and B. F. Nesbit, *Polymer*, 1960, **1**, 384.
195. K. P. McAlea, J. M. Schultz, K. H. Gardner and G. D. Wignall, *Polymer*, 1986, **27**, 1581.
196. J. Kugler, J. W. Gilmer, D. Wiswe, H. G. Zachmann, K. Hanh and E. W. Fischer, *Macromolecules*, 1987, **20**, 1116.
197. H. G. Ramjit and R. D. Sedgwick, *J. Macromol. Sci., Chem.*, 1976, **A10**, 815.
198. H. G. Ramjit, *J. Macromol. Sci., Chem.*, 1983, **A19**, 41.
199. H. G. Ramjit, *J. Macromol. Sci., Chem.*, 1983, **A20**, 659.
200. W. A. Smith, J. W. Barlow and D. R. Paul, *J. Appl. Polym. Sci.*, 1981, **26**, 4233.
201. S. M. Livengood, *US Pat.* 2 567 076 (1951) (*Chem. Abstr.*, 1951, **45**, 10 675f); P. J. Flory and F. S. Leutner, *US Pat.* 2 589 687, 2 589 688 (1952) (*Chem. Abstr.*, 1952, **46**, 5891 a and f); S. V. Kantor and F. F. Holub, *US Pat.* 3 160 605, 3 160 604 (1964) (*Chem. Abstr.*, 1965, **62**, 5416a, 7468i); *US Pat.* 3 036 990, 3 036 991, 3 036 992 (1962) (*Chem. Abstr.*, 1962, **57**, 7468b, f, i); I. Goodman, J. E. McIntyre and D. H. Aldred, *Br. Pat.* 993 272 (1965) (*Chem. Abstr.*, 1965, **63**, 8521d); I. Goodman and D. H. Aldred, *Br. Pat.* 1 000 200 (1965) (*Chem. Abstr.*, 1965, **63**, 13 448f).
202. M. Droscher and G. Wegner, *Polymer*, 1978, **19**, 43.
203. M. Droscher, *J. Appl. Polym. Sci., Appl. Polym. Symp.*, 1981, **36**, 217.
204. W. M. Eareckson, *J. Polym. Sci.*, 1959, **40**, 399.
205. A. J. Conix, *Ind. Chim. Belge*, 1957, **22**, 1457.
206. A. J. Conix, *Ind. Eng. Chem.*, 1959, **51**, 147.
207. H. Schnell, *Angew. Chem.*, 1956, **68**, 633.
208. H. Nakamura, Imanishi, K. Sanui and N. Ogata, *Polym. J.*, 1974, **11**, 661.
209. N. Ogata, K. Sanui, T. Onozaki and S. Imanishi, in 'Interfacial Synthesis', ed. C. E. Carraher, Jr. and J. Preston, Dekker, New York, 1982, vol. 3, p. 379.
210. W. Deits and O. Vogl, *J. Macromol. Sci., Chem.*, 1981, **A16**, 1145.
211. P. W. Morgan, *Polym. Rev.*, 1965, **10**, 48.
212. L. B. Sokolov, *Polym. Sci. USSR (Engl. Transl.)*, 1970, **12**, 1097.

213. E. Turska, *Polym. Sci. USSR (Engl. Transl.)*, 1973, **15**, 448.
214. V. A. Vasnev, A. I. Tarasov, S. V. Vinogradova and V. V. Korshak, *Polym. Sci. USSR (Engl. Transl.)*, 1975, **17**, 828.
215. V. V. Korshak, S. V. Vinogradova, V. A. Vasnev and T. I. Mitaishvili, *Polym. Sci. USSR (Engl. Transl.)*, 1969, **11**, 89.
216. V. V. Korshak, S. V. Vinogradova, V. A. Vasnev and A. I. Tarasov, *Polym. Sci. USSR (Engl. Transl.)*, 1975, **17**, 1388.
217. V. V. Korshak, S. V. Vinogradova, V. A. Vasnev, Yu. I. Perfilov and P. O. Okulevich, *Polym. Sci. USSR (Engl. Transl.)*, 1974, **16**, 2852.
218. V. V. Korshak, S. V. Vinogradova, V. A. Vasnev, Yu. I. Perfilov and P. O. Okulevich, *J. Polym. Sci., Polym. Chem. Ed.*, 1973, **11**, 2209.
219. E. Turska, S. Boryniec and L. Pietrzak, *J. Appl. Polym. Sci.*, 1974, **18**, 667.
220. E. Turska, S. Boryniec and A. Dems, *J. Appl. Polym. Sci.*, 1974, **18**, 671.
221. P. A. Curnuck and M. E. B. Michael, *Br. Polym. J.*, 1973, **5**, 21.
222. S. V. Vinogradova, V. V. Korshak, G. Sh. Papava, N. A. Maisuradze and P. D. Tsiskarishvili, *Polym. Sci. USSR (Engl. Transl.)*, 1972, **14**, 2685.
223. V. V. Korshak and S. V. Vinogradova, in 'Polyesters', ed. J. Burdon, Pergamon Press, Oxford, 1965, p. 178.
224. W. Kern, P. Munk, A. Sabel and K. H. Schmidt, *Makromol. Chem.*, 1956, **17**, 201.
225. W. Kern, P. Munk and K. H. Schmidt, *Makromol. Chem.*, 1956, **17**, 219.
226. W. E. Elwell and D. C. McGowan, *US Pat.* 2 585 323 (1952) (*Chem. Abstr.*, 1953, **47**, 2775c).
227. R. F. Fischer, *J. Polym. Sci.*, 1960, **44**, 155.
228. E. Schwenk, K. Gulbins, M. Roth, G. Benzing, R. Maysenholder and K. Hamann, *Makromol. Chem.*, 1962, **51**, 53.
229. D. M. Young and F. Hostettler, *Br. Pat.* 859 642 (1961) (*Chem. Abstr.*, 1961, **55**, 25 355g).
230. G. L. Brode and J. V. Koleske, *J. Macromol. Sci., Chem.*, 1972, **A6**, 1109.
231. A. Schindler, Y. M. Hibionada and G. C. Pitt, *J. Polym. Sci., Polym. Chem. Ed.*, 1982, **20**, 319.
232. H. K. Hall, Jr. and A. K. Schneider, *J. Am. Chem. Soc.*, 1958, **80**, 6409.
233. H. K. Hall, Jr. and R. Zbinden, *J. Am. Chem. Soc.*, 1958, **80**, 6428.
234. H. K. Hall, Jr., M. K. Brandt and R. M. Mason, *J. Am. Chem. Soc.*, 1958, **80**, 6420.
235. F. Korte and W. Glet, *J. Polym. Sci., Part B.*, 1966, **4**, 685.
236. K. Saotome and Y. Kodaira, *Makromol. Chem.*, 1965, **82**, 41.
237. R. Huisgen and H. Ott, *Tetrahedron*, 1959, **6**, 253.
238. D. L. Ann and H. E. Shi Ping, *Eur. Pat. Appl.* 117 538 (1984) (*Chem. Abstr.*, 1985, **102**, 25 551).
239. Y. Yamanaka, I. Suzuki, J. Omura and H. Iwashita, *Jpn. Pat.* 5910 (1968) (*Chem. Abstr.*, 1968, **69**, 11 346).
240. F. Hostettler and D. M. Young, *US Pat.* 3 169 945 (1965) (*Chem. Abstr.*, 1965, **62**, 11 936h).
241. F. Hostettler, G. Magnus and H. Vineyard, *Fr. Pat.* 1 418 360 (1965) (*Chem. Abstr.*, 1966, **65**, 2435g).
242. D. M. Young, F. Hostettler and C. F. Horn, *US Pat.* 2 890 208 (1959) (*Chem. Abstr.*, 1959, **53**, 18 546h).
243. E. A. Dawes and P. J. Senior, *Adv. Microb. Physiol.*, 1973, **10**, 135.
244. M. Doudoroff and R. Y. Stainer, *Nature (London)*, 1959, **183**, 1440.
245. D. G. Lundgren, R. E. Alper, C. Schnaitman and R. H. Marchessault, *J. Bacteriol.*, 1965, **89**, 245.
246. J. M. Merrick, *J. Photosynth. Bacteriol.*, 1978, **199**, 219.
247. J. N. Baptist *US Pat.* 3 036 959, 3 044 942 (1962) (*Chem. Abstr.*, 1962, **57**, 17 219a, 12 725d).
248. D. Ellar, D. G. Lundgren, K. Okamura and R. H. Marchessault, *J. Mol. Biol.*, 1968, **35**, 489.
249. D. H. Williamson and J. F. Wilkinson, *J. Gen. Microbiol.*, 1958, **19**, 198.
250. J. Merrick and M. Doudoroff, *J. Bacteriol.*, 1964, **88**, 60.
251. S. Akita, Y. Einaga, Y. Miyaki and H. Fujita, *Macromolecules*, 1976, **9**, 774.
252. S. Akita, Y. Einaga, Y. Miyaki and H. Fujita, *Macromolecules*, 1977, **10**, 1356.
253. R. E. Alper, D. G. Lundgren, R. H. Marchessault and W. A. Coté, *Biopolymers*, 1963, **1**, 545.
254. J. Cornibert and R. H. Marchessault, *J. Mol. Biol.*, 1972, **71**, 735.
255. R. H. Marchessault, S. Coulombe, H. Morikawa, K. Okamura and J. F. Revol, *Can. J. Chem.*, 1981, **59**, 38.
256. L. L. Wallen and W. K. Rohwedder, *Environ. Sci. Technol.*, 1974, **8**, 576.
257. R. H. Findlay and D. C. White. *Appl. Environ. Microbiol.*, 1983, **45**, 71.
258. Y. Doi, M. Kunioca and K. Soga, *Macromolecules*, 1986, **19**, 2860.
259. S. H. Collins, *Eur. Pat.* 111 231, (1982) (*Chem. Abstr.*, 1986, **102**, 156 780).
260. J. T. Dickson, H. P. W. Huggill and J. C. Welch, *Br. Pat.* 590 451 (1947) (*Chem. Abstr.*, 1948, **42**, 414h).
261. S. Bezzi, *Gazz. Chim. Ital.*, 1949, **79**, 219.
262. J. D. Doedens and E. H. Rosenbrock, *Belg. Pat.* 622 054 (1963) (*Chem. Abstr.*, 1963, **59**, 1813f).
263. E. P. Goldberg, *Br. Pat.* 870 096 (1962) (*Chem. Abstr.*, 1962, **56**, 14 173h).
264. E. P. Goldberg, *Polym. Prepr., Am. Chem. Soc., Div. Polym. Chem.*, 1964, **5**, 233.
265. BASF, *Ger. Pat.* 890 792 (1955) (*Chem. Abstr.*, 1956, **50**, 12 099a).
266. D. S. Tarbel and N. A. Leister, *J. Org. Chem.*, 1958, **23**, 1149.
267. D. S. Tarbel and N. J. Longosz, *J. Org. Chem.*, 1959, **24**, 774.
268. T. B. Windhloz, *J. Org. Chem.*, 1958, **23**, 2044.
269. T. B. Windhloz, *J. Org. Chem.*, 1960, **25**, 1703.
270. P. W. Morgan, *Polym. Rev.*, 1965, **10**, 362.
271. H. Harada and H. Higashi, *Kogyo Kagaku Zasshi*, 1965, **68**, 1980 (*Chem. Abstr.*, 1966, **64**, 6761e).
272. G. Natta, G. Mazzanti, G. Pregaglia and M. Biraghi, *J. Am. Chem. Soc.*, 1960, **82**, 5511.
273. G. Natta, G. Mazzanti, G. Pregaglia and G. Pozzi, *J. Polym. Sci.*, 1962, **58**, 1201.
274. R. G. J. Miller, E. Nield and A. Turner-Jones, *Chem. Ind. (London)*, 1962, 181.
275. E. S. Shreiner, V. P. Zubov, V. A. Kabanov and V. A. Kargin, *Dokl. Akad. Nauk SSSR*, 1964, **156**, 396 (*Chem. Abstr.*, 1964, **61**, 4495f).
276. G. Natta, G. Mazzanti, G. Pregaglia, M. Biraghi and M. Peraldo, *J. Am. Chem. Soc.*, 1960, **82**, 4742.
277. G. Natta, G. Mazzanti, G. Pregaglia and M. Biraghi, *Makromol. Chem.*, 1961, **44–46**, 537.
278. Y. Yamashita and S. Nunomoto, *Makromol. Chem.*, 1963, **67**, 10.
279. H. J. Hagemeyer, Jr., *US Pat.* 2 533 455 (1950) (*Chem. Abstr.*, 1951, **45**, 2265b).
280. J. G. N. Drewitt and J. Lincoln, *Br. Pat.* 707 913 (1954) (*Chem. Abstr.*, 1954, **48**, 12 465a).
281. Chemstrand Corp., *Br. Pat.* 769 700 (1957) (*Chem. Abstr.*, 1957, **51**, 12 552f).

282. W. Sweeny, *J. Appl. Polym. Sci.*, 1963, **7**, 1983.
283. I. Goodman and E. Haddock, *Br. Pat.* 968 390 (1964) (*Chem. Abstr.*, 1964, **61**, 14 809g).
284. *US Pat.* 3 053 810 (1962) (*Chem. Abstr.*, 1963, **58**, 1560c).
285. J. L. R. Williams, J. M. Carlson and G. A. Reynolds, *Makromol. Chem.*, 1963, **65**, 54.
286. J. G. Erikson, *J. Polym. Sci., Part A-1*, 1966, **4**, 519.
287. F. D. Trischler and J. Hollander, *J. Polym. Sci., Part A-1*, 1969, **7**, 1971.
288. I. Luderwald, in 'Developments in Polymer Degradation', ed. N. Grassie, Applied Science, London, 1979, vol. 2, p. 77.
289. D. J. Crisson, in 'Encyclopedia of Polymer Science and Engineering', 2nd edn., ed. H. F. Mark, N. M. Bikales, C. G. Overberger and G. Menges, Wiley, New York, 1986, vol. 4, p. 677.
290. V. Passalacqua, F. Pilati, V. Zamboni, B. Fortunato and P. Manaresi, *Polymer*, 1976, **17**, 1044.
291. J. Devaux, P. Godard and J. P. Mercier, *Makromol. Chem.*, 1978, **179**, 2201.
292. H. A. Pohl, *J. Am. Chem. Soc.*, 1951, **73**, 5660.
293. I. Marshall and A. Todd, *Trans. Faraday Soc.*, 1953, **49**, 67.
294. H. Zimmermann and Dao duy Chu, *Faserforsch. Textiltech.*, 1973, **24**, 445.
295. H. Zimmermann, *Lenzinger Ber.*, 1964, **36**, 64.
296. I. Luderwald and M. Urrutia, *Makromol. Chem.*, 1976, **177**, 2079.
297. R. M. Lum, *J. Polym. Sci., Polym. Chem. Ed.*, 1979, **17**, 203.
298. G. Rafler, J. Blaesche, H. Gajewski and K. Zacharias, *Acta Polym.*, 1980, **31**, 633.
299. H. Zimmerman and P. Lohmann, *Acta Polym.*, 1980, **31**, 686.
300. G. Rafler and J. Blaesche, *Acta Polym.*, 1982, **33**, 472.
301. H. Zimmerman and E. Leibnitz, *Faserforsch. Textiltech.*, 1965, **16**, 282.
302. K. Tomita, *Polymer*, 1977, **18**, 295.
303. K. Tomita, *Polymer*, 1977, **18**, 1295.
304. K. Yoda, A. Tsuboi, M. Wada and R. Yamadera, *J. Appl. Polym. Sci.*, 1970, **14**, 2357.
305. I. Luderwald and M. Urrutia, *Makromol. Chem.*, 1976, **177**, 2093.
306. V. V. Korshak and S. V. Vinogradova, in 'Polyesters', ed. J. Burdon, Pergamon Press, Oxford, 1965, p. 25.
307. J. K. Stille and T. W. Campbell, in 'High Polymers', ed. H. Mark, C. S. Marvel and H. W. Melville, Interscience, New York, 1972, vol. 27, p. 21.
308. P. W. Morgan, *Polym. Rev.*, 1965, **10**, 320.
309. C. S. Fuller and C. L. Erickson, *J. Am. Chem. Soc.*, 1937, **59**, 344.
310. C. S. Fuller and C. J. Frosh, *J. Am. Chem. Soc.*, 1939, **61**, 2575.
311. C. S. Fuller et al., *J. Am. Chem. Soc.*, 1942, **64**, 154.
312. B. F. Cinadr and E. G. Bobalek, *J. Appl. Polym. Sci.*, 1962, **6**, 32.
313. P. Morison and J. E. Hutchins, *Pap. Meet.—Am. Chem. Soc., Div. Org. Coat. Plast. Chem.*, 1961, **21:1**, 159.
314. H. J. Hagemeyer, *US Pat.* 3 043 808 (1962) (*Chem. Abstr.*, 1962, **57**, 12 725g).
315. J. K. Stille and T. W. Campbell, in 'High Polymers', ed. H. Mark, C. S. Marvel and H. W. Melville, Interscience, New York, 1972, vol. 27, p. 355.
316. B. J. Tighe, *Chem. Ind. (London)*, 1968, 1837.
317. I. J. Smith and B. J. Tighe, *Chem. Ind. (London)*, 1973, 695.
318. G. P. Blackbourn and B. J. Tighe, *J. Chem. Soc. C*, 1971, 257.
319. B. W. Evans, D. J. Fenn and B. J. Tighe, *J. Chem. Soc. B*, 1970, 1049.
320. M. D. Thomas and B. J. Tighe, *J. Chem. Soc. B*, 1970, 1039.
321. D. J. Fenn, M. D. Thomas and B. J. Tighe, *J. Chem. Soc. B*, 1970, 1044.
322. I. J. Smith and B. J. Tighe, *J. Polym. Sci., Polym. Chem. Ed.*, 1976, **14**, 2293.
323. D. G. H. Ballard and B. J. Tighe, *J. Chem. Soc. B*, 1967, 702.
324. G. P. Blackbourn and B. J. Tighe, *J. Chem. Soc., Perkin Trans. 2*, 1972, 1263.
325. G. P. Blackbourn and B. J. Tighe, *J. Chem. Soc. B*, 1971, 1384.
326. I. J. Smith and B. J. Tighe, *Chem. Ind. (London)*, 1969, 170.
327. I. J. Smith and B. J. Tighe, *J. Polym. Sci., Polym. Chem. Ed.*, 1976, **14**, 949.
328. I. J. Smith and B. J. Tighe, *Makromol. Chem.*, 1981, **182**, 313.
329. R. D. Gilbert, V. Stannett, G. C. Pitt and A. Schinder, in 'Developments in Polymer Degradation', ed. N. Grassie, Applied Science, London, 1982, vol. 4, p. 259.
330. B. J. Tighe, in 'Macromolecular Chemistry', ed. A. D. Jenkins and J. F. Kennedy, Royal Society of Chemistry, London, 1980.
331. W. F. H. Borman and M. Kramer, *Coat. Plast. Prepr. Pap. Meet. (Am. Chem. Soc., Div. Org. Coat. Plast. Chem)*, 1974, **34**, 77.
332. B. G. Jaquiss, W. F. H. Borman and R. W. Campbell, in 'Kirk-Othmer Encyclopedia of Chemical Technology', 3rd edn., Wiley, New York, 1982, vol. 18, p. 549.
333. D. C. Clagett, in 'Encyclopedia of Polymer Science and Engineering', 2nd edn., ed. H. F. Mark, N. M. Bikales, C. G. Overberger and G. Menges, Wiley, New York, 1986, vol. 6, p. 111.
334. H. Zimmerman and D. Becker, *Faserforsch. Textiltech.* 1973, **24**, 479.
335. W. L. Hergenrother, *J. Polym. Sci., Polym. Chem. Ed.*, 1974, **12**, 875.
336. V. Hornoff and L. Zeman, *Can. J. Chem.*, 1978, **56**, 2703.
337. H. Zimmerman and N. T. Kim, *Polym. Eng. Sci.*, 1980, **20**, 680.
338. M. Droscher, *J. Appl. Polym. Sci., Appl. Polym. Symp.*, 1981, **36**, 217.
339. T. Yu, H. Bu, J. Chen, J. Mei and J. Hu, *Makromol. Chem.*, 1986, **187**, 2697.
340. S. Chang, M. F. Sheu and N. H. Chang, *J. Polym. Sci., Polym. Chem. Ed.*, 1982, **20**, 2053.
341. H. Kamatani, S. Konagaya and Y. Nakamura, *Polym. J.*, 1980, **12**, 125.
342. S. M. Aharoni, C. E. Forbes, W. B. Hammond, D. M. Hindenlang, F. Mares, K. O'Brien and R. D. Sedgwick, *J. Polym. Sci., Polym. Chem. Ed.*, 1986, **24**, 1281.
343. H. K. Reimschuessel, *Ind. Eng. Chem. Prod. Res. Dev.*, 1980, **19**, 117.
344. G. W. Halek, *J. Polym. Sci., Polym. Symp.*, 1986, **74**, 83.
345. S. A. Jabarin and E. A. Lofgren, *J. Appl. Polym. Sci.*, 1986, **32**, 5315.

346. V. Hornof, *J. Macromol. Sci., Chem.*, 1981, **A15**, 503.
347. H. Renwen, Y. Fehg, H. Tinzheng and G. Shiming, *Angew. Makromol. Chem.*, 1983, **119**, 159.
348. J. R. Kirby, A. J. Baldwin and R. H. Heidner, *Anal. Chem.*, 1965, **37**, 1306.
349. S. G. Hovenkamp and J. P. Munting, *J. Polym. Sci., Part A-1*, 1970, **8**, 679.
350. H. K. Reimschuessel, T. DeBona and A. K. S. Murthy, *J. Polym. Sci., Polym. Chem. Ed.*, 1979, **17**, 3217.
351. H. K. Reimschuessel and T. DeBona, *J. Polym. Sci., Polym. Chem. Ed.*, 1979, **17**, 3241.
352. K. Ravindranath and R. A. Mashelkar, *J. Appl. Polym. Sci.*, 1981, **26**, 3179.
353. K. Ravindranath and R. A. Mashelkar, *J. Appl. Polym. Sci.*, 1982, **27**, 471.
354. K. Ravindranath and R. A. Mashelkar, *Polym. Eng. Sci.*, 1982, **22**, 610.
355. K. Ravindranath and R. A. Mashelkar, *J. Appl. Polym. Sci.*, 1982, **27**, 2625.
356. K. Ravindranath and R. A. Mashelkar, *Polym. Eng. Sci.*, 1982, **22**, 619.
357. A. Kumar, S. K. Gupta, B. Gupta and D. Kunzru, *J. Appl. Polym. Sci.*, 1982, **27**, 4421.
358. A. Kumar, S. N. Sharma and S. K. Gupta, *J. Appl. Polym. Sci.*, 1984, **29**, 1045.
359. K. Ravindranath and R. A. Mashelkar, *AIChE. J.*, 1984, **30**, 415.
360. J. R. Whinfield and J. T. Dickson, *US Pat.* 2 465 319 (1949) (*Chem. Abstr.*, 1949, **43**, 4896g).
361. B. G. Hudson and R. Barker, *J. Org. Chem.*, 1967, **32**, 3650.
362. A. Buyle Padias and H. K. Hall, Jr., *J. Polym. Sci., Polym. Chem. Ed.*, 1981, **19**, 1021.
363. A. Roviello and A. Sirigu, *Makromol. Chem.*, 1982, **183**, 895.
364. L. Strzelecki and D. van Luyen, *Eur. Polym. J.*, 1980, **16**, 299.
365. A. Blumstein, K. N. Sivramakrishnan, R. B. Blumstein and S. B. Clough, *Polymer.*, 1982, **23**, 47.
366. P. W. Morgan, *Polym. Rev.*, 1965, **10**, 319.
367. W. R. Krigbaum, A. Ciferri, J. Asrar, H. Toriumi and J. Preston, *Mol. Cryst. Liq. Cryst.*, 1981, **76**, 99.
368. A. Roviello and A. Sirigu, *Eur. Polym. J.*, 1979, **15**, 61.
369. A. Blumstein, K. N. Sivramakrishnan, S. B. Clough and R. B. Blumstein, *Mol. Cryst. Liq. Cryst. Lett.*, 1979, **49**, 255.
370. J. Asrar, H. Toriumi, J. Watanabe, W. R. Krigbaum, A. Ciferri and J. Preston, *J. Polym. Sci., Polym. Phys. Ed.*, 1983, **21**, 1119.
371. A. Blumstein, S. Vilasagar, S. Ponrathnam, S. B. Clough and R. B. Blumstein, *J. Polym. Sci., Polym. Phys. Ed.*, 1982, **20**, 877.
372. A. Blumstein and O. Thomas, *Macromolecules*, 1982, **15**, 1264.
373. H. K. Hall, Jr., T. Kuo, R. W. Lenz and T. M. Leslie, *Macromolecules*, 1987, **20**, 2041.

18
Polyarylates

BARRY D. DEAN, MARKUS MATZNER and JAMES M. TIBBITT
Amoco Performance Products Inc., Bound Brook, NJ, USA

18.1 INTRODUCTION	317
18.2 SYNTHETIC METHODS	318
18.2.1 The Acid Chloride Route	318
18.2.1.1 The interfacial reaction	318
18.2.1.2 The solution polycondensation	319
18.2.2 The Diacetate Route	320
18.2.3 The Diphenate Process	321
18.3 COMPARISON OF POLYARYLATE PROCESSES	322
18.3.1 Process Economics	322
18.3.2 Process Complexity and Reliability	323
18.3.3 Process Flexibility and Product Quality	323
18.4 PROPERTIES OF POLYARYLATES	324
18.4.1 Mechanical and Thermal	324
18.4.2 Thermal/Color Stabilization	324
18.4.3 Photo-Fries Rearrangement (UV Stability)	325
18.4.4 Hydrolytic Stability	326
18.4.5 Polyarylate Blends	327
18.5 REFERENCES	327

18.1 INTRODUCTION

Polyarylates are an emerging class of new engineering polymers. These materials are wholly aromatic polyesters derived from aromatic dicarboxylic acids and diphenols. The earliest references date back almost 30 years.[1-3] Commercial introduction, however, only began in 1970. At that time, the highly crystalline (liquid crystalline, LC) polyesters based on *p*-hydroxybenzoic acid were commercialized by the Carborundum Company. A few years later (about 1975–1978) amorphous polyarylates, made from bisphenol A and a mixture of isophthalic and terephthalic acids, were made available by Unitika in Japan (U-polymer) and by the Union Carbide Corporation ('ARDEL polyarylate') in the United States.

Given the number of available diphenols, aromatic dicarboxylic acids and aromatic hydroxycarboxylic acids, a host of polyarylate compositions are possible.[4] The class encompasses the highly oriented liquid crystalline materials mentioned above, the crystalline and semi-crystalline polyarylates derived from symmetrical diphenols such as 4,4'-biphenol and hydroquinone, and all the way to the amorphous bisphenol A polymers. As would be expected, the literature on polyarylates is very extensive. This may be illustrated by the fact that in a book by Korshak and Vinogradova[5] issued in 1965, about 80 pages are devoted to polyarylates. In a later book by Korshak, published in 1971,[6] one finds already 140 different polyarylate formulae and 279 references. The literature continued to grow and there are now thousands of references.

This chapter will discuss the synthesis and the properties of the commercial polyarylates such as the bisphenol A/isoterephthalate of structure (**1**). It is to be noted, however, that the methods useful for the preparation of (**1**) are in most cases quite general and applicable to the synthesis of almost any other aromatic polyester.

$$\left[-O -\underset{Me}{\overset{Me}{\underset{|}{\overset{|}{C}}}}- OCO-\bigcirc-CO- \right]_n$$

(1)

18.2 SYNTHETIC METHODS

The direct uncataylzed reaction of an aromatic diacid with a diphenol is inadequate as a general synthetic procedure for the preparation of high molecular weight polyarylates.[7] Two references[8,9] describe the direct preparation of high polymer using antimony, tin, titanium, and germanium catalysts. Typical monomers used were hydroquinone, substituted hydroquinones, various 4,4'-biphenols, isophthalic, terephthalic and naphthalenedicarboxylic acids, as well as hydroxyacids such as p-hydroxybenzoic acid and 6-hydroxy-2-naphthoic acid. According to these references the method is probably not applicable to the preparation of polyesters from the relatively less stable alkylidene and cycloalkylidene diphenols such as bisphenol A. Along similar lines, a disclosure[10] has indicated the possibility of preparing polyarylates from a mixture of a diphenol, a dicarboxylic acid and a lower aliphatic monocarboxylic acid in the presence of a cobalt, nickel or manganese organic/inorganic mixed salt. This latter indication, however, was not substantiated by any example of the subject disclosure.

Thus, the literature data regarding the direct phenol–acid reaction are very scarce. At this stage the approach appears to be commercially impractical for producing high molecular weight linear polyarylates. Other polycondensation techniques, yielding high quality polymer, were therefore developed over the years. In these techniques, at least one of the monomers is derivatized to enhance its reactivity. The existing polyarylate processes can be subdivided into three categories: the acid chloride route, the diacetate route and the diphenate route. The acid chloride polymerizations are generally solution reactions, whereas the diacetate and the diphenate methods can be run in solution, in the melt or as a slurry.

18.2.1 The Acid Chloride Route

This synthetic route is shown in equation (1). The condensation of an acid chloride with a diphenol can be accomplished in a variety of ways. The method most often used, which has achieved commercial status in the production of U-polymer, is interfacial polymerization. Other methods, i.e. the low temperature and the high temperature solution polycondensations, were also developed. A brief discussion of each follows.

$$nHO-\bigcirc-\underset{Me}{\overset{Me}{\underset{|}{\overset{|}{C}}}}-\bigcirc-OH + nClCO-\bigcirc-COCl \xrightarrow{-2nHCl}$$

$$\left[-O-\bigcirc-\underset{Me}{\overset{Me}{\underset{|}{\overset{|}{C}}}}-\bigcirc-OCO-\bigcirc-CO- \right]_n \quad (1)$$

18.2.1.1 The interfacial reaction

The interfacial process for the preparation of polyarylates was first described by Eareckson[3] and Conix.[1,11,12] Detailed descriptions of this process have since been published.[13,14] Basically, the interfacial polymerization involves the reaction of an aqueous solution of the dialkali metal salt of a diphenol with a solution of the acid chloride(s) in a water immiscible solvent (e.g. methylene chloride or chlorobenzene) which is inert under the reaction condition. It is advantageous that the polyester formed be soluble in the organic phase. As the two starting immiscible solutions are brought into contact, the polymerization occurs at the interface where a film of the polymer is formed. Diffusion through this film, affected by the extent to which the solvent can swell or dissolve the polyarylate, can be an important factor in the process. Obviously, this effect will be minimized if the polymer is soluble in the organic phase and/or if the agitation is sufficiently intense to break it up into small particles. Phase transfer catalysts may be used and are typically tertiary amines as well as quaternary ammonium or phosphonium salts. The reactions take place at temperatures that are generally within the range of about 0 to 35 °C and proceed at relatively high rates. Thus, a series of high molecular weight polyarylates was prepared by Eareckson[3] by stirring the two solutions in a home blender for five minutes only. Molecular weights can be controlled by adding appropriate chain stoppers such as a monophenol or a monocarboxylic acid chloride. After the polymerization is complete, the two phases are separated, the organic copolymer solution is washed, and the polymer is isolated by coagulation in a non-solvent or by devolatilization.

The interfacial reaction is quite general and allows for the preparation of a wide range of polyarylates. It tends to proceed sluggishly when the more acidic bisphenols (*e.g.* the 4,4'-dihydroxydiphenyl sulfone, *etc.*) are used. Also, the method is of rather limited usefulness when the diacid chloride is hydrolytically unstable, and/or when the polyester formed is only slightly soluble and precipitates from the reaction medium at the oligomer stage.

A variant, which could, in principle, be classed as an interfacial polymerization was described in the patent literature.[15] It involves contacting a solution of bisphenol A and iso/terephthaloyl chlorides with calcium hydroxide; methylene chloride is the preferred solvent and triethylamine is the phase transfer catalyst. When the polymerization is completed, the inorganic salts ($CaCl_2/Ca(OH)_2$) are removed *via* filtration and the polymer is isolated as described above.

An interfacial synthesis of polyarylates using acid chlorides that were prepared *via* the chlorination of the corresponding dimethyl esters was recently described.[16] The process reportedly yields materials having excellent color, as well as improved thermal and hydrolytic stability. Along related lines, the possibility of using diacid chlorides prepared *in situ*, *i.e.* without any purification, to make high molecular weight polymer was also demonstrated.[17]

18.2.1.2 *The solution polycondensation*

(i) *Low temperature solution polycondensation*

Polyarylates can be prepared using the classical Schotten–Baumann ester-forming conditions.[18–21] Polymer is produced by reacting essentially equivalent amounts of the diacid chloride and dihydroxy compound in an inert solvent, in the presence of a stoichiometric amount of an acid acceptor. The reaction is typically conducted in solvents such as tetrahydrofuran, in the presence of pyridine or triethylamine. The reactions are generally run at room temperature or below (-10 °C to $+30$ °C). With soluble polyarylates, material of excellent quality can be made *via* this route.

The direct room temperature reaction of phosgene with a stoichiometric mixture of a diacid and a diphenol in pyridine yields the corresponding polyesters.[22,23] If an excess of diphenol is employed, copoly(ester/carbonates) result. The polymerization can be performed in the tertiary base only, which then acts as both the solvent and catalyst, or in mixtures of the tertiary base and an inert solvent such as methylene chloride, xylene, chlorobenzene, *etc.* Various tertiary bases were suggested to be useful in this process. Although several mechanisms may explain the formation of the ester linkage, that shown in equation (2) was favored.[23]

$$\begin{array}{c} \text{—ArOH} + COCl_2 \xrightarrow{\text{base}} \text{—ArOCCl} \xrightarrow{-Ar^1CO_2H} \\ \text{—ArOCOCAR}^1 \xrightarrow{-CO_2} \text{—ArOCAr}^1 \\ \text{ester} \end{array} \quad (2)$$

(ii) *High temperature solution polycondensation*

The high temperature reaction of diacid chlorides with diphenols in an inert solvent yields polyester and hydrogen chloride by-product. The non-catalyzed polycondensations generally require temperatures above 200 °C to give high polymer. Thus, the reaction of isophthaloyl and terephthaloyl chlorides with bisphenol A was reported to yield good quality material at temperatures of 215–220 °C in dichloroethylbenzene.[24] Another solvent, used widely by Soviet workers, is ditolylmethane.[25] A wide variety of other media are useful and include chlorinated benzenes (*e.g.* tetrachlorobenzene), chlorinated diphenyls or diphenyl ethers, chlorinated naphthalenes, as well as non-chlorinated aromatics such as terphenyl, benzophenones, dibenzylbenzenes, and the like. The process was found useful for the preparation of hydroquinone and 4,4'-biphenol iso- and terephthalates;[7,26] these latter polymers are highly crystalline with limited solubilities and high temperatures are required to keep them in solution in order to build up their molecular weights into the useful range.

The high temperature diacid chloride/diphenol polymerization can be catalyzed with a large number of Lewis acids and bases.[27-30] Aluminum, titanium and magnesium salts are typical catalysts. The catalyzed reactions can be conducted at temperatures as low as 130 °C. Solvents such as chlorobenzene, *sym*-tetrachloroethane, and dichlorobenzene are preferred.

The high temperature solution polycondensation is useful for the preparation of crystalline polyarylates (*vide supra*). As far as amorphous polyesters are concerned, the non-catalyzed version is of interest since it yields a product that is free of catalyst, which may have a deleterious effect on stability, *etc*. Generally, however, the high temperature acid chloride reactions yield polymer of poorer color than their low temperature counterparts.

18.2.2 The Diacetate Route

The diacetate process involves reaction of stoichiometric amounts of an aromatic dicarboxylic acid and the diacetate derivative of an aromatic diphenol at high temperature under an inert atmosphere (equation 3). The diacetate process first described by Conix[1] and Levine and Temin[2] is essentially a reversible melt process, which requires continuous removal of acetic acid in order to achieve high molecular weight polymer. It was postulated that aromatic ester formation occurs by the reaction of a phenol end group and a mixed anhydride with concurrent elimination of a molecule of acetic acid (equation 4).[31]

While the chemistry of the process is relatively straightforward, there are a number of variables which must be taken into consideration to assure aromatic polyesters exhibiting good mechanical properties, low color and good melt stability.

The diacetate process can be conducted neat[32,33] but as such may suffer from stoichiometric imbalance due to aromatic diacid sublimation and slowed reaction rates at the prepolymer stage due to the reaction becoming diffusion controlled as a result of the high reaction medium viscosity. Both of these factors render the preparation of high molecular weight polyarylate resins in a neat, melt process relatively difficult. A more direct route involves the use of a diluent that is inert under the reaction conditions. Depending on the aromatic polyester composition and reaction conditions, the process is conducted homogeneously[34-36,40-42] or heterogeneously.[37-40,42] The diluent could, of course, also act as an acetic acid entraining agent. Typical diluents which have been disclosed in the literature are diphenyl ether, diphenyl ether–biphenyl eutectic mixture, chlorinated biphenyls, aromatic sulfones and γ-lactones. Under reaction conditions where the aromatic polyester is in solution (or in a swollen state) optimum molecular weight is achieved *in situ*. Comparatively, crystalline aromatic polyesters are made using slurry processes; in which a prepolymer is formed then isolated as a finely divided powder. Also, a prepolymer may be prepared as a neat melt. The molecular weight of the crystalline, aromatic polyester is advanced in the solid state by heating the powdered, crystalline polyester at a temperature below the melting temperature under high vacuum. Alternatively, it may be advanced in the molten state, such as conveying the prepolymer material through a vented extruder at a temperature above the melting point of the polyester.

The diacetate process reaction rate may be enhanced by using low levels (50–150 p.p.m.) of esterification catalysts in the presence or absence of a diluent.[32-35,37-39,41,42] Catalysts most often cited for the diacetate process are based on antimony, titanium, magnesium, manganese and zinc. The use and selection of a catalyst should be weighed against possible deleterious effects on the color and hydrolytic stability of the molded polyarylate resulting from catalyzed side reactions.

The diacetate derivative of the aromatic diphenol required for this polyarylate process can be prepared in a second reactor and subsequently charged to the polymerization reactor, or *in situ* in the polymerization reactor in the presence of the aromatic diacid(s), diluent and catalyst (optional). In order to obtain polyarylate having good color and good thermal stability, it is important to remove the residual acetic anhydride (present from the derivatization of the aromatic diphenol) from the starting diacetate.[42] Physical removal (distillation) or chemical sequestering (additional diphenol) of residual acetic anhydride are effective means for controlling the amount of free anhydride present during the high temperature condensation reaction.

One variation of the diacetate process involves the reaction of a diphenol diacetate and the dimethyl ester of the aromatic diacid. In this sequence, condensation would occur with the liberation of methyl acetate; however, the reaction proceeds only to very low molecular weight (catalyzed or uncatalyzed).[43]

Thus, the diacetate process can be applied broadly to give good quality product with the only real chemistry limitation (from a commercial viewpoint) being the solubility/reactivity of the diacid(s) component and the subsequent effect on reaction time to achieve practical molecular weight.

18.2.3 The Diphenate Process

A process which has origins in polycarbonate synthesis (see also Volume 5, Chapter 20) is the diphenyl ester route to aromatic polyesters, herein referred to as the diphenate process.[44] The reaction between a diaryl ester of an aromatic dicarboxylic acid(s) and an aromatic diphenol(s) results in high molecular weight aromatic polyester(s) (equation 5).

$$\text{Ph-O-CO-C}_6\text{H}_4\text{-CO-O-Ph} + \text{HO-C}_6\text{H}_4\text{-C(Me)}_2\text{-C}_6\text{H}_4\text{-OH} \xrightarrow[>230°C]{\Delta}$$

$$\text{+}[\text{O-C}_6\text{H}_4\text{-C(Me)}_2\text{-C}_6\text{H}_4\text{-O-CO-C}_6\text{H}_4\text{-CO}]_n + \text{Ph-OH} \quad (5)$$

First described by Blaschke and Ludwig[45] the diphenate process employs a slight excess of diphenyl carbonate in combination with the diphenyl ester of the aromatic diacid and the aromatic diphenol. The slight excess of diphenyl carbonate reportedly ensures high molecular weight polymer by compensating for any stoichiometric imbalance which might occur between the diaryl ester of the aromatic diacid and the diphenol coreactants as a result of material loss due to sublimation. As the molecular weight increases, any unreacted diphenyl carbonate can be removed downstream in a vented extruder. The ester interchange reaction in the diphenate process can proceed in the absence of a catalyst[45,46] but the rate enchancement achieved catalytically is commercially attractive.[47-50] Lithium phenoxide, sodium phenoxide and potassium borophenoxide are the preferred nominal catalysts.

The diaryl ester derivatives required for the ester interchange can be synthesized by a variety of routes (Figure 1).

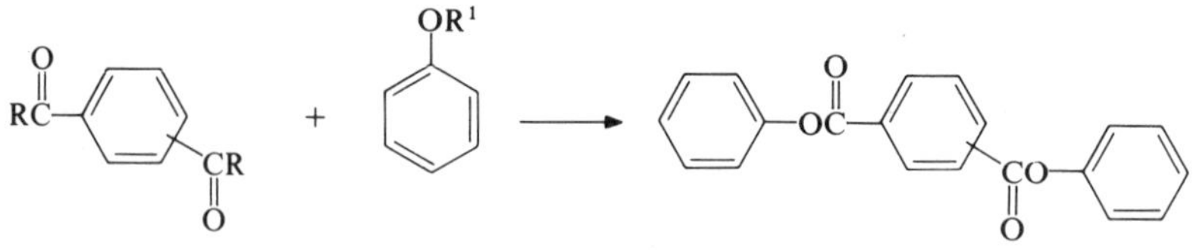

R	R^1	Conditions
Cl	H	Low temperature, no catalyst
OH	H	Δ, o-phosphoric acid option
OMe	H	Δ, titanium catalyst
OH	CO_2Me	Δ, catalyst optional
OMe	CO_2Me	Δ, titanium, antimony, tin

Figure 1 Synthesis of diaryl ester derivatives

On the occasion that more than one aromatic diacid is to be used, the complexity of raw material (diphenyl ester) availability increases in that esterification of each diacid component becomes an additional step in the overall polyarylate reaction process. The single exception is a eutectic mixture consisting of 70–75% diphenyl isophthalate and 25–30% diphenyl terephthalate which can be made in a single step. While eliminating a second esterification procedure the compositional variation offered to aromatic polyesters is quite limited. Simplification of the diphenate process is accomplished by the catalyzed, *in situ* generation of the diaryl ester of the aromatic diacid with a diaryl carbonate in the presence of the aromatic diphenol.[47]

The reversible nature of the diphenate process dictates continuous removal of phenol from the reaction system to ensure obtaining high molecular weight. As the ester interchange proceeds, the increasing melt viscosity becomes prohibitive to the efficient removal of phenol necessary for a reasonable rate of molecular weight advancement. Application of a reduced pressure environment to aid in the removal of phenol is beneficial, and in combination with (i) use of a devolatilizing extrusion apparatus to enhance polymer mass surface area exposure[48,50] and/or (ii) use of a solid state polymerization technique[51] (see also diacetate process) yields optimum molecular weight advancement.

As was the case with the diacetate process, use of an inert solvent as a slurry agent in the diphenate process has been demonstrated to aid in the melt viscosity control enhancing the rate of polymerization[52,53] in the later stages.

The diphenate process compares favorably with the diacetate process with respect to product quality (mechanical properties, color, hydrolytic stability). The limiting aspect of the diphenate process is the preparation (availability) of the diphenyl ester derivatives and in this regard each diphenate process case will be unique with specific process chemistry concerns.

18.3 COMPARISON OF POLYARYLATE PROCESSES

The three basic process options for the commercial manufacture of polyarylate polymers have been described thoroughly in the preceeding sections of this paper. At this time a critical comparison of the commercial potential of these process options will be made. The bases of comparison will be process economics, process complexity/reliability and process flexibility/product quality.

18.3.1 Process Economics

A first point of economic comparison is the variable cost requirements of each process. Here, variable costs are defined as the sum of all raw materials costs plus the utilities cost for conversion of raw materials to product. All labor, overheads and depreciation costs are not included. On a variable cost basis, both the diacetate and diphenate routes show a distinct advantage over the acid chloride route. The largest component of the cost differential results from the high cost of the acid chloride monomers relative to the free acids. The second largest component arises because the acid chloride process inherently uses greater solvent volumes than the other two routes. Solvent losses which invariably occur contribute to increased variable cost as the solvent recovery processes are not completely efficient. Variable cost differences between the diacetate and diphenate processes are not very large. Both processes can be thought of as variations to reacting free diphenol with the free diacids. In the diacetate variation, acetic anhydride is consumed in forming the diacetate, but some of this cost is recouped by selling acetic acid — the process by-product. In the diphenate route, phenol is first consumed in monomer preparation, then recovered during the polymerization. The variable cost of the diacetate route may be slightly higher than that of the diphenate route due to the conversion of anhydride to acetic acid, but this disadvantage can be mitigated depending on the phenol recovery/recycle efficiencies in the diphenate process.

Secondly, the capital investment requirement required to construct facilities to practice each of the three process technologies can be compared. The acid chloride process is a low temperature, atmospheric pressure process and process fluid viscosities are low. Thus, standard design reaction equipment with low cost supporting utilities are used in the reaction area. However, polymer recovery would generally be accomplished by precipitation, washing and drying followed by extruder pelletization — operations which are capital intensive. Also, extensive solvent recovery is required in the acid chloride process, again leading to increased capital cost. Both the melt or solution diacetate and diphenate processes on the other hand are high temperature, high vacuum processes where process fluid viscosities reach very high values. For these processes, polymer reactors will require some special design features particularly with respect to agitation and heat transfer. Supporting utilities will be rather capital intensive. To balance these costs, however, product recovery is expected to be relatively simple, requiring only one or two melt processing operations most likely using a thin film polymer processor followed by an extruder. Solvent recovery requirements would be modest for the diacetate process but somewhat more costly for the diphenate process where large quantities of phenol (especially from monomer production) will require purification prior to recycle. Some difference in capital investment required for monomer production in the diacetate and diphenate processes is also expected. Diphenyl ester production is less attractive due to the more extreme reaction conditions required and the large phenol recycle streams. However, even with the noted differences, it is estimated that any of the three described processes could be built for approximately the same dollar amount per annual pound of polymer capacity at the 15 Mlb year^{-1} scale (1 kg = 2.2 lb).

18.3.2 Process Complexity and Reliability

By analogy with other highly similar commercial chemical and polymer technologies, all three of the polyarylate processes discussed here must be regarded as reliable. As far as complexity is concerned, an advantage may be argued for the diacetate route since the monomer preparation is facile and operates under mild conditions; moreover, the polymer process uses conventional manufacturing equipment. The diphenate polycondensation is likewise straightforward, but the phenyl ester monomer preparation and in some cases recovery is complex relative to the diacetate monomer preparation and the entire plant must be engineered to contain the large quantities of phenol inherent to the process. Finally, though the acid chloride process is not particularly complex to engineer or operate, it is certainly relatively cumbersome due to the multiplicity of unit operations and nature of the process streams (especially slurries) which are present.

18.3.3 Process Flexibility and Product Quality

In the polyarylate processes under discussion, process flexibility is often measured in terms of the range of isophthalic to terephthalic acid ratio which can be accommodated. Bisphenol A compositions near 100% isophthalate and above 75% terephthalate units crystallize readily and exhibit crystalline melting points in excess of 280 °C. Therefore, processes which do not polycondense the monomers at temperatures above the melting point of certain compositional variants cannot produce these variants reliably. Thus, the acid chloride process in which the polymerization is conducted in the room temperature range is constrained to avoid the very high isophthalate and terephthalate variants. It is suggested that the compositional limits of this process are from about

85/15 = IPA/TPA to about 60/40 = TPA/IPA (IPA = isophthalate; TPA = terephthalate), although emerging technologies, *e.g.* proper end group control, have been claimed to extend this range up to 100% TPA.[54] Greater flexibility is afforded by the diacetate and diphenate routes where the polymerization is conducted in the 270–300 °C range, possibly in the presence of solvents which may reduce the polymer melting temperature. It is expected that both these processes can operate within varied IPA/TPA molar ratios, up to nearly 100% of either isomeric acid.

Product quality is of course a multifaceted subject. Here, mechanical properties, thermal stability and optical quality (color/clarity) will be used as the criteria; the tensile and toughness characteristics exhibited by the polymer, which are key mechanical properties, are most strongly affected by polymer linearity and average molecular weight. The acid chloride process, operated at low temperature, generally yields the most linear polyarylates and consequently exhibits good mechanical properties at a given average molecular weight.

Since polyarylates exhibit very high melt viscosity, processing is conducted at temperatures in excess of 370 °C and thus excellent thermal stability is imperative. The thermal stability correlates with the chemical nature of the polymer chain terminal groups and the linearity of the polymer. Acid chloride polyarylate, which is generally the most linear and is intentionally end capped with thermally stable moieties, is expected to exhibit good thermal stability. Diacetate and diphenate polyarylate, depending on process control, can exhibit microstructure defects in the backbone. Thus, the diacetate process yields polyarylate having good color and good thermal stability provided that the level of the residual acetic anhydride in the starting diacetate is low.[42] Moreover, acid terminal groups which may be present may additionally reduce the thermal stability of the material.

Polymer optical quality will be compared in terms of inherent polymer color and light transmission or clarity. Polyarylates tend to develop yellow or amber color due to quinoid structures in the matrix resulting from thermal or light-activated Fries rearrangement (see Volume 5, Chapter 20). Acid chloride polymer is least susceptible to this color development during manufacture due to the low polymerization temperature. Diacetate and diphenate polymer color is adversely affected, to approximately the same extent, by the higher process temperatures, although, as mentioned before, these temperatures can be moderated by using proper diluents.

Clarity will be reduced as color increases, but can also be negatively affected by the presence of particulate contamination (haze). With respect to haze, acid chloride polymer which will contain some trace residue of insoluble inorganic reaction by-products may have a slight disadvantage relative to diacetate and diphenate polymers.

18.4 PROPERTIES OF POLYARYLATES

18.4.1 Mechanical and Thermal

Polyarylates are tough materials having excellent mechanical and thermal properties. The amorphous bisphenol A based poly(iso/terephthalates) are somewhat similar to aromatic polycarbonates but exhibit significantly higher heat resistance. Typical properties are listed in Table 1.[55]

At the other end of the spectrum of possible products we find the highly crystalline LC polyesters. Some mechanical properties of an injection moldable material (XYDAR™ SRT-300) are shown in Table 2.[56] As with other liquid crystalline materials, the properties are better in the direction of flow than transverse to the direction of flow. It is noteworthy that even at 575 °F (302 °C) the material still shows a tensile strength of 3800 psi (26 MPa), a flexural modulus of 800 000 psi (5.5 GPa), and a flexural strength of 4100 psi (28 MPa).[56]

Based on their structural features, polyarylates are expected to display exceptional heat resistance characteristics. The high heat distortion temperature of ARDEL D-100 (Table 1) is one measure of the outstanding temperature performance. The material has an Underwriter's Laboratory continuous use temperature of 130 °C for electrical and mechanical properties, and 120 °C for mechanical properties with impact. The retention of mechanical properties is shown by the long term aging data of Table 3.[57] Very high continuous use temperatures are also observed with the liquid crystalline polyesters. Thus, XYDAR SRT-300 has a continuous use temperature of 240 °C for mechanical, impact and electrical properties.[58]

18.4.2 Thermal/Color Stabilization

The enhancement of color in polyarylate polymers during thermal processing has been related to degradation catalyzed by oxygen, the type and amount of polymer end group as well as the presence

Table 1 Relevant Properties of Amorphous Polyarylate (ARDEL D-100)

Glass transition temperature (°C)	190
Heat distortion temperature, 1/4 in[a] (°C)	174
Density (g ml^{-1})	1.21
Tensile modulus (psi)[b]	295 000
Tensile strength at yield (psi)[b]	9500
Tensile elongation at yield (%)	8.0
Tensile elongation at rupture (%)	50.0
Flexural modulus (psi)[b]	310 000
Flexural strength at 5% strain (psi)	11 000
Notched Izod impact, 1/8 in[a] (ft lb in^{-1} of notch)[c]	4.2
Tensile impact strength (ft lb in^{-2})[c]	140
Refractive index	1.61
Water absorption (%) immersion for:	
24 h at 72 °F (22 °C)	0.27
30 d at 72 °F (22 °C)	0.71
Dielectric constant (60 Hz)	2.73
Dissipation factor (60 Hz)	0.0008
Dielectric strength (V mil^{-1})[d]	400
Arc resistance (s)	125
Volume resistivity (Ω s)	1.2×10^{16}

[a] 1 in = 2.54 cm. [b] 1 psi = 6.89 × 10^3 Pa. [c] 1 ft lb = 1.36 Nm. [d] 1 mil = 10^{-3} in.

Table 2 Mechanical Properties of XYDAR SRT-300[2]

Tensile modulus (psi)[a,e]	2 400 000
Tensile strength (psi)[a,e]	20 000
Elongation (%)[a]	4.9
Flexural modulus (psi)[b,e]	2 000 000
Flexural strength (psi)[b,e]	19 000
DTUL at 264 psi[c,e]	671 °F
Impact strength (ft lb in^{-1})[d,f]	355 °C
notched	2.4
reverse	7.3

[a] ASTM Test D-638, data at 73 °F (23 °C). [b] ASTM Test D-790, data at 73 °F (23 °C). [c] ASTM Test D-648. [d] ASTM Test D-256. [e] 1 psi = 6.89 × 10^3 Pa. [f] 1 ft lb = 1.36 Nm; 1 in = 2.54 cm.

Table 3 ARDEL D-100: Heat Aging Studies at 300 °F (149 °C)

Property	Hours aged [tested at 72 °F (22 °C)]		
	0	500	2500
Tensile modulus (psi)[a]	290 000	275 000	310 000
Tensile strength (psi)[a]	9500	10 700	11 400
Tensile elongation at rupture (%)	50	12	10.5
Tensile elongation at yield (%)	8	7.6	7.5
Notched Izod impact, 1/8 inch (ft lb in^{-1} notch)[b]	4.2	3.6	2.0
Falling dart impact, 1/8 inch plaque (ft lb)[b]	>68	>68	>68
Tensile impact (ft lb in^{-2})[b]	130	117	182

[a] 1 psi = 6.89 × 10^3 Pa. [b] 1 ft lb = 1.36 Nm; 1 in = 2.54 cm.

of residual catalysts. Color-stabilized aromatic polyesters have been achieved using low levels of phosphites, phosphonates and/or phosphates, and in some instances a high boiling epoxide-containing compound.[59,60] Presumably, the phosphorus-based components and the epoxide-containing compounds interfere with and inactivate the degradative pathway.

18.4.3 Photo-Fries Rearrangement (UV Stability)

Aromatic polyesters undergo a solid state rearrangement when exposed to sunlight or an artificial source of UV radiation. The rearranged aromatic polyester exhibits an enhanced absorption in the

IR at 2.8 μm, a new carbonyl absorption at 6.1 μm and a shift of the UV cutoff from approximately 310 nm to approximately 350 nm. The spectral data is consistent with the formation of the o-hydroxybenzophenone structure. The photo-initiated Fries rearrangement for aromatic polyesters was first reported by Okawara, Tani and Imoto (see also Volume 5, Chapter 20).[61] Additional studies confirmed the general nature of the photo-Fries rearrangement in polyarylates and an observed photo-yellowing associated with the rearrangement.[62,63]

Two possible mechanisms can be envisioned for the rearrangement: (i) a cyclic transition state; and (ii) one involving free radical intermediates (Scheme 1). The free radical intermediate process has been confirmed for aromatic esters in the vapor state.[64] A free radical intermediate process would also seem reasonable for a solid state rearrangement due to a caged radical diffusion-controlled process.

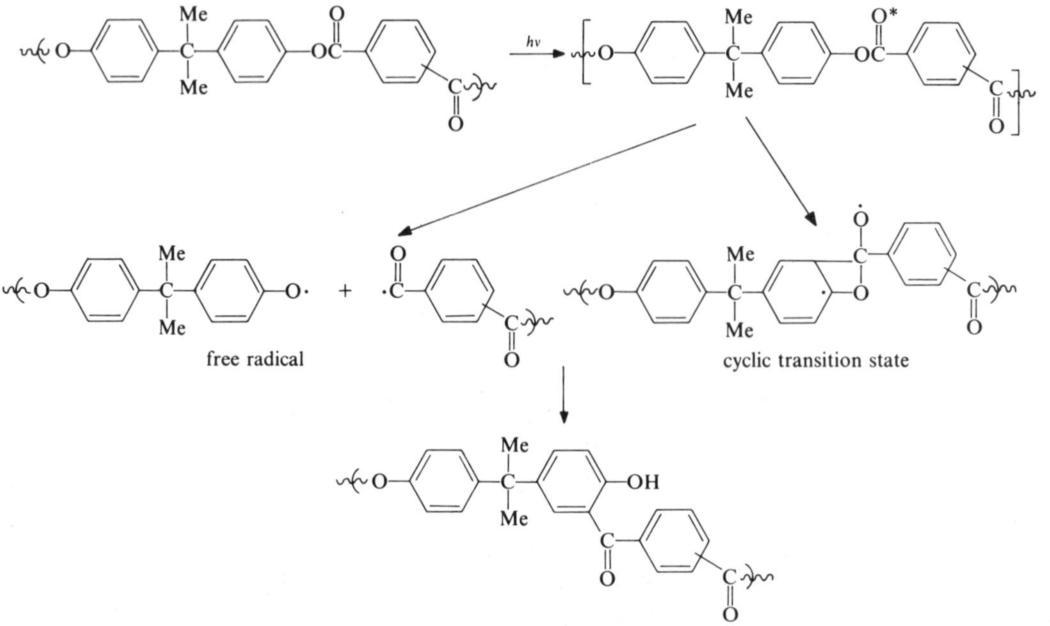

Scheme 1

The Fries rearrangement and associated yellowing can also be catalyzed with Lewis acids such as titanium and tin tetrachlorides.[63] Thus the concern relating to low color polyarylates from catalyzed melt processes (see diacetate and diphenate process, Sections 18.2.2 and 18.2.3).

The o-hydroxybenzophenone moieties are efficient UV light absorbers based on a tautomeric enol–keto equilibrium, which dissipates the energy of absorbed incident light in a radiationless manner.[65] Therefore, once sufficient rearrangement product has formed, the aromatic polyester becomes inherently UV stable.

The photo-yellowing which accompanies the rearrangement has been ascribed to the o-hydroxybenzophenone functionality. One study has indicated that the degree of yellowing may be very dependent on the structure of the aromatic polyester and thus on the hydroxybenzophenone formed and therefore at least to some extent controllable.[4]

The photo-Fries rearrangement will not occur with aromatic polyesters where the *ortho* positions of the aromatic diphenol have been completely substituted with methyl or chloro groups or similar groups.[4,66]

18.4.4 Hydrolytic Stability

Polyarylates are susceptible to hydrolytic degradation. This fact has been alluded to by Bier[43] and Freitag and Reinking[67] in terms of the necessity for thoroughly drying the polyarylate resin prior to any high temperature processing to avoid loss of mechanical properties. The hydrolysis reaction is a function of (i) the diffusion of water into the polyarylate matrix, and (ii) the heterogeneous reaction of water and the aromatic ester linkage. There have been numerous polymer blend approaches to enhancing the hydrolysis resistance of polyarylates,[68-71] the mechanisms of which have not been espoused. Presumably the effect of the various polymer additives is to influence (decrease) the rate of diffusion of water into the aromatic polyester matrix, and/or scavenging the water absorbed by the polymer.

18.4.5 Polyarylate Blends

As polymer technologists turn to the use of polymer blends to address many market/property profile needs, a chapter on the synthesis of polyarylates would not be complete without a review of polymer blends of polyarylates with other commercially available thermoplastic polymers. The blend approach is in effect an extension of polyarylate chemistry.

The properties which make polyarylates attractive blend components are the high heat resistance, ductility and inherent UV stability. The blends cited (Table 4) demonstrate the broad product ranges that have been prepared with polyarylate polymers.

Table 4 Polyarylate Blends

Other component	Refs.
Styrenic polymers	75–78
Poly(alkyl methacrylates), ionomers *etc.*	79–81
Aliphatic–aromatic polyesters	82–95
Polycarbonates and polyester carbonates	96–103
Polyamides	104–119
Poly(arylethers), poly(arylether sulfones) and poly(arylether ketones)	120–121
Poly(arylimides) and other imide-containing materials	122–124
Poly(phenylene sulfide)	125–131
Liquid crystalline polymers	132–135

Obviously, the type of polymer added to the polyarylate has a very important bearing on the properties of the resulting blend. Thus, for example, blends with nylons are more water sensitive but display an improved solvent resistance. Alloys with poly(phenylene sulfide) have been described as having improved hydrolytic stability, improved dielectric properties and good fire retardant characteristics. Addition of aromatic polycarbonates was claimed to improve color and ductility without significantly depressing the heat distortion temperatures.

Of particular interest are blends of polyarylate resins with other polyesters where the possibility of ester interchange exists. Blends of two polyesters can result in one of three situations: (i) an immiscible polymer blend where each blend component maintains chemical integrity and the free energy of mixing is positive; (ii) a miscible polymer blend where each blend component maintains chemical integrity and the free energy of mixing is negative; or (iii) a random copolymer resulting from an ester interchange reaction. This possibility is intriguing because one essentially creates a new material which can be tailored for specific applications by the addition of traditional impact modifiers, reinforcing fillers, fire retardants, *etc.*

Semicrystalline, aliphatic polyesters such as poly(ethylene terephthalate), poly(butylene terephthalate) and polyesters derived from 1,4-cyclohexane dimethanol readily transesterify with polyarylates at processing temperatures greater than 270 °C due to the presence of residual catalyst (*e.g.* titanium tetraalkoxide) in the aliphatic polyesters. However, it was claimed that if the catalyst is sequestered with arsenic oxides,[72] *o*-hydroxyaromatic acids[73] or *o*-acylphenolic compounds,[74] the rate of transesterification is slowed, and the reaction requires higher temperature and longer reaction times to achieve the same degree of randomization. Thus, polymer blend morphology can be varied by the selection of the polymer processing time/temperature profile and the use of a catalyst sequestering agent.

18.5 REFERENCES

1. A. J. Conix, *Ind. Chim. Belg*, 1957, **22**, 1457.
2. M. Levine and S. S. Temin, *J. Polym. Sci.*, 1958, **28**, 179.
3. W. M. Eareckson, *J. Polym. Sci.*, 1959, **40**, 399.

4. S. M. Cohen, R. H. Young and A. H. Markhart, *J. Polym. Sci., Part A-1*, 1971, **9**, 3263.
5. V. V. Korshak and S. V. Vinogradova, 'Polyesters', Pergamon Press, Oxford, 1965.
6. V. V. Korshak, 'The Chemical Structure and Thermal Characterization of Polymers', Israel Program for Scientific Translations, Keter, London, 1971.
7. S. W. Kantor and F. F. Holub (General Electric Co.), *US Pat.* 3 160 602 (1964).
8. S. P. Elliott (E. I. DuPont de Nemours and Co.), *US Pat.* 4 093 595 (1978).
9. A. J. East (Celanese Corp.), *Eur. Pat. Appl.* 88 546 (1983).
10. R. W. Stackman (Celanese Corp.), *US Pat.* 3 948 856 (1976).
11. A. J. Conix (Gevaert Photo Production N. V.), *US Pat.* 3 216 970 (1965).
12. A. J. Conix, *Ind. Eng. Chem.*, 1959, **51**, 147.
13. S. C. Temin, in 'Interfacial Synthesis', ed. F. Mellich and C. E. Carraher, Dekker, New York, 1977, vol. II, p. 27.
14. H. -B. Tsai and Y. -D. Lee, *J. Polym. Sci., Polym. Chem. Edn.*, 1987, **25**, 1505.
15. R. Ueno (Sumitomo Chemical Co. Ltd.), *US Pat.* 3 939 117 (1976).
16. D. Freitag, L. Bottenbruch and U. Hucks (Bayer A. G.), *US Pat.* 4 617 368 (1986).
17. H. C. Gardner and M. Matzner (Union Carbide Corp.), *US Pat.* 4 229 565 (1980).
18. W. A. Hare (E. I. DuPont de Nemours and Co.), *US Pat.* 3 234 168 (1966).
19. J. A. Pawlak, A. L. Lemper and V. A. Pattison (Hooker Chemicals and Plastics Corp.), *US Pat.* 4 049 629 (1977).
20. J. A. Pawlak, A. L. Lemper and V. A. Pattison (Hooker Chemicals and Plastics Corp.), *US Pat.* 4 051 107 (1977).
21. E. V. Gouinlock and J. C. Rosenfeld (Hooker Chemicals and Plastics Corp.), *US Pat.* 4 051 106 (1977).
22. E. P. Goldberg (General Electric Co.), *US Pat.* 3 220 976 (1965).
23. E. P. Goldberg, S. F. Strause and H. E. Munro, *Polym. Prepr., Am. Chem. Soc., Div. Polym. Chem.*, 1964, **5**, 233.
24. M. H. Keck (Goodyear Tire and Rubber Co.), *US Pat.* 3 133 898 (1964).
25. See, for example, V. V. Korshak and S. V. Vinogradova, *Vysokomol. Soedin.*, 1959, **1**, 1482.
26. S. W. Kantor and F. F. Holub (General Electric Co.), *US Pat.* 3 160 605 (1964).
27. D. R. Wilson (E. I. DuPont de Nemours and Co.), *US Pat.* 3 702 838 (1972).
28. M. Matzner and R. Barclay, Jr., *J. Appl. Polym. Sci.*, 1965, **9**, 3321.
29. W. Wolfer and E. Behr (Dynamit Nobel, A. G.), *US Pat.* 3 733 306 (1973).
30. W. Wolfer and E. Behr (Dynamit Nobel, A. G.), *Ger. Pat.* 1 933 687 (1971).
31. E. E. Riecke and F. L. Hamb, *J. Polym. Sci.*, 1977, **15**, 593.
32. A. J. Conix (Gevaert Photo Producten N. V.), *US Pat.* 3 317 464 (1967).
33. S. C. Cottis, J. Economy and L. C. Wohrer (The Carborundun Co.), *US Pat.* 3 975 487 (1976).
34. R. W. Stackman (Celanese Corp.), *US Pat.* 3 824 213 (1974); 3 948 856 (1976).
35. M. C. Yu (Phillips Petroleum Co.), *US Pat.* 4 533 720 (1985).
36. M. H. Berger, P. J. Gill and L. M. Maresca, (Union Carbide Corp.), *US Pat.* 4 294 956 (1981).
37. W. J. Jackson, H. F. Kuhfuss and J. R. Caldwell (Eastman Kodak Co.), *US Pat.* 3 780 148 (1973).
38. G. W. Calundann (Celanese Corp.), *US Pat.* 4 067 852 (1978).
39. M. H. Berger, J. M. Tibbitt and M. Matzner (Union Carbide Corp.), *US Pat.* 4 314 051 (1982).
40. M. H. Berger, L. M. Maresca and M. Matzner (Union Carbide Corp.), *US Pat.* 4 294 957 (1981).
41. L. M. Maresca, P. J. Gill and M. H. Berger (Union Carbide Corp.), *US Pat.* 4 296 232 (1981).
42. L. M. Maresca, M. Matzner and B. See (Union Carbide Corp.), *US Pat.* 4 321 355 (1982).
43. G. Bier, *Polymer*, 1974, **15**, 527.
44. H. Schnell, *Angew. Chem.*, 1956, **68**, 633.
45. F. Blaschke and W. Ludwig (Chemische·Werke Witten G. m. b. H.), *US Pat.* 3 395 119 (1968).
46. T. Urasaki, T. Yoshida, I. Inata and Y. Iraayashi (Teijin Ltd.), *Eur. Pat.* 45 499 (1982).
47. H. Schnell, V. Boilert and G. Fritz (Farbenfabriken Bayer Aktiengesellschaft), *US Pat.* 3 553 167 (1971).
48. K. Eise, R. Friedrich, H. Goemar, G. Schade and W. Wolfes (Werner Pfleiderer and Dynamit Nobel A. G.), *Ger. Pat.* 2 232 877 (1974).
49. H. Inata, S. Kawase and T. Shima (Teijin Ltd.) *US Pat.* 3 972 852 (1974).
50. G. M. Kosanovich and G. Salee (Occidental Chemical Corp.), *US Pat.* 4 465 819 (1984).
51. T. Uraski, Y. Hirabayoshi, T. Yoshida and H. Inata (Teijin Ltd.), *US Pat.* 4 436 894 (1984).
52. L. M. Maresca and M. Matzner (Union Carbide Corp.), *US Pat.* 4 459 384 (1984).
53. J. Economy and B. E. Novak (The Carborundum Co.), *US Pat.* 3 845 099 (1974).
54. D. Frietag, L. Bottenbruch, D. Rathmann and P. Tacke (Bayer A. G.) *Ger. Pat.* 3 442 125.
55. Courtesy of Amoco Performance Products, Inc.; see also L. M. Maresca and L. M. Robeson, in 'Engineering Thermoplastics: Properties and Applications', ed. J. M. Margolis, Dekker, New York, 1985, p. 255.
56. Source: Dartco Mfg., Inc., A Dart and Kraft Company, Augusta, GA, USA.
57. Source: Amoco Performance Products, Inc., Bound Brook, NJ, USA.
58. 'Modern Plastics Encyclopedia', McGraw-Hill, New York, 1980–1981, vol. 57, part 10A, p. 643.
59. J. Spanswick (Standard Oil Co.), *US Pat.* 4 302 382 (1981).
60. S. G. Gottis (Dart Industries Inc.), *US Pat.* 4 639 504 (1987).
61. M. Okawara, S. Tani and E. Imoto, *Kogyo Kagaku Zasshi*, 1965, **68**, 223.
62. S. R. Maerov, *J. Polym. Sci.*, 1965, **3**, 487.
63. D. Bellus, Z. Manasek, P. Hrdlovic and P. Slama, *J. Polym. Sci., Part C*, 1967, **16**, 267.
64. J. W. Meyer and G. S. Hammond, *J. Am. Chem. Soc.*, 1972, **94**, 2219.
65. J. E. A. Otterstedt, *J. Chem. Phys.*, 1973, **58**, 5716.
66. W. Sweeney (E. I. DuPont de Nemours and Co.), *US Pat.* 3 234 167 (1966).
67. D. Freitag and K. Reinking, *Kunststoffe*, 1981, **71**, 46.
68. L. M. Robeson (Union Carbide Corp.), *US Pat.* 4 324 869 (1982).
69. G. Salee (Hooker Chemicals and Plastics Corp.), *US Pat.* 4 327 012 (1982).
70. L. M. Robeson and T. T. Szabo (Union Carbide Corp.), *US Pat.* 4 348 500 (1982).
71. B. D. Dean (Atlantic Richfield Co.), *US Pat.* 4 542 187 (1985).
72. W. A. Smith, J. W. Barlow and D. R. Paul, *J. Appl. Polym. Sci.*, 1981, **26**, 4233.
73. R. J. McCready (General Electric Co.), *US Pat.* 4 452 933 (1984).

74. D. J. Brunelle (General Electric Co.), *US Pat.* 4 452 932 (1984).
75. A. Koshimo, H. Sakata, T. Okamoto and H. Hasegawa (Unitika Ltd.) *Jpn. Pat.* 48 51 049 (1983).
76. H. Kishikawa, K. Yasuno, S. Kitamura, K. Ueno, H. Inoue and N. Toyoda (Sumitomo Chemical Co.) *US Pat.* 3 792 118 (1974).
77. G. Salee (Hooker Chemicals and Plastics Corp.), *US Pat.* 4 126 602 (1978).
78. G. Salee (Hooker Chemicals and Plastics Corp.), *US Pat.* 4 327 012 (1982).
79. Unitika Ltd., *Jpn. Pat.* 53 51 244 (1978).
80. G. Salee (Hooker Chemicals and Plastics Corp.), *US Pat.* 4 187 259 (1980).
81. G. Salee (Hooker Chemicals and Plastics Corp.), *US Pat.* 4 304 709 (1981).
82. N. V. Onderzoekingsiustitut Research, *Fr. Pat.* 1 392 883 (1965).
83. H. Sakata, T. Okamoto and H. Haugawa (Unitika Ltd.), *US Pat.* 3 946 091 (1976).
84. Unitika Ltd., *Jpn. Pat.* 53 51 247 (1978).
85. Unitika Ltd., *Jpn. Pat.* 53 64 261 (1978).
86. Unitika Ltd., *Jpn. Pat.* 53 7875 (1980).
87. G. Salee (Hooker Chemicals and Plastics Corp.), *US Pat.* 4 221 694 (1980).
88. L. M. Robeson (Union Carbide Corp.), *US Pat.* 4 246 381 (1981).
89. D. W. Fox, B. A. Kaduk and J. B. Starr (General Electric Co.), *US Pat.* 4 358 568 (1982).
90. M. Kimura, R. S. Porter and G. Salee, *J. Polym. Sci., Polym. Phys. Ed.*, 1983, **21**, 367.
91. M. M. Schwartz, J. R. Knox and E. E. Paschke (Standard Oil Co.), *US Pat.* 4 433 118 (1984).
92. D. W. Fox, B. A. Kaduk and J. B. Starr (General Electric Co.), *US Pat.* 4 440 912 (1984).
93. M. Kimura, G. Salee and R. S. Porter, *J. Appl. Polym. Sci.*, 1984, **29**, 1629.
94. L. M. Robeson, *J. Appl. Polym. Sci.*, 1985, **30**, 4081.
95. J. I. Eguiazabal, G. Ucar, M. Cortazar and J. J. Irvin, *Polymer*, 1986, **27**, 2013.
96. H. Kishikawa, K. Yasuno, S. Kitamara, K. Ueno, H. Inoue and N. Toyoda (Sumitomo Chemical Co.), *US Pat.* 3 792 118 (1974).
97. L. M. Robeson and G. A. Skoler (Union Carbide Corp.), *US Pat.* 4 286 075 (1981).
98. Unitika Ltd., *Jpn. Pat.* 58 83 050 (1983).
99. I. Mondragan, M. Cortazar and G. M. Guzman, *Makromol. Chem.*, 1983, **184**, 1741.
100. J. Devaux, P. Devaux and P. Godard, *Makromol. Chem.*, 1985, **186**, 1227.
101. I. Mondragon and J. Nozabal, *J. Appl. Polym. Sci.*, 1986, **32**, 6191.
102. L. M. Robeson and D. M. Papuga (Union Carbide Corp.), *US Pat.* 4 598 130 (1986).
103. D. L. Allen (General Electric Co.), *Eur. Pat. Appl.* 208 940 (1987).
104. Unitika Ltd., *Jpn. Pat.* 52 98 765 (1977).
105. Unitika Ltd., *US. Pat.* 4 052 481 (1977).
106. Unitika Ltd., *Jpn. Pat.* 53 80 457 (1978).
107. K. Kyo and K. U. Yasue (Unitika Ltd.), *Ger. Pat.* 2 813 479 (1978).
108. Unitika Ltd., *Jpn. Pat.* 53 121 047 (1978).
109. Unitika Ltd., *Jpn. Pat.* 54 56 652 (1979).
110. Unitika Ltd., *Jpn. Pat.* 54 101 852 (1979).
111. K. Kyo, K. House and K. Yasue (Unitika Ltd.) *US Pat.* 4 171 330 (1974).
112. Unitika Ltd., *Jpn. Pat.* 55 13 766 (1980).
113. K. Kyo, Y. Arai and S. Kato (Unitika Ltd.), *US Pat.* 4 187 358 (1980).
114. Unitika Ltd., *Jpn. Pat.* 55 50 057 (1980).
115. K. Kyo, N. Asahara and Y. Asai (Unitika Ltd.) *US Pat.* 4 206 100 (1980).
116. K. Kyo, Y. Asai and K. Kohyama (Unitika Ltd.), *US Pat.* 4 254 242 (1981).
117. K. Kyo, N. Asahara and Y. Asai (Unitika Ltd.), *US Pat* 4 258 154 (1981).
118. Unitika Ltd., *Jpn. Pat.* 58 67 749 (1980).
119. Y. Toyoda, K. Yasue and Y. Okabayashi (Unitika Ltd.), *Jpn. Pat.* 61 183 353 (1986).
120. E. Nield (ICI Ltd.), *US Pat.* 3 742 087 (1973).
121. L. M. Robeson and J. E. Harris (Union Carbide Corp.), *Eur. Pat. Appl.* 170 067 (1986).
122. L. M. Robeson, M. Matzner and L. M. Maresca (Union Carbide Corp.), *US Pat.* 4 250 279 (1981).
123. F. F. Holub and W. R. Schlich (General Electric Co.), *Eur. Pat. Appl.* 117 326 (1984).
124. F. F. Holub (General Electric Co.), *Eur. Pat. Appl.* 117 327 (1984).
125. Unitika Ltd. and Hodogaya Chem. Ind. KK, *Jpn. Pat.* 52 25 852 (1977).
126. Unitika Ltd., *Jpn. Pat.* 53 64 265 (1978).
127. Unitika Ltd., *Jpn. Pat.* 53 88 054 (1978).
128. G. Salee (Hooker Chemicals and Plastics Corp.), *US Pat.* 4 251 429 (1981).
129. G. Salee (Hooker Chemicals and Plastics Corp.), *US Pat.* 4 284 549 (1981).
130. G. Salee (Hooker Chemicals and Plastics Corp.), *US Pat.* 4 305 862 (1981).
131. Y. Toyoda and R. Yasue (Unitika Ltd.), *Jpn. Pat.* 61 171 759 (1986).
132. Unitika Ltd., *US Pat.* 3 884 990 (1975).
133. Unitika Ltd., *Jpn. Pat.* 82 45 783 (1982).
134. Sumitomo Chemical Co. Ltd., *Jpn. Pat.* 60 44 544 (1985).
135. G. D. Kiss (Celanese Corp.), *Eur. Pat. Appl.* 169 947 (1986).

19
Unsaturated Polyesters

ALAIN FRADET
Université Pierre et Marie Curie, Paris, France
and
PATRICK ARLAUD
Norsolor, Drocourt, France

19.1 INTRODUCTION	331
19.2 CHEMISTRY OF UNSATURATED POLYESTERS	332
19.2.1 Synthesis	332
19.2.1.1 Reaction between dicarboxylic acids or anhydrides and diols	332
19.2.1.2 Reaction between anhydrides and oxiranes	334
19.2.1.3 Miscellaneous synthesis	335
19.2.1.4 Side reactions	335
19.2.2 Thickening of Polyester Resins	337
19.2.3 Crosslinking	338
19.3 STRUCTURE–PROPERTIES RELATIONSHIPS	339
19.3.1 Influence of the Diacid	339
19.3.2 Influence of the Diol	340
19.3.3 Influence of the Vinyl Monomer	340
19.4 APPLICATIONS	340
19.4.1 Casting	341
19.4.2 Gel Coats	342
19.4.3 Lamination	342
19.4.4 Molding	342
19.5 REFERENCES	342

19.1 INTRODUCTION

Since the first work on the crosslinking properties of maleic acid polyesters in the late 1930s, unsaturated polyesters have been used in a remarkably wide range of applications, making them a thermosetting system of major importance.

Unsaturated polyesters are macromolecules obtained by polycondensation of unsaturated and saturated acids or anhydrides with diols. For almost all applications, they are diluted with vinyl monomers capable of free radical copolymerization with the unsaturations in polyester chains. These solutions are called polyester resins. A typical polyester resin composition is a 60:40 (wt) mixture of a polyester based on maleic anhydride, ortho- or iso-phthalic acid, and 1,2-propanediol, with a crosslinking monomer such as styrene. Polyester resins can be compounded with fillers or glass fibers, and cured in the presence of free radical initiators to yield thermoset articles with a very wide range of mechanical and chemical properties, depending on the choice of diacids, diols, crosslinking monomers, initiators and additives.

This versatility in the properties of the final product, associated with a relatively low cost, has renewed interest in unsaturated polyester resins as matrixes for glass-fiber-reinforced composites extensively used in a large number of industries, such as the transport, electric and electronic industries, building and public works, as well as in the manufacture of sport and leisure articles, furniture, and sanitary and domestic appliances.

In order to achieve the control of properties and reproducibility of unsaturated polyesters required for automated processes, further fundamental research has been carried out over the last decade, on: (i) the kinetics and mechanisms of the polyesterification reaction; (ii) the nature of the side reactions occurring during the polyesterification and their influence on the structure and properties of the final product; (iii) the reaction of unsaturated polyesters with the metal oxide additives used for the viscosity control of molding compounds; (iv) initiation and crosslinking kinetics and the tridimensional network structure; and (v) the structure of the blends of unsaturated polyester resin and thermoplastic polymers used as low shrink compositions.

New grades of polyester resins are now available, which can compete with some thermoplastics, such as fiber-reinforced polyamides, poly(butylene terephthalate) or acrylonitrile–butadiene–styrene (ABS)/polycarbonate blends.

19.2 CHEMISTRY OF UNSATURATED POLYESTERS

19.2.1 Synthesis

19.2.1.1 Reaction between dicarboxylic acids or anhydrides and diols

The synthesis of unsaturated polyesters usually involves a bulk reaction at elevated temperatures between dibasic acids or anhydrides and diols. A general reaction scheme for maleic anhydride and 1,2-ethanediol is illustrated in Scheme 1.

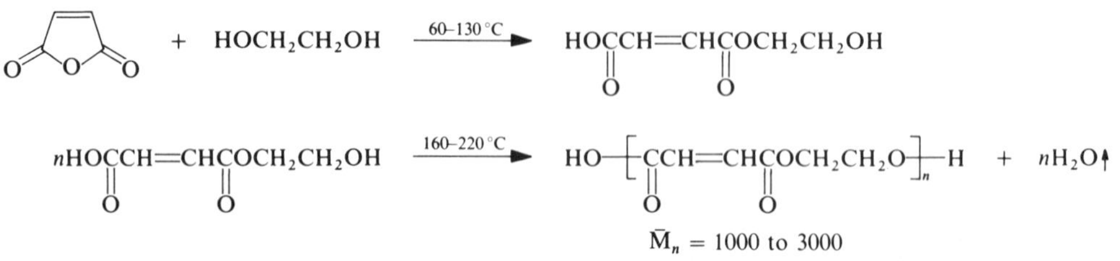

$$\text{maleic anhydride} + HOCH_2CH_2OH \xrightarrow{60-130\,°C} HOCCH{=}CHCOCH_2CH_2OH$$

$$n\,HOCCH{=}CHCOCH_2CH_2OH \xrightarrow{160-220\,°C} HO{-}[CCH{=}CHCOCH_2CH_2O]_n{-}H + nH_2O\uparrow$$

$$\bar{M}_n = 1000 \text{ to } 3000$$

Scheme 1

During this reaction most of the maleate groups are isomerized into fumarate groups. Since esterification is a reversible process, reaction water must be efficiently removed, especially in the last stages of the reaction where the decrease in carboxyl group concentration is slow and the increase in viscosity is fast. These last stages are usually carried out under vacuum. However, in order to avoid losses of volatile reactants, an azeotropic distillation of reaction water in the presence of added organic solvent such as toluene or xylene may be used.[1] The main drawbacks of this process are the longer reaction time and the difficulty in removing the last traces of solvent.

(i) Kinetics and mechanisms

The theoretical analysis of the kinetic data for bulk polyesterification reactions is difficult because of the high concentrations of reactive end groups at the beginning of the reaction and because of the changes in the dielectric constant of the medium during the reaction.[2] According to Flory,[3–7] only the experimental results obtained for extents of reaction above 0.8 should be considered, that is when the polarity no longer changes and when the reactive groups form a dilute solution in the polyester. Within these limits, experimental data show that both mono- and poly-esterifications are third-order reactions,[3–9] second order in acid and first order in alcohol. A reasonable mechanism involves non-dissociated ion pairs and can be described as in Scheme 2, and with a protonic catalyst, as in Scheme 3.[10]

$$2RCO_2H \xrightleftharpoons{K} RC(OH)_2^+ \, RCO_2^-$$

$$RC(OH)_2^+ \, RCO_2^- + R'OH \xrightarrow[\text{slow}]{k} R{-}\underset{\underset{H}{OH}}{\overset{OH}{\underset{|}{C}}}{-}\overset{+}{O}{-}R' \, RCO_2^- \xrightarrow{\text{fast}} \text{products}$$

R, R' = polyester chains $-d[RCO_2H]/dt = kK[RCO_2H]^2[R'OH]$

Scheme 2

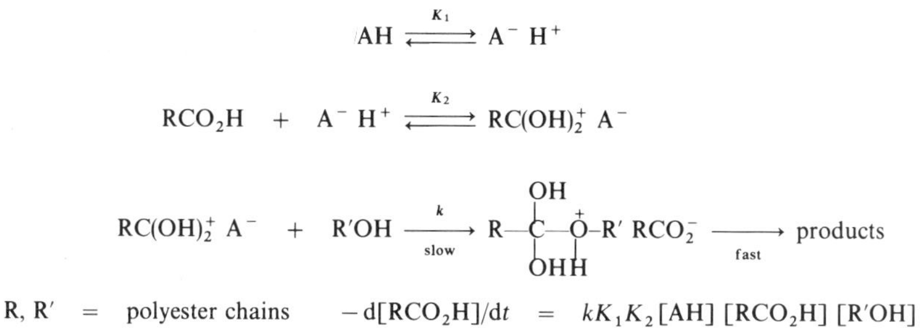

$$AH \xrightleftharpoons{K_1} A^- H^+$$

$$RCO_2H + A^- H^+ \xrightleftharpoons{K_2} RC(OH)_2^+ A^-$$

$$RC(OH)_2^+ A^- + R'OH \xrightarrow[slow]{k} R-\underset{\underset{H}{OH}}{\overset{OH}{\underset{|}{C}}}-\overset{+}{O}-R'\ RCO_2^- \xrightarrow{fast} products$$

R, R' = polyester chains $-d[RCO_2H]/dt = kK_1K_2[AH][RCO_2H][R'OH]$

Scheme 3

A number of kinetic studies are relative to the whole course of polyesterification.[11-13] In these cases, the results cannot be used for mechanistic interpretation. However they are useful for comparison purposes. Some kinetic data relative to reactants which are widely used in the synthesis of unstaturated polyesters are summarized in Table 1.

Table 1 Kinetic Parameters of Polyesterification Reactions Without Added Catalyst[a]

Diol	Diacid	r	Temperature range (°C)	Overall order d	$k \times 10^5$ $(mol^{1-d} kg^{d-1} s^{-1})$	Ref.
1,2-Ethanediol	Maleic	1	160–170	2	3.88–5.78	11
1,2-Ethanediol	Fumaric	1	170–180	2	5.35–7.65	11
1,2-Ethanediol	Orthophthalic	1	150–190	2	0.661–4.22	12
1,2-Ethanediol	Isophthalic	2–8	130–235	2.5	3.52–9.22[b]	131
1,2-Ethanediol	Hexanedioic	1	160	2.5	1.25	13
1,2-Propanediol	Maleic	1	200–210	3	2.20–3.97	74
Diethylene glycol	Maleic	1	140–170	2	1.58–5.05	11
Diethylene glycol	Fumaric	1	150–200	2	2.58–15.5	11
Diethylene glycol	Hexanedioic	1	166–202	3	6.83[c]	5
1,6-Hexanediol	Maleic	1	160–170	2	7.38–10.6	11
1,6-Hexanediol	Fumaric	1	170	2	10.6	11
1,6-Hexanediol	Hexanedioic	1	160	2.5	3.67	13

[a] r = initial molar ration of diol to diacid; k = reaction rate constant. If esterifications are carried out over a temperature range, the listed k values correspond to the lowest and highest temperatures. [b] In $mol^{-1.5} l^{1.5} s^{-1}$. [c] At 202 °C.

(ii) Catalysis

Although no catalyst is added for most industrial processes, esterification catalysts may be used in the synthesis of unsaturated polyesters. The most common are tin compounds, *e.g.* butyltin oxide hydroxide and dibutyltin diacetate, titanium or zirconium derivatives, *e.g.* tetrabutoxytitanium, sodium acetate or phosphate, zinc salts, and strong organic acids, *e.g.* p-toluenesulfonic acid. Metal derivatives are superior to protonic catalysts as the latter compounds are also efficient catalysts for side reactions, which lead to deeply colored and branched products.

The mechanisms of polyesterifications catalyzed by metal derivatives are not as well known as non-catalyzed or acid-catalyzed ones.[2] According to some studies on models,[2,14-17] the catalyst is effective only when the initial ligands of the metal can be exchanged with the hydroxylic or carboxylic species present in the medium. The existence of a complex of the acid with the metal glycolate[15] (**1**) could explain the catalytic effect by formation of a carbonyl–metal bond which induces a positive charge on the carbon atom of the carboxyl group and increases the reactivity of the system (**2**).[17-19]

$$(RCO_2)_x M[O(CH_2)_n OH]_2 \qquad\qquad M\overset{\overset{\delta-}{O}}{\underset{\underset{\delta+}{O}}{\cdots}}C-$$

(**1**) (**2**)

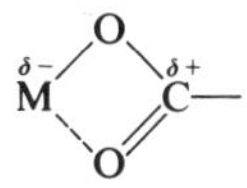

The efficiency of the various catalysts depends strongly on the nature of the metal ion and can be classified as follows: $Ti^{IV} \geqslant Sn^{II} \gg Zn^{II} > Pb^{II} > Mn^{II} \approx Co^{II} \approx Cd^{II} \approx Al^{III}$. Titanium and tin derivatives exhibit the highest catalytic activities. The presence of water in the reaction medium seems to play an important role in the kinetics of polyesterifications catalyzed by Zr or Ti alkoxides,[17,19,20] as these compounds give condensed species of lower catalytic activity after hydrolysis.

(iii) Manufacture

The traditional method of processing unsaturated polyesters is a one-stage reaction. However, when the reaction is complicated by solubility problems or when the reaction rates between reactants are very different, a two stage reaction may be used. This is typical of isophthalic acid-based resins. The reaction is carried out in stainless steel kettles equipped with efficient heating and stirring systems, and under an inert gas flow in order to prevent color-producing oxidation reactions and to make it easier to remove the reaction water. The kettles are provided with packed and steam-jacketed partial condensers whose role is to reduce the loss of glycols and other volatile components. Because of the diol-consuming side reactions described below, a slight excess of diol is used. Accurate and judicious control of polyesterification and side reactions through the determination of the carboxyl and hydroxyl group concentration, amount of evolved water and viscosity determines the reproducibility of the synthesis. When the predetermined end point is reached, the unsaturated polyester is diluted with styrene and the reactivity and other parameters are adjusted with inhibitors and other additives.

19.2.1.2 Reaction between anhydrides and oxiranes

The copolymerization reaction between dicarboxylic anhydrides and oxiranes provides a much shorter reaction time at lower temperatures (*ca.* 160 °C) than the previously described reaction between diols and dicarboxylic acids and anhydrides, (see equation 1).[21]

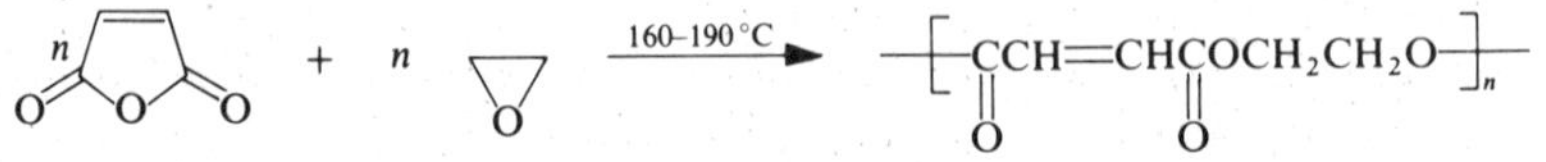

$$n \text{(maleic anhydride)} + n \text{(ethylene oxide)} \xrightarrow{160-190\,°C} -[\text{CCH}=\text{CHCOCH}_2\text{CH}_2\text{O}]_n- \quad (1)$$

Some α-diol or dicarboxylic acid is added to limit the molecular weight to the values ($M_n \approx 1000$ to 3000) commonly used in polyester resin manufacture.[22] Although the reaction can proceed without added catalyst,[23-25] a catalyst is usually used. Basic catalysts such as tertiary amines are efficient catalysts for the reaction of epoxides with carboxylic acids[26] or anhydrides.[27] However maleic anhydride becomes very dark in the presence of tertiary amines, and crosslinking by the carboxylate anion to double bonds has also been reported in these media.[27] The process is normally catalyzed with compounds such as lithium chloride,[28,29] zinc chloride,[30] aluminium chloride[31] or aluminum chloride complexes,[32] sodium acetate[33] or ethylzinc derivatives.[34-39] The isomerization of maleate units into fumarate units is achieved by further heating to higher temperatures in the presence of isomerization catalysts.

According to Schechter and Wynstra,[40] the reaction is a polyaddition. Small amounts of alcohol initiate the reaction; after reaction with anhydride, the carboxy ester formed reacts with epoxide, giving a hydroxy ester, which in turn reacts with anhydride leading to chain growth. Dicarboxylic acids or water can promote the reaction in the same way (Scheme 4). Although any hydroxyl group or carboxyl group present in the medium can start chains, they do not seem to be strictly necessary for polyester formation since the rate of reaction was found to be independent of hydroxyl group concentration.[27]

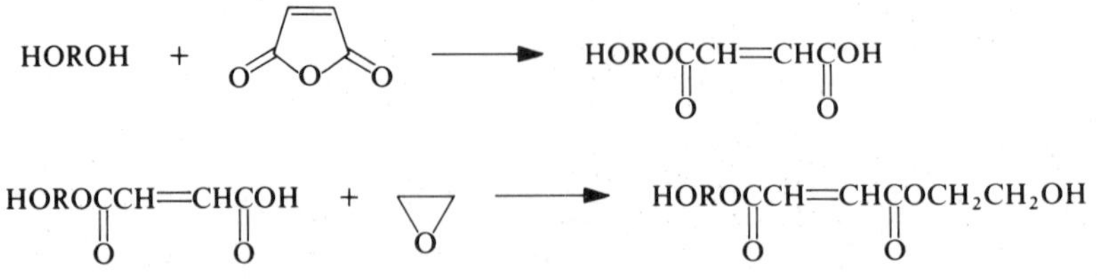

$$\text{HOROH} + \text{(maleic anhydride)} \longrightarrow \text{HOROCCH}=\text{CHCOH}$$

$$\text{HOROCCH}=\text{CHCOH} + \text{(ethylene oxide)} \longrightarrow \text{HOROCCH}=\text{CHCOCH}_2\text{CH}_2\text{OH}$$

Scheme 4

In the case of the reaction catalyzed by tertiary amines, an anionic propagation mechanism has been put forward (Scheme 5).[27] An anionic mechanism was also postulated for the catalysis by ethylzinc derivatives, with the existence of coordinated intermediate species (Scheme 6).[38] In fact, it seems that both anionic copolymerization and polyaddition coexist, as shown for the LiCl-catalyzed reaction.[29]

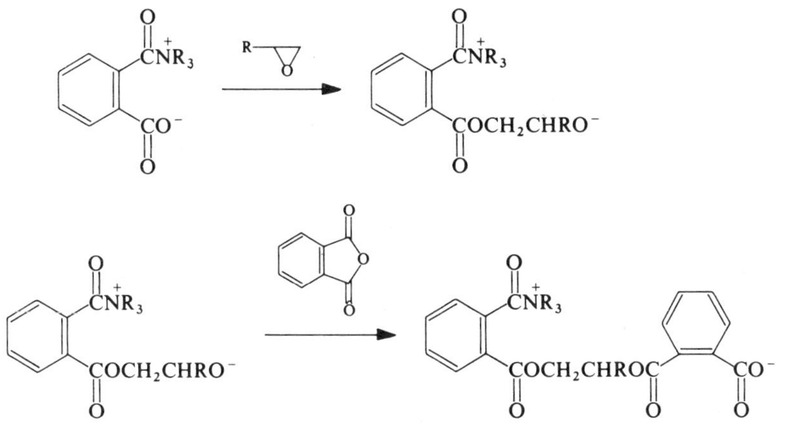

Scheme 5

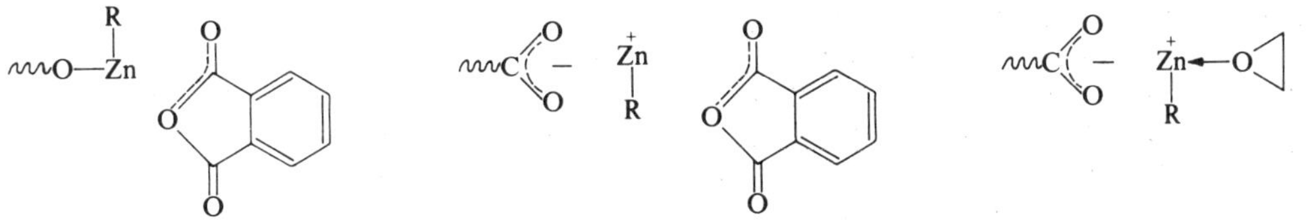

Scheme 6

Since alkylene oxides can homopolymerize in the same conditions, the copolymer obtained contains ether linkages. The amount of these linkages depends on the catalyst used and on the experimental conditions and increases product toughness and flexibility. Other disadvantages of this process are the limited range of chemical compositions available, as practically only maleic and phthalic anhydride, and ethylene and propylene oxide have been extensively used, and the more pronounced color and odor of the resins obtained. Nevertheless, the oxide method may be preferred to the diol method as it may be cheaper, depending on the relative cost of the oxide and diol.

19.2.1.3 Miscellaneous syntheses

The reaction of acid dichlorides, extensively used in the synthesis of aromatic polyesters, has been applied to the synthesis of some particular unsaturated polyesters such as (3).[41] Metal salts of maleic or fumaric acids can be reacted with halogenated aromatic[42] or aliphatic[43] compounds in the presence or absence[43] of phase transfer catalysts (equation 2).

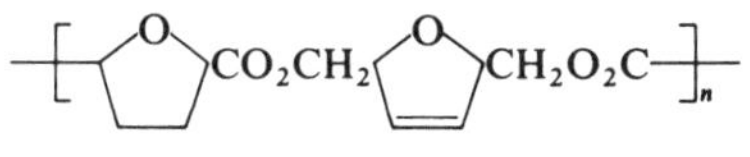

(3)

$$nM^+\ {}^-O_2CRCO_2^-\ {}^+M\ +\ n\,XR'X\ \longrightarrow\ {+\!\!\{O_2CRCO_2R'\}_n\!\!+}\ +\ 2n\widetilde{MX}\!\downarrow \qquad (2)$$

$$M\ =\ Cs,\,K;\ X\ =\ Br$$

19.2.1.4 Side reactions

The chemical structure of unsaturated polyesters is more complex than expected in view of the chemistry described above. The ^{13}C NMR spectra of unsaturated polyesters present many small peaks which cannot be assigned to carboxylic or hydroxylic end groups alone. These are due to a number of side reactions which are more or less controlled.

(i) Double bond saturation

The addition of hydroxyl groups to double bonds is one of the most important side reactions in the synthesis of unsaturated polyesters by polycondensation (equation 3). It leads to the formation of side chains and a modification of the stoichiometry due to diol consumption. Moreover, the resulting alkoxysuccinate may act as a chain-limiting agent by formation of the corresponding lactone. This reaction was studied by Ordelt et al.[44-51] who drew the following conclusions. (a) The extent of double bond saturation may reach 10 to 20%, and this very high value explains why the properties of the resin and resulting cured compound may be modified; (b) the extent of saturation increases with the reaction temperature, diol concentration, strength and concentration of the acid; and (c) the saturation reaction behaves as an equilibrium[51] when the resulting alkoxysuccinate has non-esterified carboxyl group in the α position to the alkoxy group or when a strong acid catalyst is present. On the other hand it was shown[52] that a double bond saturation occurs during alkaline hydrolysis of polyesters. However the mechanism of this addition is most probably anionic and does not concern the synthesis of unsaturated polyesters by polycondensation which is carried out in acidic medium.

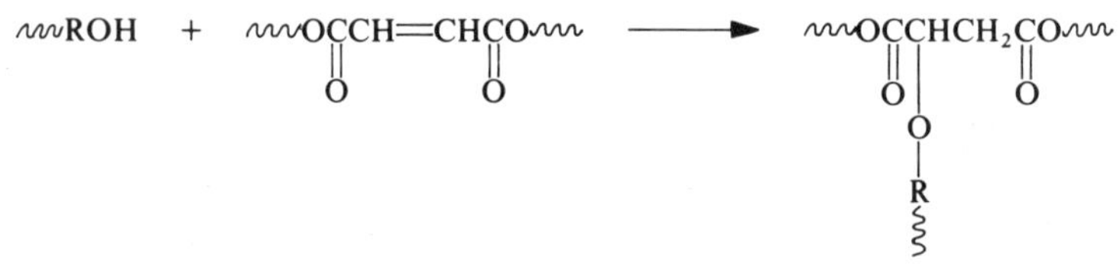

$$\sim\sim ROH + \sim\sim OCCH=CHCO\sim\sim \longrightarrow \sim\sim OCCHCH_2CO\sim\sim \qquad (3)$$

Studies on non-polycondensable models[53] showed that in polyesterification conditions, the reaction is an intermolecular reversible acid-catalyzed reaction (Scheme 7). The peaks relative to the branched structures have been assigned, both in ^{13}C and ^{1}H NMR spectra.[53-56]

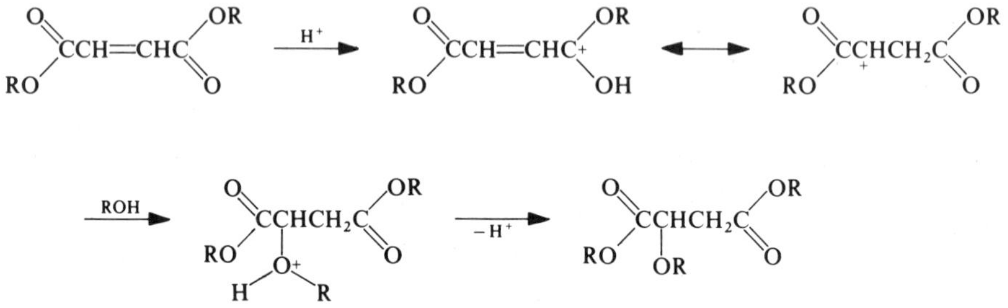

Scheme 7

(ii) Ester interchange

High temperature polyesterifications are also accompanied by ester interchange, alcoholysis or acidolysis of the chains by the monomers still present in the reaction medium or by chain end groups. Thus even if the various comonomers are introduced at different reaction times, a statistical distribution of comonomer units will be obtained within the chains of the final product if heating is continued long enough. Such reactions affect the hydroxy or carboxy end groups and result in some perturbations such as phthalic anhydride elimination[57] or terminal primary ester/secondary ester isomerization.[58]

(iii) Maleate–fumarate isomerization

Maleate double bonds are partly or wholly converted to fumarate double bonds during polyesterification, and as the content of fumarate units determines many properties of unsaturated polyesters, it must be carefully controlled.

The degree of isomerization of maleate units is determined by the esterification conditions, the reaction time and temperature,[59] the acidity of the system, and the nature of the reactants and catalysts.[60] The rate of maleate to fumarate conversion depends on the nature of the α-diol used: the longer the chain of the diol, the lower the isomerization rate.[61-63] After a certain time, the fumarate/double bonds isomerization ratio approaches a constant value which depends on the chemical composition of the polyester and on the polycondensation temperature. Increasing the bulkiness of the units associated with the maleate units in the polyester chains or using unsymmetrical diols or aromatic acids increases the conversion of maleate into fumarate.[60,63-66]

At the beginning of the reaction, non-catalyzed isomerization exhibits second-order kinetics,[60,63,67] indicating an autocatalytic effect of the protons of maleic acid on isomerization

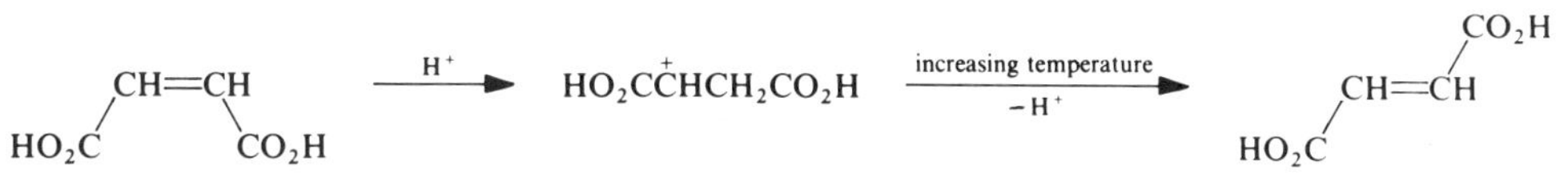

$$\underset{HO_2C}{\overset{CH=CH}{\diagdown}}\underset{CO_2H}{\diagup} \xrightarrow{H^+} HO_2C\overset{+}{C}HCH_2CO_2H \xrightarrow[-H^+]{\text{increasing temperature}} \underset{HO_2C}{\overset{CH=CH}{\diagdown}}\underset{CO_2H}{\diagup}$$

Scheme 8

(Scheme 8). The activation energies calculated from the Arrhenius equation are in the range 60–80 kJ mol^{-1}.

Some isomerization catalysts have been described, including acids, amines,[68-71] alkali metal halides[72] and sulfides.[69] They are mainly used to complete the isomerization of polyesters obtained from the epoxide–anhydride process.

(iv) Dehydration of α-diols

It is known that under experimental polyesterification conditions, dehydration of α-diol also takes place. These reactions lead to the following. (a) The formation of diethylene glycol or dipropylene glycol from 1,2-ethanediol or 1,2-propanediol, as shown by ^{1}H NMR spectroscopy.[73] Since these are dihydroxy compounds, they react with the carboxylic species and are included in the polyester chains, which may account for modifications in the mechanical properties of the final product. (b) The formation of low boiling point compounds such as tetrahydrofuran in the case of 1,4-butanediol or cyclic ethers and propanal in the case of 1,2-propanediol, particularly when strong acid catalysts are used (equation 4).[53,74]

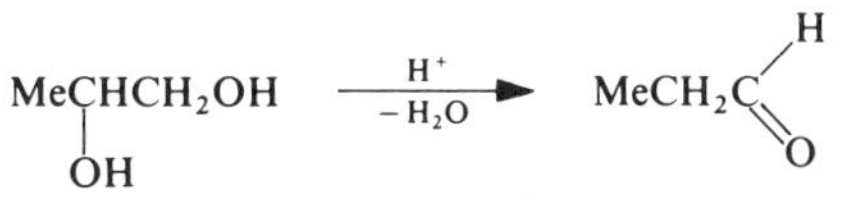

$$\underset{OH}{\text{MeCHCH}_2\text{OH}} \xrightarrow[-H_2O]{H^+} \text{MeCH}_2\text{C}\underset{O}{\overset{H}{\diagup}} \qquad (4)$$

19.2.2 Thickening of Polyester Resins

The addition of metal oxides such as MgO or CaO to polyester resins containing carboxyl groups leads to an increase in the viscosity up to a semisolid state. This reaction is referred to as the thickening reaction and is the basis of the process employed to form sheet-molding compounds (SMC). The resulting tack free, easily handled compounds are still able to undergo free radical curing while molding.[75] A stable molding viscosity must be reached as rapidly as possible, after an induction time at low viscosity to allow fillers and glass fibers to wet out. However the control of the viscosity increase is difficult to achieve, since small changes in factors such as thickener concentration, water content of the medium or amount of magnesium stearate used as mold release agent cause considerable changes in viscosity.[75,76] The important parameters of the thickening process were shown to be: (a) the molar ratio of metal oxide or hydroxide to carboxyl end groups;[77] (b) the water content of the medium;[77,78] and (c) the molecular weight of the resin.[79]

Different mechanisms have been put forward. According to Vancso-Szmercsanyi et al.,[80-84] the mechanism involves a two step reaction between carboxyl end groups and the metal oxide, with initial formation of basic or neutral salts, followed by the complexation of these salts by the ester groups in the chains and/or the hydroxyl end groups. This 'complexation reticulation' may explain the increase in viscosity. Other authors[79,85,86] described the thickening reaction as the formation of a polymeric neutral salt. However, these mechanisms cannot account for certain findings such as the viscosity increases observed with monovalent cations[87] or with neutralized carboxylic polybutadienes.[88,89] Another interpretation involving the ionic aggregation of magnesium carboxylates as in ionomers has been more recently proposed.[90-92] SAXS difractograms of magnesium oxide-neutralized polyesters[91] or their solutions in styrene[92] showed the existence of the so-called ionomer peak, which is characteristic of the existence of ion aggregation.

The changes in viscosity during thickening may be divided into three stages. (1) A low viscosity stage during which a heterogeneous reaction between carboxyl end groups and insoluble metal oxide takes place, facilitated by the presence of small amounts of water as solvating agent. This reaction is concomitant with the aggregation of the resulting magnesium oxide. As the medium is heterogeneous, the reaction rate mainly depends on the available surface of metal oxide, and on the oxide particle size. (2) An increasing viscosity stage where carboxylates undergo rearrangement to ionic aggregates, with entanglement of attached polyester chains. The rate of aggregation is directly related to the viscosity increase. (3) A high viscosity stage when the aggregation of carboxylates is completed. The plateau value of the viscosity depends, at least in part, on the degree of magnesium carboxylate group solvation by water, since it is known that water bonds easily to magnesium.[93]

To overcome the difficulties encountered in the oxide or hydroxide method, some other processes have been reported to increase the viscosity of thermosetting polyester compositions, such as the addition of crystalline polyesters insoluble in the resin at room temperature,[94] and the use of diisocyanates for the chain extension of dihydroxy unsaturated polyesters or for the formation of interspersed networks.[95]

19.2.3 Crosslinking

The crosslinking of unsaturated polyesters is a free radical polymerization initiated with compounds often referred to as 'catalysts', although they are in fact consumed during the reaction. The most widely used initiators are peroxides, diacyl peroxides, peresters, perketals and ketones, dialkyl peroxides and hydroperoxides.[66] Some peroxide free initiators have also been reported.[96,97]

The decomposition of initiators is induced by heat in the case of molding compounds or by accelerators at temperatures below the decomposition temperature of the initiator in the case of cast polyester resins. Two types of accelerator are used, metal salts — mainly cobalt salts — and amines.

The oxidoreduction of metal salts by peroxides produces free radicals. The process is very efficient since both lower and higher valencies of cations participate in the reaction (Scheme 9). Peroxide decomposition is promoted by tertiary amines following Scheme 10, where free radicals are formed from both benzoyl and amino moieties.[98]

$$RO_2H + Co^{2+} \longrightarrow Co^{3+} + RO\cdot + OH^-$$
$$RO_2H + Co^{3+} \longrightarrow Co^{2+} + RO_2\cdot + H^+$$
$$Co^{3+} + OH^- \longrightarrow Co^{2+} + OH\cdot$$

Scheme 9

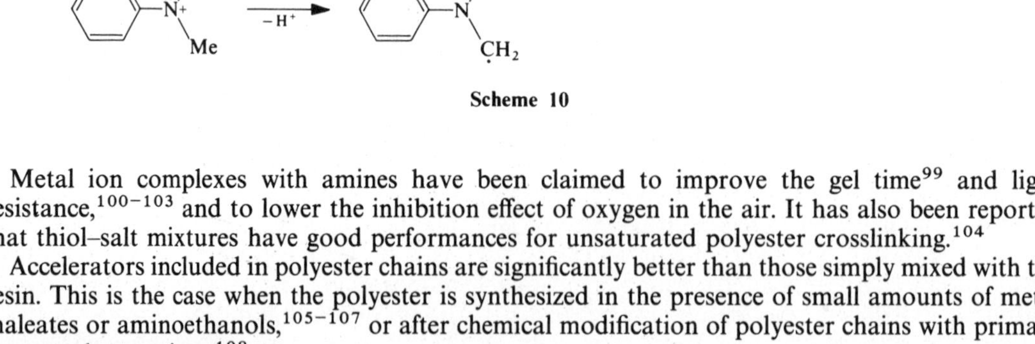

Scheme 10

Metal ion complexes with amines have been claimed to improve the gel time[99] and light resistance,[100–103] and to lower the inhibition effect of oxygen in the air. It has also been reported that thiol–salt mixtures have good performances for unsaturated polyester crosslinking.[104]

Accelerators included in polyester chains are significantly better than those simply mixed with the resin. This is the case when the polyester is synthesized in the presence of small amounts of metal maleates or aminoethanols,[105–107] or after chemical modification of polyester chains with primary or secondary amines.[108]

Unsaturated polyesters may also be cured by microwaves,[109] high frequency electric fields[110] or by UV or visible irradiation, using photoinitiators.[111,112]

The kinetics of the cure of thermosetting unsaturated polyester resins, studied by DSC[113–116] or by spin lattice relaxation time T_1 measurements using NMR spectroscopy,[117] were shown to obey the following expression[118]

$$-d\alpha/dt = (K_1 + K_2\alpha^m)(1 - \alpha^n) \tag{5}$$

where $m + n$ is generally considered equal to two.

After copolymerization between the unsaturated monomer present in the resin (most often styrene) and the fumarate or maleate double bonds of the polyester, a tridimensional network is formed, made up of polyester chains and poly(styrene-co-fumarate) chains connected with each other through the fumarate or maleate units. The properties of the cured resin strongly depend on the degree of crosslinking and on the average length of polystyrene crosslinks between the polyester chains. Crosslink density and the length of crosslinks can be controlled by varying the ratio of unsaturated diacid to saturated diacid in the polyester backbone, and the ratio of styrene to unsaturated acid double bonds respectively. A 2:1 styrene to fumarate ratio is generally used to obtain a sufficient degree of conversion of the two unsaturated species. The structure of the poly(styrene-co-fumarate) chains can be studied after chemical degradation of polyester chains by hydrolysis,[119–121] or by means of MAS-CP ^{13}C NMR[122] directly on cured products. An average value of the length of polystyrene crosslinks —(styrene)$_{\bar{n}}$— has been estimated to be $n \approx 4.5$.[121] The styrene to fumarate ratio is always higher than the original ratio, indicating that not all fumarate groups have reacted. Even for ratios lower than one, crosslinks with $n \geqslant 2$ are present.[120–122]

19.3 STRUCTURE–PROPERTIES RELATIONSHIPS

For a given polyester formulation, the properties of the final compound are a function of its condensation, e.g. carboxyl and hydroxyl group concentration, viscosity and molecular weight distribution, and of the structural features of the three-dimensional network obtained after free radical copolymerization.

An increase in the molecular weight of an unsaturated polyester improves its hardness, tensile and flexural strength, and its heat distortion temperature (HDT) until a plateau value is reached. Carboxyl end groups impart higher viscosities and better physical properties to polyesters than hydroxyl end groups.

Generally, both the physical and chemical properties of polyesters are affected by the ratio and type of the acid and diol components and of the copolymerizable monomer.

19.3.1 Influence of the Diacid

Because of economic considerations, maleic anhydride is used as the unsaturated condensation monomer for most commercial resins. Fumaric acid may be employed for some specific pruposes although it reacts more slowly with diols. These compounds can be replaced by itaconic or citraconic acid or anhydride, or mesaconic acid, but they are more expensive.

Because the double bonds in fumarate polyesters are more highly reactive, the cured resin has a greater density of crosslinks, leading to an increase in hardness and heat resistance and a simultaneous decrease in the flexibility and impact strength. On the other hand, the presence of fumarate units in polyester chains is preferable to the presence of maleate units in view of the longer pot life and better thermal properties and photostability.[123]

The following are some of the dibasic aromatic and saturated aliphatic acids used to modify the chemical and mechanical properties of the product.

Because of its low cost, phthalic anhydric is the most commonly used aromatic acid. However, it is difficult to obtain high molecular weight polyesters from this compound due to its tendency to reform and sublime from its half esters during polyesterification and also on storage. This can be avoided by using a two-step reaction where maleic anhydride is added after total reaction of phthalic anhydride.[124] Higher molecular weight polyesters with excellent physical and chemical properties can be obtained from isophthalic acid.

Tetrahydrophthalic anhydride improves hardness and glass-wetting properties and brings a refractive index close to that of glass fibers, which is important for clear-panelling applications.

Flame retardant polyester resins are obtained by replacing part of the aromatic acids or anhydrides with halogenated compounds, such as hexachloro-endo-methylene tetrahydrophthalic acid (chlorendic acid), which is widespread but expensive, or tetrabromo- or tetrachloro-phthalic anhydrides.

Saturated aliphatic acids are used to flexibilize the polymer backbone. The most commonly used is adipic acid, but succinic, azelaic and sebacic acids are also employed. On the other hand, these acids decrease water and weathering resistance.

19.3.2 Influence of the Diol

A wide range of diols or diol mixtures may be used. The most popular is 1,2-propanediol (propylene glycol). 1,2-Ethanediol (ethylene glycol) may reduce the solubility of the unsaturated polyester in the copolymerizable monomer. Diethylene glycol, dipropylene glycol and poly(oxyalkylene) glycols impart flexibility to the polyester resins, but strongly reduce their water resistance. The greater the number of ether linkages between hydroxyl end groups, the greater the flexibility, but the poorer the water resistance of the resin, due to the hydrophilicity induced by the polarity of ether oxygens.

The impact strength is highest when long chain unbranched glycols are used. On the other hand, the presence of branched diols in the polyester backbone leads to very good corrosion resistance because of the steric hindrance of these diols, which prevent the attack of ester linkages, Hydrogenated bisphenol A and its ethylene or propylene oxide derivatives are the most efficient. Neopentyl glycol and 2,2,4-trimethyl-1,3-pentanediol are also used.

Dicyclopentadiene derivatives have been reported to give polyesters with low shrinkage during curing[125] and good tensile and impact strength, useful for the preparation of laminates[126] or coatings.[127]

Fire retardant resins may be obtained from halogenated diols, *e.g.* 2,2-dibromomethyl-1,3-propanediol.

19.3.3 Influence of the Vinyl Monomer

Vinyl monomers act as solvents of the unsaturated polyester and as crosslinking agents. The degree of crosslinking can be controlled by modifying the concentrations of unsaturated acids and of the vinyl monomer. The length of crosslinks can be controlled to a certain extent by modifying the concentration and type of vinyl monomer employed.

Styrene is the most commonly used monomer, because of its good compatibility with unsaturated polyesters, high boiling point, good reactivity with polyester unsaturations and low cost. Large amounts of styrene yield a high exotherm peak temperature and a hard cured resin, with better chemical, water and solvent resistance, and with low amounts of unreacted polyester unsaturations. However, the physical properties are optimum when the amount of styrene ranges from 30 to 45%, depending on the starting unsaturated polyester. Other vinyl monomers may be used in conjunction with styrene.[128]

In contrast to styrene, α-methylstyrene copolymerizes slowly, decreases the exotherm peak temperature and imparts flexibility. The addition of such a monomer in the resin improves the control of the reactivity and shrinkage of the resulting compound.

Vinyltoluene leads to shorter cure times, higher exotherm peak temperature and less shrinkage than styrene. Because of its higher boiling point it may replace styrene in some specific applications. The reactivity, hardness and heat resistance of the resin may be enhanced with divinylbenzene, but the cured compounds are more brittle.

Weathering resistance may be obtained by using methyl methacrylate in a 50/50 mixture with styrene. Other advantages of this monomer are the lower viscosity of the resin which provides a better wetting of glass fibers and a refractive index close to that of glass fibers. On the other hand it does not copolymerize as easily with polyester double bonds as styrene and has a more pronounced odor.

Among the other vinyl monomers used in the manufacture of unsaturated polyester resin are diallyl phthalate, which brings a relatively low exotherm peak temperature, and triallyl cyanurate, which requires longer gel times but improves the heat stability of the cured resin.

Some additives are employed in low concentration in order to prevent free radical copolymerization of vinyl monomers in polyester resins during manufacture or storage. These are mainly hydroquinone or its derivatives (mono- or di-*t*-butylhydroquinone, methylhydroquinone), and also benzoquinone, *para*-*t*-butylcatechol, di-*t*-butyl-2,6-paracresol and some quaternary ammonium salts.

19.4 APPLICATIONS

To yield thermoset articles, unsaturated polyester resins are cured in the presence of a free radical initiator ('catalyst') eventually coupled with accelerators, depending on the application. The initiator is usually an organic peroxide at 0.5–2% weight concentration. These compounds are preferred to the azo initiators because of their good solubility in styrene and their high reactivity allowing curing at lower temperatures, and because they do not release gaseous by-products on curing. The polymerization kinetics may be modified to fit the application by selecting a suitable initiator and accelerator. Commonly used initiators are reported in Tables 2 and 3 for different application temperatures.

Table 2 Initiators Used for Low to Medium Application Temperatures (15 to 80 °C)

Class	Examples	Accelerator	Pot life of the catalyzed resin	Gel time	Exotherm peak time	Application temperature
Ketone peroxide	Methyl ethyl ketone peroxide	No[a]	Short	Medium	Medium	Ambient
	Acetylacetone peroxide	No[a]	Medium	Medium	Short	Ambient
	Methyl isobutyl ketone peroxide	No[a]	Short	Medium	Short	Medium
	Cyclohexanone peroxide	No[a]	Medium	Medium	Medium	Ambient
Diacyl peroxide	Benzoyl peroxide	Dimethylaniline	Long	Medium	Medium	Ambient
		Diethylaniline	Long	Long	Long	Ambient
		Dimethyl-p-toluidine	Long	Short	Very long	Ambient
	Lauroyl peroxide	No	Medium to long	Medium at 80 °C	Medium	Medium

[a] Cobalt naphthenate or octoate may be used in order to modify the reactivity depending on the application.

Table 3 Initiators Used for Medium to Very High Application Temperatures (80–160 °C)

Class	Examples	Application temperature
Perester[a]	t-Butyl peroctoate	Medium or high
	t-Butyl perbenzoate	Medium or high
	t-Butyl peracetate	Medium or high
Alkyl hydroperoxide[b]	t-Butyl hydroperoxide	Medium or high
	Cumyl hydroperoxide	Medium or high
Dialkyl peroxide[b]	t-Butyl peroxide	High or very high
	Dicumyl peroxide	High or very high

[a] May be used with cobalt octoate or naphthenate.
[b] Used without accelerator.

Unsaturated polyester resins may be formulated with or without fillers and glass fibers and cured at low or high temperature, eventually under pressure. The main fillers are calcium carbonate, in various particle sizes, and types of clay, antimony oxide, chlorinated alkanes, aluminium oxide trihydrate and glass microspheres.

Four principal application techniques can be distinguished between, depending on the presence or absence of glass fibers, and on the process temperature and pressure.

19.4.1 Casting

Cast polyester resins are not reinforced with glass fibers and the mechanical properties of the heat-cured final compound are only of secondary importance. Among the applications which use no fillers are the embedding of parts and the manufacture of buttons and varnishes. However most applications employ large amounts (up to 90%) of low cost fillers: polyester cements, simulated marble and wood, mastic, anchor bolts and various encapsulation compounds.

19.4.2 Gel Coats

To produce molded parts affording surfaces with an attractive appearance, high gloss and environmental resistance, pigmented, filled and prepromoted polyester resin paints, called gel coats, are sprayed with a peroxide initiator onto the mold before reinforcement with glass and laminating resins. Generally these gel coats contain thixotropic additives such as pyrogenated silica in order to avoid sagging on vertical surfaces. Their thicknesses range from 0.2 to 0.7 mm. Gel coats are used primarily in the marine industry, as the exterior paint layer for boats.

19.4.3 Lamination

Laminating resins are reinforced with glass fibers and are usually not cured under pressure. There are four principal processes.

Hand lay up or spray up is used to reinforce gel coats. The resin/initiator mixture is sprayed with glass fiber strands or laid on a glass cloth, then squeezed into the glass strand with a roller to ensure that the glass fibers are completely wetted; continuous lamination is used for clear or opaque panelling, filament winding is used to manufacture large tanks or ductwork, and pultrusion is used to produce various profiles.

19.4.4 Molding

Reinforced unsaturated polyester resins may be cured at high temperature and/or pressure using matched metal dies. This process is referred to as molding. The use of semifinished compounds that may be molded (sheet-molding compounds, 'SMC') or injected (bulk-molding compounds, 'BMC') has become increasingly important, particularly in automotive applications for weight reduction. SMC and BMC contain highly reactive unsaturated polyester resins, fillers, high temperature initiators, mold release agents (zinc or calcium stearate), pigments, thermoplastics to prevent shrinkage, and, for SMC, a thickening agent. In the case of SMC, two layers of this mixture are used to sandwich chopped glass strands or continuous roving between two polyethylene sheets. The roll of compound is then sliced and molded under pressure and at high temperatures. In the case of BMC, all the ingredients are blended with short glass fibers in an appropriate mixture and injection molded. Mechanical properties are low compared to those obtained from SMC.

In order to replace metals in automotive exterior parts by cured unsaturated polyester resins, excellent dimensional stability, accurate mold surface reproduction and an excellent surface appearance must be achieved. Some modified polyester resins associated with reactive initiators have such properties without additives.[129] However 'low profile' or 'low shrink' additives for surface finish and for dimensional stability respectively are nearly always used to compensate for the chemical shrinkage occurring during the copolymerization of the unsaturated polyester resin with the crosslinking monomer.[75,130] These additives are thermoplastic polymers, such as polystyrene, polyethylene, poly(vinyl acetate), poly(methyl methacrylate), poly(butadiene–styrene) elastomers, polycaprolactone and others. Most are incompatible with the thermosetting system or become incompatible with it on curing. As the temperature increases, the thermoplastic expands and compensates for the shrinkage. Thermoplastic additives usually affect the mechanical and thermal properties and the pigmentation of the final product. Further work is necessary in this field to improve our understanding of the mechanisms of the phase separation between thermoplastic additives and the thermosetting polyester and to obtain low profile compositions affording better temperature resistance without modifying any of the other properties of the resin.

19.5 REFERENCES

1. F. Lesek, J. Kitzler, K. Hajek, J. Novak, J. Drabek, V. Macku, A. Rada, J. Sedivy, L. Klancik and A. Kocian, *Czech. Pat.* 216 091 (1984) (*Chem. Abstr.*, 1985, **102**, 25 566).
2. A. Fradet and E. Maréchal, *Adv. Polym. Sci.*, 1982, **43**, 51.
3. P. J. Flory, *J. Am. Chem. Soc.*, 1936, **58**, 1877.
4. P. J. Flory, *J. Am. Chem. Soc.*, 1937, **59**, 466.
5. P. J. Flory, *J. Am. Chem. Soc.*, 1939, **61**, 3334.
6. P. J. Flory, *J. Am. Chem. Soc.*, 1940, **62**, 2261.
7. P. J. Flory, *Chem. Rev.*, 1946, **39**, 154.
8. A. Fradet and E. Maréchal, *J. Polym. Sci., Polym. Chem. Ed.*, 1981, **19**, 2905.

9. S. D. Hamann, D. H. Solomon and J. D. Swift, *J. Macromol. Sci., Chem.*, 1968, **2**, 153.
10. A. Fradet and E. Maréchal, *J. Macromol. Sci., Chem.*, 1982, **17**, 859.
11. I. Vancso-Szmercsanyi and E. Makai-Bodi, *J. Polym. Sci.*, 1968, **16**, 3709.
12. K. Matsuzaki and K. Mitani, *Kogyo Kagaku Zasshi*, 1967, **70**, 470.
13. E. Makai-Bodi and I. Vancso-Szmercsanyi, *Eur. Polym. J.*, 1969, **5**, 145.
14. O. M. O. Habib and J. Malek, *Collect. Czech. Chem. Commun.*, 1976, **41**, 3077.
15. O. M. O. Habib and J. Malek, *Collect. Czech. Chem. Commun.*, 1976, **41**, 2724.
16. L. Nondek and J. Malek, *Makromol. Chem.*, 1977, **178**, 2211.
17. F. Leverd, A. Fradet and E. Maréchal, *Eur. Polym. J.*, 1987, **23**, 695.
18. A. Fradet and E. Maréchal, *Eur. Polym. J.*, 1978, **14**, 761.
19. A. Fradet and E. Maréchal, *J. Macromol. Sci., Chem.*, 1982, **17**, 881.
20. P. Laporte, A. Fradet and E. Maréchal, *J. Macromol. Sci., Chem.*, 1987, **A24**, 1269.
21. L. N. Sedov, D. Ya. Filippenko and E. A. Khromova, *Itogi Nauki Tech.: Khim. Teknol. Vysokomol. Soedin.*, 1977, **11**, 5.
22. J. G. Milligan and H. G. Waddill (Jefferson Chemical Co., Inc.), *US Pat.* 3 355 434 (1967) (*Chem. Abstr.*, 1968, **68**, 13 711).
23. W. Fisch and W. Hoffmann, *J. Polym. Sci.*, 1954, **12**, 497.
24. W. Fisch, W. Hoffmann and J. Koskikallio, *J. Appl. Chem.*, 1956, 429.
25. W. Fisch, W. Hoffmann and J. Koskikallio, *Chem. Ind.*, 1956, 756.
26. P. J. Madec and E. Maréchal, *Adv. Polym. Sci.*, 1985, **71**, 153.
27. R. F. Fischer, *J. Polym. Sci.*, 1960, **44**, 155.
28. E. Schwenk, K. Gulbins, M. Roth, C. Benzing, R. Maysenhölder and K. Hamann, *Makromol. Chem.*, 1962, **51**, 53.
29. M. A. Bulgakova, I. A. Rrasnova, O. F. Alkaeva and M. I. Siling, *Proizvod. Pererab. Plastmass Sint. Smol.*, 1978, 12 (*Chem. Abstr.*, 1979, **91**, 5495).
30. J. A. Seiner and E. E. Parker (PPG Co.), *Fr. Pat.* 1 376 810 (1964) (*Chem. Abstr.*, 1965, **63**, 705d).
31. H. Matsuda, T. Okamoto, H. Doi and K. Kaneoka (Okura Industrial Co., Ltd.), *Jpn. Pat.* 78 30 690 (1978) (*Chem. Abstr.*, 1978, **89**, 111 203).
32. S. Inoue, T. Aida, K. Sanuki and M. Ishikawa (Hitachi Chemical Co.), *US Pat.* 4 565 845 (1986) (*Chem. Abstr.*, 1986, **104**, 207 923).
33. A. Hilt, K. H. Reichert and K. Hamann, *Makromol. Chem.*, 1967, **101**, 246.
34. W. Kuran and A. Nieslochowski, *J. Macromol. Sci., Chem.*, 1981, **15**, 1567.
35. V. A. Dodonov, R. F. Galliulina, Yu. N. Krasnov, T. R. Shnol, E. V. Chistova and E. P. Udalova, *USSR Pat.* 789 532 (1981) (*Chem. Abstr.*, 1981, **94**, 122 301).
36. T. Hasegawa, M. Miki and Y. Taniguchi (Takeda Chemical Industies Ltd.), *Fr. Pat.* 2 480 761 (1981) (*Chem. Abstr.*, 1982, **96**, 69 901).
37. W. Kuran and A. Nieslochowski (Politechnika Warsawska), *Pol. Pat.* 121 261 (1983) (*Chem. Abstr.*, 1984, **100**, 157 165).
38. S. Inoue, K. Kitamura and T. Tsuruta, *Makromol. Chem.*, 1969, **126**, 250.
39. T. Tsuruta, K. Matsura and S. Inoue, *Makromol. Chem.*, 1964, **75**, 211.
40. L. Schechter and J. Wynstra, *Ind. Eng. Chem.*, 1956, **48**, 86.
41. J. A. Moore and J. E. Kelly, *J. Polym. Sci., Polym. Chem. Ed.*, 1978, **16**, 2407.
42. J. Kielkiewicz and W. Kuran, *Makromol. Chem., Rapid Commun.*, 1981, **2**, 255.
43. N. Lacoudre, A. Leborgne, M. Sepulchre, N. Spasski, J. Djonlagic and M. S. Jacovic, *Makromol. Chem.*, 1986, **187**, 341.
44. Z. Ordelt and F. Ciganek, *Chem. Prum.*, 1964, **14**, 141 (*Chem. Abstr.*, 1964, **60**, 15 987).
45. Z. Ordelt, *Vysokomol. Soedin.*, 1962, **4**, 1110.
46. Z. Ordelt, *Chem. Prum.*, 1966, **16**, 22 (*Chem. Abstr.*, 1966, **64**, 12 814).
47. Z. Ordelt, *Chem. Prum.*, 1966, **16**, 97 (*Chem. Abstr.*, 1967, **66**, 28 18).
48. Z. Ordelt, *Makromol. Chem.*, 1963, **63**, 153.
49. Z. Ordelt, *Makromol. Chem.*, 1963, **68**, 166.
50. J. Klaban and Z. Ordelt, *Plast. Kautsch.*, 1965, **12**, 210.
51. Z. Ordelt, V. Novak and B. Kratky, *Collect. Czech. Chem. Commun.*, 1968, **33**, 405.
52. M. Bohdanecky, J. Mleziva, A. Sternschuss and V. Zvonar, *Makromol. Chem.*, 1961, **47**, 201.
53. A. Fradet and E. Maréchal, *Makromol. Chem.*, 1982, **183**, 319.
54. W. Sauer, P. Kuzay, W. Kimmer and H. Jahn, *Plast. Kautsch.*, 1976, **23**, 331.
55. M. Paci, V. Crescenzi, N. Supino and F. Campana, *Makromol. Chem.*, 1982, **183**, 377.
56. D. Judas, A. Fradet and E. Maréchal, *Makromol. Chem.*, 1984, **185**, 2583.
57. D. Judas, A. Fradet and E. Maréchal, *Makromol. Chem.*, 1983, **184**, 1129.
58. A. Fradet, M. Brigodiot and E. Maréchal, *Makromol. Chem.*, 1979, **180**, 1149.
59. V. Zvonar and A. Sternscuss, *Plast. Kautsch.*, 1960, **7**, 228.
60. O. I. Savicheva, L. N. Sedov, T. S. Khramova, Ya. G. Urman and I. Ya. Slonim, *Vysokomol. Soedin., Ser. A.*, 1973, **15**, 96.
61. I. Vancso-Szmercsanyi, K. Maros-Greger and E. Makai-Bodi, *J. Polym. Sci.*, 1961, **53**, 241.
62. L. G. Curtis, D. L. Edwards, R. M. Simons, P. J. Trend and P. T. V. Barner, *Ind. Eng. Chem., Prod. Res. Dev.*, 1964, **3**, 218.
63. I. Vancso-Szmercsanyi, L. Maros and A. Zharan, *J. Appl. Polym. Sci.*, 1966, **10**, 513.
64. S. K. Gupta and R. T. Thampy, *Makromol. Chem.*, 1970, **139**, 103.
65. V. V. Vasnev, I. N. Konkina, V. V. Korshak, S. V. Vinigradova, J. J. Lindberg, P. Jääskeläinen and K. Piiroinen, *Makromol. Chem.*, 1987, **188**, 683.
66. J. Makhlouf, in 'Encyclopedia of Chemical Technology', 3rd edn., Wiley, New York, 1982, vol. 18, p. 575.
67. M. Andreis, Z. Veksli and Z. Meic, *Polymer*, 1985, **26**, 1099.
68. V. N. Klyuchnikov, I. Ya. Slonim, Ya. G. Urman, D. Ya. Filippenko and A. Kh. Bulai, *Vysokomol. Soedin., Ser. A.*, 1980, **22**, 2058.
69. D. M. J. Filippienko, P. Penczek, E. Kicko-Walczak, V. M. Diomkin and L. V. Koval, *Pol. Pat.*, 124 395 (1984) (*Chem. Abstr.*, 1985, **103**, 37 902).
70. D. M. Filippenko, Ya. G. Urman, V. N. Klyuchnikov, L. N. Sedov, E. A. Khromova and I. Ya. Slonim, *USSR Pat.* 474 251 (1979) (*Chem. Abstr.*, 1980, **92**, 42 826).
71. J. Weiss and A. Zalmansky (PCUK), *Ger. Pat.* 2 301 159 (1973) (*Chem. Abstr.*, 1974, **80**, 15 627).

72. T. A. Smirnova, B. Ya. Eryshev, D. F. Kutepov, V. P. Pshenitsyna, B. P. Yatsenko and Yu. P. Brysin, *Z. Prikl. Khim.*, 1980, **53**, 2303.
73. D. Herrmann, W. Kimmer and W. Sauer, *Acta Polym.*, 1984, **35**, 277.
74. R. G. Robins, *Aust. J. Appl. Sci.*, 1954, **5**, 187.
75. R. Burns, 'Polyester Molding Compounds', Dekker, New York, 1982.
76. P. Arlaud, P. Blondeaux, P. Canard and R. Seury, *Composites*, 1985, 17.
77. F. B. Alvey, *J. Polym. Sci., Part A-1*, 1971, **9**, 2233.
78. I. Vancso-Szmercsanyi and A. Szilagyi, *J. Polym. Sci., Polym. Chem. Ed.*, 1974, **12**, 2155.
79. R. Burns, K. S. Gandhi, A. G. Hankin and B. M. Lynskey, *Plast. Polym.*, 1975, **12**, 228.
80. I. Vancso-Szmercsanyi and E. Noo, *Kunststoffe*, 1968, **58**, 907.
81. I. Vancso-Szmercsanyi, *Vysokomol. Soedin.*, 1973, **15**, 380.
82. I. Vancso-Szmercsanyi and A. Szilagyi, *J. Polym. Sci., Polym. Chem. Ed.*, 1974, **12**, 2155.
83. I. Vancso-Szmercsanyi and P. Hirschberg, *Acta Chim. Acad. Sci. Hung.*, 1974, **83**, 79.
84. A. Szilagyi, V. Izvekov and I. Vancso-Szmercsanyi, *J. Polym. Sci., Polym. Chem. Ed.*, 1980, **18**, 2803.
85. K. S. Gandhi and R. Burns, *J. Polym. Sci., Polym. Chem. Ed.*, 1976, **14**, 793.
86. M. Gruskiewicz and J. Collister, *Polym. Compos.*, 1982, **3**, 6.
87. B. Alt, *Kunststoffe*, 1976, 66, 786.
88. G. Broze, R. Jerome and Ph. Teyssié, *J. Polym. Sci., Polym. Phys. Ed.*, 1983, **21**, 2205.
89. G. Broze, R. Jerome, Ph. Teyssié and C. Marco, *Macromolecules*, 1983, **16**, 1771.
90. D. Judas, A. Fradet and E. Maréchal, *J. Polym. Sci., Polym. Chem. Ed.*, 1984, **22**, 3309.
91. K. Balakoteswara Rao and K. S. Gandhi, *J. Polym. Sci., Polym. Chem. Ed.*, 1985, **23**, 2135.
92. C. Habassi, M. Brigodiot, A. Fradet and C. Williams, to be published.
93. D. Judas, A. Fradet and E. Maréchal, *Polym. Bull. (Berlin)*, 1986, **16**, 13.
94. J. S. Thompson (Scott Bader Co. Ltd.), *Eur. Pat. Appl.*, 83 837 (1983) (*Chem. Abstr.*, 1983, **99**, 106 353).
95. J. L. Ferrarini, Jr. and E. Kuehn (ICI United States Inc.), *Ger. Pat.* 2 715 294 (1977) (*Chem. Abstr.*, 1978, **88**, 38 578).
96. H. J. Rosenkranz and H. Wolfers (Bayer A. G.), *US Pat.* 4 219 629 (1980) (*Chem. Abstr.*, 1981, **94**, 16 372).
97. D. Braun and R. Rengel, *Angew. Makromol. Chem.*, 1981, **98**, 265.
98. K. Demmler and J. Schlag, *Farbe Lack.*, 1971, **77**, 224.
99. U. Waiblinger, G. Baessler and H. D. Ullrich, *East Ger. Pat.* 121 524 (1976) (*Chem. Abstr.*, 1977, **86**, 172 572).
100. K. Binder, W. Edl and H. Twittenhoff (Peroxide Chemie GmbH.) *Ger. Pat.* 3 016 051 (1981) (*Chem. Abstr.*, 1982, **96**, 53 241).
101. N. G. Shtan'ko, N. M. Gol'da, T. V. Todorova and O. Kh. Aveskina, *USSR Pat.* 358 328 (1973) (*Chem. Abstr.*, 1973, **78**, 161 020).
102. C. H. Stapfer and R. W. D'Andrea (Cincinnati Milacron Chemicals Inc.), *Ger. Pat.* 2 111 882 (1971) (*Chem. Abstr.*, 1972, **76**, 101 334).
103. M. W. Uffner (Air Products and Chemicals Inc.), *Ger. Pat.* 2 327 131 (1974) (*Chem. Abstr.*, 1974, **81**, 121 833).
104. V. R. Kamath, M. F. Novits and R. B. Gallagher, *Mod. Plast.*, 1982, **59**, 90, 92, 94.
105. S. S. Jada, *Polym. Prepr, Am. Chem. Soc., Div. Polym. Chem.*, 1981, **22**, 119.
106. S. S. Jada, *Ind. Eng. Chem., Prod. Res. Dev.*, 1983, **22**, 14.
107. C. U. Pittman, Jr. and S. S. Jada, *Ind. Eng. Chem., Prod. Res. Dev.*, 1982, **21**, 281.
108. S. S. Dzhada, A. D. Valgin, D. F. Kutepov, A. I. Doskin and I. I. Gerasimova, *Polim. Stroit. Mater.*, 1977, 137 (*Chem. Abstr.*, 1979, **90**, 152 987).
109. A. J. Berteaud, H. Jullien and H. Valot, *RGE, Rev. Gen. Electr.*, 1981, 826.
110. R. E. Haven and N. P. Suh (Massachusetts Institute of Technology), *US Pat.* 4 423 191 (1983) (*Chem. Abstr.*, 1984, **100**, 86 765).
111. W. Nicolaus, A. Hesse, D. Scholz and W. Koser, *Plast. Renf., Fibres Verre Text.*, 1981, 4.
112. V. Cermak and J. Mleziva, *Polimery (Warsaw)*, 1979, **24**, 401.
113. M. E. Ryan and A. Dutta, *J. Appl. Polym. Sci.*, 1979, **24**, 635.
114. S. Y. Pusatcioglu, A. L. Fricke and J. C. Hassler, *J. Appl. Polym. Sci.*, 1979, **24**, 937.
115. C. D. Han and K. W. Lem, *J. Appl. Polym. Sci.*, 1983, **28**, 3155.
116. R. F. Storey, S. Sudhakar and M. L. Hogue, *Polym. Prepr., Am. Chem. Soc., Div. Polym. Chem.*, 1986, **27**, 167.
117. M. Paci and F. Campana, *Polymer*, 1985, **26**, 1885.
118. M. R. Kamal and S. Sourour, *Polym. Eng. Sci.*, 1973, **13**, 59.
119. D. K. Hamann, W. Funke and H. Gilch, *Angew. Chem.*, 1959, **72**, 596.
120. A. W. Birley, J. W. Dawkins, D. Kyriacos and A. Bunn, *Polymer*, 1981, **22**, 812.
121. M. Paci and F. Campana, *Eur. Polym. J.*, 1985, **21**, 717.
122. M. Paci, V. Crescenzi and F. Campana, *Polym. Bull. (Berlin)*, 1982, **7**, 59.
123. H. V. Boenig, 'Unsaturated polyesters, structures and properties', Elsevier, Amsterdam, 1964.
124. P. Canard (Société Chimique des Charbonnages), *Eur. Pat.* 179 702 (1986) (*Chem. Abstr.*, 1986, **105**, 24 834).
125. S. L. Libina, N. M. Nikitskaya, N. V. Khrenova, S. G. Alekseeva and L. N. Sedov, *Plast. Massy*, 1977, 17.
126. Dainippon Ink and Chemicals Inc., *Jpn. Pat.* 58 108 218 (1983) (*Chem. Abstr.*, 1983, **99**, 195 934).
127. Y. Matsumoto, K. Wake, T. Okada and S. Nogami (Asahi Chemical Industry Co., Ltd.), *Jpn. Pat.* 80 27 307 (1980) (*Chem. Abstr.*, 1980, **93**, 27 859).
128. H. V. Boenig, in 'Encyclopedia of Polymer Science', Wiley, New York, 1969, vol. 11, p. 129.
129. P. Canard and D. Judas, *Composites*, 1983, 71.
130. K. E. Atkins and J. F. Rocky, Jr., *Plast. Compd.*, 1982, **5**, 31.
131. M. Sumoto and Y. Hasegawa, *Kogyo Kagaku Zasshi*, 1965, **68**, 1900.

20
Polycarbonates

DONALD C. CLAGETT and SHELDON J. SHAFER
General Electric Company, Pittsfield, MA, USA

20.1	INTRODUCTION	345
20.2	CHEMISTRY	346
	20.2.1 Interfacial Resin Chemistry	347
	20.2.2 Transesterification Resin Chemistry	347
	20.2.3 Solution Polycarbonate Resin Chemistry	348
20.3	POLYCARBONATE RESIN SYNTHESIS	349
	20.3.1 Interfacial Synthetic Processes	349
	20.3.1.1 Batch interfacial synthesis	349
	20.3.1.2 Batch continuous interfacial synthesis	349
	20.3.1.3 Continuous interfacial synthesis	350
	20.3.2 Transesterification Polycarbonate Resin Synthesis	351
	20.3.3 Solution Polycarbonate Resin Synthesis	351
20.4	POLYCARBONATE RESIN PURIFICATION	352
	20.4.1 Interfacial Process Resin Purification	352
	20.4.2 Solution Process Resin Purification	352
	20.4.3 Transesterification Process Resin Purification	352
20.5	RESIN ISOLATION	352
	20.5.1 Interfacial/Solution Process Resin Isolation	352
	20.5.1.1 Concentration	352
	20.5.1.2 Spray drying	353
	20.5.1.3 Steam distillation	353
	20.5.1.4 Antisolvent precipitation	353
	20.5.1.5 Mechanical devolatilization	353
	20.5.2 Transesterification Polycarbonate Resin Isolation	353
20.6	RESIN FINISHING	353
	20.6.1 Powder, Granules and Pellets	354
	20.6.2 Melt Stream	354
20.7	SUMMARY AND TRENDS	354
20.8	REFERENCES	355

20.1 INTRODUCTION

The existence of polycarbonate resins has been known for nearly a century.[1] The literature contains numerous references to preparations or reactions which can be rationalized as having made polycarbonate resins. However, the real beginnings of commercial polycarbonate resin technology did not occur until the late 1950s when two groups at two major international chemical firms actively pursued research which led to patents that are the basis of the current 600 million pound per year industry.[2] Fox of General Electric Company and Schnell of Bayer AG led teams of researchers who announced their inventions almost simultaneously. The resin technology involved the reaction of phosgene with the salt of bisphenol A (BPA) to produce the bisphenol A polycarbonate resin. Many alternative formulations based on innumerable dihydroxy compounds have been reported in the literature and have received patent protection. But no resin has had the combination of performance, manufacturing reliability, and economy that the original bisphenol A polycarbonate resin has achieved.[3,4]

This chapter details many of the available options for polycarbonate synthesis. Indeed, the chemistry is extremely flexible. But the reader should keep in mind that the big volume polycarbonate resin remains bisphenol A based polymer and that nothing is on the horizon at this time that seems destined to supplant it. On the other hand, the ways to make bisphenol A based polycarbonate resin and to purify and finish the resin are various. This chapter will attempt to outline the majority of those options being practiced now or being actively considered. Motivations for implementing the options include the ability to convert raw polymer to finished resin products and environmental considerations. These factors will be included in the discussion.

20.2 CHEMISTRY

The chemistries employed to make polycarbonate resins include interfacial, transesterification, and solution-based methods. Although the methods differ significantly in medium of reaction, monomeric raw materials, catalysts, temperature history, and special promotors such as phase transfer agents, there is one raw material common to all methods: phosgene. In all cases phosgene provides the source of carbonate carbonyl moiety at some stage of monomer or polymer synthesis. Many man-years of effort have been expended in seeking ways to avoid use of phosgene, which is a highly toxic material. However, none have progressed beyond the laboratory curiosity stage because of yield, energy efficiency, or reagent costs. A cursory review of research efforts will be outlined in subsequent sections.

The general structure of polycarbonate resins is denoted by structure (1). This structure can be viewed as a condensation polymer of carbonic acid and a diol, endcapped with monohydroxyl moieties. Indeed, hundreds of polycarbonate resins have been produced over the years from alkyl-, alkylaryl- and aryl-dihydroxy diols and a similar range of monohydroxy chain terminators. For typical engineering plastic resins only the aromatic diols have proved useful monomers. The basic reason is thermal stability. Most alkyl diols have hydrogens *beta* to hydroxy groups making them subject to elimination mechanisms (2). Such eliminations are common to molecules with aromatically activated β-hydrogens[5] and can even be described in terms of concerted, anchiomerically assisted mechanisms.[6] Exceptions include alcohols with tertiary alkyl substitutions on the β-carbon.

Because of the low symmetry of bisphenol A (3) which promotes amorphous resin formation, and its preparation from the cheap, readily available raw materials acetone and phenol (equation 1), over 99% of the world's commercial polycarbonate resins are based on this chemical.

A few other diols have also been used commercially, either alone or in combination with bisphenol A. These include tetramethylbisphenol A (4) and tetrabromobisphenol A (5). The tetramethylbisphenol A has been used to produce resins with higher heat deflection test (HDT) values than conventional bisphenol A based resins (Table 1). Tetrabromobisphenol A is a component of flame retardant polycarbonate resins.

Table 1 Polycarbonate Resin Properties

Property	ASTM Method	Lexan® 1011	Experimental tetramethylene bisphenol A polycarbonate resin
HDT at 1.82 N mm^{-2} (°C)	D648	132	173
Notched Izod at 20 °C, 3.2 mm thickness (J M^{-1})	D256	600	30
Flex modulus at 20 °C (N mm^{-2})	D790	2300	n.d.
Specific gravity (g cm^{-3})	D792	1.21	1.09

Chain termination agents are usually aromatic. Most common are phenol and *p-t*-butylphenol. A few other chain terminators are used to adjust certain resin properties, especially flow characteristics. These are usually various alkylphenols or polycyclic aromatics.

20.2.1 Interfacial Resin Chemistry[7,8]

Interfacial synthesis involves reaction at the boundary between two immiscible solvents. These are protic and aprotic respectively. Some reactants are dissolved in the protic aqueous layer and the other reagents are dissolved in the aprotic organic layer. During the reaction the monomers react at the interface with the polymerizing resin growing into and remaining dissolved in the aprotic solvent. Theoretically the polymer chains could grow to infinite lengths which would result in coagulation of the preparation or, at best, high molecular weight resins with intractable rheologies. In practice chain termination agents are added to control molecular weight. The higher the chain termination concentration the lower will be the resultant resin molecular weight.

The typical solvents used in industry are methylene chloride and aqueous caustic. The caustic dissolves the aromatic diol and the phenolic chain terminators. The methylene chloride layer dissolves the carbonate source which invariably is phosgene. In all interfacial procedures vigorous agitation is necessary to promote practical reaction rates. The reaction is exothermic and provisions must be made to control reflux of the volatile methylene chloride. The ultimate concentration of the resin in the methylene chloride solvent is also very important. Concentrations above 26% can result in metastable solutions whose coagulation or high viscosities can wreak havoc in a production facility. The overall interfacial reaction between bisphenol A in caustic and phosgene is shown in equation (2).

$$O\text{-}C_6H_4\text{-}C(Me)_2\text{-}C_6H_4\text{-}OH + Cl\text{-}CO\text{-}Cl \xrightarrow{NaOH} NaO\text{-}C_6H_4\text{-}C(Me)_2\text{-}C_6H_4\text{-}O\text{-}CO\text{-}Cl + NaCl \quad (2)$$

(3)

20.2.2 Transesterification Resin Chemistry[9-12]

Typically transesterification reactions are carried out in the melt phase. The carbonyl is provided by a carbonate ester. The other monomers are aromatic diols. Because of the high temperatures used to attain melt conditions, only aromatic diols have the thermal stability to survive the polymerization reaction. The reaction is shown in equation (3). Here a true condensation reaction results in the exchange of hydroxylic reagents with the diol releasing the monohydroxylic agent from the reactant carbonate as a condensate. In practice the condensate is removed overhead leaving behind the polymeric resin.

$$HO\text{-}R\text{-}OH + R'O\text{-}\underset{\Delta}{\overset{\text{catalyst}}{C}}\text{-}OR' \rightleftharpoons HO\text{-}R\text{-}O\text{-}\overset{O}{\underset{\|}{C}}\text{-}OR' + R'OH \quad (3)$$

The reaction conditions employed for a transesterification reaction range from 150 °C to 320 °C and pressures from atmospheric to less than 1 mm of mercury. Typical catalysts are bases such as lithium, sodium or potassium hydroxide.

The resin molecular weight is controlled by manipulating the residence time of the reactant system at high temperature and high vacuum in the melt phase. The resulting polymer remains as a 'living' polymer, *i.e.* a polymer which on further heating can continue to polymerize or branch. An alternative method of chemical control uses small amounts of high boiling reactive monofunctional phenol as chain terminator. These are usually added in the initial formulation.

The reaction rate can be adjusted by the use of various catalysts. Typical catalysts are alkali metal hydroxides such as sodium or potassium hydroxide. These basic species apparently act by inducing the aromatic diols to form diolates prior to reaction with the carbonate ester.

The literature has shown that a number of carbonate esters have successfully produced polycarbonate resins.[11,12] By far the most common reference is to the simplest aromatic carbonate, diphenyl carbonate (6). The reaction to produce a typical transesterified polycarbonate resin is shown in equation (4).

$$(4)$$

As was noted before, all polycarbonate resins are ultimately made *via* reactions with phosgene. In the case of transesterified polycarbonate resins the phosgene reaction is with a phenolate salt to produce the condensate salt and the product diaryl carbonate. Equation (5) shows the reaction of sodium phenolate with phosgene to produce diphenyl carbonate.

$$(5)$$

It is appropriate here to address the issue of alternate routes to diaryl carbonates. The reaction of aromatic diols with diaryl carbonates is thermodynamically favorable for polycarbonate synthesis whereas reaction of aromatic diols with dialkyl carbonates is thermodynamically unfavorable. This means that the use of chemicals such as dimethyl carbonate, which is readily produced by non-phosgene routes, has not so far been shown to be practical for transesterification polymerization to produce aromatic polycarbonates. Production of diaryl carbonate from dimethyl carbonate has had more success. A number of patents have been issued for processes to convert dimethyl carbonate to diphenyl carbonate.[13,14]

Attempts have been made to produce diaryl carbonates by the direct oxidative coupling of phenols to carbon monoxide or by direct condensation with carbon dioxide.[15,16,17] The expensive catalysts, low yields and low catalyst turnover efficiencies have precluded a commercial process. The catalysts are typically Group VIIIB metals with palladium being favored. It is generally acknowledged that the discovery of a convenient and cheap direct production of diaryl carbonates without the agency of phosgene would be a revolutionary development. This would probably catapult transesterification polycarbonate synthesis into more than contention with alternate synthetic methods based on safety and environmental sensitivities alone.

20.2.3 Solution Polycarbonate Resin Chemistry[18,19]

Solution polycarbonate synthesis was the first process pursued commercially. In solution polycarbonate synthesis all the reactants are soluble in the reaction matrix as is the synthesized polycarbonate resin. The polycarbonate resin is then isolated by a variety of means which will be discussed in subsequent sections.

The usual solution matrices are an organic solvent such as methylene chloride and an organic base such as pyridine. Pyridine is sufficiently basic to succeed in dissolving the aromatic diol to produce reactive pyridinium salts.

Phosgene is bubbled into the stirred reaction mixture with the rapid polymerization being accompanied by the production of pyridinium hydrochloride. At the conclusion of the reaction the pyridinium hydrochloride must be removed from the reaction solution prior to polymer isolation. The removal of the pyridinium hydrochloride is usually accomplished by multiple aqueous acid and water washes. The pyridine can be recovered by neutralization and subsequent purification. The major disadvantages of the solution process are in the difficulty in removing all traces of pyridine/pyridinium hydrochloride from the polymer solution and in the cost of recovery/purification of the pyridine. Equation (6) depicts the solution process chemistry.

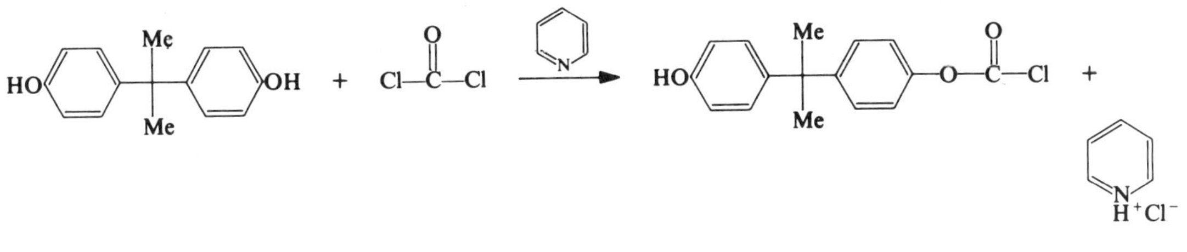

(6)

20.3 POLYCARBONATE RESIN SYNTHESIS

20.3.1 Interfacial Synthetic Processes

Interfacial resin synthesis is performed commercially by three techniques: batch, batch continuous, and continuous synthesis. These methods will be discussed in turn.

20.3.1.1 Batch Interfacial Synthesis[20-23]

Batch interfacial synthesis is accomplished as the title implies. The reactants, reaction media, *etc.* are charged to a single vessel where the polymerization reaction takes place. A typical procedure involves charging a stirred reaction vessel with caustic, methylene chloride, bisphenol A, chain terminator, phase transfer agent, and any other modifying monomers or other proprietary agents. Phosgene is bubbled into the rapidly stirred vessel at a rate which is tailored to the parameters of the system insuring maximum phosgene uptake. Because pH decreases due to caustic consumption and concurrent brine production, additional caustic is added as the reaction progresses. The exothermic reaction that ensues generates heat that must be moderated by cooling coils or by an efficient reflux system. At the end of the reaction the aqueous brine solution is separated from the polycarbonate/methylene chloride solution. This solution is then purified and the resin isolated per procedures which will be discussed in subsequent sections.

The polycarbonate resin that results from batch interfacial synthesis can be produced with dispersities as low as low as 2.5. Because chain terminators are added in the initial charge diaryl carbonates and low molecular weight oligomers are produced and are retained in the polymer at the 100 to 1000+ p.p.m. levels. Batch processes are also plagued with the problems of batch to batch variability which requires very excellent process controls and sophisticated blending/rework procedures to cope with off-specification product. **Scheme 1** shows a typical batch process.

Single stirred tank reactor

NaOH, BPA, phase transfer agent, phosgene, phenol in CH$_2$Cl$_2$ and H$_2$O

↓

15–20% solids polycarbonate resin in CH$_2$Cl$_2$

Scheme 1 Batch Interface Synthesis

20.3.1.2 Batch continuous interfacial synthesis[24-29]

Batch continuous operations use a sequence of stirred tank reactors to promote the polycarbonate resin polymerization. Commercial systems can include two to four reactors. Three series reactors are a common process configuration. This three reactor configuration allows sequential programming of pH, addition of chain terminators, and phase transfer agents. The first reactor is charged similarly to the single batch interfacial reactor. Methylene chloride, caustic and bisphenol A are added and are subsequently treated with phosgene gas. Oligomeric bisphenol A chloroformates are formed and are dissolved in the methylene chloride solvent. During this reaction the aqueous layer is maintained at a pH of around ten. At this point there exist a variety of process options. In general the whole reaction mixture is transferred to a second reactor where a phase transfer agent ($[R_4N^+]Cl^-$) is added and the polymer is advanced in molecular weight. This transfer and reaction can be done in one or a number of steps. The resulting methylene chloride solution is separated and is ready for purification and resin isolation.

The batch continuous reaction polycarbonate resin is generally produced with a molecular weight dispersity similar to that for batch reactions. If chain-terminating agent is added in the second reactor after bisphenol A oligomer is produced, the levels of low molecular weight oligomers and diaryl carbonate contaminates are reduced relative to the batch process. The batch continuous process thus offers certain quality and semicontinuous process advantages. However, control systems must be precise in order to avoid product variability. Scheme 2 indicates the batch-continuous process sequence.

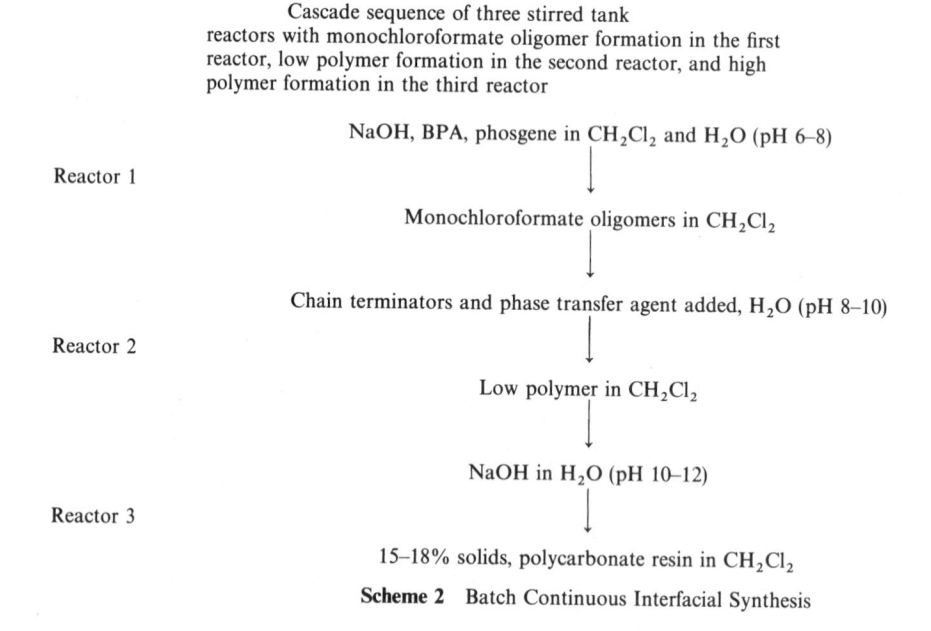

Cascade sequence of three stirred tank reactors with monochloroformate oligomer formation in the first reactor, low polymer formation in the second reactor, and high polymer formation in the third reactor

Reactor 1

NaOH, BPA, phosgene in CH_2Cl_2 and H_2O (pH 6–8)
↓
Monochloroformate oligomers in CH_2Cl_2
↓
Chain terminators and phase transfer agent added, H_2O (pH 8–10)

Reactor 2
↓
Low polymer in CH_2Cl_2
↓
NaOH in H_2O (pH 10–12)

Reactor 3
↓
15–18% solids, polycarbonate resin in CH_2Cl_2

Scheme 2 Batch Continuous Interfacial Synthesis

20.3.1.3 *Continuous interfacial synthesis*[26,30-32]

Continuous polycarbonate processes are really only variants of the batch continuous processes. Here the functions of the first reactor are performed by tubular continuous mixers. The reactants are fed into the system with all reactant concentrations carefully set and the rates of addition precisely maintained. The bisphenol A is added as a preprepared basic aqueous salt solution. Methylene chloride and phosgene are concurrently introduced resulting in a bisphenol A chloroformate oligomer solution. Either in a continuing tubular mixing system or in a series of subsequent stirred tank reactors, phase transfer agent, chain terminators, and additional caustic and bisphenol A salt are introduced. At the end of the train the methylene chloride polycarbonate resin is separated from the brine for purification and isolation.

The polycarbonate resin produced has similar dispersity and purity characteristics to those that are possessed by batch continuous resins. Product resin will be produced with consistent properties provided that strict controls are maintained throughout the process sequence. A typical reaction sequence is shown in Scheme 3.

Polycarbonates 351

```
                        Tube reactors to make oligomers, followed by a series of stirred tank
                        reactors to build high polymer

Tube reactor            NaOH, BPA, phosgene, reducing agent in CH₂Cl₂ and H₂O
                                               │
                                               ↓
                        ? % Solids monochloroformate oligomers in CH₂Cl₂
                                               │
                                               ↓
                                         Decant brine

Subsequent stirred reactors  NaOH, chain terminator and phase transfer agent added
                                               │
                                               ↓
                        12–16% solids polycarbonate resin in CH₂Cl₂
```

Scheme 3 Continuous Interfacial Synthesis

20.3.2 Transesterification Polycarbonate Resin Synthesis[9–12]

Transesterification processes usually involve preliminary melting of precursor diaryl carbonates and bisphenol A. The melted reagents, along with carefully metered catalysts and chain terminators, are fed to a heated mixer under reduced pressure. During the entire process the temperature will be elevated to as high as 290 °C. Oligomeric polycarbonate resin is formed at the same time that over 90% of the aryl hydroxyl condensate is removed overhead. The molten oligomer is then treated under still higher vacuum and heat with equipment which maximizes melt surface exposure. Wiped film evaporators, vented extruders, and other polymer mixer/reaction equipment have been successfully employed. The high molecular weight polymer is produced as a molten stream. This is usually stranded and chopped to produce pellets of virgin polycarbonate resin.

The properties of the transesterified polycarbonate resin are similar to interfacial resins. Molecular weight dispersity ranges between 2–3. Because vacuum is employed during the polymerization process the transesterified resins tend to be low in volatile contaminants such as diaryl carbonates. However unless the catalysts decompose during the polymerization they remain in the polymer. Although proof is sketchy it is suspected that this residual catalyst will result in the polymer being subject to molecular weight alteration during subsequent heat histories. The transesterification process is shown in Scheme 4.

```
              BPA, DPC melted in stirred tank reactor, transferred to
              a flash evaporation unit and subsequently to high
              vacuum polymerization/devolatilization equipment

                            BPA, DPC, catalyst melted
                                      │
                                      ↓
90% of reaction to oligomer with evolution of
phenol in flasher                     │
                                      ↓
100% reaction completion in wiped film evaporator,
vented extruder with evolution of phenol
                                      │
                                      ↓
                          Polycarbonate resin melt stream
```

Scheme 4 Transesterification Synthesis

20.3.3 Solution Polycarbonate Resin Synthesis[18,19]

Commercial solution polycarbonate resin processes have been batch processes. The reactants are placed in a stirred tank reactor. Typically, methylene chloride, pyridine, bisphenol A and chain terminator are added to the reaction vessel. Phosgene is bubbled into the stirred mixture. Purification and resin isolation follows.

Solution process resin is similar to batch interfacial resin. The process for solution polycarbonate resin is outlined in Scheme 5.

$$\text{BPA, CH}_2\text{Cl}_2\text{, pyridine, chain terminator. Phosgene is added}$$
$$\downarrow$$
$$\text{10–20\% solids in CH}_2\text{Cl}_2\text{, and contains pyridinium hydrochloride}$$

Scheme 5

20.4 POLYCARBONATE RESIN PURIFICATION

20.4.1 Interfacial Process Resin Purification[33-36]

The interfacial processes result in chlorocarbon solvent based solutions of polycarbonate resin which also contain residual monomer, phase transfer agent and small amounts of inorganic salts. Before isolation of the polymer these impurities must be removed, otherwise the performance of the isolated resin will be degraded. The methods most frequently used involve washing the resin with sequences of acid and demineralized water. The plant units employed are washing extraction columns, liquid–liquid centrifuges, or some series combination of both kinds of units. At the end of the wash sequence the solution may be filtered to remove particulates.

Different configurations of the various manufacturers' interfacial reaction processes produce chlorocarbon–polycarbonate resin solutions of rather high viscosities. In order to use the washing extraction columns and the centrifuges efficiently the viscosities are usually adjusted by dilution. In order to process properly the real solution viscosities should not exceed 1 poise ($0.1\,\text{kg}\,\text{m}^{-1}\,\text{s}^{-1}$). This means that the solids levels of the solution may end up as low as 10%.

20.4.2 Solution Process Resin Purification[37]

Solution process polycarbonate resin solutions, like those produced in interfacial systems, need to be washed to clean the solution prior to resin isolation. The wash media must be tailored to the solvent. If chlorocarbon solvent is used in combination with organic bases, then acid/demineralized water combinations are employed similar to those for the interfacial processes. However, the large quantity of pyridinium hydrochloride generated in the solution process usually requires greater quantities of washing media than in the interfacial process. Filtration of the resulting solution can be employed to remove particulates prior to isolation.

20.4.3 Transesterification Process Resin Purification

The transesterification process resin product stream is a molten polymer stream. In large part the cleanliness of the resin product is determined at the outset by raw materials purity and process control. However, particulates can be removed by melt filtration. Many-on line techniques are available ranging from regenerable cannisters to screen packs.[38]

20.5 RESIN ISOLATION

20.5.1 Interfacial/Solution Process Resin Isolation

In both process cases the process stream afforded for isolation is polycarbonate resin dissolved in an organic solvent. Many different isolation means are theoretically possible. However in industrial practice those isolation means most commonly encountered include concentration, spray drying, steam distillation, antisolvent precipitation, or mechanical devolatilization.

20.5.1.1 Concentration[39-44]

Polycarbonate resin can be isolated as a friable solid gel which remains rich in solvent, perhaps as much as 50% by weight. Typical units used for concentration include plate heaters, rising film evaporators, or pump-around vaporizers. The resultant solvent rich gel can then be freed from the residual solvent using vented extrusion, drying towers, *etc*. The resin at the completion of the solvent removal is a granular particle which can be transported and blended conveniently.

20.5.1.2 Spray drying[45,46]

The polycarbonate resin solution is fed to a nozzle which sprays the solution into a stream of hot gas. The solvent evaporates first, forming a small droplet of concentrated polymer. As solvent removal nears completion the droplet is transformed into a powder particle. The solvent rich stream is treated to recover the process solvents and the powder is air-conveyed to storage bins to await finishing.

20.5.1.3 Steam distillation[47,48]

Steam distillation removes the solvent from the polycarbonate resin by a combination of volatilization, azeotroping and gaseous sweeping. The polycarbonate solution is agitated during the steam distillation. The polycarbonate resin produced is a solid, porous particulate which contains both water and the original solvents. These particles must undergo additional drying and desolventization prior to resin finishing.

20.5.1.4 Antisolvent precipitation[19,29,49]

In antisolvent precipitation, a solvent which is miscible with the polycarbonate solvent solution but cannot dissolve the polycarbonate resin is added to the polycarbonate resin solution. For example, for methylene chloride, solvents like anisole, acetone or hexane may be used. The polycarbonate resin precipitates as a powder. This is separated centrifugally, crushed and further treated to remove residual solvent.

20.5.1.5 Mechanical devolatilization[50]

Mechanical devolatilization is accomplished by introduction of a concentrated solution of polycarbonate resin to a device which will develop high surface area in the presence of heat and vacuum. If the interfacial or solvent process resin solution is made up with a volatile solvent like methylene chloride then prior to the devolatilization a solvent exchange is accomplished with a higher boiling solvent. An example is chlorobenzene in place of methylene chloride.

Concentration is also effected so that about 50% solids solution can be fed to the devolatilization equipment.

There are a number of devolatilization units that can be used. These include wiped film evaporators, vented extruders[51] and disc ring evaporators. For all of these systems the solvent is removed for recycle and the polycarbonate is extruded as a molten stream. Typically this is stranded and pelletized.

20.5.2 Transesterification Polycarbonate Resin Isolation

As was noted in Section 20.4.3, transesterified polycarbonate resin is produced as a molten stream. This stream is stranded and the strand is chopped to make pellets. This pelletized product is similar to that produced from mechanical devolatilization processes.

20.6 RESIN FINISHING

Very little polycarbonate resin is sold and used to make articles as a pure polymeric chemical. To achieve engineering specifications, compounding with modifying agents is almost always employed. This is referred to in the industry as finishing. Finishing operations are critical to making resins that yield viable products with consistent quality.

20.6.1 Powder, Granules and Pellets

Polycarbonate resin isolated as powder, granules or pellets can be blended with thermal stabilizers, UV stabilizers, hydrolytic stabilizers, pigments, fillers, reinforcers and other agents in blending equipment. The blend can then be extruded to produce the homogeneous polycarbonate resin product. There are obviously differences in the blendabilities of powder, granules or pellets. Very small amounts of blending agents can readily be distributed throughout a powder matrix. However with pellets channeling can result in uneven distribution. Pellet blending then usually employs pellet concentrates of low concentration additives. Granule blending falls somewhere in between powder and pellet-blending mechanisms.

The blended premix is then extruded to produce the pelletized product.

20.6.2 Melt Stream

Polycarbonate resin produced by transesterification or by mechanical devolatilization is a melt stream. This can be stranded and pelletized for finishing as described in Section 20.6.1 or it can be finished on-line using extruder blending techniques. Thus the molten polycarbonate resin product can be introduced directly into down-line extruder equipment, which can receive inputs of solution borne or side stream molten additive streams metered to assure constant product composition. Such operations were not practical in previous years unless a single large volume product was being made. The advent of on-line extruder controls using rheometry, spectrophotometry and density with appropriate Boolean feed-back loops has made what was a fairly inflexible process much more useful in multiproduct production environments.[52]

20.7 SUMMARY AND TRENDS

The future of polycarbonate resin appears to be very bright. As an engineering resin it brings the advantages of moldability and structural strength that promote its use as a substitute for traditional materials ranging from glass to steel. Additionally, the coming of age of engineering plastics is seeing new products using polycarbonate resin not just as a substitute material, but as the first material of choice and the basis for the product design. For the foreseeable future the majority of the polycarbonate resin used will be bisphenol A based. This is due to the unique blend of economics and properties that is provided by the bisphenol A monomer.

New polycarbonate related materials are, however, reaching the marketplace. These are the poly(ester carbonate)s which afford higher HDT performance. These materials are typically based on isophthalic and terephthalic acid as the ester-forming monomers, and bisphenol A. The ratio of both isophthalate to terephthalate and ester to carbonate can be varied, however; some compositions will result in crystalline resins.[53,54] A typical example of this type of resin is General Electric's LEXAN® PPC 4701 (7) which has an HDT of 325 °F (163 °C). These resins can be produced by interfacial, solvent or transesterification means. Thus current technology can easily be extended to accommodate these resins' production in existing facilities.

(7)

Possible governmental requirements to reduce use of chlorocarbon solvents could result in control requirements and retrofits that could seriously affect solution and interfacial technologies economics. Also, a breakthrough in production of diaryl carbonate monomer precursors for the transesterification process could greatly enhance transesterification desirability.

In conclusion we can only say that polycarbonate resin manufacture is certain to increase over the next decade and that the current manufacturing processes are likely to be those used as models for the next production expansions. We do however caution that the world concerns for environment and resultant economic effects should not be ignored. The polycarbonate resin process picture could be altered instantaneously by governmental regulatory actions targeted at solvent emissions, bisphenol-A, phosgene utilization, and/or residual process-related resin contaminants.

DISCLAIMER

Inasmuch as General Electric Company has no control over the use to which others may put this material, it does not guarantee that the same results as those described herein will be obtained. Nor does General Electric Company guarantee the effectiveness or safety of any possible or suggested design for article of manufacture as illustrated herein by any photographs, technical drawings and the like. Each user of the material or design or both should make his own tests to determine the suitability of the material or any material for the design, as well as the suitability of the material or design or both for his own particular use. Statements concerning possible or suggested uses of the materials or designs described herein are not to be construed as constituting a license under any General Electric patent covering such use or as recommendations for use of such materials or designs in the infringement of any patent.

20.8 REFERENCES

1. H. Einhorn, *Justus Liebigs Ann. Chem.*, 1898, **300**, 135.
2. *Plast. Focus*, 1984, **16**, no. 49, 1.
3. H. Schnell, 'Chemistry and Physics of Polycarbonates', Interscience, New York, 1964.
4. W. F. Christopher and D. W. Fox, 'Polycarbonates', Reinhold, New York, 1962.
5. Depuy and King, *Chem. Rev.*, 1960, **60**, 432.
6. For a brief review of β-eliminations, see J. March, 'Advanced Organic Chemistry', McGraw-Hill, New York, 1977, 895.
7. (a) H. Vernalekein, in 'Interfacial Synthesis', ed. F. Millich and C. Carreher, Jr., Dekker, New York, 1977, vol. II, chap. 3.
   (b) Plastics World, 1984 Plastics Directory, Plastics Property Guide, 1984, no. 54.
8. M. Rabinovitz, Y. Cohen and M. Halpern, *Angew. Chem., Int. Ed. Engl.*, 1986, **25**, 960; D. Landini, A. Maia and A. Rampoldi, *J. Org. Chem.*, 1986, **51**, 5474.
9. U. Curtius, L. Bottenbruch and H. Schnell (Bayer AG), *US Pat.* 3 442 854 (1969).
10. H. Yamana, T. Kuni, T. Furusawa, Y. Sugimura, H. Nakai and Y. Hiro (Mitsubishi Gas Chemical Co.), *US Pat.* 3 888 826 (1975).
11. D. Brunelle (General Electric Co.), *US Pat.* 4 321 356 (1982).
12. J. B. Starr and A. Ko (General Electric Co.), *US Pat.* 4 383 092 (1983).
13. G. Illuminati, V. Romanu and R. Tesei (Snamprogetti SpA.), *US Pat.* 4 182 726 (1977).
14. J. E. Hallgren (General Electric Co.), *US Pat.* 4 410 464 (1983).
15. J. E. Hallgren (General Electric Co.), *US Pat.* 4 096 168 (1978).
16. A. J. Chalk (General Electric Co.), *US Pat.* 4 096 169 (1978).
17. J. E. Hallgren (General Electric Co.), *US Pat.* 4 349 485 (1982).
18. N. P. Chopey, *Chem. Eng.*, 1960, Nov. 14, 174.
19. K. Dick, G. Ham and J. Gross (Dow Chemical Co.), *US Pat.* 4 378 454 (1983).
20. (a) J. H. Vestergaard (General Electric Co.), *US Pat.* 3 989 672 (1976).
    (b) W. Alewelt, G. Jacobs, D. Margotte and E. Lax (Bayer AG), *US Pat.* 4 127 561 (1978).
21. V. Mark (General Electric Co.), *US Pat.* 4 277 597 (1981).
22. W. Alewelt, D. Margotle, C. Wulft and H. Vernaleren (Bayer AG), *US Pat.* 4 346 210 (1982).
23. J. Campbell and J. Lynch (General Electric Co.), *US Pat.* 4 384 108 (1983).
24. Y. Narita, H. Konuma, H. Nishitani, N. Komori and K. Ogishima (Idemitsu Kosan Co.), *US Pat.* 3 947 126 (1976).
25. T. Megumi and S. Kondo (Mitsubishi Gas Chemical Co.), *US Pat.* 4 097 457 (1978).
26. H. Koda, T. Megumi and H. Yoshizahi (Mitsubishi Gas Chemical Co.), *US Pat.* 4 122 112 (1978).
27. K. Katsuhisa, N. Katsuhiko and M. Akira (Mitsubishi Chemical Industries), *US Pat.* 4 413 103 (1983).
28. S. Matsuo and M. Itoi (Idemitsu Kosan Co.), *US Pat.* 4 452 966 (1984).
29. S. Glass (Dow Chemical Co.), *US Pat.* 4 529 791 (1985).
30. H. Vernaleken, O. Court and K. Weirauch (Bayer AG), *US Pat.* 3 674 740 (1972).
31. P. Horn and H. Kuerten (BASF), *US Pat.* 3 945 969 (1976).
32. H. Vernaleken and U. Hucks (Bayer AG), *US Pat.* 4 038 252 (1977).
33. K. Morgenstern, L. Bottenbruch, H. Schwarz, H. Vernaleken, H. Schnell and O. Court (Bayer AG), *US Pat.* 3 666 719 (1970).
34. A. Horbuck and H. Vernalehon (Bayer AG), *US Pat.* 3 939 118 (1976).
35. R. Rinaldi, G. Govoni and F. Visani (Montedison SpA.), *US Pat.* 4 316 009 (1982).
36. H. Mori, K. Kohyama, K. Nakamura and S. Tahamatsu (Mitsubishi Chemical Industries), *US Pat.* 4 323 519 (1982).
37. C. Pannell (Dow Chemical Co.), *US Pat.* 4 177 343 (1979).
38. C. D. Morland and B. D. Mitchell, 'Guide to the Selection of Polymer Filtration Systems', Purolator Technologies, Inc., 1981.
39. H. Koda, K. Hamaya, H. Yoshizahi, Y. Kojima, S. Tsuchiga, S. Fukuda and O. Tikeno (Mitsubishi Gas Chemical Co. and Hitachi Shipbuilding & Engineering Co.), *US Pat.* 4 184 911 (1980).
40. G. Govoni, G. DiDrusco, C. Corazzari and P. Guardigl (Montedison SpA.), *US Pat.* 4 212 967 (1980).
41. J. W. Flock and S. L. Matson (General Electric Co.), *US Pat.* 4 408 040 (1983).
42. J. W. Flock, S. L. Matson and P. H. Bollenbeck (General Electric Co.), *US Pat.* 4 423 207 (1983).
43. F. L. Rubin, H. A. Moak, A. D. Holt, F. C. Standiford and D. Stuhlbarg, in 'Perry's Chemical Engineers Handbook', ed. R. H. Perry and D. Green, McGraw-Hill, New York, 1984, pp. 11–31.
44. K. Kohyama, A. Matsuno and K. Tsuruhara (Mitsubishi Chemical Industries), *US Pat.* 4 546 172 (1985).
45. A. Clementi (Snamprogetti SpA.), *US Pat.* 3 772 262 (1973).
46. K. Masters, 'Spray Drying', 2nd edn., Wiley, New York, 1976.

47. H. Schnell and H. Schwarz (Bayer AG), *US Pat.* 3 427 370 (1969).
48. K. Kohyama, A. Matsumo and K. Tsuruhara (Mitsubishi Chemical Industries), *US Pat.* 4 546 172 (1985).
49. C. Marshall and R. Staffanoon (Dow Chemical Co.), *US Pat.* 4 182 850 (1980).
50. L. Bottenbruch, H. Lotter and H. Schnell (Bayer AG), *US Pat.* 3 437 638 (1969).
51. D. Gras and K. Eise, 'The Use of Multiscrew Extruders for Devolatilizing Low Solids Polymer Solutions', Pfleiderer Corp., code 28-7-1.
52. M. D. Bertolucci and D. E. Delaney, *1983 Tech. Conf. Soc. Plast. Eng. Inc.*, 1983, Sept. 20–22.
53. C. Quinn (General Electric Co.), *US Pat.* 4 238 596 (1980).
54. R. Markezich and C. Quinn (General Electric Co.), *US Pat.* 4 238 597 (1980).

21
Aliphatic Polyamides

REINOUD J. GAYMANS
Technische Hogeschool Twente, Enschede, Netherlands
and
DOETZE J. SIKKEMA
Akzo Fibres and Polymers Research, Arnhem, Netherlands

21.1	INTRODUCTION	357
	21.1.1 Equilibria and Kinetics	359
	21.1.2 Condensation and Ring Opening	359
	21.1.3 Amide Interchange and Ring Addition	361
	21.1.4 Molecular Weight Control and Molecular Weight Distribution	361
21.2	POLYAMIDE 4	361
21.3	POLYAMIDE 6	361
	21.3.1 Hydrolytic Process	361
	21.3.2 Random Copolymers	365
21.4	POLYAMIDES 11 AND 12	366
21.5	POLYAMIDE 4,6	366
21.6	POLYAMIDE 6,6	366
	21.6.1 Polyamide Salt	367
	21.6.2 Batch Process	367
	21.6.3 Continuous Process	368
	21.6.4 Laboratory Synthesis	368
21.7	POLYAMIDES 6,10 AND 6,12	368
21.8	PARTIALLY AROMATIC POLYAMIDES	369
21.9	BLOCK COPOLYMERS	369
21.10	OTHER ROUTES FOR THE SYNTHESIS OF ALIPHATIC POLYAMIDES	370
21.11	REFERENCES	371

21.1 INTRODUCTION

Many excellent reviews have been written on polyamides and their synthesis,[1-13] and comprehensive (private) literature surveys are published by the Stanford Research Institute. Polyamides are mostly prepared by a stoichiometric stepwise reaction and belong to the group of condensation polymers (see also Volume 5, Chapter 1). The polymers are formed by successive reactions between a difunctional reactant A–A with a difunctional reactant B–B or by reaction of the difunctional monomer A–B with itself. In order to obtain high molecular weight polymers, these two functional groups must be present in stoichiometric amounts. Of course, the stoichiometry is automatically obtained with a difunctional monomer of the type A–B.

Polyamides are an important group of polymers, frequently referred to as nylons, which was the trade name of the first commercial aliphatic polyamide. The different types are named according to the number of carbon atoms in the main chain. The polyamides $-(AABB)_n$ can be prepared from diamines and diacids (equation 1) and are called polyamide or nylon $x,(y + 2)$, *e.g.* polyamides 4,6, 6,6, 6,10 and 6,12. Polyamides of the type $-(AB)_n$ can either be prepared from amino acids by the self-condensation reaction in equation (2) or from cyclic amides (lactams) by an addition process (equation 3).

$$n H_2N{-}(CH_2)_x{-}NH_2 \;+\; n HO_2C{-}(CH_2)_y{-}CO_2H \longrightarrow$$

$$-[-N(H)-(CH_2)_x-N(H)-C(=O)-(CH_2)_y-C(=O)-]_n\; +\; 2n H_2O \tag{1}$$

$$n H_2N{-}(CH_2)_x{-}CO_2H \longrightarrow -[-N(H){-}(CH_2)_x{-}C(=O)-]_n\; +\; n H_2O \tag{2}$$

$$-NH_2 \;+\; \text{lactam}[(CH_2)_x, N(H){-}C(=O)] \longrightarrow -N(H){-}C(=O){-}(CH_2)_x{-}NH_2 \tag{3}$$

To initiate the hydrolytic polymerization of lactams, a fraction of the rings has first to be opened (ring-opening reaction, equation 4). These A–B polymers are named polyamide or nylon $(x+1)$, and important examples are polyamides 4, 6, 11 and 12. Another way of synthesizing polyamides from cyclic lactams is anionic polymerization, and this is discussed in Volume 3, Chapter 35.

$$\text{lactam}[(CH_2)_x, N(H){-}C(=O)] \;+\; H_2O \longrightarrow H_2N{-}(CH_2)_x{-}CO_2H \tag{4}$$

Polyamides can also be prepared from other starting materials, such as acyl chlorides. These more reactive compounds are particularly important for the preparation of aromatic polyamides, such as poly(p-phenylene terephthalamide) (equation 5).

$$n H_2N{-}C_6H_4{-}NH_2 \;+\; n ClC(=O){-}C_6H_4{-}C(=O)Cl \longrightarrow -[-N(H){-}C_6H_4{-}N(H){-}C(=O){-}C_6H_4{-}C(=O)-]_n \;+\; 2n\,HCl \tag{5}$$

Another synthetic route uses interchange reactions, which involve at least one amide group (equations 6 and 7). The total number of end groups is not changed by these reactions, but the chain lengths are continuously redistributed. In equation (7), a diamine is formed which can, if the conditions are right, be stripped from the system. The lactam ring addition step (equation 3) can also be regarded as an interchange reaction.

$$-R^1C(=O){-}NR^2(H){-} \;+\; -R^3C(=O){-}NR^4(H){-} \longrightarrow -R^1C(=O){-}NR^4(H){-} \;+\; -R^3C(=O){-}NR^2(H){-} \tag{6}$$

$$-R^1C(=O){-}NR^2(H)NH_2 \;+\; H_2NR^3 \longrightarrow -R^1C(=O){-}NR^3(H) \;+\; H_2NR^2NH_2 \tag{7}$$

Industrial processes for the synthesis of aliphatic polyamides are high temperature (200–300 °C) melt polymerizations, which can be followed by a solid-state postcondensation (see also Volume 5, Chapters 9 and 13).

21.1.1 Equilibria and Kinetics

The general aspects of hydrolytic polymerization will be discussed in this chapter.

Hydrolytic polyamidation involves proton transfers and therefore several equilibria have to be considered (equations 8–14). Equation (8) shows polyamide salt formation and polyamidation is either from the salt (equation 9) or *via* the acid-catalyzed reaction (equation 14). As can be seen, the amount of water and the pH of the system must also have an effect.

$$—NH_2 + HO_2C— \rightleftharpoons —NH_3^+ ^-O_2C— \tag{8}$$

$$—NH_3^+ ^-O_2C— \rightleftharpoons \underset{\underset{O}{\|}}{—N—C—} + H_2O \tag{9}$$
$$\phantom{—NH_3^+ ^-O_2C— \rightleftharpoons \ }\overset{H}{|}$$

$$—NH_2 + H_2O \rightleftharpoons —NH_3^+ + OH^- \tag{10}$$

$$—CO_2H + H_2O \rightleftharpoons —CO_2^- + H^+ \tag{11}$$

$$2H_2O \rightleftharpoons H_3O^+ + OH^- \tag{12}$$

$$—CO_2H + H^+ \rightleftharpoons —C(OH)_2^+ \tag{13}$$

$$—C(OH)_2^+ + H_2N— \rightleftharpoons \underset{\underset{H}{|}}{\overset{\overset{O}{\|}}{—C—N—}} + H^+ + H_2O \tag{14}$$

The reactivity of the functional groups has little dependence on either the length of the aliphatic group or the length of the polymer chain (see Volume 5, Chapter 3).[13,14] Therefore the polyamidation kinetics of the different polyamides are comparable. In contrast, for lactam ring structures, the reaction equilibria depend on the ring size.[11] This effect of the strain energy in cyclic lactams is discussed in Volume 5, Chapter 6.

The main amide equilibria are: (a) condensation and ring opening and (b) amide interchange and ring addition. These will be discussed in the following sections.

21.1.2 Condensation and Ring Opening

The equilibria of the hydrolytic polyamidation reaction have been thoroughly investigated.[9-11,15] The equilibrium for these reactions can be expressed as equation (15), where B is the temperature-independent equilibrium constant and ΔH_a is the enthalpy change of the reaction.

$$\frac{[—CONH—][H_2O]}{[—CO_2H][—NH_2]} = K = B \exp\frac{-\Delta H_a}{RT} \tag{15}$$

The equilibrium constant K (also called K_{III} in the synthesis of polyamide 6) has also been found to be dependent on the water concentration.[16-20] Giori and Hayes[21] found (Figure 1) that, up to 0.5 mol water per kg reaction mass (0.9 wt %), K increases linearly, and at water concentrations higher than 1 mol kg^{-1} (1.8 wt %), K decreases again. At very high water concentrations, K seems to be independent of the water concentration.

With increasing water content, the dielectric constant of the medium and the degree of end-group ionization will increase.[22] This is likely to influence the end-group activity coefficients, depending on whether the polycondensation reaction involves predominantly the condensation of neutral or ionized species.

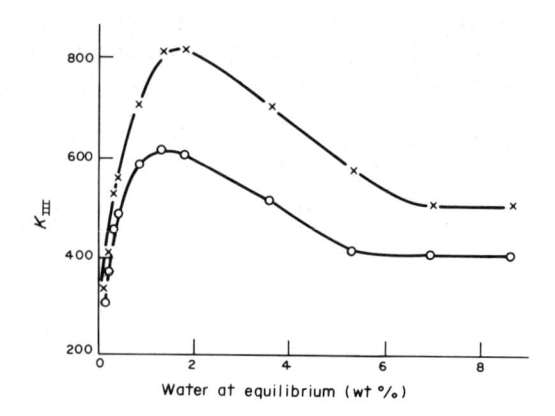

Figure 1 Equilibrium constants for polycondensation K_{III} vs. equilibrium water content: ($\times$) 240 °C ($\bigcirc$) 260 °C (reproduced with permission of J. Wiley & Sons from ref. 21)

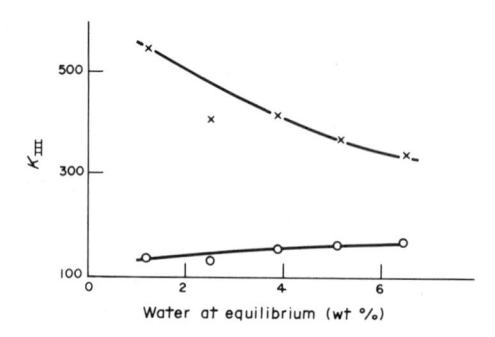

Figure 2 Equilibrium constants for nylon 6 polycondensation K_{III} vs. equilibrium water content: ($\times$) apparent equilibrium constants; ($\bigcirc$) equilibrium constants corrected for the water activity coefficient. $T = 270$ °C (reproduced with permission of J. Wiley & Sons from ref. 22)

If the equilibrium constant is calculated with activity constants from Raoult's law instead of concentrations (which can be obtained from vapor–liquid equilibria as studied by Giori and Hayes[22]), then K is virtually independent of the water concentration (Figure 2).

The reported values of ΔH_a for hydrolytic polyamidation are in the order of 25–29 kJ mol^{-1}.[11,23–25] This means that on decreasing the temperature, the equilibrium molecular weight shifts to higher values at constant water concentration.

When steric effects are present, such as in lactam ring structures, this equilibrium constant can change.[11]

The kinetics of the hydrolytic polyamidation-type reaction have the form shown in equation (16). The kinetics of the polyamidation have been a matter of discussion for many years. One of the reasons for this is that the equilibrium constant is dependent on the water concentration (Figure 1).

$$-\frac{d(-CO_2H)}{dt} = k[-CO_2H][-NH_2] - \frac{K}{[amide][H_2O]} \quad (16)$$

In aqueous solution, second-order kinetics are found for the condensation of simple carboxylic acids and amines.[25–26] For studies using high water concentrations (5–10 mol kg^{-1}), a second-order reaction was observed with an activation energy of approximately 85 kJ mol^{-1}.[26–29] In experiments using low water concentrations (0.4–1.2 mol kg^{-1}), a mixed uncatalyzed second- and acid-catalyzed third-order reaction was observed and the rate constant can then be written as shown in equation (17).

$$k = k' + ck'' \quad (17)$$

In this region, the K value decreases with water concentration. The calculated activation energies were 96 kJ mol^{-1} for the uncatalyzed second-order part and 88 kJ mol^{-1} for the third-order part.[2]

From hydrolysis data using very low water concentrations (0.005–0.1 mol kg^{-1}), the reaction was found to be second order, but showed a dependence of water concentration on the rate constants, this being due to the linear increase in K value with water concentration.

For the caprolactam hydrolysis with 1.1 mol kg^{-1} water, catalyzed second- and acid-catalyzed third-order reactions were observed, with activation energies of 109 and 63 kJ mol^{-1}, respectively.[6]

In the solid-state polymerization of (pre)polymers, the reaction rate is not only dependent on the end-group concentrations and the temperature, but also at which end-group concentration the postpolymerization started, and this reaction is limited by diffusion of the catalyzing groups.[30,31] The reaction is also influenced by the presence of the crystalline phase.[32]

The acid catalyst can be composed of carboxylic acids, as well as boric and phosphoric acids.[2,31]

21.1.3 Amide Interchange and Ring Addition

During the polymerization, the amide-interchange reaction also takes place, but its equilibrium value is not usually important.

However, for the ring-addition reaction (equation 3), the equilibrium value under polymerization conditions determines the minimum final content of lactam, which is 7.8% for polyamide 6 at 250 °C and even higher for some other lactam polymers.[9-11] These unfavorable values for some lactam polymers (polyamides 4 and 5) make these polymers thermally unstable at higher temperatures.

When two polyamides of the same type but different molecular weights are mixed, the initial bimodal distribution changes in 3 min at 260 °C to a single distribution with an almost normal width.

If two different polyamides are mixed for 3 min at 260 °C, 5% of the amide groups (as measured by NMR) have already undergone amide interchange and a block copolymer is formed. By reacting for longer periods of time (120 min at 260 °C), a completely random copolyamide can be formed.[33-36]

The amide-interchange reaction has also been found to be acid catalyzed.[2,34]

21.1.4 Molecular Weight Control and Molecular Weight Distribution

The molecular weight of a material can be stabilized by creating non-reactive end-groups, or by adding monofunctional acids or amines, or adding excess acid or amine (see Volume 5, Chapters 2 and 4).

In some instances, the residual monomers can be stripped by an amide-interchange reaction (equation 7). Volatile diamines particularly are stripped in this way and are therefore often added in excess at the beginning of the reaction.

The molecular weight distributions of melt-polymerized polyamides are in the order of two and these values are in good agreement with the calculated distribution for condensation polymers by Flory,[37-43] assuming the polymers to be linear and the reactivity of the end groups to be independent of the degree of polymerization.

The usual method used nowadays for studying the molecular weight distribution is high performance liquid chromatography (HPLC) [or, alternatively, gel-permeation chromatography (GPC) or size-exclusion chromatography (SEC)].[44-47] Two systems of analysis are used: (i) solutions of polyamides in m-cresol at high temperatures (100 °C)[44] and (ii) trifluoroacetylated polyamides dissolved in solvents such as tetrahydrofuran (THF) or chlorinated hydrocarbons.[45,46] Besides the molecular weight distribution of the polymer, the monomer and higher cyclic oligomer content can also be determined in this way.

21.2 POLYAMIDE 4

Polyamide 4 can only be synthesized to high molecular weights by the anionic ring-opening polymerization of pyrrolidone at low polymerization temperatures,[3] and it depolymerizes when heated close to its melting temperature.[48-50]

21.3 POLYAMIDE 6

21.3.1 Hydrolytic Process

The hydrolytic process for nylon 6 preparation (Scheme 1) was the subject of a recent review,[51] which stressed the kinetics, including the side reactions which form cyclic oligomers,[52,53] and pointed out our relative lack of understanding of the physical transport processes that ultimately limit the kinetics.[54] The simpler model of Jacobs and Schweigman[55] takes advantage of the fact that, in industrial practice, residence times are sufficiently long to allow the process to essentially reach chemical equilibrium at the end of a continuous reactor. A detailed simulation of an industrial reactor has recently appeared.[56] A classical review of nylon 6 polymerization is the paper by Reimschuessel.[9]

Ring opening: $C_1 + W \underset{k'_1 = k_1/K_1}{\overset{k_1}{\rightleftharpoons}} P_1$

Polycondensation: $P_n + P_m \underset{k'_2 = k_2/K_2}{\overset{k_2}{\rightleftharpoons}} P_{n+m} + W;\quad n, m = 1, 2, 3, \ldots$

Polyaddition: $P_n + C_1 \underset{k'_3 = k_3/K_3}{\overset{k_3}{\rightleftharpoons}} P_{n+1};\quad n = 1, 2, \ldots$

Ring opening of cyclic dimer: $C_2 + W \underset{k'_4 = k_4/K_4}{\overset{k_4}{\rightleftharpoons}} P_2$

Polyaddition of cyclic dimer: $P_n + C_2 \underset{k'_5 = k_5/K_5}{\overset{k_5}{\rightleftharpoons}} P_{n+2};\quad n = 1, 2, \ldots$

Reaction with monofunctional acid: $P_n + P_{mx} \underset{k'_2 = k_2/K_2}{\overset{k_2}{\rightleftharpoons}} P_{m+n,x} + W;\quad n, m = 1, 2, \ldots$

W = water; C_1 = caprolactam; P_m = polymer chain of chain length m; P_{mx} = polymer chain of chain length m with one end capped with a monofunctional acid; cyclic dimer =

Scheme 1 Kinetic scheme for nylon 6 polymerization (reproduced with permission of Marcel Dekker from ref. 51)

The reaction is carried out at between 220 and 280 °C. The process is acid catalyzed, which means that carboxylic end-groups increase the rate. Early procedures used a simple tubular reactor, the so-called VK-column (Vereinfacht Kontinuierlich; Figure 3) where caprolactam and water were added continuously to the top at atmospheric pressure under reflux. No water was removed in the downstream sections of the reactor. The actual water content in the initiation section (the reactor top), which translates to final water content and end-group concentrations *via* chemical equilibration (Figure 4), is a function of the temperature only[55] in the top section (Figure 5). In order to achieve high molecular weights, one needs high temperatures; in order to induce a rapid rate of production, one needs lower temperatures, correlating with higher end-group contents. This

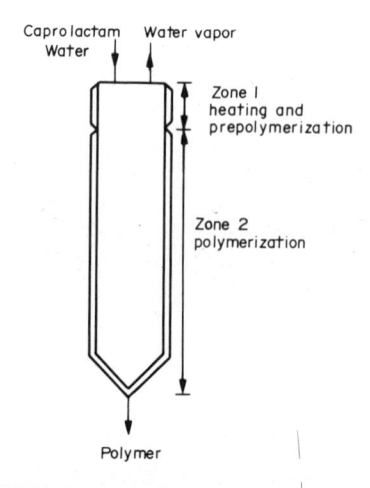

Figure 3 Conventional VK-column (reproduced with permission of Elsevier from ref. 55)

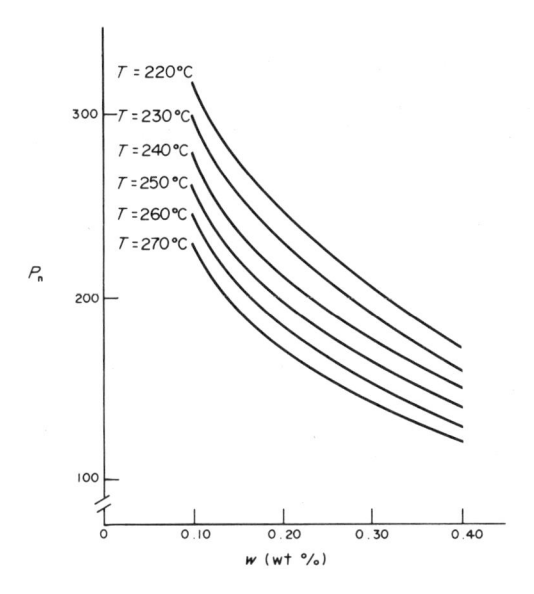

Figure 4 Influence of the water content w on the degree of polymerization P_n at chemical equilibrium (reproduced with permission of Elsevier from ref. 55)

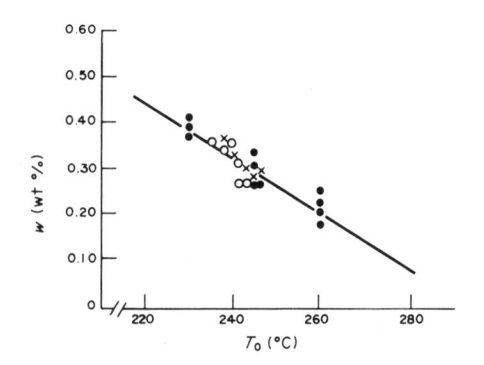

Figure 5 The water concentration w in the melt at the top of continuous reactors as a function of the temperature T_0 in the melt: (●) pilot plant; industrial reactors, (○) Glanzstoff A. G., Germany, (+) Enka N. V. Emmen, Netherlands, (×) Enka N. V. Emmen, Netherlands. Equation for straight line is $w = 1.76 - 0.0060\, T_0$ (reproduced with permission of Elsevier from ref. 55)

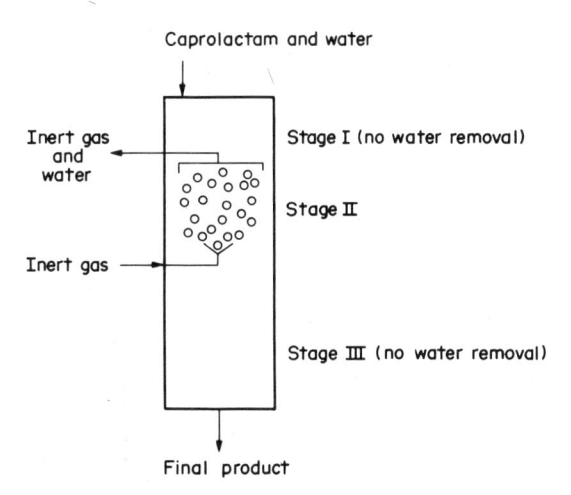

Figure 6 A three-stage reactor with no water removal in stages I and III (due to high pressures) but water removal by inert-gas sparging in stage II

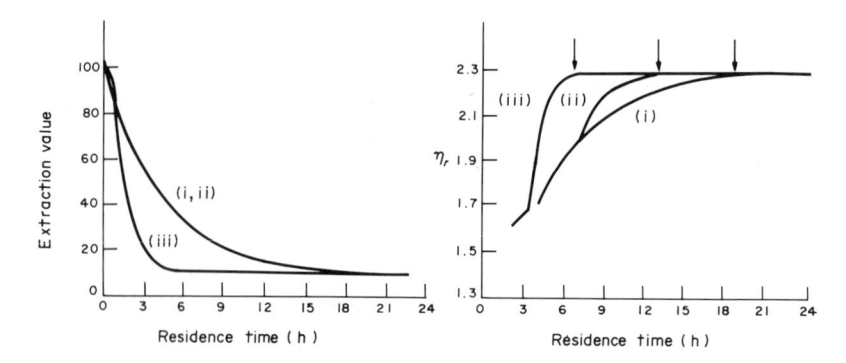

Figure 7 Comparison of polymerization techniques: (i) conventional VK-column; (ii) prepolymerization at atmospheric pressure followed by water removal; (iii) prepolymerization at elevated pressure followed by water removal (reproduced with permission of Elsevier from ref. 55)

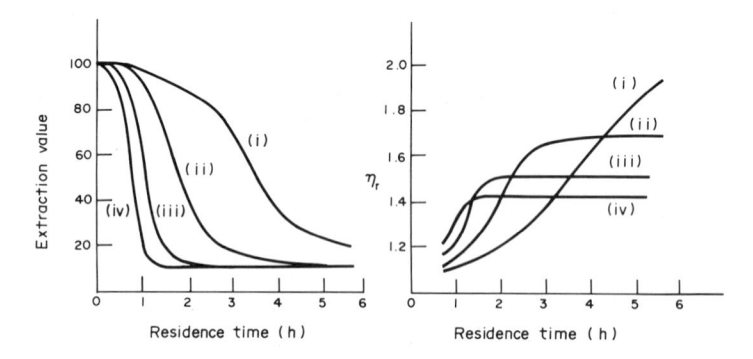

Figure 8 Prepolymerization at elevated pressure: curve (i) 0.5% water in lactam feed; curve (ii) 1.0% water in lactam feed; curve (iii) 2.0% water in lactam feed; curve (iv) 3.0% water in lactam feed (reproduced with permission of Elsevier from ref. 55)

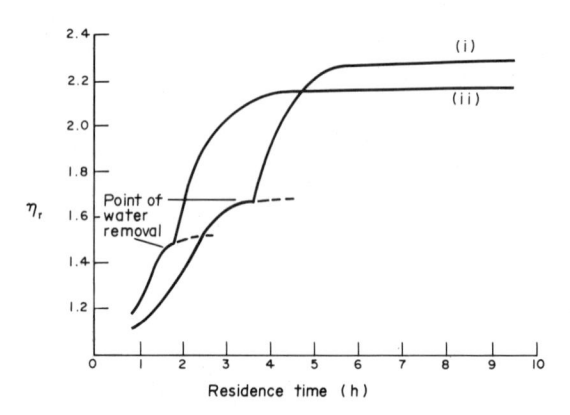

Figure 9 Water removal after prepolymerization at elevated pressure: curve (i) prepolymerization with 1.0% water, water removal up to 0.1% water; curve (ii) prepolymerization with 2.0% water, water removal up to 0.1% water (reproduced with permission of Elsevier from ref. 55)

problem can be solved by aiming for a higher water content at the start of the process, using a lower temperature or higher pressure, and removing water from the product later in the process, either by flashing in a separate reactor or by sparging with inert gas (Figure 6). In batch processes, it is customary to use fairly large amounts of water for initiation under autogenous pressure for about an hour, and polymerize at atmospheric pressure (Figures 7, 8 and 9).

Kinetic data, including the relatively high-energy cyclic dimer side-product, are presented in Table 1.[51]

Further discussions on side reactions center on the formation of unwanted end-groups.[57] Various catalysts have been recommended for increasing the capacity of a given facility;[58] however, in

Table 1 Rate and Equilibrium Constants for Nylon 6 Polymerization[a,b]

i	A_i° (kg mol^{-1} h^{-1})	E_i° (cal mol^{-1})	A_i^c (kg^2 mol^{-2} h^{-1})	E_i^c (cal mol^{-1})	ΔH_i (cal mol^{-1})	ΔS_i (cal K^{-1} mol^{-1})
1	5.9874×10^5	1.9880×10^4	4.3075×10^7	1.8806×10^4	1.9180×10^3	-7.8846
2	1.8942×10^{10}	2.3271×10^4	1.2114×10^{10}	2.0670×10^4	-5.9458×10^3	9.4374×10^{-1}
3	2.8558×10^9	2.2845×10^4	1.6377×10^{10}	2.0107×10^4	-4.0438×10^3	-6.9457
4	8.5778×10^{11}	4.2000×10^4	2.3307×10^{12}	3.7400×10^4	-9.6000×10^3	-1.4520×10^1
5	2.5701×10^8	2.1300×10^4	3.0110×10^9	2.0400×10^4	-3.1691×10^3	5.8265×10^{-1}

[a] Reproduced with permission of the American Chemical Society from K. Tai and T. Tagawa, *Ind. Eng. Chem. Prod. Dev.*, 1983, **22**, 192.

[b] $k_i = k_i^\circ + k_i^c[\text{—CO}_2\text{H}] = A_i^\circ \exp(-E_i^\circ/RT) + A_i^c \exp(-E_i^c/RT)[\text{—CO}_2\text{H}]$; $K_i = \exp\left\{\dfrac{\Delta S_i}{R} - \dfrac{\Delta H_i}{RT}\right\}$; $i = 1, 2, \ldots, 5$.

practice, nylon 6 is normally produced without a catalyst, although phosphoric acid is used in cases where a high molecular weight rather than exact reproducibility and constancy of quality is desired. Acid catalysts increase the formation of amidine end-groups,[57] which deactivate the catalyst (Scheme 2). This side reaction is of significance, especially in the anhydrous cationic polymerization of caprolactam, for instance initiated by its complex with orthophosphoric acid.[59–61] The anhydrous cationic process is not used industrially, in contrast with the anhydrous anionic caprolactam polymerization (*cf.* Volume 3, Chapter 35).

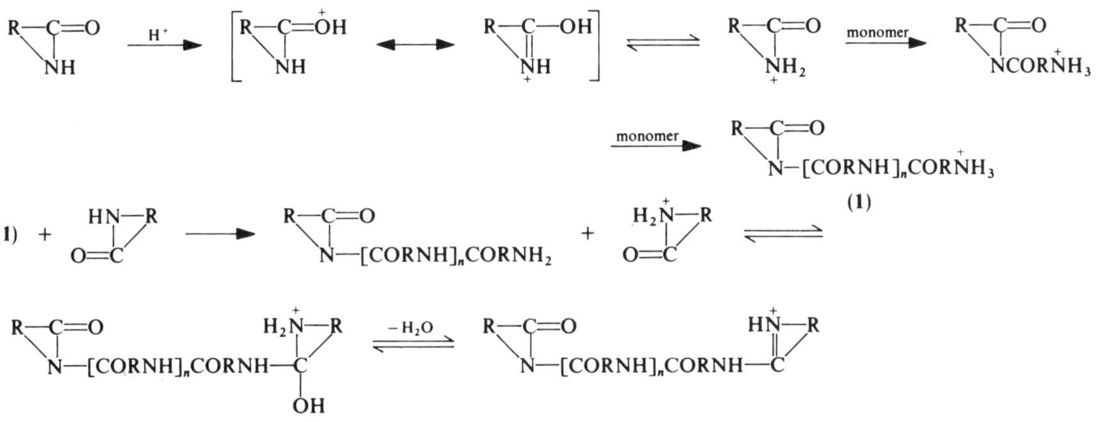

Scheme 2

For laboratory experiments, it is usually convenient to start with caprolactam and a small amount of 6-aminohexanoic acid instead of water. Great care must be taken to exclude all oxygen.

The significant amounts of extractable product (mainly monomer) remaining after polymerization (*cf.* Figure 7) necessitates extraction of the virgin polymer.

For many applications, where a high molecular weight is desired, a solid-phase postcondensation is carried out, typically for 5–20 h at 160–180 °C *in vacuo*, to increase [η] from 1.0–1.3 to 1.5–5, as desired. Solution viscosities are usually determined in formic acid or *m*-cresol.

21.3.2 Random Copolymers

Copolymerization of caprolactam with other lactams, notably 12-dodecanelactam, by hydrolytic schemes has the same advantages and disadvantages as homopolymerization. Copolymerization with nylon salts can be performed without the addition of water, eliminating the hydrolysis step and retaining only the ring-opening and polycondensation steps. Compared with the homopolymer, the random copolymers showing depressed crystallinity are produced on a much smaller scale. Small amounts of comonomer are incorporated in order to influence the dyeing behavior of textile or carpet fibers (using diamine rather than nylon salt to arrive at excess amino end-groups for increased affinity for acid dyestuffs and 5-sulfoisophthalic acid, for example, to affect the affinity for basic dyestuffs). Higher levels of comonomer systems (*e.g.* HMD–isophthalic acid salt) are used to prepare (largely) amorphous products which are valued for their transparency.[62]

21.4 POLYAMIDES 11 AND 12

Nylons 4, 8, 11 and 12 are known; nylons 11 and 12 are produced commercially for plastics applications. Their lower melting point, lower moisture absorption and lower modulus of elasticity make them useful in a different set of applications to those of nylon 6. Nylon 11 is prepared from 11-aminoundecanoic acid by a melt polycondensation; nylon 12 is produced from 12-dodecanelactam by a process similar to the nylon 6 process. These polymers contain very low levels of monomer and cyclic oligomers in the virgin product[63] and need not be extracted before use.

21.5 POLYAMIDE 4,6

Polyamide 4,6 is a newly marketed polyamide. It has a high amide content and a high chain regularity, thus its T_m is quite high (290 °C).[64-68]

Melt polymerization from its polyamide salt gives low molecular weight polymers.[64,65] This is due to a side reaction, the cyclization of tetramethylenediamine (TMDA) to pyrrolidine (equation 18). High molecular weight polymers can be obtained by preparing a prepolymer from its polyamide salt at 210 °C and 1.5×10^6 Pa steam pressure and postcondensing this polymer in its solid state.[65-68] The cyclization of TMDA is a first-order process of amine-terminated end-groups with an activation energy of 137 kJ mol^{-1}.[68]

$$-\overset{O}{\underset{}{C}}-\underset{H}{N}-(CH_2)_4-NH_2 \longrightarrow -\overset{O}{\underset{}{C}}-N\text{\textlangle}\rangle + NH_3 \qquad (18)$$

The solid-state process gives high molecular weight polymers for several reasons: (a) the polyamidation equilibrium constant K is higher at lower temperatures; (b) the cyclization of TMDA, which it is desirable to avoid, has a higher activation energy than the polymerization reaction, thus a lower reaction temperature is an advantage; and (c) the thermal degradation process also has a high activation energy.

In the polymerization, an excess of TMDA from the prepolymerization seems to be preferable,[66-68] probably because of the stripping of pyrrolidine end-groups by aminolysis.

21.6 POLYAMIDE 6,6

The synthesis of polyamide 6,6 has been studied extensively and several good review articles have appeared.[1,2] Polyamide 6,6 can be prepared from 1,6-diaminohexane and adipic acid to form an AA–BB type polymer with even–even numbers of methylene units. This gives the polymer a high chain regularity and interesting properties.

The polymer is prepared by a melt-polymerization method starting with a polyamide salt solution in water.[1,2,69-71] A pressurized prepolymerization and a polycondensation step above the melting

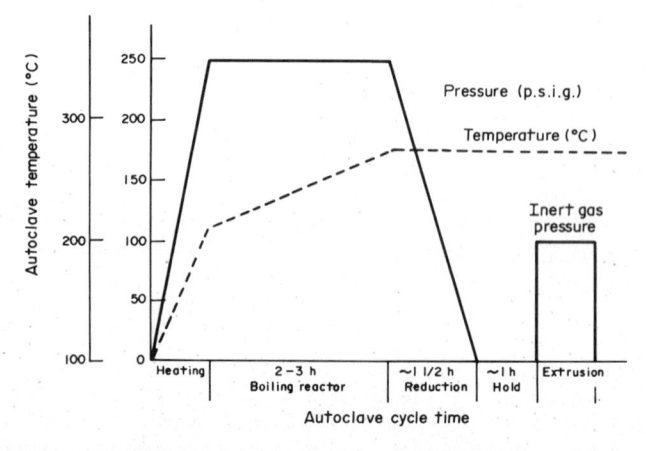

Figure 10 A typical autoclave cycle useful for batch preparation of nylon 6,6 (Du Pont Technical Laboratory, Seaford, Delaware) (reproduced with permission of Wiley-Interscience from ref. 2)

temperature of the polymer is carried out. Typical process conditions are given in Figure 10. In this process there are the following stages: (i) prepolymerization at 210–275 °C and 1.8×10^6 Pa; (ii) flashing stage, reducing the pressure to the atmospheric value; and (iii) polymerization at 275–290 °C and 1×10^5 Pa.

To increase the molecular weight further, the polymer can be postcondensed by either a melt-finishing or solid-state postcondensation method.

21.6.1 Polyamide Salt

As soon as the diamines and the diacids are mixed, the polyamide salt is formed. If this is done with the starting materials in methanol or ethanol solvents, the partially soluble salt is precipitated. As the equimolar salt is least soluble, it is an easy way of obtaining balanced salts with, at the same time, a purification step. A 50% solution in water is made with this salt, which has to be kept at 50 °C to prevent crystallization. With pure starting materials, it is possible to make a concentrated salt solution in water directly, but as the salt formation is an exothermic reaction, and at this stage no temperature increase is wanted, the mass has to be cooled. The composition of the salt can be determined by the pH of a 1% solution in water (the pH–composition relationship is given in Figure 11).

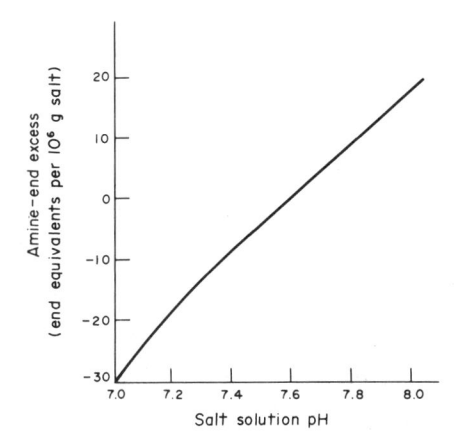

Figure 11 Relation of free amine group concentration of growing polyamide to the pH of 6,6-salt aqueous solution (Du Pont Technical Laboratory, Seaford, Delaware) (reproduced with permission of Wiley-Interscience from ref. 2)

The 50% salt solution is an easy way of transporting the salt, but before polymerizing it is usually first concentrated at 150–180 °C.[71]

21.6.2 Batch Process

The synthesis of polyamide 6, 6 is mainly achieved by a batch melt-polymerization method. An example of an autoclave cycle for polyamide 6,6 synthesis is given in Figure 10. The autoclave is charged with a concentrated solution of polyamide salt and heated to 210 °C. During the prepolymerization the pressure is maintained at 1.8×10^6 Pa and the temperature is raised from 210 to 275 °C. Pressure is employed for two reasons; firstly, to keep water in the system, so that the mass remains liquid below its melting temperature (the T_m of polyamide is 265 °C) and, secondly, to prevent the more volatile diamine from evaporating.[2,71] With these high water concentrations only prepolymers are produced.

During the polymerization step, at atmospheric or reduced pressure, a high molecular weight polymer is formed and the viscosity of the reaction mixture is considerably increased. With the very high viscosities, it is difficult to keep the reaction mass well-stirred, but this stirring is necessary in order to prevent local overheating and to allow efficient evaporation of the condensation water.

As polyamide 6, 6 is susceptible to gelation, long reaction times at high temperature and reduced pressure should be avoided.[72,73] The reaction which triggers this gelation is thought to be dimerization of the diamine to a triamine.[72,73]

$$2H_2NRNH_2 \longrightarrow H_2NR\overset{H}{\underset{|}{N}}RNH_2 + NH_3 \quad (19)$$

At the end of the polymerization cycle, the polymer is extruded out of the reactor.

For higher molecular weight materials, this autoclave reactor process can be followed by a postcondensation step in the solid state.[1,2,74,75] The solid-state polymerization is usually performed on granular material in tumble dryers at reduced pressure well below the melting temperature of the polymer and for many hours.[74,75]

21.6.3 Continuous Process

Continuous melt processes for polyamide 6,6 have been reported for over 20 years.[76-85] In a typical process (Figure 12), a concentrated salt solution is pumped into a set of heating tubes. These tubes have several heating zones and the diameter is gradually increased in size.[1,76,78] The process starts with a one-phase flow of a low viscosity polyamide salt solution and ends with a two-phase flow with the viscous polyamide melt along the tube wall and the steam through the middle. The steering and flexibility of this continuous process seems to be quite a problem. Coupled to the polymerization tubes is a steam separator and a finishing reactor.

For very high molecular weights, these materials are also postcondensed in the solid state. This postcondensation is usually a batch process.

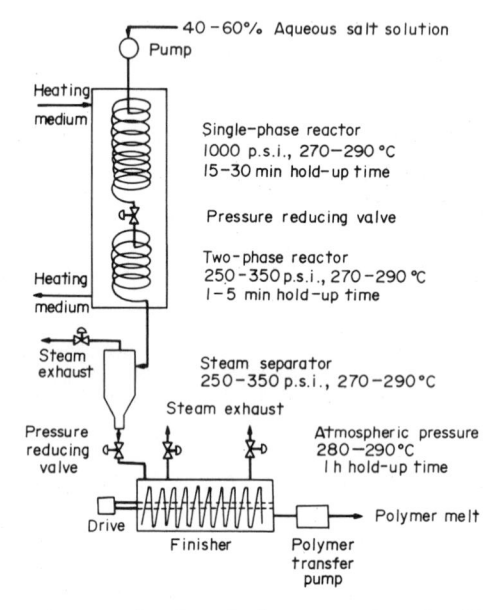

Figure 12 Diagram of an early continuous melt-polymerization process for nylon 6,6 (1 p.s.i. = 6.9 kN m^{-2}; reproduced from ref. 76)

21.6.4 Laboratory Synthesis

A good method for melt polymerization in a laboratory is given in Macromolecular Syntheses[86] and other literature sources. The first step is a pressurized reaction with a Carius tube; handling Carius tubes is somewhat hazardous. It is preferable to use an autoclave or a capsule in an autoclave.[65,87]

21.7 POLYAMIDE 6,10 AND 6,12

The syntheses of polyamide 6,10 with 1,6-diaminohexane and sebacic acid and polyamide 6,12 with 1,6-diaminohexane and dodecanedioic acid are very similar to that of polyamide 6,6; however, their susceptibility to thermal degradation is less than that of polyamide 6,6.[1,2,5]

21.8 PARTIALLY AROMATIC POLYAMIDES

Partially aromatic amorphous polyamides are of interest, as they combine a high T_g with transparency and a particularly good solvent resistance to hydrocarbons.

Partially aromatic semi-crystalline polyamides have a high T_g and a high T_m. The melting temperatures are usually well above 300 °C, a temperature at which most organic polymers start to degrade; therefore melt processing without degradation is not possible for these polymers. However, with the structural units of semi-crystalline polymers, copolymers can be formed with interesting combinations of properties.[88-92]

The partially aromatic polyamides can be synthesized from aromatic diamines and aliphatic diacids or from aliphatic diamines and aromatic diacids.

Aromatic diamines are very unreactive and susceptible to oxidation with color formation.[8,93] Partially aromatic polymers based on these aromatic diamines are usually prepared with acyl chlorides by a low temperature solution or interfacial method.[94,95] Partially aromatic polyamides with aromatic diacids can be synthesized from solution,[96] or interfacially,[88,97-99] but are mostly prepared from their salts.[92,100-102] Although the reactivities of the aromatic acids are lower than those of aliphatic acids, high molecular weight polymers can be synthesized in this way.[92]

A number of important partially aromatic polyamides are given in Table 2.[103-108]

21.9 BLOCK COPOLYMERS

Poly(ether amide) block copolymers have been investigated as potential antistatic compounds. High molecular weight materials were proposed as elastomers[109] and they are used as elastomeric

Table 2 Important Amorphous Partially Aromatic Polyamides[a]

	Structure of starting materials			Ref.
phthalic acid (CO₂H, CO₂H)	H₂NCH₂C(Me)CH₂C(Me)(H)CH₂CH₂NH₂ with (Me) alt			103
isophthalic acid (CO₂H, CO₂H)	H₂N–⟨Me-cyclohexyl⟩–CH₂–⟨cyclohexyl-Me⟩–NH₂	(CH₂)₁₁ C=O / NH		104
	H₂N–(CH₂)₆–NH₂			
HO₂C–(CH₂)₄–CO₂H	H₂N–⟨C₆H₄⟩–C(Me)(Me)–⟨C₆H₄⟩–NH₂			105
isophthalic acid (CO₂H, CO₂H)	H₂N–(CH₂)₆–NH₂			106
HO₂C–(CH₂)₄–CO₂H / HO₂C–(CH₂)₇–CO₂H	H₂N–⟨C₆H₄⟩–CH(Me)(Me)–⟨C₆H₄⟩–NH₂			107
phthalic acid (CO₂H, CO₂H)	H₂NH₂C–⟨norbornane⟩–CH₂NH₂	(CH₂)₅ C=O / NH		108

[a] Reproduced with permission of Huethig lund Wepf Verlag from ref. 101.

additives in nylon plastics to prepare impact-resistant grades.[110] Preparation is straightforward,[109,110] given the availability of α,ω-diaminopolyethers.[111] (Polyether)diols, which are more widely available, can be used for the preparation of poly(ether ester amide)s by esterification of the diol prepolymer with carboxyl-terminated polyamide prepolymer.[112] This route is relatively difficult because prepolymers of elevated molecular weights will give phase separation and esterification is much slower than amidation or caprolactam ring opening. Most research on the synthesis of poly(ether ester amide)s has been directed towards nylon 11 or 12/poly(oxytetramethylene)diol (polyTHF) systems, where the phase separation problem is less severe (equations 20 and 21).[113-115]

$$\text{Poly-THF diol} + [\text{nylon 11 diacid prepolymer}] \xrightarrow{\text{esterify}} \text{'Pebax' poly(ether ester amide) (Ato Chimie)} \quad (20)$$

$$\text{Poly(oxypropylene)diamine} + \text{caprolactam} + \text{diacid} \longrightarrow \text{'Arnitel A' poly(ether amide) (Akzo)} \quad (21)$$

Using amine/acyl chloride solutions or interface polymerization technology one can prepare various block copolymers.[116] Nylon 6/6,6/aromatic polyamide block copolymers[117] have been proposed as dispersion adjuvants in mixed solutions of aliphatic and aromatic nylons that are studied for making composites.[118] The miscibility of different polymers, even in solution, is low at best and negligible in the case where one forms an isotropic solution and the other a liquid-crystalline solution.

21.10 OTHER ROUTES FOR THE SYNTHESIS OF ALIPHATIC POLYAMIDES

As discussed above, the industrial process used for manufacturing aliphatic polyamides is hydrolytic polymerization in the melt; however, several other methods have been described in the literature.[8,119-126]

With activated acids, the reaction with diamine is much faster, but the condensation product might react with the free diamine, thus disturbing the stoichiometry of the reaction.

$$n\text{H}_2\text{NR}^1\text{NH}_2 + n\text{XCR}^2\text{CX} \longrightarrow \underset{\underset{\text{H}}{|}}{+}\text{NR}^1\text{N}\underset{\text{H}}{|}\text{CR}^2\overset{\text{O}}{\underset{}{\text{C}}}\underset{n}{)} + 2n\text{XH} \quad (22)$$

$$\text{H}_2\text{NR}^1\text{NH}_2 + 2n\text{XH} \longrightarrow \text{X}^-\overset{+}{\text{H}_3\text{N}}\text{R}^1\overset{+}{\text{NH}_3}\text{X}^- \quad (23)$$

Acyl chlorides have been used for studies on the synthesis of aliphatic polyamides. The condensation product of this reaction (HCl) reacts readily with free diamine and only a strong acid acceptor can prevent this.

Solution polymerization has been studied with solvents such as THF and dimethylacetamide (DMAc) and with tertiary amines (such as triethylamine[8,119-121] and diisopropylbenzylamine[121]), metal hydroxides[8] and excess diamines[8] as acid acceptors. The molecular weights obtained are often low.[8]

An elegant way to overcome the problem of the disturbing effect of the condensation product is interfacial polymerization. In interfacial polymerization, the acyl chloride is dissolved in an apolar solvent (*e.g.* toluene) and the diamine is dissolved in water; the acid acceptor (*e.g.* sodium hydroxide) is added to the diamine phase.

The reaction takes place at the interface of these solutions and the polymer precipitates as it is formed. In a static process, a fiber can be pulled out (Figure 13). In a vigorously stirred process, the molecular weights obtained are usually somewhat higher.[8,112-122] The polymers prepared in this way have a broad molecular weight distribution.

Not all polyamides can be prepared equally well by this method. The lower the length of the aliphatic sections, the lower the maximum obtainable molecular weight.[8,122] There are several reasons for this.

One of the side reactions of this process is the hydrolysis of the acyl chloride. For the reaction to take place, the diacyl chloride has to react in the organic layer with the diffused diamine, before it diffuses into the aqueous layer and is hydrolyzed. The solubility ratios of the diacyl chloride and the diamine in the apolar/water phases are the controlling parameters for these diffusion processes. As the length of the aliphatic chain in the acyl chloride is shortened, the molecule becomes more polar;

Aliphatic Polyamides

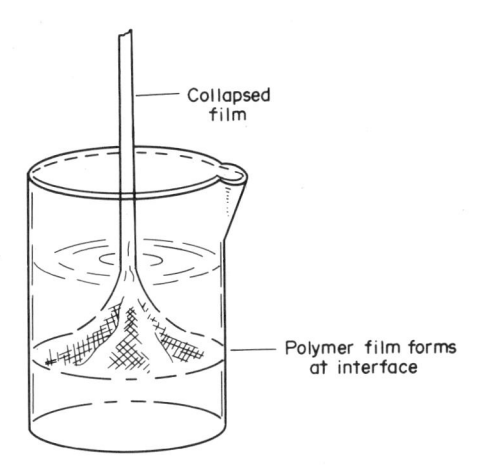

Figure 13 Sketch of an interfacial polymerization with the collapsed polymer film being withdrawn from the surface between the immiscible phases (reproduced with permission of the American Chemical Society from P. W. Morgan and S. L. Kwolek, *J. Chem. Educ.*, 1959, **36**, 182)

diffusion into the water phase becomes faster with more chance of hydrolysis, and the maximum obtainable molecular weight is decreased.

Similarly, as the length of the aliphatic chain in the diamine is shortened, the diamine also becomes more polar and the relative diffusion rate into the organic phase is reduced, with more chance of hydrolysis.

The solubility of the polymer also plays a role. As the solubility of the polymer is decreased, so is the maximum obtainable molecular weight.[8]

Esters have been used as starting materials for high temperature polymerization and sometimes even higher molecular weights have been obtained in this way than with polyamide salts.[123,124]

The low temperature solution method has been studied with triphenyl phosphite and other catalytic systems,[125,126] but this method produces mainly low viscosity type polyamides.

21.11 REFERENCES

1. H. Hopff, in 'Man-Made Fibers, Science and Technology', ed. H. F. Mark, S. M. Atlas and E. Cernia, Wiley-Interscience, New York, 1968, vol. 2, p. 181.
2. D. B. Jacobs and J. Zimmerman, in 'High Polymers', ed. C. E. Schieldknecht and I. Skeist, Wiley-Interscience, New York, 1977, vol. XXIX, chap. 12.
3. D. C. Jones and T. R. White, 'Kinetics and Mechanisms of Polymerization Series', Dekker, New York, 1972, vol. 3, p. 41.
4. M. I. Kohan, 'Nylon Plastics', Wiley-Interscience, New York, 1973, chap. 2.
5. J. Zimmerman, in 'Encyclopedia of Polymer Science and Engineering', 2nd edn., ed. H. F. Mark, N. M. Bikales, C. G. Overberger and G. Menges, Wiley, New York, 1988, vol. 11, p. 315.
6. R. E. Putscher, in 'Kirk-Othmer Encyclopedia of Chemical Technology', ed. H. F. Mark, D. F. Othmer, C. G. Overberger and G. T. Seaborg, Wiley, New York, 1987, vol. 18, p. 328.
7. I. K. Miller and J. Zimmerman, *ACS Symp. Ser.*, 1985, **285**, 159.
8. P. W. Morgan, 'Condensation Polymers: by Interfacial and Solution Methods', Interscience, New York, 1965.
9. H. K. Reimschuessel, *Macromol, Rev.*, 1977, **12**, 65.
10. W. Sbrolli, in 'Man-Made Fibers, Science and Technology', ed. H. F. Mark, S. M. Atlas and E. Cernia, Wiley-Interscience, New York, 1968, vol. 2, p. 227.
11. H. Sekiguchi, in 'Ring-opening Polymerization', ed. K. J. Ivin and T. Saegusa, Elsevier, London, 1984, vol. 2, chap. 12.
12. L. B. Sokolov, in 'Synthesis of Polymers by Polycondensation', Israel Program for Scientific Translations, Jerusalem, 1968.
13. E. Schwartz, in 'Kunststoff Handbuch', ed. E. H. R. Vieweg and A. Muller, Hanser, Munchen, 1966, vol. 6, chap. 2.
14. P. J. Flory, 'Principles of Polymer Chemistry', Cornell University Press, Ithaca, NY, 1953, p. 102.
15. D. Heikens, *J. Polym. Sci.*, 1956, **22**, 65.
16. P. H. Hermans, *J. Appl. Chem.*, 1955, **5**, 493.
17. P. F. van Velden, G. M. van der Want, D. Heikens, Ch. A. Kruissink, P. H. Hermans and A. J. Staverman, *Recl. Trav. Chim. Pays-Bas*, 1955, **74**, 1376.
18. F. Wiloth, *Z. Phys. Chem. (Frankfurt)*, 1955, **5**, 66.
19. K. Tai, H. Teranishi, Y. Arai and T. Tagawa, *J. Appl. Polym. Sci.*, 1979, **24**, 211.
20. K. Tai, H. Teranishi, Y. Arai and T. Tagawa, *J. Appl. Polym. Sci.*, 1980, **25**, 77.
21. C. Giori and B. T. Hayes, *J. Polym. Sci., Part A-1*, 1970, **8**, 335.
22. C. Giori and B. T. Hayes, *J. Polym. Sci., Part A-1*, 1970, **8**, 351.
23. O. Fukumoto, *J. Polym. Sci.*, 1956, **22**, 263.
24. F. Wiloth, *Makromol. Chem.*, 1956, **15**, 98.

25. H. Morawetz and P. S. Otaki, *J. Am. Chem. Soc.*, 1963, **85**, 463.
26. P. J. Flory, *Chem. Rev.*, 1964, **39**, 137.
27. D. H. Solomon (ed.), 'Kinetics and Mechanisms of Polymerization', Dekker, New York, 1972, vol. 3.
28. Ch. A. Kruissink, G. M. van der Want and A. J. Staverman, *J. Polym. Sci.*, 1958, **30**, 67.
29. E. Roerdink and J. M. M. Warnier, *Polymer*, 1985, **26**, 1582.
30. G. C. Monroe, Jr. (Du Pont), *US Pat.* 3 031 433 (1962) (*Chem. Abstr.*, 1963, **58**, 588g).
31. R. J. Gaymans, J. Amirtharaj and H. Kamp, *J. Appl. Polym. Sci.*, 1982, **27**, 2513.
32. J. Zimmerman, *J. Polym. Sci., Part B*, 1964, **2**, 955.
33. C. Ayers, *J. Appl. Chem.*, 1954, **4**, 444.
34. L. F. Beste and R. C. Houtz, *J. Polym. Sci.*, 1952, **8**, 395.
35. P. J. Flory, *J. Am. Chem. Soc.*, 1942, **64**, 2205.
36. R. J. Gaymans, unpublished results.
37. P. J. Flory, *J. Am. Chem. Soc.*, 1936, **58**, 1877.
38. G. B. Taylor, *J. Am. Chem. Soc.*, 1947, **69**, 638.
39. P. H. Hermans, D. Heikens and P. F. van der Velden, *J. Polym. Sci.*, 1955, **16**, 451.
40. H. C. Beachell and D. W. Carson, *J. Polym. Sci.*, 1955, **16**, 451.
41. J. J. Burke and T. A. Orofino, *J. Polym. Sci., Part A-2*, 1969, **7**, 1.
42. Z. Tuzar, P. Kratochvil and M. Bohdanecky, *Adv. Polym. Sci.*, 1979, **30**, 117.
43. W. Conti and E. Sorta, *Eur. Polym. J.*, 1972, **8**, 475.
44. M. A. Dudley, *J. Appl. Polym. Sci.*, 1972, **16**, 493.
45. E. Jacobi, H. Schuttenberg and R. C. Schulz, *Makromol. Chem., Rapid Commun.*, 1980, **1**, 397.
46. E. Biagini, E. Gattiglia, E. Pedemonte and S. Russo, *Makromol. Chem.*, 1983, **184**, 1213.
47. G. Pastuska, U. Just and H. August, *Angew. Makromol. Chem.*, 1982, **107**, 173.
48. R. Bacskai, *Polym. Bull. (Berlin)*, 1985, **14**, 527.
49. C. E. Barnes, *US Pat.* 2 739 959 (1973) (*Chem. Abstr.*, 1972, **77**, 63 211e).
50. E. M. Peters and J. A. Gervasi, *CHEMTECH*, 1972, January, 16.
51. A. Kumar and S. K. Gupta, *J. Macromol. Sci., Rev. Macromol. Chem.*, 1986, **C26**, 183.
52. P. J. Flory, Yu. W. Suter and M. Mutter, *J. Am. Chem. Soc.*, 1976, **98**, 5733.
53. M. Mutter, Yu. W. Suter and P. J. Flory, *J. Am. Chem. Soc.*, 1976, **98**, 5745.
54. K. Tai, Y. Arai and T. Tagawa, *J. Appl. Polym. Sci.*, 1982, **27**, 731.
55. H. Jacobs and C. Schweigman, in 'Proceedings of the 5th European Symposium on Chemical Reaction Engineering', Elsevier, London, 1972, vol. B7, p. 1.
56. S. K. Gupta and M. Tjahjadi, *J. Appl. Polym. Sci.*, 1987, **33**, 933.
57. G. Bertalan, T. T. Nagy, I. Rusznak, L. Toeke, P. Anna and G. Marosi, *Makromol. Chem.*, 1987, **188**, 317; see also ref. 60, p. 17.
58. S. M. Aharoni, *Polym. Bull. (Berlin)*, 1983, **10**, 210.
59. R. Gomola, J. Kondelikova and J. Kralicek, *Angew. Makromol. Chem.*, 1986, **143**, 49.
60. Z. Csueroes, I. Rusznak, G. Bertalan, L. Trezl and J. Koeroesi, *Makromol. Chem.*, 1970, **137**, 9, 17.
61. S. Doubravzky and F. Gelcji, *Makromol. Chem.*, 1968, **113**, 270.
62. See, for example, J. G. Dolden, *Polymer*, 1976, **17**, 875.
63. R. Feldmann and R. Feinauer, *Angew. Makromol. Chem.*, 1973, **34**, 9.
64. R. G. Beaman and F. B. Cramer, *J. Polym. Sci.*, 1956, **21**, 223.
65. R. J. Gaymans, T. E. C. van Utteren, J. W. A. van den Berg and J. Schuyer, *J. Polym. Sci., Polym. Chem. Ed.*, 1977, **15**, 537.
66. R. J. Gaymans and E. H. J. P. Bour (DSM), *Eur. Pat.* 38 094 (1981) (*Chem. Abstr.*, 1982, **96**, 96 636m).
67. R. J. Gaymans and E. H. J. P. Bour (DSM), *Eur. Pat.* 39 524 (1981) (*Chem. Abstr.*, 1982, **96**, 69 637n).
68. E. Roerdink and J. M. M. Warnier, *Polymer*, 1985, **26**, 1582.
69. C. Shaw (ICI), *Br. Pat.* 854 223 (1960) (*Chem. Abstr.*, 1961, **55**, 9953d).
70. F. Wiloth (Glanzstoff), *US Pat.* 3 130 180 (1962) (*Chem. Abstr.*, 1963, **58**, 1596a).
71. W. H. Carothers and G. D. Graves (Du Pont), *US Pat.* 2 163 584 (1939) (*Chem. Abstr.*, 1939, **33**, 7816[7]).
72. G. H. Kroes, Thesis, Technical University Delft, 1963.
73. L. H. Peebles, Jr. and M. W. Huffman, *J. Polym. Sci., Part A-1*, 1971, **9**, 1807.
74. G. C. Monroe, Jr. (Du Pont), *US Pat.* 3 031 433 (1962) (*Chem. Abstr.*, 1964, **58**, 588g).
75. F. Wiloth (Glanszoff), *US Pat.* 3 379 696 (1968) (*Chem. Abstr.*, 1963, **62**, 10 553d).
76. G. B. Taylor (Du Pont), *US Pat.* 2 361 717 (1944) (*Chem. Abstr.*, 1945, **39**, 2672[4]).
77. H. Taul and F. Wiloth (Glanzstoff), *US Pat.* 3 027 355 (1962) (*Chem. Abstr.*, 1962, **57**, 13 992i).
78. J. A. Carter (British Nylon), *Belg. Pat.* 614 387 (1962) (*Chem. Abstr.*, 1963, **58**, 8081a).
79. Onderzoekingsinstituut, *Br. Pat.* 856 915 (1960) (*Chem. Abstr.*, 1961, **55**, 14 999a).
80. J. C. Bryan (Du Pont), *US Pat.* 3 357 955 (1967) (*Chem. Abstr.*, 1968, **68**, 30 977v).
81. W. K. Kwok (Du Pont), *US Pat.* 3 846 381 (1974) (*Chem. Abstr.*, 1975, **82**, 44 468f).
82. J. M. Iwasyk (Du Pont), *US Pat.* 3 948 862 (1974) (*Chem. Abstr.*, 1976, **84**, 60 274m).
83. J. T. Rich (ICI), *US Pat.* 3 868 352 (1975) (*Chem. Abstr.*, 1974, **81**, 106 379t).
84. B. M. Pinney (Du Pont), *US Pat.* 3 960 820 (1976) (*Chem. Abstr.*, 1975, **83**, 194 053b).
85. F. Mertes (BASF), *US Pat.* 4 060 517 (1977) (*Chem. Abstr.*, 1976, **84**, 45 228f).
86. P. F. Beck and E. E. Magat, *Macromol. Synth.*, 1966, **3**, 101.
87. D. M. W. van der Ham. *Chem. Ind. (London)*, 1972, 730.
88. V. E. Shashoua and W. M. Eareckson, III, *J. Polym. Sci.*, 1959, **40**, 343.
89. G. Roques and J. Neel, *C. R. Hebd. Seances Acad. Sci., Ser. C*, 1967, **264**, 63, 178.
90. C. Aubineau, R. Audebert and G. Champetier, *Bull. Soc. Chim. Fr.*, 1970, 533, 1404.
91. J. S. Ridgway, *J. Polym. Sci., Part A-1*, 1970, **8**, 3089.
92. R. D. Chapman, D. A. Holmer, O. A. Pickett, K. R. Lea and J. H. Saunders, *Text. Res. J.*, 1981, **51**, 564.
93. P. W. Morgan and S. L. Kwolek, *J. Polym. Sci.*, 1959, **40**, 299.
94. D. A. Holmer, O. A. Pickett, Jr. and J. H. Saunders, *J. Polym. Sci., Polym. Chem. Ed.*, 1972, **10**, 1547.

95. P. W. Morgan and S. L. Kwolek, *Macromolecules*, 1975, **8**, 104.
96. R. J. Gaymans, *J. Polym. Sci., Polym. Chem. Ed.*, 1985, **23**, 1599.
97. R. G. Beaman, P. W. Morgan, C. R. Koller, E. L. Wittbecker and E. E. Magat, *J. Polym. Sci.*, 1959, **40**, 329.
98. W. H. Bonner, Jr., *US Pat.* 3 088 794 (1963) (*Chem. Abstr.*, 1963, **59**, 1796b).
99. F. M. Silver and F. Dobinson, *J. Polym. Sci., Polym. Chem. Ed.*, 1977, **15**, 2535.
100. J. G. Dolden, *Polymer*, 1976, **17**, 875.
101. W. Nierlinger, B. Brassat and D. Neuray, *Angew. Makromol. Chem.*, 1981, **98**, 225.
102. R. J. Gaymans and A. G. J. van der Ham, *Polymer*, 1984, **25**, 1755.
103. J. Schneider, *Kunststoffe*, 1974, **64**, 365.
104. E. Schmid and W. Griecht (Inventa), *Ger. Pat.* 2 642 244 (1977) (*Chem. Abstr.*, 1977, **86**, 190 656g).
105. H. G. Peine, G. Falkenstein, H. Doerfel, P. Raff and L. Schuster (BASF), *Ger. Pat.* 1 595 354 (1966) (*Chem. Abstr.*, 1969, **71**, 13 679u).
106. F. G. Lum and E. F. Carlston (California Research Corp.), *US Pat.* 2 742 496 (1952) (*Chem. Abstr.*, 1956, **50**, 12 536d).
107. H. W. Hill, Jr., R. W. Campbell and R. S. Shue, *Polym. Eng. Sci.*, 1978, **18**, 36.
108. E. Reske, L. Brinkmann, H. Frischer and F. Rohnscheid (Hoechst), *Ger. Pat.* 2 156 723 (1971) (*Chem. Abstr.*, 1973, **79**, 67 155m).
109. D. J. Sikkema, *J. Appl. Polym. Sci.*, unpublished results.
110. J. L. Cohen, D. J. Sikkema and R. W. M. van Berkel, *Neth. Pat.* 8 101 262 (1981) (*Chem. Abstr.*, 1983, **98**, 5124).
111. E. J. Yeakey, *US Pat.* 3 654 370 (1970) (*Chem. Abstr.*, 1969, **71**, 50 809).
112. Ato Chimie, *Neth. Pat.* 7 809 006 (1970) (*Chem. Abstr.*, 1979, **90**, 205 147).
113. M. Xie and Y. Camberlin, *Makromol. Chem.*, 1986, **187**, 383.
114. G. Deleens, P. Foy and E. Marechal, *Eur. Polym. J.*, 1977, **13**, 353.
115. S. W. Shalaby, E. M. Pearce and H. K. Reimschuessel, *Pap. Meet.— Am. Chem. Soc., Div. Org. Coat. Plast. Chem.*, 1973, **33** (1), 640.
116. D. Acierno, A. Ciaperoni, F. P. La Mantia and G. Polizzotti, *Acta Polym.*, 1986, **37**, 278.
117. R. Martin, W. Goetz and B. Vollmert, *Angew. Makromol. Chem.*, 1985, **132**, 91.
118. M. Takayanagi and T. Kajiyama, *Br. Pat.* 2 008 598 (1980) (*Chem. Abstr.*, 1980, **92**, 23 547).
119. Y. Arai, M. Watanabe, K. Sanui and N. Ogata, *J. Polym. Sci., Polym. Chem. Ed.*, 1985, **23**, 3081.
120. T. Konomi, S. Endo and O. Sawaguchi, *Jpn. Pat.* 76 131 595 (1976) (*Chem. Abstr.*, 1977, **86**, 73 685n).
121. L. B. Sokolov, V. M. Savinov and V. S. Petruchin, *Ger. Pat.* 2 632 416 (1978) (*Chem. Abstr.*, 1978, **88**, 106 699s).
122. P. W. Morgan, *SPE J.*, 1959, **15**, 485.
123. N. Ogata, K. Sanui, T. Ohtake and H. Nakamura, *Polym. J.*, 1979, **11**, 827.
124. W. Nielinger and D. Neuray (Bayer), *Ger. Pat.* 2 743 483 (1977) (*Chem. Abstr.*, 1979, **91**, 5670g).
125. N. Yamazaki, F. Higashi and J. Kawabata, *J. Polym. Sci., Polym. Chem. Ed.*, 1974, **12**, 2149.
126. N. Ogata and H. Tanaka, *Polym. J.*, 1975, **7**, 412.

22
Aromatic Polyamides

LEO VOLLBRACHT

AKZO Fibres and Polymers Research, Arnhem, Netherlands

22.1 WHOLLY AROMATIC HOMOPOLYAMIDES	375
22.1.1 Introduction	375
22.1.2 Meta-coupled Aromatic Polyamides	377
22.1.2.1 Poly(m-phenyleneisophthalamide) Nomex® (Du Pont) and Conex® (Teijin)	377
22.1.3 Para-coupled Aromatic Polyamides	377
22.1.3.1 General	377
22.1.3.2 Poly(p-benzamide) (PBA)	378
22.1.3.3 Poly(p-phenyleneterephthalamide) (PPDT)	378
22.1.4 Future Prospects	382
22.1.5 Attempted Alternative Preparation Methods	382
22.1.5.1 Phosphorylation reaction	382
22.1.5.2 Polymerization from the vapour phase	383
22.1.5.3 Polymerization with thionyl chloride as activating agent	383
22.1.5.4 Polymerization in liquid sulfur trioxide	383
22.1.5.5 Polymerization using silicon tetrachloride	383
22.1.5.6 Polymerization of PPDT using formadinium salts	383
22.1.5.7 Polymerization of p-aminobenzoic acid using carbon disulfide	384
22.1.5.8 Polymerization using phenyl esters of aromatic acids	384
22.1.5.9 High temperature polymerization in the presence of a catalyst	384
22.1.5.10 Polymerization using N-silylated aromatic diamines and aromatic diacid chlorides	384
22.2 AROMATIC COPOLYAMIDES	384
22.3 AROMATIC HYDRAZIDES AND AMIDE–HYDRAZIDES	385
22.4 REFERENCES	385

22.1 WHOLLY AROMATIC HOMOPOLYAMIDES

22.1.1 Introduction

Some 30 years ago, methods were found which permitted the preparation of high molecular weight fully aromatic polyamides. Special solvent systems were required to obtain high molecular weights. For these polyamides to be spun into fibres, the existing spinning techniques had to be adapted to the type of polymer used, but there was no need for the development of new spinning processes. For example, the *para*-coupled aromatic polyamides were dissolved in hot 100% sulfuric acid and the solution thus prepared was spun using the air-gap technique. As a result, the desired high-strength high-modulus fibres could be obtained in a commercially viable process. The first commercially produced wholly aromatic polyamide fibre was Nomex®, introduced in 1961.[1] It is a fully *meta*-oriented polymer, viz. poly(*m*-phenyleneisophthalamide). It showed excellent thermal and flame-resistant properties and, unlike the conventional synthetic fibres, it did not drip. This material was followed, some years later, by development of the preparation and processing of *para*-coupled aromatic polyamides (Section 22.1.3).

Both the *meta* and *para* polymers have become of great commercial importance. Since the wholly aromatic polyamides show a tendency to decompose during or even before melting, they must be spun from solution. The *meta* product [Nomex® (Du Pont) and Conex® (Teijin)] is used in cases where flame resistance and thermal stability are required. The *para* products [Kevlar® (Du Pont) and Twaron® (Akzo)] are not only flame resistant and thermally stable, but also display exceptional strength and modulus values. Because of the special fibre properties which are unique and unknown for commercially available synthetic and chemical fibres, these products form a class apart, denoted by the generic standard name of aramides.[2] Reviews which in part or entirely cover the subject of aromatic polyamides have been published in the literature.[2-8]

There are a large number of wholly aromatic polyamides if all the possible *ortho-*, *meta-* and *para-*oriented monomers are combined. However, only two of them are of commercial importance: poly(*m*-phenyleneisophthalamide) (**1**; Nomex®) and poly(*p*-phenyleneterephthalamide) (**2**; PPDT, Kelvar®, Twaron®). PPDT is a combination of the abbreviations used for the monomers, *p*-phenylenediamine (PPD) and terephthalic acid (T). Initially, Du Pont used poly(*p*-benzamide) (PBA), but later on they changed over to PPDT.[2]

$$\left(\text{NH}-\bigcirc-\text{NHC}-\bigcirc-\text{C}\right)_n \qquad \left(\text{NH}-\bigcirc-\text{NHC}-\bigcirc-\text{C}\right)_n$$

(1)　　　　　　　　　　　　　　　　(2)

The wholly aromatic polyamides cannot be prepared by melt polycondensation. The potential polyamides either do not melt or melt at high temperatures with decomposition. Many aromatic diacids will decarboxylate[9] and the aromatic diamines are readily oxidized and have a tendency to sublime.[5,10] Consequently, use is made of low temperature polymerization methods. Since the reactivity of aromatic diamines and aromatic dicarboxylic acids is too low, it is generally necessary to employ aromatic carboxylic acid chlorides (equation 1). An acid acceptor is needed to bind the hydrochloric acid evolved. Usually, the polymerization proceeds very rapidly. Two methods are in use, viz. the interfacial polymerization method (a two-phase process) and the solution polymerization method (a single-phase process). In interfacial polymerization, the reaction takes place at the interface between a solution of the diacid chloride in an organic solvent (immiscible with water under the conditions applied) and an aqueous solution of the diamine.[11] In the solution polymerization method, the reaction simply takes place in a homogeneous solution. For the solution polymerization of aromatic polyamides, it is advantageous to use aprotic polar amide solvents, such as dimethylacetamide (DMAc), N-methylpyrrolidone (NMP) and hexamethylphosphoramide [HMPA; phosphoric tris(dimethylamide)]. Dimethylformamide (DMF) is unsuitable because it reacts with acid chlorides, nor is dimethyl sulfoxide (DMSO) of any use because of its violent reaction with acid chlorides. HMPA, a powerful solvent, must not be used, since it is a carcinogen in rats.[12] The dissolving power of the amide solvents can be increased by addition of salts, such as lithium chloride or calcium chloride. The amide solvents further serve as acid acceptors, which is why such compounds do not need to be added. Though these solvents are far weaker bases than the primary aromatic diamines, the huge excess of solvent brings the polymerization to completion without loss of diamine as an inactive hydrochloride. A further advantage of the amide solvents may appear in the case of the more soluble aromatic polyamides, such as Nomex®. When the hydrochloric acid formed in the polymerization is neutralized with calcium hydroxide, calcium chloride, a solvating agent, is formed, which leads to dissolution of the polymer so that the reaction mass can be spun directly (combined process).

$$\text{ArNH}_2 + \text{Ar}'\overset{\text{O}}{\underset{\|}{\text{C}}}\text{Cl} \longrightarrow \text{ArNH}\overset{\text{O}}{\underset{\|}{\text{C}}}\text{Ar}' + \text{HCl} \qquad (1)$$

Suggestions for Nomex® manufacture are given in refs. 2 and 7. Proposals for a mechanism accounting for the role of lithium chloride as a solvating agent for aromatic polyamides in DMAc can be found in refs. 13 and 14. In general, high molecular weight wholly aromatic polyamides can be obtained more consistently by low temperature solution polycondensation than by the interfacial method.[2] The polymer prepared is usually isolated by coagulation and neutralization with aqueous alkali, filtration, washing with water or an organic solvent and drying. The dry polymer is very liable to be electrostatically charged. The two aromatic polyamides which are of commercial interest (**1** and **2**) will now be discussed separately. A section will also be devoted to PBA, the forerunner of PPDT. The same procedures can be followed for other aromatic polyamides and copolyamides. The aromatic copolyamides are usually easier to prepare because of their better solubility. They will be discussed in Section 22.1.4.

22.1.2 Meta-coupled Aromatic Polyamides

22.1.2.1 Poly(m-phenyleneisophthalamide) Nomex® (Du Pont) and Conex® (Teijin)[2]

This polymer is prepared from *m*-phenylenediamine and isophthaloyl dichloride (equation 2). It can be prepared by interfacial polymerization or by solution polymerization. Examples of the interfacial method can be found in refs. 15 and 16, and of solution polymerization in refs. 4 and 16. Products of sufficiently high molecular weight are formed.

$$n\ H_2N{-}C_6H_4{-}NH_2 + n\ ClOC{-}C_6H_4{-}COCl \longrightarrow {+}NH{-}C_6H_4{-}NHCO{-}C_6H_4{-}CO{+}_n + 2nHCl \quad (2)$$

Nomex® is used when heat and flame resistance are essential. Because of its attractive electrical properties, which are retained at high temperatures, Nomex® is also used as an electrical insulating material, in the form of fibre or paper.

22.1.3 Para-coupled Aromatic Polyamides

22.1.3.1 General

The *para* orientation leads to less solubility, generally. For this reason, only a few solvent systems have been found that are capable of producing polyamides with a molecular weight sufficiently high for spinning purposes. If interfacial polymerization is used to produce these rigid-rod extended-chain polymers, it results only in low molecular weights.[17,18]

The view that *para*-coupled aromatic polyamides, because of their rigid chains, will give materials with exceptional properties, such as high melting points, thermal stability and high strength and modulus, dates back to the 1960s. (See, for example, a review by Mark on polymer applications, published in 1965[19]). In 1961, the PPDT polymer itself was mentioned and exemplified in a Du Pont aramid patent.[16] The solvent HMPA was used and an inherent viscosity of 1.9 dl g^{-1} was attained. In a subsequent patent,[20] dealing with the heat treatment of aramids, the PPDT polymer is also mentioned specifically. A publication by Preston and Smith in 1966 revealed an initial modulus of 245 g denier^{-1} (denier = weight in g of 9000 m of fibre) for an all-*para*-ordered aromatic copolymer.[21] The spinning of various poly(terephthalamide)s, dissolved in high concentrations in sulfuric acid, is described in a Celanese patent published in 1965.[22] The use of air-gap spinning in the case of wholly aromatic polyamides, in order to improve the spinning process and the fibre properties, was disclosed in a Monsanto patent of 1968.[23] High polymer concentrations of 10–30% were employed. Finally, they were used commercially by Du Pont in 1970. The first all-*para* homopolymer to be used was PBA, which was replaced later on by PPDT.[24,34] Akzo started pilot-plant production of aramid fibres in 1977.

The polymerization of *para*-coupled polymers is sometimes associated with the peculiar phenomenon of liquid crystallinity, also called mesomorphism or solution anisotropy. We shall discuss this briefly.

The all-*para* coupling causes the polymer units to be linked so as to extend the chain in the same direction, while the amide bonds are presumed to be predominantly in the *trans* configuration. These rigid extended-chain polyamides, which have a rod-like shape, lead to mesomorphic (liquid crystalline, anisotropic) solutions beyond a certain concentration. This was predicted by Flory in 1956[25] and proved experimentally by Hermans in 1962.[26] As the concentration of rod-like molecules is increased, the system may simply become a saturated solution with excess polymer or, to accommodate more dissolved polymer in a particular space, is forced to form regions or arrays in which the polymer chains are arranged in parallel. Continued addition of polymer now enlarges the regions of the ordered state, which for aromatic polyamides is the nematic phase. A schematic representation of polymer states in solution, including that of random coils, is given in Figure 1.[4,26] Some polyamides with stiff chains in the extended state which have been shown to yield liquid-crystalline solutions are given in Table 1.[4] Note that a fully aromatic system is not required *per se*. The occurrence of a liquid-crystalline phase is revealed by some special phenomena, such as opalescence upon stirring and variation in solution viscosity. Upon increasing the polymer concentration in the isotropic region, the viscosity first exhibits a normal increase (see Figure 2[4]) up to a critical point where an anisotropic liquid-crystalline phase is formed, after which the solution viscosity *decreases* with *increasing* polymer concentration. Beyond a certain point, the viscosity increases again. This viscosity behaviour of the anisotropic solution has been demonstrated by

Hermans[26] for a different polymer, *viz.* poly(γ-benzyl-L-glutamate). For spinning purposes, PPDT in a high concentration is dissolved in 100% sulfuric acid, which gives rise to anisotropy of the solution. The effect of the anisotropy of spinning solutions on yarn properties is smaller than generally assumed. Weyland[27] has elaborated some hypotheses on this subject.

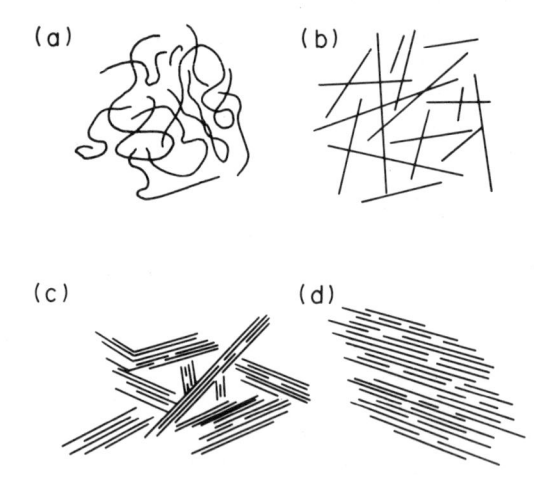

Figure 1 Schematic representation of polymer states in solution: (a) random coils; (b) random rods; (c) rods in liquid-crystalline arrays; and (d) nematic state for one array (reproduced with permission from ref. 4)

At a certain molecular weight, the rod-like character of the all-*para* polyamides cause the inherent viscosity η_{inh} to be higher than for the corresponding *meta* polymers. According to ref. 28, an η_{inh} value of $0.4 \, dl \, g^{-1}$ for poly(*m*-benzamide) approximately corresponds to an η_{inh} value of $1.7 \, dl \, g^{-1}$ for poly(*p*-benzamide). Furthermore, it makes a difference whether the inherent viscosity is determined in amide solvents containing lithium chloride or in concentrated sulfuric acid. The considerable difference observed is attributed to the formation of aggregates in the amide solvent.[28,29]

In ref. 29, molecular weight data as functions of the inherent viscosity are given for PBA and PPDT. For spinning purposes, the inherent viscosity of the all-*para*-coupled aromatic polyamides is reported to have a value of $5 \, dl \, g^{-1}$ or higher.[30]

22.1.3.2 Poly(p-benzamide) (PBA)

A comprehensive review of the synthesis of this polymer has been given by Kwolek and co-workers.[31] The monomer (3) must be prepared separately (Scheme 1). Note that the amino group is masked as the hydrochloric acid salt in order to prevent spontaneous polymerization. Even so, the hydrochloric acid salt of *p*-aminobenzoyl chloride is unstable, as it is sensitive to moisture, light and heat. A brief summary of the special problems encountered in using this type of AB monomer for polymerization is given by Morgan.[32] To start the polymerization, the salt is dissolved in one of the aprotic polar amide solvents mentioned previously, thereby releasing the amine group from the hydrochloride, so that the polymerization can proceed. The polymer is isolated as described before. Another possibility is to transform the reaction mixture into a spinning dope by the addition of lithium hydroxide, thus producing the solvating agent lithium chloride *in situ* (see introduction). The liquid anisotropic dope can be dry spun or wet spun.[31,33]

Despite the disadvantages of using monomers such as (3), Du Pont started with PBA as a fibre material (first generation Fibre B), probably with a view to using a combined polymerization–spinning process. However, a smooth particle-free spinning solution of high molecular weight PBA is difficult to prepare by a combined polymerization–spinning process and will certainly be attended by spinning problems. Later on, Du Pont switched over to the use of PPDT as a fibre starting material, which does not permit the use of a combined process.[34] The fibres (second generation Fibre B), which are air-gap spun from sulfuric acid, have a considerably higher strength than those of PBA.

22.1.3.3 Poly(p-phenyleneterephthalamide) (PPDT)

This polymer is prepared from two monomers in a typical AA–BB polymerization, *viz.* *p*-phenylenediamine (PPD) and terephthaloyl dichloride (TDC) (equation 3). PPD is dissolved in the

Table 1 Some Polyamides Yielding Liquid-crystalline Solutions[a]

[a] Reproduced with permission from ref. 4.

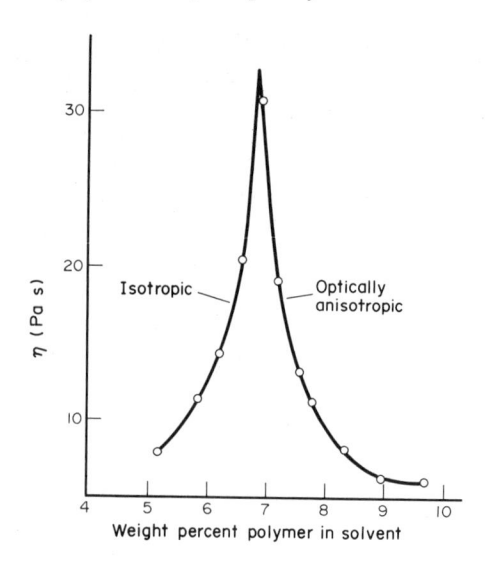

Figure 2 Effect of polymer concentration and the formation of a liquid-crystalline solution on the bulk viscosity of solutions of poly(chloro-1,4-phenylene/1,4-phenylene-2,6-naphthalamide) (70/30) in hexamethylphosphoramide–N-methylpyrrolidone (1–1, v/v) containing 2.9% LiCl at 27 °C; polymer η_{inh} = 2.0. Bulk viscosity was determined with a Brookfield Model RV viscometer with a no. 6 spindle at 10 r.p.m. (reproduced with permission from ref. 4)

$$HCl \cdot H_2N-C_6H_4-COCl \quad (3)$$

Scheme 1 Poly(p-benzamide) preparation (reproduced with permission from ref. 31)

amide solvent and TDC is added with cooling in the molten state or as a solid material (powder or flakes). The polymerization is very fast. The kinetics in HMPA have been studied by Borkent and co-workers[35] by applying stopped-flow spectrophotometry.

$$nH_2N-C_6H_4-NH_2 + nClOC-C_6H_4-COCl \xrightarrow{\text{amide solvent}} +NH-C_6H_4-NHC(O)-C_6H_4-C(O)+_n + 2n\,\text{amide}\cdot HCl \quad (3)$$

In a Celanese patent, the premixing of the solid reactants PPD and TDC is described.[36] Care should be taken that the solid reactants do not contain water, oxygen or carbon dioxide. This method does not seem to offer special advantages, however.

PPDT is even less soluble than PBA.[37] So far, only two solvent systems have been found suitable for solution polymerization on an industrial scale in which polymer of a sufficiently high molecular weight is obtained. These systems are a solvent mixture of HMPA and NMP in the volume ratio 2:1 and an NMP–calcium chloride mixture. The use of a mixture of HMPA and NMP leads to higher molecular weight polymers than the use of the neat solvents separately.[38,39] However, in a continuous process, using a special technique of premixing the reactants and extrusion, use is made of neat HMPA.[40-42] The HMPA system was used by Du Pont until 1982. The NMP/CaCl$_2$ system was discovered and patented by Akzo.[43] As HMPA was found to be a carcinogen in rats,[12] the NMP/CaCl$_2$ system has become of special importance, especially since, after many years, no commercial alternative has yet been found. As HMPA is no longer permitted for use in commercial operations, we shall discuss briefly the use of the NMP/CaCl$_2$ system. Before it became known that HMPA was a carcinogen in rats, Akzo had already found the NMP/CaCl$_2$ system as a result of a special study to find an alternative to HMPA. The study was started because the use of HMPA on a plant scale would present serious difficulties. The compound is thermally and chemically unstable, especially under acidic conditions, which results in incomplete recovery. Another objectionable aspect is the formation (during polymerization, coagulation and recovery) of phosphorus-containing degradation products. At the time (1974) the prospects of finding an alternative were not great. World-wide, the prevailing idea was that amide solvent/salt systems were inferior to the mixed solvent system HMPA/NMP, a conclusion which had been clearly expressed by Fedorov et al., who had tried in vain to find an alternative.[38] Sokolov had the same view.[44] Of the unattractive amide solvent/salt systems, DMAc/LiCl was considered to be the best.[14,38,44] Of the many possible combinations, the NMP/CaCl$_2$ system was not considered by anybody; apparently[14,38,44] because it was expected to be even less attractive than the DMAc/LiCl system. Yet, it turned out that in this system the same results can be achieved as in HMPA/NMP, but, to that end, it is necessary to use a large amount of calcium chloride, preferably surpassing the solubility limit of 6% for calcium chloride in NMP at room temperature. This seems strange and illogical, simply because of the apparent uselessness of non-dissolved calcium chloride particles. However, when the polymerization is started by the introduction of terephthaloyl dichloride, the calcium chloride particles quickly dissolve, probably due to the complexation of calcium chloride with the amide groups which have just formed, as a result of which calcium chloride is removed from the solution, which in turn results in dissolution of the non-dissolved particles. Therefore, the amount of calcium chloride present should be larger than the amount which dissolves at room temperature. The excess of calcium chloride may then be present in a solid or supersaturated state. In our opinion, the complexation renders the PPDT polymer more soluble, keeping it in a swollen state when a high degree of polymerization is reached, as in the case of HMPA/NMP. In this swollen state, the polymerization is apt to continue, because it is still accessible to reactive species. Finally, a polymer is obtained that is no different from the polymer produced in HMPA/NMP. The amount of calcium chloride is related to the amount of polymer. For satisfactory results, the weight of calcium chloride should not be lower than the weight of polymer to be formed (*i.e.* approximately equimolar), but preferably higher. The polymer concentration should be as high as possible, of course, but the concentration to be obtained depends on the equipment used and should be determined experimentally. The same applies to the amount of calcium chloride. Amazingly, the shape of the apparatus and the mixing efficiency can have a tremendous effect on the inherent viscosity. This is illustrated by the data in Table 2. In this connection, a role is also played by the high speed of the polymerization reaction and the resulting rapid onset of gelation.

Table 2 Influence of Apparatus on η_{inh}[a]

Apparatus	η_{inh} (dl g^{-1})
Drais kneader 1½ l[b]	0.7
Glass flask with anchor stirrer	2.3
Waring Blendor 1 l	3.8
160 l Drais TS[c]	5.2

[a] Conditions: polymer concentration 8%, CaCl$_2$ 12% (both calculated on 100 parts by weight of NMP); initial temperature 0–20 °C.
[b] Planetary mixer from Drais Ltd., Mannheim, West Germany. [c] Turbulent Schnellmischer.

The polymerization is carried out with cooling and exclusion of moisture. Under proper conditions, the viscosity is fairly independent of the monomer (polymer) concentration applied. For HMPA/NMP, this dependency has been reported.[45]

The effect of the calcium chloride concentration in NMP on the inherent viscosity of the polymer produced differs strongly from that of lithium chloride in, for example, DMAc. At a low lithium chloride concentration in DMAc, a peak is found in the curves of the inherent viscosity as a function of lithium chloride concentration.[14] Up to higher lithium chloride concentrations, the inherent viscosity remains at a constantly low level. In contrast, after an initial steep rise, the inherent viscosity of the polyamide in the NMP/$CaCl_2$ system will continue to increase with the calcium chloride concentration and will remain far above the peak viscosity in the DMAc/LiCl system. The absence of a peak viscosity is of importance in commercial production, because it makes the process less sensitive to slight variations in the salt concentration. At a certain stage, depending on the equipment used, the calcium chloride content has risen to the extent that it prevents proper mixing. The inherent viscosity will then level off and begin to decrease. The function of calcium chloride in PPDT polymerization in NMP is unique. An important point is the recovery of NMP and calcium chloride, which proceeds without any problem.

Northolt[46,47] has determined the crystal and molecular structure of PPDT by X-ray diffraction. Further studies by the same author on the elastic properties and the stress–strain behaviour in relation to structure can be found in the literature.[48–51]

Kevlar® and Twaron® are used when properties such as high strength, high modulus and low density are of advantage, viz. ropes, cables, reinforcement of rubber (tyres) and plastics.[52,53] An interesting application may be the replacement of asbestos in automotive brake linings.

22.1.4 Future Prospects

It seems most probable that Nomex® will maintain its established position. No major changes can be foreseen in its synthesis and spinning.

In the field of *para*-coupled polyamides, PBA seems to have hardly any future. The complicated monomer synthesis needs to be simplified and the yarn properties improved. This is not the case for PPDT, the polymer on which Kevlar® and Twaron® are based. Its future is promising. The monomer synthesis is attractive, the polymerization may be performed very well in the almost harmless and easily recoverable NMP/$CaCl_2$ system and the spinning from 100% sulfuric acid using the air-gap technique offers no basic problems. Improvement of the already exceptional yarn properties are to be expected as a result of further optimization of the process.

22.1.5 Attempted Alternative Preparation Methods

Several investigations into finding different ways of preparing PPDT and PBA have been made by various authors. They are described below.

22.1.5.1 *Phosphorylation reaction*

This method involves the direct polycondensation of aromatic amino acids, or of aromatic diamines and aromatic diacids, in the presence of an aryl phosphite and an organic base. It has been the subject of several publications. In 1974,[54] Yamazaki, Higashi and Kawabata published an article on a procedure for the synthesis of, *inter alia*, aromatic polyamides using dicarboxylic acids and diamines in the presence of triphenyl phosphite in NMP with pyridine as base. A feature of this method is that the use of acid chlorides is avoided. Low molecular weight products were obtained. In 1975, Yamazaki, Matsumoto and Higashi improved the process, predominantly for PBA, by the addition of lithium chloride.[55] The molecular weights obtained were still too low. Over the years, the process was further improved. In 1978, Preston and Hofferbert stressed the importance of temperature control.[28] In 1982, Higashi, Ogata and Aoki managed to increase the inherent viscosity of PPDT to $4.5\,dl\,g^{-1}$ by adding calcium chloride to the NMP/pyridine/LiCl medium.[56] Finally, in 1984,[57,58] Krigbaum, Kotek, Mihara and Preston arrived at an inherent viscosity of $6.2\,dl\,g^{-1}$ for PPDT by increasing the reaction temperature from 100 °C to 115 °C. Strangely enough, these improvements do not apply to PBA; since 1975 the maximum inherent viscosity has remained on the $1.8\,dl\,g^{-1}$ level. These results show that unexpected differences may be encountered in the preparation of two very similar polymers. Though for PPDT an inherent viscosity of 6.2 dl g^{-1} has been obtained, a rather complex reaction medium is required. Furthermore, the concentration of reactants is low,[57] 0.08 mol l^{-1}, vs. a value of 0.25 mol l^{-1} for the HMPA/NMP system[45] and ≥0.25 mol kg^{-1} for the NMP/CaCl$_2$ system. Application on an industrial scale is not expected for the time being.

22.1.5.2 *Polymerization from the vapour phase*

Demonstrating a skillful experimentation technique, Shin,[59] and Ikeda and co-workers[60,61] were able to react PPD and TDC in the vapour phase in a temperature range of 200–300 °C. After its formation, the polymer deposited on specially provided foils and films. While the inherent viscosites usually ranged from 2 to 3.6 dl g^{-1},[61] a maximum value of 5.3 dl g^{-1} was obtained.[59] However, this high value is only obtained in a small proportion of the polymer produced, widely different values being found at the various sampling spots. Most of the polymer has an inherent viscosity ≤3 dl g^{-1}.

22.1.5.3 *Polymerization with thionyl chloride as activating agent*

The direct synthesis of wholly aromatic polyamides using thionyl chloride as an activating agent has been reported,[62–64] with NMP as the preferred solvent.[62] The monomers used included PPD and terephthalic acid.[64] The inherent viscosities are low. For PPDT, an η_{inh} value of 0.67 dl g^{-1} is mentioned.[64] For the reaction of *p*-aminobenzoic acid with thionyl chloride, see Scheme 1.

22.1.5.4 *Polymerization in liquid sulfur trioxide*

It has been found possible to apply the direct condensation of terephthalic acid and PPD sulfate in sulfur trioxide.[65–67] A major side-reaction is the sulfonation of the aromatic ring. Solution anisotropy was found to take place in the reaction mass. The reaction mass could be spun, but the process was impeded by the tremendous evolution of heat during coagulation of the yarn in water as well as by the weakness of the gelatinous fibres produced. The latter phenomenon has been attributed to the hydrophilicity brought about by the introduction of the sulfonic acid group. The intrinsic viscosities were low (≤2.2 dl g^{-1}). The synthesis of sulfonated PPDT has also been performed by reacting 1,4-bis(trichloromethyl)benzene and PPD sulfate in sulfur trioxide.[68] A low viscosity polymer resulted (η_{inh} ≤ 1.2 dl g^{-1}).

22.1.5.5 *Polymerization using silicon tetrachloride*

p-Aminobenzoic acid has been polymerized to PBA, and *p*-phenylenediamine with terephthalic acid to PPDT, in the presence of silicon tetrachloride as a condensation agent.[69] The reaction was carried out under reflux in pyridine or a mixture of pyridine and chlorobenzene. For PBA, an intrinsic viscosity of 1.4 dl g^{-1} was found, for PPDT a value of 0.6 dl g^{-1}.

22.1.5.6 *Polymerization of PPDT by using formamidinium salts*

In this process, *p*-phenylenediamine is converted first into *N*,*N*-*p*-phenylenebis(*N'*,*N'*-dimethylformamidine), which forms a salt with terephthalic acid after its addition. Upon heating this salt to above 225 °C, PPDT is formed with evolution of dimethylformamide (DMF; equation 4). The intrinsic viscosity is low, ≤0.94 dl g^{-1}.[70]

22.1.5.7 Polymerization of p-aminobenzoic acid by using carbon disulfide

p-Aminobenzoic acid can be reacted with carbon disulfide to give an isothiocyanate which condenses spontaneously to an oligomeric product. Upon heating the oligomer, carbon oxysulfide is released with formation of PBA (equation 5).[71] The conversion is low, 34%. A maximum inherent viscosity of 4.8 dl g^{-1} has been reported.[71]

$$n\text{HO}_2\text{C}-\text{C}_6\text{H}_4-\text{NH}_2 + n\text{CS}_2 \longrightarrow n\text{H}_2\text{S} + n\text{HO}_2\text{C}-\text{C}_6\text{H}_4-\text{N}=\text{C}=\text{S}$$

$$\longrightarrow -(\text{O}_2\text{C}-\text{C}_6\text{H}_4-\text{NHC(S)})_n- \xrightarrow{\Delta} -(\text{C(O)}-\text{C}_6\text{H}_4-\text{NH})_n- + n\text{COS} \quad (5)$$

22.1.5.8 Polymerization using phenyl esters of aromatic acids

Heating phenyl p-aminobenzoate to 325 °C produces oligomeric PBA with evolution of phenol, which is removed by distillation. The solid mass is subsequently heated to 425 °C (solid-state polymerization). PBA with an inherent viscosity of 1.9 dl g^{-1} is formed.[72] The same procedure is followed to obtain PPDT by reacting PPD with diphenyl terephthalate. The polymer has an inherent viscosity of 2.5 dl g^{-1}.[73]

22.1.5.9 High temperature polymerization in the presence of a catalyst

PPD and terephthalic acid are heated in a mixture of diphenyl sulfone and p-xylene in an inert atmosphere.[74] The catalyst consists of a mixture of p-toluenesulfonic acid and boric acid. The temperature is gradually increased to 375 °C. p-Xylene and water are distilled off. An inherent viscosity of 2.09 dl g^{-1} is obtained.

22.1.5.10 Polymerization using N-silylated aromatic diamines and aromatic diacid chlorides

This recently reported[75] reaction proceeds as shown in equation (6). Silylated m- and p-phenylenediamine have been used; they were prepared by reaction of the diamines with trimethylsilyl chloride in the presence of triethylamine. The polymerization took place at −10 °C in the system HMPA/NMP containing approximately 5% of lithium chloride. High inherent viscosities were obtained, also in the case of PPDT, for which a value of 7.4 dl g^{-1} was found. It is clear, however, that this reaction is only useful in the laboratory.

$$n\text{Me}_3\text{SiNHArNHSiMe}_3 + n\text{ClCAr'CCl(O,O)} \longrightarrow 2n\text{Me}_3\text{SiCl} + -(\text{NHArNHCAr'C(O,O)})_n- \quad (6)$$

22.2 AROMATIC COPOLYAMIDES

By following the same procedures that apply to the manufacture of aromatic homopolymers,[43] viz. amine/acid chloride reaction in a solution of NMP/CaCl$_2$ (or weaker solvent systems in the case of easily soluble polymers), one can, of course, prepare copolymers (equation 7). Various systems have been proposed for solution polycondensation of the amine/carboxylic acid system using condensing agents. Typically, such schemes (notably triphenyl phosphite/pyridine[55,76]) are more effective for aromatic polyesters than for polyamides, and, among the polyamides, are more useful for the more soluble polymers than for the poorly soluble products (equation 8). By carrying out the reaction at high temperature, however, acceptable molecular weights can be attained.[56,57]

$$\text{ArCOCl} + \text{Ar'NH}_2 \xrightarrow{\text{solvent}} \text{ArCONHAr'} + \text{HCl} \quad (7)$$

$$\text{ArCO}_2\text{H} + \text{Ar'NH}_2 \xrightarrow[\text{(PhO)}_3\text{P, pyridine}]{\text{solvent}} \text{ArCONHAr'} + \text{PhOH} + (\text{PhO})_2\text{POH} \quad (8)$$

A third route[77] is applicable to copolyamides where the molten state is accessible, *viz.* acidolysis (equation 9). This reaction type has been studied intensely in connection with all-aromatic polyesters,[78] but aromatic poly(ester amide)s can be made along the same lines.[79]

$$ArCO_2H + Ar'NHAc \xrightarrow{heat} ArCONHAr' + AcOH \uparrow \qquad (9)$$

22.3 AROMATIC HYDRAZIDES AND AMIDE–HYDRAZIDES

Condensation of diacid chlorides with hydrazine or with preformed bishydrazides can be performed in solution (DMAc, NMP with salts if needed). Copolymers will stay in solution: the solution can be spun as such. Most attention in this field has centred on the semi-regular copoly(amide–hydrazide) from 4-aminobenzhydrazide and terephthaloyl chloride. Attractive yarns have been obtained after hot drawing.[3,80] Intense dehydration of hydrazides, either by heating after isolation[81] or in solution in a dehydrating agent, such as polyphosphoric acid or oleum, produces poly(oxadiazole)s.[82,83]

22.4 REFERENCES

1. M. S. Reisch, *Chem. Eng. News*, 1987, **65** (5), 9.
2. J. Preston, in 'Kirk-Othmer Encyclopedia of Chemical Technology', 3rd edn., ed. M. Grayson and D. Eckroth, Wiley-Interscience, New York, 1978, vol. 3, p. 213.
3. W. Bruce-Black and J. Preston, 'High-modulus Wholly Aromatic Fibres', Dekker, New York, 1973.
4. P. W. Morgan, *Macromolecules*, 1977, **10**, 1381.
5. P. W. Morgan, *CHEMTECH*, 1979, 316.
6. P. W. Morgan, *J. Polym. Sci., Polym. Symp.*, 1985, **72**, 27.
7. T. Skwarski, *Lenzinger Ber.*, 1978, **45**, 28.
8. M. Jaffe and R. S. Jones, in 'Handbook of Fiber Science and Technology', ed. M. Lewin and J. Preston, Dekker, New York, 1985, vol. III, part A, p. 349.
9. R. C. Blume (Du Pont Co.), *US Pat.* 3 558 571 (1971) (*Chem. Abstr.*, 1971, **74**, 88 382n).
10. P. W. Morgan and S. L. Kwolek, *Macromolecules*, 1975, **8**, 104.
11. P. W. Morgan, 'Condensation Polymers: By Interfacial and Solution Methods', Interscience, New York, 1965.
12. J. A. Zapp, *Science*, 1975, **190**, 422.
13. J. R. Schaefgen, T. I. Bair, J. W. Ballou, S. L. Kwolek, P. W. Morgan, M. Panar and J. Zimmerman, in 'Ultra-high Modulus Polymers', ed. A. Ciferri and I. M. Ward, Applied Science, Barking, 1979, p. 173.
14. A. A. Fedorov, V. M. Savinov and L. B. Sokolov, *Polym. Sci. USSR (Engl. Transl.)*, 1971, **12**, 2745.
15. P. W. Morgan, 'Condensation Polymers: By Interfacial and Solution Methods', Interscience, New York, 1965, p. 493.
16. Du Pont de Nemours Co., *Br. Pat.* 871 581 (1961) (*Chem. Abstr.*, 1961, **55**, 24 113e).
17. H. F. Mark, S. M. Atlas and N. Ogata, *J. Polym. Sci.*, 1962, **61**, S49.
18. J. Preston and R. W. Smith (Monsanto Co.), *US Pat.* 3 225 011 (1965) (*Chem. Abstr.*, 1965, **62**, 14 856h).
19. H. F. Mark, *J. Polym. Sci., Part C*, 1965, **9**, 1.
20. Du Pont de Nemours Co., *Br. Pat.* 877 885 (1961) (*Chem. Abstr.*, 1962, **56**, 11 844h).
21. J. Preston and R. W. Smith, *J. Polym. Sci., Part B*, 1966, **4**, 1033.
22. Celanese Corp., *Br. Pat.* 979 342 (1965).
23. H. S. Morgan, Jr. (Monsanto Co.), *US Pat.* 3 414 645 (1968).
24. R. E. Wilfong and J. Zimmerman, *J. Appl. Polym. Sci.*, 1973, **17**, 2039.
25. P. J. Flory, *Proc. R. Soc. London, Ser. A*, 1956, **234**, 73.
26. J. Hermans, Jr., *J. Colloid Sci.*, 1962, **17**, 638.
27. H. G. Weyland, *Polym. Bull. (Berlin)*, 1980, **3**, 331.
28. J. Preston and W. L. Hofferbert, Jr., *J. Polym. Sci., Polym. Symp.*, 1978, **65**, 13.
29. J. R. Schaefgen V. S. Foldi, F. M. Logullo, V. H. Good, L. W. Gulrich and F. L. Killian, *Polym. Prepr., Am. Chem. Soc., Div. Polym. Chem.*, 1976, **17** (1), 69.
30. M. Jaffe and R. S. Jones, in 'Handbook of Fiber Science and Technology', ed. M. Lewin and J. Preston, Dekker, New York, 1985, vol. III, part A, p. 361.
31. S. L. Kwolek P. W. Morgan, J. R. Schaefgen and L. W. Gulrich, *Macromolecules*, 1977, **10**, 1390.
32. P. W. Morgan, 'Condensation Polymers: By Interfacial and Solution Methods', Interscience, New York, 1965, p. 193.
33. S. L. Kwolek (Du Pont Co.), *US Pat.* 3 600 350 (1971) (*Fr. Pat.* 1 526 745, 1968; *Chem. Abstr.*, 1969, **71**, 4426g).
34. J. Preston, in 'Kirk-Othmer Encyclopedia of Chemical Technology', 3rd edn., ed. M. Grayson and D. Eckroth, Wiley-Interscience, New York, 1978, vol. 3, p. 214.
35. G. Borkent, P. A. T. Tijssen, J. P. Roos and J. J. van Aartsen, *Recl. Trav. Chim. Pays-Bas*, 1976, **95**, 84.
36. R. S. Jones (Celanese Corp.), *US Pat.* 4 057 536 (1977).
37. M. Jaffe and R. S. Jones, in 'Handbook of Fiber Science and Technology', ed. M. Lewin and J. Preston, Dekker, New York, 1985, vol. III, part A, p. 359.
38. A. A. Fedorov, V. M. Savinov, L. B. Sokolov, M. H. Zlatogorsky and V. S. Greciskin, *Sow. Beitr. Faserforsch. Textiltech.*, 1973, **10** (5), 217.
39. B. Jingsheng, Y. Anji, Z. Shengquing, Z. Shufan and H. Chang, *J. Appl. Polym. Sci.*, 1981, **26**, 1211.
40. T. F. Ficklinger, W. N. Jeter and K. K. Likhyani (Du Pont Co.), *US Pat.* 3 849 074 (1974).

41. J. A. Fitzgerald and K. K. Likhyani (Du Pont Co.), *US Pat.* 3 850 888 (1974) (*Chem. Abstr.*, 1975, **82**, 99 881r).
42. A. R. Bice, J. A. Fitzgerald and A. E. Hoover (Du Pont Co.), *US Pat.* 3 884 881 (1975) (*Chem. Abstr.*, 1975, **83**, 115 383h).
43. L. Vollbracht and T. J. Veerman (Akzo), *US. Pat.* 4 308 374 (1981) (*Chem. Abstr.*, 1976, **85**, 160 859v).
44. L. B. Sokolov, V. M. Savinov and V. S. Petrukhin, *US Pat.* 4 169 932 (1979).
45. T. I. Bair, P. W. Morgan and F. L. Killian, *Macromolecules*, 1977, **10**, 1396.
46. M. G. Northolt and J. J. van Aartsen, *J. Polym. Sci., Polym. Lett. Ed.*, 1973, **11**, 333.
47. M. G. Northolt, *Eur. Polym. J.*, 1974, **10**, 799.
48. M. G. Northolt and J. J. van Aartsen, *J. Polym. Sci., Polym. Symp.*, 1977, **58**, 283.
49. M. G. Northolt, *Polymer*, 1980, **21**, 1119.
50. M. G. Northolt, *Br. Polym. J.*, 1981, **13**, 64.
51. M. G. Northolt and R. van der Hout, *Polymer*, 1985, **26**, 310.
52. Kh. Hillermeier and H. G. Weyland, *Plastica*, 1977, **30**, 374.
53. R. E. Wilfong and J. Zimmerman, *J. Appl. Polym. Sci.: Appl. Polym. Symp.*, 1977, **31**, 1.
54. N. Yamazaki, F. Higashi and J. Kawabata, *J. Polym. Sci., Polym. Chem. Ed.*, 1974, **12**, 2149.
55. N. Yamazaki, M. Matsumoto and F. Higashi, *J. Polym. Sci., Polym. Chem. Ed.*, 1975, **13**, 1373.
56. F. Higashi, S. Ogata and Y. Aoki, *J. Polym. Sci., Polym. Chem. Ed.*, 1982, **20**, 2081.
57. W. R. Krigbaum, R. Kotek, Y. Mihara and J. Preston, *J. Polym. Sci., Polym. Chem. Ed.*, 1984, **22**, 4045.
58. W. R. Krigbaum, R. Kotek, Y. Mihara and J. Preston, *J. Polym. Sci., Polym. Chem. Ed.*, 1985, **23**, 1907.
59. H. Shin (Du Pont Co.), *US Pat.* 4 009 153 (1977) (*Chem. Abstr.*, 1977, **86**, 141 532n).
60. R. J. Angelo, R. N. Blomberg, F. P. Boettcher, R. M. Ikeda, M. R. Samuels (Du Pont Co.), *US. Pat.* 4 104 438 (1978).
61. R. M. Ikeda, R. J. Angelo, F. P. Boettcher, R. N. Blomberg and M. R. Samuels, *J. Appl. Polym. Sci.*, 1980, **25**, 1391.
62. M. Ueda, S. Aoyama, M. Konno and Y. Imai, *Makromol. Chem.*, 1978, **179**, 2089.
63. Y. Imai, S. Aoyama, T. Nguyen and M. Ueda, *Makromol Chem., Rapid Commun.*, 1980, **1**, 655.
64. F. Higashi and T. Nishi, *J. Polym. Sci., Polym. Chem. Ed.*, 1986, **24**, 701.
65. F. M. Silver, *Polym. Prepr., Am. Chem. Soc., Div. Polym. Chem.*, 1979, **20** (1), 207.
66. F. M. Silver, *J. Polym. Sci., Polym. Chem. Ed.*, 1979, **17**, 3535.
67. F. M. Silver and W. Bruce-Black, *J. Polym. Sci., Polym. Chem. Ed.*, 1979, **17**, 3543.
68. F. M. Silver, *J. Polym. Sci., Polym. Chem. Ed.*, 1980, **18**, 1787.
69. P. Strohriegl and W. Heitz, *Makromol. Chem., Rapid Commun.*, 1985, **6**, 111.
70. J. W. Spiewak, *J. Polym. Sci., Polym. Chem. Ed.*, 1978, **16**, 2303.
71. W. Memeger, *Macromolecules*, 1976, **9**, 195.
72. R. S. Jones (Celanese Corp.), *US Pat.* 3 753 957 (1973). (*Ger. Pat.* 2 258 567, 1973; *Chem. Abstr.*, 1973, **79**, 79 486a).
73. R. S. Jones (Celanese Corp.), *US Pat.* 3 901 854 (1975).
74. R. S. Jones (Celanese Corp.), *US Pat.* 3 870 685 (1975) (*Chem. Abstr.*, 1975, **82**, 171 747w).
75. Y. Oishi, M. Kakimoto and Y. Imai, *Macromolecules*, 1987, **20**, 703.
76. Y. Imai, M. Kajiyama, S. Ogata and M. Kakimoto, *Polym. J.*, 1985, **17**, 1173.
77. D. J. Sikkema, *Neth. Pat.* 8 600 436 (1986).
78. G. W. Calundann, *US Pat.* 4 161 470 (1979) (*Ger. Pat.* 2 844 817, 1979; *Chem. Abstr.*, 1979, **91**, 58 578q).
79. J. E. McIntyre and A. H. Milburn, *Br. Polym. J.*, 1981, **13**, 5.
80. F. Dobinson, C. A. Pelezo, W. B. Black, K. R. Lea and J. H. Saunders, *J. Appl. Polym. Sci.*, 1979, **23**, 2189.
81. I. P. Bragina, V. V. Korshak, G. L. Berestneva, S. V. Vinogradova, D. R. Tur, V. A. Khomutov and V. V. Krylova, *Polym. Sci. USSR (Engl. Transl.)*, 1976, **18**, 2652.
82. *Belg. Pat.* 819 043 (1974) (*Ger. Pat.* 2 439 948, 1975, *Chem. Abstr.*, 1975, **82**, 172 487y).
83. Y. Iwakura, K. Uno and S. Hara, *J. Polym. Sci., Part A*, 1965, **3**, 45.

23

Isocyanate-derived Polymers

FABRIZIO PARODI
EniChem SpA, Milano, Italy

23.1 INTRODUCTION	387
23.2 CHEMISTRY OF ORGANIC ISOCYANATES	388
23.2.1 Reactions with Protic Nucleophiles	390
23.2.2 Mechanism, Kinetics and Catalysis of Reactions between Isocyanates and Protic Nucleophiles	394
23.2.3 Cycloaddition Reactions	396
23.2.4 Dimerization, Trimerization and Polymerization of Isocyanates	399
23.3 ADDITION POLYMERS	403
23.3.1 Polyurethanes	403
23.3.2 Polyureas	404
23.3.3 Miscellaneous Polymers from Isocyanates and Protic Nucleophiles	405
23.3.4 Poly(2-oxazolidone)s	405
23.3.5 Poly(isocyanurate)s	406
23.4 CONDENSATION POLYMERS	407
23.4.1 Poly(carbodiimide)s	407
23.4.2 Polyamides and Polyimides	408
23.5 ADDITION–CYCLIZATION POLYMERS	409
23.5.1 Poly(5-iminoimidazolidione)s and Poly(parabanic acid)s	409
23.5.2 Polyhydantoins, Poly(quinazolinedione)s and Poly(hydrouracils)	409
23.5.3 Poly(oxazolidinedione)s and Poly(benzoxazinedione)s	410
23.6 REFERENCES	411

23.1 INTRODUCTION

In the second half of the 1930s, polymers of a new family, polyurethanes, were obtained by O. Bayer and co-workers at Farbenfabriken Bayer by polyaddition from organic diisocyanates and diols.

Since their first appearance, polyurethanes have revealed an unusual versatility; their chemistry, as well as the chemistry of related intermediates (diisocyanates among them), has been enormously developed and polyurethanes have one of the widest ranges of polymer applications throughout the world: fibres, elastomers, foams, skins, adhesives, coatings, *etc.*

Polyurethanes were the first examples of polymers whose synthesis involves organic isocyanates, but the exceptional versatility of the isocyanate moiety has been exploited so that an entire series of different polymers has arisen.

Macromolecular products which are synthesized by means of mono-, di- or poly-isocyanates can be grouped together under the broad definition of isocyanate-derived polymers, even though great differences of structure, properties, polymerization reactions and polymer growth mechanism exist among them: moreover, most of them can also be prepared from reactants other than isocyanates.

The most relevant polymers of such an 'isocyanate family' are polyurethanes and polyureas, to which the next two chapters of this volume (Chapters 24 and 25) are devoted, attainable by step polyaddition of diisocyanates with diols or diamines respectively. However, the intrinsic features of the isocyanate moiety imply that mono- or poly-isocyanates can undergo a plethora of different reactions, as well as a variety of polymerization processes following either a chain- or a step-growth mechanism.

Mono- or poly-isocyanates can behave as monomers in chain polymerization processes: homopolymerization yields 'nylon-1' polymers; copolymerization of monoisocyanates with various comonomers and cyclopolymerization of certain diisocyanates are also possible.

Conversely, di- or poly-isocyanates can act as di- or poly-functional reactants in step polymerizations.

Step-polyadditions are possible with at least difunctional protic reactants (e.g. affording polyurethanes or polyureas, as mentioned above), with various reactants such as diepoxides giving poly(2-oxazolidone)s, etc; polyaddition processes also result from isocyanate dimerization and trimerization reactions, the latter giving poly(isocyanurate)s.

Step polycondensations are feasible with, e.g., dicarboxylic acids or cyclic dianhydrides leading to polyamides or polyimides, respectively, or also directly by isocyanate–isocyanate condensation which yields poly(carbodiimide)s.

Two-stage addition–cyclization polymerizations are finally possible with reactants such as hydrogen cyanide, di(glycinate ester)s, bis(α-hydroxy acid)s, etc., affording poly(5-iminoimidazolidinone)s, poly(oxazolidinedione)s, polyhydantoins and so on, respectively.

Although far less relevant, the polymers deriving from isothiocyanates are worth mentioning. Isothiocyanate chemistry is largely parallel to that of isocyanates and leads, in many cases, to polymers which are thio-analogues of those arising from the corresponding isocyanates.

23.2 CHEMISTRY OF ORGANIC ISOCYANATES

Isocyanates, the esters of isocyanic acid, are those compounds which bear one or several —N=C=O groups, and the entire spectrum of polymerization processes involving them arises from the peculiar features of this moiety. A general description of the —NCO group chemistry can simply be drawn considering monoisocyanates, since di- or poly-functional compounds behave similarly, except in terms of reactivity when two or more groups on the same molecule influence each other by inductive and/or conjugative effects (as is often the case, especially in aromatic isocyanates).

Several reviews deal with synthesis,[1-3] physical properties[4] or with both the synthesis and chemistry of isocyanates.[5-8] However, their synthesis is beyond the scope of this chapter, with the exception of the thermolysis of certain compounds which yield isocyanates under simple heating or act as isocyanate equivalents in some reactions, as shown below.

Isocyanates belong to the class of heterocumulenes,[9] compounds containing a system of cumulated double bonds between atoms, one of which, at least, is of an element other than carbon (N, O, S, P, etc.), such as ketenes, isothiocyanates, carbodiimides, etc. Heterocumulenes are generally characterized by a great and versatile reactivity; isocyanates, in particular, are able to participate in such a variety of reactions that only a brief summary can be given here of those most important in polymerization processes.

The reactions of isocyanates, reviewed in several monographs,[5-8,10,11] arise mainly from the resonance hybrids of the —NCO group, as in equation (1), which determine the pronounced electrophilic character of the C atom, further enhanced by conjugation with aromatic residues (R = Ar; last form in equation 1) in aromatic isocyanates. The —NCO group is particularly prone to undergo attack of nucleophilic agents; most of its reactions, indeed, occur (at least formally) by addition of nucleophilic reactants (equation 2), or are highly catalyzed (or initiated) through a Lewis base (or an anion) attack, exemplified in equations (3a) and (3b).

$$R-\ddot{N}=C=\ddot{O} \longleftrightarrow R-\ddot{N}=\overset{+}{C}-\ddot{O}{:}^{-} \longleftrightarrow R-\overset{-}{\ddot{N}}-\overset{+}{C}=\ddot{O} \quad (\longleftrightarrow \quad {}^{-}Ar=\ddot{N}-\overset{+}{C}=\ddot{O}) \qquad (1)$$

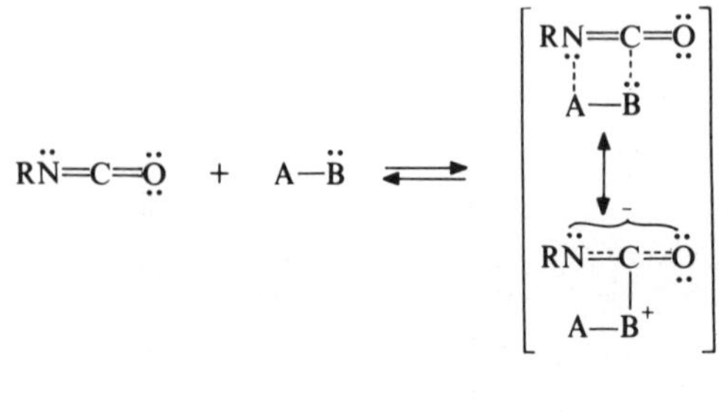

$$R\ddot{N}=C=\ddot{O} + A-\ddot{B} \rightleftharpoons \begin{bmatrix} RN=C=\ddot{O} \\ | \\ A-\overset{+}{B} \\ \uparrow \\ \underbrace{} \\ R\ddot{N}=C=\ddot{O} \\ | \\ A-B^{+} \end{bmatrix} \rightleftharpoons \begin{matrix} R\ddot{N}-C=\ddot{O} \\ | \quad | \\ A \quad B \end{matrix} \qquad (2)$$

$$RN=C=O \; + \; B \; \rightleftharpoons \; RN=C=O \atop \underset{B^+}{|} \qquad (3a)$$

$$RN=C=O \; + \; B^- \; \rightleftharpoons \; \left[RN=C=O \atop \underset{B}{|} \right]^- \qquad (3b)$$

On the other hand, catalysis by protic acids, Lewis acids and metal compounds can also be expected by interaction with the nucleophilic N and O atoms. In practice, however, acids display feeble catalytic effects on isocyanate reactions.

Electron-withdrawing substrates bonded to the —NCO group enhance the C atom electrophilicity and thereby the isocyanate reactivity both in addition of nucleophiles and in base-promoted processes. Electron-releasing substrates act in an opposite sense.

Inductive, conjugative, as well as steric effects must be considered.

Conjugation with aromatic nuclei makes aromatic isocyanates particularly reactive, especially if electron-withdrawing ring substituents are present (and even more when they are in the *ortho* or *para* positions to the —NCO), as, for example, in phenyl and *p*-nitrophenyl isocyanate, (**1**) and (**2**) respectively. Conversely, electron-releasing substituents depress the reactivity.

Most of the kinetic evaluations of isocyanate reactivity have dealt with the addition reaction to alcohols as protic nucleophiles, giving urethanes (equation 4), because of the relevance of this reaction to polyurethane syntheses.

$$RNCO \; + \; HOR' \; \underset{\Delta}{\rightleftharpoons} \; RNHCO_2R' \qquad (4)$$

Although such data cannot be extended to all isocyanate reactions, a general order of reactivity may be given as follows: chlorosulfonyl > *p*-nitrophenyl > phenyl > benzyl > *n*-alkyl > cyclohexyl > *t*-butyl isocyanate.

The effect of ring substituents on aromatic isocyanate reactivity can be depicted, for example, by the reaction rates of variously monosubstituted phenyl isocyanates with 2-ethylhexanol, reported in Table 1, which fit a correlation through the Hammett equation. The —NCO group itself and, to a lesser extent, the urethane moiety —NHCO$_2$R exhibit an activating effect on isocyanate reactivity,

Table 1 Relative Reaction Rates for Monosubstituted Phenyl Isocyanates: Reaction with 2-Ethylhexanol;[a,b] Catalyzed Trimerization[c,d]

Substituent	Relative reaction rate + 2-Ethylhexanol	Trimerization	Substituent	Relative reaction rate + 2-Ethylhexanol	Trimerization
p-NO$_2$	41.7	Too fast	*p*-Br	—	144.0
m-NO$_2$	33.3	Too fast	*p*-NHCO$_2$R[e]	1.43	—
m-CF$_3$	9.9	—	*p*-Ph	1.36	—
m-Cl	7.0	70.0	*m*-NHCO$_2$R[e]	1.35	—
o-NO$_2$	—	20.5	*m*-OMe	1.27	—
o-Cl	—	0.09	None	1.00	10.0
m-NCO	4.7	—	*m*-Me	0.64	1.9
p-NCO	3.6	—	*p*-Me	0.61	3.9
p-Cl	3.4	128.0	*p*-OMe	0.51	—

[a] 2-Ethylhexanol in excess, benzene, at 28 °C.
[b] M. Kaplan, *J. Chem. Eng. Data*, 1961, **6**, 272.
[c] In chlorobenzene; catalyst: *N,N',N''*-tris(3-dimethylaminopropyl)hexahydro-*s*-triazine (TDHT).
[d] L. N. Nicholas and G. T. Gmitter, *J. Cell. Plast.*, 1965, **1**, 85.
[e] R = 2-ethylhexyl.

mainly in aromatic isocyanates (see the data of Table 1), but also in aliphatic and cycloaliphatic ones by their inductive electron withdrawal.

Steric hindrance reduces the —NCO reactivity, so that primary alkyl isocyanates display faster kinetics than secondary, and these are faster than tertiary ones. Analogous effects are exerted by *ortho* substitutions in aromatic isocyanates.

The combination of such electronic and steric effects can explain the differences of reactivity existing between two —NCO groups in diisocyanates, particularly pronounced in aromatic diisocyanates, as shown in Table 2. This accounts, in turn, for the marked variations of overall reaction rates during polymerization processes between diisocyanates and difunctional reactants.

Table 2 Relative Rates of Reaction of Mono- and Di-isocyanates with Ethanol[a,b]

Isocyanate	Relative reaction rate		Activation energy (kJ mol^{-1})	
	First NCO	Second NCO	First NCO	Second NCO
Phenyl isocyanate	14	—	39.8	—
Toluene-2,4-diisocyanate	42 (4-NCO)	1.7 (2-NCO)	46	—
Diphenylmethane 4,4'-diisocyanate	32	10	30.1	39.3
α-Naphthyl isocyanate	5.8	—	38.1	—
Naphthalene-1,5-diisocyanate	38	11	21.3	39
Benzyl isocyanate	1.4	—	30.1	—
1,3-Xylylene diisocyanate	2.8	1.1	52.7	49.8
1,4-Xylylene diisocyanate	2.5	1.3	58.2	46.9
5-*t*-Butyl-1,3-xylylene diisocyanate	2.7	1.0	49.4	49.8

[a] Ethanol in excess, in toluene at 30 °C.
[b] L. L. Ferstanding and R. A. Scherrer, *J. Am. Chem. Soc.*, 1959, **81**, 4838.

Most of the reactions of isocyanates occur across their C=N bond; the following paragraphs, in fact, deal with C=N reactions, except for the catalyzed formation of carbodiimides from isocyanates across their C=O bonds.

The isocyanate and isothiocyanate groups —NCO and —NCS have an almost identical structure, so that the reactivity criteria exposed above are in general valid also for isothiocyanates, for both the synthesis and chemistry of which several reviews are available.[3,12,13]

23.2.1 Reactions with Protic Nucleophiles

The addition of protic nucleophiles belongs to the class of so-called 'insertion reactions' of isocyanates[6] and follows, when uncatalyzed, the scheme of equations (5a)–(5c), which show three of the proposed mechanisms (alternative or coexisting): *via* an intermediate ionic complex or *via* a cyclic (four-centre) transition state; interaction with a second molecule of protic nucleophile is also proposed. A is the nucleophilic atom of the reactant.

$$RN=C=O + H-A \rightleftharpoons \begin{cases} RN-C=O \\ H-A^+ \end{cases} \quad (5a)$$

$$RN=C-O^- \rightleftharpoons RN=C \xrightarrow{+HA} RN=C \xrightarrow{-HA} RNHCO \quad (5b)$$

$$\begin{cases} RN=C=O \\ H-A \end{cases} \quad (5c)$$

A broad spectrum of hydrogen-containing nucleophiles react with both aromatic and aliphatic isocyanates: compounds containing OH groups (H_2O, alcohols, oximes, phenols, acids); SH groups

(H_2S, mercaptans, thiophenols); NH groups (NH_3, amines, hydrazines, amides, ureas, urethanes, etc.); acidic CH (enolizable compounds such as malonic and acetoacetic esters, nitroalkanes, HCN, etc.); and PH groups (phosphines, hydrogenphosphites).

A survey of some relevant reactions, most of which are in principle suitable for the synthesis of polymers from at least difunctional compounds and diisocyanates, is given in Scheme 1.

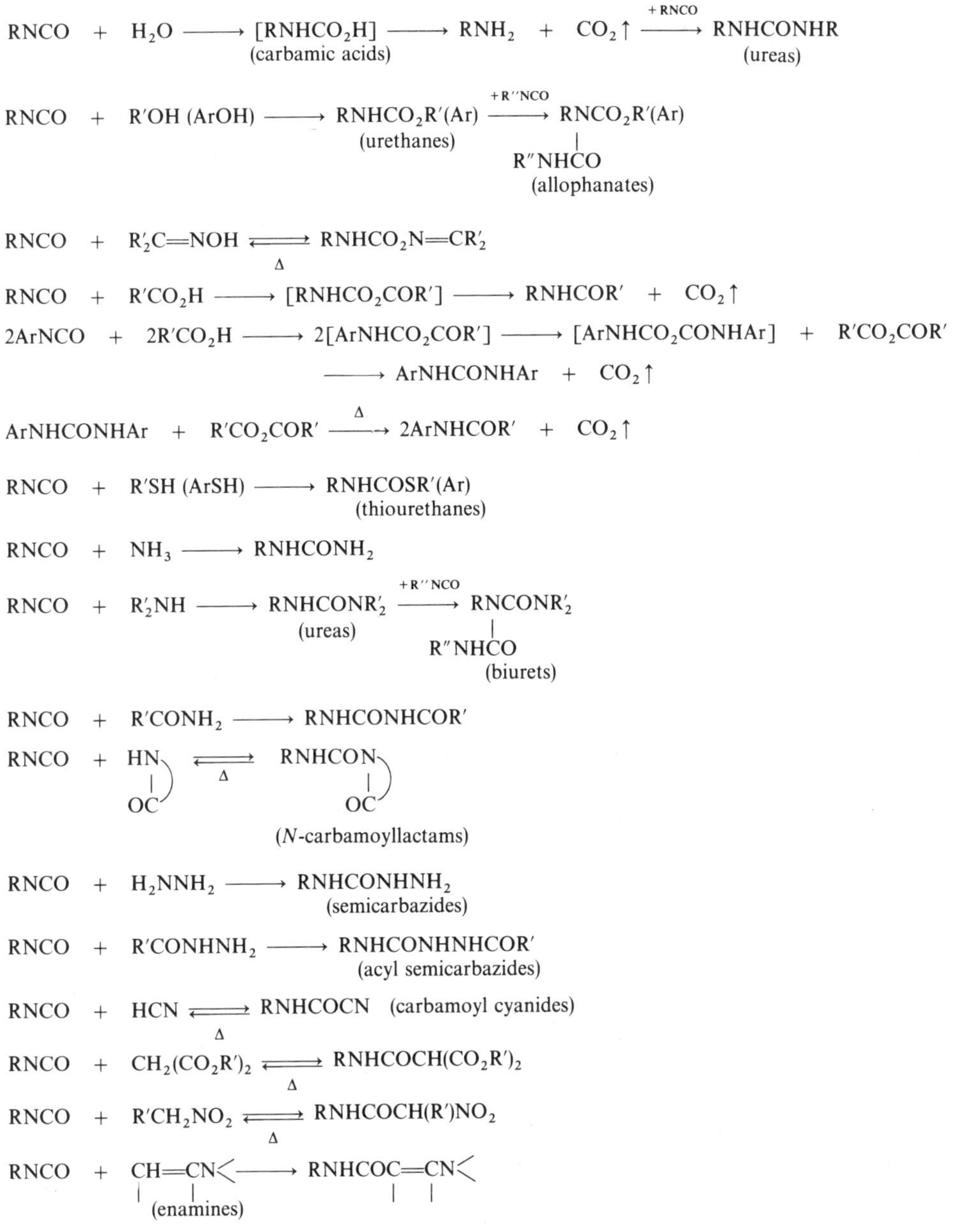

$$RNCO + H_2O \longrightarrow [RNHCO_2H] \longrightarrow RNH_2 + CO_2\uparrow \xrightarrow{+RNCO} RNHCONHR$$
$$\text{(carbamic acids)} \qquad \text{(ureas)}$$

$$RNCO + R'OH\,(ArOH) \longrightarrow RNHCO_2R'(Ar) \xrightarrow{+R''NCO} RNCO_2R'(Ar)$$
$$\text{(urethanes)} \qquad\qquad |$$
$$R''NHCO$$
$$\text{(allophanates)}$$

$$RNCO + R'_2C{=}NOH \underset{\Delta}{\rightleftharpoons} RNHCO_2N{=}CR'_2$$

$$RNCO + R'CO_2H \longrightarrow [RNHCO_2COR'] \longrightarrow RNHCOR' + CO_2\uparrow$$

$$2ArNCO + 2R'CO_2H \longrightarrow 2[ArNHCO_2COR'] \longrightarrow [ArNHCO_2CONHAr] + R'CO_2COR'$$
$$\longrightarrow ArNHCONHAr + CO_2\uparrow$$

$$ArNHCONHAr + R'CO_2COR' \xrightarrow{\Delta} 2ArNHCOR' + CO_2\uparrow$$

$$RNCO + R'SH\,(ArSH) \longrightarrow RNHCOSR'(Ar)$$
$$\text{(thiourethanes)}$$

$$RNCO + NH_3 \longrightarrow RNHCONH_2$$

$$RNCO + R'_2NH \longrightarrow RNHCONR'_2 \xrightarrow{+R''NCO} RNCONR'_2$$
$$\text{(ureas)} \qquad |$$
$$R''NHCO$$
$$\text{(biurets)}$$

$$RNCO + R'CONH_2 \longrightarrow RNHCONHCOR'$$

$$RNCO + HN{\bigg)} \underset{\Delta}{\rightleftharpoons} RNHCON{\bigg)}$$
$$\quad\; |\qquad\qquad\qquad |$$
$$\;\; OC \qquad\qquad\quad OC$$
$$\text{(N-carbamoyllactams)}$$

$$RNCO + H_2NNH_2 \longrightarrow RNHCONHNH_2$$
$$\text{(semicarbazides)}$$

$$RNCO + R'CONHNH_2 \longrightarrow RNHCONHNHCOR'$$
$$\text{(acyl semicarbazides)}$$

$$RNCO + HCN \underset{\Delta}{\rightleftharpoons} RNHCOCN \quad \text{(carbamoyl cyanides)}$$

$$RNCO + CH_2(CO_2R')_2 \underset{\Delta}{\rightleftharpoons} RNHCOCH(CO_2R')_2$$

$$RNCO + R'CH_2NO_2 \underset{\Delta}{\rightleftharpoons} RNHCOCH(R')NO_2$$

$$RNCO + CH{=}CN{<} \longrightarrow RNHCOC{=}CN{<}$$
$$\quad\;\; |\quad |\qquad\qquad\qquad |\quad |$$
$$\text{(enamines)}$$

Scheme 1

Reaction kinetics with the various reagents become faster as their nucleophilicity is increased, *i.e.* their protic acidity is decreased. Strong bases such as primary and secondary aliphatic and cycloaliphatic amines react so promptly and quantitatively that they may be used to titrate the —NCO moiety itself. Contrary to isocyanates, protic nucleophile reactivity is restricted by electron-withdrawing groups, which reduce basicity inductively and/or conjugatively, and in any case by steric hindrance.

The following order of reactivity can be given: primary aliphatic amines > primary aromatic amines > primary alcohols > secondary alcohols > tertiary alcohols > phenols > thiophenols.

Velocity constant values for reaction of phenyl isocyanate and several diisocyanates with different protic nucleophiles are reported in Tables 3 and 4.

Table 3 Velocity Constant Values for Reaction of Phenyl Isocyanate with Various Protic Nucleophiles[a,b,c]

Reagent	$k \times 10^4$ (l mol^{-1} s^{-1})	
	at 25 °C	at 80 °C
PhNH$_2$	10–20	—
Bu^{n}OH	2–4	30
EtCH(OH)Me	1	15
Me$_3$COH	0.01	—
H$\pm$O(CH$_2$)$_2$O(CH$_2$)$_2$OCO(CH$_2$)$_4$CO$\pm_n$O(CH$_2$)$_2$O(CH$_2$)$_2$OH	0.5	5
H$_2$O	0.4	6
Bu^{n}SH	0.005	—
PhOH	0.01	—
PhNHCONHPh	—	2
Pr^{n}CO$_2$H	—	2
EtOCONH$_2$	—	0.02

[a] Reactants in equivalent amounts, in toluene.
[b] J. H. Saunders and K. C. Frisch, 'Polyurethanes: Chemistry and Technology—Part 1. Chemistry', Interscience, New York, 1962, p. 208.
[c] L. Thiele, *Acta Polym.*, 1979, **30**, 323.

Table 4 Velocity Constant Values for the Overall Reaction of Various Diisocyanates with Protic Nucleophiles, at 100 °C[a]

Diisocyanate	$k \times 10^4$ (l mol^{-1} s^{-1})			
	Hydroxyl[b]	Amine[c]	Urea[d]	H$_2$O
p-Phenylene	36.0 (9.0)[e]	17.0	13.0	7.8
2-Chloro-p-phenylene	38.0	23.0	13.0	3.6
Toluene-2,4	21.0	36.0	2.2	5.8
Toluene-2,6	7.4	6.9	6.3	4.2
Naphthalene-1,5	4.0	7.1	8.7	0.7
1,6-Hexamethylene	8.3	2.4	1.1	0.5

[a] W. Cooper, R. W. Pearson and S. Darke, *Ind. Chem.*, 1960, **36**, 121.
[b] Dihydroxy-terminated poly(ethylene adipate).
[c] 3,3′-dichlorobenzidine.
[d] N,N′-diphenylurea.
[e] 1,4-butanediol.

Isothiocyanates react with protic nucleophiles[13] similarly to isocyanates as shown, for example, in Scheme 2.

$$RNCS + 2H_2O \xrightarrow{H^+, \Delta} RNH_2 + H_2S\uparrow + CO_2\uparrow$$

$$RNCS + R'OH (ArOH) \longrightarrow RNHCSOR'(Ar)$$
(thiourethanes)

$$RNCS + R'SH (ArSH) \longrightarrow RNHCSSR'(Ar)$$
(dithiourethanes)

$$RNCS + R'NH_2 \longrightarrow RNHCSNHR'$$
(thioureas)

$$RNCS + H_2NNH_2 \longrightarrow RNHCSNHNH_2$$
(thiosemicarbazides)

Scheme 2

Compounds containing two different protic nucleophile groups, such as hydroxy acids or amino acids, add one mole of isocyanate through the more reactive group. If their structure is suitable, as shown in Scheme 3, intramolecular condensations and ring closure may occur under the influence of heating and/or catalysts, and heterocyclic products are generated.

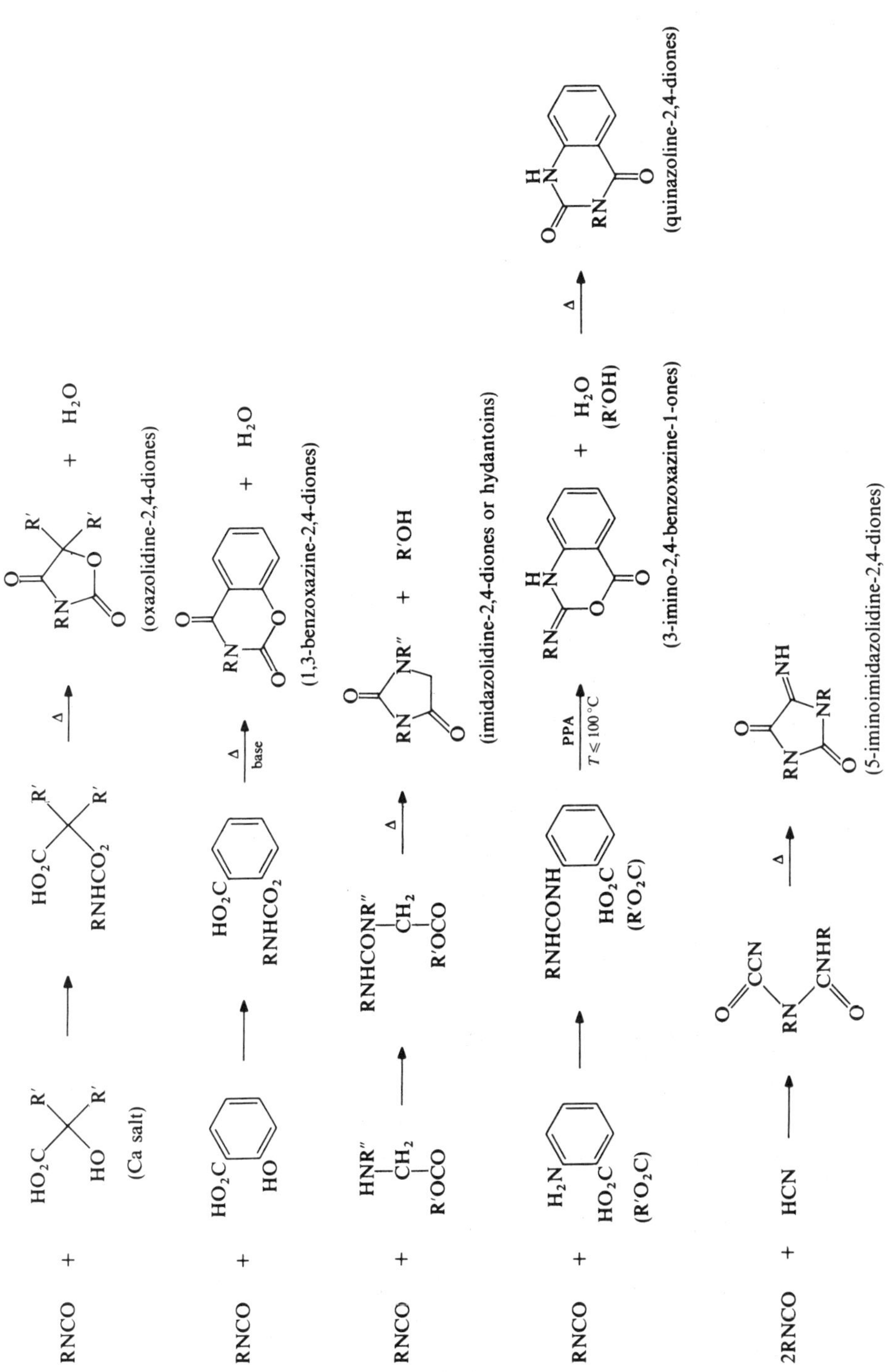
(oxazolidine-2,4-diones)
(1,3-benzoxazine-2,4-diones)
(imidazolidine-2,4-diones or hydantoins)
(3-imino-2,4-benzoxazine-1-ones)
(quinazoline-2,4-diones)
(5-iminoimidazolidine-2,4-diones)
(Ca salt)
PPA
T ≤ 100 °C

Scheme 3

As indicated, for instance, in equation (4), the isocyanate–protic nucleophile reactions are characterized by an intrinsic reversibility under heating and the thermal decomposition of the resulting products has been studied.[10] The thermostability of typical urethanes can be roughly given as: n-alkyl-$NHCO_2$-n-alkyl (stable up to $\sim 250\,°C$); aryl-$NHCO_2$-n-alkyl ($\sim 200\,°C$); n-alkyl-$NHCO_2$-aryl ($\sim 180\,°C$); and aryl-$NHCO_2$-aryl ($\sim 120\,°C$). Products as such tend to decompose[10] to isocyanate and alcohol or phenol in higher yields the lower the temperature needed for decomposition is. Many other compounds resulting from isocyanate–protic nucleophile addition, on the contrary, display different thermodecomposition pathways: this is the case, e.g. of urethanes of tertiary alcohols (which may yield alkene, primary amine and CO_2 at temperatures as low as 50–100 °C) or urethanes of α-hydroxy acids (which can give cyclization as seen in Scheme 3).

Isocyanates which are reacted with truly acidic nucleophiles (such as phenols, oximes, dialkyl malonates, lactams, hydrogen cyanide, etc.) are usually termed capped or blocked isocyanates.[10,14] They decompose at temperatures which decrease roughly as the isocyanate reactivity and the nucleophile acidity increase (50–150 °C for aromatic isocyanates, 120–180 °C and over for aliphatic ones) and give back, often in almost quantitative yields, the starting isocyanate and the 'blocking agent'. Due to a nearly perfect reversibility of reaction (5), blocked di- or poly-isocyanates are employed to generate isocyanates *in situ* as crosslinking agents for many industrial applications.

23.2.2 Mechanism, Kinetics and Catalysis of Reactions between Isocyanates and Protic Nucleophiles

A variety of mechanisms and kinetic schemes have been proposed for these reactions; several reviews devoted to this subject[11,15–19] deal principally with the extensively studied —NCO addition to alcohols and amines. Because of the multiplicity of the reactions given by isocyanates (see Scheme 1), any formally simple process often implies further addition of the isocyanate itself to the first reaction product; noticeable cases are formation of allophanates or biurets by reaction with urethanes or ureas, produced from alcohols or amines respectively. Moreover, the reversibility of some reactions with acidic compounds such as those known as 'blocking agents' (phenols, oximes, dialkyl malonates, etc.) leads, at temperatures appropriately high, to equilibrium processes. Again, the variety of species catalytically affecting the —NCO reactions will imply catalytic effects exerted by the reactants and the reaction products themselves, by solvents, impurities, etc. The catalytic activity of reactants and products varies, on the other hand, during the overall course of the process along with their change in concentration. Complex kinetics result and neat kinetic schemes usually are valid only for the earlier stages of one process. Further complications arise in polymerization processes from the different reactivity of the —NCO groups in diisocyanates and from diffusion-controlled kinetics in highly viscous systems (polymerizations performed in bulk or in concentrated solutions).

Reaction with alcohols. When uncatalyzed, this reaction should involve the general mechanism depicted by equations (5a)–(5c), more likely via the pathways (5a) and (5c). The alcohol–isocyanate reaction was found to follow second-order kinetics (reaction rate = $k[\text{—NCO}][\text{—OH}]$); however, the second-order rate constant k almost invariably increases during the process due to a moderate catalytic effect of the formed urethane, which behaves as a weak base. The alcohol, which is a very weak base, was found to slightly catalyze the reaction, whose kinetics have an apparent reaction order higher than one with respect to [—OH] (reaction rate = $k[\text{—NCO}][\text{—OH}]^{n>1}$), via a mechanism involving a second molecule of protic nucleophile, as in equation (5b). Kinetic schemes have been proposed in order to account for these 'internal' catalytic effects; it is questionable if a truly uncatalyzed —NCO/—OH reaction does exist at all.

Solvents of increasing dielectric constant (or more powerfully hydrogen bonded) tend to slow down the reaction by virtue of a higher association with the hydroxyl group, which opposes the —NCO/—OH interaction; the reaction order with respect to the alcohol concentration progressively decreases towards unity as the solvent dielectric constant rises.

The isocyanate–alcohol reaction is promoted by Lewis bases (such as tertiary amines) and the pathway of such a catalyzed reaction can be written simply as in equation (6a), where the ionic intermediate is attacked by the alcohol and the base is regenerated.

$$R\ddot{N}{=}C{=}\ddot{O} \; + \; \ddot{B} \; \rightleftharpoons \; R\ddot{N}{=}\overset{-}{C}{=}\ddot{O} \underset{B^+}{|} \; \xrightarrow{+R'OH} \; RNHCO_2R' \; + \; \ddot{B} \quad (6a)$$

$$R\ddot{N}{=}C{=}\ddot{O} \; + \; R'OH \; \longrightarrow \; RNHCO_2R' \quad (6b)$$

The uncatalyzed reaction (formally given by equation 6b), the alcohol-catalyzed, the urethane-catalyzed and the base-catalyzed reactions run in parallel. If an individual rate constant is attributed to each one of them, an overall kinetic expression (rate of urethane formation) can be given by equation (7), where i, a, x and c are the initial concentration of —NCO and —OH, the instantaneous concentration of urethane and the catalyst concentration respectively.

$$\frac{dx}{dt} = k_0(i - x)(a - x) + k_a(i - x)(a - x)^2 + k_x(x)(i - x)(a - x)$$
$$+ k_c(c)(i - x)(a - x) \qquad (7)$$

The various terms of equation (7), with their rate constants, refer to the contributions of the uncatalyzed reaction, alcohol-catalyzed, urethane-catalyzed and base-catalyzed reactions respectively to the overall rate. Highly dipolar solvents reduce, by interaction with the different catalytic species, the k_a, k_x and k_c values. The promoting effect of bases is roughly proportional to their nucleophilicity and highly basic tertiary amines are powerful catalysts widely used in polyurethane syntheses. The most sterically unhindered amines (with respect to their tertiary N atoms) are particularly efficient catalysts, for example 1,4-diazabicyclo[2.2.2]octane (DABCO), somehow irrespective of their basicity.

The effectiveness of some tertiary amines as catalysts for isocyanate–alcohol, isocyanate–urea and isocyanate–water reactions are given in Table 5.

Table 5 Relative Rates of Reaction of Phenyl Isocyanate with Protic Nucleophiles in the Presence of Various Catalysts[a,b]

Catalyst	pK_b	Relative reaction rate		
		n-Butanol	N,N'-Diphenylurea	H_2O
None	—	1.0	2.2	1.1
N-Methylmorpholine	6.60	40	10	25
Triethylamine	3.36	86	4	47
N,N'-Tetramethyl-1,3-butanediamine		260	12	100
1,4-Diazabicyclo[2.2.2]octane	8.60	1200	90	380
Tributyltin acetate		80 000	8000	14 000
Dibutyltin diacetate		600 000	12 000	100 000

[a] In dioxane, at 70 °C; isocyanate and protic nucleophile: 0.25 mol l^{-1}; catalyst: 0.025 mol l^{-1}.
[b] F. Hostettler and E. F. Cox, *Ind. Eng. Chem.*, 1960, **52**, 609.

Metal compounds soluble in organic media (such as metal acetylacetonates, 2-ethylhexanoates, alkylmetal salts of fatty acids, *etc.*) exert an outstanding catalytic effect on the isocyanate–alcohol reaction. Many different compounds of transition metals as well as those of some non-transition elements have been tested and references to papers concerned with their effectiveness and mechanism of action can be found in several reviews.[5,11,15,16,18,19] Although their intrinsic mechanism is still incompletely elucidated, metal compounds seem to act by forming complexes which involve the metal catalyst itself, the isocyanate and the hydroxyl compound. Complex formation leads to an increased electrophilicity of the isocyanate C atom and 'brings closer' the —NCO and —OH moieties. One formal mechanism can be given by equation (8).

$$\text{RN=C=O} + \text{MX}_m \rightleftharpoons \text{RN=C=O} \cdots \text{MX}_m \xrightleftharpoons{+R'OH} \text{RN=C=O} \cdots \text{MX}_m \cdots \text{H—O—R'} \longrightarrow$$

$$\text{RN—C=O} \cdots \text{MX}_m \cdots \text{H—O—R'} \rightleftharpoons \text{RN—C=O} \cdots \text{H—O—R'} + \text{MX}_m \qquad (8)$$

The reaction rate between the ligands (—NCO and —OH), after the complex has been rapidly formed, depends on the coordination number of the metal, the complex configuration and the ionic radius of the metal. The effect of a metal catalyst on the reaction kinetics can be expressed by a term (the last one in equation 7) such as $k_c(c)^n(i - x)(a - x)$, where n depends on the parameters mentioned above.

Many metals (Bi, Pb, Sn, Fe, Co, Zn, Ni, Cu, Mn, *etc.*) have shown a marked activity which, however, largely depends on the chemical structure of each specific metal compound. Tin compounds (typically dibutyltin diacetate (DBTDA) or dibutyltin dilaurate (DBTDL) and Sn^{II} 'octoate') are among the most effective and widely adopted catalysts. The combined use of metal compounds and tertiary amines shows high efficiency levels due to complex synergisms. Data on the catalytic activity of tin compounds as compared to tertiary amines have been reported in Table 5.

Reaction of isocyanates with amines. This is characterized by a pronounced catalytic effect of the amine itself as well as by the formed substituted urea, which acts as a weak base; this 'internal' amine catalysis further increases the already high reactivity of aliphatic and cycloaliphatic amines. These two catalytic actions clearly become stronger as the amine reactant and/or the product itself become more basic. The direct and indirect pathways depicted by equations (5a)–(5c) are recognized to be followed simultaneously.

Reaction of isocyanates with phenols. This is highly catalyzed by bases, not only by tertiary amines but also by bases such as sodium carbonate, acetate, *etc.* It is thought that the phenoxide anion (equation 9) attacks the —NCO group, and the reaction can proceed as in equation (10).

$$ArOH + \ddot{B} \rightleftharpoons Ar\ddot{O}:^- + HB^+ \quad (9)$$

$$R\dot{N}{=}C{=}\ddot{O} + Ar\ddot{O}:^- + HB \rightleftharpoons \left[\begin{array}{c} R\dot{N}{=}C{=}O \\ | \\ OAr \end{array} \right]^- HB^+ \longrightarrow RNHCO_2Ar + \ddot{B} \quad (10)$$

Other acidic compounds such as dialkyl malonates, acetoacetic esters, nitroalkanes, thiophenols, *etc.* react well with isocyanates after a salification step (alkali metal alkoxides are also used) probably similarly to phenols. This procedure allows the increase in the reaction rate at temperatures lower than those at which thermal splitting of the products occurs, as seen above for these reactants and for phenols (all widely used as 'blocking agents' for isocyanates). However, side-reactions may take place, promoted by a strongly basic environment: allophanate or thioallophanate formation, isocyanate trimerization, *etc.* Metal compounds can effectively be used as promoters, as well as tertiary amines or their combinations.

Reactions with urethanes and ureas. The formation of allophanates and biurets (see Scheme 1) often accompanies the synthesis of urethanes or ureas from isocyanates and alcohols or amines respectively, by further reaction of the urethanes or ureas with the isocyanates themselves.

Allophanate and biuret formation is favoured as more drastic reaction conditions are adopted, by higher concentration and reactivity of the isocyanates involved, and the more catalytically prone the system is.

The catalysts which accelerate the isocyanate–alcohol or isocyanate–amine reactions generally promote in the same way the isocyanate–urethane and isocyanate–urea ones. This is the case of the widely used tertiary amines and also of metal compounds in general. Particularly efficient catalysts for allophanates are lead, cobalt and zinc compounds.[11]

Both allophanates and biurets tend to undergo thermal dissociation to the isocyanate plus the urethane or the urea respectively, so that their formation tends to become an equilibrium process at elevated temperatures.

23.2.3 Cycloaddition Reactions

Isocyanates,[5,6,20] as well as isothiocyanates,[21] display the general feature of heterocumulenes of undergoing cycloaddition reactions with a broad spectrum of reagents.

Isocyanates afford [2+2] cycloadducts (by 1,2-cycloaddition) with C=C bond-containing compounds (alkenes, dienes, allenes, ketenes, ketenimines), with C=O bond-containing reactants (aldehydes, ketones, anhydrides, *N,N*-disubstituted formamides, *N*-substituted and *N,N*-disubstituted amides of carboxylic acids, isocyanates), with C=N bond-containing compounds (azomethines, carbodiimides), nitroso compounds, etc.

[2+3] Cycloadducts (by 1,3-cycloaddition) are given by reaction of isocyanates with 1,3-dipolar compounds (*N*-oxides of imines or heteroamines, azomethine-*N*-arylimines, nitrile-*N*-arylimines, acyl azides, azines, epoxides, aziridines, alkylene carbonates, *etc.*), acetylenes, carbenes and their precursors, *etc.*

[2+4] Cycloadducts (by 1,4-cycloaddition) are formed by reaction with C=C bond-containing products (alkenes, 1,3-dienes and 1,3-heterodienes, cumulenes), C=O group-containing compounds (CO_2, isocyanates, ketones, aldehydes), with C=N bonds of imines, carbodiimides, *etc.*

Some relevant 1,2- and 1,3-cycloadditions are reported in Schemes 4 and 5 respectively.

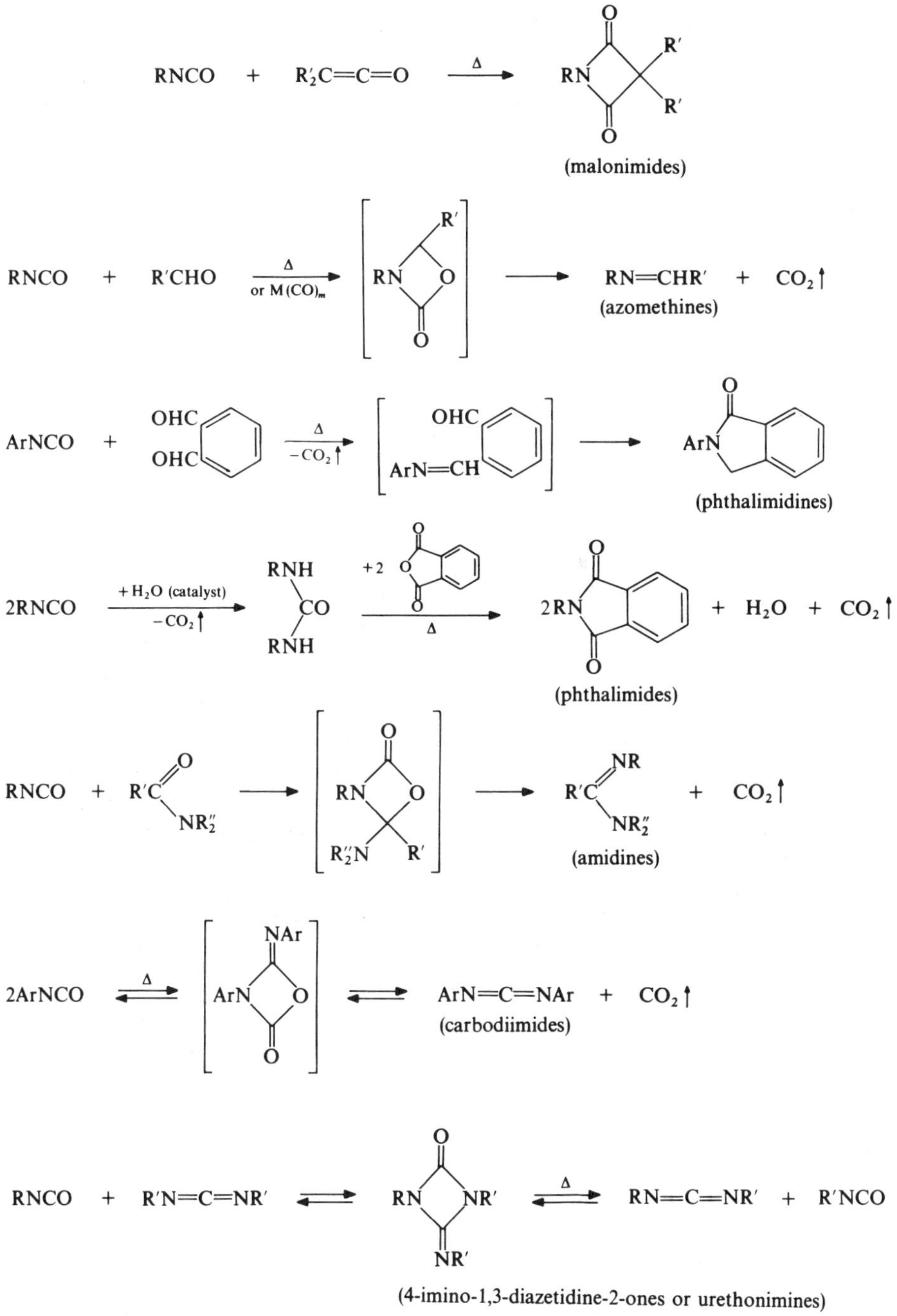

Scheme 4

A number of these reactions are suitable (and have indeed been used) to synthesize polymers from difunctional reactants, *e.g.* from diisocyanates and cyclic dianhydrides, diisocyanates and diepoxides or diaziridines, *etc.* Some other reactions, on the contrary, must or may be considered as responsible for side-processes which take place during polymer syntheses from diisocyanates if certain potential reactants or catalysts are present, or also under suitable reaction conditions (especially at high temperature and/or in bulk polymerizations). Typical examples are the isocyanate consumption by

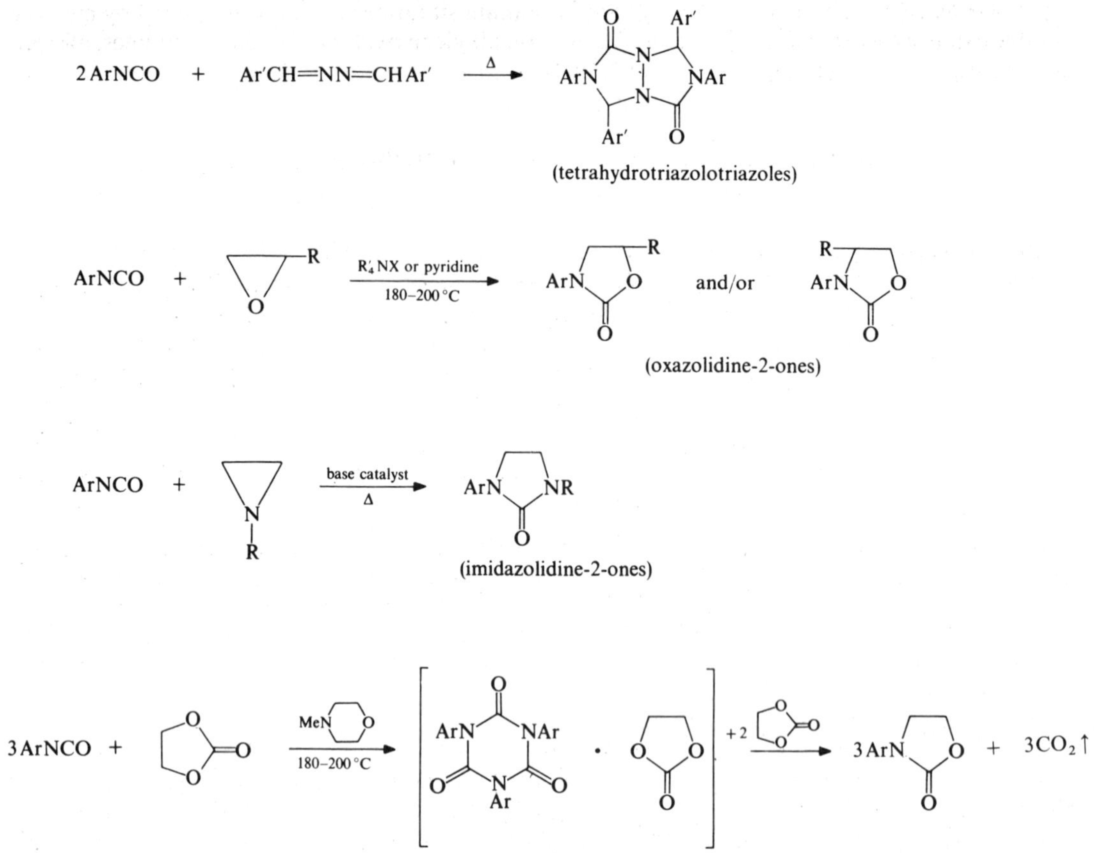

Scheme 5

reaction with amides (DMF or other amides are commonly employed as solvents in polyurethane and polyurea syntheses) and the uncatalyzed thermal formation of carbodiimides from aryl isocyanates.

Cycloadditions which involve isocyanates consist of the addition of dipolar compounds (or dipolar intermediates) to the *dipolarophile* isocyanate moiety and are generally catalyzed by bases.

The formal mechanism through which 1,2-, 1,3- and 1,4-cycloadditions may be considered to occur are outlined in equations (11a), (11b) and (11c) respectively. A, B, D and E represent single carbon atoms or heteroatoms, which may bear hydrogen atoms or various side-groups, and may also be linked to, or be part of, cyclic systems.

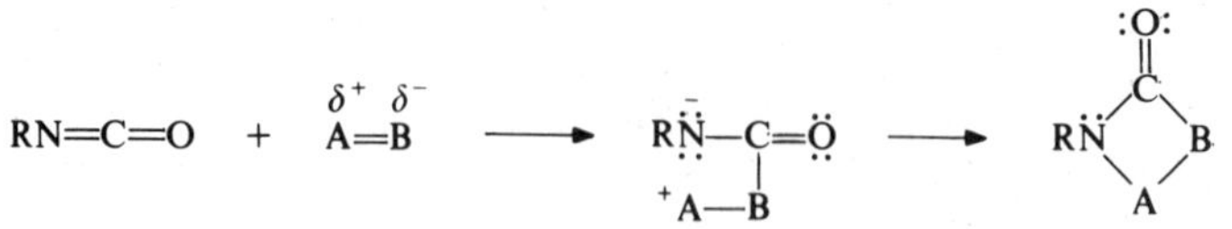

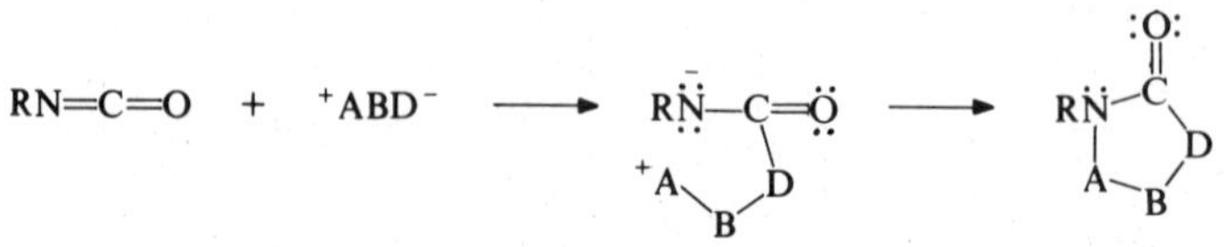

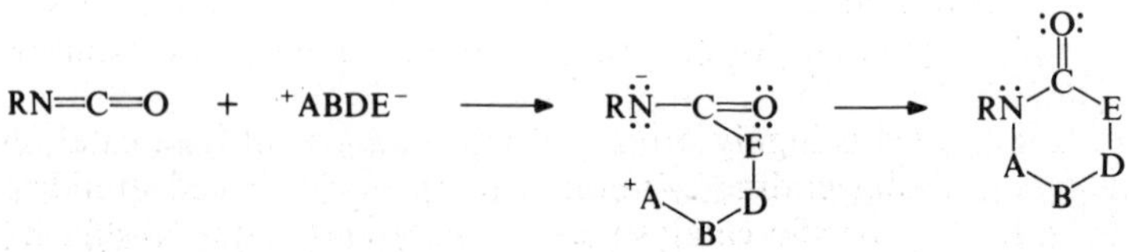

Dipolar reactants can exist as such (the most common situation) or can be generated by means of suitable catalysts, *e.g.* by action of tertiary amines on alkylene oxides or sulfides, aziridines, alkylene carbonates, *etc.*, as in equation (12) for epoxides.

$$\triangle_{O} + R_3\overset{..}{N} \longrightarrow R_3\overset{+}{N}CH_2CH_2\overset{..}{\underset{..}{O}}{:}^- \tag{12}$$

Alkylene carbonates give dipolar intermediates and lose CO_2 with tertiary amines, but also in their absence by simple thermal decomposition, as in equations (13a) and (13b).

$$\underset{O}{\overset{O}{\diagdown}}C{=}O \;\; \overset{+R_3N}{\underset{\Delta}{\diagdown\!\!\!\nearrow}} \;\; \begin{array}{l} R_3\overset{+}{N}CH_2CH_2\overset{..}{\underset{..}{O}}{:}^- + CO_2\uparrow \quad (13a) \\ \\ {}^+CH_2CH_2\overset{..}{\underset{..}{O}}{:}^- + CO_2\uparrow \quad (13b) \end{array}$$

The dipolarophile–dipolar compound mechanism for cycloadditions is in agreement with the general criteria given in Section 23.2.1 for isocyanate reactivity. More reactive isocyanates (*i.e.* with —NCO groups more powerfully electron-withdrawn by the substrate) are able to react even with feebly dipolar compounds, display faster reaction kinetics and higher cycloaddition yields; the highest reactivity is exhibited by sulfonyl and acyl isocyanates. Moreover, the isocyanate group itself can behave as a dipolarophile and as a dipolar moiety in the same reaction (*e.g.* in the uncatalyzed formation of carbodiimides; see Scheme 4).

The catalyzed formation of carbodiimides from isocyanates is another example of a 1,2-cycloaddition-involving process. As shown in equations (14a) and (14b), in fact, isocyanates react with phosphine oxides and the resulting phosphinimines give carbodiimides with further isocyanate. The second stage (equation 14b) occurs across the C=O bond of the isocyanate and regenerates the phosphine oxide, which acts as the catalyst, even if present in low amounts. Many phosphine oxides, and mainly those of cyclic phosphines, have proved to be excellent catalysts for conversion of both aryl and alkyl isocyanates into carbodiimides.

$$RNCO + R'_3PO \rightleftharpoons \left[\begin{array}{c} O \\ RN \diagup \diagdown O \\ \diagdown P \diagup \\ R'_3 \end{array} \right] \rightleftharpoons R\overset{-}{\underset{..}{N}}\overset{+}{P}R'_3 + CO_2\uparrow \tag{14a}$$

$$R\overset{-}{\underset{..}{N}}\overset{+}{P}R'_3 + RNCO \rightleftharpoons \left[\begin{array}{c} R'_3 \\ P \\ RN \diagup \diagdown O \\ \diagdown \diagup \\ \| \\ NR \end{array} \right] \rightleftharpoons RN{=}C{=}NR + R'_3PO \tag{14b}$$

Isothiocyanates also give rise to some cycloaddition reactions similarly to isocyanates and following analogous criteria.[21] The spectrum of these cycloadditions is less broad and less extensively investigated than for isocyanates. Some reactions are reported in Scheme 6.

23.2.4 Dimerization, Trimerization and Polymerization of Isocyanates

Dimerization, trimerization and homopolymerization of isocyanates, reviewed in several monographs,[5–7,20,22] are a small family of interrelated processes based on addition reactions of the isocyanate moiety with itself.

Dimerization of isocyanates is highly promoted by a number of base catalysts and occurs with formation of symmetric 1,3-diazetidine-2,4-diones (often simply called uretidine diones or uretediones) as in equation (15). Specific catalysts are trialkylphosphines, aryldialkylphosphines, tris-

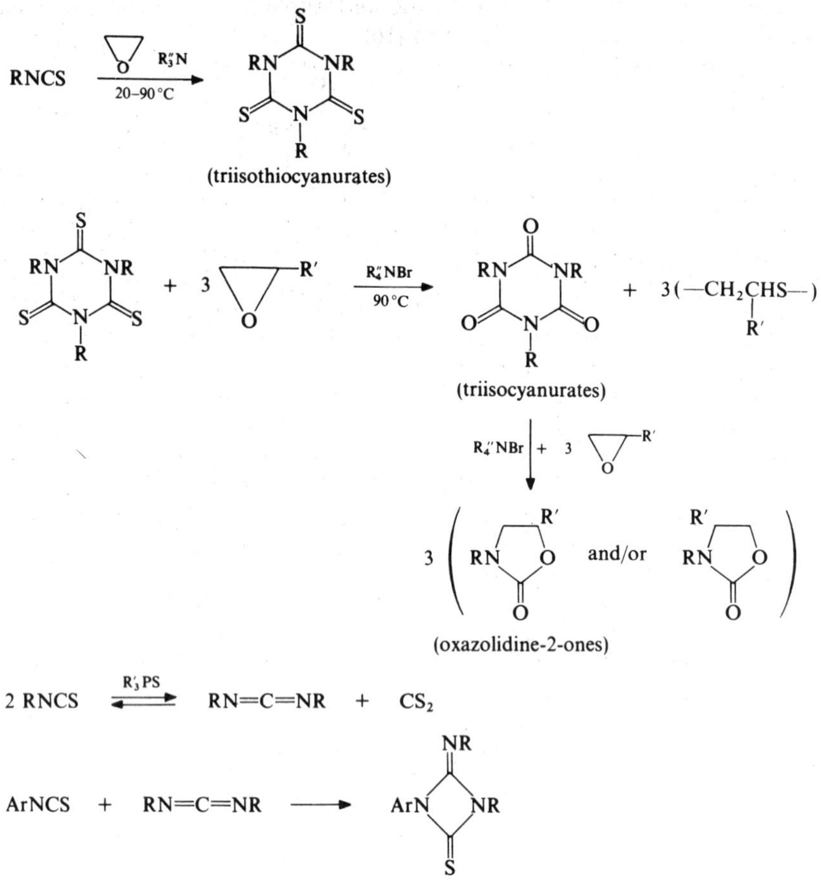

Scheme 6

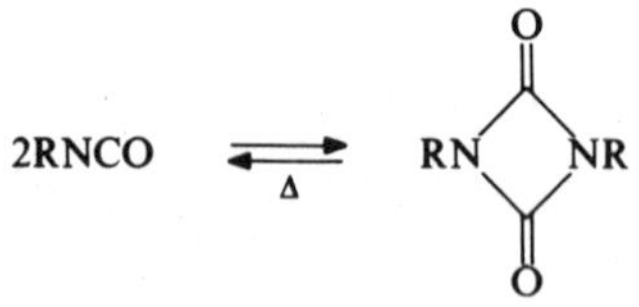

$$2\,RNCO \underset{\Delta}{\rightleftharpoons} \text{(uretedione)} \tag{15}$$

dialkylcarbamoyl phosphites, hexaalkylphosphortriamides, pyridines, N-substituted guanidines, etc.

The dimerization reaction is *per se* rather unsuitable for polymer syntheses from diisocyanates because of the almost perfect reversibility of reaction (15) at even moderate temperatures; for example thermal splitting of dimers of toluene 2,4-diisocyanate (TDI) and of diphenylmethane 4,4'-diisocyanate (MDI) occurs completely at 175 °C and ≥200 °C respectively. In this respect, dimers can be employed as stable sources of isocyanates to be generated *in situ* at elevated temperature (*e.g.* for crosslinking of hydroxylated polymers in coatings).

Isocyanates of increasing reactivity are progressively more prone to undergo dimerization; aromatic isocyanates and diisocyanates (*e.g.* MDI and *p*-phenylene diisocyanate) tend to give uretediones spontaneously even during their storage at room temperature in the solid state and in the absence of catalysts. Diisocyanate dimers, which are high melting and slightly soluble compounds,[22] are hard to react under standard polymerization conditions (*e.g.* in polyurethane synthesis) and their formation in any case alters the calculated stoichiometry. Moreover, uretedione linkage formation must be considered as a side-reaction in polymer syntheses, in that it introduces thermally weak points in polymer backbones. Aliphatic and cycloaliphatic isocyanates, also in the presence of specific catalysts, have a low tendency to afford dimers; their formation has been claimed, however, *e.g.* from hexamethylene 1,6-diisocyanate and from 3-isocyanatomethyl-3,5,5-trimethylcyclohexyl isocyanate (or 'isophorone diisocyanate', IPDI).

Trimerization is readily given by both aromatic and aliphatic isocyanates; six-membered isocyanurate structures are produced as in equation (16).

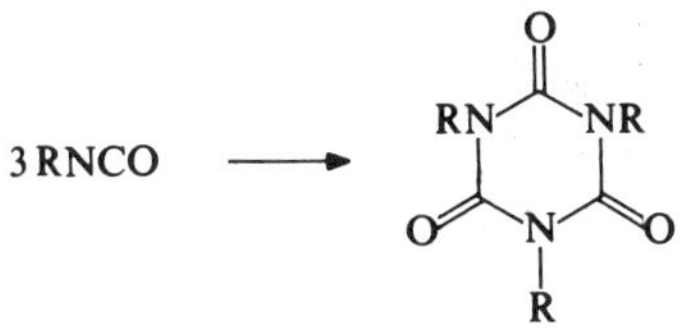

$$3\,RNCO \longrightarrow \text{(isocyanurate ring)} \qquad (16)$$

Analogously to dimerization and addition reactions in general, isocyanates of higher reactivity are more prone to trimerize (see Table 1); thereby, aromatic isocyanates give isocyanurates far more easily than aliphatic ones. Base catalysts promote this reaction and a large number of compounds have been found to be efficient: alkali metal carbonates, alkoxides, oxides and acetates, various lead salts, some metal naphthenates, tertiary amines in general, hydroxides of quaternary ammonium or phosphonium, *etc.* Mixtures of an alkylene oxide and pyridine act as an excellent trimerization catalyst and acidic catalysis has also been shown to be effective. Trimerization is further favoured in solvents of higher dielectric constant (such as DMF, DMSO, *etc.*).

Triisocyanurates are characterized by an outstanding thermal stability, so that their formation has been exploited to prepare many branched or crosslinked polymers of commercial relevance [poly(isocyanurate)s]. Moreover, because of the broad spectrum of effective catalysts, their formation often takes place as a side-reaction in processes and in polymer syntheses which involve isocyanates; trimerization is especially favoured under drastic reaction conditions, in the presence of strong bases, if highly reactive isocyanates are employed, and is frequently accompanied by dimerization as well as by carbodiimide generation.

Dimerization and trimerization generally occur together, the latter being preferred at higher temperatures; dimer–trimer mixtures result even though specific catalysts are used.

The same mechanism, outlined in Scheme 7, is proposed for base-catalyzed dimerizations and trimerizations; both are initiated by the standard nucleophilic attack already shown in equations (3a) and (3b). Dimers or trimers are afforded by ring closure after two or three steps of the same process respectively.

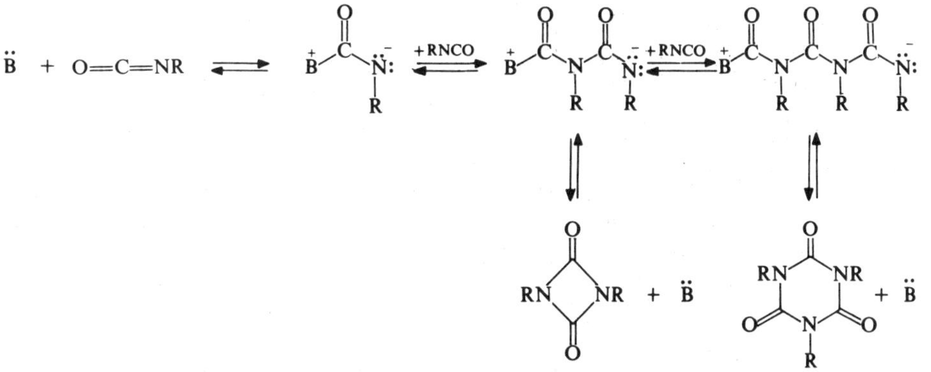

Scheme 7

Homopolymerization of monoisocyanates (equation 17) can be assumed to follow a mechanism similar to base-catalyzed dimerization and trimerization; it occurs in place of them when stronger bases are added as the catalysts (sodium cyanide, sodium naphthalene, sodium in DMF, lithium alkyls, *etc.*) and low temperatures are maintained (below $-40\,°C$). Polymers with a nylon-1 structure are afforded. The low temperature polymerization of diisocyanates yields analogous but crosslinked high polymers.

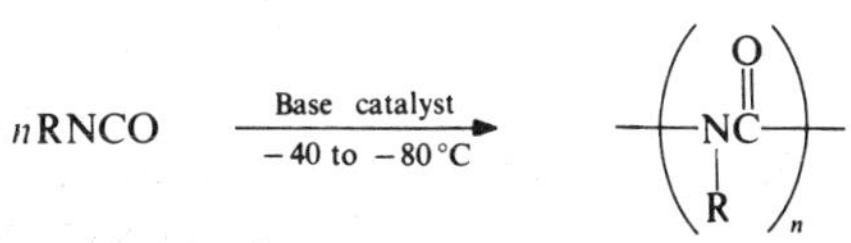

$$n\,RNCO \xrightarrow[-40 \text{ to } -80\,°C]{\text{Base catalyst}} \left(\!\!-N-C-\!\!\right)_n \qquad (17)$$

The main features of this type of polymerization are reported in several monographs.[6,7,20,23–25] Its mechanism is typical of anionic chain polymerizations *via* initiation and propagation stages which may be written, though probably in an oversimplified form, as in equations (18a) and (18b); termination can be caused, as usual, by reaction with protic compounds, as in equation (18c).

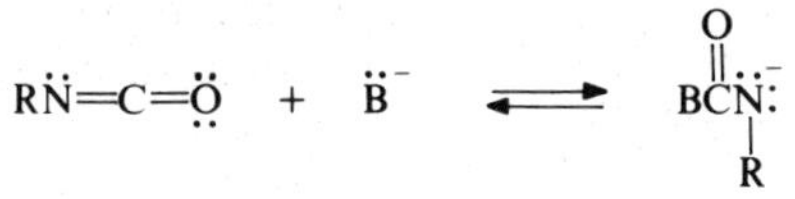

$$R\dot{N}=C=\ddot{O} + \ddot{B}^- \rightleftharpoons B\overset{O}{\underset{R}{C}}\dot{N}:^- \quad (18a)$$

$$B\overset{O}{\underset{R}{C}}\dot{N}:^- + n\overset{..}{C}=\dot{N} \longrightarrow BC(NC)_n\dot{N}:^- \quad (18b)$$

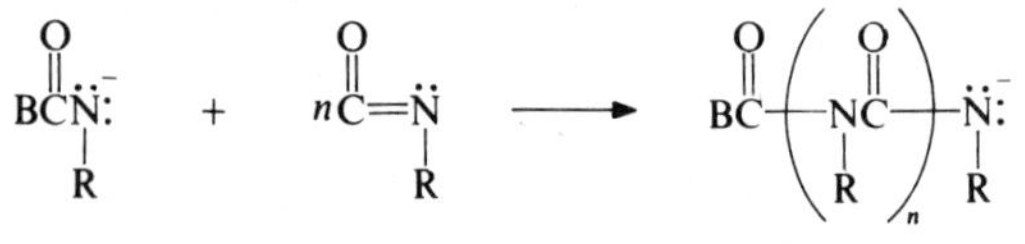

$$BC(NC)_n\dot{N}:^- + HA \longrightarrow BC(NC)_n NH + \ddot{A} \quad (18c)$$

Isocyanate dimerization/trimerization and linear polymerization are competing processes; the latter becomes more relevant as lower reaction temperatures are adopted.

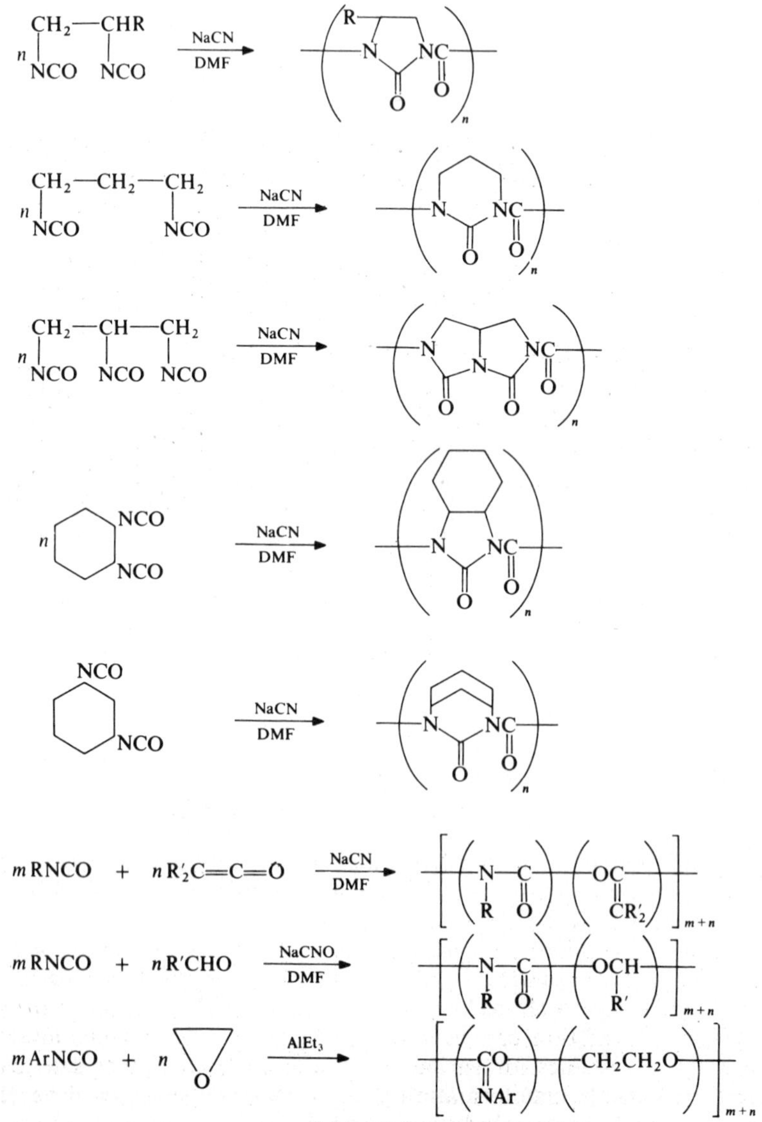

Scheme 8

High-melting cyclopolymers have been obtained by anionic polymerization of 1,2- and 1,3-diisocyanato-alkanes and -cycloalkanes, as summarized in Scheme 8, under conditions similar to those effective for monoisocyanates.[26]

At temperatures as low as $-70\,°C$ to $-80\,°C$ and in the presence of the catalysts (initiators) which promote the homopolymerization, several cycloaddition reactions (among those discussed in Section 23.2.3) can become true anionic chain copolymerizations of isocyanates with various reactants (comonomers). Copolymerizations with alkenes, ketenes, aldehydes and ethylene oxide have been briefly reviewed[26] and are surveyed in Scheme 8.

23.3 ADDITION POLYMERS

23.3.1 Polyurethanes

All linear, branched or crosslinked polymers whose structure contains urethane linkages —NHCO$_2$— or >NCO$_2$— are often considered to belong to the family of polyurethanes; polymeric products of widely different composition and including many other linking groups besides urethane (ether, ester, carbonate, urea, semicarbazide, *etc.*) are described under this heading.

In such a broad context, a large number of reviews and books have been published, which deal with synthesis, properties and applications of polyurethanes.[27-40] Chapter 24 of this volume is devoted to them. However, more properly, one should consider as polyurethanes those polymers whose spectrum of properties largely depends on physicochemical contributions brought by the urethane groupings themselves and/or those whose synthesis is accomplished *via* their formation.

The isocyanate–alcohol addition is by far the most widely followed route both for urethane[41] and polyurethane synthesis. The simple scheme of difunctional polymerization is depicted by equation (19), where Y and Z are residues chosen from an extremely wide range, including polymeric sequences which lead to block copolymers.

$$n\,\text{OCNYNCO} + n\,\text{HOZOH} \longrightarrow {\left(\!\!-\text{YNH}\overset{\overset{\text{O}}{\|}}{\text{C}}\text{OZO}\overset{\overset{\text{O}}{\|}}{\text{C}}\text{NH}-\!\!\right)}_n \qquad (19)$$

Obviously, only —NHCO$_2$— urethanes can be obtained by —NCO/—OH reaction; the nitrogen-substituted ones >NCO$_2$— can be prepared from chloroformates and secondary amines.

One of the major problems encountered in polyurethane preparation (as well as for other isocyanate-derived polymers) is the wide spectrum of side-reactions caused by the intrinsic reactivity and versatility of the —NCO moiety itself (surveyed in Section 23.2), especially when more reactive aromatic diisocyanates are involved.

Kinetics and catalysis of reactions between isocyanates and protic nucleophiles have been described in Section 23.2.2; allophanate formation by addition of —NCO groups to urethane linkages has been briefly examined. Mainly aromatic and also aliphatic diisocyanates (much more than the cycloaliphatic ones) are quite prone to give allophanate groupings, which may cause branching and even crosslinking of otherwise linear polymers, as in equation (20).

$$\left(\!\!-\text{YNH}\overset{\overset{\text{O}}{\|}}{\text{C}}\text{OZO}\overset{\overset{\text{O}}{\|}}{\text{C}}\text{NH}-\!\!\right)_n + \text{OCN}- \longrightarrow {\left(\!\!-\text{YN}\underset{\underset{\text{O=CNH}\sim\!\sim}{|}}{\overset{\overset{\text{O}}{\|}}{\text{C}}}\text{OZO}\overset{\overset{\text{O}}{\|}}{\text{C}}\text{NH}-\!\!\right)}_n \qquad (20)$$

Branching, up to complete gelation, frequently occurs during the synthesis of polyurethanes, especially when carried out in bulk or in solution in the presence of tertiary amines or certain metal catalysts added in order to speed up the urethane formation.

The preparation of thermoplastic polyurethanes requires allophanate inhibition; it can be accomplished by adding, for example, small amounts of acyl chlorides (such as benzoyl chloride or *p*-nitrobenzoyl chloride) which act as 'scavengers' of any basicity present in the system.[32] On the contrary, allophanate formation can be promoted intentionally and employed in order to get 'vulcanization' of polyurethanes for various applications. Several other side-reactions may take place: trimerization, carbodiimide formation, reactions with solvents, *etc.* They can be limited or avoided by adopting mild process conditions and/or by excluding the presence of specific catalysts.

Various di- and tri-isocyanates,[7,26] their dimers and their blocked derivatives are suitable for polyurethane synthesis. Some other compounds must be considered: cyclic dinitrile carbonates (3), sulfites (4) and oxalates (5). They can directly generate isocyanates *in situ* by thermolysis, or react in the presence of DABCO and/or tin catalysts with diols or triols (or diamines, *etc.*) giving polymers which decompose to yield polyurethanes, as in equations (21)–(23).[7,42]

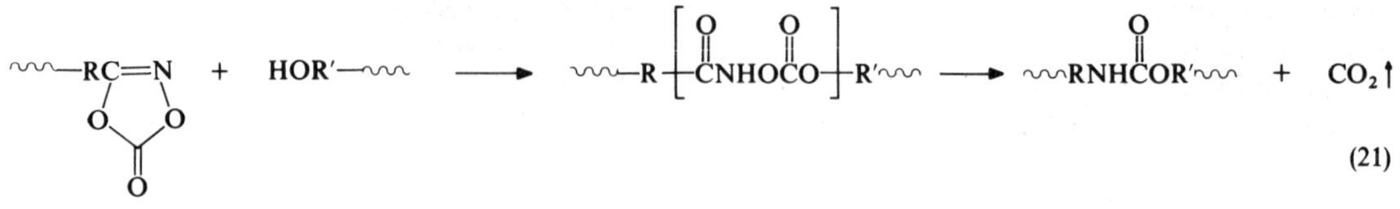

$$\sim\sim\text{RC}=\text{N} + \text{HOR}'\sim\sim \longrightarrow \sim\sim\text{R}\left[\overset{O}{\underset{\|}{\text{C}}}\text{NHO}\overset{O}{\underset{\|}{\text{C}}}\text{O}\right]\text{R}'\sim\sim \longrightarrow \sim\sim\text{RNH}\overset{O}{\underset{\|}{\text{C}}}\text{OR}'\sim\sim + \text{CO}_2\uparrow \quad (21)$$

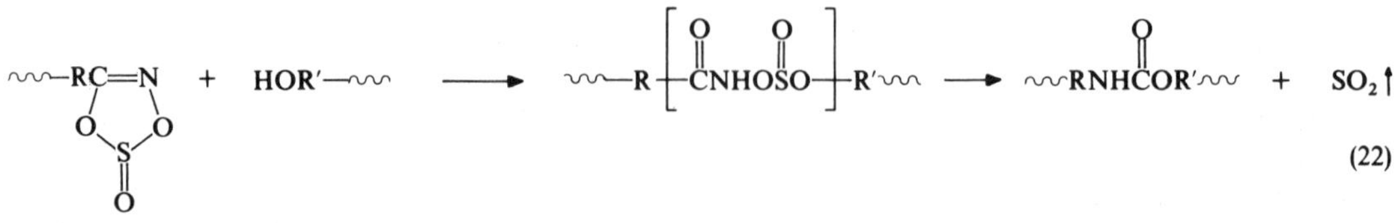

(3)

$$\sim\sim\text{RC}=\text{N} + \text{HOR}'\sim\sim \longrightarrow \sim\sim\text{R}\left[\overset{O}{\underset{\|}{\text{C}}}\text{NHOSO}\right]\text{R}'\sim\sim \longrightarrow \sim\sim\text{RNH}\overset{O}{\underset{\|}{\text{C}}}\text{OR}'\sim\sim + \text{SO}_2\uparrow \quad (22)$$

(4)

$$\sim\sim\text{RC}=\text{N} + \text{HOR}'\sim\sim \longrightarrow \sim\sim\text{R}\left[\overset{O}{\underset{\|}{\text{C}}}\text{NHO}\overset{O}{\underset{\|}{\text{C}}}\text{—CO}\right]\text{R}'\sim\sim \longrightarrow \sim\sim\text{RNH}\overset{O}{\underset{\|}{\text{C}}}\text{OR}'\sim\sim + \text{CO}_2\uparrow$$
$$+ \text{CO}\uparrow \quad (23)$$

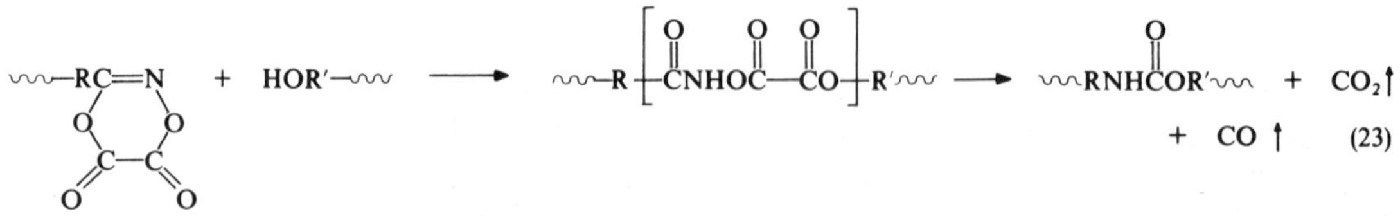

(5)

Other interesting reactants are aminimides, which yield isocyanates by thermal splitting,[7,43] as in equation (24).

$$\text{R}'_3\text{N}^+\text{N}^-\text{CRCN}^-\text{N}^+\text{R}'_3 \xrightarrow{\Delta} \text{OCNRNCO} + 2\text{R}'_3\text{N}\uparrow \quad (24)$$

Polyurethanes have also been obtained by heating acyl azides of hydroxy acids, as in equation (25).

$$n\text{HORCON}_3 \xrightarrow[-n\text{N}_2\uparrow]{\Delta} n\left[\text{HORNCO}\right]^- \longrightarrow \left(\text{RNHCO}_2\right)_n \quad (25)$$

23.3.2 Polyureas

N-substituted ureas can be synthesized by a variety of methods,[44] so that several routes are suitable to obtain polymers whose structural unit contains N,N'-di-, -tri- or -tetra-substituted urea linkages (—NHCONH—, —NHCON< or >NCON<).[45–47]

Chapter 25 deals specifically with polyureas. Only di- or tri-substituted ureas are attainable from isocyanates, reacted with primary or secondary amines respectively. Difunctional schemes for isocyanate/amine polymerizations are given in equations (26a) and (26b).

$$n\text{OCNYNCO} + n\text{H}_2\text{NZNH}_2 \longrightarrow \left(\text{YNH}\overset{O}{\underset{\|}{\text{C}}}\text{NHZNH}\overset{O}{\underset{\|}{\text{C}}}\text{NH}\right)_n \quad (26a)$$

$$n\text{OCNYNCO} + n\text{HNZNH} \longrightarrow \left(\text{YNH}\overset{O}{\underset{\|}{\text{C}}}\text{NZN}\overset{O}{\underset{\|}{\text{C}}}\text{NH}\right)_n \quad (26b)$$

Side-reactions may take place during such polyurea syntheses, analogously to polyurethanes. Most commonly, biuret formation (see Section 23.2.2) may lead to branching or crosslinking, though amines are generally far more reactive than ureas. The isocyanate consumption proceeds rapidly by addition to amines and free —NCO groups are not left present for a long time for isocyanate–urea reaction, if equivalent amounts of reactants are used. Syntheses are normally carried out in highly dipolar aprotic solvents such as DMF, DMSO, N-methylpyrrolidone (NMP), N,N-dimethylacetamide, etc., or their mixtures with ketones, esters, etc. Interfacial methods are also applicable.

23.3.3 Miscellaneous Polymers from Isocyanates and Protic Nucleophiles

Linear polymers have been prepared by polyaddition of diisocyanates with several diprotic nucleophiles,[5,48] generally in dipolar aprotic solvents: with diketooximes as in equation (27); with hydrazine or substituted hydrazines (polyureylenes; **6**); with dihydrazides (poly(acyl semicarbazide)s; **7**); and with enamines (*e.g.* enamines of ketones as in equation 28) which can give cycloadducts with isocyanates but can behave also in a fashion 'equivalent' to a protic nucleophile.

$$\text{(27)}$$

$$\text{(6)} \quad \text{(7)}$$

$$\text{(28)}$$

23.3.4 Poly(2-oxazolidone)s

As seen in Section 23.2.3 (Scheme 5), isocyanates, and especially aromatic ones, undergo, with alkylene oxides, a 1,3-cycloaddition with alkylene oxides which yields oxazolidine-2-ones (simply called 2-oxazolidones), potentially as mixtures of pairs of isomeric products.

The synthesis of polymers whose backbones contain 2-oxazolidone rings from diisocyanates and diepoxides is described briefly in some reviews.[49,50] The polymerization scheme is shown by equation (29), where the diepoxide is a diglycidyl ether (*e.g.* diglycidyl ethers of various bisphenols,

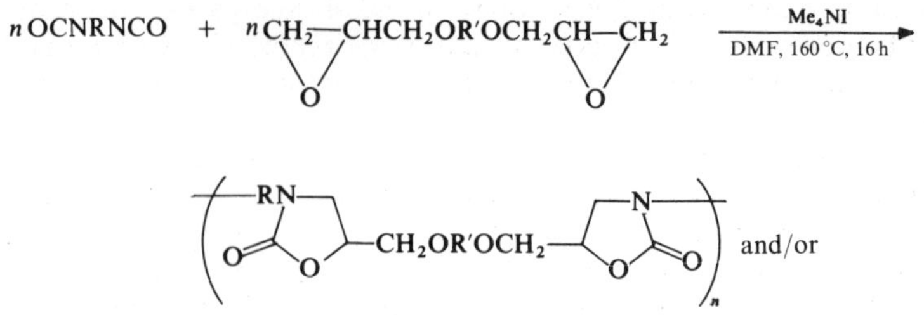

$$n\,\text{OCNRNCO} + n\,\text{CH}_2\text{—CHCH}_2\text{OR'OCH}_2\text{CH—CH}_2 \xrightarrow[\text{DMF, 160 °C, 16 h}]{\text{Me}_4\text{NI}}$$

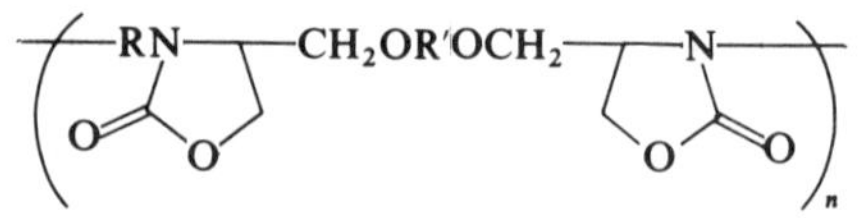

(structure **8**) and/or

(structure **9**) (29)

which are conventional liquid epoxy resins). Both isomeric oxazolidone structures, (**8**) and (**9**), have been reported.

Polymerizations are successful, for example, in DMF at 120–160 °C and in the presence of various catalysts (tetramethylammonium iodide, pyridine, $ZnBr_2$, *etc.*).

23.3.5 Poly(isocyanurate)s

Trimerization of isocyanates to isocyanurates has been described above. This reaction can be used for preparation of polymer networks starting both from diisocyanates and from isocyanate-terminated polymers, as in equation (30).

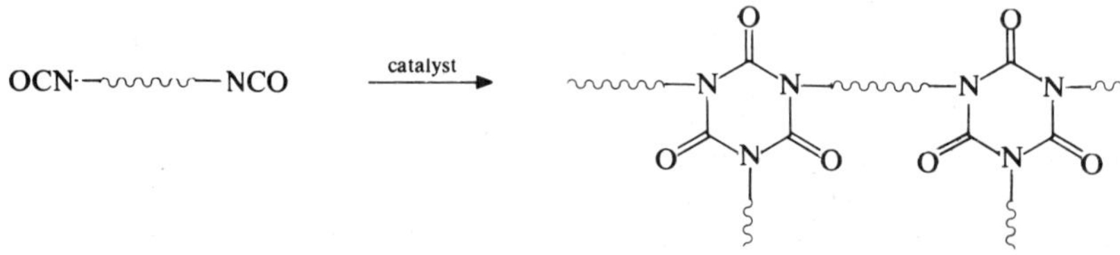

$$\text{OCN—}\sim\sim\sim\text{—NCO} \xrightarrow{\text{catalyst}} \text{(isocyanurate network)} \quad (30)$$

The isocyanurate ring displays a fairly good thermal and hydrolytical stability and these polymers have found wide applications for rigid foams, heat resistant adhesives, coatings, *etc.* If isocyanate-terminated polymers are used, a great variability of properties is possible depending on the length and structure of the polymeric segments (polyether, polyester, polyurethane, *etc.*) between the pairs of —NCO groups. Rather recent reviews deal with the chemistry of poly(isocyanurate)s or isocyanurate-modified polymers.[16,50-54]

The mechanism of base-catalyzed trimerization has been discussed in Section 23.2.4, where several families of catalysts were mentioned. On the other hand, an enormous number of compounds have been tested and claimed to be effective through a variety of mechanisms.

Protic acids (such as hydrochloric or oxalic) and also Lewis acids (Friedel–Crafts catalysts in general) promote isocyanurate formation, as in equation (31), though usually with slow kinetics.

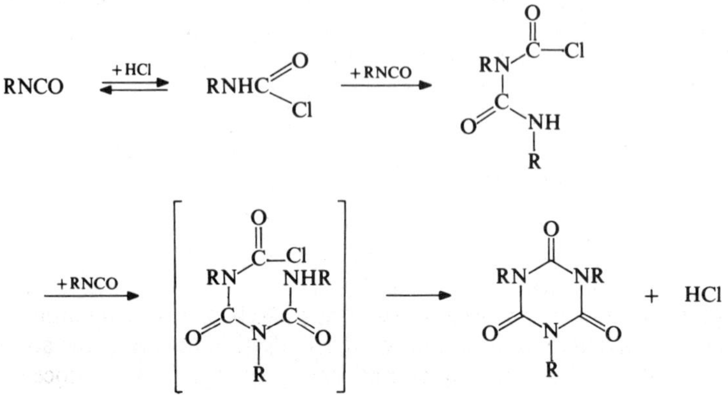

$$\text{RNCO} \underset{}{\overset{+\text{HCl}}{\rightleftarrows}} \text{RNHC(O)Cl} \xrightarrow{+\text{RNCO}} \text{(intermediate)} \xrightarrow{+\text{RNCO}} [\text{cyclic intermediate}] \longrightarrow \text{(isocyanurate)} + \text{HCl} \quad (31)$$

Many metal compounds with highly polarized metal–oxygen, metal–nitrogen, or metal–sulfur bonds (R_3SnOR', $R_3Sb(OR')_2$, $R_3PbNR'_2$, etc.) can add to the N=C bond of the isocyanate as in equation (2).[5,20] They act as trimerization catalysts *via* further reaction of the resulting adducts with isocyanates.

Several mixed catalysts must also be mentioned: combinations of tertiary amines with alkylene oxides, alkylene sulfides, alkylenimines, alkylene carbonates, alcohols, phenols, *etc.* have displayed a pronounced efficiency. For example, the action of amine/epoxide catalysts may be explained as in equation (32), which shows a mechanism similar to that involved in 1,3-cycloaddition of isocyanates and alkylene oxides to yield 2-oxazolidones (see Section 23.2.3). Alkylene oxide alone is uneffective and acts as the cocatalyst.

$$\text{(equation 32)} \tag{32}$$

Mixed isocyanurate/2-oxazolidone polymers result from combinations of diisocyanates (or isocyanate-terminated polymers) and epoxy resins, which in turn cocatalyze the isocyanurate formation itself. Alkylene sulfides, alkylenimines and alkylene carbonates act analogously to epoxides (see equations 12 and 13).

23.4 CONDENSATION POLYMERS

23.4.1 Poly(carbodiimide)s

Carbodiimides can be synthesized by a variety of methods.[55,56] The uncatalyzed condensation of isocyanates with themselves to give carbodiimides (see Scheme 4) is possible (slowly) only under a stream of nitrogen bubbling through boiling aromatic isocyanates. Nitrogen carries out the CO_2 and overcomes the reversibility of reaction (33).

$$2\,RNCO \rightleftharpoons RN{=}C{=}NR + CO_2 \tag{33}$$

Up to almost quantitative yields of carbodiimides are obtained in a few hours at moderate temperatures in the presence of catalysts. The catalyzed formation of carbodiimides has been described briefly in Section 23.2.3. Phosphine oxides are among the most efficient catalysts, but many oxides of elements of groups VB and VIB are effective too (*e.g.* oxides of tertiary amines, of triarylarsines, of triarylstibines, dimethyl sulfoxide, *etc.*), as well as various alkyl or aryl phosphoramides. At present, the preferred catalysts for carbodiimide preparations are 1-ethyl-3-methyl-3-phospholene-1-oxide (**10a**) and 1-phenyl-3-methyl-3-phospholene 1-oxide; other similar compounds, such as (**10b**), (**10c**) *etc.*, have shown lower activity.

(10a)   (10b)   (10c)

All these compounds induce the mechanism of catalysis given in equations (14a) and (14b). Aliphatic and cycloaliphatic isocyanates require higher reaction temperatures and longer times than aromatic ones. Exactly the same principles and catalysts and similar procedures are valid for synthesis of poly(carbodiimide)s from diisocyanates (equation 34), on which topic some reviews have been published.[50,57,58]

$$n\text{OCNRNCO} \xrightarrow{\text{catalyst}} -(\text{RN}=\text{C}=\text{N})_n- + n\text{CO}_2\uparrow \quad (34)$$

Polymerizations are usually carried out in various inert solvents, though precipitation normally occurs as molecular weight values increase, due to the generally small solubility of poly(carbodiimide)s. Carbodiimide linkages are often formed during poly(isocyanurate) production.

As typical heterocumulenes, carbodiimides can undergo an interesting spectrum of reactions, including homopolymerization, across their double bonds.[9,59,60] Poly(carbodiimide)s must thereby be considered reactive polymers, which can add protic nucleophiles similarly to isocyanates (as in equation 35) and are used, for example, as acid scavengers for polyester stabilization.

$$-\text{RN}=\text{C}=\text{N}- + \text{HA} \longrightarrow -\text{RNHC}=\text{N}-\underset{A}{|} \quad (35)$$

23.4.2 Polyamides and Polyimides

The reactions of isocyanates with carboxylic acids have been outlined in Scheme 1 and proceed through several steps which differ from aliphatic–aliphatic to aliphatic–aromatic, aromatic–aliphatic or aromatic–aromatic isocyanate–acid, but whose overall result is in each case amide production with elimination of carbon dioxide.

Polyamides of commercial relevance are obtained *via* such a condensation, carried out in inert dipolar solvents, from diisocyanates and dicarboxylic acids as in equation (36).[26]

$$n\text{OCNRNCO} + n\text{HO}_2\text{CR'CO}_2\text{H} \xrightarrow{\Delta} -(\text{RNHCR'CNH})_n- + 2n\text{CO}_2\uparrow \quad (36)$$

Aliphatic dicarboxylic acids, such as adipic or azelaic acid, and diphenylmethane 4,4'-diisocyanate (MDI) are currently employed; aromatic dicarboxylic acids can also be used in combination with aliphatic ones. Block copolyamides have been prepared from MDI and acid-terminated low polymers (*e.g.* polyesters), by virtue of the absence of interchange reactions as compared to analogous polyamidations from dicarboxylic acids and diamines.

Polyamides have also been synthesized from diisocyanates and enamines as reported in Section 23.3.3.

As seen in Section 23.2.3, isocyanates formally undergo a 1,2-cycloaddition with cyclic anhydrides; imides are afforded and CO_2 is evolved. This reaction is best performed in solution in the presence of small amounts of water, which act as the catalyst (see Scheme 4, where the most probable mechanism is shown.

Polyimides are attainable under similar conditions in highly dipolar solvents, such as NMP or *m*-cresol. Aromatic polyimides, as in equation (37), get the major attention for their high thermal resistance. These syntheses are reviewed in a monograph.[61]

$$n\text{OCN}\text{-Ar-CH}_2\text{-Ar-NCO} + n\text{(dianhydride)} \xrightarrow{\Delta} \text{polyimide} + 2n\text{CO}_2\uparrow \quad (37)$$

Poly(amide–imide)s are prepared in solution by condensation of diisocyanates with monoanhydrides of tricarboxylic acids,[26,62] as shown in equation (38).

$$n\,\text{OCNRNCO} + n\;\underset{\text{HO}_2\text{C}}{\text{[phthalic anhydride-COOH]}} \xrightarrow[80-160\,°C]{\text{NMP}} +\!\!\left(\text{RNHC(O)}\text{-[phthalimide]}\right)\!\!\frac{}{n} + 2n\,\text{CO}_2\uparrow \quad (38)$$

DMF or NMP are the solvents most commonly employed, but o-dichlorobenzene, ethylene glycol monomethyl ether acetate and cresols have been used as well. Traces of water accelerate the reaction.

23.5 ADDITION–CYCLIZATION POLYMERS

Several heterocyclic polymers can be obtained from diisocyanates and suitable reactants, *via* a choice of those cyclization reactions mentioned in Section 23.2.1 and summarized in Scheme 3. Their synthesis is accomplished in two stages: addition of the isocyanate to a protic nucleophile and then intramolecular cyclization (the latter occurring by an addition or a condensation reaction). Some of these polymers have become of certain interest by virtue of their good thermal stability and reviews of their syntheses are available.[26,61]

23.5.1 Poly(5-iminoimidazolidione)s and Poly(parabanic acid)s

Poly(5-iminoimidazolidine-2,4-dione)s can be synthesized from diisocyanates and hydrogen cyanide[63] (see Scheme 3). The direct cyclization of di(carbamoyl cyanide)s as in equation (39) fails to give high polymers, which, on the contrary, are obtained by polyaddition of the same intermediates and diisocyanates (equation 40).

$$n\,\text{OCNRNCO} + 2n\,\text{HCN} \longrightarrow n\;\text{NCC(O)NHRNHC(O)CN} \xrightarrow{-n\text{HCN}}$$

$$\left(\!\!\text{[RN-iminoimidazolidinone ring with NH]}\!\!\right)_{\!n} \xrightarrow{\text{H}_2\text{O}} \left(\!\!\text{[RN-parabanic acid ring]}\!\!\right)_{\!n} \quad (39)$$

$$n\,\text{NCC(O)NHRNHC(O)CN} + n\,\text{OCN-C}_6\text{H}_4\text{-CH}_2\text{-C}_6\text{H}_4\text{-NCO} \longrightarrow$$

$$\left(\!\!\text{[RN-iminoimidazolidinone]-C}_6\text{H}_4\text{-CH}_2\text{-C}_6\text{H}_4\text{-[iminoimidazolidinone]}\!\!\right)_{\!n} \quad (40)$$

Polymerizations are performed at 90–160 °C in NMP or DMSO with pyridine as a catalyst. Poly(5-iminoimidazolidione)s can be hydrolyzed to poly(parabanic acid)s as in equation (39).

23.5.2 Polyhydantoins, Poly(quinazolinedione)s and Poly(hydrouracils)

Diisocyanates polycondense with di(glycinate ester)s affording poly(imidazolidine-2,4-dione)s (simply called polyhydantoins) in two steps, as shown in equation (41).[61,63]

Poly(4-iminoimidazolidine-2-one)s, prepared by polyaddition and cyclization from diisocyanates and bis(α-aminonitriles) can be hydrolyzed to polyhydantoins (equation 42).

$$n\text{OCNRNCO} + n\text{R''OCCH}_2\text{NHR'NHCH}_2\text{COR''} \xrightarrow[50°C]{m\text{-cresol}}$$

$$\left(-\text{RNHCN}-\text{R'}-\text{NCNH}-\right)_n \xrightarrow{150°C} \left(-\text{RN}\underset{O}{\overset{O}{\diagup}}\text{N}-\text{R'}-\text{N}\underset{O}{\overset{O}{\diagup}}\text{N}-\right)_n + 2n\text{ R''OH} \quad (41)$$

with side groups R''OCCH$_2$ and CH$_2$COR'' on the first structure.

$$n\text{ OCNRNCO} + n\text{ NCCH}_2\text{NHR'NHCH}_2\text{CN} \xrightarrow{\text{DMF}}$$

$$\left(-\text{RNHCN}-\text{R'}-\text{NCNH}-\right)_n \longrightarrow \left(-\text{RN}\underset{HN}{\overset{O}{\diagup}}\text{N}-\text{R'}-\text{N}\underset{NH}{\overset{O}{\diagup}}\text{N}-\right)_n \xrightarrow{\text{H}_2\text{O}}$$

$$\left(-\text{RN}\underset{O}{\overset{O}{\diagup}}\text{N}-\text{R'}-\text{N}\underset{O}{\overset{O}{\diagup}}\text{N}-\right)_n \quad (42)$$

Poly(quinazoline-2,4-dione)s are attainable from bisanthranilic acids (equation 43). A sequence of reactions similar to that of equation (41) gives poly(3-imino-2,4-benzoxazine-1-one)s which are then converted by heating to the final polymers.

$$n\text{ OCNRNCO} + n \begin{array}{c} \text{H}_2\text{N}-\text{Ar}-\text{NH}_2 \\ \text{HO}_2\text{C}\phantom{-\text{Ar}-}\text{CO}_2\text{H} \end{array} \xrightarrow[-2n\text{H}_2\text{O}]{160-180°C} \text{[poly(3-imino-2,4-benzoxazine-1-one)]}$$

$$\xrightarrow{230-350°C} \text{[poly(quinazoline-2,4-dione)]} \quad (43)$$

Poly(5,6-dihydrouracil)s can be synthesized in dipolar solvents from aromatic diisocyanates and esters of bis(β-aminocarboxylic acid)s as in equation (44). The cyclization is accomplished by heating in polyphosphoric acid (PPA).

$$n\text{OCN}-\text{R}-\text{NCO} + n\text{ MeOCCH}_2\text{CH}_2\text{NHR'NHCH}_2\text{CH}_2\text{COMe}\longrightarrow$$

$$\left(-\text{RNHCN}-\text{R'}-\text{N}-\text{C}-\text{NH}-\right)_n \xrightarrow[\Delta]{\text{PPA}} \left(-\text{RN}\underset{O}{\overset{O}{\diagup}}\text{N}-\text{R'}-\text{N}\underset{O}{\overset{O}{\diagup}}\text{N}-\right)_n + 2n\text{ MeOH} \quad (44)$$

23.5.3 Poly(oxazolidinedione)s and Poly(benzoxazinedione)s

Poly(oxazolidine-2,4-dione)s are obtained from diisocyanates and bis(α-hydroxy acid)s or their esters,[61] as shown in equation (45).

$$n\,OCNRNCO + n\,HOCHR'CHOH(R''O_2C)(CO_2R'') \longrightarrow \left[RNHCOCHR'CHOCNH(R''O_2C)(CO_2R'')\right]_n$$

$$\xrightarrow{\Delta} \left[\text{RN-cyclic(O,O,C=O,C=O)-R'-cyclic-N}\right]_n + 2n\,R''OH \quad (45)$$

Thermally and hydrolytically stable poly(1,3-benzoxazine-2,4-dione)s result from aromatic diisocyanates and diphenyl esters of aromatic bis(o-hydroxycarboxylic acid)s (equation 45).[62] Aliphatic esters can be used under more drastic conditions.

$$n\,OCNRNCO + n\,\text{(biphenyl with PhO}_2C, CO_2Ph, HO, OH\text{)} \xrightarrow[100\,°C]{\text{DMSO, tertiary amine}}$$

$$\left[\text{biphenyl(PhO}_2C, CO_2Ph)\text{-RNHCO, OCNH}\right]_n \xrightarrow[100\,C]{\text{DMSO}} \left[\text{-RN-benzoxazinedione-benzoxazinedione-N-}\right]_n + 2n\,PhOH \quad (46)$$

23.6 REFERENCES

1. J. H. Saunders and K. C. Frisch, 'Polyurethanes: Chemistry and Technology—Part 1. Chemistry', Interscience, New York, 1962, p. 17.
2. D. J. David and H. B. Staley, 'Analytical Chemistry of the Polyurethanes', Wiley, New York, 1969, chaps. 2 and 3.
3. S. R. Sandler and W. Karo, 'Organic Functional Group Preparations', 2nd edn., Academic Press, New York, 1983, vol. 1, chap. 12.
4. Ref. 2, chap. 7.
5. A. A. R. Sayigh, H. Ulrich and W. J. Farrissey, Jr., in 'Condensation Monomers', ed. J. K. Stille and T. W. Campbell, Wiley, New York, 1972, chap. 5.
6. R. Richter and H. Ulrich, in 'The Chemistry of Cyanates and their Thio Derivatives', ed. S. Patai, Wiley, New York, 1977, part 2, chap. 17.
7. D. H. Chadwick and T. H. Cleveland, in 'Kirk–Othmer Encyclopedia of Chemical Technology', 3rd edn., ed. M. Grayson and D. Eckroth, Wiley, New York, 1981, vol. 13, p. 789.
8. K. Findeisen, K. König and R. Sundermann, in 'Houben–Weyl—Methoden der Organischen Chemie', 4th edn., ed. H. Hagemann, Thieme Verlag, Stuttgart, 1983, band E4, p. 738.
9. G. Tennant, in 'Comprehensive Organic Chemistry', ed. D. H. R. Barton and W. D. Ollis, Pergamon Press, Oxford, 1979, vol. 2, p. 513.
10. Ref. 1, chap. 3.
11. Ref. 1, chap. 4.
12. L. Drobnica, P. Kristián and J. Augustin, in 'The Chemistry of Cyanates and their Thio Derivatives', ed. S. Patai, Wiley, New York, 1977, part 2, chap. 22.
13. A. Hartmann, in 'Houben–Weyl—Methoden der Organischen Chemie', 4th edn., ed. H. Hagemann, Thieme Verlag, Stuttgart, 1983, band E4, p. 834.
14. K. C. Frisch, in 'Polyurethane Technology', ed. P. F. Bruins, Interscience, New York, 1969, chap. 1.
15. A. Farkas and G. A. Mills, *Adv. Catal.*, 1962, **13**, 6.
16. S. L. Reegen and K. C. Frisch, *Adv. Urethane Sci. Technol.*, 1971, **1**, 1.
17. G. Borkent, *Adv. Urethane Sci. Technol.*, 1974, **3**, 1.
18. T. E. Lipatova, *Adv. Urethane Sci. Technol.*, 1976, **4**, 34.
19. D. E. Giles, in 'The Chemistry of Cyanates and their Thio Derivatives', ed. S. Patai, Wiley, New York, 1977, part 2, chap. 12; T. E. Lipatova and Yu. Nizel'sky, *Adv. Urethane Sci. Technol.*, 1981, **8**, 217.
20. H. Ulrich, 'Cycloaddition Reactions of Heterocumulenes', Academic Press, New York, 1967, chap. 4.
21. Ref. 20, chap. 5.
22. Ref. 2, p. 112.

23. C. G. Overberger and J. A. Moore, in 'Encyclopedia of Polymer Science and Technology', ed. N. M. Bikales, H. F. Mark and N. G. Gaylord, Wiley, New York, 1967, vol. 7, p. 743.
24. W. Sweeny and J. Zimmerman, in ref. 23, 1969, vol. 10, p. 577.
25. A. J. Bur and L. J. Fetters, *Chem. Rev.*, 1976, **76**, 727.
26. H. Ulrich, in 'Encyclopedia of Polymer Science and Engineering', 2nd edn., ed. H. F. Mark, N. M. Bikales, C. G. Overberger and G. Menges, Wiley, New York, 1987, vol. 8, p. 448.
27. J. H. Saunders and K. C. Frisch, 'Polyurethanes: Chemistry and Technology—Part 2. Technology', Interscience, New York, 1964.
28. B. A. Dombrow, 'Polyurethanes', 2nd edn., Reinhold, New York, 1965.
29. R. Vieweg and A. Höchtlen (eds.), 'Kunststoff-Handbuch', Hanser Verlag, München, 1966, band 7.
30. A. Damusis, in 'Sealants', ed. A. Damusis, Reinhold, New York, 1967, chap. 5.
31. K. A. Pigott, in ref. 23, 1969, vol. 11, p. 506.
32. P. Wright and A. P. C. Cumming, 'Solid Polyurethane Elastomers', MacLaren, London, 1969.
33. D. C. Allport and A. A. Mohajer, in 'Block Copolymers', ed. D. C. Allport and W. H. Janes, Applied Science, London, 1973, chap. 5.
34. S. R. Sandler and W. Karo, 'Polymer Syntheses', Academic Press, New York, 1974, vol. 1, chap. 8.
35. K. Uhlig and D. Dieterich, in 'Ullmanns Encyklopaedie der Technischen Chemie', 4th edn., ed. E. Bartholome, E. Biekert, H. Hellmann, H. Ley, M. Weigert and E. Weise, Verlag Chemie, Weinheim, 1980, band 19, p. 301.
36. C. Hepburn, 'Polyurethane Elastomers', Applied Science, London, 1982.
37. H. Ulrich, in ref. 7, 1983, vol. 23, p. 576.
38. G. Oertel (ed.), 'Kunststoff-Handbuch', Hanser Verlag, München, 1983; band 7 (Engl. edn.: 'Polyurethane Handbook', Hanser, Munich, 1985).
39. S. K. Gupta and A. Kumar, 'Reaction Engineering of Step Growth Polymerization', Plenum Press, New York, 1987, chap. 9.
40. D. Dieterich, in 'Houben–Weyl—Methoden der Organischen Chemie', 4th edn., ed. H. Bartl and J. Falbe, Thieme Verlag, Stuttgart, 1987, band E20, teil 2, p. 1561.
41. A. F. Hegarty, in 'Comprehensive Organic Chemistry', ed. D. H. R. Barton and W. D. Ollis, Pergamon Press, Oxford, 1979, vol. 2, p. 1083.
42. K. C. Frisch, *Adv. Urethane Sci. Technol.*, 1973, **2**, 1.
43. W. J. McKillip, *Adv. Urethane Sci. Technol.*, 1974, **3**, 81.
44. S. R. Sandler and W. Karo, 'Organic Functional Group Preparations', Academic Press, New York, 1971, vol. 2, chap. 6.
45. H. G. J. Overmars, in ref. 23, 1969, vol. 11, p. 464.
46. Ref. 34, chap. 7.
47. Ref. 40, p. 1721.
48. Ref. 27, chap. 12.
49. R. J. Cotter and M. Matzner, 'Ring-Forming Polymerizations', Academic Press, New York, 1972, part B, vol. 1, p. 194, 272.
50. S. R. Sandler and W. Karo, 'Polymer Syntheses', Academic Press, New York, 1980, vol. 3, chap. 4.
51. A. A. R. Sayigh, *Adv. Urethane Sci. Technol.*, 1974, **3**, 141.
52. H. E. Reymore, Jr., P. S. Carleton, R. A. Kolakowski and A. A. R. Sayigh, *J. Cell. Plast.*, 1975, **11**, 328.
53. K. C. Frisch and J. E. Kresta, *Int. Prog. Urethanes*, 1977, **1**, 191; S. G. Entelis and R. P. Tiger, *Adv. Urethane Sci. Technol.*, 1981, **8**, 19.
54. Ref. 40, p. 1739.
55. Ref. 44, chap. 9.
56. W. Rasshofer, in 'Houben–Weyl—Methoden der Organischen Chemie', 4th edn., ed. H. Hagemann, Thieme Verlag, Stuttgart, 1983, band E4, p. 883.
57. Ref. 49, p. 14, 18, 71.
58. Ref. 40, p. 1753.
59. Ref. 2, p. 146.
60. Ref. 20, chap. 6.
61. R. Merten, *Angew. Chem.*, 1971, **83**, 339; *Adv. Urethane Sci. Technol.*, 1973, **2**, 123.
62. R. J. Cotter and M. Matzner, 'Ring-Forming Polymerizations', Academic Press, New York, 1972, part B, vol. 2, chap. 1.
63. Ref. 49, p. 230.

24
Polyurethanes

KURT C. FRISCH and DANIEL KLEMPNER
University of Detroit, MI, USA

24.1 SYNTHESIS OF POLYURETHANES	413
24.2 CURING REACTIONS OF POLYURETHANES	415
24.2.1 *General*	415
24.2.2 *Elastomers, Sealants and Elastoplastics*	416
24.2.3 *Flexible Foams*	418
24.2.4 *Rigid Foams*	419
24.2.5 *Coatings and Adhesives*	419
24.2.5.1 Blocked isocyanates	419
24.2.5.2 Coating systems	420
24.3 REFERENCES	425

24.1 SYNTHESIS OF POLYURETHANES

The synthesis of polyurethanes can be carried out by a variety of methods, although the most widely used production method is the reaction of di- or poly-functional hydroxycompounds (colloquially referred to as polyols) with di- or poly-functional isocyanates. A schematic representation of the preparation of linear polyurethanes by the polyaddition procedure is shown in equation (1).

$$\text{HOROH} + \text{OCNR'NCO} \rightarrow +\!\!\left(\text{OROCNHR'NHC}\right)\!\!\underset{\text{polyurethane}}{\overset{\text{O}\ \ \ \ \ \ \ \text{O}}{\underset{\|\ \ \ \ \ \ \ \|}{}}}\!\!\!_n \qquad (1)$$

If the functionality of the hydroxy-containing compounds and/or of the isocyanate is increased beyond two, branched or crosslinked polymers are produced. Due to the fact that the nature of the polyol and isocyanate components can be widely varied, polyurethanes are among the most versatile polymers, producing flexible and rigid foams, soft and hard elastomers (elastoplastics), coatings for soft and hard substrates, adhesives, fibers, films, plastics and composites, among others.

Although polyurethanes are of relatively recent origin, as far as synthetic polymers are concerned, the chemistry of organic isocyanates goes back almost 140 years, when Wurtz[1] synthesized the first aliphatic isocyanates in 1849. He also described a number of important isocyanate reactions, such as the reaction of ethyl isocyanate with ethanol to form ethyl carbamate, and with a secondary amine to form a substituted urea derivative.

However, it was not until 1937 that Bayer of the German company I.G. Farbenindustre developed the 'diisocyanate polyaddition' procedure which resulted in the production of polyurethanes and other isocyanate-based polymers.[2,3] The first efforts by the Bayer team to produce fibers which could rival polyamide (nylon) fibers without infringing the Du Pont patents led to the synthesis of polyureas from 1,6-hexamethylene diisocyanate and 1,6-hexamethylenediamine, shown in equation (2). However, these polyureas were infusible and were not suitable for the preparation of fibers or plastics.

$$\text{OCN(CH}_2)_6\text{NCO} + \text{H}_2\text{N(CH}_2)_6\text{NH}_2 \rightarrow +\!\!\left(\overset{\text{O}}{\overset{\|}{\text{C}}}\text{NH(CH}_2)_6\text{NH}\overset{\text{O}}{\overset{\|}{\text{C}}}\text{NH(CH}_2)_6\text{NH}\right)\!\!_{\overline{n}} \qquad (2)$$

polyurea

On the other hand, the reaction of 1,4-butanediol with 1,6-hexamethylene diisocyanate led to the formation of a polyurethane possessing properties which made it of interest in the production of plastics and fibers, especially of bristles (equation 3).

$$HO(CH_2)_4OH + OCN(CH_2)_6NCO \rightarrow -\!\!-\!\!O(CH_2)_4O\overset{O}{\overset{\|}{C}}NH(CH_2)_6NH\overset{O}{\overset{\|}{C}}\!\!-\!\!\!\!)_{\overline{n}} \quad (3)$$
$$(1)$$

The polyurethane (1) was marked commercially under the tradename of 'Perlon U' for plastics. However, these products attained only a limited sales volume in Germany and found even less acceptance in the US.

Besides the 'polyaddition' procedure for the production of polyurethanes, there are many other syntheses which are of lesser importance. Perhaps one of the most versatile alternative routes is the reaction of bis(chloroformate)s with diamines, which was disclosed as early as 1945[4] (equation 4).

$$Cl\overset{O}{\overset{\|}{C}}OROC\overset{O}{\overset{\|}{C}}Cl + H_2NR'NH_2 \rightarrow -\!(\overset{O}{\overset{\|}{C}}OROC\overset{O}{\overset{\|}{C}}NHR'NH)_{\overline{n}} \quad (4)$$

A very convenient way of producing polyurethanes by the above route is *via* 'interfacial polycondensation'. The principle of this method consists in the formation of a polymer at the interface of the two reactants which are dissolved in water while the chloroplastic derivative is dissolved in an organic solvent. This method lends itself to the preparation of a great variety of polyurethanes. Many investigations of the synthesis of polyurethanes as well as of copolymers, such as poly(urethane-*co*-urea)s, poly(urethane-*co*-amide)s, poly(urethane-*co*-carbonate)s, *etc.*, by interfacial condensation were made, notably by Morgan,[5,6] Wittbecker *et al.*,[7] Malichenko *et. al.*[8,9] and Carraher *et al.*[10] A very good summary of polyurethanes and copolymers prepared by interfacial polycondensation has been given by Tanaka and Yokoyama.[11]

There are a number of 'indirect' reactions leading to the formulation of polyurethanes. Some of these methods were used to a limited degree in commercial applications.

One of these consists of the reaction of 3,3'-tetramethylenedi(1,4,2-dioxazol-5-one) (2; adiponitrile carbonate, ADNC) with diols or polyols in the presence of certain metal catalysts, *e.g.* tin catalysts, either alone or in combination with tertiary-amine catalysts, which leads to the formation of polyurethanes with evolution of carbon dioxide, as shown in equation (5).[12]

$$\underset{(2)\ \text{ADNC}}{\text{[structure]}} + HOROH \xrightarrow{\Delta} -\!(O\overset{O}{\overset{\|}{C}}NH(CH_2)_4NH\overset{O}{\overset{\|}{C}}OR)_{\overline{n}} + CO_2 \quad (5)$$

The preparation of ADNC and other aliphatic and aromatic dinitrile carbonates, as well as the technology associated with nitrile carbonates, are given in a number of patents.[13,14]

In addition to dinitrile carbonates, Burk and co-workers[14] have shown that cyclic sulfites (3) and oxalates (4) can also be used for the preparation of polyurethanes. These cyclic derivatives yield hydroxamates by reaction with active-hydrogen compounds. In the case of diols or polyols, these hydroxamates can be decomposed to yield polyurethanes in a manner similar to that for nitrile carbonates (Scheme 1).

(3) (4)

Another 'indirect' method for the generation of urethane linkages in polycarbonates has been reported by Foti *et al.*[15] The first step is the reaction of piperazine with an aromatic polycarbonate to yield a urethane diphenol (Scheme 2). The second step is the repolymerization with phosgene to produce an alternating copolymer (equation 6).

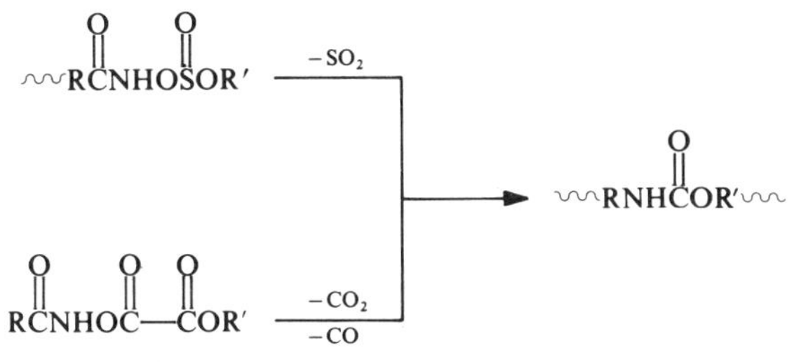

Scheme 1

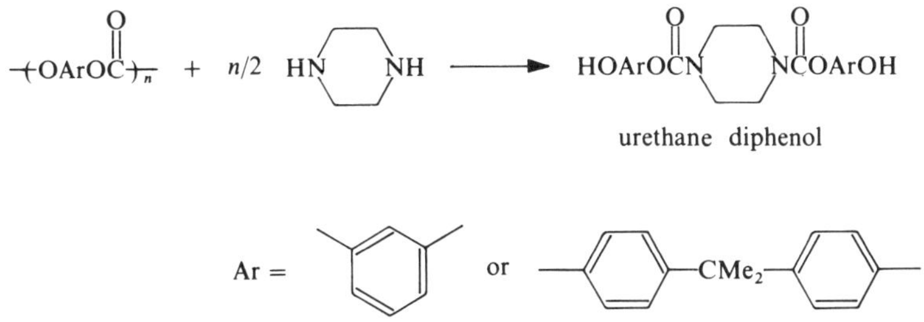

$$\text{\textapprox}RC(O)NHOSO_2R' \xrightarrow{-SO_2}$$

$$\text{\textapprox}RC(O)NHOC(O)-C(O)R' \xrightarrow[-CO]{-CO_2} \text{\textapprox}RNHC(O)R'\text{\textapprox}$$

hydroxamates

Scheme 2

$$+(OArOC(O))_n + n/2\ HN\underset{}{\overset{}{\bigcirc}}NH \longrightarrow HOArOC(O)N\underset{}{\overset{}{\bigcirc}}NC(O)ArOH$$

urethane diphenol

Ar = *m*-xylylene or $-\text{C}_6\text{H}_4-\text{CMe}_2-\text{C}_6\text{H}_4-$

$$HOArOC(O)N\underset{}{\overset{}{\bigcirc}}NC(O)ArOH + COCl_2 \longrightarrow +(OArOC(O)OArOC(O)N\underset{}{\overset{}{\bigcirc}}NC(O))_n \quad (6)$$

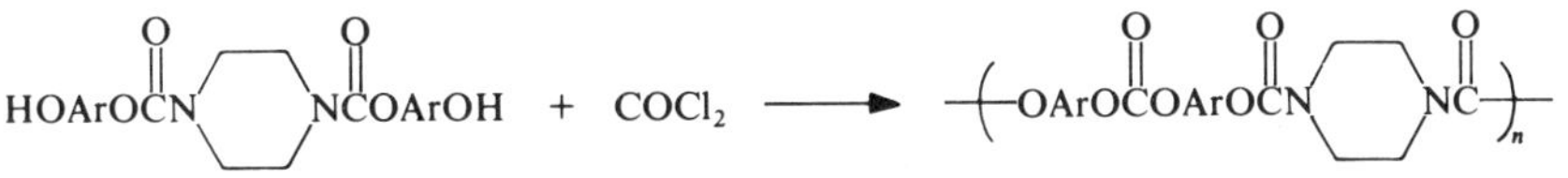

poly(carbonate-*co*-urethane)

An interesting synthesis of polyurethanes uses aminimides containing two hydroxy groups (5), formed by reaction of an ester, *e.g* methyl methacrylate, with 1,1-dimethylhydrazines and 2-hydroxymethyloxirane.[16] Homopolymerization of this monomer, with azobis(isobutyronitrile) (AIBN) as initiator in either water or methanol, leads to a polymer which, on pyrolysis at 170 °C for 2 h, yields a crosslinked product containing urethane linkages, as determined by IR. Crosslinking presumably takes place during the pyrolysis, as shown in Scheme 3.

$$CH_2=CMeC(O)\overset{-}{N}\overset{+}{N}Me_2CH_2CHOHCH_2OH$$

(5)

$$\text{\textapprox}C(O)\overset{+}{N}Me_2CH_2CHOHCH_2OH \rightarrow \text{\textapprox}N=C=O + Me_2NCH_2CHOHCH_2OH \xrightarrow{2\text{\textapprox}N=C=O}$$

$$+(NHC(O)OCH_2CHOC(O)NH)_n$$
$$|$$
$$CH_2NMe_2$$

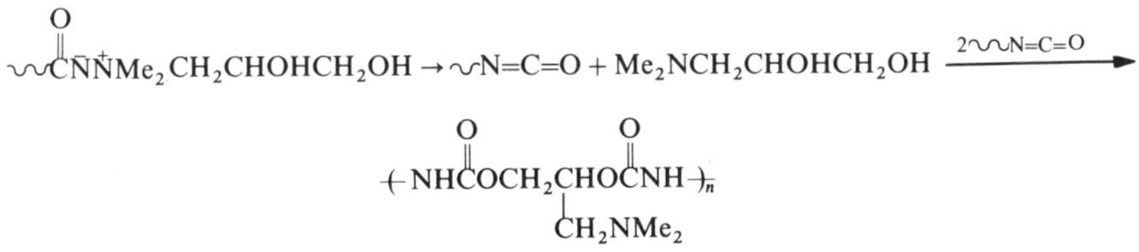

Scheme 3

Due to the fact that the synthesis of polyurethanes and urethane copolymers is determined, to a large extent, by the type of curing reaction used for the formation of foams, elastomers, sealants, coatings and adhesives, as well as other polyurethane products, the rest of this chapter is devoted to the chemistry of the curing reactions.

24.2 CURING REACTIONS OF POLYURETHANES

24.2.1 General

The curing of urethane polymers and prepolymers involves the reaction of isocyanate groups (either monomers, oligomers or prepolymers) with compounds (again either monomeric or oligomeric) containing hydroxy groups.[17] The functionality of all materials must be greater than or equal to two. If thermoplastic (linear) polyurethanes are desired, a functionality of two is required for both the isocyanate and the active-hydrogen-containing material. If thermosetting polyurethanes are the goal, functionalities greater than two are required for either the polyisocyanate or the active-hydrogen compound or both to provide crosslinking.

The foregoing holds for all types of urethane products, including elastomers of all types, coatings, elastoplastics, sealants, adhesives and foams (flexible, semi-flexible and rigid) of all densities. In general, foams of all types are crosslinked, the more rigid foams being more highly crosslinked. The other products can be either thermoplastic or thermosetting, although the latter is somewhat more prevalent.

Suitable active-hydrogen-containing materials include those with hydroxy groups (polyols), including water, amine groups (primary or secondary) and already-formed urethane and urea groups. Another group which can result in curing of isocyanate-containing materials but contains no active hydrogen is the epoxy group. In addition, isocyanate groups can react with themselves and thus participate in the curing of urethane materials with no other active-hydrogen-containing compounds.

The curing reaction can use an isocyanate-terminated prepolymer (NCO/OH = 2) or quasi-prepolymers (NCO/OH > 2) or it can be a one-shot approach in which no preformed urethane is involved. Catalysis is often required to obtain complete curing, although the type and amount depends on the type of isocyanate and active-hydrogen-containing material. The most common types of catalysts used are organometallic catalysts, notably organotin compounds and tertiary amines. If aliphatic isocyanates are used, catalysis must generally be employed, since these materials react much more slowly than aromatic isocyanates. Amines (in particular primary aliphatic amines) react exceptionally rapidly with isocyanates (aromatic amines are slower), so that no catalyst is needed. In fact, this reaction is so fast that hindered amines must often be used to allow for a controllable reaction. Primary hydroxy groups react much faster than secondary hydroxy groups and generally do not need catalysis (depending, of course, on the isocyanate used). Reaction with epoxy groups is relatively slow and needs specific catalysis as well as higher temperatures to result in a cure to yield oxazolidones.

24.2.2 Elastomers, Sealants and Elastoplastics

Polyurethane elastomers, sealants and elastoplastics (high-modulus elastomers) may be prepared by either the one-shot or the prepolymer approach. In the former, all ingredients (isocyanates, polyols with short-chain diol or amine chain-extender and/or polyamines) are mixed simultaneously, together with catalysts, fillers, plasticizer and any other additives. They are then cast or molded and (generally) thermally cured, although room-temperature curing systems can be prepared by using high levels of mixed catalysts or appropriately fast-reacting isocyanates and active-hydrogen-containing materials. Such systems are generally two components, consisting of an A component, being the di- or poly-isocyanate, and a B component, being the polyol(s) and/or polyamine(s), with a chain-extender (generally a low molecular weight polyol or polyamine), any other desired active-hydrogen material, catalyst(s) and any other ingredients. Such one-shot systems include reaction injection-molding (RIM) systems as well as some cast elastomers. Typical reactions are shown in Scheme 4.

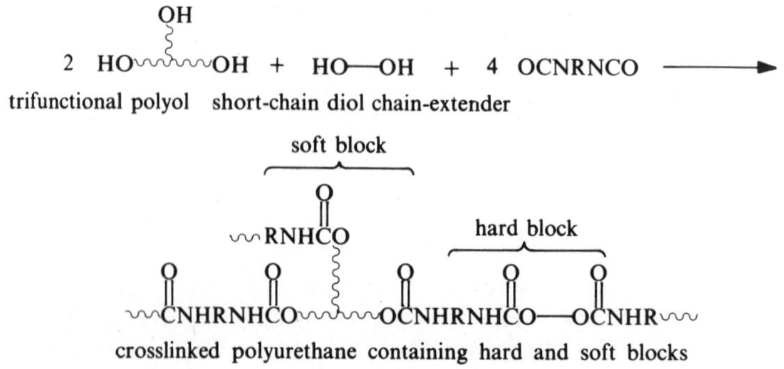

Scheme 4

The polyols are generally hydroxy-containing polyesters (caprolactones, adipates or, to a lesser extent, castor oil and transesterification derivatives thereof), polyethers [poly(oxypropylene), poly(oxypropylene-co-oxyethylene), or poly(1,4-oxybutylene)] or, to a lesser extent, hydroxy-containing hydrocarbon polymers (hydroxy-containing butadiene homopolymers and copolymers). If oxypropylene-based polyethers are used, they are sometimes capped with oxirane (ethylene oxide) in order to create primary hydroxy-groups, which are more reactive than the secondary oxypropylene hydroxy groups. When linear polyurethanes are desired, difunctional polyols are used. Typical short-chain diol chain-extenders include 1,4-butanediol, ethylene glycol, bisethanol ether, bispropanol ether, 1,6-hexanediol, 2,2-dimethyl-1,3-propanediol, the bis(hydroxyethyl) derivative of hydroquinone (HQEE) and modified 2,2-bis(4-hydroxyphenyl)propane (bisphenol *A*). Short triols, such as tri(hydroxymethyl)propane, may also be used when additional crosslinking is desired.

Diamines may be introduced as short chain-extenders, including hindered diamines, such as 4,4'-methylenedi(*o*-chloroaniline) or the dimethyl ester of 4,4'-methylenedianthranilic acid, primary aromatic diamines (where excessive reactivity is not a problem), such as phenylenediamine, tolylenediamines and methylenedianilines, and (more recently) aromatic secondary diamines, such as 4,4'-di(*s*-butylamino)diphenylmethane. Longer-chain diamines are also used [amine-terminated poly(oxypropylene)s] especially in RIM polyurea systems. These reactions are shown in equation (7) (everything is shown as difunctional for convenience).

$$HO\sim\sim OH + H_2N\sim\sim NH_2 + 2\ OCNRNCO \longrightarrow$$
polyol, short- or long-chain diamine

$$\sim\sim RNHCO\sim\sim OCNHRNHCNH\sim\sim NHCNH\sim\sim \quad (7)$$
poly(urethane-*co*-urea)

If no polyol is being used and only amine-terminated polyethers, such as poly(oxypropylene), diols and triols together with a diamine chain-extender such as diethyltoluenediamine (DETDA) are employed, a polyurea RIM system is obtained as shown in equation (8).

$$H_2N\sim\sim O\sim\sim O\sim\sim NH_2 + H_2N-NH_2 + OCNRNCO \longrightarrow$$
polyether amine, short hindered diamine

$$\sim\sim RNHCNH\sim\sim O\sim\sim ONHCNHR-NHCNH\sim\sim O\sim\sim \quad (8)$$

When a more structured polyurethane or poly(urethane-*co*-urea) is desired, the prepolymer approach is used. An isocyanate-terminated prepolymer is first prepared by reacting excess diisocyanate with a polyol. The curing involves the reaction of the prepolymer (A component) with a chain extender (B component), generally a low molecular weight polyol or polyamine (the functionality depending on the desired amount of crosslinking, if any), such as those described earlier. The curing reactions for polyurethanes and poly(urethane-*co*-urea)s are shown in equations (9) and (10).

$$OCNRNHCO\sim\sim OCNHRNCO + HOR'OH \longrightarrow \left(CNHRNHCO\sim\sim OCNHRNHCOR'O\right)_n \quad (9)$$
prepolymer, polyurethane

$$OCNRNHCO\sim\sim OCNHRNCO + H_2NR'NH_2 \longrightarrow$$
prepolymer

$$\left(CNHRNHCO\sim\sim OCNHRNHRNHCNHR'NH\right)_n \quad (10)$$
poly(urethane-*co*-urea)

The isocyanates used are primarily aromatic (tolylene diisocyanate or 4,4'-diphenylmethyl diisocyanate, pure or modified); aliphatic isocyanates are also occasionally employed.

Sealants are low-modulus polyurethane elastomers which involve higher molecular weight polyols, low levels of crosslinking (if any) and very little (if any) chain extension. Elastoplastics are high-modulus (rigid and semi-rigid) elastomers prepared using higher levels of crosslinking, lower equivalent-weight polyols, and higher amounts of short chain-extenders, the latter resulting in greater amounts of hard segments.

Additional reactions which can occur in the above systems to result in further crosslinking include allophanate (**6**) formation (reaction of isocyanate with urethane groups), and biuret (**7**) formation (reaction of isocyanate with urea groups) shown in equations (11) and (12).

$$\text{\textasciitilde\textasciitilde NHCO\textasciitilde\textasciitilde} + \text{OCN\textasciitilde\textasciitilde} \longrightarrow \text{allophanate (6)} \quad (11)$$

$$\text{\textasciitilde\textasciitilde NHCNH\textasciitilde\textasciitilde} + \text{OCN\textasciitilde\textasciitilde} \longrightarrow \text{biuret (7)} \quad (12)$$

When increased rigidity and/or high-temperature performance is desired, further crosslinking to form heterocyclic rings may be accomplished via 1,3,5-trisubstituted-1,3,5-triazine-2,4,6(1H,3H,5H)-trione (**8**; isocyanurate) formation. This trimerization reaction, which will take place with the employment of excess isocyanate and selective catalysts, is shown in equation (13).

$$3 \text{ OCN}\text{---}\text{NCO} \xrightarrow{\text{trimerization catalyst}} \text{isocyanurate (8)} \quad (13)$$

Another method for obtaining polyurethanes with high-temperature properties is curing with epoxy-containing materials to form oxazolidones, shown in equation (14).

$$\text{OCN\textasciitilde\textasciitilde NCO} + \text{epoxide} \xrightarrow{\text{oxazolidone catalyst}} \text{bis(1,3,2-oxazolidone)} \quad (14)$$

Polyurethane elastomers with enhanced energy-absorbing abilities (both mechanical and acoustical) represent a more recent application. These materials are made by either underindexing (NCO/OH < 1), using chain stoppers (monofunctional alcohols) or both. This results in a broad distribution of molecular weights, which gives rise to a very broad glass transition.

24.2.3 Flexible Foams

Flexible polyurethane foams are prepared from basically the same raw materials as polyurethane elastomers. Polyether or polyester polyols are used, generally with a functionality of about three. No short chain-extenders are used. In general, the one-shot or quasi-prepolymer approach is used. The curing reactions involved are those of polyisocyanates with polyols, as previously described, and are catalyzed primarily by a combination of tertiary amines and organotin compounds. The main difference between flexible foams and elastomers is the presence of a blowing agent, which, for most flexible polyurethane foams, is water. Excess isocyanate [more than that needed to react with the polyol(s)] is used in the A component. The B component, in addition to polyols, contains water and a surfactant as a cell control agent. Water reacts with the excess isocyanate to form an amine and carbon dioxide, which results in the foaming. This blowing reaction is catalyzed by tertiary amines.

The diamine thus created can then react with additional isocyanate to result in polyurea formation, so that a water-blown flexible polyurethane foam is actually a poly(urethane-*co*-urea). The isocyanate–water reaction is shown schematically in Scheme 5. The overall reaction can be represented as equation (15).

$$\sim\!NCO + H_2O \rightarrow \left[\sim\!NH\overset{O}{\overset{\|}{C}}OH\right] \rightarrow \sim\!NH_2 + CO_2$$

$$\sim\!NCO + \sim\!NH_2 \rightarrow \sim\!NH\overset{O}{\overset{\|}{C}}NH\sim$$

Scheme 5

$$2\sim\!NCO + H_2O \rightarrow \sim\!NH\overset{O}{\overset{\|}{C}}NH\sim + CO_2\uparrow \tag{15}$$

In some cases, some fluorocarbons or dichloromethane are also used as physical (auxiliary) blowing agents in addition to water (they simply evaporate due to the exothermic gelling reaction). It is critical that a proper balance of catalysts be employed in order to assure balanced gelling, curing and blowing reactions.

24.2.4 Rigid Foams

Rigid polyurethane foams are prepared from polyisocyanates and polyols of higher functionalities (in general four or more), so that high degrees of crosslinking result. Short chain-extenders or amine reactivity (other than in tertiary nitrogen-containing polyols) are not generally used. The one-shot or quasi-prepolymer approach is most often used, with the former being most common. The gelling (and curing) reactions are the same as those involved in urethane elastomers. Blowing is generally accomplished with fluorocarbons (physical blowing) although sometimes small amounts of water are used. Water-blown foams are generally employed in the manufacture of high-density foams or foam composites. Trimerization of the isocyanate (the reaction shown earlier) is also often employed in curing rigid foams to achieve a more thermally stable lower-combustible foam (urethane–isocyanurate foam).

24.2.5 Coatings and Adhesives

The basic chemistry of polyurethane coatings and adhesives is similar to that already described for elastomers and foams. Solvents may or may not be employed. Both aqueous and non-aqueous systems are used. Aliphatic as well as aromatic isocyanates are used, although there is a trend toward increasing use of aliphatic isocyanates because of their greatly superior light and color stability.

The diisocyanate and polyols can be reacted in various combinations by using diols, triols and tetrols, or combinations of these, in which the NCO/OH ratio is lower than two.

24.2.5.1 *Blocked isocyanates*

For certain applications, such as wire enamels, coil coatings, fabric coatings, powder coatings, cationic electrodeposition coatings, *etc.*, 'blocked' isocyanates are being used in order to provide one-component coatings with hydroxy-containing or other active-hydrogen-containing components. The 'blocking' reaction is reversible according to the classic reaction of the adduct from tri(hydroxymethyl)propane and tolylene diisocyanate (TDI) with phenol (equation 16). Application of heat ($\sim 150\,°C$) regenerates the free isocyanate, which is then capable of reacting with the hydroxy component during cure.

$$EtC(CH_2OCONHRNCO)_3 + 3\,PhOH \rightarrow EtC(CH_2OCONHRNHCO_2Ph)_3 \tag{16}$$

Both aliphatic and aromatic isocyanates can be blocked by a variety of blocking agents. These include alcohols, phenols, oximes, lactams, dicarbonyl compounds, hydroxamates, bisulfite addition compounds, hydroxylamines and esters of *p*-hydroxybenzoic acid and salicylic acid. Excellent reviews of blocked isocyanates have been written by Wicks.[18] Perhaps the most widely used blocking agents at present are phenol, branched alcohols, 2-butanone oxime and ε-caprolactam.

Catalysts play an important role in the deblocking or thermal dissociation of the blocked isocyanates. Notably, organometallic compounds and tertiary amines are capable of lowering both the deblocking temperature and time as compared to the uncatalyzed system.

In addition to tin compounds, such as dibutyltin dilaurate and dibutyltin diacetate, various other metal compounds have been claimed to be effective deblocking catalysts, such as zinc naphthenate,[19] lead naphthenate, bismuth salts and titanates,[20] and calcium, magnesium, strontium or barium salts of hexanoic, octanoic, naphthenic or (Z,Z,Z)-9,12,15-octadecatrienoic (linolenic) acid.[21]

The use of blocked isocyanates for aqueous systems is of special interest. The sodium bisulfite blocking of 1,6-hexamethylene diisocyanate was described in Petersen's classic paper on blocked isocyanates.[22] The stability of bisulfite-blocked aromatic isocyanates in water can present a difficulty, and various stabilizer systems have been proposed.

Sodium-bisulfite-blocked isocyanate-terminated prepolymers are extensively used as shrink-resistant finishes for wool. Poly(carbamoylsulfonate)s (**10**; PCS) are produced from isocyanate-terminated prepolymers (**9**) with bisulfite salts in aqueous alcohol as the solvent, shown schematically in equation (17).[23-25] The principal curing reaction of PCS is the hydrolysis to form urea crosslinks (equation 18). Other water-soluble blocked isocyanate crosslinkers have been prepared by reaction of diisocyanate with 1,2-di(hydroxymethyl)propionic acid followed by blocking with caprolactam and subsequent solubilization in water by means of a tertiary amine[26] or 2-ethylimidazole.[27]

$$R(NCO)_n + NaHSO_3 \xrightarrow[\text{EtOH}]{H_2O} R(NHCOSO_3^- Na^+)_n \qquad (17)$$
$$(\mathbf{9}) \qquad\qquad\qquad (\mathbf{10})$$

$$2\, RNHCOSO_3^- Na^+ \xrightarrow{H_2O} RNHCONHR + CO_2 + 2NaHSO_3 \qquad (18)$$

24.2.5.2 Coating systems

The polyols and chain extenders used in coatings and adhesives are similar to those used for elastomers previously described. In addition, hydroxy-terminated urethane polyethers (polyether polyols extended with diisocyanates) are often used.

Historically, urethane coatings were first grouped into five American Society for Testing and Materials (ASTM) categories. Because these coatings have been reveiwed in detail in many articles and chapters,[28-31] they are listed below together with some brief comments.

(i) One-package urethane alkyd (oil-modified urethanes) coatings (ASTM type 1)

One-package urethane alkyd coatings, also referred to as uralkyds or urethane oils, are made by transesterification of drying oils with polyols, followed by the reaction of the resulting di- or monoglycerides with a diisocyanate. Curing of the uralkyds occurs by oxidation of the double bonds present in the fatty-acid portion of the coating system.

(ii) One-package moisture-cure urethane coatings (ASTM type 2)

One-package moisture-cure urethane coatings consist of isocyanate-terminated polymers cured by atmospheric moisture. The rate of cure depends upon the humidity in the air and the presence of certain tertiary-amine catalysts that accelerate the NCO–water reaction. A major difficulty with this type of coating system is pigmentation.

(iii) Single-package blocked adduct urethane coatings (ASTM type 3)

Single-package blocked adduct urethane coatings are based on blocked isocyanate components in a blend with hydroxy-bearing materials, which are stable in a single package and which cure after deblocking at elevated temperatures, about 150–160 °C, and at lower temperatures and shorter curing times when catalyzed.

(iv) Two-package catalyst urethane coatings (ASTM type 4)

Two-package catalyst urethane coatings are similar to type 2, consisting of an isocyanate-terminated polymer to which a catalyst is added as a second component prior to application. Two groups of catalyst are usually employed: (1) reactive catalysts, containing hydroxy groups such as alkanolamines: or (2) non-reactive catalysts such as tertiary amines and metal salts of carboxylic acids.

(v) Two-package polyol urethane coatings (ASTM type 5)

Two-package polyol urethane coatings consist of isocyanate-terminated adducts of polymers that are cured by reaction with di- or poly-functional hydroxy-containing materials. The latter may consist of low- to medium-weight polyols with a polyester, polyether, polyether urethane or castor oil backbone. When the two components (OH– and NCO–) are mixed together, they have only a limited pot-life. Therefore, the components are mixed prior to application. Catalysts may be used to speed up the cure either for room temperature or oven cure.

In addition to the five groups of urethane coatings described above, other coating systems have been developed that are detailed in the following sections.

(i) Urethane lacquers

These include polymerized thermoplastic coatings of relatively high molecular weight, dissolved in suitable solvents. The polymers are prepared by reaction of the polyisocyanates with a polyol until no free isocyanate remains. The resulting polymer is then dissolved in solvents such as N-methylpyrrolidone, dimethylformamide, tetrahydrofuran or solvent blends, such as methyl cellosolve/cellosolve acetate/xylene. The films are formed by mere solvent evaporation. One serious disadvantage is the generally low solids-content of these coating systems. These lacquers find extensive use as coatings over flexible substrates such as textiles, flexible foams, microcellular foams, rubber and other elastomer products.

(ii) Two-package coatings: isocyanate-terminated prepolymers cured with polyamines

Due to the fact that isocyanates react very rapidly with amines to form urea groups, the success of preparing suitable coating systems depends upon the following alternatives: (1) the use of hindered aromatic amines to slow down the urea formation; (2) the use of aliphatic isocyanates, for example 4,4′-methylenebis(cyclohexyl diisocyanate) ($H_{12}MDI$) in the prepolymer formation, followed by reaction with an aromatic diamine (aliphatic isocyanate groups react considerably slower than aromatic ones); or (3) the use of aromatic diamines in spray systems where short pot-lives can be used.

In all of the above-described cases, urethane–urea block copolymers are formed that are generally characterized by higher strength properties and higher hardness when compared to coatings made from the same prepolymer but cured with polyols. A disadvantage of these types of coatings is the tendency toward yellowing upon weathering.[30] The use of methylenebis(o-chloroaniline) (MOCA) and other hindered diamines (as a replacement for MOCA) leads to a coating with a pot life of 0.5 to 1 h. 4,4′-Methylenebisaniline, due to its greter reactivity, is used either in spray systems with aromatic diisocyanate-based prepolymers or with prepolymers made from aliphatic isocyanates.

In addition to diamines, hydrazine[32,33] has also been reported to give coatings with good abrasion resistance. Likewise, guanidine[34] has been used to cure isocyanate-terminated prepolymers, yielding highly crosslinked coatings due to the functionality of guanidine.

(iii) Two-package coatings: isocyanate-terminated prepolymers cured with ketimines

Instead of diamines, ketimines can be employed to cure aromatic or aliphatic polyisocyanates. Ketimines are generally produced from primary amines and a ketone. Since ketimines are not very reactive with aliphatic isocyanates, the presence of moisture is necessary in order to hydrolyze the ketimines to form amines, which further react with the isocyanate groups to yield urea linkages. The addition of drying agents, that act as scavengers for moisture, to prevent premature unblocking of the ketimines has been reported.[35,36] The curing mechanism between ketimines and polyisocyanates depends upon the amount of moisture to bring about hydrolysis of the ketimines; several reactions are possible, as shown by Furukawa et al.[37] Hence, side reactions may cause the formation of coatings having inferior properties and a tendency to yellow.

(iv) Two-package coatings with amino–formaldehyde cure

One component consists of a hydroxy-terminated urethane prepolymer while the other component is an alkyl ether (usually the methyl or butyl ether of formaldehyde condensation products or derivatives thereof).[38] Various catalysts may be employed to accelerate the heat cure (*ca.* 125 °C) and to lower the curing temperature. Curing occurs *via* splitting-off of the respective alcohol, as shown in equation (19). The storage stability of these coatings systems is excellent, and a wide range of properties can be obtained through variation in the raw materials.

$$>\!NCH_2OR + HOR^1OCONHR^2\sim\!\sim \rightarrow >\!NCH_2OR^1OCONHNHR^2\sim\!\sim + ROH \tag{19}$$

(v) Aqueous urethane coating systems

Published literature on the composition and applications of water-based urethane systems, *e.g.* emulsions or latices, has been rather limited until a few years ago when the use of these systems began to increase substantially. Both non-ionic and ionic systems have been prepared. Altscher[39] described the use of emulsions of blocked urethane systems containing 5–20% solvent for use in various textile applications.

Urethane–urea latices, consisting of small particles of high molecular weight polymers dispersed in water, are prepared by chain extension of isocyanate-terminated prepolymers in aqueous diamine solutions, employing either non-ionic or ionic surfactants.[40–42]

Ionic polyurethanes include both urethane polyelectrolytes and urethane ionomers. The principal difference between ionomers and polyelectrolytes is that the latter consist essentially of ionic monomer units, while ionomers contain only a relatively small number of ionic monomer units (*e.g.* acid groups) that can be crosslinked by means of metal salts. It is interesting to note that when polyurethanes assume positive or negative charges significant changes occur in their mechanical properties and solubility characteristics. In general, the strength properties are increased and the usually hydrophobic urethane-polymers may become highly hydrophilic at a high concentration of charges. Most properties of ionic polyurethanes may be accounted for by (a) the primary structure, *e.g.* the main chain, the nature and sequence of the segments and the concentration of ionic groups; (b) hydrogen bonding; (c) ionic associations; and (d) association of hydrophobic parts of polyurethane segments due to van der Waals' forces.[43]

Dieterich et al.[44,45] reviewed, in depth, aqueous ionic and non-ionic polyurethane systems and described different methods for the preparation of these polymers.

Suskind[46] described the formation of film-forming cationic urethane latices. The isocyanate-terminated prepolymer derived from either a polyester or polyether diol and tolylene diisocyanate was first chain-extended with an *N*-alkyldiethanolamine to yield a relatively low molecular weight urethane capable of further chain-extended reactions (equation 20).

$$3\,OCN\sim\!\sim NCO + 2\,RN(CH_2CH_2OH)_2 \rightarrow OCN\sim\!\sim\overset{R}{\underset{|}{N}}\sim\!\sim\overset{R}{\underset{|}{N}}\sim\!\sim NCO \tag{20}$$

Emulsification occurs when the partially extended urethane is added to 3% aqueous acetic acid with high-speed mixing. Curing of the latex takes place either by reaction of water[47,48] with the terminal isocyanate groups or by reaction with water-soluble diamines (equations 21 and 22).

$$OCN\sim\!\sim\overset{R}{\underset{H}{N^+}}\sim\!\sim\overset{R}{\underset{H}{N^+}}\sim\!\sim NCO \xrightarrow{H_2O} \sim\!\sim\overset{R}{\underset{H}{N^+}}\sim\!\sim NH\overset{O}{\overset{\|}{C}}NH\sim\!\sim\overset{R}{\underset{H}{N^+}}\sim\!\sim + CO_2 \tag{21}$$

$$OCN\sim\!\sim\overset{R}{\underset{H}{N^+}}\sim\!\sim\overset{R}{\underset{H}{N^+}}\sim\!\sim NCO \xrightarrow{H_2NR'NH_2} \sim\!\sim\overset{R}{\underset{H}{N^+}}\sim\!\sim NH\overset{O}{\overset{\|}{C}}NHR'NH\overset{O}{\overset{\|}{C}}NH\sim\!\sim\overset{R}{\underset{H}{N^+}}\sim\!\sim \tag{22}$$

When triethanolamine is used as a third component in the preliminary chain-extension step, at an NCO/OH ratio of 1.5, the resulting product consists of both linear and branched units containing terminal isocyanate groups. Further chain extension, as shown, results in crosslinked structures (**11**).

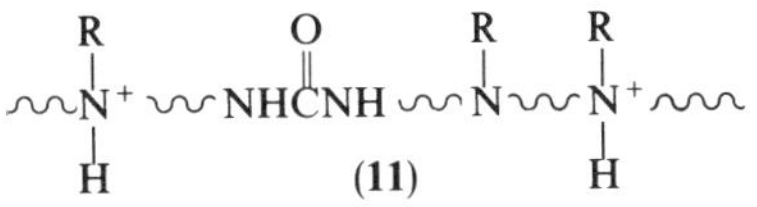

$$\sim\sim\underset{\underset{H}{|}}{\overset{\overset{R}{|}}{N^+}}\sim\sim NH\overset{O}{\overset{\|}{C}}NH\sim\sim\overset{\overset{R}{|}}{N}\sim\sim\underset{\underset{H}{|}}{\overset{\overset{R}{|}}{N^+}}\sim\sim$$
(11)

It has been assumed[46] that volatilization of acetic acid occurs during the drying cycle of the urethane latices, which then no longer contain hydrophilic amine acetate groups. Rembaum and co-workers[49–53] also reported the preparation of cationic polyurethanes and various applications of these polymers, which were termed 'ionene polymers' because they contained ionic amines.

Dieterich et al.[43] and Taft and Mohar[54] have prepared excellent reviews of urethane ionomers. One of the most important characteristics of urethane ionomers is the ease with which they form stable water-dispersions without the use of emulsifiers.[55–61] These dispersions consist of colloidal two-phase systems that can be readily prepared by adding a solution of the urethane ionomer in polar solvents, such as methyl ethyl ketone or tetrahydrofuran, to water with strong agitation, followed by removal of the solvent by distillation. When chemical crosslinking is also carried out during or after the emulsion process, precipitation of the dispersion occurs. However, in many cases, the solid particles may be redispersed by simple agitation. Dispersible polyurethane powders have thus been obtained.[62]

Dieterich and Reiff[63] have described the formation of aqueous urethane dispersions by the dispersion of ionomer melts with subsequent polycondensation in two-phase systems. The principle of this procedure consists of reacting molten ionic modified polyester or polyether prepolymers containing isocyanate groups with urea to yield bisbiurets, followed by hydroxymethylation by means of aqueous formaldehyde in a homogeneous phase, and the resulting plasticized melt of hydroxymethylated ionic urethane bisbiurets is dispersed in water at 50–130 °C. These steps can be represented schematically as shown in Scheme 6. The above-described method can be simplified by masking the diamine with a ketone to form a diketimine. Ketimines can be mixed with isocyanate prepolymers without a reaction occurring. This mixture can be mixed with water to form a dispersion; subsequently the imine hydrolyzes to form a diamine that reacts with the isocyanate prepolymer to form a chain-extended high molecular weight polyurethane dispersion. Similarly, isocyanate prepolymers can be mixed with a ketazine, aldazine or hydrazone before dispersing in water. Hydrolysis of these compounds produces hydrazine, whose chain extends the isocyanate prepolymer.

OCNNCO + HO—OH + OCNNCO + HON(|)OH + OCNNCO ⟶
(NCO/OH = 1.5)

OCN—◯⌇◯—◯—N(|)—◯—NCO $\xrightarrow{NH_2CONH_2}$

H₂NCONHCONH—◯⌇◯—◯—N(|)—◯—NHCONHCONH₂ $\xrightarrow{ClCH_2CONH_2}$

H₂NCONHCONH—◯⌇◯—◯—N⁺(|CH₂CONH₂)—◯—NHCONHCONH₂ $\xrightarrow{2-4\ mol\ CH_2O}$

Hydroxymethylated product $\xrightarrow[acidic\ pH]{H_2O\ 50-130\ °C}$ Hydrophobic polymer (high molecular weight)

—◯— = urethane group

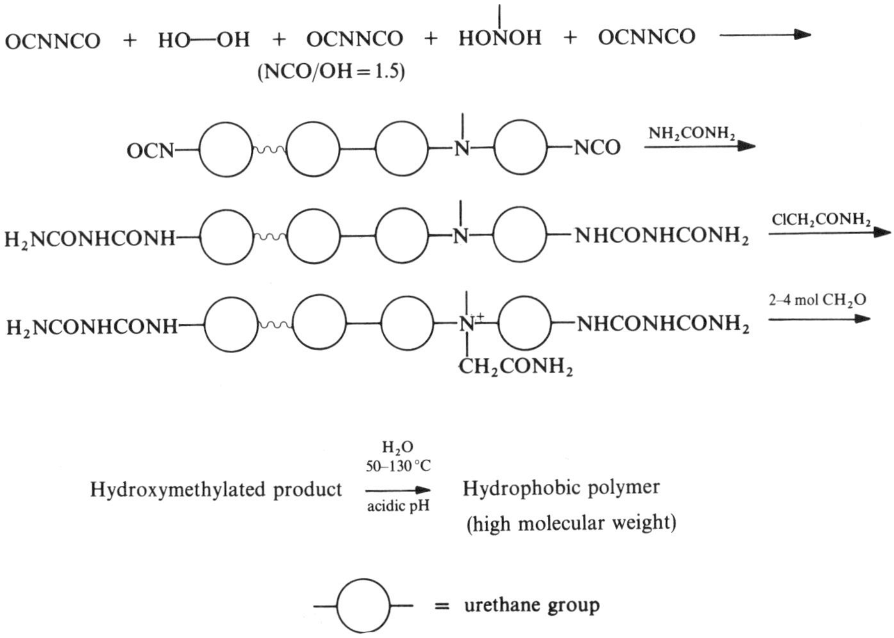

Scheme 6

Perhaps the aqueous urethane coating-systems most widely used at present are the anionic polyurethane dispersions. They are usually prepared by reaction between a difunctional polyol, an acidic solubilizing group and a diisocyanate to form a low molecular weight prepolymer. The prepolymer is neutralized with a suitable base and then chain extended with a diamine. The resulting polyurethanes consist generally of colloidal dispersions of small particle size (less than 0.1 μm). Taub[64] has reviewed in detail the various components and formulations for anionic polyurethane dispersions. Many different acidic solubilizing compounds, such as sulfonic acids and hydroxy-containing carboxylic acids, have been evaluated in anionic urethane dispersions. Di(hydroxymethyl)propionic acid finds wide application, and the level of the acid content has considerable influence on the colloidal nature of the system. Because it is desirable for the dispersions to have a relatively low viscosity, a highly polar water-soluble solvent is usually employed, such as N-methylpyrrolidone, tetrahydrofuran, methyl ethyl ketone, etc. The solvent may either be left in the final system or vacuum-stripped to yield a solvent-free product. Neutralization of carboxy-containing prepolymer is carried out with bases such as sodium hydroxide or tertiary amines. Triethylamine is usually employed for air-drying systems, while tertiary amines such as N,N-diethylethanolamine are preferred for systems that are baked after application. The amount of water used in the dispersion is very important both from a processing standpoint and for final consistency of the product and particle size. The final step in the preparation of the anionic dispersions is the chain extension of neutralized isocyanate prepolymer in water with a diamine such as ethylenediamine, hydrazine, diamino-3,5,5-trimethyl-2-cyclohexen-1-one and methylenebis (4-cyclohexylamine).[64]

An excellent review of water-borne polyurethanes has recently been published by Rosthauser and Nachtkamp.[65]

Relatively few urethane powder-coating systems have become commercially available. Urethane powder-coatings based on polyester polyols and ε-caprolactam-blocked diisocyanate have yielded good films when baked at 200 °C for 30 min.[66] Both ε-caprolactam-blocked aliphatic and aromatic isocyanates together with appropriate solid polyester systems are now available.

(vi) Radiation-cured urethane coatings

Urethane polymers containing allyl or vinyl groups can be cured by ultraviolet (UV) or electron-beam irradiation. Johnson and Labana[67] employed electron-beam irradiation to cure the reaction product of 2 mol of hydroxyethyl acrylate with TDI (equation 23).

$$2\,CH_2{=}CHCO_2CH_2CH_2OH + TDI \rightarrow (CH_2{=}CHCO_2CH_2O_2CNH)_2C_6H_3Me \quad (23)$$

Other types of curing systems for allyl-terminated urethane polymers were described by Kehr and Wszolek.[68] Polyether and polyester urethane–ene systems were prepared by reaction of polyester or polyether diols with allyl isocyanate or by reaction of polyester or polyether diols with allyl isocyanate or by reaction of isocyanate-terminated prepolymers with allyl alcohol. These urethane–ene systems can be cured by branched monomeric polythiols such as the tetrakis(thioglycolate) ester of pentaerythritol $C(CH_2OCOCH_2SH)_4$.

Another curing method employing photocuring rate-accelerators used the same mixtures of polymeric allylic diene and tetrathiol in the presence of 1 part of a phenolic antioxidant. UV radiation of about 3600 Å is optimum for curing the thiol–ene system. Thus films or coatings can be cured in a few seconds depending upon the photosensitizer and the intensity of the radiation source.[68] Photopolymers consisting of urethane–cinnamates have been described by Thomas.[69]

An excellent review of radiation curing including the use of urethane systems was presented recently by Tu.[70]

As previously described, high-temperature resistant coatings and adhesives can be prepared: by

$$OCN{\sim\!\sim}OX{\sim\!\sim}NCO \xrightarrow{\text{termination catalyst}} \left[\,{\sim\!\sim}OX{-}N\overset{\displaystyle \underset{OX}{N}}{\underset{}{\bigcirc}}N{\sim\!\sim}OX{\sim\!\sim}\,\right] \quad (24)$$

poly(oxazolidone-*co*-isocyanurate)

trimerizing the isocyanate (isocyanurate); by reacting it with an epoxide to form an oxazolidone (OX); or combinations of the two (equation 24).[71]

24.3 REFERENCES

1. A. Wurtz, *Ann. Chem. Pharm.*, 1849, **71**, 326.
2. O. Bayer, *Angew. Chem.*, 1947, **A59**, 275.
3. O. Bayer, H. Rinke, W. Siefken, L. Ortner and H. Schild, (I.G. Farbenindustrie), *Ger. Pat.* 728 981 (1942).
4. G. P. Hoff and D. B. Wicker, 'Perlon U. Polyurethanes at I.G. Farben, Boringen, Augsburg', P.B. Report 1122, Sept. 12, 1945.
5. P. W. Morgan, 'Condensation Polymers by Interfacial and Solution Methods', Wiley, New York, 1965, p. 278.
6. P. W. Morgan, *US Pat.* 3 373 139 (1968) (*Chem. Abstr.*, 1968, **68**, 96 752); *US Pat.* 3 295 122 (1968).
7. E. L. Wittbecker, W. S. Spliethoff and C. R. Stine, *J. Appl. Polym. Sci.*, 1965, **9**, 213.
8. B. F. Malichenko, O. N. Tsypina and A. E. Nesteron, *Vysokomol. Soedin., Ser. B.*, 1969, **11**, 67.
9. B. F. Malichenko, Y. V. Shelud'ke, Yu. Yu. Kercha and R. L. Savchenko, *Polym. Sci. USSR (Engl. Transl.)*, 1969, **11**, 423; *Polym. Sci. USSR (Engl. Transl.)*, 1969, **11**, 1721.
10. C. E. Carraher, Jr., *Inorg. Macromol. Rev.*, 1972, **1**, 287.
11. T. Tanaka and T. Yokoyama, in 'Polymer Applications and Technology', ed. F. Millich and C. E. Carraher, Jr., Dekker, New York, 1977, vol. II.
12. J. A. Dieter, K. C. Frisch and L. G. Wolgemuth, *J. Paint Technol.*, 1975, **47** (603), 65.
13. E. H. Burk, Jr., H. W. Kutts and L. B. Wolgenuth (Sinclair Research), *US Pat.* 3 531 425 (1970) (*Chem. Abstr.*, 1971, **71**, 31 023).
14. E. H. Burk, Jr. and D. D. Carlos (Sinclair Research), *US Pat.* 3 423 449 (1969) (*Chem. Abstr.*, 1969, **70**, 67 581); *US Pat.* 3 560 518 (1971) (*Chem. Abstr.*, 1971, **74**, 64 249).
15. S. Foti, M. Giuffrida, P. Maravigna and G. Montaudo, *J. Polym. Sci., Polym. Chem. Ed.*, 1983, **21**, 1599.
16. B. M. Culbertson, E. A. Sedor and R. C. Slagel, *Macromolecules*, 1968, **1**, 254.
17. J. H. Saunders and K. C. Frisch, 'Polyurethanes', Wiley, New York, 1962, part I, 1964, part II.
18. Z. W. Wicks, Jr., *Prog. Org. Coat.*, 1975, **3**, 73.
19. W. J. Mijs, J. B. Reesink, J. C. Groenenboom and J. P. Volmer, *J. Coat. Technol.*, 1978, **50**, 58.
20. Wyandotte Chemicals Corp., *Br. Pat.* 994 348 (1965)(*Chem. Abstr.*, 1965, **63**, 5904).
21. I. A. Altynbaer, S. F. Borisov, N. G. Rogov, E. G. Romanova and M. S. Fedoseev, *USSR Pat.* 324 250 (1971) (*Chem. Abstr.*, 1972, **77**, 35 556).
22. S. Petersen, *Justus Liebigs Ann. Chem.*, 1949, **562**, 205.
23. G. B. Guise, M. B. Jackson and J. A. MacLaren, *Aust. J. Chem.*, 1972, **25**, 2583.
24. G. B. Guise, *J. Appl. Polym. Sci.*, 1977, **21**, 3427; *Polym. News*, 1981, **7**, 149.
25. A. G. DeBoos and M. A. White, *Melliand Textilber.*, 1980, **61**, 267.
26. S. Murayama, H. Nishisawa, K. Ebisawa, T. Tanuma and S. Tanaka, *Jpn. Pat.* 75 139 829 (1976) (*Chem. Abstr.*, 1976, **84**, 61 331).
27. H. Nishisawa, S. Tanaka, and K. Hashiya, *Jpn. Pat.* 78 37 663 (1978) (*Chem. Abstr.*, 1978, **89**, 112 584).
28. A. Damusis and K. C. Frisch, in 'Film Forming Compositions', ed. R. R. Myers and J. S. Long, Dekker, New York, 1967, part I.
29. J. H. Saunders and K. C. Frisch, 'Polyurethanes', Wiley, New York, 1964, part II, chap. X.
30. J. M. Buist and H. Gudgeon, 'Advances in Polyurethane Technology', Wiley, New York, 1968.
31. W. Chang, R. Scriven, J. Pfeffer and S. Porter, *Ind. Eng. Chem. Prod. Res. Dev.*, 1973, **12** (4), 278.
32. J. P. Brennan, *Br. Pat.* 1 215 922 (1970) (*Chem. Abstr.*, 1971, **74**, 65 292).
33. Farbenfabriken Bayer, *Neth. Pat.* 7 104 911 (1971) (*Chem. Abstr.*, 1972, **77**, 49 973).
34. L. Trepasso, *US Pat.* 3 493 543 (1970) (*Chem. Abstr.*, 1970, **72**, 91 583).
35. A. S. Scheibelhoffer, *US Pat.* 3 547 127 (1971) (*Chem. Abstr.*, 1975, **83**, 180 939).
36. W. E. Jackson, *US Pat.* 3 645 907 (1972) (*Chem. Abstr.*, 1972, **77**, 6566).
37. K. Harada, Y. Mizol, J. Furukawa and S. Yamashita, *Makromol. Chem.*, 1970, **132**, 77.
38. M. Kaplan, in 'Advances in Urethane Science and Technology', ed. K. C. Frisch and S. L. Reegen, Technomic, Westport, Conn., 1971, vol. 1, chap. 8.
39. S. Altscher, *Am. Dyest. Rep.*, 1965, **54** (24), 32.
40. S. L. Axelrood, *US Pat.* 3 148 173 (1964) (*Chem. Abstr.*, 1962, **57**, 4883); *US Pat.* 3 281 397 (1966) (*Chem. Abstr.*, 1967, **66**, 19 145); *US Pat.* 3 294 724 (1966) (*Chem. Abstr.*, 1967, **66**, 46 912).
41. J. M. McClellan and I. C. MacGugan, *Rubber Age (N.Y.)*, 1967, **100** (3), 66.
42. J. M. McClellan, S. L. Axelrood and O. M. Grace, *US Pat.* 3 410 817 (1968) (*Chem. Abstr.*, 1967, **67**, 109 294).
43. D. Dieterich, W. Keberle and H. Witt, *Angew. Chem.*, 1970, **82** (2), 53.
44. D. Dieterich, W. Keberle and R. J. Wuest, *J. Oil. Colour Chem. Assoc.*, 1970, **53**, 363.
45. D. Dieterich, *Angew. Makromol. Chem.*, 1981, **98**, 133.
46. S. P. Suskind, *J. Appl. Polym. Sci.*, 1965, **9**, 2457.
47. W. Chang, R. Scriven, J. Peffer and S. Porter, *Mol. Eng. Chem. Prod. Res. Dev.*, 1973, **12** (4), 278.
48. J. E. Mallonee, *US Pat.* 2 968 575 (1961).
49. A. Rembaum, *J. Macromol. Sci., Chem.*, 1969, **3** (1), 87.
50. A. Rembaum, W. Baumgarten and A. Eisenberg, *J. Polym. Sci., Part B*, 1968, **6**, 159.
51. R. Somoano, S. P. S. Yen and A. Rembaum, *J. Polym. Sci., Part B*, 1970, **8**, 467.
52. A. Rembaum, H. Rile and R. Somoano, *J. Polym. Sci., Part B*, 1970, **8**, 457.
53. A. Rembaum, S. P. S. Yen, R. F. Lendel and M. Shen, *J. Macromol. Sci., Chem.*, 1970, **4** (3), 715.
54. D. D. Taft and A. F. Mohar, *J. Paint Technol.*, 1970, **42**, 674.
55. D. Dieterich, O. Bayer and J. Peter, *Ger. Pat.* 1 184 946 (1962) (*Chem. Abstr.*, 1965, **62**, 7984).

56. D. Dieterich and O. Bayer, *Br. Pat.* 1 078 202 (1965) (*Chem. Abstr.*, 1965, **63**, 5868).
57. W. Keberle and D. Dieterich, *Br. Pat.* 1 076 688 (1965) (*Chem. Abstr.*, 1966, **65**, 17 170).
58. W. Keberle, D. Dieterich and O. Bayer, *Ger. Pat.* 1 237 306 (1964) (*Chem. Abstr.*, 1966, **65**, 9134).
59. W. Keberle and D. Dieterich, *Br. Pat.* 1 076 909 (1966) (*Chem. Abstr.*, 1967, **66**, 95 788).
60. D. Dieterich, E. Mueller and O. Bayer, *Ger. Pat.* 1 178 586 (1962) (*Chem. Abstr.*, 1965, **62**, 6633); *Ger. Pat.* 1 179 363 (1963) (*Chem. Abstr.*, 1965, **62**, 2900).
61. W. Keberle and E. Mueller, *Br. Pat.* 1 146 890 (1969) (*Chem. Abstr.*, 1969, **70**, 107 169).
62. H. Witt and D. Dieterich, *Ger. Pat.* 1 282 962 (1966) (*Chem. Abstr.*, 1969, **70**, 29 769).
63. D. Dieterich and H. Reiff, *Angew. Makromol. Chem.*, 1972, **26**, 85; in 'Advances in Urethane Science and Technology', ed. K. C. Frisch and D. Klempner, Technomic, Westport, Conn., 1976, vol. 4, p. 112.
64. B. Taub, 'Recent Advances in Coatings', Conference, State University of New York in New Paltz and Polymer Institute, University of Detroit, Lake Mohonk, NY, 1982.
65. J. W. Rosthauser and K. Nachtkamp, in 'Advances in Urethane Science and Technology', ed. K. C. Frisch and D. Klempner, Technomic, Westport, Conn., 1987, vol. 10, p. 121.
66. R. Dhein, H. Rudolph, H. Kreuder and R. Gebauer, *Ger. Pat.* 1 957 483 (1971) (*Chem. Abstr.*, 1971, **75**, 65 379).
67. O. B. Johnson and S. S. Labana, *US Pat.* 3 660 143 (1972) (*Chem. Abstr.*, 1972, **77**, 50 280).
68. C. L. Kehr and W. R. Wzolek, *Pap. Meet. — Am. Chem. Soc., Div. Org. Coat. Plast. Chem.*, 1973, **33** (1), 295.
69. D. C. Thomas, *US Pat.* 3 655 625 (1972) (*Chem. Abstr.*, 1972, **77**, 20 881).
70. S. T. Tu, 'Recent Advances in Radiation Curing', The 1978 Modern Engineering Technology Seminar, Taiwan, 1978.
71. P. Kordomenos, J. E. Kresta and K. C. Frisch, *J. Coat. Technol.*, 1983, **55**, 49; *J. Coat. Technol.*, 1953, **55**, 59.

25
Polyureas

ANTHONY J. RYAN and JOHN L. STANFORD
University of Manchester Institute of Science and Technology, UK

25.1 INTRODUCTION	427
25.2 POLYUREAS FROM DIAMINES AND VARIOUS CARBOXY COMPOUNDS	428
25.2.1 Carbon Dioxide	428
25.2.2 Carbonates	428
25.2.3 Carboxysulfide, Phosgene or Bis(carbamoyl chlorides)	433
25.2.4 Diurethanes, Urea or Urea Derivatives	435
25.2.5 Carbon Monoxide or Silane Derivatives	435
25.3 POLYUREAS FROM POLYAMINES AND DIISOCYANATES	436
25.3.1 Kinetics of Amine–Isocyanate Reactions	437
25.3.2 Solution Polymerizations	441
25.3.3 Interfacial Polymerizations	446
25.3.4 Bulk Polymerizations by Reaction Injection Moulding (RIM)	447
25.4 REFERENCES	452

25.1 INTRODUCTION

Polyureas are polycondensation or step-wise addition products in which the urea linking group (1) occurs periodically in oligomeric or polymeric chains of otherwise different chemical structure: the chains may be linear or non-linear and comprise aliphatic, cycloaliphatic, aromatic or heteroatomic structural repeat units. Much of the early work[1] on the synthesis of polyureas was carried out by Bayer (1938–1942) in order to produce alternative fibre-forming polymers analogous to the polyamides originally patented by Carothers.[2] These linear polyureas were formed from aliphatic diamines and possessed intrinsically high melting points (240–340 °C). Consequently, the polymers were unstable at the elevated temperatures required for melt spinning, due mainly to the decomposition by residual —NH_2 groups at chain ends. Thus, only by 1960 and in one case, namely poly(nonamethyleneurea) or polyurea-9,1, was commercial production achieved.[3] A comprehensive survey of the synthesis, characterization, structure–property relations and fibre-forming technology of polyureas up to 1968 is contained in the excellent reviews of Overmars[4] and of Snider and Richardson.[5]

$$-\text{N}-\overset{\overset{\displaystyle O}{\|}}{\text{C}}-\text{N}-$$
$$\underset{\text{H}}{|}\qquad\underset{\text{H}}{|}$$
(1)

There is now a much wider range of commercial polymers containing urea groups, such as urea–formaldehyde resins,[6] poly(amide–urea)s,[7] poly(urethane–urea)s used as cast elastomers[8] and poly(urethane–urea–acrylate)s used as surface coatings. These polymers, however, are strictly speaking block copolymers containing additional periodic linking groups and as such are not included in this review. Most of the polyureas reported are formed from reactions involving diamines and carboxy-containing reactants such as carbon dioxide, metal carbonates, bis-(carbamoyl chloride)s and urethanes, ureas and thioureas. These synthesis routes are discussed in the first part of this chapter. The second part considers the syntheses of polyureas based on diamine and/or polyamine–diisocyanate reactions, which are becoming increasingly important in the production of bulk elastomers and rigid plastics, particularly those formed using the process known as reaction injection moulding (RIM).[9]

25.2 POLYUREAS FROM DIAMINES AND VARIOUS CARBOXY COMPOUNDS

Of the diamine–carboxy compound reactions referred to in the introduction, only those involving carbon dioxide, carbonates and bis(carbamoyl chloride)s are of major significance: the remainder are of minor importance and are discussed only briefly for the sake of completeness.

25.2.1 Carbon Dioxide

The earliest documented attempts[10] to synthesize polyureas were based on the facile reaction between tetramethylenediamine and carbon dioxide. Reactions were carried out for 36 h under pressure at 220 °C to yield products of low molecular weight but with melting points in excess of 320 °C. It was found later that pressures above 100 atm and temperatures up to 350 °C were more effective.[11] Thus, the ultimate degree of polymerization was shown to depend mainly on the CO_2 pressure, and to obtain polyureas of sufficiently high molecular weight for spinning fibres, pressures from 500 to 2000 atm for reactions in silver-lined autoclaves were essential.[12,13] In the reaction shown in equation (1), water is formed as a condensation by-product and has to be removed to produce high molecular weight polyurea, either by blowing off the CO_2 (which then has to be replenished) or by adsorbing the water using a drying agent contained in a vessel attached to the main reactor. Not surprisingly, this synthesis method was regarded to be of little practical importance because of the very high pressures and temperatures, and processing problems involved.

$$H_2N-R-NH_2 \; + \; CO_2 \; \rightleftharpoons \; \left(\begin{matrix} -N-R-N-\overset{O}{\underset{}{C}}- \\ \underset{H}{|} \quad \underset{H}{|} \end{matrix} \right)_n \; + \; H_2O \quad (1)$$

In more recent studies[14–20] on reactions of diamines with CO_2, however, the use of combinations of phosphorus-containing compounds and pyridine salts as promoters has resulted in lower temperatures (40 °C) and pressures (20 atm) during polyurea formation. Initially,[14,15] work focussed on the use of diphenyl phosphite in N-methylpyrrolidinone (NMP) and pyridine to form model urea compounds as well as polyureas. For example, quantitative yields of diphenylurea[15] and polyurea were obtained from reactions involving CO_2 with, respectively, aniline and methylene(bisaniline). Further studies[18] showed that the use of triphenyl phosphite (2) and catalytic amounts of pyridine hydrochloride (3), as shown in Scheme 1, resulted in polyureas having significantly higher intrinsic viscosities. The reaction in Scheme 1 may proceed via the phosphonium salt (4) followed by the formation of the carbamoyloxy-N-phosphonium salt (5) of pyridine in a manner similar to that proposed for the diphenyl phosphite-promoted reactions.[14–17] It is believed that the final urea-forming stage in Scheme 1 involves the formation of a four-centred transition state (6), as shown in Scheme 2. A number of diamines[18] were used in polycondensation reactions carried out for 4 h at 60 °C and under 50 atm of CO_2 to give the wide range of polyureas having inherent viscosities ranging from 0.27 to 2.23 dl g^{-1} shown in Table 1.

In a modification of this procedure,[21,22] ethylene chlorophosphite was used with pyridine, which acted both as a solvent and as an acid acceptor. The product having the highest molecular weight was obtained using 4,4′-diaminodiphenyl ether (DADPE) as the diamine at temperatures less than 60 °C and with CO_2 pressures between 20 and 25 atm. However, it was shown that in contrast to previous work,[14–17] the use of NMP or DMAc as solvent resulted in products of lower yields and molecular weights.[20] Such observations have been attributed[23] to interactions between the ethylene chlorophosphite and these solvents.

25.2.2 Carbonates

Original work[24] showed that polyureas can be formed from reactions, involving diamines and carbonic esters such as dibutyl or diphenyl carbonates, carried out in a solvent such as m-cresol[25] at

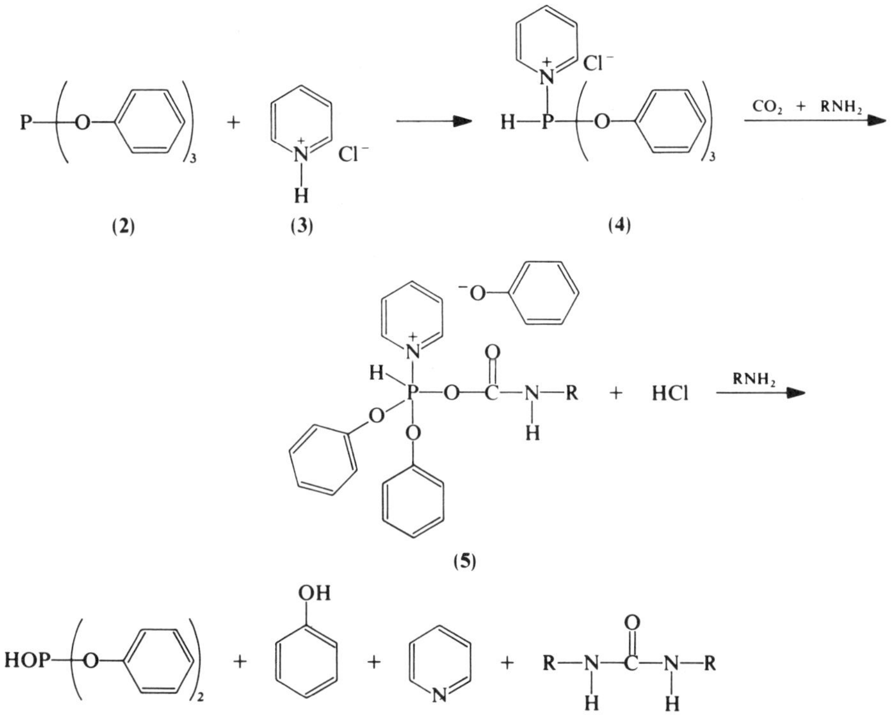

Scheme 1

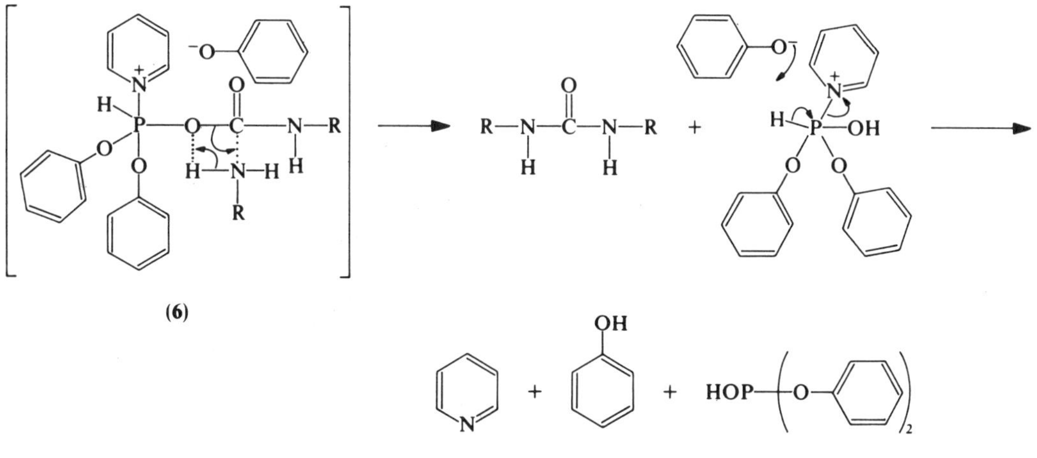

Scheme 2

255 °C for several hours. The general reaction is shown in equation (2). Stoichiometric equivalence of reactants was essential to produce high molecular weight polyureas, but in so doing the melt viscosities experienced made it impossible to remove completely the solvent and in the case of diphenyl carbonate, the phenol condensation by-product. More recently,[26] this synthesis has been improved by using the more volatile DMAc as solvent and catalyst, which allows for a lower reaction temperature (150 °C) and easier solvent removal.

$$nH_2N-R-NH_2 + nR'O-\overset{O}{\underset{}{C}}-OR' \rightleftharpoons \sim\!\!\left(\!\!N-R-N-\overset{O}{\underset{}{C}}\!\!\right)_{\!n}\!\!\sim + 2nR'OH \qquad (2)$$

In a very recent development of the carbonate route,[27,28] polyureas have been synthesized by the direct polycondensation of diamines with alkali metal carbonates. This new synthesis is an adaption of that first used to prepare polyamides and polyesters[29-31] by reactions involving the formation of an active acyl group from carboxylic acid, triphenylphosphine (Ph$_3$P) and pyridine, followed by

Table 1 Aromatic Polyureas[c] Synthesized from Various Diamines and Carbon Dioxide using Triphenyl Phosphite

Diamine	η_{inh} (dl g^{-1})			
	DMAc	Ref.	NMP	Ref.
NH_2—⟨⟩—O—⟨⟩—NH_2	1.21	a	1.10	b
NH_2—⟨⟩—CH_2—⟨⟩—NH_2	2.23	a	0.47	b
NH_2—⟨⟩—S—⟨⟩—NH_2	1.10	a	—	—
NH_2—⟨⟩·(HCl)$_2$ with NH_2	0.99	a	0.23	b
NH_2—⟨⟩(MeO)—CH_2—⟨⟩(OMe)—NH_2	0.92	a	—	—
NH_2—⟨⟩—SO_2—⟨⟩—NH_2	0.27	a	—	—

[a] F. Higashi, T. Murukami and Y. Taguchi, *J. Polym. Sci., Polym. Chem. Ed.*, 1982, **20**, 103.
[b] C. I. Chiriac, *Polym. Bull. (Berlin)*, 1986, **15**, 65.
[c] Inherent viscosities were measured using 0.5 g dl^{-1} solutions in DMAc with 5% w/w LiCl (30 °C) and NMP (20 °C).

reaction with amino or hydroxy groups to yield, respectively, amide or ester linkages. In the case of polyureas, an alkali metal carbonate, Ph$_3$P and a polyhalo compound are reacted in pyridine as shown in Scheme 3 to form the active acyl intermediate (**7**), which then undergoes aminolysis with the requisite diamine. Using DADPE as the diamine, the effects of monomer concentration, in relation to the molar ratios of Ph$_3$P and C$_2$Cl$_6$, on both polyurea yield and intrinsic viscosity have been studied.[28] In two series of experiments,[27] polyureas were synthesized in pyridine solution from DADPE using the different carbonates shown in Table 2(a), and from lithium carbonate Li$_2$CO$_3$ using the different diamines in Table 2(b), to study the effects of carbonate structure and diamine structure, respectively, on polyurea yield and molecular weight. Identical reaction conditions (10 °C for 4 h) were used with the concentrations of diamine and carbonate (0.35 mol dm^{-3}) and Ph$_3$P and C$_2$Cl$_6$ (0.70 mol dm^{-3}) constant in each case. The results are given in Table 2 and in general, these studies showed that the reaction times and temperatures required for the polyureas were greater than those used in the preparation of analogous aromatic polyamides or polyesters.[28,30,31] It is thought that the more severe reaction conditions may be required as a consequence of the lower rate of formation of the acyl intermediate in the polyurea synthesis.

In a small modification of the Li$_2$CO$_3$ route,[32,33] the range of polyhalo compounds used with Ph$_3$P has been extended to include other chlorides and bromides of general structure C$_n$X$_{2n+2}$, where X = Cl or Br and n = 1, 2 and 3. Thus, the reaction steps and conditions were essentially the same as those in Scheme 3 except that the use of C$_2$Cl$_6$ was specifically excluded. These more recent developments[27,28,32,33] in polyurea synthesis are attempts to produce highly aromatic, rigid-rod macromolecules and are a clear indication of the strong interest currently being shown in producing polymers with liquid crystalline structures.

A synthetic route using similar activated ester groups has been described[34] for the preparation of aliphatic polyureas. For example, the activated ester (**8**) of hexamethylenediamine was reacted

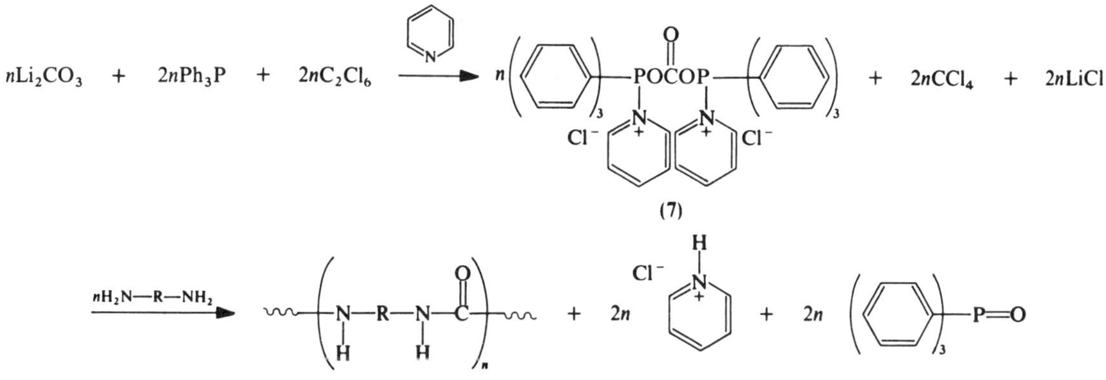

Scheme 3

Table 2 Polyureas[a] Synthesized from (a) 4,4'-Diaminodiphenyl Ether and Various Carbonates and (b) Lithium Carbonate and Various Diamines

(a) Carbonate	Yield (%)	η_{inh} (dl g^{-1})	(b) Diamine	Yield (%)	η_{inh} (dl g^{-1})
Li$_2$CO$_3$	98.0	0.23	NH$_2$—(CH$_2$)$_6$—NH$_2$	0	—
Na$_2$, K$_2$ and CaCO$_3$	0	—	NH$_2$—Ph—NH$_2$	Trace	—
(NH$_4$)$_2$CO$_3$	Trace	—	NH$_2$—PhCH$_2$Ph—NH$_2$	50.6	0.12
NaHCO$_3$	94.9	0.15	NH$_2$—PhOPh—NH$_2$	98.0	0.23
KHCO$_3$	97.5	0.23			

[a] Inherent viscosities were measured using 1.8 g dl^{-1} solutions in DMAc with 7% w/w LiCl (30 °C).

with equimolar amounts of each of the diamines, **(9)**, **(10)** or **(11)** in dipolar aprotic solvents at temperatures between 25 and 90 °C to form higher molecular weight polyureas. The polycondensation reaction shown in Scheme 4 probably involves the urethane groups of **(8)** in a combination of dissociation, to give nitrophenol and diisocyanate, and direct aminolysis to give the four-centred intermediate **(12)**.

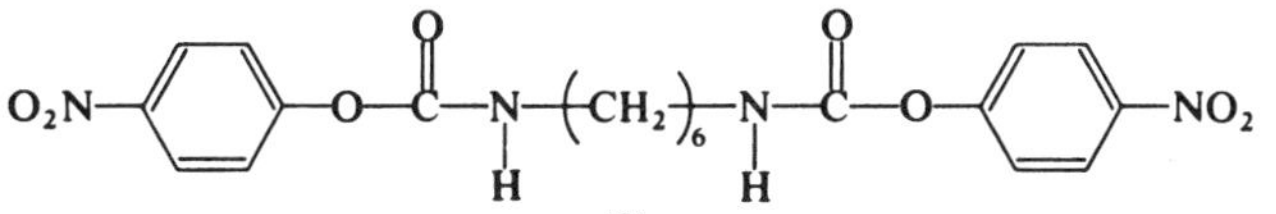

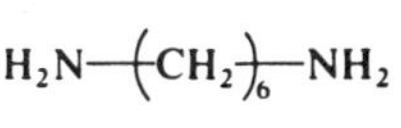

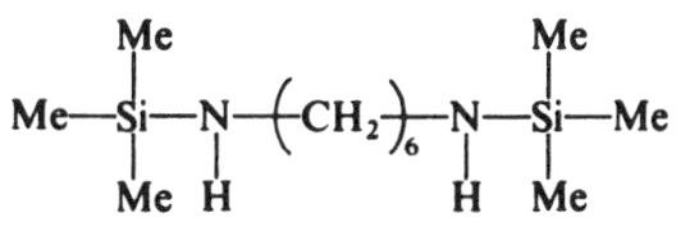

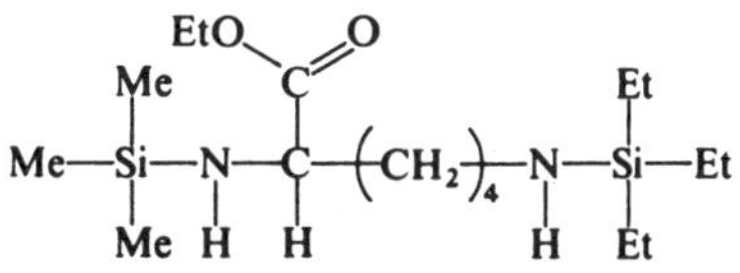

Poly(ethyleneurea) has also been synthesized[26] using the activated ester approach. The reaction which is shown in equation (3) involved *N,N'*-bisphenyloxycarbonyl-1,2-diaminoethane **(13)** and ethylenediamine and was carried out in DMAc. However, the activated ester **(13)** is less reactive than the ester **(8)** used to prepare the polyureas in Scheme 4.

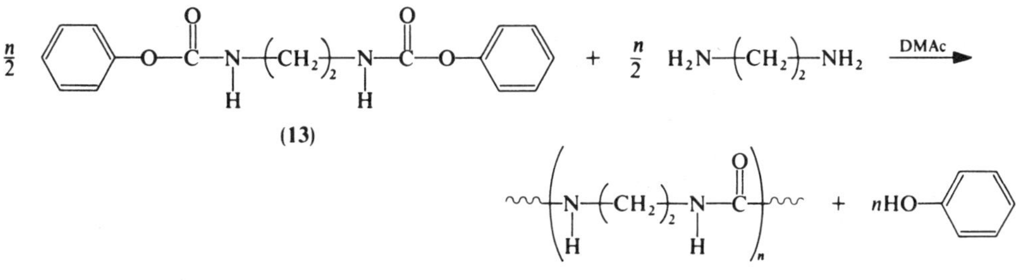

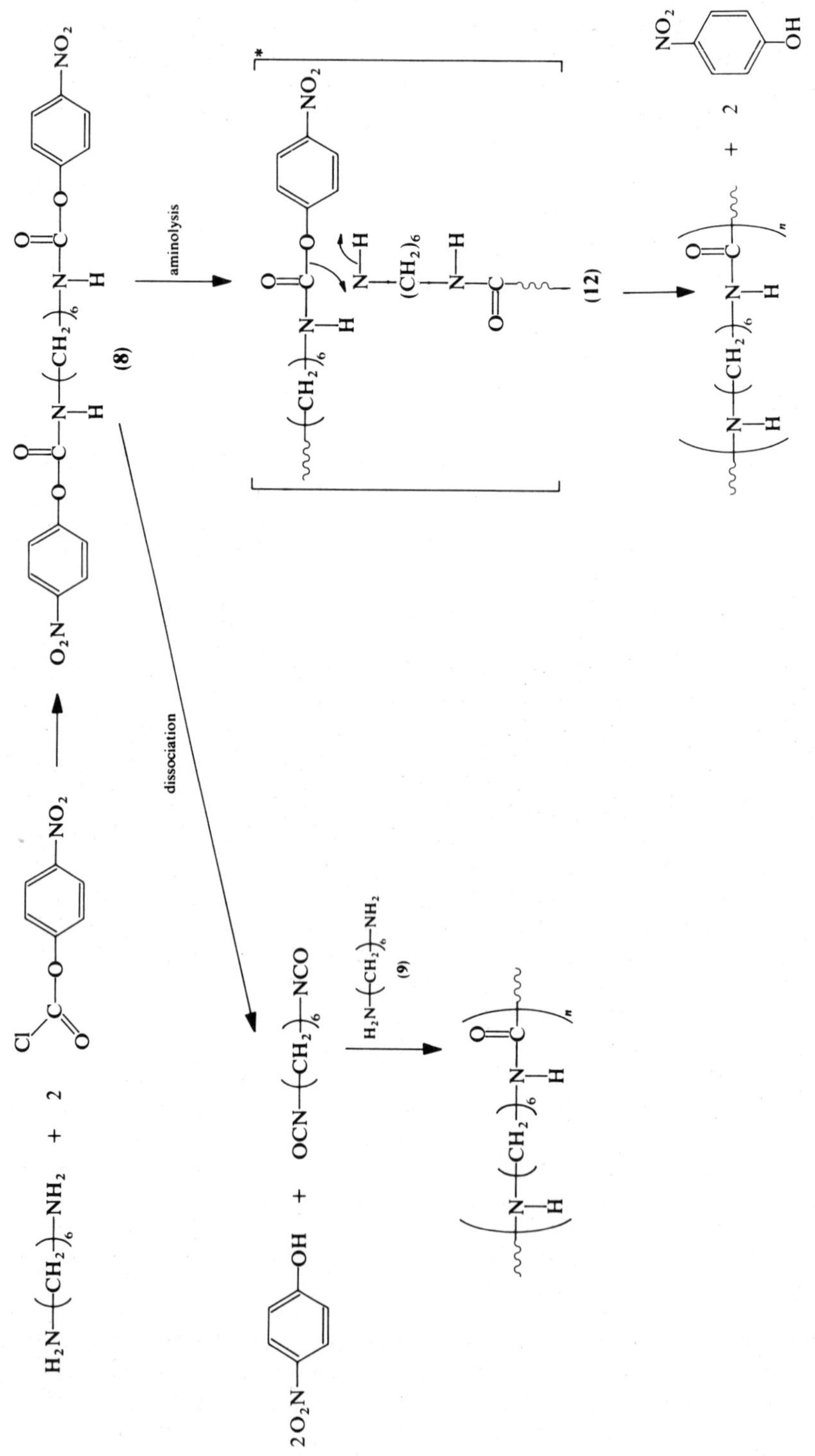

Scheme 4

25.2.3 Carboxysulfide, Phosgene or Bis(carbamoyl chlorides)

Thiocarbamate salts are formed by the reactions between COS and a diamine in solution, and due to the salt-like nature of the products, equivalency of functional groups is ensured in subsequent self-condensation reactions to form polyureas. A detailed description of the synthesis of poly(nonamethyleneurea) by this route has been reported[35] in which a two-stage preparation was used involving the formation of a sulfur-free prepolymer. In this method, hydrogen sulfide is eliminated quantitatively by the controlled heating (80 °C *in vacuo*) of the thiocarbamate salt, which is followed by further heating (180 to 200 °C) to form linear polyurea. The polymerization stage has the advantage of taking place entirely in the solid phase without the melting of any intermediate products or of the polyurea.

$$COS + H_2N-R-NH_2 \longrightarrow H_3\overset{+}{N}R-\underset{H}{N}-\overset{O}{\underset{\|}{C}}-S^- \xrightarrow{\Delta} H_2NR\left(\underset{H}{N}-\overset{O}{\underset{\|}{C}}-\underset{H}{N}-R\right)NH_2 + H_2S$$

$$\xrightarrow{\Delta} \sim\sim\left(\underset{H}{N}-\overset{O}{\underset{\|}{C}}-\underset{H}{N}-R\right)_n\sim\sim + H_2N-R-NH_2 \qquad (4)$$

The reactions of diamines with phosgene ($COCl_2$) using interfacial techniques, either in the liquid/liquid[36] or the liquid/gas phase,[37] have been used to prepare aliphatic polyureas. These polycondensations involve two immiscible solutions, namely an aqueous solution of a diamine or mixture of diamines and an organic solution of phosgene. An alkali, normally NaOH, is added to the aqueous diamine solution to act as an HCl acceptor and the general reaction is shown in equation (5). When a mixture of diamines is used, the phosgenation route results in the formation of copolymers having a random distribution of structural repeat units along the backbone, and in the particular case of benzidine and hexamethylenediamine, the effects of reaction variables on polyurea yield and molecular weight have been reported in detail.[38] A review[39] of the interfacial preparation of polyureas highlights the advantages of this route in terms of the low temperatures used, which virtually eliminate the thermal instability problems encountered in other methods. The interfacial technique has been particularly successful in the rapid preparation of high molecular weight poly(hexamethyleneurea) for use as finishing agents for wool textiles and in microencapsulation applications.

$$H_2N-R-NH_2 + Cl-\overset{O}{\underset{\|}{C}}-Cl \xrightarrow{NaOH} \sim\sim\left(\underset{H}{N}-R-\underset{H}{N}-\overset{O}{\underset{\|}{C}}\right)_n\sim\sim + NaCl \qquad (5)$$

Bis(carbamoyl chloride)s, formed from a diamine and an excess of phosgene, can also be used as precursors to polyureas. The precursors can be polymerized with a second diamine of different structure to that used for the precursor to yield alternating copolymers[40-42] according to the reactions shown in equation (6). Most of the copolyureas formed in this way are based on piperazine-1,4-di(carbamoyl chloride), which, in organic solution, is interfacially polymerized with an aqueous mixture of diamine and base at room temperature followed by heating to reflux for about one hour to complete polymerization. By this method, a series of high molecular weight copolyureas were produced in quantitative yields using stoichiometric amounts of the aliphatic diamines, $NH_2-R'-NH_2$, in which $-R'-$ was varied as shown in Table 3. These piperazine-based, alternating copolyureas of general structure (14) can be compression moulded into films or extruded as fibres and were highly crystalline with melting points much higher than the melt temperatures of corresponding random copolyureas. Attempts have also been made[43] to prepare commercial fibre-forming copolyureas using bipiperidyl bis(carbamoyl chloride)s (15), but because of their lower reactivity produced materials of lower molecular weight and crystallinity than those based on piperazine. A comparative study[44] of alternating aromatic–aliphatic, heterocyclic–aliphatic and aliphatic–aliphatic polyureas, formed using the different bis(carbamoyl chloride)s and diamines listed in Table 4, showed quite dramatic changes in materials behaviour from, respectively, stiff but brittle, crystalline glasses to soft and highly extensible elastomers.

$$H_2N-R-NH_2 \xrightarrow{xs.\ COCl_2} Cl-\overset{O}{\underset{}{C}}-\underset{H}{N}-R-\underset{H}{N}-\overset{O}{\underset{}{C}}-Cl \xrightarrow{H_2N-R-NH_2}{NaOH}$$

$$\leftaligned\sim\sim\left(\overset{O}{\underset{}{C}}-\underset{H}{N}-R-\underset{H}{N}-\overset{O}{\underset{}{C}}-\underset{H}{N}-R-\underset{H}{N}\right)\sim\sim + NaCl \qquad (6)$$

Table 3 Alternating Copolyureas (14)[a] Synthesized from Piperazine-1,4-di(carbamoyl chloride) and the Various Diamines $NH_2-R'-NH_2$

$NH_2-R'-NH_2$	η_{inh} (dl g^{-1})	T_m (°C)
$-(CH_2)_2-$	0.41	250
$-(CH_2)_3-$	0.70	285
$-(CH_2)_6-$	0.79	230
$-(CH_2)_7-$	0.53	220
$-(CH_2)_9-$	0.87	215
$-(CH_2)_{10}-$	1.45	215
$-(CH_2)_3-O-(CH_2)_3-$	0.54	200

[a] Inherent viscosities were measured using 0.2 g dl^{-1} solutions in p-chlorophenol (25 °C).

(14)   (15)

Table 4 Alternating Copolyureas Synthesized from Various Aromatic, Heterocyclic and Aliphatic Bis(carbamoyl chloride)s and Aliphatic Diamines

Bis(carbamoyl chloride)	Diamine	η_{inh}[a] (dl g^{-1})	T_m (°C)	E (GPa)	σ_u (MPa)	ε_u (%)
Piperazine-1,4-di(carbamoyl chloride)	$NH_2-(CH_2)_{10}-NH_2$	1.45	215	1.45	46	57
N,N'-diphenyl-p-phenylene bis(carbamoyl chloride)	$NH_2-(CH_2)_{10}-NH_2$	0.57	150	1.78	38	8
N,N'-diphenyl-p-phenylene bis(carbamoyl chloride)	$NH_2-(CH_2)_6-NH_2$	0.53	170	1.71	55	5
N,N'-dimethyl-hexamethylene bis(carbamoyl chloride)	$NH_2-(CH_2)_6-NH_2$	0.49	—	0.02	12	670

[a] 0.2 g dl^{-1} solutions in p-chlorophenol at 25 °C. E, σ_u and ε_u are tensile modulus, strength and elongation.

25.2.4 Diurethanes, Urea or Urea Derivatives

The reactions of diamines with diurethanes (equation 7),[45] have been carried out between 200 and 250 °C in bulk or in the presence of a solvent such as *m*-cresol and required an oxygen-free atmosphere. For these reasons, this route to polyureas is no longer of much interest.

$$n\text{R'O-CO-NH-R-NH-CO-OR'} + n\text{H}_2\text{N-R''-NH}_2 \longrightarrow \left(\text{-CO-NH-R-NH-CO-NH-R''-NH-}\right)_n + 2n\text{R'OH} \quad (7)$$

However, the reaction between a diamine and urea has received more attention, being one of the first methods[46] for preparing polyureas, and the fundamental aspects of the reaction have already been reviewed.[4] Heating stoichiometric amounts of diamine and urea in the absence of air produces a polyurea with the evolution of ammonia (NH_3), as shown by equation (8). The best results are achieved if the reaction is carried out in two stages. Firstly, the ω-aminoalkylurea is formed (equation 9) by heating to 130 °C and maintaining the temperature until the evolution of NH_3 ceases. Secondly, the temperature is raised above the melting point of the final polymer and vacuum applied to remove the final traces of NH_3 in order to yield high molecular weight polyurea (equation 10). The mechanism of the first stage has been studied[47] by measuring the formation of NH_3 and was observed to be first order in urea with an activation energy of 25–30 kcal mol^{-1} (1 kcal = 4.18 kJ). The rate-determining step was shown to be the dissociation of urea to isocyanic acid and NH_3 (equation 11). In a similar kinetics study[48] on the formation of the polyurea from dodecamethylene-diurea and diamine, the rate of NH_3 evolution was observed to be slower compared with that in the reaction involving urea. The preparations of polyureas using the diurea were carried out in solution using either *m*-cresol or water. In the latter case, palmitic acid was used as a stabilizer to convert unreacted —NH_2 end groups to more stable amide groups, which enhanced the oxidative degradation resistance of the products and enabled the melt-spinning of polyurea fibres.[48]

$$n\text{H}_2\text{N-R-NH}_2 + n\text{H}_2\text{N-CO-NH}_2 \longrightarrow \left(\text{-NH-CO-NH-R-}\right)_n + 2n\text{NH}_3\uparrow \quad (8)$$

$$\text{H}_2\text{N-R-NH}_2 + \text{H}_2\text{N-CO-NH}_2 \xrightleftharpoons{130\,°C} \text{H}_2\text{N-R-NH-CO-NH}_2 + \text{NH}_3\uparrow \quad (9)$$

$$n\text{H}_2\text{N-R-NH-CO-NH}_2 \xrightleftharpoons{180\text{ to }200\,°C} \left(\text{-NH-CO-NH-R-}\right)_n + n\text{NH}_3\uparrow \quad (10)$$

$$\text{H}_2\text{N-CO-NH}_2 \rightleftharpoons \text{H-N=C=O} + \text{NH}_3\uparrow \quad (11)$$

25.2.5 Carbon Monoxide or Silane Derivatives

The reactions of diamines with carbon monoxide (CO) have been shown[49] to yield polyureas when mercury(II) oxide or mercury(II) salts such as mercury(II) acetate, propionate or sulfate are used as catalysts in solution. The reactions were carried out at moderate temperatures (25–75 °C) but require CO pressures of up to 70 atm to form addition products, which were then decomposed by further heating (150–300 °C) to give polyureas. This route has been significantly improved recently[50] using selenium compounds as catalysts in solutions of DMAc containing substituted pyridines. Much lower temperatures (30 °C) and pressures (1 atm) for 2–3 h were required to produce high molecular weight polyureas.

The use of silane derivatives with hexamethylenediamine or diurea provides an interesting method for the synthesis[51] of poly(hexamethyleneurea). The two routes to the polyurea are shown in Scheme 5 and involve either the reaction of the diamine (16) with trimethylisocyanatosilane (17) or the reaction of the diurea (18) with hexamethyldisilazane (19). Both reactions are carried out in the bulk and proceed *via* the disilylated diurea intermediate (20) with the evolution of NH_3. Polyureas were obtained in good yield from both routes but no other characterization data were reported.

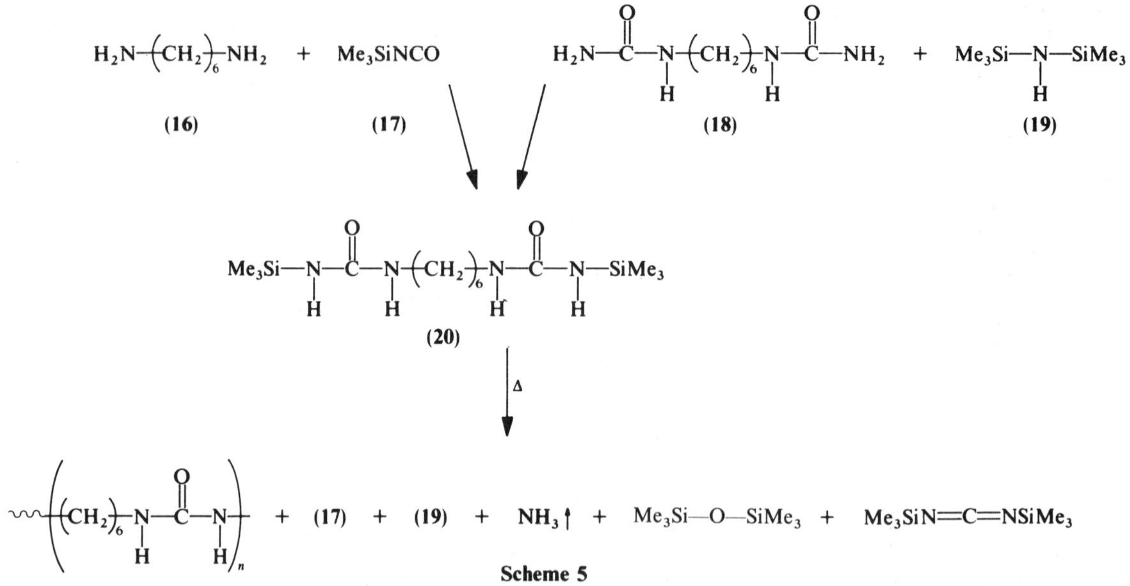

Scheme 5

25.3 POLYUREAS FROM POLYAMINES AND DIISOCYANATES

Technically, the most important method used for polyurea formation is based on the reactions between polyamines and polyisocyanates. One immediate advantage of these polyaddition reactions is that polyureas are formed without the evolution of condensation by-products. In the simplest case involving difunctional reactants as shown in equation (12), linear homopolyureas are obtained when R = R', and copolyureas when R ≠ R': the use of higher functionality reactants produces correspondingly crosslinked polyureas. Of the many diisocyanate compounds available, the four which appear to have received most attention are hexamethylene diisocyanate (HDI; 21), toluene diisocyanate (TDI), used as a mixture of 2,4- and 2,6-isomers (22, 23), 4,4'-methylenediphenylene diisocyanate (MDI; 24) and naphthalene diisocyanate (NDI; 25). The synthesis and characterization of these diisocyanates (and others) are well documented.[52]

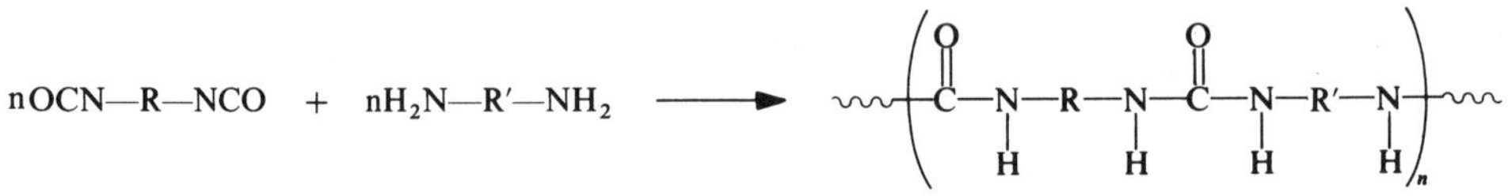

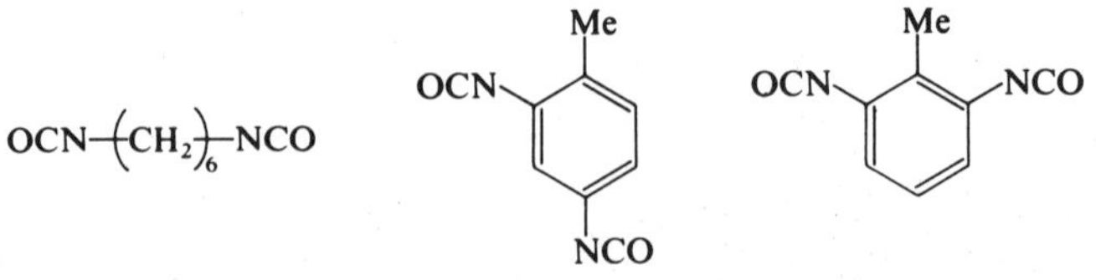

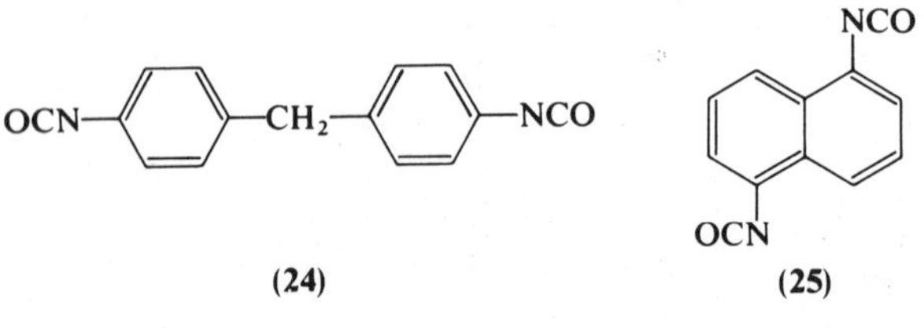

25.3.1 Kinetics of Amine–Isocyanate Reactions

A knowledge of the kinetics and mechanisms involved in the synthesis of polyureas is essential, particularly for the bulk polymerization of copolyureas from mixtures of amine-functionalized reactants having —NH_2 groups with significantly different rates of reaction toward —NCO groups. The effects of relative reactivities on the structure, morphology and properties of copolyureas formed using the bulk synthesis process, reaction injection moulding (RIM), are discussed in a later section. In this section, the kinetics and mechanisms of the reactions between amines and isocyanates are reviewed.

Independent studies[53,55,56] of model urea-forming systems led to the proposal of two different mechanisms. In the first,[53] the reactions of phenyl, o-tolyl and p-tolyl isocyanates (26)–(28) with aniline (29), o-toluidine (30), p-toluidine (31) and o-chloroaniline (32) in dioxane solution were studied. The relative reactivities of these systems followed the classical electronic trend with electron-donating groups that increase nucleophilicity or base strength increasing the reactivity of the amine (and the converse for the isocyanate). Substituents in the *ortho* position, particularly on the isocyanate, cause steric hindrance and reduce reactivity, and whereas bases such as pyridine and other tertiary amines showed little catalytic action, ureas were found to accelerate the reaction. Furthermore, the initial rate was found to be more dependent on the concentration of the amine than on the isocyanate and the order with respect to amine was close to two.

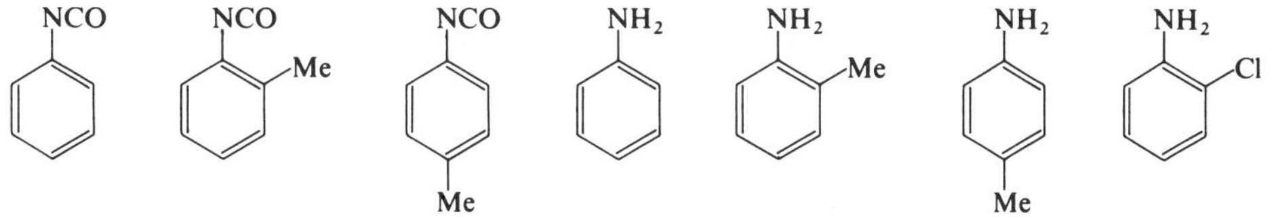

(26) (27) (28) (29) (30) (31) (32)

On the basis of these observations, a mechanism involving the formation of a single complex was proposed (Scheme 6). For the *initial* stages of the reaction, the rate, r_0, is given by equation (13) which, with respect to amine concentration, reduces to second order when $k_3[ArNH_2] \ll k_2$, or first order when $k_3[ArNH_2] \gg k_2$. As the reaction proceeds, the concentration of urea, [ArNHCONHAr], increases, whilst [$ArNH_2$] decreases, and the product-catalyzed reaction becomes increasingly dominant. Kinetics data[53] were found to be in accord with this scheme and led to the suggestion that complex formation was fast and that the product-induced reaction, represented by the term k'_3[ArNHCONHAr], occurred with greater efficiency than the amine-induced reaction, that is $k'_3/k_3 > 1$.

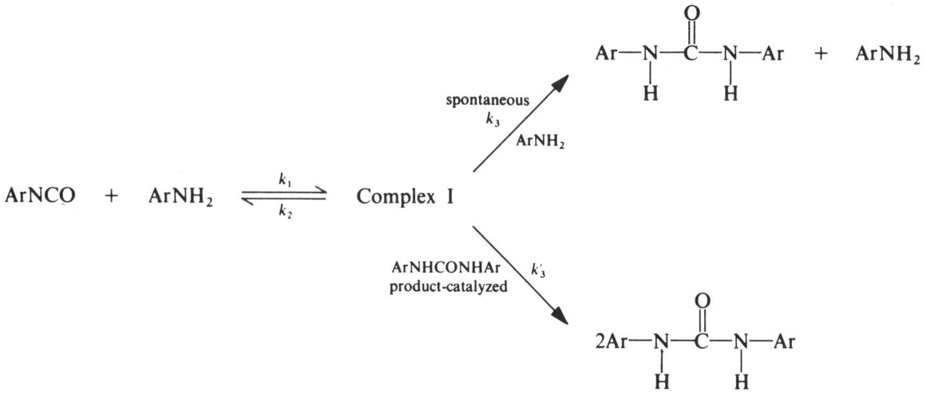

Scheme 6

$$r_0 = \frac{k_1 k_3 [ArNCO][ArNH_2]^2}{k_2 + k_3[ArNH_2]} \tag{13}$$

A somewhat different mechanism was proposed on the basis of independent studies[54–56] on reactions between phenyl isocyanate (26) and ethyl p-aminobenzoate (33), cyclohexyl isocyanate (34) and aniline (29), phenyl isocyanate and aniline, and p-methoxyphenyl isocyanate (35) and aniline in benzene solutions. The mechanism comprised two complexes,[55] both involving the isocyanate: one with the amine (complex II) and the other with the product urea (complex III). Thus in Scheme 7, the truely spontaneous reaction involves complex II, whereas the product-catalyzed reaction involves complex III. Assuming stationary state conditions for the formation and decomposition of the two complexes[55] leads to equations (14) and (15) for k_s and k_c, the second-order rate constants for the spontaneous and product-catalyzed reactions. Constants k_s and k_c are related to the overall second-order rate constant, k, according to equation (16).

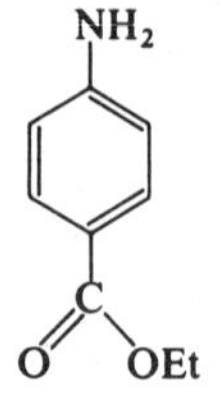

(33) 4-aminobenzoic acid ethyl ester

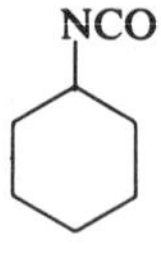

(34) cyclohexyl isocyanate

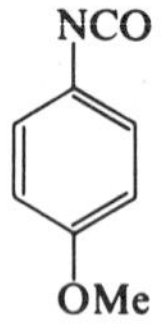

(35) 4-methoxyphenyl isocyanate

$$\text{ArNCO} + \text{ArNH}_2 \underset{k_5}{\overset{k_4}{\rightleftharpoons}} \text{complex II} \xrightarrow[\text{ArNH}_2]{k_6} \text{Ar-NH-CO-NH-Ar}$$

$$\text{ArNCO} + \text{Ar-NH-CO-NH-Ar} \underset{k_8}{\overset{k_7}{\rightleftharpoons}} \text{Complex III} \xrightarrow[\text{ArNH}_2]{k_9} 2\,\text{Ar-NH-CO-NH-Ar}$$

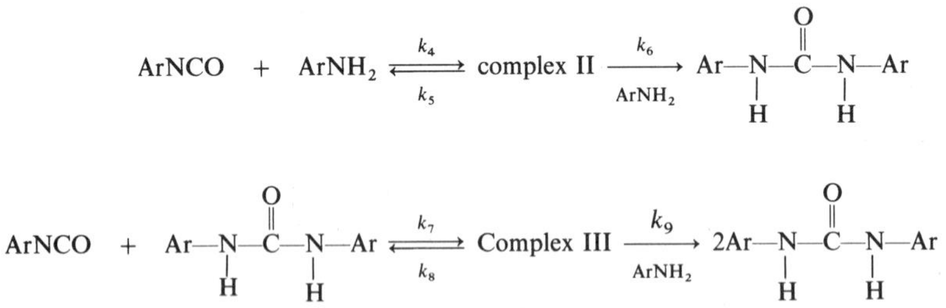

Scheme 7

$$k_s = \frac{k_4 k_6 [\text{ArNH}_2]}{(k_5 + k_6)[\text{ArNH}_2]} \quad (14)$$

$$k_c = \frac{k_7 k_9}{k_8 + k_9 [\text{ArNH}_2]} \quad (15)$$

$$k = k_s + k_c [\text{ArNHCONHAr}] \quad (16)$$

In the presence of an added catalyst[56] such as a tertiary base (:B), a base-catalyzed reaction between amine and isocyanate is superimposed on the previous reactions, to give the overall reaction scheme shown in Scheme 8. Initially, reaction involves the isocyanate–base complex, complex IV, and the amine, and the overall second-order rate constant is k'_c. In addition, reaction between isocyanate, added base and product urea yields a ternary complex, complex V, which on subsequent reaction with amine produces two urea molecules and regenerates the base: the overall second-order rate constant for these reaction steps is k''_c. Thus, the complete base-catalyzed reaction comprises contributions from spontaneous, product-catalyzed, added base-catalyzed and ternary complex-forming reactions for which the combined rate constant, k_e, is given by equation (17). The reactions of amines and isocyanates studied in benzene were further complicated because the product urea had only limited solubility in benzene[54] and precipitated out of solution. However, in most cases,[55,56] complete second-order behaviour was observed after the reaction medium became saturated with urea product and [ArNHCONHAr] became constant.

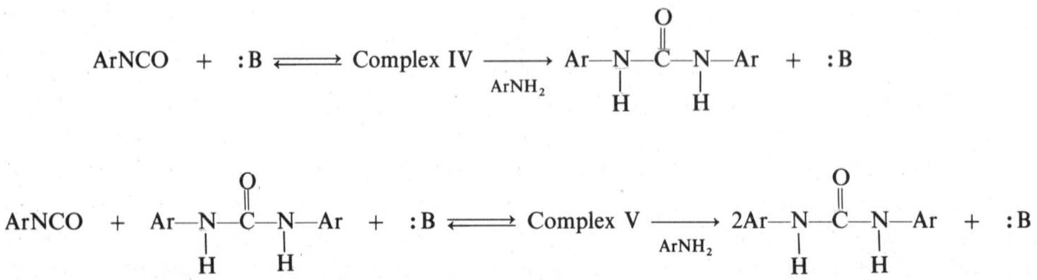

$$\text{ArNCO} + :\text{B} \rightleftharpoons \text{Complex IV} \xrightarrow[\text{ArNH}_2]{} \text{Ar-NH-CO-NH-Ar} + :\text{B}$$

$$\text{ArNCO} + \text{Ar-NH-CO-NH-Ar} + :\text{B} \rightleftharpoons \text{Complex V} \xrightarrow[\text{ArNH}_2]{} 2\,\text{Ar-NH-CO-NH-Ar} + :\text{B}$$

Scheme 8

$$k_e = k_s + k_c[\text{ArNHCONHAr}] + k'_c[:\text{B}] + k''_c[:\text{B}][\text{ArNHCONHAr}] \qquad (17)$$

The kinetics of the reaction between o-cyanoaniline (36) and phenyl isocyanate (26), derived from titrimetric data,[57] were used as a model for the polymerization of 3,3'-dicyano-4,4'-oxydianiline (37) with MDI (24) as shown in Scheme 9. The model reaction in either sulfolane or NMP was observed to be second order overall and first order with respect to amine up to a high degree of conversion. The activation energy, E_a, of the spontaneous reaction was low (39.5 kJ mol^{-1}) and decreases in the presence of added catalyst, triethylamine. The rate of reaction, however, increased and the reaction was first order with respect to [catalyst]. For both the model reactions and the polymerization, the rates of spontaneous reactions were reported to be low ($\sim 13 \times 10^4$ l mol^{-1} s^{-1}). These observations indicate that product catalysis is not an important feature in these reactions. Thus the electron-withdrawing nature and steric hindrance of the —C≡N groups reduce the reactivity of the amine functional groups, which results in slow reactions with isocyanate groups yielding low molecular weight polyureas. Changing the position of the nitrile group from the amine to the isocyanate molecule increases the rate of urea formation. The rate of reaction of o-cyanophenyl isocyanate (38) with aniline (29) in Scheme 9 is 20 times faster than that of phenyl isocyanate (26) with o-cyanoaniline (36). An increase in reaction rate is also observed for bifunctional compounds and leads to the formation of polyureas with much higher molecular weights.

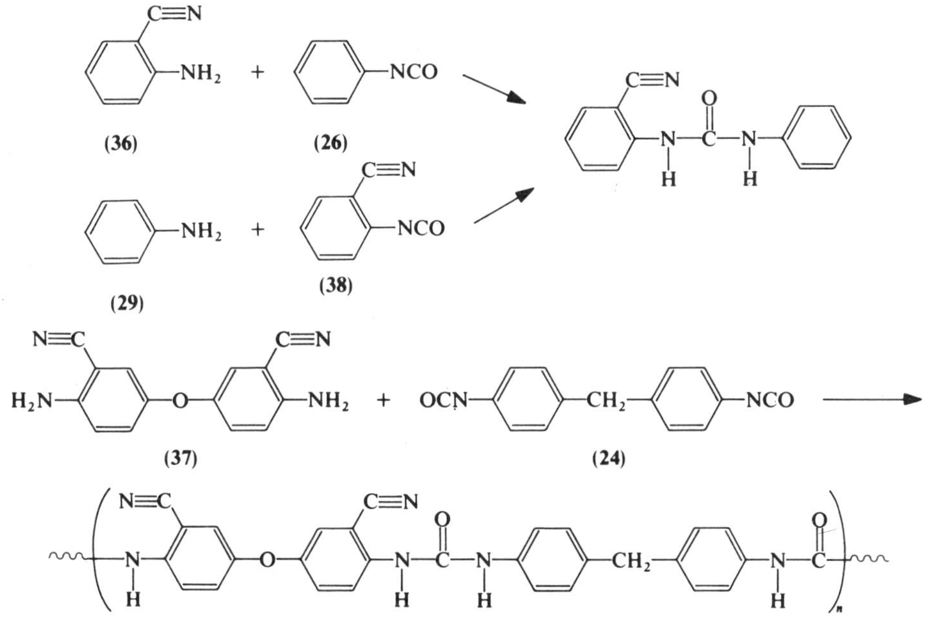

Scheme 9

In a more recent study,[58] the adiabatic temperature rise technique was used to investigate the reaction kinetics of amines and isocyanates. The fractional conversion of functional groups was related to the adiabatic temperature rise, ΔT_{ad}, for a series of model reactions in DMAc solution. The reactions of aliphatic and aromatic amines with phenyl isocyanate (26) were investigated and the reaction times for 25, 50 and 75% conversion of —NCO groups are reported in Table 5. The importance of comparing reaction times at increasing levels of conversion in order to assess relative reactivities was demonstrated by the ratio of reaction times for n-butanol and o-toluidine, which decreased from 5 to 2 at 25% at 75% conversions, respectively. In contrast, the relative ratios for aniline and o-toluidine remained essentially constant over the same conversion range. Numerical analysis of the ΔT_{ad} vs. time data using an Arrhenius expression gave different activation energies, E_a, and reaction orders for n-butanol and o-toluidine reactions, whereas those for aniline and o-toluidine were similar with only values of the pre-exponential factor showing any significant differences.

The data in Table 5 show quite clearly the steric hindrance effect on reactivity of the methyl group in o-toluidine as shown by comparing the reaction times of o-toluidine and aniline. The competition between steric and electronic effects is indicated by the greater initial reactivity of diethyltoluenediamine, DETDA (39) and (40), compared with o-toluidine (30) and shows that in this case, the inductive effect dominates. However, the influence of steric hindrance on reaction rate is readily seen by comparing pure 2,6-DETDA (40) with the less reactive isomeric mixture of DETDA. In the

Table 5 Conversion Times and Activation Energies for Adiabatic Reactions[a] Involving Phenyl Isocyanate (0.5 M in DMAc) and Various Reactants

Reactant	Time (s) for —NCO Conversions (%)			Activation energy (kJ mol^{-1})
	25	50	75	
Aniline	< 3.5	10.0	40.0	17.2
o-Toluidine	19.0	59.0	271	16.8
n-Butanol	92.0	242	510	61.1
DETDA (2,6)	11.0	56.0	470	—
DETDA (2,4/2,6)	13.0	63.0	490	—
M-600	—	$\sim 2 \times 10^{-3}$	—	—

[a] Initial reaction temperature, $T_0 = 23\,°C$.

former, each —NH$_2$ group is sterically hindered by Me and Et substituents, whereas 40% of —NH$_2$ groups in the mixture are more sterically hindered by the two Et substituents. Also apparent is the difference in reactivities between unreacted and monoreacted DETDA which reach 25% conversion before the less sterically hindered o-toluidine, but require 1.75 times longer to reach 75% conversion: this was ascribed to the urea formed from the reaction of the first amine on DETDA withdrawing electrons from the ring, thereby deactivating the second amine. The initial —NH$_2$ groups are, however, *meta* to each other and any deactivation occurs in the *ortho* and *para* positions. In addition, the possibility of an inductive effect occurring across three phenyl C atoms needs to be considered, although its magnitude may not be sufficient to account for the high reactivity ratio of 3.2 ± 0.8 determined[58] from product analysis. Reduced reactivity also arises from steric hindrance of the second —NH$_2$ group due to the bulkiness of the phenylurea substituent on the ring, which reduces the number and efficiency of subsequent reactant encounters of the monoreacted DETDA molecule.

(39)   (40)

As discussed earlier, previous studies on the mechanisms of model amine–isocyanate reactions showed the overall order of reactions to be between two[55–57] and three,[53] and first order in isocyanate.[53–57] The reaction order of the phenyl isocyanate–aniline system was determined[58] using ΔT_{ad} data from reactions conducted using different molar ratios. A value of 2.05 ± 0.19 with respect to amine was determined from a double logarithmic plot of reaction rate, r_a, at 50% conversion, divided by initial [PhNCO] vs. initial [PhNH$_2$], assuming the isocyanate reaction to be first order. With an overall order of three based on an Arrhenius kinetics model, as represented by equation (18), numerical fitting of reaction rate data with fixed values of $n = 1$ and $m = 2$, gave calculated values 542.6 (l mol^{-1})^2 s^{-1} and 17.2 kJ mol^{-1}, respectively, for the pre-exponential factor, A, and activation energy, E_a. Using m in equation (18) as a regression variable, values of 654.6 (l mol^{-1})$^{2.02}$ s^{-1} and 17.6 kJ mol^{-1} were obtained for A and E_a, and the value of m was 2.02, in good agreement with the third-order model. The high order and low activation energy were thus shown to account for the high initial reactivity of this system. However, attempts to correlate ΔT_{ad} data for o-toluidine and DETDA reactions with the third-order kinetics model proved unsuccessful.

$$r_a = A e^{(-E_a/RT)}[\text{—NCO}]^n[\text{—NH}_2]^m \tag{18}$$

The significantly higher reactivity toward isocyanate of aliphatic compared with aromatic amine is clearly demonstrated by the result shown in Table 5 for M-600, a monofunctional, poly-(oxypropylene) amine. Limited ΔT_{ad} data were obtained using a continuous-flow[58] apparatus and the exotherms were simulated using both second- and third-order models. However, the shortest time after mixing within which a temperature measurement could be made was 1.92×10^{-3} s, and the lowest conversion was 57%, even at the highest flow rate. Simulation of the data was therefore stated to be uncertain because of the variability in measurements. Although the Arrhenius model does not take into account the changing concentration and efficient autocatalysis of the product

25.3.2 Solution Polymerizations

The formation of polyureas in solution using the amine–isocyanate route is long established.[1] Indeed, in the formation of linear aliphatic polyureas, the use of solvents is essential if branching and eventual crosslinking by biuret side reactions[59] are to be avoided. Aromatic polyurea formation is less prone to crosslinking because of steric hindrance of active chain atoms and functional groups afforded by the benzene rings. Dipolar, aprotic solvents such as tetramethylene sulfone (TMS), DMSO, DMAc, DMF and NMP are good solvents not only for the reactants but also for the polyureas formed, so that premature precipitation of low molecular weight products during polymerization is prevented, which would otherwise cause an imbalance of the functional group stoichiometry.

The solution polymerization of aromatic diamines and diisocyanates has been successfully employed[60] to produce the series of high melting point linear polyureas (41)–(44) shown in Table 6. Stoichiometric amounts of reactants were polymerized at temperatures below 100 °C using some of the non-corrosive, spinning solvents listed above. The polyurea fibres obtained had excellent properties (Table 6) with high values of tensile modulus (E) and strength (σ_u), together with reasonable levels of toughness as reflected by the values of ultimate elongation (ε_u). As expected, replacing the aromatic MDI in (43) by its hydrogenated version, HMDI, in (44) resulted in fibres of lower strength and stiffness. In a more recent study,[61] a series of linear aliphatic–aromatic and heterocyclic–aromatic polyureas were formed in high yields from, respectively, hexamethylenediamine with either MDI or TDI, and piperazine with HDI. These polyureas, (45)–(47) in Table 7, make an interesting comparison with the previous, wholly-aromatic polyureas since the former, as expected, have lower melting points, whereas their molecular weights overall are significantly higher. The higher molecular weights probably result from the use of DMAc containing polyacrylonitrile as the solvent[61] during polyurea formation. Also apparent from Table 7 is the dramatic effect on polyurea structure caused by using the asymmetric TDI (22) rather than MDI (24), seen by comparing the properties of the highly crystalline polyurea (45) with the amorphous polyurea (46).

A wide range of homopolyureas and alternating and random copolyureas, synthesized by the amine–isocyanate solution route, have been characterized by NMR spectroscopy.[62] Both ^{13}C NMR and natural abundance ^{15}N NMR spectra were obtained from trifluoroacetic acid solutions. In contrast to analogous polyamides, the carbonyl signals of polyureas are insensitive to neighbouring

Table 6 Linear Aromatic, Alternating Copolyureas Synthesized from Diamines and Diisocyanates in Solution

Polyurea	η_{inh}^a (dl g^{-1})	T_m (°C)	E (GPa)	σ_u (GPa)	ε_u (%)
(41)	0.41	321	4.12	0.32	25
(42)	0.63	307	3.92	0.31	37
(43)	0.76	—	5.09	0.35	18
(44)	0.63	—	3.29	0.26	32

^{a} DMF containing 5% w/w LiCl (temperature not reported).

Table 7 Linear Aromatic– and Heterocyclic–Aliphatic Copolyureas Synthesized from Diamines and Diisocyanates in DMAc/Acrylonitrile Solution

Polyurea	Yield (%)	η_{inh}^a (dl g^{-1})	T_g (°C)	T_m (°C)
(45)	94	0.68	200	266
(46)	97	1.34	210	Amorphous
(47)	96	1.40	200	270

^{a} 0.2 g dl^{-1} solutions in 95% H$_2$SO$_4$ at 25 °C.

group effects. The ^{15}N NMR signals of the urea groups, however, are sensitive to substituents and, from structure–shift relationships derived from spectra, information on the sequence structure in the polyureas was obtained.

There are severe problems associated with the processing of wholly aromatic polyureas in terms of their limited solubility and high melting points. In an attempt to overcome these problems, N-substituted aromatic diamines were used[63] with MDI, TDI and m-xylene α,α'-diisocyanate in 1,1,2,2-tetrachloroethane solution to form polyureas in which H bonding between urea groups was eliminated. Thus, by using the N-phenylated diamines (48) and (49) completely amorphous and readily soluble polyureas were obtained, which had significantly lower glass transition temperatures (T_g) compared with corresponding unsubstituted polyureas, although data for the latter were not reported. Nevertheless, the data shown in Table 8[63] for the two substituted polyureas (50) and (51) can be compared with those in Table 6 to show significant physical property differences, with the former being more flexible and having much lower strengths and elongations.

Another use of substituted phenylenediamines with diisocyanates has been reported[64] in the solution polymerization of phosphorus-containing polyureas. The substituted phenylenediamines (52) and (53) in which the R group was either Et or CH$_2$CH$_2$Cl, together with the unsubstituted

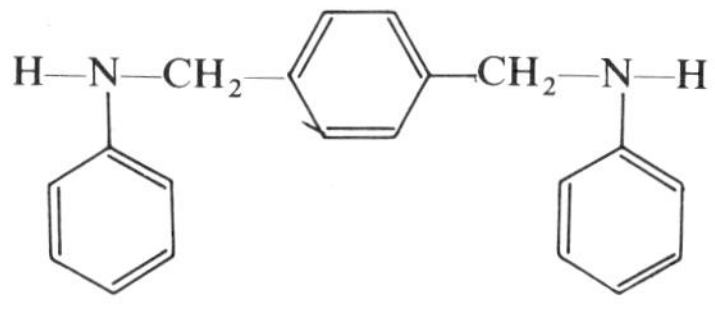

(48) 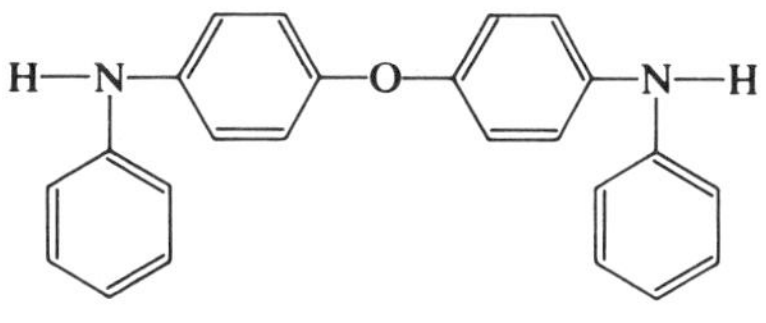   **(49)**

Table 8 Substituted Aromatic Polyureas Synthesized from the N-Substituted Diamine (48) and either MDI or TDI in 1,1′,2,2′-Tetrachloroethane Solution

Polyurea	η_{inh}^a (dl g^{-1})	T_g (°C)	E (GPa)	σ_u (MPa)	ε_u (%)
(50)	0.54	122	2.0	84	7
(51)	0.30	117	2.0	48	3

^{a} 0.5 g dl^{-1} solutions in DMAc at 30 °C.

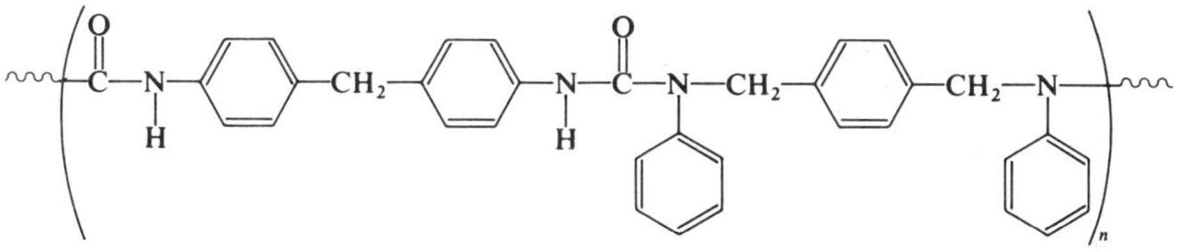

(50)

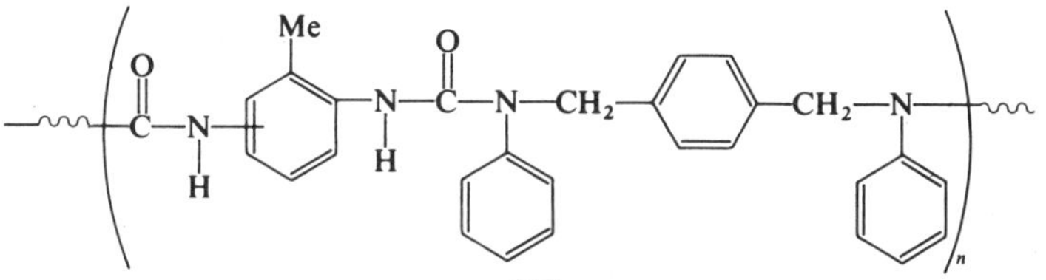

(51)

m-phenylenediamine (54), were reacted with different diisocyanates, MDI, TDI and HDI, to produce a comparative series of polyureas. In terms of reactivity towards the diisocyanates, the substituted diamines were less reactive because of the electron-withdrawing nature of the phosphenyl groups. Furthermore, despite the incorporation of phosphorus residues on the chains, the substituted polyureas were shown to have inferior thermal stability but higher char yields compared with unsubstituted polyurea.

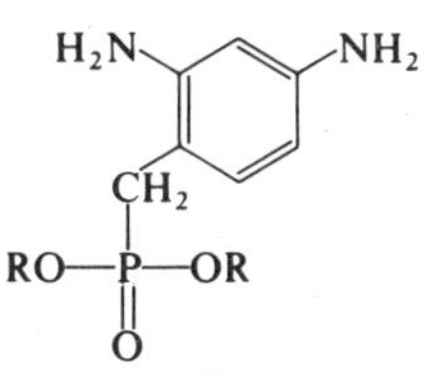

(52) 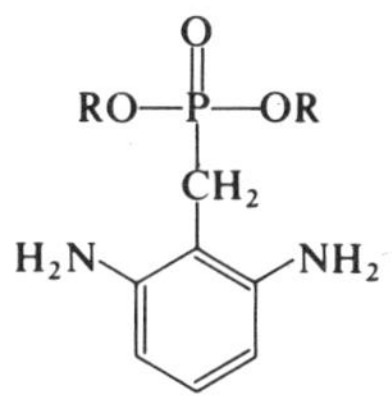   **(53)** 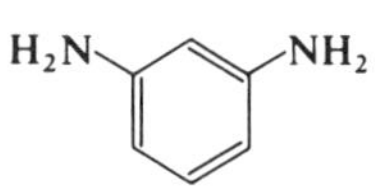   **(54)**

In a more recent development,[65] sulfur-containing polyureas have been produced which exhibit much improved thermal stability. A wide range of polyureas, all based on thiophene 2,5-diisocyanate (55), were synthesized using the various aliphatic and aromatic diamines (56) to (60). The polyureas were characterized using thermogravimetry (TG) to give the weight loss data in Table 9. It is interesting to note that in the aliphatic materials containing methylene sequences $+CH_2+_n$, thermal stability appears to depend more on whether n is even or odd than on its magnitude: increasing n from 2 to 6 causes only a small increase, whereas changing n from 2 to 3 produces a reduction in weight loss which is much bigger. As expected, using aromatic diamines results in polyureas with much enhanced thermal stability. Overall, the impressively high char yields (500 °C) suggest that these polyureas could be of potential use in ablative applications.

Table 9 Sulfur-containing Polyureas Synthesized from Thiophene 2,5-diisocyanate (55) and the Various Diamines (56)–(60)

Diamine	Yield (%)	$\eta_{inh}{}^a$ (dl g^{-1})	Temperature (°C)		500 °C residual weight (%)
			5% weight loss	10% weight loss	
(56)	99	0.52	202	215	38.6
(57)	99	0.49	196	205	39.2
(58)	98	0.62	205	220	42.5
(59)	98	0.43	245	300	36.4
(60)	98	0.68	245	320	46.3

^{a} 0.5 g dl^{-1} solutions in DMSO at 30 °C.
^{b} From TG data conducted in N_2 at 10 °C per minute.

An interesting class of polyureas containing sugar residues has recently been reported.[66] The diamine, 2,6-diamino-2,6-dideoxy-D-glucose (61) was reacted with different diisocyanates, OCN—R—NCO, in DMSO to produce linear polyureas as suggested by equation (19). There is a clear difference in reactivities of the —NH$_2$ and —OH functional groups on (61) toward —NCO groups, and the difference was shown to be unusually temperature dependent. When polyurea synthesis was carried out at 25 °C, the selectivity between —NH$_2$ and —OH groups was reduced and branching and gelation occurred. However, at 5 °C, essentially linear polyureas were obtained in which the almost complete absence of urethane linkages was confirmed by IR spectroscopy.

The amine–isocyanate solution polymerization route has also been used for the synthesis of siloxane–urea copolymers.[67] Specially synthesized, oligomeric α,ω-bis(aminopropyl)poly(dimethylsiloxane)s (PSX; 62) with molecular weights 1000 to 4000, were reacted with various diisocyanates (63) as shown in Scheme 10. Reactions were carried out in either THF or bis(2-ethoxyethyl) ether at room temperature, although heating at 50–60 °C for several hours was required to develop high molecular weights. In some cases, diamine chain extenders, NH$_2$—R′—NH$_2$ (64) were included in polymerizations which required the use of NMP as cosolvent. Attempts to produce copolyureas with high levels of chain extender and MDI proved unsuccessful, however, due to the premature phase separation of oligomeric hard segments. The general structure of the siloxane–urea copolymers produced is represented by (65) in which A and B represent hard and soft segments formed, respectively, from reactants (**64**) and (**62**). (The special case of $A = 0$ represents the polyurea formed without chain extender.) Although the siloxane amine–isocyanate reaction is slow relative to corresponding reactions involving alkane- or alkoxy-amines, the siloxane–urea copolymers were obtained in high yields and were readily soluble in various common organic solvents. The polyureas based on HMDI and PSX with molecular weight 1150, were shown to have the highest strength and toughness compared with corresponding polyureas based on MDI or TDI. Even without chain extenders, the thermoplastic copolyureas possessed good elastomeric properties, which infers that some phase separation between the phenyl- or cyclohexyl-urea and PDMS moieties must have occurred.

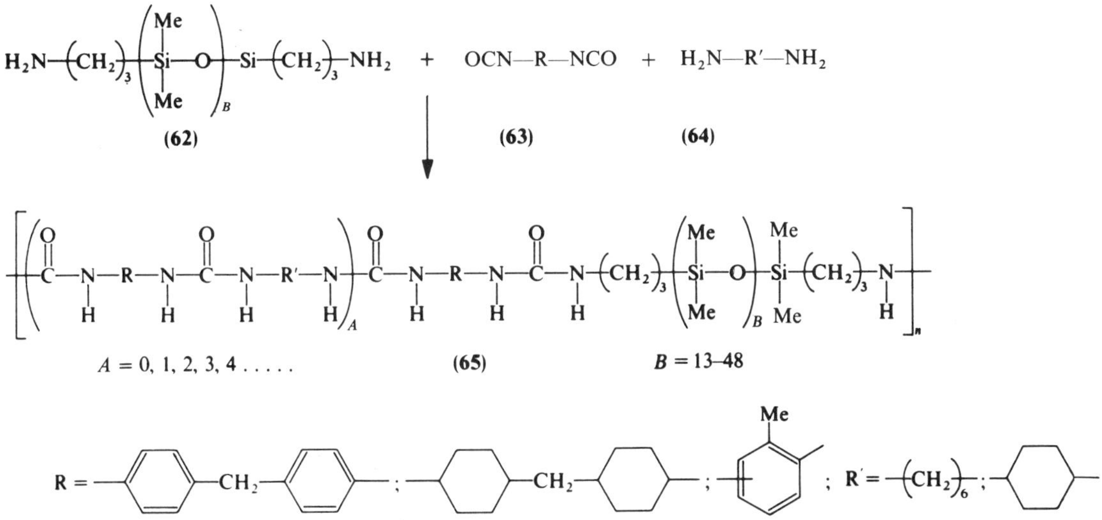

Scheme 10

As a special case of the amine–isocyanate reaction, novel reactions between tertiary alcohols and isocyanates have been investigated[68,69] for the synthesis of polyureas and elastomeric block copolymers. The reaction route involves several steps in which amine groups are generated *in situ* and react with free diisocyanate to form polyurea in the final step. The synthesis is based on a previous observation[70] that heating tertiary alcohols with phenyl isocyanate yields a carbanilide and the corresponding alkene. Scheme 11 represents the reaction route to polyurea in terms of a simplified monofunctional scheme in which cumyl alcohol (**66**) and phenyl isocyanate (**26**) are used as the initial reactants. The urethane intermediate (**67**) formed dissociates at elevated temperature yielding the alkene, α-methylstyrene (**68**) and the amine, which in this case is aniline (**29**). Finally, the reaction between the amine and phenyl isocyanate produces the carbanilide, N,N'-diphenylurea (**69**).

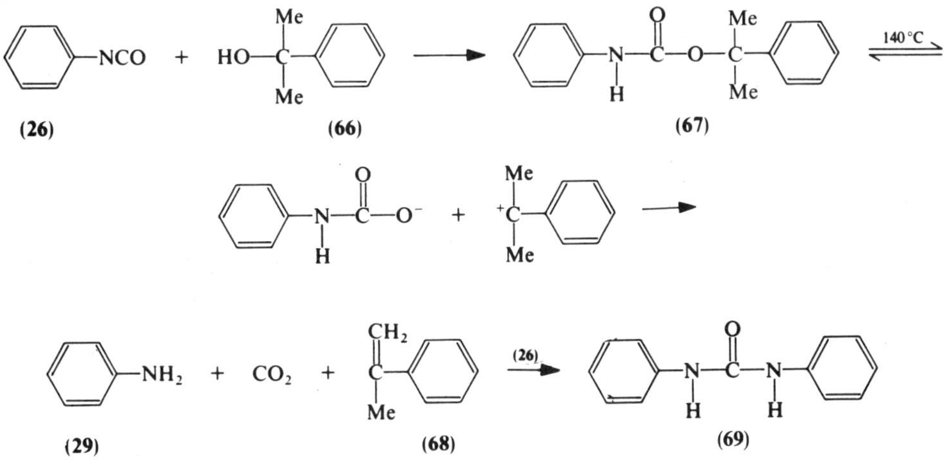

Scheme 11

In the same study,[68,69] the formation of polyureas starting from silyl diols and diisocyanates was also investigated. The synthesis is similar to that shown in Scheme 11 in that amine groups are generated *in situ* and react with free diisocyanate in the final step to form polyurea. However, the route to amine groups is different, as shown in Scheme 12, which again represents the synthesis of polyurea in terms of simple monofunctional species. Heating the diol (**70**) with a catalyst produces the siloxane (**71**) and liberates water. The water, being in a molecularly dispersed form in the reaction medium, is therefore more miscible with the isocyanate, thereby facilitating the formation of the carbamic acid intermediate (**72**). Dissociation of (**72**) yields the amine which as before reacts with isocyanate to form the urea product (**69**).

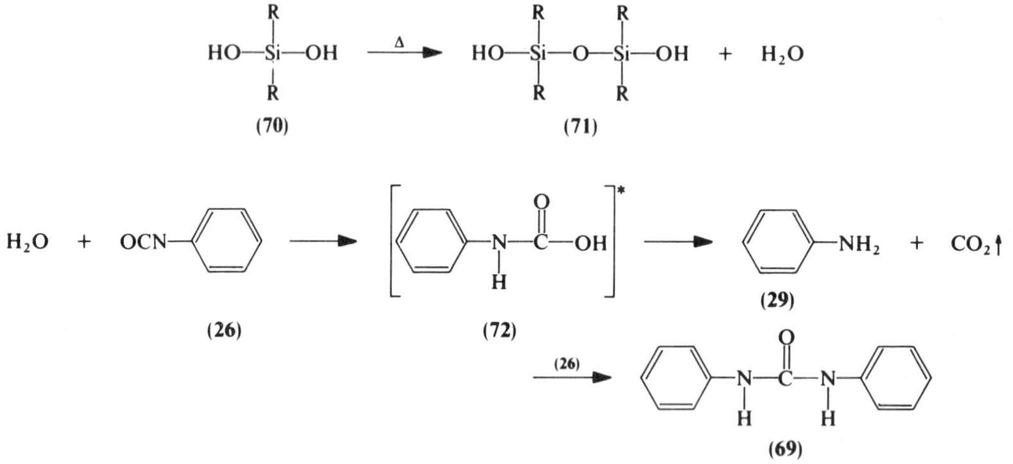

Scheme 12

25.3.3 Interfacial Polymerizations

Developments in this form of polymerization have been reviewed previously[39] and more detailed experimental information is contained in some of the earlier patent literature.[41,42,44,71] The main advantage of interfacial polymerization is that it offers a straightforward route to alternating copolyureas which are completely linear. In terms of the amine–isocyanate route, there is the additional advantage that troublesome by-products are not formed which would otherwise remain in the system as is the case in the direct phosgenation and bis(carbamoyl chloride) interfacial routes discussed previously in Section 25.2.3. In interfacial amine–isocyanate reactions, polyureas are formed at the interface of an aqueous solution of diamine and an organic solution of diisocyanate. Mechanistically, it has been shown[72] that polymerization occurs on the organic side of the interface. For example, in the formation of poly(hexamethyleneurea) from a benzene solution of HDI and an aqueous solution of HDA, polymerization occurs when diamine diffuses into the organic phase and reacts with HDI, and not *vice versa*. Indeed, no hydrolysis or polymer formation was observed even after several hours, when the HDI solution was left in contact with water (that is the aqueous phase without diamine present).

In a recent application of this method, optically active polyureas were synthesized[73] using a catalyst-modified version of a route reported previously.[71] The polyureas were synthesized from L-lysine diisocyanate methyl ester (**73**) and various racemic and other optically inactive diamines. Initially the conditions for interfacial polymerization were studied using HDI or MDI and 1,2-diaminoethane. In contrast to previous observations,[72] problems were encountered with side reactions in the aqueous phase involving diisocyanate which resulted in the formation of polyurea and CO_2. It was observed that large amounts of side reaction occurred when the diisocyanate was present in stoichiometric excess over the diamine. However, the more reactive MDI (compared with HDI) polymerizes with diamine without any side reaction, provided that within the two phases, the diamine concentration exceeds that of the diisocyanate: the optimum monomer ratio of $NH_2:NCO$ was shown to be 2:1. Using these interfacial reaction conditions, a series of polyureas based on (**73**) were synthesized using α,ω-diaminoalkanes containing methylene sequences $-(CH_2)_n-$, in which n was increased from 2 to 9. Polyurea yield increased with the value of n, leading to the conclusion that the longest chain diamine was the most reactive. This presumably is due to the increased hydrophobicity of the amine in the aqueous phase as the hydrocarbon chain length increases. However, the inductive effect of $\text{-(CH}_2\text{)}_n\text{-}$ causing an increase in the electronegativity of the amino groups as n increases appears to have been overlooked. In the particular case of the racemic diamine 1,2-diaminopropane (**74**) reaction by asymmetric selective polyaddition did not occur in the desired manner and the polyurea obtained had optical properties similar to those of the polyurea formed using the optically inactive 1,2-diaminoethane.

$$\text{OCN—}\overset{*}{\text{C}}\text{H-(CH}_2\text{)}_4\text{-NCO} \qquad \text{H}_2\text{NCH}_2\text{—}\overset{*}{\text{C}}\text{H—Me}$$
$$\underset{\text{OMe}}{\underset{|}{\overset{|}{\text{C}}=\text{O}}} \qquad\qquad\qquad \underset{}{\overset{|}{\text{NH}_2}}$$

(73) (74)

In an interesting development of the interfacial process involving liquid and gas phases, polyurea materials were produced[74] in bead form. Polyurea colloids were formed as spherical particles with a narrow size distribution by spraying aerosol droplets of liquid diisocynate into ethylenediamine vapour. Particle sizes were varied by changing the temperature of the aerosol generator and by the flow rate of the carrier gas used with the diamine. Mixed polyurea–metal oxide particles were also produced by exposing a polymer colloid to alkoxide vapours with subsequent hydrolysis in a humid atmosphere. In this way, metal oxides were incorporated within the polyurea beads rather than forming a surface layer.

25.3.4 Bulk Polymerizations by Reaction Injection Moulding (RIM)

The process known as reaction injection moulding (RIM) has been used in the formation of a wide variety of block copolymers,[75, 79] including polyurethanes, poly(urethane–urea)s and, most recently, polyureas. In effect, RIM provides a bulk synthesis route in which polymerization and moulding (of large and complex components) occur simultaneously. In the process, low viscosity monomer or oligomer reactants are molecularly combined by impingement mixing at high pressures (100–200 bar; 1 bar = 10^5 Pa) immediately prior to laminar flow under low pressure (~ 10 bar) into a mould. A schematic diagram of the RIM process is shown in Figure 1. The flow rate ratio of the two reactant streams, A and B, must be carefully controlled to ensure that the correct stoichiometry of the reactants is achieved and maintained during mould filling. The key element in the process is the small volume (~ 10 cm^3) mixing chamber in the mix-head in which the reactant streams impinge at high velocity (~ 200 m s^{-1}) and polymerization commences and continues during mould filling: complete polymerization is often effected by postcuring.

The RIM process was originally developed for the production of large, block copolyurethane mouldings. The first chemical formulations comprised a liquid polyisocyanate (based essentially on MDI, **24**) as one reactant stream, and a mixture of a high molecular weight polyol (usually

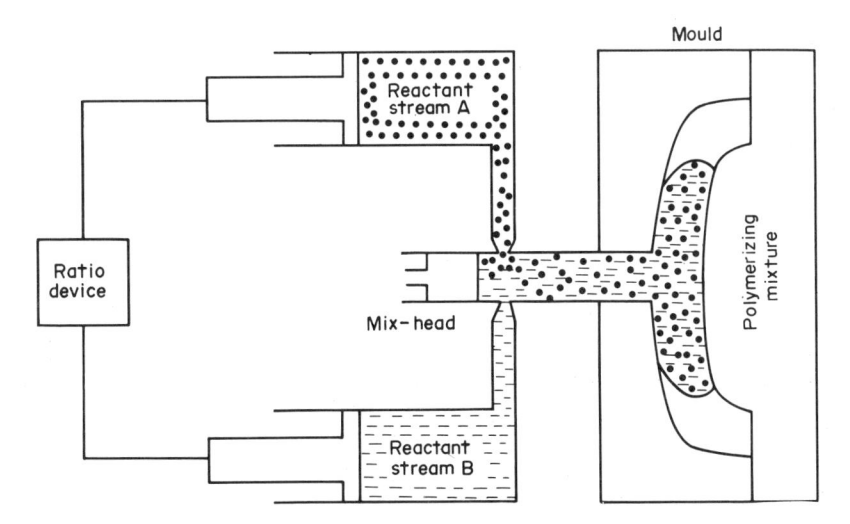

Figure 1 Schematic representation of the reaction injection moulding (RIM) process used for the bulk synthesis of copolyureas

poly(oxypropylene), tipped with poly(oxyethylene) units), a diol chain extender and catalysts as the second reactant stream. The copolymers formed,[77-79] therefore, contained only urethane linking groups. Subsequently, chemical formulations were developed, in which the diol chain extender was replaced by an aromatic diamine (usually an isomeric mixture of DETDA **39** and **40**) to form block copoly(urethane–urea)s. In this case, the copolymers contain both urethane and urea linking groups.[80-85] As before, the polymerizations require catalysts, with the isocyanate–polyol reactions being preferentially catalyzed[82,84,85] because of the much lower intrinsic reactivity of —OH compared with —NH_2 functional groups. Overall, the copoly(urethane–urea) formulations are much faster than those of corresponding copolyurethanes under similar processing conditions.

The most recent developments in RIM have been in the bulk synthesis of block copolyureas. Formulations for these comprise the liquid, MDI-based polyisocyanate reacting with a mixture of a high molecular weight, amine-functionalized polyether (**75**)[86,87] and an aromatic diamine such as DETDA as shown in Scheme 13. The copolymers formed contain only urea linking groups and the amine–isocyanate reactions are so fast compared with the previous two cases that under equivalent polymer-forming conditions, the addition of catalysts is unnecessary. The polymerization rates are such that the RIM process provides the only bulk synthesis route to these block copolyureas. The polyureas thus formed have the $-(AB)_n-$ type of block copolymer structure (**76**) shown in Scheme 13, in which the A and B blocks comprise aromatic urea, oligomeric hard segments (HS) and urea-linked, polyether soft segments (SS). Most of the initial work reported on novel RIM copolyureas is contained in commercially based literature[88-91] with little detailed information on the chemistry and structure involved in these block copolymers.

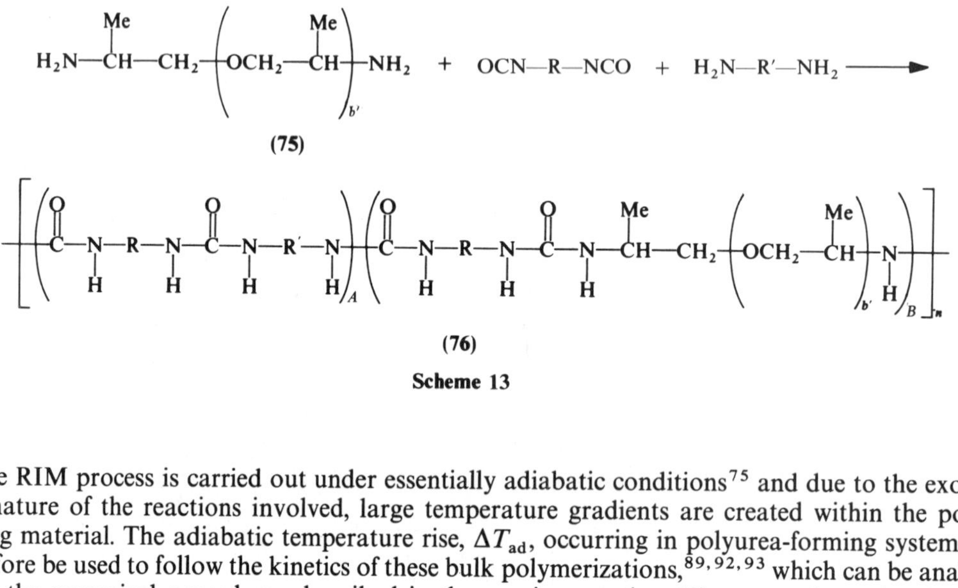

Scheme 13

The RIM process is carried out under essentially adiabatic conditions[75] and due to the exothermic nature of the reactions involved, large temperature gradients are created within the polymerizing material. The adiabatic temperature rise, ΔT_{ad}, occurring in polyurea-forming systems can therefore be used to follow the kinetics of these bulk polymerizations,[89,92,93] which can be analyzed using the numerical procedures described in the previous section. The rates of reaction between aliphatic amines (on polyether prepolymers) and aromatic isocyanates are extremely high. For example,[89] during a RIM process using stoichiometric amounts of MDI and a poly(oxypropylene) (POP) triamine of molecular weight 5000, the liquid mixture was extruded from the mixing head as a rubbery solid. Thus, in less than 0.1 s, at least 70% conversion of functional groups (~ 0.5 mol l^{-1} overall concentration) had occurred, even without added catalyst and at an initial temperature of $\sim 30\,°C$. In a separate study[94] involving the same polyether triamine and MDI in very dilute solution (~ 0.05 M in toluene or 90% by weight of solvent), gelation was observed to occur within the time required for mixing (<1 s), whereas in previous studies[95] on polyurethane network formation in such high dilutions, gelation was difficult to observe and took several weeks.

A number of attempts have been made to quantify the reactivity differences between aromatic and aliphatic amines, which are fundamental to the final structure and properties of RIM copolyureas. The bulk kinetics of a crosslinking copolyurea system have been investigated[89] using a RIM formulation comprising a polyether triamine (5000 molecular weight), DETDA and a uretonimine-modified diisocyanate, u-MDI. Independent ΔT_{ad} measurements were made on polymerizing mixtures in either the mould runner close to the outlet of the mix-head or in a separate adiabatic reactor. A second-order kinetics model based on the simple equation (20) was used to predict the temperature rise profile which was shown to be in good agreement with the experimental ΔT_{ad} data.

$$\text{RNCO} + \text{R'NH}_2 \xrightarrow{k} \text{R}-\underset{H}{\overset{}{N}}-\overset{O}{\underset{}{\overset{\|}{C}}}-\underset{H}{\overset{}{N}}-\text{R'} \qquad (20)$$

In a similar kinetics study[93] of the synthesis of a linear copolyurea based on a polyether diamine (2000 molecular weight), DETDA and u-MDI, separate temperature measurements were made using two thermocouples to give ΔT_{ad} data at 0.025 and 0.403 s after impingment mixing. The almost instantaneous temperature rise recorded was associated with the polyether diamine–u-MDI reaction, an observation which is substantiated by the reaction half-time of 2×10^{-3} s reported in Table 5. A more detailed study of the slower u-MDI–DETDA reaction, in isolation, was carried out in a 'solution' of non-reactive polyether. The ΔT_{ad} data indicated that not all of the functional groups had reacted, which was attributed to premature phase separation and precipitation of the aromatic polyurea from the polyether. Experimental data were compared with temperature rise profiles predicted from the Arrhenius-based kinetics model discussed in the previous section, with a first-order reaction in isocyanate and second-order reaction in amine. Good agreement between the model and experiment was obtained for chemical conversions (of functional groups) in the range 35 to 70%: below 35%, consistent data were difficult to obtain due to the high rate of temperature rise. Above 70% conversion, the reaction rate continued to decrease, which was explained in terms of possible differences in reactivity of the two amine groups on DETDA. However, the effects causing a reduction in reaction rate due to the large increase in viscosity and the vitrification of the polyurea HS associated with phase separation, also need to be considered. In summary, both kinetics studies[89,93] of the bulk synthesis of copolyureas showed good correlations between experiment and predicted behaviour, based respectively on second-order[89] and third-order[93] kinetics models.

In recent studies,[9,96] comparative series of non-linear copolyureas were synthesized in which the chemical structure of the diamine chain extender was varied. The copolyureas were based on reactants (**24**) and (**77**) to (**80**), and were produced by the RIM process[77] using a stoichiometric ratio of isocyanate to amine groups of 1.03. Hard segment contents, defined as the mass of diamine and the stoichiometric equivalent of isocyanate relative to the total reactant mass, were varied from 35 to 70% by increasing the relative proportion of aromatic diamine to polyether triamine in the formulations. Initial reactant and mould temperatures were 35 and 117 °C, respectively, and mould filling was achieved within 1 s. Gelation and vitrification times of < 2 s were typical for the systems, which again exemplify the extremely fast rates of polymerization encountered in bulk copolyurea synthesis.

(**77**) (**24**) (**78**)

(**79**) (**80**)

The properties of the copolyureas are dominated by the essentially two-phase structure comprising urea end-linked poly(oxypropylene) soft segments and amorphous, rigid and extensively hydrogen-bonded hard segments. The much faster rate of reaction of the aliphatic polyether amine groups compared with aromatic amine groups (see Table 5), leads to the rapid formation of isocyanate-tipped polyether oligomers. Competitive consecutive reactions between the diamine chain extender and the oligomers and free isocyanate then occur, leading to gross gelation/vitrification and ultimate formation of a heterogeneous macronetwork with well-developed connectivity between hard and soft segment phases. Copolyurea properties were shown to be dependent on urea group density and thus hydrogen-bonding potential, as well as copolymer compositon, by the use of the different diamines (77) to (79). Copolymer composition in terms of hard segment contents had significant effects on the dynamic mechanical–thermal behaviour, as shown in Figure 2 for the DETDA-based copolyureas.[9] Two major relaxations are observed, associated with soft and hard segment glass transitions occurring, respectively, at temperatures T_g^S ($\sim -40\,°C$) and T_g^H (210 to 240 °C). Less-pronounced relaxations arising from mixed-phase transitions are also evident at temperatures around 50 °C. Further evidence of the phase-separated structure was obtained from DSC data. Changes in heat capacity at T_{gs} were analyzed[85] to give the percentage degree of phase separation in these copolyureas, and the value of 65% obtained is relatively low considering the high thermodynamic driving force for phase separation in these bulk copolymer-forming systems.

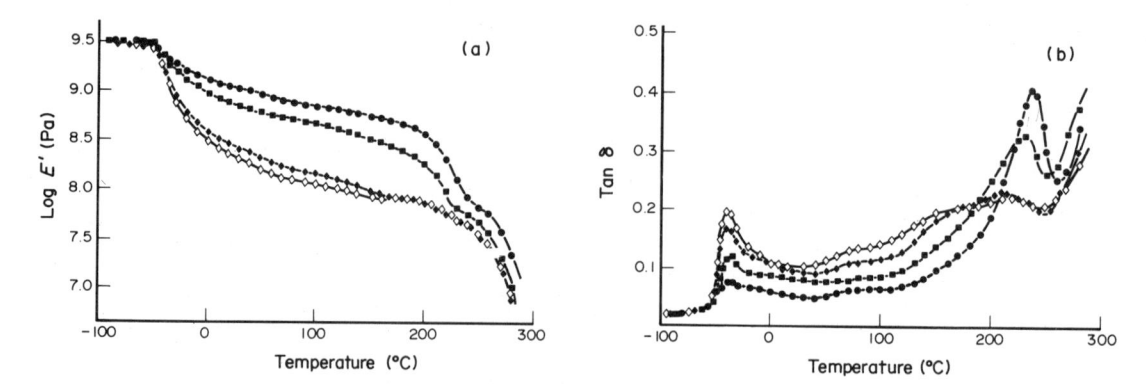

Figure 2 Dynamic mechanical–thermal behaviour of postcured (100 °C/18 h) segmented copolyureas, synthesized in bulk using the RIM process.[9] (a) Flexural modulus, E'; (b) mechanical damping, $\tan\delta$; ♦, R35; ■, R50; ●, R65. Data for non-postcured R35 ◇ are also included. Numbers in sample codes refer to HS contents (%)

A summary of the properties of the most representative copolyureas reported in the literature[9,97-99] are given in Table 10 in which the numbers in the sample code refer to the percentage of hard segment contents. The tensile properties show clearly the effects of increasing HS content in changing the copolyureas from soft, extensible elastomers to stiff, relatively brittle plastics. In addition the modulus-temperature dependence is significantly reduced as shown by the decreasing values of $E'(-30)/E'(65)$, the ratio of moduli at -30 and $65\,°C$. The excellent high temperature stability is further exemplified by the very low values (<2) of the ratio $E'(65)/E'(160)$, and by the ability of these copolyureas to retain physical integrity at temperatures in excess of 250 °C (see Figure 2).

Table 10 Comparative Physical Properties of Copolyureas Synthesized in Bulk using the RIM Process

Copolyurea	Ref.	E (MPa)	σ_u (MPa)	ε_u (%)	$E(-30)/E(65)$	$E(65)/E(160)$
R35	a	276	15.9	276	4.9	2.0
R50	a	617	25.0	119	2.9	1.8
R65	a	1144	37.6	17	2.2	1.5
D35	b	225[e]	21.2	200	2.7	—
D42	b	409[e]	3.47	190	2.8	1.0
D52	b	650[e]	28.7	110	2.1	1.1
N45	c	580[e]	27.0	300	2.9	—
W22	d	20	4.6	250	4.2[f]	—
W44	d	243	18.3	89	3.7[f]	—
W66	d	750	38.2	4	2.0[f]	—

[a] A. J. Ryan, J. L. Stanford and A. N. Wilkinson, *Polym. Bull. (Berlin)*, 1987, **18**, 517.
[b] R. J. G. Dominguez, 'New Product Data', Texaco Chemical Company, NPD-027, 1985.
[c] D. Nissen, W. Schonleben, M. Marx, K. H. Illers and P. Simak (BASF), *Ger. Pat.* 3 215 909 (1983) (*Chem. Abstr.*, 1983, **100**, 8213a).
[d] N. R. Willkomm, Z. S. Chen, C. W. Macosko, D. Gobran and E. L. Thomas, *Polym. Eng. Sci.*, 1988, in press.
[e] Flexural moduli.
[f] Ratio of shear moduli at -30 and $70\,°C$.

Immediately after processing, the high HS content copolyureas are difficult to remove from the mould because of their brittleness, ascribed to the relatively low degree of polymerization in the HS resulting from the onset of vitrification during copolymer formation. This was confirmed by GPC data from quenched, bulk-polymerized copolyurea, which showed evidence of discrete, low molecular weight oligomers.[97]

An interesting comparison of the bulk and solution syntheses of copolyureas has recently been reported.[97] The linear copolyureas listed in Table 11 were based on a poly(oxypropylene) diamine (molecular weight 2000), DETDA and MDI, and were synthesized either in DMAc solution (S22, S44, S66) or in bulk (W22, W44, W66). The solution copolyureas had molecular weights (GPC) more than double those of their bulk counterparts and their GPC traces each showed a smooth low molecular weight tail compared with the oligomer peaks referred to previously. The anomalously high values of MWD (~ 4) for both sets of copolyureas were attributed either to the polymerizations not yielding the most probable distribution (for which MWD is 2) or to inaccuracies in GPC data arising from the PS calibrant which does not adequately represent the structure of the block copolymers.

Table 11 Molecular Weight Data for Copolyureas Synthesized in Solution and Synthesized in Bulk using the RIM Process

Copolyurea	$10^3 \times \overline{M}_n$	$10^3 \times \overline{M}_w$	MWD
S22	34.0	128.0	3.7
S44	32.0	113.0	3.6
S66	21.0	76.0	3.6
W22	13.6	53.1	3.9
W44	7.2	28.2	4.0
W66	6.6	23.1	3.5

Compared with copolymers formed in bulk, those cast from solution have purer phases as shown by DSC, and larger domain sizes as shown by small angle X-ray scattering. These differences are due possibly to the slow rate of domain formation in copolyurea solutions, which become more concentrated as solvent is removed, whereas in the bulk process there is kinetic competition between spinodal phase separation and polymerization. The tensile properties of both sets of copolyureas were also reported, and those of the bulk copolyureas are included in Table 10. Values of Young's modulus (E) and tensile strength (σ_u) are similar for both types of copolymer with 22 and 44% HS contents. However, copolyureas formed in bulk have ultimate elongations (ε_u) almost double the values of those formed in solution, despite having approximately half the molecular weight. The tensile properties of polymers are molecular weight dependent and reach limiting values at high molecular weight. It would appear that these limits have been reached by the copolymers formed in bulk and that the highly hydrogen-bonded, network-like structure of the glassy aromatic HS is the dominant morphological feature characterizing these copolyureas.

Hydrogen bonding in polyureas has been investigated,[100] in comparison with polyurethanes, using crystallographic, DSC and IR techniques on model compounds. Thus the bisurea (**81**) was shown to have a higher melting point and a lower carbonyl-stretching wavenumber than the corresponding bisurethane (**82**). Crystallographic data were analyzed using a computer model to give the detailed structures of the compounds shown in Figure 3. The hydrogen bonds in the case of the bisurethane are singular and the carbonyl oxygen is bonded to the —N—H dipole of the opposite urethane group, whereas the hydrogen bonds in the bisurea are bifurcated, that is each carbonyl oxygen is bonded to both the —N—H dipoles of the opposite urea group. The enhanced hydrogen bonding thus accounts for the superior physical and thermal properties of polyureas, compared with analogous polyurethanes.

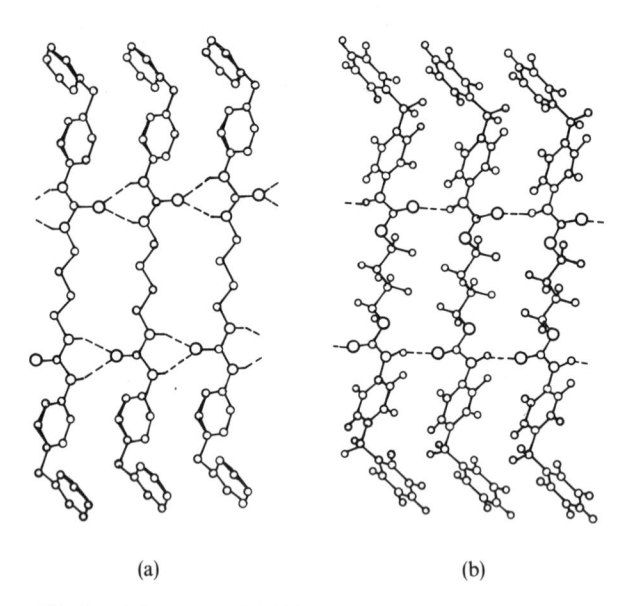

Figure 3 Molecular structures[100] of model compounds (a) bisurea (**81**) and (b) bisurethane (**82**) showing the arrangement of adjacent molecules with hydrogen bonds (-----) in the plane of the paper (reproduced by permission of Springer-Verlag from *Colloid Polym. Sci.*, 1985, **236**, 335).

25.4 REFERENCES

1. O. Bayer, *Angew. Chem., Ausg. A*, 1947, **59**, 257.
2. H. Mark and G. S. Whitby, 'Collected Papers of Wallace Hulme Carothers of High Polymeric Substances', 'High Polymer', Interscience, New York, 1940, vol. 1.
3. Toyo Koatsu Industries, *Br. Pat.* 981 813 (1965).
4. H. G. J. Overmars, in 'Encyclopaedia of Polymer Science and Technology', ed. H. F. Mark, N. Bikales, C. G. Overberger and G. Merges, Elsevier, Amsterdam, 1969, vol. 11, p. 464.
5. O. E. Snider and R. J. Richardson, in 'Encyclopaedia of Polymer Science and Technology', ed. H. F. Mark, N. Bikales, C. G. Overberger and G. Merges, Elsevier, Amsterdam, 1969, vol. 11, p. 495.
6. B. Meyer, 'Urea–Formaldehyde Resins', Addison-Wesley, New York, 1979.
7. K. Hayashi, S. Harny and Y. Iwakura, *Makromol. Chem.*, 1965, **86**, 64.
8. C. Hepburn, 'Polyurethane Elastomers', Applied Science, London, 1982.
9. A. J. Ryan, J. L. Stanford and A. N. Wilkinson, *Polym. Bull. (Berlin)*, 1987, **18**, 517.
10. E. Fischer, *Chem. Ber.*, 1913, **46**, 2504.
11. G. D. Buckley and N. H. Ray (ICI Ltd.), *US Pat.* 2 550 767 (1951) (*Chem. Abstr.*, 1951, **45**, 7821i).
12. M. Katz (Du Pont), *US Pat.* 2 975 157 (1961) (*Chem. Abstr.*, 1961, **55**, 13 921i).
13. W. W. Mosely and R. G. Parrish (Du Pont), *US Pat.* 3 318 849 (1967) (*Chem. Abstr.*, 1967, **67**, 22 832h).
14. N. Yamazaki and F. Higashi, *Polym. Prepr., Am. Chem. Soc., Div. Polym. Chem.*, 1976, **17**, 157.
15. N. Yamazaki, T. Iguchi and F. Higashi, *J. Polym. Sci., Polym. Chem. Ed.*, 1975, **13**, 785.
16. N. Yamazaki, T. Iguchi and F. Higashi, *Tetrahedron*, 1975, **31**, 3031.
17. N. Yamazaki and F. Higashi, *Adv. Polym. Sci.*, 1981, **38**, 1.
18. F. Higashi, T. Murukami and Y. Taguchi, *J. Polym. Sci., Polym. Chem. Ed.*, 1982, **20**, 103.
19. C. I. Chiriac, *Rev. Roum. Chim.*, 1984, **29**, 329.
20. C. I. Chiriac, *Polym. Bull. (Berlin)*, 1986, **15**, 65.
21. C. I. Chiriac and J. K. Stille, *Macromolecules*, 1977, **10**, 710.
22. C. I. Chiriac and J. K. Stille, *Macromolecules*, 1977, **10**, 712.
23. S. Kluchko and R. I. Meltzer, *J. Org. Chem.*, 1965, **30**, 3454.
24. W. H. Carothers (Du Pont), *US Pat.* 2 071 250, 2 071 251 (1937) (*Chem. Abstr.*, 1937, **31**, 27 149, 27 151); 2 190 770 (1940) (*Chem. Abstr.*, 1940, **34**, 41 878).
25. I. G. Farbenindustrie, *Ger. Pat.* 745 684 (1944, prior 1935) (*Chem. Abstr.*, 1935, **40**, 7702).
26. H. R. Kricheldorf, *J. Macromol. Sci., Chem.*, 1980, **A14**, 959.
27. N. Ogata, K. Sanui, M. Watanabe and Y. Kosaka, *J. Polym. Sci., Polym Lett. Ed.*, 1986, **24**, 65.
28. Y. Kosaka, M. Watanabe, K. Saniu and N. Ogata, *J. Polym. Sci., Polym. Chem. Ed.*, 1986, **24**, 1915.

29. G. Wu, H. Tanaka, K. Sanui and N. Ogata, *Polym. J.*, 1982, **14**, 571.
30. S. Yasuda, G. Wu, H. Tanaka, K. Sanui and N. Ogata, *J. Polym. Sci., Polym. Chem. Ed.*, 1983, **21**, 2609.
31. G. Wu, H. Tanaka, K. Sanui and N. Ogata, *Polym. J.*, 1982, **14**, 979.
32. O. Ohara (Teijin Chemicals Ltd.), *Jpn. Pat.* 61 203 129 (1986) (*Chem. Abstr.*, 1986, **106**, 85 278b).
33. M. Takamatsu and O. Ohara (Teijin Chemicals Ltd.), *Jpn. Pat.* 61 203 130 (1986) (*Chem. Abstr.*, 1986, **106**, 85 234).
34. R. D. Katsarava, T. M. Kartvelishvli and M. M. Zaalishvli, *Soobshch. Akad. Nauk. Gruz. SSR*, 1984, **113**, 533 (*Chem. Abstr.*, 1984, **101**, 171 861q).
35. H. G. J. Overmars, *Macromol. Synth.*, 1968, **3**, 8.
36. L. Alexandru and L. Dascaw, *J. Polym. Sci.*, 1961, **52**, 331.
37. L. B. Sokolov and L. V. Turetskii, *Vysokomol. Soedin., Ser. A.*, 1964, **6**, 346.
38. E. A. Hassan, *U. A. R. J. Chem.*, 1970, **13**, 213 (*Chem. Abstr.*, 1970, **75**, 49 701s).
39. K. C. Stueben and A. E. Barnabeo, in 'Interfacial Synthesis', ed. F. Millich and C. E. Carraher Jr., Dekker, New York, 1977, vol. 2, p. 269.
40. D. J. Lyman and S. L. Jung, *J. Polym. Sci.*, 1959, **40**, 407.
41. R. J. Cotter (Union Carbide), *US Pat.* 3 130 179 (1964) (*Chem. Abstr.*, 1964, **61**, 4514e).
42. R. J. Cotter (Union Carbide), *US Pat.* 3 131 167 (1964) (*Chem. Abstr.*, 1964, **61**, 4514b).
43. D. C. Allport (ICI Ltd.), *Br. Pat.* 1 115 003 (1968) (*Chem. Abstr.*, 1968, **76**, 15 169e).
44. R. J. Cotter (Union Carbide), *US Pat.* 3 164 571 (1965).
45. H. Dreyfus (American Celanese), *US Pat.* 2 568 885 (1951) (*Chem. Abstr.*, 1951, **46**, 1808h).
46. O. V. Schickh, R. Baümler and F. Ebel, *Ger. Pat.* 896 412 (1948).
47. H. Iiyama, M. Asakura and K. Kimoto, *Kogyo Kagaku Zasshi*, 1966, **68**, 236 (*Chem. Abstr.*, 1965, **63**, 490).
48. G. S. Kolesnikov, O. Ya. Fedotova and Thai-Suong-Tinh, *Vysokomol. Soedin., Ser. A*, 1969, **11**, 2682.
49. D. M. Fenton (Union Oil Company), *US Pat.* 3 277 061 (1966).
50. J. J. Herman and A. Lecloux (Solvay et Cie), *Eur. Pat.* 193 982 (1986).
51. A. L. DiSalvo, *J. Polym. Sci., Polym. Lett. Ed.*, 1974, **12**, 641.
52. G. Oertl, 'Polyurethane Handbook', Hanser, Berlin, 1985.
53. R. L. Craven, in 'Proceedings of the American Chemical Society Division of Paints, Plastics and Printing Ink Chemistry, Atlantic City, September 1956', ACS, Washington, DC, 1956, paper 33.
54. J. W. Baker and D. N. Bailey, *J. Chem. Soc.*, 1957, 4649.
55. J. W. Baker and D. N. Bailey, *J. Chem. Soc.*, 1957, 4562.
56. J. W. Baker and D. N. Bailey, *J. Chem. Soc.*, 1957, 4663.
57. N. N. Barashkov, E. N. Teleshov and A. N. Pravendnikov, *Vysokomol. Soedin., Ser. A*, 1978, **20**, 2749.
58. M. C. Pannone and C. W. Macosko, *J. Appl. Polym. Sci.*, 1987, **34**, 2409.
59. O. Bayer, *Angew. Makromol. Chem.*, 1947, **59**, 263.
60. M. Katz (Du Pont), *US Pat.* 2 888 438 (1959) (*Chem. Abstr.*, 1959, **53**, 17 582b).
61. E. L. Lawton, T. Murayama, V. F. Holland and D. C. Felty, *J. Appl. Polym. Sci.*, 1980, **25**, 187.
62. H. R. Kricheldorf, *J. Macromol. Sci., Chem.*, 1980, **A14**, 959.
63. Y. Oishi, M.-A. Kakimoto and Y. Imai, *J. Polym. Sci., Polym. Chem. Ed.*, 1987, **25**, 2185.
64. J. A. Mikroyannidis, *J. Polym. Sci., Polym. Chem. Ed.*, 1984, **22**, 3423.
65. S. K. Kwon, K. L. Choi and S. K. Choi, *J. Polym. Sci., Polym. Chem. Ed.*, 1987, **25**, 1781.
66. K. Kurita, K. Murakami, K. Kobayashi, M. Takahashi and Y. Koyana, *Makromol. Chem.*, 1986, **187**, 1359.
67. I. Yilgor, A. K. Sha'aban, W. P. Steckle, Jr., D. Tyagi, G. L. Wilkes and J. E. McGrath, *Polymer*, 1984, **25**, 1800.
68. D. Tyagi, J. P. Armistead, G. L. Wilkes, B. Lee and J. E. McGrath, *Polym. Prepr., Am. Chem. Soc., Div. Polym. Chem.*, 1985, **26**, 12.
69. B. Lee, J. E. McGrath, D. Tyagi and G. L. Wilkes, *Polym. Prepr., Am. Chem. Soc., Div. Polym. Chem.*, 1986, **27**, 100.
70. J. H. Saunders and R. J. Slocombe, *Chem. Rev.*, 1948, **43**, 203.
71. W. Lehman and H. Rinke (Bayer, A. G.), *US Pat.* 2 852 494 (1958) (*Chem. Abstr.*, 1959, **53**, 1829b).
72. P. W. Morgen and S. L. Kwolek, *J. Polym. Sci.*, 1959, **40**, 299.
73. T. Yasuzawa, H. Yamaguchi and Y. Minoura, *J. Polym. Sci., Polym. Chem. Ed.*, 1979, **17**, 3387.
74. R. E. Partch, K. Nakamura, K. J. Wolfe and E. Matijevic, *J. Colloid Interface Sci.*, 1985, **105**, 560.
75. L. J. Lee, *Rubber Chem. Technol.*, 1980, **53**, 542.
76. R. E. Camargo, C. W. Macosko, M. Tirrell and S. T. Wellinghoff, *Polymer*, 1986, **26**, 1145.
77. N. Barksby, D. Dunn, A. Kaye, J. L. Stanford and R. F. T. Stepto, *ACS Symp. Ser.*, 1985, **270**, 27.
78. R. B. Turner, H. L. Spell and J. A. Vanderhider, *Polym. Sci. Technol.*, 1982, **18**, 63.
79. C. W. Macosko, 'RIM Fundamentals', Hanser, Munich, FDR, 1988, in press.
80. D. Nissen and R. A. Markovs, *J. Elastomers Plast.*, 1983, **15**, 96.
81. R. A. Markovs, *J. Cell. Plast.*, 1985, **21**, 326.
82. R. E. Camargo, J. S. Andrews, C. W. Macosko and S. T. Wellinghoff, *Polym. Prepr., Am. Chem. Soc., Div. Polym. Chem.*, 1984, **25**, 294.
83. R. B. Turner, *Polym. Comp.*, 1984, **5**, 151.
84. A. J. Ryan, J. L. Stanford and R. H. Still, in 'Integration of Fundamental Polymer Science and Technology—2', ed. P. J. Lemstra and L. A. Kleintjens, Elsevier, Applied Science, London; 1988, part 7, p. 515.
85. A. J. Ryan, J. L. Stanford and R. H. Still, *Br. Polym. J.*, 1988, **20**, 77.
86. D. M. Rice and R. J. G. Dominguez (Texaco Inc.), *US Pat.* 4 371 160 (1984) (*Chem. Abstr.*, 1984, **100**, 104 874a).
87. R. J. G. Dominguez, *J. Cell. Plast.*, 1984, **20**, 433.
88. J. H. Ewen, *J. Elastomers Plast.*, 1985, **17**, 281.
89. N. P. Vespoli, L. M. Alberino, A. A. Peterson and J. H. Ewen, *J. Elastomers Plast.*, 1986, **18**, 159.
90. R. A. Grigsby and D. M. Rice, *J. Cell. Plast.*, 1986, **22**, 484.
91. R. J. G. Dominguez, D. M. Rice and R. A. Grigsby, *Plast. Eng.*, 1987, November, 41.
92. R. E. Camargo, C. W. Macosko, V. M. Gonzalez and M. Tirrell, *Rubber Chem. Technol.*, 1983, **56**, 774.
93. M. C. Pannone and C. W. Macosko, *Polym. Eng. Sci.*, 1988, **28**, 660.
94. A. J. Ryan and J. L. Stanford, unpublished results.
95. J. L. Stanford and R. F. T. Stepto, *Br. Polym. J.*, 1977, **9**, 124.

96. D.-K. Lee, A. J. Ryan, J. L. Stanford and A. N. Wilkinson (UMIST), *Br. Pat. Appl.*, 88 00281 (1988).
97. W. R. Willkomm, Z. S. Chen, C. W. Macosko, D. Gobran and E. L. Thomas, *Polym. Eng. Sci.*, 1988, in press.
98. R. J. G. Dominguez, 'New Product Data, NPD-027', Texaco Chemical Company, 1985.
99. D. Nissen, W. Schonleben, M. Marx, K. H. Illers and P. Simak (BASF), *Ger. Pat.*, 3 215 909 (1983) (*Chem. Abstr.*, 1984, **100**, 82 146).
100. L. Borne and H. Hespe, *Colloid. Polym. Sci.*, 1985, **263**, 335.

26
Polybenzyls

CHRISTOS P. TSONIS
University of Petroleum and Minerals, Dhahran, Saudi Arabia

26.1	INTRODUCTION	455
26.2	POLYMERIZATION OF BENZYL CHLORIDE	456
26.3	POLYMERIZATION OF BENZYL DERIVATIVES	456
26.4	KINETICS AND MECHANISM	458
26.5	STRUCTURE CHARACTERIZATION	461
26.6	BRANCHING	462
26.7	REFERENCES	463

26.1 INTRODUCTION

Linkage of one benzene or arene ring to another by a methylene group with the elimination of hydrogen halide, water, or any other good leaving group in the presence of Friedel–Crafts catalysts normally gives isomers of benzyl dimers. The condensation can be repeated producing tribenzyls, tetrabenzyls and higher polymers known as polybenzyls (Scheme 1).

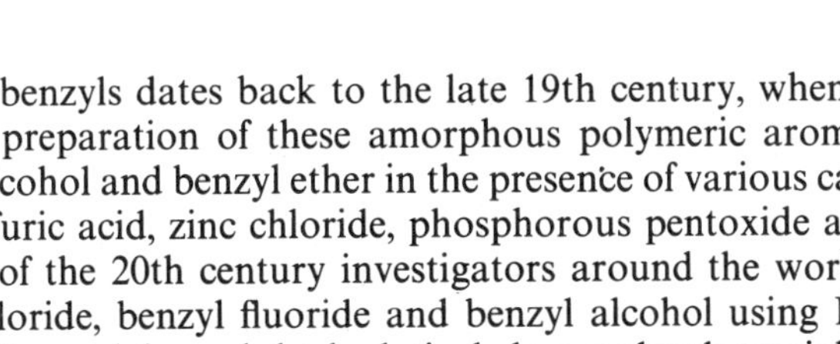

Scheme 1

The synthesis of polybenzyls dates back to the late 19th century, when a number of prominent chemists described the preparation of these amorphous polymeric aromatic hydrocarbons from benzyl halides, benzyl alcohol and benzyl ether in the presence of various catalytic systems including aluminum chloride, sulfuric acid, zinc chloride, phosphorous pentoxide and many others.

During the first half of the 20th century investigators around the world undertook the task of polymerizing benzyl chloride, benzyl fluoride and benzyl alcohol using Lewis and Brönsted acid type catalysts. The resulting polybenzyls had relatively low molecular weight (2000–3000) and varied in appearance, color, texture, solubility and melting point. Structural characterization of these polybenzyls was generally poor, mainly due to the lack of proper instrumentation. Some early good summaries and discussions on the subject can be found in the chemical literature.[1-3]

During the past three decades a more systematic approach has been taken to both the synthesis and characterization of polybenzyls. The prime objective of the effort has been to prepare linear high molecular weight polymers which may lead to high temperature application materials. It should be emphasized at this point that the incorporation of phenyl groups into linear polymer chains is of paramount importance in polymer synthesis because they control rotation of the chemical bonds along the chain, which increases rigidity, crystallinity and hence the thermal stability of the polymer.

26.2 POLYMERIZATION OF BENZYL CHLORIDE

The polymerization of benzyl chloride has been studied to a greater extent than polymerization of its derivatives, mainly because it is readily available and relatively inexpensive. Its polymerization has been investigated extensively with systems that involve conventional Friedel–Crafts catalysts[4–9] such as $AlCl_3$, $SnCl_4$, SbF_5, $TiCl_4$, $FeCl_3$ and PCL_5, some metal oxides,[10,11] and transition metal carbonyl complexes,[12–14] including $Mo(CO)_6$, $Cr(CO)_6$, $W(CO)_6$, $ArCr(CO)_3$, $Re(CO)_5Cl$, $Re(CO)_5Br$, $Mn(CO)_5Cl$ and $Mn(CO)_5Br$.

Theoretically, for every mole of benzyl chloride reacted, one mole of hydrochloric acid is released. The conventional Friedel–Crafts catalysts usually require low temperature (-130 to $+25\,°C$) conditions due to their high reactivity and low selectivity. The solvents are carefully selected to have relatively high polarity and to be inert, but at the same time the reaction solution must remain homogeneous during polymerization at such low temperatures. Some examples of good solvents that meet these requirements are dichloromethane, chloroform, trichloroethane, sulfur dioxide and nitromethane. Despite these low temperature conditions, the resulting polybenzyls prepared with the Friedel–Crafts Lewis acid catalysts are generally low melting, amorphous materials with a great deal of branching or crosslinking. The soluble polybenzyls have normally an average degree of polymerization (DP) in the range 20–50. However, the synthesis of a linear crystalline unsubstituted oligobenzyl can be achieved[5] from the condensation of 4,4'-bis(chloromethyl)diphenylmethane and diphenylmethane (1:15 mole ratio) in chloroethane catalyzed by $AlCl_3$ at $10\,°C$. The product has a melting point around $135\,°C$ and a molecular weight of 610.

In contrast to most conventional Friedel–Crafts catalysts, which require low temperature polymerization conditions, polybenzyls may also be prepared by relatively high temperature polycondensation of benzyl chloride using group IIA metal oxides,[10] group VIB metal carbonyls,[13] arenetricarbonylchromium[12] and group VIIB metal carbonyl halides.[14] These reaction systems are carried out at temperatures above $100\,°C$ and often in bulk. The relatively activity of group IIA metal oxides, $MgO > CaO > BaO$, follows that of their Lewis acidity and ability to accept electrons by interacting with the chloride part of the monomer.

The relatively high temperature catalytic properties of group VIB metal carbonyls, arenetricarbonylchromium and VIIB metal carbonyl halides for the polymerization of benzyl chloride have recently come under investigation. Complexing the metal with carbonyl and arene ligands allows the catalyst to dissolve in the reaction medium, which remains homogeneous throughout. As the temperature of the reaction mixture rises, the organic substituent usually dissociates partially from the metal, exposing the vacant metal d orbitals. This allows the benzyl chloride to interact with these vacant sites and polymerization occurs. For example, arenechromiumtricarbonyls catalyze the self-polycondensation of benzyl chloride when they are thermally activated, and their activity may be controlled by the nature of the arene attached to the metal. For group VIB metal hexacarbonyls relative activities are $Mo \gg W > Cr$, but phosphine ligands attached to the metal do not easily dissociate, and thus poison its catalytic properties.

Group VIIB rhenium and manganese carbonyl halides also induce the thermal polymerization of benzyl chloride, with rhenium having higher activity than its manganese counterpart.

Even though the group IIA, VIB and VIIB metal-based homogeneous catalysts are milder than the conventional Friedel–Crafts metal halides, it still appears impossible to polycondense benzyl chloride to a linear p-substituted, high molecular weight polymer.

26.3 POLYMERIZATION OF BENZYL DERIVATIVES

Although the polymerization of benzyl chloride has been investigated extensively, attention has also been focused on other benzyl derivatives. In the polymerization of benzyl halides,[1] the fluorides appear to be the most reactive. In general, even though reactivity decreases in the order $F > Cl > Br > I$, benzyl chlorides and bromides are most frequently used because of low cost. Benzyl iodides are seldom prepared and polymerized because of their instability and tendency to enter into side reactions. All benzyl halides are self-polycondensed in a similar manner to that described for benzyl chloride. Benzyl alcohol can be polymerized[1] easily when it comes into contact with small amounts of concentrated sulfuric acid or with other acids. The polybenzyls formed are basically similar in structure and properties to those obtained from benzyl halides.

Certain alkyl substituents attached to the aromatic ring or on the benzyl carbon of the monomer may bring about drastic changes in the structure of the resulting polybenzyls. Some elegant examples are the monomers α-methylbenzyl chloride,[4,6,15,16,19] α,α-dimethylbenzyl chloride,[17,18] α-ethylbenzyl chloride[20] and 2,5-dimethylbenzyl chloride,[4,5,16,21] which upon homopolymerization produce the corresponding polybenzyls (1)–(4) with high linearity and crystallinity. This phenomenon is normally attributed to steric effects of the methyl substituents so that during polycondensation substitution on the *para* position is highly favored.

When all the aromatic carbons on the benzyl ring are substituted except the *para* position, as is the case with 2,3,5,6-tetramethylbenzyl chloride,[2,5] the polybenzyl (5; $n=4$) is linear and high melting, but with very low molecular weight. The low degree of polymerization may be explained by steric effects of the methyl substituents of the aromatic ring, or by the fact that growing oligobenzyl chains become insoluble in the reaction medium and precipitate out of solution before having a chance to grow.

Linear low molecular weight oligobenzyls can also be synthesized if one carefully designs the monomers and reaction conditions. For example, a series of 2,5-dimethylbenzyl and 2,3,5,6-tetramethylbenzyl compounds have been prepared from the condensation of a number of chloromethylated aromatic and other substituted benzenes by carefully monitoring the reaction parameters such as comonomer mole ratio. The resulting oligomers (5; $n=1$–4) and (6) are linear and exhibit crystallinity.

The effect of monomer structure on polymer properties may also be illustrated in a number of polybenzyl copolymers. For example, α,α'-dichloro-*p*-xylene can be copolymerized with a series of arenes, including 1,2-dimethylbenzene, 1,3-dimethylbenzene, 1,4-dimethylbenzene, tetrahydronaphthalene and 1,2,4,5-tetramethylbenzene. Their corresponding copolymers shown in (7)–(10) are usually branched with a great deal of crosslinking; (11), prepared using 1,2,4,5-tetramethylbenzene as comonomer, exhibits high linearity and crystallinity.

A linear crystalline polybenzyl of structure (12) is also obtained[22] from the copolymerization of 2,5-dimethyl-α,α'-dichloro-*p*-xylene with 1,4-dimethylbenzene.

In a parallel investigation,[22] the 2,3,4,5-tetramethyl-α,α'-dichloro-*p*-xylene is copolymerized with di-, tri- and tetra-substituted arenes, namely 1,3-dimethylbenzene, 1,4-dimethylbenzene, 1,2,4-

trimethylbenzene, 1,3,5-trimethylbenzene and 1,2,4,5-tetramethylbenzene. The copolymers (13)–(17) which are usually produced from these comonomers are generally crystalline, of low molecular weight and high melting point.

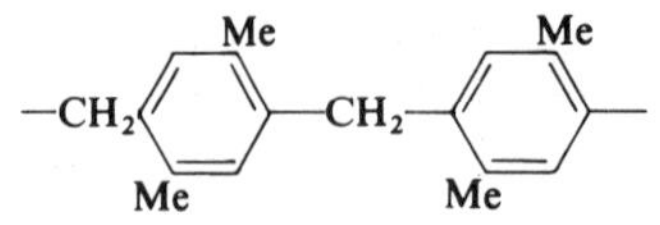

(12) linear

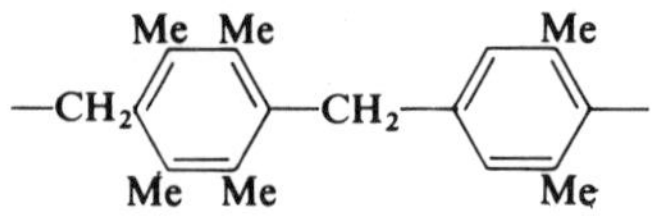

(13) linear

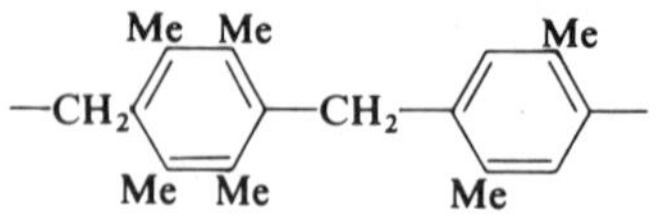

(14) linear

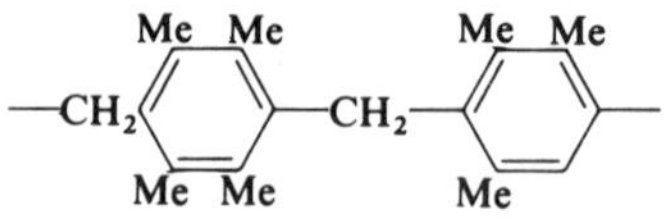

(15) linear

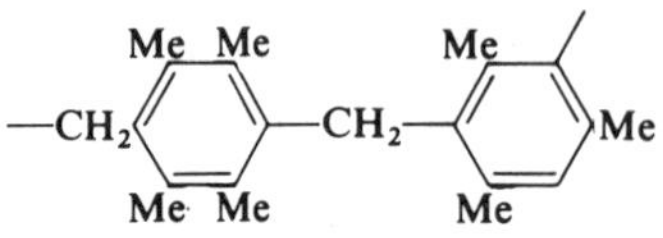

(16) linear

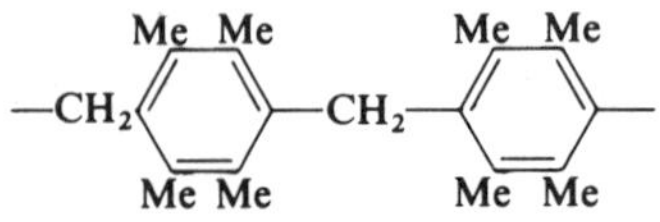

(17) linear

All these observations suggest that linear disubstituted polybenzyl copolymers could be formed if the total number of methyl substituents on both monomers under copolymerization is greater than or equal to four. However, when these monomer systems have less than four methyl groups, the resulting copolymers are usually branched and amorphous with some crosslinking.

So far, from the present discussion on the structure–property relationships of polybenzyls it can be concluded that those polymers derived from unsubstituted benzyl halides are generally amorphous, low melting, low molecular weight and have good solubility, while the incorporation of methyl substituents makes these polymers usually crystalline and reduces to some extent their molecular weight, increases their melting point, and decreases their solubility sharply. However, the incorporation of anthracene into the backbone of these polybenzyls can improve markedly their melting point and solubility. These anthracene-based alternating copolymers may be synthesized[23] by two general reactions using $SnCl_4$ catalyst in the nitroethane at about 65 °C. In the first type (equation 1) 9,10-bis(chloromethyl)anthracene reacts with a series of aromatic monomers such as anthracene, p-xylene, mesitylene, 1,2,4,5-tetramethylbenzene and bis(2,3,5,6-tetramethylphenyl)methane. The resulting polybenzyls were characterized by NMR and UV, and compared with model compounds; they were found to contain 9,10-disubstituted anthracene units. The polymers of the second type (equation 2) were prepared from the copolymerization of anthracene with a number of chloromethylated aromatics, including α,α′-dichloro-p-xylene, 1,3-bis(chloromethyl)-2,4,6-trimethyl benzene, 1,4-bis(chloromethyl)-2,3,5,6-tetramethyl benzene, and 4,4′-bis(chloromethyl)diphenylmethane and its octamethyl derivative. All these alternating copolymers possess 1,4- or a mixture of 1,4- and 9,10-disubstituted anthracene units.

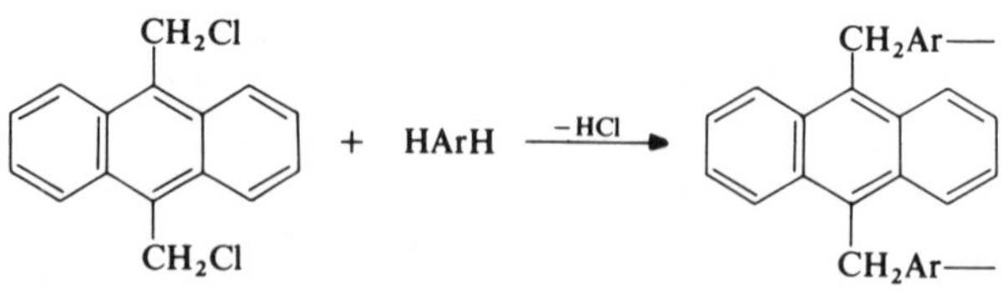

(1)

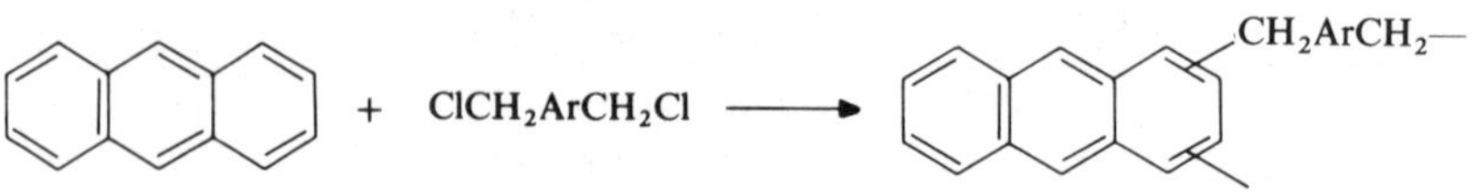

(2)

26.4 KINETICS AND MECHANISM

The self-polycondensation of benzyl chloride and its derivatives in the presence of a suitable catalyst is undoubtedly an electrophilic aromatic substitution process. It can thus be considered analogous to Friedel–Crafts benzylation of toluene or diphenylmethane. The kinetics and mechanism of these and other alkylations are very complex and there is no single scheme which is appropriate to all systems.

The polycondensation of benzyl chloride and its derivatives does not appear to follow the conventional step growth polymerization process, because conversion often rises rapidly and accelerates until its rate reaches first-order kinetics[21,24,25] with respect to the monomer. The molecular weights of the resulting polybenzyls generally increase with increasing conversion and later level off. At high conversions, temperature and catalyst concentration do not seriously affect the degree of polymerization. The polybenzyls formed at any time during polymerization have similar structures, which are mainly branched due to their polysubstitution pattern on the aromatic ring.

Although at low conversion the polymerization usually follows first-order kinetics with respect to the monomer, at high conversions the rate slows down and does not obey first-order kinetics. The kinetics of these systems may be reliable only at low conversions and short reaction times if one wishes to avoid considerable catalyst decomposition and interference from high viscosity. The reaction rate can be followed easily by determining the HCl evolution at different time intervals, since for every mole of gas released one mole of benzyl halide monomer is consumed. A rate expression that correlates well with all temperatures and catalyst concentrations for the polymerization of benzyl chloride catalyzed by tin(IV) chloride in bulk is represented[24] by equation (3). In another study[6] the polymerization of benzyl chloride, α-chloroethylbenzene, and diphenylchloromethane were catalyzed by $AlCl_3$ or $SnCl_4$ in nitrobenzene. The kinetics for the first two reactions were second-order with respect to the monomer and first-order with respect to the catalyst, whereas the polycondensation of diphenylchloromethane under the same conditions was found to be first order with respect to both monomer and catalyst.

$$d[HCl]/dt = K_1[SnCl_4][PhCH_2Cl]/(1 + K_2[SnCl_4]^2) \qquad (3)$$

Interaction of the catalyst with the leaving group attached to the methylene carbon, often a halide or hydroxyl, may form a highly polarized complex, or a carbonium ion, which is usually stabilized *via* resonance on the aromatic ring. The counterion generated alongside the benzyl ion is often stable since the catalysts used for the polymerization have strong affinity for halide and hydroxyl ions. The rate of polymerization is usually influenced by the type of catalyst employed, which probably suggests that when carbonium ions are involved, they are not always kinetically free but rather coexist in equilibrium with ion pairs. Generally one would expect that the greater the polarity of the medium (solvent), the higher the concentration of the free carbonium ion. Thus the solvents employed in these polymerizations should not react with the catalysts or the carbonium ions that might be present in solution (*e.g.* H_2O, MeOH) because they will inhibit the polymerization. In addition, solvents that are easily alkylated, such as toluene, should be avoided. Those with high dielectric constant (*e.g.* polar solvents) and low freezing point are generally preferred, since polarity increases ion solvation and enhances polymerization rate. When Lewis acid catalysts of the $AlCl_3$ type are used, the polymerization is often carried out at low temperatures and very low freezing point solvents are required. These reaction conditions are necessary in order to suppress side reactions because these catalysts are highly reactive and less selective.

Although, numerous Friedel–Crafts catalysts have been used for the polymerization of benzyl chloride and its derivatives, it is probable that no single order of reactivity of the catalysts can be drawn up since the substituent on the benzyl monomer, the nature of the leaving group present on the methylene carbon and the reaction conditions might change the sequence for the polymerization of different monomers. A sequence suggested[26] for some metal halide catalysts is $ZrCl_4 > TiCl_4 > SnCl_4 \geq BF_3 \gg ZnCl_2$.

The effect of a substituent on the reactivity of a benzyl monomer in polymerization depends largely on whether it can stabilize the benzyl complex formed from the interaction of the catalyst with the group present on the methylene carbon (usually a halide). In general, electron-donating substituents attached to the *ortho* or *para* positions of the benzene ring release electrons *via* resonance toward the intermediate (*e.g.* carbonium ion) and stabilize it by reducing its positive character. The electron-withdrawing substituents usually have the opposite effect. Thus substituents increase the reactivity of benzyl chloride in the approximate order[3,25] *p*-OMe > *p*-CHMe$_2$ > H > *m*-Cl > *p*-Cl > *p*-NO$_2$, which follows the order of their electron-releasing effect. The substituted benzyl bromides basically follow a similar order of reactivity.[27]

The effect of the leaving group upon the polymerization rate generally depends on whether the leaving group can stabilize the negative charge it has acquired. This is exemplified[21] by three dimethylbenzyl derivatives (**18**), namely 2,5-dimethylbenzyl chloride, 2,5-dimethylbenzyl acetate and 2,5-dimethylbenzyl methyl ether. Their polymerization rate in the highly ionizing acetic acid, catalyzed by sulfuric acid, is controlled by the nature of the leaving group. Among the three examined the acetoxy group gives the fastest polymerization and chloride the slowest.

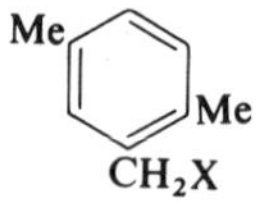

(**18**) X = OCOMe > OMe > Cl

All the above experimental evidence discussed supports the fact that the polymerization of benzyl chloride and its derivatives follow the Friedel–Crafts aromatic polyalkylation process. Thus the interaction of the catalyst with the organic monomer can form a strongly polarized complex or a carbonium ion, which can be in equilibrium with its ion pair, depending on solvent polarity, type of catalyst, the nature of the substituents on the benzyl ring and reaction conditions. It then reacts with the aromatic nucleus forming a π and/or σ complex, which can lose a proton to some Lewis base, usually the halide, to give benzyl isomers of diphenyl methane (Scheme 2). As the aromatic polyalkylation continues, the dimers, trimers and oligomers generated in solution have higher reactivity than the monomer because they have more potential phenyls for attack by the electrophilic species. Furthermore, the benzyl group in the polybenzyl chains activates the ring to substitution, whereas the chloromethyl deactivates the monomer. The resulting polybenzyl has a highly random multisubstituted branched structure (**19**).

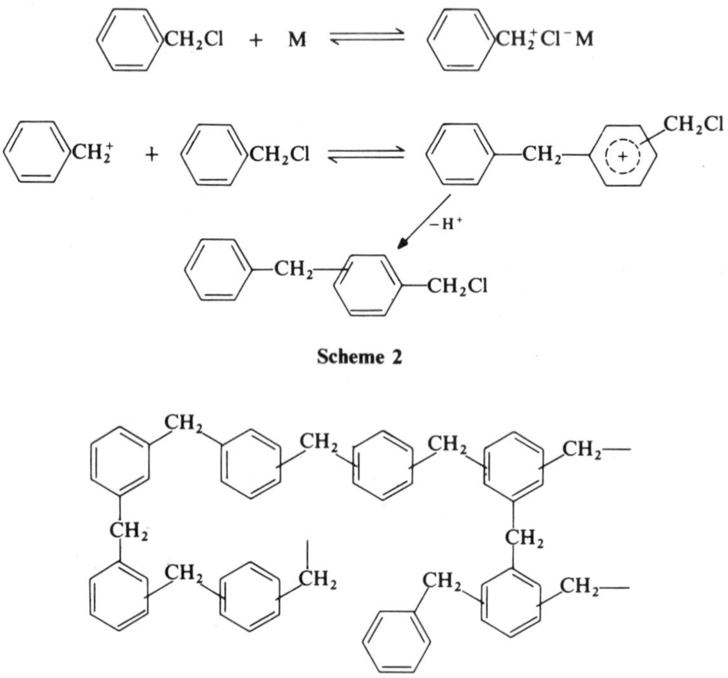

Scheme 2

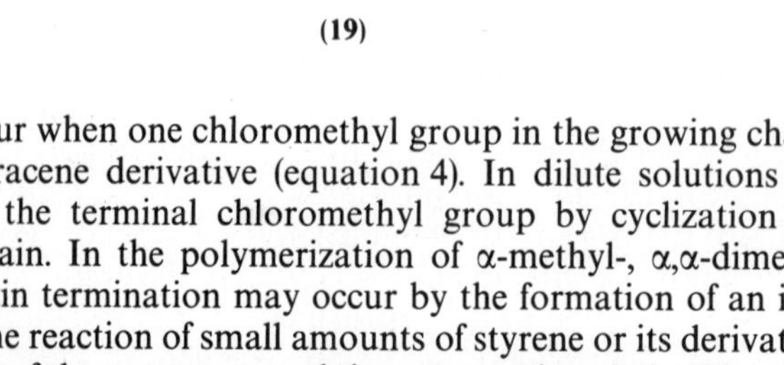

(**19**)

Termination may occur when one chloromethyl group in the growing chains reacts with another forming a dihydroanthracene derivative (equation 4). In dilute solutions the dihydroanthracene could be formed from the terminal chloromethyl group by cyclization (back biting) with the oligobenzyl growing chain. In the polymerization of α-methyl-, α,α-dimethyl-, or α-ethyl-benzyl chloride monomers, chain termination may occur by the formation of an indanyl group (**20**). This may come about from the reaction of small amounts of styrene or its derivatives generated *in situ* by the dehydrochlorination of the monomer, and the propagating chain. The polymerization will cease when all the stereochemically and electronically possible phenyl groups on the chain have been substituted. There is also the possibility that the growing chloromethyl units become increasingly buried at high conversions in highly viscous reaction media. In fact, the purified polybenzyls

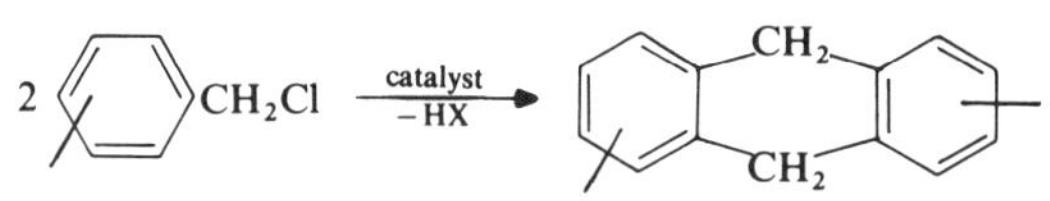

$$2 \text{ PhCH}_2\text{Cl} \xrightarrow[-HX]{\text{catalyst}} \text{(structure)} \quad (4)$$

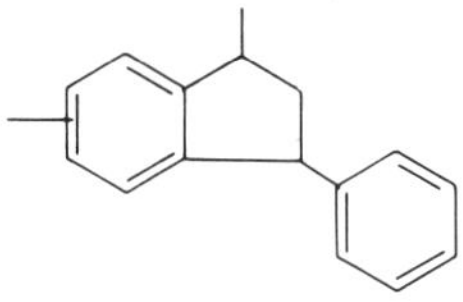

(20)

prepared under various conditions contain small amounts of chlorine and conversions based on HCl evolution seldom go over 95%. These observations suggest that chloromethyl groups are present on the polybenzyl chains.

26.5 STRUCTURE CHARACTERIZATION

Early studies on the structure determination of polybenzyls were generally poor, mainly because of inadequate chemical spectroscopic methods. The structures originally proposed for the polybenzyls derived from the polymerization of benzyl halides and benzyl alcohol were linear, containing polystilbene (21)[28] or poly(p-benzyl) (22)[1] repeating units. However, later studies by IR, oxidation and degradation methods ruled out the possibility that these linear structures exist and indicated that the polymers contain essentially benzene rings polysubstituted in a random fashion.

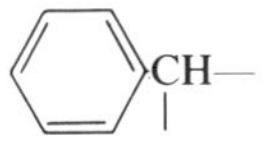

(21) 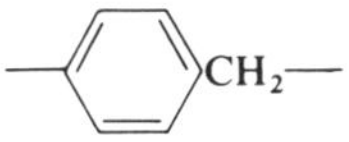   (22)

Theoretically, the monosubstituted phenyl group can provide five sites for aromatic substitution (23). The polybenzyl disubstitution pattern could have any or all of the three possible isomers as shown in structures (24)–(26) and one would expect the *ortho* and *para* isomers to be more prevalent than the *meta* isomer, since the methylene group is an *ortho, para* director. In fact, model Friedel–Crafts low temperature alkylations such as the reaction of benzyl chloride with toluene or diphenylmethane have demonstrated[6,8] that all possible *ortho*, *meta* and *para* isomers are formed. In both reactions the *para* isomer was the major product and *meta* the minor.

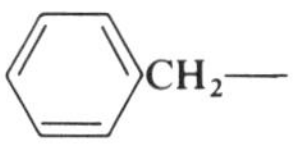

(23) 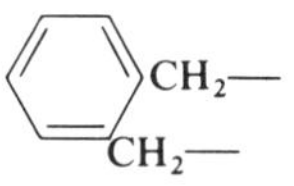   (24) 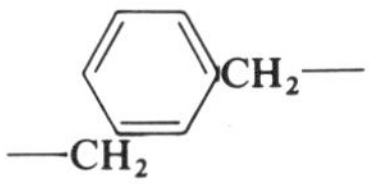   (25)    (26)

If any trisubstitution occurs, the disubstituted polybenzyls should produce three trisubstituted isomeric derivatives (27)–(29). The most likely isomer is (27), since the trisubstituted polybenzyl with (28) is sterically crowded and (29) is not expected to exist at all or in very low concentration because its precursor, the *m*-disubstituted isomer, is probably present only in small amounts.

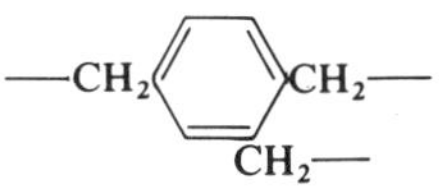

(27) 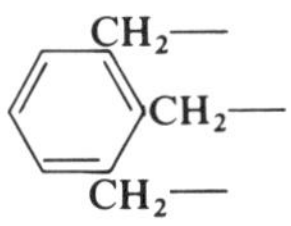   (28) 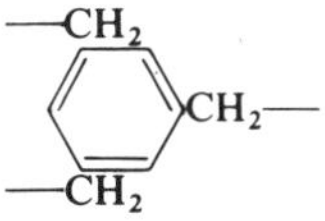   (29)

There is also the possibility that some tetrasubstitution will occur on the polybenzyl chains producing all or any of the three tetrasubstituted isomers (30)–(32). The major isomer should be (30) because of less steric hindrance relative to (31) and (32). It is very unlikely that any further substitution will take place on the already crowded tetrasubstituted polybenzyl chain due to steric effects; however, such isomers, if formed at all, would be of very low concentration and will probably have negligible effect on the structure of the whole polybenzyl.

NMR, UV, X-ray and degradation studies have shown[2,8,12,29,30] that polybenzyls derived from benzyl halides are mainly polysubstituted and highly branched. Furthermore, model compounds such as o-, m- and p-dibenzylbenzene have been synthesized, and their IR and ^{13}C NMR spectra have been analyzed.[8,32] Monosubstituted benzenes show two strong absorption bands in the 675–700 cm^{-1} and 725–775 cm^{-1} regions, while p-substituted derivatives exhibit a strong peak in the 800–860 cm^{-1} range. The o-, m- and p-dibenzylbenzenes give IR peaks at 1050, 1090 and 1023 cm^{-1}, which are specific to the *ortho*, *meta* and *para* isomers of dibenzylbenzene. The band at 1032 cm^{-1} is characteristic of a monosubstituted benzyl group. The information provided by IR alone for the characterization of polybenzyls is limited since specific absorptions characteristic of polysubstituted polybenzyls are not known. Furthermore, the p-substitution pattern in the polybenzyl may be absent due to the conformational factors of the polymers.

The application of ^{13}C NMR to the elucidation of the structure of these polybenzyls is more reliable and informative. The ^{13}C NMR chemical shifts for o-, m- and p-dibenzylbenzene appear at 139.0, 141.0, and 138.8 p.p.m., respectively. For the polybenzyls the spectrum[33] may be divided into three main regions. The aliphatic resonance usually appears in the 34–44 p.p.m. region and reveals a number of peaks arising from the presence of the methylene groups in the polymer. The 124–144 p.p.m. range represents the aromatic carbons having hydrogens present. The 134–144 p.p.m. region may be assigned to the substituted aromatic carbons. The chemical shifts of the mono-, di-, tri- and tetra-substituted phenyl units are predicted using ^{13}C additivity parameters.

Although the first two regions provide practically no useful information about the multisubstitution pattern of polybenzyls due to signal overlapping problems, the chemical shifts of the substituted aromatic carbons are very informative. Thus the chemical shifts 141.0–142.0, 137.6–139.6, 139.6–141.0, 136.0–136.4 and 136.7–137.6 p.p.m. may be assigned to mono-, di-, tri- and tetra-substituted aromatic polybenzyl carbons and their intensities should give the relative number of carbon atoms. A number of polybenzyls examined by ^{13}C NMR reveal[33] that approximately 24% of the phenyl rings are monosubstituted, 51% disubstituted, 17% trisubstituted and 8% tetrasubstituted. The pentasubstituted and hexasubstituted patterns were not observed, probably due to steric and electronic effects.

26.6 BRANCHING

Branching plays an important role in the polymer properties. In general, as random branching increases, melting point decreases, solubility increases, crystallinity decreases and the thermal stability of the polymer decreases.

In polybenzyls, the relative branching may be estimated if one considers the ratios of IR or NMR peak areas of: (i) monosubstituted phenyls (M) to that of methylene groups (B); (ii) polysubstituted phenyls (P) to that of methylene groups (B); and (iii) polysubstituted to monosubstituted phenyls. In general, the larger the M:B ratio, the higher the degree of branching, since the terminal groups of all short branches consist of monosubstituted benzene rings. Of course, one has to assume that no chloromethyl, dihydroanthracene, or other units are present as end groups on the chains. For the other two cases the opposite is true; thus, branching increases as the P:M and P:B ratios decrease.

Quantitatively, the number of branches in a polybenzyl could be determined from equation (5). The term X represents the number of branches in a polybenzyl and $X+1$ the monosubstituted benzene rings. Hence, the total number of protons on the terminal benzyls (monosubstituted phenyls) is $5(X+1)$, whereas the number of hydrogens on multisubstituted benzene rings will be $4[\text{DP}-(X+1)-X]$, since for every branch generated, one hydrogen is lost from the chain. The

$$\frac{P}{M} = \frac{4[\text{DP}-(X+1)-X]}{5(X+1)} \tag{5}$$

degree of polymerization is denoted by DP. Thus, the number of hydrogens on all terminal benzyls (HTB) can be calculated from equation (6).

$$\text{HTB (\%)} = \frac{5(X+1)}{5(X+1)+4[\text{DP}-(X+1)]-X} \times 100 \tag{6}$$

The degree of branching in polybenzyls derived from the polymerization of benzyl chloride is usually dependent upon molecular weight and reaction temperature. The higher the molecular weight, the higher the degree of branching, since longer polybenzyl chains provide more sites for substitution. Branching may be reduced at low temperature because the reaction rate is often slower; alkylation becomes more selective and favors the *para* position of the polybenzyl aromatic rings. The effect of catalyst and solvent does not appear to play a major role in branching for the benzyl halide polymers.

Polymer branching can be controlled by the presence of substituents on either the aromatic or the benzyl carbon. Thus, polymers prepared from the polymerization of α-ethylbenzyl chloride, 2,5-dimethylbenzyl chloride and 2,3,5,6-tetramethylbenzyl chloride are highly linear. Also, temperature and catalyst are important parameters in determining the degree of branching in polybenzyls derived from the polymerization of substituted benzyl halide monomers. For example,[19] all poly-(α-methylbenzyl) polymers synthesized from the polycondensation of α-methylbenzyl chloride in chloroethane catalyzed by $AlCl_3$ at $-125\,°C$ are linear, stereoregular and crystalline, while the same reaction system at $-78\,°C$ gives branched amorphous polybenzyls. Branching increases with increasing molecular weight and temperature. The catalyst to monomer ratio also affects the degree of polymerization and branching; the highest ratio gives the lowest molecular weight and the highest degree of branching.

Although the $AlCl_3$ catalyst can be employed to synthesize linear poly(α-methylbenzyl) at temperatures below $-80\,°C$, the introduction of an $MeCH_2NO_2 \cdot AlCl_3$ complex allows the polymerization to take place at higher temperature without the polymer losing its linearity. In fact, the model reaction of α-chloroethylbenzene with α,α'-diphenylethane gives the *p*-substituted bis(α-phenylethyl)benzene isomer exclusively when a catalyst ratio of $MeCH_2NO_2/AlCl_3 = 6$ in CH_3CH_2Cl at $-65\,°C$ is employed.[19] This behavior could be attributed to the complexation effect of nitroethane, which reduces the effective concentration of $AlCl_3$ and makes it less reactive, hence more selective.

26.7 REFERENCES

1. R. L. Shriver and L. Berger, *J. Org. Chem.*, 1941, **6**, 305.
2. H. C. Haas, D. I. Livingston and M. Saunders, *J. Polym. Sci.* 1955, **15**, 503.
3. R. W. Lenz, 'Organic Chemistry of Synthetic High Polymers', Interscience, New York, 1967, p. 228.
4. J. P. Kennedy and R. B. Isaacson, *J. Macromol. Chem.*, 1966, **1**, 541.
5. G. Montaudo, F. Bottino, S. Caccamese, P. Finocchiaro and G. Bruno, *J. Polym. Sci., Part A-1*, 1970, **8**, 2453.
6. G. Montaudo, P. Finocchiaro, S. Caccamese and F. Bottino, *J. Polym. Sci., Part A-1*, 1970, **8**, 2475.
7. N. Grassie and I. G. Meldrum, *Eur. Polym. J.*, 1971, **7**, 629.
8. J. Kuo and R. W. Lenz, *J. Polym. Sci., Polym. Chem. Ed.*, 1976, **14**, 2749.
9. A. G. Pinkus and W. H. Lin, *J. Macromol. Sci., Chem.* 1979, **A13**, 133.
10. E. J. Spanier and F. E. Caropreso, *J. Polym. Sci., Part A-1*, 1969, **7**, 2679.
11. M. Hiro and K. Arata, *Chem. Lett.* 1979, **9**, 1141.
12. C. P. Tsonis and M. U. Hasan, *Polymer*, 1983, **24**, 707.
13. C. P. Tsonis, *Polym. Bull. (Berlin)*, 1983, **9**, 349.
14. C. P. Tsonis, *J. Mol. Catal.*, 1985, **33**, 61.
15. G. Bruno, G. Montaudo, N. V. Hien, R. H. Marchessault, P. R. Sundararajan, J. E. Chandler and R. W. Lenz, *J. Polym. Sci., Polym. Lett. Ed.*, 1975, **13**, 559.
16. J. Kuo and R. W. Lenz, *J. Polym. Sci., Polym. Chem. Ed.*, 1977, **15**, 119.
17. A. Fritz and R. W. Rees, *J. Polym. Sci., Part A-1*, 1972, **10**, 2365.
18. A. Fritz, *Chem. Technol.*, 1972, **2**, 687.
19. J. E. Chandler, B. H. Johnson and R. W. Lenz, *Macromolecules*, 1980, **13**, 377.
20. J. Skura, R. W. Lenz, *Polym. Bull. (Berlin)*, 1980, **2**, 31.
21. N. A. Peppas and G. N. Valkanas, *J. Polym. Sci., Polym. Chem. Ed.*, 1974, **12**, 2567.
22. W. C. Overhults and A. D. Ketley, *Makromol. Chem.*, 1966, **95**, 143.
23. G. Montaudo, P. Finocchiaro and S. Caccamese, *J. Polym. Sci., Part A-1*, 1971, **9**, 3627.
24. L. Valentine and R. W. Winter, *J. Chem. Soc.*, 1956, 4768.
25. D. B. V. Parker, W. G. Davies and K. D. South, *J. Chem. Soc. (B)*, 1967, 471.
26. D. B. V. Parker, *Eur. Polym. J.*, 1969, **5**, 93.
27. P. F. G. Praill, *J. Chem. Soc.*, 1957, 3162.
28. A. L. Henne and H. M. Leicester, *J. Am. Chem. Soc.*, 1938, **60**, 864.

29. J. H. Lady, I. Kesse and R. E. Adams, *J. Appl. Polym. Sci.*, 1960, **3**, 71.
30. R. T. Conley, *J. Appl. Polym. Sci.*, 1965, **9**, 1107.
31. M. U. Hasan and S. A. Ali, *J. Polym. Sci., Polym. Chem. Ed.*, 1985, **23**, 1847.
32. P. R. Young and J. E. Fernandez, *J. Polym. Sci., Part A-1*, 1971, **9**, 1771.
33. M. U. Hasan and C. P. Tsonis, *J. Polym. Sci., Polym. Chem. Ed.*, 1984, **22**, 1349.

27
Polyphenylenes

MARTIN B. JONES
University of North Dakota, Grand Forks, ND, USA
and
PETER KOVACIC
University of Wisconsin-Milwaukee, Milwaukee, WI, USA

27.1 INTRODUCTION	465
27.2 OXIDATIVE COUPLING	465
27.2.1 *Chemical Means*	465
27.2.2 *Electrochemical Means*	466
27.3 ORGANOMETALLIC COUPLING	467
27.3.1 *Ullmann Type*	467
27.3.2 *Grignard Type*	467
27.4 DEHYDROGENATION OF POLYCYCLOHEXYLENES	468
27.5 CYCLOADDITION REACTIONS	469
27.6 OTHER SYNTHETIC ROUTES	470
27.7 REFERENCES	471

27.1 INTRODUCTION

Polyphenylenes are macromolecules which comprise benzenoid aromatic nuclei directly joined to one another by C–C bonds. These materials have been known for many years. Until 1979, the major interest in polyphenylenes stemmed from their characteristic thermal and thermo-oxidative stabilities. More recent continuing interest in them has arisen from the finding that one member of this class of polymers, poly (*p*-phenylene) (PPP; **1**), can be transformed from an electrical insulator into an electrical conductor upon doping with electron acceptors or donors.[1]

(1)

A comprehensive review of synthetic routes to polyphenylenes last appeared in 1971.[2] The nomenclature which will be used throughout this discussion is also found in that paper. More recent related reviews have focused on oxidative polymerizations[3,4] in general and PPP prepared by oxidative polymerization.[5] The present treatment will concentrate primarily on synthetic and mechanistic developments since 1971. Comparisons of PPPs prepared by different methods, including those reported in this chapter are given in a prior review.[5]

27.2 OXIDATIVE COUPLING

27.2.1 Chemical Means

Perhaps the most widely employed method for the preparation of polyphenylenes involves oxidative coupling of benzene and substituted benzenes *via* treatment with a Lewis acid catalyst–

oxidant system. For example, equation (1) illustrates the synthesis of PPP by stirring benzene with anhydrous $AlCl_3$ and anhydrous $CuCl_2$ for two hours at mild temperatures.[6] Recent studies[7] suggest that the mechanism of this reaction involves an initial one-electron oxidation of benzene to its radical cation, followed by association of the radical cation with several neutral benzene molecules in a propagative fashion (Scheme 1). Viewed from the side, the chain would resemble a flight of stairs and hence this is called the stair-step mechanism. When the radical cation character becomes too delocalized to further promote propagation, covalent bond formation takes place to give an oligomeric radical cation. Subsequently, a second one-electron oxidation followed by loss of two protons aromatizes the terminal rings. Oxidative rearomatization of the dihydro structures by $CuCl_2$ gives PPP.

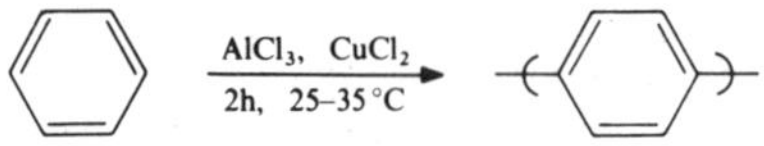

Scheme 1

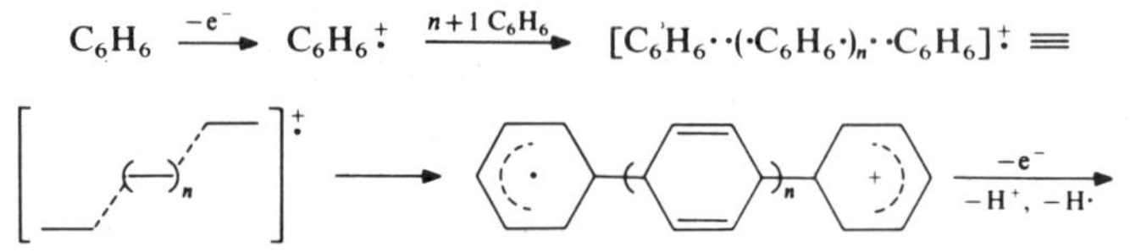

Numerous other catalysts and oxidants may also be employed for the synthesis of PPP, although the exact nature of the polymer obtained depends to some extent on the identity of the reagents.[5] Among the more novel reagents which convert benzene to PPP are Cu^{2+} and Ru^{3+} ion-exchanged montmorillonite clays,[8,9] and AsF_3/AsF_5,[10] which yields a homogeneous solution of PPP. AsF_5 may also be used to convert biphenyl and p-terphenyl to PPP.[11] The latter reaction combines the polymerization step with a doping step to give conductive PPP directly. Structural features and physical properties of PPP prepared *via* oxidative polymerization have been reviewed.[5]

Polyphenylenes with *ortho* linkages (2) are obtained from oxidative polymerization of monosubstituted benzenes such as toluene or chlorobenzene under conditions analogous to those employed for the preparation of PPP.[5] Polyphenylenes with *meta* linkages (3) are synthesized by treatment of *m*-terphenyl or mixtures of *m*-terphenyl and biphenyl with $AlCl_3/CuCl_2$ at temperatures of 85–180 °C.[12,13] In general, the *ortho* and *meta* isomers are more processable than the *para* isomer, because of the *para* polymer's propensity to form intermolecular stacks.

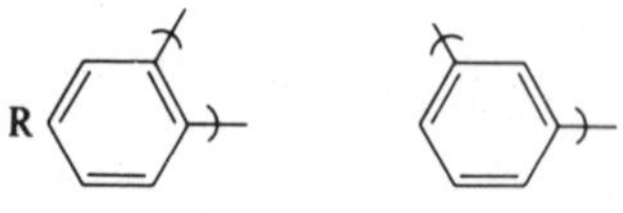

R = Me, Cl
(2)   (3)

27.2.2 Electrochemical Means

Anodic oxidation of benzene or biphenyl in liquid sulfur dioxide on a platinum electrode yields passivating films with quaternary ammonium perchlorate as the electrolyte or conductive dendritic deposits with quaternary ammonium tetrafluoroborate as the electrolyte.[14] In either case, the polyphenylene which is obtained largely consists of *para* linkages and contains variable amounts of oxygen, present as phenolic groups. Electrochemical oxidation of benzene in a two phase, HF–benzene system on a platinum electrode gives free-standing films of polyphenylene which contain *ortho*, *meta* and *para* links.[15] Electrochemical polymerization of benzene in nitromethane with aluminum chloride and water or an amine as additives results in the formation of black deposits of polyphenylene on a platinum anode.[16] Flexible, electrically conductive films of PPP are obtained *via* anodic oxidation of benzene in a nitrobenzene solution containing a composite electrolyte of $CuCl_2$ and $LiAsF_6$.[17,18] The latter product is structurally similar to chemically produced PPP, as determined by IR and elemental analyses. Likewise, anodic oxidation of benzene in the presence of $BF_3 \cdot OEt_2$ gives flexible, highly conducting films of PPP.[19] Electrochemical oxidation of *p*-terphenyl in acetonitrile solution with tetrabutylammonium perchlorate as the electrolyte gives films whose dominant structure is PPP-like.[20] Under analogous conditions, *o*-terphenyl initially dehydrocyclizes to triphenylene, then couples to give an electroactive polymeric film. The structural features of films obtained from anodic oxidation of biphenyl and *m*-terphenyl are more disordered but are dominated by triphenylene-like moieties.

27.3 ORGANOMETALLIC COUPLING

27.3.1 Ullmann Type

A number of methods for preparing oligophenylenes and polyphenylenes involve the use of metals. Condensation of dihaloaromatics with copper (Ullmann reaction) or with alkali metals (Wurtz–Fittig reaction) has been used extensively in the past to prepare biphenyl and terphenyl species.[2] However, these reactions suffer from deficiencies such as low yield and significant formation of byproducts and have usually been employed for synthesis of oligomers with specific structures. Linear polyphenylenes with both *p*-phenylene units (1–4 consecutive) and *o*- or *m*-phenylene units have been prepared *via* Ullmann crosscoupling of iodobiphenyl with diiodobenzene, diiodobiphenyl or iodoterphenyl.[21,22] For example, the reaction of 2-iodobiphenyl with *p*-diiodobenzene in the presence of Cu powder at 250–265 °C for two hours affords, after chromatographic separation, 2,2″-diphenyl-*p*-terphenyl (**4a**, 13.8%), 2,2‴-diphenyl-*p*-quaterphenyl (**4b**, 9.8%), 2,2⁗-diphenyl-*p*-quinquephenyl (**4c**, 7.4%) and 2,2⁗′-diphenyl-*p*-sexiphenyl (**4d**, 2.4%).[21] Similarly, highly branched polyphenylenes containing *p*-phenylene units (*e.g.* **5**) have been synthesized by Ullmann coupling reactions.[23]

(4)
a: *n* = 1
b: *n* = 2
c: *n* = 3
d: *n* = 4

(5)

27.3.2 Grignard Type

Of greater utility for polymer preparation is the coupling of Grignard reagents. A discussion of early research toward the preparation of polyphenylenes *via* Grignard reactions may be found in a prior review.[2] More recently, poly(*p*-phenylene) has been synthesized in good yield *via* coupling of the mono-Grignard reagent of *p*-dihalobenzene in the presence of organometallic promoters such as (2,2′-bipyridyl-*N*,*N*′)dichloronickel [NiCl$_2$(bipy)][24,25] and bis(acetylacetonato)nickel [Ni(acac)$_2$] (equation 2).[26] In each case the resulting product is lighter in color but otherwise similar in physical and IR spectral characteristics to PPP prepared *via* oxidative polymerization. Poly(*m*-phenylene) may be synthesized in analogous fashion from *m*-dichlorobenzene.[25] The proposed mechanism for the Ni-catalyzed coupling is illustrated in Scheme 2.

Organic promoters have also been gainfully employed in the synthesis of polyphenylenes from Grignard reagents. Reaction of the mono-Grignard reagent of *p*-dibromobenzene with CCl$_2$FCClF$_2$ in THF at 30–50 °C gives PPP as a light yellow, insoluble powder.[27] Treatment of the bis-Grignard reagent of *p*-dibromobenzene with either *cis*- or *trans*-1,4-dichloro-2-butene in THF yields PPP which is similar in spectral and physical properties to the polymer prepared *via* oxidative polymerization (equation 3).[28] The mechanism of the latter reaction is thought to involve electron transfer from the Grignard reagent to the promoter, resulting in phenyl-like radicals which then couple to give the polymer (Scheme 3).

Crosscoupling of aryl Grignard reagents with aryl halides and dihalobenzenes catalyzed by Ni(acac)$_2$ has been employed for the synthesis of specific polyphenylenes.[29,30] For example, equation (4) illustrates the preparation of a linear sexiphenyl with both *para* and *meta* linkages in 84% yield.[30] When the reactants are sterically crowded, the yield of the reaction is diminished (*e.g.* 1% of the corresponding sexiphenyl was obtained from reaction of 2,2′-diiodobiphenyl with the

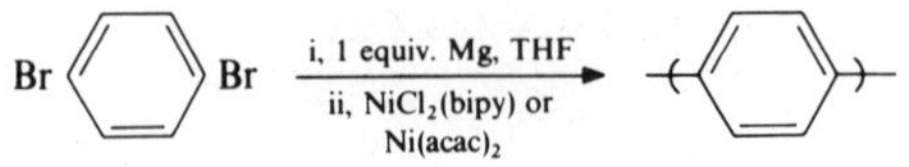
Br
Br
i, 1 equiv. Mg, THF
ii, NiCl2(bipy) or Ni(acac)2

(2)

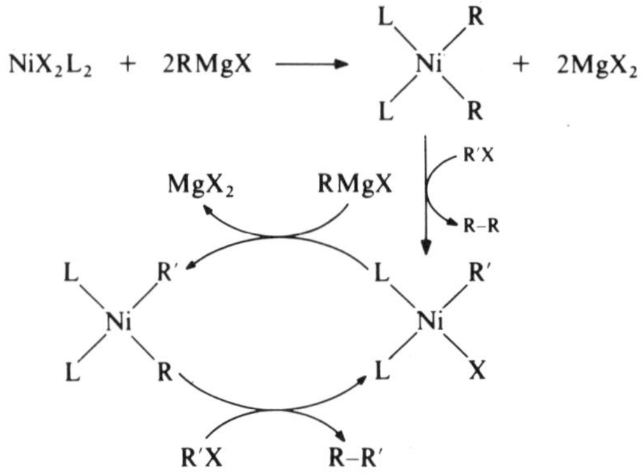
NiX2L2 + 2RMgX
Ni
+ 2MgX2
R'X
R–R
MgX2
RMgX
R'X
R–R'

Scheme 2

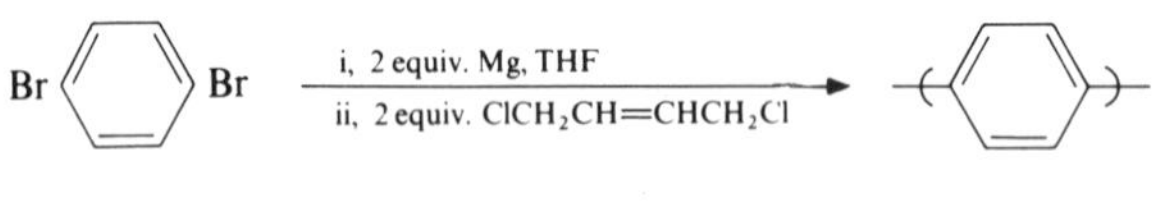
Br
Br
i, 2 equiv. Mg, THF
ii, 2 equiv. ClCH2CH=CHCH2Cl

(3)

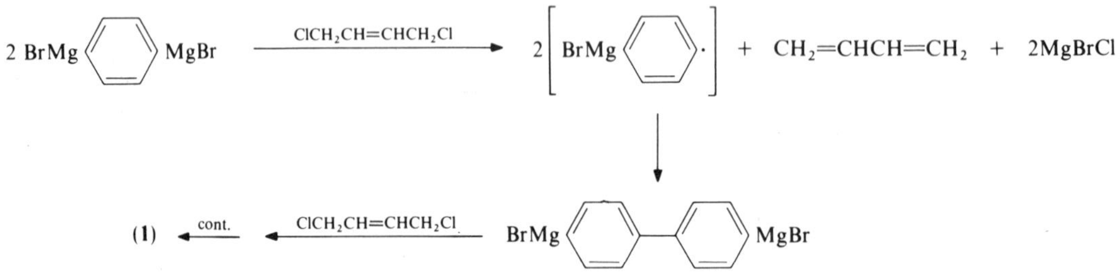
2 BrMg
MgBr
ClCH2CH=CHCH2Cl
2 BrMg
+ CH2=CHCH=CH2 + 2MgBrCl
(1)
cont.
ClCH2CH=CHCH2Cl
BrMg
MgBr

Scheme 3

Grignard reagent derived from 2-bromobiphenyl[30]). In general, the Grignard crosscoupling reactions give higher yields and fewer side-products than Ullmann preparations of the same compounds.

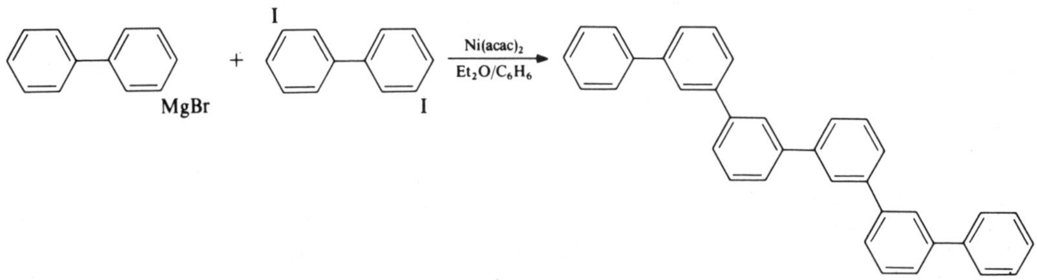
MgBr
I
I
Ni(acac)2
Et2O/C6H6

(4)

27.4 DEHYDROGENATION OF POLYCYCLOHEXYLENES

Polymerization of 1,3-cyclohexadiene by treatment with various Ziegler type catalysts[2,31] or *n*-butyllithium[32] affords poly(1,3-cyclohexadiene) with regular 1,4-linked mer units. Subsequent dehydrogenation with chloranil or *via* halogenation/pyrolysis gives PPP (Scheme 4).[31] Cationic polymerization of 1,3-cyclohexadiene produces a polymer with a mixture of 1,4- and 1,2-bonded structures, which can be converted by halogenation/pyrolysis to a polyphenylene with *ortho* and *para* linkages.[31]

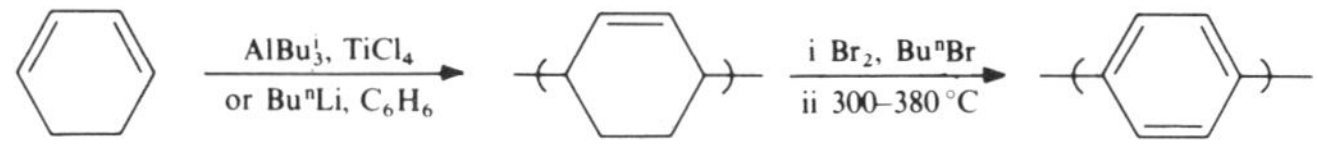

Scheme 4

Diester derivatives of 5,6-dihydroxy-1,3-cyclohexadiene have been polymerized under radical conditions using benzoyl peroxide or AIBN and the resulting polycyclohexylenes converted to polyphenylenes by pyrolysis (Scheme 5).[33,34] The *cis*-diacetate derivative bears particular mention as a starting material, since it is prepared *via* bacterial oxidation of benzene by the genetically modified bacteria *Pseudomonas putida*.[33] Proton and ^{13}C NMR spectra of the polycyclohexylenes suggest that these materials are predominantly, but not exclusively, of the 1,4-structure.[34] Small amounts (<15%) of the 1,2-structure are also present. The molecular weights of the polycylohexylenes and their thermal stabilities depend on the relative stereochemistry and the size of the ester substituents.[34] The *trans*-diacetate polymer has the lowest M_w (6300 g mol^{-1}) and has a glass transition temperature (T_g) of 190 °C, while the *cis*-dipivalate polycyclohexylene possesses the highest M_w (68 000 g mol^{-1}) and a T_g of 238 °C. Pyrolysis of each of the *cis*-diesters proceeds smoothly with loss of both esters, resulting in aromatization of the polymer.[34] The *trans*-diacetate, however, loses more weight than could be accounted for by the expected aromatization reaction. Each of the pyrolyzed polymers contains residual carbonyl absorption in the IR, indicating that complete aromatization does not occur during the thermal treatment.

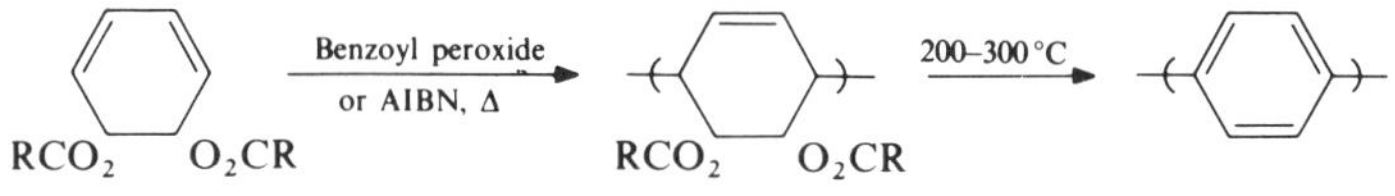

Scheme 5

27.5 CYCLOADDITION REACTIONS

Phenylated polyphenylenes of high molecular weight[2,35] and unsubstituted poly(*p*-phenylene)[36] have been obtained from 1,4-cycloadditions of biscyclopentadieneones or bispyrones with bisacetylenes. The phenyl-substituted polymers, which have approximately equivalent amounts of *m*- and *p*-phenylene units, are light yellow, amorphous and soluble in typical organic solvents. The unsubstituted PPP prepared *via* cycloaddition of 5,5'-*p*-phenylenebis-2-pyrone with *p*-diethynylbenzene (equation 5) shares the same physical properties of insolubility and infusibility as PPP prepared by the previously mentioned routes. However, the polymer from cycloaddition is light yellow in color, compared to dark brown for the product of oxidative polymerization of benzene with AlCl$_3$/CuCl$_2$. Increased amounts of *m*-phenylene units in the polyphenylenes leads to decreased melting points and increased solubility.[37]

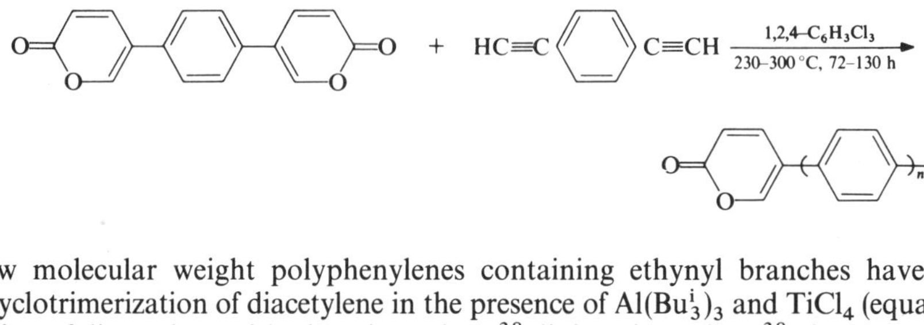

Low molecular weight polyphenylenes containing ethynyl branches have been prepared by polycyclotrimerization of diacetylene in the presence of Al(Bu$^i_3$)$_3$ and TiCl$_4$ (equation 6).[38] Copolymerization of diacetylene with phenylacetylene,[38] diphenylacetylene,[38] vinylacetylene,[38] or various 1-alkynes[39] under analogous conditions gives polyphenylenes with phenyl, vinyl, or alkyl branches respectively. Thermal treatment of the homopolymer of diacetylene or of the copolymer of diacetylene and phenylacetylene results in crosslinking to yield materials with high thermal stabilities.[38]

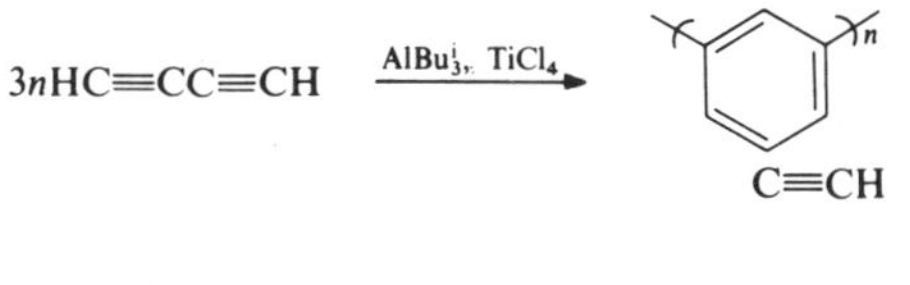

1,4-Diphenyl-*N,N*-dimethylaminohex-5-en-1-yne, obtained from the base-catalyzed Stevens rearrangement of the corresponding allylpropynylammonium cation, has been shown to undergo thermal ring closure to give terphenyl in good yield (Scheme 6).[40] The mechanism of the thermal cyclization is postulated to proceed through a [3,3] sigmatropic rearrangement of the *N,N*-dimethylaminohex-5-en-1-yne to give an alleneamine, which then undergoes a [1,3]H shift to give the corresponding *N,N*-dimethylamino-substituted hexatriene. Electrocyclic ring closure of the hexatriene yields a 1,3-cyclohexadiene which then eliminates dimethylamine to yield the final product.[40]

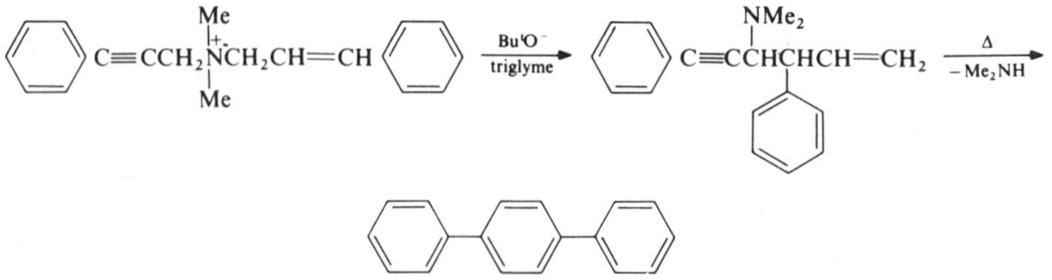

Scheme 6

Extension of this procedure to poly(*N,N*-dimethylaminohex-5-en-1-yne)s has successfully afforded phenyl-substituted PPPs (Scheme 7).[41] Thus alkylation of 4,4'-bis(3-*N,N*-dimethylamino-1-propynyl)-1,1'-biphenyl with 1,4-bis(2-bromomethyl-2-phenylethenyl)benzene furnishes an excellent yield of the corresponding polymeric ammonium bromide. Treatment with anhydrous potassium *t*-butoxide gives an isomeric mixture of the Stevens rearrangement products. Finally, pyrolysis of the polymeric dimethylamino-substituted ene–yne produces the polyphenylene. Support for the previously mentioned thermal cyclization mechanism is given by a differential scanning calorimetry (DSC) thermogram of the poly(*N,N*-dimethylaminohex-5-en-1-yne), which exhibits a major exotherm corresponding to the cyclization at 195–220 °C and a shoulder on the high temperature side of this exotherm which corresponds to the loss of dimethylamine and concomitant aromatization.[41]

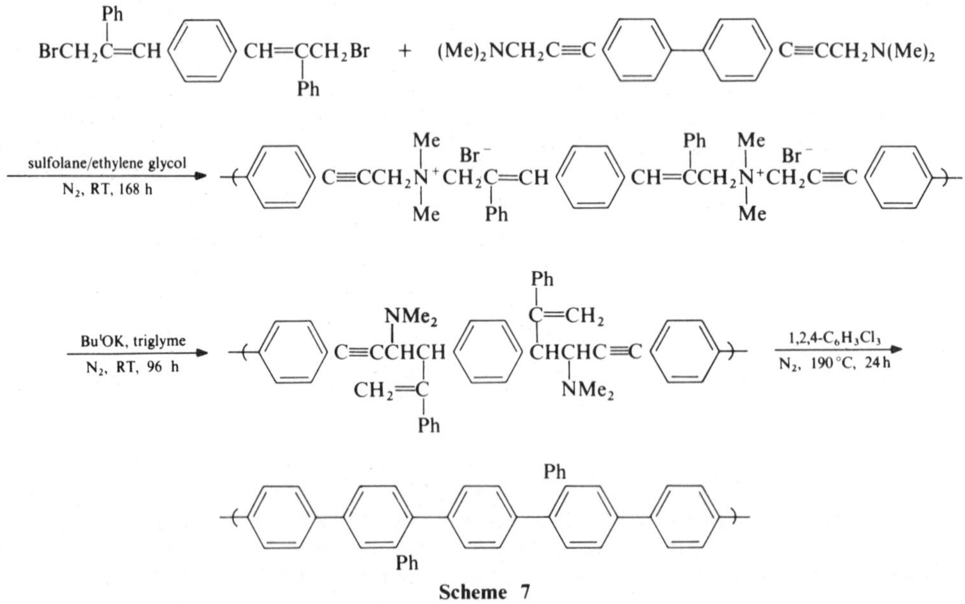

Scheme 7

27.6 OTHER SYNTHETIC ROUTES

Poly(*p*-phenylene) containing small amounts of nitrogen in the form of bridging diazo groups has been prepared by heating of the bisdiazonium salt of benzidine in ammoniacal copper(I) ion (equation 7).[2,42] The polymer obtained from this reaction is largely insoluble. Formation and subsequent heating of the bisdiazonium salt/CuCl complex in pure water rather than in base increases the yield of benzene-soluble polymer.[43] $FeCl_2$ can be employed in place of CuCl, with similar results. The soluble PPP, which has molecular weights from 1000 to 16 000, contains one azo group and two to three chlorines for each ten phenyl rings. The polymerization mechanism is thought to proceed *via* homolytic cleavage of the C–N≡N bonds, with formation of N_2 and biphenyl radicals (diradicals). Homocoupling of the radicals yields the polymer.

$$^+N_2\text{-C}_6H_4\text{-}N_2^+ \xrightarrow[\Delta]{CuCl, NH_3} -(\text{C}_6H_4)- \quad (7)$$

Electrochemical reduction of 1,4-dibromobenzene in THF/HMPA on a mercury pool electrode in the presence of Ni^0 complexes as catalysts and lithium perchlorate as the electrolyte gives good yields of insoluble PPP as a pale yellow powder (equation 8).[44,45] Similar product is obtained from 4,4′-dibromobiphenyl.[45] The Ni^0 complexes are generated *in situ* by electrochemical reduction of NiX_2L_2 (X = Cl or Br; L = PPh_3 or $Ph_2PCH_2CH_2PPh_2$). The IR spectrum of this polymer is nearly identical to that of PPP prepared by oxidative polymerization of benzene.

$$Br\text{-C}_6H_4\text{-}Br + 2ne^- \xrightarrow[THF, HMPA]{Ni^0} -(\text{C}_6H_4)- + 2nBr^- \quad (8)$$

In related work, an acetonitrile solution of $Ni(PPh_3)_2(p\text{-}C_6H_4Br)Br$ (prepared from $Ni(PPh_3)_4$ and *p*-dibromobenzene) has been reduced on Pt or glassy carbon electrodes with tetrabutylammonium perchlorate as the electrolyte.[46] The resulting electroactive polymeric coating is proposed to be of the *p*-phenylene type, but containing one Ni (bridging between adjacent phenylene rings) per 6–7 phenylene moieties.

Soluble prepolymers are obtained from polycyclocondensation of the bis(ethyl ketal) of 4,4′-diacetylbiphenyl with the ethyl ketal of acetophenone in the presence of HCl.[47,48] Upon heating from 300–450 °C, the prepolymers are converted to crosslinked polyphenylenes.

27.7 REFERENCES

1. R. L. Elsenbaumer and L. W. Shacklette, in 'Handbook of Conducting Polymers', ed. T. A. Skotheim, Dekker, New York, 1986, vol. 1, chap. 7.
2. J. G. Speight, P. Kovacic and F. W. Koch, *J. Macromol. Sci., Rev. Macromol. Chem.*, 1971, **5**, 275.
3. M. B. Jones and P. Kovacic, in 'Encyclopedia of Polymer Science and Engineering', ed. J. I. Kroschwitz, Wiley, New York, 1987, vol. 10, p. 670.
4. H. Naarmann, M. Beaujean, R. Merenyi and H. G. Viehe, *Polym. Bull. (Berlin)*, 1980, **2**, 683.
5. P. Kovacic and M. B. Jones, *Chem. Rev.*, 1987, **87**, 357.
6. P. Kovacic and A. Kyriakis, *J. Am. Chem. Soc.*, 1963, **85**, 454.
7. S. Milosevich, K. Saichek, L. Hinchey, W. B. England and P. Kovacic, *J. Am. Chem. Soc.*, 1983, **105**, 1088.
8. F. Stoessel, J. L. Guth and R. Wey, *Clay Miner.*, 1977, **12**, 255.
9. Y. Soma, M. Soma and I. Harada, *Chem. Phys. Lett.*, 1983, **99**, 153.
10. M. Aldissi and R. Liepins, *J. Chem. Soc., Chem. Commun.*, 1984, 255.
11. L. W. Shacklette, H. Eckhardt, R. R. Chance, G. G. Miller, D. M. Ivory and R. H. Baughman, *J. Chem. Phys.*, 1980, **73**, 4098.
12. N. Bilow and J. B. Rust (Hughes Aircraft Co.), US Pat. 3 582 498 (1971) (*Chem. Abstr.*, 1971, **75**, 49 859).
13. N. Bilow and L. J. Miller, *J. Macromol. Sci., Chem.*, 1967, **1**, 183.
14. M. Delamare, P.-C. Lacaze, J.-Y. Dumousseau and J.-E. Dubois, *Electrochim. Acta*, 1982, **27**, 61.
15. I. Rubinstein, *J. Polym. Sci., Polym. Chem. Ed.*, 1983, **21**, 3035.
16. K. Kaeriyama, M. Sato, K. Someno and S. Tanaka, *J. Chem. Soc., Chem. Commun.*, 1984, 1199.
17. M. Satoh, K. Kaneto and K. Yoshino, *J. Chem. Soc., Chem. Commun.*, 1985, 1629.
18. M. Satoh, M. Tabata, K. Kaneto and K. Yoshino, *Polym. Commun.*, 1985, **26**, 356.
19. T. Ohsawa, T. Inoue, S. Takeda, K. Kaneto and K. Yoshino, *Polym. Commun.*, 1986, **27**, 246.
20. G. Schiavon, S. Zecchin, G. Zotti and S. Cattarin, *J. Electroanal. Chem. Interfacial Chem.*, 1986, **213**, 53.
21. S. Ozasa, N. Hatada, Y. Fujioka and E. Ibuki, *Bull. Chem. Soc. Jpn.*, 1980, **53**, 2610.
22. S. Ozasa, Y. Fujioka and E. Ibuki, *Chem. Pharm. Bull.*, 1982, **30**, 2698.
23. Y. Fujioka, S. Ozasa, K. Sato and E. Ibuki, *Chem. Pharm. Bull.*, 1985, **33**, 22.
24. T. Yamamoto and A. Yamamoto, *Chem. Lett.*, 1977, 353.
25. T. Yamamoto, Y. Hayashi and A. Yamamoto, *Bull. Chem. Soc. Jpn.*, 1978, **51**, 2091.
26. M. Sato, K. Kaeriyama and K. Someno, *Makromol. Chem.*, 1983, **184**, 2241.
27. Toshiba Corp., *Jpn. Pat.* 84 58 029 (1984) (*Chem. Abstr.*, 1984, **101**, 73 279).

28. S. K. Taylor, S. G. Bennett, I. Khoury and P. Kovacic, *J. Polym. Sci., Polym. Lett. Ed.*, 1981, **19**, 85.
29. E. Ibuki, S. Ozasa, Y. Fujioka and Y. Yanagihara, *Chem. Pharm. Bull.*, 1982, **30**, 802.
30. E. Ibuki, S. Ozasa, Y. Fujioka, M. Okada and Y. Yanagihara, *Chem. Pharm. Bull.*, 1982, **30**, 2369.
31. D. A. Frey, M. Hasegawa and C. S. Marvel, *J. Polym. Sci., Part A*, 1963, **1**, 2057.
32. P. E. Cassidy and C. S. Marvel, *Macromol. Synth.*, 1972, **4**, 7.
33. D. G. H. Ballard, A. Courtis, I. M. Shirley and S. C. Taylor, *Macromolecules*, 1988, **21**, 294.
34. D. R. McKean and J. K. Stille, *Macromolecules*, 1987, **20**, 1787.
35. J. K. Stille, F. W. Harris, H. Mukamal, R. O. Rakutis, C. L. Schilling, G. K. Noren and J. A. Reed, *Adv. Chem. Ser.*, 1969, **91**, 625.
36. H. F. vanKerckhoven, Y. K. Gilliams and J. K. Stille, *Macromolecules*, 1972, **5**, 541.
37. J. N. Braham, T. Hodgins, T. Katto, R. T. Kohl and J. K. Stille, *Macromolecules*, 1978, **11**, 343.
38. V. V. Korshak, V. A. Sergeev and Yu. A. Chernomordik, *Vysokomol. Soedin., Ser. B*, 1977, **19**, 493 (*Chem. Abstr.*, 1977, **87**, 136677).
39. V. V. Korshak, V. A. Sergeev, Yu. A. Chernomordik and S. B. Alaev, *Izv. Akad. Nauk SSSR, Ser. Khim.*, 1977, 1645 (*Chem. Abstr.*, 1977, **87**, 118498).
40. T. Laird, W. D. Ollis and I. O. Sutherland, *J. Chem. Soc., Perkin Trans. 1*, 1980, 1473.
41. M. R. Unroe, B. A. Reinhardt and E. J. Soloski, *Polym. Prepr., Am. Chem. Soc., Div. Polym. Chem.*, 1987, **28** (1), 183.
42. A. A. Berlin, *J. Polym. Sci.*, 1961, **55**, 621.
43. S. Hayama and S. Niino, *J. Polym. Sci., Polym. Chem. Ed.*, 1974, **12**, 357.
44. J.-F. Fauvarque, M.-A, Petit, F. Pfluger, A. Jutand, C. Chevrot and M. Troupel, *Makromol. Chem., Rapid Commun.*, 1983, **4**, 455.
45. J.-F. Fauvarque, A. Digua, M.-A. Petit and J. Savard, *Makromol. Chem.*, 1985, **186**, 2415.
46. G. Schiavon, G. Zotti and G. Bontempelli, *J. Electroanal. Chem. Interfacial Chem.*, 1984, **161**, 323.
47. V. V. Korshak, M. M. Teplyakov, D. M. Kakauridze, D. A. Shapiro and E. L. Vulakh, *Vysokomol. Soedin., Ser. A*, 1976, **18**, 1831 (*Chem. Abstr.*, 1977, **86**, 44247).
48. V. V. Korshak, M. M. Teplyakov and R. A. Dvorikova, *Vysokomol. Soedin., Ser. A*, 1982, **24**, 277 (*Chem. Abstr.*, 1982, **92**, 181664).

28
Poly(phenylene oxide)s

DWAIN M. WHITE
General Electric Co., Schenectady, NY, USA

28.1 INTRODUCTION	473
28.2 OXIDATIVE COUPLING POLYMERIZATION	473
28.3 DISPLACEMENT POLYMERIZATION	478
28.4 ULLMAN ETHER SYNTHESIS	478
28.5 ELECTROCHEMICAL SYNTHESES	478
28.6 SUBSTITUTION REACTIONS	479
28.7 POLYMER STRUCTURE	479
28.8 REFERENCES	480

28.1 INTRODUCTION

Poly(phenylene oxide)s [poly(oxyphenylene)s] are useful materials for engineering thermoplastic applications because of their thermal, oxidative and chemical stability. The polymers are readily prepared from phenols in good yields by a variety of oxidative coupling procedures and other types of step-growth polymerization reactions. Poly(2,6-dimethyl-1,4-phenylene oxide), PPO® resin, is the most important commercial poly(phenylene oxide) and is sold as a blend with polystyrene and other additives as Noryl® resin by the General Electric Co. The preparation of PPO and other poly(phenylene oxide)s is described in this chapter, with a major emphasis on PPO.

28.2 OXIDATIVE COUPLING POLYMERIZATION

Oxidative coupling polymerization is often the preferred route to poly(phenylene oxide)s. The copper-catalyzed oxidation of 2,6-dimethylphenol to PPO was first reported by Hay in 1959 (equation 1).[1] He found that the polymer (**1**) forms in high yield along with a small amount of a by-product, 3,3′,5,5′-tetramethyl-4,4′-diphenoquinone (**2**). Since then, a large number of other catalyst systems have been reported. A number of transition metal catalyzed systems and also a variety of uncatalyzed systems are described below.

Many phenols can be used in oxidative coupling polymerization reactions. These phenols usually have substituents in the two *ortho* positions. Most phenols with *ortho* alkyl, aryl, chloro or bromo substituents produce poly(phenylene oxide) in high yield. When the groups are bulky (*e.g. t*-butyl or isopropyl), the diphenoquinone becomes the main product and the yield of polymer is low. The diphenoquinone is also formed in significant quantities from 2,6-dimethoxyphenol. Examples of product yields are shown in Table 1.

Table 1 Effect of Substituents on Product Yields

Monomer	Polymer (%)	Diphenoquinone (%)	Ref.
2,6-Dimethylphenol	85	3	2
2-Ethyl-6-methylphenol	82	—	2
2-Isopropyl-6-methylphenol	62	—	2
2,6-Diethylphenol	81	—	2
2-Chloro-6-methylphenol	83	—	2
2-Bromo-6-methylphenol	18	—	2
2-Methyl-6-methoxyphenol	60	—	2
2-Methyl-6-phenylphenol	93	5	3
2,6-Diphenylphenol	96	3	4
2-t-Butyl-6-methylphenol	—	45	2
2,6-Diisopropylphenol	—	53	2
2,6-Di-t-butylphenol	—	97	2
2,6-Dimethoxyphenol	—	74	2

Phenols with only one *ortho* substituent and an open *para* position can be oxidatively coupled but the polymer is usually highly branched and colored. Presumably, oxidative coupling at the open *ortho* position leads to branching while other types of oxidation produce quinones and other colored moieties. Less branching is achieved by using a bulky amine as part of the catalyst to block the open *ortho* position.[5] 2-Amylpyridine and 2,5-dinonylpyridine have been used instead of pyridine in the catalyst to improve both the yield and viscosity properties of the polymer formed from o-cresol. Small amounts of phenol and o-cresol can be incorporated into PPO when they are copolymerized with 2,6-dimethylphenol. The resultant polymers are not highly branched even when the catalyst systems do not contain bulky amines.

2,3,6-Trisubstituted phenols can also be polymerized. In the case of 2,3,6-trimethylphenol, the polymer crystallizes and becomes insoluble in the reaction medium.[6] Copolymers from 2,6-dimethylphenol and 2,3,6-trimethylphenol are soluble when the 2,3,6-trimethylphenol is the minor component.[6,7] 2,3,5,6-Tetrasubstituted phenols usually form low molecular weight products.[6]

Phenols with *para* substitution often act as chain stoppers and limit the molecular weight of the polymer when used in small amounts. Under some conditions, however, the *para* substituent is removed during polymerization and the phenol becomes a typical 1,4-oxyphenylene repeat unit. For example, the 4-methyl group in 2,4,6-trimethylphenol is converted to formaldehyde during oxidation with a copper-catalyst system[8] or with manganese oxide as the oxidizing agent.[9] In another example, the 4-halo group in 4-halo-2,6-dimethylphenol is displaced by phenoxide by a process described in Section 28.3.

Catalysts for oxidative coupling polymerizations are usually composed of a transition metal salt and a base. Molecular oxygen is normally the oxidizing agent. Most liquids that dissolve the polymer, such as benzene or toluene, can be used as solvents for the reaction. A desiccant or a polar liquid is often added to prevent water, a by-product of the polymerization, from forming a separate phase which may interact with the catalyst. The transition metal is usually copper or manganese, although other metals, such as cobalt, are also active.

Copper-catalyzed systems often contain a copper halide and either an aliphatic or heterocyclic amine.[1,10,11] When diamines such as N,N,N',N'-tetramethylethylenediamine (TMEDA) are used, they are often combined with an aliphatic monoamine (*e.g.* trimethylamine).[10] N,N'-Di-t-butylethylenediamine forms a very active catalyst that is not easily hydrolyzed during polymerization.[11] Di-n-butylamine also forms a catalyst with copper that is not sufficiently affected by water to require addition of a drying agent.[12] The types of heterocyclic amines that are used include pyridine,[1] morpholine, 4-N,N-(dimethylamino)pyridine,[13] and diazabicyclononene.[14] Many polymeric amines are also reported to function as ligands, including poly(vinylpyridine),[15,16] imidazole polymers,[17] poly(N,N-dialkylaminostyrene)s,[18] poly(iminotrimethylene)s[19] and oligomeric amines.[16] Copper catalysts using insoluble copper species,[20] copper(II) chloride with potassium hydroxide,[21] and mixtures of copper(I) and copper(II) ions[22] are also known.

A typical polymerization procedure based on a copper-catalyst system[12] that produces high molecular weight PPO is as follows. Oxygen is bubbled into a vigorously stirred solution of 30 cm³ toluene, 2.5 cm³ di-n-butylamine, 2 cm³ methanol and 0.096 g copper(I) bromide at 25 °C. Recrystallized 2,6-dimethylphenol (6.1 g in 18 cm³ toluene) is added over a 20 min period. An exothermic reaction occurs during the monomer addition. When the temperature approaches 40 °C, it is maintained at that temperature with a water bath. The formation of diphenoquinone (**2**) during the reaction results in a red-brown color and eventually the formation of red crystals of (2). Approximately 75 min after the start of the monomer addition, the reaction mixture is viscous and the polymerization is terminated by the addition of 2 cm^3 acetic acid. The PPO is precipitated by adding 200 cm^3 methanol. After washing the precipitate with methanol, dissolving the polymer in toluene and reprecipitating with methanol, the polymer is dried in a vacuum oven at 80 °C. The yield is 5 g and the intrinsic viscosity (measured in chloroform at 25 °C) is approximately 5×10^{-5} m^3 g^{-1}.

A variety of manganese-catalyst systems are described in the literature. Many are similar to the copper systems since they require a ligand and a base. Some of the catalysts are effective at very low manganese concentrations. An example is a combination of manganese chloride, benzoin oxime and sodium hydroxide in toluene and methanol.[23] High molecular weight polymer is formed with a manganese to monomer ratio of 1:2000. Other systems use ligands such as triethanolamine[24] and o-hydroxyazo compounds.[25] Thiols have also been used in manganese-catalyzed systems.[26]

Poly(phenylene oxide)s are also prepared with non-catalytic systems by using metal oxides and other inorganic compounds as the oxidizing agent instead of molecular oxygen. Manganese dioxide,[27] lead dioxide,[28] silver oxide[29] and sodium bismuthate[30] produce high molecular weight polymer from many 2,6-disubstituted phenols including 2,6-dimethylphenol, 2,6-diphenylphenol and 2-benzyl-6-methylphenol.[31] Addition of an excess of the solid oxidizing agent to the phenol in benzene or toluene at 25 °C causes an exothermic reaction and rapid formation of polymer. The polymer is isolated by filtering off the metal salt residues and then precipitating the polymer with methanol from the filtrate solution. Sometimes amines are added to the reaction mixture to enhance the activity.[28,31]

Low molecular weight PPO with a very narrow molecular weight distribution can be made by redistributing low molecular weight oligomers in methylene chloride at 0 °C (equation 2). Either the dimer (3; $x = 2$), the trimer (3; $x = 3$) or a mixture of these and other low oligomers redistribute in the presence of a small amount of an initiator such as the diphenoquinone (2) to produce 2,6-dimethylphenol and higher molecular weight oligomers. The high molecular weight oligomers form a complex with methylene chloride and precipitate from the solution. The isolated polymer has a molecular weight of approximately 1800 and a dispersity (M_w/M_n) of 1.08.[32]

A variety of mechanisms have been proposed for the oxidative coupling polymerization of phenols.[33] The role of catalysts and the kinetics of polymerization have been studied extensively.[34] The pathways for the progression of the oxidized phenols to polymer are described here. Oxidation of the phenol by the oxidized form of the catalyst generates an aryloxy radical which dimerizes to form a cyclohexadienone that can undergo enolization and redistribution reactions. With 2,6-dimethylphenol as the monomer, 2,6-dimethylphenoxy radicals are generated. They couple to form a quinol ether (4) which enolizes to produce the dimer (5) as shown in equation (3). Oxidation of the dimer produces a dimer radical, which couples with other radicals that are present. When it couples with a 2,6-dimethoxy radical to form a quinol ether, enolization produces the trimer, as shown in equation (4). Higher molecular weight oligomeric radicals are converted to the next higher oligomers in an analogous manner and eventually even higher molecular weights are attained (equation 5).

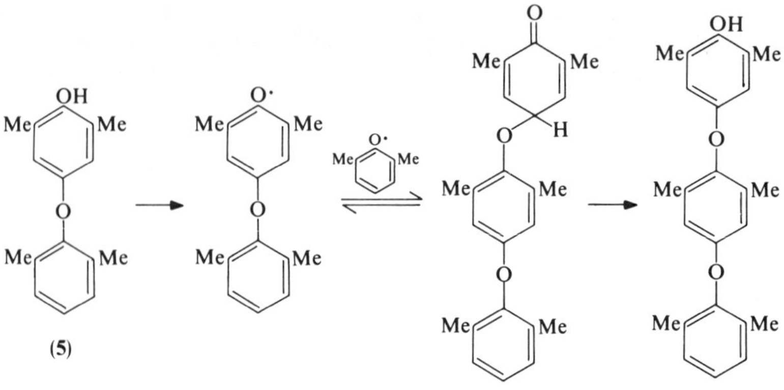

(4)

(5)

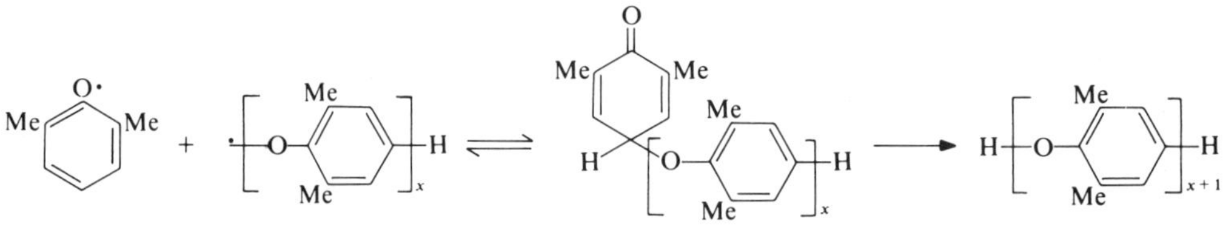

(5)

The polymerization does not proceed only by this process of adding one unit at a time to the growing chain, however, since coupling can also occur between other combinations of oligomeric radicals. For example, when the dimer radical couples with another dimer radical, ketal (**6**) is formed (equation 6). Or if the dimer radical couples with a larger oligomeric radical, an analogous higher molecular weight ketal is formed. The quinone ketals are in equilibrium with the aryloxy radicals and can dissociate either back to their precursor radicals or to a pair of new oligomeric radicals. Eventually, these equilibration processes produce the 2,6-dimethylphenoxy radical which couples with another oligomeric radical to form the quinol ether and tautomerizes to the phenol. For example (equation 7), trimer radical (**7**) is produced when (**6**) dissociates. A tetramer is produced whenever (**7**) couples with a 2,6-dimethylphenoxy radical (the other product of the dissociation of **6**) to form the intermediate quinol ether (**8**) and then the tetramer. The 2,6-dimethylphenoxy radical can also couple with other radicals in the system in an analogous manner.

The processes above show how the redistribution reactions of oligomers, the formation of ketals and their subsequent dissociations eventually generate quinol ethers which enolize to the more stable phenol. The enolization drives the reaction to high molecular weight since it produces a phenol-terminated oligomer that is more stable than either the precursor radicals or the quinol ether. The phenolic group that is generated can be oxidized to generate a new aryloxy radical which can then participate in the redistribution–enolization scheme. Each combination of oxidation, radical coupling, dissociation and enolization effectively doubles the average molecular weight of the system.

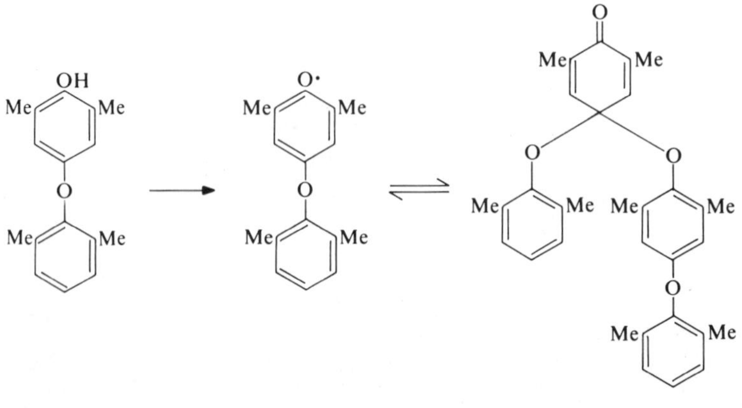

(6)

(6)

A second process that occurs along with the dissociation/redistribution scheme is an intramolecular rearrangement in which the cyclohexadienone acetal groups move to an adjacent unit on a polymer chain.[35,36] The rearrangement may proceed *via* two consecutive sigmatropic rearrangements with an intermediate bis(cyclohexadienone) (equation 8). The rearrangement allows an acetal group to move along the chain randomly until it reaches a terminal position. The enolization converts it to a phenol (equation 9). This process, which predominates at low temperatures, leads to migration of the internal acetal structure to the end of the chain without requiring the dissociation that occurs in the redistribution process. At higher temperatures, the quinone ketal may dissociate before reaching the end of the chain and both mechanisms become important.

A large body of evidence supports these pathways and rules out several alternative mechanisms.[35-43]

28.3 DISPLACEMENT POLYMERIZATION

Another general method for preparing poly(phenylene oxide)s is by a displacement of halogen from a 4-halo-2,6-disubstituted phenol to produce an ether linkage. With 4-bromo-2,6-dimethylphenol, aqueous sodium hydroxide and a catalytic amount of an initiator, high molecular weight PPO forms rapidly at 25 °C (equation 10).[44,45,46] Unlike oxidative coupling polymerization, which requires a stoichiometric amount of oxidizing agent, the displacement polymerization requires only a catalytic amount of an oxidizing agent to serve as an initiator. Typical initiators include potassium ferricyanide, lead dioxide and other metal oxides, 2,4,6-tri-t-butylphenoxy radical, iodine and sodium hypochlorite.[44,47]

$$n\ HO\text{-}C_6H_2(Me)_2\text{-}Br\ +\ n\ KOH\ \xrightarrow[H_2O,\ C_6H_6]{K_3Fe(CN)_6}\ [\text{-}O\text{-}C_6H_2(Me)_2\text{-}]_n\ +\ nKBr \tag{10}$$

Moderately hindered phenols can be polymerized. For example, 3,4-dibromo-2,6-dimethylphenol is converted to high molecular weight poly(3-bromo-2,6-dimethyl-1,4-phenylene oxide).[48] Only the bromine in the 4-position is displaced. With 3,4,5-tribromo-2,6-dimethylphenol, the polymeric product is too insoluble to attain high molecular weight. High molecular weight soluble copolymers containing 3,5-dibromoPPO units can be prepared, however. As much as 70% 3,5-dibromo units can be introduced when the other starting phenol is 3,4-dibromo-3,5-dimethylphenol.

2,4,6-Trihalophenols are polymerized under similar conditions. The sodium salt of 4-bromo-2,6-dichlorophenol in dimethyl sulfoxide with an initiator is converted to linear poly(2,6-dichloro-1,4-phenylene oxide).[49] When either a copper or silver salt of 2,4,6-trichlorophenol is heated, a branched poly(dichlorophenylene oxide) is formed.[50] These reactions may be intermediate in behavior between halogen displacement polymerizations and the Ullman ether synthesis type of process described in the next section.

Substituents in the *ortho* positions are required for these reactions. The starting phenol is usually recovered unchanged in high yield if the *ortho* position is open.[51] Apparently, oxidation at an open *ortho* position consumes initiator and possibly interferes with the propagation step in the chain process.

The mechanism of the displacement polymerization has been described as a coupling reaction between an aryloxy radical (initially generated by the initiator) and a phenoxide anion (generated by an excess of base) to form a radical-anion intermediate that loses bromide ion to form a new aryloxy radical.[45] Recently, evidence has been presented that shows that the reaction has the characteristics of a single electron transfer (SET) mechanism.[52]

Since radical intermediates are involved in both displacement polymerization and in oxidative coupling polymerization, it is not surprising that mixtures of 2,6-dimethylphenol and its halogenated derivatives (3,4-dibromo and 3,4,5-tribromo) can form random copolymers.[47] When the base used for these copolymerizations is an excess of amine, the amine serves both as a ligand in the oxidative coupling process and as the base for the displacement process.

28.4 ULLMAN ETHER SYNTHESIS

The Ullman ether synthesis provides a route to poly(phenylene oxide)s that cannot be prepared by the methods described above since it is not necessary to have *ortho* substituted phenols. An example of the reaction is the synthesis of poly(1,4-phenylene oxide) from the sodium salt of 4-bromophenol (equation 11).[53] The reaction differs considerably from the displacement polymerizations described above. Not only are *ortho* substituents not a requirement but oxidizing agents such as oxygen must also be avoided and high reaction temperatures are needed. The solvent requirement is fairly specific; materials such as the dimethoxybenzenes are effective solvents.

$$n\ NaO\text{-}C_6H_4\text{-}Br\ \xrightarrow[\text{pyridine}]{CuCl}\ [\text{-}O\text{-}C_6H_4\text{-}]_n\ +\ n\ NaBr \tag{11}$$

28.5 ELECTROCHEMICAL SYNTHESES

Recently, there has been increased activity in applying electrochemical techniques to poly(phenylene oxide) synthesis. Many of the procedures produce polymeric films that are insoluble and coat the electrode. This provides a method of coating a metal with an insoluble polymeric coating. Monomers polymerized include *o*-cresol, allylmethylphenols, catechol and resorcinol.[54] In many cases, the polymers have not been thoroughly characterized.

An electrochemical method that produces soluble poly(phenylene oxide)s uses tetramethylammonium bromide as the supporting electrolyte and methylene chloride as the solvent.[55] Monomers that have been polymerized include phenol and 2,6-dimethylphenol.

28.6 SUBSTITUTION REACTIONS

Another approach for the preparation of specific poly(phenylene oxide)s is through chemical derivitization of an existing poly(phenylene oxide). Halogenation,[56,57] alkylation,[58] carboxylation,[59] sulfonation,[60,61] nitration[62] and amination of halogenated methyl groups[63] produce the corresponding derivatized polymers. For example, bromination under ionic conditions introduces bromine into one of the open aryl positions on each repeat unit to produce poly(3-bromo-2,6-dimethyl-1,4-phenylene oxide). With high bromine concentrations, dibromination of each ring is possible. The product, poly(3,5-dibromo-2,6-dimethyl-1,4-phenylene oxide) appears to be highly crystalline since it is insoluble in organic solvents. Under free radical conditions, the methyl group can be brominated. With an excess of *N*-bromosuccinimide and with light as an initiator, poly(2-bromomethyl-6-methyl-1,4-phenylene oxide) is formed.[57] When lithiation[58,59] of PPO is followed by carboxylation[59] or alkylation, complex side-chains (*e.g.* comb-like structures[64] and polycyclic ring systems[65]) are possible end-products.

28.7 POLYMER STRUCTURE

When PPO is made in a typical copper-catalyzed polymerization system, the polymer is composed of 2,6-dimethyl-1,4-oxyphenylene units.[66] Lower molecular weight polymers have 4-hydroxy-3,5-dimethylphenyl 'head' end-groups and 2,6-dimethylphenoxy 'tail' end-groups.[67] Side reactions reduce the relative number of 4-hydroxy-3,5-dimethylphenyl end-groups in higher molecular weight polymers.

Polymers that have been heated in solution with biphenoquinone (2) chemically incorporate the diphenoquinone by a redox process and, as a result, contain a biphenyl moiety (9) either at the end of the chain ($x = 0$) or internally ($x = 1, 2, etc.$).[68] The resultant polymer then has phenolic hydroxyl groups at both ends of the molecule. Redistribution[42] of PPO or copolymerization[69] of 2,6-dimethylphenol with a bisphenol such as 2,2-bis(3,5-dimethyl-4-hydroxyphenyl)propane incorporates the bisphenol and also generates a polymer with hydroxyl groups on both ends of the molecule.

(9) $x = 0, 1, 2, \ldots$

When PPO is made in the presence of a secondary amine, the amine is incorporated into the polymer as an aminomethylphenol (10) at the head end of the polymer chain. Smaller amounts of the amine are found internally on backbone methyl groups. The end-group units are thermally unstable and are not present in thermally processed polymers.[70]

(10)   (11)

After PPO has been heated to temperatures at the high end of the thermal processing range (approximately 350 °C), a small number of internal phenolic units with structure (11) are present along the backbone. They arise from a free-radical induced rearrangement of backbone units.[71] Usually, there is less than one of these units per PPO molecule.

28.8 REFERENCES

1. A. S. Hay, H. S. Blanchard, G. F. Endres and J. W. Eustance, *J. Am. Chem. Soc.*, 1959, **81**, 6335.
2. A. S. Hay, *J. Polym. Sci.*, 1962, **58**. 581.
3. D. M. White and H. J. Klopfer, *J. Polym. Sci., Part A-1*, 1970, **8**, 1427.
4. D. M. White and H. J. Klopfer, *J. Polym. Sci., Part A-1*, 1972, **10**, 1565.
5. A. S. Hay and G. F. Endres, *J. Polym. Sci., Polym. Lett. Ed.*, 1965, **3**, 887.
6. D. M. White, unpublished results.
7. E. Yonemitsu, A. Sugio and T. Kawaki, US Pat. 4 011 200 (1977) (*Chem. Abstr.*, 1980, **92**, 90 966).
8. G. D. Cooper, US Pat. 3 749 693 (1973) (*Chem. Abstr.*, 1973, **79**, 146 989).
9. E. M. McNelis, *J. Am. Chem. Soc.*, 1966, **88**, 1074.
10. A. S. Hay, US Pat. 3 306 875 (1967) (*Chem. Abstr.*, 1965, **62**, 708).
11. A. S. Hay, US Pat. 4 028 341 (1977) (*Chem. Abstr.*, 1976, **84**, 5674).
12. J. G. Bennett and G. D. Cooper, US Pat. 4 092 294 (1978) (*Chem. Abstr.*, 1978, **89**, 25 103).
13. J. P. J. Verlaan, P. J. T. Alferink and G. Challa, *J. Mol. Catal.*, 1984, **24**, 235.
14. H. Wieden and U. Bahr, Br. Pat. 1 134 613 (1969) (*Chem. Abstr.*, 1969, **70**, 29 589).
15. E. Tsuchida, M. Kaneko and H. Nishide, *Makromol. Chem.*, 1972, **151**, 221.
16. E. Tsuchida, H. Nishikawa and E. Terada, *J. Polym. Sci., Polym. Chem. Ed.*, 1976, **14**, 825.
17. E. Tsuchida, 'New Frontiers in Organometallic and Inorganic Chemistry, 2nd Meeting', Science Press, Beijing, 1982, p. 207 (*Chem. Abstr.*, 1985, **102**, 185 554z).
18. E. Tsuchida, H. Nishide and T. Nishiyama, *Makromol. Chem.*, 1974, **175**, 3047.
19. C. E. Konin, B. L. Hiemstra, G. Challa, M. van de Velde and E. J. Goethals, *J. Mol. Catal.*, 1985, **32**, 309.
20. H. Nishide, Y. Suzuki and E. Tsuchida, *Eur. Polym. J.*, 1981, **17**, 573.
21. S. Tsuruya, K. Kinumi, K. Hagi and M. Masai, *J. Mol. Catal.*, 1983, **22**, 47; S. Tsuruya, N. Nakamae and T. Yonezawa, *J. Catal.*, 1976, **44**, 40.
22. T. Yoshimura, W. Storck and G. Manecke, *Makromol. Chem.*, 1977, **178**, 75, 97.
23. W. K. Olander, US Pat 4 054 553 (1977) (*Chem. Abstr.*, 1977, **86**, 73 412).
24. H. Komoto and K. Ohmura, *Makromol. Chem.*, 1973, **166**, 57.
25. K. Ueno, T. Maruyama, H. Inoue, Y. Tatsukami and M. Isobe, *Eur. Pat. Appl.* 0 059 958 (1983) (*Chem. Abstr.*, 1983, **98**, 35 144).
26. M. Kaneko and A. Yamada, *Makromol. Chem.*, 1981, **182**, 101.
27. E. J. McNelis, *J. Org. Chem.*, 1966, **31**, 1255.
28. H. M. van Dort, C. R. H. I. de Jonge and W. J. Mijs, *J. Polym. Sci., Part C*, 1968, **22**, 431.
29. B. O. Lindgren, *Acta Chem. Scand.*, 1960, **14**, 1203.
30. E. Kon and E. McNelis, *J. Org. Chem.*, 1975, **40**, 1515.
31. T. Sugie, K. Ishihara and K. Honda, *Jpn. Pat.* 51 30 899 (1976) (*Chem. Abstr.*, 1976, **85**, 33 707u).
32. D. M. White, *Macromolecules*, 1979, **12**, 1008.
33. H. L. Finkbeiner, A. S. Hay and D. M. White, in 'Polymer Processes, High Polymers', ed. C. E. Schildknecht and I. S. Skeist, Wiley, New York, 1977, vol. 29, p. 537.
34. G. F. Endres, A. S. Hay, J. W. Eustance, *J. Org. Chem.*, 1963, **28**, 1300; E. Ochiai, *Tetrahedron*, 1964, **20**, 1831; M. Kaneko, H. Nishide and E. Tsuchida, *Kogyo Kagaku Zasshi*, 1971, **74**, 1194; E. Tsuchida, H. Nishide and T. Nishiyama, *Makromol. Chem.*, 1974, **175**, 3047; A. J. Schouten, N. Prak and G. Challa, *Makromol. Chem.*, 1977, **178**, 401; D. P. Mobley, *J. Polym. Sci., Polym. Chem. Ed.*, 1984, **22**, 3203; C. C. Price and K. Nakaoka, *Macromolecules*, 1971, **4**, 363.
35. H. L. Finkbeiner, G. F. Endres, H. S. Blanchard and J. W. Eustance, *SPE Trans.*, 1962, **2**, 110; *Polym. Prep., Am. Chem. Soc., Div. Polym. Chem.*, 1968, **2**, 340.
36. D. M. White, *Polym. Prepr., Am. Chem. Soc., Div. Polym. Chem.*, 1968, **9** (1), 663.
37. G. F. Endres and J. Kwiatek, *J. Polym. Sci.*, 1962, **58**, 593; G. D. Cooper, H. S. Blanchard, G. F. Endres and H. L. Finkbeiner, *J. Am. Chem. Soc.*, 1965, **87**, 3996.
38. W. J. Mijs, O. E. van Lohuizen, J. Bussink and L. Vollbracht, *Tetrahedron*, 1967, **23**, 2253.
39. D. J. Williams and R. Kreilick, *J. Am. Chem. Soc.*, 1967, **89**, 3408.
40. G. D. Cooper, A. R. Gilbert and H. L. Finkbeiner, *Polym. Prep., Am. Chem. Soc., Div. Polym. Chem.*, 1966, **7**, 166.
41. D. A. Bolon, *J. Org. Chem.*, 1967, **32**, 1584.
42. D. M. White, *J. Org. Chem.*, 1969, **34**, 297.
43. G. D. Cooper and J. G. Bennett, *J. Org. Chem.*, 1972, **37**, 441.
44. G. D. Staffin and C. C. Price, *J. Am. Chem. Soc.*, 1960, **82**, 3632.
45. C. C. Price, *J. Paint Technol.*, 1966, **38**, 705.
46. C. C. Price and N. S. Chu, *J. Polym. Sci.*, 1962, **61**, 135.
47. J. R. Hall, *J. Polym. Sci., Polym. Lett. Ed.*, 1966, **4**, 463.
48. D. M. White, *Polym. Prepr., Am. Chem. Soc., Div. Polym. Chem.*, 1974, **15** (1), 210; D. M. White and H. J. Klopfer, *ACS Symp. Ser.*, 1975, **6**, 169.
49. G. S. Stamatoff, US Pat. 3 236 807 (1966) (*Chem. Abstr.*, 1966, **64**, 14 386).
50. H. S. Blanchard, H. L. Finkbeiner and G. A. Russell, *J. Polym. Sci.*, 1962, **58**, 469; J. F. Harrod, *Can. J. Chem.*, 1969, **47**, 637; B. Carr and J. F. Harrod, *J. Am. Chem. Soc.*, 1973, **95**, 5707.
51. D. M. White, unpublished results.
52. V. Percec and T. D. Shaffer, *J. Polym. Sci., Polym. Lett. Ed.*, 1986, **24**, 439.
53. H. M. van Dort, C. A. M. Hoefs, E. P. Magre, A. J. Schopf and K. Yntema, *Eur. Polym. J.*, 1968, **4**, 275.
54. M. Musiani, G. Pagura and G. Mengoli, *Electrochim. Acta*, 1985, **30**, 501; S. Azim and S. Guruvich, *Bull. Electrochem.*, 1986, **2**, 353; M. Langmuir and V. R. Koch, 'Materials Research Society, Symposium Proceedings', 1986, p. 55 (*Biomed. Mater.*, 1986, 293; *Chem. Abstr.*, 1987, **106**, 72 860); M. Vijayan, S. Pitchumani and V. Krishnan, *Bull. Electrochem.*, 1986, **2**, 349.
55. K. Yamamoto, H. Nishide and E. Tsuchida, *Makromol. Chem., Rapid Commun.*, 1987, **8**, 11; E. Tsuchida, H. Nishide and T. Maekawa, *ACS Symp. Ser.*, 1985, **282**, 175; *J. Macromol. Sci., Chem.*, 1984, **A21**, 1081.
56. A. S. Hay, US Pat 3 528 858 (1968) (*Chem. Abstr.*, 1971, **74**, 4222).

57. D. M. White and C. M. Orlando, *ACS Symp. Ser.*, 1975, **6**, 178.
58. A. S. Hay and A. J. Chalk, *J. Polym. Sci., Polym. Lett. Ed.*, 1968, **6**, 105.
59. A. S. Hay and A. J. Chalk, *J. Polym. Sci., Part A-1*, 1969, **7**, 691.
60. Y. Huang, G. Cong and W. J. MacKnight, *Macromolecules*, 1986, **19**, 2267.
61. A. S. Hay, *US Pat.* 3 528 858 (1970) (*Chem. Abstr.*, 1971, **74**, 4222).
62. J. V. Crivello, *J. Org. Chem.*, 1981, **46**, 3056.
63. A. S. Hay, *US Pat.* 3 378 505 (1968) (*Chem. Abstr.*, 1968, **68**, 115 414).
64. C. Pugh and V. Percec, *Polym. Bull. (Berlin)*, 1986, **16** (6), 513.
65. J. F. Klebe, *Polym. Prep., Am. Chem. Soc., Div. Polym. Chem.*, 1971, **12** (1), 43.
66. F. Laupretre and L. Monnerie, *Eur. Polym. J.*, 1975, **11**, 845.
67. D. M. White, *Polym. Prepr., Am. Chem. Soc., Div. Polym. Chem.*, 1972, **13**, 373.
68. D. M. White, *J. Polym. Sci., Polym. Chem. Ed.*, 1981, **19**, 1367.
69. W. Heitz and W. Risse, *US Pat.* 4 521 584 (1985) (*Chem. Abstr.*, 1984, **101**, 192 700); H. Nava and V. Percec, *J. Polym. Sci., Polym. Chem. Ed.*, 1986, **24**, 965.
70. D. M. White and S. A. Nye, in press.
71. A. Factor, *J. Polym. Sci., Part A-1*, 1969, **7**, 363; A. Factor, H. L. Finkbeiner, R. A. Jerussi and D. M. White, *J. Org. Chem.*, 1970, **35**, 57.

29
Poly(ether ketone)s

PHILIP A. STANILAND
ICI PLC, Wilton, Cleveland, UK

29.1	INTRODUCTION	483
29.2	BACKGROUND	484
29.3	ELECTROPHILIC ROUTE	484
	29.3.1 *Early Work*	484
	29.3.2 *AlCl₃ Route*	486
	29.3.3 *HF Process*	487
	29.3.4 *Use of Phosgene*	488
	29.3.5 *Other Catalytic Systems*	488
29.4	NUCLEOPHILIC ROUTE	489
	29.4.1 *Stoichiometry and Molecular Weight Control*	490
	29.4.2 *Carbonate Process*	490
	29.4.3 *Poly(ether ether ketone) PEEK*	491
	29.4.4 *Other Polymers Derived From 4,4'-Difluorobenzophenone (BDF)*	492
	29.4.5 *Ether Cleavage*	492
	29.4.6 *Variants of the Nucleophilic Process*	493
29.5	PROPERTIES OF POLY(ETHER KETONES)	494
	29.5.1 *Miscibility and Isomorphism*	494
29.6	CHEMISTRY OF POLY(ETHER KETONES)	495
	29.6.1 *Characterization*	496
	29.6.2 *Stabilization*	496
	29.6.3 *Thermal Decomposition*	496
	29.6.4 *Sulfonation*	496
	29.6.5 *Crosslinking*	496
29.7	REFERENCES	496

29.1 INTRODUCTION

Poly(ether ketones) comprise the class of polymers in which arylene groups are linked by ether and carbonyl groups. Since alkylene linkages are generally not present, and are undesirable, the acronym PAEK [poly(aryl ether ketone)] may be used[1] to describe the general class.

A polymer marketed as Stilan (RTM) by the Raychem Corporation had structure (**1**), usually referred to as PEK, while the polymer with structure (**2**) is PEEK, produced commercially by Imperial Chemical Industries as Victrex PEEK (RTM). The systematic name of (**2**) is poly(oxy-1,4-phenyleneoxy-1,4-phenylenecarbonyl-1,4-phenylene), which is sufficiently cumbersome to justify the use of the acronym.

In these cases 1,4-phenylene groups have been assumed, but other poly(ether ketones) may contain such groups as 1,3-phenylene, biphenylene, naphthylene, *etc.*

29.2 BACKGROUND

The poly(ether ketones) are a desirable class of polymers which are attracting increasing interest at the present time. Their development has been the subject of recent reviews;[2,3] however, new publications, mainly in the form of patents, are becoming ever more frequent. Their desirability stems from their extremely high thermal stability, and in this they resemble the polysulfones whose synthesis and development has also been reviewed.[2,3] However, the polysulfones are, with few exceptions, amorphous and subject to attack by solvents. Liquids which may cause cracking, softening, *etc.* are present among those which occur in or around aircraft, *e.g.* dichloromethane (paint stripper), isopropanol (de-icing fluid), hydraulic fluid, kerosene, *etc.* Poly(ether ketones) are usually crystalline and are therefore resistant to attack by solvent, which is especially important when they are used, as is commonly the case, in an aerospace environment. The only common room-temperature solvent known for PEK or PEEK is concentrated sulfuric acid.

While crystallinity gives the poly(ether ketones) a unique set of properties among thermoplastics, it is this property, coupled with melting points generally above 300 °C, which makes their synthesis so difficult. The problem is how to keep the polymers in solution and so obtain high molecular weights. This can be achieved by working at very high temperatures[4] (the 'nucleophilic' process) or by working in strongly acidic media[5] (the 'electrophilic' process) in which the carbonyl groups are protonated, allowing the polymer to remain in solution.

The first generally acknowledged attempt[6] to produce a poly(ether ketone) (Scheme 1) failed to produce high molecular weight polymer because the product came out of solution prematurely. This was an example of the electrophilic approach (or Friedel–Crafts acylation) to the synthesis, as was the first attempt[7] to make PEK (Scheme 2); again, the molecular weight was too low to give useful mechanical properties. An inherent viscosity (IV) of about 0.6 (or a reduced viscosity (RV) of about 0.8), measured in sulfuric acid, is required before useful mechanical properties, particularly toughness, are obtained. This is very approximate, as the solution viscosity required will vary according to structure and the presence of structural defects such as branching.[8]

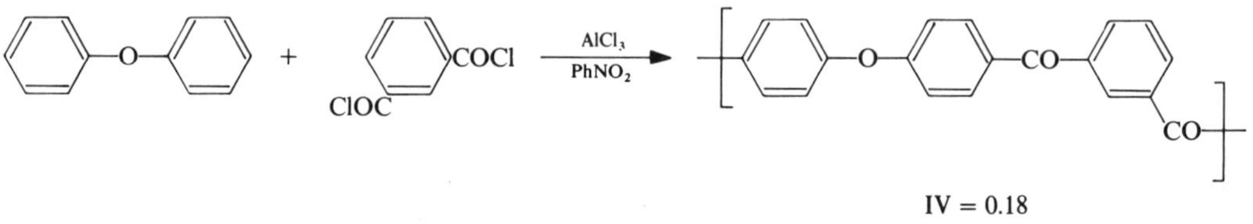

Scheme 1

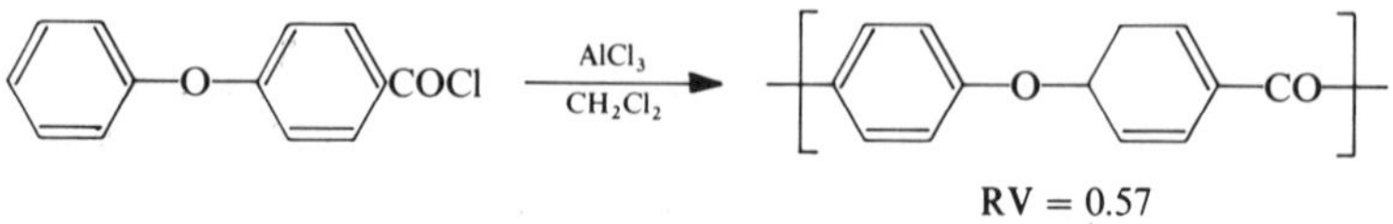

Scheme 2

The properties of the PAEKs are such that high selling prices are possible, allowing the use of costly raw materials and/or costly processes. However, there is undoubted pressure to utilize cheap and available starting materials, such as phosgene, diphenyl ether, terephthaloyl chloride, *etc.*, to produce another generation of these polymers, and in this respect the first attempts at synthesizing PAEKs are instructive.

29.3 ELECTROPHILIC ROUTE

29.3.1 Early Work

The first attempt[6] to produce poly(ether ketones) used aluminum chloride (1.5 mols per mol of –COCl) in nitrobenzene to condense isophthaloyl chloride (IC) or terephthaloyl chloride (TC) with diphenyl ether (DPO) or biphenyl. Poly(ether ketone ketone) (PEKK) with IV = 0.18 was produced (Scheme 1) at 65 °C from IC/DPO, from which fibres could be drawn. The material 'softened' at 250 °C and was soluble in *m*-cresol and tetrachloroethylene. The first PEK was produced[7] in a similar manner using aluminum chloride (1.4 mols per mol of –COCl) in dichloromethane to condense *p*-phenoxybenzoyl chloride at room temperature (Scheme 2). The polymer was soluble in dichloroacetic acid and had a viscosity ratio (VR) of 1.57 (equivalent to a reduced viscosity of 0.57) at a concentration of 1% in that solvent. Among others, the polymers (3)–(5) were also synthesized (Scheme 3). Polymer (4) yielded tough coherent film on pressing at 270 °C, while (5) was said not to melt below 310 °C.

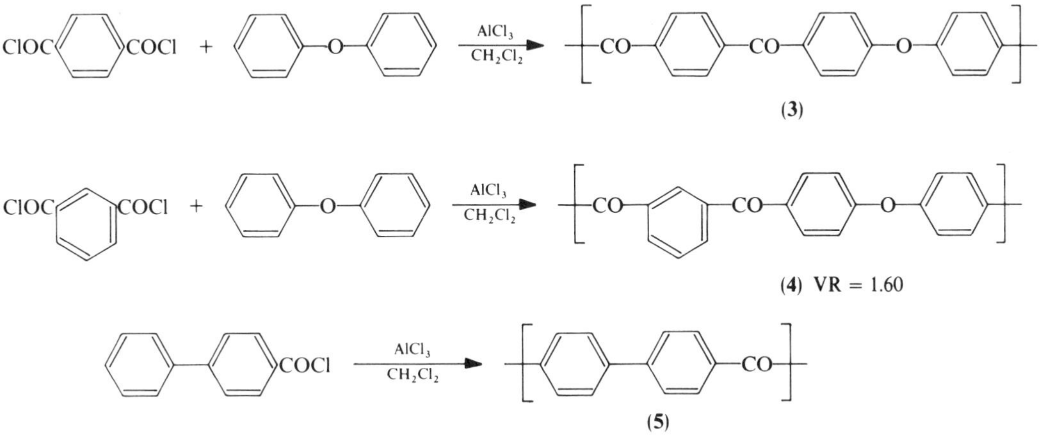

Scheme 3

It is clear that, in general, the polymers made in these first experiments were too low in molecular weight to give really useful properties, and in some cases must have been of low crystallinity, as shown by their solubility, this possibly being due to the presence of aberrant structures.

PEK was prepared from *p*-phenoxybenzoic acid, using polyphosphoric acid (PPA)[9] as solvent and catalyst (equation 1).

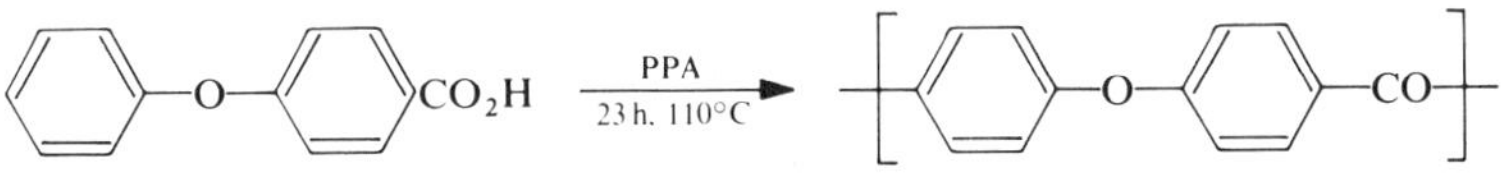

The polymer was reported to have a melting point (T_m) of 360 °C and to be insoluble in organic solvents, but, surprisingly, was described as 'essentially amorphous'. *m*-Phenoxybenzoic acid was also polymerized to give a polymer of IV = 0.45 and T_m = 350 °C. A more effective acidic system, which led to the first truly high molecular weight poly(ether ketone), was the use of a mixed HF/BF$_3$ solvent/catalyst combination.[5,30] *p*-Phenoxybenzoyl chloride was polymerized in a stainless steel reactor to give a 97% yield of PEK of IV = 2.76, glass transition temperature (T_g) = 163 °C and T_m = 361 °C. The latter properties are essentially those now accepted for PEK polymer.

BF$_3$ was used at 2 or 3 mol per carbonyl group and HF at 2–10 mol per mol of BF$_3$. Molecular weight, it was suggested, could be controlled by varying reaction time, with IV values as high as 7–8 being accessible. Polymer (5) was synthesized with IV = 1.7, which could be moulded at 530 °C to a hard rigid chip. In all, some 63 examples of homopolymers and copolymers were reported, but not in detail.

An attempt to produce PEK by reaction of phosgene with diphenyl ether under these conditions, gave only a low yield of low molecular weight material (equation 2). The main interest appears to have been in film- and fibre-forming materials. It was shown that a cast film of PEK could be biaxially drawn to give oriented films of high modulus and tensile strength and capable of elongations up to 82%.

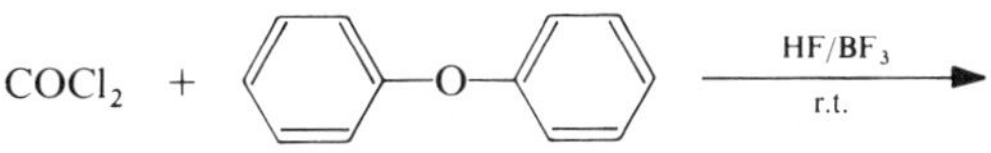

29.3.2 AlCl₃ Route

The crystalline poly(ether ketones) are insoluble in the commonly used Friedel–Crafts reaction solvents such as dichloromethane, 1,2-dichloroethane, nitrobenzene, carbon disulfide, *etc.*, and it is this which prevents attainment of high molecular weight. Nevertheless, because of its ease as a laboratory technique, the AlCl₃-catalyzed process has been extensively used,[10–15] to produce research quantities of, usually, low molecular weight polymers.

The realization that the polymer/catalyst complex is solubilized if large excesses of AlCl₃ are used has enabled the production of high molecular weight polymers.[16–20]

The use of AlCl₃ in the ratio of 2.6–5.6 mol per mol of reacting acid chloride group, in the presence of a Lewis base (which is believed to reduce alkylation by the reaction solvent), has yielded a number of high molecular weight compositions (Scheme 4).[16]

Scheme 4

A wide variety of Lewis bases may be employed, but DMF, *n*-butyronitrile, tetramethylammonium chloride and lithium chloride are favoured.

The effect of varying the ratio of AlCl₃ to monomer is shown in Figure 1. The process has seen further development with the use of dispersants[19] (*e.g.* lithium stearate) and liquefaction agents[20] (*e.g.* liquid HCl).

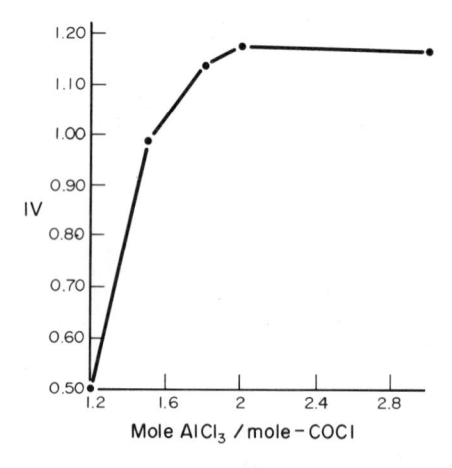

Figure 1 Effect of AlCl₃/*p*-phenoxybenzoyl chloride ratio on IV of PEK produced

The route appears now to be commercially viable, although requiring the use and disposal of large amounts of aluminum chloride. It is versatile, and has been used[22] to prepare complex copoly(ether ketones) containing imide, amide, ester, azo, quinoxaline, benzimidazole, benzoxazole and benzothiazole structures, of which (**6**) is an example.

(6)

29.3.3 HF Process

Polymerization in HF/BF$_3$ has also seen considerable development[5,23-26] and has been used commercially. When *p*-phenoxybenzoyl chloride is polycondensed in a non-metallic reactor, *e.g.* of poly(tetrafluoroethylene) (PTFE) construction, PEK of improved colour and melt stability, capable of elongations of 50% or more, is obtained.[24] The molecular weight is controlled in an IV range of 0.8 to 1.65 by inclusion of a capping agent, such as biphenyl or diphenyl ether.

The mechanical properties of PEK made by this process are improved when low polymer is removed by acetone extraction.[26] Further improvements in colour, stability and reduced branching are achieved by 'double end capping',[27] in which a nucleophilic and an electrophilic reagent are included in the recipe, as shown in Scheme 5. The process is versatile and has been used to prepare the polymers (**7**)[28] and (**8**) to (**11**)[31] (Scheme 6).

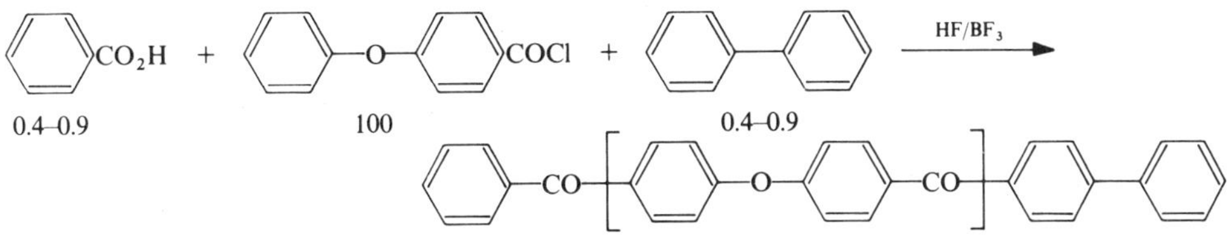

Scheme 5

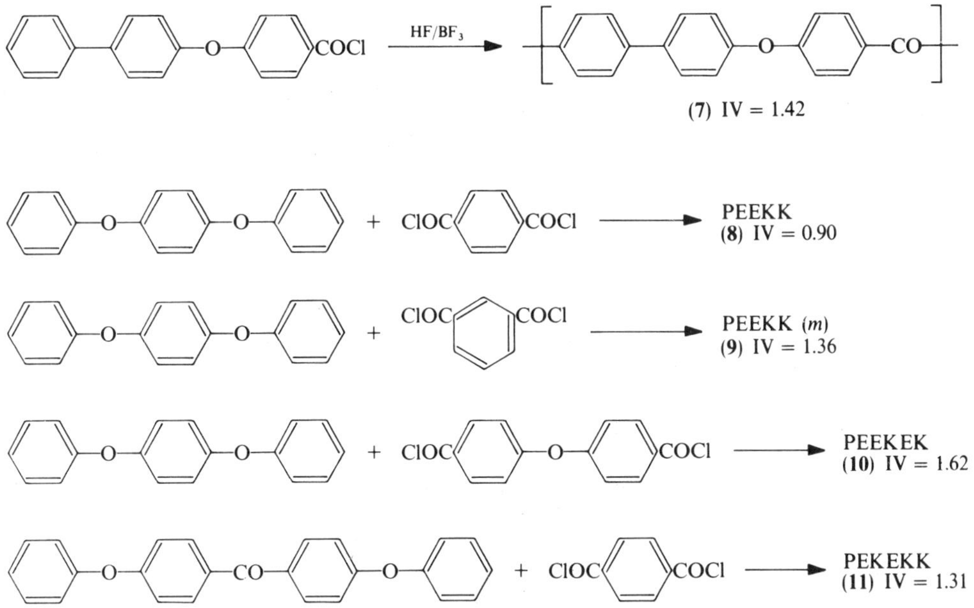

Scheme 6

Polymers prepared from terephthaloyl and isophthaloyl chloride with diphenyl ether contain 0.6–1.0% of 9-phenylenexanthydrol end groups[35] (**12**), which limit molecular weight and cause melt instability. These are detectable by absorbance in the UV at 455 nm, and can be chemically reduced[35] to the corresponding 9-phenylenexanthene (**13**), thereby improving melt stability.

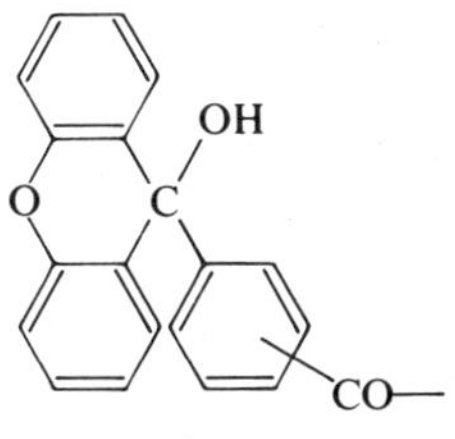

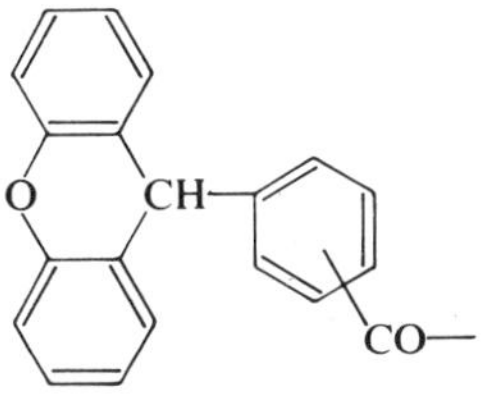

By incorporating an appropriate comonomer, such as 1,4-diphenoxybenzene, the amount of xanthydrol end groups is reduced, and an IV of up to 2.55 is attained.[21]

A variant of the usual process[29] uses derivatives of thio- or dithio-carbonic acids in combination with a difunctional nucleophilic substrate (Scheme 7). This process is claimed to use cheap starting materials but does not appear to have been used commercially.

Scheme 7

29.3.4 Use of Phosgene

The synthesis of useful (tough, high molecular weight) PEK from the cheap reagents phosgene and diphenyl ether (Scheme 8) has long been a desirable target, but one not so far achieved. Generally, low yields of low molecular weight material have been obtained;[18,30] no systematic study of the reasons for this has been carried out, but instability of diphenyl ether under certain conditions has been postulated,[31] and clearly there is increased scope for non-*para* structures.

Scheme 8

Polymerization in 1,2-dichloroethane with $AlCl_3$/LiCl[16] yields PEK with IV = 0.60, but this was brittle on annealing. Polymerization in carbon disulfide[17] gave IV values of 0.59–0.73, while $AlCl_3$ in trifluoromethanesulfonic acid gave a polymer of IV = 0.95.

29.3.5 Other Catalytic Systems

Systems other than those based on $AlCl_3$ or HF have been used on the laboratory scale. Trifluoromethanesulfonic acid (triflic acid), a 'superacid', has been used as solvent and catalyst to polycondense acid chlorides;[32] for example, those shown in Scheme 9. Triflic acid also allows condensation of aromatic carboxylic acids;[33,34] some examples are shown in Scheme 10.

The polymers were shown by ^{13}C NMR to have all *para* structures. Terephthalic acid fails to react with aryl ethers under these conditions, and, in a systematic study, it was found that electron-withdrawing substituents *para* to the carboxyl group inhibit acylation. This must also explain the failure to produce high polymers in other systems using terephthaloyl chloride.[6,7,14,30,35,36]

Triflic acid does not catalyze the polycondensation of 4-phenoxybenzoic acid; this, and work with model compounds, shows that both rings in diphenyl ether are deactivated following monoacylation. This is presumed to be the result of partial positive charge induction on the bridging oxygen (Scheme 11).

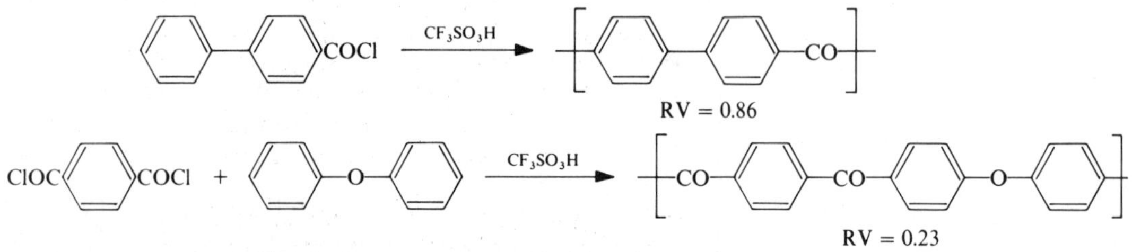

Scheme 9

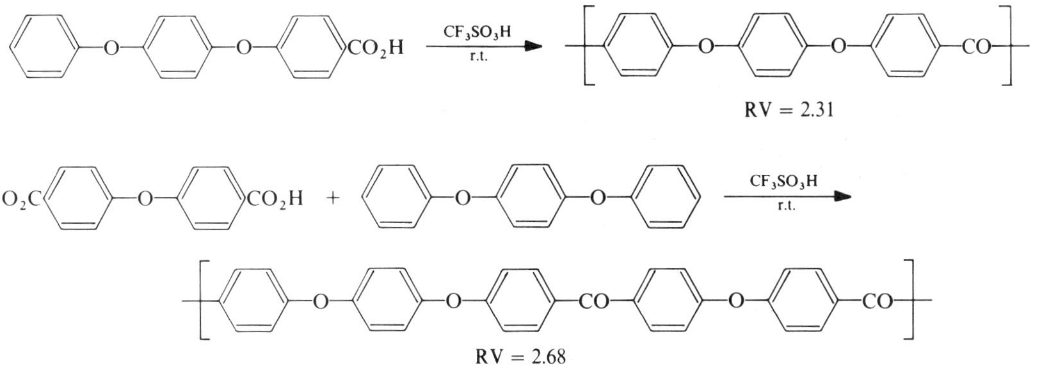

[Scheme 10 structures]

Scheme 10

[Scheme 11 resonance structures]

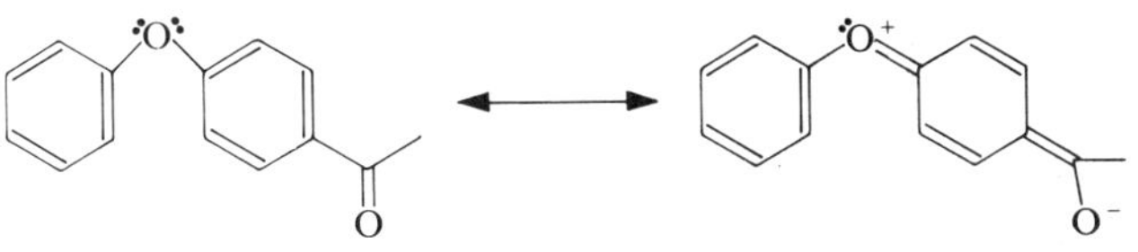

Scheme 11

A 1:10 combination of phosphorus pentoxide in methanesulfonic acid (PPMA) has been proposed as a substitute for polyphosphoric acid.[37] PPMA has been used to synthesize polyketones[38] (Scheme 12). PPMA was the solvent for the production of polyketones containing dibenzo[18]crown-6[39] and of thermotropic polyketones.[40]

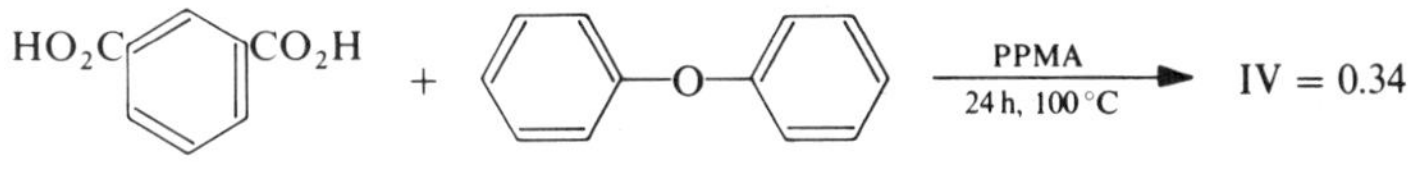

[Scheme 12 structure] PPMA, 24 h, 100 °C, IV = 0.34

Scheme 12

29.4 NUCLEOPHILIC ROUTE

The polymerization process by nucleophilic displacement, originally developed for the production of polysulfones,[4] has been successfully adapted by ICI to the production of poly(ether ketones). Essential features in the reaction (Scheme 13) of aromatic halides and alkali metal salts of phenols, to produce sulfone and ketone polymers, are the use of a dipolar aprotic solvent and activation of the halide X by an electron-withdrawing group Q in the *ortho* or *para* position.

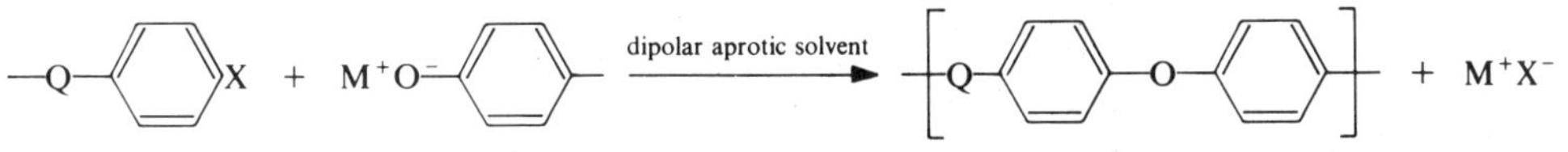

[Scheme 13 structure] —Q—C$_6$H$_4$—X + M$^+$O—C$_6$H$_4$— → —[Q—C$_6$H$_4$—O—C$_6$H$_4$]— + M$^+$X$^-$

Scheme 13

The nature of the electron-withdrawing group, the halide, the cation and the solvent all have a marked influence on the course and rate of reaction. The power of the electron-withdrawing group lies in the order $NO_2 \sim SO_2 > C=O > N=N$; for the halogens, $F \gg Cl > Br > I$; and for the alkali metal cations in the order $Cs > K > Na > Li$. The choice of solvent is complicated, with dimethyl sulfoxide (b.p. 180 °C) being favoured for polymers which will form and remain in solution below 170 °C.

Crystalline poly(ether ketones) generally require a solvent which can be used at higher temperatures without decomposition which may have a reduced catalytic effect; diphenyl sulfone (DPS) is a favoured example.

Poly(ether ketones) will usually be prepared from a halide activated by the carbonyl group, and can be produced from one-monomer or two-monomer reactions, of which equations (3) and (4) are examples.

$$M^+O-\text{Ar}-CO-\text{Ar}-X \longrightarrow [-O-\text{Ar}-CO-\text{Ar}-] + MX \qquad (3)$$

$$M^+O-\text{Ar}-CO-\text{Ar}-O^-M^+ + X-\text{Ar}-CO-\text{Ar}-X \longrightarrow [(-O-\text{Ar}-CO-\text{Ar}-)_2] + 2MX \qquad (4)$$

While chlorine is sufficiently activated by a sulfone group it is usually not sufficiently activated by a carbonyl group, and it may be necessary to employ fluorine.[42] While an adequate molecular weight can sometimes be achieved using chloro monomers the polymer may be brittle and of high melt viscosity, due, it is believed, to the presence of aberrant structures, branching in particular.

In practice, it was found that fluorine was approximately 100 times more reactive than chlorine in a model synthesis[43] (equation 5). Although diaryl sulfones have been found to be the most effective solvents in the preparation of high molecular weight crystalline poly(ether ketones),[9] benzophenone has also been employed.[44]

$$\text{Ph}-CO-\text{Ar}-O-\text{Ar}-OK + X-\text{Ar}-CO-\text{Ph} \longrightarrow$$

$$\text{Ph}-CO-\text{Ar}-O-\text{Ar}-O-\text{Ar}-CO-\text{Ph} \qquad (5)$$

$$X = Cl, F$$

29.4.1 Stoichiometry and Molecular Weight Control

The attainment of high molecular weight is dependent upon accurate control of stoichiometry, i.e. the molar ratios of reactants (which may also include the base) undergoing polycondensation.

Since it is usually possible to obtain a polymer of undesirably high molecular weight when the monomers are stoichiometrically in balance, it is usual[9] to arrange for a reactant, most often the dihalide, to be in a slight excess. In the case of the one-monomer reaction (equation 3) it is necessary to incorporate a small amount of another reactant, wich could be a mono- or di-halide.

Molecular weight control can also be achieved by monitoring the course of the polymerization and, at the desired point, making an addition of a monofunctional reagent, which may be a phenol or phenoxide or an active halogen compound. This technique not only stops the polymerization but may be used to convert unstable end groups, such as phenoxide or phenol, to the more stable ether end group (equation 6).

$$\sim\text{Ar}-O^-M^+ + RX \longrightarrow \sim\text{Ar}-OR + MX \qquad (6)$$

29.4.2 Carbonate Process

Controlling stoichiometry on an industrial scale can be problematic. The use of alkali metal hydroxides to prepare phenoxide monomers requires careful stoichiometric control in addition to that required for the balance with dihalide. If the salt formation is carried out *in situ* by reaction of the phenol with alkali metal carbonate[45,46] it is found that the amount of the latter is not critical, since excess carbonate (unlike hydroxide) does not lead to hydrolysis of the halo monomer.

The problems of insolubility and instability of the salts, most marked in the case of hydroquinone, are alleviated, because disalts do not appear to be formed. Monosalt, as soon as it is formed, reacts with dihalo compound to form a soluble phenol ether, which goes on to react with more carbonate, as shown in Scheme 14.

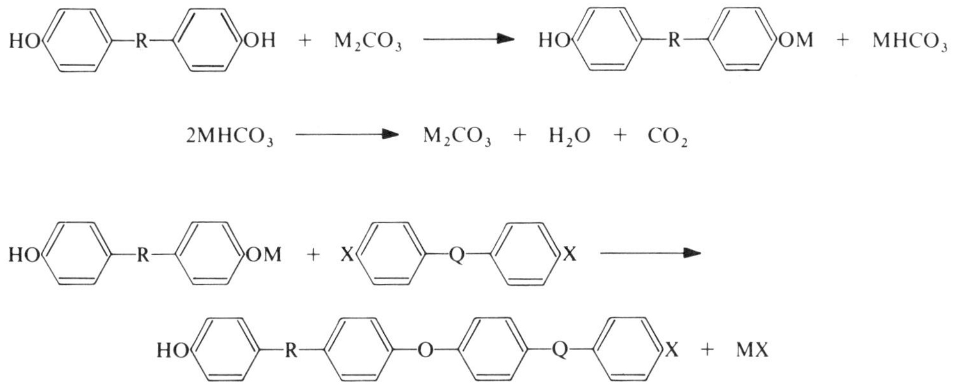

Scheme 14

While the carbonate route has many advantages, it does suffer from some disadvantages. There is present throughout the polymerization a finite concentration of phenolic hydroxyl groups not yet converted to phenoxide. These acidic conditions may cause decomposition of DMSO,[47] and the bisphenol itself may undergo decomposition, particularly if the reaction temperature is raised too rapidly. The evolution of water vapour and carbon dioxide may cause foaming problems; this is avoided by extending the reaction at specific hold temperatures.

29.4.3 Poly(ether ether ketone) PEEK

The nucleophilic process and its difficulties can be illustrated by the specific example of PEEK, a polymer produced commercially[48] by the chemistry of equation (7).[42] Hydroquinone (HQ) is commerically available; its salts are very unstable, being subject to oxidation, and only the carbonate route is workable. In this instance 4,4'-difluorobenzophenone (BDF) must be used; this monomer is not a commodity material but can be produced via a variety of Friedel–Crafts processes, such as those shown in Scheme 15.

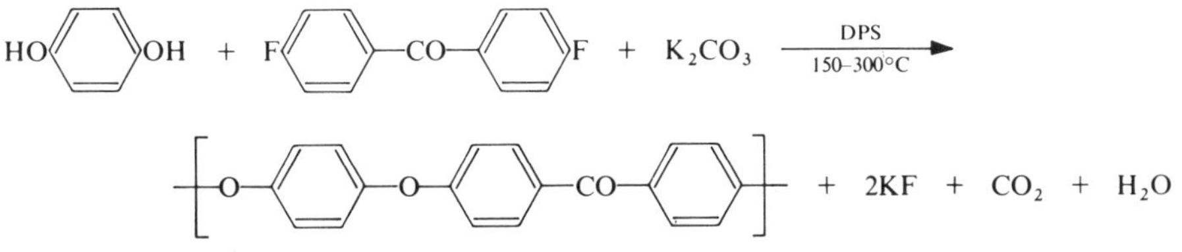

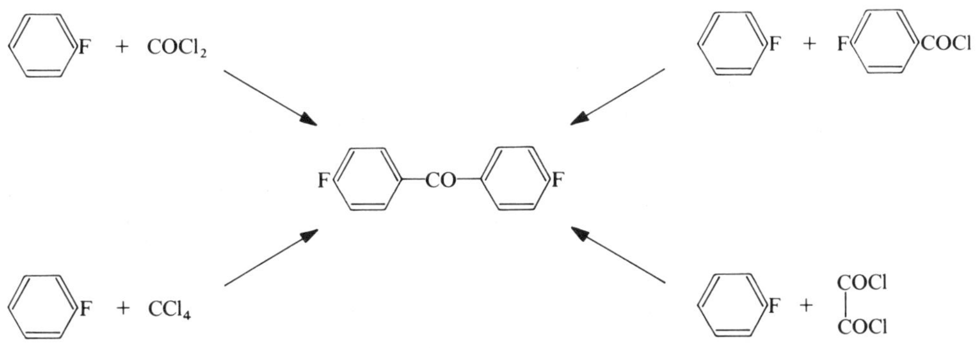

Scheme 15

Alternatively, the synthesis starts from 4,4'-diaminodiphenylmethane (DADM), an intermediate derived from aniline and formaldehyde, used in methylene diisocyanate manufacture on a large scale (Scheme 16).

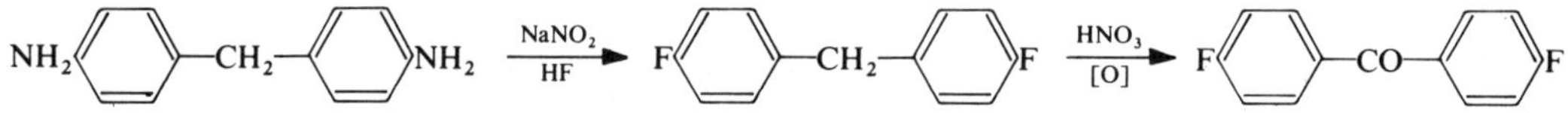

Scheme 16

Sodium carbonate cannot be used alone to produce PEEK, but can be used in the presence of the carbonate of an alkali metal of higher atomic weight.[49]

PEEK has a melting point (T_m) of 334 °C, and, in order to keep the polymer in solution, the final reaction temperature must be not less than about 300 °C. PEEK and its composites have been reviewed.[41]

29.4.4 Other Polymers Derived from 4,4'-Difluorobenzophenone (BDF)

Early workers considered difluoro monomers to be too expensive for commercial use,[47] but the current availability of BDF makes possible a number of polymers derived from commercially available bisphenols, such as those shown in Scheme 17.

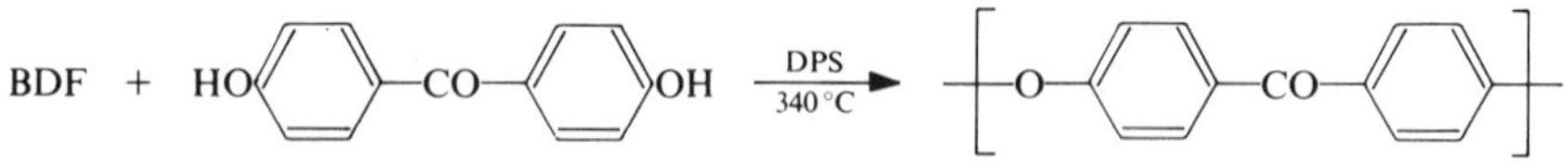

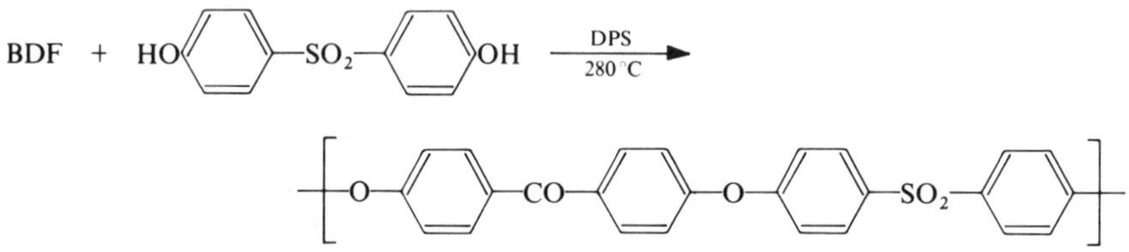

Scheme 17

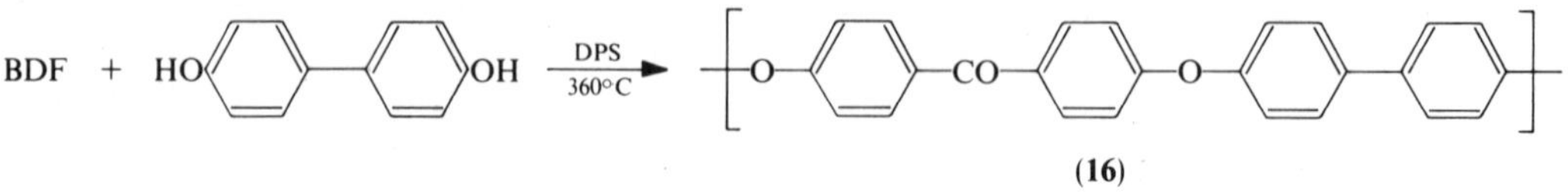

The same sort of considerations apply to these syntheses as for PEEK, but where the crystalline polymers have a higher T_m, such as (14) and (16), then higher reaction temperatures are required. In the case of (15), the polymer is amorphous and lower temperatures may be employed. In the case of (14) and (16), sodium carbonate may be used alone. At the present state of knowledge, conditions for individual syntheses must be determined experimentally. Other fluoro ketones, such as (17)[8] and (18),[50] have been polymerized with bisphenols to provide additional classes of poly(ether ketones).

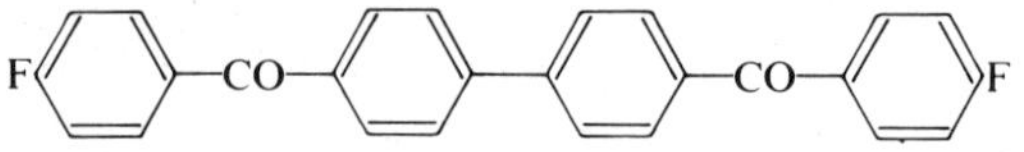

29.4.5 Ether Cleavage

Under certain conditions, activated ether links in poly(aryl ethers) can be broken in a process which is the reverse of the nucleophilic polycondensation.[1] If KF is present, for example, the polymerization reaches an equilibrium portrayed in Scheme 18. This process may be of nuisance value in the case where an end-capping agent RX is present in excess, since the phenoxide end group will react with RX instead of re-forming polymer.[51] The process may be inhibited by the presence of akali or alkaline earth metal salts of a non-oxidizing anion; lithium chloride is preferred.[1] On the other hand, the ether cleavage process may be turned to advantage in the preparation of block copolymers.[52,53] The block precursors (**19**) may be synthesized by the electrophilic route and then subjected to ether interchange in a nucleophilic process, as shown in Scheme 19.

Where transetherification is the route to block-polymer formation, the oligomer need not possess functional end groups.

29.4.6 Variants of the Nucleophilic Process

Nitro groups may be displaced by nucleophilic attack to form polymers;[54,55] see, for example, equation (8).

$$RV = 0.18, T_m = 236-255\,°C \tag{8}$$

This process gave a low molecular weight polymer of dubious melting point. Production of high melting poly(ether ketones) requiring high reaction temperatures is likely to fail, as a result of oxidation reactions involving nitrite by-product.

BDF has been condensed with trimethylsilyl derivatives of bisphenols[56] in the presence of caesium fluoride catalyst (Scheme 20).

The advantage claimed for this melt process is that the polymer is obtained in a pure form, not requiring separation from solvents and by-product salts. However, the process still requires the relatively costly difluoro compound.

The problem of insolubility with the crystalline poly(ether ketones), which necessitates use of high temperatures, can be avoided if the polymer is first produced in an amorphous form, which can subsequently be converted into the crystalline form. This may be achieved for PEK by converting 4,4'-dihydroxybenzophenone into an acetal derivative (**20**),[57,58] as illustrated in Scheme 21.

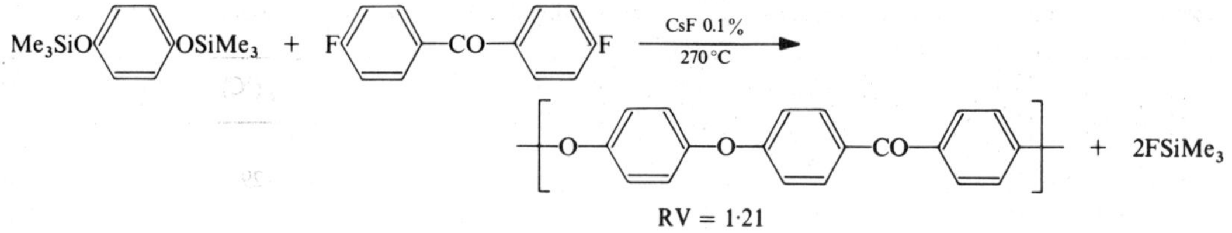

Scheme 20

Me₃SiO–C₆H₄–OSiMe₃ + F–C₆H₄–CO–C₆H₄–F $\xrightarrow[270\,^\circ\text{C}]{\text{CsF 0.1\%}}$ [–O–C₆H₄–O–C₆H₄–CO–C₆H₄–]ₙ + 2FSiMe₃

RV = 1·21

Scheme 20

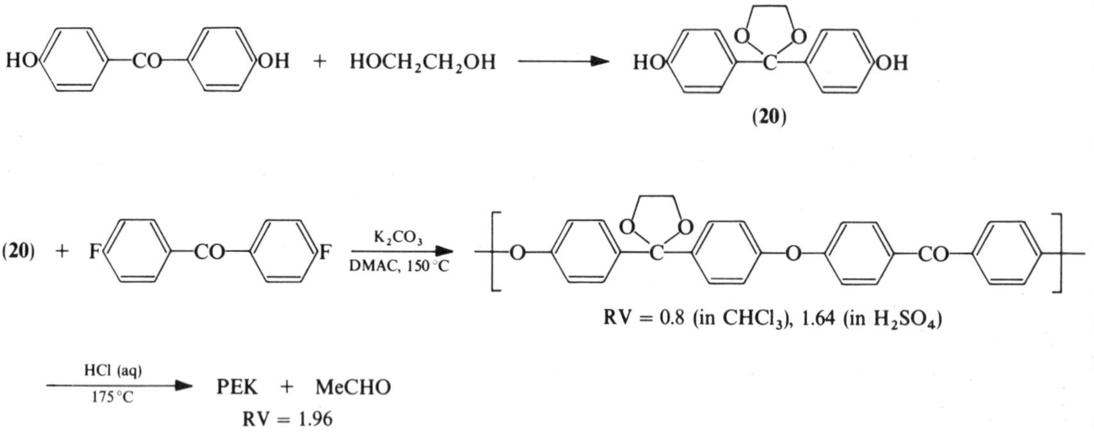

Scheme 21

HO–C₆H₄–CO–C₆H₄–OH + HOCH₂CH₂OH ⟶ HO–C₆H₄–C(OCH₂CH₂O)–C₆H₄–OH

(20)

(20) + F–C₆H₄–CO–C₆H₄–F $\xrightarrow[\text{DMAC, 150}\,^\circ\text{C}]{\text{K}_2\text{CO}_3}$ [–O–C₆H₄–C(OCH₂CH₂O)–C₆H₄–O–C₆H₄–CO–C₆H₄–]ₙ

RV = 0.8 (in CHCl₃), 1.64 (in H₂SO₄)

$\xrightarrow[175\,^\circ\text{C}]{\text{HCl (aq)}}$ PEK + MeCHO

RV = 1.96

Scheme 21

As shown, the amorphous poly(acetal ketone) can be prepared at low temperatures and is soluble in solvents such as chloroform. However, the process still requires the relatively costly difluorobenzophenone, and complete hydrolysis to the polyketone may prove impractical on a commercial scale.

29.5 PROPERTIES OF POLY(ETHER KETONES)

The crystalline poly(ether ketones) are stiff and tough, and resist wear, abrasion and fatigue. They exhibit high temperature performance; that is the polymers withstand the high (400 °C) temperatures required for processing and, subsequently, can be used at high temperatures ($\geq 200\,^\circ$C) without oxidation and loss in properties.

They possess low flammability and, when burning, give low levels of smoke and toxic gas. They are solvent resistant and resistant to radiation.

Detailed property data are available in review articles, papers and trade literature referred to therein.[2-4,8,9,41,59] However, the important properties of glass transition temperature, T_g, and crystalline melting point, T_m, obtained by differential scanning calorimetry, are shown for a number of polymer structures in Table 1.

It can be seen that the T_g values can range from about 100 °C to over 200 °C and T_m values from about 300 °C to well over 400 °C. Relatively small changes in structure, such as substitution of methylene for carbonyl in the chain, can prevent crystallization from occurring.

29.5.1 Miscibility and Isomorphism

An interesting and unusual property of the poly(ether ketones) is that blends may exhibit miscibility and isomorphic behaviour.[60]

Blends are either miscible and isomorphic or immiscible and not isomorphic. The proposed explanation is that the unit cells of the PAEKs are nearly identical, i.e. the units PhO– and PhCO– are interchangeable in the crystalline lattice,[61] and, if a blend is miscible in the melt, the two types of chains can be in close proximity during crystallization and will be isomorphic. PEEK is miscible with PEK but not with PEKK, while PEKK is miscible with PEK and PEEKK.

Table 1 Glass Transition Temperature (T_g) and Crystalline Melting Point (T_m) for some Representative Poly(ether ketones)

Structure	T_g (°C)	T_m (°C)	Ref.
—O—⟨⟩—O—⟨⟩—O—⟨⟩—CO—⟨⟩—	129	324	a
—O—⟨⟩—O—⟨⟩—CO—⟨⟩—	144	335	b
—O—⟨⟩—CO—⟨⟩—	154	367	b
—O—⟨⟩—CO—⟨⟩—CO—⟨⟩—	165	391	a
—O—⟨⟩—O—⟨⟩—CO—⟨⟩—CO—⟨⟩—	150	365	c
—O—⟨⟩—⟨⟩—O—⟨⟩—CO—⟨⟩—	167	416	b
—O—⟨⟩—⟨⟩—CO—⟨⟩—	210	440	d
—O—⟨⟩—CH$_2$—⟨⟩—O—⟨⟩—CO—⟨⟩—	123	f	e
—O—⟨⟩—C(Me)$_2$—⟨⟩—O—⟨⟩—CO—⟨⟩—	155	f	e
—O—⟨⟩—SO$_2$—⟨⟩—O—⟨⟩—CO—⟨⟩—	181	f	e
—O—(naphthyl)—O—⟨⟩—CO—⟨⟩—	155	f	e

[a] J. E. Harris and L. M. Robeson, *J. Polym. Sci., Polym. Phys. Ed.*, 1987, **25**, 311.
[b] T. E. Attwood, P. C. Dawson, J. L. Freeman, L. R. J. Hoy, J. B. Rose and P. A. Staniland, *Polym. J.*, 1981, **22**, 1096.
[c] K. J. Dahl and V. Jansons (Raychem Corp.), US Pat. 3 956 240 (1976) (*Chem. Abstr.*, 1976, **85**, 63 655).
[d] K. J. Dahl (Raychem Corp.), Br. Pat. 1 383 393 (1975) (*Chem. Abstr.*, 1973, **78**, 98 766).
[e] Author's unpublished results.
[f] Does not crystallize on cooling at 20 °C min^{-1} from the melt.

29.6 CHEMISTRY OF POLY(ETHER KETONES)

Analysis and characterization of PAEKs is made difficult by the absence of room temperature solvents; in spite of this, many of the laboratory techniques applicable to polymers have been successfully employed. IR spectroscopy is possible using powder or film samples, NMR spectroscopy may be carried out in sulfuric or triflic acids, *etc.*

29.6.1 Characterization

Crystallinity and crystal structure have been elucidated by means of wide and small angle X-ray diffraction.[61-63] Morphology has been studied by electron microscopy[64] and light-polarizing microscopy.[62]

Molecular weights have been measured by light scattering and by gel permeation chromatography (GPC), run in a mixed phenol/trichlorobenzene solvent at 115 °C.[65]

29.6.2 Stabilization

The poly(ether ketones) are inherently very stable, but their high melting points lead to elevated processing temperatures, generally above those at which conventional antioxidants can function, and under these conditions oxidative changes can occur. It has been found[66] that certain amphoteric oxides can function as antioxidants for PAEKs, with γ-alumina apparently being most effective.

29.6.3 Thermal Decomposition

Isothermal decomposition of PEEK and PEK has been studied by thermogravimetry.[67] Volatile decomposition products were analyzed by mass spectroscopy and were found to contain phenol and dibenzofuran.

29.6.4 Sulfonation

Poly(ether ketones) containing phenylene groups unprotected by an electron-withdrawing group, such as in PEEK, may be sulfonated in sulfuric acid.[36,68,69] The sulfonated polymer can be converted into its sodium salts[70] by neutralization with sodium acetate. The T_g increases from 143 °C to 415 °C for 100% sodium sulfonate PEEK; ionic clustering is believed to occur at below 25–30% sodium sulfonate. Fully sulfonated PEEK polymer is water soluble.

29.6.5 Crosslinking

Crosslinking may be deliberately induced to improve high temperature performance. This has been achieved using elemental sulfur,[71] by insertion of biphenylene into the chain,[10,11] by incorporation of alkynic groups[12] and by incorporation of butadiene moieties.[72]

29.7 REFERENCES

1. D. R. Kelsey (Union Carbide Corp.), *Eur. Pat.* 211 693 (1987) (*Chem. Abstr.*, 1987, **107**, 7846).
2. J. P. Critchley, G. J. Knight and W. W. Wright, 'Heat Resistant Polymers', Plenum Press, New York, 1983.
3. J. B. Rose, in 'High Performance Polymers, Their Origin and Development', ed. R. B. Seymour and G. S. Kirshenbaum, Elsevier, New York, 1986.
4. R. N. Johnson, A. G. Farnham, R. A. Clendinning, W. F. Hale and C. N. Merriam, *J. Polym. Sci., Part A-1*, 1967, **5**, 2375.
5. B. M. Marks (Du Pont Ltd.), *US Pat.* 3 441 538 (1969) (*Chem. Abstr.*, 1967, **67**, 44 371).
6. W. H. Bonner (Du Pont Ltd.), *US Pat.* 3 065 205 (1962) (*Chem. Abstr.*, 1963, **58**, 5806f).
7. I. Goodman, J. E. McIntyre and W. Russell (ICI plc), *Br. Pat.* 971 227 (1964) (*Chem. Abstr.*, 1964, **61**, 14 805b).
8. T. E. Attwood, P. C. Dawson, J. L. Freeman, L. R. J. Hoy, J. B. Rose and P. A. Staniland, *Polym. J.*, 1981, **22**, 1096.
9. Y. Iwakura, K. Uno and T. Takiguchi, *J. Polym. Sci., Part A-1*, 1968, **6**, 3345.
10. R. J. Swedo and C. S. Marvel, *J. Polym. Sci., Polym. Lett. Ed.*, 1977, **15**, 683.
11. A. Sutter, P. Schmutz and C. S. Marvel, *J. Polym. Sci., Polym. Chem. Ed.*, 1982, **20**, 609.
12. C. S. Marvel, in 'Contemporary Topics in Polymer Science', ed. M. Shen, Plenum Press, New York, 1979.
13. V. Sankaran and C. S. Marvel, *J. Polym. Sci., Polym. Chem. Ed.*, 1979, **17**, 3949.
14. J. Lee and C. S. Marvel, *J. Polym. Sci., Polym. Chem. Ed.*, 1983, **21**, 2189.
15. K. Niume, F. Toda, K. Uno, M. Hasegawa and Y. Iwakura, *J. Polym. Sci., Polym. Chem. Ed.*, 1982, **20**, 1965.
16. V. Jansons and H. C. Gors (Raychem Corp.), *World Pat.* 84 03 891 (1984) (*Chem. Abstr.*, 1985, **102**, 204 469).
17. S. Nozawa and M. Nakata (Mitsubishi Chemical Co. Ltd.), *Eur. Pat.* 135 938 (1985) (*Chem. Abstr.*, 1985, **103**, 71 829).
18. M. I. Litter and C. S. Marvel, *J. Polym. Sci., Polym. Chem. Ed.*, 1985, **23**, 2205.
19. V. Jansons, H. C. Gors, S. Moore, R. H. Reamey and P. Becker (Raychem Corp.), *Eur. Pat.* 174 207 (1986) (*Chem. Abstr.*, 1986, **105**, 115 584).

20. R. H. Reamey (Raychem Corp.), *Eur. Pat.* 173 408 (1986) (*Chem. Abstr.*, 1986, **104**, 207 906).*
21. S. Moore, V. Jansons and K. J. Dahl (Raychem Corp.), *World Pat.* 86 01 199 (1986) (*Chem. Abstr.*, 1986, **105**, 79 565).
22. K. J. Dahl, P. J. Horner, H. C. Gors, V. Jansons and R. H. Whitely (Raychem Corp.), *World Pat.* 86 02 368 (1986) (*Chem. Abstr.*, 1987, **106**, 33 661).
23. C. E. Berr (Du Pont Ltd.), *US Pat.* 3 637 592 (1972) (*Chem. Abstr.*, 1969, **71**, 92 082).
24. K. J. Dahl (Raychem Corp.), *US Pat.* 3 953 400 (1976) (*Chem. Abstr.*, 1976, **85**, 64 968).
25. K. J. Dahl (Raychem Corp.), *US Pat.* 3 751 398 (1973) (*Chem. Abstr.*, 1973, **78**, 30 770).*
26. K. J. Dahl (Raychem Corp.), *US Pat.* 4 024 314 (1977) (*Chem. Abstr.*, 1977, **87**, 54 036).
27. K. J. Dahl (Raychem Corp.), *US Pat.* 4 247 682 (1981) (*Chem. Abstr.*, 1977, **87**, 40 068).*
28. K. J. Dahl (Raychem Corp.), *US Pat.* 3 914 298 (1975) (*Chem. Abstr.*, 1976, **84**, 60 173).
29. V. Jansons (Raychem Corp.), *Eur. Pat.* 70 147 (1983) (*Chem. Abstr.*, 1983, **98**, 90 127).*
30. C. E. Berr (Du Pont Ltd.), *US Pat.* 3 516 966 (1970) (*Chem. Abstr.*, 1970, **73**, 35 968).
31. K. J. Dahl and V. Jansons (Raychem Corp.), *US Pat.* 3 956 240 (1976) (*Chem. Abstr.*, 1976, **85**, 63 655).
32. J. B. Rose (ICI plc), *Eur Pat.* 63 874 (1982) (*Chem. Abstr.*, 1983, **98**, 180 081).
33. H. M. Colquhoun and D. F. Lewis (ICI plc), *Br. Pat.* 2 116 990 (1985) (*Chem. Abstr.*, 1984, **100**, 7426).*
34. H. M. Colquhoun, *Polym. Prepr., Am. Chem. Soc., Div. Polym. Chem.*, 1984, **25**, 17.
35. R. J. Angelo, R. Darms and R. D. Wysong (Du Pont Ltd.), *US Pat.* 3 767 620 (1973) (*Chem. Abstr.*, 1973, **79**, 67 280).*
36. T. Ogawa and C. S. Marvel, *J. Polym. Sci., Polym. Chem. Ed.*, 1985, **23**, 1231.
37. P. E. Eaton, G. R. Carlson and J. T. Lee, *J. Org. Chem.*, 1973, **38**, 4071.
38. M. Ueda and T. Kano, *Makromol. Chem., Rapid Commun.*, 1985, **5**, 833.
39. M. Ueda, T. Kano, T. Waragai and H. Sugita, *Makromol. Chem., Rapid Commun.*, 1985, **6**, 847.
40. T. D. Shaffer and V. Percec, *Polym. Bull. (Berlin)*, 1985, **14**, 367.
41. H. X. Nguyen and H. Ishida, *Polym. Compos. Eng.*, 1987, **8**, 57.
42. J. B. Rose and P. A. Staniland (ICI plc), *Eur. Pat.* 879 (1979) (*Chem. Abstr.*, 1982, **96**, 200 397).*
43. J. R. Lovering, Ph.D. Thesis, University College, London, 1986.
44. I. Fukawa and T. Tanabe (Asahi), *Eur. Pat.* 193 187 (1986) (*Chem. Abstr.*, 1986, **105**, 227 514).
45. R. A. Clendinning, A. G. Farnham, N. L. Zutty and D. C. Priest (Union Carbide Corp.), *Can. Pat.* 847 963 (1970).
46. Celanese Corp., *Br. Pat.* 1 264 900 (1972) (*Chem. Abstr.*, 1970, **73**, 46 063).*
47. R. A. Clendinning, A. G. Farnham and R. N. Johnson, in 'High Performance Polymers, Their Origin and Development', ed. R. B. Seymour and G. S. Kirshenbaum, Elsevier, New York, 1986.
48. C. P. Smith, *Swiss Plastics*, 1981, **3**, 37.
49. M. B. Cinderey and J. B. Rose (ICI plc), *Br. Pat.* 1 586 972 (1981) (*Chem. Abstr.*, 1978, **89**, 147 398).*
50. J. B. Rose, E. Nield, P. T. McGrail and H. M. Colquhoun (ICI plc), *Eur. Pat.* 194 062 (1986) (*Chem. Abstr.*, 1987, **106**, 120 420).
51. T. E. Attwood, A. B. Newton and J. B. Rose, *Br. Polym. J.*, 1972, **4**, 391.
52. L. R. J. Hoy and J. B. Rose (ICI plc), *Br. Pat.* 1 541 568 (1979) (*Chem. Abstr.*, 1977, **86**, 156 205).*
53. R. A. Clendinning, J. E. Harris, D. R. Kelsey, M. Matzner, L. M. Robeson, P. A. Winslow and L. M. Maresca (Amoco Corp.), *World Pat.* 86 06 389 (1986) (*Chem. Abstr.*, 1987, **106**, 214 908).
54. T. Takekoshi, *Polym. J.*, 1987, **19**, 191.
55. V. E. Radlmann, W. Schmidt and G. E. Nischk, *Makromol. Chem.*, 1969, **130**, 45.
56. H. R. Kricheldorf and G. Bier, *Polymer*, 1984, **25**, 1151.
57. D. R. Kelsey (Union Carbide Corp.), *Eur. Pat.* 148 633 (1985) (*Chem. Abstr.*, 1985, **103**, 215 983).
58. D. R. Kelsey, L. M. Robeson, R. A. Clendinning and C. S. Blackwell, *Macromolecules*, 1987, **20**, 1204.
59. D. Sek and H. Zak, *Eur. Polym. J.*, 1981, **17**, 1193; W. R. Heslop and L. J. Frisco, 'Technical Paper, Wire and Cable Symposium, Atlantic City', 1973; P. M. Hergenrother, B. J. Jensen and S. J. Havens, *Polymer*, 1988, **29**, 358.
60. J. E. Harris and L. M. Robeson, *J. Polym. Sci., Polym. Phys. Ed.*, 1987, **25**, 311.
61. P. C. Dawson and D. J. Blundell, *Polymer*, 1980, **21**, 577.
62. D. J. Blundell and B. N. Osborn, *Polymer*, 1983, **24**, 953.
63. D. R. Rueda, F. Ania, A. Richardson, I. M. Ward and F. J. Balta Calleja, *Polym. Commun.*, 1983, **24**, 258; J. N. Hay, D. J. Kemmish, J. I. Langford and A. I. M. Rae, *Polym. Commun.*, 1984, **25**, 175.
64. A. J. Lovinger and D. D. Davis, *J. Appl. Phys.*, 1985, **58**, 2843.
65. J. Devaux, D. Delimoy, D. Daoust, R. Legras, J. P. Mercier, C. Strazielle and E. Nield, *Polymer*, 1985, **26**, 1994.
66. K. J. Dahl and F. M. Kameda (Raychem Corp.), *US Pat.* 3 925 307 (1975) (*Chem. Abstr.*, 1975, **82**, 99 335).*
67. J. N. Hay and D. J. Kemmish, *Polymer*, 1987, **28**, 2047; R. B. Prime and J. C. Seferis, *J. Polym. Sci., Polym. Lett. Ed.*, 1986, **24**, 641.
68. X. Jin, M. T. Bishop, T. S. Ellis and F. E. Karasz, *Br. Polym. J.*, 1985, **17**, 4.
69. J. B. Rose (ICI plc), *US Pat.* 4 268 650 (1981) (*Chem. Abstr.*, 1980, **93**, 240 291).*
70. C. Bailly, D. J. Williams, F. E. Karasz and W. J. MacKnight, *Polymer*, 1987, **28**, 1009.
71. C. M. Chan and S. Venkatramen, *J. Appl. Polym. Sci.*, 1986, **32**, 5933.
72. V. Sankaran and C. S. Marvel, *J. Polym. Sci., Polym. Chem. Ed.*, 1979, **17**, 3949.

* In some instances the *Chem. Abstr.* references cited are *not* for the patents quoted, but for equivalent patents.

30

Polyimides and Other Heteroaromatic Polymers

BERNARD SILLION
CEMOTA, Vernaison, France

30.1	INTRODUCTION	499
30.2	GENERAL FEATURES FOR THE SYNTHESIS OF HETEROCYCLIC POLYMERS	500
	30.2.1 *Direct Formation of High Molecular Weight Rod-like Cyclized Heterocyclic Polymers*	500
	30.2.2 *Intermediate Isolation of High Molecular Weight Non-cyclized Heterocyclic Precursors*	500
	30.2.3 *Intermediate Isolation of Low Molecular Weight Prepolymers*	500
	30.2.4 *Direct Formation of High Molecular Weight Cyclized Soluble and/or Fusible Polyheterocyclic Thermoplastics*	501
	30.2.5 *Thermosetting Resins*	501
30.3	ROD-LIKE HETEROCYCLIC POLYMERS AND COPOLYMERS	502
	30.3.1 *Polyquinolines*	502
	30.3.2 *Polybenzimidazoles, Polybenzoxazoles and Polybenzothiazoles*	503
30.4	POLYIMIDES	505
	30.4.1 *Polycondensation Linear Polyimides*	505
	30.4.1.1 *Polymerization of dianhydrides and diamines*	505
	30.4.1.2 *Polymerization of a dianhydride with a diisocyanate*	510
	30.4.1.3 *Polymerization of a dianhydride with silylated diamines*	511
	30.4.1.4 *Polycondensation of diester diacids with diamines*	512
	30.4.1.5 *Formation of polyimides via nucleophilic nitro group displacement reactions*	513
	30.4.1.6 *Chemistry of special linear polyimides and derivatives*	514
	30.4.2 *Thermosetting Polyimides*	518
	30.4.2.1 *The chemistry of 1,1'-arylenebis-1H-pyrrole-2,5-diones (bismaleimides)*	518
	30.4.2.2 *The chemistry of 3a,4,7,7a-tetrahydro-4,7-methano-1H-isoindole-1,3(2H)-diones (bisnadimides)*	522
	30.4.2.3 *Acetylene-terminated imides*	523
30.5	POLYQUINOXALINES AND POLY(PHENYLQUINOXALINE)S	525
	30.5.1 *Linear Polyquinoxalines and Poly(phenylquinoxaline)s*	525
	30.5.2 *Cross-linked Poly(phenylquinoxaline)s*	527
	30.5.2.1 *Cross-linking of high molecular weight poly(phenylquinoxaline)s*	527
	30.5.2.2 *Phenylquinoxaline oligomers with reactive terminal unsaturation*	528
30.6	REFERENCES	529

30.1 INTRODUCTION

Heteroaromatic polymers were introduced at the beginning of the 1960s to meet new demands for heat-resistant plastics for space and military applications. Speciality polymers now find many other applications, such as adhesives, matrices for laminates, membranes for gas separation, dielectrics used in the production of integrated circuits, and films, coatings and resins for assembly and packaging in microelectronics.

The chemistry of these heterocyclic polymers could be included within the general chemistry of condensation polymers. However, a new concept is required, because heterocycle formation is used for chain growth, following the general scheme illustrated in equation (1).

$$n\text{A–Ar}^1\text{–A} + n\text{B–Ar}^2\text{–B} \rightarrow \text{A}(\text{Ar}^1\text{–(heterocycle–Ar}^2)_n\text{–B} + \text{by-products coming from the condensation and the cyclization} \quad (1)$$

Generally, the condensation is not a simple reversible reaction, as in a conventional condensation, but a multistep process involving a polycondensation followed by a final irreversible cyclization. More than 40 different heterocycles have been introduced into macromolecules using the so-called polyheterocyclization process; all the main heterocyclic and aromatic polymers were discovered between 1962 and 1972[1] and many reviews discuss the synthesis and properties of heat-resistant heterocyclic polymers.[2-4]

The direct linkage between aromatic and heterocyclic nuclei makes the chain very rigid and ensures high glass transition temperatures (T_g) or crystalline melting points (T_m), leading to good mechanical properties at high temperatures. Another consequence of the aromatic nature of the chain is a high bonding energy, so that the macromolecules are very resistant to chemical agents. Depending on the more or less symmetrical structure of the heterocyclic macromolecule, it is possible to obtain crystalline or amorphous polymers. However, the high T_g or T_m and very poor solubilities are important drawbacks to their use, therefore the synthetic strategy used will depend strongly on the final application for which they are intended.

When a polymer is used for example as a film, coating or membrane, either special solvents have to be found for its synthesis, to keep the cyclized high molecular weight heterocycles in solution, or the synthesis of soluble high molecular weight precursors must be used. The precursor undergoes a thermal cyclization in the solid state after solvent removal.

When polymers are used as adhesives or laminates, one synthetic route is to prepare a special thermoplastic either with flexible linkages between the aromatics and heterocycles, or carrying bulky substituents. A second possibility is the synthesis of low molecular weight heterocyclic precursors, followed by an increase in the molecular weight by thermal treatment during processing. More recently, a third type of product has been studied. These products are telechelic heterocyclic oligomers carrying two reactive functions, such as maleimide, nadimide, acetylene, cyanate, phthalonitrile, *etc.* These reactive oligomers polymerize when they are heated, leading to a cross-linked network.

30.2 GENERAL FEATURES FOR THE SYNTHESIS OF HETEROCYCLIC POLYMERS

The five main synthetic pathways used to obtain different types of heterocyclic polymers can be summarized according to the general features described in the following sections.

30.2.1 Direct Formation of High Molecular Weight Rod-like Cyclized Heterocyclic Polymers

The polymers are prepared by polycondensation at low concentrations (<10%) in special solvents, such as polyphosphoric acid containing P_2O_5 or a cresol solution containing P_2O_5. The polymers exhibit lyotropic properties. The solutions are spun, and produce a fiber with good mechanical properties. Polyquinolines, polybenzoxazoles, polybenzothiazoles and polybenzimidazoles have been obtained using this synthetic route (see Section 30.3).

30.2.2 Intermediate Isolation of High Molecular Weight Non-cyclized Heterocyclic Precursors

Raw materials with high reactivities (*e.g.* acids, chlorides, anhydrides, isocyanates and amines) are usually condensed at room temperature in low concentrations (20%) in an aprotic dipolar solvent (dimethylformamide, dimethylacetamide, *N*-methylpyrrolidinone, dimethyl sulfoxide, *etc.*) or in an ether solvent (diglyme, *etc.*), and the precursor is spin coated, cast, dip coated or spun. The solvent is removed by heating and the final insoluble and non-fusible heterocyclic polymer is obtained either by thermal treatment or by chemical cyclization. This general process has been used to obtain polyimides from poly(amic acid)s[5] as well as polyquinazolinediones from poly(urea acid)s,[6] polybenzoxazinones from poly(amide ortho acid)s[7] and polyisoindoloquinazolinediones from poly(amide ortho amide)s.[8]

30.2.3 Intermediate Isolation of Low Molecular Weight Prepolymers

Using monomers with low reactivities (esters instead of anhydrides), the condensation is performed in bulk or at very high concentrations (50% dry matter) in an aprotic solvent. The polycondensation degree is kept below 10. The prepolymer is used to prepare prepregs or adhesive films, and the high molecular weight polymer is obtained by thermal curing during the processing. For example, polybenzimidazoles have been obtained from diphenyl esters and tetraamines,[9] and polyimides from diester diacids and diamines,[10] using this process.

30.2.4 Direct Formation of High Molecular Weight Cyclized Soluble and/or Fusible Polyheterocyclic Thermoplastics

The reaction is performed with a special reactive monomer having a flexible linkage, such as an aromatic ether, sulfone, *meta* linkage, *etc.*, or carrying bulky substituents such as *gem*-dimethyl. phenyl, cardo (cyclic pendant) substituents, *etc.* The cyclization is performed during the polycondensation, and the high molecular weight product obtained in solution exhibits thermoplastic behavior in the solid state. The following polymers are typical of this family: the polyimide from a dianhydride with an aromatic ether linkage, Ultem TR, used commercially by General Electric;[11] the polyimide from a diaminosiloxane developed by M & T;[12] the polyimide from 3,3'-diaminobenzophenone, LARC TPI from NASA;[13] the polyimide from diaminotrimethylphenylindane (CIBA XU 218);[14] polyhydantoins;[15] and poly(phenylquinoxaline)s.[16]

30.2.5 Thermosetting Resins

Low molecular weight telechelic oligomers with terminal functions are able to react by an addition mechanism. The characteristics of these resins are shown schematically in Figure 1. Block A is an oligomer with a number-average molecular weight ($\bar{M}_n$) ranging from 1000 to 2000. The main types of chemical compounds used for the synthesis of block A are aromatic ether sulfones, aromatic esters and heterocyclic compounds such as pyridine, quinoxaline or an imide.[4,17] The chemical structure and the molecular weight of the central moiety A will govern the initial T_g or T_m. The telechelic groups Z react by an addition polymerization without the evolution of volatiles. This point is important for structural applications. The range of the thermal reaction temperature for the different reactive groups is given in Table 1.

As can be seen from Table 1, the polymerization temperature can be very different for the same type of reactive group. Actually, the polymerization temperature is strongly dependent on the physical state of the resin. Thermal polymerization starts, in fact, when the resin melts; this point was

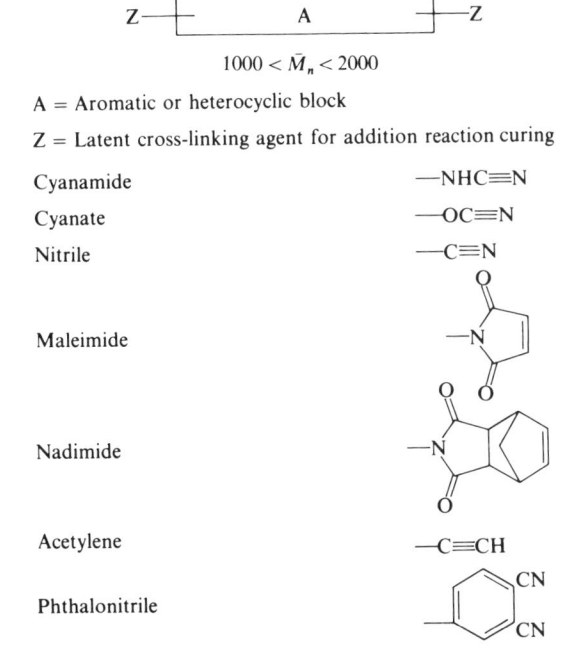

$$Z \!-\!\boxed{\quad A \quad}\!-\! Z$$

$$1000 < \bar{M}_n < 2000$$

A = Aromatic or heterocyclic block

Z = Latent cross-linking agent for addition reaction curing

Cyanamide	—NHC≡N
Cyanate	—OC≡N
Nitrile	—C≡N
Maleimide	(maleimide structure)
Nadimide	(nadimide structure)
Acetylene	—C≡CH
Phthalonitrile	(phthalonitrile structure)

Figure 1 Telechelic oligomers that can be cross-linked by addition reactions

Table 1 Reaction Temperature of Reactive End-groups

Function	Catalyst	Temperature (°C)	Ref.
Nitrile	PTSA	350	a
Cyanate	Alkylphenol	177	b
Cyanamide	—	150–200	c
Maleimide	—	177–286	d
Nadimide	—	250–275	e
Acetylene	—	130–140, 200	f, g
Phthalonitrile	Redox	220	h

[a] Li Chen Hsu and W. Phillips, *ACS Symp. Ser.*, 1982, **195**, 285.
[b] D. A. Shimp, *Polym. Mater. Sci. Eng.*, 1986, **54**, 107.
[c] G. E. Shukurov, V. A. Pandratov and V. V. Korshak, *Plast. Massy*, 1985, **9**, 5.
[d] I. K. Varma, G. M. Fohlen and J. A. Parker, *J. Polym. Sci., Polym. Chem. Ed.*, 1982, **20**, 283.
[e] H. R. Lubowitz, *Polym. Prepr., Am. Chem. Soc., Div. Polym. Chem.*, 1971, **12**, 329.
[f] A. Dussart–Lermusiaux, M. Senneron, M. Bartholin and B. Sillion, *Proc. Congr. Annu. G. F. P. Pau*, 1986, vii.
[g] C. Y. C. Lee, *Dev. Reinf. Plast.*, 1986, **5**, 121.
[h] T. Pascal, J. Malinge, B. Sillion, P. Claudy and J. M. Letoffé, *J. Polym. Sci., Polym. Chem. Ed.*, (in press).

clearly demonstrated in the case of maleimides. Resins with a low melting point polymerize at 180 °C and resins with a high melting point react only at the melting point. The same observation has been made with alkynic resins.

Three main points have to be considered: the melting point, T_m, or the glass transition temperature, T_g, of the resin, the polymerization temperature and the T_g of the final network. The difference between the T_m or T_g of the resin and the reaction temperature determines the 'processability window' of the resin considered. At first, it seems rather advantageous to use the oligomer with the lowest T_m or T_g. However, if the initial T_g is too low, the T_g of the final network will also be too low for high temperature applications. It is also very important to take into account the fact that the final T_g is only reached after a postcure at high temperature, because the total extent of reaction cannot be reached in the glassy state.

30.3 ROD-LIKE HETEROCYCLIC POLYMERS AND COPOLYMERS

30.3.1 Polyquinolines

Polyquinolines have been synthesized in the Stille Laboratory *via* a modification of the Friedlander synthesis (equation 2). The polymerization medium used was a mixture of *m*-cresol and dimethylcresyl phosphate at 135–140 °C for 24–48 h in nitrogen.[18] By this process, flexible polymers containing 2,6- or 2,4-quinoline units and aromatic ether linkages were obtained.[19–21] These polymers show T_g ranging from 200 °C to 300 °C, and good oxidative and thermal properties, but they lose their mechanical properties when the T_g is reached, well below the decomposition temperature (550 °C).

$$\text{ArCOR}^1(NH_2) + R^2CH_2COR^3 \longrightarrow \text{quinoline} + 2H_2O \qquad (2)$$

The self condensations of 1-(4-amino-3-benzoylphenyl)ethanone (**1**; X = nil) and 1-(4′-amino-3′-benzoyl[1,1′-biphenyl]-4-yl)ethanone (**1**; X = *p*-C_6H_4) lead to AB polymers (**2**; equation 3)[22] which

$$\text{MeCOX–Ar(COPh)(NH}_2) \xrightarrow{-2nH_2O} [\text{quinoline–X}]_n \qquad (3)$$

(**1**)   (**2**)

exhibit flow birefringence in a solution containing 1.5% solid. These polymers show a very high degree of crystallinity in the solid state.

In the same way, as shown in equation (4), the condensation of (4,4'-diamino[1,1'-biphenyl]-3,3'-diyl)bis[phenylmethanone] (3) with 1,1'-[1,1'-biphenyl]-4,4'-diylbisethanone (4) gives AABB rod-like polyquinoline (5).[22] Polymer (5) also exhibits birefringence domains and can be spun in fibers with a high degree of crystallinity. The introduction of pendant aryl ether groups improves the solubility in the polymerization medium.[23] The excellent thermal and oxidative stability of polyquinoline, associated with its crystallinity, ensures retention of its mechanical properties up to about 500 °C.

30.3.2 Polybenzimidazoles, Polybenzoxazoles and Polybenzothiazoles

Polybenzimidazoles (PBI) and polybenzoxazoles (PBO) were patented in 1958 by Du Pont.[24] Polybenzothiazoles (PBT) were studied at the same time by Japanese and US scientists.[25,26] The general synthesis shown in equation (5) is based on the reaction of aromatic bis(o-diamines) (6; X = NH), bis(o-aminophenols) (6; X = O) and bis(o-aminothiols) (6; X = S), with dicarboxylic acids (7; Y = OH), chlorides (7; Y = Cl) or phenyl esters (7; Y = OPh).

Thermal condensation with free diacids needs very high temperatures of 200–350 °C in order to occur rapidly.[24] Two other syntheses were developed for industrial applications. The first one is based on the use of diphenyl dicarboxylates, using Marvel's pioneering work[9] for PBI, and extending it to PBO[27] and PBT.[25] The two-step procedure developed by Marvel involves an initial melt polymerization at 250–290 °C in an inert atmosphere. The prepolymer is then ground and reheated to 385 °C in a high vacuum in order to avoid cross-linking.

Using diphenyl 1,3-benzenedicarboxylate and [1,1'-biphenyl]-3,3',4,4'-tetraamine as raw materials, Celanese markets PBI fibers[28] obtained by wet spinning from dimethylacetamide solutions. From the same starting materials, low molecular weight prepolymers were isolated and used as an adhesive or matrix for laminates which retain their mechanical properties, after processing, for a short time at temperatures of up to 650 °C.[30,31]

Recently, Hoechst, Celanese and Alpha Precision Plastics have developed a sintering process starting from PBI (inherent viscosity = 0.55 dl g^{-1}) and leading to molded PBI. This material exhibits a 124 MPa tensile strength, 5.8 GPa tensile modulus, 220 MPa flexural strength and 5.8 GPa flexural modulus. PBI could be used from −100 °C up to 400 °C.

In order to improve the mechanical properties of PBI, PBT and PBO, Arnold developed a new type of rigid-rod condensed heterocyclic system. Two polymers have been synthesized, using a polycondensation method in polyphosphoric acid (PPA). Poly(benzo[1,2-d:4,5-d']bisthiazole-2,6-diyl-1,4-phenylene) (11) was prepared by the condensation of 2,5-diamino-1,4-benzenedithiol dihydrochloride (9) with 1,4-benzenedicarboxylic acid (10; equation 6) in PPA.[32] Under the same conditions, the reaction of 1,4-benzenedicarboxylic acid (10) with 4,6-diamino-1,3-benzenediol (12) yielded poly(benzo[1,2-d:5,4-d']bisoxazole-2,6-diyl-1,4-phenylene) (13; equation 7).[33]

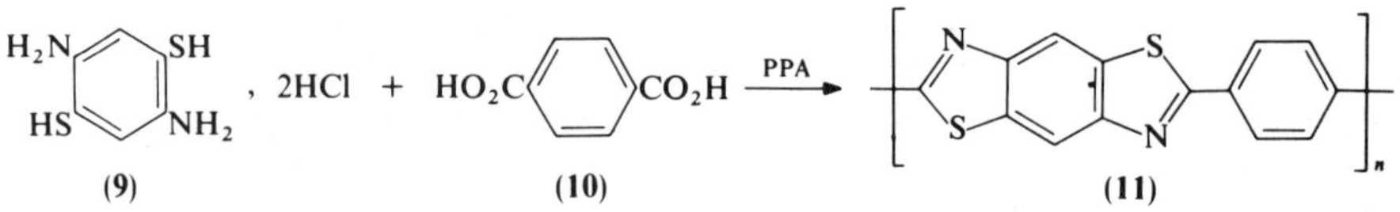

(6) [structure: compound (9) H₂N/SH, HS/NH₂, 2HCl + HO₂C-C₆H₄-CO₂H (10) → PPA → polymer (11)]

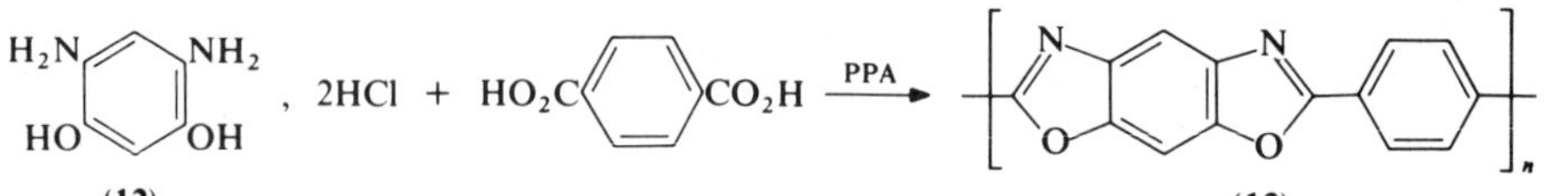

(7) [structure: compound (12) H₂N/NH₂, HO/OH, 2HCl + HO₂C-C₆H₄-CO₂H → PPA → polymer (13)]

Both polymers form liquid crystalline phases during polymerization when the concentration reaches 5%. In order to obtain a high molecular weight, it was found that the P_2O_5 concentration in PPA would be higher than 82% at the end of polycondensation; the new process can create polymer concentrations as high as 21%.[34,35]

The same polycondensation procedure has been used to synthesize AB-PBT (15) and AB-PBO (17) polymers from 4-amino-3-mercaptobenzoic acid (14) and 3-amino-4-hydroxybenzoic acid (16; equations 8–9). Both polymers (15) and (17) form liquid crystal phases during polycondensation at concentrations of 14%.

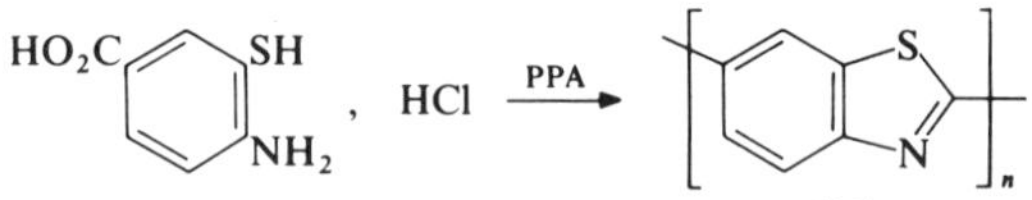

(8) [structure: compound (14) HO₂C/SH/NH₂, HCl → PPA → polymer (15)]

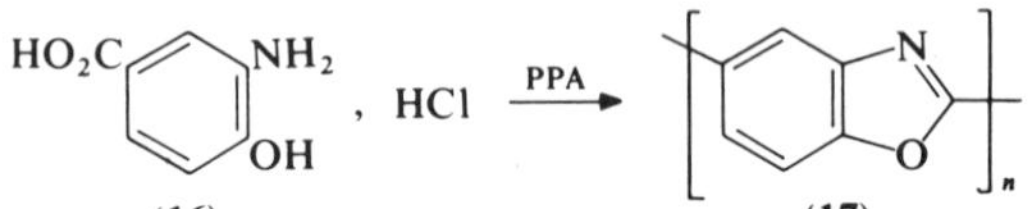

(9) [structure: compound (16) HO₂C/NH₂/OH, HCl → PPA → polymer (17)]

The comparison of AABB and AB polymers, from the point of view of molecular rigidity, indicates that AABB polymers are characterized by a catenation angle of 180°, and a coefficient a in the Mark–Houwink equation of 1.8. In the case of AB polymers, the catenation angle and coefficient a are, respectively, 150° and 1.02 for AB-PBO and 162° and 1.00 for AB-PBT[36] (Figure 2). These results are in agreement with a semiflexible polymer chain for AB-PBO and AB-PBT.

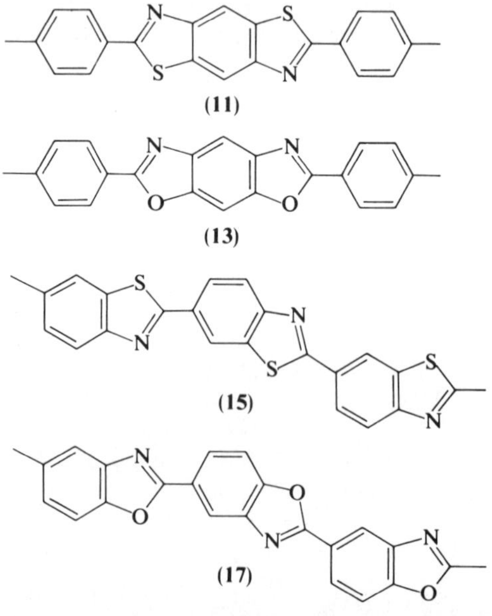

Structure	Catenation angle
(11)	180°
(13)	180°
(15)	162°
(17)	150°

Figure 2 Catenation angles of rigid rod-like heterocyclic polymers

Rod-like rigid PBT exhibits very high mechanical properties. Some experimental fibers show a tenacity of 5.2 GPa, a modulus of 380 GPa and an elongation of 11%; after 66 h at 525 °C in a nitrogen atmosphere, the measured tenacity was 4 GPa, the modulus was 260 GPa and the elongation was increased slightly by 1.5%. After 200 h at 326 °C in air, PBT retains about 90% of its modulus and 75% of its tenacity.[37]

Polybenzodiimidazoles were synthesized using similar methods. For example, N,N'-(4,6-diamino-1,3-phenylene)bis[4-methylbenzenesulfonamide] (**18**) reacts with 1,4-benzenedicarboxylic acid in PPA after heating at 90 °C for 10 h, then at 190 °C for 1 h, to give poly[(1,5-dihydrobenzo[1,2-d: 4,5-d']diimidazole-2,6-diyl)-1,4-phenylene] (**19**),[38,39] soluble in methanesulfonic acid or in polyphosphoric acid (equation 10).

$$\text{(18)} + \text{HO}_2\text{C-C}_6\text{H}_4\text{-CO}_2\text{H} \xrightarrow[90-190\,°C]{\text{PPA}} \text{(19)} \qquad (10)$$

Polycondensation in PPA produces ABA block copolymers with rigid and semiflexible sequences. For example, heating 2,5-diamino-1,4-benzenedithiol with an excess of diacid (**10**) in polyphosphoric acid leads to an α,ω-dicarboxy PBT which reacts with 3,4-diaminobenzoic acid dihydrochloride. The copolymer was spun into fibers, which had a Young's modulus of 95–116 GPa, tensile strength of 1565–1700 MPa and an elongation at break of 1.4–2.3%.[40]

30.4 POLYIMIDES

Polyimides are the most popular heat-resistant polymers. They are used in various industries, such as electrical insulation, space, aviation, electronics and industries using gas permeation. The annual production in 1984 was about 1600 tonnes and the growth rate is estimated at 20% per year. There are a lot of papers devoted to the syntheses and applications of polyimides.[5,41,42]

Two types of polyimides have been developed: (i) condensation polyimides (**20**), linear polymers coming from the reaction between bis(*o*-dicarboxylic acid) derivatives and diamines; and (ii) addition polyimides, mainly bismaleimide (**21**) and bisnadimide (**22**), which react during the processing to give cross-linked networks.

(**20**)   (**21**)   (**22**)

30.4.1 Polycondensation Linear Polyimides

According to IUPAC rules, cyclic anhydrides and imides must be named as heterocyclic derivatives, 1,3-benzofurandiones for the former and isoindolediones for the latter. This convention will be applied in the following discussion to well-known compounds, but the more comprehensive terms anhydride and imide will be used for naming general products.

30.4.1.1 Polymerization of dianhydrides and diamines

(i) Formation of polyamic acids

As illustrated in Scheme 1, the polycondensation reaction of $1H,3H$-benzo[1,2-c-4,5-c']difuran-1,3,5,7-tetrone (pyromellitic dianhydride, PMDA; **23**) with an aromatic diamine (**24**) is a two-stage reaction leading to poly(amic acid) (PAA; **25**) and then to polyimide (**26**).[43,44]

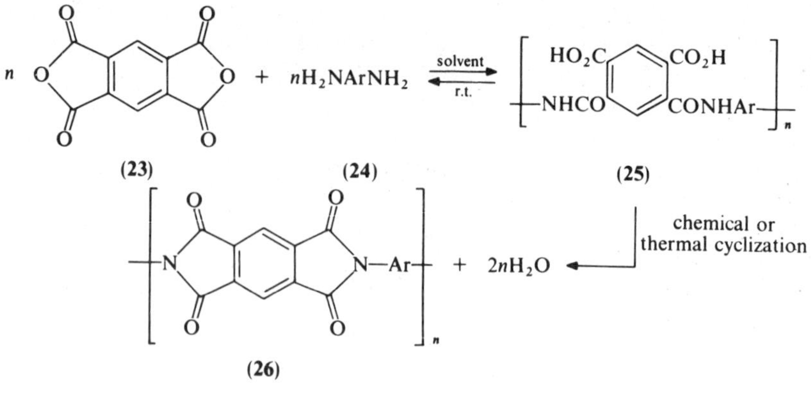

Scheme 1

(a) Poly(amic acid) formation in aprotic polar solvents. Various solvents can be used to perform the first step of the reaction; dimethylformamide (DMF), dimethylacetamide (DMAC), dimethyl sulfoxide (DMSO) and N-methylpyrrolidone (NMP) are the most commonly used. At first, a complexation between the anhydride and the solvent occurs, as was observed in the case of PMDA.[45] Then, the nucleophilic attack of anhydride takes place, leading to PAA.

As would be expected, water has very deleterious effects. First, water can hydrolyze the anhydride function of either the starting monomer or the growing chain. Indeed, it was observed that higher molecular weight PAA was obtained when PMDA was added in a solid form to the solution of diamine[46] because solid anhydrides are less sensitive to moisture than dissolved ones. Another effect of water is the hydrolysis of PAA. As a matter of fact, an ortho amide acid such as 2-[(phenylamino)-carbonyl]benzoic acid (**27**) is hydrolyzed 10^5 times faster than the corresponding unsubstituted N-phenylbenzamide (**28**), by a cooperative effect[47] (Scheme 2).

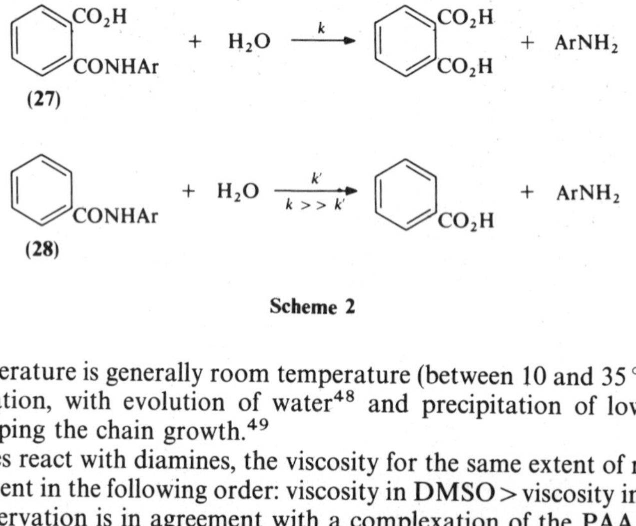

Scheme 2

The reaction temperature is generally room temperature (between 10 and 35 °C); higher temperatures induce imidization, with evolution of water[48] and precipitation of low molecular weight polyimide, thus stopping the chain growth.[49]

When dianhydrides react with diamines, the viscosity for the same extent of reaction depends on the nature of the solvent in the following order: viscosity in DMSO > viscosity in DMAC > viscosity in DMF.[50] This observation is in agreement with a complexation of the PAA by the solvent.

The problem of the polyelectrolyte behavior of PAA has been studied. A polyelectrolyte effect was observed in the determination of $\bar{M}_w$ and $\bar{M}_n$,[51] but it was clearly established, for the case of PAA prepared from PMDA and 4,4'-oxybisbenzeneamine (oxydianiline, ODA) in NMP, that the polyelectrolyte behavior was due to triethylamine, which is an impurity in NMP. In pure NMP distilled on P_2O_5, or in pure NMP to which water has been added, the variation of the specific viscosity to concentration ratio (η_{sp}/C) vs. concentration is a straight line.[52] Another aspect of the PAA synthesis is the reversibility of the reaction; Cotts and Volksen[53] have polymerized very pure PMDA and ODA in different molecular ratios by addition of PMDA in a solid state or in solution to a solution of ODA. The results are summarized in Table 2.

When PMDA is added in solution there is good agreement between the experimental and the calculated $\bar{M}_w$. When PMDA is added in a solid state, the higher $\bar{M}_w$ decreases as a function of time. The question was whether this phenomenon occurred because of degradation of the highest

Table 2 Results from Polymerization of Very Pure PMDA and ODA in Different Molecular Ratios

PMDA/ODA	% solids	M_w (calc.)	M_w (exp.)
Addition of solid PMDA to ODA solution			
0.8182	25.6	2000	6700
0.9231	17.5	10 000	22 000 (10 000)[a]
0.9725	18.6	30 000	43 000 (29 000)[b]
Addition of PMDA solution to ODA solution			
0.8182	20	2000	2400
0.9259	20.4	10 000	11 000
0.9717	16.7	28 000	26 000

[a] After aging at 0 °C for 60 days.
[b] After aging at 25 °C for 10 days.

molecular weight or because of equilibration; the answer was given by Walker,[54] who showed by size exclusion chromatography (SEC) that, when $\bar{M}_w$ decreases, $\bar{M}_n$ is kept constant, thus clearly demonstrating equilibration.

The reversibility of the reaction is mainly dependent on proton acidity, the basicity of the amide and the nature of the solvent. Kinetic aspects have been studied in the case of ODA with 1,3-isobenzofurandione and with an alkylene bis(1,3-isobenzofurandione).[55] Considering the Mark–Houwink relationship in the case of PAA, Cotts and Wolksen[56] found, respectively, in NMP and in a mixture of NMP and dioxane, 0.74 and 0.70 for the *a* coefficient; such values are in agreement with a flexible chain for the polymers.

(b) Synthesis of poly(amic acid)s in other solvents. The adhesive properties of the PAA prepared from 5,5'-carbonylbis(1,3-isobenzofurandione) (benzophenone tetracarboxylic dianhydride, BTDA) and 3,3'-carbonylbisbenzeneamine are improved if a diglyme solution is used.[57] This solvent was also used to prepare PAA from BTDA and a mixture of ODA and 1,3-benzenediamine[58] or from 5,5'-thiobis(1,3-isobenzofurandione) and the same diamines.[59] In the case of a diamine with a low reactivity, such as 4,4'-sulfonylbisbenzeneamine (**29**), the solvent enhances the basicity of the amine by complexation, and the polymer obtained exhibits higher viscosity than the same polymer prepared in NMP, DMAC or DMF.[60]

(i) Cyclization of poly(amic acid)s

(a) Thermal cyclization of poly(amic acid)s. This reaction is generally performed by heating a cast solution of PAA with a low solid content (15–25% solid) *in vacuo* or in an inert atmosphere. Different cure cycles have been studied as a function of the nature of the main chain (flexibility) and of the application required. Generally, the solvent is removed at a temperature between 100 °C and 150 °C and the polymer is heated to 300 °C or 350 °C.[50] The loss of weight observed with PAA, by thermogravimetric methods at 170 °C, 185 °C and 200 °C, was interpreted as a two-stage imidization *via* a complex between PAA and the solvent (DMAC).[61] A process of complexation/decomplexation between amic acid and NMP was also demonstrated by Feger.[62] In the case of PAA derived from PMDA and ODA, the rate of thermal cyclization at 161 °C was increased by adding a tertiary amine,[61] according to Scheme 3.

The correlation between the basicity of the starting diamine and the rate of cyclization has been mentioned by Soviet workers;[63] the higher the pK_a of the diamine, the higher the rate constant of ring closure.

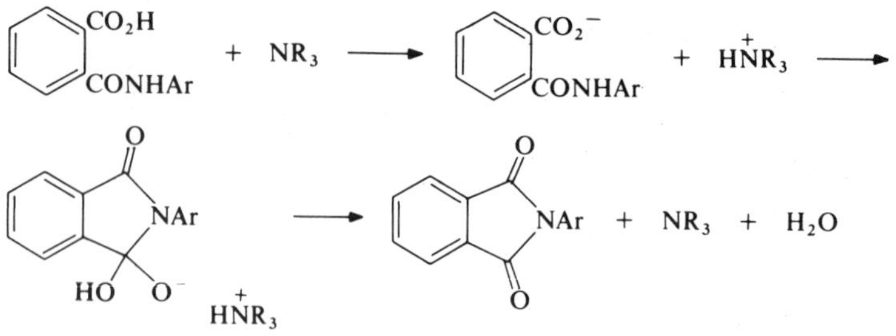

Scheme 3

Competition between cyclization and depolymerization was observed in the case of polymers obtained from condensation of PMDA with 4,4'-, 3,4'- and 3,3'-carbonylbisbenzeneamines. These polymers exhibit brittle behavior between 175 °C and 225 °C, but are flexible at 300 °C.[64] The same observation was made for other polyimides.[65] The variation of $\bar{M}_n$ as a function of the course of imidization at 150 °C shows a decrease in $\bar{M}_n$ during the first stage of the reaction and then a partial recovery of $\bar{M}_n$,[66] in the case of PAA from 4,4'-(fluorene-9,9-diyl)bisbenzeneamine and 5,5'-[1-methylethylidene]bis(1,3-isobenzofurandione).

Using the Fourier transform infrared (FTIR) technique, Young[67] observed, in the case of PMDA–ODA poly(amic acid) cyclization, an anhydride band at 1850 cm^{-1} which reached a maximum intensity between 150 and 200 °C and diminished in intensity at higher temperatures. This phenomenon was also observed for the soluble polymer (**30**) carrying flexible linkages (Scheme 4).

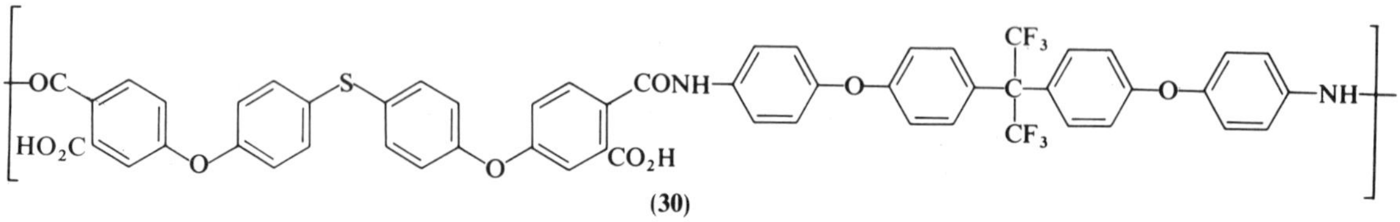

(**30**)

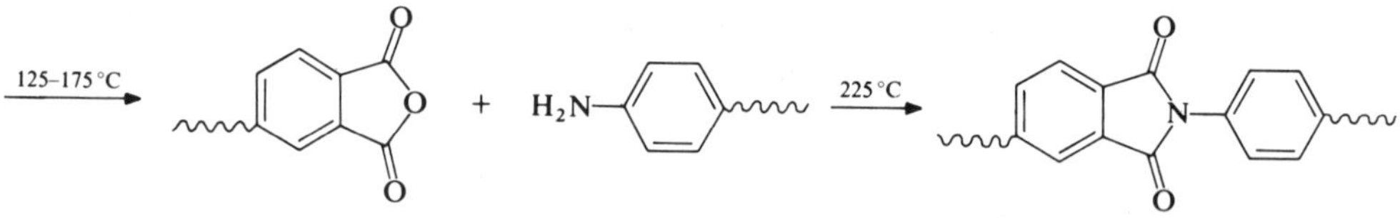

Scheme 4

The anhydride band appeared when the film was heated to 152 °C, showed a maximum intensity at 175 °C and was not detectable at 225 °C. The solubility of these polymers made SEC and viscosity characterization possible, which agreed with an initial reduction in molecular weight before final solid-state polycondensation. The general thermal cyclization of a PAA solution can be summarized according to Scheme 5.

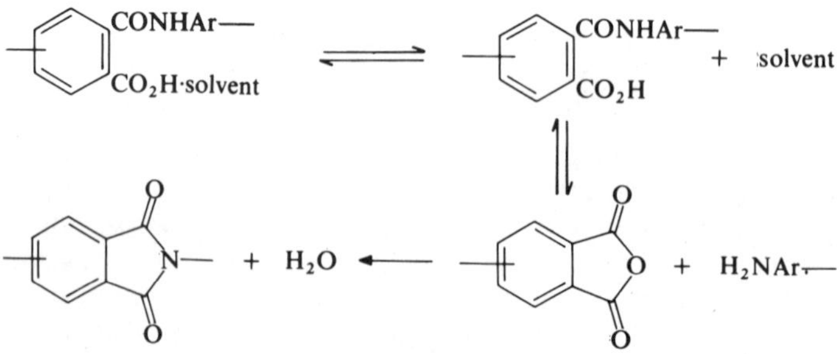

Scheme 5

(b) Chemical cyclization of poly(amic acid). PAA can be dehydrated by chemical processes. Using dicyclohexylcarbodiimide, poly[3-phenylimino-1(3*H*)-isobenzofuranone]s (polyisoimides; **31**)

were obtained,[68] as shown in Scheme 6. Isoimides rearrange into imides on being heated or on being chemically treated with acetate ions.

Another chemical dehydration extensively studied involves the use of an anhydride in the presence of organic bases such as pyridine or trialkylamine.[69] The mechanism was recently published[70] from a study on a model compound and on PAA from PMDA–ODA and PMDA–diaminocumene. When dehydration was performed with acetic anhydride using triethylamine, only imide was obtained, but pyridine gave a mixture of predominantly imide and isoimide; tertiary amine plays a catalytic role. The suggested mechanism is given in Scheme 7.

The first step is the formation of a mixed anhydride, then two independent routes can give imide or isoimide. Isoimide formation is kinetically favored *via* intramolecular *O*-acylation through an ion pair in equilibrium with the mixed anhydride. The imide form is thermodynamically favored. In the case of poly(amic acid) cyclization, viscosity measurements show that no degradation and no cross-linking occur during the cyclization. Chemical cyclization with trifluoroacetic anhydride gives only the isoimide product.

(iii) Polymerization of a dianhydride with a diamine — reaction without solvent

In microelectronics, polyimides are used as dielectric coatings. The solutions of PAA are spin coated and, after solvent removal, thermally cyclized. In order to avoid problems due to solvent evaporation, PMDA and ODA are coevaporated at 200 °C under a vacuum on a substrate. The effusion rate of each reactant is controlled in order to keep the stoichiometry better than 1% molar; the substrate is kept at 25–50 °C. The deposited film contains 30–50% of unreacted product and the PAA. The final polymerization is obtained by heating for 30 min at 175 °C and for 30 min at 300 °C in nitrogen, but the molecular weight is rather low ($M_w = 13\,000$). Thermal properties of cured vapor-deposited monomer are very similar to the properties of polyimides coming from PAA, but the dielectric properties are better ($\varepsilon = 2.9$, tan $\delta = 0.008$ vs. $\varepsilon = 3.2$, tan $\delta = 0.01$).[71]

(iv) Polymerization of a dianhydride with a diamine in phenolic solvents: poly(ether imide)s

Dianhydrides containing diaromatic ethers were prepared by the aromatic nucleophilic substitution of nitroimide (32) by phenoxide dianions (33),[72] followed by alkaline hydrolysis and cyclodehydration to dianhydrides (34; Scheme 8).

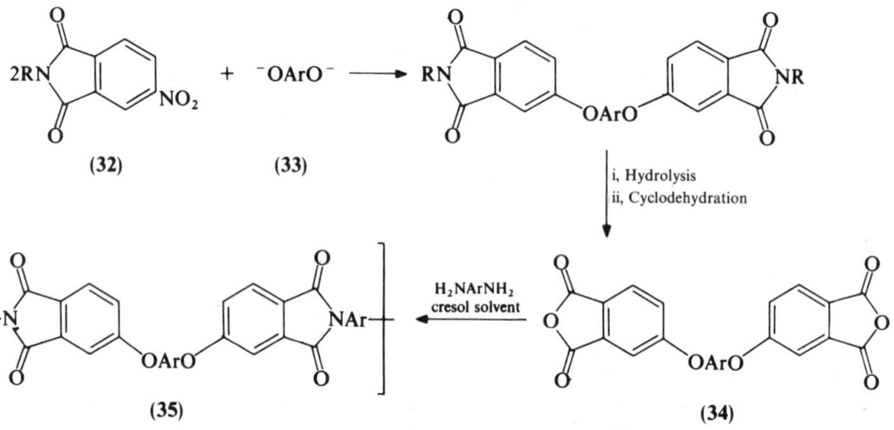

Scheme 8

The reaction between the dianhydride and diamines was conducted in phenol, cresol or chlorinated solvents, in combination with inert diluents such as toluene or chlorobenzene. The reaction mixture was heated to 160–180 °C and the water was removed azeotropically; the polymer was recovered by precipitation in methanol. Poly(ether imide)s (35) were generally amorphous, T_g 210–265 °C, except in the case of dianhydrides coming from diphenols with *para* linkages, such as 1,4-benzenediol and [1,1'-biphenyl]-4,4'-diol.[73]

The scope of reaction for formation of aromatic polymers *via* nitro group displacement has been reviewed by Takekoshi.[74]

30.4.1.2 Polymerization of a dianhydride with a diisocyanate

In 1967 it was observed that aromatic and aliphatic diisocyanates react with PMDA to yield polyimides.[75] Using the reaction of a mixture of diisocyanates with a dianhydride, moldable polyimides were obtained.[76] The reaction is catalyzed by water[77] but also by metallic alkoxides[78] or alkali-metal lactamates.[79] From a mechanistic point of view, the catalytic effect of water is interpreted as arising from a partial hydrolysis of the isocyanate,[80] followed by reaction of the aromatic amine with the anhydride (Scheme 9).

Meyers[81] was able to isolate a soluble seven-membered ring intermediate (37) besides polyimide (38) in the reaction of PMDA with 1,1'-methylenebis(4-isocyanatobenzene) (36) in DMF at 130 °C (Scheme 10). The intermediate loses carbon dioxide on heating. This mechanism was supported by IR spectra and by the observation that the yield of the seven-membered intermediates was increased when the reaction was carried out under CO_2 pressure.

This mechanism was confirmed later from the reaction of a diisocyanate with a mixture of dianhydride and tetraacids[82] or with a mixture of dianhydride and diester diacid.[83] Higher

$$ArNCO + H_2O \longrightarrow ArNH_2 + CO_2$$

$$ArNCO + ArNH_2 \longrightarrow ArNHCONHAr$$

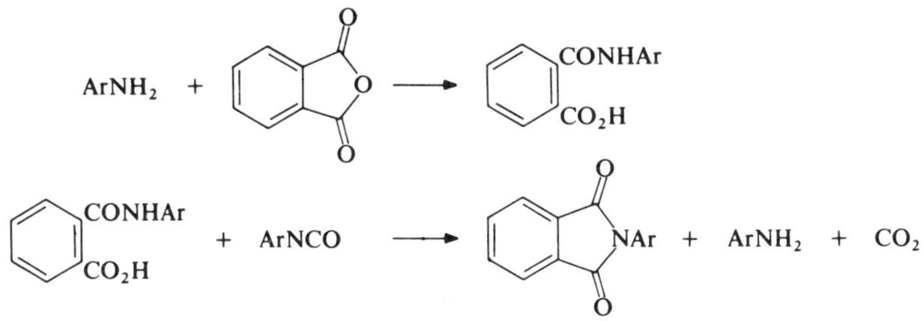

Scheme 9

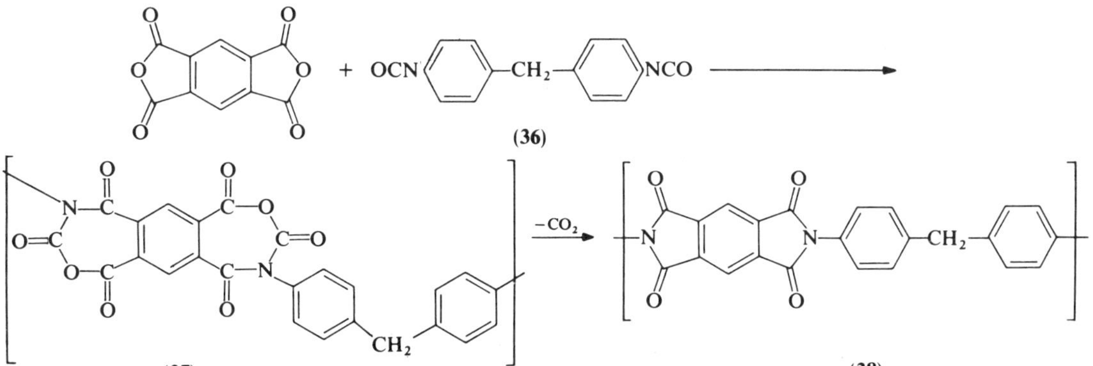

(36) (37) (38)

Scheme 10

molecular weights were obtained with a mixture of tetraacid and dianhydride in a molar ratio ranging from 1:7 to 1:4. 1,1′-sulfonylbis(4-isocyanatobenzene),[84] 1,1′-methylenebis(3-chloro-4-isocyanatobenzene) and 1,1′-azobis(4-isocyanatobenzene)[85] give only low molecular weight polyimides by reaction with PMDA and BTDA in DMAC. Recently, a tractable polyimide was obtained by reaction of a mixture of 1,1′-methylenebis(4-isocyanatobenzene) and 1,1′-methylenebis(4-isocyanatocyclohexane) (mixture of isomers) with PMDA and BTDA in DMSO and DMF and NMP.[86]

Although the synthesis of polyimides from isocyanates gives a rather low molecular weight, the reaction is used industrially for the production of Polyimide 2080 by Upjohn. The condensation of 5-carboxy-1,3-isobenzofurandione with diisocyanates was an early patent[87] and has been said[88] to give a soluble high polymer by polymerization in NMP. The poly(amide–imide) prepared by this process is used for wire insulation.

30.4.1.3 Polymerization of a dianhydride with silylated diamines

Silylated diamines (**39**) are obtained by reaction of trimethylchlorosilane on the diamine. They react with the dianhydride in solvents such as tetrahydrofuran[89,90] at room temperature, giving a

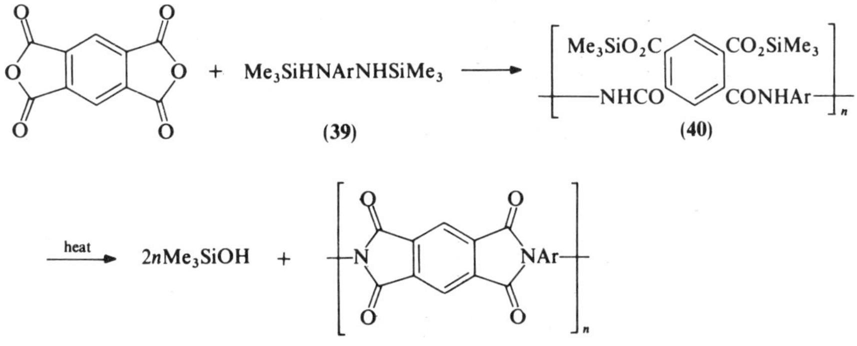

Scheme 11

poly(amide trimethylsilyl ester) (**40**), by transfer of the silyl protective group on the carboxylic group, which is subsequently cyclized to polyimide, with elimination of trimethylsilanol, by heating at 150 °C for 30 min (Scheme 11).

It was reported that diamines (**41**), fully substituted by trimethylsilyl groups, also react with anhydrides to form an imide, giving a disiloxane as a by-product (equation 11).[91]

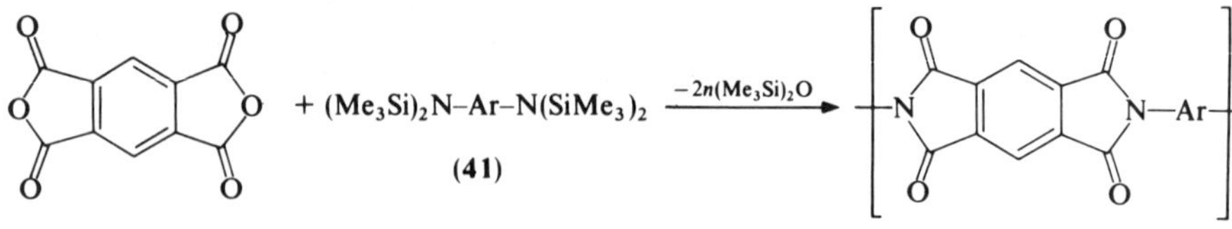

$$\text{pyromellitic dianhydride} + (Me_3Si)_2N\text{-}Ar\text{-}N(SiMe_3)_2 \xrightarrow{-2n(Me_3Si)_2O} \left[-N\underset{O}{\overset{O}{\diamond}}N\text{-}Ar- \right]_n \quad (11)$$

(**41**)

30.4.1.4 Polycondensation of diester diacids with diamines

Many applications, such as adhesives, matrices for composites and coatings in electronics, need high solid-content solutions. The reactivity of diester diacids is lower than the reactivity of the corresponding dianhydrides, so it is possible to obtain soluble low molecular weight chain-extendable polyimides by reaction with diamines.

Aliphatic polyimides were prepared in 1955 from dialkyl tetracarboxylates (**42**) and aliphatic diamines by a two-stage melt condensation of the resulting organic salts (**43**) at 110–140 °C, then at 250–300 °C[92] (Scheme 12).

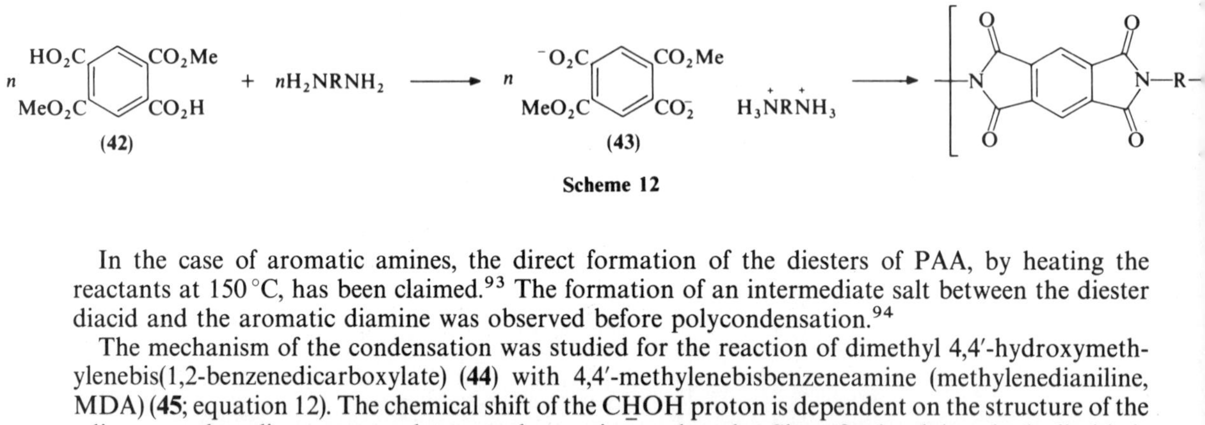

Scheme 12

In the case of aromatic amines, the direct formation of the diesters of PAA, by heating the reactants at 150 °C, has been claimed.[93] The formation of an intermediate salt between the diester diacid and the aromatic diamine was observed before polycondensation.[94]

The mechanism of the condensation was studied for the reaction of dimethyl 4,4'-hydroxymethylenebis(1,2-benzenedicarboxylate) (**44**) with 4,4'-methylenebisbenzeneamine (methylenedianiline, MDA) (**45**; equation 12). The chemical shift of the C$\underline{H}$OH proton is dependent on the structure of the adjacent carboxylic groups, and was used as an internal probe. Since final polybenzhydrolimide is soluble in NMP, it is possible to examine the course of polycondensation and the extent of reaction as a function of time and temperature.[95] It was concluded that the intermediate was the amic acid, which can be observed up to 160 °C. At 180 °C, only polyimide (**46**) was detected; high molecular weights were obtained by heating to 200 °C.

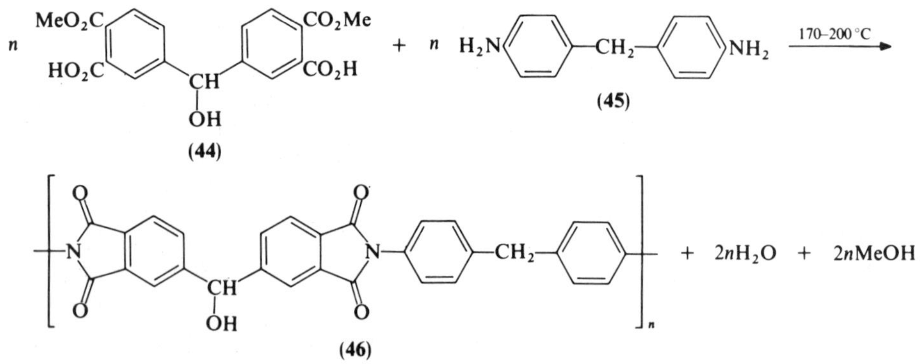

Another study based on dimethyl 4,4'-carbonylbis(1,2-benzenedicarboxylate) and MDA, using nuclear magnetic resonance (NMR) and FTIR, focused on the mechanism of the substitution of the ester group by the amino group.[96] The formation of an intermediate anhydride via proton abstraction from the carboxylic group by the amine was demonstrated (Scheme 13).

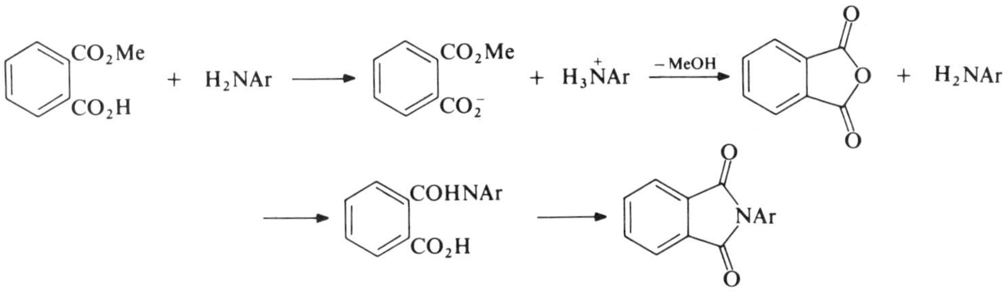

Polyimides and Other Heteroaromatic Polymers 513

Scheme 13

The reactions of more acidic alcohols, such as chloro- or fluoro-propanols or ethyl glycolate, with anhydrides give ester acids with better leaving groups than is the case for alkyl ester acids.[97] The reactivity with aromatic amines is greatly increased, and it has been demonstrated that the reaction is due to the fast elimination of alcohol followed by reaction of amine with anhydride (Scheme 14).

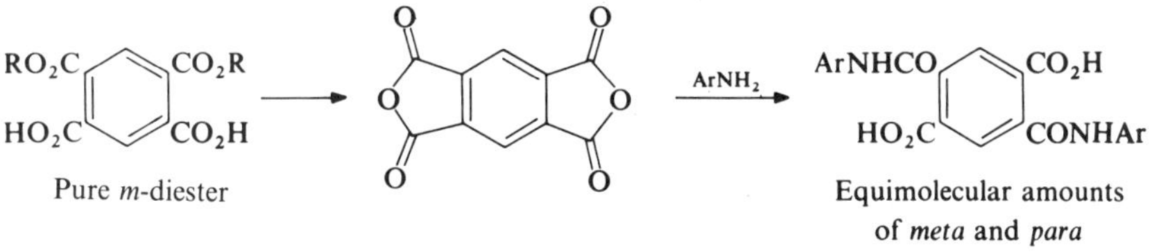

Scheme 14

The reactions of diester diacids have been used to prepare high concentration PAA by reaction of dianhydrides partially diesterified with an alcohol. The intermediate PAA was end-capped with the diester diacid, which only reacted at higher temperatures[98] (Scheme 15).

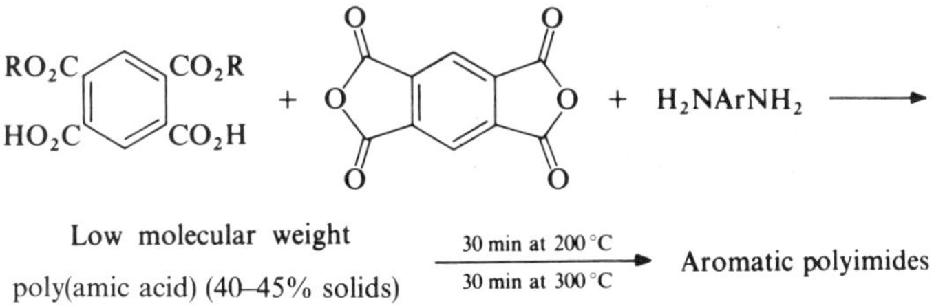

Scheme 15

30.4.1.5 Formation of polyimides via nucleophilic nitro group displacement reactions

The application of aromatic nucleophilic substitution to polymer synthesis is now well documented. Industrial polymers such as polysulfone, poly(phenylene sulfide) and poly(ether ketone) are produced by this synthetic route.[99] As was shown in Section 30.4.1.4, nitro groups activated by the imide function react very smoothly with phenolate anions. This reaction has been extended to polycondensation,[100] using the nucleophilic displacement of 2,2′-arylenebis(nitroisoindoledione) (**47**) by bisphenol salts (equation 13).

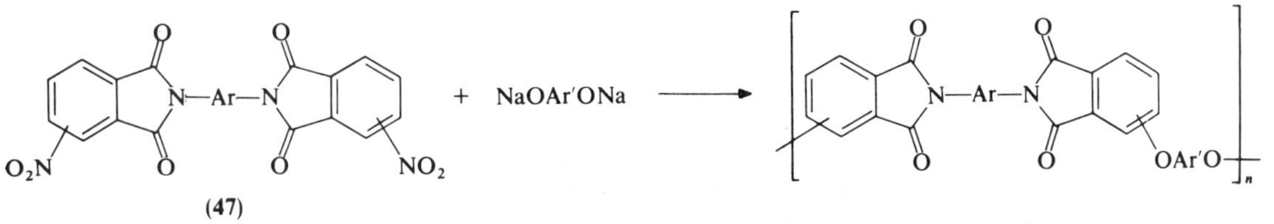

(13)

The reaction takes place in anhydrous conditions in DMSO or DMF after the formation of the bisphenolate. Moisture reacts with the phenolate, leading to an equilibrium with phenol, and the formation of an OH⁻ ion which can open the imide ring. The carboxylate (**48**) created deactivates the nitro group and the phenol (**49**) does not behave like a nucleophile, so the polymerization is stopped. Even under anhydrous conditions, the molecular weight seems rather moderate from the intrinsic viscosity observed, ranging from 0.2 to 0.5 dl g.$^{-1}$

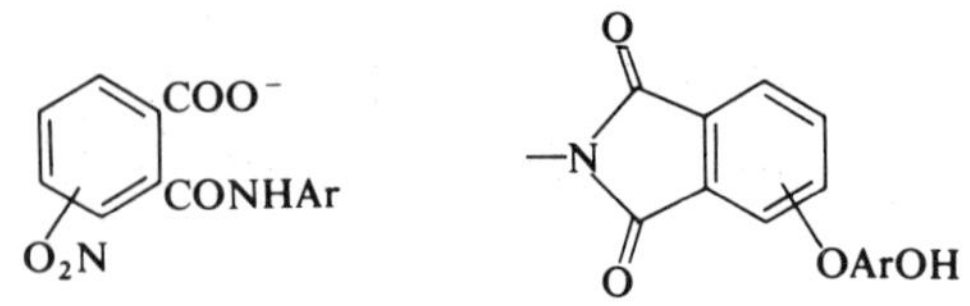

Deactivated nitro group (**48**)

Non-active phenol (**49**)

The 4- and 5-nitro-1,3-isobenzofurandiones (**50**) have been opened by the amino groups of 3- and 4-aminophenols (**51**) at low temperature in DMF. On heating to 140–145 °C, the cyclization took place with the formation of an AB monomer (**52**; Scheme 16), which was polymerized after formation of phenolate. These AB poly(ether imide)s (**53**) also showed low viscosities.[101]

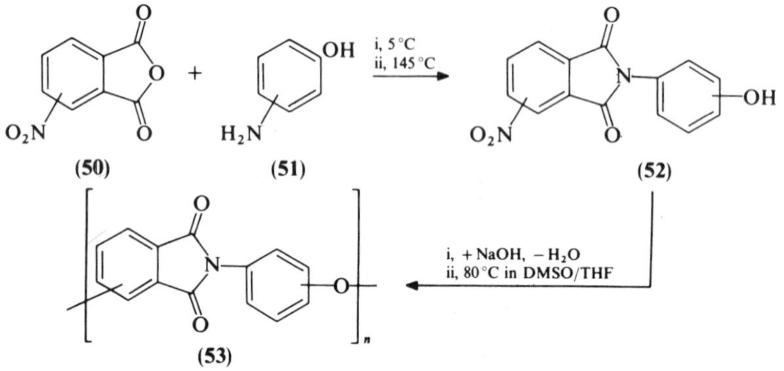

Scheme 16

30.4.1.6 Chemistry of special linear polyimides and derivatives

An important part of recent work on polyimides has been related to the development of polymers starting from new raw materials, dianhydrides or diamines, in order to improve processability and high temperature durability, thus lowering moisture resistance, dielectric constant and loss factor, and lowering the thermal expansion coefficient, increasing the toughness, etc.[102]

(i) Thermoplastic soluble polyimides

The main synthetic strategy starts from dianhydrides or diamines carrying flexible linkages, such as SO$_2$, O, etc., and *meta* bonding between the aromatic rings.[13] Some of these modified polyimides are either commercially available or currently being developed. Table 3 summarizes the chemical structure and T_g values of the most extensively studied thermoplastics.

Two types of thermoplastic material appear in this table. Polyimides with a high content of *meta* linkages or flexible heteroatoms exhibit rather low T_g values. The other type of material was synthesized with bulky groups such as indane derivatives (XU 218)[103] or with groups which tend to restrict rotation, such as hexafluoroisopropylidene, and their T_g values are rather high.[104,105]

Crystallinity has been observed in some linear polyimides[5,106,107] and it was mentioned recently that the imidized LARC TPI could exhibit a transient form of crystallinity.[108] Hergenrother was able to prepare thermoplastic crystalline poly(imide ether ketone) (**54**) using the concept developed for the crystalline poly(ether ketone) (Scheme 17).[109]

(ii) Polyimides with low thermal expansion coefficients

The relationship between chemical structure and thermal expansion coefficients (TEC) was studied for the case of polyimides prepared by the reaction of dianhydrides, such as PMDA and

Table 3 Thermoplastic Soluble Polyimides

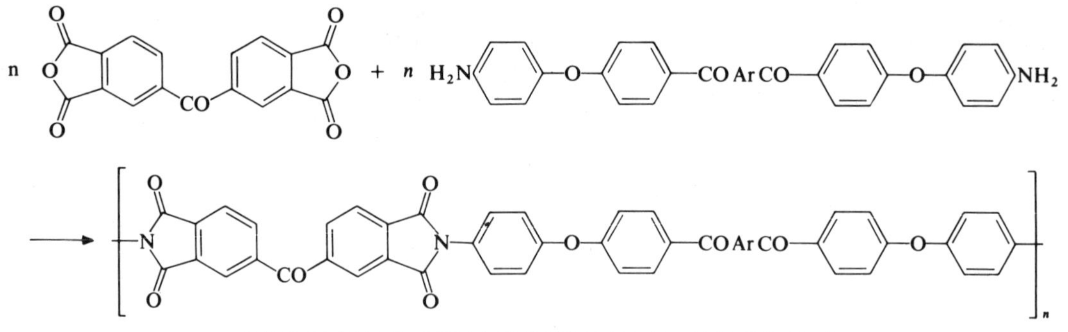

Scheme 17

Ar	T_g (°C)	T_m (°C)
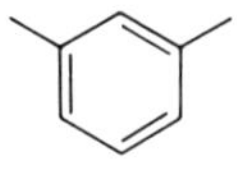 (meta-phenylene)	222	350
 (para-phenylene)	233	427
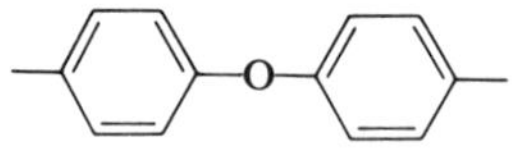 (4,4′-oxydiphenylene)	215	418

(5,5′-bisisobenzofuran)-1,1′,3,3′-tetrone (4,4′-biphthalic dianhydride, DPDA), with rigid diamines, *via* PAA in NMP. The lowest TEC values were observed with benzene or pyridine diamines with *para* linkages. The linearity of the conformation is responsible for the low TEC, which is associated with crystallinity, good mechanical properties and excellent thermal stability.[110,111] TEC values ranging from 0.4×10^{-6} K^{-1} to 1×10^{-5} K^{-1} can be obtained. These polymers find intermetallic dielectric applications in electronics.

(iii) Photosensitive polyimides

Photosensitive polymers are classified into two groups according to their behavior when they are irradiated: the negative resins cross-link and the positive ones are cleaved into soluble fragments.

The first type of negative polyimide was developed by Asahi, Ciba-Geigy, Du Pont and Merck from the pioneering work of the Siemens team.[112–114] PMDA reacts with 2-hydroxyethyl 2-methyl-2-propenoate (**55**), then the diester diacid is condensed with ODA. The poly(amic ester) (**56**) is

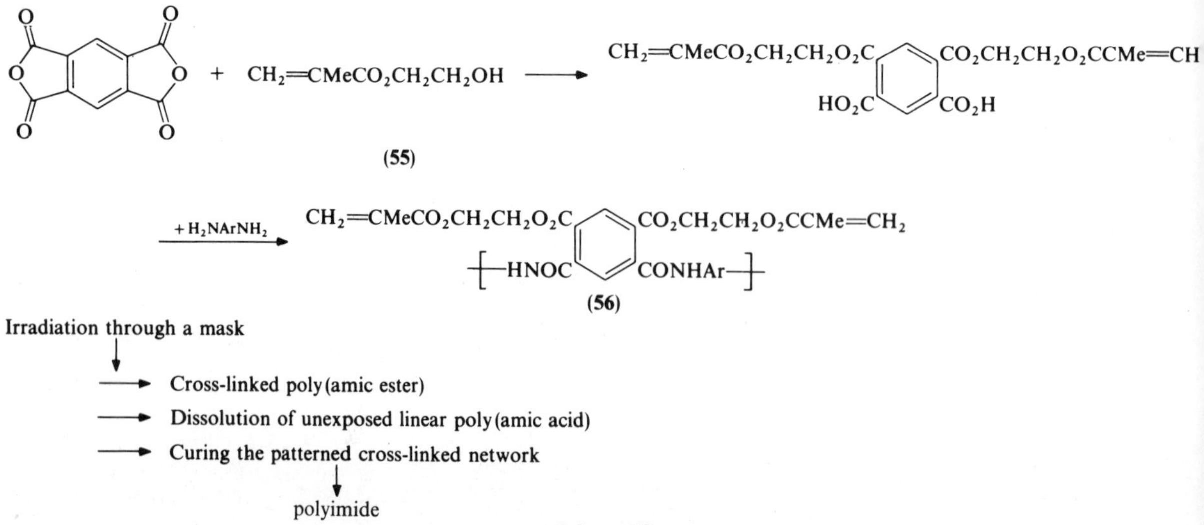

Scheme 18

photopatterned, and the non-irradiated polymer is removed by dissolution. Curing at 300–350 °C induces the degradation of the acrylic network and the formation of polyimide (Scheme 18).

A second approach is based on the modification of PAA with acrylic or cinnamic derivatives, carrying functions which are able to react with free carboxylic groups (Scheme 19). The process includes irradiation development and formation of patterned polyimide by heating to 300–350 °C.[115]

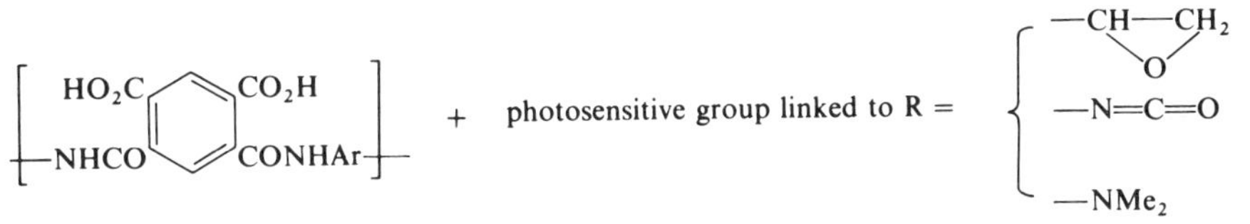

Scheme 19

The third type of photosensitive polyimide was developed using the concept of a soluble preimidized polymer (57), obtained from the condensation of DPDA and N-[2-amino-4-(4-aminophenoxyl)phenyl]-2-propenamide in NMP followed by chemical cyclization. This type of fully cyclized polymer has two main advantages: the stability of the solution and a low shrinkage during the final curing due to the precyclized structure.[116]

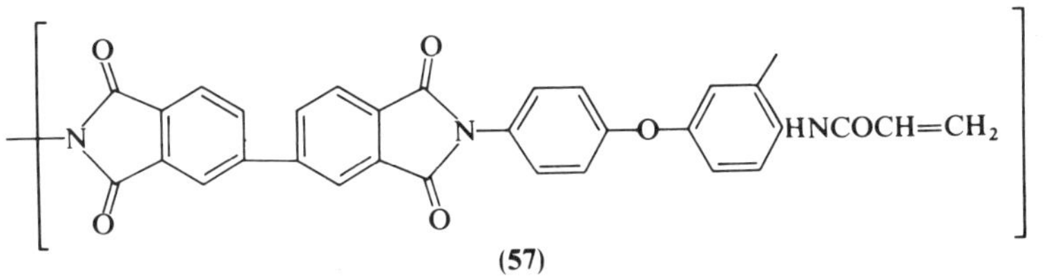

(57)

(iv) Copolyimide isoindoloquinazolinedione

Poly(amic acid) synthesis was applied to aromatic amino compounds (58) carrying carbonamido groups in the *ortho* position (Scheme 20). The thermal cyclization steps can be performed by heating or by chemical reaction, with the formation of isoindoloquinazoline fused rings (59). Copolymers with diamines have also been prepared[118,119] and marketed by Hitachi for heat-resistant dielectric applications.

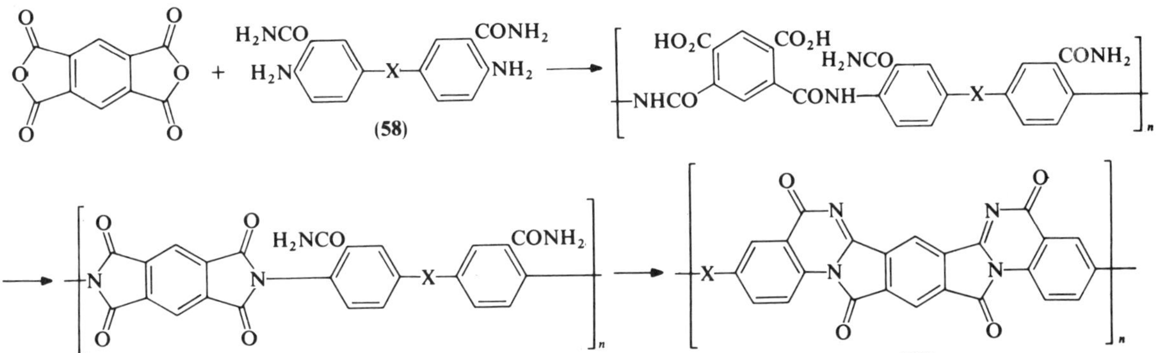

Scheme 20

(v) Preparation of mono- and multi-layers of polyimides by the Langmuir–Blodgett (LB) technique

LB films are prepared from straight-chain amphiphilic compounds, such as fatty acids, by spreading them on water and forming a condensed phase by increasing the surface pressure. Two chemical approaches have been tried, in order to obtain very thin films of organized polyimides.

The first one is based on the transformation of PMDA in distearyl ester (60), followed by condensation of the corresponding diacid chloride (61) into polyamide (62) by reaction with diamine

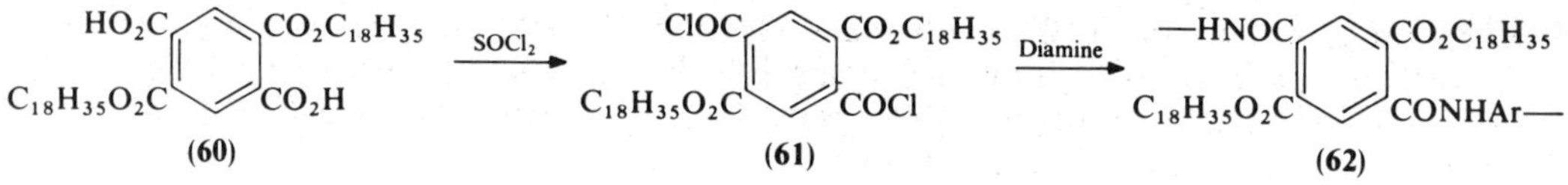

(Scheme 21). The amphiphilic poly(amide ester) (**62**) was spread onto the water surface, and, when surface pressure reached 55 dyn cm^{-1} (dyne = 10^{-5} N), a condensed film with a thickness of about 1800 Å was obtained. The poly(amide ester) film was cyclized at 400 °C with evolution of stearyl alcohol.[120]

Similar results were obtained by direct salification of the PAA by dimethylcetylamine. The multilayer film obtained with the salt was chemically cyclized by acetic anhydride and pyridine. The thickness of the 100-layer polyimide film was 400 Å.[121]

30.4.2 Thermosetting Polyimides

Three main types of thermosetting resins have been studied: bismaleimides, bisnadimides, and acetylene-terminated imides.

30.4.2.1 The chemistry of 1,1'-arylenbis-1H-pyrrole-2,5-diones (bismaleimides)

(i) Synthesis

The ring opening of 2,5-furandione (**63**) by diamines in chloroform, acetone or toluene gives a precipitated maleamic acid (**64**; Scheme 22). The amic acid is soluble in aprotic polar solvents. Cyclization to bismaleimide (**65**) takes place either by a thermal method, by heating the unisolated amic acid in DMF or acetic acid,[122-124] or by chemical cyclization, carried out with sodium acetate and acetic anhydride;[125-128] the latter method is the most generally used. Recently, some patents have claimed a chemical cyclization by P_2O_5 in NMP[129] or by methanesulfonic acid in DMF–toluene.[130] Aliphatic and aromatic maleimides have rather high melting points.[123,124]

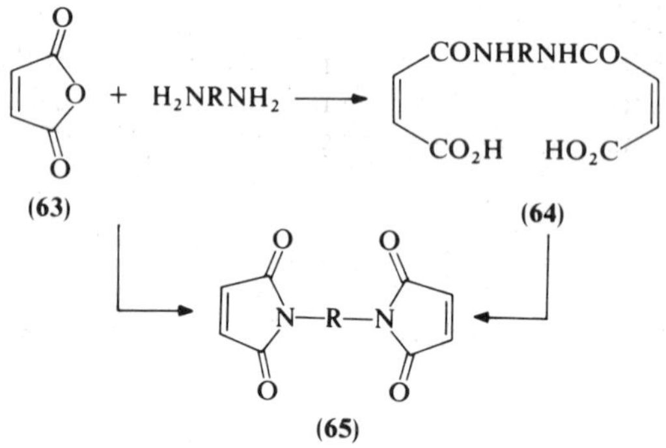

Scheme 22

(ii) Linear polymerization by nucleophilic addition

The maleimide carbon–carbon double bond is strongly electrophilic and can react with nucleophilic reagents such as amines and thiols, as shown in equation (14). This type of polymerization was recently reviewed.[131] The condensation with thiols has to be carried out in protic solvents in order to avoid the side reactions of intermediate anionic maleimido species with maleimide, giving crosslinking.

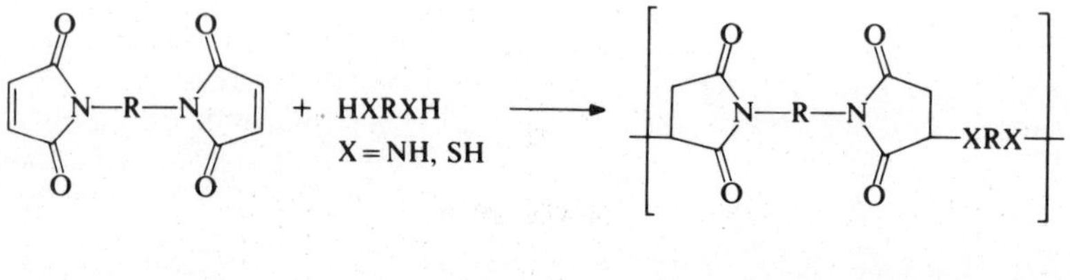

Condensation with aromatic diamines takes place in *m*-cresol solution and high molecular weights can be obtained. With aliphatic diamines side reactions can occur; for example, 1,2-ethanediamine and 1,3-propanediamine (67) open 1,1'-(1,3-phenylene)bis-1*H*-pyrrole-2,5-dione (66) to yield the corresponding polymaleamide (68; equation 15).

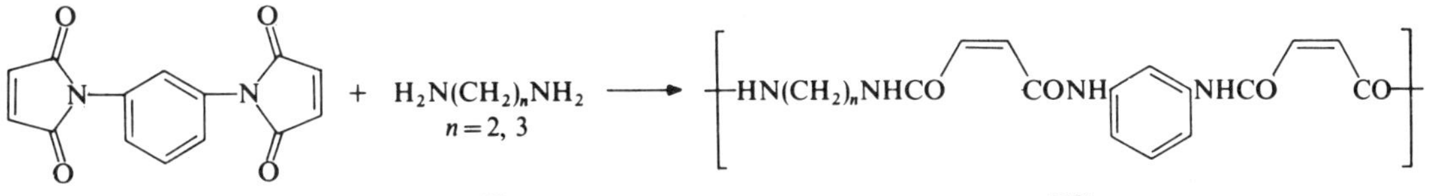

$$\text{(66)} + \text{H}_2\text{N(CH}_2)_n\text{NH}_2 \longrightarrow \text{(68)} \quad n = 2, 3 \tag{15}$$

(iii) Linear polymerization by Diels–Alder reaction

The maleimide carbon–carbon double bond can undergo a Diels–Alder reaction with dienes. When a bis aromatic maleimide (69) and an arylenebis(2,3,5-triphenyl-2,4-cyclopentadien-1-one) (70) are heated in chloronaphthalene at reflux, a polyaddition takes place. The ketonic adduct (71) is unstable, loses CO_2 and gives a diene, but the ensuing aromatization is difficult to perform quantitatively[132] (Scheme 23).

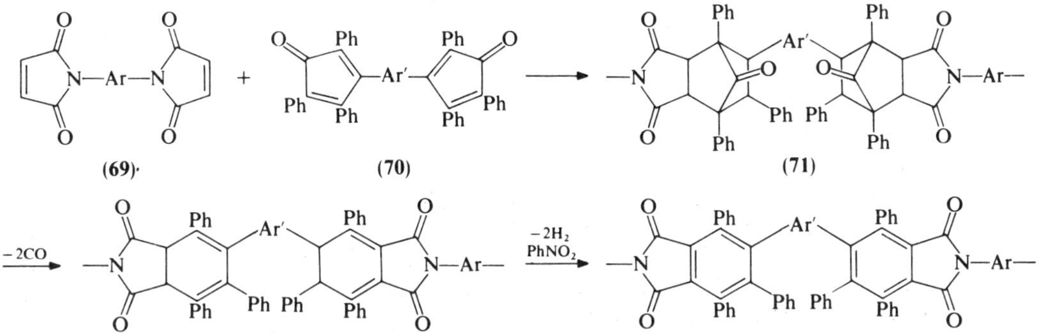

Scheme 23

Furan end-capped siloxanes (72) also react with bismaleimide in THF at 70 °C. The dienic reaction is followed by aromatization in acetic anhydride (Scheme 24). The molecular weights of both unaromatized and aromatized polymers ranged from 20 000 to 50 000.[133]

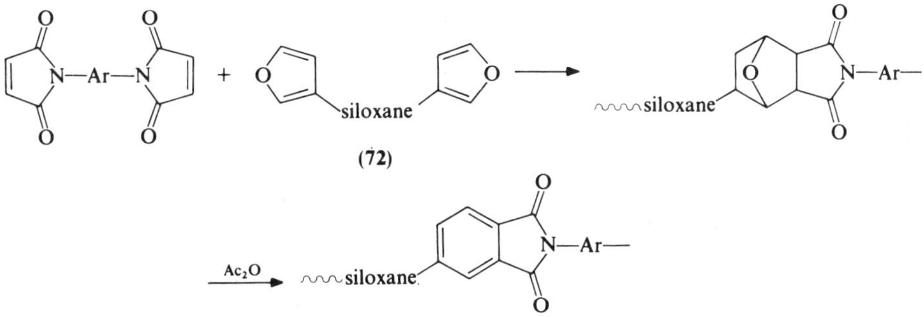

Scheme 24

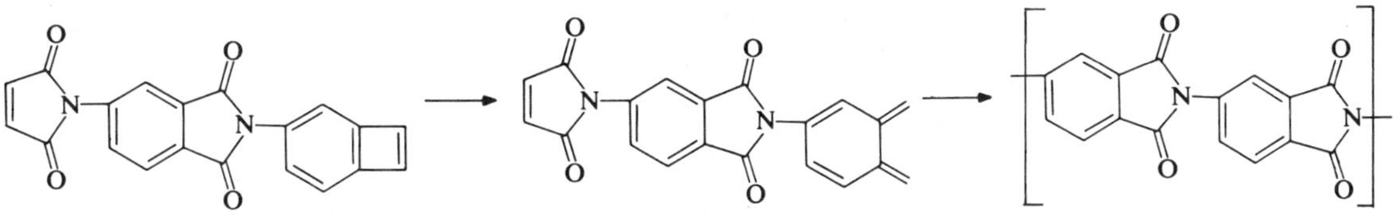

Scheme 25

The problem of the aromatization of the Diels–Alder adduct was neatly solved by Arnold, using the ring opening of benzocyclobutene end-capped resin (**73**) as shown in Scheme 25. The heated benzocyclobutene is in equilibrium with the dienic system (**74**) which directly gives the aromatic imide (**75**) by Diels–Alder reaction with maleimide.[134]

(iv) Cross-linking of bismaleimides

When bismaleimides are heated to 210 °C, they polymerize by an addition mechanism with the formation of a cross-linked network (**76**; equation 16).[135]

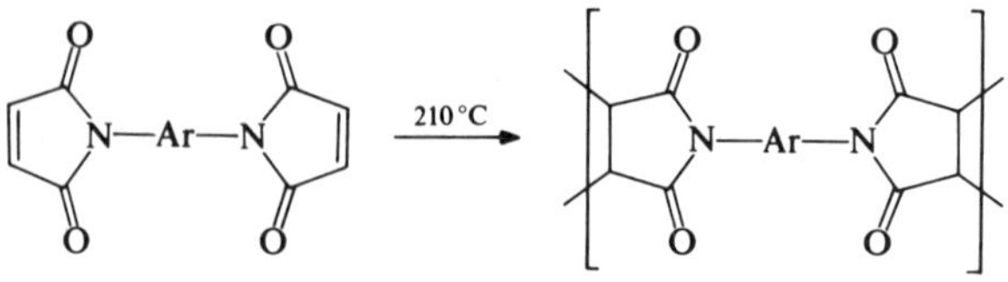

$$\text{(16)}$$

(**76**)

In the case of rigid bismaleimides with high melting points, the correlation between structure and polymerization temperature shows that polymerization has to be carried out above 230 °C.[136] When R is a diphenyl sulfone or a diphenyl ether sulfone oligomer (**77**), the more flexible linkages make it possible to obtain a B-stage by heating to 180–200 °C. A postcure at 270 °C gives the final network by a radical polymerization mechanism.[137]

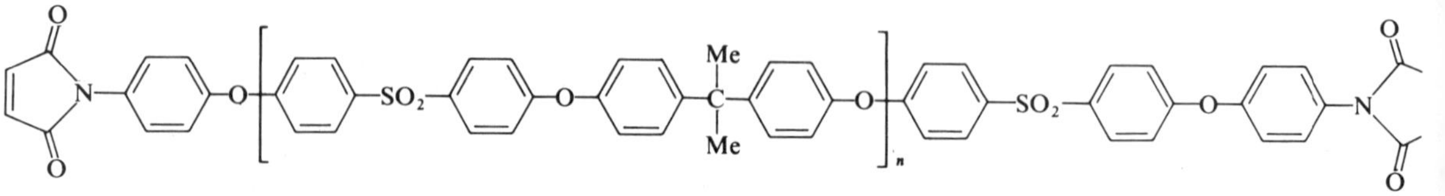

(**77**)

When R is aliphatic, it was shown by IR spectroscopy that, up to a conversion of 20–30%, polymerization was pseudo first order. There is a relationship between the length of the methylene sequence ($6 < n < 12$) and the activation energy, which decreases with decreasing n.[123] Bismaleimides can also be polymerized and cross-linked with a basic catalyst such as 2-methylimidazole or diazabicyclooctane, by an anionic mechanism.[138]

(v) Polymerization of maleimides by combination of a nucleophilic process and an addition process

On reacting a molar excess of bismaleimide with a diamino compound, linear chain growth occurs by nucleophilic addition. Maleimide end-capped low molecular weight aspartimide (**78**) can be cross-linked by further heating[139] (Scheme 26).

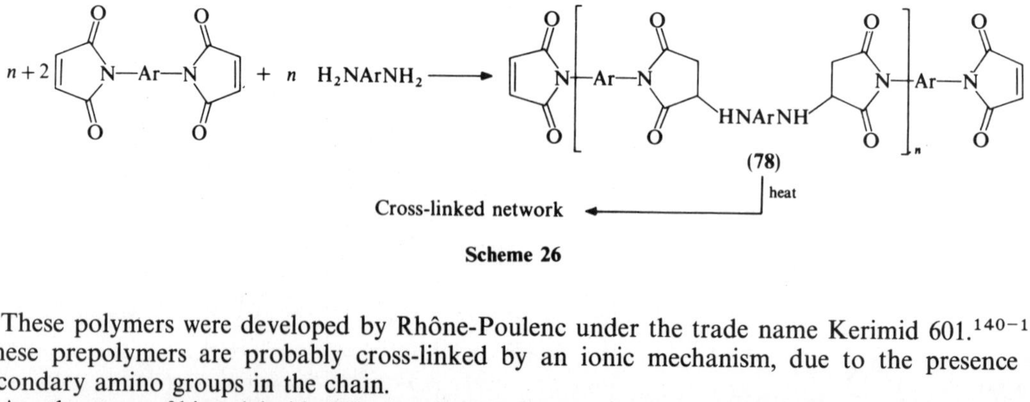

(**78**)

Cross-linked network ⟵ heat

Scheme 26

These polymers were developed by Rhône-Poulenc under the trade name Kerimid 601.[140–142] These prepolymers are probably cross-linked by an ionic mechanism, due to the presence of secondary amino groups in the chain.

Another type of bismaleimide (**80**), described by Technochemie as H795,[143] is formed by reaction of 3-aminobenzoic acid hydrazide (**79**) on bismaleimide (Scheme 27). The linear oligomer (**80**) is soluble in low boiling solvents such as toluene and oxygenated solvents.

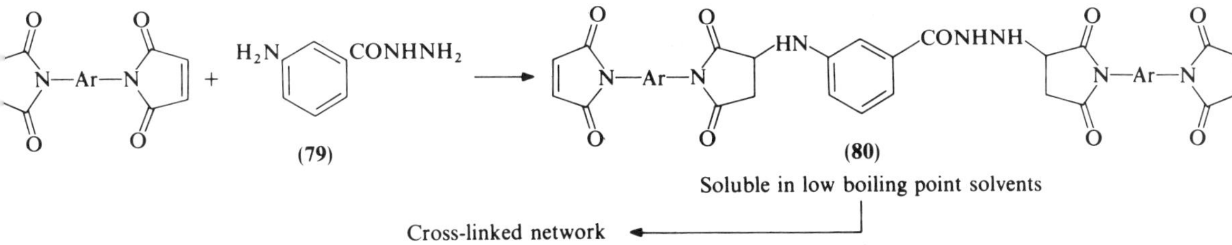

Polyimides and Other Heteroaromatic Polymers 521

Scheme 27

The cross-linking behavior of end-capped aspartimides has been studied by differential scanning calorimetry. Temperatures of polymerization and polymerization enthalpies are strongly dependent on the basicity of the amines: the more basic the amine, the lower the temperature of polymerization.[144]

(vi) Miscellaneous copolymerization of bismaleimides

Copolymerization of bismaleimides with other unsaturated monomers has been extensively investigated in order to develop more processable resins. Reaction with *o,o'*-diallylphenols is thought to proceed initially by ene-synthesis, followed by Diels–Alder reaction[145,146] (Scheme 28). Aromatic bismaleimides are soluble in liquid 4,4'-(1-methylethylidene)bis[2-(2-propenyl)phenol] and this system is the base of the XU 292 Ciba-Geigy resin.

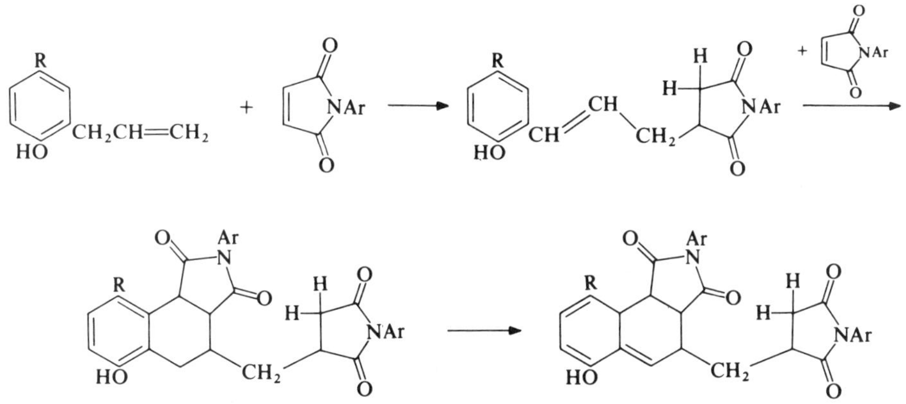

Scheme 28

By copolymerization of bis[4-(2-[2-propenyl]phenoxy)phenyl]methanone (**81**) with bismaleimides (BMI), Stenzenberger[147] (equation 17) obtained resins with improved fracture toughness compared to resins coming from pure bismaleimide.

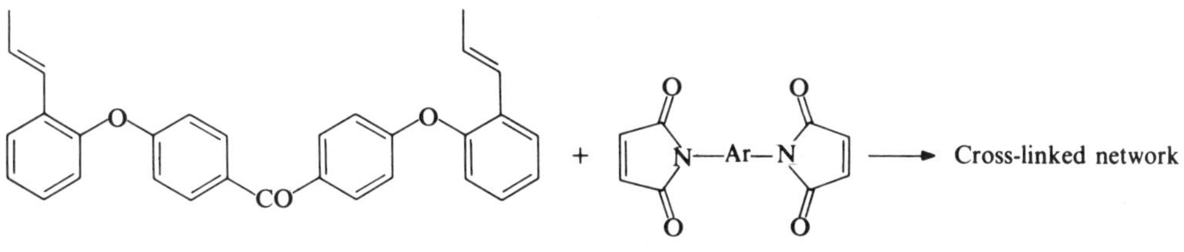

The copolymerization of bismaleimides and the biscyanate of 4,4'-(1-methylethylidene)bisphenol (**82**) can produce an oligomeric B-stage resin that is soluble in ketones and other oxygenated solvents and is compatible with epoxy resins.[148] The chemical structure of these bismaleimidetriazine (BT) resins has not yet been elucidated. The formation of an intermediate pyrimidine cycle (**83**) coming from cycloaddition between two cyanate functions and one maleimide has been claimed (equation 18). An excellent review of bismaleimide resins and their formulation has recently been published.[149]

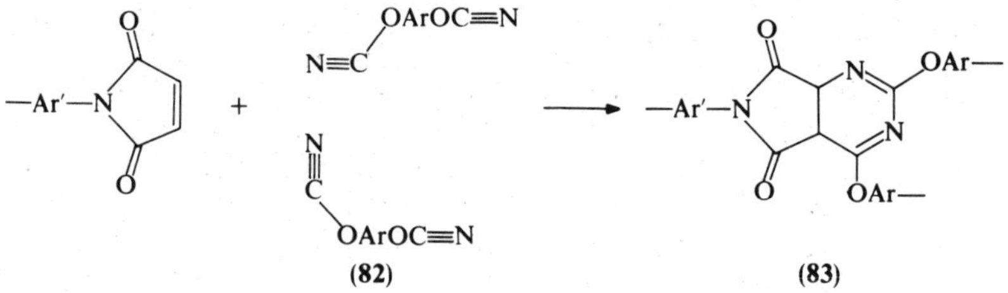

30.4.2.2 The chemistry of 3a,4,7,7a-tetrahydro-4,7-methano-1H-isoindole-1,3(2H)-diones (bisnadimides)

The first attempt to develop nadimide end-capped polyimides was carried out using the standard poly(amic acid) synthesis between BTDA (**84**) and MDA (**85**) with 3a,4,7,7a-tetrahydro-4,7-methanoisobenzofuran-1,3-dione (nadic anhydride); (**86**) for chain termination (Scheme 29).

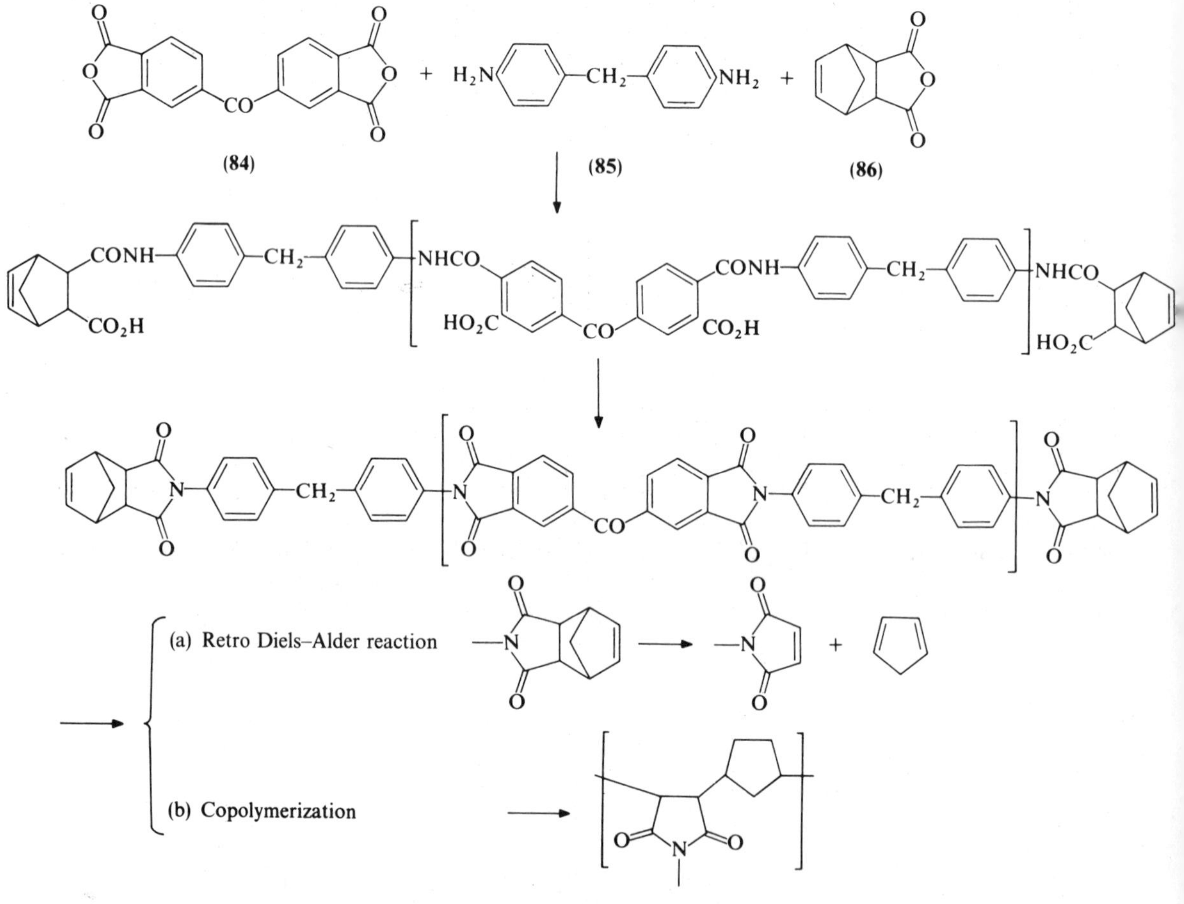

Scheme 29

The molecular weight limited amic acid solution is used to impregnate fiber reinforcement. Thermal cyclization is followed by a retro Diels–Alder reaction and a further copolymerization between cyclopentadiene and maleimide units.[150,151] End-capped oligoamic acid was marketed by Ciba-Geigy as P13N resin ($M_n = 1300$). However, this type of resin has various shortcomings, such as instability of PAA, poor resin flow, residual solvent and water evolution, which create porosity in composites. The processability was improved by a new synthetic route using a mixture of the dimethyl 4,4′-carbonylbis(1,2-benzenedicarboxylate), methyl bicyclo[2.2.1]hept-5-ene-2,3-dicarboxylate and diamine (Scheme 30).[151]

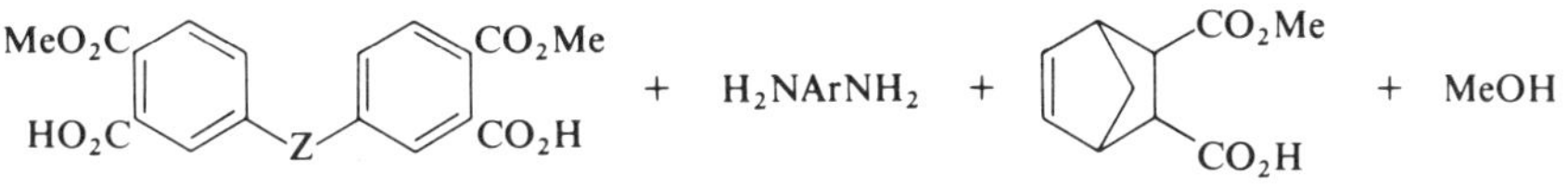

$$\text{MeO}_2\text{C} \underset{\text{HO}_2\text{C}}{\diagup\!\!\!\bigcirc\!\!\!-\!\!Z\!\!-\!\!\bigcirc\!\!\!\diagdown} \text{CO}_2\text{Me} \atop \text{CO}_2\text{H} \quad + \quad \text{H}_2\text{NArNH}_2 \quad + \quad \underset{\text{CO}_2\text{H}}{\bigcirc\!\!\!\diagdown}^{\text{CO}_2\text{Me}} \quad + \quad \text{MeOH}$$

(1) Mixture of reactants

    Z = CO   Ar = p-C$_6$H$_4$CH$_2$C$_6$H$_4$— $\bar{M}_n$ = 1500 (PMR 15)

    Z = C(CF$_3$)$_2$ Ar = p-C$_6$H$_4$CH$_2$C$_6$H$_4$— $\bar{M}_n$ = 1100 (PMR 11)

    Z = CO   Ar = Jeffamine $\bar{M}_n$ = 1600 (LARC 160)

(2) Impregnation

(3) Cure cycle   ⟶   Cross-linked network

Scheme 30

This new concept has been called polymerization of monomeric reactants (PMR). The stoichiometry of the reactants was adjusted to obtain an oligomer with $M_n = 1500$. In practice, BTDA, nadic anhydride and diamine are dissolved in methanol, and the reinforts are impregnated before condensation. A residual solvent (11%) is needed to provide drapability to the prepreg. The cure cycle includes thermal cyclization (water and methanol elimination) then retro Diels–Alder reaction and polymerization (Scheme 30).

PMR15 is now marketed as prepreg by various companies. It has been observed[152] that, in methanolic solution, BTDA can give tri- or tetra-esters that are less reactive than the diester. If 5,5'-[2,2,2-trifluoro-1-(trifluoromethyl)ethylidene]bis(1,3-isobenzofurandione) (6FDA) is used instead of BTDA, a fluorinated PMR11 with better oxidative stability is obtained.[153] The tack and drape of the prepreg were improved by replacing the rigid diamines with more flexible ones, such as Jeffamine (LARC 160, $\bar{M}_n$ 1600).[154]

The mechanism of thermal cross-linking for nadimide resins still remains unclear. The first report postulated an initial partial retro Diels–Alder dissociation followed by the reaction of cyclopentadiene with maleimide, giving a diradical. This diradical should be able to promote copolymerization of the nadimide with maleimide and cyclopentadiene, giving a partially unsaturated copolymer.[151]

Gailord and Martan[155] studied the radical initiated polymerization of model 3a,4,7,7a-tetrahydro-2-phenyl-4,7-methano-1H-isoindole-1,3(2H)-dione, and pointed out that *endo*→*exo* isomerization takes place at 260 °C. Polymerization gives a saturated polymer with both *endo* and *exo* configurations. The thermal polymerization of model compounds and of norbornene end-capped polyimides, studied by NMR, confirms isomerization above 200 °C. The Diels–Alder dissociation occurs at 275 °C. The polymer formed at 285 °C is mainly saturated, with units derived from *endo* and *exo* structures, from maleimide and from the cyclopentadiene nadimide adduct.[156]

30.4.2.3 *Acetylene-terminated imides*

Although the mechanism of the thermal polymerization of arylacetylene has not yet been elucidated,[157,158] the chemistry of acetylene end-capped resin is comprehensively documented.[159,160] Acetylene-terminated imides were introduced in 1974, *via* the reaction pathway represented in Scheme 31.

These oligomers were obtained by the reaction of two moles of dianhydride (**88**) with one mole of 3,3'-[1,3-phenylenebis(oxy)]bisbenzeneamine (**89**) and two moles of 3-ethynylbenzeneamine (**87**). It has been demonstrated that 4-ethynylbenzeneamine is hydrolyzed into an acetyl group during the imide formation,[161] so the *meta* isomer is used. Thermal cyclization of oligomer (**90**) leads to an imide oligomer (**92**), but the isoimide resin (**91**) obtained by chemical dehydration exhibits better solubility and a lower melting point.[162] Isoimide is thermally isomerized into imide; this point has been studied with a thermoplastic polyisoimide, where it was observed that the isomerization is a single body reaction, non-diffusion controlled, but sensitive to the onset of glass transition.[163] Acetylene polymerization starts at 200 °C with a maximum rate at 280 °C. Both imides and isoimides obtained with BTDA are marketed by National Starch.

Oligoimide obtained with 6FDA has a softening point around 170 °C and good solubility in many solvents.[164] Other oligomers have been investigated, and 5-ethynyl-1,3-isobenzofurandione has been used for end-capping diamino telechelic oligo(amic acid)s.[165] Graphite-fiber reinforced lami-

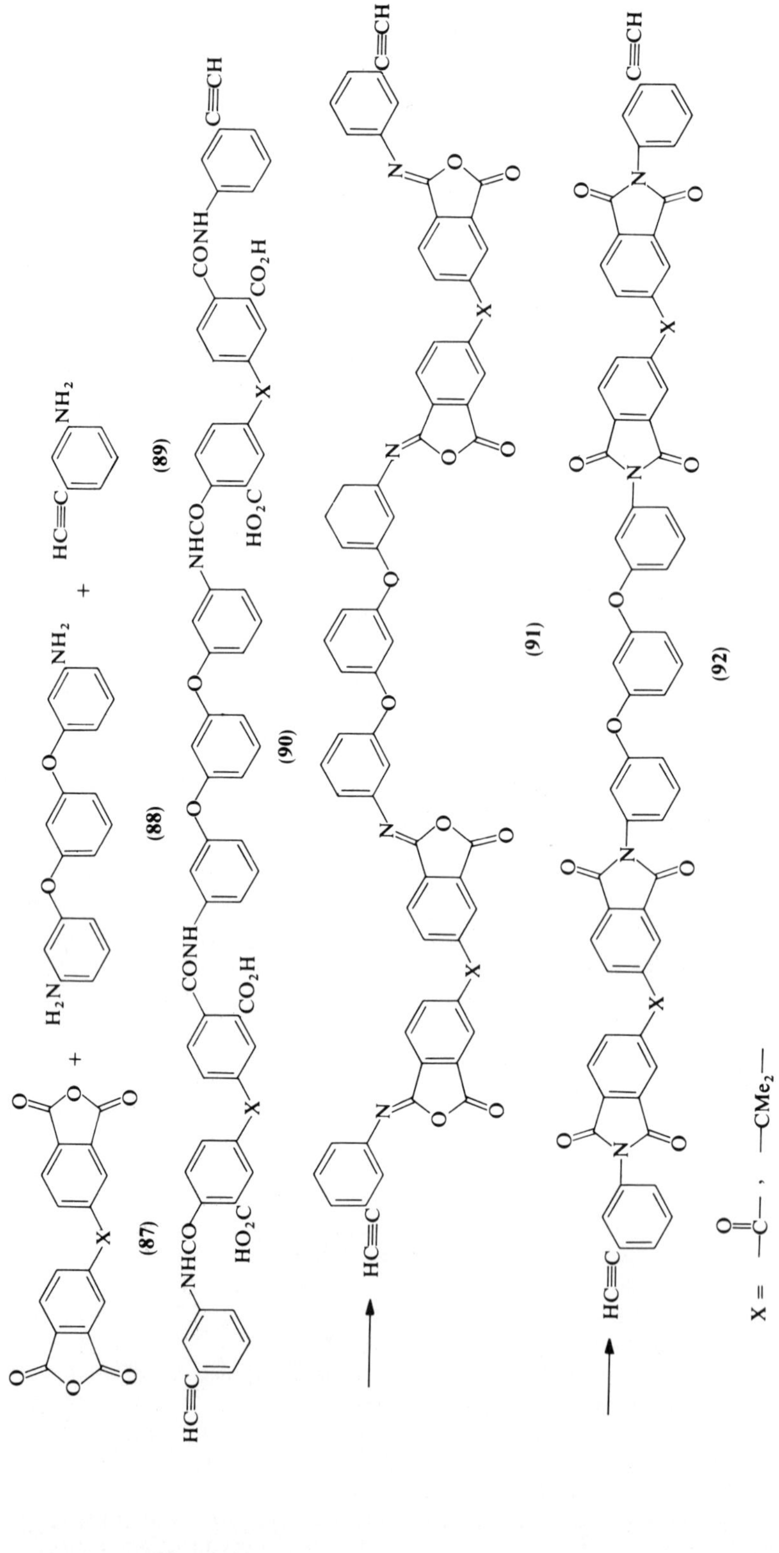

Scheme 31

nates have been prepared with chemically different ethynyl end-capped polyimide resins. Laminates retain their mechanical properties after 1000 h at 316 °C.[166]

A new type of thermostable semiinterpenetrated network was studied, by reacting an acetylene-terminated imide sulfone oligomer with an equal weight of thermoplastic poly(imide sulfone).[167] This network was tested as an adhesive and exhibits good properties after 1000 h at 230 °C in air.

30.5 POLYQUINOXALINES AND POLY(PHENYLQUINOXALINE)S

The general synthesis of quinoxalines is based on the old Hinsberg reaction[168] using the condensation of 1,2-benzenediamine with α-oxoaldehydes. The reaction takes place at low temperatures with high yields.

30.5.1 Linear Polyquinoxalines and Poly(phenylquinoxaline)s

The first papers that were published dealt with unsubstituted polyquinoxalines[169,170] obtained by reaction of α,α′-dioxoarylenediacetaldehydes (**93**; X = H) with aromatic tetraamines (**94**; equation 19). Tetracarbonyl compounds were, in fact, obtained in the dihydrate form.[169] The reaction was carried out in the solid state or in a solvent, such as dimethylacetamide, dimethylaniline, polyphosphonic acid or cresol.[171]

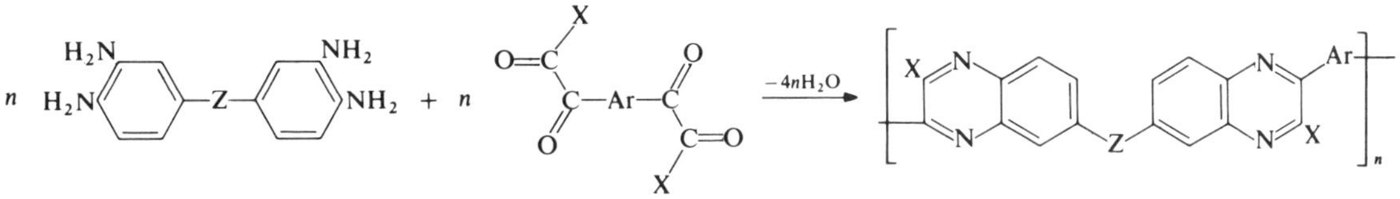

X = H, α-ketoaldehyde (dihydrate) ⟶ polyquinoxalines

X = Ph, α-diketone ⟶ poly(phenylquinoxaline)s

(19)

Poly(phenylquinoxaline)s were reported later,[172] using 1,1′-arylenebis(2-phenylethanedione) (**93**; X = Ph) instead of α-oxoaldehydes. The kinetic study by UV of the formation of poly(phenylquinoxaline)s in cresol solution is in agreement with data concerning the formation of the model compound 2,3-diphenylquinoxaline. The rate constant in cresol is about 100 times more than that in chloroform.[173] The catalytic effects of the proton donor, and the correlation between the pK_a of the acid and the rate constant, was demonstrated,[174] and it was suggested that the m-cresol gives a molecular complex with tetraamine, followed by the simultaneous reaction of both o-amino groups with the α-diketone group.

The effect of the reaction medium on the structure of the condensation product was studied using model compounds.[175] In a non-protic solution, it was possible to isolate some azomethine intermediates (**96**), coming from the reaction between one amino group and one of the α-diketone groups (equation 20).

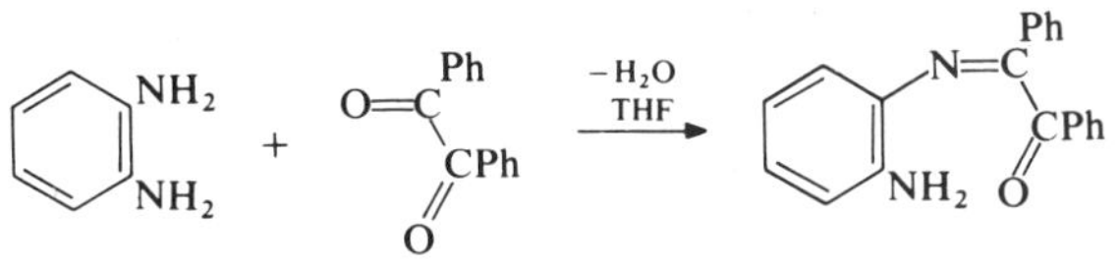

(20)

We shall now consider the macromolecular structure of the bisquinoxaline repetitive units; the three positional isomers represented in Scheme 32 can be formed during the polycondensation reaction.

For unsubstituted polyquinoxaline, it has been suggested, from UV data obtained with model compounds, that the main chain resulted from 2,2′-linkages. This regioselectivity could explain the crystallinity observed by X-ray diffraction.[176] For poly(phenylquinoxaline), ^{13}C NMR shows the presence of all three types of linkages,[177] which explains the amorphous nature of the polymers, which are soluble in chlorinated and phenolic solvents. The variation of the glass transition

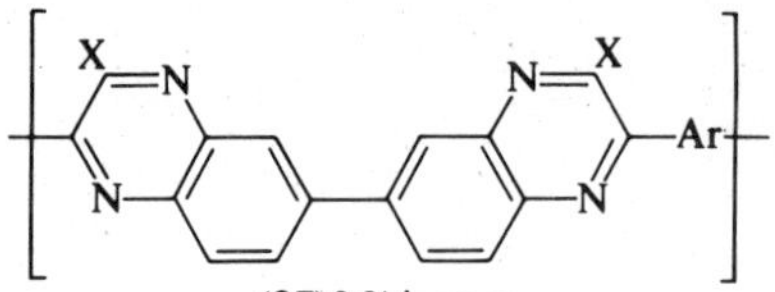

Scheme 32

(97) 2,2'-isomer
(98) 2,3'-isomer 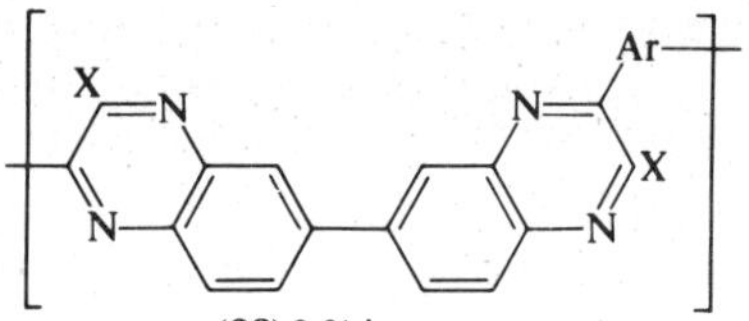
(99) X = H, Ph 3,3'-isomer 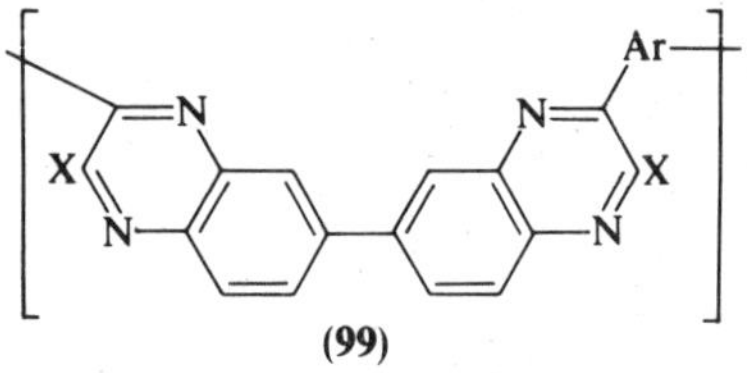

Table 4 Glass Transition Temperature of Poly(phenylquinoxaline)s

Ar	T_g (°C)
 –C$_6$H$_4$– (para)	351
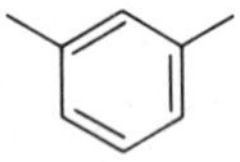 –C$_6$H$_4$– (meta)	319
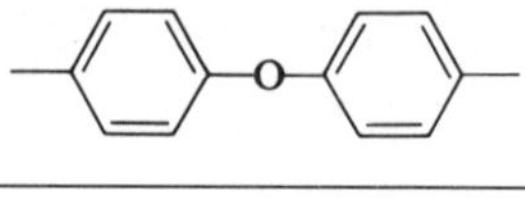 –C$_6$H$_4$–O–C$_6$H$_4$–	290

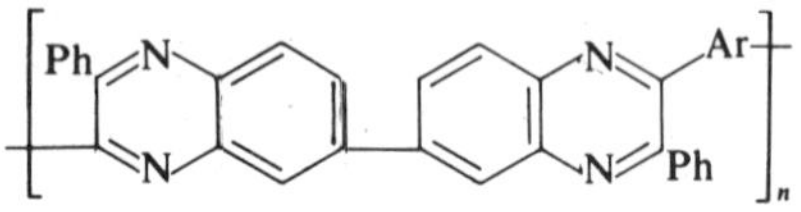

temperature was studied as a function of structure;[178] Table 4 shows the effect for poly(phenylquinoxaline)s of groups that add flexibility.

The properties of the solutions were studied and various authors have found a value of between 0.6 and 0.7 for the a coefficient of the Mark–Houwink equation,[176] corresponding to a rather flexible chain.

The thermooxidative stability is strongly dependent on substitution (phenylated quinoxalines are more stable than unphenylated).[179] It has also been shown that the flexible oxygen linkages, used to increase the solubility and to decrease the T_g,[178] also drastically decrease the thermooxidative behavior.[179] The effect of chain termination on thermal stability has been clearly demonstrated in the case of poly(phenylquinoxaline) (PPQ; equation 19; X = Ph, Ar = 1,4-C$_6$H$_4$). The loss of weight at 400 °C is in the following order: amine-terminated polymer > ketone-terminated polymer > phenyl-terminated polymer.[180] The activation energy for the breakdown of the thermal PPQ was about 60 kcal mol^{-1}.[176,180]

Polyquinoxalines are chemically inert; no degradation was observed after a treatment of 6 h in boiling 40% potash,[181] or after 70 h in sulfuric acid.[182]

30.5.2 Cross-linked Poly(phenylquinoxaline)s

30.5.2.1 Cross-linking of high molecular weight poly(phenylquinoxaline)s

PPQs exhibit relatively low T_g values probably due to their isomerism and the presence of bulky pendant phenyl groups. Several different approaches have been used to try to improve their mechanical properties at high temperatures.

(i) Thermal cross-linking occurring in the main chain

By heating at 477 °C for 120 min, a great increase of T_g was observed for PPQ. It has been suggested that the cross-link occurs *via* two pathways: loss of phenyl or loss of hydrogen followed by radical recombination.[183] Similar observations were made with PPQ having *m*-terphenyl groups adding flexibility in the main chain.[184] The cross-linking reaction can take place more easily in the case of poly(tolylquinoxaline) due to the reactivity of the methyl group.[185]

(ii) Thermal cross-linking of modified main chain

The strained biphenylene ring (**100**) is opened at 350 °C, giving an intermediate biradical (**101**) which recombines (Scheme 33). The introduction of a biphenylene moiety into the backbone (**102**) makes PPQ cross-linkable at 350–380 °C.[186] Insulating varnish for copper wire has been prepared by this process.[187]

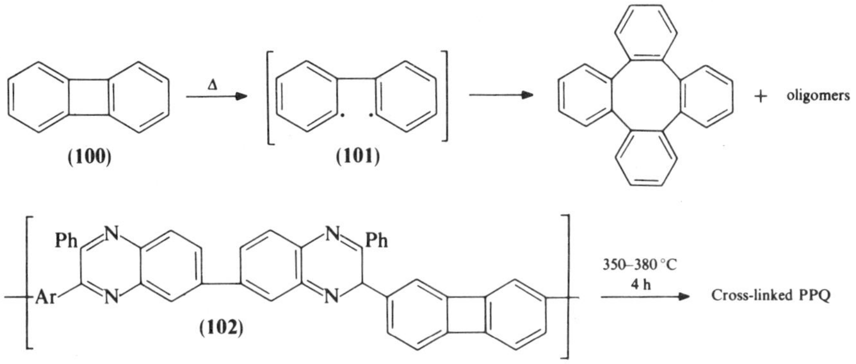

Scheme 33

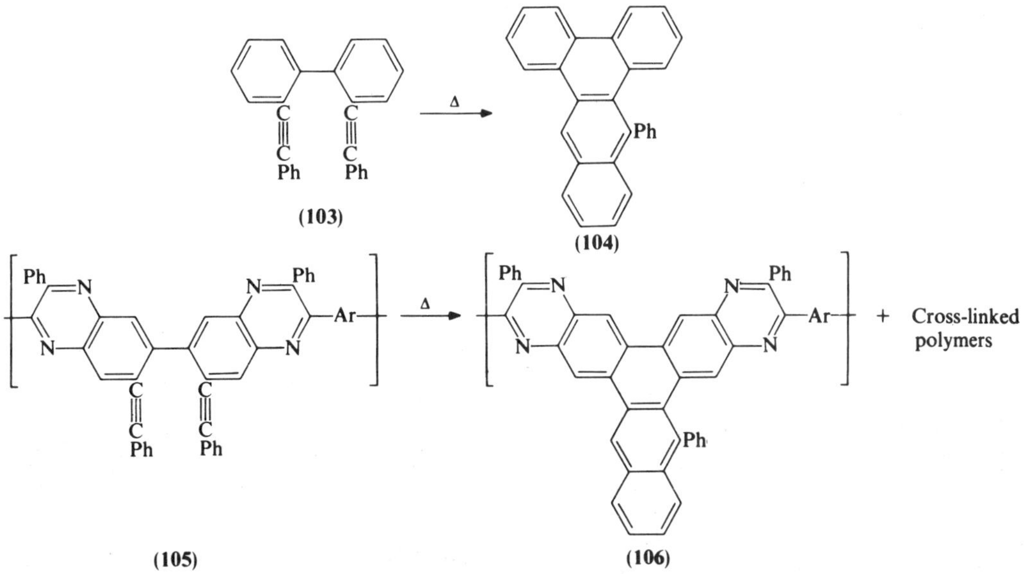

Scheme 34

(iii) Intramolecular cycloaddition

2,2′-Bis(phenylethynyl)-1,1′-biphenyl has been reported to undergo an intramolecular cycloaddition to yield a dibenzanthracene structure (Scheme 34). The corresponding tetraamino compound was used to prepare a phenylethynyl substituted PPQ (**105**) showing an initial T_g of 215 °C. After curing at 245 °C, the T_g reached 365 °C. However, this phenomenon is not due only to the intramolecular cyclization giving a more rigid structure (**106**), as some cross-linking was observed.[188]

(iv) Cross-linking by reactive groups on the side chain

Cyano- and cyanato-substituted PPQs react at 400 °C, giving cross-linked resins with increased T_g values.[178,189] The ethynyl and phenylethynyl side chains were introduced *via* tetraketone derivatives (**107** and **108**; equation 21).[190,191] The cross-link occurs by heating over a temperature range of 375–445 °C. A decrease in the thermal stability of resins cross-linked by acetylenic groups when compared with linear PPQ has been reported.

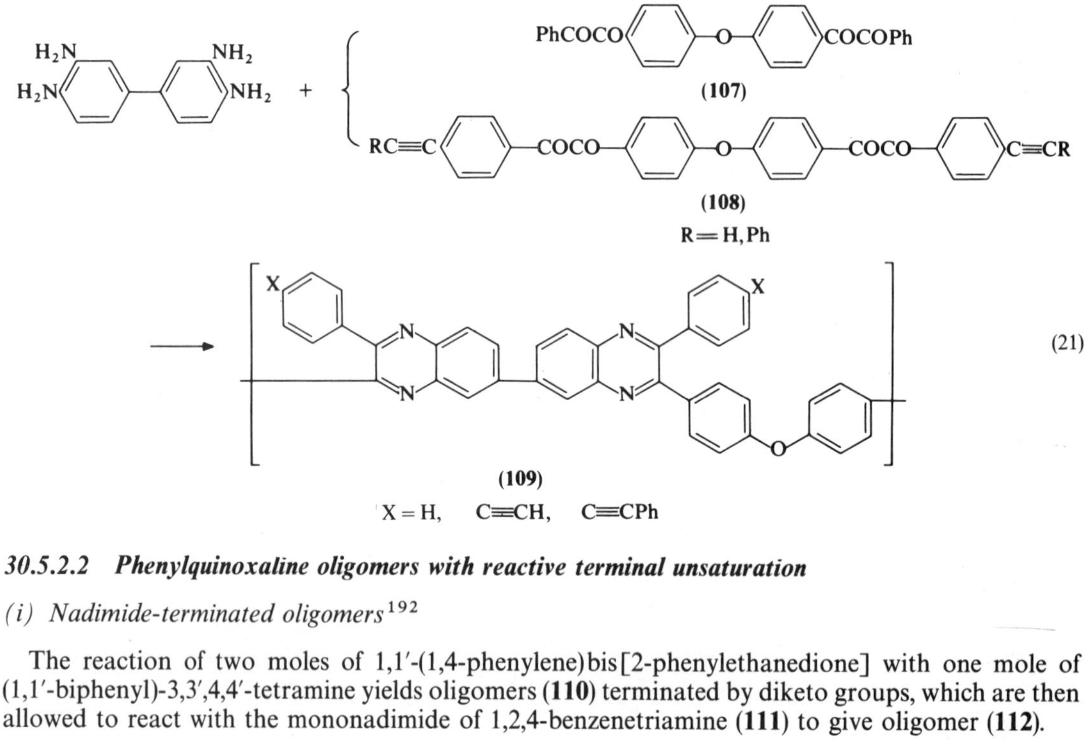

(107) (108) R=H, Ph

(109) X = H, C≡CH, C≡CPh (21)

30.5.2.2 Phenylquinoxaline oligomers with reactive terminal unsaturation

(i) Nadimide-terminated oligomers[192]

The reaction of two moles of 1,1′-(1,4-phenylene)bis[2-phenylethanedione] with one mole of (1,1′-biphenyl)-3,3′,4,4′-tetramine yields oligomers (**110**) terminated by diketo groups, which are then allowed to react with the mononadimide of 1,2,4-benzenetriamine (**111**) to give oligomer (**112**).

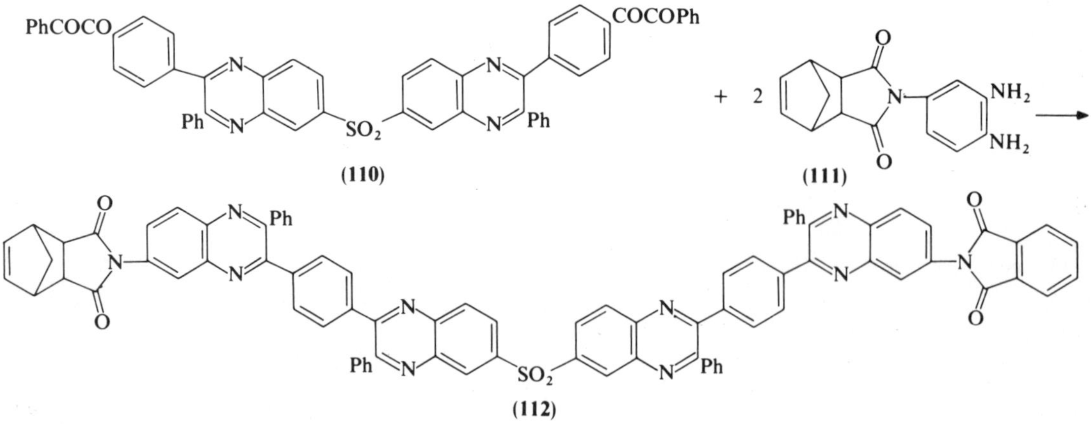

(110) (111)

(112)

$\bar{M}_n = 1900$

$\xrightarrow{\Delta}$ Cross-linked polymer (T_g after cure = 347 °C)

Scheme 35

(ii) Acetylene-terminated oligomers

Two synthetic approaches have been tested, differing in the nature of the end-capping agent. In the first case, oligomers obtained with an excess of tetraketone were end-capped with 4-(3-ethynylphenoxy)-1,2-benzenediamine.[193] As can be seen in Table 5, the softening temperature was not strongly influenced by the Ar group, but the very different thermooxidative behavior is unexpected.

Table 5 Thermal Properties of Acetylene-terminated Quinoxaline (ATQ) Resins

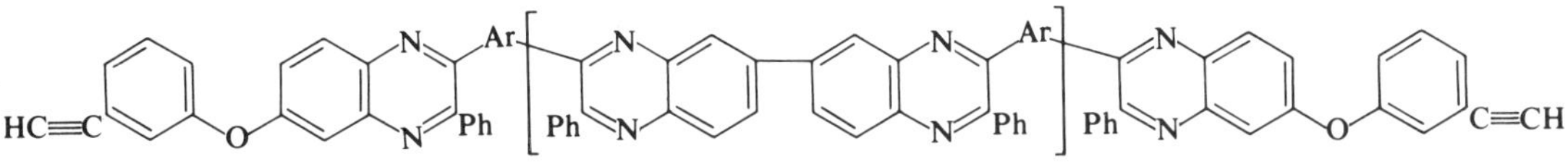

(113)

Ar	Oligomer	Softening temperature (°C)	Cure rate maximum (°C)	T_g^a (°C)	Weight loss[b] (%)
p-C$_6$H$_4$OC$_6$H$_4$–	ATQ-O	159	274	321	12
p-C$_6$H$_4$SC$_6$H$_4$–	ATQ-S	144	277	331	30
p-C$_6$H$_4$–	ATQ-PP	161	277	340	22

[a] After curing for 3 hours at 280 °C.
[b] After 150 hours at 316 °C.

The second approach is based on the synthesis of diamino-terminated oligomer end-capped agents with [4-(3-ethynylphenoxy)phenyl]phenylethanedione, giving so-called bis-acetylene-terminated quinoxaline (BATQ) resins[194] (Table 6). The softening point, cure temperature and final T_g are in the same temperature range as for ATQ and BATQ. An attempt to decrease the softening temperature was made by mixing BATQ with low molecular weight, low melting, diethynyl compounds;[195] but the final T_g of cured resins is always depressed.

Table 6 Thermal Properties of BATQ Resins

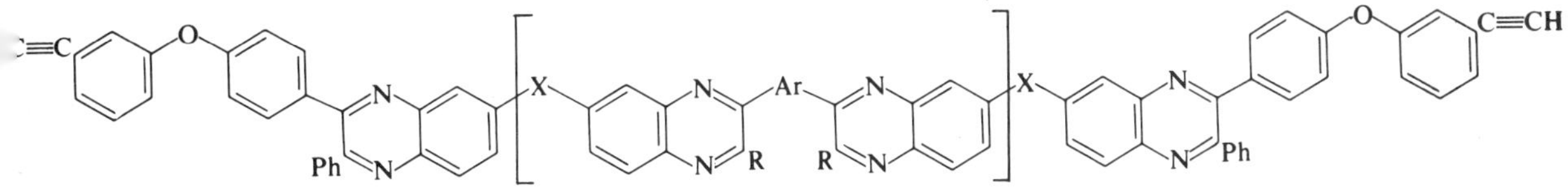

(114)

Oligomer	X	R	Ar	Softening temperature (°C)	Cure rate maximum (°C)	T_g^a (°C)
BATQ-O	—	Ph	p-C$_6$H$_4$OC$_6$H$_4$–	145	276	310
BATQ-M	—	Ph	m-C$_6$H$_4$–	145	274	311
BATQ-S	—	Ph	p-C$_6$H$_4$SC$_6$H$_4$–	160	273	309
BATQ-P	—	p-PhOC$_6$H$_4$–	m-C$_6$H$_4$–	145	283	275
BATQ-H	O	Ph	p-C$_6$H$_4$(OC$_6$H$_4$)$_2$–	150	280	322
BATQ-O,O	O	Ph	p-C$_6$H$_4$OC$_6$H$_4$–	145	280	298
BATQ-O,P	O	p-PhOC$_6$H$_4$–	m-C$_6$H$_4$–	145	285	290
BATQ-O,M	O	Ph	m-C$_6$H$_4$–	125	293	257

[a] After curing for 6 hours at 280 °C.

30.6 REFERENCES

1. R. J. Cotter and M. Matzner, 'Ring Forming Polymerization', Academic Press, New York, 1972, vol. 13 B1 and 13 B2.
2. P. E. Cassidy, 'Thermally Stable Polymers', Dekker, New York, 1980.
3. J. P. Critchley, C. J. Knight and W. W. Wright, 'Heat Resistant Polymers', Plenum Press, London, 1983.
4. J. P. Critchley, *Angew. Makromol. Chem.*, 1982, **109/110**, 41.
5. C. E. Sroog, in 'Encyclopedia of Polymer Science and Technology', ed. N. M. Bikales, H. F. Mark and N. G. Gaylord, Interscience, New York, 1969, vol. 11, p. 247.

6. N. Yoda, R. Nakanishi, M. Kurihara, Y. Bamba, S. Tohyama and K. Ikeda, *J. Polym. Sci., Part B*, 1966, **4**, 11.
7. N. Yoda, in 'Encyclopedia of Polymer Science and Technology', ed. N. M. Bikales, H. F. Mark and N. G. Gaylord, Interscience, New York, 1969, vol. 10, p. 682.
8. G. Rabilloud, B. Sillion and G. de Gaudemaris, *Makromol. Chem.*, 1967, **18**, 108.
9. H. A. Vogel and C. S. Marvel, *J. Polym. Sci.*, 1961, **50**, 511.
10. S. A. Johnson and N. K. Roberts, *Int. SAMPE Symp.*, 1987, **32**, 868.
11. I. W. Serfaty, in 'Polyimides', ed. K. L. Mittal, Plenum Press, New York, 1984, p. 149.
12. R. Edelman, *Proc. Int. Conf. Polyimides, 2nd*, 1985, 182.
13. V. L. Bell, B. L. Stump and H. Gager, *J. Polym. Sci., Polym. Chem. Ed.*, 1976, **14**, 2275.
14. J. H. Bateman, W. Geresy and A. D. Neiditch, *Org. Coat. Plast. Chem.*, 1975, **35** (2), 72.
15. P. Sallé, B. Sillion and G. de Gaudemaris, *Bull. Soc. Chim. Fr.*, 1968, 3378.
16. P. M. Hergenrother, H. H. Levine, *J. Polym. Sci., Part A-1*, 1967, **5**, 1453.
17. P. M. Hergenrother, *ACS Symp. Ser.*, 1985, **282**, 1.
18. W. H. Beever and J. K. Stille, *J. Polym. Sci., Polym. Symp.*, 1978, **65**, 41.
19. J. F. Wolfe and J. K. Stille, *Macromolecules*, 1976, **9**, 489.
20. S. O. Norris and J. K. Stille, *Macromolecules*, 1976, **9**, 496.
21. W. Wrasidlo, S. O. Norris and J. K. Stille, *Macromolecules*, 1976, **9**, 512.
22. P. D. Sybert, W. H. Beever and J. K. Stille, *Macromolecules*, 1981, **14**, 493.
23. D. M. Sutherlin and J. K. Stille, *Macromolecules*, 1985, **18**, 2669.
24. K. C. Brinker and I. M. Robinson (Du Pont Co.), *Br. Pat.* 798 004 (1958).
25. P. M. Hergenrother, W. Wrasidlo and H. H. Levine, *Polym. Prepr., Am. Chem. Soc., Div. Polym. Chem.*, 1964, **5**, 153.
26. Y. Imai, I. Taoka, K. Uno and Y. Iwakura, *Makromol. Chem.*, 1965, **83**, 167.
27. T. Anios and W. W. Moyer, *US Pat.* 3 423 483 (1969).
28. G. M. Moelter, R. F. Tetreault and M. J. Hefferon, *Polym. News*, 1983, **9** (5), **134**.
29. L. H. Lee, *Polym. Mater. Sci. Eng.*, 1987, **56**, 198.
30. B. C. Ward, *Int. SAMPE Symp.*, 1987, **32**, 853.
31. M. E. Harb, J. W. Treat and B. C. Ward, *Int. SAMPE Symp.*, 1987, **32**, 795.
32. J. F. Wolfe, B. H. Loo and F. E. Arnold, *Macromolecules*, 1981, **14**, 915.
33. J. F. Wolfe and F. E. Arnold, *Macromolecules*, 1981, **14**, 909.
34. J. F. Wolfe, P. D. Sybert and J. R. Sybert, *US Pat.* 4 533 693 (1985).
35. J. F. Wolfe, *Polym. Mater. Sci. Eng.*, 1986, **54**, 99.
36. A. W. Chow, P. E. Penwell, S. P. Bitler and J. F. Wolfe, *Polym. Prepr., Am. Chem. Soc., Div. Polym. Chem.*, 1987, **28** (1), 50.
37. F. E. Arnold, 'Other High Performance, High Temperature Polymers; Interdisciplinary Workshop on the Chemistry and Properties of Polyimides', Reno, American Chemical Society, Division of Polymer Chemistry, 1985.
38. T. E. Helminiak, *US Pat. Appl.* 638 211 (1975).
39. R. F. Kovar and F. E. Arnold, *J. Polym. Sci., Polym. Chem. Ed.*, 1976, **14**, 2807.
40. T. T. Tsai, F. E. Arnold and H. W. Hwang, *Polym. Prepr., Am. Chem. Soc., Div. Polym. Chem.*, 1985, **26**, 144.
41. M. I. Bessonov, M. M. Koton, V. V. Kudryatsev and L. A. Laius 'Polyimides Thermally Stable Polymers', Consultant Bureau, New York, 1987.
42. K. L. Mittal, 'Polyimides, Synthesis, Characterization and Applications', Plenum Press, New York, 1982, vol. 1 and vol. 2.
43. C. E. Sroog, A. L. Endrew, S. V. Abramo, C. E. Berr, M. W. Edwards and K. L. Oliver, *J. Polym. Sci., Part A.*, **3**, 1965, 1374.
44. S. R. Sandler and W. Karo, 'Polymer Synthesis', Academic Press, New York, 1974, vol. 1, p. 216.
45. A. Ya. Ardasnikov, I. E. Kardash, B. V. Kotov and A. N. Pravednikov, *Dokl. Chem. (Engl. Transl.)*, 1965, **164**, 1006.
46. R. A. Dine–Hart and W. W. Wright, *J. Appl. Polym. Sci.*, 1967, **11**, 609.
47. S. A. Zakoshchikov, G. M. Zubareva and G. M. Zolotareva, *Sov. Plast. (Engl. Transl.)*, 1967, **4**, 13.
48. I. W. Frost and I. Kesse, *J. Appl. Polym. Sci.*, 1964, **8**, 1039.
49. C. E. Sroog, *J. Polym. Sci., Part C*, 1967, **16**, 1191.
50. G. M. Bower and L. W. Frost, *J. Polym. Sci., Part A*, 1965, **3**, 3135.
51. M. L. Wallach, *J. Polym. Sci., Part A-2*, **5**, 1967, 653.
52. P. M. Cotts, in 'Polyimides', ed. K. L. Mittal, Plenum Press, New York, 1984, p. 223.
53. W. Wolksen and P. M. Cotts, in 'Polyimides', ed. K. L. Mittal, Plenum Press, New York, 1984, p. 163.
54. C. C. Walker, *Proc. Int. Conf. Polyimides, 2nd,* 1985, 426.
55. R. Kaas, *J. Polym. Sci., Polym. Chem. Ed.*, 1981, **19**, 2255.
56. P. M. Cotts and W. Wolksen, *ACS Symp. Ser.*, 1984, **242**, 227.
57. D. J. Progar and T. L. Saint–Clair, *Natl. SAMPE Tech. Conf.*, 1975, **7**, 53; D. J. Progar, V. L. Bell and T. L. Saint-Clair, *US Pat.* 4 065 345 (1977).
58. D. J. Progar and T. L. Saint–Clair, *Proc. Int. Conf. Polyimides, 2nd*, 1985, 575.
59. H. Burks, T. L. Saint–Clair and D. J. Progar, *Proc. Int. Conf. Polyimides, 2nd*, 1985, 529.
60. A. H. Egli and T. L. Saint–Clair, *Proc. Int. Conf. Polyimides, 2nd*, 1985, 395.
61. J. A. Kreuz, A. L. Endrey, F. P. Gay and C. E. Sroog, *J. Polym. Sci., Part A-1*, 1966, **4**, 2607.
62. M. J. Brekner and C. Feger, *J. Polym. Sci., Polym. Chem. Ed.*, 1987, **25**, 2005.
63. S. V. Lavrov, *Vysokomol. Soedin., Ser. A*, 1977, **19**, 2374.
64. V. L. Bell, B. L. Stump and N. Gager, *J. Polym. Sci., Polym. Chem. Ed.*, 1976, **14**, 2275.
65. R. A. Dine–Hart and W. W. Wright, *J. Appl. Polym. Sci.*, 1968, **6**, 1935.
66. P. P. Nechayer, Ya. S. Vygodskii, G. E. Zaikov and S. V. Vinogradova, *Polym. Sci. USSR (Engl. Transl.)*, 1976, **18**, 1903.
67. P. R. Young and A. C. Chang, *Natl. SAMPE Symp. Exhib.*, [*Proc.*], 1985, **30**, 889.
68. R. J. Angelo, *US Pat.* 3 073 785 (1963); *US Pat.* 3 282 878 (1967).
69. R. J. Angelo, *US Pat.* 3 420 795 (1969).
70. R. J. Angelo, R. C. Golike, W. E. Tatum and J. A. Kreuz, *Proc. Int. Conf. Polyimides, 2nd*, 1985, 631.
71. R. J. Salem, F. O. Lequeda, J. Duran, W. Y. Lee, R. M. Yang, *J. Vac. Sci. Technol.*, 1986, **A4** (3), 369.

72. T. Takekoshi, J. E. Kachanovski, J. S. Mantello and J. M. Weber, *J. Polym. Sci., Polym. Chem. Ed.*, 1985, **23**, 1759.
73. T. Takekoshi, J. E. Kochanovski, J. S. Mantello and M. J. Weber, *J. Polym. Sci., Polym. Symp.*, 1986, **74**, 93.
74. T. Takekoshi, *Polym. J.*, 1987, **19**, 191.
75. Soc. Rhodiaceta, *Neth. Pat.* 66 90 214 (1967).
76. L. M. Alberino, W. J. Farrisey and J. S. Rose, *US Pat.* 3 708 458 (1973).
77. P. S. Carleton, W. J. Farrisey and J. S. Rose, *J. Appl. Polym. Sci.*, 1972, **16**, 2983.
78. The Upjohn Co., *US Pat.* 4 001 186 (1977).
79. The Upjohn Co., *US Pat.* 4 021 412 (1977).
80. W. J. Farrisey, L. M. Alberino and A. A. Sayig, *J. Elastomers Plast.*, 1975, **7**, 285.
81. R. A. Meyers, *J. Polym. Sci., Part A-1*, 1969, **7**, 2757.
82. W. M. Alvino and L. E. Edelman, *J. Appl. Polym. Sci.*, 1975, **19**, 2961.
83. W. M. Alvino and L. E. Edelman, *J. Appl. Polym. Sci.*, 1978, **22**, 1983.
84. N. D. Ghatge and D. K. Dandge, *Angew. Makromol. Chem.*, 1976, **56**, 163.
85. B. M. Shinde, N. D. Ghatge, N. Patil, *J. Appl. Polym. Sci.*, 1985, **30**, 3505.
86. S. Kilic, D. K. Mohanty, I. Yilgor and J. E. McGrath, *Polym. Prepr., Am. Chem. Soc., Div. Polym. Chem.*, 1986, **27** (1), 318.
87. Soc. Rhodiaceta, *Fr. Pat.* 1 498 015 (1967).
88. S. Terney, J. Keating, J. Zielenski, J. Hakala and H. Sheffer, *J. Polym. Sci., Part A-1*, 1970, **8**, 683.
89. E. M. Boldebuck and J. F. Klebe, *US Pat.* 3 303 157 (1967).
90. G. Greber and D. Lohmann, *Angew. Chem., Int. Ed. Engl.*, 1969, **8**, 899.
91. J. R. Pratt and S. F. Thames, *Synthesis*, 1973, 233.
92. W. M. Edwards and I. M. Robinson, *US Pat.* 2 710 853 (1955).
93. R. E. de Brunner and J. K. Fincke, *US Pat.* 3 423 366 (1969).
94. W. M. Edwards, I. M. Robinson and E. N. Squire, *US Pat.* 3 867 609 (1959).
95. M. E. Quenneson, J. Garapon, M. Bartholin and B. Sillion, *Proc. Int. Conf. Polyimides, 2nd*, 1985, 74.
96. J. C. Johnston, M. A. Meador and W. B. Alston, *J. Polym. Sci., Polym. Chem. Ed.*, 1987, **25**, 2175.
97. W. Wolksen, R. Diller and D. Young, *Polym. Mater. Sci. Eng.*, 1986, **54**, 94.
98. M. Fryd and B. T. Merriman, *US Pat.* 4 562 100 (1985).
99. S. Maiti and B. K. Mandal, *Prog. Polym. Sci.*, 1986, **12**, 111.
100. D. M. White, T. Takekoshi, F. J. Williams, H. M. Relles, P. M. Donahue, H. J. Klopper, G. R. Louks, J. S. Mantello, R. O. Matthews and R. W. Schluenz, *J. Polym. Sci., Polym. Chem. Ed.*, 1981, **19**, 1635.
101. B. K. Mandal and S. Maiti, *J. Polym. Mater.*, 1985, **2**, 115.
102. P. M. Hergenrother, *Polym. J.*, 1987, **19**, 73.
103. J. H. Bateman, W. Geresy and D. S. Neiditch, *Org. Coat. Plast. Chem.*, 1975, **35** (2), 72.
104. W. B. Alston and R. F. Gratz, *Proc. Int. Conf. Polyimides, 2nd*, 1985, 30.
105. H. H. Gibbs, *J. Appl. Polym. Sci.: Appl. Polym. Symp.*, 1979, **35**, 207.
106. C. E. Sroog, *Macromol. Rev.*, 1976, **11**, 161.
107. S. Isoda, M. Kochi and H. Kambe, *J. Polym. Sci., Polym. Phys. Ed.*, 1982, **20**, 837.
108. T. L. Saint-Clair, H. D. Burks, N. T. Wakelyn and T. H. Hou, *Polym. Prepr., Am. Chem. Soc., Div. Polym. Chem.*, 1987, **28**, 90.
109. P. M. Hergenrother, N. T. Wakelyn and S. Havens, *J. Polym. Sci., Polym. Chem. Ed.*, 1987, **25**, 1093.
110. S. Numata, K. Fujisaki, D. Makino and N. Kinjo, *Proc. Int. Conf. Polyimides, 2nd*, 1985, 492.
111. D. Hofer, *Polym. Mater. Sci. Eng.*, 1986, **54**, 85.
112. R. Rubner, W. Bartel and G. Bald, *Siemens Forsch.-Entnicklungsber.*, 1976, **5**, 235.
113. H. Ahne, H. Kruger, E. Pammer and R. Rubner, in 'Polyimides', ed. K. L. Mittal, Plenum Press, New York, 1984, p. 905.
114. H. Merrem, R. Klug and H. Martner, in 'Polyimides', ed. K. L. Mittal, Plenum Press, New York, 1984, p. 919.
115. N. Yoda and H. Hiramoto, *J. Macromol. Sci., Chem.*, 1984, **A21**, 1641.
116. T. Nakano, *Proc. Int. Conf. Polyimides, 2nd*, 1985, 163.
117. G. Rabilloud, B. Sillion and G. de Gaudemaris, *Makromol. Chem.*, 1967, **108**, 18.
118. G. Rabilloud, B. Sillion and G. de Gaudemaris, *US Pat.* 3 678 005 (1972).
119. A. Saiki, K. Mukai, S. Harada and Y. Miyadera, *Org. Coat. Plast. Chem.*, 1980, **43**, 459.
120. M. Uekita and H. Awaji, *Eur. Pat. Appl.* 0 209 114 (1987).
121. M. Kakimoto, M. Suzuki, T. Konishi, Y. Imaio, M. Iwamoto and T. Hino, *Chem. Lett.*, 1986, 823.
122. H. N. Cole and W. F. Gruber, *US Pat.* 3 127 414 (1964).
123. R. O. Hummel, K. U. Heinen, H. D. Stenzenberger and H. Siesler, *J. Appl. Polym. Sci.*, 1974, **18**, 2015.
124. H. D. Stenzenberger, D. V. Heinen and D. O. Hummel, *J. Polym. Sci., Polym. Chem. Ed.*, 1976, **14**, 2911.
125. J. V. Crivello, *J. Polym. Sci., Polym. Chem. Ed.*, 1973, **11**, 1885.
126. J. E. White, M. D. Scaïa and D. A. Snider, *J. Appl. Polym. Sci.*, 1984, **29**, 891.
127. J. E. White, D. A. Snider and M. D. Scaïa, *J. Polym. Sci., Polym. Chem. Ed.*, 1984, **22**, 589.
128. Y. R. Elliot (ed.), 'Macromolecular Synthesis', Wiley, New York, 1966, vol. 2, p. 211.
129. S. Kadoi and T. Aya, *US Pat.* 4 535 392 (1985).
130. H. Hirobumi, O. Yuji and A. Koichi, *Jpn. Pat.* 61 229 863 (*Chem. Abstr.*, 1987, **106**, 156 995).
131. J. E. White, *Ind. Eng. Chem. Prod. Res. Dev.*, 1986, **25**, 395.
132. F. W. Harris and S. O. Norris, *J. Polym. Sci., Polym. Chem. Ed.*, 1973, **11**, 2143.
133. G. C. Tesoro and V. R. Sastri, *Ind. Eng. Chem. Prod. Res. Dev.*, 1986, **25**, 444.
134. L. S. Tan, E. J. Solovski and F. E. Arnold, *Polym. Prepr, Am. Chem. Soc., Div. Polym. Chem.*, 1986, **27** (2), 240; L. S. Tan, E. J. Solovski and F. E. Arnold, *Polym. Mater. Sci. Eng.*, 1987, **56**, 650; L. R. Denny, I. J. Goldfarb and M. Farr, *Polym. Mater. Sci. Eng.*, 1987, **56**, 656.
135. F. Grundschober and J. Sambeth, *US Pat.* 3 380 964 (1968).
136. I. K. Varma, G. M. Fohlen and J. A. Parker, *J. Polym. Sci., Polym. Chem. Ed.*, 1982, **20**, 283.
137. G. T. Kwiatkovski, L. M. Robeson, G. L. Brude and A. W. Bedwin, *J. Polym. Sci., Polym. Chem. Ed.*, 1975, **13**, 961.
138. H. D. Stenzenberger, M. Herzog, W. Romer and R. Scheiblich, *Natl. SAMPE Symp. Exhib., [Proc.]*, 1985, **30**, 1568.
139. F. Grundschober, *US Pat.* 3 533 966 (1970).

140. F. P. Darmory, *Natl. SAMPE Symp. Exhib., [Proc.]*, 1974, **19**, 693.
141. M. A. Mallet and F. P. Darmory, *ACS Symp. Ser.*, 1974, **4**, 112.
142. R. T. Alvarez and F. P. Darmory, *Natl. SAMPE Symp. Exhib., [Proc.]*, 1975, **20**, 253.
143. H. D. Stenzenberger, M. Herzog, W. Roemer and R. Scheiblich, *Natl. SAMPE Symp. Exhib., [Proc.]*, 1984, **29**, 1043.
144. I. K. Varma–Sangita and D. S. Varma, *J. Polym. Sci., Polym. Chem. Ed.*, 1984, **22**, 1419.
145. J. J. King, M. Chaudari and S. Zahir, *Natl. SAMPE Symp. Exhib., [Proc.]*, 1984, **29**, 392.
146. Th. Wagner–Jauregy, *Synthesis*, 1980, **3**, 772.
147. H. D. Stenzenberger, P. König, M. Herzog, W. Roemer, S. Pierce and M. S. Canning, *Natl. SAMPE Symp. Exhib., [Proc.]*, 1987, **32**, 44.
148. M. Gaba, K. Suzuki and N. Ikegushi, *Proc. Electr./Electron. Insul. Conf.*, 1977, **13**, 11.
149. D. Landman, *Dev. Reinf. Plast.*, 1987, **5**, 39.
150. H. R. Lubowitz, *Polym. Prepr., Am. Chem. Soc., Div. Polym. Chem.*, 1971, **12**, 329.
151. T. T. Serafini, P. Delvigs and G. R. Lightsey, *J. Appl. Polym. Sci.*, 1972, **16**, 905.
152. R. W. Lauwer and R. D. Vannucci, *NASA Tech. Memo.*, 1979, NASA TM-X-79068.
153. T. T. Serafini, R. D. Vannucci and W. B. Alston, *NASA Tech. Memo.*, 1975, NASA TM-X-71682.
154. T. L. Saint–Clair and R. A. Jewel, *Natl. SAMPE Tech. Conf.*, 1976, **8**, 82.
155. N. G. Gailord and M. Martan, *ACS Symp. Ser.*, 1982, **195**, 97.
156. A. L. Wong and W. M. Ritchey, *Macromolecules*, 1981, **14**, 825; A. L. Wong, A. N. Garroway and W. M. Ritchey, *Macromolecules*, 1981, **14**, 832.
157. M. D. Seifik, E. O. Stejshal, R. A. McKay and J. Shaefer, *Macromolecules*, 1979, **12**, 423.
158. C. Y. C. Lee, *Dev. Reinf. Plast.*, 1986, **5**, 121.
159. P. M. Hergenrother, in 'Encyclopedia of Polymer Science and Engineering', ed. J. I. Kroschwitz, 2nd edn., Wiley, New York, 1986, vol. 1, p. 61.
160. L. R. Denny, I. J. Goldfarb and C. Y. C. Lee, *Int. SAMPE Symp.*, 1986, **31**, 153.
161. P. M. Hergenrother, *J. Heterocycl. Chem.*, 1980, **17**, 5.
162. A. L. Landis and A. B. Baselow, *Natl. SAMPE Tech. Conf.*, 1982, **14**, 236.
163. C. Y. C. Lee, *Natl. SAMPE Symp. Exhib., [Proc.]*, 1985, **30**, 1665.
164. D. J. Capo and J. Schoenberg, *Int. SAMPE Tech. Conf.*, 1986, **18**, 710.
165. P. M. Hergenrother, *Polym. Prepr., Am. Chem. Soc., Div. Polym. Chem.*, 1980, **21** (1), 81.
166. N. Bilow, L. B. Keller, A. L. Landis, R. H. Boshan and A. A. Castillo, *Natl. SAMPE Symp. Exhib., [Proc.]*, 1978, **23**, 791.
167. A. Hanky and T. L. Saint–Clair, *SAMPE J.*, 1985, **21** (July–August), 40.
168. O. Hinsberg, *Ber. Dtsch. Chem. Ges.*, 1884, **17**, 320.
169. G. P. de Gaudemaris and B. Sillion, *J. Polym. Sci., Part B*, 1964, **2**, 203.
170. J. K. Stille and J. R. Williamson, *J. Polym. Sci., Part B*, 1964, **2**, 209.
171. G. de Gaudemaris, B. Sillion and J. Prévé, *Bull. Soc. Chim. Fr.*, 1964, **8**, 1793.
172. P. M. Hergenrother and H. H. Levine, *J. Polym. Sci., Part A-1*, 1967, **5**, 1453.
173. G. L. Hagnauer and G. D. Mulligan, *Macromolecules*, 1973, **6**, 477.
174. V. V. Korshak, E. S. Krongauz and A. P. Travnikova, *Vysokomol. Soedin., Ser. A*, 1980, **22**, 1450.
175. A. I. Kuzaev, V. V. Korshak, E. S. Krongauz, G. A. Mirontseva and A. P. Travnikova, *Vysokomol. Soedin., Ser. A*, 1983, **25**, 396.
176. J. K. Stille and J. R. Williamson, *J. Polym. Sci., Part A*, 1964, **2**, 3867.
177. Y. G. Urman, N. S. Zabel'nikov, S. G. Alekseyeva, V. D. Vorobev and I. Y. Slonin, *Vysokomol. Soedin., Ser. A*, 1978, **20**, 2236.
178. P. M. Hergenrother, *Macromolecules*, 1974, **7**, 575.
179. W. J. Wrasidlo, *J. Polym. Sci., Part A-1*, 1970, **8**, 1107.
180. J. M. Augl, *J. Polym. Sci., Part A-1*, 1972, **10**, 2403.
181. P. M. Hergenrother and H. H. Levine, *Polym. Prepr., Am. Chem. Soc., Div. Polym. Chem.*, 1967, **8**, 501.
182. V. V. Korshak, E. S. Krongauz, A. M. Berlin, O. Z. Neiland and S. Skuya, *Vysokomol. Soedin., Ser. A*, 1974, **16**, 1770.
183. R. Becker and H. Raubach, *Faserforsch. Textiletech.*, 1975, **26** (8), 406.
184. G. Rabilloud and B. Sillion, *J. Polym. Sci., Polym. Chem. Ed.*, 1978, **16**, 2093.
185. S. E. Wentworth and G. D. Mulligan, *Polym. Prepr., Am Chem. Soc., Div. Polym. Chem.*, 1974, **15**, 697.
186. A. Recca and J. K. Stille, *Macromolecules*, 1978, **11**, 479.
187. G. Rabilloud and B. Sillion, *Fr. Pat.* 2 453 873 (1979).
188. F. L. Hedberg and F. E. Arnold, *J. Polym. Sci., Polym. Chem. Ed.*, 1976, **14**, 2607.
189. Lu Fengcal, W. Beiging and C. Jinbiao, *Org. Coat. Appl., Polym. Sci. Proc.*, 1981, **46**, 208.
190. P. M. Hergenrother, *Macromolecules*, 1981, **14**, 898.
191. P. M. Hergenrother, *J. Appl. Polym. Sci.*, 1983, **28**, 355.
192. N. Odagiri, T. Yamashita and K. Tobokura, 'Proceedings of the Japan–US CCM-III', ed. K. Kawata, S. Umekawa and A. Kobayashi, Japan Society for Composite Materials, Tokyo, 1986, p. 53.
193. R. F. Kovar, G. F. Etilers and F. E. Arnold, *J. Polym. Sci., Polym. Chem. Ed.*, 1977, **15**, 1081.
194. F. L. Hedberg and F. E. Arnold, *J. Appl. Polym. Sci.*, 1979, **24**, 763.
195. S. S. Sikka and I. J. Goldfarb, *Org. Coat. Plast. Chem.*, 1981, **45**, 138.

31
Polysulfides

DAVID E. VIETTI
Morton Thiokol Inc., Woodstock, IL, USA

31.1 INTRODUCTION	533
31.2 INORGANIC POLYSULFIDES	534
31.2.1 Synthesis of Inorganic Polysulfides	534
31.2.2 Equilibria in Sodium Polysulfide Solutions	534
31.3 SYNTHESIS OF POLYSULFIDE POLYMERS	535
31.3.1 From Sodium Polysulfide and Organic Dihalides	535
31.3.1.1 'Toughening'	536
31.3.1.2 Polymerization process	536
31.3.1.3 Preparation of liquid polysulfides	537
31.3.1.4 Choice of halide monomers	537
31.3.2 Other Methods of Polysulfide Synthesis	539
31.3.2.1 From reactions with elemental sulfur	539
31.3.2.2 From vinyl-terminated polysulfides	540
31.3.2.3 From reactions with sulfur chlorides	540
31.3.2.4 From oxidation of dithiols	540
31.3.2.5 Disulfide interchange	540
31.3.2.6 From cyclic disulfides	541
31.3.2.7 Resin modification	541
31.4 REFERENCES	541

31.1 INTRODUCTION

Polysulfide polymers have the general structure $(-R-S_x-)_n$ (**1**) where x is referred to as the 'rank' and represents the average number of sulfur atoms in the polysulfide unit. The rank of polysulfide polymers usually ranges from slightly less than two to about four. The origin of polysulfide polymers dates back to the infancy of organic chemistry. As early as 1838, there were reports of rubbery, intractable, high sulfur, semisolid products from the reaction of potassium sulfides with a halogenated product designated 'chloraetherin'. 'Chloraetherin' was a mixture of products obtained from the addition of chlorine to ethylene in aqueous suspension.[1-3] Subsequently, there were a number of reports of similar products obtained by various methods,[4-6] but the first useful products were developed from Patrick's studies of the reaction of sodium polysulfide with ethylene dichloride. In 1927, the first of an extensive series of patents on the reaction of organic dihalides and inorganic polysulfides was issued to Patrick and Mnookin.[7] Shortly thereafter, production of the ethylene tetrasulfide polymer under the trade name Thiokol A was started. This was the first synthetic elastomer manufactured commercially in the United States. These new synthetic rubbers, available from potentially low cost raw materials, excited considerable interest and various modifications of the polysulfide elastomers appeared all over the world. However, these original polymers were difficult to processs, evolved irritating fumes during compounding, and possessed some undesirable physical properties. Gradual improvements in the products and in methods of processing solved some of these problems but the most significant improvement came in the early 1940's when a method for preparing thiol-terminated liquid polysulfides was developed. Cure of the liquid polysulfides could be accomplished by oxidative coupling of the terminal thiol groups. The liquid polysulfides were particularly useful because now, in effect, a rubber could be compounded without the need of heavy mixing equipment. The liquid polysulfides are the predominant form of polysulfides produced today. Polysulfide rubbers have unusually good resistance to solvents and the environment and good low temperature properties. Because of their unique properties, they have found use in a variety of applications and a number of review articles have been published on the preparation, compounding and performance of polysulfide polymers.[8-16]

31.2 INORGANIC POLYSULFIDES

The inorganic polysulfide polymers are generally prepared from aqueous solutions of alkali polysulfides. Since many aspects of the synthesis and chemistry of inorganic polysulfides are relevant to the preparation of polysulfide polymers, a brief description of inorganic polysulfide chemistry is a good starting point. The focus will be on sodium polysulfides because they are the most commonly prepared and the chemistry of other alkali metal polysulfides is closely related.

31.2.1 Synthesis of Inorganic Polysulfides

Sodium polysulfides have been prepared by a number of techniques.[17,18] Reaction of the appropriate proportions of anhydrous sodium sulfide with molten sulfur (equation 1) has given pure crystalline Na_2S_2, Na_2S_4 and Na_2S_5 which were characterized by X-ray powder diffraction and Raman spectroscopy. The di- and penta-sulfides were found to exist in more than one crystalline form.[19-21] Attempts to prepare Na_2S_3 by this method were unsuccessful and a mixture of Na_2S_4 and Na_2S_2 was obtained. Apparently, Na_2S_3 is unstable with respect to the di- and tetra-sulfide.[20] Sodium pentasulfide is the highest polysulfide that can be obtained in pure form by fusing sodium monosulfide and sulfur. Sodium polysulfides have also been prepared directly from the elements sodium and sulfur (equation 2). This may be accomplished by combining the molten elements[22] or the reaction may be carried out in liquid ammonia.[23,24] Anhydrous alkali polysulfides have also been prepared from the reaction of Na_2S or NaSH with sulfur in anhydrous ethanol.[25,26]

Sodium polysulfides can also be prepared in aqueous solution from the reaction of Na_2S or NaSH and sulfur (equations 1 and 3). However, the product is always a mixture of species Na_2S_x where $x = 1-5$. In addition, sodium polysulfides are hydrolyzed to some extent in aqueous solution and thiosulfate can be produced as a by-product as shown in equations (4)–(6). When NaSH is used the hydrogen sulfide evolved (equation 3) represses the hydrolysis and relatively pure high rank polysulfides can be prepared. The most widely used process for the commercial production of inorganic polysulfide solutions is the reaction of caustic with sulfur. This well-known reaction dates to 1798 and the process can be visualized as taking place in two steps as shown in equations (5) and (6). The reaction is carried out at elevated temperatures and under these conditions, the second reaction (6) takes place very rapidly.[27] However, it is possible to carry out the reaction at a concentration sufficiently high to cause the precipitation of the sodium sulfite formed and sodium polysulfide solutions are produced which contain low levels of thiosulfate.[28] The sodium sulfite crystals are collected by filtration and are useful in a later step of the polysulfide polymer synthesis.

$$Na_2S + (x-1)S \rightarrow Na_2S_x \tag{1}$$

$$2Na + xS \rightarrow Na_2S_x \tag{2}$$

$$2NaSH + (x-1)S \rightarrow Na_2S_x + H_2S \tag{3}$$

$$Na_2S_x + H_2O \rightarrow NaS_xH + NaOH \tag{4}$$

$$6NaOH + (2x+1)S \rightarrow 2Na_2S_x + 3H_2O + Na_2SO_3 \tag{5}$$

$$Na_2SO_3 + S \rightarrow Na_2S_2O_3 \tag{6}$$

31.2.2 Equilibria in Sodium Polysulfide Solutions

Because of hydrolysis, disproportionation to thiosulfate, and equilibria between polysulfide species (equations 4–7), aqueous polysulfide solutions contain a complex mixture of ions. One approach to analyzing these solutions has been to determine the concentration of 'sulfidic sulfur' (S^{2-}) and the 'polysulfidic' or zero valent sulfur present in the solution. This information is needed in order to work with polysulfide solutions and wet,[29,30] spectrophotometric[31] and recently thermometric[32] and electrochemical[33] methods of analysis have been developed.

$$(x-2)S_x^{2-} + HS^- + OH^- \rightleftharpoons (x-1)S_{x-1}^{2-} + H_2O \tag{7}$$

There have also been a number of approaches taken to try to determine the actual equilibrium distribution of the various polysulfide species present in aqueous solution. Early work based on electrode potential measurements suggested that all ions S_x^{2-} with $x = 1-5$ were present in significant amounts within the pH range of 10–13.[34-36] This was disproved by later results based on rapid-mixing acidimetric and spectroscopic studies which showed that the tetra- and penta-sulfide ions predominated in solutions of intermediate alkalinity (pH 9–14).[37-39] The absence of di- and tri-sulfides was attributed to their rapid disproportionation in aqueous solution as shown in equations (8) and (9). High rank polysulfide solutions were also found to be inhomogeneous due to the disproportionation of tetrasulfide as in equation (10). At one time, the pentasulfide ion was considered to be the only species to dissolve in water without disproportionation.[37] However, solubility measurements in the system sulfur–sodium sulfide showed that the rank of the polysulfide solution did not exceed 4.8 in solutions saturated with sulfur at 25 °C.[40] This would preclude the possibility of the formation of pure pentasulfide solutions. Also, Raman studies of aqueous solutions of high-rank polysulfide indicate the presence of HS^-, S_4^{2-} and S_5^{2-}.[20]

$$3S_2^{2-} + H_2O \rightarrow 2HS^- + 2OH^- + S_4^{2-} \tag{8}$$

$$3S_3^{2-} + H_2O \rightarrow HS^- + OH^- + 2S_4^{2-} \tag{9}$$

$$4S_4^{2-} + H_2O \rightarrow HS^- + OH^- + 3S_5^{2-} \tag{10}$$

Studies of the near UV spectra of polysulfide solutions have given much useful information on their composition.[27,38,39,41,42] However, the determination of precise peak locations and widths was complicated by the strong overlap between absorptions of the various species. Recently, a computer iterative technique has been applied to analysis of the spectra and has provided a more precise means of measuring equilibrium constants and calculating the distributions of the species in solution. Upon input of sulfur, alkali metal sulfide and alkali metal hydroxide concentration, the distribution of OH^-, H^+, H_2S, HS^-, S_x^{2-}, ($x = 1-5$) water and alkali metal cation can be calculated. The calculation does not consider the protonated species HS_x^-, because previous studies indicated that it is not present in alkaline polysulfide solutions. Although these species do form on lowering the pH of polysulfide solutions they decompose rapidly accordingly to equation (11). This recent interpretation of polysulfide spectra provides the most accurate description of the species in polysulfide solutions. Although spectral and equilibrium constants calculated by this technique vary by as much as an order of magnitude from previous values, the general conclusions drawn from the earlier spectroscopic and acidimetric studies are still supported. That is, polysulfide solutions at pH 9–14 are largely a mixture of mono-, tetra- and penta-sulfide and significant concentrations of the di- and tri-sulfides exist only in very alkaline polysulfide solutions. Also, aqueous polysulfide solutions are thermodynamically unstable and disproportionate on heating to form thiosulfate and monosulfide with the rate and degree of disproportionation proportional to the hydroxide ion concentration.[27,43]

$$HS_x^- \rightleftharpoons (x-1)S + SH^- \tag{11}$$

31.3 SYNTHESIS OF POLYSULFIDE POLYMERS

31.3.1 From Sodium Polysulfide and Organic Dihalides

The most important commercial method for producing polysulfide polymers is the reaction of aqueous sodium polysulfide with a dichloroalkane as in equation (12). The reaction is pseudo first order because the dichloroalkane is less than 1% soluble in the sodium polysulfide solution and thus its concentration in the aqueous phase remains constant. Because of this reaction, polysulfide polymers have been classified as condensation polymers. However, polysulfide polymerization differs markedly from a typical condensation polymerization. In order to obtain high molecular weight condensation polymers, complete purity and precise proportions of reactants are required along with an absolute avoidance of side reactions. Even under the best conditions, condensation polymers are rarely made with molecular weights of more than thirty thousand. In contrast, polysulfide polymers can be readily prepared with molecular weights estimated as greater than

200 000 and this is accomplished by using an excess of sodium polysulfide. The excess polysulfide functions in several ways to increase the molecular weight of the polymer. First of all, it helps drive the reaction to completion and eliminate chloride terminals in the polymer. More importantly, it acts as an oxidizing or coupling agent to bind two polymer chains as in equation (13). In addition, the excess polysulfide provides for a process referred to as 'toughening'.

$$n\text{Cl–R–Cl} + n\text{Na}_2\text{S}_x \rightarrow (-\text{RS}_x-)_n + 2n\text{NaCl} \tag{12}$$

$$2\text{RSNa} + \text{Na}_2\text{S}_x \rightarrow \text{RSSR} + 2\text{Na}_2\text{S}_{x/2} \tag{13}$$

31.3.1.1 'Toughening'

As discussed earlier, aqueous sodium polysulfide solutions are hydrolyzed to a certain extent (equation 4). This leads to some displacement of chloro groups by hydroxyl ions to form hydroxy terminated oligomers as in equations (14) and (15). These reactions result in chain termination and when equimolar quantities of dihalide and sodium polysulfide are used, the resulting polymer usually has a molecular weight of less than 5000. However, it has been established that polysulfide polymers in the presence of sodium polysulfide are constantly rearranging as the aliphatic disulfide group is reversibly cleaved by inorganic sulfides as in equation (16). When this cleavage occurs near the end of a hydroxy terminated oligomer (equation 17), the low molecular weight fragment can be solubilized in the mother liquor. This removes the chain terminating hydroxyl fragment and allows the thiol-terminated fragment to react and continue to build molecular weight. This phenomenon is referred to as 'toughening' and in the process of polysulfide manufacture sufficient time is allowed for this to occur. Support for this theory of 'toughening' comes from experiments where hydrophobic groups such as butyl or amyl were deliberately placed at the end of a polysulfide chain. In this case treatment with sodium polysulfide was not very effective in increasing the molecular weight. On the other hand, when hydroxy terminals were deliberately introduced, treatment with sodium polysulfide quickly increased the molecular weight of the products.[44] Because of the equilibrium in equation (16) the excess polysulfide has some effect in limiting the molecular weight of the polymer. However, during the process of washing the polymer, the excess polysulfide is removed and equilibrium is shifted back in favour of the high molecular weight product.

$$\text{Cl–R–Cl} + \text{OH}^- \rightarrow \text{Cl–R–OH} + \text{Cl}^- \tag{14}$$

$$\text{Cl–R–OH} + -\text{RS}_x\text{Na} \rightarrow -\text{RS}_x\text{ROH} + \text{NaCl} \tag{15}$$

$$-\text{RS}_x\text{R} + \text{Na}_2\text{S}_x \rightarrow 2-\text{RS}_x\text{Na} \tag{16}$$

$$-\text{RSSROH} + \text{Na}_2\text{S}_2 \rightarrow -\text{RSSNa} + \text{NaSSROH} \tag{17}$$

31.3.1.2 Polymerization process

The first commercial alkyl polysulfide polymer was prepared from the reaction of ethylene dichloride with aqueous sodium tetrasulfide. The product was a rubbery solid which could be worked up only by cutting it out of the reactor after the preliminary washing process followed by washing on a corrugated rubber mill. The product was of poor quality due to difficulties in washing it free of impurities. The process was improved significantly with the addition of surfactant to the reaction medium. This gave a dispersion of discrete polymer particles which reduced occlusion of partly reacted material and produced polymers with improved physical properties. However, it did not solve the purification problem. If sufficient dispersing agents were used to prepare the desired stable latex, then the resulting dispersion settled so slowly that it could not be washed free of impurities by decantation. If flocculating agents such as calcium chloride or aluminum sulfate were used there was a risk of coagulation and the agents had to be removed in a second treatment with a suitable complexing agent. The discovery that a freshly precipitated magnesium hydroxide dispersion could serve as a nucleating system proved to be the development that made feasible the large scale production of polysulfide polymers.[45,46] The coarse dispersions that were formed still allowed for the preparation of high molecular weight products but the polymer particles settled readily and could easily be washed by decantation. The dispersion was stable at pH 8–10 but on acidification the coagulated rubber could be obtained.

31.3.1.3 Preparation of liquid polysulfides

Although pure high molecular weight polymers could be obtained by the methods described, there were still some deficiencies in their properties. One problem was their thermoplastic character which was due to the absence of effective crosslinking. Also, the polymer properties are significantly affected by the rank of the sulfur linkages and generally, it was desirable to remove excess sulfur to obtain a polymer with a rank approaching two. This desulfurization or 'stripping' can be accomplished by treating a suspension of the washed polymer with alkali hydroxides, alkali sulfides, hydrosulfides and sulfites as shown in equations (18)–(20) and alkali carbonates are also effective. In the 'stripping' with sulfides an equilibrium is established and a sufficient excess of sulfide is required for removal of sulfur to the ultimate rank. The reaction with caustic is irreversible, but complications arise from undesirable side reactions. The use of sulfite is the preferred method as it is cheap, easy to handle and quite effective. Furthermore, if NaSH is also added to the solution during the 'stripping' process, the polymer is cleaved or 'split' to give a lower molecular weight polymer by means of the reactions shown in equations (20)–(22). After neutralization with bisulfite (equation 23) and washing and drying, the thiol-terminated polymers can be obtained as a liquid. By adjustment of the amount of NaSH used, polymers of molecular weight ranging from 400 to crude rubbers of 25 000 or more can be prepared. More recently, alternative methods have been developed which involve stripping with caustic followed by treatment with hydrazine or dithionite.[47,48]

$$-RS_4R- + 2NaOH \rightarrow -RS_2R- + 4/3Na_2S + 1/3Na_2S_2O_3 \tag{18}$$

$$-RS_4R- + 2Na_2S \rightarrow -RS_2R- + 2Na_2S_2 \tag{19}$$

$$-RS_xR- + (x-2)Na_2SO_3 \rightarrow -RS_2R- + (x-2)Na_2S_2O_3 \tag{20}$$

$$-RSSR- + NaSH \rightarrow -RSH + NaSSR- \tag{21}$$

$$-RSSNa + Na_2SO_3 \rightarrow -RSNa + Na_2S_2O_3 \tag{22}$$

$$-RSNa + NaHSO_3 \rightarrow -RSH + Na_2SO_3 \tag{23}$$

These processes also provided a suitable means of introducing crosslinking into the polymers. Previous efforts to introduce network formation by adding polyhalides with functionality greater than two were unsuccessful because the high molecular weight products could not be processed without introduction of molecular weight modifiers which destroyed the desired properties. However, with the advent of the splitting and stripping process, crosslinked networks could be introduced and the polymer reduced to bring the viscosity down to ranges suitable for processing and compounding. The original polysulfide link could then be reconstituted by curing with oxidizing agents which couple the terminal thiol groups. The process for preparing liquid polysulfides greatly expanded the utility of the polysulfide polymers because now, in effect, a rubber can be compounded without the need of heavy mixing equipment. The compounded liquid polymers can be poured into casting molds, fed from caulking guns, spread into thin sheets, *etc.* and allowed to cure to form a solid elastomeric product. Although the solid polysulfide elastomers and aqueous dispersions are still used in some applications the liquid polysulfides are by far the most widely used form of the polysulfide polymers.

31.3.1.4 Choice of halide monomers

The basic polymerization reaction is an aliphatic nucleophilic substitution reaction and a variety of difunctional reagents such as tosylates, sulfates and thiosulfates and even *gem*-dinitrates have been reacted with alkali polysulfides to form polymers. In some cases there are advantages to using these reactants, but the availability and reactivity of dihalides, especially dichlorides, has generally made them the reagents of choice for preparing polysulfide polymers. The relative reactivities of the halides with alkali polysulfides follow the pattern expected for nucleophilic substitution reactions. Alkyl bromides are more reactive than alkyl chlorides, and alkyl fluorides are generally unreactive. Aromatic and vinyl halides, unless strongly activated by substituent groups, are too unreactive to form high polymers under normal reaction conditions. The order of reactivity for the alkyl halides is primary > secondary > tertiary and neopentyl halides are unreactive. Secondary and tertiary halides are generally unsuitable as monomers because of their tendency to undergo dehydrohalogenation reactions; however, some vicinal dihalides such as 2,3-dibromopropyl-*n*-butyl ether have given polymers of suitable molecular weight. With the exception of dichloro- and dibromo-methane, *gem*-dihalides do not give polymeric products under normal reaction conditions.[13,49] With regard to the reactivity of the sodium polysulfide species, there is evidence to suggest that the nucleophilicity of the species increases with increasing polysulfide rank (i.e. $S_5^{2-} > S_4^{2-} > \ldots S_1^{2-}$).[50]

The ease of formation, yield, and physical and chemical properties of the polysulfide polymers depend to some degree on the rank of the alkali polysulfide solution used for the polymerization, but are primarily determined by the structure of the dihalide monomer used. Methylene chloride and methylene bromide are the simplest dihalides and these react with sodium polysulfide to give a rubbery polymer. An identical product can also be formed from the reaction of formaldehyde with alkali polysulfide (equation (24), R = H). Substituted methylene polymers were also prepared in a similar manner.[51] These polymers were among the first polysulfides to be studied but they were not very useful because they have strong odors and poor physical properties. Polymers derived from ethylene and propylene dichlorides had good physical properties and were the first commercially successful polysulfide elastomers in spite of the fact that these also were malodorous and evolved irritating fumes during processing. However, it was found that by copolymerization with monomers which had greater than five atom spacing between reactive terminals, the evolution of irritating gases was eliminated and the odor was reduced.

$$\text{RCHO} + \text{Na}_2\text{S}_x + \text{H}_2\text{O} \rightarrow (-\text{RCHS}_x-)_n + 2\text{NaOH} \tag{24}$$

Another problem arises, however, when the spacing between the reactive halide functions permits the formation of a stable heterocyclic product. Cyclic monosulfides are obtained in substantial amounts when dihalide monomers with four or five atoms between the halogen terminals are used (equation 25). Both the rank and concentration of the inorganic sulfide are significant in determining the proportion of cyclic compound formed. The higher the rank and the higher the concentration of the polysulfide, the less the amount of ring compound obtained. Cyclic disulfides are also formed when the polysulfide polymers undergo thermal decomposition. The mechanism of formation of these compounds appears to be a tail-biting-backbone (back-biting) depolymerization as in equation (26). This is a reversible process so in time an equilibrium concentration of the cyclic disulfide is achieved. Steam distillation disturbs the equilibrium by removing the cyclic disulfide as it is formed and a steady rate of formation of these cyclic compounds over a period of months has been observed. Dihalides with spacing that allows formation of 5–7 membered cyclic disulfides depolymerize most rapidly. Introduction of chain stoppers reduces the rate of cyclic compound formation and the addition of sodium hydroxide or increasing the number of thiol terminals accelerates the process. Because of the odors and loss of yield associated with these side reactions, it is advantageous to choose a dihalide with spacing that will minimize the formation of these cyclic products.

$$\text{XCH}_2\text{CH}_2\text{CH}_2\text{CH}_2\text{X} + \text{S}^{2-} \longrightarrow \text{[cyclic sulfide]} \tag{25}$$

$$-\text{SSCH}_2\text{CH}_2\text{OCH}_2\text{CH}_2\text{S}-\text{S}-[\text{ring}] \longrightarrow -\text{SSCH}_2\text{CH}_2\text{OCH}_2\text{CH}_2\text{S}^- + [\text{cyclic disulfide}] \tag{26}$$

In order to improve the properties and processability of the polysulfide polymers, quite a variety of halide monomers have been investigated.[13] For example, unsaturated dihalides such as 1,4-dichloro-2-butene have been used to prepare polysulfide polymers. This monomer is particularly reactive because the halides are allylic. It was hoped that the unsaturation would provide a site for vulcanization but attempts to cure the polymer by sulfur and accelerator combinations were not successful. Polychloromethylated benzene compounds react to give stiff waxy products with very little odor but which are not suitable as elastomers. Under drastic conditions, polymers have also been obtained from aromatic dihalides such as p-dichlorobenzene. Sulfur containing dihalides, primarily bis(2-chloroethyl) sulfide ('mustard gas'), have also been reacted to form polysulfides but the products never became commercially available because of the obvious difficulties in handling the highly vesicant monomers. Hydroxy containing monomers such as glycerol dichlorohydrin or the essentially equivalent epichlorohydrin were found to give polysulfides with relatively low odor, but they were rather moisture sensitive because of the high hydroxyl content. Ester containing dihalides such as chloroacetate derivatives of glycerol have been studied. These materials are not easily polymerized because they are susceptible to hydrolysis in the alkaline polysulfide solution. However, the dicyclopentadiene dichloroacetic acid adduct was successfully polymerized with sodium tetrasulfide in DMF.[52]

Polysulfides

Polymers and copolymers based on ether-containing dihalides provided a substantial improvement in handling properties of the polysulfide elastomers. A variety of these compounds have been examined and bis(2-chloroethyl) formal has been found to be the most useful. It has sufficient spacing between chloro groups to minimize ring formation and is readily prepared from inexpensive starting materials. For these reasons, and for the fact that it imparts desirable physical properties to the polysulfide polymer, it is the dihalide used in nearly all of the polysulfides prepared commercially today.

31.3.2 Other Methods of Polysulfide Synthesis

Although the reaction of organic dihalides with inorganic polysulfides is the most important commercial method for preparing polysulfide polymers, there are a number of alternative methods which have been developed. Many of these methods have been described in earlier reviews,[13,15] so only selected examples of the different approaches will be discussed here and emphasis will be on recent developments.

31.3.2.1 From reactions with elemental sulfur

A number of polysulfides have been prepared by reacting alkenes or dialkenes directly with molten sulfur. The general objective has been to prepare systems with a large proportion of sulfur (over 90%) and these compositions are often termed 'modified sulfur'. The purpose is to obtain a stable form of flexible or 'plastic' sulfur which is useful in a variety of applications.[53,54] The most useful product appears to be that obtained from the reaction of sulfur with dicyclopentadiene.[55]

Another approach to preparing polysulfide polymers from alkenes and sulfur has been developed which involves anionic polymerization of vinyl compounds in the presence of sulfur. Adding butadiene to a mixture of sulfur, lithium and THF solvent at $-40\,°C$ gave products (equation 27) which could be converted to thiol terminated oligomers by Clemmensen reduction. The rank, x, of the polysulfide was found to decrease as the ratio of initiator concentration over sulfur concentration increased and the chain length n depended mainly on monomer concentration.[56,57]

$$Li + CH_2=CH-CH=CH_2 + S \rightarrow [-(-C_4H_6)_n-S_x-]-_p \qquad (27)$$

Elemental sulfur can also be copolymerized with cyclic sulfides such as propylene sulfide as in equation (28). High molecular weight polymers may be prepared in this manner and polysulfides with a rank as high as eight were obtained.[58,59]

$$\text{(propylene sulfide)} + S_8 \longrightarrow -CH_2-\overset{Me}{\underset{H}{C}}-S_x- \qquad (28)$$

Epoxy compounds also react with hydrogen sulfide and sulfur in the presence of basic catalysts to form mono- and poly-thiodiglycols and bis(2-hydroxyethyl)polysulfide (**3**, $x=2$) may be prepared in this manner. These diols will yield polyformals when reacted with formaldehyde and an acid catalyst. The product has the same chain structure as the polysulfide obtained from the reaction of bis(2-chloroethyl) formal with sodium polysulfide. However, the polymer is terminated with a hydroxyl group rather than a thiol. A thiol-terminated polysulfide may be obtained by treating an aqueous dispersion of (**3**) with sodium tetrasulfide by way of a process reminiscent of the 'toughening' procedure. Because of the enhanced reactivity of a hydroxyl group on a carbon β to sulfur, (**2**) can be readily converted to polyether by heating in the presence of an acid catalyst and azeotropic removal of water (equation 30). It is also possible to prepare copolymers by adding unactivated glycols such as ethylene or diethylene glycols.[13,15]

$$HOCH_2CH_2S_xCH_2CH_2OH + CH_2O \rightarrow HO(CH_2CH_2S_xCH_2CH_2OCH_2O)_nH \qquad (29)$$
$$(\mathbf{2}) \qquad\qquad\qquad\qquad (\mathbf{3})$$

$$(\mathbf{2}) + H^+ \rightarrow HO(CH_2CH_2S_xCH_2CH_2O)_nH \qquad (30)$$

31.3.2.2 From vinyl terminated polysulfides

Recently, reaction (31) was reported to give divinyl polysulfides (4) from vinyl epoxy ethers and sodium polysulfide. The reaction provides a means of introducing highly active polymerizable double bonds into thiodiglycols and the products (4) polymerize on heating.[60] Also, (vinylaryl)alkyl-terminated polysulfides have been prepared by reacting sodium polysulfides with vinylbenzyl chloride or with mixtures of vinylbenzyl chloride, and other aliphatic dihalides.[61] For example divinylbenzyltetrasulfide (5, $x=4$) was prepared in this manner (equation 32). These vinyl-terminated polysulfides can be cured with heat or with free radical initiators. The polymers are reported to have low odor and when blended or copolymerized with other polymers may impart many of the beneficial properties of polysulfide polymers such as resistance to oxygen permeation, water, UV light and solvent.[62-65]

$$(2) \ CH_2{=}CHOROCH{-}CH_2 \ (epoxide) + Na_2S_x \longrightarrow (CH_2{=}CHOROCHCH_2{-})_2{-}S_x \quad (31)$$
$$\underset{OH}{} \quad (4)$$

$$Na_2S_x + 2CH_2{=}CH{-}C_6H_4{-}CH_2Cl \rightarrow (CH_2{=}CH{-}C_6H_4{-}CH_2{-})_2{-}S_x \quad (32)$$
$$(5)$$

31.3.2.3 From reactions with sulfur chlorides

Polysulfide polymers have been prepared by the reaction of dialkenes with S_2Cl_2 as in equation (33), but the product was unstable because of the labile chlorines present.[13] However, dechlorination of (6) with sodium sulfide gives products which are useful as additives for oils.[66] Reaction of S_2Cl_2 with dithiols, diols and diamines also gives polysulfides (equation 34).[67] Polyamine disulfides prepared in this manner exhibited semiconductor properties.[68]

$$nCH_2{=}CH{-}CH{=}CH_2 + nS_2Cl_2 \rightarrow ({-}SCH_2{-}\underset{Cl}{CH}{-}\underset{Cl}{CH}{-}CH_2S{-})_n \quad (33)$$
$$(6)$$

$$HX{-}R{-}XH + S_2Cl_2 \rightarrow ({-}X{-}R{-}X{-}S_2{-})_n \quad (34)$$
$$X = S, O, NH, RN$$

31.3.2.4 From oxidation of dithiols

The oxidation of thiols to disulfides is well known and there have been a few instances where this method was used to form polysulfide polymers. There are problems with this method, however, because it is difficult to prepare dithiols in a pure state so that no unreactive terminal groups are present in the monomer and because of side reactions in the oxidation process. Only low molecular weight polymers have been obtained from this method. Also, in most cases the dithiol is made from the dihalide so it is easier to produce a polysulfide directly from the dihalide. However, this approach has proven useful in the preparation of the semiconducting polymer, poly(m-phenylene disulfide) by oxidation of 1,3-dimercaptobenzene with DMSO.[69]

31.3.2.5 Disulfide interchange

Thiols are capable of cleaving disulfide bonds and the interchange reaction (35) occurs readily when catalyzed by base.[70] This reaction can be used to modify the structure of polysulfide polymers. When an organic dithiol is added to a liquid polysulfide it is incorporated into the polymer. Liquid polysulfides prepared from bis(2-chloroethyl) formal have been modified in this manner and are claimed to give products with improved properties for certain applications.[71]

$$HSR'SH + {-}SSRSSRSS{-} \rightarrow {-}SSRSSR'SH + HSRSS{-} \quad (35)$$

31.3.2.6 From cyclic disulfides

The ring opening polymerization of cyclic disulfides as in equation (36) is another way of preparing polysulfide polymers. The polymerization may be initiated by β radiation, UV light and radical or ionic catalysts. The cyclic disulfides also copolymerize with vinyl monomers when initiated by free radicals.[15,72] Unfortunately, the cyclic disulfides are usually difficult to prepare so this is not a very useful route to polysulfide polymers.

$$\text{(cyclic disulfide with O and S—S)} \longrightarrow (-SCH_2CH_2OCH_2CH_2S-)_n \qquad (36)$$

31.3.2.7 Resin modification

Polysulfide polymers, particularly thiol-terminated liquid polysulfides are capable of reacting chemically with other resins and modifying their properties. Resin modifications and processes where polysulfide polymers are formed simultaneously with the formation of another resin are discussed in earlier reviews.[10,13] For example, phenolic resins and, to a certain extent, furfuryl alcohol resins can be flexibilized by reaction with polysulfides.

Thiols also react with epoxides and some useful modified resin compositions have been prepared by combining liquid polysulfides with epoxide resins. The products retain much of the hardness of an epoxy while the introduction of the elastomeric polysulfide greatly improves impact strength and imparts some flexibility. Amine catalysts are usually used to cure these resins. Anhydride cures also give satisfactory products, but strong acids cause degradation of the polysulfide.[73-77]

31.4 REFERENCES

1. C. Lowig and S. Wiedmann, *Pogg. Ann. Physik*, 1839, **46**, 81.
2. C. Lowig and S. Wiedmann, *Pogg. Ann. Physik*, 1842, **49**, 123.
3. C. Lowig and S. Weidman, *Ann. Chem. Pharm.*, 1840, **36**, 320.
4. V. Meyer, *Ber. Dtsch. Chem. Ges.*, 1886, **19**, 3259.
5. H. Fasbender, *Ber. Dtsch. Chem. Ges.*, 1887, **20**, 460.
6. J. Baer, *Kautschuk*, 1934, **10**, 55.
7. J. C. Patrick and N. M. Mnookin, *Br. Pat.*, 302 270 (1927) (*Chem. Abstr.*, 1929, **23**, 4307).
8. S. M. Ellerstein, in 'Encyclopedia of Polymer Science and Engineering', ed. J. I. Kroschwitz, Wiley, New York, 1988, vol. 13, p. 186.
9. S. M. Ellerstein and E. R. Bertozzi, in 'Kirk-Othmer Encyclopedia of Chemical Technology', ed. M. Grayson, Wiley, New York, 1978, vol. 18, p. 814.
10. E. R. Bertozzi, *Rubber Chem. Technol.*, 1968, **41**, 114.
11. A. V. Tobolsky and W. J. MacKnight, 'Polymeric Sulfur and Related Polymers', Wiley, New York, 1965.
12. A. V. Tobolsky, 'The Chemistry of Sulfides', Interscience, New York, 1968.
13. M. B. Berenbaum, in 'Polyethers, Part III', ed. N. G. Gaylord, Interscience, New York, 1962, vol. 13, p. 43.
14. J. R. Panek, in 'Polyethers, Part III', ed. N. G. Gaylord, Interscience, New York, 1962, vol. 13, p. 115.
15. E. M. Fettes, in 'Organic Sulfur Compounds', ed. N. Kharasch, Pergamon, London, 1961, vol. 1, p. 266.
16. E. A. Peterson and A. D. Yazujian, *Adhes. Age*, 1987, 6.
17. O. Erametsa and K. Karlsson, *Acta Polytech. Scand., Chem. Incl. Metall. Ser.*, 1961, **15**, 5.
18. J. W. Mellor, 'A Comprehensive Treatise on Inorganic and Theoretical Chemistry', Longman, Green and Co., New York, 1946, vol. 2, p. 629.
19. E. Rosen and R. Tegman, *Acta Chem. Scand.*, 1971, **25**, 3329.
20. G. J. Janz, J. R. Downey, Jr., E. Roduner, G. J. Wasilczyk, J. W. Coutts and A. Eluard, *Inorg. Chem.*, 1976, **15**, 1759.
21. G. J. Janz, E. Roduner, J. W. Coutts and J. R. Downey, Jr., *Inorg. Chem.*, 1976, **15**, 1751.
22. F. Bittner, W. Hinrichs, H. Hovestadt and L. Lange, *US Pat.* 4 640 832 (1987) (*Chem. Abstr.*, 1986, **104**, 227 187f).
23. J. Letoffe, J. Blanchard and J. Bousquet, *Bull. Soc. Chim. Fr.*, 1976, 395.
24. F. Feher and H. J. Berthold, *Z. Anorg. Allg. Chem.*, 1953, **273**, 144.
25. A. Rule and J. S. Thomas, *J. Chem. Soc.*, 1914, **105**, 177.
26. J. S. Thomas and A. Rule, *J. Chem. Soc.*, 1917, **111**, 1063.
27. W. F. Giggenbach, *Inorg. Chem.*, 1974, **13**, 1724.
28. E. R. Bertozzi, *US Pat.* 2 796 325 (1957) (*Chem. Abstr.*, 1957, **51**, 14 220h).
29. J. Karchmer, 'The Analytical Chemistry of Sulfur and Its Compounds', Wiley, New York, 1970, Part I.
30. J. Boulegue, *J. Geochem. Explor.*, 1981, **15**, 21.
31. A. Teder, *Sven. Papperstidn.*, 1967, **70**, 500.
32. J. W. Stahl and J. Jordan, *Anal. Chem.*, 1987, **59**, 1222.
33. J. E. Yakupkovic and J. Jordan, *J. Electroanal. Chem. Interfacial. Electrochem.*, in press.
34. D. Peschanski and G. Valensi, *C.R. Hebd. Seances. Acad. Sci.*, 1948, **227**, 845.

35. D. Peschanski and G. Valensi, *J. Chim. Phys. Phys. Chim. Biol.*, 1949, **46**, 602.
36. C. Maronny, *J. Chim. Phys. Phys.-Chim. Biol.*, 1959, **56**, 140.
37. G. Schwarzenbach and A. Fischer, *Helv. Chim. Acta*, 1960, **43**, 1365.
38. A. Teder, *Ark. Kemi*, 1969, **30**, 379.
39. A. Teder, *Ark. Kemi*, 1969, **31**, 173.
40. R. H. Arntson, F. W. Dickson and G. Tunell, *Science*, 1958, **128**, 716.
41. W. Giggenbach, *Inorg. Chem.*, 1972, **11**, 1201.
42. A. Teder, *Sven. Papperstidn.*, 1969, **72**, 245.
43. S. Licht, G. Hodes and J. Manassen, *Inorg. Chem.*, 1986, **25**, 2486.
44. E. M. Fettes, presented at 136th Meeting of the American Chemical Society, Atlantic City New Jersey, Sept., 1979.
45. J. C. Patrick, *Br. Pat.* 359 000 (1929) (*Chem. Abstr.*, 1932, **26**, 4981).
46. J. C. Patrick, *US Pat.* 1 950 744 (1934) (*Chem. Abstr.*, 1934, **28**, 3541).
47. V. J. Rekalic, M. E. Tenc–Popovic and S. D. Radosavljevic, *J. Polym. Sci., Polym. Chem. Ed.*, 1980, **18**, 2033.
48. M. E. Tenc–Popovic, S. Popov, S. D. Radosavljevic and V. J. Rekalic, *J. Polym. Sci., Part A-1*, 1972, **10**, 2583.
49. R. W. Lenz, 'Organic Chemistry of Synthetic High Polymers', Interscience, New York, 1972, p. 154.
50. L. A. Averko–Antonovich, G. V. Ronamova and R. A. Ibragimova, *Khim. Tekhnol. Elementoorg. Soedin. Polim.*, 1972, 120 (*Chem. Abstr.*, 1974, **81**, 154 076p).
51. J. De la Pena, M. C. Luciv, J. Guzman and E. Riande, *Macromolecules*, 1986, **19**, 486.
52. K. M. McCreedy and K. R. Hilton, *US Pat.* 4 493 927 (1985) (*Chem. Abstr.*, 1985, **102**, 2 044 492w).
53. B. R. Currell, A. J. Williams, A. J. Mooney and B. J. Nash, in 'Advances in Chemistry Series', ed. J. R. West, American Chemical Society, Washington D.C., 1975, vol. 140. p. 1.
54. L. Blight, B. R. Currell, B. J. Nash, R. A. M. Scott and C. Stillo, in 'Advances in Chemistry Series', ed. D. J. Bourne, American Chemical Society, Washington D.C., 1978, vol. 165, p. 13.
55. B. K. Bordoloi and E. M. Pearce, in 'Advances in Chemistry Series', ed. D. J. Bourne, American Chemical Society, Washington D.C., 1978, vol. 165, p. 31.
56. J. M. Catala, J. M. Pujol and J. Brossas, *Polym. Prepr., Am. Chem. Soc., Div. Polym. Chem.*, 1986, **27**, 198 (*Chem. Abstr.*, 1986, **105**, 6900b).
57. J. M. Catala, J. M. Pujol and J. Brossas, *Fr. Demande* FR 2 577 932 (1985) (*Chem. Abstr.*, 1987, **106**, 102 849h).
58. A. Duda and S. Penczek, *Macromolecules*, 1982, **15**, 36.
59. A. Duda and S. Penczek, *Makromol. Chem.*, 1980, **181**, 995.
60. B. A. Trofimov, N. A. Nedolya, V. I. Komel'kova and M. Ya. Khil'ko, *Zh. Pr. Khim.*, 1986, **59**, 2382.
61. G. Wulff and I. Schulze, *Angew. Chem. Int. Ed. Engl.*, 1978, **17**, 537.
62. V. E. Meyer and T. E. Dergazarian, *US Pat.* 4 438 259 (1984) (*Chem. Abstr.*, 1984, **100**, 211 825u).
63. T. E. Dergazarian, *US Pat.* 4 607 078 (1986) (*Chem. Abstr.*, 1986, **105**, 2105a).
64. V. E. Meyer, T. E. Dergazarian, *US Pat.* 4 608 433 (1986) (*Chem. Abstr.*, 1987, **106**, 85 275y).
65. R. E. Hefner, Jr., *US Pat.* 4 692 500 (1987) (*Chem. Abstr.*, 1988, **108**, 7042x).
66. J. Bolle and A. Dabir, *US Pat.* 4 284 520 (1981) (*Chem. Abstr.*, 1982, **96**, 38171b).
67. L. Tokarzewski and Z. Szymik, *Plaste Kautsch.*, 1965, **7**, 387.
68. Y. P. Losev, V. N. Isakovich, Y. M. Pushkin, L. P. Loseva, M. A. Ksenofontov and L. V. Volod'ko, *Vysokomol. Soedin., Ser. A.*, 1980, **22**, 607 (*Chem. Abstr.*, 1980, **92**, 198 828s).
69. C. D. Casa, P. C. Bizzarri and S. Nuzziello, *J. Polym. Sci., Polym. Lett. Ed.*, 1985, **23**, 323.
70. M. B. Berenbaum in 'Chemical Reactions of Polymers', ed. E. M. Fettes, Interscience, New York, 1964, p. 528.
71. L. Morris and H. Singh, *US Pat.* 4 623 711 (1986) (*Chem. Abstr.*, 1987, **106**, 68 558a).
72. Japan Synthetic Rubber Co., *Jpn. Pat.* 164 181 (1981) (*Chem. Abstr.*, 1982, **96**, 162 752f).
73. H. Kuramoto and K. Watanabe, *Jpn. Pat.* 148 280 (1986) (*Chem. Abstr.*, 1986, **105**, 174 060w).
74. T. C. Lee and T. M. Rees, *Eur. Pat. Appl.* 171 198 (1986) (*Chem. Abstr.*, 1986, **104**, 226 445b).
75. K. Rushland and B. Winkler, *East Ger. Pat.* 226 001 (1985) (*Chem. Abstr.*, 1986, **104**, 169 751x).
76. L. Popescu and M. Banta, *Rom. Pat.* 82 410 (1981) (*Chem. Abstr.*, 1984, **103**, 54 977n).
77. Dainippon Toryo Co., *Jpn. Pat.* 74 156 (1981) (*Chem. Abstr.*, 1981, **95**, 152 270n).

32
Poly(phenylene sulfide)s

J. F. GEIBEL and R. W. CAMPBELL
Phillips Petroleum Co., Bartlesville, OK, USA

32.1	INTRODUCTION	543
32.2	HISTORY OF POLY(PHENYLENE SULFIDE)	543
32.3	CHEMISTRY OF POLY(PHENYLENE SULFIDE)	545
	32.3.1 Polymerization	545
	32.3.2 Cure	547
32.4	PROCESSES FOR POLY(PHENYLENE SULFIDE)	547
	32.4.1 Polymerization	547
	32.4.2 Cure	550
32.5	PROPERTIES OF POLY(PHENYLENE SULFIDE)	551
	32.5.1 Thermal Properties	551
	32.5.2 Chemical Resistance	554
	32.5.3 Flame Resistance	554
	32.5.4 Electrical Properties	555
	32.5.5 Melt and Solution Properties	556
	32.5.6 Mechanical Properties	557
32.6	OTHER POLY(ARYLENE SULFIDE)S	557
32.7	REFERENCES	559

32.1 INTRODUCTION

Poly(phenylene sulfide), PPS, has a long history in the chemical literature. Reported syntheses of PPS or PPS-related resins date back almost 100 years. Commercial development activities began approximately 40 years ago, ultimately leading to the first commercial production of PPS in 1973.

The structure of PPS consists of a series of alternating disubstituted aromatic rings (*p*-phenylene units) and divalent sulfur atoms (sulfide linkages), as shown in (**1**). PPS is a semi-crystalline thermoplastic polymer which possesses excellent mechanical, electrical, thermal and chemical resistance properties in addition to the polymer being inherently self-extinguishing. This unique combination of properties explains why polymers such as PPS comprise a class of materials referred to as 'engineering thermoplastics'. These materials have found uses in applications such as coatings, injection molding, film, fiber and advanced composites. New applications are being discovered as designers are gaining familiarity with these materials.

(**1**) Poly(*p*-phenylene sulfide)

32.2 HISTORY OF POLY(PHENYLENE SULFIDE)

Prior to the commercialization of a poly(phenylene sulfide) process, developments relating to the synthesis of PPS could be classified as either: (1) electrophilic substitution reactions, or (2) nucleophilic substitution reactions. The earliest reference to reactions leading to poly(phenylene sulfide) structures is that of Friedel and Crafts in 1888[1] in which they studied reactions of benzene with sulfur sources (equation 1, X = S, S_2Cl_2, SCl_2). These reactions were low yield (50–80%) and consisted of crude products containing other products such as thianthrene. Analogous results were observed for the self-condensation of benzenethiol in the presence of aluminum chloride or sulfuric acid.[2] There is no mention of any polymer formation in this early report.

$$\bigcirc + X \xrightarrow{AlCl_3} +\!\!\left[\bigcirc\!-\!S\right]\!\!+ + \text{by-products} \qquad (1)$$

Phenylene sulfide structures

Genvresse in 1897 was the first to assign the PPS structure to an amorphous, insoluble resin which was prepared from the reaction of benzene and sulfur in the presence of aluminum chloride.[3] The products produced were very poorly characterized by today's standards. Genvresse's products were low molecular weight and contained too much sulfur to be pure PPS. He isolated thianthrene and an amorphous material which melted at 295 °C. The early synthetic efforts have been reviewed in greater detail elsewhere.[4–6] Other researchers,[7–10] attempting to synthesize diphenyl sulfide and related aromatic sulfur compounds, also directed their efforts toward phenylene sulfide-type structures in the years subsequent to the initial reports of Genvresse and Friedel and Crafts.

In 1984, the electrophilic substitution reactions of benzene and sulfur in the presence of aluminum chloride were reinvestigated.[11] Improved analytical methods allowed the detection of structures which contained thianthrylene sulfide linkages, phenylene sulfide linkages and polysulfides, and were unstable to the reaction conditions. It was concluded that the formation of high molecular weight poly(arylene sulfide)s in the presence of aluminum chloride is, therefore, highly unlikely.

The 60 years following the work of Friedel and Crafts saw no significant improvements in the synthesis of PPS-type polymers. In 1948 A. D. Macallum discovered the first nucleophilic displacement route to poly(arylene sulfide) resins.[12] Although the electrophilic reactions just discussed are straightforward, they display a lack of selectivity and produce only low molecular weight products. Typical molecular weights for polymers from benzene and sulfur in the presence of aluminum chloride[11] were only 3500, too low to have useful engineering properties. It is the nucleophilic substitution route to PPS that yields products with usable properties.

The Macallum process involves the reaction of elemental sulfur, sodium carbonate and dichlorobenzene at 275–300 °C in the melt in a sealed vessel (equation 2). Elemental analysis of polymers prepared by this route generally contain more sulfur than is predicted by the structure of the repeat unit (i.e. X is in the range 1.2–2.3).

$$Cl-\bigcirc-Cl + S + Na_2CO_3 \longrightarrow \left(\bigcirc-S_x\right)_n \qquad (2)$$

Although the polymerization reaction was highly exothermic and difficult to control, Macallum recognized that the product had unusual thermal stability and the potential to be a high performance engineering plastic. PPS analogs were prepared from difunctional and trifunctional chlorinated aromatic compounds and mixtures thereof. Macallum was the first to observe that the higher molecular weight resins were suitable molding materials which possessed good mechanical properties.[13]

The Macallum PPS synthesis is versatile, with several modifications being reported. The sulfur source can be comprised of a mixture of sodium carbonate and sulfur which has been prereacted prior to reaction with dichlorobenzene or entirely replaced with sodium sulfide which contains a small amount of elemental sulfur.[2] Another analogous synthesis is to utilize an arenebis(diazonium) salt in place of dichlorobenzene which is reacted with sodium sulfide.[14] This procedure, however, typically results in polymers containing nitrogen.

The Macallum PPS process generated interest within the industrial sector, eventually leading to the sale of his process patents[15] in 1954 to Dow Chemical Company. Researchers in the Dow laboratories conducted during the late 1950s and early 1960s a detailed study of the Macallum polymerization as well as the structures and characteristics of the polymers.[4,16,17] Despite these efforts, the initial problems associated with the severe reaction conditions, control of the exothermic reaction and unpredictable polymer yields and properties remained largely unsolved.[18]

Because PPS remained as an attractive polymer with an excellent combination of properties, the Dow researchers explored alternate synthetic routes to PPS. An improved synthetic procedure consisted of the self-condensation reaction of copper(I) p-bromobenzenethiolate.[17] This nucleophilic substitution reaction was carried out at 200–250 °C in the solid state or, alternatively, in the presence of materials such as pyridine as a reaction medium (equation 3).

$$Br-\langle\bigcirc\rangle-SCu \longrightarrow -(\langle\bigcirc\rangle-S)_n + CuBr \quad (3)$$

The resultant polymer possessed a linear structure. However, the monomer and process were costly and they experienced considerable difficulty in removing the by-product, copper(I) bromide, from PPS made *via* this process.[19] Scale-up only made these problems worse. Dow actually test marketed a phenylene sulfide polymer in the early and mid-1960s; however, their PPS product never reached commercialization because of these difficulties.

Studying model coupling reactions involving nucleophilic displacement of aryl halides by sulfur nucleophiles, researchers at Monsanto found unexpectedly good suitability of amide solvents.[20] At the same time, researchers at Phillips Petroleum Company developed a new synthetic route to PPS. This new process involves the production of PPS from p-dichlorobenzene and sodium sulfide in a polar organic solvent at elevated temperature. In 1967, Phillips Petroleum Company was granted a US patent for this process.[21]

Between 1967 and 1973, workers at Phillips optimized the new process, defined and characterized the polymer properties, constructed and operated a pilot plant, established market demand for this new polymer, and constructed a full-scale commercial plant. In 1973 the world's first PPS plant came on stream in Phillips Petroleum Company's facility in Borger, Texas. The product is marketed under the trade name of Ryton® Polyphenylene Sulfide.

32.3 CHEMISTRY OF POLY(PHENYLENE SULFIDE)

32.3.1 Polymerization

Two synthetic routes for the formation of PPS have been discussed in the preceding section. The earliest references to phenylene sulfide structures describe electrophilic substitution reactions. A variety of these reactions have been reported. Product characterization was lacking and the reaction specificity clearly indicated that a complex collection of reactions was taking place. The product distribution reported in recent reinvestigations[6,11] of electrophilic substitution processes directed towards PPS and the fact that PPS itself is unstable to aluminum chloride[6,22] indicate that a detailed mechanistic understanding of this process would be difficult. This complexity, coupled with the fact that the nucleophilic substitution process has more commercial significance, has made nucleophilic routes the subject of greater investigation.

A series of nucleophilic displacements can be envisioned leading to poly(phenylene sulfide) (equations 4 and 5). Reaction (4) depicts the initiation reaction of inorganic sulfide with the monomer, p-dichlorobenzene. Reaction (5) shows the beginning of polymer chain growth as the product of initiation reacts with additional p-dichlorobenzene. These reactions may be extended to include the reaction of sodium sulfide and sodium p-chlorobenzenethiolate with chlorophenyl end groups on polymeric species. Analogously, thiolate nucleophile end groups on polymer molecules can react with any other aryl chloride. In this way, the growth of high molecular weight PPS proceeds, resulting in the consumption of the reactive moieties originally present in the monomers.

$$Na_2S + Cl-\langle\bigcirc\rangle-Cl \longrightarrow \left[Cl-\langle\bigcirc\rangle\genfrac{}{}{0pt}{}{Cl}{SNa}\right]^- Na^+ \xrightarrow{-NaCl} Cl-\langle\bigcirc\rangle-SNa \quad (4)$$

$$Cl-\langle\bigcirc\rangle-SNa + Cl-\langle\bigcirc\rangle-Cl \xrightarrow{-NaCl} Cl-\langle\bigcirc\rangle-S-\langle\bigcirc\rangle-Cl \quad (5)$$

As further mechanistic alternatives to the nucleophilic process just described, single electron transfer (SET) chain processes can be envisioned. Recently, Koch and Heitz published a mechanistic study[23] in which they concluded that the PPS polycondensation proceeds *via* a SET process with radical cations as reactive intermediates. Initiation is proposed *via* oxidation of hydrosulfide *via* SET to a hydrosulfenyl radical, which reacts with *p*-dichlorobenzene, ultimately leading to 4-chlorophenylsulfenyl radical (**2**; equation 6). Sulfenyl radical (**2**) can then react with additional dichlorobenzene in a chain propagation series of reactions as shown in equations (7) and (8).

$$HS^- \xrightarrow{-e} HS\cdot \xrightarrow{Cl-C_6H_4-Cl} Cl-C_6H_4(Cl)(SH)\cdot \xrightarrow{-Cl} Cl-C_6H_4-SH\cdot \xrightarrow{-H} Cl-C_6H_4-S\cdot \quad (6)$$

$$(2)$$

$$Cl-C_6H_4-Cl + \cdot S-C_6H_4-Cl \longrightarrow Cl-C_6H_4(Cl)(S-C_6H_4-Cl)\cdot \xrightarrow{-Cl}$$

$$(3)$$

$$Cl-C_6H_4\cdot^+-S-C_6H_4-Cl \quad (7)$$

$$(4)$$

$$Cl-C_6H_4\cdot^+-S-C_6H_4-Cl + HS^- \longrightarrow Cl-C_6H_4-S-C_6H_4-Cl + HS\cdot \quad (8)$$

$$(5)$$

Reactions (7) and (8) show the production of the first oligomer of PPS (**5**) and hydrosulfenyl radical, which is available for reentry into reaction (6). It is presumed that species such as (**5**) can also participate in reactions (6) and (7) in place of dichlorobenzene, resulting in still higher oligomers of PPS and ultimately producing polymer. Other reactions which produce aryl radicals, such as (**3**), or aryl radical cations, such as (**4**), would also provide entry to the chain mechanism. Presumably, radical coupling and hydrogen abstraction processes would account for chain termination reactions.

In contrast to the SET radical cation mechanism just discussed, an $S_{RN}1$ mechanism, analogous to that described by Bunnett,[24] may also be operative in the PPS polymerization. Low molecular weight PPS has been synthesized from *p*-halobenzenethiolates by an $S_{RN}1$ mechanism by Novi *et al.*[25] An initiation step proceeds *via* a one electron reduction of dichlorobenzene with subsequent expulsion of a chloride ion from the dichlorobenzene radical anion to produce an aryl radical (**6**) shown in reaction (9).

$$Cl-C_6H_4-Cl + e^- \longrightarrow [Cl-C_6H_4-Cl]^{\cdot -} \xrightarrow{-Cl} Cl-C_6H_4\cdot \quad (9)$$

$$(6)$$

Aryl radical (**6**) is an electron deficient species which can react with hydrosulfide to produce chlorobenzenethiol radical anion which can chain transfer (reaction 10a) or lose a chloride ion to produce another aryl radical (reaction 10b).

$$Cl-C_6H_4\cdot + HS^- \longrightarrow [Cl-C_6H_4-SH]^{\cdot -} \xrightarrow{-e} Cl-C_6H_4-SH \quad (10a)$$

$$\xrightarrow{-Cl} \cdot C_6H_4-SH \quad (10b)$$

Under the basic PPS polymerization conditions, species such as *p*-chlorobenzenethiol produced in reaction (10a) will rapidly deprotonate and be available for reactions with the variety of aryl radicals, such as (**6**), to produce higher oligomers (**7**) of PPS, which can continue the chain process (equation 11).

$$\text{Cl-C}_6\text{H}_4\cdot + {}^-\text{S-C}_6\text{H}_4\text{-Cl} \longrightarrow [\text{Cl-C}_6\text{H}_4\text{-S-C}_6\text{H}_4\text{-Cl}]^{\overline{\cdot}} \xrightarrow{-\text{Cl}^-} \text{Cl-C}_6\text{H}_4\text{-S-C}_6\text{H}_4\cdot \quad (11)$$

(**7**)

Whether the PPS polymerization proceeds by a purely nucleophilic (S_N2) process or by an electron transfer radical ($S_{RN}1$) process has not been established.

32.3.2 Cure

The ability of PPS to undergo 'change' upon heating was recognized early in the development of the polymer. Smith and Handlovits[18] reported that a polymer with useful properties could be obtained from a thermal treatment of PPS. PPS combines attributes of both thermosetting and thermoplastic polymers. Under normal processing conditions, PPS behaves as a true thermoplastic. It can be repeatedly processed with only minor changes in its rheology. However, under appropriate conditions, PPS can be changed in an incompletely understood 'curing' process.[26] Attempts have been made to elucidate the reactions which take place when PPS is cured. Since the polymers are not readily analyzed by conventional techniques, this task is difficult.

Several studies have been published describing the curing, degradation, annealing and thermal stability of PPS in which mechanisms of the curing reactions have been postulated.[16,17,19,27-35] Smith and Handlovits[19] cite evidence for homolysis of C—S bonds in the PPS backbone and subsequent reactions of the thiyl and aryl free radicals to produce crosslinking and coupling reactions. Black and coworkers[27] found evidence for chain scission, crosslinking and oxidation processes. These workers observed chain extension, loss of both hydrogen and sulfur, as well as oxygen uptake during the aging studies. They concluded that PPS curing involves crosslinking by loss of hydrogen and creation of new carbon–sulfur bonds, the evolution of low molecular weight chain fragments (*e.g.* phenyl sulfide), and the formation of oxidation products of both the sulfur and aromatic moieties. Ehlers and coworkers[29] confirmed the work of Black, citing evidence for homolysis of carbon–sulfur bonds leading to crosslinking and recombination. They, too, observed evolution of volatile dimer, trimer and other sulfur compounds during the curing process.

Port and Still[34] reported on the influence of cure temperature, atmosphere and polymer structure. They found the curing reactions of PPS to be complex and depend upon the above-mentioned variables. For PPS cured below the crystalline melting point of the polymer, reactions occurred which increased the solution viscosity as well as reduced the crystallizability of the polymer. This confirms earlier results of Brady[31] in which PPS cured in air below the crystalline melting point showed no significant change in percent crystallinity, although the ability to crystallize after melting and annealing was dramatically reduced. Port and Still demonstrated that curing reactions are more rapid in air than nitrogen. At curing temperatures above the crystalline melting point, crosslinking occurred with little weight loss yielding an insoluble network structure.

32.4 PROCESSES FOR POLY(PHENYLENE SULFIDE)

32.4.1 Polymerization

The Phillips process for preparing PPS consists of the reaction of sodium sulfide with *p*-dichlorobenzene carried out in a polar organic solvent (equation 12). Polymer formation is accompanied by the production of sodium chloride as a by-product.[13] The feedstocks for this process are relatively inexpensive, large volume chemicals. This feature was of critical importance in the commercialization of PPS at a reasonable price.

$$Cl-\langle\bigcirc\rangle-Cl + Na_2S \xrightarrow[\text{solvent}]{\text{polar organic}} -(\langle\bigcirc\rangle-S)_n + 2NaCl \qquad (12)$$

The key process steps for producing PPS (Figure 1) are: (1) preparation of sodium sulfide from aqueous sodium hydrosulfide (or alternatively hydrogen sulfide) and aqueous sodium hydroxide; (2) dehydration of the sodium sulfide feed stream in the presence of the polar organic solvent; (3) reaction of sodium sulfide and dichlorobenzene in the polar solvent, yielding a slurry of PPS and sodium chloride in the solvent; (4) polymer recovery; (5) polymer washing to remove the sodium chloride and residual solvent; (6) polymer drying; and (7) packaging.

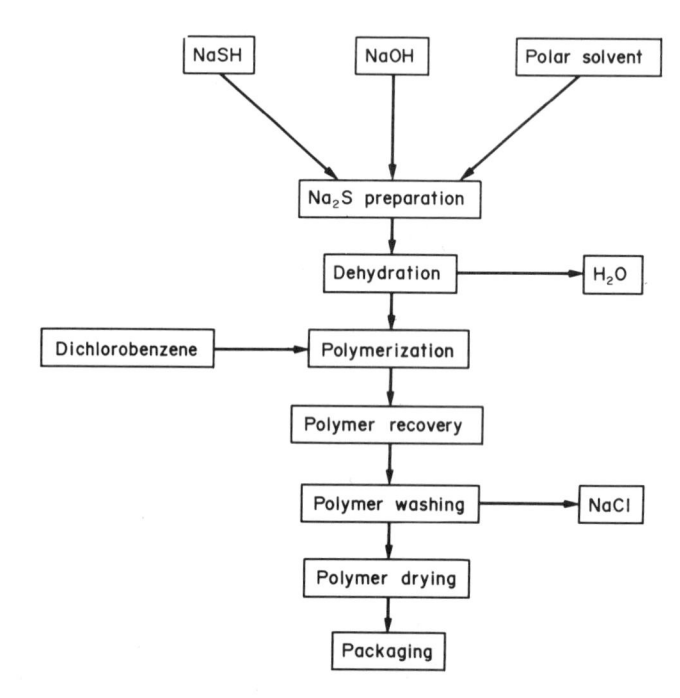

Figure 1 Polymerization process for Ryton PPS

The polymer produced by the Phillips process is an off-white powder which has a linear structure containing approximately 150–200 repeat units. This translates to a molecular weight of 16 000–22 000 Daltons. Conventional molecular weight measurement techniques are complicated by the solubility characteristics of PPS.[36] Dilute solution light scattering and gel permeation chromatography studies (performed in 1-chloronaphthalene at 220 °C) indicate a molecular weight of approximately 18 000.[37,38] The inherent viscosity, measured at 206 °C in 1-chloronaphthalene, is typically 0.16 dl g^{-1}.

Advances in the understanding of the polymerization chemistry of PPS from the laboratories of Phillips Petroleum Company, have made possible the synthesis of high molecular weight PPS directly in the polymerization vessel.[39] These high molecular weight polymers, frequently referred to as modified polymers, do not require a curing step. They are used as extrusion resins for production of fiber and film as well as the resin feedstock for injection molding compounds and composites.

This new polymerization process is based upon use of an alkali metal carboxylate as a polymerization modifier. The process steps for producing high molecular weight PPS are shown in Figure 2. The process is similar to the unmodified process described earlier. The major difference is that an alkali metal carboxylate is included in the sodium sulfide preparation step. The remainder of the process steps (dehydration, polymerization, recovery, washing, drying and packaging) are similar to those described previously. Polymer obtained from the Phillips modified process is an off-white powder with a linear structure. The molecular weight of this modified polymer is approximately 35 000 Daltons,[40] indicating a structure with approximately 300 to 400 repeat units. Even higher molecular weight polymer can be produced by the copolymerization of a tri- or tetra-functional comonomer, introducing long chain branches into the polymer structure. Small amounts of 1,2,4-

Poly(phenylene sulfide)s 549

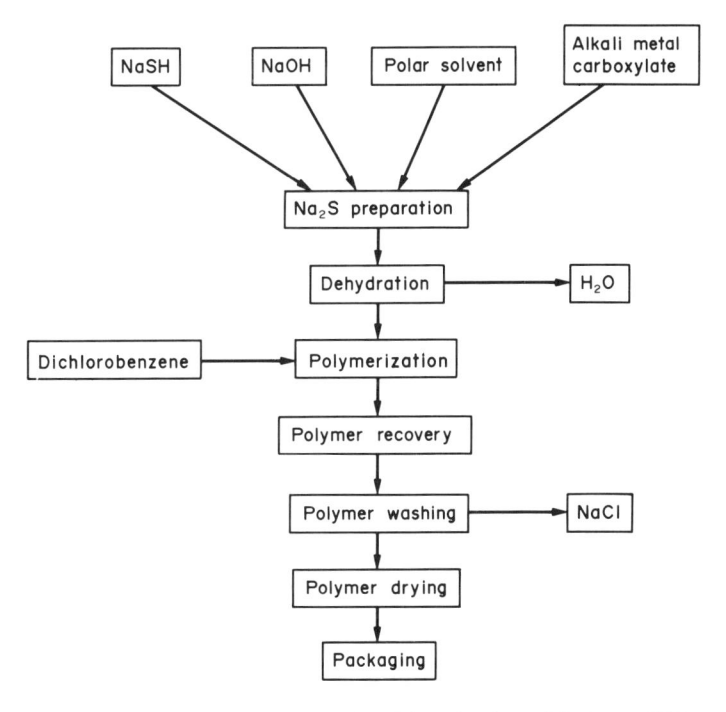

Figure 2 Polymerization process for high molecular weight Ryton PPS

trichlorobenzene can be used to produce high molecular weight PPS polymers in the Phillips processes.[21]

Kureha Chemical Industry has disclosed a multistep process for the preparation of high molecular weight PPS.[41] The key process steps involved in the Kureha process are: (1) dehydration of sodium sulfide in N-methyl-2-pyrrolidone (NMP); (2) polymerization of the dehydrated sodium sulfide with dichlorobenzene at a low temperature for an extended period of time; (3) addition of water to the 'prepolymer' formed during the low temperature polymerization step; (4) a second, higher temperature polymerization step for an extended period of time; and (5) polymer recovery. Figure 3 shows a block diagram of the Kureha process. The patent literature indicates that an additional process step can be inserted between the first (low temperature) polymerization step and the water

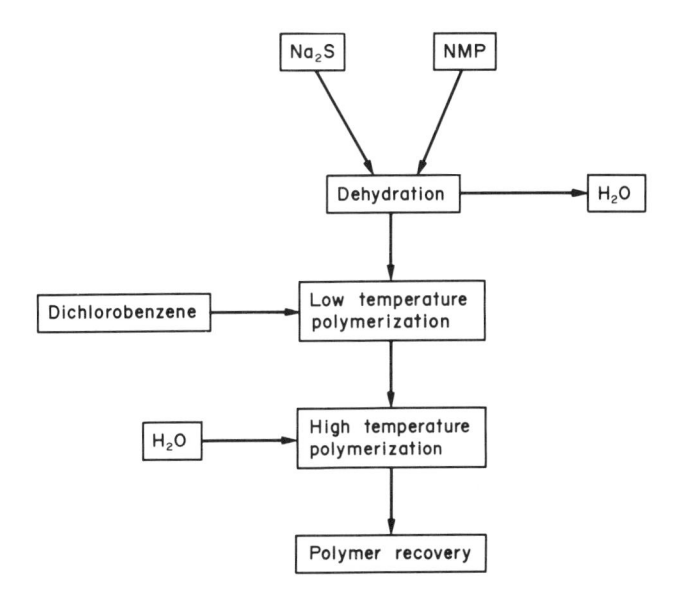

Figure 3 Polymerization process for Fortron PPS (Y. Iizuka, T. Iwasaki, T. Katto and Z. Shiiki (Kureha Kagaku Kabushiki Kaisha), US Pat. 4 645 826, 1987)

charge. This step involves washing the prepolymer with a basic solution, termed a 'purification' of the prepolymer. When the 'purified polymer' is subjected to the second, high temperature polymerization period in an NMP–water solvent mixture, even higher molecular weight PPS is obtained. While examples in the patent literature indicate some flexibility in the choice of times and temperatures for the two polymerization periods, the highest molecular weight polymers are reported for long polymerization cycles. The use of water added to the PPS polymerization mixture in order to effect polymer stabilization[42] and improvements in molecular weight[42,43] have been reported previously.

Processes for the production of PPS have also been reported by Bayer.[44] Key aspects of the Bayer procedures are the use of N-methylcaprolactam as the polymerization solvent, moderate to large amounts of a branching reagent (e.g. 1,2,4-trichlorobenzene), polymerization cosolvents (e.g. carboxylic amides,[44] carboxylic anhydrides,[45] carboxylic esters,[45] amino acids[46]) and various inorganic salts as polymerization modifiers.[47,48] The carboxylic acid derivatives used by Bayer are reactive under the PPS polymerization conditions described and produce carboxylates, whose effect on PPS polymerization has been well documented.[39]

Another process for the preparation of PPS has been published by Idemitsu.[49] This process involves reaction of an alkali metal sulfide with a dihaloaromatic compound (e.g. p-dichlorobenzene) in a two phase solvent system comprised of high molecular weight polyethylene glycol and water. A lower molecular weight polyethylene glycol is reported to function as a phase transfer catalyst for the reactant sodium sulfide. PPS is isolated from the polyethylene glycol phase.

32.4.2 Cure

PPS can be converted to a tougher material by a thermal treatment. The melt-curing process is used for the production of tough, intractable coatings from PPS. The curing reactions of PPS represent one of the most commercially useful characteristics of low to moderate molecular weight PPS. Enhancement of the properties of PPS by curing is the basis for a variety of products and applications derived from a single starting material. This gives PPS unprecedented versatility compared to conventional polymer systems.

Curing of PPS can be accomplished in one of two ways. The first involves a melt process[35] in which the polymer is heated above its crystalline melting point in the presence of air. Upon continued heating, the viscosity of the molten polymer increases and the color of the polymer darkens. Extensive heating causes continued increases in viscosity and eventual gelation and solidification yielding a dark infusible solid. The solid polymer is believed to be crosslinked and is insoluble in all organic solvents, even at elevated temperatures.

The second curing process is a solid state process in which the polymer powder is heated to just below its melting point in the presence of air.[50] Solid state curing is a practical process for converting large quantities of polymer in bulk. The process is generally carried out at 175–280 °C. The progress of the solid state curing reaction is conveniently followed by measuring changes in the melt viscosity, determined by a slightly modified ASTM melt flow procedure. As curing occurs, molecular weight and melt viscosity increase, as evidenced by a decrease in the melt flow compared to that for the uncured polymer. Figure 4 shows the rate of decrease of melt flow for PPS cured in the solid state at several temperatures. The rate of cure is a strong function of the curing temperature, with faster decreases in melt flow occurring at higher curing temperatures. This process allows easy control of molecular weight.

Solid state curing of PPS may be conducted in an air-circulating oven. Adequate contact between polymer particles and air must be maintained. If curing is performed in pans or trays in the oven, care should be exercised not to have more than one inch of polymer in the pan. This insures adequate air contact and helps avoid thermal differences between the bulk polymer temperature and the oven temperature. If 'hot spots' in the oven exist, the polymer may fuse and aggregate, producing large particles which may not cure uniformly. The solid state cure of PPS may also be performed in an agitated bed with an air stream passing through the polymer or in an aqueous slurry in the presence of oxygen.[51]

Several important properties of PPS change when the polymer is cured: (1) molecular weight increases; (2) an insoluble fraction is produced; (3) toughness increases; (4) melt viscosity increases; (5) kinetics and extent of crystallization are decreased; and (6) the color of the polymer changes from off-white to tan/brown/black. The extent to which these changes are observed depends directly on the extent of cure, a function of time and temperature of the cure.

Poly(phenylene sulfide)s 551

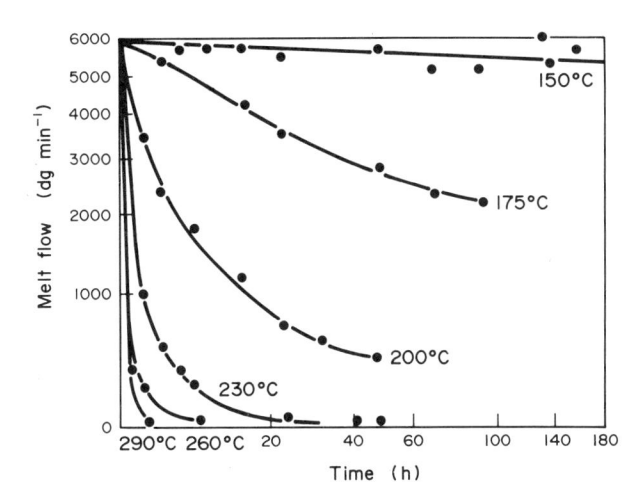

Figure 4 Cure rate of PPS as a function of solid cure temperature (reproduced with permission of the publisher from ref. 36)

A single uncured feedstock is used to produce a family of cured polymers, each with melt flow properties appropriate for different applications. The data in Table 1 show typical melt flow values for different grades of PPS.[36] The uncured feedstock has a very high melt flow value, typically in the range of 3000–8000 g/10 min. Powder coating grade PPS resins are relatively lightly cured, while the injection molding grade PPS resins have melt flow values of approximately 60 to 600 g/10 min, resulting from a more extensive curing cycle. The range of melt flows for the injection molding grade PPS resins accommodates variations in the nature and amounts of fillers and reinforcing agents added during the production of compounds. Heavily cured PPS resins used for compression molding and free sintering will soften, but do not flow under usual melt flow measurement conditions.

Table 1 Typical Melt Flow Values for PPS Resins

Resin	Melt flow (g/10 min)
Uncured PPS	3000–8000
Powder coating PPS	1000
PPS for mineral/glass-filled compounds	600
PPS for glass-filled compounds	60
Compression molding PPS	0

32.5 PROPERTIES OF POLY(PHENYLENE SULFIDE)

PPS is recognized among thermoplastic resins for its remarkable combination of inherent properties and physical characteristics. These include: unusual thermal stability, excellent chemical resistance, inherent flame resistance, high electrical insulation properties and good mechanical properties of molded parts.

32.5.1 Thermal Properties

The thermal stability of PPS results from the structure of the polymer, involving only carbon–carbon, carbon–hydrogen and carbon–sulfur bonds. All are quite thermodynamically stable. Table 2 lists the dissociation energies for bonds present in PPS.

The implication of the thermodynamic stability of the bonds in PPS is that it will take a large input of energy (high temperatures) in order to dissociate any of the bonds (induce thermal degradation). PPS is well suited for use at high temperatures for extended periods of time.

A standard test for measuring the thermal stability of a material is thermogravimetric analysis. In this test, the percent weight loss is recorded as a function of the temperature of the material. Figure 5

Table 2 Bond Dissociation Energy in PPS

Bond	Bond energy (kcal mol^{-1} at 25 °C)[a]
C—C	114
C—H	99
C—S	66

[a] 1 kcal = 4.18 kJ

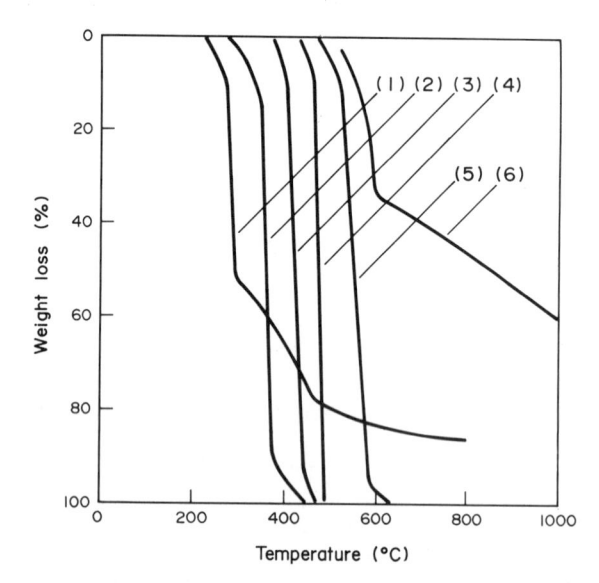

Figure 5 Comparative thermogravimetric analyses of polymers in nitrogen: (1) poly(vinyl chloride); (2) poly(methyl methacrylate); (3) polystyrene; (4) polyethylene; (5) poly(tetrafluoroethylene); (6) PPS (reproduced with permission of the publisher from Figure 2, ref. 30)

shows PPS in comparison to several other polymers. PPS retains approximately 40% of its original weight at temperatures as high as 1000 °C in nitrogen.

PPS is a semicrystalline polymer. The thermal transitions of PPS are appropriate for melt processing of material into finished parts. As defined by differential scanning calorimetry (DSC), PPS possesses a high crystalline melting point (T_m, 285 °C) and a modest glass transition temperature (T_g, 85 °C). The polymer crystallizes rapidly above T_g as indicated by an exothermic crystallization peak located at 120–130 °C (T_{cc}, heating rate of 10 °C/min).[30] The polymer also crystallizes rapidly when cooled from the melt. PPS displays a melt crystallization exotherm at approximately 220 °C (T_{mc}, cooling rate of 10 °C/min).[36] By proper choice of molding and annealing conditions, a crystalline part can be obtained which is dimensionally stable and possesses good mechanical properties. Typical heats of fusion for crystalline samples of PPS are in the range of 10–12 cal g^{-1} (42–50 J kg^{-1}).[36] The ability to mold PPS in a crystalline form is critical in order to achieve dimensionally stable parts and to allow those parts to be used at elevated temperatures. According to an X-ray diffraction method described by Brady, fully crystallized PPS has a crystallinity index of about 65%.[31] In this publication, the mechanical properties of PPS were correlated with the degree of crystallinity.

The thermal history of PPS can influence the crystallizability and ultimate level of crystallinity attainable, as described previously. Curing PPS below T_m does not reduce the percent crystallinity. However, after melting and recrystallizing the cured PPS, a dramatic reduction in the level of crystallinity has been reported.[31] This observation was recently reconfirmed and expanded upon,[52] in a study of the effect of curing temperature on the crystallinity of PPS coatings. Figure 6 shows a plot of percent crystallinity as a function of curing temperature for PPS films cured for 30 min. The percent crystallinity remains approximately constant for samples cured well below the melting point of the polymer, but rapidly approaches zero for polymers cured above the melting point. Extensive annealing (3–4 hours at 200 °C) of PPS films cured at 300 °C for 30 min induced low levels of crystallinity (15%). Samples cured at 370 °C displayed no crystallinity after annealing.

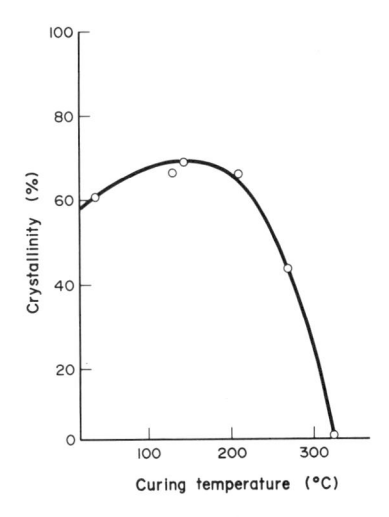

Figure 6 Percent crystallinity as a function of curing temperature for PPS (reproduced with permission of the publisher from ref. 52)

Another useful thermal characteristic of engineering materials is the ASTM heat deflection temperature under high load. This test (ASTM method D648) measures short term property retention at elevated temperature. The high load heat deflection temperatures for a variety of glass-reinforced engineering polymers are compared in Table 3.

Long term performance at high temperatures is best described by the Underwriters Laboratory's (UL) Temperature Index, defined as the maximum temperature at which a molded part can be exposed continuously for a period of 10 years with no more than a 50% loss in properties. Table 4 summarizes the UL Temperature Indices for a variety of engineering thermoplastics. Based on its high UL Temperature Index, PPS is a proven high temperature engineering thermoplastic.

Table 3 ASTM High Load Heat Deflection Temperatures for Glass-reinforced Engineering Thermoplastics[a]

Polymer	ASTM Heat Deflection Temperature at 264 psi (°C)
Poly(ether ether ketone) (PEEK)	312
Poly(amideimide) (Torlon)	274
PPS	263
Nylon 6,6	254
Poly(butylene terephthalate)	220
Poly(etherimide) (Ultem)	216
Polyarylate	179
Polysulfone	177
Polycarbonate	149

[a] Taken from ref. 53. Higher value used when range was reported.

Table 4 UL Temperature Indices for Engineering Thermoplastics[a]

Polymer	UL temperature index (°C)
PPS	220
Poly(ether sulfone)	170
Polysulfone	140
Poly(butylene terephthalate)	140
Nylon 6,6	140
Polycarbonate	125

[a] Taken from ref. 26.

32.5.2 Chemical Resistance

Poly(phenylene sulfide) has excellent chemical resistance. There are no known solvents for the polymer below 200 °C. Parts molded from PPS find application in a variety of hostile chemical environments where inertness is required. A chemically resistant polymer will maintain its dimensional stability, physical properties, weight and appearance when exposed to hostile environments.[54] PPS is affected by high temperature exposure to only a few organic solvents, strong mineral acids and strong oxidizing reagents.[55] The action of these reagents is due to the expected chemistry for the structure of PPS. Oxidizing reagents can convert sulfides present in PPS to sulfoxides or sulfones. The aromatic ring can be halogenated, nitrated or sulfonated under appropriate conditions to yield derivatized polymers. Table 5 documents a comparison of PPS with selected engineering polymers for short term exposure to a variety of reagents.

Table 5 Comparative Chemical Resistance of Various Polymers at 93 °C/24 h: Percent Tensile Strength Retained[a]

Chemical	Nylon 6,6	Polycarbonate	Polysulfone	Modified PPO	PPS
37% HCl	0	0	100	100	100
10% HNO_3	0	100	100	100	96
30% H_2SO_4	0	100	100	100	100
85% H_3PO_4	0	100	100	100	100
30% NaOH	89	7	100	100	100
28% NH_4OH	85	0	100	100	100
H_2O	66	100	100	100	100
$FeCl_3$	13	100	100	100	100
NaOCl	44	100	100	100	84
Bromine	8	48	92	87	64
Butyl alcohol	87	94	100	84	100
Phenol	0	0	0	0	100
Butylamine	90	0	0	0	50
Aniline	85	0	0	0	96
2-Butanone	87	0	0	0	100
Benzaldehyde	98	0	0	0	84
Chlorobenzene	73	0	0	0	100
Chloroform	57	0	0	0	87
Ethyl acetate	89	0	0	0	100
Butyl phthalate	90	46	63	19	100
p-Dioxane	96	0	0	0	88
Butyl ether	100	61	100	0	100
Gasoline	80	99	100	0	100
Diesel fuel	87	100	100	36	100
Toluene	76	0	0	0	98
Benzonitrile	88	0	0	0	100
Nitrobenzene	100	0	0	0	100

[a] From ref. 55.

Long term exposure of PPS to hostile environments also shows excellent retention of properties. Table 6 shows the retention of tensile strength after exposure to a variety of reagents for 24 hours and three months.

32.5.3 Flame Resistance

Poly(phenylene sulfide) possesses outstanding fire retardancy. This inherent flame resistance can be attributed to its aromatic structure and to its tendency to char. When exposed to a flame, PPS will burn with a yellow-orange flame until the flame source is removed. The oxygen index for neat PPS resin is 44[36] and compounds of PPS have oxygen indices from 44 to above 60.[13] Moldings exhibit very low radiant panel flame spread index, low smoke generation, and are classified as self-extinguishing and non-dripping. Table 7 summarizes the flammability characteristics of PPS.

Table 6 Long Term Chemical Resistance of PPS Molding at 93 °C[a]

Chemical	Percent tensile retained	
	24 h	3 months
37% HCl	100	56
10% HNO_3	96	31
30% H_2SO_4	100	100
85% H_3PO_4	100	100
30% NaOH	100	100
H_2O	100	100
5% NaOCl	84	90
Butyl alcohol	100	100
Cyclohexanol	96	81
Butylamine	50	0
Aniline	96	87
2-Butanone	100	70
Benzaldehyde	84	90
CCl_4	100	77
Chloroform	87	79
Ethyl acetate	100	100
Butyl ether	100	100
p-Dioxane	88	75
Gasoline	100	87
Toluene	98	74
Benzonitrile	100	77
Nitrobenzene	100	85
Phenol	100	100
N-Methylpyrrolidone	94	74
Air	—	90

[a] From ref. 55.

Table 7 Flammability Characteristics of Poly(phenylene sulfide)

Test	Test method	Value
Oxygen index	ASTM D 2863	44–62[a]
Flammability	UL 94	V-O/5V
Flame spread index	ASTM E 162	50.8 mm
Autoignition temperature[b]	—	540 °C
Smoke density[c]	NBS test chamber	
obscuration time, smoldering	—	15.5 min
obscuration time, flaming	—	3.2 min

[a] Unfilled PPS has oxygen index = 44. Filled compositions vary with filler and composition.
[b] Ref. 36.
[c] Time required to reach specific optical density of 16. Results from ref. 56.

32.5.4 Electrical Properties

Poly(phenylene sulfide) resin and compounds display properties characteristic of good insulators. Typical properties are high dielectric strength, low dielectric constant, low dissipation factor and high volume resistivity. Table 8 summarizes the electrical properties of two major types of PPS molding compounds used for electronic and electrical applications.

PPS compounds which contain only glass fillers are better insulators than those containing glass and mineral fillers. Alternatively, compounds containing both glass and mineral fillers provide better arc resistance and tracking index. Both types of compounds, however, provide an excellent balance of electrical properties. The dissipation factor and dielectric constant for a 40% glass-filled PPS compound display excellent stability over a wide frequency range and at elevated temperatures.

PPS (40% glass-filled) compounds also display excellent retention of electrical properties after exposure to humid environments. For example, the volume resistivity decreases by only one decade (10^{16} ohm cm to 10^{15} ohm cm) after immersion in 60 °C water for 12 d. The unusual combination of good electrical properties and high temperature resistance accounts for the Underwriters Laboratories approval of the use of PPS compounds at high temperatures (200–240 °C[36]). These properties make PPS compounds well suited for electronic and electrical applications.

Table 8 Electrical Properties of PPS Compounds

Property	40% glass filled	Glass and mineral filled	Test method
Dielectric strength (1 kHz, V mil^{-1})[c]	450	340–400	D149[a]
Dielectric constant (1 MHz)	3.8	4.6	D150[a]
Dissipation factor (1 MHz)	0.0013	0.016	D150[a]
Volume resistivity (2 min, Ω cm)	4.5×10^{16}	2.0×10^{15}	D257[a]
Arc resistance (s)	35	200	D495[a]
Comparative tracking index (V)	180	235	UL 746A[b]
Insulation resistance (Ω)	10^{11}	10^{9}	—

[a] ASTM.
[b] Underwriters Laboratory.
[c] 1 mil = 10^{-3} in.

Despite the fact that PPS is an excellent insulator, it can be rendered electrically conducting by the addition of dopants such as arsenic pentafluoride,[57,58] antimony pentafluoride,[59] sulfur trioxide[59,60] and nitrosyl hexafluorophosphate.[61] Doped PPS is not unique as a conducting polymer. Many other polymers with π-conjugated structures in the backbone of the polymer have been doped with strong electron acceptors to yield conducting polymers. Polyacetylene,[62] poly(p-phenylene),[63] and polypyrrole[64] are examples of such polymers. These polymers all suffer, however, from processing difficulties in that they are intractable. Since PPS does not present processing difficulties in its predoped state, it has been the object of significant efforts directed toward conductive polymers. The reaction of arsenic pentafluoride with PPS appears to result in structural modification of the polymer.[65] Dibenzothiophene moieties are produced by the coupling of adjacent phenyl rings in the PPS backbone. These aromatic rings are forced into a coplanar conformation not found in PPS crystal structures[66] or amorphous state studies,[67] presumably facilitating π-electron overlap between the aromatic rings. A further consequence of doping PPS with arsenic pentafluoride is the loss of crystallinity,[68] due to the above-mentioned dopant-induced disorders. The solvent resistance of PPS is due in part to its crystallinity. Thus, doping PPS with arsenic pentafluoride in the presence of arsenic trifluoride (the solvent for this process) resulted in rapid dissolution of the polymer.[69] The resulting solution was intensely blue, conductive and indefinitely stable in dry air. Evaporation of the solvent produced film which had conductivities ranging from 5×10^{-3} to 200 S cm^{-1} and which possessed good strength and flexibility. The extension of these technologies to practical electronic and electrical applications is attractive and awaiting further development.

32.5.5 Melt and Solution Properties

Rheological measurements with PPS melts are complicated by the high temperatures required (T_m 285 °C) and the propensity of the polymer toward curing reactions. Kraus and Whitte[40] described the melt rheology of three commercially important classes of PPS: (1) linear and uncured; (2) branched and uncured; and (3) linear and cured. They reported that the Newtonian viscosity of linear, uncured PPS increases approximately as the 4.9th power of the weight average molecular weight, much more steeply than the usual 3.4th power relationship. There appears to be no viscosity enhancement in randomly branched versions of PPS. The Newtonian viscosity of branched PPS was determined to be less than that for the corresponding linear PPS (equal M_w). The increasingly branched forms of PPS display increasingly non-Newtonian behavior. This observation was rationalized on the basis of the branched polymers having greater weight average molecular weight and a broader molecular weight distribution. The molecular weight for entanglement of linear PPS chains was estimated to be 20 000 Daltons.

More common melt viscosity determinations are made using a modification of the ASTM test method D1238 for melt flow determination. This is an isoshear stress test and has been described for the various feedstocks used in manufacturing PPS compounds (see Table 1).

Solution viscosity measurements on PPS are made difficult by the insolubility of PPS. Inherent viscosity measurements are performed in 1-chloronaphthalene at 206 °C at a concentration of 0.4 g dl^{-1}. Table 9 shows the relationship of inherent viscosity and melt flow for linear PPS.

Advances have been made in developing gel permeation chromatography (GPC) techniques which are compatible with the limited solubility of PPS. Commercial instruments do not have the

Table 9 Inherent Viscosities of PPS[a]

Polymerization process	Inherent viscosity (dl g^{-1})	Melt flow (g/10 min)
Unmodified	0.18	>6000
Carboxylate modification	0.24	1295
Carboxylate modification	0.28	665
Carboxylate modification	0.35	93

[a] Data taken from examples in ref. 39.

temperature capabilities required for PPS solubility. Researchers at Phillips[38,70] and Toray[71] constructed custom built high temperature gel permeation chromatographs. These studies have shown the heterogeneity index (M_w/M_n) to be 1.7 for PPS, lower than the value of 2.0 expected for a theoretical distribution from a condensation polymer.[72] The significance of the deviation is an area of current investigation.

32.5.6 Mechanical Properties

Parts molded from PPS are generally characterized by high strength and stiffness and moderate impact resistance. An unusual property possessed by PPS moldings is their 'metallic ring' when given a sharp tap. Fillers and reinforcing fibers can be easily added to PPS, yielding reinforced compounds with enhanced properties. These compounds are useful for compression and injection moldings, displaying high tensile strength, high flexural modulus, good flexural strength, high heat deflection temperature, low elongation and moderate impact strength. Compounds can be custom-tailored by inclusion of other mineral fillers, pigments or fiber reinforcements to suit specific applications. Table 10 summarizes the properties of neat PPS and a variety of filled compounds.

Table 10 Properties of PPS Resin and Compounds[a]

Property	Neat[b]	40% glass reinforced	Glass and mineral filled	Test method[c]
Density (g cm^{-3})	1.35	1.6	1.8–2.0	D1505
Tensile strength (p.s.i. × 10^{-3})[e]	9.5	17.5	10.8	D638
Elongation (%)	1.6	1.2	0.54	D638
Flexural modulus (p.s.i. × 10^{-6})[e]	0.56	1.7	2.2	D790
Flexural strength (p.s.i. × 10^{-3})[e]	14.0	26.0	14.5	D790
Izod impact (ft lbf in^{-1})[e]				D256
notched	0.3	1.3	0.6	
unnotched	1.9	4.5	1.9	
Compressive strength (p.s.i. × 10^{-3})[e]	16.0	21.0	16.0	D695
Heat deflection temperature (°C)[d]	135	>260	>260	D648
Rockwell hardness	R-120	R-123	R-121	D785
Coefficient of linear expansion (10^{-5} °C^{-1})	4.9	4.0	2.8	—

[a] Injection molded into a 135 °C mold, samples unannealed unless specified.
[b] Cured feedstock.
[c] ASTM test methods.
[d] Samples annealed at 260 °C for four hours.
[e] 1 p.s.i. = 0.00689 MPa; 1 lb ft in^{-1} = 53.38 J m^{-1}.

32.6 OTHER POLY(ARYLENE SULFIDE)S

Although PPS is the poly(arylene sulfide) which has received the most attention, a variety of other structures have been synthesized.

Copolymers have been prepared[73] from mixtures of *p*- and *m*-dichlorobenzene and sodium sulfide (Table 11). Incorporation of *meta* units results in disruption of crystallinity. Copolymers containing 50–100 mol % *meta* structure are amorphous. The 50 and 75 mol % *meta* copolymers are quite soluble in tetrahydrofuran (THF).

Table 11 Poly(phenylene sulfide) Copolymers[73,74]

Para/meta ratio	T_g (°C)	T_m (°C)	Solubility[a]
1/0	83	283	Insoluble
3/1	68	205	Insoluble
1/1	49	—[b]	Soluble
1/3	27	—[b]	Soluble
0/1	15	—[b]	Nearly insoluble
1/1[c]	39	271	Soluble fraction

[a] In refluxing tetrahydrofuran.
[b] Amorphous polymers softening at 90–100 °C.
[c] Block copolymer.

The *meta* homopolymer is nearly insoluble in refluxing THF but is soluble in NMP at 100 °C. Molecular weight in this series decreases as *meta* content increases, contributing to the lowering of T_g. The last entry in Table 11 is a block copolymer prepared from a mixture of *meta* and *para* homopolymers.[74] It consisted of soluble and insoluble fractions (NMP at 100 °C) which contained 75 and 25 mol % *meta* units, respectively.

Physical properties of other poly(phenylene sulfide)s are summarized in Table 12. Variation of diad structure (head-to-head and head-to-tail) for the 2-methyl derivative accounts for its lack of crystallinity. The addition of symmetrically placed substituents in the 2,6-dimethyl and 2,3,5,6-tetramethyl derivatives does not significantly reduce crystallinity.

Physical properties of other poly(arylene sulfide)s are summarized in Table 13. Structural irregularities in the polynuclear derivatives and the bulky sulfone group disrupt crystallinity. Replacement of sulfide groups with ether groups lowers crystallinity. Biphenyl and benzophenone repeat units give high-melting, crystalline products. Both nucleophilic and electrophilic synthetic procedures were used in producing these polymers.

Table 12 Other Poly(phenylene sulfide)s

Structure	Softening temperature (°C)	Crystallinity	Solubility	Ref.
–(C$_6$H$_4$–S)$_n$– (para)	280	Some	ClC$_{10}$H$_7$	21
–(C$_6$H$_4$–S)$_n$– (meta)	100	None	NMP	73
–(2-Me-C$_6$H$_3$–S)$_n$–	100 (140)	None	C$_6$H$_6$	75 (76)
–(2,6-Me$_2$-C$_6$H$_2$–S)$_n$–	170 (220)	Slight	C$_6$H$_6$	75 (76)
–(2,3,5,6-Me$_4$-C$_6$–S)$_n$–	290	Some	o-Cl$_2$C$_6$H$_4$	77

Table 13 Other Poly(arylene sulfide)s

Structure	Softening temperature (°C)	Crystallinity	Solubility	Ref.
(naphthalene-S)ₙ	130	None	CHCl₃	78
(anthracene-S)ₙ	105	None	CHCl₃	78
(biphenyl-S)ₙ	430	High	None	14
(C₆H₄-S-C₆H₄-O-C₆H₄-S)ₙ	160	Some		79
(C₆H₄-O-C₆H₄-S)ₙ	160	Some	DMSO	80
(C₆H₄-CO-C₆H₄-S)ₙ	340	Some	H₂SO₄	81
(C₆H₄-SO₂-C₆H₄-S)ₙ	275	None	NMP	82

32.7 REFERENCES

1. C. Friedel and J. M. Crafts, *Ann. Chim. Phys.*, 1888, **14** (6), 433.
2. N. G. Gaylord, 'Polyethers', in 'High Polymers Series', Interscience, New York, 1962, vol. XIII, part III, p. 31.
3. P. Genvresse, *Bull. Soc. Chim. Fr.*, 1897, **17**, 599.
4. R. W. Lenz, C. E. Handlovits and W. K. Carrington, *J. Polym. Sci.*, 1959, **41**, 333.
5. H. A. Smith, in 'Encyclopedia of Polymer Science and Technology', ed. H. F. Mark, Interscience, New York, 1969, vol. 10, p. 653.
6. J. W. Cleary, *Adv. Polym. Synth.*, 1985, **31**, 159.
7. J. J. B. Deuss, *Recl. Trav. Chim. Pays-Bas.*, 1909, **28**, 136.
8. H. O. Jones and H. S. Tasker, *Proc. Chem. Soc.*, 1909, **25**, 24.
9. T. P. Hilditch, *J. Chem. Soc.*, 1910, **97**, 2579.
10. H. B. Glass and E. E. Reid, *J. Am. Chem. Soc.*, 1929, **51**, 3428.
11. J. W. Cleary, *Polym. Prepr., Am. Chem. Soc., Div. Polym. Chem.*, 1984, **25** (2), 36.
12. A. D. Macallum, *J. Org. Chem.*, 1948, **13**, 154.
13. D. G. Brady, *J. Appl. Polym. Sci.: Appl. Polym. Symp.*, 1981, **36**, 231.
14. M. Poninski and M. Kryszewski, *Bull. Acad. Pol. Sci., Ser. Sci. Chim.*, 1965, **13** (1), 49.
15. A. D. Macallum, US Pat. 2 513 188 (1950) (*Chem. Abstr.*, 1950, **44**, 8165a); US Pat. 2 538 941 (1951) (*Chem. Abstr.*, 1951, **45**, 5193c).
16. R. W. Lenz and C. E. Handlovits, *J. Polym. Sci.*, 1960, **43**, 167.
17. R. W. Lenz, C. E. Handlovits and H. A. Smith, *J. Polym. Sci.*, 1962, **58**, 351.
18. H. A. Smith and C. E. Handlovits, ASD-TDR-62-372, Report on Conference on High Temperature Polymer and Fluid Research, Dayton, Ohio, 1962, p. 123.
19. H. A. Smith and C. E. Handlovits, ASD-TDR-62-322, Part II, Phenylene Sulfide Polymers, 1962, p. 18.
20. J. R. Campbell, *J. Org. Chem.*, 1964, **29**, 1830.
21. J. T. Edmonds, Jr. and H. W. Hill, Jr. (Phillips Petroleum Co.), US Pat. 3 354 129 (1967) (*Chem. Abstr.*, 1967, **68**, 13 598e).
22. T. W. Johnson (Phillips Petroleum Co.), US Pat. 4 426 500 (1984) (*Chem. Abstr.*, 1984, **100**, 104 565).
23. W. Koch and W. Heitz, *Makromol. Chem.*, 1983, **184**, 779.
24. J. F. Bunnett, *Acc. Chem. Res.*, 1978, **11**, 413.

25. M. Novi, G. Petrillo and M. L. Sartirana, *Tetrahedron Lett.*, 1986, **27**, 6129.
26. R. S. Shue, *Dev. Plast. Technol.*, 1985, **2**, 259.
27. R. M. Black, C. F. List and R. J. Wells, *J. Appl. Chem.*, 1967, **17**, 269.
28. N. S. J. Christopher, J. L. Cotter, G. J. Knight and W. W. Wright, *J. Appl. Polym. Sci.*, 1968, **12**, 863.
29. G. F. L. Ehlers, K. R. Fisch and W. R. Powell, *J. Polym. Sci.*, 1969, **7**, 2955.
30. J. N. Short and H. W. Hill, *Chem. Technol.*, 1972, **2**, 481.
31. D. G. Brady, *J. Appl. Polym. Sci.*, 1976, **20**, 2541.
32. R. T. Hawkins, *Macromolecules*, 1976, **9**, 189.
33. V. A. Sergeyev, V. K. Shitikov and V. I. Nedelkin, *Vysokomol. Soedin.*, 1977, **19**, 396.
34. A. B. Port and R. H. Still, *Polym. Degradation Stab.*, 1980, **2**, 1.
35. J. T. Edmonds, Jr. and H. W. Hill, Jr. (Phillips Petroleum Co.), *US Pat.* 3 524 835 (1970) (*Chem. Abstr.*, 1970, **73**, 121 201g).
36. H. W. Hill, Jr. and D. G. Brady, 'Polymers Containing Sulfur, Poly(Phenylene Sulfide),' in 'Kirk-Othmer Encyclopedia of Chemical Technology', 3rd edn., ed. M. Grayson, Wiley, New York, 1982, vol. 18, p. 793.
37. H. W. Hill, Jr., *Ind. Eng. Chem. Prod. Res. Dev.*, 1979, **18**, 252.
38. C. J. Stacy, *Polym. Prepr., Am. Chem. Soc., Div. Polym. Chem.*, 1985, **26** (1), 180.
39. R. W. Campbell (Phillips Petroleum Co.), *US Pat.* 3 919 177 (1975) (*Chem. Abstr.*, 1975, **83**, 115 380e).
40. G. Kraus and W. M. Whitte, 28th Macromolecular Symposium of the IUPAC, Amherst, MA, July 12, 1982 (*Chem. Abstr.*, 1983, **99**, 123 454c).
41. (a) Y. Iizuka, T. Iwasaki, T. Katto and Z. Shiiki (Kureha Kagaku Kogyo Kabushiki Kaisha), *US Pat.* 4 645 826 (1987) (*Chem. Abstr.*, 1986, **104**, 169 104p); (b) Y. Iizuka, T. Iwasaki, T. Katto and Z. Shiiki (Kureha Kagaku Kogyo Kabushiki Kaisha), *Eur. Pat.* 166 368 (1986) (*Chem. Abstr.*, 1986, **104**, 169 104p).
42. J. T. Edmonds, Jr. (Phillips Petroleum Co.), *US Pat.* 4 071 509 (1978) (*Chem. Abstr.*, 1978, **88**, 122 161r).
43. J. T. Edmonds, Jr. and L. E. Scoggins (Phillips Petroleum Co.), *US Pat.* 4 116 947 (*Chem. Abstr.*, 1979, **90**, 39 424z).
44. (a) K. Idel, D. Freitag and L. Bottenbruch, (Bayer Aktiengesellschaft), *US Pat.* 4 433 138 (1984) (*Chem. Abstr.*, 1982, **98**, 72 943f); (b) E. Ostlinning and K. Idel (Bayer Aktiengesellschaft), *US Pat.* 4 663 430 (1987) (*Chem. Abstr.*, 1984, **102**, 79 463i).
45. K. Idel, E. Ostlinning and D. Freitag (Bayer Aktiengesellschaft), *Ger. Pat.* 3 428 986 (1986) (*Chem. Abstr.*, 1986, **105**, 115 583n).
46. K. Idel, E. Ostlinning, D. Freitag and W. Alewelt (Bayer Aktiengesellschaft), *Ger. Pat.* 3 428 984 (1986) (*Chem. Abstr.*, 1986, **104**, 225 452g).
47. W. Ebert, R. Meyer, K. Idel and R. Schubert (Bayer Aktiengesellschaft), *Ger. Pat.* 3 317 820 (1984) (*Chem. Abstr.*, 1985, **102**, 114 127b).
48. K. Idel, D. Freitag, L. Bottenbruch and J. Merten (Bayer Aktiengesellschaft), *Ger. Pat.* 3 019 732 (1981) (*Chem. Abstr.*, 1981, **96**, 52 936m).
49. R. G. Sinclair, H. B. Benekay and S. Sowell (Idemitsu Petrochemicals Co.), *Jpn. Pat.* 61 145 226 (1986) (*Chem. Abstr.*, 1986, **105**, 173 275h).
50. R. G. Rohlfing (Phillips Petroleum Co.), *US Pat.* 3 717 620 (1973) (*Chem. Abstr.*, 1973, **78**, 112 277n).
51. F. T. Sherk and J. T. Edmonds, Jr. (Phillips Petroleum Co.), *US Pat.* 4 376 196 (1983) (*Chem. Abstr.*, 1982, **97**, 73 384r)
52. S. G. Joshi and S. Radhakrishnan, *Thin Solid Films*, 1986, **142**, 213.
53. J. Agranoff (ed.), 'Modern Plastics Encyclopedia,' McGraw-Hill, New York, 1985–1986 edn, 1985, vol. 62, no. 10A.
54. D. G. Brady and H. W. Hill, *Mod. Plast.*, 1974, **51** (5), 60.
55. H. W. Hill, Jr. and D. G. Brady, *Polym. Eng. Sci.*, 1976, **16**, 831.
56. C. J. Hilado, 'Flammability Handbook for Plastics,' 2nd edn., Technomic Publishing Co., Westport, CT, 1974, p. 60.
57. J. F. Rabolt, T. C. Clarke, K. K. Kanazawa, J. R. Reynolds and G. B. Street, *J. Chem. Soc., Chem. Commun.*, 1980, 347.
58. R. R. Chance, L. W. Shacklette, G. G. Miller, D. M. Ivory, J. M. Sowa, R. L. Elsenbaumer and R. H. Baughman, *J. Chem. Soc., Chem. Commun.*, 1980, 348.
59. H. Shimizu, Y. Tanabe and H. Kanetsuna, *Polym. J.*, 1986, **18**, 367.
60. K. F. Schoch, Jr., *Polym. Prepr., Am. Chem. Soc., Div. Polym. Chem.*, 1984, **25** (2), 278.
61. M. Rubner, P. Cukor, H. Jopson and W. Deits, *J. Electron. Mater.*, 1982, **11**, 261.
62. Y. W. Park, M. A. Druy, C. K. Chiang, A. G. MacDiarmid, A. J. Heeger, H. Shirakawa and S. Ikeda, *J. Polym. Sci., Polym. Lett. Ed.*, 1979, **17**, 195.
63. D. M. Ivory, G. G. Miller, J. M. Sowa, L. W. Shacklette, R. R. Chance and R. H. Baughman, *J. Chem. Phys.*, 1979, **71**, 1506.
64. K. K. Kanazawa, A. F. Diaz, R. H. Geiss, W. D. Gill, J. F. Kwak, J. A. Logan, J. F. Rabolt and G. B. Street, *J. Chem. Soc., Chem. Commun.*, 1979, 854.
65. L. W. Shacklette, R. L. Elsenbaumer, R. R. Chance, H. Eckhardt, J. E. Frommer and R. H. Baughman, *J. Chem. Phys.*, 1981, **75**, 1919.
66. B. J. Tabor, E. P. Magré and J. Boon, *Eur. Polym. J.*, 1971, **7**, 1127.
67. T. P. H. Jones, G. R. Mitchell and A. H. Windle, *Colloid Polym. Sci.*, 1983, **261**, 110.
68. N. S. Murthy, R. L. Elsenbaumer, J. E. Frommer and R. H. Baughman, *Synth. Met.*, 1984, **9** (1), 91.
69. J. E. Frommer, R. L. Elsenbaumer and R. R. Chance, *Org. Coat. Appl. Polym. Sci. Proc.*, 1983, **48**, 552.
70. C. J. Stacy, *J. Appl. Polym. Sci.*, 1986, **32**, 3959.
71. A. Kinugawa, *Kobunshi Ronbunshu*, 1987, **44** (2), 139.
72. P. J. Flory, 'Principles of Polymer Chemistry', Cornell University Press, Ithaca, NY, 1953, p. 325.
73. R. W. Campbell and L. E. Scoggins (Phillips Petroleum Co.), *US Pat.* 3 869 434 (1975) (*Chem. Abstr.*, 1975, **83**, 11 382r).
74. R. W. Campbell (Phillips Petroleum Co.), *US Pat.* 3 966 688 (1976) (*Chem. Abstr.*, 1976, **85**, 94 921f).
75. S. Tsunawaki and C. C. Price, *J. Polym. Sci., Part A*, 1964, **2**, 1511.
76. A. B. Port and R. H. Still, *J. Appl. Polym. Sci.*, 1979, **24**, 1145.
77. G. Montaudo, G. Bruno, P. Maravigna and F. Bottino, *J. Polym. Sci., Polym. Chem. Ed.*, 1974, **12**, 2881.
78. Z. Binenfeld and A. F. Damanski, *Bull. Soc. Chim. Fr.*, 1961, 679.
79. B. Hortling, M. Söder and J. J. Lindberg, *Angew. Makromol. Chem.*, 1982, **107**, 163.
80. T. Fujisawa and M. Kakutani, *J. Polym. Sci., Polym. Lett. Ed.*, 1970, **8**, 19.
81. D. Mukherjee and P. Pramanik, *Indian J. Chem., Sect. A*, 1982, **21** (5), 501.
82. R. W. Campbell (Phillips Petroleum Co.), *US Pat.* 4 016 145 (1977) (*Chem. Abstr.*, 1977, **86**, 190 850).

33
Polysulfones

FABRIZIO PARODI

EniChem SpA, Milano, Italy

33.1 INTRODUCTION	561
33.2 A SURVEY OF SYNTHETIC ROUTES TO POLYSULFONES	562
33.3 POLYCONDENSATION *VIA* SULFONYL LINKAGE (POLYSULFONYLATION)	564
33.3.1 Polycondensation of Arenesulfonyl Halides	564
33.3.1.1 Friedel–Crafts sulfonylation and polysulfonylation	564
33.3.1.2 Polysulfonylation by high catalyst-concentration	565
33.3.1.3 Polysulfonylation by low catalyst-concentration	568
33.3.2 Polycondensation of Arenesulfonic Acids	569
33.3.3 Polycondensation of Arenesulfinates and Related Synthetic Routes	571
33.4 POLYCONDENSATION OF MONOMERS BEARING SULFONYL GROUPS	572
33.4.1 Poly(arylene ether sulfone)s by Polyetherification	572
33.4.1.1 Synthesis of diaryl ethers	572
33.4.1.2 Polyetherification	573
33.4.1.3 Synthesis of bis(4-chlorophenyl) sulfone–bisphenol A poly(ether sulfone)	576
33.4.1.4 Mechanism and kinetics of polyetherification	577
33.4.1.5 Self-polycondensation of (halophenylsulfonyl)phenoxides	580
33.4.1.6 Side reactions	581
33.4.1.7 Choice of solvents	584
33.4.1.8 Choice of monomers	584
33.4.1.9 Alternative and miscellaneous methods of polyetherification	584
33.4.2 Aliphatic–Aromatic Poly(ether sulfone)s	585
33.4.3 Miscellaneous Polysulfones by Various Step-polymerization Reactions	585
33.4.3.1 Poly(arylene sulfide sulfone)s and poly(arylene ether sulfide sulfone)s	585
33.4.3.2 Poly(amide sulfone)s	586
33.4.3.3 Poly(arylene ester sulfone)s	587
33.4.3.4 Poly(arylene ether ketone sulfone)s	587
33.4.3.5 Polymers from divinyl sulfones	588
33.5 REFERENCES	588

33.1 INTRODUCTION

During the last 15 years, aromatic polysulfones have achieved a remarkable position among other thermoplastic polymers by virtue of their excellent properties, such as thermal stability, high distortion-temperature, chemical inertness, electrical performance and flame retardancy.

The products currently known as aromatic polysulfones [poly(sulfonylarylene)s], and manufactured world-wide in large amounts, are amorphous polymers of the general formula (**1**). Most of them belong to the sub-family of poly(arylene ether sulfone)s [**2**; poly(oxyarylenesulfonylarylene)s], or are based on such a structural unit. At present, many grades of these polymers are marketed under different trade names, *e.g.* 'Victrex' (powders/granules) and 'Stabar S' (films) by ICI; 'Udel' and 'Radel' (powders/granules) (previous trademarks of Union Carbide) by Amoco Chem.; 'Ultrason-E' by BASF; 'Talpa 1000' films by Mitsui Toatsu Chem. Ind.; and 'Sumilite FS-1300' (films) by Sumitomo Chem. Ind.

$$\mathrm{+(ArSO_2)_{\mathit{n}}} \qquad \mathrm{+(Ar'SO_2Ar''O)_{\mathit{n}}}$$

$$(\mathbf{1}) \qquad\qquad (\mathbf{2})$$

The properties, applications, processing and preparation of aromatic polysulfones, and especially poly(arylene ether sulfone)s, have been extensively reviewed,[1-5] particularly where commercial products are concerned.

From a general point of view, two distinct sub-families of polysulfones can be identified. The most important one has the sulfonyl group, $-SO_2-$, as a linking unit of the main chain and has the general formula (3); this obviously includes the aromatic polysulfones (1) and (2). The other sub-family, with the sulfonyl as a side group, includes those polymers with the general formulae (4) and (5). Y and Z represent aliphatic, cycloaliphatic, aromatic, heterocyclic or mixed residues, with or without additional functions as main-chain elements or side groups.

$$-(YSO_2)_n- \qquad -(Y)_n- \atop ZSO_2 \qquad -(Y)_n- \atop SO_2$$

(3) (4) (5)

Polymers containing the sulfonyl unit together with different functional groups (*e.g.* ether, sulfide, amide and ester) combine the peculiar features of each moiety in the same macromolecular structure. Such polymers belong to different families at the same time (polysulfones, polyethers, polysulfides, *etc.*). They can be treated more properly as polysulfones when the $-SO_2-$ group plays a 'predominant' role in the polymer structure by its intrinsic features[6] (electron-withdrawing character, high polarity contribution) and/or its concentration, thus strongly affecting the polymer end-properties.

In addition to the fundamental characteristics of the hydrocarbon skeleton of the macromolecule, in terms of segmental mobility, further linking functions, other than sulfonyl, can affect both polymer-backbone flexibility and intermolecular interactions, and hence the range of viscoelastic, physico-mechanical and rheological properties. In this respect, polysulfones with suitably balanced toughness, strength, melt viscosity and glass transition temperature T_g may be precisely tailored.

Because of the high 'stiffening' effect of the $-SO_2-$ unit, the polymers richest in sulfonyl groups may give materials ranging from the scarcely mouldable up to the intractable (although intrinsically thermoplastic); thus, the presence of 'softening' groups in the repeat unit is often desired. For instance, the introduction of ether linkages is used to lower both T_g and the melt viscosity of poly(arylene sulfone)s (1), resulting in poly(arylene ether sulfone)s (2), whose thermoplasticity and processability are considerably enhanced.

Moreover, chemical groups additional to the sulfonyl function can be used *per se* as linking units; their formation from suitable monomers can be the synthetic route of choice, as in the case of polyetherification commonly adopted to prepare poly(ether sulfone)s.

33.2 A SURVEY OF SYNTHETIC ROUTES TO POLYSULFONES

Rarely have so many different routes been devised to obtain polymers of a single family as in the case of polysulfones. Dealing with their synthesis implies investigation of chain polymerization, polymer oxidation and, above all, step polymerization; this latter will be treated in detail in this chapter.

Polysulfones of type (3) can be synthesized by the well-reviewed[7,8] free-radical copolymerization of unsaturated compounds, mainly monoalkenes or dienes, with sulfur dioxide, following the general scheme of equation (1). Typical chain-reactions are also the homopolymerization of vinyl- or unsaturated cyclic-sulfones, as well as their copolymerization with various vinyl monomers; polysulfones of general formula (4), as in equation (2), or (5) are thus afforded.[9]

$$n \, \mathrm{C{=}C} + n\mathrm{SO_2} \longrightarrow -(\mathrm{C{-}CSO_2})_n- \qquad (1)$$

$$n \, \mathrm{C{=}C} \atop ZSO_2 \longrightarrow -(\mathrm{C{-}C})_n- \atop ZSO_2 \qquad (2)$$

As shown by equation (3), polysulfones can also be obtained by oxidizing polysulfides or polysulfoxides with oxidants, such as H_2O_2 in $MeCO_2H$ or *t*-butyl hydroperoxide.[10-14,148] This rather unusual route, mostly applied to aromatic polymers, can provide variously mixed sulfide–sulfoxide–sulfone polymers, since reaction (3) can be stopped at any desired stage.

$$\text{\textendash(YS\textendash)}_n \xrightarrow{\text{oxidant}} \text{\textendash(YSO\textendash)}_n \xrightarrow{\text{oxidant}} \text{\textendash(YSO}_2\text{\textendash)}_n \qquad (3)$$

The above-mentioned methods, although interesting and efficient, should be regarded as complementary to step-polymerization routes, mainly by polycondensation processes. Due to their industrial relevance, thermoplastic aromatic polysulfones (and polycondensation reactions leading to them) have been investigated in detail and many alternative synthetic pathways have been devised.

Polycondensations affording polysulfones may be of two general types: polycondensations *via* sulfonyl-linkage formation (polysulfonylations) and polycondensations of reactants containing sulfonyl groups.

Three different routes, mostly used to prepare wholly aromatic polysulfones, are possible for polysulfonylation: Friedel–Crafts polycondensation (equation 4) of arenedisulfonyl halides with compounds bearing at least two aromatic hydrogens; polycondensation (equation 5) of arenedisulfonic acids with compounds having at least two aromatic hydrogens; polycondensation (equation 6) of alkali metal arenedisulfinates with dihalides (active towards nucleophilic displacement of halogen). Similar results can also be achieved by self-polycondensation of arenesulfonyl halides (equation 7) or arenesulfonic acids (equation 8).

$$n\text{XO}_2\text{SArSO}_2\text{X} + n\text{HAr'H} \xrightarrow{-2n\text{HX}} \text{\textendash(ArSO}_2\text{Ar'SO}_2\text{\textendash)}_n \qquad (4)$$

$$n\text{HO}_3\text{SArSO}_3\text{H} + n\text{HAr'H} \xrightarrow{-2n\text{H}_2\text{O}} \text{\textendash(ArSO}_2\text{Ar'SO}_2\text{\textendash)}_n \qquad (5)$$

$$n\text{MO}_2\text{SArSO}_2\text{M} + n\text{XYX} \xrightarrow{-2n\text{MX}} \text{\textendash(ArSO}_2\text{ArYSO}_2\text{\textendash)}_n \qquad (6)$$

$$n\text{HArSO}_2\text{X} \xrightarrow{-n\text{HX}} \text{\textendash(ArSO}_2\text{\textendash)}_n \qquad (7)$$

$$n\text{HArSO}_3\text{H} \xrightarrow{-n\text{H}_2\text{O}} \text{\textendash(ArSO}_2\text{\textendash)}_n \qquad (8)$$

Pairs of reactants, one or both of them bearing a sulfonyl group, can be polycondensed by suitable chemical groups A and B, able to react with each other, giving a linking group E plus condensation by-products. A and B must be present on different coreactants (AABB-type polycondensation, equation 9) or as a couple AB on the same monomer (self-polycondensation, equation 10).

$$n\text{AYSO}_2\text{Y'A} + n\text{BZB} \longrightarrow \text{\textendash(YSO}_2\text{Y'EZE\textendash)}_n + \text{by-products} \qquad (9)$$

$$n\text{AYSO}_2\text{Y'B} \longrightarrow \text{\textendash(YSO}_2\text{Y'E\textendash)}_n + \text{by-products} \qquad (10)$$

A wide series of condensation reactions are applicable, such as, for instance, alkali metal phenoxides or thiophenoxides with aryl or alkyl halides, acyl halides or aryl haloformates with amino or hydroxy groups, thus leading to ether, sulfide, amide, urethane, ester or carbonate linkages, respectively.

Accordingly, suitable polycondensations have the features of polyetherification, polythioetherification, polyamidation, *etc.*, and afford hybrid polymers such as poly(ether sulfone)s, poly(sulfide sulfone)s, poly(amide sulfone)s, and so on.

Polysulfonylation can give the highest densities of sulfonyl linkages and can lead to 'true' polysulfones; these polymers have a backbone free from any extra function not brought by the reactants themselves.

However, additional linkages are quite often desired, for different reasons, and can be provided by the same reactants selected for polysulfonylation or built in by alternative polycondensation of reactants containing sulfonyl groups. Although several polymers are attainable by only one of the two processes, a large number can be prepared by both methods. Polysulfonylation and polyetherification may be interchangeable, *e.g.* in the synthesis of certain poly(arylene ether sulfone)s, as shown in Scheme 1.

Finally, a brief mention must be made of alternative step-polymerization routes, the most notable one being the addition of divinyl sulfones to compounds containing acidic hydrogens.

$$nXO_2SArOAr'SO_2X + nHAr''OAr'''H \qquad nXAr'''SO_2ArX + nMOAr'SO_2Ar''OM$$

$$\searrow_{-2nHX} \qquad \swarrow_{-2nMX}$$

$$-\!\!\!+\!\!ArOAr'SO_2Ar''OAr'''SO_2\!\!+\!\!\!_n-$$

Scheme 1

Besides general summaries of polymerization processes giving polysulfones,[9,15] several reviews have been devoted to the industrially most relevant polycondensations leading to aromatic polysulfones[1,3,18] and, more specifically, to poly(arylene ether sulfone)s.[2,16,17]

33.3 POLYCONDENSATION *VIA* SULFONYL LINKAGE (POLYSULFONYLATION)

Sulfones can be prepared by a variety of extensively reviewed methods.[19-22]

The traditional routes to diaryl sulfones by a Friedel–Crafts reaction (equation 11) from arenesulfonyl halides and aromatic hydrocarbons, by condensation (equation 12) of arenesulfonic acids with aromatic hydrocarbons and by condensation (equation 13) of alkali metal arenesulfinates with activated aryl halides, represent the fundamental pathways suitable for synthesis of poly(arylene sulfone)s through sulfonyl-linkage formation.

$$ArSO_2X + Ar'H \xrightarrow[\text{catalyst}]{\text{Friedel-Crafts}} ArSO_2Ar' + HX\uparrow \qquad (11)$$

$$ArSO_3H + Ar'H \xrightarrow[\text{or } \Delta]{\text{drying agent}} ArSO_2Ar' + H_2O \qquad (12)$$

$$ArSO_2M + Ar'X \longrightarrow ArSO_2Ar' + MX \qquad (13)$$

33.3.1 Polycondensation of Arenesulfonyl Halides

33.3.1.1 *Friedel–Crafts sulfonylation and polysulfonylation*

According to the mechanism and activation/orientation criteria valid in general for electrophilic aromatic substitutions,[23,24] the Friedel–Crafts sulfonylation (as well as the acylation) of aromatic compounds with sulfonyl (or acyl) halides[25,26] differs from the corresponding alkylation in some respects.

Electron-releasing alkyl groups activate the aromatic nucleus towards further electrophilic attack of the alkylating agent, whereas electron-withdrawing sulfonyl and acyl groups tend to hinder further electrophilic attack of the acid halide by an opposite effect. Multiple sulfonylations and acylations are, therefore, almost suppressed by the first-linked sulfonyl or acyl residue, unless drastic reaction-conditions or activated aromatic substrates are involved.

Moreover, both sulfonylation and acylation traditionally require far larger amounts of Friedel–Crafts catalyst (at least one mole of metal halide per mole of acid halide) than alkylation does (only a small fraction per mole of alkylating agent).

For the sulfonylation (equation 11) by arenesulfonyl halides,[27,28a] $AlCl_3$, $FeCl_3$, $SbCl_5$, $AlBr_3$ and BF_3 were proved to be efficient catalysts,[29] with $FeCl_3$ claimed as the best among them.[30]

The sulfonylation mechanism involves a two-stage reaction, as shown in equation (14): attack of sulfonylium cation $ArSO_2^+$ on the aromatic nucleus with formation of an intermediate complex and its successive split. It is assumed that the effective sulfonylating agent is the sulfonylium salt generated by action of the Lewis-acid catalyst on the sulfonyl halide.[28b]

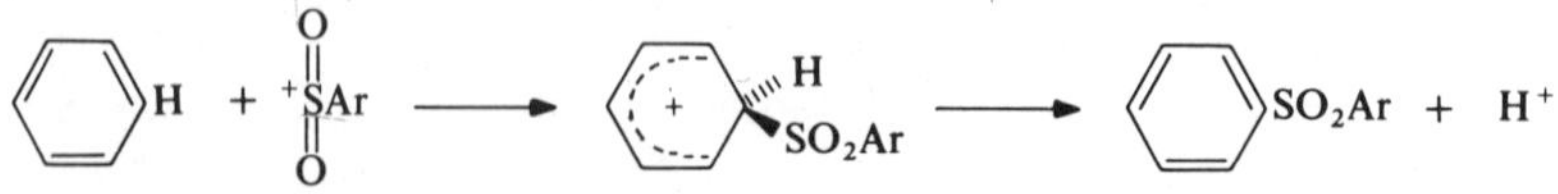

$$\text{Ph-H} + {}^+SO_2Ar \longrightarrow [\text{complex}] \longrightarrow \text{Ph-}SO_2Ar + H^+ \qquad (14)$$

Although traditionally carried out by high catalyst-concentrations, sulfonylation can be better accomplished at elevated temperatures and with small concentrations (catalytic amounts) of $FeCl_3$.[31] For example, arenesulfonyl chlorides react smoothly in the molten state at 120–140 °C with aromatics in the presence of 1–5 mol % of $FeCl_3$ (or $SbCl_5$, $InCl_3$, iron(II) or iron(III) acetylacetonate (acetylacetone is 2,4-pentanedione) or $BiCl_3$), giving, in a few hours, the corresponding sulfones in high yields.[32,33]

Described in preceding reviews,[34,35] polysulfonylation routes to poly(arylene sulfone)s, which obey such general criteria, follow two different reaction routes depicted by equations (15) and (16). They are equivalent in principle, and lead to the same polymer when the substituted or unsubstituted arylene groups, Ar and Ar′, are equal. X is a halogen (most commonly chlorine) and MX_m a Friedel–Crafts metal-halide catalyst. Either the two-component polycondensation reaction (equation 15) of disulfonyl halides with compounds containing at least two aromatic hydrogens, or the self-polycondensation (equation 16) of monosulfonyl halides having at least one aromatic hydrogen, can be successfully adopted by using a wide selection of reactants under well-established experimental conditions.

$$n\text{XO}_2\text{SArSO}_2\text{X} + n\text{HAr}'\text{H} \xrightarrow{MX_m} (\text{ArSO}_2\text{Ar}'\text{SO}_2)_n + 2n\text{HX}\uparrow \qquad (15)$$

$$2n\text{HArSO}_2\text{X} \xrightarrow{MX_m} (\text{ArSO}_2\text{ArSO}_2)_n + 2n\text{HX}\uparrow \qquad (16)$$

Since more than one aromatic hydrogen (variously prone to electrophilic substitution) is commonly present in the reacting molecules, different repeating units can arise from the same formal polycondensation.

Structural irregularities due to possible sulfonylation at various ring-positions, disulfonylations on one aromatic nucleus (which can cause branching or even crosslinking), as well as side reactions directly promoted by the Friedel–Crafts catalyst, have been reported.

33.3.1.2 Polysulfonylation by high catalyst-concentration

By adopting high amounts of Friedel–Crafts catalyst in inert high-boiling solvents, the self-polycondensation of aromatic monosulfonyl chlorides (many hours at 140–225 °C in dimethyl sulfone, in the presence of $FeCl_3$) has been patented.[36] Using one mole (or a slight excess) of $AlCl_3$ or $FeCl_3$ per g equivalent of $-SO_2Cl$ in nitrobenzene solution, both the polycondensation (Scheme 2, path a) of diphenyl ether with 4,4′-oxydibenzenesulfonyl chloride[37] and the self-polycondensation (Scheme 2, path b) of 4-phenoxybenzenesulfonyl chloride[38] have been reported to produce soluble poly(diphenylene ether sulfone) [6; poly(sulfonylphenyleneoxyphenylene)]. Deviations from the theoretical stoichiometry were nevertheless pointed out. In both cases, more than the stoichiometric amount of HCl is evolved (1.2 to 1.5 times more), probably because of the attack of anhydrous metal halides on the aromatic nuclei, a known reaction for $FeCl_3$[39] and $AlCl_3$.[40] The polysulfonylation (Scheme 2, path a) has indeed been reported to yield polymers containing chemically linked aluminum and chlorine atoms, and it has been proposed that, in addition to the standard repeating units (6), units such as (7) or (8) could be generated (equation 17).[37]

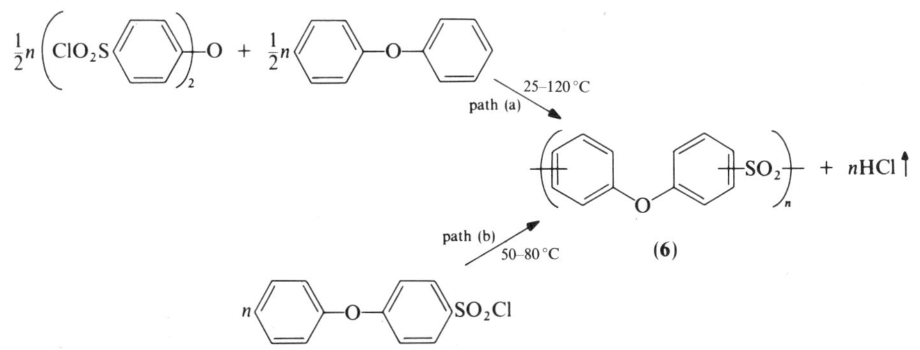

Scheme 2

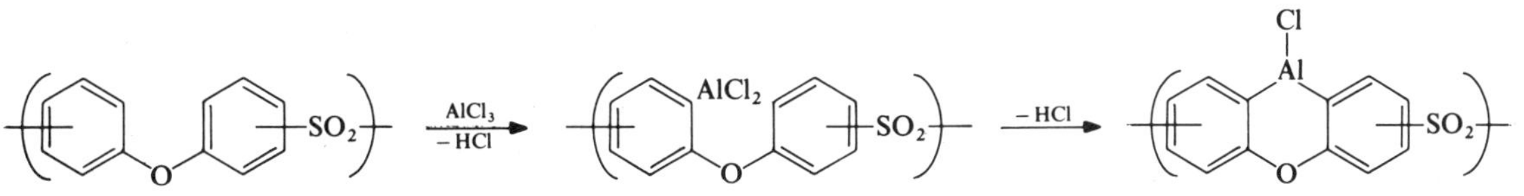

Table 1 Polycondensation of Aromatics with Disulfonyl Chlorides and Self-polycondensation of Monosulfonyl Chlorides, in Bulk or in Solution

Reactant(s)	Solvent	Reaction conditions $FeCl_3$ (wt %)	Final temperature (°C)	Time (h)	Yield (%)	Polymer insoluble product (%)	RV_1^a	Ref.
ClO_2S–Ph + Ph	—	1.2	300	4	90	25	0.09	33
ClO_2S–Ph + Ph–O–Ph	—	0.6	320	4	92	30	0.11	33
(ClO$_2$S–Ph)$_2$ + Ph–Ph	—	0.7	230	4	92	10	0.15	33
(ClO$_2$S–Ph)$_2$ + Ph–O–Ph	—	≤0.3	150	4	>85	<20	≤0.6	33
(ClO$_2$S–Ph)$_2$ + Ph–O–Ph	PhNO$_2$	0.08	130	3.5		0	0.7	47
(ClO$_2$S–Ph)$_2$ + Ph–O–Ph	PhNO$_2$	0.14	120	22	~100	0	1.07	42
(ClO$_2$S–Ph–O–Ph–SO$_2$)$_2$	—	≤0.3	150	4	>85	<20	≤0.4	33

Structure	Solvent		Temp	Time	Yield (%)		Reduced viscosity[a]	Ref
⌬–⌬–SO₂Cl	—	1.3	250	3.5	95	100	—	41
⌬–O–⌬–SO₂Cl	—	≤4	260	3.5	>90	<10	≤2.0	41
⌬–O–⌬–SO₂Cl	PhNO₂	0.14	120	22	~100	0	0.81	42
⌬–S–⌬–SO₂Cl	—	1.2	240	3.5	>90	<10	0.56	41
naphthyl–SO₂Cl	—	≤40	≤290	3.5	90	0	<0.1	41

[a] Reduced viscosity $= (\eta_{\text{solution}} - \eta_{\text{solvent}})/\eta_{\text{solvent}}$ for a 1% solution of polymer in DMF at 25 °C.

33.3.1.3 Polysulfonylation by low catalyst-concentration

Far more relevant, and, to a large extent, free from side reactions such as those mentioned above, is the polysulfonylation by low concentrations of Friedel–Crafts catalyst (almost invariably $FeCl_3$, 0.1–4 wt %), performed in bulk or in solution, *e.g.* in nitrobenzene, dimethyl sulfone, or chlorinated biphenyls.

General criteria for this method of polysulfonylation have been drawn from different bulk polycondensations of aromatics with disulfonyl chlorides[33] and bulk self-polycondensations of aromatic monosulfonyl chlorides,[41] as well as from detailed studies on the reaction mechanism and the polymers afforded by the polycondensations of Scheme 2 in nitrobenzene solution in the presence of $FeCl_3$.[38,42]

Electron-releasing groups strongly activate (by conjugation sulfonylation at positions *ortho* or *para* to themselves; a moderate inductive effect additionally activates any position of the aromatic nucleus. *Para* sulfonylation is most favoured, due to steric effects hindering *ortho* substitution. An opposite effect is exerted by deactivating (and *meta* orienting) electron-withdrawing groups, including $-SO_2Cl$ and $-SO_2Ar$ themselves.

Highly reactive monomers and elevated reaction-temperatures (as well as high catalyst-amounts) promote high-polymer formation and high conversion, but at the same time also promote sulfonylation in less-favoured ring positions or even multiple sulfonylations. Thus, structurally irregular polysulfones can arise and multiple sulfonylations on the same aromatic nucleus can, in turn, cause polymer branching and crosslinking, up to completely insoluble products.

Therefore, polysulfones with higher molecular weights and yields, as well as more regular structures with lower formation of insoluble products, derive from a proper balance between the structure and reactivity of the monomers on one hand, and polymerization conditions on the other.

Reaction conditions and yields for several bulk and solution polysulfonylations are collected in Table 1.

Polycondensation of dichlorosulfonyl compounds with aromatics (equation 15) and self-polycondensation of monosulfonyl chlorides (equation 16) behave quite differently one from the other, even though designed to give the same polymer. Self-polysulfonylation is, by definition, free from preferential vaporization losses of more volatile reactants, and generally provides more reproducible results as well as higher molecular weights and yields. This arises from a better control of the desired equimolar ratio of coreacting functions (the sulfonyl chloride and the aromatic hydrogen), an intrinsic feature of the monosulfonyl chloride employed, which depends only on its own purity. On the other hand, the electron-withdrawing sulfonyl chloride group makes the aromatic hydrogens of the monosulfonyl chlorides less reactive, so that self-polysulfonylation normally requires, even at the beginning, more drastic reaction-conditions, or is successful only by using activated monosulfonyl chlorides (in most cases, activated by ether linkages).

The two-component polycondensation (equation 15) actually occurs by two different condensation reactions: sulfonylation (equation 18) of the very reactive aromatic compound $Ar'H_2$ (the prevalent reaction at the beginning of the process) and sulfonylation (equation 19) of the same compound deactivated by monosulfonylation, which becomes progressively more important as the reaction proceeds. From the early stages of the process to the subsequent ones, there is a continuously changing tendency to undergo sulfonylation among the different ring positions, which leads to the formation of structurally irregular or branched polymers.

$$\sim\sim SO_2ArSO_2Cl\ +\ HAr'H\ \xrightarrow{-HCl}\ \sim\sim SO_2ArSO_2Ar'H \qquad (18)$$

$$\sim\sim SO_2ArSO_2Cl\ +\ HAr'SO_2\sim\sim\ \xrightarrow{-HCl}\ \sim\sim SO_2ArSO_2Ar'SO_2\sim\sim \qquad (19)$$

Self-polysulfonylation, by contrast, is always influenced at any stage by the deactivating effect of $-SO_2Cl$ or $-SO_2Ar$ groups. Sulfonylation occurs on one preferential ring-position of the monosulfonyl chloride (equation 20) or of the growing-chain ends (equation 21) and more regular polysulfones are thus produced. For example, the reactions in Scheme 2, carried out in nitrobenzene solution, have displayed approximate ratios of *para* to *ortho* to *ortho,para* linking (structures **9, 10** and **11**, respectively, Scheme 3) or 90:10:1.6 for the two-component polymerization and 99: $\approx$ 0:0.7 for the self-polycondensation, respectively.[42]

$$\sim\sim ArSO_2Cl\ +\ HArSO_2Cl\ \xrightarrow{-HCl}\ \sim\sim ArSO_2ArSO_2Cl \qquad (20)$$

$$\sim\sim ArSO_2Cl\ +\ HArSO_2Ar\sim\sim\ \xrightarrow{-HCl}\ \sim\sim ArSO_2ArSO_2Ar\sim\sim \qquad (21)$$

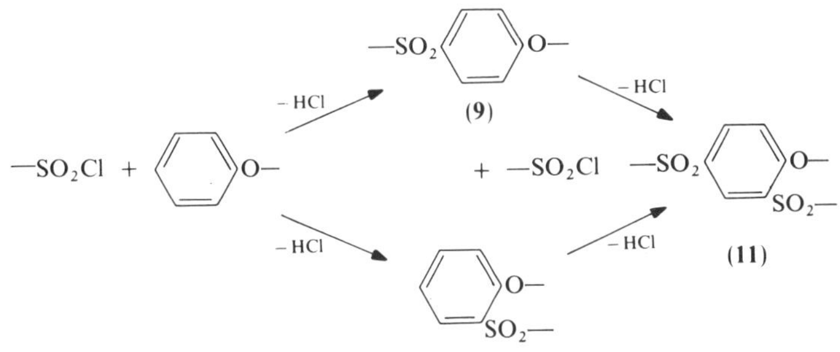

Polysulfones

Scheme 3

In Scheme 2, path (a), the activating and *ortho,para*-orienting ether group mainly promotes *para* but also, significantly, *ortho* sulfonylation of the highly reactive diphenyl ether in the initial stages of polymerization. In the subsequent stages of this reaction (and during the overall course of the reaction in Schemes 2, path b) the activating effect of the ether group is greatly reduced, by conjugation, by —SO_2Cl or —SO_2Ar present on one adjacent ring (**12**); the least-hindered *para* sulfonylation is thus largely preferred, and an almost perfectly linear polymer is obtained.

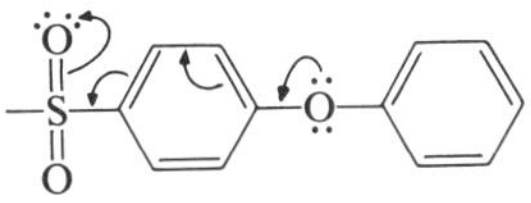

(12)

Ortho sulfonylation and *ortho,para* disulfonylation represent structural irregularities. *Ortho* units adversely affect the toughness of otherwise straight-chain poly(*p*-diphenylene ether sulfone) [poly(oxy-1,4-phenylenesulfonyl-1,4-phenylene],[16,119] whilst *ortho,para* branching can result in crosslinking, unless mild solution-polymerization conditions are adopted.

Bulk polysulfonylation is initially performed in the molten state and the resulting solid low-polymer is ground; further polymerization follows with powder sintering. The reaction is carried out in 3–20 h in the presence of 0.6–4 wt % of $FeCl_3$, with final temperatures from 150 to 320 °C, under a slow stream of dry nitrogen and final vacuum to help HCl elimination[33,41] (as also reported in several patents[43–45]).

Since sulfonyl chlorides generally decompose above 250 °C by a radical mechanism,[46] final temperatures of 230–250 °C represent the best choice to get high molecular weights and moderate crosslinking.

Whilst bulk polymerization invariably leads to some (or even total) insoluble products, solution polysulfonylation[38,42,44,47] (under rigorously anhydrous conditions, with $FeCl_3$ amounts as low as 0.05 wt % and at an optimum temperature of 120–140 °C) gives soluble high polymers smoothly in almost quantitative yields with stoichiometric HCl evolution. Multiple sulfonylations are indeed largely suppressed by the lower process-temperature as compared to that necessary for completion of the reaction in bulk polymerizations.

All synthesized polysulfones must be freed from the metal-halide catalyst, unreacted intermediates and high-boiling solvents by repeated dissolution and reprecipitation; residual catalyst can be thoroughly eliminated by washing with hot aqueous HCl or chelating-agent solutions.[43]

Bulk polymerizations at low temperatures (−20 to +5 °C), by using $BF_3 \cdot HF$ as the catalyst, were developed;[48] the polycondensation between aromatics and SO_2Cl_2 was also claimed.[48]

A complete process has been patented[49] that starts from disulfonic acids, which are converted into sulfonyl halides by phosphoryl halides and then polycondensed *in situ* with aromatics by a standard Friedel–Crafts procedure.

33.3.2 Polycondensation of Arenesulfonic Acids

It is well known that sulfonation and halosulfonation of aromatic compounds by sulfur trioxide or oleums (disulfuric acids) and halosulfonic acids, respectively, is normally accompanied by sulfone formation, sometimes in large amounts.[50] Diaryl sulfones are formally produced by water elimination between aromatic hydrogens and sulfonic acid groups generated by sulfonation or as the first step of halosulfonation. Following the general criteria of electrophilic aromatic substitution, the old scheme of direct attack of the sulfonic acid on the aromatic nucleus can be assumed, where the electrophilic agent is the sulfonylium ion $ArSO_2^+$; mechanisms involving pyrosulfonic acid ($ArSO_2OSO_3H$) or sulfonic anhydride ($ArSO_2OSO_2Ar$) intermediates are, however, possible.[50,51]

Sulfone formation is promoted by drastic reaction-conditions, by low molar ratios of sulfonating agent to aromatic compound, by dehydrating media and on aromatics activated by electron-releasing substituents.

Selective sulfonating agents (SO_3-complexes, acyl sulfates, *etc.*), mild reaction-conditions and empirical 'sulfone inhibitors' (such as carboxylic acids and anhydrides, Na_2SO_4, $CHCl_3$, ketones and clay) have been devised in order to hinder sulfone formation.[50]

However, the condensation of arenesulfonic acids with aromatics, exemplified by equation (12), can become, if suitably promoted, a procedure of choice for producing diaryl sulfones instead of a side reaction. Thus, they can be obtained in high yields, under rather mild conditions (60–70 °C, 4 h), in $(CF_3CO)_2O$ (trifluoroacetic anhydride)[52] or, under slightly more drastic conditions, in polyphosphoric acid (PPA)[53] (80 °C, 8h) or, in much higher yields, in PPA/P_2O_5 mixtures.[54] Either strongly acidic or dehydrating media are beneficial to this sort of sulfonylation, and both favour the two following different mechanisms:[37] *via* mixed sulfonic phosphoric[53,55] or sulfonic carboxylic[52,56] anhydride formation (and attack by sulfonylium ions $ArSO_2^+$), or by direct attack of sulfonylium ions. Arenes can be efficiently converted into diaryl sulfones by one-step treatment with $(CF_3CO)_2O/H_2SO_4$ 2:1 molar mixture,[57] the formed bis(trifluoroacetyl) sulfate being the active agent.

As already seen for polysulfonylation with sulfonyl halides, either polycondensation (equation 5) of arenedisulfonic acids with aromatic compounds or self-polycondensation (equation 8) of arenesulfonic acids are effective routes to poly(arylene sulfone)s.

Polymers with an inherent viscosity of up to 0.52 and yields ranging between 74 and 88% have been obtained, for instance, by reaction (22) carried out in PPA.[37] The sulfonation of the aromatic compound can be accomplished by a stoichiometric amount of H_2SO_4, with simple azeotropic distillation of the resulting water, and can be directly followed by polycondensation, without any isolation and purification of the sulfonic acid intermediate.[37]

$$\tfrac{1}{2}n\left(HO_3S\text{-}\langle\rangle\text{-}\right)_2 O + \tfrac{1}{2}n\,\langle\rangle\text{-}O\text{-}\langle\rangle \xrightarrow[240\,°C,\,4\text{-}8\,h]{PPA} \left(\text{-}\langle\rangle\text{-}\langle\rangle\text{-}O\text{-}\langle\rangle\text{-}SO_2\text{-}\right)_n + nH_2O \quad (22)$$

Soluble and high molecular weight polysulfones were indeed claimed[58] to have been obtained by polycondensation of diphenyl ether or diphenyl sulfide with the corresponding 4,4'-disulfonic acids; such disulfonic acids are obtained previously by sulfonation, azeotropically distilling the resulting water with CCl_4, and then polycondensed *in situ*. After refluxing the monomer mixture for many hours at 125–145 °C in *o*-dichlorobenzene/CCl_4, the resulting oligomers (freed from solvents) are further polymerized in PPA or benzoic anhydride, as shown in equation (22).

Under milder reaction-conditions, high-yield polysulfonylations of both types equation (5) and equation (8) (sodium salts may be used in place of the free sulfonic acids) have been claimed in liquid mixtures of alkanesulfonic acids and P_2O_5 as reaction media.[59]

The polycondensation (equation 23) in a 10:1 molar mixture of methanesulfonic acid and P_2O_5 affords 99% of poly(*p*-diphenylene ether sulfone) of intrinsic viscosity 0.46.

$$n\,\langle\rangle\text{-}O\text{-}\langle\rangle\text{-}SO_3Na \xrightarrow[120\,°C,\,12\,h]{MeSO_3H/P_2O_5} \left(\text{-}\langle\rangle\text{-}O\text{-}\langle\rangle\text{-}SO_2\text{-}\right)_n + Na\ salts \quad (23)$$

Simultaneous sulfonation and polysulfonylation can be accomplished under mild conditions by a one-step process in fluoroalkanesulfonic acids;[60] for instance, diphenyl ether is reacted with an equimolecular amount of H_2SO_4 in trifluoromethanesulfonic acid, as in equation (24).

$$n\,\langle\rangle\text{-}O\text{-}\langle\rangle + nH_2SO_4 \xrightarrow[100\,°C,\,3.5\,h]{CF_3SO_3H} \left(\text{-}\langle\rangle\text{-}O\text{-}\langle\rangle\text{-}SO_2\text{-}\right)_n + 2nH_2O \quad (24)$$

33.3.3 Polycondensation of Arenesulfinates and Related Synthetic Routes

As shown in equation (13), sodium arenesulfinates can be condensed either with alkyl halides, giving alkyl aryl sulfones by nucleophilic substitution, or with reactive aryl halides (activated by electron-withdrawing substituents *ortho* or *para* to the halogen on the aromatic nucleus), affording diaryl sulfones by nucleophilic aromatic substitution.

Because of the double character of the sulfinate anion as a nucleophile (equation 25), *O*-alkylation (-arylation) largely competes with *S*-alkylation (-arylation); the former leads to sulfinate esters (due to sulfinyloxy —SO—O— group formation), the latter produces sulfones,[21] often in much lower yields.[61]

$$\text{(25)}$$

As is generally true in nucleophilic substitutions, slow reaction kinetics and side reactions (in this case O-alkylation or -arylation, halide hydrolysis or dehydrohalogenation) result in moderate yields of sulfones, mainly when halides are reacted with sodium arenesulfinates in highly dipolar solvents (alcohols, water, DMF, DMSO, etc.), in which these salts are soluble enough. Significantly higher yields of sulfones under milder conditions can be achieved by using tetrabutylammonium sulfinates in common solvents, such as THF, in which they are completely soluble.[62] Similar results are attainable, in principle, with sodium sulfinates in two-phase systems (organic solvent/water) by aid of phase-transfer catalysts and/or chelating agents, which is a method of general relevance for nucleophilic substitutions.[63,64]

Only recently, the halogen-displacement reaction of sodium arenesulfinates with alkyl or activated-arylhalides has been tentatively adopted for the synthesis of polysulfones from difunctional reagents in heterogeneous systems, under phase-transfer conditions.

Polycondensation of disodium 4,4′-oxydibenzenesulfinate with the highly activated bis(4-chloro-3-nitrophenyl) sulfone, as in equation (26), has afforded soluble polymers of sufficiently high molecular weight and acceptable thermal stability.[65]

$$\text{(26)}$$

This polysulfonylation has been carried out on equimolecular mixtures of reactants in nitrobenzene, acetonitrile or, optimally, in water/nitrobenzene mixtures; in the presence of different phase-transfer catalysts (tetrabutyl-ammonium or -phosphonium halides, half mole per mole of sulfinate). The yields range from 71 to 93% after 24 h at 80–100 °C.

Under similar conditions, disodium 4,4′-oxydibenzenesulfinate has been polycondensed with aliphatic dihalides, such as dibromo-*p*-xylene[66] and 1,3-dichloroacetone,[67] with yields of 53–85%. According to their general performances, these two polycondensations, based on alkylation of disodium arenedisulfinates, have given low yields and polymers containing appreciable amounts of sulfinate ester units, caused by O-alkylation. Thermally and hydrolytically weak sulfinyloxy-linkages are believed responsible for both the much poorer thermal stability and the depolymerization in strong acids exhibited by such polymers[66,67] in comparison with commercial polysulfones prepared by different synthetic routes.[35] To a lesser extent, similar deficiencies have been also displayed by wholly aromatic polysulfones synthesized by sulfinate arylation as in equation (26).[65]

Several years ago, the reaction of benzyl chloride with zinc sulfoxylate ($ZnSO_2$) was mentioned as being capable of producing dibenzyl sulfone.[68] Only about 20 years later, the same sulfone and the homologous polymer were synthesized (though in low yields) by condensation of sodium dithionite (hydrosulfite) with benzyl chloride and dichloro-*p*-xylene, respectively, as in equation (27).[69]

$$n\text{ClCH}_2\text{-C}_6\text{H}_4\text{-CH}_2\text{Cl} + n\text{Na}_2\text{S}_2\text{O}_4 \xrightarrow[110\,°C,\,8\,h]{\text{DMF}} \text{+}(\text{CH}_2\text{-C}_6\text{H}_4\text{-CH}_2\text{SO}_2)_n\text{+} + n\text{SO}_2\uparrow + 2n\text{NaCl} \quad (27)$$

This sort of reaction probably follows a nucleophilic displacement mechanism (perhaps S_N1[69]) by virtue of the peculiar character of the dithionite anion ($^-$O—SOSO—O$^-$) present in solutions of $Na_2S_2O_4$.[70,71] Such an ion is active as a nucleophilic species, similarly to the sulfinyl ion $ArSO_2^-$, and, indeed, can give either sulfones or sulfinyl esters.[69]

More recently, poly(p-xylene sulfone) [poly(sulfonyl-p-xylylene)] has been obtained in varying yields directly from dichloro-p-xylene, an equivalent amount of alkali metal and liquid sulfur dioxide (large excess), in the presence of crown ethers as catalysts.[72] The highest yields have been achieved following equation (28), but similar results have been reported by using K_2CO_3 in place of potassium metal. Although some doubts exist that the reaction proceeds through intermediate formation of akali metal dithionite (SO_2 is known to give dithionites by reaction with electropositive metals),[71] the true mechanism has still to be adequately elucidated.

$$n\text{ClCH}_2\text{-C}_6\text{H}_4\text{-CH}_2\text{Cl} + n\text{SO}_2(l) + 2n\text{K} \xrightarrow[80°C, 24h]{SO_2(l), \text{dicyclohexyl-[18]-crown-6 ether}} +(\text{CH}_2\text{-C}_6\text{H}_4\text{-CH}_2\text{SO}_2)_n + 2n\text{KCl} \quad (28)$$

33.4 POLYCONDENSATION OF MONOMERS BEARING SULFONYL GROUPS

As shown in Section 33.2, a wide series of condensation reactions are available for synthesizing hybrid polymers from reactants containing sulfonyl groups, *via* the two general reaction schemes of equations (9) and (10). Such condensations, as compared to sulfonylations, are, in most cases, feasible under milder conditions and can give little or no structural irregularity and branching[16] or crosslinking. They often represent the best synthetic routes to certain polymers which can also be attained by polysulfonylation, as in the case of polyetherification to produce poly(ether sulfone)s.

33.4.1 Poly(arylene ether sulfone)s by Polyetherification

33.4.1.1 *Synthesis of diaryl ethers*

Aryl halides activated by electron-withdrawing substituents (—NO_2, —CN, —SO_2Ar, —SO_2R, —COAr, —COR, *etc.*) *ortho* or *para* to the halogen on the aromatic nucleus, can undergo, in highly dipolar solvents, halogen substitution by nucleophilic reagents. Aromatic nucleophilic substitutions[73,74] operated by reagents of weak to medium basicity (such as ArO^-, RO^-, HO^-, $ArSO_2^-$ and CN^-) follow a bimolecular addition–elimination mechanism, through intermediate complexes with highly delocalized negative charges. The general scheme of this nucleophilic displacement on aryl halides is shown by equation (29), where G, X and Z represent the electron-withdrawing substituent, the halogen and the anion of the nucleophilic reagent, respectively. The intermediate adducts (**11**), often called Meisenheimer complexes, can become stable enough (and vividly coloured), by virtue of strong conjugation, as shown for etherification (equation 30) of *p*-nitrohalobenzenes by alkali metal methoxides.

$$G\text{-C}_6\text{H}_4\text{-X} + Z^- \xrightarrow{\text{slow}} [G\text{-C}_6\text{H}_4(X)(Z)]^- \xrightarrow{\text{fast}} G\text{-C}_6\text{H}_4\text{-Z} + X^- \quad (29)$$

(13)

$$O_2N\text{-C}_6\text{H}_4\text{-X} + {}^-OMe \longrightarrow [\text{red Meisenheimer complex}] \longrightarrow O_2N\text{-C}_6\text{H}_4\text{-OMe} + X^- \quad (30)$$

red

Electron-withdrawing substituents cause strong stabilization of the intermediate complex by conjugation with the negative charge when they are *ortho* or *para* to the halogen, and feeble stabilization by induction from any ring position; electron-withdrawing groups therefore promote the slowest (rate determining) initial reaction-step and the overall reaction-path (equation 29).

Like the nitro group in equation (30), the sulfonyl moiety powerfully activates halogen displacement, as in equation (31), by stabilization of the intermediate adduct resulting from attack of

$$\text{ArS}\overset{O}{\underset{O}{\|}}\!\!\!\diagdown\!\!\!\text{—}\!\!\!\diagup\!\!\!\text{X} + {}^-\text{OAr}' \;\text{Na}^+ \longrightarrow \left(\text{ArS}\overset{O}{\underset{O}{\|}}\!\!\!\diagdown\!\!\!\text{—}\!\!\!\diagup\!\!\!\overset{X}{\underset{\text{OAr}'}{}}\right) \longleftrightarrow \left(\text{ArS}\overset{\overset{:\ddot{O}:^-}{\|}}{\underset{O}{}}\!\!\!\diagdown\!\!\!\text{—}\!\!\!\diagup\!\!\!\overset{X}{\underset{\text{OAr}'}{}}\right)\text{Na}^+ \longrightarrow$$

$$\text{Ar}\!\!\diagdown\!\!\!\text{—}\!\!\!\diagup\!\!\text{OAr}' + \text{NaX} \qquad (31)$$

phenoxide anion, and hence promotes the etherification; analogous conjugative delocalization of the negative charge is accomplished by sulfonyl as the *ortho* substituent.

As for nucleophilic substitutions in general, these reactions are successfully carried out in high-boiling, dipolar aprotic solvents {such as DMF, DMSO, N,N-dimethylacetamide, N-methylpyrrolidone, sulfolane (tetrahydrothiophene-1,1-dioxide) and hexamethylphosphoramide [phosphoric tris(dimethylamide)]} which enhance the active concentration of the attacking base by marked counter-cation solvation, and assist the bimolecular addition step. Similarly, aromatic nucleophilic substitutions are accelerated in the presence of crown ethers or cryptates,[75] still by an addition–elimination mechanism.

Aryl chlorides, bromides and iodides display nearly the same reactivity, whilst those of aryl fluorides are normally several orders of magnitude greater.[76] The ease of displacement of the different halogens by nucleophiles can be expressed as F(312) $\gg$ Cl (1) $\geq$ Br (0.74) $\geq$ I (0.36), the numbers in parentheses being, for the sake of comparison, the relative reaction rates for the etherifications of equation (30).[77]

It is well known that diaryl ethers can be obtained in high yields by condensation of aryl halides (substituted by electron-withdrawing groups) with alkali metal phenoxides,[78,79] as in equation (31), following the standard mechanism of aromatic nucleophilic substitution. Such syntheses are carried out in the aprotic solvents above-mentioned at moderate to high temperatures, under an inert atmosphere to protect the readily oxidizable phenoxides; the selection of alkali metal (in some instances an alkaline earth metal, but usually Na or K) depends on phenoxide solubility.

Moreover, strongly activated halides, such as 2,4-dinitrofluorobenzene, react so promptly and quantitatively under mild conditions in ordinary solvents, that they can be used to characterize alcohols[80] and phenols[81] by converting them into the corresponding 2,4-dinitrophenyl ethers.

Although far less reactive, unactivated aryl halides can be successfully condensed with alkali metal phenoxides under the catalytic conditions of the Ullmann reaction,[82,83] as in equation (32). This 'modified Ullmann condensation' must not be confused[84] with the Ullmann synthesis of biaryls by coupling of aryl halides, induced by copper, copper(I) compounds, *etc.* Similar catalysts and reaction conditions activate the aforementioned etherification. High yields of diaryl ethers are achieved, even from feebly reactive aromatic halides, in pyridine, DMF, DMSO or tetramethylurea as solvents, in the presence of activated copper powder,[85] Cu_2O,[86] Cu_2Cl_2[87] or, more recently, pentafluorophenylcopper.[88]

$$\text{ArX} + \text{KOAr}' \xrightarrow[180\,°\text{C}]{\text{copper catalyst}} \text{ArOAr}' + \text{KX} \qquad (32)$$

A basic mechanism of aromatic nucleophilic substitution has been admitted for the Ullmann etherification, but a reverse order of mobility for the different halogens (I $\approx$ Br > Cl $\gg$ F) has been shown.[89]

Finally, aryl ethers can also be obtained from aryl (or alkyl) halides condensed with silver[90] or copper(I)[91] phenoxides.

33.4.1.2 *Polyetherification*

The aforementioned syntheses of diaryl ethers may become routes to poly(arylene ether)s (polyoxyarylenes) as in equations (33) and (34), and hence to poly(arylene ether sulfone)s if the reactants bear sulfonyl linkages, as in equations (35) and (36). Two general pathways can be followed: polyetherification (equations 33, 35) and self-polyetherification (equations 34, 36); the metal M is almost always Na or K.

$$n\text{XArX} + n\text{MOAr'OM} \xrightarrow{-2n\text{MX}} \text{(ArOAr'O)}_n \qquad (33)$$

$$n\text{MOArX} \xrightarrow{-n\text{MX}} \text{(ArO)}_n \qquad (34)$$

Table 2 Polycondensation of Aromatic Halides with Alkali Metal Phenoxides in DMSO Solution

Reactant(s)	Catalyst	Reactants conc. (wt %)	Temperature (°C)	Time (h)	Polymer RV_1 [a]	Ref.
NaO–C₆H₄–Br	Cu_2O	10	150	22	0.04	33
	Cu_2Cl_2–pyridine[b]	25	200	various	0.1–1.1[c]	97
Br–C₆H₄–Br + (NaO–C₆H₄–)₂CMe₂	Cu_2O	12	150	22	0.09	33
(Br–C₆H₄–)₂O + (NaO–C₆H₄–)₂CMe₂	Cu_2O	14	150	22	0.05	33
(Br–C₆H₄–)₂CO + (NaO–C₆H₄–)₂CMe₂	Cu_2O	12	150	22	0.20	33
(Cl–C₆H₄–)₂SO₂ + (NaO–C₆H₄–)₂CMe₂	Cu_2O	15	150	22	0.36	33
Br–C₆H₄–SO₂–C₆H₄–ONa	Cu_2O	15	150	22	0.16	33
(Cl–C₆H₄–)₂CO + (KO–C₆H₄–)₂CMe₂	none	10	135	18	0.16	98

Monomers	Catalyst	Temp (°C)	Time (h)	RV	Yield (%)
(Cl-C6H4-SO2)2 + (NaO-C6H4-CMe2)2	none	20 / 40	135 / 160	4 / 1	0.71 / 1.35	98 / 98
(F-C6H4-SO2)2 + (KO-C6H4-CMe2)2	none	20	25–145	0.5	0.94	98
(Cl-C6H4-SO2)2 + (KO-C6H4-CMe2)2	none	25	165	4	0.7	98
(Cl-C6H4-SO2)2 + (KO-C6H4-CO)2	none	20	130	3.5	0.5	98
(Cl-C6H4-SO2)2 + KO-C6H4-OK	none	20	135	6	0.4	98
(Cl-C6H4-SO2)2 + (KO-C6H4-SO2)2	none	20	150	29	0.2	98
KO-C6H4-SO2-C6H4-Cl	none[d]	40	230	24	0.38	102
KO-C6H4-SO2-C6H4-F	none[d]	30	180	24	0.88	102

[a] Reduced viscosity = $(\eta_{solution} - \eta_{solvent})/\eta_{solvent}$ for a 1% solution of polymer in various solvents at 25 °C.
[b] Reaction carried out in 1,4-dimethoxybenzene.
[c] RV_1 in nitrobenzene at 140 °C.
[d] Reaction carried out in sulfolane.

$$n x \text{ArSO}_2\text{Ar'X} + n\text{MOAr''OM} \xrightarrow{-2n\text{MX}} \text{+ArSO}_2\text{Ar'OAr''O+}_n \quad (35)$$

$$n\text{MOArSO}_2\text{Ar'X} \xrightarrow{-n\text{MX}} \text{+ArSO}_2\text{Ar'O+}_n \quad (36)$$

Polyetherifications which involve scarcely activated or unactivated aromatic halides were reported as rather unsuccessful in affording high polymers. For example, self-condensation of sodium bromophenoxides[92] or silver 4-bromo-2,6-dimethylphenoxide[93] gives only oligomers. By contrast, high molecular weight polyethers are easily synthesized, in the presence of atmospheric oxygen or traces of oxidizing agents, by condensation of 2,6-disubstituted alkali metal 4-halophenoxides by a free-radical mechanism.[94,95]

However, both the oxidation reactions occurring in solutions of alkali metal phenoxides and the catalyzed oxidative polymerization of phenols (the so-called 'oxidative coupling', effective only with 2,6-disubstituted phenols) proceed as free-radical processes through formation of quinones and pertain to the synthesis of poly(arylene ether)s.[94,95]

The catalyzed Ullmann etherification could represent, in principle, a good general route for successful polycondensation of feebly reactive aromatic halides with phenoxides. Several condensations of dialkali metal bisphenoxides with aromatic dihalides and the self-condensation of different alkali metal halophenoxides (carried out at 150 °C in DMSO, in the presence of 5–10 mol % Cu_2O), were nevertheless reported[33] as generally failing to yield high polymers.

Indeed, as shown in Table 2, polymers with low reduced viscosity RV_1 [$R V_1 = (\eta_{solution} - \eta_{solvent})/\eta_{solvent}$ for a 1% solution of polymer], from 0.03 to 0.36, were achieved.[33] Polyethers with low to moderate molecular weights were obtained only from halides activated by electron-withdrawing groups; bromides, apparently, were more reactive than chlorides, according to the above-depicted features of the Ullmann condensation.

Copper-catalyzed self-polycondensation of alkali metal m-chlorophenoxide[96] or sodium p-bromophenoxide[97] showed a general irreproducibility, due to a series of side reactions which further reduce the potential effectiveness of the Ullmann polyetherification.

Polycondensations of a wide series of similar monomers, mainly bis(4-halophenyl) sulfones condensed with disodium or dipotassium bisphenoxides, as in equation (35), were reported as reliably successful in affording high polymers in DMSO (or sulfolane, diphenyl sulfone, etc.) in the absence of copper catalysts.[98] As summarized in Table 2, moderate to high molecular weights (RV_1 values ranging from 0.16 to 1.4) show the effectiveness of uncatalyzed polyetherification, provided that aromatic halides powerfully activated by electron-withdrawing substituents are employed.

After further investigations and experimental refinements on such uncatalyzed polyetherification, the two-component polycondensation (equation 35) of bis(haloaryl) sulfones with dialkali metal bisphenoxides and the self-polycondensation (equation 36) of alkali metal (haloarylsulfonyl)phenoxides have become the two routes of choice for the synthesis of poly(arylene ether sulfone)s.

33.4.1.3 Synthesis of bis(4-chlorophenyl) sulfone–bisphenol A poly(ether sulfone)

This speciality polymer has been produced since 1965 by the Union Carbide Corporation by polycondensation of bis(4-chlorophenyl) sulfone (p,p'-dichlorodiphenyl sulfone, DCDPS) with 2,2-bis(4-hydroxyphenyl)propane (bisphenol A) dialkali metal salt, as in equation (37).

$$n\text{Cl}\text{-}\langle\text{C}_6\text{H}_4\rangle\text{-}SO_2\text{-}\langle\text{C}_6\text{H}_4\rangle\text{-}\text{Cl} + n\text{NaO}\text{-}\langle\text{C}_6\text{H}_4\rangle\text{-}C(\text{Me})_2\text{-}\langle\text{C}_6\text{H}_4\rangle\text{-}\text{ONa} \xrightarrow{\text{DMSO}}_{160\,°C}$$

$$\text{+}\langle\text{C}_6\text{H}_4\rangle\text{-}SO_2\text{-}\langle\text{C}_6\text{H}_4\rangle\text{-}O\text{-}\langle\text{C}_6\text{H}_4\rangle\text{-}C(\text{Me})_2\text{-}\langle\text{C}_6\text{H}_4\rangle\text{-}O\text{+}_n + 2n\text{NaCl} \quad (37)$$

A description of the experimental procedure for this synthesis[98] may represent a valid example for poly(ether sulfone)s in general, either in laboratory- or industrial-scale preparations. It summarizes, in fact, the general features of polyetherifications, and appropriate experimental conditions, in order to get high yields, reproducibly pure and high molecular weight polymers.

Typical reaction conditions[98] are reported here. One mole of high purity bisphenol A, dissolved in about 2 kg of a 1:2 DMSO/chlorobenzene mixture, is converted into its disodium salt by adding 2 mol of 50% aqueous NaOH. Water is thoroughly removed by azeotropic distillation with chlorobenzene. One mole of DCDPS as a 50% solution in dry chlorobenzene is added, and the temperature is kept constant at 160 °C until the desired molecular weight is reached (1 h or longer). The whole process has to be carried out under an inert gas atmosphere (nitrogen or argon) in order to avoid phenoxide oxidations, which cause polymer discolouration, *etc*. Addition of the sulfone makes the reaction mixture brightly coloured, yellow to orange but often green (most probably because of the aforementioned formation of Meisenheimer complexes). When the proper RV has been reached, polyetherification can be stopped in many ways, *e.g.* by destroying the alkali metal phenoxide terminals of the growing polymer chains by adding acids or bubbling gaseous methyl chloride (for a few minutes at 120–160 °C). The intensely coloured solution turns to light amber, showing the reaction end-point.

The cooled viscous solution is then diluted with about 3 kg of chlorobenzene, freed from the by-product NaCl by filtration and coagulated in three or four volumes of ethanol. The resulting fluffy polymer is dried under vacuum at 135 °C (yield ≥ 90%).

33.4.1.4 *Mechanism and kinetics of polyetherification*

The polyetherification mechanism and kinetics have been investigated in detail (also by means of model compounds) for the polycondensation (equation 35) of several bis(halophenyl) sulfones with alkali metal bisphenoxides[98,99,101] and for the self-polycondensation (equation 36) of different alkali metal (halophenylsulfonyl)phenoxides.[100–102] The influence of various solvents, cations and reaction temperatures has been considered as well.

Such polyetherifications invariably occur by the well-described bimolecular addition–elimination path of aromatic nucleophilic substitution.

Regularly decreasing molecular weights as functions of increasing deviations from the 1:1 stoichiometry of the coreacting groups,[98] as well as polydispersity index values of 2–2.3[98] (as also reported in an accurate study of dilute solution properties of DCDPS–bisphenol A polymer[103]) and rather symmetrical curves of molecular weight GPC distribution,[16] are fully consistent with the principles of condensation polymerization.

Higher overall polymerization rates and higher molecular weights result from more reactive halides and phenoxides, as the data of Table 2 show for a series of polycondensations.

Such data are in good agreement with the known general order of ease of displacement exhibited by the different halogens (F ≫ Cl ≥ Br ≥ I) in aromatic nucleophilic substitutions, the condensation of fluorides being extremely fast. It can also be noted how various bisphenoxides of increasing basicity (and, conversely, bisphenols of decreasing acidity) display higher reaction rates.

Aromatic halide reactivity and phenoxide basicity are strongly influenced, in opposite senses, by both electron affinity and the position of aromatic-ring substituents, and by their conjugative, inductive and steric effects. Electron-withdrawing groups (and, primarily, sulfonyl moieties among them) enhance the halide reactivity (as seen in detail in Section 33.4.1.1) and *vice versa* depress the phenoxide basicity by delocalization of the negative charge of the phenoxide anion, as in equation (38). Indeed, the strongly acidic bis(4-hydroxyphenyl) sulfone (bisphenol S) behaves as one of the least reactive bisphenols, and thus requires very drastic etherification conditions.

$$-\overset{O}{\underset{O}{S}}-\!\!\left\langle\!\!\bigcirc\!\!\right\rangle\!\!-O^-\ K^+\ \longleftrightarrow\ -\overset{O^-}{\underset{O}{S}}-\!\!\left\langle\!\!\bigcirc\!\!\right\rangle\!\!=\!O\ \ K^+ \qquad (38)$$

A good general correlation has recently been found[104] for the polycondensation rate constants of bis(4-halophenyl) sulfones with a number of bisphenoxides, in terms of Hammett substituent constants[105] and bisphenol pK values.

The second-order rate constants k_2 for model etherification reactions,[100] reported in Table 3, are in agreement with the general criteria described above. However, actual polyetherification kinetics are much more complex, and the second-order rate 'constants' do change as the polymer chains grow; this indicates remarkable variations in reactivity of the functional groups during the polymerization. Initial rate constants and the whole polymerization course are shown in Table 4 and Figure 1, respectively, for several solution polycondensations.

Table 3 Second-order Rate Constants for Halogen Displacement from Aromatic Halides by Potassium Phenoxides, in DMSO at 120°C[100]

Reactants	$k_2 \times 10^5 (\text{l mol}^{-1}\text{s}^{-1})$
Ph–SO$_2$–C$_6$H$_4$–Cl (para) + KO–Ph	10.5
Ph–SO$_2$–C$_6$H$_4$–Cl (ortho) + KO–Ph	100
Ph–SO$_2$–C$_6$H$_4$–Cl + KO–Ph (F instead) — see figure	3024
Ph–SO$_2$–C$_6$H$_4$–Cl + KO–C$_6$H$_4$–CMe$_2$–C$_6$H$_4$–OK	3740[a]
Ph–SO$_2$–C$_6$H$_4$–Cl + KO–C$_6$H$_4$–SO$_2$–C$_6$H$_4$–OK	5.1[a]
Ph–SO$_2$–C$_6$H$_4$–Cl + KO–C$_6$H$_4$–SO$_2$–Ph	1.08

[a] For each —OK group.

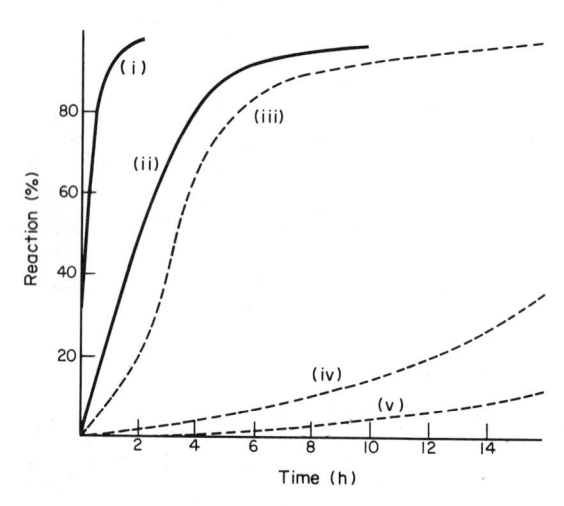

Figure 1 Polycondensation of aromatic halides with potassium phenoxides (from refs. 100–101):

———, in sulfolane at 158 °C, (i) (F–C$_6$H$_4$–)$_2$SO$_2$ + (KO–C$_6$H$_4$–)$_2$SO$_2$, (ii) KO–C$_6$H$_4$–SO$_2$–C$_6$H$_4$–F;

- - - - in DMSO at 178 °C, monomers 40 wt%, (iii) KO–C$_6$H$_4$–SO$_2$–C$_6$H$_4$–Cl,

(iv) KO–C$_6$H$_4$–SO$_2$–C$_6$H$_4$–Cl (Cl ortho), (v) C$_6$H$_5$–SO$_2$–C$_6$H$_3$(OK)–Cl

Table 4 Initial Second-order Rate Constants for Several Polyetherifications[100,101]

Reactant(s)	in DMSO at 120 °C	$k_2 \times 10^5 (\text{l mol}^{-1} \text{s}^{-1})$ in DMSO at 178 °C	in sulfolane at 158 °C
KO–C₆H₄–SO₂–C₆H₄–F	1.45	—	~25
KO–C₆H₄–SO₂–C₆H₄–F (ortho-OK)	0.07	—	—
KO–C₆H₄–SO₂–C₆H₄–Cl	0.01	1.28	—
KO–C₆H₄–SO₂–C₆H₄–Cl (meta)	0.58	—	—
KO–C₆H₄–SO₂–C₆H₄–Cl (ortho-OK)	—	0.07	—
KO–C₆H₄–SO₂–C₆H₄–Cl (ortho-Cl)	—	0.13	—
KO–C₆H₄–SO₂–C₆H₄–O–C₆H₄–SO₂–C₆H₄–F	—	—	~1150
F–C₆H₄–SO₂–C₆H₄–F + KO–C₆H₄–SO₂–C₆H₄–OK	—	—	~1150
Cl–C₆H₄–SO₂–C₆H₄–Cl + KO–C₆H₄–SO₂–C₆H₄–OK	16.0	—	—
Cl–C₆H₄–SO₂–C₆H₄–Cl + KO–C₆H₄–CMe₂–C₆H₄–OK	6100	—	—

The activating effect of the sulfonyl group on the halogen in one ring (and, conversely, its deactivating effect on the phenoxide) can be further enhanced or attenuated by electron-withdrawing or electron-releasing substituents in a sulfonyl-linked adjacent nucleus, mainly by conjugation when groups are in the *ortho* or *para* positions. In a certain sense, the sulfonyl group can play a 'bridge effect', in that it transmits, by its 'modulated' electron affinity, the influence of substituents to an adjacent ring.[100]

For these reasons (as can be seen comparatively by examining the kinetic data of Tables 3 and 4) halogen atoms in bis(halophenyl) sulfones activate each other. Halogen atoms inductively enhance, from any position of the aromatic nucleus, the electron withdrawal of the sulfonyl bridge, although they slightly attenuate it by conjugative electron-release from *ortho* or *para* positions, as in (14). Both inductively and conjugatively, electron-releasing phenoxide anions of bis(hydroxyphenyl) sulfones, such as (15), attenuate the electron withdrawal of the sulfonyl group and mutually activate themselves.

(14)   (15)

In contrast, but by an identical mechanism, the halogen and phenoxide anion deactivate each other in (halophenylsulfonyl)phenoxides such as (16), and therefore have self-polyetherification rates which are comparatively much lower than those of the corresponding two-component polycondensation,[101,106] as shown in Table 4. In such self-polyetherifications, however, the 'bridged' mutual deactivations of halide and phenoxide progressively disappears as their reciprocal distance increases on going from monomer to dimer, trimer, *etc.*; the increasing reactivity of functional groups at the ends of the growing polymer chains promotes higher polymerization rates after the earliest stages of the process (see Figure 1). After the oligomerization stage, the reactivity of the functional groups levels off; chemical rate constants then remain unchanged, unless the mobility of halide and phenoxide terminal-groups is limited (with far slower kinetics) by the increasing viscosity of the reaction medium. The process becomes diffusion controlled above certain viscosity values, which depend on both concentrations and average molecular weights.

(16)

In this respect, the polycondensation of DCDPS with bisphenol *A* disodium salt in DMSO has recently been investigated[107] and a correlation has been found between rate constants (in dilute solution) and rheokinetic constants for polyetherification in concentrated solutions (50–80 wt % of reactants). With monomer concentrations of 70 wt% and over, a transition from a 'kinetic' to a diffusion-controlled stage (accompanied by a sharp deceleration of the polymerization process) invariably takes place when the activation energy for the viscous flow increases until it equals the activation energy for the etherification reaction itself.[107]

33.4.1.5 *Self-polycondensation of (halophenylsulfonyl)phenoxides*

Self-polyetherification represents, in spite of intrinsically low reaction rates, a successful alternative method to the two-component polyetherification, similarly to self-polysulfonylation. The 1:1 stoichiometric ratio of the coreacting functions, halide and phenoxide, is, of course, precisely controlled.

High-purity alkali metal (halophenylsulfonyl)phenoxides can be easily prepared by partial hydrolysis (half of reaction 39) of bis(halophenyl) sulfones by alkali metal hydroxides and then directly polymerized[108] without extraction of the (halophenylsulfonyl)phenol. Polymerization can even be carried out *in situ* after thorough elimination of water by azeotropic distillation.[109–111]

$$\text{X}\underset{\text{X}}{\bigcirc}-\text{SO}_2-\underset{\text{X}}{\bigcirc}\text{X} \xrightarrow[-\text{KX}]{+2\text{KOH}} \text{KO}\bigcirc-\text{SO}_2-\underset{\text{X}}{\bigcirc}\text{X} \xrightarrow[-\text{KX}]{+2\text{KOH}} \text{KO}\bigcirc-\text{SO}_2-\bigcirc\text{OK}$$

(39)

The halogen displacement by KOH in aqueous DMSO was investigated both from DCDPS[112] and, later, from a number of different bis(halophenyl) sulfones.[100] Because of the feeble reactivity of the halogen in (halophenylsulfonyl)phenoxides towards nucleophilic attack, the reaction (39) stops halfway, in most cases, and affords the intermediate product in high yield. Indeed, these features can be exemplified by the second-order rate-constant values ($10^5 \times k_2$ l mol^{-1} s^{-1}) displayed over the arrows in equation (40), for the reaction carried out in 80.5:19.5 DMSO/H$_2$O at 120 °C.[100]

$$\text{Cl}\bigcirc-\text{SO}_2-\bigcirc\text{Cl} \xrightarrow[\substack{+2\text{KOH}\\-\text{KCl}\\-\text{H}_2\text{O}}]{358\,°\text{C}} \text{KO}\bigcirc-\text{SO}_2-\bigcirc\text{Cl} \xrightarrow[\substack{+2\text{KOH}\\-\text{KCl}\\-\text{H}_2\text{O}}]{3.24} \text{KO}\bigcirc-\text{SO}_2-\bigcirc\text{OK}$$

(40)

Only under comparatively drastic polymerization conditions (elevated temperatures, long reaction times), can high molecular weight poly(arylene ether sulfone)s be obtained by solution self-polyetherification, as shown in Table 2 and pointed out in a detailed study.[102]

In (halophenylsulfonyl)phenoxides, the reactivity order for halogen (*p*-F > *p*-Cl > *o*-Cl ≫ *m*-Cl) and phenoxide (*m*-OK > *p*-OK > *o*-OK), in agreement with the kinetic data of Table 4, implies that compounds such as (**17**) have very high reactivities. Being characterized by the lowest conjugative deactivation of phenoxide and the highest conjugative activation of halogen, such compounds may lead to polymers with unusually high molecular weights.[102] Very high molecular weights have been obtained from (**17**),[113] *e.g.* when polycondensed for 16 h at 150 °C in DMSO or heated *in vacuo* for 15 min at 300 °C.

$$\text{KO}\bigcirc-\text{SO}_2-\bigcirc\text{Cl}$$

(**17**)

Alkali metal (halophenylsulfonyl)phenoxides can be melt polymerized[102] best under vacuum at high temperature (not exceeding 300 °C, in order to minimize formation of small amounts of gel), overcoming the need for long reaction times (typical for self-polyetherifications), as the data of Table 5 show.

33.4.1.6 *Side reactions*

Even in the earliest studies of polyetherification routes to poly(arylene ether sulfone)s, a series of side reactions (falling under the general name of hydrolytic side-reactions) were pointed out[98] and investigated.[112] Such reactions either directly hinder high-polymer synthesis or cause cleavage of the formed polysulfone.

In two-component polycondensation, even under anhydrous starting conditions, an excess of base (alkali metal hydroxide) over the bisphenol implies irreversible halogen displacement from part of the bis(halophenyl)sulfone and yields a certain amount of alkali metal (halophenylsulfonyl) phenoxide. Although it is still able to self-polycondensate under more drastic conditions, the required 1:1 halide/phenoxide stoichiometry is altered, an excess of free alkali-metal phenoxide is left present and lower molecular weights result.[112]

On the other hand, residual water or moisture present during the polymerization, though unable to react with the halide, reversibly hydrolyzes the alkali metal phenoxide (equation 41) and yields free alkali-metal hydroxide, the equilibrium concentration of which depends mainly on the phenol acidity. The hydroxide produced, in turn, leads again to irreversible halogen-displacement from the dihalide.

$$\text{ArOK} + \text{H}_2\text{O} \rightleftharpoons \text{ArOH} + \text{KOH} \tag{41}$$

Table 5 Melt[a] and Solution[b] Self-polycondensation of (Halophenylsulfonyl)phenoxides[102]

Monomer	Monomer conc. (wt%)	Reaction conditions Temperature (°C)	Time (h)	Polymer RV_1[c]
KO–C₆H₄–SO₂–C₆H₄–F (para)	30[b]	180	24	0.88
KO–C₆H₄–SO₂–C₆H₄–F (para)	100	280	0.75	0.79–3.48
KO–C₆H₄–SO₂–C₆H₄–Cl (para)	40[b]	230	24	0.38
KO–C₆H₄–SO₂–C₆H₄–Cl (para)	70[b]	230	24	0.45
KO–C₆H₄–SO₂–C₆H₄–Cl (para)	100	280	0.75	0.50–0.84
KO–C₆H₄–SO₂–C₆H₄–Cl (meta-OK, para-Cl)	100	300	0.25	2.3
OK–C₆H₄–SO₂–C₆H₄–Cl (ortho-OK)	100	300	3	0.12
KO–C₆H₄–SO₂–C₆H₄–Cl (meta-Cl)	100	300	0.5	0.09

[a] Under vacuum.
[b] In sulfolane solution.
[c] Reduced viscosity = $(\eta_{solution} - \eta_{solvent})/\eta_{solvent}$ for a 1% solution of polymer at 25 °C.

Although diaryl ethers are known to be very stable towards attack by base, ethers activated by electron-withdrawing groups can be cleaved rather readily by nucleophilic reagents.[79] The formed poly(ether sulfone) can thus be cleaved by nucleophilic fission of ether linkages which are 'weakened' by the electron withdrawal of a sulfonyl group through an aromatic nucleus. For the model compound (**18**), second-order rate constants ($k_2 \times 10^5$ l mol^{-1} s^{-1}) for bond fission by KOH in aqueous DMSO at 120 °C are indicated.[100]

$$\text{Cl}\overset{64}{\downarrow}\text{–C}_6\text{H}_4\text{–SO}_2\text{–C}_6\text{H}_4\overset{530}{\downarrow}\text{–O}\overset{530}{\downarrow}\text{–C}_6\text{H}_4\text{–SO}_2\text{–C}_6\text{H}_4\overset{64}{\downarrow}\text{–Cl}$$

(**18**)

Free alkali-metal hydroxides, accidentally present in excess or coming from alkali metal phenoxide hydrolysis, cause drastic lowering[112] of the molecular weight via reaction (42). Moreover, under proper anhydrous conditions, the phenoxide end-groups of the growing polymer can attack ether linkages, causing ether interchange, as shown in equilibrium (43), similarly to ester interchange in polyesterification.[112]

$$\sim\sim SO_2ArOArSO_2Ar\sim\sim + {}^-OH \rightleftharpoons \sim\sim SO_2ArO^- + HOArSO_2Ar\sim\sim \quad (42)$$

$$\sim\sim SO_2ArOArSO_2Ar\sim\sim + {}^-OAr' \sim\sim \rightleftharpoons \sim\sim SO_2ArOAr'\sim\sim + {}^-OArSO_2Ar\sim\sim \quad (43)$$

Ether interchange proceeds without any influence on molecular weight values, but leads, in some cases, to polymers with an irregular repeating structure, as shown for the polycondensation of dihalide (**18**) with bisphenol A dialkali metal salt.[16] By the mechanism of reaction (43), in contrast, spurious non-polymeric phenoxides, exceeding the 1:1 stoichiometry, can cause a rather abrupt molecular weight decrease.[101,112]

Rigorously equivalent amounts of reactants and perfectly pure monomers and solvents must be used. Water must be thoroughly eliminated, usually by azeotropic distillation, after salification of the bisphenol which is to be condensed with the dihalide, or after partial hydrolysis of the dihalide before its self-condensation. Moisture and oxygen are excluded by inert gas (N_2) streaming during the polymerization (vacuum pumping being the best condition for bulk polymerization). Excess of alkali metal hydroxide must be absolutely avoided, and its adverse effects can be almost completely suppressed if the bisphenol salification is carried out with weaker bases, *e.g.* sodium or potassium carbonate or bicarbonate, as often reported.[114-117,127,128] Such weak bases are unable to displace halogen from bis(halophenyl) sulfones in aprotic solvents and allow addition of all reactants at the same time; the polymerization is then carried out after the water has been azeotropically distilled off.

Halide ions are generally poor nucleophilic reagents and are much poorer than phenoxide anions. Alkali metal halides, which are weak bases and are always present as condensation by-products, can nevertheless attack activated ether-linkages of poly(ether sulfone)s during the polyetherification itself.[101]

Ether fission by chlorides, bromides or iodides has been shown to be very slow, in accordance with their feeble basicity. Fluoride ion, by contrast, is a rather powerful nucleophile, and KF (as well as CsF or RbF and much more than NaF or LiF) behaves as a strong base, especially in dipolar aprotic solvents.[118] In spite of the low solubility of KF in such solvents, a fast decrease in molecular weights has been shown, *e.g.* on dissolving a poly(ether sulfone) in sulfolane saturated with KF at 200°C; stable molecular weight values are asymptotically reached over longer times.[101]

The equilibrium (44) can therefore be proposed for ether-linkage fission by F^-.

$$\sim\!\!\text{ArOArSO}_2\!\!\sim + \text{KF} \underset{k_1}{\overset{k_2}{\rightleftharpoons}} \sim\!\!\text{ArOK} + \text{FArSO}_2\!\!\sim \tag{44}$$

The synthesis of poly(ether sulfone)s from aromatic fluorides easily affords high polymers and is characterized by high reaction rates, but, in addition to the general ether interchange (equation 43), is also controlled by the equilibrium (44). This equilibrium, between polymer cleavage and polyetherification, follows the same kinetic scheme in both directions.

The reaction proceeds towards the polyetherification by removal of KF from the system (because of its low solubility), and the achieved molecular weights are equilibrium values. At equilibrium, $k_1[-\text{F}][-\text{OK}] = k_2[-\text{O}-][\text{KF}]$, and [KF] can be considered constant after solution saturation by KF produced in the earliest stages of the process. Higher molecular weight values (higher $[-\text{O}-]$) can be obtained by increasing the initial concentration of the reactants $[-\text{F}]$ and $[-\text{OK}]$, as clearly shown by experimental data.[101] Similar favourable effects of high monomer concentrations on the resulting molecular weights can also be seen[102] for halides other than fluorides, but to a far lesser extent (some data are given in Table 5).

The GPC chromatograms of polysulfones synthesized by polyetherification almost always show, besides the symmetrical major peak, a secondary peak at lower elution volumes; the relative area of this secondary peak increases as more elevated reaction temperatures are adopted.[16,119] This component, with larger hydrodynamic volumes, has been identified as a slightly branched fraction of the polymer, originating from traces of trifunctional reactants (which has been reproduced, for the sake of investigation[102]) or from the side reactions (45) and (46), which involve attack of phenoxide anion on the activated aromatic nuclei.[119]

This undesired reaction may be promoted by high temperature, and is most probably responsible for the slight gelling in melt polyetherifications carried out at temperatures above 300 °C.

Chain branching implies no adverse influence on the mechanical properties of the polymers, unless high branching-levels are achieved, e.g. by carrying out the polymerizations in the presence of appreciable amounts of trihalodiphenyl sulfones. Such polysulfones have shown significantly reduced toughness, analogously to poly(diphenylene ether sulfone)s, which deviate from the 'toughest' straight *para,para* chain-structure. In fact, lower toughness has been displayed by poly-(*p*-diphenylene ether sulfone)s bearing *ortho* or *meta* linkages deliberately introduced by increasing the amounts of *ortho* or *meta* monomers, such as those indicated in Table 5.[120]

33.4.1.7 Choice of solvents

The solubility and reactivity of alkali metal phenoxides, and hence the overall rates of reactions involving them, are greatly enhanced in highly dipolar aprotic solvents, by virtue of their high dielectric constant and preferential cation solvation. The choice of such solvents is determined by process temperature (higher for less reactive monomers), solubility requirements of monomers and resulting polymers and solution viscosity.[98] The preferred solvents (and polymerization temperatures) are as follows: DMSO (135–180 °C), dimethyl sulfone (180–200 °C), sulfolane (200–240 °C) and diphenyl sulfone (230–260 °C and over),[98] most frequently adopting DMSO or sulfolane and sometimes using *N,N*-dimethylacetamide[127] or *N*-methylpyrrolidone (NMP).[117] A scarcely reactive system such as DCDPS–bisphenol S can afford high polymer only under drastic conditions allowed by an appropriate solvent, e.g. 6–8 h at 220–240 °C in sulfolane,[98,121] 6 h at 300 °C in biphenyl[122] or, in bulk, 35 h at 200–290 °C.[123]

Cosolvents forming azeotropes with water (chlorobenzene, toluene or xylene) are currently added at the initial stage of the process, in order to accomplish a thorough elimination of water after the phenol-salification step; cosolvents are then usually distilled off before polymerization.

33.4.1.8 Choice of monomers

Viscoelastic properties, rheological behaviour, T_g, electrical properties and thermal stability of polysulfones are determined by the structure of their repeating unit and primarily, therefore, by the monomers selected.

Increasingly bulky and dipolar monomers result in polymers with higher T_g values, as has been shown for a series of poly(ether sulfone)s[16,35] and for polymers obtained by polycondensation of DCDPS with a number of bisphenols.[98]

Polycondensation of DCDPS with hydroquinone or perfluoroalkylbisphenol A affords crystallizable and high-melting products.[98] Specific attention and various patent claims have been devoted to polymers with high T_g values derived from the condensation of 4,4'-bis(4-chlorophenylsulfonyl)biphenyl with hydroquinone[124] and from the condensation of DCDPS with hydroquinone/bisphenol A mixtures,[125,126] with 3,3',5,5'-tetramethyl-4,4'-dihydroxybiphenyl[127] and with 3,3',5,5'-tetramethyl bisphenol S.[128]

Other polysulfones worth mentioning are those derived from the condensation of DCDPS with 1,5-dihydroxynaphthalene[129] or with α,α'-bis(4-hydroxyphenyl)-*p*-diisopropylbenzene[130,131] (for which polymer melt viscosities significantly lower than those for bisphenol-*A*-based polysulfones have been claimed).

33.4.1.9 Alternative and miscellaneous methods of polyetherification

As is well known, halide anions, and particularly the fluoride anion, which is a rather powerful Lewis-base,[118] exert a catalytic action in aromatic nucleophilic substitutions in dipolar aprotic solvents.[132] For instance, protic nucleophiles are directly and efficiently alkylated by alkyl halides in the presence of KF[133] and, more specifically, phenols are alkylated in the presence of KF (or CsF) absorbed on Celite[134,135] or Et_4NF.[136]

Following this criterion, halophenols and dihalides with bisphenols have been condensed (in sulfolane at 220–280 °C) by using KF as the base,[102] with successful though slow polymerizations. Even the KF-promoted polycondensation (equation 47) of highly activated 3,3'-dinitroDCDPS with bisphenol A has been carried out,[137] again with low reaction rates. Only in the presence of a large

$$\text{equation shown in image}\quad(47)$$

excess of KF (at least 5 mol per 1 mol of bisphenol) and long reaction times (48 h at 100 °C in sulfolane) does the condensation proceed successfully in terms of yields (up to 99%) and molecular weights.

Tetrabutylammonium fluoride is known to be an effective reagent for the cleavage of silyl ethers under mild conditions in aprotic solvents.[138] A series of reactions (*e.g.* alkylation) on trimethylsilyl ethers, esters, *etc.*, have been reported as proceeding with fluoride-catalyzed desilylation with the aid of KF, CsF or $(C_4H_9)_4NF$.[139]

Based on such general features, the bulk polycondensation of bis(trimethylsilyl ether)s of various bisphenols with bis(4-fluorophenyl)sulfone has been investigated.[140] Fast polymerization kinetics and high yields have been obtained under the catalytic influence of KF or CsF, as in equation (48). Highly thermostable polysulfones, with $\bar{M}_n$ ranging from 2500 to 17 500 and $\bar{M}_w/\bar{M}_n$ values reasonably close to 2, have been afforded.

$$\text{equation shown in image}\quad(48)$$

By using phase-transfer catalysts, phenols (as alkali metal phenoxides) can be smoothly alkylated with alkyl halides[141] or arylated with activated aryl halides[142] to the corresponding ethers, under mild conditions in two-phase water/organic-solvent mixtures.

Slow but successful polycondensations of 3,3′-dinitroDCDPS with bisphenol *A* dipotassium salt have indeed been accomplished, at room temperature, in two-phase water/organic-solvent systems, in the presence of phase-transfer catalysts [*e.g.* tetraalkylammonium halides, crown ethers and poly(ethylene glycol)s].[143] The best results have been obtained in H_2O/CH_2Cl_2, in the presence of dicyclohexyl-[18]-crown-6 ether.

Although the reported molecular weights are generally lower than those attainable by homogeneous polyetherification, such phase-transfer methods look promising,[144] affording high-enough poly(ether sulfone)s under very mild conditions in ordinary solvents.

33.4.2 Aliphatic–Aromatic Poly(ether sulfone)s

Usual molecular weight regulation of poly(arylene ether sulfone)s, by rapid reaction of the alkali metal phenoxide end-groups with MeCl, produces alkyl aryl ether terminal-units.

Alkyl aryl ether linkages and, hence, poly(alkylene arylene ether sulfone)s can be specifically synthesized by condensing dialkali metal bisphenoxides with aliphatic dihalides. This etherification simply follows a mechanism of aliphatic nucleophilic substitution and requires milder reaction conditions, since aliphatic halides are generally more reactive than aromatic ones. The polymerization of bisphenol *S* dipotassium salt with bis[4-(chloromethyl)phenyl]methane[145] and the chain extension of sodium-phenoxide-terminated 'prepolymers' DCDPS–bisphenol *A* with CH_2Cl_2 or CH_2Br_2,[146] may be given as examples.

33.4.3 Miscellaneous Polysulfones by Various Step-polymerization Reactions

33.4.3.1 *Poly(arylene sulfide sulfone)s and poly(arylene ether sulfide sulfone)s*

Some of these highly thermostable polymers have been synthesized by polysulfonylation (self-polycondensation of 4-(phenylthio)benzenesulfonyl chloride,[41,43] or polycondensation of diphenyl sulfide with 4,4′-oxydibenzenesulfonic acid or 4,4′-thiodibenzenesulfonic acid[58]) or by polyetherification (polycondensation of DCDPS with 4,4′-dihydroxydiphenyl sulfide[98]).

Several patents specifically claim, moreover, the preparation of these polymers by two general polythioetherification methods, which are among those suitable for the synthesis of poly(arylene sulfide)s:[147] condensation of an activated aromatic halide with either an alkali metal thiophenoxide or an alkali metal sulfide, under the same experimental conditions as used for polyetherification.

Bis(haloaryl) sulfones can indeed be polycondensed with dialkali metal bisthiophenoxides[148-150] as in equation (49),[149] and alkali metal (haloarylsulfonyl)thiophenoxides can be self-polycondensed as in equation (50);[151] in some cases, the polymerization of unsalified (haloarylsulfonyl)thiophenol is promoted by KF.[152,153] Bis(haloaryl) sulfones may also be polycondensed with alkali metal sulfides in the presence of $MeCO_2Li$ or heterocyclic amines,[154] as in equation (51), or, additionally, copolycondensed with dihalobenzenes.[155]

$$n\text{Cl-Ar-SO}_2\text{-Ar-Cl} + n\text{KS-Ar-Ar-SK} \xrightarrow[150°C, 24h]{DMF} \text{-(-Ar-SO}_2\text{-Ar-S-Ar-Ar-S-)}_n\text{-} + 2n\text{KCl} \quad (49)$$

$$n\text{KS-Ar-SO}_2\text{-Ar-Cl} \xrightarrow[180°C, 5h]{NMP} \text{-(-Ar-SO}_2\text{-Ar-S-)}_n\text{-} + n\text{KCl} \quad (50)$$

$$n\text{Cl-Ar-SO}_2\text{-Ar-Cl} + n\text{Na}_2\text{S} \longrightarrow \text{-(-Ar-SO}_2\text{-Ar-S-)}_n\text{-} + 2n\text{NaCl} \quad (51)$$

Besides the use of monomers containing ether links,[58,98,149,150] poly(arylene ether sulfide sulfone)s can result from a combination of polyetherification and polythioetherification, e.g. by reacting bis(haloaryl) sulfones with mixtures of the dialkali metal salts of bisphenols and bisthiophenols,[156] or with alkali metal sulfides and bisphenoxides, respectively.[157-159]

33.4.3.2 Poly(amide sulfone)s

Aromatic polymers of this sub-family, which are characterized by high softening- or melting-points, fair thermal stability, high toughness and good solubility in solvents such as DMF or NMP, have always been synthesized by polyamidation reactions.

Using the reaction between amine and acyl chloride, high polymers were initially prepared from the interfacial condensation of 4,4′-sulfonyldibenzoyl chloride with a series of non-aromatic diamines.[160] More recently, several aromatic diamines have been condensed with two different acyl dichlorides,[161,162] as in equation (52).[161]

$$n\left(\text{ClCO-Ar-O-Ar-}\right)_2\text{SO}_2 + n\text{H}_2\text{NArNH}_2 \xrightarrow{-2n\text{HCl}}$$

$$\text{-(-Ar-O-Ar-SO}_2\text{-Ar-O-Ar-CONHArNHCO-)}_n\text{-} \quad (52)$$

Direct polyamidation from diamines and dicarboxylic acids has been reported, e.g. condensing bis(4-aminophenyl) sulfone with azelaic acid or azelaic/isophthalic acid mixtures in bulk at high temperature[163] or, at moderate temperature, azeotropically distilling off the water, as in equation (53).[164] Diamino-terminated polysulfones have been further polycondensed with dianhydrides.[165]

$$n\left(\text{HOCO}-\phenyl-\text{O}-\phenyl-\text{SO}_2-\phenyl\right)_2\text{O} + n\text{H}_2\text{N}-\phenyl-\text{NH}_2 \xrightarrow[-2n\text{H}_2\text{O}]{\text{DMSO/toluene}}$$

$$\left(-\phenyl-\text{O}-\phenyl-\text{SO}_2-\phenyl-\text{O}-\phenyl-\text{SO}_2-\phenyl-\text{O}-\phenyl-\text{CONH}-\phenyl-\text{NHCO}-\right)_n \quad (53)$$

33.4.3.3 Poly(arylene ester sulfone)s

Poly(phenylene ester sulfone)s have been prepared by interfacial condensation of bisphenol S disodium salt with iso- and tere-phthaloyl chloride mixtures (or with 4,4′-sulfonyldibenzoyl chloride) in the presence of surfactants, as in equation (54).[166] Similarly, various mixtures of bisphenol S and bisphenol A have been used.[167]

$$n\text{NaO}-\phenyl-\text{SO}_2-\phenyl-\text{ONa} + n\text{ClCO}-\phenyl-\text{COCl} \xrightarrow[\text{room temperature}]{\text{H}_2\text{O/CH}_2\text{Cl}_2, \text{ surfactant}}$$

$$\left(-\phenyl-\text{SO}_2-\phenyl-\text{O}_2\text{C}-\phenyl-\text{CO}_2-\right)_n + 2n\text{NaCl} \quad (54)$$

Dihydroxy-terminated poly(ether sulfone)s (prepared by standard polyetherification from $(m + 1)$ mol of bisphenol A and m mol of DCDPS) have been condensed with iso- and tere-phthaloyl chloride mixtures, either interfacially in the presence of phase-transfer catalysts[168] or in solution.[169] Acetyl derivatives of dihydroxy-terminated poly(arylene ether sulfone)s and poly(bisphenol A terephthalate) have been copolycondensed by transesterification, giving poly(ether sulfone-block-ester) copolymers.[170]

Sulfone-containing polycarbonates have been synthesized by direct reaction of bisphenol S[171] or bisphenol S/bisphenol A mixtures with phosgene,[172] under phase-transfer conditions.

33.4.3.4 Poly(arylene ether ketone sulfone)s

Some examples of these polymers have been obtained by Friedel–Crafts condensation of mixtures of aromatic sulfonyl and carbonyl halides under the same experimental conditions adopted for polysulfonylations,[43] and from 4,4′-diphenoxydiphenyl sulfone condensed with iso- and tere-phthaloyl chlorides.[173]

The use of a moderately acidic catalyst, constituted of mixtures of $AlCl_3$ (Lewis acid) with LiCl (Lewis base), prevents, as specifically claimed,[174] any *ortho* or *meta* linking in polycondensations such as that of equation (55).

$$m\,\phenyl-\text{O}-\phenyl-\text{COCl} + n\,\phenyl-\text{O}-\phenyl-\text{SO}_2\text{Cl} \xrightarrow{\text{AlCl}_3/\text{LiCl}}$$

$$\left[\left(-\phenyl-\text{O}-\phenyl-\text{CO}-\right)_m \left(-\phenyl-\text{O}-\phenyl-\text{SO}_2-\right)_n\right] + (m+n)\text{HCl}\uparrow \quad (55)$$

Similar polymers have also been synthesized by polyetherification, *e.g.* by condensation of DCDPS with 4,4′-dihydroxybenzophenone.[98,114,175] Synthesis and properties of high polymers (RV values up to 2.5) have been described for the copolyetherification in equation (56),[176] which yields random structures due to the well-known ether-interchange reaction.

$$m\text{KO}\langle\text{C}_6\text{H}_4\rangle\text{-CO-}\langle\text{C}_6\text{H}_4\rangle\text{OK} + n\text{Cl}\langle\text{C}_6\text{H}_4\rangle\text{-CO-}\langle\text{C}_6\text{H}_4\rangle\text{Cl} + q\text{Cl}\langle\text{C}_6\text{H}_4\rangle\text{-SO}_2\text{-}\langle\text{C}_6\text{H}_4\rangle\text{Cl}$$

$$\xrightarrow[300-355\,°C]{\text{diphenyl sulfone}} \left[\begin{array}{c}(\langle\rangle\text{-CO-}\langle\rangle\text{-O})_{(m+n)} \\ (\langle\rangle\text{-SO}_2\text{-}\langle\rangle\text{-O})_q\end{array}\right] + 2m\text{KCl} \quad (56)$$

$$m = n + q$$

Polymers of this sub-family are semicrystalline and exhibit elevated melting temperatures (300–360 °C), increasingly high as the molar ratio of crystallizable ether–ketone units vs. amorphous ether–sulfone ones increases.[176] Only very high process-temperatures can prevent polymer crystallization and separation from the reacting system, allowing high molecular weights to be reached. Similar synthetic conditions and a variety of such high-melting polymers (300–430 °C) have been patented,[177] e.g. from bis(haloaryl) ketones and bisphenol S dialkali metal salts.

33.4.3.5 Polymers from divinyl sulfones

Due to the strong electron-withdrawal exerted by the sulfonyl group, α,β-unsaturated sulfones are prone to add nucleophiles (forming Michael-type adducts) or can behave as powerful dienophiles.[6]

Divinyl sulfones of the general formula (**19**) (including the variously substituted members of the family, the first one of which is divinyl sulfone **20**) represent difunctional monomers able to polymerize with difunctional protic nucleophiles under the catalytic influence of bases, as shown with a diol in equation (57).

$$\text{CH}_2\text{=CHSO}_2\text{RSO}_2\text{CH=CH}_2 \qquad \text{CH}_2\text{=CHSO}_2\text{CH=CH}_2$$
$$(\textbf{19}) \qquad\qquad\qquad (\textbf{20})$$

$$n\text{CH}_2\text{=CHSO}_2\text{CH=CH}_2 + n\text{HOROH} \xrightarrow[\text{catalyst}]{\text{base}} \text{-(CH}_2\text{CH}_2\text{SO}_2\text{CH}_2\text{CH}_2\text{ORO)}_{\overline{n}} \quad (57)$$

Polymerization of divinyl sulfones has been reported with diols,[178] dithiols,[178,179] compounds containing an active methylene group (e.g. dialkyl malonates)[180,181] and urea.[182]

Finally, by virtue of their fairly good reactivity with such functional groups under mild conditions, divinyl sulfones have been claimed as crosslinking agents of proteins,[183,184] poly(vinyl alcohol) in textiles treatment[185] and also as surface modifiers for cellulosic fibres.[186,187]

33.5 REFERENCES

1. R. N. Johnson, in 'Encyclopedia of Polymer Science and Technology', ed. N. M. Bikales, H. F. Mark and N. G. Gaylord, Wiley, New York, 1969, vol. 11, p. 447.
2. D. M. White and G. D. Cooper, in 'Kirk-Othmer Encyclopedia of Chemical Technology', 3rd edn., ed. M. Grayson and D. Eckroth, Wiley, New York, 1982, vol. 18, p. 605.
3. N. J. Ballintyn, in 'Kirk-Othmer Encyclopedia of Chemical Technology', 3rd edn., ed. M. Grayson and D. Eckroth, Wiley, New York, 1982, vol. 18, p. 832.
4. H. Buchert and G. Blinne, *Synthetic*, 1985, **16**, 41.
5. O. B. Searle and R. H. Pfeiffer, *Polym. Eng. Sci.*, 1985, **25**, 474.
6. J. Strating, in 'Organic Sulfur Compounds', ed. N. Kharasch, Pergamon Press, Oxford, 1961, vol. 1, chap. 15.
7. N. Tokura, in 'Encyclopedia of Polymer Science and Technology', ed. N. M. Bikales, H. F. Mark and N. G. Gaylord, Wiley, New York, 1968, vol. 9, p. 460.
8. S. R. Sandler and W. Karo, 'Polymer Syntheses', Academic Press, New York, 1980, vol. 3, chap. 1.
9. E. J. Goethals, in 'Encyclopedia of Polymer Science and Technology', ed. N. M. Bikales, H. F. Mark and N. G. Gaylord, Wiley, New York, 1970, vol. 13, p. 460.
10. J. Studinka and R. Gabler (Inventa A. G.), *Swiss Pat.* 491 981 (1970) (*Chem. Abstr.*, 1970, **73**, 99 414).
11. I. C. Taylor, *Def. Publ. U.S. Pat. Off.* 911 007 (1973) (*Chem. Abstr.*, 1973, **79**, 54 779).
12. T. Fujisawa, M. Sumitani and N. Kobayashi (Sagami Chem. Res. Cent.), *Jpn. Pat.* 73 23 560 (1973) (*Chem. Abstr.*, 1974, **80**, 134 087).
13. R. Gabler and J. Studinka, *Chimia*, 1974, **28**, 567.
14. Y. Imai, M. Ueda, A. Kato and Y. Kano, *Kobunshi Ronbunshu*, 1980, **37**, 445 (*Chem. Abstr.*, 1980, **93**, 133 025).
15. K. J. Ivin and J. B. Rose, *Adv. Macromol. Chem.*, 1968, **1**, 335.

16. J. B. Rose, *Polymer*, 1974, **15**, 456.
17. J. B. Rose, *Chimia*, 1974, **28**, 561.
18. Z. K. Brzozowski and G. Rokicki, *Przem. Chem.*, 1974, **53**, 526 (*Chem. Abstr.*, 1975, **82**, 58 611).
19. C. M. Suter, 'The Organic Chemistry of Sulfur', Wiley, New York, 1944, chap. 7.
20. W. E. Truce, T. C. Klingler and W. W. Brand, in 'Organic Chemistry of Sulfur', ed. S. Oae, Plenum Press, New York, 1977, chap. 10.
21. T. Durst, in 'Comprehensive Organic Chemistry', ed. D. H. R. Barton and W. D. Ollis, Pergamon Press, Oxford, 1979, vol. 3, part 11.8.
22. S. R. Sandler and W. Karo, 'Organic Functional Group Preparations', 2nd edn., Academic Press, New York, 1983, vol. 1, chap. 20.
23. R. O. C. Norman and R. Taylor, 'Electrophilic Substitution in Benzenoid Compounds', Elsevier, Amsterdam, 1965.
24. F. A. Carey and R. J. Sundberg, 'Advanced Organic Chemistry', 2nd edn., Plenum Press, New York, 1984, part A, p. 481.
25. G. A. Olah, 'Friedel–Crafts Chemistry', Wiley, New York, 1973, p. 91.
26. F. A. Carey and R. J. Sundberg, 'Advanced Organic Chemistry', 2nd edn., Plenum Press, New York, 1983, part B, p. 380.
27. C. M. Suter, 'The Organic Chemistry of Sulfur', Wiley, New York, 1944, p. 673.
28. (a) G. A. Olah, 'Friedel–Crafts Chemistry', Wiley, New York, 1973, p. 122; (b) p. 488.
29. W. E. Truce and C. W. Vriesen, *J. Am. Chem. Soc.*, 1953, **75**, 5032.
30. S. F. Cox and K. G. Neill (ICI Ltd.), *Aust. Pat.* 242 187 (1962) (*Chem. Abstr.*, 1965, **63**, 4 208g).
31. J. Huismann (I. G. Farbenind A. G.), *Ger. Pat.* 701 954 (1941) (*Chem. Abstr.*, 1942, **36**, 987).
32. M. E. B. Jones (ICI Ltd.), *Br. Pat.* 979 111 (1965) (*Chem. Abstr.*, 1965, **62**, 9065h).
33. B. E. Jennings, M. E. B. Jones and J. B. Rose, *J. Polym. Sci., Part C*, 1967, **16**, 715.
34. J. B. Rose, *Chem. Ind.*, **1968**, 461.
35. H. A. Vogel, *J. Polym. Sci., Polym. Chem. Ed.*, 1970, **8**, 2035.
36. H. A. Vogel (3M Co.), *Br. Pat.* 1 163 975 (1969) (*Chem. Abstr.*, 1969, **71**, 113 498).
37. S. M. Cohen and R. H. Young, *J. Polym. Sci., Polym. Chem. Ed.*, 1966, **4**, 722.
38. M. E. A. Cudby, R. G. Feasey, S. Gaskin, M. E. B. Jones and J. B. Rose, *J. Polym. Sci., Part C*, 1969, **22**, 747.
39. P. Kovacic and C. Wu, *J. Org. Chem.*, 1961, **26**, 759, 762.
40. C. A. Thomas, 'Anhydrous Aluminum Chloride in Organic Chemistry', Reinhold, New York, 1961, p. 648.
41. M. E. A. Cudby, R. G. Feasey, B. E. Jennings, M. E. B. Jones and J. B. Rose, *Polymer*, 1965, **6**, 589.
42. M. E. A. Cudby, R. G. Feasey, S. Gaskin, V. Kendall and J. B. Rose, *Polymer*, 1968, **9**, 265.
43. ICI Ltd., *Belg. Pat.* 639 634 (1964) (*Chem. Abstr.*, 1965, **63**, 700f).
44. H. A. Vogel (3M Co.), *Ger. Pat.* 1 745 085 (1973) (*Chem. Abstr.*, 1974, **80**, 121 522).
45. M. E. B. Jones (ICI Ltd.), *US Pat.* 4 008 203 (1977) (*Chem. Abstr.*, 1977, **86**, 172 380).
46. P. J. Bain, E. J. Blackman, W. Cummings, S. A. Hughes, E. R. Lynch, E. B. McCall and R. J. Roberts, *Proc. Chem. Soc., London*, 1962, 86.
47. ICI Ltd., *Belg. Pat.* 667 401 (1966) (*Chem. Abstr.*, 1966, **65**, 15 544c).
48. E. I. du Pont de Nemours & Co., *Neth. Pat.* 6 611 019 (1967) (*Chem. Abstr.*, 1967, **67**, 44 371).
49. M. E. B. Jones (ICI Ltd.), *Br. Pat.* 1 166 624 (1969) (*Chem. Abstr.*, 1969, **71**, 125 233).
50. E. E. Gilbert, 'Sulfonation and Related Reactions', Interscience, New York, 1965, chap. 2.
51. N. H. Christensen, *Acta Chem. Scand.*, 1964, **18**, 954.
52. E. J. Bourne, M. Stacey, J. C. Tatlow and J. M. Tedder, *J. Chem. Soc.*, 1951, 718.
53. B. M. Graybill, *J. Org. Chem.*, 1967, **32**, 2931.
54. H. J. Sipe, Jr., D. W. Clary and S. B. White, *Synthesis*, 1984, 283.
55. L. Field, *J. Am. Chem. Soc.*, 1952, **74**, 394.
56. E. J. Bourne, M. Stacey, J. C. Tatlow and J. M. Tedder, *J. Chem. Soc.*, 1949, 2976.
57. T. E. Tyobeka, R. A. Hancock and H. Weigel, *J. Chem. Soc., Chem. Commun.*, 1980, 114.
58. S. M. Cohen and R. H. Young, Jr. (Monsanto Co.), *US Pat.* 3 418 277 (1968) (*Chem. Abstr.*, 1969, **70**, 48 045).
59. M. Ueda (Idemitsu Kosan Co.), *Jpn. Pat.* 85 228 541 (1985) (*Chem. Abstr.*, 1986, **104**, 187 100).
60. J. B. Rose (ICI plc), *Eur. Pat. Appl.* 49 070 (1982) (*Chem. Abstr.*, 1982, **97**, 6999).
61. R. J. Mulder, A. M. van Leusen and J. Strating, *Tetrahedron Lett.*, 1967, 3061.
62. G. E. Vennstra and B. Zwaneburg, *Synthesis*, 1975, 519.
63. C. M. Starks and C. Liotta, 'Phase-transfer Catalysis', Academic Press, New York, 1978.
64. W. E. Keller, 'Phase-transfer Reactions — Fluka Compendium', Thieme, Stuttgart, 1986, vol. 1.
65. M. Sato, H. Kondo and M. Yokoyama, *Makromol. Chem., Rapid Commun.*, 1982, **3**, 821.
66. M. Sato and M. Yokoyama, *Makromol. Chem.*, 1984, **185**, 629.
67. M. Sato, *Makromol. Chem., Rapid Commun.*, 1984, **5**, 151.
68. P. C. L. Thorne and E. R. Roberts, in 'Ephraim's Inorganic Chemistry', 4th edn., Oliver and Boyd, Edinburgh, 1943, p. 544; 5th edn., 1948. p. 555.
69. E. Wellisch, E. Gipstein and O. J. Sweeting, *J. Polym. Sci., Polym. Lett. Ed.*, 1964, **2**, 35.
70. A. F. Wells, 'Structural Inorganic Chemistry', 3rd edn., Oxford University Press, Oxford, 1962, p. 422.
71. M. Schmidt and W. Siebert, in 'Comprehensive Inorganic Chemistry', ed. J. C. Bailar, Jr., H. J. Emeleus, R. Nyholm and A. F. Trotman-Dickenson, Pergamon Press, Oxford, 1973, vol. 2, p. 881.
72. K. Soga, K. Nagata, I. Hattori and S. Ikeda, *Makromol. Chem.*, 1980, **181**, 2019.
73. C. M. Suter, 'The Organic Chemistry of Sulfur', Wiley, New York, 1944, p. 400.
74. J. Miller, 'Aromatic Nucleophilic Substitution', Elsevier, New York, 1968.
75. D. J. Sam and H. E. Simmons, *J. Am. Chem. Soc.*, 1974, **96**, 2253.
76. A. E. Pavlath and A. J. Leffler, 'Aromatic Fluorine Compounds', Reinhold, New York, 1962, p. 29.
77. G. P. Briner, J. Miller, M. Liveris and P. G. Lutz, *J. Chem. Soc.*, 1954, 1265.
78. R. W. Bost and F. Nicholson, *J. Am. Chem. Soc.*, 1935, **57**, 2368.
79. J. F. Bunnett and R. E. Zahler, *Chem. Rev.*, 1951, **49**, 273.
80. W. B. Whalley, *J. Chem. Soc.*, 1950, 2241.
81. J. D. Reinheimer, J. P. Douglass, H. Leister and M. B. Voelkel, *J. Org. Chem.*, 1957, **22**, 1743.
82. P. E. Fanta, *Synthesis*, 1974, 9.

83. M. F. Semmelhack, P. Helquist, L. D. Jones, L. Keller, L. Mendelson, L. S. Ryono, J. G. Smith and R. D. Stauffer, *J. Am. Chem. Soc.*, 1981, **103**, 6460.
84. C. A. Buehler and D. E. Pearson, 'Survey of Organic Syntheses', Wiley, New York, 1970, vol. 1, p. 292.
85. H. E. Ungnade and E. F. Orwoll, in 'Organic Syntheses', ed. E. C. Horning, Wiley, New York, 1955, collective vol. 3, p. 566.
86. R. G. R. Bacon and O. J. Stewart, *J. Chem. Soc.*, 1965, 4953.
87. A. L. Williams, R. E. Kinney and R. F. Bridger, *J. Org. Chem.*, 1967, **32**, 2501.
88. M. P. Cava and A. Afzali, *J. Org. Chem.*, 1975, **40**, 1553.
89. H. Weingarten, *J. Org. Chem.*, 1964, **29**, 977, 3624.
90. G. F. Woods and I. W. Tucker, *J. Am. Chem. Soc.*, 1948, **70**, 2174.
91. G. M. Whitesides, J. S. Sadowski and J. Lilburn, *J. Am. Chem. Soc.*, 1974, **96**, 2829.
92. G. P. Brown and A. Goldman, *Polym. Prepr., Am. Chem. Soc., Div. Polym. Chem.*, 1964, **5**, 195.
93. H. S. Blanchard, H. L. Finkbeiner and G. A. Russell, *J. Polym. Sci.*, 1962, **58**, 469.
94. S. R. Sandler and W. Karo, 'Polymer Syntheses', Academic Press, New York, 1974, vol. 1, p. 239.
95. D. M. White and C. D. Cooper, in 'Kirk-Othmer Encyclopedia of Chemical Technology', 3rd edn., ed. M. Grayson and D. Eckroth, Wiley, New York, 1982, vol. 18, p. 594.
96. J. H. Beeson and R. E. Pecsar, *Anal. Chem.*, 1969, **41**, 1678.
97. H. M. van Dort, C. A. M. Hoefs, E. P. Magré, A. J. Schöpf and K. Yntema, *Eur. Polym. J.*, 1968, **4**, 275.
98. R. N. Johnson, A. G. Farnham, R. A. Clendinning, W. F. Hale and C. N. Merriam, *J. Polym. Sci., Polym. Chem. Ed.*, 1967, **5**, 2375.
99. S. R. Shultze and A. L. Baron, *Adv. Chem. Ser.*, 1969, **91**, 692.
100. A. B. Newton and J. B. Rose, *Polymer*, 1972, **13**, 465.
101. T. E. Attwood, A. B. Newton and J. B. Rose, *Br. Polym. J.*, 1972, **4**, 391.
102. T. E. Attwood, D. A. Barr, T. King, A. B. Newton and J. B. Rose, *Polymer*, 1977, **18**, 359.
103. G. Allen, J. McAinsh and C. Strazielle, *Eur. Polym. J.*, 1969, **5**, 319.
104. A. K. Mikitaev, *Acta Polym.*, 1981, **32**, 453.
105. F. A. Carey and R. J. Sundberg, 'Advanced Organic Chemistry', 2nd edn., Plenum Press, New York, 1984, part A, p. 179.
106. M. J. Wang, Y. L. Yang, J. F. Yao and K. Z. Liu, *Kao Fen Tzu T'ung Hsun*, 1979, 257 (*Chem. Abstr.*, 1980, **93**, 205 072).
107. A. Kh. Bulai, V. N. Klyuchnikov, Ya. G. Urman, I. Ya. Slonim, L. M. Bolotina, V. A. Kozhina, M. M. Gol'der, S. G. Kulichikhin, V. P. Beghishev and A. Ya. Malkin, *Polymer*, 1987, **28**, 1349.
108. ICI Ltd., *Neth. Pat. Appl.* 6 613 475 (1967) (*Chem. Abstr.*, 1967, **67**, 44 283).
109. D. A. Barr and J. B. Rose (ICI Ltd.), *Br. Pat.* 1 153 528 (1969) (*Chem. Abstr.*, 1969, **71**, 39 671).
110. D. A. Barr and J. B. Rose (ICI Ltd.), *US Pat.* 3 634 355 (1972) (*Chem. Abstr.*, 1972, **76**, 127 774).
111. D. A. Barr and J. B. Rose (ICI Ltd.), *US Pat.* 4 232 142 (1980) (*Chem. Abstr.*, 1981, **94**, 66 428).
112. R. N. Johnson and A. G. Farnham, *J. Polym. Sci., Polym. Chem. Ed.*, 1967, **5**, 2415.
113. A. B. Newton (ICI Ltd.) *Ger. Pat.* 1 914 324 (1969) (*Chem. Abstr.*, 1970, **72**, 3941); *Br. Pat.* 1 265 145 (1972) (*Chem. Abstr.*, 1972, **77**, 20 299).
114. ICI Ltd., *Jpn. Pat.* 78 12 991 (1978) (*Chem. Abstr.*, 1978, **89**, 44 411); *Jpn. Pat.* 78 10 696 (1978) (*Chem. Abstr.*, 1978, **89**, 110 791).
115. G. Blinne and C. Cordes (BASF A. G.), *Ger. Pat.* 2 731 816 (1979) (*Chem. Abstr.*, 1979, **90**, 138 421).
116. J. P. Xu, Y. S. Ni, Y. M. Wen, J. L. Zhang, Z. Y. Zhang, X. Z. Qu and J. P. Shu, *Kao Fen Tzu T'ung Hsun*, 1979, 266 (*Chem. Abstr.*, 1980, **93**, 186 851).
117. D. K. Mohanty, J. L. Hedrick, K. Gobetz, B. C. Johnson, L. Yilgor, E. Yilgor, R. Yang and J. E. McGrath, *Polym. Prepr., Am. Chem. Soc., Div. Polym. Chem.*, 1982, **23**, 284.
118. L. F. Fieser and M. Fieser, 'Reagents for Organic Synthesis', Wiley, New York, 1967, vol. 1, p. 933.
119. T. E. Attwood, T. King, I. D. McKenzie and J. B. Rose, *Polymer*, 1977, **18**, 365.
120. T. E. Attwood, T. King, V. J. Leslie and J. B. Rose, *Polymer*, 1977, **18**, 369.
121. Kirin University, *Chi Lin Ta Hsueh, Tsu Jan K'o Hsueh Hsueh Pao*, 1979, 73 (*Chem. Abstr.*, 1980, **93**, 47 234).
122. T. Aoyagi, N. Yagi, H. Matsumura and I. Kishi (Denki Kagaku Kogyo KK), *Jpn. Pat.* 79 18 899 (1979) (*Chem. Abstr.*, 1979, **90**, 205 156).
123. N. Yagi, H. Okai, M. Fukuda and I. Kishi (Denki Kagaku Kogyo KK), *Ger. Pat.* 2 433 400 (1975) (*Chem. Abstr.*, 1975, **82**, 171 705).
124. T. King and J. B. Rose (ICI Ltd.), *Ger. Pat.* 2 355 927 (1974) (*Chem. Abstr.*, 1975, **82**, 58 489).
125. R. Viswanathan and J. E. McGrath, *Polym. Prepr., Am. Chem. Soc., Div. Polym. Chem.*, 1979, **20**, 365.
126. Y. Ding and X. Wang, *Gaodeng Xuexiao Huaxue Xuebao*, 1985, **6**, 177 (*Chem. Abstr.*, 1985, **103**, 142 473).
127. L. A. Hartmann (ICI Americas Inc.), *US Pat.* 4 156 068 (1979) (*Chem. Abstr.*, 1979, **91**, 75 088).
128. L. M. Maresca and H. S. Chao (Union Carbide Corp.) *US Pat.* 4 473 684 (1984) (*Chem. Abstr.*, 1984, **101**, 231 205).
129. T. E. Attwood, J. B. Rose and A. F. Lennox (ICI Ltd.) *Def. Publ. U.S. Pat. Off.* 905 009 (1972) (*Chem. Abstr.*, 1973, **78**, 44 699).
130. R. J. Cornell, *Polym. Prepr., Am. Chem. Soc., Div. Polym. Chem.*, 1972, **13**, 607.
131. R. J. Cornell, *Adv. Chem. Ser.*, 1973, **129**, 131.
132. G. C. Finger and C. W. Kruse, *J. Am. Chem. Soc.*, 1956, **78**, 6034.
133. J. H. Clark and J. M. Miller, *J. Chem. Soc., Chem. Commun.*, 1976, 229; *Tetrahedron Lett.*, **1977**, 139.
134. T. Ando and J. Yamawaki, *Chem. Lett.*, 1979, 45.
135. T. Ando, J. Yamawaki, T. Kawate, S. Sumi and T. Hanafusa, *Bull. Chem. Soc. Jpn.*, 1982, **55**, 2504.
136. J. M. Miller, K. H. So and J. H. Clark, *Can. J. Chem.*, 1979, **57**, 1887.
137. Y. Imai, M. Ueda and M. Ii, *Makromol. Chem.*, 1978, **179**, 2989.
138. E. J. Corey and B. B. Snider, *J. Am. Chem. Soc.*, 1972, **94**, 2549.
139. M. Fieser, 'Reagents for Organic Synthesis', Wiley, New York, 1982, vol. 10, p. 81, 325, 378.
140. H. R. Kricheldorf and G. Bier, *J. Polym. Sci., Polym. Chem. Ed.*, 1983, **21**, 2283.
141. A. McKillop, J. C. Fiaud and R. P. Hug, *Tetrahedron*, 1974, **30**, 1379.
142. H. Alsaidi, R. Gallo and J. Metzger, *Synthesis*, 1980, 921.
143. Y. Imai, M. Ueda and M. Ii, *J. Polym. Sci., Polym. Lett. Ed.*, 1979, **17**, 85.

144. D. J. Gerbi, R. F. Williams, R. Kellman and J. L. Morgan, *Polym. Prepr., Am. Chem. Soc., Div. Polym. Chem.*, 1981, **22**, 385.
145. W. Podkoscielny, M. Dethloff, J. Dethloff and A. Dawidowicz, *Angew. Makromol. Chem.*, 1978, **69**, 67.
146. Idemitsu Kosan Co. Ltd., *Jpn. Pat.* 85 32 826 (1985) (*Chem. Abstr.*, 1985, **103**, 37 886).
147. S. R. Sandler and W. Karo, 'Polymer Syntheses', Academic Press, New York, 1980, vol. 3, p. 98.
148. A. Kreuchunas, *US Pat.* 2 822 351 (1958) (*Chem. Abstr.*, 1958, **52**, 7778b).
149. A. L. Baron and D. R. Blank, *Makromol. Chem.*, 1970, **140**, 83.
150. R. Gabler (Inventa A. G.), *Ger. Pat.* 2 009 323 (1970) (*Chem. Abstr.*, 1970, **73**, 110 331).
151. R. G. Feasey (ICI Ltd.), *US Pat.* 3 819 582 (1974) (*Chem. Abstr.*, 1975, **82**, 157 207).
152. R. G. Feasey and J. B. Rose (ICI Ltd.), *Ger. Pat.* 2 156 343 (1972) (*Chem. Abstr.*, 1972, **77**, 75 774).
153. M. S. Fortuin (ICI Ltd.), *Br. Pat.*, 1 369 217 (1974) (*Chem. Abstr.*, 1975, **82**, 98 741).
154. R. W. Campbell (Phillips Petroleum Co.), *Ger. Pat.* 2 726 862 (1977) (*Chem. Abstr.*, 1978, **88**, 137 178); *US Pat.* 4 070 349 (1978) (*Chem. Abstr.*, 1978, **89**, 25 071); *US Pat.* 4 125 525 (1978) (*Chem. Abstr.*, 1979, **90**, 55 528).
155. K. Idel, E. Ostlinning, W. Koch and W. Heitz (Bayer A. G.), *Ger. Pat* 3 312 254 (1984) (*Chem. Abstr.*, 1985, **102**, 25 584).
156. R. Gabler and J. Studinka (Inventa A. G.), *Ger. Pat.* 1 909 441 (1969) (*Chem. Abstr.*, 1970, **72**, 32 479).
157. H. Kawahara, I. Otsuka and K. Kashiwame (Asahi Glass Co.), *Jpn. Pat.* 86 72 020 (1986) (*Chem. Abstr.*, 1986, **105**, 173 292).
158. H. Kawahara, I. Otsuka, K. Kashiwame and T. Yasuda (Asahi Chem. Ind. Co.), *Jpn. Pat.* 86 76 523 (1986) (*Chem. Abstr.*, 1986, **105**, 153 738).
159. K. Kashiwame, I. Otsuka and S. Ozawa (Asahi Chem. Ind. Co.), *Jpn Pat.* 86 168 629 (1986) (*Chem. Abstr.*, 1987, **106**, 19 206).
160. C. W. Stephens, *J. Polym. Sci.*, 1959, **40**, 359.
161. C. Chiriac and J. K. Stille, *Macromolecules*, 1977, **10**, 712.
162. K. J. Scariah, V. N. Krishnamurthy, K. V. C. Rao and M. Srinivasan, *Makromol. Chem.*, 1985, **186**, 2427.
163. I. Thomas and J. R. Traynor (ICI Ltd.), *Ger. Pat.* 2 252 933 (1973) (*Chem. Abstr.*, 1973, **79**, 126 914).
164. G. F. D'Alelio (Plastics Eng. Co.), *US Pat.* 4 517 354 (1985) (*Chem. Abstr.*, 1985, **103**, 88 375).
165. R. H. Lubowitz and C. H. Sheppard (Boeing Co.; TRW Inc.), *US Pat.* 4 584 364 (1986) (*Chem. Abstr.*, 1986, **105**, 79 886).
166. W. M. Eareckson, III, *J. Polym. Sci.*, 1959, **40**, 399.
167. Mitsui Toatsu Chem. Inc., *Jpn. Pat.* 83 79 014 (1983) (*Chem. Abstr.*, 1983, **99**, 195 620).
168. A. K. Banthia, D. C. Webster and J. E. McGrath, *Org. Coat. Plast. Chem.*, 1980, **42**, 127.
169. T. C. Chiang and S. L. Ng, *Polymer*, 1981, **22**, 3.
170. J. M. Lambert, E. Yilgor, I. Yilgor, G. L. Wilkes and J. E. McGrath, *Polym. Prepr., Am. Chem. Soc., Div. Polym. Chem.*, 1985, **26**, 275.
171. H. Schnell, *Ind. Eng. Chem.*, 1959, **51**, 157.
172. R. L. Price, M. W. Witman and S. Krishnan (Mobay Chem. Corp.), *US Pat.* 4 535 143 (1985) (*Chem. Abstr.*, 1985, **103**, 161 004).
173. K. P. Sivaramakrishnan and C. S. Marvel, *J. Polym. Sci., Polym. Chem. Ed.*, 1974, **12**, 1945.
174. V. Jansons and H. C. Heinrich (Raychem Corp.), *Eur. Pat. Appl.* 178 183 (1986) (*Chem. Abstr.*, 1986, **105**, 43 579).
175. D. A. Barr and J. B. Rose (ICI Ltd.) *US Pat. Reissue* 28 252 (1974) (*Chem. Abstr.*, 1975, **82**, 125 810).
176. T. E. Attwood, P. C. Dawson, J. L. Freeman, L. R. J. Hoy, J. B. Rose and P. A. Staniland, *Polymer*, 1981, **22**, 1096.
177. J. B. Rose, E. Nield, P. T. McGrail and H. M. Colquhoun (ICI), *Eur. Pat. Appl.* 194 062 (1986) (*Chem. Abstr.*, 1987, **106**, 120 420).
178. D. L. Schoene (United States Rubber Co.), *US Pat.* 2 505 366 (1950) (*Chem. Abstr.*, 1950, **44**, 6676i).
179. Y. Imai, Y. Asamidori, T. Inoue and M. Ueda, *J. Polym. Sci., Polym. Chem. Ed.*, 1981, **19**, 583.
180. D. L. Schoene (United States Rubber Co.), *US Pat.* 2 493 364 (1950) (*Chem. Abstr.*, 1950, **44**, 5643c).
181. United States Rubber Co., *Br. Pat.* 650 742 (1951) (*Chem. Abstr.*, 1951, **45**, 6430a).
182. J. W. Schappel (American Viscose Corp.), *US Pat.* 2 623 035 (1952) (*Chem. Abstr.*, 1953, **47**, 3038a).
183. Fuji Photo Film Co., *Brit. Pat.* 1 534 495 (1978) (*Chem. Abstr.*, 1979, **91**, 66 257).
184. J. Heyna, L. Berlin and E. Schinzel (Farbw. Hoechst A. G.), *Ger. Pat.* 1 100 942 (1961) (*Chem. Abstr.*, 1961, **55**, 24 064e).
185. G. C. Tesoro, *US Pat.* 3 031 435 (1962) (*Chem. Abstr.*, 1962, **57**, 2459d).
186. J. P. Stevens & Co., *Br. Pat.* 985 150 (1965) (*Chem. Abstr.*, 1965, **62**, 11 962c).
187. H. Mark and S. M. Atlas, in 'Chemical Aftertreatment of Textiles', ed. H. F. Mark, Interscience, New York, 1971, p. 411.

34
Polysiloxanes

J. J. LEBRUN and H. PORTE
Rhône Poulenc, Saint Fons, France

34.1 INTRODUCTION	593
34.2 POLYCONDENSATION OF SILANOLS AND SILOXANOLS	594
34.2.1 Reactivity of Silanols	594
34.2.2 Condensation Reactions of Silanols	594
34.2.2.1 Thermal condensation	594
34.2.2.2 Homogeneous catalysis	594
34.2.2.3 Other polycondensation catalysts	598
34.3 CONDENSATION OF SILANOLS WITH ORGANOSILANES	598
34.3.1 Introduction to the Preparation of Elastomers	598
34.3.2 Chemistry of Crosslinking	600
34.3.2.1 Crosslinking via acyloxy groups	600
34.3.2.2 Crosslinking via alkoxy groups	602
34.3.2.3 Crosslinking via alkenyloxy groups	603
34.3.2.4 Crosslinking via amino groups	603
34.3.2.5 Crosslinking via ketiminoxy groups	604
34.3.2.6 Crosslinking via amido groups	604
34.3.2.7 Crosslinking via aminoxy groups	605
34.3.3 Conclusion	606
34.4 REFERENCES	606

34.1 INTRODUCTION

The synthesis of poly(dimethylsiloxane) macromolecules having $\equiv$SiOH groups at the end of the chain may be carried out by the following methods.

(i) Either by ring-opening polymerization (equation 1), for example in a basic medium using water as a transfer agent during polymerization or as a deactivating agent at the end.

$$Mt^+OH^- + (SiO)_x \xrightarrow{\text{Initiation}} HOSi\sim\sim\sim SiO^-Mt^+ \xrightarrow[\text{Propagation}]{O_x} HOSi\sim\sim\sim SiO^-Mt^+ \xrightarrow{\text{Deactivation}} HOSi\sim\sim\sim SiOH \quad (1)$$

$X = 3\text{--}6$

(ii) Or by polycondensation catalyzed by acids or bases (equation 2), starting with α,ω-siloxanediol oligomers resulting from the hydrolysis of Me_2SiCl_2.

$$n\text{HO}\!-\!(SiO)_y\!-\!H \xrightarrow{H^+ \text{ or } OH^-} HO\!-\!(SiO)_{ny}\!-\!H + (n-1)\,H_2O \quad (2)$$

By either of these processes, average degrees of polymerization of a few thousands may be obtained.

Although there have been many studies relating to the anionic or cationic polymerization of cyclosiloxanes, the polycondensation of linear α,ω-siloxanediol derivatives has formed the subject of only a few published fundamental investigations. Nonetheless, this reaction is widely utilized in silicone chemistry on an industrial scale for the production of macromolecules which form part of elastomeric formulations.

In the latter case, the reactions mainly employed involve the use of silanol or siloxanol groups with polyfunctional organosilicon derivatives R_nSiX_{4-n}, where $n=0$, 1 or 2, the group X being chosen so as to have a ≡Si—X bond which is easily hydrolyzable: *i.e.* acyloxy, alkoxy, amino, amido and the like.

34.2 POLYCONDENSATION OF SILANOLS AND SILOXANOLS

34.2.1 Reactivity of Silanols

Condensation reactions involving silanols are governed by the polarity of the Si—O and O—H bonds. This depends on the electronic nature of the substituents of silicon.

Silanols are stronger acids than their hydrocarbon homologues; this is demonstrated by the formation and the characterization of complexes of silanol and electron-donating compounds: ethers,[1-8] amines,[8,9] ketones[3] and the like. The bond between the silicon and the oxygen is σ and $p\pi$–$d\pi$ in nature, which increases the polarity of the O—H bond.[1,4,5,9-11]

This property of the silanol bond depends on the nature of the silicon substituents. For example, the acidity of silanol decreases when the inductive effect +I of the silicon substituents increases.

In contrast to the chemistry of hydrocarbon compounds, the basicity of silanols is not inversely proportional to their acidity.[7,12]

34.2.2 Condensation Reactions of Silanols

Like all reactions of this type, the condensation of silanols is an equilibrium reaction (3).

$$\equiv\text{SiOH} + \text{HOSi}\equiv \rightleftharpoons \equiv\text{SiOSi}\equiv + H_2O \qquad (3)$$

It will therefore be possible to promote the formation of a siloxane bond by decreasing the partial pressure of water in the medium.

34.2.2.1 *Thermal condensation*

Because of their acid nature, silanols may be made to undergo homocondensation by heat treatment. This is so, for example, in the case of trimethylsilanol which undergoes condensation from room temperature upwards.[14] The heat stability of tris(organo)silanols increases with the steric hindrance of the silicon substituents.

Thus, tris(isoamyl)silanol undergoes condensation from 270 °C upwards,[15] whereas triphenylsilanol is stable up to 300 °C.[16]

The reactivity of silanols increases with the number of hydroxyl groups on the silicon.[13]

Diorganosilanediols are much more heat unstable than triorganosilanols: dimethylsilanediol must be stored at low temperature[17-18] and the impurities present in the glass catalyze its condensation.[19]

As in the case of the triorgano derivatives, the stability of diorganosilanediol increases with the size of the substituents: diethylsilanediol is stable at room temperature,[20] diisopropylsilanediol undergoes condensation from 130 °C upwards,[21] and dicyclohexylsilanediol at 200 °C.[21]

Linear oligomers of formulae $HO[Me_2SiO]_nH$ (with $n=3$, 4 or 5) are much more stable and do not react even at 200 °C;[22] therefore the condensation reaction must be catalyzed.

34.2.2.2 *Homogeneous catalysis*

(i) Catalysis by acids

Many cationic catalysts bring about the condensation of silanols, among which are those mentioned in Table 1.

Table 1

Acids	Refs.
H_2SO_4	23–32
HCl	26, 33–42
HBr	43
H_3PO_4	23, 32, 44
H_3BO_4	45–47
HNO_3	33
CF_3CO_2H	48
Aryl SO_3H	49–51
$MeSO_3H$	52
CF_3SO_3H	52–54

Studies on the kinetics of cationic polycondensation of silanediols were carried out in the 1960s.[55] The results obtained depend on: the nature of the silanediol employed; the type of catalyst chosen; and the possible presence of additives.

Taking the condensation rate of dimethylsilanediol as reference, the reactivity of different monomers in the decreasing order is as follows: $Me_2Si(OH)_2$ > $Me(CH{=}CH_2)Si(OH)_2$ > $(ClCH_2)MeSi(OH)_2$ > $(Cl_2CH)MeSi(OH)_2$; and in the aromatic series: $Me_2Si(OH)_2$ > $PhMeSi(OH)_2$ > $Ph_2Si(OH)_2$.

Increasing the electronegative groups on the silicon causes a decrease in the electron density on the oxygen of the silanol bond and renders attack by a proton more difficult.[55]

An increase in the size of the substituent causes a decrease in condensation rate by a steric effect.[58]

Many catalysts have been studied: HCl, HBr, H_2SO_4 and $HClO_4$.[56,67]

It is possible to demonstrate the following common trends. (i) Second order kinetics with respect to silanediol and first order kinetics with respect to the catalyst.[56-67] (ii) Existence of an induction period.[60-62,64] (iii) Significant modification of the progress of reaction by the introduction of water: elimination of the induction period and acceleration of the rate of condensation catalyzed by HCl and H_2SO_4;[60,61,63] decrease in the rate of condensation catalyzed by HBr and $HClO_4$;[63] and increase in activation energy in some cases (Table 2). (iv) Similarity in behaviour of condensations in the presence of water and of alcohol.[64]

Table 2

$H_2O(M\,l^{-1} \times 10^4)$	E_a(kcal mol^{-1}) H_2SO_4	HBr	$HClO_4$
0.15	4.5	6.1	14.1
0.35	—	11.1	14.2
0.50	7.8	—	14.0

Monomer = $Et_2Si(OH)_2$ and solvent = dioxane (1 cal = 4.19 J).

The mechanism proposed[59] is the sum total of two elementary processes: (a) production of silanol to give a primary oxonium ion; and (b) reaction of a silanol with the oxonium ion formed (Scheme 1).

$\equiv$SiOH + HX $\rightleftharpoons$ $\equiv$SiOH$_2^+$X$^-$ (a)

$\equiv$SiOH + $\equiv$SiOH$_2^+$X$^-$ $\rightleftharpoons$ $\equiv$SiOSi$\equiv$ + HX + H_2O (b)

2 $\equiv$SiOH $\rightleftharpoons$ $\equiv$SiOSi$\equiv$ + H_2O

Scheme 1

However, it should be noted that there is no experimental evidence for the existence of this oxonium ion.

It is only very recently that cationic polycondensation of siloxanes has been studied by using models for the formation of reaction products.[52] The monomer employed is decamethylpentasiloxane-1,9-diol and the catalysts are methanesulfonic and trifluoromethanesulfonic acids. The condensation leads to the formation of two categories of products (Scheme 2).

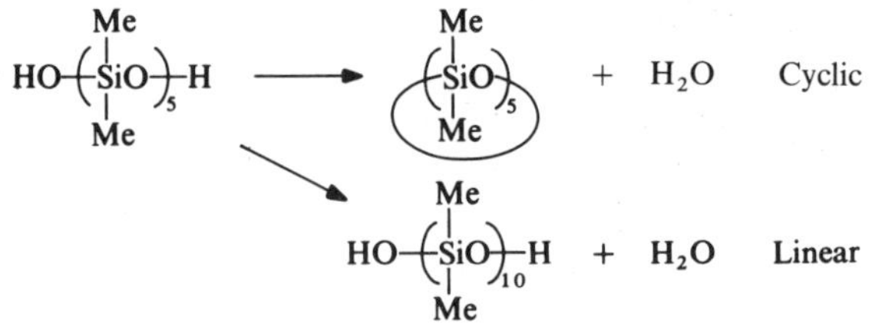

$$HO\text{-}(SiMeO)_5\text{-}H \longrightarrow (SiMeO)_5 + H_2O \quad \text{Cyclic}$$

$$\longrightarrow HO\text{-}(SiMeO)_{10}\text{-}H + H_2O \quad \text{Linear}$$

Scheme 2

The nature of solvent employed has a significant effect on the nature of products formed: in dioxane, cyclic oligomers are predominantly formed and a difference in order with respect to the monomer is observed (first order for the formation of cyclic molecules and second order for the formation of linear oligomers); whilst in dichloromethane, the main compounds formed are linear.

The author proposes a reaction mechanism which involves a nucleophilic assistance of dioxane towards silanols, a role which cannot be fulfilled by dichloromethane.

(ii) Catalysis by bases

The polycondensation reaction of silanediols and of α,ω-siloxanediols may also be catalyzed by adding strong bases: KOH, NaOH or LiOH, at concentrations of the order of 10^{-3} mol l^{-1}.[68-70]

A base-catalyzed condensation passes through a preliminary stage of silanolate formation (equation 4).

$$\equiv\text{SiOH} + \text{Mt}^+\text{OH}^- \rightleftharpoons \equiv\text{SiO}^-\text{Mt}^+ + H_2O \quad (4)$$

Considering the respective concentrations of silanol and the base, it can be assumed that the silanolate concentration in the medium is equal to the initial concentration of the catalyst, when $\text{Mt}^+ = \text{Li}^+$, Na^+ or K^+ in a polar solvent.

The propagation stage may be approximated to a nucleophilic attack by the silanolate on a silanol with hydroxyl regeneration, as in equation (5), which determines the kinetics of the polycondensation.

$$\equiv\text{SiO}^-\text{Mt}^+ + \equiv\text{SiOH} \rightleftharpoons \equiv\text{SiOSi}\equiv + \text{OH}^-\text{Mt}^+ \quad (5)$$

From a comparison of the condensation by acid catalysis with that by basic catalysis, it is observed that the reactivity of the different silanediols is reversed: $Me_2Si(OH)_2 < (CH_2=CH)\text{-}MeSi(OH)_2 < (ClCH_2)MeSi(OH)_2 < (Cl_2CH)MeSi(OH)_2$ and $(Me)_2Si(OH)_2 < MePhSi(OH)_2 < Ph_2Si(OH)_2$.

The nucleophilic attack by the base on the silicon atom is facilitated by the electronegativity of the substituents. It is also interesting to note that in the case of basic catalysis, the electronegativity prevails over the steric effect of the substituent.[72]

The condensation rate of trimethylsilanol in a methanolic medium is 500 times faster with hydrochloric acid than with potassium hydroxide.[70]

Kinetic studies on polycondensation in methanol have been carried out with dimethylsilanediol or with methylphenylsilanediol, restricting the conversion level to 10%.[71] Beyond this value, the formation of ≡SiOMe groups interferes with the determinations. The condensation rate of silanols is proportional to the catalyst concentration, but does not depend on the nature of the cation. The limiting stage of the reaction would be the action of the anion on the silanol[70,73] and its kinetics are first order with respect to the silanol.[71] When the silanol is condensed in acetone using di-(n-hexyl)ammonium 2-ethylhexanoate, the reaction is first order with respect to silanol, but it is of second order with acetonitrile as solvent.[74] A study of the kinetics of condensation of 1,4-bis-(dimethylhydroxysilyl)benzene and 1,1,3,3-tetramethyl-1,3-dihydroxydisiloxane in toluene under conditions of azeotropic distillation of water, in the presence of potassium hydroxide, indicates a second order with respect to the monomer and an order of 0.5 with respect to the catalyst.[75] In this case, it is possible that the silanolate in a solvent of low polarity might be in the form of an associated ion pair (**1**), and that propagation occurs predominantly on non-associated silanolate.

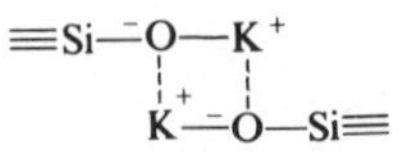

$$\equiv\text{Si}\text{-}\text{O}\text{-}\text{K}^+$$
$$\text{K}^+\text{-}\text{O}\text{-}\text{Si}\equiv$$

(**1**)

More recently, Chojnowski[76] has proposed a polycondensation mechanism involving an elimination reaction with the intermediate formation of a silicon-containing derivative carrying a double bond (Scheme 3, where R = organic residue on polymer chain).

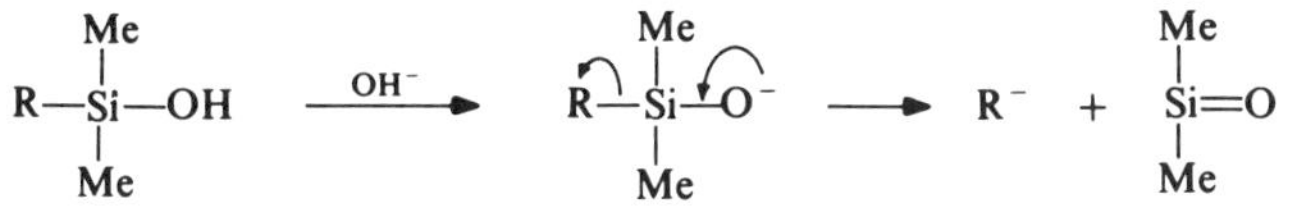

Scheme 3

The formation of such a silanone had been proposed previously.[77] This intermediate is unstable and undergoes rapid change, as in equation (6).

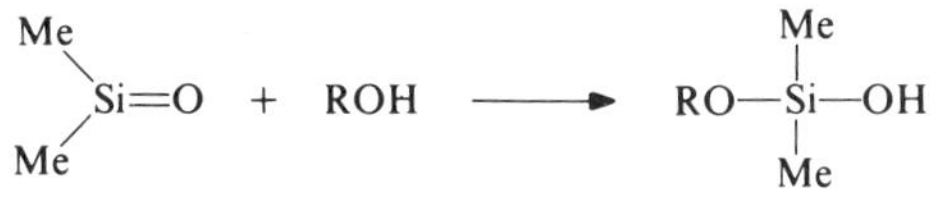

$$\underset{Me}{\overset{Me}{Si}}=O + ROH \longrightarrow RO-\underset{Me}{\overset{Me}{Si}}-OH \qquad (6)$$

Thus, the anionic polymerization of cyclosiloxanes and the polycondensation catalyzed by bases could involve this reaction intermediate, the formation of which would explain the results obtained for the model polycondensation of decamethylpentasiloxane-1,9-diol. This study was carried out in 1,4-dioxane in the presence of 2% of water and of NaOH. The reactions expected are as in Scheme (4).

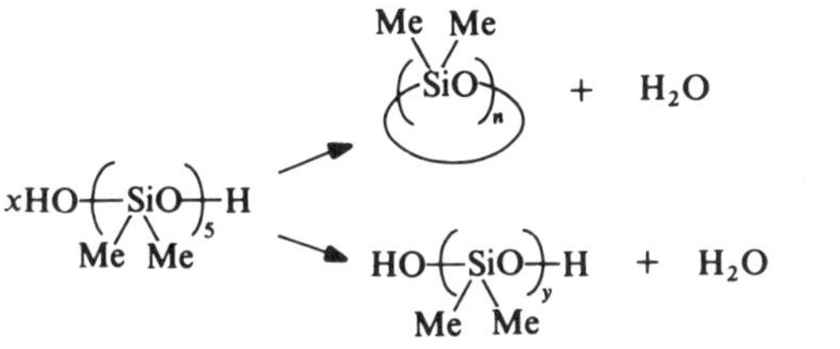

Scheme 4

Logically, at the beginning of the reaction, the production of compounds in which $n=5$ and $y=10$ may be expected. However, at low conversion levels (10%), (2) and (3) are essentially formed, but no cyclic derivative D_5 such as (4) nor (5).

(2) $HO{\text{-}}(SiO)_4{\text{-}}H$ with Me Me
(3) $HO{\text{-}}(SiO)_6{\text{-}}H$ with Me Me
(4) cyclic $(SiO)_5$ with Me Me
(5) $HO{\text{-}}(SiO)_{10}{\text{-}}H$ with Me Me

This is confirmed even for a high conversion level (80%). However, from the kinetic determinations, it is not possible to decide between the conventionally proposed mechanisms (Scheme 5) and the mechanism put forward by Chojnowski (Scheme 6).

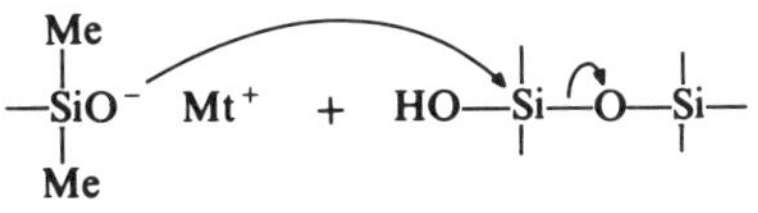

Scheme 5

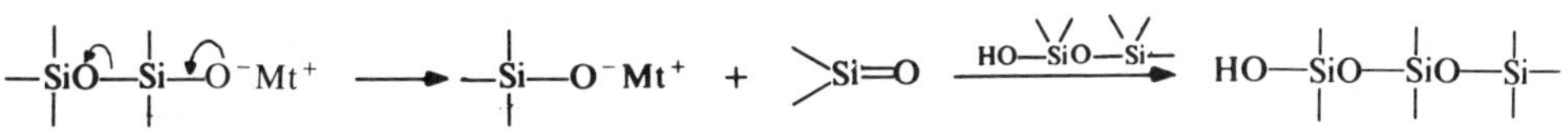

Scheme 6

Nevertheless, assuming the passage through a silanone, it would be possible to explain some secondary reactions encountered in anionic polymerization and written previously (Scheme 7; R = organic residue or polymer chain; Mt = Li, Na, K and the like). This mechanism can hardly be considered because of the proximity of the negative charges.

$$-\underset{Me}{\underset{|}{Si}}-O-\underset{Me}{\underset{|}{Si}}-O^- Mt^+ \;+\; R^- Mt^+ \longrightarrow -\underset{Me}{\underset{|}{Si}}-O^- Mt^+ \;+\; R-\underset{Me}{\underset{|}{Si}}-O^- Mt^+$$

Scheme 7

On the other hand, the mechanism in Scheme 8 appears to be more realistic.

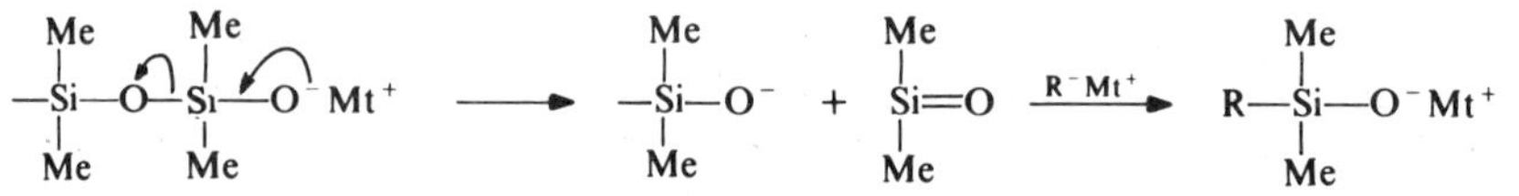

$$-\underset{Me}{\underset{|}{Si}}-O-\underset{Me}{\underset{|}{Si}}-O^- Mt^+ \longrightarrow -\underset{Me}{\underset{|}{Si}}-O^- \;+\; \underset{Me}{Si}=O \xrightarrow{R^- Mt^+} R-\underset{Me}{\underset{|}{Si}}-O^- Mt^+$$

Scheme 8

In fact, irrespective of whether it is by the formation of a silanone or by the intramolecular or intermolecular reaction of the siloxanolate with the polysiloxane chain, it is difficult to restrict the redistribution reactions taking place during polycondensation performed to high conversion levels. These secondary reactions result in an increase of the fraction of volatile compounds with low molecular weights.

Additionally, the activation of the polycondensation reaction using suitable catalysts generally results in an acceleration of the secondary reactions as well. However, the use of agents which sequester the cations and free the anion enables this mechanism to be avoided from a kinetic point of view.

In the presence of cryptands,[78] polyheteromacropolycyclic compounds such as (**6**) or of the sequestering agent[79] tris(oxaalkylamine) (**7**), it is observed that the rate of polycondensation catalyzed by strong bases is greatly accelerated while the formation of volatile compounds of low masses is restricted at the same time.

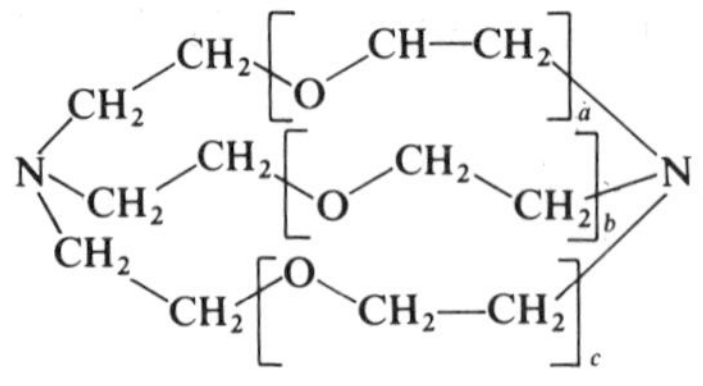

(**6**)

$$N\text{\textendash}[CHR_1\text{\textendash}CHR_2\text{\textendash}O\text{\textendash}(CHR_3\text{\textendash}CHR_4\text{\textendash}O)_n R_5]_3$$

(**7**)

34.2.2.3 Other Polycondensation catalysts

Many acid, basic and metal catalysts have been described and claimed for the polycondensation of silanols. Table 3 gives a non-exhaustive list thereof.

34.3 CONDENSATION OF SILANOLS WITH ORGANOSILANES

34.3.1 Introduction to the Preparation of Elastomers

Reactions involving the use of a silanol group with another silicon derivative containing the labile group ≡SiX have especially been studied in the formulation of silicone elastomers.

Silicone macromolecules have very specific properties, among which a weakness of their cohesive forces and a high freedom of rotation of the SiOSi bonds are noteworthy; these result in a relatively low value for their glass transition temperature ($< -100\,°C$). Therefore, the preparation of elastomeric materials having superior mechanical properties requires the crosslinking of linear polysiloxanes into which reinforcing fillers have been incorporated.

Table 3

Catalyst	Ref.
Amines	80–82
Amine salts + carboxylic acid	82–84
Amine salts + sulfonic acid	85
Amine salts + phosphoric acid	82, 86
BF_3, $MeCO_2H$ + pyridine	87
Alkali metal alkylarylsulfonate	88
Carboxylate of Na, K, Li, Pb, Hg and the like	89, 90
Iron or tin octoate	91, 92
$(RO)_3VO$	93
Alcoholate of Al, Mg, Na and the like	94, 95
Phenolate	96, 97
Metal halides	98
Ion exchange resins	99
Zeolites and the like	100

Elastomers are generally classified according to the method of crosslinking employed: hot vulcanizable elastomers (crosslinking by radical reactions); single component cold vulcanizable elastomers (crosslinking by condensation reactions); and two-component cold vulcanizable elastomers (crosslinking by hydrosilylation or by condensation).

Most of the elastomeric compositions which can be vulcanized at room temperature contain the following components: an α,ω-hydroxylated polysiloxane (**8**); a crosslinking agent of the $RSiX_3$ or SiX_4 type, where X is a group which can be hydrolyzed (acyloxy, amino, ketiminoxy, enoxy and the like); a filler (silica); a catalyst *e.g.* a metal salt having the property of a Lewis acid or a base; various additives *e.g.* colouring agents, fungicides and the like.

$$HO\text{-}(\underset{\underset{Me}{|}}{\overset{\overset{Me}{|}}{Si}}O)_y\text{-}H$$

(**8**)

The structure of the chemical network of the elastomer depends on a number of stages which may be sequential or simultaneous:

(i) Functionalization of the linear chains (equation 7).

$$HOSi\text{—}SiOH + 2RSiX_3 \longrightarrow X_2RSiOSi\text{—}SiOSiX_2R + 2HX \quad (7)$$

(ii) Hydrolysis (equation 8).

$$X_2RSiOSi\text{—}SiOSiX_2R + H_2O \longrightarrow HOXRSiOSi\text{—}SiOSiX_2R + HX \quad (8)$$

(iii) Crosslinking by condensation (equation 9).

$$\sim\!\!SiX + HOSi\!\!\sim \longrightarrow \sim\!\!SiOSi\!\!\sim + HX \quad (9)$$

Among the most commonly employed organosilicon groups ≡SiX, those in Table 4 may be mentioned.

Cold vulcanizable elastomers produced by single component or two-component formulations find their uses as flowing materials for applications in coating, adhesive bonding, moulding and electrical insulation or as non-flowing thixotropic materials in the fields of adhesive bonding and sealing (building and public works).

The economic investment of this class of material is increasingly important. For thirty years, there has been a succession of innovations which are described exclusively in the form of patents.

Table 4

Organosilicon Groups	Formula
Acyloxy groups	≡SiOCOR
Alkoxy groups	≡SiOR
Amino groups	≡SiN(R$_1$)(R$_2$)
Amido groups	≡SiN(R$_1$)—C(=O)—R$_2$
Alkenyloxy (or enoxy) groups	≡Si—O—C(R$_1$)=CHR$_2$
Aminoxy groups	≡Si—O—N(R$_1$)(R$_2$)
Ketiminoxy groups	≡Si—O—N=C(R$_1$)(R$_2$)

The first single component formulations provided have employed an acyloxysilane-based crosslinking system. The elimination of the undesirable acid release has led to the production of a second generation of elastomers. The use of silane carrying hydrolyzable groups, such as alkoxy, amino and the like, has enabled this problem to be overcome.

The current trend consists of improving the environment for the crosslinking system, for instance by adding: 'scavengers' which increase the stability during storage; and catalysts and crosslinking agent (but the problem of their toxicity becomes very important).

The analysis which follows is not exhaustive, but tends to illustrate each category of elastomer. The classification is carried out according to the type of crosslinking employed.

34.3.2 Chemistry of Crosslinking

34.3.2.1 *Crosslinking* via *acyloxy groups*

The principle of this crosslinking is illustrated by equations (10) and (11).

Hydrolysis   ≡SiOCOR + H$_2$O ⇌ ≡SiOH + RCO$_2$H   (10)

Condensation   ≡SiOCOR + ≡SiOH ⇌ ≡SiOH≡ + RCO$_2$H   (11)

The reaction between a polysiloxane carrying silanol ends and a methyltriacetoxysilane type crosslinking agent was described as early as in 1957 by Rhône Poulenc.[101] The composition is stable during storage for a few months in a confined atmosphere and crosslinks by atmospheric moisture without catalyst in a few hours.[102-103]

The condensation reaction may be catalyzed by metal salts and more particularly by tin salts,[104,105] titanium salts[108] or even by a mixture of the two.[107] Catalysis by titanium requires much lower quantities than the quantities of tin employed.[108]

The fillers employed are generally silicas[137] which have optionally undergone a treatment designed to make them hydrophobic.

The production of functional macromolecules is carried out *in situ* or beforehand Section (34.2.2.) Different categories of materials have been employed (Table 5).

It is worth noting that other types of functionalization may be employed, although they do not currently have significant industrial outlets (equation 12).[106,110]

Table 5

Category	Formula	Ref.
Acyloxysilane	$MeSi(OAc)_3$ and $EtSi(OAc)_3$	104
	$MeSi(OAc)_3 + Me_2Si(OAc)_2$	114
	$ViSi(OAc)_3$,[a] $PhSi(OAc)_3$	
	$MeSi(OCOPh)_3$	140–142
Acyloxy-/alkoxy-silane mixture	$MeSi(OAc)_3 + Si(OEt)_4$	109–112
	$MeSi(OAc)_3$ + alkyl polysilicate	
Mixed acyloxy–alkoxy compounds	$(Bu^tO)_2Si(OAc)_2$	111
	$(EtO)_xSi(OAc)_{4-x}$	115
	$R_{4-a-b}Si(OR)_a(OCOR)_b$	116
Others	$(AcO)_2\underset{\underset{Me}{\mid}}{Si}O\underset{\underset{Me}{\mid}}{Si}(OAc)_2$	113
	$(AcO)_3Si(O\underset{\underset{Me}{\mid}}{Si})_3OAc$	117
	$(AcO)_3SiCH_2CH_2\underset{\underset{Me}{\mid}}{Si}(OAc)_2$	118, 119
	Silyl 2-ethylhexanoate	138

[a] Vi = Vinyl group.

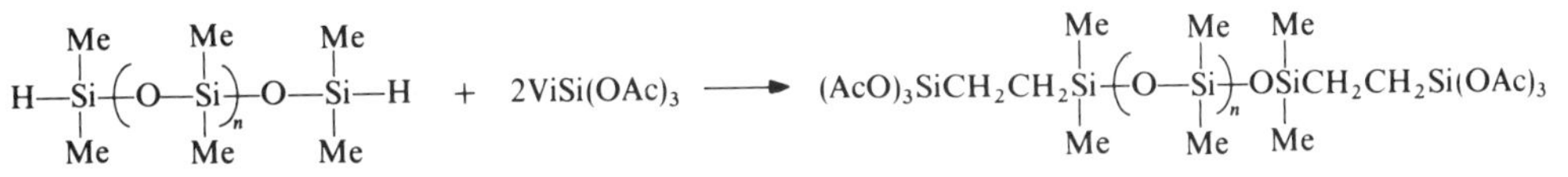

$$H-\underset{\underset{Me}{\mid}}{\overset{\overset{Me}{\mid}}{Si}}{\left(O-\underset{\underset{Me}{\mid}}{\overset{\overset{Me}{\mid}}{Si}}\right)}_n O-\underset{\underset{Me}{\mid}}{\overset{\overset{Me}{\mid}}{Si}}-H + 2ViSi(OAc)_3 \longrightarrow (AcO)_3SiCH_2CH_2\underset{\underset{Me}{\mid}}{\overset{\overset{Me}{\mid}}{Si}}{\left(O-\underset{\underset{Me}{\mid}}{\overset{\overset{Me}{\mid}}{Si}}\right)}_n OSiCH_2CH_2Si(OAc)_3 \quad (12)$$

The crosslinking of the functionalized linear chains occurs through the hydrolysis of the acyloxysilanes under the influence of atmospheric moisture. The silanols formed condense with other acyloxysilanes.

The crosslinking therefore takes place through the surface in contact with ambient air and spreads into the mass by diffusion of water vapour. This causes problems in the case of crosslinking in thick film. An improvement consisted of incorporating lime into the elastomer, which enables water to be released within the mass.[120-124] This enables setting speeds to be accelerated very substantially (equations 13 and 14).

$$\equiv SiOAc + H_2O(atm) \longrightarrow \equiv SiOH + AcOH \quad (13)$$

$$Ca(OH)_2 + 2AcOH \longrightarrow Ca(OAc)_2 + 2H_2O \quad (14)$$

The intrinsic properties of silicone elastomers may be modified or improved by adding specific additives. One of the properties most commonly sought is the adhesiveness to supports of different kinds. Thus, a number of additives may be grouped together (Table 6).

The use of silicone elastomers in a humid environment requires the addition of antifungal substances such as solvents,[135] alkyl benzimidazolylcarbamate[136] or thiurams.[139]

Table 6

Additive	Support	Ref.
$Si(OR)_x(OCOR')_x$	Aluminum	115
Zirconium salt	Aluminum, steel	133
Epoxysilane	Glasses, metals	125, 131
Methyl ethyl silicate	Metals, PVC, concrete	134
Silicone resin	Metals	129, 132

34.3.2.2 Crosslinking via *alkoxy groups*

The principle of this crosslinking is illustrated by equations (15) and (16).

Hydrolysis $\quad\quad\quad\quad \equiv SiOR + H_2O \longrightarrow \equiv SiOH + ROH \quad\quad\quad\quad (15)$

Condensation $\quad\quad \equiv SiOR + \equiv SiOH \longrightarrow \equiv SiOSi\equiv + ROH \quad\quad\quad (16)$

Alkoxysilane-based cold vulcanizable elastomers have undergone a great deal of development over the past 20 years and currently continue to form the subject of many investigations. The composition may be provided in a single component or a two-component form, with formulations which are very close.

The first single component elastomers contained a functionalized oil, a crosslinking agent and a catalyst.

The synthesis of functionalized oil, which has been known for a long time,[143,144] has been improved by the use of functionalization catalysts, such as organic derivatives of titanium,[145] alkoxyaluminum chelates[146] and *N,N*-disubstituted hydroxylamines.[147] The most efficient currently available catalysts are amine based.[148]

The crosslinking and/or functionalization agents employed are alkoxysilanes such as: $Si(OMe)_4$, $MeSi(OMe)_3$, $MeSi(OCH_2CH_2OMe)_3$, $ViSi(OMe)_3$, $PhSi(OMe)_3$ and $PhSi(OCH_2CH_2OMe)_3$.

In two-component elastomeric compositions, it is possible to employ partially hydrolyzed substances called alkyl polysilicates.[186]

The crosslinking catalysts are chosen from amongst the following: (a) aliphatic or arylaliphatic amines; (b) organic compounds of titanium[149-155] such as: titanium alkoxides, *e.g.* $Ti(OEt)_4$, $Ti(OPr^n)_4$, $Ti(OBu^n)_4$, $Ti(OCH_2CH_2OMe)_4$. and $Ti(OSiMe_3)_4$; and titanium chelates, *e.g.* (9), (10) and (11). (It is worth noting that other metals catalyze the crosslinking, *e.g.* Zr,[156] Al[157] and the like.) (c) Organic compounds of tin;[158] among the many compounds that have been employed are tin carboxylates,[159-160] tin chelates[161-163] and tin oxides.[164]

The choice of the catalytic system will depend on the application aimed at (setting time). Thus, the catalysts will most frequently be employed in the form of mixtures, and tin is often included in the compositions in order to accelerate the crosslinking.

This two-stage process (functionalization followed by crosslinking) has been simplified by the use of a direct mixture[165] of: α,ω-dihydroxypolysiloxane oil; an excess of crosslinking agent; a functionalization catalyst (an amine); and a crosslinking catalyst.

The main problem encountered in this type of composition relates to stability during storage. The principle of 'scavengers' consists in blocking the non-functionalized silanols, which are responsible for the instability of the mixture, by the crosslinking agent.

A very large number of patents aimed at improving this stability by using 'scavengers' of silanol groups,[166-174] the main categories claimed being silazanes,[168,169,172,174] enoxysilanes,[166,171] and isocyanates.

The amines play a part important in elastomer composition, such as: catalysts for functionalization of oils; crosslinking catalysts; and adhesive agents.

The amines employed for the functionalization of oils are generally aliphatic amines (*n*-butylamine and amylamine), aliphatic polyamines or arylaliphatic amines (benzylamine and phenylethylamine).[175-177] One of the improvements consisted of the use of organoaminosilanes,[178-181] which are capable of taking part in the crosslinking,[182-185] *e.g.* $NH_2(CH_2)_3Si(OMe)_3$, $NH_2(CH_2)_3Si(OCH_2CH_2OMe)_3$, $NH_2(CH_2)_2NH(CH_2)_3Si(OMe)_3$, $NH_2(CH_2)_3O(Me)_2CCH=CHSi(OMe)_3$ and the like.

Another role of the amine consists in providing adhesiveness to the supports.[186-189] Many other compounds may improve adhesive properties: isocyanate, hydroxyl, mercapto,[190,191] carbamate[192] and epoxy.[193]

Among the different additives required to modify the intrinsic properties of elastomers, plasticizers (trimethylsilyl-blocked silicone oils[158] and organic polymers)[194,195] and heat stabilizers (rare earth hydroxides)[158] may be mentioned.

34.3.2.3 Crosslinking via alkenyloxy groups

The principle of this crosslinking is illustrated by equations (17) and (18).

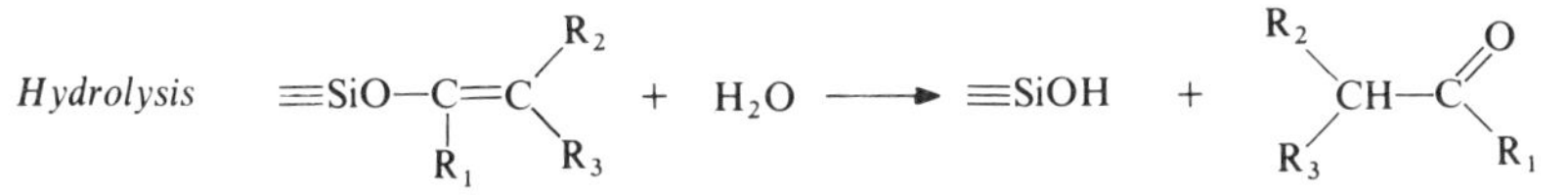

$$\text{Hydrolysis} \quad \equiv\!\text{SiO}-\underset{R_1}{\overset{R_2}{C}}=\underset{R_3}{C} + H_2O \longrightarrow \equiv\!\text{SiOH} + \underset{R_3}{\overset{R_2}{C}H}-\underset{R_1}{\overset{O}{C}} \tag{17}$$

$$\text{Condensation} \quad \equiv\!\text{SiOH} + \equiv\!\text{SiO}\underset{R_1}{\overset{R_2}{C}}=\underset{R_3}{C} \longrightarrow \equiv\!\text{SiOSi}\!\equiv + \underset{R_3}{\overset{R_2}{C}H}-\underset{R_1}{\overset{O}{C}} \tag{18}$$

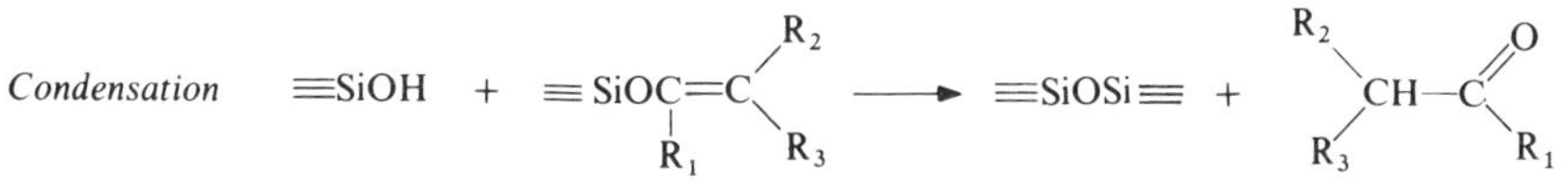

The crosslinking of α,ω-dihydroxysiloxane oils using tri- or tetra-functional alkenyloxysilane derivatives has been described by Shinetsu.[196] The reaction is carried out in the presence of a catalyst such as a metal salt of an acid, an Al alcoholate, a titanium ester and the like. The reagent systems consisting of alkenyloxysilane may comprise a curing accelerator containing the units (12).[197]

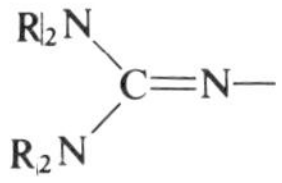

$$\underset{R_2N}{\overset{R_2N}{>}}C=N-$$

(12)

Instead of reacting alkenyloxysilane in contact with atmospheric moisture with α,ω-dihydroxy oils at the time of using the composition, it is possible to functionalize the oils in a previous stage using a tin-based catalyst (equation 19).[198]

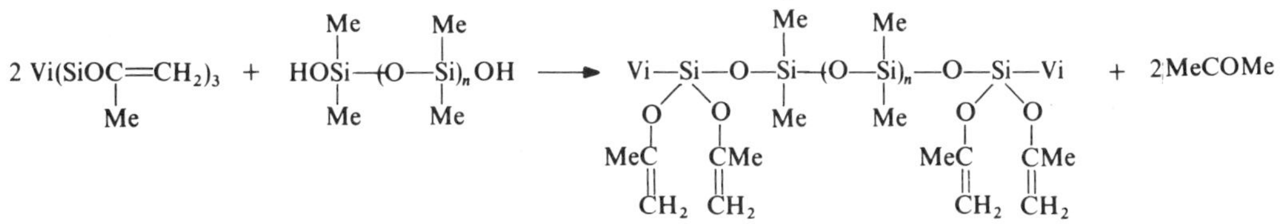

$$2\,\text{Vi}(\text{SiOC}=CH_2)_3 + \text{HOSi}(\text{O}-\text{Si})_n\text{OH} \longrightarrow \text{Vi}-\text{Si}-\text{O}-\text{Si}-(\text{O}-\text{Si})_n-\text{O}-\text{Si}-\text{Vi} + 2\,\text{MeCOMe} \tag{19}$$

34.3.2.4 Crosslinking via amino groups

The principle of this crosslinking is illustrated by equations (20) and (21).

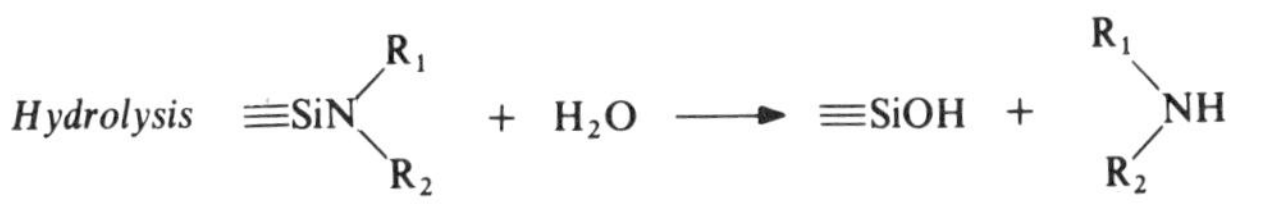

$$\text{Hydrolysis} \quad \equiv\!\text{SiN}\underset{R_2}{\overset{R_1}{<}} + H_2O \longrightarrow \equiv\!\text{SiOH} + \underset{R_2}{\overset{R_1}{>}}NH \tag{20}$$

$$\text{Condensation} \quad \equiv\!\text{SiOH} + \equiv\!\text{SiN}\underset{R_2}{\overset{R_1}{<}} \longrightarrow \equiv\!\text{SiOSi}\!\equiv + \underset{R_2}{\overset{R_1}{>}}NH \tag{21}$$

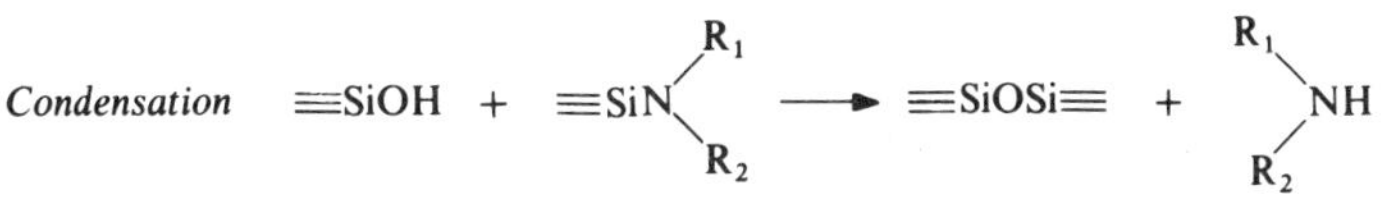

The first patents were filed by Wacker.[199] The aminosilanes claimed as crosslinking agents in single component formulae are of the type $R_n\text{Si}(NR'R'')_{4-n}$ or (13), where R' and R'' = H, ethyl or butyl.

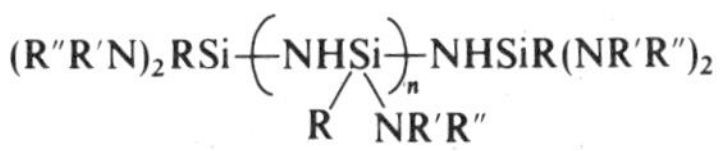

$$(R''R'N)_2 RSi(NHSi)_n NHSiR(NR'R'')_2$$
$$\qquad\qquad\quad R\ \ NR'R''$$

(13)

By using tricyclohexylamine alkylsilane as the crosslinking agent, the adhesiveness of the material is improved and the amine released during the crosslinking has the advantage of being not very toxic. An identical result may be obtained by using dicyclohexylamine dimethylsilane[200] as the chain extender. Bayer[201] proposes a mixed crosslinking system consisting of alkoxylated and aminated silanes, for example (14) or (15).

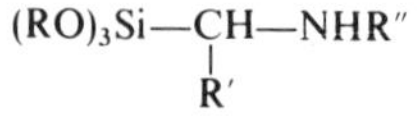

$$(RO)_3Si-CH-NHR''$$
$$\quad\quad\quad |$$
$$\quad\quad\quad R'$$

(14)

$$(RO)_3SiCH_2$$
$$\quad\quad\quad\quad\backslash NR'$$
$$(RO)_3SiCH_2 \nearrow$$

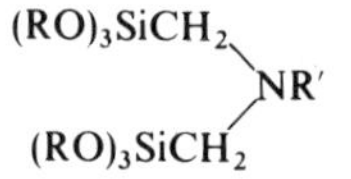

(15)

The presence of aminosilane or of aminosilazane would enable the surface curing of conventional single component CVE composition (where CVE = Cold Vulcanized Elastomer), in the presence of carbon dioxide, to be accelerated.[202]

34.3.2.5 Crosslinking via ketiminoxy groups

The principle of this crosslinking is illustrated by equations (22) and (23).

$$\text{Hydrolysis} \quad \equiv SiON=C{\stackrel{R_1}{\diagdown}}_{R_2} + H_2O \longrightarrow \equiv SiOH + {\stackrel{R_1}{\diagdown}}_{R_2}C=NOH \quad\quad (22)$$

$$\text{Condensation} \quad \equiv SiOH + \equiv SiON=C{\stackrel{R_1}{\diagdown}}_{R_2} \longrightarrow \equiv SiOSi\equiv + {\stackrel{R_1}{\diagdown}}_{R_2}C=NOH \quad\quad (23)$$

As crosslinking agent for α,ω-dihydroxy oils, Wacker[203] uses a mixed system based on aminosilane and ketiminoxysilane (equation 24).

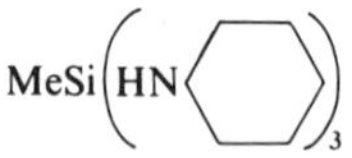

$$\text{MeSi}\left(HN\!-\!\!\bigcirc\right)_3 + \text{MeSi[ON=C(Me)}_2]_3 \quad\quad (24)$$

In particular, the use of organotri(ketiminoxy)silane is proposed in the following: (a) a flame resistant CVE composition by Dow Corning,[204] (b) a cold vulcanizable formulation by Toray[205] based on an α,ω-silaxanediol oil, a crosslinking agent which may be a ketiminoxysilane, a silicate or polysilicate, a zeolite and a metal salt; (c) a CVE formulation incorporating a fungicide, by Bayer;[206] (d) a CVE composition by Rhône Poulenc, which employs a mixed ketiminoxysilane/alkoxysilane crosslinking system;[207] (e) a tin-catalyzed elastomeric composition to form oil resistant low modulus seals;[218] and (f) single component systems catalyzed by organotitanium esters (Rhône Poulenc[108] and Toray[219]).

34.3.2.6 Crosslinking via amido groups

The principle of this reaction is illustrated by equations (25) and (26).

$$\text{Hydrolysis} \quad \equiv Si-N-C-R_2 + H_2O \longrightarrow \equiv SiOH + R_1NHCOR_2 \quad\quad (25)$$
$$\quad\quad\quad\quad\quad\quad\quad | \quad ||$$
$$\quad\quad\quad\quad\quad\quad R_1 \; O$$

$$\text{Condensation} \quad \equiv SiOH + \equiv Si-N-C-R_2 \longrightarrow \equiv SiOSi\equiv + R_1NHCOR_2 \quad\quad (26)$$
$$\quad\quad\quad\quad\quad\quad\quad\quad\quad | \quad ||$$
$$\quad\quad\quad\quad\quad\quad\quad\quad R_1 \; O$$

The first patents date from 1964 and were filed by Bayer.[208] The compositions are single component formulations based on oils blocked by silanols or alkoxysilanes, filler and other additives. The amido derivatives may be a mixed alkoxyamidosilane of the type (16).

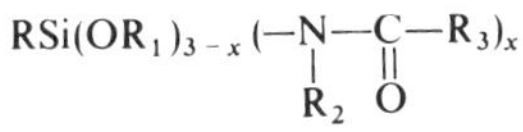

$$RSi(OR_1)_{3-x}(-\underset{R_2}{\underset{|}{N}}-\underset{O}{\underset{\|}{C}}-R_3)_x$$

(16)

In these compositions, the amidosilane derivative plays the role of crosslinking agent ($x = 3$) but more frequently the role of chain extender ($x = 2$). For example, in a low modulus composition, Dow Corning[209] employs a diorganodiacetamidosilane (17) in the presence of an oil having silanol end groups, and a polyfunctional aminoxysilane (18).

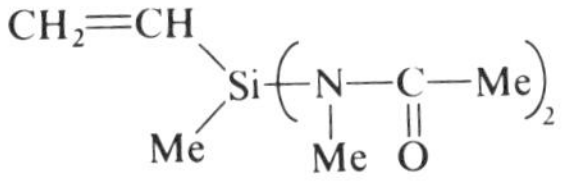

$$\underset{Me}{\overset{CH_2=CH}{\diagdown}}Si(-\underset{Me}{\underset{|}{N}}-\underset{O}{\underset{\|}{C}}-Me)_2 \qquad Me_3SiO(\underset{Me}{\underset{|}{Si}}-O)_x(\underset{Me}{\underset{|}{Si}}-O)_y SiMe_3$$

(17) 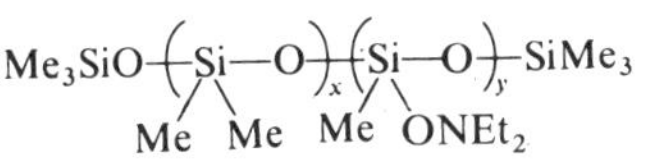 (18)

The acetamidosilane is obtained by reacting an alkali metal salt of an N-alkyl acetamide with an organohalosilane. Similar compositions were proposed by Toray[210,211] and Toshiba[212] with bis(N-methylacetamido)methylvinylsilane as the chain extender and a crosslinking agent based on aminoxysilane groups, also proposed by Shinetsu.[216,217] The presence of acetamidosilane promotes the rapid formation of the surface crosslinking.

Another method for the synthesis of amidosilane and amidosiloxanes consists of reacting an amide containing at least one active hydrogen with H_nSiR_{4-n} ($n = 2$, 3 or 4) or $R_mH_pSiO_{(4-m-p)/2}$ where $m = 0-3$ and $p = 0.005-2$ in the presence of Pt.[215]

34.3.2.7 Crosslinking via aminoxy groups

The principle of this reaction is illustrated in equations (27) and (28).

Hydrolysis $\equiv SiON\underset{R_2}{\overset{R_1}{\diagdown}} + H_2O \longrightarrow \equiv SiOH + R_1R_2NOH$ (27)

Condensation $\equiv SiON\underset{R_2}{\overset{R_1}{\diagdown}} + \equiv SiOH \longrightarrow \equiv SiOSi\equiv + R_1R_2NOH$ (28)

The early synthesis of aminoxysiloxanes and aminoxysilanes were described by Rhône Poulenc,[220] Stauffer[221] and General Electric.[222] The products were obtained by the action of the desired hydroxylamine, most frequently Et_2NOH, on a halosilane, an alkoxysilane or a silicon-containing derivative carrying the bond $\equiv Si-H$.[222]

Aminoxysilanes are employed as chain extenders (bifunctional) or as crosslinking agents (functionality ≥ 3) in elastomeric compositions vulcanizing by atmospheric moisture.

Among the compounds most commonly encountered, cyclosiloxane derivatives may be mentioned, such as (19) and (20) (where R = $-(CH_2)_4Me$,[223] $-(CH_2)_7Me$,[225] $-(CH_2)_3Me$,[226] $-Me$),[227-230] and also organoaminoxysilane derivatives, such as $R-Si(ONEt_2)_3$ or (21).[230-232]

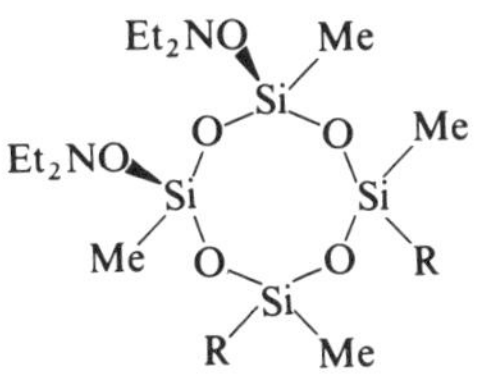

(19) 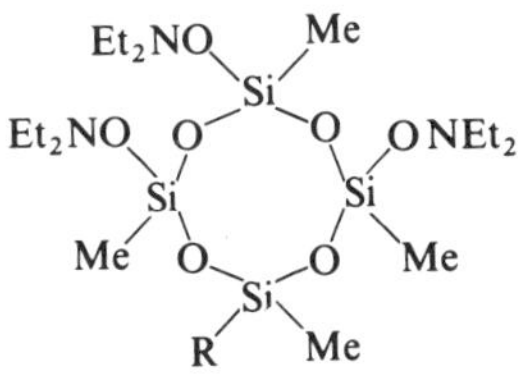 (20) 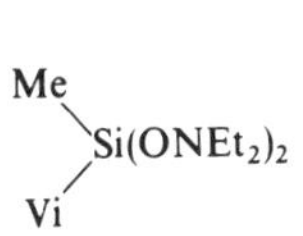 (21)

The condensation reactions between aminoxysilanes or aminoxysiloxanes and silanols may be catalyzed by: dibutyltin dilaurate;[224] butyl titanate;[228] or a boric acid ester.[229]

34.3.3 Conclusion

The choice of the elastomeric composition, related to the crosslinking agents, is made according to several important parameters for the desired applications.

Among these parameters, one may list: the adhesion to different substrates; the kinetics of crosslinking; and the toxicity and the corrosion generated by the elimination of organic compounds.

Table 7 summarizes the advantages and the drawbacks of the most common types of elastomeric compositions.

Table 7

Crosslinking system of the composition	Advantages	Drawbacks
Alkoxy	Odourless Low toxicity Excellent adhesion properties No corrosion of substrates	Rather long crosslinking time
Acyloxy	Good storage properties Good adhesion on glass Good thermal stability Short crosslinking time	Strong odour Corrosion of metallic substrates
Alkenyloxy	No corrosion Good properties for electronic applications	
Amino	Good adhesion properties	Unpleasant odour Long crosslinking time
Ketiminoxy	Short crosslinking time Low toxicity Good adhesion	Corrosion of metallic substrates (Cu)
Amido		Long crosslinking time
Aminoxy		Rather low adhesion properties Rather long crosslinking time

34.4 REFERENCES

1. N. A. Natwiyoff and R. S. Drage, *J. Organomet. Chem.*, 1965, **3**, 393.
2. T. Kagiya, Y. Sumida, T. Watanabe and T. Tachi, *Bull. Chem. Soc. Jpn.*, 1971, **44**, 923.
3. T. Kagiya, Y. Sumida and T. Watanabe, *Bull. Chem. Soc. Jpn.*, 1970, **43**, 3716.
4. R. West and R. H. Baney, *J. Am. Chem. Soc.*, 1959, **81**, 6145.
5. R. West, R. H. Baney and D. L. Powele, *J. Am. Chem. Soc.*, 1960, **82**, 6269.
6. G. J. Harris, *J. Chem. Soc.*, 1963, 5978.
7. G. J. Harris, *J. Chem. Soc. (B)*, 1971, 2083.
8. Ya. I. Ryskin, *Opt. Spektrosk.*, 1958, **4**, 532.
9. Ya. I. Ryskin, *Opt. Spektrosk.*, 1959, **7**, 278.
10. J. T. Wang and C. H. Van Dyke, *Inorg. Chem.*, 1967, **6**, 1741.
11. N. Viswanathan and C. H. Van Dyke, *J. Chem. Soc.*, 1968, 487.
12. L. P. Kuhn, *J. Am. Chem. Soc.*, 1952, **74**, 2492.
13. M. G. Voronkov, V. P. Mileshkevich and Yu. A. Yuzhelevskii, 'The Siloxane Bond', Plenum, New York, 1978.
14. K. Licht and H. Kriegsmann, *Z. Anorg. Allg. Chem.*, 1964, **330**, 151.
15. F. Taurke, *Chem. Ber.*, 1905, **38**, 1661.
16. B. Dolgov and Yu. Vol'nov, *Zh. Obshch. Khim.*, 1931, **1**, 91.
17. T. Takiguchi, *J. Am. Chem. Soc.*, 1959, **81**, 2359.
18. T. Takiguchi and F. Hitara, *Kogyo Kagaku Zasshi*, 1959, **62**, 484.
19. S. W. Kantor, *J. Am. Chem. Soc.*, 1953, **75**, 2712.
20. K. A. Andrianov and N. V. Sokolov, *Dokl. Akad. Nauk SSSR*, 1955, **101**, 81.
21. C. Eaborn, *J. Chem. Soc.*, 1952, 2840.
22. K. A. Andrianov, V. V. Astakhin and V. K. Pyzhov, *Izv. Akad. Nauk SSSR, Ser. Khim.*, 1962, 2243.
23. General Electric, *US Pat.* 2 371 068 (1940) (*Chem. Abstr.*, 19??, **39**, 4889).
24. L. H. Sommer, E. W. Pietrusza and F. C. Whitrore, *J. Am. Chem. Soc.*, 1946, **68**, 2282.
25. A. G. Kuznetsova and V. I. Ivanov, *Plast. Massy*, 1963, **10**, 17.
26. J. B. Gangi and F. A. Bettelheim, *J. Polym. Sci., Part A*, 1964, **2**, 4011.
27. Dow Corning, *US Pat.* 2 605 274 (1952) (*Chem. Abstr.*, 1952, **46**, 11 770).
28. Allegemeine Elektricitats, *Ger. Offen.*, 926 811 (1955).
29. F. S. Kipping and G. Martin, *J. Chem. Soc.*, 1909, **95**, 489.
30. H. Marsden and F. S. Kipping, *J. Chem. Soc.*, 1908, **93**, 198.

31. Milton Yusem, *US Pat.* 3 094 507 (1963); *Br. Pat.* 1 053 550 (*Chem. Abstr.*, 1963, **59**, 7741a).
32. Corning Glass Works, *US Pat.* 2 467 976 (1949) (*Chem. Abstr.*, 1949, **43**, 5639).
33. F. S. Kipping and L. L. Lloyd, *J. Chem. Soc.*, 1901, **79**, 449.
34. K. A. Andrianov and V. E. Nikitenkov, *Izv. Akad. Nauk SSSR, Ser. Khim.*, 1961, 441.
35. F. S. Kipping and R. Robinson, *J. Chem. Soc.*, 1914, **105**, 484.
36. C. A. Burkhard, *J. Am. Chem. Soc.*, 1945, **67**, 2173.
37. J. F. Hyde and R. C. Delong, *J. Am. Chem. Soc.*, 1941, **63**, 1194.
38. T. Takiguchi, *J. Org. Chem.*, 1959, **24**, 989.
39. R. Robinson and F. S. Kipping, *Proc. Chem. Soc., London*, 1912, **28**, 45.
40. Yu. A. Yuzhelevskii, E. G. Kagan, A. L. Klebanskii, A. V. Kharlamova and I. A. Zevakin, *Vysokomol. Soedin., Ser. B*, 1969, **11**, 854.
41. V. E. Nikitenkov, *USSR Pat.* 162 138 (1964) (*Chem. Abstr.*, 1964, **61**, 16 094a).
42. R. Robinson and F. S. Kipping, *J. Chem. Soc.*, 1912, **101**, 2156.
43. M. Hunt and V. Weinmayer, *US Pat.* 2 467 132 (1945) (*Chem. Abstr.*, 1949, **43**, 6232e).
44. Corning Glass Works, *US Pat.* 2 435 147 (1948) (*Chem. Abstr.*, 1948, **42**, 2819).
45. United States Rubber *US Pat.* 2 609 201 (1952) (*Chem. Abstr.*, 1953, **47**, 347).
46. V. D. Lobkov, A. L. Klebanskii and E. V. Kogan, *Vysokomol Soedin.*, 1965, **7**, 1535.
47. General Electric, *US Pat.* 2 644 805 (1953) (*Chem. Abstr.*, 1954, **48**, 310).
48. Union Carbide, *Fr. Pat.* 1 275 384 (1959).
49. Dow Corning, *Br. Pat.* 1 024 024 (1966), *Belg. Pat.* 545 219 (*Chem. Abstr.*, 1965, **63**, 10 093).
50. Union Carbide, *Br. Pat.* 910 544 (1962) (*Chem. Abstr.*, 1963, **59**, 245 F).
51. Ciba, *Ger. Offen.* 1 928 610 (1970) (*Chem. Abstr.*, 1970, **72**, 68 171z).
52. J. Chojnowski, S. Rubinsztajn and L. Wilczek, *J. Chem. Soc., Chem. Commun.*, 1984, **2**, 69.
53. Dow Corning, *Eur. Pat.* 120 645 (1983) (*Chem. Abstr.*, 1984, **101**, 39 699d).
54. Dow Corning, *Eur. Pat.* 119 816 (1983) (*Chem. Abstr.*, 1985, **102**, 8049T).
55. W. Noll, 'Chemistry and Technology of Silicones', Academic Press, New York, 1968.
56. Z. Lasocki and Z. Michalska, *Bull. Acad. Pol. Sci., Ser. Sci. Chim.*, 1961, **9**, 589.
57. Z. Lasocki and Z. Michalska, *Bull. Acad. Pol. Sci., Ser. Sci. Chim.*, 1961, **9**, 591.
58. Z. Lasocki and Z. Michalska, *Bull. Acad. Pol. Sci., Ser. Sci. Chim.*, 1964, **12**, 223; 1963, **11**, 637.
59. Z. Lasocki and Z. Michalska, *Bull. Acad. Pol. Sci., Ser. Sci. Chim.*, 1964, **12**, 227.
60. Z. Lasocki and Z. Michalska, *Bull. Acad. Pol. Sci., Ser. Sci. Chim.*, 1965, **13**, 261.
61. Z. Lasocki and Z. Michalska, *Bull. Acad. Pol. Sci., Ser. Sci. Chim.*, 1965, **13**, 267.
62. Z. Lasocki and Z. Michalska, *Bull. Acad. Pol. Sci., Ser. Sci. Chim.*, 1965, **13**, 597.
63. Z. Lasocki and Z. Michalska, *Bull. Acad. Pol. Sci., Ser. Sci. Chim.*, 1966, **14**, 819.
64. Z. Lasocki and Z. Michalska, *Bull. Acad. Pol. Sci., Ser. Sci. Chim.*, 1966, **14**, 825.
65. Z. Lasocki, B. Dejak and A. Mogilmskiw, *Bull. Acad. Pol. Sci., Ser. Sci. Chim.*, 1969, **17**, 7.
66. Z. Lasocki, B. Dejak and A. Mogilmskiw, *Bull. Acad. Pol. Sci., Ser. Sci. Chim.*, 1969, **17**, 571.
67. J. Chojnowski and S. Chrzczonowich, *Bull. Acad. Pol. Sci., Ser. Sci. Chim.*, 1962, **10**, 161.
68. S. Chrzczonowich and Z. Lasocki, *Bull. Acad. Pol. Sci., Ser. Sci. Chim.*, 1961, **9**, 591.
69. S. Chrzczonowich and Z. Lasocki, *J. Polym. Sci.*, 1962, **59**, 259.
70. W. T. Grubb, *J. Am. Chem. Soc.*, 1954, **76**, 3408.
71. B. Dejak, Z. Lasocki and W. Mogilnicki, *Bull. Acad. Pol. Sci., Ser. Sci. Chim.*, 1969, **17**, 7.
72. J. Chojnowski and S. Chrzczonowich, *Int. Symp. Organosilicon Chem., Sci. Commun.*, Prague, 1965, 41.
73. S. Chrzczonowich and Z. Lasocki, *Rocz. Chim.*, 1962, **35**, 433.
74. R. H. Baney and J. Lipowitz, *Int. Symp. Organosilicon Chem.*, 3rd, Madison, 1972, 19.
75. N. I. Martyakova, C. B. Dolgoplosk and E. G. Kagan, *Vysokomol. Soedin.*, 1971, **13**, 579.
76. J. Chojnowski, S. Rubinsztajn, W. Stanczyk and M. Scibiorek, *Makromol. Chem. Rapid Commun.*, 1983, **4**, 703.
77. C. Eaborn and W. Stanczyk, *J. Chem. Soc, Perkin Trans. 2*, 1984, 2099.
78. Rhône Poulenc, *Fr. Pat.* 2 461 730 (1979): *Eur. Pat.* 23 187 (*Chem. Abstr.*, 1981, **94**, 175 859e).
79. Rhône Poulenc, *Fr. Pat.* 2 571 731 (1984) (*Chem. Abstr.*, 1986, **105**, 153 595s).
80. General Electric, *US Pat.* 3 294 737 (1966) (*Chem. Abstr.*, 1967, **66**, 46 908t).
81. Dow Corning, *US Pat.* 3 202 634 (1965): *Br. Pat.* 925 433 (*Chem. Abstr.*, 1963, **59**, 8950d).
82. Dow Corning, *Fr. Pat.* 1 523 068 (1968) (*Chem. Abstr.*, 1969, **71**, 13 514m).
83. S. B. Dolgoplosk, E. G. Kagan, and A. D. Skonorovskaya, *Vysokomol. Soedin., Ser. B*, 1969, **11**, 635.
84. O. R. Pierce, Y. K. Kim and D. B. Bourrie, *Polym. Prepr., Am. Chem. Soc., Div. Polym. Chem.*, 1971, **12**, 497.
85. Union Carbide, *US Pat.* 3 036 035; *Br. Pat.* 942 479 (*Chem. Abstr.*, 1962, **57**, 6093a).
86. Dow Corning, *Br. Pat.* 1 163 029 (1969); *Fr. Pat.* 1 523 068 (1968) (*Chem. Abstr.*, 1969, **71**, 13 514m).
87. R. C. Smith and G. F. Kellum, *Anal. Chem.*, 1967, **39**, 338.
88. Ciba, *Ger. Pat.* 1 928 610 (1970) (*Chem. Abstr.*, 1970, **72**, 68 172z).
89. Union Carbide, *Br. Pat.* 910 334 (1963) (*Chem. Abstr.*, 1963, **58**, 9142h).
90. Union Carbide, *Ger. Pat.* 1 243 395 (1967): *Br. Pat.* 910 334 (*Chem. Abstr.*, 1963, **58**, 9142h).
91. General Electric, *Fr. Pat.* 1 504 367 (1976): *Br. Pat.* 1 125 170 (*Chem. Abstr.*, 1968, **69**, 97 584d).
92. Dow Corning, *US Pat.* 3 294 718 (1966) (*Chem. Abstr.*, 1967, **66**, 38 459g).
93. General Electric, *Fr. Pat.* 1 497 896 (1967): *US Pat.* 3 334 066 (*Chem. Abstr.*, 1968, **68**, 50 601w).
94. Milton Yusem, *US Pat.* 3 094 507 (1963) (*Chem. Abstr.*, 1963, **59**, 7141a).
95. Distillers, *Br. Pat.* 889 125 (1962) (*Chem. Abstr.*, 1962, **56**, 14 472h).
96. Dow Corning, *Fr. Pat.* 1 352 529 (1964). (*Chem. Abstr.*, 1964, **61**, 9653c).
97. Dow Corning, *Ger. Pat.* 1 301 137 (1969).
98. N. I. Martyakova, C. B. Dolgoplosk, E. G. Kagan and V. P. Mileshkevich, *Vysokomol. Soedin., Ser. B*, 1971, **13**, 579.
99. Kali-Chemie, *Ger. Pat* 1 069 388 (1964).
100. Union Carbide, *Br. Pat.* 943 841 (1964) (*Chem. Abstr.*, 1964, **61**, 5816a).
101. L. F. Ceyzeriat and P. Dumont, Rhône Poulenc, *Ger. Offen.* 1 121 329 (1957); *Fr. Pat.* 1 188 495 (1954) (*Chem. Abstr.*, 1960, **54**, 20 278g).

102. L. F. Ceyzeriat and P. Dumont, Rhône Pôulenc, *Ger. Offen.* 1 121 803 (1958): *Fr. Pat.* 1 198 749 (1955) (*Chem. Abstr.*, 1961, **55**, 7888f).
103. L. B. Bruner, Dow Corning, *US Pat.* 3 035 126 (1958) (*Chem. Abstr.*, 1963, **59**, 1545f).
104. L. B. Bruner, Dow Corning, *US Pat.* 3 077 465 (1959) (*Chem. Abstr.*, 1963, **58**, 14 275d).
105. J. R. Russel, Dow Corning, *US Pat.* 3 061 575 (1960) (*Chem. Abstr.*, 1963, **58**, 11 559d).
106. Dow Corning, *US Pat.* 3 109 013 (1960): *Belg. Pat.* 611 888 (*Chem. Abstr.*, 1963, **58**, 9312c).
107. Rhône Poulenc, *Fr. Pat.* 1 392 648 (1964) (*Chem. Abstr.*, 1965, **63**, 1816d).
108. Rhône Poulenc, *Eur. Pat.* 102 268 (1982) (*Chem. Abstr.*, 1984, **100**, 211 446w).
109. Dow Corning, *US Pat.* 3 642 692 (1970): *Ger. Pat.* 2 117 026 (*Chem. Abstr.*, 1972, **76**, 73 499y).
110. Bayer, *US Pat.* 3 819 674 (1971) (*Chem. Abstr.*, 1975, **83**, 30 092n).
111. General Electric, *US Pat.* 3 334 067 (1966) (*Chem. Abstr.*, 1967, **67**, 74 337v).
112. Dow Corning, *US Pat.* 3 754 967 (1966) (*Chem. Abstr.*, 1974, **80**, 4683z).
113. Toray, *Fr. Pat.* 2 362 189 (1976) (*Chem. Abstr.*, 1978, **89**, 198 904b).
114. Wacker, *Fr. Pat.* 1 359 578 (1962): *Ger. Pat.* 1 157 020 (*Chem. Abstr.*, 1964, **61**, 9643f).
115. Dow Corning, *US Pat.* 3 647 917 (1970): *Ger. Pat.* 2 116 816 (*Chem. Abstr.*, 1972, **76**, 35 056u).
116. General Electric, *US Pat.* 3 296 195 (1963) (*Chem. Abstr.*, 1967, **66**, 47 088n).
117. General Electric, *Fr. Pat.* 1 495 984 (1965): *Br. Pat.* 1 130 074 (*Chem. Abstr.*, 1969, **70**, 12 444f).
118. Wacker, *US Pat.* 3 661 816 (1970): *Ger. Pat.* 1 812 039 (*Chem. Abstr.*, 1969, **71**, 91 630k).
119. Rhône Poulenc, *Fr. Pat.* 1 198 749 (1958) (*Chem. Abstr.*, 1955, **61**, 7888f).
120. Rhône Poulenc, *Fr. Pat.* 2 540 128 (1983) (*Chem. Abstr.*, 1985, **102**, 8078b).
121. Rhône Poulenc, *Eur. Pat.* 117 772 (1983): *Fr. Pat.* 2 540 129 (*Chem. Abstr.*, 1985, **102**, 8077a).
122. Rhône Poulenc, *Eur. Pat.* 140 770 (1983) (*Chem. Abstr.*, 1985, **103**, 38 552q).
123. Rhône Poulenc, *Eur. Pat.* 204 641 (1985) (*Chem. Abstr.*, 1987, **107**, 41 515t).
124. Triplex Safety Glass, *Br. Pat.* 1 308 985 (*Chem. Abstr.*, 1973, **78**, 160 761n).
125. Dow Corning, *US Pat.* 4 115 356 (1977) (*Chem. Abstr.*, 1979, **90**, 170 016k).
126. Dow Corning, *US Pat.* 3 719 635 (1971): *Ger. Pat.* 2 229 574 (*Chem. Abstr.*, 1973, **78**, 112 520m).
127. Dow Corning, *US Pat.* 3 772 240 (1971) (*Chem. Abstr.*, 1974, **80**, 109 611c).
128. Dow Corning, *US Pat.* 3 836 506 (1973): *Ger. Pat.* 2 153 602 (*Chem. Abstr.*, 1972, **77**, 49 156j).
129. Dow Corning, *US Pat.* 3 960 800 (1974): *Belg. Pat.* 826 836 (1976) (*Chem. Abstr.*, 1976, **84**, 165 728r).
130. Dow Corning, *US Pat.* 3 957 714 (1975) (*Chem. Abstr.*, 1976, **85**, 47 964m).
131. Dow Corning, *US Pat.* 4 115 356 (1977) (*Chem. Abstr.*, 1979, **90**, 170 016k).
132. General Electric, *US Pat.* 3 382 205 (1963); *Fr. Pat.* 1 408 662; *Br. Pat.* 1 080 785 (*Chem. Abstr.*, 1966, **65**, 5640a).
133. General Electric, *Fr. Pat.* 2 433 558 (1978): *Belg. Pat.* 877 845 (*Chem. Abstr.*, 1980, **92**, 165 004s).
134. Bayer, *Eur. Pat.* 22 976 (1979) (*Chem. Abstr.*, 1981, **94**, 149 980k).
135. Bayer, *Fr. Pat.* 2 235 981 (1973); *Ger. Pat.* 2 333 966; *Belg. Pat.* 817 203 (*Chem. Abstr.*, 1975, **82**, 157 568q).
136. Bayer, *Ger. Pat.* 2 737 405 (1977) (*Chem. Abstr.*, 1979, **90**, 181 581s).
137. Degussa, *Ger. Pat.* 2 929 587 (1979) (*Chem. Abstr.*, 1981, **94**, 4827p).
138. General Electric, *US Pat.* 4 247 445 (1979) (*Chem. Abstr.*, 1981, **94**, 158 512u).
139. F. Perrin, Faure (Rhône Poulenc) *Eur. Pat.* 141 685 (1983) (*Chem. Abstr.*, 1981, **94**, 179 313v).
140. Rhône Poulenc, *Fr. Pat.* 1 198 749 (1958) (*Chem. Abstr.*, 1961, **55**, 7888f).
141. Dow Corning, *Fr. Pat.* 1 220 348.
142. General Electric, *Fr. Pat.* 2 464 288; *Belg. Pat.* 877 267 (*Chem. Abstr.*, 1980, **92**, 60 183t).
143. Dow Corning, *US Pat.* 3 122 522 (*Chem. Abstr.*, 1964, **61**, 32 846).
144. Dow Corning, *US Pat.* 3 161 614; *Belg. Pat.* 623 603 (*Chem. Abstr.*, 1964, **60**, 12 138a).
145. SWS Silicones, *US Pat.* 4 111 890 (1979) (*Chem. Abstr.*, 1979, **90**, 88 576j).
146. General Electric, *Br. Pat.* 2 144 758 (*Chem. Abstr.*, 1985, **102**, 186 000c).
147. Rhône Poulenc, *Fr. Pat.* 2 508 467 (*Chem. Abstr.*, 1983, **99**, 23 843k).
148. Midland Silicones, *US Pat.* 3 542 901 (*Chem. Abstr.*, 1971, **74**, 54 893q).
149. Rhône Poulenc, *Fr. Pat.* 1 266 528 (1963) (*Chem. Abstr.*, 1962, **56**, 11 771t).
150. Dow Corning, *US Pat.* 3 294 739 (1961) (*Chem. Abstr.*, 1967, **66**, 38 695f).
151. Dow Corning, *Fr. Pat.* 1 330 625 (1961).
152. General Electric, *Fr. Pat.* 2 121 289 (*Chem. Abstr.*, 1973, **78**, 98 803x).
153. General Electric, *Fr. Pat.* 2 121 631; *Ger. Pat.* 2 200 346 (*Chem. Abstr.*, 1972, **77**, 127 718k).
154. Dow Corning, *Fr. Pat.* 1 359 396 (*Chem. Abstr.*, 1965, **62**, 4197b).
155. General Electric, *Fr. Pat.* 2 251 602; *Ger. Pat.* 2 454 408 (*Chem. Abstr.*, 1975, **83**, 117 337v).
156. General Electric, *US Pat.* 4 357 443 (*Chem. Abstr.*, 1983, **98**, 5136g).
157. General Electric, *US Pat.* 4 533 503 (*Chem. Abstr.*, 1985, **102**, 96 785j).
158. W. Noll, 'Chemistry and Technology of Silicones', Academic Press, New York, 1968.
159. Midland Silicones, *US Pat.* 3 186 963.
160. Wacker-Chemie, *US Pat.* 3 862 919; *Ger. Pat.* 2 259 802 (*Chem. Abstr.*, 1974, **81**, 154 803y).
161. Rhône Poulenc, S. C., *Eur. Pat.* 147 323 (*Chem. Abstr.*, 1986, **104**, 6999v).
162. General Electric, *US Pat.* 4 517 337 (*Chem. Abstr.*, 1985, **103**, 38 553r).
163. General Electric, *US Pat.* 4 554 310 (*Chem. Abstr.*, 1986, **104**, 110 834y).
164. General Electric, *Fr. Pat.* 2 240 263; *US Pat.* 3 839 246 (*Chem. Abstr.*, 1975, **82**, 32 166t).
165. Rhône Poulenc, *Eur. Pat.* 21 859 (*Chem. Abstr.*, 1981, **94**, 158 142y).
166. General Electric, *US Pat.* 4 377 706 (1981) (*Chem. Abstr.*, 1983, **99**, 5814x).
167. General Electric, *US Pat.* 4 424 157 (1982) (*Chem. Abstr.*, 1984, **100**, 210 124c).
168. General Electric, *Eur. Pat.* 110 251 (1982) (*Chem. Abstr.*, 1984, **101**, 92 599g).
169. General Electric, *US Pat.* 4 417 042 (1982) (*Chem. Abstr.*, 1984, **100**, 8241h).
170. General Electric, *Fr. Pat.* 2 546 396 (1983).
171. General Electric, *Fr. Pat.* 2 546 525 (1983) (*Chem. Abstr.*, 1985, **102**, 96 790f).
172. General Electric, *Eur. Pat.* 139 064 (*Chem. Abstr.*, 1985, **102**, 114 957x).
173. General Electric, *Fr. Pat.* 2 543 562 (1983) (*Chem. Abstr.*, 1985, **102**, 47 185d).

174. General Electric, *US Pat.* 4 536 540 (*Chem. Abstr.*, 1985, **103**, 55 578v).
175. Dow Corning, *Fr. Pat.* 1 302 035.
176. Dow Corning, *Fr. Pat.* 1 330 623.
177. Dow Corning, *US Pat.* 3 122 522 (*Chem. Abstr.*, 1964, **61**, 3284g).
178. General Electric, *US Pat.* 2 754 311 (*Chem. Abstr.*, 1957, **51**, 4423i).
179. Union Carbide, *US Pat.* 2 832 754 (*Chem. Abstr.*, 1958, **52**, 14 652h).
180. Union Carbide, *US Pat.* 2 930 809 (*Chem. Abstr.*, 1960, **54**, 19 484d).
181. Dow Corning, *US Pat.* 2 971 864 (*Chem. Abstr.*, 1961, **55**, 14 310d).
182. Midland Silicones, *Fr. Pat.* 1 443 657.
183. C. A. Cheeseman, *US Pat.* 3 686 357 (*Chem. Abstr.*, 1972, **77**, 153 552).
184. Midland Silicones, *Belg. Pat.* 774 830.
185. General Electric, *Fr. Pat.* 2 152 908 (*Chem. Abstr.*, 1973, **79**, 32 617d).
186. Rhône Poulenc, *Eur. Pat.* 184 966 (*Chem. Abstr.*, 1986, **105**, 192 684g).
187. Wacker-Chemie, *Fr. Pat.* 2 074 144.
188. Shinetsu, *US Pat.* 4 180 642 (*Chem. Abstr.*, 1979, **90**, 153 306n).
189. Bayer, *Eur. Pat.* 74 001 (*Chem. Abstr.*, 1983, **99**, 70 979u).
190. *Austria Pat.* 271 665.
191. *Austria Pat.* 271 666.
192. General Electric, *Fr. Pat.* 2 333 022 (*Chem. Abstr.*, 1977, **87**, 69 524n).
193. Dow Corning, *US Pat.* 4 115 356 (*Chem. Abstr.*, 1979, **90**, 170 016k).
194. N. K. F. Groep, *Fr. Pat.* 2 392 476 (*Chem. Abstr.*, 1979, **90**, 187 950h).
195. Krafft S. A., *Fr. Pat.* 2 446 849 (*Chem. Abstr.*, 1980, **93**, 240 978p).
196. Shinetsu, *US Pat.* 3 819 563 (1972) (*Chem. Abstr.*, 1974, **81**, 92 467f).
197. Shinetsu, *US Pat.* 4 180 642 (1977).
198. Shinetsu, *Ger. Pat.* 3 032 625 (1979) (*Chem. Abstr.*, 1981, **94**, 209 796z).
199. Wacker-Chemie, *Fr. Pat.* 1 248 826.
200. Wacker-Chemie, *Fr. Pat.* 1 510 778 and 1 510 779; *Ger. Pat.* 1 255 924 and 1 260 140 (*Chem. Abstr.*, 1968, **68**, 22 622k and **68**, 60 373n).
201. Bayer, *Fr. Pat.* 1 588 712 (*Chem. Abstr.*, 1968, **68**, 6399 or **71**, 82 405p).
202. SWS Silicones, *US Pat.* 4 170 700 (1977) (*Chem. Abstr.*, 1980, **92**, 23 611p).
203. Wacker-Chemie, *US Pat.* 3 678 003 (1969): *Ger. Pat.* 1 964 502 (*Chem. Abstr.*, 1971, **75**, 141 625x).
204. Dow Corning, *Fr. Pat.* 2 145 700 (1971): *Ger. Pat.* 2 234 790 (*Chem. Abstr.*, 1973, **78**, 137 668d).
205. Sato, *Jpn. Pat.* 51 030 584 (1974).
206. Bayer, *Fr. Pat.* 2 235 981 (1973): *Ger. Pat.* 2 333 966 (*Chem. Abstr.*, 1975, **82**, 157 568g).
207. Rhône Poulenc, *Fr. Pat.* 2 067 636 (1969): *Ger. Pat.* 2 055 712 (*Chem. Abstr.*, 1971, **75**, 89 238c).
208. Bayer, *Fr. Pat.* 1 423 477; *Neth. Pat.* 6 501 494 (*Chem. Abstr.*, 1968, **68**, 87 593w).
209. Dow Corning, *US Pat.* 3 766 128 (1972) (*Chem. Abstr.*, 1974, **80**, 84 497).
210. Toray, *Neth. Pat.* 80 1904 (1980) (*Chem. Abstr.*, 1982, **96**, 87 198).
211. Toray, *Jpn. Pat.* 80 66 983 (1980) (*Chem. Abstr.*, 1980, **93**, 134 018).
212. Toshiba, *Jpn. Pat.* 81 38 349 (1981) (*Chem. Abstr.*, 1981, **95**, 63 900).
213. Dow Corning, *US Pat.* 3 776 933 (1972) (*Chem. Abstr.*, 1974, **80**, 48 160).
214. Dow Corning, *US Pat.* 3 776 934 (1972) (*Chem. Abstr.*, 1974, **80**, 134 562).
215. General Electric, *US Pat.* 4 602 094 (1985) (*Chem. Abstr.*, 1986, **105**, 191 387).
216. Shinetsu, *Jpn. Pat.* 81 93 755 (1981) (*Chem. Abstr.*, 1981, **95**, 221 133).
217. Shinetsu, *Jpn. Pat.* 76 08 864 (1976) (*Chem. Abstr.*, 1976, **84**, 165 963).
218. Loctite, *Eur. Pat.* 157 580 (*Chem. Abstr.*, 1985, **103**, 78 302).
219. Toray, *Eur. Pat.* 135 293 (1983) (*Chem. Abstr.*, 1980, **92**, 115 230s).
220. Rhône Poulenc, *Fr. Pat.* 1 359 240 (1963).
221. Stauffer Chemical Co., *Fr. Pat.* 1 506 185 (1965) (*Chem. Abstr.*, 1968, **69**, 106 868).
222. General Electric, *Fr. Pat.* 1 462 725 (1965) (*Chem. Abstr.*, 1967, **67**, 54 258).
223. General Electric, *Fr. Pat.* 1 462 728 (1965) (*Chem. Abstr.*, 1967, **67**, 44 627).
224. Dow Corning, *Ger. Pat.* 2 308 806 (1972) (*Chem. Abstr.*, 1974, **81**, 154 287).
225. Toshiba Silicone Co., *Jpn. Pat.* 79 90 349 (1979) (*Chem. Abstr.*, 1980, **92**, 24 001).
226. Toshiba Silicone Co., *Jpn. Pat.* 81 43 365 (1981) (*Chem. Abstr.*, 1981, **95**, 99 445).
227. Toray, *Eur. Pat.* 87 013 (1983) (*Chem. Abstr.*, 1983, **99**, 196 728).
228. Toshiba, *Jpn., Pat.* 84 155 483 (1984) (*Chem. Abstr.*, 1985, **102**, 26 566).
229. Toshiba, *Jpn. Pat.* 86 159 463 (1986) (*Chem. Abstr.*, 1986, **105**, 228 654).
230. Toray, *Eur. Pat.* 36 262 (1981) (*Chem. Abstr.*, 1982, **96**, 87 192).
231. Dow Corning, *Ger. Pat.* 3 207 336 (1982) (*Chem. Abstr.*, 1983, **98**, 5309).
232. Toray, *US Pat.* 4 387 177 (1983) (*Chem. Abstr.*, 1983, **99**, 89 476).

35

Phenol–Formaldehyde Polymers

ANDRE KNOP
Rütgerswerke AG, Frankfurt, Federal Republic of Germany
VOLKER BÖHMER
Johannes-Gutenberg-Universität, Mainz, Federal Republic of Germany
and
LOUIS A. PILATO
Temecon Group International Inc., Edison, NJ, USA

35.1 INTRODUCTION	612
35.1.1 Historical Development	612
35.1.2 Commercial Status and Applications	612
35.2 SYNTHESIS MATERIALS	613
35.2.1 Phenols	613
35.2.2 Aldehydes	614
35.3 REACTION MECHANISM	615
35.3.1 Molecular Structure and Reactivity	615
35.3.2 Phenol–Formaldehyde Reactions under Alkaline Conditions	616
35.3.3 Phenol–Formaldehyde Reactions under Acidic Conditions	619
35.3.4 Phenol–Formaldehyde Reactions under Special Conditions	620
35.3.5 Crosslinking Reactions and Curing	621
35.3.6 Statistical Approach and Computer Simulation	623
35.4 STRUCTURALLY UNIFORM OLIGOMERS	623
35.4.1 Synthesis of Linear Oligomers	624
35.4.2 Synthesis of Cyclic Oligomers	625
35.4.3 Solid-state Properties	626
35.4.4 Solution Properties	628
35.5 SYNTHESIS OF INDUSTRIAL-GRADE RESINS	630
35.5.1 Synthesis of Novolak Resins	630
35.5.2 Synthesis of Resol Resins	632
35.5.3 Synthesis of High-ortho Resins	632
35.5.4 Synthesis of Aqueous Dispersion Resins	633
35.5.5 Resin Characterization According to Application	634
35.5.5.1 Particle boards, plywood and fiber boards	634
35.5.5.2 Molding compounds	634
35.5.5.3 Heat and sound insulation materials	634
35.5.5.4 Bonded textile felts	635
35.5.5.5 Industrial laminates and paper impregnation	635
35.5.5.6 Coatings	635
35.5.5.7 Grinding materials	636
35.5.5.8 Friction materials	636
35.6 TOXICOLOGY OF MONOMERS, PREPOLYMERS AND RESINS	636
35.6.1 Toxicology of Phenol	636
35.6.2 Toxicology of Formaldehyde	637
35.6.3 Toxicology of Prepolymers and Cured Resins	637
35.6.4 Microbial Degradation and Environmental Protection Methods	637
35.7 ANALYTICAL CHARACTERIZATION	638
35.7.1 Spectroscopic/Spectrometric Methods	638
35.7.2 Chromatographic Methods	638

35.7.3 Thermal Characterization	642
35.7.4 Other Methods	642
35.8 REFERENCES	643

35.1 INTRODUCTION

35.1.1 Historical Development

In 1872, von Bayer[1] obtained a colorless noncrystallizing resinous product from the reaction of phenol with formaldehyde, while he was investigating phenol-based dyes. The occurrence of similar intractible materials in an acidic medium was also reported by ter Mer,[2] Claus and Trainer,[3] Claisen[4] and others. The reaction between formaldehyde and phenol in the alkaline pH range was first recorded in 1894 by Lederer[5] and Manasse.[6] This reaction is generally referred to as the Lederer–Manasse reaction. These early investigators did not perceive any practical use for the ill-defined products. Speyer[7] and Luft[8] were the first to recognize the technical significance and practical use of curable phenolic resins.

In 1907, Baekeland developed an economical method to convert these resins into moldable compositions which could be transformed by heat and pressure to hard and heat-resistant molded products.[9] Recognizing the commercial importance of his invention, he concurrently disclosed in numerous patents an extensive number of applications[10] for this new composition.

The first commercial phenolic resin plant, Bakelite GmbH, was started on 25 May 1910 by Rütgerswerke AG at Erkner near Berlin. Excellent review articles and books by Hultzsch,[11] Martin,[12] Megson[13] and others have provided a summary of chemical research and existing phenolics technology from that time to the present day.

35.1.2 Commercial Status and Applications

The most important markets[14] for phenolic resins are those that relate to the wood-working industry, thermal insulation and molding compounds. About 75% of all phenolic resins are consumed in these market areas (Table 1).

Table 1 Use of Phenolic Resins in Western Europe (1987)

Use	%
Wood-working industry	32
Thermal insulation	18
Molding compounds	16
Decorative laminates	6
Foundry	5
Coatings	4
Electrical laminates	4
Filters and separators	3
Abrasives	3
Felt bonding	3
Friction	2
Rubber and adhesives	2
Others	3

Virtually all the early applications initially established by Baekeland continue to use phenolic resin in significant volumes. The favorable development of the automotive and electronics industries created a greater phenolic resin demand in foundry, abrasives, filters, felt bonding, friction and laminate applications.

Relatively new and promising applications for phenolic resins[14] are phenolic foams, precursors for polymeric carbon (glassy carbon), ablation systems, low-smoke laminating resins, photoresists, color developers for carbonless copying paper (CCP) and polymeric phenolic composites for aircraft interiors, among others.

All these diverse application areas require specially designed resins differing in molecular weight distribution (MWD), cure characteristics, melting point (m.p.), solids content and other selected properties.[14]

In 1985, the world production of phenolic resins amounted to 2.5 million tons (Table 2), with the US producing 48%, Western Europe 20%, and Japan 13%. The phenolic resin consumption *per capita* in 1985 amounted to 2.9 kg in Japan, 2.8 kg in West Germany, 2.2 kg in the US (without considering wood products), 1.8 kg in France and Italy and 1.1 kg in Great Britain. By volume, phenolic resins comprise about 4% of the world-wide production of plastics today (including cellulosics but excluding elastomers).

Table 2 Phenolic Resin Production[15-17] (Captive Production Excluded)

Country	1000 t
USA	1258[c]
USSR	399[a]
Japan	313[c]
West Germany	195[b]
France	65[a]

[a] 1985. [b] 1986. [c] 1987.

In total resin production, phenolic resins are comparable to polyesters or polyurethanes (in volume). In most applications, phenolics are combined with reinforcing fillers or fibers to function as the adherent or the critical binder of the composition. Key properties,[14] such as high temperature resistance, intractibility and flame retardance, are recognizable features that are expected to contribute to further market growth. Because of the multitude of application areas, phenolic resin consumption closely parallels the gross national product growth. The average annual resin production growth-rate between 1974 and 1986 in the US was 1.4%, and between 1982 and 1986 it was 6.7%.

Major phenolic resin producers are, for example, Georgia Pacific, Borden, Plastics Engineering Co. and Occidental in the US, Bakelite and Perstorp in Western Europe, and Sumitomo Bakelite and Mitsui Toatsu in Japan, among others.

35.2 SYNTHESIS MATERIALS

Phenolic resins are produced by the reaction of phenols with aldehydes. The simplest representatives of these types of compounds, phenol and formaldehyde, are by far the most important. Additives, modifiers and fillers are usually introduced after polymerization to provide selected properties. Physical properties of phenols[18-23] are listed in Table 3.

35.2.1 Phenols

The cumene process (Scheme 1), developed by Hock and Lang,[24] is by far the most important synthetic process[19,20] for the production of phenol, and accounts for more than 95% of the synthetic phenol capacity in the Western world today. In addition to the cumene synthesis process, about 3% of phenol is derived from coal.

Cumene (isopropylbenzene), required for the Hock process, is produced[25] by alkylation of benzene with propylene over a solid phosphoric acid catalyst (UOP-process). Cumene is oxidized

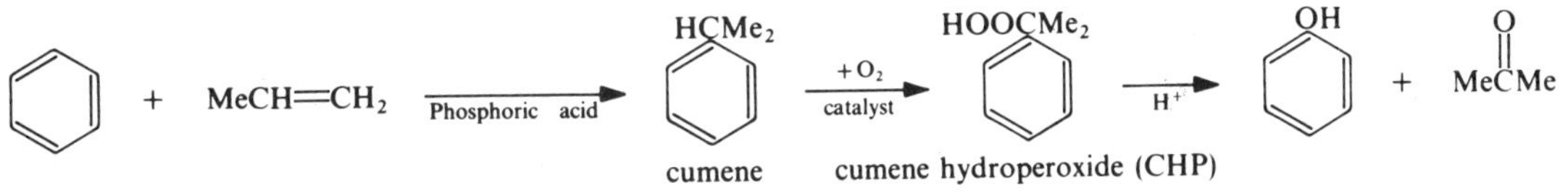

Scheme 1

Table 3 Physical Properties of Phenols[18-23]

Name		Molecular weight	m.p. (°C)	b.p. (°C)	pK_a (25 °C)
Phenol	(hydroxybenzene)	94.1	40.9	181.8	10.00
o-Cresol	(1-methyl-2-hydroxybenzene)	108.1	30.9	191.0	10.33
m-Cresol	(1-methyl-3-hydroxybenzene)	108.1	12.2	202.2	10.10
p-Cresol	(1-methyl-4-hydroxybenzene)	108.1	34.7	201.9	10.28
p-t-Butylphenol	(1-t-butyl-4-hydroxybenzene)	150.2	98.4	239.7	10.25
p-t-Octylphenol	(1-t-octyl-4-hydroxybenzene)	206.3	85	290	—
p-Nonylphenol	(1-nonyl-4-hydroxybenzene)	220.2	—	295	—
2,3-Xylenol	(1,2-dimethyl-3-hydroxybenzene)	122.2	75.0	218.0	10.51
2,4-Xylenol	(1,3-dimethyl-4-hydroxybenzene)	122.2	27.0	211.5	10.60
2,5-Xylenol	(1,4-dimethyl-2-hydroxybenzene)	122.2	74.5	211.5	10.40
2,6-Xylenol	(1,3-dimethyl-2-hydroxybenzene)	122.2	49.0	212.0	10.62
3,4-Xylenol	(1,2-dimethyl-4-hydroxybenzene)	122.2	62.5	226.0	10.36
3,5-Xylenol	(1,3-dimethyl-5-hydroxybenzene)	122.2	63.2	219.5	10.20
Resorcinol	(1,3-dihydroxybenzene)	110.1	110.8	281.0	—
Bisphenol A	(2,2-bis(4-hydroxyphenyl)propane)	228.3	157.3	—	—

with oxygen in air in the liquid phase to cumene hydroperoxide (CHP). The mechanism for the acid-catalyzed peroxide decomposition was investigated by Seubold and Vaugham.[26] CHP decomposes very rapidly under acidic conditions and elevated temperatures. Finally, phenol is obtained and purified by distillation.

The largest use of phenol is the production of phenol–formaldehyde resins (over 30%), followed by the production of bisphenol A, ε-caprolactam and other compounds. Cresols, xylenols and alkylphenols are less important compared to phenol in commercial resin production. They are commonly used in resin formulations for electrical laminates, coatings, friction linings, adhesives and antioxidants. Cresol and xylenol resins exhibit solubility in nonpolar solvents, increased flexibility and compatibility with different modifying materials. Resorcinol, as a dihydric phenol (1,3-dihydroxybenzene), exhibits a high reaction rate with formaldehyde and is used preferentially for adhesive formulations that cure at ambient temperature. Its broader application is limited by its high price.

35.2.2 Aldehydes

Formaldehyde[27] is virtually the only carbonyl component for the synthesis of technically important phenolic resins. Special resin properties can be obtained with other aldehydes, e.g. acetaldehyde, 2-furaldehyde or glyoxal, but these aldehydes (Table 4) have not achieved significant commercial importance.

Formaldehyde is produced by dehydrogenation of methanol, over either an iron oxide/molybdenum oxide catalyst or over a silver catalyst (equation 1). Because of the hazards in handling mixtures of pure oxygen and methanol, air is used as the oxidizing agent. Oxygen is used to burn the hydrogen by-product.

$$\text{MeOH} + \tfrac{1}{2}O_2 \xrightarrow{\text{catalyst}} H_2C{=}O + H_2O \tag{1}$$

Table 4 Aldehydes Suitable for Resin Modification[18,28,29]

Type	Formula	m.p. (°C)	b.p. (°C)
Formaldehyde	$CH_2{=}O$	−118	−19
Acetaldehyde	$MeCH{=}O$	−123	20.8
Propionaldehyde	$MeCH_2CH{=}O$	−81	48.8
n-Butyraldehyde	$Me(CH_2)_2CH{=}O$	−97	74.7
Isobutyraldehyde	$Me_2CHCH{=}O$	−66	61
Glyoxal	$O{=}CHCH{=}O$	−15	50.4
2-Furaldehyde	(furan)CH=O	−31	162

Formaldehyde, a colorless pungent irritating gas (Table 5), exists in aqueous solution almost exclusively as a mixture of oligomers of poly(oxymethylene).[29,30] The portion of monomeric formaldehyde CH_2=O in aqueous solution[27] is very low (0.01%). Further equilibria exist in the presence of methanol,[28] which aids in stabilization by forming a hemiformal terminus. Trace amounts (0.05%) of formic acid found in commercial products (Table 6) are due to the Cannizzaro reaction (equation 2).

$$2CH_2\!=\!O \;+\; NaOH \;\rightleftharpoons\; MeOH \;+\; MeCO_2Na \tag{2}$$

Table 5 Properties of Formaldehyde[29,30]

Molecular weight	30.03
Boiling point	$-19.2\,°C$
Melting point	$-118\,°C$
Enthalpy of solution in H_2O	$62\,kJ\,mol^{-1}$
Enthalpy of formation (25 °C)	$-116\pm 6\,kJ\,mol^{-1}$
Ignition temperature	$430\,°C$
Flammability limits in air	7–73 vol % CH_2=O
Dissocation constant in H_2O at 0 °C	1.4×10^{-14}
Dissocation constant in H_2O at 50 °C	3.3×10^{-13}

Table 6 Formaldehyde Solution Specification, Resin Grade[14]

Formaldehyde		$37.0\pm 0.1\%$
Methanol	max.	0.5%
Formic acid	max.	0.02%
Chloride	max.	0.5 p.p.m.
Fe	max.	0.12 p.p.m.
Al	max.	0.25 p.p.m.

The major use of formaldehyde is in the production of thermosetting resins based on phenol, urea and melamine. Other uses comprise, for example, the production of polyacetal resins, pentaerythritol [2,2-bis(hydroxymethyl)-1,3-propanediol] and hexamethylenetetramine (HMTA, 1,3,5,7-tetraazatricyclo[3.3.1.1]decane). HMTA, used as a formaldehyde source to crosslink novolak resins, is prepared from formaldehyde and ammonia (equation 3). The reaction is reversible. HMTA decomposes at elevated temperatures, generally above 250 °C. In aqueous solutions, HMTA is easily hydrolyzed to amines and is also used as a catalyst in the resol-formation reaction, rather than ammonia, with equivalent results.

$$6CH_2\!=\!O \;+\; 4NH_3 \;\rightleftharpoons\; (CH_2)_6N_4 \;+\; 6H_2O \tag{3}$$

35.3 REACTION MECHANISM

35.3.1 Molecular Structure and Reactivity

Phenolic resins are obtained by step-growth polymerization of difunctional monomers (mainly formaldehyde) with monomers of functionality of two or higher. Under usual reaction conditions, substitution occurs only in the *ortho* or *para* position relative to the phenolic hydroxyl group. Thus phenol itself and *m*-cresol are trifunctional, *o*- and *p*-cresol are difunctional, and 2,4- and 2,6-dimethylphenol are monofunctional. Those phenols of lower functionality are used to incorporate special properties in the resin (Table 3). In practice, a decreasing 'functionality' is found with increasing molecular weight, due to molecular shielding effects, leading, for instance, to an experimentally determined average phenol novolak functionality of 2.31.[31]

In solution as well as in the solid state phenol shows a strong propensity to form hydrogen bonds. Hydrogen-bonded chains in the form of a threefold spiral are found by X-ray analysis.[32,33] In solution, *e.g.* in benzene containing small amounts of water, trimolecular species Z_3, $Z_2 \cdot H_2O$ and $Z \cdot 2H_2O$ (Z = phenol) were identified.[34] For Z_3 a cyclic structure was proposed.

Phenol is a weak acid and therefore the phenolate anion is the reactive species under alkaline conditions. Alkylphenols are slightly less acidic than phenol, while hydroxymethylphenols are more acidic.[35] These differences of the dissociation equilibria must be considered, if the reaction is not carried out with an excess of base.

Formaldehyde is present in aqueous solution to the extent of less than 0.01% in its monomeric nonhydrated form.[27,30] Methanediol (equation 4) is observed as the key monomeric species which, in higher concentrations, is in equilibrium with oligomers (equation 5).[29,30,36] In the presence of alcohols, hemiformal equilibria (equation 6) must be considered; formals are not formed under these conditions. The phenolic hydroxyl group and hydroxymethyl groups formed by substitution (see Section 35.3.2) may react in a similar manner (equation 7). Peaks for $m/n = 0, 1, 2, 3$ have been identified by high-resolution nuclear magnetic resonance (NMR). The overall competitive characteristics of the hydroxyl group of methanol:benzyl alcohol:phenol is 70:20:1 for hemiformal formation.[37]

$$CH_2=O + H_2O \rightleftharpoons HOCH_2OH \quad (K = 0.7 \times 10^{14}) \tag{4}$$

$$nHOCH_2OH \rightleftharpoons HO(CH_2O)_nH + (n-1)H_2O \tag{5}$$

$$ROH + nHOCH_2OH \rightleftharpoons RO(CH_2O)_nH + nH_2O \tag{6}$$

$$\text{(phenol-CH}_2\text{OH)} + (m+n) HOCH_2OH \rightleftharpoons \text{(phenol with O(CH}_2\text{O)}_m\text{H and CH}_2\text{O(OCH}_2)_n\text{H)} + (m+n)H_2O \tag{7}$$

Three reactions must be considered in the formation of phenol–formaldehyde polymers: formaldehyde addition to phenol; chain growth or prepolymer formation; and finally the crosslinking or curing reaction. The rates of the first two reactions, which lead to the formation of prepolymers, are pH dependent. Thus mainly two prepolymer types are obtained, dependent on pH.

Novolaks are obtained by the reaction of phenol and formaldehyde in the strongly acidic pH region. The reaction is usually carried out at a molar ratio of 1 mol (phenol) to 0.75–0.85 mol (formaldehyde). Novolaks are linear or slightly branched condensation products, linked by methylene bridges of a relatively low molecular weight up to approximately 2000. These resins are soluble and permanently fusible, *i.e.* thermoplastic, and are cured only by addition of a hardener, almost exclusively formaldehyde supplied as HMTA, to insoluble and infusible products.

Resols are obtained by reaction of phenols and aldehydes under alkaline conditions, where the aldehyde is used in excess. Phenol to formaldehyde ratios of between 1:1.0 to 1:3.0 are customary. They consist of mono- or poly-nuclear hydroxymethylphenols which are stable at room temperature, but are transformed into three-dimensional crosslinked insoluble and infusible polymers by the application of heat (or, less frequently, with acids).

The progressive or finite polymerization, which is commonly referred to as curing, is distinguished by crosslinking of mainly linear chains with the occurrence of gelation at some intermediate stage in the polymerization reaction. The methylene bridge is thermodynamically the most stable crosslink site and therefore it is prevalent in cured phenolic resins.

35.3.2 Phenol–Formaldehyde Reactions under Alkaline Conditions

The main products formed from phenol and formaldehyde at room temperature are a variety of mono- and poly-hydroxymethyl phenols (Table 8), many of which have been isolated as pure compounds. Sodium hydroxide, sodium carbonate, calcium, magnesium and barium hydroxide and tertiary amines are used as catalysts in this alkaline hydroxymethylation. The reactive species is the phenolate anion, which is substituted *via* the electrophilic formaldehyde species. C-alkylation occurs almost exclusively in the *ortho* and *para* positions. *Meta* substitution is not detected (equations 8 and 9).

$$\text{(phenolate)} + \overset{\delta+}{CH_2}=\overset{\delta-}{O} \longrightarrow \text{(cyclohexadienone intermediate with CH}_2\text{O}^-\text{)} \longrightarrow \text{(o-hydroxymethylphenolate)} \tag{8}$$

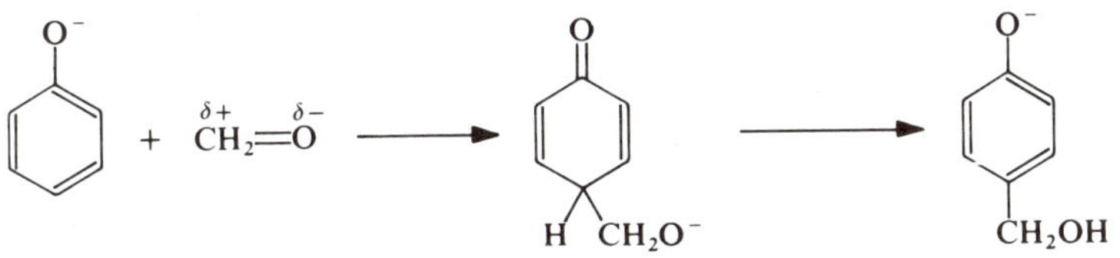

$$\text{(9)}$$

The overall reaction rate of the base-catalyzed phenol–formaldehyde reaction may be expressed by a second-order rate equation,[38–44] which is similarly observed for several alkylphenols[45–47] (equation 10).

$$\text{rate} = k[\text{phenolate anion}][\text{formaldehyde}] \tag{10}$$

Initial studies, which determined all the individual rate constants for the various hydroxymethylation steps in the general reaction scheme (Scheme 2), were carried out by Freeman and Lewis, using quantitative paper chromatography.[38] Similar studies were reported later by different laboratories using various techniques to separate and quantify different products.[39–44] A direct comparison of the rate constants is quite difficult. Even a comparison of the relative reactivities is challenging, because all the reaction conditions were varied to some extent by the different investigators.

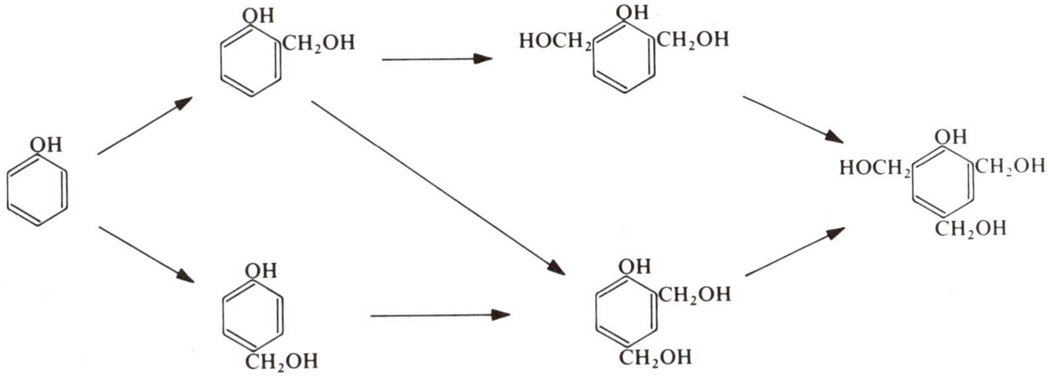

Scheme 2

Obviously, the *para* position in phenol shows a slightly higher reactivity than the *ortho* position, but the results differ for *o*-hydroxymethylphenol. Introduction of the *p*-hydroxymethyl group decreases the reactivity of phenol, while the effect is not clear for *o*-hydroxymethylphenol. An enhanced rate is found, however, for the *o*-hydroxymethyl group. This effect, which is especially pronounced for two hydroxymethyl groups, suggests an interpretation in terms of an interaction of the hydroxymethyl group and the phenolic hydroxyl group. For experiments with catalytic amounts of base only, the higher acidity of the *o*-hydroxymethylphenols,[35] caused by the stabilization of the anion by an intramolecular hydrogen bond, provides a plausible explanation.

The reaction mechanism of *C* alkylation of phenolate by formaldehyde (equations 8 and 9) is oversimplified, because it does not explain changes in the *ortho/para* substitution ratio observed for different pH values, and especially for different metal hydroxides[48,49] used as catalysts. Ortho substitution is favored by divalent metal and mainly by transition metal cations, which can be explained by chelated transition states (compare Section 35.3.4). The *ortho*-directing effect is increased with decreasing polarity of the reaction medium, which is analogous to other electrophilic substitutions of phenol.[50,51]

The mechanism also neglects the formation of hemiformals (see equations 6 and 7) and formaldehyde oligomers (equation 5). These reactions were included in the treatment of the kinetic scheme by Zavitsas.[41] Yet, similar positional reactivities were noted, and compare favorably with other studies. Figure 1 shows several hydroxymethyl intermediates, including hemiformals, as a function of time, as determined recently by Perrin et al.[52]

The hydroxymethyl compounds formed in the first reaction sequence undergo further condensation reactions (see later in Table 8) especially at higher temperatures (70–100 °C). Analysis of the complicated mixtures thus formed has become more and more feasible by the combination of modern chromatographic and spectroscopic techniques. Mechin et al. have separated and isolated by high performance liquid chromatography (HPLC) a variety of compounds consisting of two to

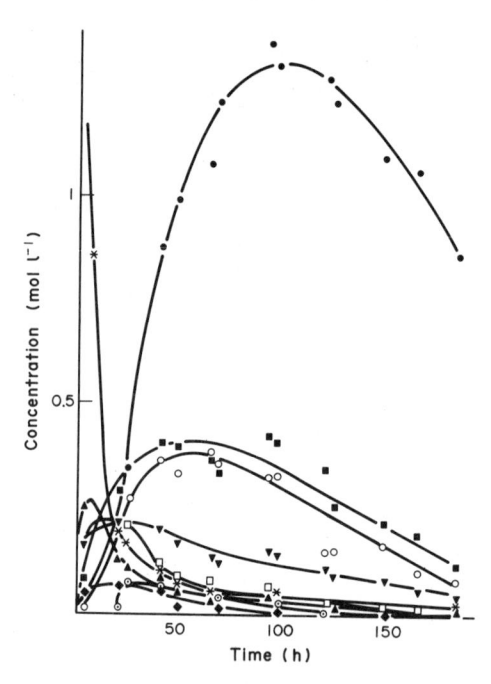

Figure 1 Variation of products during the sodium hydroxide-catalyzed reaction of phenol and formaldehyde: ∗, phenol; ▼, 4-hydroxymethylphenol; ▲, 2-hydroxymethylphenol; ■, 2,4-dihydroxymethylphenol; □, hemiformal of 2,4-dihydroxymethylphenol; ◆, 2,6-dihydroxymethylphenol; ●, 2,4,6-trihydroxymethylphenol; ○, ⊙, hemiformals of 2,4,6-trihydroxymethylphenol. A small amount of 3,3′,5,5′-tetrahydroxymethyl-4,4′-dihydroxydiphenylmethane is also detected[52]

five phenolic units from resins obtained with sodium hydroxide catalysis.[53,54] They were identified by ^{13}C and ^{1}H NMR spectroscopy. With the exception of one trimer containing a —CH$_2$OCH$_2$— bridge, only compounds with methylene bridges were found, and only methylene bridges connecting two *para* positions or an *ortho* with a *para* position, but never two *ortho* positions. On the other hand, Prokai claims the presence of 2,2′-dihydroxydiphenylmethane derivatives in resins obtained under similar conditions.[55] They were separated by gas chromatography (GC) after trimethylsilylation; however, product identification by electron impact mass spectrometry alone is not sufficiently reliable compared to other characterization methods.

The absence of *ortho–ortho* bridges is in accordance with a recent kinetic study of the self-condensation of tri(hydroxymethyl)phenol.[56] A first-order reaction in alkaline solution was interpreted as the formation of quinone methides in the rate-determining step (equation 11). The greater tendency to form the *p*-quinone methide, which is also noted in other reactions,[57] would account for the observed substitution pattern.

$$\text{HOCH}_2\!\!-\!\!\underset{\text{CH}_2\text{OH}}{\underset{|}{\text{C}_6\text{H}_2(\text{OH})}}\!\!-\!\!\text{CH}_2\text{OH} \xrightarrow{-\text{H}_2\text{O}} \text{HOCH}_2\!\!-\!\!\underset{\text{CH}_2}{\underset{\|}{\text{C}_6\text{H}_2(\text{O})}}\!\!-\!\!\text{CH}_2\text{OH} \quad \left(\text{HOCH}_2\!\!-\!\!\underset{\text{CH}_2\text{OH}}{\underset{|}{\text{C}_6\text{H}_2(\text{O})}}\!\!=\!\!\text{CH}_2\right) \quad (11)$$

Quinone methides[58] formed as intermediates (see also Section 35.3.5) may react with free *ortho* or *para* positions to form methylene bridges, or with hydroxymethyl groups to form dimethylene ether links, which again may lose formaldehyde to form methylene bridges. A direct displacement of formaldehyde (*ipso* substitution) seems possible too (Scheme 3).

Although it was shown that self-condensation of hydroxymethylphenols is faster than their condensation with phenols,[59] in general there is no convincing experimental evidence in favor of one of the reaction pathways leading to the most stable and therefore abundantly observed methylene bridge.

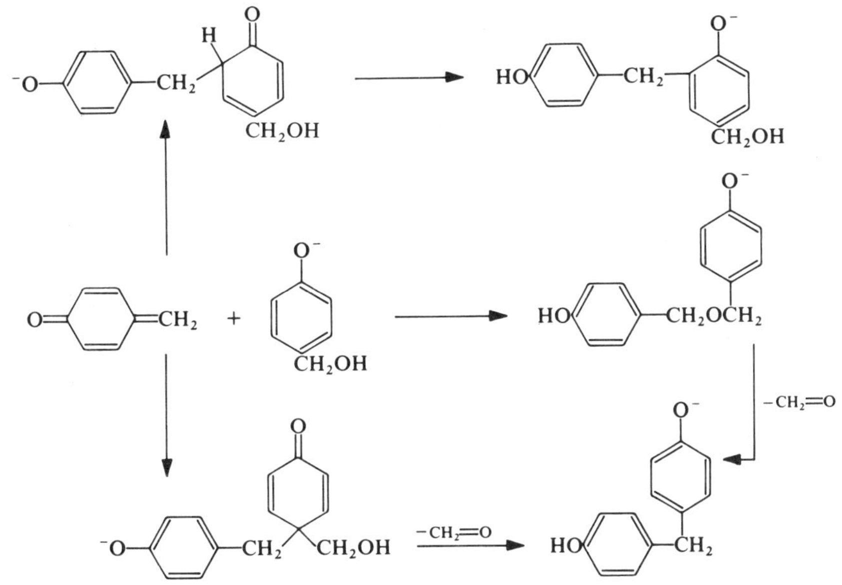

Scheme 3

35.3.3 Phenol–Formaldehyde Reactions under Acidic Conditions

Catalysts useful for the acid-catalyzed condensation of phenol and formaldehyde are, for example, oxalic acid, hydrochloric acid, sulfuric acid or *p*-toluenesulfonic acid. The overall reaction rate was found to be second order in most cases.[60] The reaction rate increases with decreasing pH. An increase in the activation energy and a change from negative to positive values for the activation entropy was observed[61] when the pH was increased from 1.14 to 3.00.

The first step of the reaction is again an electrophilic hydroxymethylation, the carbonium ion formed from methanediol being the reacting species (equations 12 and 13).

$$HOCH_2OH + H^+ \rightleftharpoons {}^+CH_2OH + H_2O \qquad (12)$$

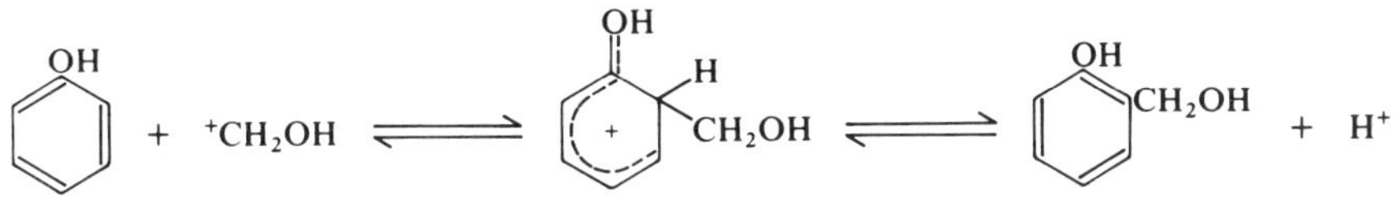

$$(13)$$

In contrast to the reaction under alkaline conditions, hydroxymethylated phenols cannot be isolated in the acid-catalyzed reaction. However, their existence as intermediates was detected by NMR spectroscopy.[37] This is due to the rapid formation of the resonance-stabilized benzylic carbonium ion, which then reacts quite rapidly to yield dihydroxydiphenylmethanes[62] (equations 14 and 15). This second condensation step was found to be 10–13 times faster than the initial substitution by formaldehyde (equations 12 and 13).[61] Thus prepolymer can only be prepared if phenol is present in considerable excess over formaldehyde.

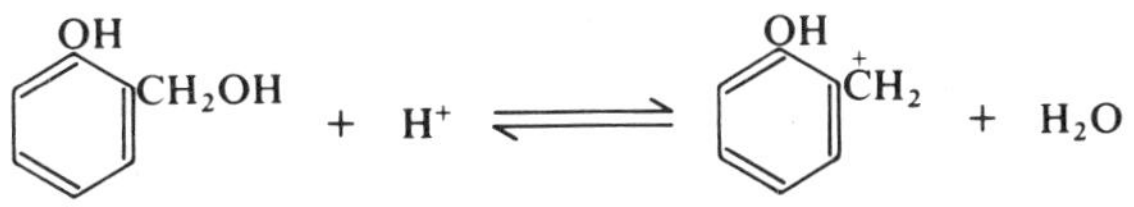

$$(14)$$

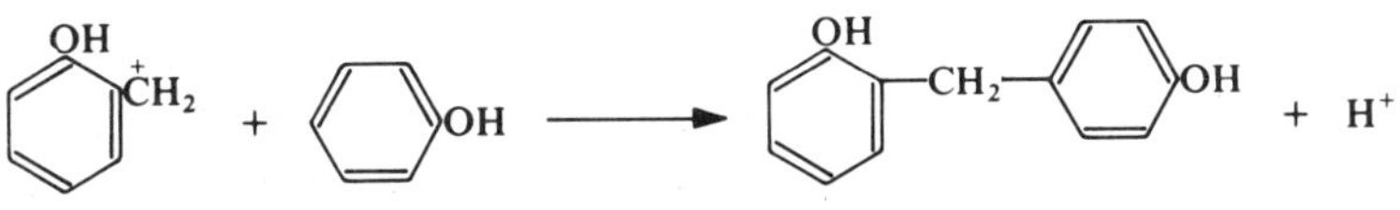

$$(15)$$

The *para*-substituted intermediates are more reactive than the *ortho*-substituted ones, analogous to alkaline conditions. Thus, in a recent *in situ* study by ^{13}C NMR, only *ortho* substitution was observed in the initial stages of an acid-catalyzed phenol–formaldehyde reaction, not as hydroxymethyl but rather in the form of hemiacetal groups. The *para*-substituted phenol entities obviously undergo immediate condensation and the first methylene bridges that appear are *para–para* bridges. The next are the *ortho–para* and finally the *ortho–ortho* bridges.[63]

In general, a further formaldehyde addition does not occur at the usual phenol/formaldehyde ratios of 1:(0.7–0.85). Mainly linear chain molecules are formed, containing 5–10 phenol units connected by methylene bridges. Branching is proposed as occurring above 10 units.[64,65]

35.3.4 Phenol–Formaldehyde Reactions under Special Conditions

Ammonia-catalyzed resols (which usually have higher average molecular weights) differ significantly from all other resols by their characteristic yellow color, which is believed to be attributable to the azomethine group —CH=N—. This color is also typical of novolaks cured with HMTA. It has been shown[37,66,67] that these prepolymers contain secondary (**1**) and tertiary amine groups (**2**) as well as benzoxazine structures (**3**). The reaction may be described as an aminomethylation or a Mannich reaction.

Formaldehyde and ammonia form HMTA in a complicated reversible reaction sequence.[68] It is not surprising that similar structures are formed if HMTA is used as a formaldehyde source. By monitoring the reaction of HMTA with phenol by ^{13}C NMR and Fourier transform infrared (FTIR), Sojka[69] has shown that there is an initial rapid decay of HMTA with concurrent benzylamine formation. The amine is predominantly secondary with some tertiary amine and trace amounts of primary amine. The substitution pattern of the amine is exclusively *ortho*, suggesting again the apparent intermediacy of benzoxazine which rearranges to the amine(s).

High *ortho–ortho* resins can be either solid novolaks or liquid resols depending on the phenol/formaldehyde ratio. They became commercially important and of academic interest as a result of early studies of Bender and Farnham,[70] who established their unusually rapid cure-rate due to the presence of vacant *para* positions. The early preparative conditions leading to high *ortho–ortho* novolaks consisted of an intermediate pH range of 4–7 and the use of divalent metal salts[49,71] such as Ca, Mg, Zn, Cd, Pb, Cu, Co and Ni with a large excess of phenol. Characterization of the *ortho–ortho* structure was corroborated by IR and NMR[72] and may be represented by structure (**4**). If the phenol/formaldehyde ratio is 1:1.5–1.8 under similar conditions (pH 4–7, divalent salts, azeotropic removal of water), high-*ortho* liquid resols are obtained,[73] with a generalized structure as shown in structure (**5**).

The mechanism of the *ortho*-selective substitution involves the formation of chelate-like complexes. By directly forming the chelated material from phenol and ethylmagnesium bromide, and then reacting it with formaldehyde (or other aldehydes) in an aprotic solvent (boiling benzene or toluene), Casiraghi et al. obtained high yields of novolaks with high-*ortho* specificity.[74,75] The method is of general use for the synthesis of a variety of *ortho*-substituted phenol derivatives. High-*ortho* novolaks with 96% regioselectivity are also obtained by the direct reaction of phenol and poly(oxymethylene) in xylene at 170–220 °C in a pressurized reactor.[76] The reaction mechanism is proposed to occur *via* quinone methide intermediates in all these examples.

35.3.5 Crosslinking Reactions and Curing

Due to the presence of reactive hydroxymethyl groups, resols may be converted to highly crosslinked products without the introduction of additional reagents. Heat curing, by far the most important crosslinking process, is conducted at temperatures between 130 °C and 200 °C. Previously, Hultzsch[77,78] and von Euler[79,80] proposed quinone methides as intermediates in the curing process. These may be formed by the elimination of water from hydroxymethyl compounds or from dibenzyl ethers. Thus, most of the crosslinking steps are similar to equation (11) and Scheme 3, although the reaction conditions are quite different.

Quinone methides are highly reactive compounds which, if derived from formaldehyde and phenols, cannot be isolated as pure monomers. If they are prepared in the absence of reactive compounds, *e.g.* by elimination of HCl from chloromethylated phenols,[81–83] these reactive species undergo linear oligomerization by forming dimers, trimers (in the case of *o*-quinone methides, *via* a Diels–Alder cycloaddition[83,84]) or other low molecular weight oligomers. Some typical structures are shown in Scheme 4, and similar structures may be present in a cured resol.

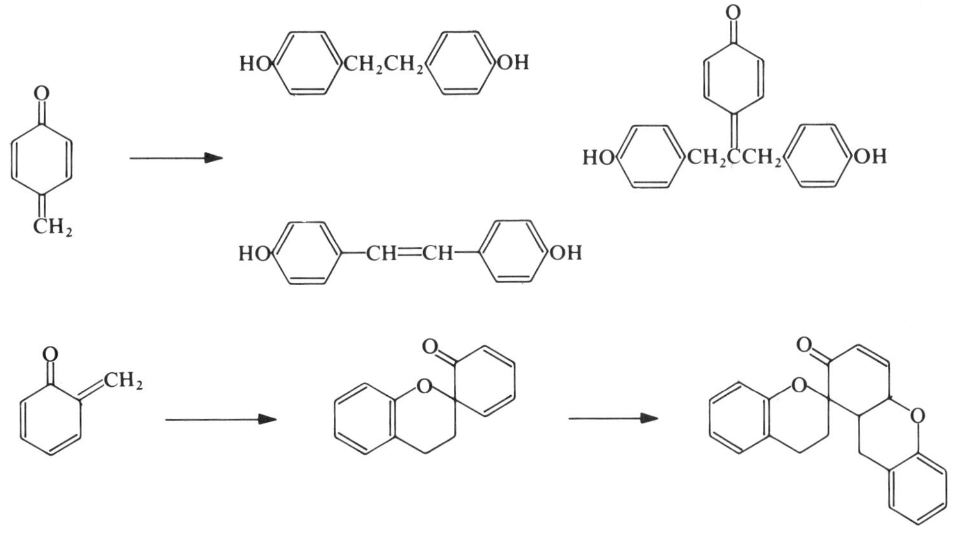

Scheme 4

Substituted quinone methides are known and have been characterized by ^{13}C NMR.[85] In some cases their structure was confirmed by X-ray analysis.[86–88] Phenol is known to add to them in the expected way.[88]

The condensation of bis(hydroxymethyl)cresols (or similar compounds) has been studied as a model reaction. Under suitable conditions only water is eliminated, leading to linear polymers in which the phenolic units are connected exclusively by dimethylene ether linkages.[89] The elimination of formaldehyde is reported in the presence of sodium cations.[90] A transition state consisting of hydrogen bonding between a phenolic hydroxyl group and a neighboring hydroxymethyl group is proposed, with the benzylic alcohol moiety present in *o*-hydroxymethylphenols susceptible to nucleophilic displacement under very mild conditions.[91] Under carefully controlled conditions only water (1.5 mol mol^{-1}) is eliminated even from 2,4,6-trihydroxymethylphenol,[92] a result which was confirmed by thermal analysis very recently.[52] Within this line of reasoning it is difficult to visualize a network in which all phenolic units are linked by exactly three —CH$_2$OCH$_2$— bridges. Sterically isolated quinone methide species which cannot undergo further reactions would also account for this result.

A solid-state ^{13}C NMR study by Maciel[93] has shown the direct involvement of the phenolic hydroxyl groups, and the condensation of methylene bridges with hydroxymethyl groups or formaldehyde liberated during the curing process (as proposed already by Zinke[94]), as shown in equations (16)–(18).

$$\text{PhOH} + \text{HOCH}_2\text{-PhOH} \longrightarrow \text{PhO-OCH}_2\text{-PhOH} + H_2O \quad (16)$$

$$\text{HO-Ph-CH}_2\text{-Ph} + \text{Ph-CH}_2\text{OH} \longrightarrow \text{HO-Ph-CH(-Ph)-CH}_2\text{-PhOH} + H_2O \quad (17)$$

$$2\,\text{HO-Ph-CH}_2\text{-Ph} + CH_2{=}O \longrightarrow \text{[bis(dihydroxydiphenylmethyl)methane]} + H_2O \quad (18)$$

Acid curing of resols may be achieved at ambient temperature by a variety of strong organic and inorganic acids. Most probably the benzyl cation is the reactive intermediate, and the mechanism corresponds to the second step in the formation of novolaks (equations 14 and 15).

Curing of (thermoplastic) novolak resins requires the addition of a crosslinking compound, which is mainly HMTA and rarely poly(oxymethylene) or 1,3,5-trioxane. For information related to the use of other crosslinking agents, such as epoxide resins, diisocyanates, urea, melamine or the corresponding formaldehyde resins, the original literature should be consulted.[14] Nitrogen-containing crosslinked resins (up to 6% of chemically bound nitrogen may be found) are obtained with the generation of ammonia. Reaction intermediates and structures deduced from model compound studies have already been described (see Section 35.3.4). Evidence for the intermediacy of benzoxazine, benzyl-substituted triazines and tribenzylamine species, as well as ammonia, in the curing reaction of novolak resins with HMTA has been recently reported by Maciel[95] through the use of ^{13}C and ^{15}N cross-polarized/magic-angle spinning (CP/MAS) NMR spectroscopy. However, no evidence was found to indicate the presence of azomethine structures. It is postulated that, initially, the HMTA becomes involved in hydrogen bonding with the hydroxyl groups of the novolak resin, and, in a concerted manner, the HMTA rings open and react with resin to form the various intermediates mentioned previously. An acid-catalyzed mechanism is also deduced from the considerable rate increase with decreasing pH.[37,96]

Comparative kinetic studies of the HMTA curing of a random novolak, a high *ortho* or 2,2′-material and an all 2,4′-novolak (prepared from the condensation of *p*-hydroxymethylphenol) showed the highest reactivity for the free *para* position of the 2,2′-material[97] which is in agreement with earlier studies related to the isomeric dihydroxydiphenylmethanes.[70,71] This is in contrast, but not a contradiction, to the reactivity found for isomeric dihydroxydiphenylmethanes towards formaldehyde in alkaline solution.[98] Yet it is difficult to rationalize benzoxazine intermediates, which require vacant *ortho* positions, for the 2,2′-novolak. The high reactivity of the 2,2′-material could be associated also with the 'hyperacidity' of *ortho*-linked novolaks *via* the intermolecular hydrogen-bonding mechanism with HMTA. However, a considerably higher reactivity of the *para* position is often found for phenols under acidic conditions.

In summary, many plausible and reasonable reaction sequences have been proposed to occur during the crosslinking process, but it is still uncertain which precise reaction occurs, the extent of it, and the conditions under which the preferred reaction sequence takes place. Possibly *meta* positions may also be involved in the curing process, since some *meta* substitution has been found under certain conditions, *e.g.* during the condensation of chloromethylated nitrophenols with alkylphenols.[99]

35.3.6 Statistical Approach and Computer Simulation

Flory treated the phenol–formaldehyde reaction numerically by assuming a statistical condensation of the trifunctional phenol with the difunctional formaldehyde by merely supplying internuclear linkages.[100] However, Flory's simplifying assumptions: (a) equal reactivity of nuclear positions and functional groups; (b) all groups react independently of one another; and (c) no intramolecular interaction occurs as molecular size increases, obviously do not apply to this condensation reaction. The unequal reactivity of the *ortho* and *para* positions of phenol, as well as of the hydroxymethyl group in methanediol and hydroxymethylphenols, and the general change in reactivity as the reaction proceeds, leads to significant differences between experimental observations and statistically derived molecular-weight-distribution functions and polydispersity index. A better agreement with experimental data was obtained with the Flory–Stockmeyer model, using an average reactivity of 2.31 for phenol in its acid-catalyzed reaction with formaldehyde.[31,101]

Recently, the mathematical modeling of the phenol–formaldehyde reaction has become more sophisticated. Two rather different procedures are being actively pursued.

(i) Rate equations may be formulated for the different reaction steps, leading to a set of differential equations which are solved numerically, assuming certain values for the single rate constants or for their ratios. Assuming different reactivities for *ortho* and *para* positions, as well as for 'internal' reactive sites and those at the end of the molecule, Kumar *et al.* have treated the formation of novolaks[102-104] and resols[105,106] under various conditions. A similar treatment, using somewhat different basic assumptions (*e.g.* equal reactivity for *ortho* and *para* positions) was reported by Steffan.[107]

(ii) The phenol–formaldehyde polycondensation reaction may also be simulated *via* a computer with a limited number of molecules in a statistical way, known as the Monte-Carlo method. Different rates are introduced as different 'reaction probabilities' and again these probabilities are varied until the statistically derived properties (*e.g.* molecular weight distribution) best fit the experimental data. In this way, Ishida *et al.* have studied several phenol–formaldehyde condensation[108,109] and cocondensation[110] reactions, as well as the crosslinking process.[111] For instance, it can be demonstrated that, under acidic conditions, a hydroxymethyl group reacts 5–12 times faster than formaldehyde (methanediol)[108] which is in reasonable agreement with kinetic studies.[81]

In the future it is anticipated that mathematical modeling or computer simulation in combination with the progress of analytical techniques will provide more detailed information on the complicated nature of phenol–formaldehyde reactions.

35.4 STRUCTURALLY UNIFORM OLIGOMERS

Model compounds as well as prototype reactions are frequently used to understand more complicated chemical systems. In the case of phenolic resins, a large number of model compounds have been synthesized. They are regarded as either components of a complicated mixture of novolaks or resols or as representative segments of the phenolic crosslinked network. These model compounds were formerly the basis for the correlation of physical properties with molecular structure. Presently, progress in analytical separations and identification methods is formidable; highly sophisticated chemical instrumentation allows the analysis of highly complicated mixtures. Nevertheless, well-defined compounds with definite structure are still required for the calibration of those analytical techniques, to test mechanistic schemes, to furnish basic data for computer simulation, *etc.*

The vast number of possible structures of oligomeric phenolic compounds is illustrated by Table 7.[112] In addition to these linear oligomers, branched and cyclic compounds are possible when more than three phenolic compounds are connected by methylene bridges. Various substituents (including OMe or OH) and different bridges besides methylene further increase the number of possible compounds.

This variety of oligomers with a definite molecular structure also makes them attractive as model compounds for fundamental studies, *e.g.* of neighboring-group effects or 'long range' cooperative

Table 7 Possible Number of Linear Oligomers which can be Obtained by Linking a Given Number n of Phenol, o-Cresol or p-Cresol Units via Methylene Groups in the Ortho- or Para-position[112]

n	2	3	4	5	6	7	8	9	10
Phenol	3	7	21	57	171	495	1485	4401	13203
o-Cresol	3	4	10	16	36	64	136	256	528
p-Cresol	1	1	1	1	1	1	1	1	1

effects. The rapid development of calixarenes[113] (cyclic phenolic oligomers, see Section 35.4.2) as potential host molecules and enzyme mimics is an important example.

35.4.1 Synthesis of Linear Oligomers

Two different reactions are used to connect phenolic units by methylene bridges: the direct condensation with formaldehyde, leading to symmetrical structures (equation 19) and the condensation of hydroxymethyl derivatives with other phenols which can also generate nonsymmetrical structures (equation 20).

Instead of hydroxymethyl compounds, which are available by reaction with formaldehyde under alkaline conditions,[114] the use of chloromethylated (or bromomethylated) compounds is convenient or necessary in special cases.[115-117]

The rational synthesis of definite oligomers usually requires suitable protective or blocking groups. In the case of phenolics, halogen (chlorine, bromine) is used to protect (or block) ortho or para positions from undergoing reaction. Dehalogenation is performed under very mild conditions (room temperature, normal pressure) with hydrogen in alkaline solution using Raney nickel as catalyst.[115,118,119] This method may be applied in the presence of nonreducible substituents, such as alkyl, aryl or —CO_2R.

As an independent protecting group, the t-butyl group[120] may be used, which can be eliminated by transalkylation in toluene in the presence of $AlCl_3$ at 50 °C.[121] Halogen atoms remain unchanged under these conditions. Bromine atoms may be selectively eliminated by Zn/NaOH while leaving chlorine atoms intact in the same molecule.[122,123] Among these possibilities, only hydrogenation is sufficiently smooth and does not affect hydroxymethyl groups.[119]

Selective coupling reactions directed to vacant positions ortho to the phenolic hydroxyl group are possible using bromomagnesium salts in refluxing benzene.[124] Obviously, magnesium acts as a coordination site for formaldehyde[125] and facilitates the formation of the quinone methide,[126] the proposed reaction intermediate,[127] which again is coordinated in a similar manner (equations 21 and 22).

The ensuing oligomerization reactions consist of a 'duplication' procedure with formaldehyde or a 'stepwise' extension with salicyl alcohol, generally with an *o*-hydroxybenzyl alcohol (compare equations 19 and 20). In this way, Casiraghi *et al.* synthesized linear *ortho*-linked oligomers with up to nine phenol units without protecting the *para* position.[128] The same compounds were initially obtained by Kämmerer *et al.* via the corresponding oligomers of *p*-chlorophenol.[115]

Under similar conditions, novolaks with alkylidene bridges (for example from acetaldehyde[129] or isobutyraldehyde[130]) can be prepared. These compounds attract considerable theoretical interest, since stereoisomers are obtained as a consequence of the unsymmetrically (ethylidene or isobutylidene) bridged phenolic units. As proposed by Casiraghi, this method not only enables regio- but also stereo-controlled synthesis. If suitable chiral alkoxy aluminum chlorides are used instead of the magnesium derivatives, enantioselective *ortho*-hydroxyalkylation[131] as well as enantiocontrolled synthesis of dinuclear compounds[132] becomes feasible.

35.4.2 Synthesis of Cyclic Oligomers

During the last decade, the interest in cyclic condensation products of *para*-substituted phenols with formaldehyde has intensified. The name for this family of cyclic compounds, calix(*n*)arenes, was coined by Gutsche[113] and is derived from calix (Greek, chalice) and arene (indicating a macrocyclic array of aromatic rings).

The formation of calix(4)arenes was first recognized by Zinke *et al.*[133] during the alkaline condensation of *p*-alkylphenols with formaldehyde in a three-stage process. Under similar conditions, Cornforth later isolated two condensation products from *p*-*t*-octylphenol with different melting points, which he believed to be stereoisomeric cyclic tetramers.[134] Later, Gutsche showed that the direct condensation of *p*-*t*-butylphenol with formaldehyde leads to a mixture of cyclic (and probably linear) compounds, containing not only methylene, but also dimethylene ether bridges[135] (equation 23).

$n = 4-8$

The composition of these mixtures is greatly dependent on reaction conditions. A recent HPLC study claims that small amounts of calix(4, 5, 6, 7 and 8)arenes are also formed in an acid-catalyzed reaction of *p*-*t*-butylphenol with formaldehyde. The mechanism for the formation of cyclic compounds is not yet well understood.[137] It has been shown, however, that under the drastic reaction conditions, cleavage of methylene bridges and restructuring of phenolic units occurs, as calix(4)arenes can be obtained from calix(8)arenes.

Synthesis conditions are now well defined to prepare calixarenes with 4, 6 or 8 *p*-*t*-butylphenol units, with yields of up to 80% in a preparative manner,[135,136] and even on an industrial scale. Calixarenes with isopropyl-,[138] *t*-pentyl-,[139] *t*-octyl-, octyl- and dodecyl-phenol[140] have been obtained in the same way. Although even-numbered oligomers are favored for, as yet, unknown reasons, small amounts of odd-numbered cyclic oligomers are also obtained.[141-143]

A stepwise synthesis of calixarenes was first described by Hayes and Hunter[144] and systematically studied and extended by Kämmerer and Happel.[145-148] Starting with *o*-bromo- or *o*-chloro-*p*-alkylphenol, it consists of subsequent hydroxymethylation and condensation steps. Finally, the halogen is eliminated and the linear monohydroxymethylated precursor is cyclized under high dilution conditions in refluxing acetic acid (Scheme 5). The ring size is unambiguously determined by the synthetic pathway, which also allows some substituent variation in the *para* position (R, R′, R″). Calixarenes with up to seven phenol units (*p*-cresol, *p*-*t*-butylphenol) were obtained in this way. The obvious disadvantages are the long reaction sequence and limited overall yields. Therefore, a more convergent synthesis for the linear precursor used in the final cyclization step was proposed by Gutsche.[149]

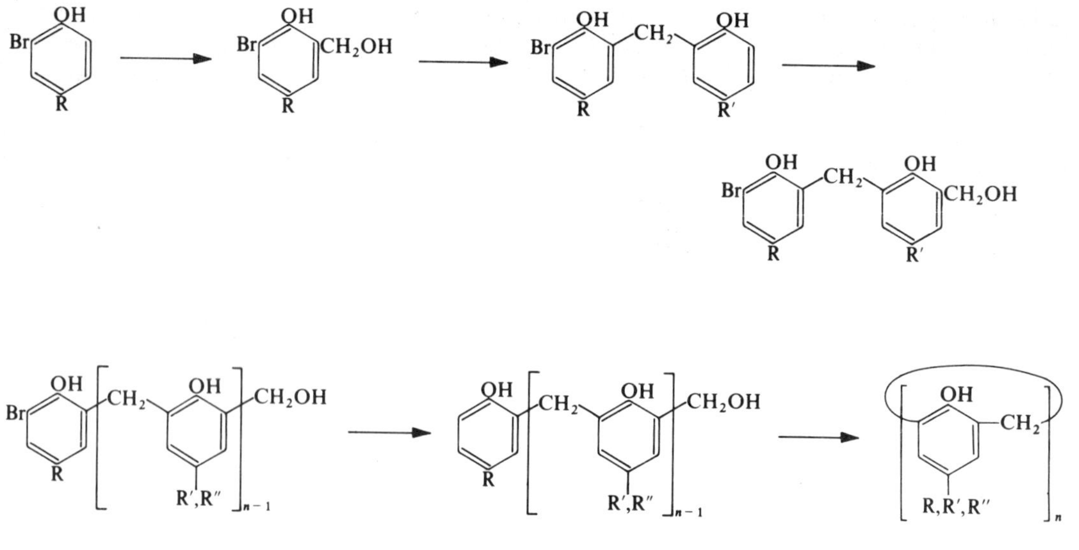

Scheme 5

A rather straightforward way to calix(4)arenes with different substituents in the *para* position consists of the condensation of a suitable linear trimer or dimer with the corresponding bis(bromomethyl)ated phenol or dimer[150-152] (Scheme 6). It allows the preparation of compounds with reducible substituents, such as nitro, azophenol or halogen. Especially interesting are asymmetrically substituted calix(4)arenes, as the basis for chiral host molecules. They are obtained with three (in the order ABCC) or four different phenolic units[151,152] or by substitution of a single *meta* position.[153]

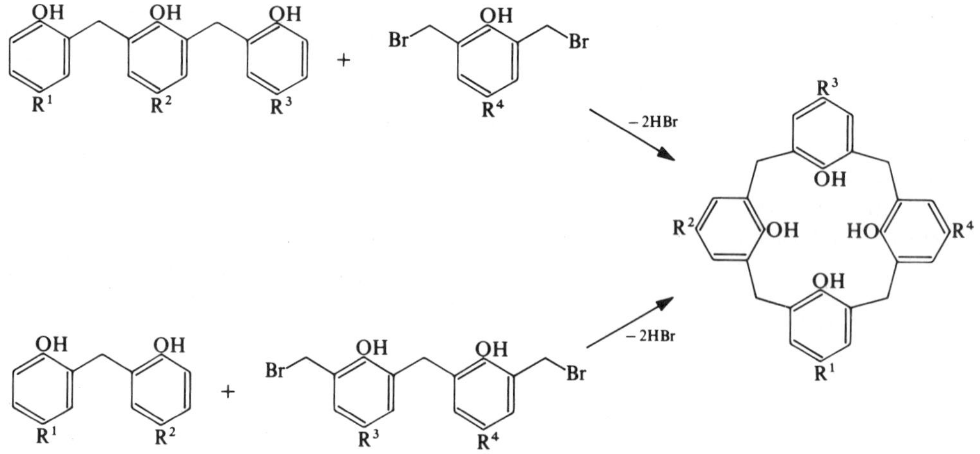

Scheme 6

In relation to Scheme 6, a series of calix(4)arenes was prepared in which two opposite phenolic units are connected *via* their *para* positions by an aliphatic chain.[154,155] Other cyclic compounds with dimethylene ether bridges,[156] with pyrocatechol[157] or resorcinol[158,159] units, can also be mentioned. In addition, calixarenes undergo chemical modification by those reactions which are known to occur with phenols.

35.4.3 Solid-state Properties

Melting points of several series of oligo[(2-hydroxy-5-methyl-1,3-phenylene)methylene]s were reported by Kämmerer and Niemann,[160] showing a minimum for $n = 5$ in all cases. In comparison to linear oligomers, the corresponding calixarenes exhibit remarkably high melting points, in some cases in excess of 400 °C. Lower melting points are found, however, for differently substituted calixarenes, and the melting point may be considerably different for isomeric calix(4)arenes.[151] In general, calixarene derivatives, such as esters or ethers, have lower melting points, with some exceptions.[135]

The first single-crystal X-ray diffraction studies in the field of phenolic oligomers were reported for *p-t*-butylcalix(4)arene[161] and for the octaacetate of *p-t*-butylcalix(8)arene.[162] They are regarded as the earliest definite proof of the ring size of these two oligomers. Recently, several crystal structures of calix(4)arenes have been published,[163-165] always exhibiting the so-called 'cone' conformation (Figure 2). Obviously, the conformation of the molecule is uniquely organized by the cyclic array of intramolecular hydrogen bonds. For all compounds, the distances between adjacent oxygen atoms are found in the range of 2.63–2.67 Å. Slight conformational differences are reflected by the inclination of the phenyl rings with respect to the normal of the plane of the methylene carbon atoms. It varies normally between 121° and 129°, but may be deformed to 115° or 137° by special packing effects.[163]

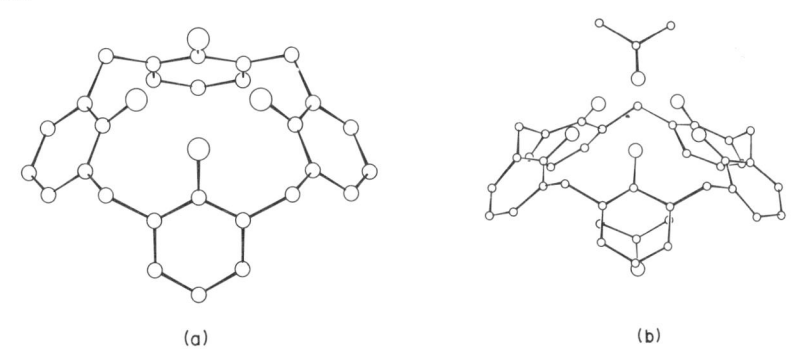

Figure 2 Conformation of (a) calix (4)- and (b) calix (5)-arene as determined by X-ray analysis;[143,163] acetone, which is incorporated in the crystal lattice in different ways, is omitted for calix(4)arene

A similar cone conformation is found for calix(5)arene (Figure 2),[143] and a complete cycle of intramolecular hydrogen bonds is present also in isopropylcalix(6)arene[166] and *t*-butylcalix(8)arene.[167]

Solvent entrapment in the crystal lattice often occurs in calixarenes although solvent-free crystals have been obtained from *p-t*-octylcalix(4)arene.[174] However, the guest molecules are not always located inside the cavity of the host molecules. The host/guest ratio may be different from unity[163,165] and different ratios have been found for the same host/guest pair.[163]

Crystal structures for linear oligomers were first reported by Casiraghi *et al.*[128,129,168] All results show that the conformation is again determined by a maximum number of intramolecular hydrogen bonds between hydroxyl groups of adjacent phenolic units.[169] In the crystal lattice, the molecules are ordered by further intermolecular hydrogen bonds either to indefinite chains (Figure 3a) or to cyclic dimers (Figure 3b).[169] In this arrangement, each hydroxyl acts alternately as a donor and an acceptor concurrently ('isodromic hydrogen bonds') and never as double acceptor. The intramolecular O—O distances for all known structures are in the range of 2.62–2.71 Å for methylene-bridged compounds (2.71–2.79 Å were found for ethylidene-bridged trimers[129]) and the intermolecular O—O distances are only slightly longer (2.68–2.77 Å). This indicates that all the hydrogen bonds should exhibit the same strength. While no special direction of hydrogen bonds is found for oligomers with a nonsubstituted *ortho* position, a methyl or *t*-butyl group in one *ortho* position obviously directs the hydrogen bonds in the opposite direction, as found for tetranuclear compounds.[169]

As shown in Figure 4, three phenolic units may be placed in a chair-like *anti* or *trans* position, or in a boat-like *syn* or *cis* position. In the case of the ethylidene-bridged oligomers, this sequence is determined by the relationship of the methyl substituent to the O—H ···· O bond system on the opposite side.

Crystallographic characterization of 2,4,6-tri(hydroxymethyl)phenol and 3,3′,5,5′-tetra(hydroxymethyl)-4,4′-dihydroxydiphenylmethane was performed by Perrin *et al.*[52] All bond distances and bond angles are consistent with literature values. Dihedral angles between the benzene-ring plane and the C—C—O planes of the hydroxymethyl groups were found to be between 3° and 143°, obviously determined by the requirements of intermolecular hydrogen bonds. Thus, a three-dimensional network is formed leading to a very compact arrangement in the crystalline state.

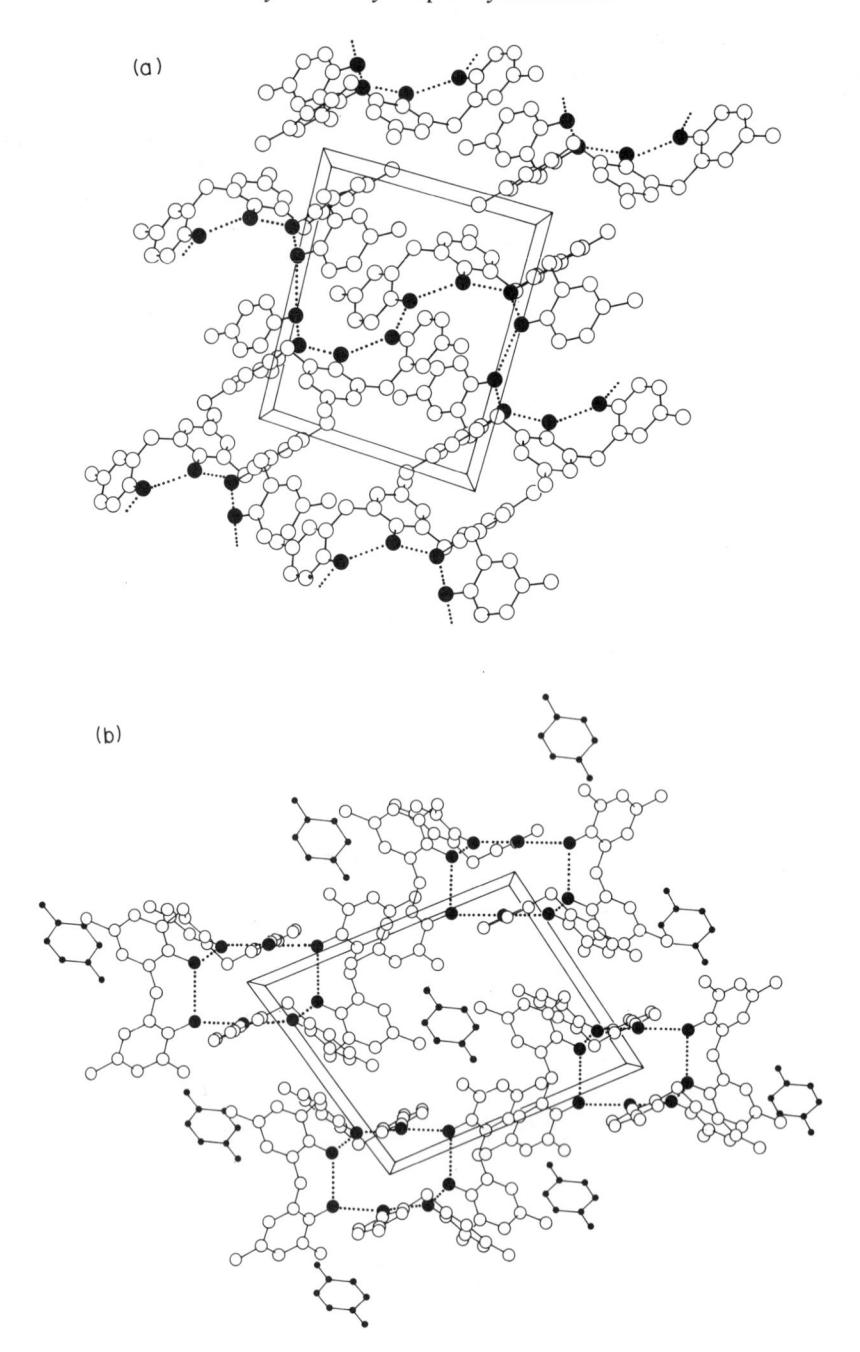

Figure 3 Molecular and crystal structure of tetranuclear compounds

Conformation is (a) *anti/anti* for R = H and (b) *syn/anti* for R = Me; hydrogen bonds are indicated by dotted lines[169]

Surprisingly, no intramolecular hydrogen bonds were identified in either compound, nor in the mono- and di-hydroxymethyl derivatives of *para*-substituted phenols.[170,171]

35.4.4 Solution Properties

The conformation of oligo[(hydroxyphenylene)methylene]s in solution (and, as a consequence, their chemical reactivity) is mainly determined by the presence of intramolecular hydrogen bonds

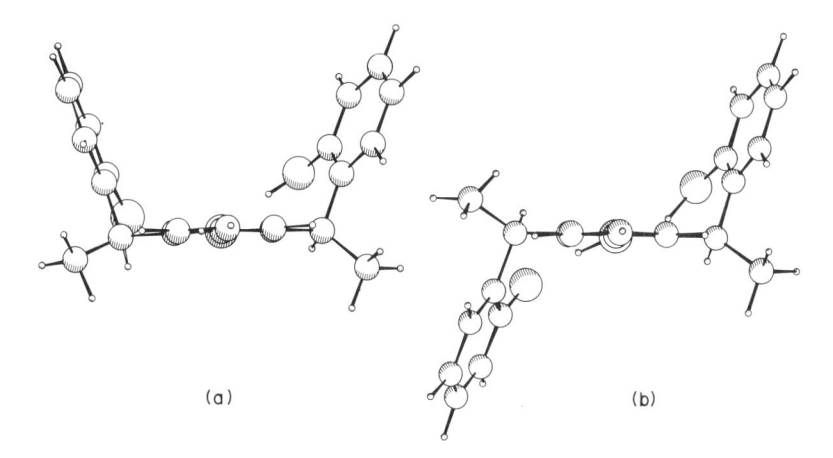

Figure 4 Molecular conformation of the diastereomeric o-ethylidene-linked phenol trimers. The *meso* compound (a) has the *syn* conformation, the racemic compound (b) the *anti* conformation[129]

between the hydroxy groups of adjacent *ortho,ortho*-linked phenolic units. In aprotic solvents, intra- as well as intermolecular hydrogen bonds were observed by IR[172–174] and ^{1}H NMR spectra.[175] Conclusions related to oligomer conformation may also be drawn from the temperature dependence of the dipole moment that has been reported for a series of di-, tri-, and tetra-nuclear compounds by Tobiason *et al.*[176]

The conformational mobility of calixarenes is limited by their cyclic structure. It was shown[177,178] that the interconversion of various calixarene conformers is rapid on the NMR time-scale at elevated temperature, while the cone conformation is 'frozen' at low temperature. As expected, cyclic penta-, hexa- and hepta-nuclear compounds are more flexible than tetranuclear compounds but, surprisingly, calix(8)arenes exhibit a similar energy barrier for the interconversion as calix(4)arenes.[179,180] This is again caused by intramolecular hydrogen bonds.

The 'hyperacidity' of *ortho,ortho*-linked di-, tri- and tetra-nuclear compounds (in comparison with similar isomers) was first noticed by Sprengling on applying potentiometric titration.[181] It is easily explained by strong intramolecular hydrogen bonds which stabilize the monoanion. An extensive study of these effects was performed,[182] using 'well designed' oligomers, consisting of alkylphenol units and one nitrophenol unit,[183] which allows an easy determination of pK_{a1} (K_{a1} = first acid dissociation constant). Decreasing values for pK_{a1} with increasing chain length are found for linear compounds with the *p*-nitrophenol unit at the terminus. This indicates an increased stabilization of the monoanion by a chain of intramolecular hydrogen bonds. Bulky substituents *ortho* to the hydroxyl group at the other end of the molecule cause a higher acidity even in tetranuclear compounds. The decrease in pK_{a1} is greater if the *p*-nitrophenol unit is placed in the interior of the chain. In calix(4)arenes, the first acid dissociation constant is obviously very sensitive to small conformational changes in the cyclic array of hydrogen bonds,[184] and a surprisingly low value for pK_{a1} in *p*-nitrocalix(4)arene was reported by Shinkai.[185]

The ability of *t*-butyl- and *t*-pentyl-calixarenes to transport metal ions from an alkaline source-phase through hydrophobic liquid membranes into a neutral receiving-phase was demonstrated by Izatt *et al.*[139,186] The effect is most pronounced for Cs$^+$ ions, where the highest transport rate is found for the cyclic octamer, while the cyclic tetramer shows the highest selectivity.

Photo-CIDNP (photochemically induced dynamic nuclear polarization) studies have shown very recently a rapid intramolecular hydrogen transfer (within 10^{-10}–10^{-9} s) from phenol to adjacent phenoxy radical units.[187,188] In contrast to an earlier interpretation,[189] they provide further evidence for the presence of intramolecular hydrogen bonds.

Efforts to correlate the reactivity of linear oligomers with the chain length have been reported by Imoto *et al.* for the acid-catalyzed reaction with formaldehyde.[190] More detailed studies are reported for the electrophilic bromination with bromine in acetic acid.[191,192] It was found that the reactivity is decreased by an intramolecular hydrogen bond between the phenolic hydroxyl groups, especially if this hydrogen bond is directed by bulky substituents from the adjacent unit towards the reacting phenolic unit. The directing effect of an *ortho* substituent is transmitted from one end of the molecule to the other end in linear *ortho*-linked oligomers even up to $n=6$,[193] which again is strong evidence for a concerted intramolecular hydrogen-bond phenomenon between all phenolic hydroxyl groups. Again the strongest effect is observed when the interior phenolic unit is influenced by both adjacent units.[194]

Unique differences were found for the reactivity of chloromethyl groups in methanolysis and aminolysis reactions for two series of isomeric dinuclear compounds.[57] The rate of methanolysis[195] leading to hydroxybenzyl methyl ethers, is nearly independent of structure, while the aminolysis[196] is strongly accelerated (by factors greater than 10^3) in *ortho,ortho*-linked dimers. Higher reaction rates for trimers and tetramers again suggest a chain of intramolecular hydrogen bonds.[197]

Electrophilic substitutions under alkaline conditions, which require the phenolate ion as substrate, may be retarded for *ortho,ortho*-linked compounds. Obviously, the intramolecular interaction of the hydroxyl groups, leading to a decrease of pK_{a1} and an increase of pK_{a2} makes a complete anionic structure of the reacting phenolic unit impossible, at least under certain conditions. The lowest reactivity attributable to the 2,2'-dihydróxydiphenylmethane in comparison to the 4,4'- and 2,4'-isomers in the alkali-catalyzed addition of formaldehyde (in spite of the presence of two more reactive *para* positions) was explained by such a hydrogen-bonded monoanion.[98] Similar results were obtained for coupling reactions with benzenediazonium chloride.[198]

35.5 SYNTHESIS OF INDUSTRIAL-GRADE RESINS

The various resin types and the final physical state of the resulting product determine the process and equipment that is used in the manufacture of phenolic resins. The key monomers, phenol and formaldehyde, are transformed into a variety of forms—flakes or powders, liquids, solutions or dispersions. Either acid (novolak) or base (resol) catalyzed conditions are used. The use of substituted phenols is significantly lower and relates to speciality applications. Most resin production is carried out by a discontinuous or batch process. Several continuous processes have been described in the literature but only the continuous process for standard novolaks is in operation. The multitude of resins with different specifications (different gel times and melt flows) required by different customers would render the continuous process uneconomical, except in those instances when a single large-volume resin can be used for a particular large volume end-use application.

The total reaction enthalpy of the phenol–formaldehyde substitution and condensation reaction was determined experimentally by Manegold and Petzold[199] and Jones[200] to be $\Delta H_0 = 81.1$ kJ mol^{-1} and 82.3 kJ mol^{-1}, respectively. Individual values of 20.1 kJ mol^{-1} for the substitution reaction, 78.1 kJ mol^{-1} for the condensation reaction and 98.2 kJ mol^{-1} for the total reaction have been calculated *via* combustion enthalpies.[201] The heat generation per unit time and peak temperature depend on production conditions, the molar ratio of reactants and catalyst concentration.

Aqueous formaldehyde (37–50%) is generally the preferred formaldehyde source, because the highly exothermic reaction of phenol and formaldehyde is more conveniently controlled by the vaporization and condensation of the excess water. Some selected liquid resol resins[202] are reportedly prepared with poly(oxymethylene) under continuous addition of poly(oxymethylene) to 90% phenol in water and maintaining the temperature below 90 °C.

35.5.1 Synthesis of Novolak Resins

The preparation of novolak resins consists of an acid-catalyzed reaction of phenol and formaldehyde at reflux. Strong acids, such as H_2SO_4 or HCl, provide a low pH resulting in a rapid exothermic reaction between phenol and formaldehyde. The reaction rate of phenol with formaldehyde increases with decreasing pH. Thus, with strong acids, a catalyst is introduced incrementally to moderate the exothermic reaction, and formaldehyde addition is carefully metered into the phenol–acid catalyst composition. These strong acids can be neutralized with caustic soda or lime at the conclusion of the reaction.

Weaker acids, oxalic acid or H_3PO_4, promote a more controllable reaction and result in novolak resins with reduced color. The preferred catalyst for the manufacture of novolaks is oxalic acid. It sublimes at about 157 °C and ambient pressure. At a higher temperature (180 °C) it decomposes to CO, CO_2 and H_2O and requires no removal step. Due to its reducing behavior very light-colored resins are obtained.

The molar ratio of phenol to formaldehyde is normally within the range of 1:(0.75–0.85). The molar ratio generally relates to novolak MWD and melting point range. Lower melting novolaks (m.p. 70–75 °C) are obtained when a molar ratio of 1:0.75 is used, whereas higher melting novolaks (m.p. 80–100 °C) result with a 1:0.85 ratio. Novolaks are flaked to an appropriate size, mixed with HMTA, pulverized into a powder and processed in this form.

In the batch process, phenol is heated to 95 °C. After catalyst addition, aqueous formaldehyde is introduced. The temperature is maintained at reflux until the formaldehyde is consumed. Then the

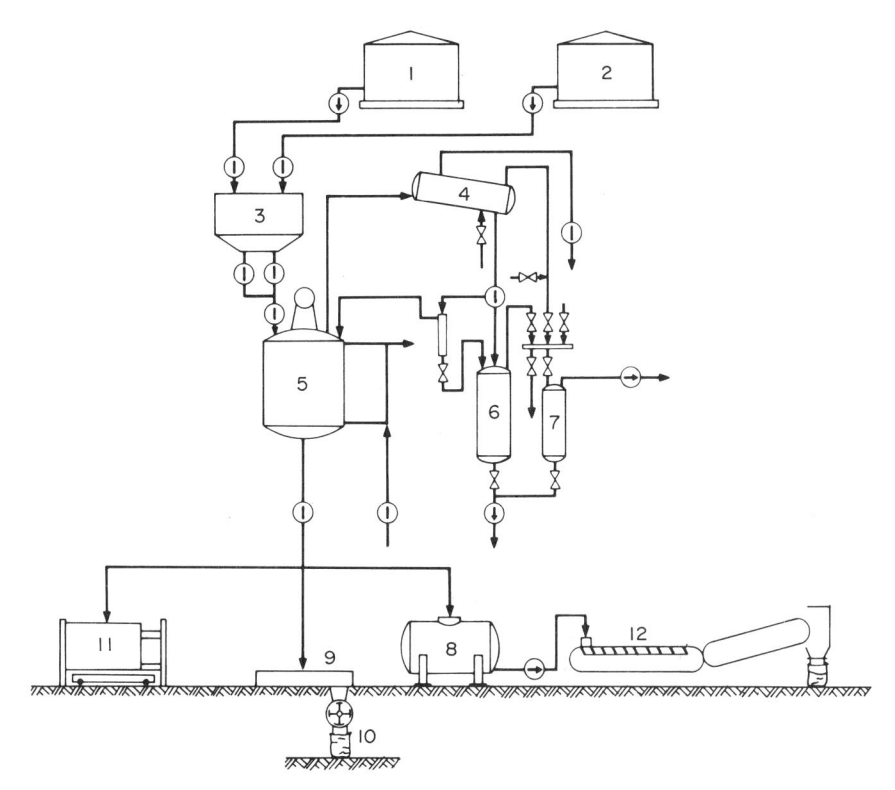

Figure 5 Phenolic-resin production plant, batch process:[14] 1, phenol; 2, formaldehyde; 3, scale; 4, condenser; 5, reactor; 6, condensate receiver; 7, vacuum; 8, resin receiver; 9, resin through; 10, mill; 11, cooling carriage; 12, cooling belt

water is removed with the unreacted phenol by distillation. As soon as the desired melting point is obtained, the resin is transferred to a heated vessel and then flaked on a continuous cooling belt flaker (Figure 5).

In the continuous process,[203,204] shown in Figure 6, formaldehyde, phenol and catalysts are transferred to the first stage reactor where the reaction commences. The reaction is completed in the

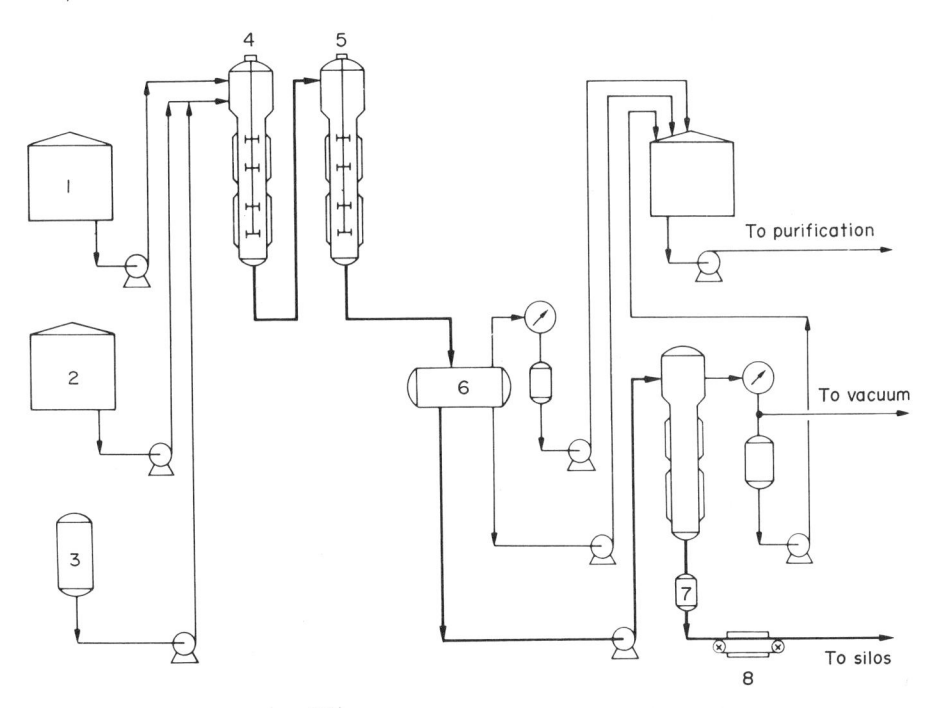

Figure 6 Continuous novolak-resin process:[203] 1, phenol; 2, formaldehyde; 3, catalyst; 4, first reactor; 5, second reactor; 6, flash drum; 7, continuous distillation; 8, cooling belt

second stage reactor at a temperature of 120–180 °C and a pressure of 7×10^5 Pa. The resulting reaction mixture is conveyed to a flash drum, which acts as a vapor–liquid separator. The flashed vapor is condensed and transferred to the purification section, while the lower liquid-phase settles into two layers, the heavier one containing resin. Resin is purified and fed to a belt flaking machine.

A novolak resin may be prepared on a laboratory scale according to the following procedure.[205] A total of 130 g of phenol (1.38 mol), 13 ml water, 92.4 g of 37% aqueous formaldehyde (1.14 mol) and 1 g of oxalic acid dihydrate were introduced into a resin kettle. Molar ratio of phenol to formaldehyde is 1:0.826. The mixture is refluxed for 30 min with constant stirring. An additional 1 g of oxalic acid dihydrate is added with continued refluxing for another hour. Then 400 ml of water are added with cooling. Water is decanted from the upper two-phase composition. Water and residual phenol is removed by vacuum distillation until the residue temperature reaches 120 °C or until a sample of the resin is brittle at room temperature. About 140 g of resin is obtained. A melting point of 80–90 °C is identified.

Yields of production batch or continuous novolaks are generally 105% based on phenol charge.

Residual water has a greater effect on resin properties and flow when compared to unreacted phenol. Resin melt viscosity can be reduced as much as 90% at elevated temperatures by the addition of 3% water. Water also facilitates the HMTA cure of novolaks, whereby the reactivity of an HMTA-cured novolak is accelerated by increased water content.

35.5.2 Synthesis of Resol Resins

The alkaline phenol–formaldehyde reaction is more versatile than the corresponding production of novolaks and results in a wide spectrum of resins consisting of low to intermediate molecular weight liquid resins and solid resins. The phenol:formaldehyde molar ratio ranges from 1:1 to 1:3. Catalyst-type influences the MWD and resin microstructure and differs significantly from the acid-catalyzed novolak reaction. Group I and IIa metal oxides and hydroxides are most prominent, with ammonia or HMTA and amines used to a lesser extent. Ammonia or HMTA becomes incorporated within the resol microstructure.[69] Catalyst performance and its removal (if necessary) as well as cost are the main criteria for catalyst selection. When necessary, ionic catalysts are removed by salt precipitation and filtration. Catalyst removal is required if dielectric properties, ageing and humidity resistance are critical requirements in the final end-use application. The molar catalyst ratio based on phenol in commercial resols is quite diverse. It ranges from 1:1 to 1:0.01.

Most resols are produced by a discontinuous or batch process, whereby single or multiple steps are involved and relate to product type. Low molecular weight water-soluble resins for paper laminates or plywood glues are examples of single-step resols, where, if resin advancement is needed, partial condensation is carried out in a second step. Usually water is removed by vacuum dehydration, and the resin is recovered as a high-solids liquid resin, or solvent is added resulting in a solution resin. High melting resols of phenol or cresol (substituted phenols) are usually catalyzed by ammonia and require accurate temperature control, in-line analysis and rapid discharge and cooling.

The following procedure is suggested for the preparation of a resol resin on a laboratory scale.[205] To a resin kettle equipped with reflux condenser, stirrer, thermometer and siphoning tube for removal of samples, 94 g (1 mol) of phenol, 123 g of 37% formaldehyde (1.5 mol) and 4.7 g of barium hydroxide octahydrate were added. Heating with constant stirring is maintained for 2 h at 70 °C. In the absence of stirring, phase separation of resin from water occurs. The pH is then adjusted to 6–7 by controlled addition of 10% sulfuric acid. Vacuum is applied by water aspirator with pressure regulated between 30–50 mmHg (1 mmHg = 133.322 38 Pa). Water is then removed by vacuum distillation. The temperature is maintained below 70 °C. Samples of resin are removed every 15 min until the desired gel time is obtained. Gel time is determined by pressing the liquid sample with a spatula on a hot plate at 150 °C and until the fluid resin becomes 'stringy'. The value is recorded in seconds. Full resol characterization involves percent solids determination (150 °C for 30 min), viscosity value at 20 °C or 25 °C, water content by Karl–Fischer titration, and percent free formaldehyde by hydroxylamine hydrochloride titration.

35.5.3 Synthesis of High-*ortho* Resins

Phenolic resins with a significantly high amount of *ortho,ortho'*-orientation are prepared at an intermediate pH range of 4–7. These *ortho,ortho'*-materials can be either novolaks or resols. The unique properties of a wholly *ortho*-substituted phenolic resin are primarily a high rate of cure with HMTA (novolaks) and attractive rheological behavior of liquid resols.

Conventional novolak equipment is used to manufacture high-*ortho* novolaks, except mildly acidic catalysts, such as divalent metal acetates (Ca, Mg, Zn, *etc.*), are used with a molar ratio of phenol to formaldehyde in the range of 1:0.8. Formaldehyde is added during the course of the reaction. Usually the reaction/condensation and high temperature distillation (to 145 °C) is conducted in two stages to minimize any exotherm or gelation that can occur due to the use of a higher concentration of formaldehyde solutions (50%). Improved processing techniques have been developed for the preparation of high-*ortho* novolaks and liquid resols through the use of toluene or xylene as an azeotropic solvent. Constant removal of water provides more favorable control of reaction exotherm and rate enhancement. Others features include dual catalyst system and split feed of formaldehyde.

The preparation of a high *o,o'*-novolak is recommended as follows.[206] A catalyst system of 0.1% ZnO, 0.2% magnesium acetate and 0.11% magnesium methyl sulfonate based on phenol content is placed in a resin reactor along with 100 g of xylene and 1000 g (10.6 mol) phenol and heated to 108 °C. 50% solution of formaldehyde (458 g, 7.62 mol) is added at a rate of 4 g per min, during which time the water concentration is increased to 10%. This concentration is maintained through distillation of a xylene–water mixture *via* a Dean–Stark trap, which allows water removal, and xylene is returned to the reactor. Upon completion of the formaldehyde addition, water is removed over a period of 3 h until the temperature is increased to 140 °C. The mixture is then distilled under vacuum to remove xylene and excess phenol at a vacuum of 6.2 kPa and 140 °C temperature. The resin is removed from the vessel and cooled. GC analysis including silylation indicated a 97% *ortho* content and a 94% yield based on phenol.

Other methods for high-*ortho* novolaks consist of conducting the reaction of phenol and poly(oxymethylene) in an azeotropic nonpolar media (xylene) at 170–220 °C for 12 h[76] or by the reaction of selected metal phenolates with poly(oxymethylene).[74]

The latter method, for the preparation of high-*ortho* novolaks by selected metal phenolates, is as follows. To a suspension of phenoxymagnesium bromide (59.2 g, 0.3 mol) in one liter of anhydrous benzene, poly(oxymethylene) (6.0 g, 0.2 mol) is added with stirring at room temperature. The resulting yellow slurry is vigorously stirred under reflux for 12 h and then quenched with an excess of 10% aqueous HCl. Triple extraction with ether (150 ml each) of the yellow solid, followed by drying over Na_2SO_4 and concentration of the dry ether extract, yields a pale yellow syrup. Further drying of the residue *via* high vacuum (10 Pa) at about 50 °C, and then cooling yielded 26.1 g (92.7% based on phenol) of a powdered colorless solid, m.p. 56–60 °C. HPLC analysis indicated 40% dimer, 34% trimer and 36% tetramer. ^1H and ^{13}C NMR indicated the oligomers were wholly *ortho–ortho'* linked.

35.5.4 Synthesis of Aqueous Dispersion Resins

A novel method for manufacturing specialty resols is the preparation of phenol–formaldehyde resins as a stable dispersion of discrete particles. The manufacture of stable aqueous resol dispersions of phenol or substituted phenols with formaldehyde requires the use of a polysaccharide-type protective colloid with a base catalyst. It is a two-stage reaction with the initial stage consisting of phenol–formaldehyde condensation to a specific molecular weight, followed by transformation of the aqueous phenolic solution to a phenolic resin in water dispersion. The protective colloid is added during the phase change and provides stabilization of the phenolic particles without agglomeration. The average particle diameter is 10–20 μm.

The synthesis[207] of a stable aqueous phenolic dispersion is as follows. In a 5 l round-bottomed flask 1200 g (12.76 mol) of phenol, 1668 g of 40% formaldehyde (22.24 mol) and 30 g of barium hydroxide octahydrate are introduced. Upon heating the mixture to 75 °C, heat is removed and the mixture is allowed to reflux atmospherically. Refluxing is continued for 45 min. After this heating period, an aqueous solution of 24 g of gum arabic, 6 g of guar gum in 720 ml of water is added. Subsequent to the 'gum solution' addition, 20% aqueous H_2SO_4 is added to neutralize the barium hydroxide. The clear solution turns cloudy and ultimately milky, indicative of the formation of a uniform dispersion. A temperature of 85 °C is maintained for about 10 min followed by cooling of the dispersion to 50 °C. Dispersion is 42% solids content by weight with an average particle size of 10 μm. It possesses a polydispersity (M_w/M_n) value of 2.1 by gel permeation chromatography (GPC) and free phenol content of 6.2%. Gel time (150 °C) was determined as 75 s.

35.5.5 Resin Characterization According to Application

The variation of the main synthesis parameters, *i.e.* composition, catalyst, phenol–formaldehyde mole ratio, MWD and modification yields an extremely broad range of resins exhibiting selective and unique properties as required in different application fields. The most important resin types according to application fields[14] are characterized below.

35.5.5.1 *Particle boards, plywood and fiber boards*

Particle-board resins are a resol-type obtained by reaction of phenol and formaldehyde in a molar ratio of 1:(1.8–3.0) in an aqueous solution with sodium hydroxyde as catalyst. Apart from its catalytic action, sodium hydroxide has other important features,[208] such as providing a low viscosity for a high average molecular weight resin and excellent water solubility. In contact with the acidic wood components, the resin is probably neutralized and exhibits a strong increase in viscosity and develops bonding properties. Therefore, sodium hydroxide is often used in an equimolar ratio based on phenol. Often special catalysts are used to accelerate the curing rate. Organic or inorganic carbonates, formamide, lactones, resorcinol and resorcinol resins and isocyanates have been proposed. As the binder represents the major cost of particle board, phenol resins are sometimes modified by addition of urea, lignosulfonates, or other cost-reducing additives.

For wafer board and oriented-strand board production, spray-dried solid phenolic resins in combination with molten wax are used.

For weather-resistant plywood, resols catalyzed by sodium hydroxide are used. The viscosity of the resins varies between 700 to 4000 mPa s according to the gluing process — dry or wet. As with PB-resins, the content of free phenol and free formaldehyde is extremely low and practically nondetectable.

Phenol resol resins, sometimes modified with natural resins such as colophonium (rosin), are used in fiber boards. In general, the dry resin content in fiber boards is only between 1 and 3% achieved by precipitation of a diluted aqueous resin solution by dilute sulfuric acid or aluminium sulfate at pH 4.

35.5.5.2 *Molding compounds*

For molding compounds, novolak-type resins produced mainly by use of oxalic acid as catalyst are applied; hydrochloric, sulfuric or phosphoric acids are seldom used.[14]

The phenol–formaldehyde molar ratio is within the range of 1:(0.75–0.85). Novolak resins with a low free-phenol content ($\leq 2\%$) and relatively high flow are used in newer formulations. HMTA is the curing agent in most cases. High-*ortho* novolaks are added in some formulations to enhance the curing rate. Resols are used for special electrical applications when high hydrolytic resistance is required.

35.5.5.3 *Heat and sound insulation materials*

Thermal and acoustical insulation is one of the largest and fastest growing markets for phenolic resins. Thermal insulation markets include the use of phenolic resins as binders for mineral and glass fibers as well as smaller volumes in the manufacture of phenolic foam.[14]

Appropriate low molecular weight aqueous resol resins for mineral and glass fiber mats are obtained by reacting phenol with an excess of formaldehyde at temperatures below 70 °C. The phenol–fomaldehyde ratio is between 2.5 and 3.5. In general, alkaline earth hydroxides, *e.g.* calcium or barium hydroxide, seldom sodium hydroxide, are used as catalysts (see Figure 8). High-quality resins are normally ash-free, the catalyst is precipitated as sulfate or carbonate and removed by filtration. A high formaldehyde ratio favors high resin efficiency. Urea is added in most formulations up to 40% (calculated on dry weight) to reduce cost. The resin is applied as a 10 to 15% aqueous solution; a high water dilutability is an important requirement. A satisfactory resin consists mainly of mononuclear polyhydroxymethylated compounds, the prevailing species being tri(hydroxymethyl)phenol. The amount of polynuclear compounds should be as low as possible.

Aqueous resols consisting of phenol and formaldehyde in a molar ratio 1:(1.5–2.5) are generally used for phenolic foam production. Sodium hydroxide or alkaline earth hydroxides are used as catalysts. Typical resol resins used in foam preparation contain about 80% solids, volatile components consist mainly of water, formaldehyde and phenol. After the reaction is completed within a temperature range of 60 to 90 °C, the percent solids and viscosity are adjusted by distillation. The resin viscosity is an important variable, among others, for foam-density adjustment. Foam density is further regulated by the type of surfactant, amount of blowing agent, temperature and acid reactivity of the resin. Crosslinking of the resol resin is promoted by the addition of a strong organic sulfonic acid such as p-toluenesulfonic acid, methanesulfonic acid or phenolsulfonic acid.

35.5.5.4 Bonded textile felts

Pulverized novolak resins with a low free-phenol content and special particle-size distribution are used for bonded felts production.[14] They are mixed with generally 10% HMTA and often with flame-retarding inorganic materials.

35.5.5.5 Industrial laminates and paper impregnation

The hydrophilic structure of noncured phenolic resins characterizes them as suitable material for the impregnation of paper and cotton fabrics used in the manufacture of electrical and decorative laminates, molded parts, filter papers and battery separators.[14] Low molecular weight components, preferably mononuclear phenol alcohols, penetrate into the capillary cavities of cellulose fibers and fill the cavities, whereas resins of higher molecular weight coat the fibers and make them water-repellent.

Phenol, cresol and xylenol are used to manufacture high quality electrical laminates, p-t-butylphenol or nonylphenol enhance flexibility. The higher compatibility of cresol resins with natural oils results in mechanically tougher and less brittle resin films. Today, phenol resols modified with synthetic flexibilizers (*i.e.* polyether, polyester, polyurethane or polybutadiene compounds) are used, which are comparable to cresol and xylenol resins.

Resins of different molecular weight are necessary, regardless of whether the impregnation is performed in one or two steps. The first (preimpregnating) resin, a low molecular weight resol resin, contains only a minor portion of dinuclear phenols ($\sim 20\%$). It mainly consists of mono-, di-, and tri-(hydroxymethyl)phenol in addition to free phenol ($\sim 10\%$). Calcium, magnesium or barium hydroxide are used as catalysts for the production of preimpregnating resins, which are precipitated at the end of the reaction by the addition of sulfuric acid. Melamine resins can be added to improve flame retardancy. Hydrophobic modified resols of medium average molecular weight are used as the 'main' impregnating resin. Mostly, they are flexibilized, *e.g.* with phosphoric acid esters, tung oil or other synthetic prepolymers.

35.5.5.6 Coatings

High metal adhesion and low water vapor/oxygen transmission are the key factors for the effectiveness of phenolic coatings. They are always used in combination with more flexible hydrophobic resins like epoxy, poly(vinylbutyral) alkyd or natural resins and maleinized oils.[14] Because of coloration or the propensity toward discoloration, phenolics are mainly employed as primers and undercoating materials. If solubility in hydrocarbons and aromatics is required, it is easily achieved by the use of alkyl phenol resins.

Water-borne or water-reducible resins are increasing in importance due to ecological and worker-safety reasons. Water solubility of the phenolic moiety is frequently achieved by introducing carboxy, sulfone or sulfomethyl groups, followed by ion formation with ammonia, amines or inorganic bases. Low molecular weight resols are water soluble while higher molecular weight resols require higher amounts of inorganic bases for water solubility. Many water-reducible phenolic coating resins are described in the literature,[209-220] including salicyclic acid or diphenolic acid modified resins.[14]

Ammonia-catalyzed resols based on cresols or phenol, which are typically yellow in coloration (gold lacquers), are mainly used in coatings for metal containers. In general, the phenolic resin must be flexibilized with other polymers, for instance, epoxy, poly(vinylbutyral) or alkyd resins.

Alkyl- and aryl-phenolic resins can be cooked with drying oils as a result of reduced self-reactivity.[221] Tung oil is preferred; sometimes linseed oil or castor oil is used. The proportion of phenolic resin is between 25% (resols) and 100% (novolaks) in relation to the oil, depending on self-condensation ability.

Rosin-modified phenolic resins[14] are used in printing inks, in oil lacquers and as additives to alkyd paints because of their good compatibility with natural oils ('art copals') in which they improve the drying and gloss.

35.5.5.7 Grinding materials

A combination of liquid and pulverized phenol resins is used for grinding-wheel manufacture.[14] The liquid resin functions as a wetting agent for the abrasive grains, powder resin and fillers. In addition, 2-furaldehyde, furfuryl alcohol, cresols and anthracene oil are also used in combination with liquid phenol resins. Liquid phenol resins and 2-furaldehyde possess very good wetting properties and good compatibility with pulverized phenol resins. The qualitative requirements of phenol resins are high, especially as far as the uniformity of the batches is concerned. Sodium hydroxide or sodium carbonate are preferred as catalysts for the production of liquid phenol–formaldehyde resols. Low viscosity resins with the highest possible dry solids content are required.

The novolak resin is mixed with HMTA and then finely ground. The HMTA portion may be between 4 and 14%, preferably 9%. Low HMTA content leads to lower crosslinking and softer bonds. High HMTA content results in higher crosslinking density and thereby higher hardness and heat resistance of the wheel.

35.5.5.8 Friction materials

A wide spectrum of phenolic resins are used as binders for friction materials.[14] These include phenol-type resins, such as novolaks, resols, novolak/resol blends, oil-modified novolaks, solubilized oil–phenol types, cresol resins and rubber or thermoplastic modified novolaks. Oil modification of phenol resins is conducted with cashew nut shell liquid (CNSL), tung, linseed or soya bean oil.

35.6 TOXICOLOGY OF MONOMERS, PREPOLYMERS AND RESINS

The risks associated with handling phenolic resins (during manufacture, formulation or fabrication into commercial products) must be distinguished between starting materials, phenol–formaldehyde oligomers or prepolymers and cured phenolic resin. Key factors must be considered, such as compositional characteristics (resol or novolak), as well as the molecular weight of the resin. These help to determine the physiological effects of various phenolic resins. Low to medium molecular weight resins (resols) possess varying amounts of free phenol and formaldehyde. High molecular weight resins, primarily novolaks, are formaldehyde-free and moderately low in free phenol, while solid resols possess small amounts of formaldehyde and phenol. Cured phenolic resins are completely innocuous. The US Food and Drug Administration (FDA) permits phenolic coatings and phenolic molded articles to come in contact with food.

35.6.1 Toxicology of Phenol

Phenol is a highly toxic compound causing protein degradation and tissue erosion. Occupational and environmental exposure to phenol is largely restricted to source-dominated areas. Ambient levels of phenol are quite low compared to the threshold limit value (TLV) of 5 p.p.m. The US Occupational Safety and Health Administration (OSHA) time-weighted average (TWA) (on skin) is 5 p.p.m. The human oral LD_{Lo} (lethal dose low) value is 140 mg kg^{-1}. Odor-recognition threshold of phenol is 0.05 p.p.m., far below levels of toxic effects. Upon skin contact, skin becomes white with a subsequent red blister-like appearance and an accompanying burning sensation. Solid or liquid phenol is rapidly absorbed by the skin and results in very severe tissue damage.

Exposure to large amounts of phenol results in death due to paralysis of the central nervous system. Minor ingestion causes damage mainly to the kidneys, liver and pancreas. If phenol is inhaled or swallowed, local cauterization occurs with headaches, dizziness, vomiting, irregular breathing, respiratory arrest and heart failure.

Excellent reviews summarizing environmental and health risks of phenol are recommended.[222,223]

35.6.2 Toxicology of Formaldehyde

Formaldehyde vapor is very irritating to the mucous membranes; the pungent odor is noticeable below 1 p.p.m. Aqueous formaldehyde is equally irritating. The preservation of medical or biological specimens by formaldehyde is well known. Formaldehyde is believed to attack bacteria by reacting with protein amino groups thus altering the nature and activity of these groups. Formaldehyde in the organism is partly exhaled or rapidly metabolized into carbon dioxide and formic acid, and the remainder is excreted in the urine. Formaldehyde is mutagenic in bacteria test systems and test systems using fungi and certain insects. It has also been shown that formaldehyde forms adducts with DNA and proteins in both *in vivo* and *in vitro* test systems.

The US Department of Labour reduced the permissible worker exposure level (PEL) to formaldehyde from 3 to 1 p.p.m. in 1987.[24] In addition, a short term exposure level (STEL) (*i.e.* exposure during any 15 minute period) of 2 p.p.m. is permitted. In Western Europe, permissible exposure levels (8 h) vary from country to country, allowing only, for example, 0.5 p.p.m. in West Germany.

Several toxicological studies have shown that formaldehyde can induce a rare type of malignant nasal tumor in both sexes of rats, in multiple inhalation experiments and in multiple species (both rats and mice).

Epidemiological studies, sponsored by the Formaldehyde Institute (US), of persons occupationally exposed to formaldehyde have shown no overall cancer excess, no nasal cancer excess and no excess respiratory cancer. The European Chemical Industry Ecology and Toxicology Center Report of 1982 claimed that the nasal cancers in experimental animals develop only at concentrations which produce chronic tissue irritation, and that there is no relationship between formaldehyde exposure and cancer in humans.[225] A recent article reviews formaldehyde as a pollutant, its occurrence and regulation.[226]

In 1987, the status of formaldehyde was changed by the US Environmental Protection Agency (EPA) from possible to probable human carcinogen (Group B-1).[224] Currently, the EPA is considering further regulatory action.

35.6.3 Toxicology of Prepolymers and Cured Resins

Prepolymers contain varying amounts of free phenol and formaldehyde. In particular, liquid resol resins are known to contain some free phenol and formaldehyde depending on the molar ratio of formaldehyde to phenol and whether a formaldehyde scavenger is used. A high ratio of formaldehyde to phenol significantly reduces free-phenol content in the resin. Inexpensive formaldehyde scavengers such as ammonia or urea are effective; more expensive resorcinol is employed under conditions which merit the expense. Solid novolak resins are formaldehyde-free and contain low amounts of free phenol (usually 2% to about 6%). Solid resols contain some formaldehyde and phenol.

Cured resins are virtually harmless and are permitted by the US FDA in food-type applications.

35.6.4 Microbial Degradation and Environmental Protection Methods

Selective methods requiring either degradation or adsorption have been developed for the reduction and disposal of solid or liquid waste that is generated during the manufacture of phenolic resins. These include microbial degradation, thermal incineration, physical or physicochemical scrubbing, chemical oxidation, resinification and adsorption methods.[14]

Microbial degradation is the most efficient process for the treatment of waste water containing phenol. Certain types of microorganisms such as pseudomonas, eubacteriales, actinomycetes and others exist in water containing up to $1000\ \text{mg l}^{-1}$ of phenol. These species are quite active at 25–35 °C. Within this aqueous environment, sufficient nutrient content (N, P) and oxygen, a pH of 7.5–8.5 and the absence of heavy metal ions ($<5\ \text{mg l}^{-1}$) are necessary conditions. Ammonium phosphate is a popular nutrient. Final effluents in the range of $0.1\ \text{mg l}^{-1}$ are achieved. Another equally effective method for phenol removed in waste water is adsorption by activated carbon.

With selective chemical oxidizing systems such as H_2O_2 with transition metal salts, ozone, sodium hypochlorite, $KMnO_4$ or $K_2Cr_2O_7$, phenol content is readily reduced to $<1\ \text{mg l}^{-1}$. Resinification and precipitation of phenolic prepolymers and/or unreacted monomers can occur by adding H_2SO_4 or ammonia and reacting at elevated temperatures. Precipitants, such as iron(III) chloride or aluminum sulfate, are recommended in the treatment of waste water emerging from plywood, particle-board and fiber-board industries.

Extraction of phenol[14] from waste water is not economically feasible, nor does it result in a low level of phenol (*i.e.* less than 1 mg l^{-1}). Phenol extraction processes, *e.g.* the benzene–caustic process according to Pott and Hilgenstock or Lurgi's Phenosolvane process, are used for the recovery of phenol in coking plants, coal liquefaction plants or those installations that manufacture phenol or phenolic resins and require subsequent biological treatment.

35.7 ANALYTICAL CHARACTERIZATION

A multitude of analytical methods are used in commercial resin production to determine, for example, monomer content, resin advancement, product quality and batch consistency. Many of these procedures have been in existence for many decades. Free-phenol determination by the Koppenschar method or colorimetric phenol determination, the hydroxylamine hydrochloride method for formaldehyde determination, resin viscosity, gel time, water content by the Karl–Fischer method, water dilutability, refractive index, melting point, flow distance and sieve analysis, as preferred analytical methods, have been compiled by Knop *et al.*[14]

35.7.1 Spectroscopic/Spectrometric Methods

Extensive IR studies have been reported by Hummel[227,228] and provide many peak assignments for resin systems. The use of FTIR has been shown to be a convenient method to monitor the reaction of phenol with HMTA.[69] The curing pattern of novolaks and resols was examined by FTIR by Pearce and co-workers[229] in a study directed to structure/property relationships and polymer flammability.

Solution and solid-state NMR (^{1}H and ^{13}C) analyses have provided extensive microstructures of prepolymers (see Section 35.3.5) and characterization of cured resin systems[14,63,69,93,95,230] including high-*ortho* resins. The use of ^{13}C NMR has provided greater resolution and identity of all nonequivalent carbon atoms without derivatization of the phenolic hydroxyl group. Sojka and co-workers[69,231] have prepared and characterized a large number of linear trimers and tetramers with definite structure. These ^{13}C NMR spectra of trimers and tetramers combined with known dimer spectra have led to a computerized structure library of novolak oligomers.

More recently Maciel[95] combined ^{15}N and ^{13}C CP/MAS NMR methods for a better understanding of the HMTA curing of a novolak resin.

The electron spectroscopy for chemical analysis (ESCA) technique has been applied to the characterization of phenolic resin surfaces that have been treated by plasma or ion beams. ESCA has also been used to distinguish the location of sodium ions of a conventional caustic-catalyzed solid resol with that of an aqueous-dispersion-type resol.[14]

Field desorption mass spectrometry (FDMS) has been applied to the identification of molecular species in novolaks prepared from phenol and alkylphenols. Commercial phenolic 'tackifying' resins with *t*-octylphenol or *t*-dodecylphenol substituents have been characterized by FDMS.[14] Prokai[232] has examined both novolaks and resols. Molecular ions (M$^+$) up to hexamer or $m/z = 624$ for novolak resin were obtained. Resol analysis was more complicated and required resin acetylation for better resolution.

35.7.2 Chromatographic Methods

Polystyrene, poly(vinyl acetate) or silica gels of very narrow sieve fractions of approximately 5 μm are used as the stationary phase in GPC of phenolic resins.[14] Molecular weight, size and distribution have been investigated[233,234] and have encompassed solvent and concentration effects as well as monomer content. The separation of resol components is incomplete due to aggregation,[235] at least in the case of the GPC of resols which differ in ability to form hydrogen bonds with solvents. High-alkali resols, *e.g.* wood binder resins,[236] are better analyzed by solubilizing these resins with trichloroacetic acid so that the MWD can be measured using conventional instrumentation and techniques.

GPC contour plots of novolak resins, particularly those for positive photoresist applications, provide an additional 'dimension' to the MWD plot.[237] The two-dimensional contour MWD plot, which is monitored by a UV detector, identifies phenolic resin and photosensitive dissolution inhibitors and allows rapid analysis of both resin and inhibitors after light exposure.

The comprehensive review by Tesarova and Pacakova[238] of GC and HPLC of phenols contains over 290 references and lists many satisfactory stationary phases, mobile-phase compositions and retention data. A review emphasizing HPLC of resols and novolaks has been published by Werner and Barber.[239]

Recognition of the highly polar structure and thermal instability of hydroxymethylphenols suggests that HPLC may be the best technique for the separation of low molecular weight prepolymers up to a molecular weight of 1000. HPLC has, in fact, developed into an extremely facile and powerful tool to completely characterize resols without the fear of decomposition or the necessity to derivatize the phenol alcohols.

The elegance of HPLC is best illustrated by examining the work of Mechin, Hanton and others (Figure 7, Table 8).[53,54] About 30 mono-, di- and tri-nuclear hydroxymethyl derivates were separated by a semipreparative HPLC method and identified by ^{13}C and ^{1}H NMR. More recently, the studies have been extended to the identification of compounds with up to five phenolic groups.[54]

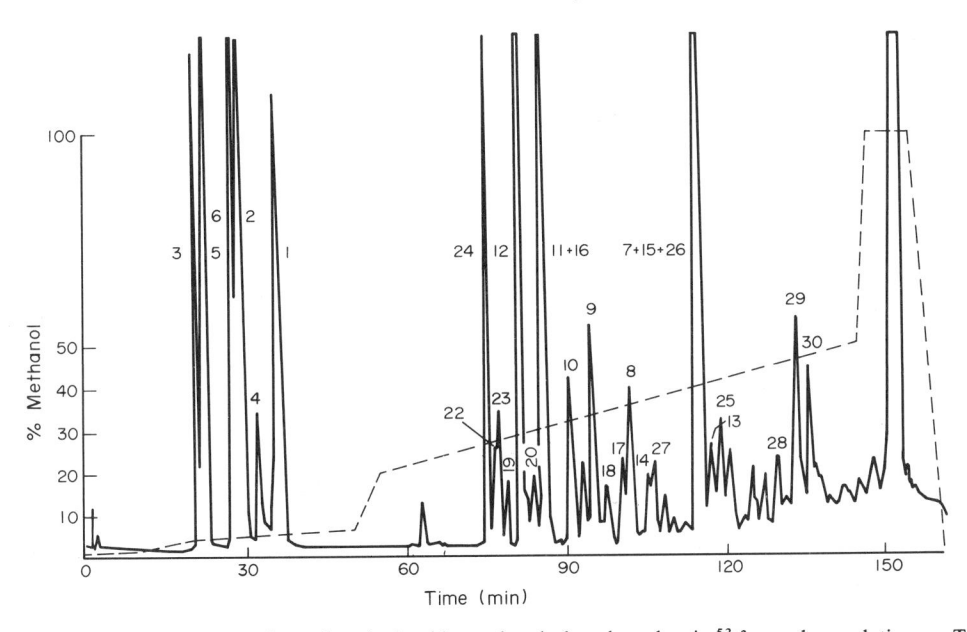

Figure 7 HPLC chromatogram of a sodium hydroxide-catalyzed phenol resol resin,[53] for peak correlation see Table 8

Table 8 Structures for Labeled Peaks in Figures 7 and 8[53]

Table 8 (*continued*)

A comprehensive review of the GC of phenols is recommended.[238] Strong hydrogen and thermal instability of phenol prepolymers requires etherification of the hydroxyl group (*e.g.* with *N,O*-bis (trimethylsilyltrifluoroacetamide), BSTFA) for effective GC separation. Haub[240] and Lindner[241] separated mono-, di- and tri-nuclear hydroxymethylphenols by silylation with BSTFA using, for example, Chromosorb WAW DMSC+10% OV-101. Significantly improved resolution was obtained by Gnauk and co-workers.[242] Combined GC/MS analysis of resols and etherified resols led to complete characterization of the individual components of sodium hydroxide-catalyzed phenol–formaldehyde resin.[243] Capillary GC facilitated separation of the trimethylsilyl derivatives. Equally effective resolution of various oligomers resulted when GC was used for both sodium hydroxide- and barium hydroxide-catalyzed resols (Figure 8).[53] Both resols are identical with the resins described in the previous section; peak identification is given in Table 8. Additional peaks which appear in the barium hydroxide-catalyzed resol are attributable to hemiformal species.

High pressure steric exclusion chromatography (HPSEC) can be effective in monitoring the progress of phenolic resin molecular weight during synthesis. The method requires the pumping of a dilute resin solution through a column packed with porous particles which separates polymer components according to size.[244] Two moderately similar procedures have been described[245–248] using HPSEC for determining the MWD of phenolic resins differing in the number of theoretical plates and time required for analysis. Recently, it was shown[249] that the MWD can be monitored in less than 5 min using a silica-based column packing (Bondagel E-125) with polar solvents such as DMF with LiCl.

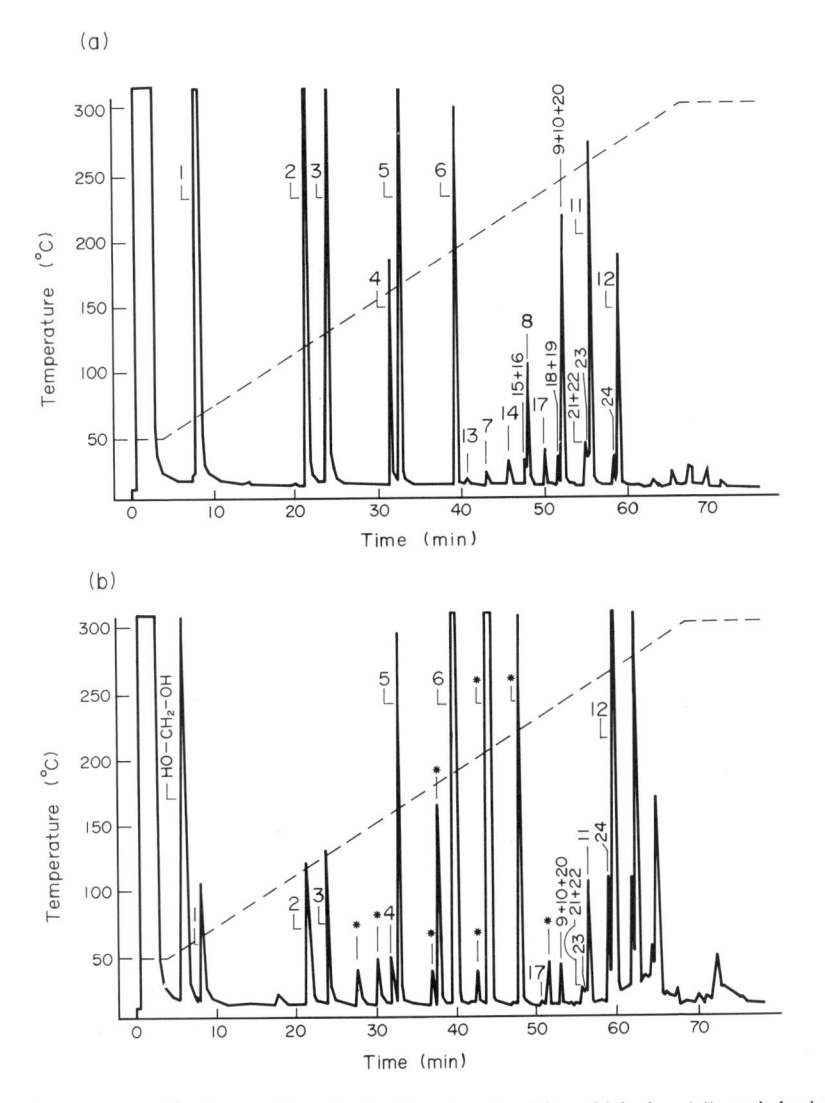

Figure 8 GC chromatograms[53] of a sodium hydroxide- (a phenol:formaldehyde = 1:1) and barium hydroxide-catalyzed resol (b, phenol:, formaldehyde = 1:3.3); peak identification in Table 8 * for hemiformal compounds

35.7.3 Thermal Characterization

Differential scanning calorimetry (DSC) has been an effective method in the determination of peak exotherm temperatures, degree of cure, cure rate and activation energies of resols and novolak composition (Figure 9).[250-254]

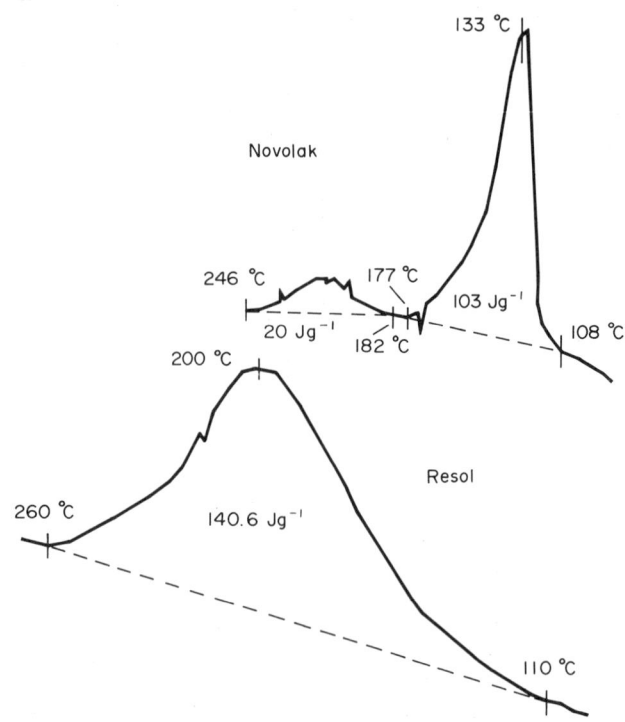

Figure 9 DSC analysis of novolak/HMTA and resol resin curing reaction[254]

Pressurized DSC cell systems are used for liquid resins to suppress the endothermic volatilization of water which completely masks the exothermic curing reaction.

Erä[255] used DSC to examine the oxalic acid-catalyzed phenol–formaldehyde reaction and reported activation energies of 155–165 kJ mol^{-1}. A heat of reaction of 94–100 kJ mol^{-1} was determined and is in accord with Vlk's value of 98.2 kJ mol^{-1}.[201] Erä's other studies consisted of several phenol–formaldehyde resols with phenol to formaldehyde ratios of 1:1.5 to 1:2.1. Heat of reaction data of resols correlated with phenol formaldehyde mole ratios, with a lower heat value corresponding to a high phenol to formaldehyde ratio.

In a study related to the reactivity of low molecular weight liquid resins,[256] DSC analyses exhibited two major exothermic peaks: one at 100–130 °C, attributable to unreacted free formaldehyde reacting with the phenolic ring; and the other peak at 140–150 °C, related to the condensation reaction of hydroxymethyl groups with an unreacted site on phenol or self-condensation of hydroxymethyl groups.

The effect of structure on cure and properties of high-*ortho* and random novolak resins cured with HTMA[257] and liquid resol compositional variables[258] were measured by torsional braid analysis (TBA).[257] A lower activation energy for gelation was noted for the high-*ortho* novolak as compared to the random material.

Dynamic mechanical analysis (DMA) allows a more rapid determination of the glass transition temperature (T_g), especially in the high temperature range. Large differences have been evaluated in T_g values of resol or novolak phenolic molding compounds.[259,260] Increased T_g values were obtained as cure temperature and cure residence times were increased. Postbaking further increased T_g to a value near 300 °C.

35.7.4 Other Methods

Results of single-crystal X-ray diffraction studies[52] are mentioned in Section 35.4.3. Linear thermal expansion coefficient, specific heat capacity and elastic constants or Young's modulus have been determined[261] for cured resol resin in the low temperature region of 6 to 100 K. Specific-heat capacity proportionality or $C_p = 0.0042T$ J K^{-1} g^{-1}, where C_p is proportional to T (temperature) has been recorded for resol resin in the 30 to 370 K temperature range.[262,263] Only specific-heat-capacity values of wood flour or asbestos-filled novolak systems (223 K to 473 K) are reported.[264]

The Mark–Houwink–Sakurada expression for novolak resin fractions of M_n from 1000 to 8050 in acetone or THF were determined by Tobiason[265] and Kamide.[266] Ishida[267,268] expressed a viscosity–diffusion MWD relationship in acetone. In his studies of random and high-*ortho* novolak resins, Kamide[266] described the molecular configuration of both resins. The random resin is somewhat coiled with two branches[64,65] while the high-*ortho* resin is completely linear within a low molecular weight regime.

35.8 REFERENCES

1. A. Bayer, *Ber. Dtsch. Chem. Ges.*, 1872, **5**, 25; *Ber. Dtsch. Chem. Ges.*, 1872, **5**, 1095.
2. E. ter Mer, *Ber. Dtsch. Chem. Ges.*, 1874, **7**, 1200.
3. A. Claus and E. Trainer, *Ber. Dtsch. Chem. Ges.*, 1886, **19**, 3009.
4. L. Claisen, *Justus Liebigs Ann. Chem.*, 1887, **237**, 261.
5. L. Lederer, *J. Prakt. Chem.*, 1894, **50**, 223.
6. O. Manasse, *Ber. Dtsch. Chem. Ges.*, 1894, **27**, 2409.
7. A. Speier, *Ger. Pat.* 99 570 (1897).
8. A. Luft, *Ger. Pat.* 140 552 (1902).
9. L. H. Baekeland, *US Pat.* 942 699 (1907); *US Pat.* 949 671 (1907).
10. A. Knop and W. Scheib, in 'Chemistry and Applications of Phenolic Resins', Springer, Berlin, 1975, p. 3.
11. K. Hultzsch, in 'Chemie der Phenolharze', Springer, Berlin, 1950.
12. R. W. Martin, in 'The Chemistry of Phenolic Resins', Wiley, New York, 1956.
13. N. J. L. Megson, in 'Phenolic Resin Chemistry', Butterworths, London, 1958.
14. A. Knop and L. A. Pilato, in 'Phenolic Resins', Springer, Berlin, 1985.
15. *Mod. Plast. Int.*, 1988, 24.
16. *Kunststoffe*, 1986, **76**, 10.
17. 'Verband Kunststofferzeugende Industrie, Frankfurt, Annual Report', 1986.
18. R. C. Weast (ed.), 'Handbook of Chemistry and Physics', 63rd edn., The Chemical Rubber Company, Cleveland, 1985.
19. C. Thurman, in 'Kirk-Othmer Encyclopedia of Chemical Technology', Wiley, New York, 1982, vol. 17, p. 373.
20. W. Jordans and B. Cornils, in 'Methodicum Chimicum', Thieme, Stuttgart, 1975, vol. 5, p. 105.
21. H. Fiege *et al.*, in 'Ullmanns Encyclopädie der technischen Chemie', 4th edn., Verlag Chemie, Weinheim, 1979, vol. 18, p. 191.
22. K. E. Clouts and R. A. McKetta, in 'Encyclopedia of Chemical Processing and Design', ed. J. J. McKetta, Dekker, New York, 1981, vol. 13, p. 212.
23. H. W. B. Reed, in 'Kirk-Othmer Encyclopedia of Chemical Technology', Wiley, New York, 1987, vol. 2, p. 72.
24. H. Hock and S. Lang, *Chem. Ber.*, 1944, **77**, 257.
25. P. R. Pujado and J. R. Salazar and C. V. Berger, *Hydrocarbon Process.*, 1976, **46**, 91.
26. F. H. Seubold and W. E. Vaugham, *J. Am. Chem. Soc.*, 1953, **75**, 3790.
27. J. F. Walker, *ACS Monogr.*, 1964, **159**.
28. J. Zabicka (ed.), 'The Chemistry of the Carbonyl Group', Interscience, New York, 1970.
29. H. Diehm and A. Hilt, in 'Ullmanns Encyclopädie der techn. Chemie', 4th edn., Verlag Chemieclo, Weinheim, 1976, vol. 11.
30. J. R. Fair and R. C. Khetz, in 'Encyclopedia of Chemical Processing and Design', ed. J. J. McKetta, Dekkar, New York, 1985, vol. 23.
31. M. F. Drumm and J. R. Le Blanc, *Kinet. Mech. Polym.*, 1972, **3**, 157.
32. C. H. Rochester, in 'The Chemistry of the Hydroxyl Group', ed. S. Patai, Interscience, London, 1971.
33. H. Gillier-Pandraud, *Bull. Soc. Chim. Fr.*, 1967, 1988.
34. M. Saunders and J. B. Hyne, *J. Chem. Phys.*, 1958, **29**, 1319.
35. G. R. Sprengling and C. W. Lewis, *J. Am. Chem. Soc.*, 1953, **75**, 5709.
36. K. Moedritzer and J. V. Wazer, *J. Phys. Chem.*, 1966, **70**, 2025.
37. P. W. Kopf and E. R. Wagner, *J. Polym. Sci., Polym. Chem. Ed.*, 1973, **11**, 939.
38. J. H. Freeman and C. W. Lewis, *J. Am. Chem. Soc.*, 1954, **76**, 2080.
39. T. Minami and T. Ando, *Kogyo Kagaku Zasshi*, 1956, **59**, 668.
40. K. C. Eapen and L. M. Yeddanapalli, *Makromol. Chem.*, 1968, **119**, 4.
41. A. A. Zavitsas, R. D. Beaulieu and J. R. Leblanc, *J. Polym. Sci., Part A-1*, 1968, **6**, 2541.
42. J. W. Aldersley and P. Hope, *Angew. Makromol. Chem.*, 1972, **24**, 137.
43. A. Sebenik and S. Lapanje, *Angew. Makromol. Chem.*, 1977, **63**, 139.
44. W. H. F. Bardey and K. H. Schmidt, *Polym. Prepr., Am. Chem. Soc., Div. Polym. Chem.*, 1983, **24** (2), 171.
45. H. C. Malhotra and V. K. Gupta, *J. Appl. Polym. Sci.*, 1978, **22**, 343.
46. H. C. Malhotra and V. Kumar, *J. Macromol. Sci., Chem.*, 1979, **A13**, 143.
47. H. C. Malhotra and V. P. Tyagi, *J. Macromol. Sci., Chem.*, 1980, **A14**, 675.
48. H. G. Peer, *Recl. Trav. Chim. Pays-Bas*, 1959, **78**, 851.
49. H. G. Peer, *Recl. Trav. Chim. Pays-Bas*, 1960, **79**, 825.
50. V. Calò, L. Lopez, G. Pesce, F. Ciminale and P. E. Todesco, *J. Chem. Soc., Perkin Trans. 2*, 1974, 1189.
51. V. Calò, L. Lopez, G. Pesce and P. E. Todesco, *J. Chem. Soc., Perkin Trans. 2*, 1974, 1192.

52. R. Perrin, R. Lamartine, J. Vicens, M. Perrin, A. Thozet, D. Hanton and R. Fugier, *Nouv. J. Chim.*, 1986, **10**, 179.
53. B. Mechin, D. Hanton, J. Le Goff and J. P. Tanneur, *Eur. Polym. J.*, 1984, **20**, 333.
54. B. Mechin, D. Hanton, J. Le Goff and J. P. Tanneur, *Eur. Polym. J.*, 1986, **22**, 115.
55. L. Prokai, *J. Chromatogr.*, 1985, **331**, 91.
56. R. T. Jones, *J. Polym. Sci., Polym. Chem. Ed.*, 1983, **21**, 1801.
57. V. Böhmer and G. Stein, *Makromol. Chem.*, 1984, **185**, 163.
58. H. U. Wagner and G. Gompper, in 'The Chemistry of The Quinoid Compounds', ed. S. Patai, Wiley, New York, 1974, vol. 2, chap. 18.
59. L. M. Yeddanapalli and D. J. Francis, *Makromol. Chem.*, 1962, **55**, 74.
60. J. I. DeJong and J. DeJonge, *Recl. Trav. Chim. Pays-Bas*, 1953, **72**, 497.
61. H. C. Malhotra and (Mrs.) Avinash, *J. Appl. Polym. Sci.*, 1976, **20**, 2461.
62. N. Kornblum, R. A. Smiley, R. K. Blackwood and D. C. Iffland, *J. Am. Chem. Soc.*, 1955, **77**, 7269.
63. R. A. Pethrick and B. Thomson, *Br. Polym. J.*, 1986, **18**, 380.
64. K. Kamide and Y. Miyakawa, *Makromol. Chem.*, 1978, **179**, 359.
65. S. A. Sojka, R. A. Wolfe, E. A. Dietz and B. F. Dannels, *Macromolecules*, 1979, **12**, 767.
66. A. Zinke and F. Hanus, *Monatsh. Chem.*, 1948, **78**, 311.
67. K. Hultzsch, *Chem. Ber.*, 1949, **82**, 16.
68. Y. Ogata and A. Kawasaki, in 'The Chemistry of the Carbonyl Group', ed. J. Zabicky, Interscience, London, 1970, vol. 2.
69. S. A. Sojka, R. A. Wolfe and G. D. Guenther, *Macromolecules*, 1981, **14**, 1539.
70. H. L. Bender, A. G. Farnham, J. W. Guyer, F. N. Apel and T. B. Gibb, *Ind. Eng. Chem.*, 1952, **44**, 1619.
71. H. L. Bender, *Mod. Plast.*, 1953, **30**, 136.
72. H. P. Higginbottom, H. M. Culbertson and J. C. Woodbrey, *J. Polym. Sci., Part A*, 1965, **3**, 1079.
73. G. L. Brode and S. W. Chow (Union Carbide Corp.), *US Pat.* 4 433 129 (1984) (*Chem. Abstr.*, 1984, **100**, 157 144f) and further patents cited in ref. 14.
74. G. Casiraghi, G. Sartori, F. Bigi, M. Cornia, E. Dradi and G. Casnati, *Makromol. Chem.*, 1981, **182**, 2151.
75. G. Casiraghi, M. Cornia, G. Ricci, G. Balduzzi and G. Casnati, *Makromol. Chem.*, 1983, **184**, 1363.
76. G. Casiraghi, G. Casnati, M. Cornia, G. Sartori and F. Bigi, *Makromol. Chem.*, 1981, **182**, 2973.
77. K. Hultzsch, *Ber. Dtsch. Chem. Ges.*, 1941, **74**, 898.
78. K. Hultzsch, *Angew. Chem.*, 1948, **A60**, 179.
79. H. von Euler, E. Adler and J. O. Cedwall, *Ark. Kemi, Mineral. Geol.*, 1941, **14A** (14).
80. H. von Euler, E. Adler, J. O. Cedwall and O. Törngren, *Ark. Kemi, Mineral Geol.*, 1942, **15A** (11).
81. N. P. Neureiter, *J. Org. Chem.*, 1963, **28**, 3486.
82. D. W. Chasar and J. C. Westfahl, *J. Org. Chem.*, 1977, **42**, 2177.
83. M. S. Chauhan, F. M. Dean, S. McDonald and M. S. Robinson, *J. Chem. Soc., Perkin Trans. 1*, 1973, 359.
84. L. Jurd, *Tetrahedron*, 1977, **33**, 163.
85. M. Benson and L. Jurd, *Org. Magn. Reson.*, 1984, **22**, 86.
86. T. W. Lewis, I. C. Paul and D. Y. Curtin, *Acta Crystallogr., Sect. B*, 1980, **36**, 70.
87. E. N. Duesler, T. W. Lewis, D. Y. Curtin and I. C. Paul, *Acta Crystallogr., Sect. B*, 1980, **36**, 166.
88. A. Arduini, A. Pochini, R. Ungaro and P. Domiano, *J. Chem. Soc., Perkin Trans. 1*, 1986, 1391.
89. H. Kämmerer, W. Kern and G. Heuser, *J. Polym. Sci.*, 1958, **28**, 331.
90. H. Kämmerer, *Makromol. Chem.*, 1952, **8**, 98.
91. S. N. Tong, Y. P. Kyung and H. G. Harwood, *Polym. Prepr., Am. Chem. Soc., Div. Polym. Chem.*, 1983, **24**, 196.
92. H. Kämmerer, M. Grossmann and G. Umsonst, *Makromol. Chem.*, 1960, **29**, 39.
93. G. E. Maciel, I. S. Chuang and L. Gollob, *Macromolecules*, 1984, **17**, 1081.
94. A. Zinke, *J. Appl. Chem.*, 1951, **1**, 257.
95. G. E. Maciel and G. R. Hatfield, *Macromolecules*, 1987, **20**, 608.
96. I. V. Kamenskii, L. N. Kuznetsov and L. N. Moisenko, *Vysokomol. Soedin., Ser. A*, 1976, **18/8**, 1787.
97. J. H. Mackey, G. Lester, L. E. Walker and J. K. Gillham, *Polym. Prepr., Am. Chem. Soc., Div. Polym. Chem.*, 1981, **22** (2), 131.
98. D. J. Francis and L. M. Yeddanapalli, *Makromol. Chem.*, 1968, **119**, 17.
99. V. Böhmer and B. Mathiasch, *Makromol. Chem.*, 1971, **148**, 41.
100. P. J. Flory, *Chem. Rev.*, 1946, **39**, 137.
101. J. Borrajo, M. I. Aranguren and R. J. J. Williams, *Polymer*, 1982, **23**, 263.
102. A. Kumar, A. K. Kulshreshtha and S. K. Gupta, *Polymer*, 1980, **21**, 317.
103. A. Kumar, U. K. Phukan and S. K. Gupta, *J. Appl. Polym. Sci.*, 1982, **27**, 3393.
104. A. Kumar, S. K. Gupta and B. Kumar, *Polymer*, 1982, **23**, 1929.
105. P. K. Pal, A. Kumar and S. K. Gupta, *Br. Polym. J.*, 1980, **12**, 121.
106. P. K. Pal, A. Kumar and S. K. Gupta, *Polymer*, 1981, **22**, 1699.
107. R. Steffan, *Angew. Makromol. Chem.*, 1985, **131**, 25.
108. S. Ishida, Y. Tsutsumi and K. Kaneko, *J. Polym. Sci., Polym. Chem. Ed.*, 1981, **19**, 1609.
109. S. Ishida, S. Wakaki, Y. Kato and Y. Nakamoto, *Ind. Eng. Chem. Prod. Res. Dev.*, 1984, **23**, 380.
110. S. Ishida, private communication.
111. S. Ishida, T. Yamagishi and Y. Nakamoto, 'Japan–U. S. Polymer Symposium, Preprints', The Society of Polymer Science, Tokyo, 1985, p. 189.
112. N. J. L. Megson, *Chem.-Ztg.*, 1972, **96**, 17.
113. C. D. Gutsche, *Top. Curr. Chem.*, 1984, **123**, 1.
114. H. Kämmerer and G. Happel, *Makromol. Chem., Rapid. Commun.*, 1980, **1**, 461.
115. H. Kämmerer and H. Lenz, *Makromol. Chem.*, 1958, **27**, 162.
116. V. Böhmer, J. Deveaux and H. Kämmerer, *Makromol. Chem.*, 1976, **177**, 1745.
117. J. S. Rodia, *J. Org. Chem.*, 1961, **26**, 2967.
118. H. Kämmerer and M. Großmann, *Chem. Ber.*, 1953, **86**, 1492.
119. H. Kämmerer, G. Happel and V. Böhmer, *Org. Prep. Proced. Int.*, 1976, **8**, 245.
120. M. Tashiro, *Synthesis*, 1979, 921.

121. V. Böhmer, D. Rathay and H. Kämmerer, *Org. Prep. Proced. Int.*, 1978, **10**, 113.
122. G. H. Hakimelahi and A. A. Mosfegh, *Helv. Chim. Acta*, 1981, **64**, 599.
123. A. A. Mosfegh, B. Mazandarani, A. Nahid and G. H. Hakimelahi, *Helv. Chim. Acta*, 1982, **65**, 1229.
124. G. Casnati, G. Casiraghi, A. Pochini, G. Sartori and R. Ungaro, *Pure Appl. Chem.*, 1983, **55**, 1677.
125. G. Casiraghi, G. Casnati, M. Cornia, A. Pochini, A. Puglia, G. Sartori and R. Ungaro, *J. Chem. Soc., Perkin Trans. 1*, 1978, 318.
126. A. Pochini and R. Ungaro, *J. Chem. Soc., Chem. Commun.*, 1976, 309.
127. G. Casnati, A. Pochini, M. G. Terenghini and R. Ungaro, *J. Org. Chem.*, 1983, **48**, 3783.
128. G. Casiraghi, M. Cornia, G. Sartori, G. Casnati, V. Bocchi and G. D. Andreetti, *Makromol. Chem.*, 1982, **183**, 2611.
129. G. Casiraghi, M. Cornia, G. Ricci, G. Casnati, G. D. Andreetti and L. Zetta, *Macromolecules*, 1984, **17**, 19.
130. G. Casiraghi, M. Cornia, G. Balduzzi and G. Casnati, *Ind. Eng. Chem. Prod. Res. Dev.*, 1984, **23**, 366.
131. F. Bigi, G. Casiraghi, G. Casnati, G. Sartori and L. Zetta, *J. Chem. Soc., Chem. Commun.*, 1983, 1210.
132. G. Casiraghi, M. Cornia, G. Casnati and L. Zetta, *Macromolecules*, 1984, **17**, 2933.
133. A. Zinke and E. Ziegler, *Chem. Ber.*, 1944, **77**, 264.
134. J. W. Cornforth, P. D. A. Hart, G. A. Nicholls, R. J. W. Rees and J. A. Stock, *Br. J. Pharmacol.*, 1955, **10**, 73.
135. C. D. Gutsche, B. Dhawan, K. H. No and R. Muthukrishnan, *J. Am. Chem. Soc.*, 1981, **103**, 3782.
136. J. Ludwig, Sr. and G. Bailie, Jr., *Anal. Chem.*, 1986, **58**, 2069.
137. B. Dhawan, S. Chen and C. D. Gutsche, *Makromol. Chem.*, 1987, **188**, 921.
138. R. Lamartine, R. Perrin, J. Vicens, D. Gamet, M. Perrin, D. Oehler and A. Thozet, *Mol. Cryst. Liq. Cryst.*, 1986, **134**, 219.
139. S. R. Izatt, R. T. Hawkins, J. J. Christensen and R. M. Izatt, *J. Am. Chem. Soc.*, 1985, **107**, 63.
140. Y. Nakamoto, T. Kozu, S. Oya and S. Ishida, *Netsu Kokasei Jushi*, 1985, **6**, 73.
141. A. Ninagawa and H. Matsuda, *Makromol. Chem., Rapid Commun.*, 1982, **3**, 65.
142. Y. Nakamoto and S. Ishida, *Makromol. Chem., Rapid Commun.*, 1982, **3**, 705.
143. M. Corruzzi, G. D. Andreetti, V. Bocchi, A. Pochini and R. Ungaro, *J. Chem. Soc., Perkin Trans. 2*, 1982, 1133.
144. B. T. Hayes and R. F. Hunter, *J. Appl. Chem.*, 1958, **8**, 743.
145. H. Kämmerer, G. Happel, V. Böhmer and D. Rathay, *Monatsh. Chem.*, 1978, **109**, 767.
146. H. Kämmerer and G. Happel, *Makromol. Chem.*, 1980, **181**, 2049.
147. H. Kämmerer, G. Happel and B. Mathiasch, *Makromol. Chem.*, 1981, **182**, 1685.
148. H. Kämmerer and G. Happel, *Monatsh. Chem.*, 1981, **112**, 759.
149. K. H. No and C. D. Gutsche, *J. Org. Chem.*, 1982, **47**, 2713.
150. V. Böhmer, P. Chhim and H. Kämmerer, *Makromol. Chem.*, 1979, **180**, 2503.
151. V. Böhmer, F. Maschollek and L. Zetta, *J. Org. Chem.*, 1987, **52**, 3200.
152. V. Böhmer, L. Merkel and U. Kunz, *J. Chem. Soc., Chem. Commun.*, 1987, 896.
153. H. Casablanca, J. Royer, A. Satrallah, A. Taty-C and J. Vicens, *Tetrahedron Lett.*, 1987, **28**, 6595.
154. V. Böhmer, H. Goldmann and W. Vogt, *J. Chem. Soc., Chem. Commun.*, 1985, 667.
155. E. Paulus, V. Böhmer, H. Goldmann and W. Vogt, *J. Chem. Soc., Perkin Trans. 2*, 1987, 1609.
156. B. Dhawan and C. D. Gutsche, *J. Org. Chem.*, 1983, **48**, 1536.
157. J. Canceill, J. Gabard and A. Collet, *J. Chem. Soc., Chem. Commun.*, 1983, 122.
158. A. G. S. Högberg, *J. Am. Chem. Soc.*, 1980, **102**, 6046.
159. A. G. S. Högberg, *J. Org. Chem.*, 1980, **45**, 4498.
160. H. Kämmerer and W. Niemann, *Makromol. Chem.*, 1973, **169**, 1.
161. G. D. Andreetti, R. Ungaro and A. Pochini, *J. Chem. Soc., Chem. Commun.*, 1979, 1005.
162. G. D. Andreetti, R. Ungaro and A. Pochini, *J. Chem. Soc., Chem. Commun.*, 1981, 533.
163. R. Ungaro, A. Pochini, G. D. Andreetti and V. Sangermano, *J. Chem. Soc., Perkin Trans. 2*, 1984, 1979.
164. G. D. Andreetti, A. Pochini and R. Ungaro, *J. Chem. Soc., Perkin Trans. 2*, 1983, 1773.
165. R. Ungaro, A. Pochini, G. D. Andreetti and P. Domiano, *J. Chem. Soc., Perkin Trans. 2*, 1985, 197.
166. R. Perrin and R. Lamartine, *Makromol. Chem., Macromol. Symp.*, 1987, **9**, 69.
167. C. D. Gutsche, A. E. Gutsche and A. I. Karaulov, *J. Inclusion Phenom.*, 1985, **3**, 447.
168. G. Casiraghi, M. Cornia, G. Balduzzi, G. Casnati and G. D. Andreetti, *Makromol. Chem.*, 1983, **184**, 1363.
169. E. Paulus and V. Böhmer, *Makromol. Chem.*, 1984, **185**, 1921.
170. D. Oehler, A. Thozet and M. Perrin, *Acta Crystallogr.*, 1985, **C41**, 1766.
171. D. Oehler, Ph. D. Thesis, University of Lyon, 1987; private communication.
172. T. Cairns and G. Eglinton, *Nature (London)*, 1962, **162**, 535.
173. T. Cairns and G. Eglinton, *J. Chem. Soc.*, 1965, 5906.
174. S. Kovacs and G. Eglinton, *Tetrahedron*, 1969, **25**, 3599.
175. L. Zetta and V. Böhmer, unpublished results.
176. F. L. Tobiason, K. Houglum, A. Shanafelt and V. Böhmer, *Polym. Prepr., Am. Chem. Soc., Div. Polym. Chem.*, 1983, **24** (2), 131.
177. G. Happel, B. Mathiasch and H. Kämmerer, *Makromol. Chem.*, 1975, **176**, 3317.
178. C. D. Gutsche, B. Dhawan, J. A. Levine, K. H. No and L. J. Bauer, *Tetrahedron*, 1983, **39**, 409.
179. C. D. Gutsche and L. J. Bauer, *Tetrahedron Lett.*, 1981, **22**, 4763.
180. C. D. Gutsche and L. J. Bauer, *J. Am. Chem. Soc.*, 1985, **107**, 6052.
181. G. R. Sprengling, *J. Am. Chem. Soc.*, 1954, **76**, 1190.
182. V. Böhmer, E. Schade, C. Antes, J. Pachta, W. Vogt and H. Kämmerer, *Makromol. Chem.*, 1983, **184**, 2361.
183. V. Böhmer, W. Lotz, J. Pachta and S. Tütüncü, *Makromol. Chem.*, 1981, **182**, 2671.
184. V. Böhmer, E. Schade and W. Vogt, *Makromol. Chem., Rapid Commun.*, 1984, **5**, 221.
185. S. Shinkai, K. Araki, H. Koreshi, T. Tsubaki and O. Manabe, *Chem. Lett.*, 1986, 1351.
186. R. M. Izatt, J. D. Lamb, R. T. Hawkins, P. R. Brown, S. R. Izatt and J. J. Christensen, *J. Am. Chem. Soc.*, 1983, **102**, 1782.
187. L. Zetta, V. Böhmer and R. Kaptein, *J. Magn. Reson.*, 1988, **76**, 587.
188. V. Böhmer, H. Goldmann, R. Kaptein and L. Zetta, *J. Chem. Soc., Chem. Commun.*, 1987, 1358.
189. L. Zetta, A. DeMarco, G. Casiraghi, M. Cornia and R. Kaptein, *Macromolecules*, 1985, **18**, 1095.

190. M. Imoto, J. Ijiichi, C. Tanaka and M. Kinoshita, *Makromol. Chem.*, 1965, **113**, 117.
191. V. Böhmer and W. Niemann, *Makromol. Chem.*, 1976, **177**, 787.
192. V. Böhmer, D. Stotz, K. Beismann and W. Niemann, *Monatsh. Chem*, 1983, **114**, 411.
193. V. Böhmer. K. Beismann, D. Stotz, W. Niemann and W. Vogt, *Makromol. Chem.*, 1983, **184**, 1793.
194. V. Böhmer, D. Stotz, K. Beismann and W. Vogt, *Monatsh. Chem.*, 1984, **115**, 65.
195. G. Stein, V. Böhmer, W. Lotz and H. Kämmerer, *Z. Naturforsch. B*, 1981, **36**, 231.
196. G. Stein, H. Kämmerer and V. Böhmer, *J. Chem. Soc., Perkin Trans. 2*, 1984, 1285.
197. V. Böhmer, G. Lempert and W. Vogt, unpublished results.
198. V. Böhmer, H. Schalla and W. Vogt, unpublished results.
199. E. Manegold and W. Petzold, *Kolloid-Z.*, 1941, **94**, 284.
200. T. T. Jones, *J. Soc. Chem. Ind., London, Trans. Commun.*, 1946, **65**, 284.
201. O. Vlk, *Plaste Kautsch.*, 1957, **4**, 127.
202. J. D. Carlson, E. W. Kifer, V. J. Wojtyna and J. P. Colton (Koppers Co.), *US Pat.* 4 478 958 (1984).
203. Euteco, S.P.A. Euteco Continuous Process, Technical Bulletin.
204. S. Vargin, U. Nistri and S. Pezzoli (Societa Italiana Resine SPA), *US Pat.* 3 687 896 (1972).
205. W. R. Sorenson and T. W. Campbell, 'Preparative Methods of Polymer Chemistry', 2nd edn., Interscience, New York, 1968.
206. H. M. Culbertson (Monsanto Co.), *US Pat.* 4 113 700 (1978).
207. J. Harding (Union Carbide Corp.), *US Pat.* 3 823 103 (1974).
208. H.-J. Deppe, *Holz Roh-Werkst.*, 1983, **403**, 41.
209. Vianova Kunstharz AG, *Austrian Pat.* 180 407 (1954); *Austrian Pat.* 299 543 (1970).
210. Reichhold-Albert-Chemie AG, *US Pat.* 1 254 528 (1913); *US Pat.* 1 254 529 (1917); *Ger. Pat.* 1 669 286 (1967); *Ger. Pat.* 2 046 458 (1971).
211. *Dtsch. Farben-Z.*, 1973, **370**, 27.
212. Kansai Paint Co. Ltd., *Ger. Pat.* 2 237 830 (1973).
213. Vianova Kunstharz AG, *US Pat.* 2 981 703 (1961) (*Chem. Abstr.*, 1961, **55**, 18 198i).
214. W. Daimer, *Dtsch. Farben-Z.*, 1973, **358**, 27.
215. Chemische Werke Hüls, *US Pat.*, 3 546 184 (1970).
216. R. Kita and A. Kimi, *J. Coat. Technol.*, 1976, **53**, 48.
217. Ford Motor Co., *Ger. Pat.* 1 794 354 (1963) (*Chem. Abstr.*, 1966, **65**, 7460e).
218. BASF AG, *Ger. Pat.* 1 949 294 (1969) (*Chem. Abstr.*, 1971, **75**, 22 685c).
219. M. R. Rifi, *J. Paint Technol.*, 1973, **73**, 45.
220. J. Anisfeld, *Farbe u. Lack*, 1975, **81**, 1024.
221. H. Kittel (ed.), 'Lehrbuch d. Lacke u. Beschichtungen', Columb, Stuttgart, 1971, vol. 1, part 1.
222. H. Babich and D. L. Davies, *Regul. Toxicol. Pharmacol.*, 1981, **1**, 90.
223. R. M. Bruce, *Gov. Rep. Announce. Index (U. S.)*, 1986, **13**, 86 (*Chem. Abstr.*, 1987, **106**, 37 659).
224. US Department of Labor, Press Release Nov. 20, 1987; USDL 87-523.
225. European Chemical Industry, Ecology and Toxicology, 'Technical Report No. 6, Formaldehyde Toxicology', September 4, 1982.
226. H. J. Nantke and W. Lohrer, *Adhaesion*, 1986, **30**, 17.
227. D. O. Hummel, 'Polymer Spectroscopy', Verlag Chemie, Weinheim, 1976.
228. D. O. Hummel and K. Scholl, in 'Infrared, Analysis of Polymers, Resins and Additives', ed. Carl Hanser, Verlag Chemie, Weinheim, 1973.
229. Y. Zaks, D. Raucher and E. M. Pearce, *J. Appl. Polym. Sci.*, 1982, **27**, 913.
230. D. D. Werstler, *Polymer*, 1986, **27**, 750.
231. L. E. Walker, E. A. Dietz, Jr., R. A. Wolfe, B. F. Dannels and S. A. Sojka, *Polym. Prepr., Am. Chem. Soc., Div. Polym. Chem.*, 1983, **24**, 177.
232. L. Prokai, *J. Polym. Sci., Polym. Lett. Ed.*, 1986, **24**, 223.
233. D. Braun and J. Arndt, *Angew. Makromol. Chem.*, 1978, **73**, 133, 143.
234. D. Braun and J. Arndt, *Fresenius' Z. Anal. Chem.*, 1979, **294**, 130.
235. A. Rudin, C. A. Fyfe and S. M. Vines, *J. Appl. Polym. Sci.*, 1983, **28**, 2611.
236. D. R. Bain and J. D. Wagner, *Polymer*, 1984, **25**, 403.
237. W. A. Dark, *J. Anal. Purif.*, 1987, **2**, 62.
238. E. Tesarova and V. Pacakova, *Chromatographia*, 1983, **17**, 269.
239. W. Werner and O. Barber, *Chromatographia*, 1982, **15**, 101.
240. H. G. Haub and H. Kämmerer, *J, Chromatogr.*, 1963, **11**, 487.
241. W. Lindner, *J. Chromatogr.*, 1978, **151**, 406.
242. R. Gnauck and D. Habisch, *Plaste Kautsch.*, 1980, **27**, 485.
243. C. M. Anthony and G. Kemp, *Angew. Makromol. Chem.*, 1983, **115**, 183.
244. W. W. Yau, J. J. Kirkland and D. D. Bly, 'Modern Size Exclusion Liquid Chromatography', Wiley, New York, 1979.
245. A. Sebenik, *J. Chromatogr.*, 1978, **160**, 205.
246. T. Takeuchi and D. Ishii, *J. Chromatogr.*, 1983, **257**, 327.
247. M. Duval, B. Bloch and S. Kohn, *J. Appl. Polym. Sci.*, 1972, **16**, 1585.
248. J. D. Wellons and L. Gollob, *Wood Sci.*, 1980, **13**, 68.
249. A. R. Walsh and A. G. Campbell, *J. Appl. Polym. Sci.*, 1986, **32**, 4291.
250. R. Kay and A. R. Westwood, *Eur. Polym. J.*, 1975, **11**, 25.
251. M. R. Kamal and S. Sourour, *Polym. Eng. Sci.*, 1973, **13**, 59.
252. L. J. Taylor and S. W. Watson, *Anal. Chem.*, 1970, **42**, 297.
253. A. Katovic and M. Stefanic, *Polym. Prepr., Am. Chem. Soc., Div. Polym. Chem.*, 1983, **24**, 191.
254. Bakelite GmbH, unpublished.
255. V. A. Erä, *J. Therm. Anal.*, 1982, **25**, 79.
256. A. W. Christiansen and L. Gollob, *J. Appl. Polym. Sci.*, 1985, **30**, 2279.
257. J. H. Mackey, G. Lester, L. E. Walker and J. K. Gillham, *Polym. Prepr., Am. Chem. Soc., Div. Polym. Chem.*, 1981, **22**, 131.

258. S. S. Kelley, L. Gollob and J. D. Wellons, *Holzforschung*, 1986, **40**, 303.
259. V. R. Landi, J. M. Mersereau and S. E. Dorman, *Polym. Compos.*, 1986, **7**, 154.
260. V. R. Landi, *Advances in Polymer Technology*, 1987, **7** (1), 209.
261. C. Harwood, H. H. Wostenholm, B. Yates and D. Badami, *J. Polym. Sci., Polym. Phys. Ed.*, 1978, **16**, 759.
262. S. S. Chang, in 'Thermal Analysis in Polymer Characterization', ed. E. A. Turi, Heyden, New York, 1981, p. 98.
263. S. S. Chang, *Polym. Prepr., Am. Chem. Soc., Div. Polym. Chem.*, 1983, **24**, 187.
264. H. Wilski, *Prog. Colloid Polym. Sci.*, 1978, **64**, 33.
265. F. L. Tobiasion, C. Chandler and F. E. Schwarz, *Macromolecules*, 1972, **5**, 321.
266. K. Kamide and Y. Miyakawa, *Makromol. Chem.*, 1978, **179**, 359.
267. S. Ishida, M. Nakagawa, H. Suda and K. Kaneko, *Kobunshi Kagaku*, 1971, **28**, 250 (*Chem. Abstr.*, 1971, **75**, 49 959).
268. S. Ishida, T. Kitagawa, Y. Makamoto and K. Kaneko, *Chem. Abstr.*, 1984, **100**, 52 413.

36
Urea–Formaldehyde and Melamine–Formaldehyde Polymers

DIETRICH BRAUN and HANS-JOSEF RITZERT
Deutsches Kunststoff-Institut, Darmstadt, Federal Republic of Germany

36.1 HISTORY	649
36.2 RAW MATERIALS	650
36.2.1 Urea	650
36.2.2 Melamine	650
36.2.3 Formaldehyde	650
36.3 REACTION OF UREA AND MELAMINE WITH FORMALDEHYDE	651
36.3.1 Urea/Formaldehyde System	651
36.3.1.1 Formation of hydroxymethylurea compounds	651
36.3.1.2 Reactions of hydroxymethylurea compounds	653
36.3.1.3 Crosslinking	654
36.3.2 Melamine/Formaldehyde System	655
36.4 COPOLYCONDENSATES	656
36.4.1 Phenol–Melamine–Formaldehyde Resins	656
36.4.2 Urea–Melamine–Formaldehyde Resins	656
36.4.3 Other Copolycondensates	657
36.5 RESIN ANALYSIS	657
36.5.1 Precondensates	657
36.5.1.1 Gel permeation chromatography	657
36.5.1.2 ^{13}C NMR spectroscopy	658
36.5.1.3 Thin-layer chromatography	658
36.5.1.4 Chemical methods	659
36.5.2 Cured Products	660
36.5.2.1 Thermoanalysis	660
36.5.2.2 Other methods	661
36.6 RESIN MANUFACTURING	661
36.7 PROPERTIES AND APPLICATIONS OF UF- AND MF-POLYMERS	662
36.7.1 Adhesives	662
36.7.2 Moulding Compounds	662
36.7.3 Coatings	662
36.7.4 Foams	663
36.8 REFERENCES	663

36.1 HISTORY

According to Meyer[1] the *history of urea–formaldehyde (UF) resins* can be divided into five different periods:

(i) Synthesis of the *raw materials* urea and formaldehyde. This period started with the synthesis of urea by Wöhler in 1824. Formaldehyde was first recognized by Butlerov in 1859 when attempting to synthesize methanediol.

(ii) Investigations of the resinous products resulting from the reaction of urea with formaldehyde. The first attempt to investigate the structure of the products of the urea–formaldehyde reaction was made by Tollens in 1884.[2] In the following years a large number of authors worked on the structure of these resins.

(iii) Commercialization of the UF-resins. This period started with Goldschmidt's patent in 1897[3] to use UF-resins as a disinfectant. In the following decades, more and more applications were described in the literature.[4-11] After about 1930 it is very difficult to comprehensively cover all the literature about the applications of UF-resins, which indicates the great importance of these materials.

(iv) Adjustment of the UF-resins. In the fourth period the properties of the UF-resins were improved (1930–1975), but a lot of applications were also lost in favour of thermoplastic materials.

(v) Adaptation of the UF-condensates to modern applications. About 10 years ago condensation products from urea and formaldehyde again became important, due to their good properties and reasonable prices.

The *history of melamine–formaldehyde (MF) resins* began with the discovery of melamine in 1834 by Liebig. The industrial production of melamine started in 1935;[12,13] in the same year the production of MF-resins began.[12] MF-resins are superior to UF-resins in heat- and water-resistance.

During the practical development of UF- and MF-resins, many investigations were done with respect to resin structure and reaction mechanisms. On the following pages the results of these investigations will be discussed.

36.2 RAW MATERIALS

36.2.1 Urea

The principle of the *industrial production of urea* is shown in Scheme 1. First, *ammonium carbamate* is formed from the gases ammonia and carbon dioxide at 1×10^7 Pa and 190 °C. In the second step, ammonium carbamate is decomposed to urea and water. The molar ratio of ammonia to carbon dioxide is higher than two, in order to increase the yield of ammonium carbamate.[14-16]

$$2NH_3(g) + CO_2(g) \rightleftharpoons H_2NCO_2NH_4(l) \quad \Delta H = -117 \text{ kJ mol}^{-1}$$

$$H_2NCO_2NH_4(l) \rightleftharpoons H_2NCONH_2(l) + H_2O(l) \quad \Delta H = +15.5 \text{ kJ mol}^{-1}$$

Scheme 1

In the 'once-through-process', there is no recovery of the ammonium carbamate. The decomposition products are worked up to ammonium sulfate and nitrate.[17,18]

In newer processes, the excess ammonium carbamate is brought back to the reaction mixture (stripping process).[19]

36.2.2 Melamine

Two processes are used to produce *melamine*. The first industrial method starts from 2-cyanoguanidine (*dicyanodiamide*).[20] Equation (1) shows the principle of the synthesis; the reaction is difficult to control because of its exothermic character. There are many variations of this process using, for example, applied pressure or by working in liquid or solid phases, *etc.*

$$3(NH_2)_2C=NCN \rightarrow 2C_3H_6N_6 \quad \Delta H = -218 \text{ kJ mol}^{-1} \tag{1}$$

Today *melamine* is produced *from urea*.[21] Equation (2) shows the principle of the reaction. The endothermic reaction needs a large supply of energy. There are high- and catalytic low-pressure procedures with yields of nearly 95% melamine.

$$6H_2NCONH_2 \rightarrow C_3H_6N_6 + 6NH_3 + 3CO_2 \quad \Delta H = +471.5 \text{ kJ mol}^{-1} \tag{2}$$

36.2.3 Formaldehyde

Formaldehyde is one of the most important compounds in the chemical industry. Therefore there are many different procedures used for its synthesis, but they can be divided into two groups.

In the *silver contact process*, methanol is partially oxidized and dehydrated to formaldehyde. The main reactions in this process are shown in Scheme 2.[22,23] The hydrogen which is formed during the dehydrogenation is burned to give water. The reaction temperature is about 600 °C.

$$MeOH \rightleftharpoons CH_2O + H_2 \qquad \Delta H = +84 \text{ kJ mol}^{-1}$$

$$H_2 + \tfrac{1}{2}O_2 \rightarrow H_2O \qquad \Delta H = -243 \text{ kJ mol}^{-1}$$

$$MeOH + \tfrac{1}{2}O_2 \rightarrow CH_2O + H_2O \qquad \Delta H = -159 \text{ kJ mol}^{-1}$$

Scheme 2

In the *Formox process*,[24,25] catalysts based on iron(III) and molybdenum oxide are used. In this process, formaldehyde is formed from methanol by oxidation. The reaction is exothermic, thus the reaction temperature can be lower (270–380 °C) and the yield of formaldehyde is nearly quantitative.

In both processes the formaldehyde is isolated by absorption in water. The formaldehyde content of these aqueous solutions (*formaline*) can be increased by distillation.[26,27]

36.3 REACTION OF UREA AND MELAMINE WITH FORMALDEHYDE

Compounds with NH groups can react with aldehydes and ketones to form addition and condensation products. Examples of these reactions are the condensations of urea and melamine with formaldehyde. It is possible to divide these reactions into three steps: (1) formation of *hydroxymethyl compounds*; (2) reaction of these hydroxymethyl compounds to form oligomers; and (3) irreversible reaction of the oligomers to form a network.

In the following sections these reactions will be treated in more detail, particularly for the urea/formaldehyde system.

36.3.1 Urea/Formaldehyde System

36.3.1.1 *Formation of hydroxymethylurea compounds*

In the first step of aminoplastic condensation, various hydroxymethylurea compounds are formed. The reaction underlies the *general acid/base catalysis* (Schemes 3 and 4).[28-43] The reactions (3) and (6), to the shown transition states, are the slowest ones (*trimolecular step*). The hydroxymethyl compounds are formed by the abstraction of a proton (reaction 5), or the base (reaction 8). Using urea as the NH compound, three hydroxymethyl groups can be formed per molecule. *Tetra(hydroxymethyl)urea* has not yet been observed in a reaction mixture of urea and formaldehyde.

However, with some catalysts (*e.g.* hydrogencarbonate) the rate of hydroxymethylurea formation cannot be explained by a trimolecular mechanism. The HCO_3^- ion is a bifunctional catalyst; Eugster and Zollinger[44] proposed the mechanism shown in Scheme 5. With a bifunctional catalyst a *double proton transfer* takes place and so the high reaction rate can be explained. The rate of hydroxymethylation of NH groups follows equation (9) (general acid/base catalysis), where [>NH],

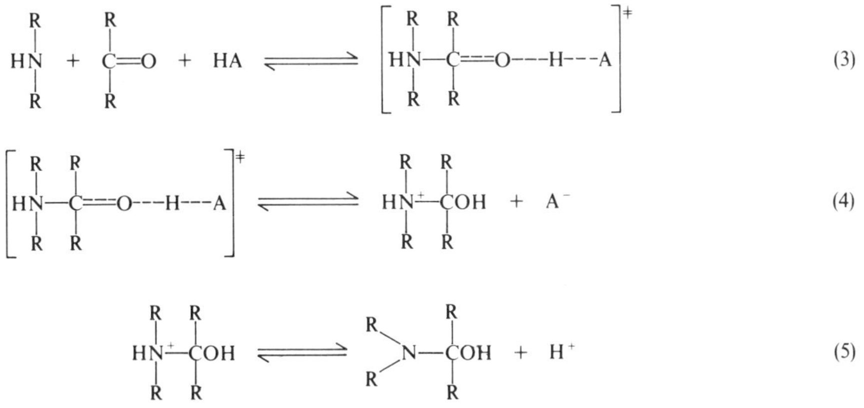

$$R_2HN + R_2C=O + HA \rightleftharpoons [R_2HN-CR_2=O---H---A]^+ \qquad (3)$$

$$[R_2HN-CR_2=O---H---A]^+ \rightleftharpoons R_2HN^+-CR_2OH + A^- \qquad (4)$$

$$R_2HN^+-CR_2OH \rightleftharpoons R_2N-CR_2OH + H^+ \qquad (5)$$

Scheme 3

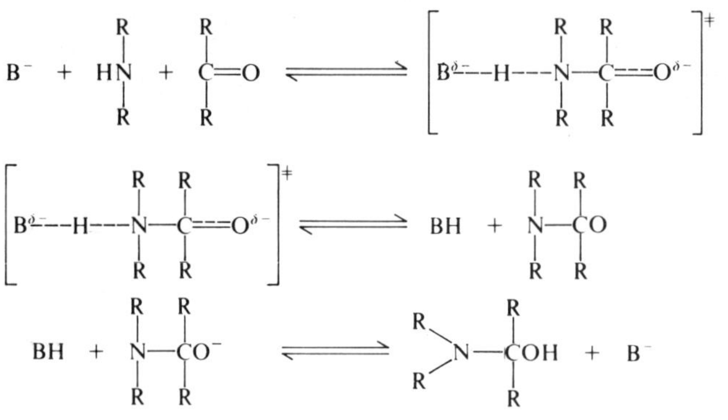

Scheme 4

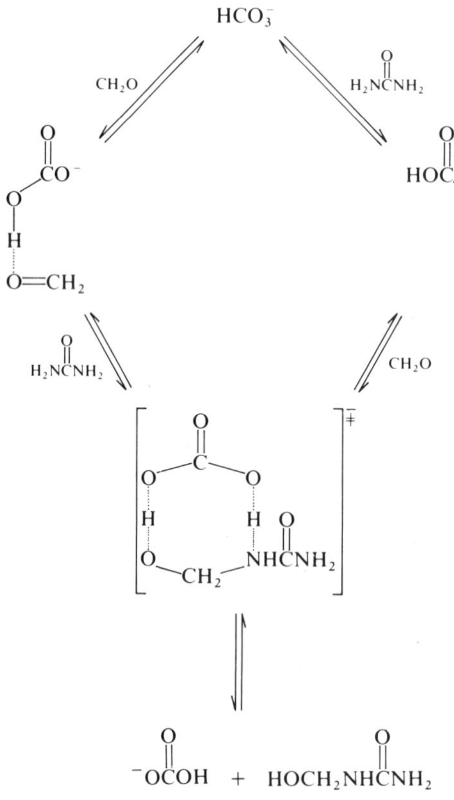

Scheme 5

[CH₂O] and [K_i] are the concentrations of NH compound, formaldehyde and catalyst, respectively. A plot of reaction rate against concentration of a buffer shows a linear relation.[37]

$$v = \sum k_i [{>}\text{NH}][\text{CH}_2\text{O}][K_i] \tag{9}$$

The *decomposition of N-hydroxymethyl compounds* to starting materials also follows acid/base catalysis. Petersen studied this dehydroxymethylation using *1,3-di(hydroxymethyl)-2-oxohexahydropyrimidine* (equation 10)[40,45] as a model compound.

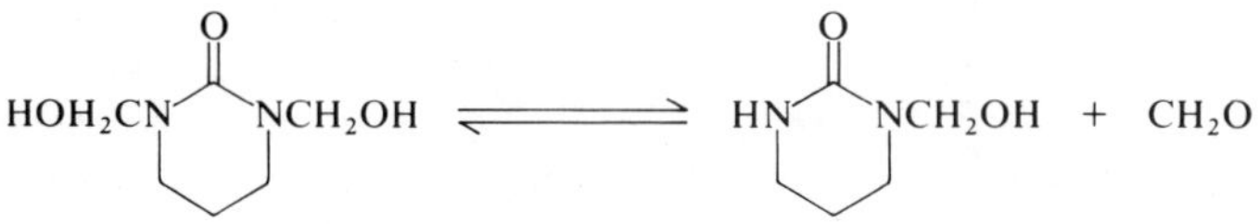

Figure 1 shows schematically the specific reaction rate as a function of pH value for the dehydroxymethylation of this cyclic model compound. The dependence of the hydroxymethylation on pH value is similar.

Similar investigations were done with other model compounds.[45-50]

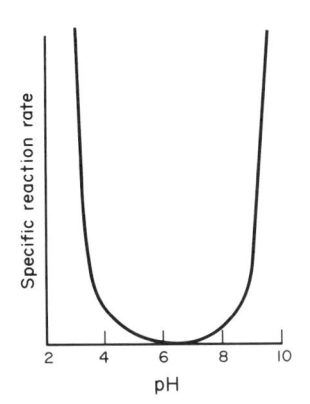

Figure 1 Specific reaction rates of the dehydroxymethylation of 1,3-di(hydroxymethyl)-2-oxohexahydropyrimidine as a function of pH value

36.3.1.2 Reactions of hydroxymethylurea compounds

The hydroxymethyl compounds of urea can undergo intermolecular condensation to form oligomers. There are two types of linkages: (a) methylene and (b) dimethylene ether bridges (structures **1** and **2**, respectively).

$$\underset{(1)}{-\overset{O}{\overset{\|}{C}}NHCH_2NH\overset{O}{\overset{\|}{C}}-} \qquad \underset{(2)}{-\overset{O}{\overset{\|}{C}}NHCH_2OCH_2NH\overset{O}{\overset{\|}{C}}-}$$

As, in principle, urea is tetrafunctional with respect to formaldehyde, a complicated reaction mixture can result from condensation of urea with formaldehyde (Scheme 6).

Tetra(hydroxymethyl)urea was not isolated. This compound reacts very quickly to form di-(hydroxymethyl)uron.[51-53] The formation of urons is only noticed at a reaction pH value above 7. The other primary products in alkaline solution are hydroxymethyl and ether compounds. Under these conditions methylene bridges are formed only very rarely.

Under acidic conditions, the condensation reactions become faster and mainly methylene bridges are formed. Figure 2 shows schematically the reaction rates of initial urea–formaldehyde reaction steps according to Jong and Jonge.[31]

The formation of the different hydroxymethylureas shows a minimum in the neutral region. The reaction rates of condensation and, as a consequence, the formation of oligomers increase with decreasing pH value. This dependence for the reaction of urea with formaldehyde is a basic rule of aminoplastic kinetics.[54-58]

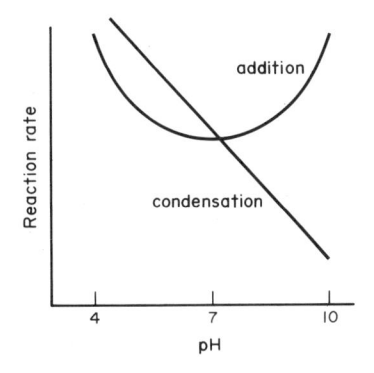

Figure 2 Reaction rates of addition and condensation reactions as functions of pH value

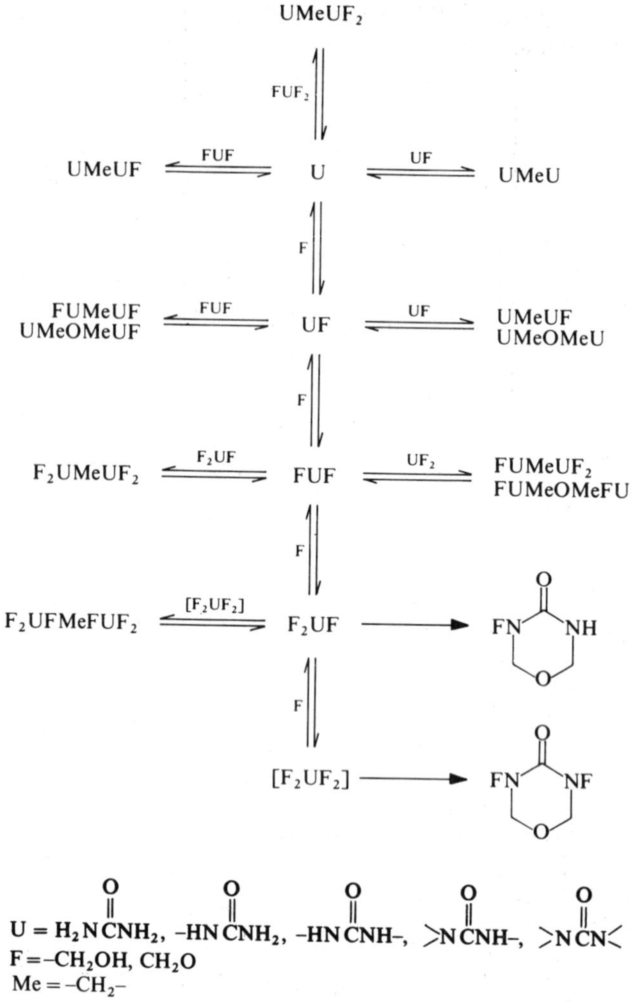

Scheme 6

```
                        UMeUF₂
                          ↕ FUF₂
  UMeUF  ⇌ FUF ⇌   U    ⇌ UF ⇌  UMeU
                          ↕ F
  FUMeUF                          UMeUF
  UMeOMeUF ⇌ FUF ⇌  UF   ⇌ UF ⇌  UMeOMeU
                          ↕ F
  F₂UMeUF₂ ⇌ F₂UF ⇌ FUF  ⇌ UF₂ ⇌ FUMeUF₂
                                  FUMeOMeFU
                          ↕ F
  F₂UFMeFUF₂ ⇌ [F₂UF₂] ⇌ F₂UF  →   FN⌒NH
                                    \_O_/  (=O)
                          ↕ F
                         [F₂UF₂]  →  FN⌒NF
                                    \_O_/  (=O)
```

$U = H_2NCNH_2$, $-HNCNH_2$, $-HNCNH-$, $>NCNH-$, $>NCN<$ (all with C=O)
$F = -CH_2OH$, CH_2O
$Me = -CH_2-$

36.3.1.3 Crosslinking

As urea has a functionality to formaldehyde higher than two, the theory of polycondensation predicts the formation of *networks*; therefore, UF-resins are able to form three-dimensional molecular networks. Scheme 7 shows the two reaction types which lead to crosslinking.

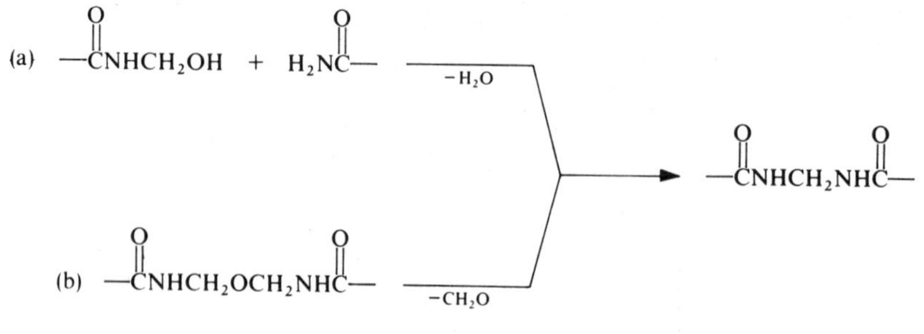

(a) $-CNHCH_2OH + H_2NC-$ $\xrightarrow{-H_2O}$
                                                    $-CNHCH_2NHC-$
(b) $-CNHCH_2OCH_2NHC-$ $\xrightarrow{-CH_2O}$

Scheme 7

First (Scheme 7, a) a hydroxymethyl compound reacts with free urea, and methylenediurea is formed by abstraction of water. The second possibility to give such methylene linkages is the abstraction of formaldehyde from a dimethylene ether linkage (Scheme 7, b). Both reactions are possible under acidic conditions or at higher temperatures.[58,59] Theoretically, a network with only methylene linkages is formed, structure (**3**). The crosslinking is an irreversible reaction and therefore UF-resins are thermosetting materials.

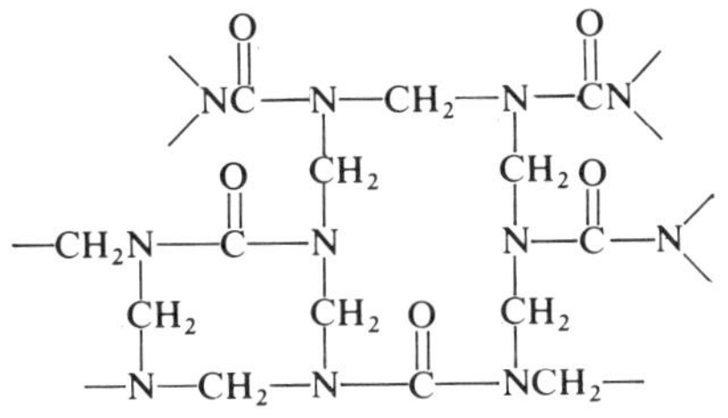

$$\text{(structure 3)}$$

(3)

The real structure, however, is very different from this theoretical network. Apart from methylene linkages, the cured products contain dimethylene ether linkages, free hydroxymethyl groups, unreacted imino groups, *etc.*

36.3.2 Melamine/Formaldehyde System

Melamine, i.e. 2,4,6-triamino-1,3,5-triazine, reacts with many different aldehydes and ketones. In principle, the reaction of melamine and formaldehyde is similar to the UF-condensation. The hydroxymethylation to *hydroxymethylmelamines* is also catalysed by acids and bases. These hydroxymethylmelamines form oligomers with methylene and dimethylene ether linkages during the first stage of the condensation. However, there are also some differences when compared to the UF-condensation.

The structure of melamine shows three amino groups, so the molecule is hexafunctional with respect to formaldehyde. In contrast to urea, it is possible to prepare the *hexa(hydroxymethyl)melamine* compound.[60] Schemes 8 and 9 show the mechanisms of the acid/base catalyzed hydroxymethylation reactions.[61-64]

$$MelNH_2 + B^- \rightarrow MelNH^- + BH$$
$$MelNH^- + CH_2O \rightarrow MelNHCH_2O^-$$
$$MelNHCH_2O^- + BH \rightarrow MelNHCH_2OH + B^-$$

Mel = (2,4,6-triamino-1,3,5-triazine structure)

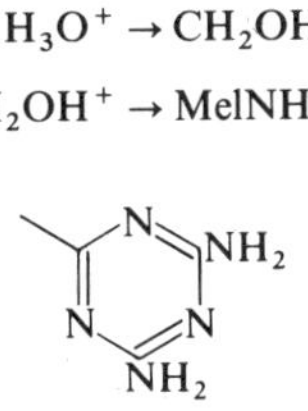

Scheme 8

$$MelNH_2 + H_3O^+ \rightarrow Mel\overset{+}{N}H_3 + H_2O$$
$$CH_2O + H_3O^+ \rightarrow CH_2OH^+ + H_2O$$
$$MelNH_2 + CH_2OH^+ \rightarrow MelNHCH_2OH + H^+$$

Mel = (2,4,6-triamino-1,3,5-triazine structure)

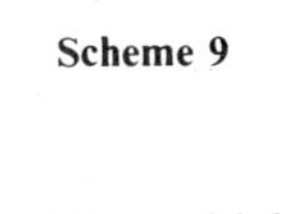

Scheme 9

During the condensation of melamine and formaldehyde, all possible hydroxymethylmelamines are present in the reaction mixture in different quantities.[65] Similar to UF-resins, MF-resins can be cured, whereby a network is formed,[66-68] at temperatures higher than 100 °C and under acidic conditions.

36.4 COPOLYCONDENSATES

Melamine–, urea– and phenol–formaldehyde (PF) resins have different properties based on the different chemical structure of the cured products. It is possible to create materials with new properties if different formaldehyde resins are combined in a suitable way. Another reason for application of cocondensates is their price; for example, UF-resins are cheaper than MF- or PF-resins.

36.4.1 Phenol–Melamine–Formaldehyde Resins

Phenol–melamine–formaldehyde resins (PMF-resins) show, in some respects, better properties than the corresponding MF- and PF-condensates.[59,69,70] The preparation of these resins is described in various patents and publications.[71–81]

Analysis of the molecular structure of PMF-resins using model compounds[82] and by spectroscopic and chromatographic methods[83,84] show that no *cocondensates of phenol and melamine compounds* are formed in the reaction mixture. A schematic formula of a possible cocondensate is shown in structure (4).

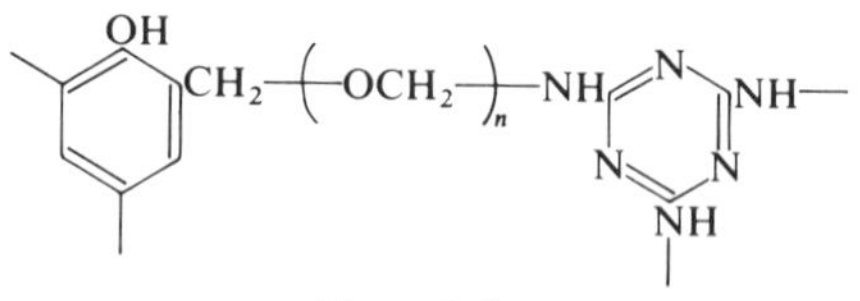

(4) $n = 0, 1$

The reason why no cocondensation takes place is the different reactivity of phenol and melamine hydroxymethyl compounds as a function of pH value.

The structure of cured crosslinked PMF-products was analyzed with the model system *p-cresol–melamine–formaldehyde*.[85] These investigations show that, even in cured networks, no cocondensate structure elements are present. So it is obvious that, in cured PMF-products, separate PF- and MF-networks form an *interpenetrating network* (IPN).

36.4.2 Urea–Melamine–Formaldehyde Resins

Urea is much cheaper than melamine and therefore many efforts were made to prepare *copolycondensates of urea, melamine and formaldehyde* (UMF-resins). Such UMF-resins are used mainly as adhesives for *pulpwoods*.[86–91]

In contrast to the PMF system, the molecular structure could be proven[92] by using reactions of model compounds of the urea- and melamine-type. During the UMF-cocondensation, copolycondensates are formed, structure (5).

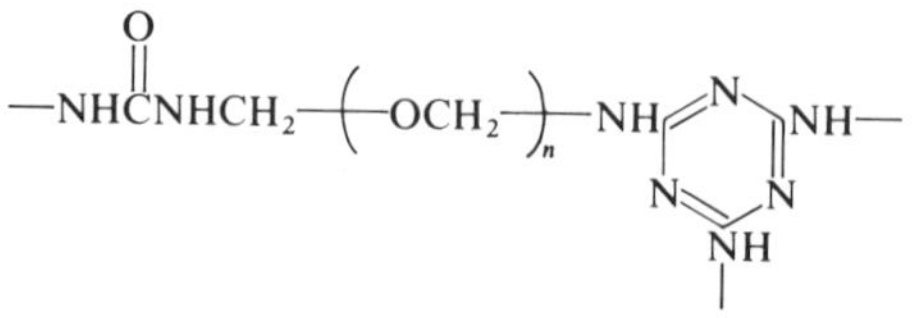

(5) $n = 0, 1$

The dependences of the reactivities of melamine and urea (with respect to formaldehyde) on pH values in mixtures are quite similar to the respective homocondensations.[93] Mechanical and electrical properties of these copolycondensates were tested on moulding compounds of UMF-precondensates after curing. The comparison with UF- and MF-resins and a mixture of these homopolycondensates shows that the pure MF-resin has the best properties.

The properties of the UMF-resin are between the UF- and MF-resins. However, UMF-resins are superior to a mixture of UF- and MF-resins, because processing of such mixtures is very difficult.[94]

36.4.3 Other Copolycondensates

For special applications, some other copolycondensates are prepared in industry, such as *phenol–urea–formaldehyde resins*[95] or *phenol–urea–melamine–formaldehyde resins*. Also, the cocondensation of the system *urea–furfuryl alcohol–formaldehyde*[96,97] and *phenol–furfuryl alcohol–formaldehyde*[98] has been described. These furfuryl resins are applied as adhesives in foundries.

36.5 RESIN ANALYSIS

36.5.1 Precondensates

Various methods of analysis were developed to investigate the structure and properties of UF- and MF-resins. The uncured precondensates are soluble in solvents such as *N,N-dimethylformamide* (DMF) or *dimethyl sulfoxide* (DMSO) and therefore analytical methods accessible to solutions can be used for precondensates.

36.5.1.1 Gel permeation chromatography

Gel permeation chromatography (GPC) gives information on the composition and *molecular weight distribution* of UF- and MF-resins.[99–105] Mostly DMF, DMSO or aqueous salt solutions are used as eluents. The gel materials are crosslinked polystyrenes (Styragels), poly(vinyl acetate)s (Fractogel PVA OR) and saccharides (Sephadex). Generally, a differential refractometer (RI detector) is used as a detector. Figure 3 shows a GPC of a typical UF-precondensate[106] and Figure 4 that of an MF-resin.[107]

In some cases it is useful to first prepare derivatives of the resins to eliminate the interactions of hydrogen bridge linkages. *Silylation reactions* are suitable for this. The hydroxymethyl groups and, in some cases, the NH groups can be substituted by several methods with trimethylsilyl groups; equation (11) shows the principle of this reaction. By this method it is also possible to separate the six different hydroxymethylmelamine compounds.[108–111]

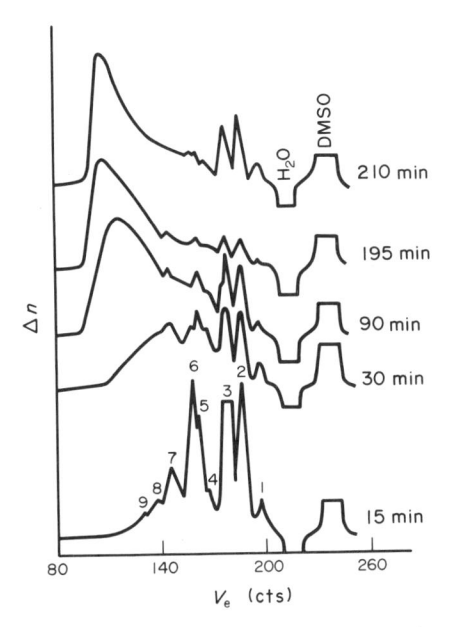

$$\text{melamine} \xrightarrow{CH_2O} \text{H}_2\text{N-triazine-NHCH}_2\text{OH} \xrightarrow{Et_2NSiMe_3} \text{H}_2\text{N-triazine-NHCH}_2\text{OSiMe}_3 \tag{11}$$

Figure 3 Typical GPC of a UF-precondensate as a function of reaction time: 1, urea; 2, mono(hydroxymethyl)urea; 3, di(hydroxymethyl)urea; 4, 5, 6, dimers; 7, trimers; 8, tetramers; and 9, pentamers. Δn is difference in refractive index; V_e is elution volume (cts = counts, arbitrary units)

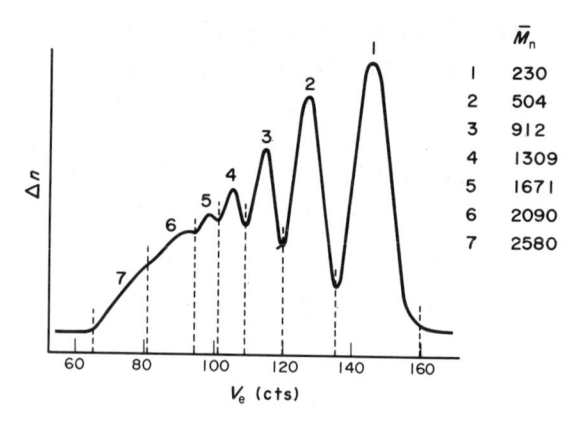

Figure 4 Typical GPC of a MF-precondensate: 1, mono-; 2, di-; 3, tri-; 4, tetra-; 5, penta-; and 6, hexa-(hydroxymethyl)melamine

	$\bar{M}_n$
1	230
2	504
3	912
4	1309
5	1671
6	2090
7	2580

36.5.1.2 ^{13}C NMR spectroscopy

In the last decade, ^{13}C *NMR spectroscopy* has been developed as a powerful new method for analyzing UF-[112-117] and MF-precondensates.[118-119] Figure 5 shows two typical ^{13}C NMR spectra of UF-resins, prepared with stabilized and non-stabilized formaline, and in Table 1 the *chemical shifts* of the most important and typical structure elements are given.

With this physical method it is possible to clearly distinguish between the signals of methylene and dimethylene ether linkages and hydroxymethyl and methoxymethyl groups. The solvent most widely used for ^{13}C NMR spectroscopy of amino resins is DMSO-d_6.

36.5.1.3 Thin-layer chromatography

A cheap and quick alternative to GPC is *thin-layer chromatography* (TLC). As shown in Figure 6, it is possible to separate the different hydroxymethylurea monomers,[120-123] whereas higher

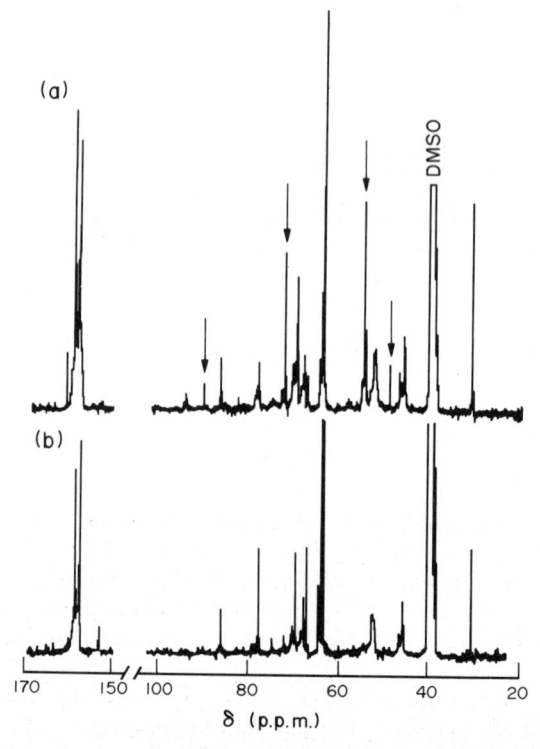

Figure 5 ^{13}C NMR spectra of UF-resins: (a) formaline with a high content of methanol (stabilized); (b) formaline with a low content of methanol (not stabilized)

Table 1 Chemical shifts of Typical Structure Elements

Structure element		Chemical shift, δ (p.p.m.)	Difference, Δ (p.p.m.)
Methylene linkages	$-NHCH_2NH-$	45.9	6.3
	$>NCH_2NH-$	52.2	5.9
	$>NCH_2N<$	58.1	
Dimethylene linkages	$-NHCH_2OCH_2NH-$	67.3	10.4
	$>NCH_2OCH_2NH-$	77.7	
Hydroxymethyl groups	$-NHCH_2OH$	63.9	5.0
	$>NCH_2OH$	68.9	
Methoxymethyl groups	$-NHCH_2OMe$	71.9	6.5
	$>NCH_2OMe$	78.4	

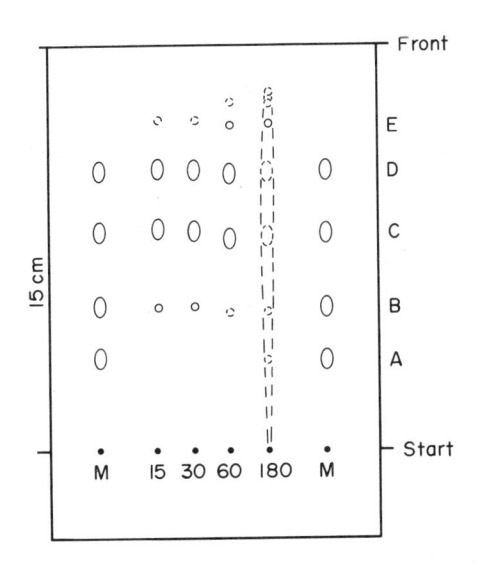

Figure 6 Thin-layer chromatogram of a UF-precondensate as a function of reaction time (min): M, model compound mixture; A, methylenediurea; B, urea; C, mono(hydroxymethyl)urea; D, di(hydroxymethyl)urea; E, tri(hydroxymethyl)urea

condensates cannot be isolated; investigations on the molecular weight distributions of the resins are not possible by TLC.

36.5.1.4 Chemical methods

One of the most important methods for analysis of amino resins is the *determination of formaldehyde*. The content of formaldehyde in pure aqueous solutions can be detected according to Walker.[124] In the presence of products with hydroxymethyl groups the free-formaldehyde content must be detected by the Peterson method.[125] Also, the alkaline splitting method can be used, which includes the determination of hydroxymethyl groups, dimethylene ether groups and free formaldehyde.[126]

In most cases, measurements are done by titration. However, in recent years, other techniques have also been developed to detect *formaldehyde* and hydroxymethyl groups; for example, *UV/VIS spectroscopy*,[127-133] *polarography*,[33,134-136] *IR and NMR spectroscopy*.[112,137,138] A summary of the different methods is available in the literature.[139,140]

A simple but practically important method to control the rate of condensation of urea- and melamine–formaldehyde resins is the measurement of *acetone- and water-dilutability* and *turbidimetric titration*.[141,142] Because formaldehyde resins are only slightly miscible with water and acetone, the resin preparation can be controlled by titration of a definite volume of the resin solution with acetone or water. Figure 7 shows the water (acetone) dilutability of a urea resin as a function of time during the condensation reaction.

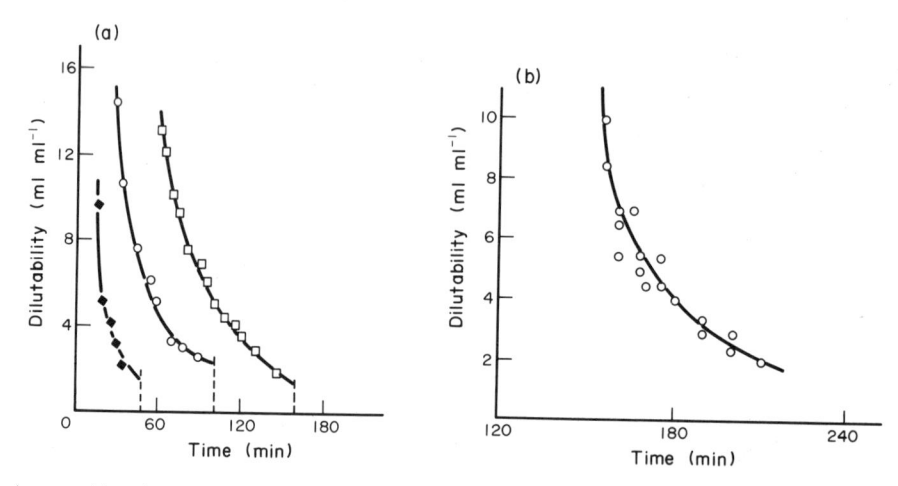

Figure 7 Acetone (a) and water (b) dilutability of a UF-resin as a function of reaction time and temperature: □, 60 °C; ○, 70 °C; ◆, 90 °C; U:F = 1:2; pH = 8

As this method is normally used at room temperature, it is unsuitable for resins which precipitate at this temperature. However, the precipitation temperature can also be used to investigate the reactions of urea or melamine with formaldehyde, mainly after long reaction times. Figure 8 shows this turbidimetric temperature as a function of reaction time.

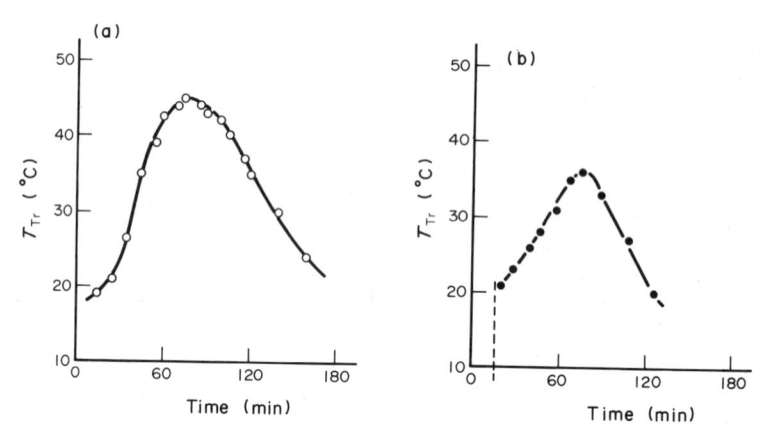

Figure 8 Turbidimetric curves of a UF-condensate as functions of reaction time (a) U:F = 1:2, T = 90 °C, pH = 8; (b) U:F = 1:2, T = 95 °C, pH = 8.5 (15 min) → 5.0 (180 min)

36.5.2 Cured Products

Cured formaldehyde resins are neither soluble nor meltable because these products are crosslinked. Therefore all methods which need a solution of the resins are not suitable for cured materials.

36.5.2.1 *Thermoanalysis*

In *thermogravimetric analysis* (TGA), the weight loss of a sample as a function of temperature is determined.[143,144] In most cases, the sample is heated with a constant heat rate under air or nitrogen. Sometimes, the differential weight-loss curve is also obtained (DTG). The thermogram of an uncured MF-resin is shown schematically in Figure 9. It can be divided into three areas:[84,145] (i) 120–190 °C, reaction of the hydroxymethyl groups and dimethylene ether bridges to form methylene bridges with emission of water and formaldehyde; (ii) 350–400 °C, oxidative degradation of the melamine rings; and (iii) from 450 °C, *degradation* of the residual fraction.

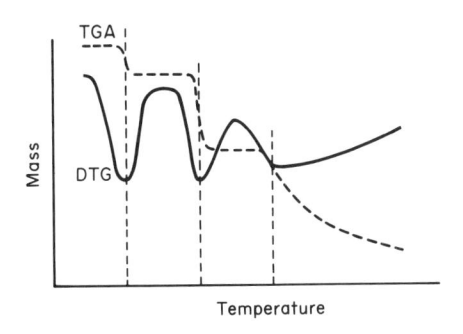

Figure 9 Typical thermogravimetric curve of a MF-precondensate under air

The thermograms of UF-resins are similar to those of MF-resins with one difference: the oxidative degradation of the urea units is observed at 250–300 °C instead of the degradation of the melamine rings.[146–148]

With TGA, both the degree of condensation and the thermal stability of aminoplastics can be investigated. The thermoanalyzer can also be used for quantitative measurements of UMF- and UF-resins.

36.5.2.2 *Other methods*

Curing times of amino resins can also be detected with the Brabender test.[149–151] This test also gives information about processing properties of such materials.

36.6 RESIN MANUFACTURING

In industry, UF- and MF-resins are mostly produced in *discontinuously stirred vessels*. Figure 10 shows the scheme of the *discontinuous industrial production* of aminoplastics.[152]

The main reasons for using discontinuous processes are the relatively small production capacities of many producers and the large number of different products.

Urea or melamine and formaldehyde are condensed at 50–100 °C. Urea and melamine are used as powders, formaldehyde as a 37–50% aqueous solution. After the condensation, the solution of the

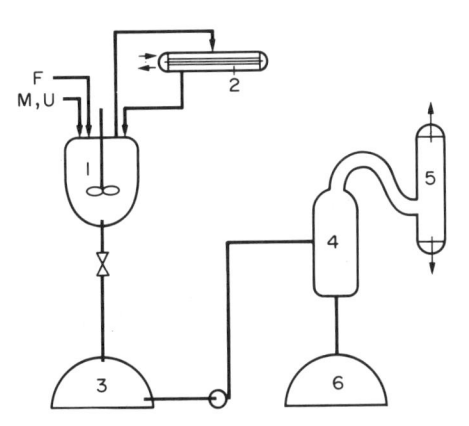

Figure 10 Industrial discontinuous manufacturing scheme of UF- and MF-resins: F, formaldehyde; M, melamine; U, urea; 1, reaction vessel; 2, reflux condenser; 3, intermediate tank; 4, evaporator; 5, condenser; 6, tank for the product

soluble reaction product is pumped into an intermediate tank. Subsequently, the aqueous UF- or MF-resins are dried with an evaporator. For the various applications, different quantities of water are evaporated. For example, a resin for a moulding compound must be completely dry, whereas a glue contains up to 50% water.

Continuous processes for UF- and MF-resin production have also been described.[153,154] A series of vessels (*cascade*) is usually used. The advantage of this procedure is the constant quality of the product, but in each reactor a minimum quantity of resin must be produced.

36.7 PROPERTIES AND APPLICATIONS OF UF- AND MF-POLYMERS

36.7.1 Adhesives

The largest quantities of UF- and MF-polymers are used for *adhesives*. Mostly, the products are applied as aqueous solutions of the resins, but powdered materials are also available. The advantage of the powdered products is their better chemical stability compared to the aqueous adhesives. Hardeners are necessary for most applications of UF-resins, whereas MF-resins can be cured without a catalyst above 120 °C. MF-resins also exhibit a better stability to water than UF-resins. The main application is the production of particle boards, where the resins are used as binders.

UF- and MF-resins can also be used for paper bonding, mainly for the combination of papers for production of boxboards, corrugated papers, *etc*.

36.7.2 Moulding Compounds

Another application of UF- and MF-resins is the production of *moulding compounds*.[155,156] Here, the dried resins in the precondensate stage are mixed with additives, such as cellulose, asbestos, wood flour, *etc*., and hardeners to give the moulding compound.

The manufacturing to the crosslinked final product is done by *pressing, casting-* and *injection-moulding*. The resulting products exhibit good thermal stabilities.

36.7.3 Coatings

Pure amino resins are not suitable for *coatings*. The bulk of aminoplastic coatings are butylated with *n-butanol or isobutanol*. For preparation of these resins, urea or melamine are first condensed with formaldehyde and then the alcohol is added to transform the hydroxymethyl groups into ethers. The water is removed during the reaction. The resulting products form brittle films but, in combination with other products, these coatings show good properties.

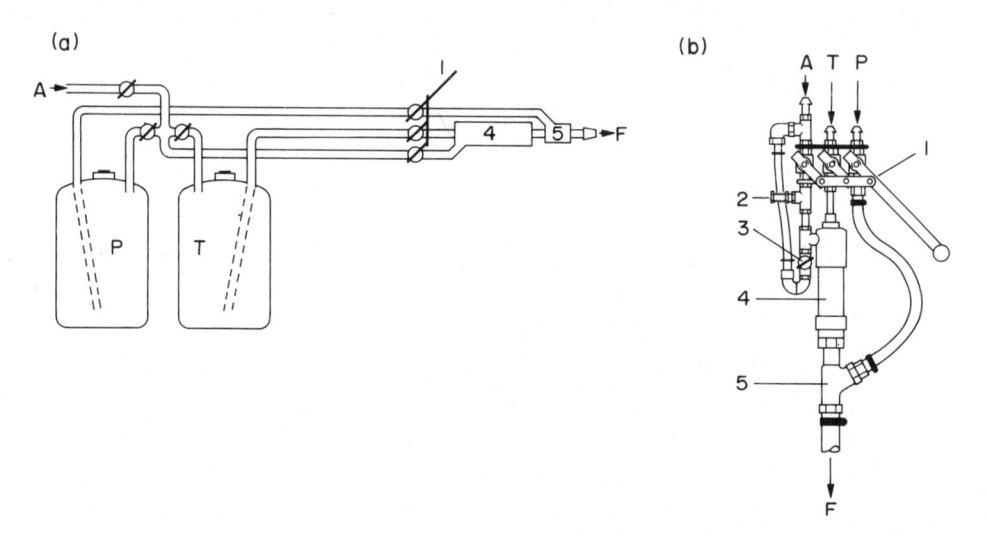

Figure 11 Foam apparatus for preparing UF-foams: (a) foaming device IG50, (b) foaming apparatus 125; A, compressed air; P, precondensate; T, surfactant solution; F, foam; 1, stop-cock; 2, micrometer for A; 3, air flush valve; 4, foam generator; 5, mixing chamber

36.7.4 Foams

It is also possible to prepare aminoplastic *foams*. Figure 11 shows a *foaming apparatus* developed by Bauer.[157] One tank of this apparatus contains a UF-precondensate. Into the other tank a mixture of an acid (*e.g.* phosphoric acid) as the hardener and an aqueous solution of a *surfactant* are brought. Both mixtures are led into a mixture pipe where first a tenside foam is formed. On the laminaes of this foam, the UF-condensate is cured and a crosslinked UF-foam is formed.[158,159] In recent years, melamine–formaldehyde foams were also prepared using low-boiling solvents, such as pentane, as gas-developing agents.

There are a lot of other applications of UF- and MF-resins, but it would be beyond the scope of this article to report on them in detail.

36.8 REFERENCES

1. B. Meyer, 'Urea–Formaldehyde Resins', Addison-Wesley, London, 1979.
2. B. Tollens, *Ber. Dtsch. Chem. Ges.*, 1884, **17**, 659.
3. C. Goldschmidt, *Ger. Pat.* 96 164 (1897).
4. A. Vosswinkel, *Ger. Pat.* 171 788 (1904).
5. R. Lauch and A. Vosswinkel, *Ger. Pat.* 180 864 (1904).
6. H. John, *Ger. Pat.* 392 183 (1918); *Austrian Pat.* 78 251 (1918); *US Pat.* 1 355 834 (1919).
7. H. Ramstetter (BASF), *Ger. Pat.* 403 645 (1922).
8. H. John, *Ger. Pat.* 394 488 (1919).
9. H. Goldschmidt and O. Neuß, *Ger. Pat.* 412 614 (1921).
10. A. Curs and H. Wolf (IG Farbenindustrie), *Ger. Pat.* 636 658 (1933).
11. A. Curs and H. Wolf (Plescon Co.), *US Pat.* 2 076 295 (1934).
12. *Br. Pat.* 455 008 (1935).
13. *Fr. Pat.* 811 804 (1936).
14. J. Mavrocvic, *Hydrocarbon Process.*, 1971, **50**, 161.
15. S. Yoshimura, *Hydrocarbon Process.*, 1970, **49**, 111.
16. S. Inoue, *Bull. Chem. Soc. Jpn.*, 1972, **45**, 1339.
17. BASF, *Ger. Pat.* 294 793 (1914).
18. BASF, *Br. Pat.* 24 117 (1914).
19. Stamicarbon, *Br. Pat.* 952 764 (1964).
20. R. Köhler, *Hydrocarbon Process. Pet. Refiner*, 1964, **43**, 177.
21. L. Reitter and R. Lihotzky, in 'Ullmanns Encyklopädie der Technischen Chemie', ed. E. Bartholome, E. Biekert, H. Hellmann, H. Ley, W. Weigert and E. Weise, Verlag Chemie, Weinheim, 1978, vol. 16, p. 507.
22. BASF, *Ger. Pat.* 2 442 231 (1976).
23. H. Diem and A. Hilt, in 'Ullmanns Encyklopädie der Technischen Chemie', ed. E. Bartholome, E. Biekert, H. Hellmann, H. Ley and W. Weigert, Verlag Chemie, Weinheim, 1976, vol. 11, p. 691.
24. M. Dinte, *Chim. Ind. (Milan)*, 1964, **46**, 1326.
25. H. Diem and A. Hilt, in 'Ullmanns Encyklopädie der Technischen Chemie', ed. E. Bartholome, E. Biekert, H. Hellmann, H. Ley and W. Weigert, Verlag Chemie, Weinheim, 1976, vol. 11, p. 693.
26. A. Ilitaco, *Chim. Ind. (Milan)*, 1954, **36**, 523.
27. Cities Service Oil Co., *US Pat.* 2 665 241 (1948).
28. J. de Jong and J. de Jonge, *Recl. Trav. Chim. Pays-Bas*, 1950, **69**, 1566.
29. J. de Jong and J. de Jonge, *Recl. Trav. Chim. Pays-Bas*, 1952, **71**, 643, 661.
30. J. de Jong and J. de Jonge, *Recl. Trav. Chim. Pays-Bas*, 1953, **72**, 38.
31. J. de Jong and J. de Jonge, *Recl. Trav. Chim. Pays-Bas*, 1953, **72**, 139.
32. J. de Jong and J. de Jonge, *Recl. Trav. Chim. Pays-Bas*, 1953, **72**, 202, 207, 213.
33. N. Landquist, *Acta Chem. Scand.*, 1955, **9**, 1127.
34. N. Landquist, *Acta Chem. Scand.*, 1955, **9**, 1459, 1466, 1477.
35. N. Landquist, *Acta Chem. Scand.*, 1956, **10**, 244.
36. N. Landquist, *Acta Chem. Scand.*, 1957, **11**, 776, 780, 792.
37. B. R. Glutz and H. Zollinger, *Helv. Chim. Acta*, 1969, **52**, 1976.
38. B. R. Glutz and H. Zollinger, *Helv. Chim. Acta*, 1969, **52**, 1985.
39. H. Petersen, *Textilveredlung*, 1967, **2**, 744.
40. H. Petersen, *Textilveredlung*, 1968, **3**, 160.
41. H. Petersen, *Textilveredlung*, 1968, **3**, 51, 353, 397, 629.
42. H. Petersen, *Textilveredlung*, 1969, **4**, 254.
43. H. Petersen, *Angew. Chem.*, 1964, **76**, 909.
44. P. Eugster and H. Zollinger, *Helv. Chim. Acta*, 1969, **52**, 1985.
45. H. Petersen, *Chem.-Ztg.*, 1971, **95**, 625.
46. L. E. Smythe, *J. Phys. Chem.*, 1947, **51**, 369.
47. L. E. Smythe, *J. Am. Chem. Soc.*, 1951, **73**, 2735.
48. L. E. Smythe, *J. Am. Chem. Soc.*, 1952, **74**, 2713.
49. L. E. Smythe, *J. Am. Chem. Soc.*, 1953, **75**, 574.
50. H. Kadowaki, *Bull. Chem. Soc. Jpn.*, 1936, **11**, 248.
51. M. T. Beachem, J. C. Oppelt, F. M. Cowen, P. D. Schickendanz and V. D. Mayer, *J. Org. Chem.*, 1963, **28**, 1876.
52. B. Tomita and Y. Hirose, *J. Polym. Sci., Polym. Chem. Ed.*, 1976, **14**, 387.

53. D. Braun and F. Bayersdorf, *Chem.-Ztg.*, 1972, **96**, 193.
54. D. Braun and F. Bayersdorf, *Angew. Makromol. Chem.*, 1979, **81**, 147.
55. D. Braun and F. Bayersdorf, *Angew. Makromol. Chem.*, 1979, **83**, 21.
56. D. Braun and F. Bayersdorf, *Angew. Makromol. Chem.*, 1980, **85**, 1.
57. D. Braun and F. Bayersdorf, *Angew. Makromol. Chem.*, 1980, **89**, 183.
58. C. P. Vale and W. G. K. Taylor, 'Aminoplastics', Iliffe Books, London, 1964, p. 31.
59. A. Bachmann and T. Bertz, 'Aminoplaste', VEB Verlag für Grundstoffindustrie, Liebzig, 1967, p. 81.
60. A. Gams, G. Widmer and W. Fisch, *Helv. Chim. Acta*, 1941, **24**, 302, 316.
61. M. Okano and Y. Ogata, *J. Am. Chem. Soc.*, 1952, **74**, 5728.
62. K. Sato and S. Ouchi, *Polym. J.*, 1978, **10**, 1.
63. D. Braun and V. Legradic, *Angew. Makromol. Chem.*, 1974, **36**, 41.
64. K. Sato, T. Konakara and M. Kawashima, *Makromol. Chem.*, 1982, **183**, 875.
65. D. Braun and V. Legradic, *Angew. Makromol. Chem.*, 1972, **25**, 193.
66. K. Sato and N. Takanobu, *Polym. J.*, 1973, **5**, 144.
67. A. Berge, *Proc. Int. Conf. Coat. Sci. Technol.*, 1977, **3**, 31.
68. D. Braun, *Angew. Makromol. Chem.*, 1979, **76/77**, 360.
69. A. Knop and W. Scheib, 'Chemistry and Application of Phenolic Resins', Springer, Berlin, 1979, p. 134.
70. K. Bruncken, in 'Kunststoffhandbuch', ed. R. Vieweg and E. Becker, Hanser, München, 1968, vol. 10, p. 352.
71. H. Scheuermann (BASF), *Br. Pat.* 557 557 (1953).
72. R. Köhler (Henkel Cie.), *Ger. Pat.* 875 568 (1953).
73. I. H. Updegraff and N. R. Segro (American Cyanamid Co.), *US Pat.* 2 826 559 (1958) (*Chem. Abstr.*, 1958, **52**, 10 652).
74. Formica Co., *Br. Pat.* 882 140 (1961) (*Chem. Abstr.*, 1962, **56**, 11 793).
75. H. Michaud, K. Scheinost and J. Seeholzer (Süddeutsche Kalkstickstoff-Werke AG), *Ger. Pat.* 1 149 902 (1963).
76. K.-H. Mädebach (VEB Leuna Werke Walter Ulbricht), *Ger. Pat.* 1 153 901 (1963).
77. K. E. S. Knutsson (Fosfatbolaget AB), *US Pat.* 3 321 551 (1967) (*Chem. Abstr.*, 1967, **67**, 33 292).
78. Ibigawa Electric Industry Co., *Br. Pat.* 1 057 400 (1967) (*Chem. Abstr.*, 1967, **66**, 66 124).
79. S. Murayama, M. Asakuno and T. Kobayashi (Sumitomo Bakelite Co.), *Jpn. Pat.* 69 00 552 (1969) (*Chem. Abstr.*, 1969, **70**, 107 101).
80. T. Nakamura (Matsushita Electric Works), *Jpn. Pat.* 74 14 147 (1974) (*Chem. Abstr.*, 1974, **81**, 170 528).
81. A. B. Abdinova, *Azerb. Khim. Zh.*, 1976, **2**, 56.
82. D. Braun and W. Krauße, *Angew. Makromol. Chem.*, 1982, **108**, 141.
83. D. Braun and W. Krauße, *Angew. Makromol. Chem.*, 1983, **118**, 165.
84. D. Braun and H.-J. Ritzert, *Angew. Makromol. Chem.*, 1984, **125**, 9.
85. D. Braun and H.-J. Ritzert, *Angew. Makromol. Chem.*, 1984, **125**, 27.
86. F. Tröger, A. Grigoriou and F. Hey, *Holz Roh-Werkst.*, 1977, **35**, 379.
87. W. Clad and C. Schmidt–Hellerau, *Holz-Zentralbl.*, 1976, **102**, 313.
88. O. Liiri and A. Kivistoe, *Holz Roh-Werkst.*, 1981, **39**, 7.
89. R. E. Kreibich, *Adhes. Age*, 1976, **19**, 27.
90. R. Marutzky, L. Ranta and E. Schriever, *Holz-Zentralbl.*, 1978, **104**, 1747.
91. H. J. Deppe and R. Stolzenburg, *J. Appl. Polym. Sci.: Appl. Polym. Symp.*, 1984, **40**, 41.
92. D. Braun and H.-J. Ritzert, *Angew. Makromol. Chem.*, 1988, **156**, 1.
93. D. Braun and H.-J. Ritzert, *Angew. Makromol. Chem.*, 1985, **135**, 193.
94. D. Braun and H.-J. Ritzert, *Kunststoffe*, 1987, **77**, 1264.
95. G. Zigeuner, *Monatsh. Chem.*, 1955, **86**, 165.
96. R. S. Bushnell and E. Parks, *Br. Foundryman*, 1962, **55**, 325.
97. R. S. Bushnell, *Giesserei-Prax.*, 1963, **4**, 83.
98. K. Roczpiak, T. Biernacka and Z. Wertz, *Polimery (Warsaw)*, 1976, **21**, 306.
99. P. Hope and P. B. Stark, *Br. Polym. J.*, 1973, **5**, 363.
100. M. Tsuge, T. Muyabayashi and S. Tanaka, *Bunseki Kagaku*, 1974, **23**, 1146.
101. K. Kamlin and R. Simonson, *Angew. Makromol. Chem.*, 1978, **68**, 175.
102. D. Braun and F. Bayersdorf, *Angew. Makromol. Chem.*, 1979, **81**, 147.
103. D. Braun and F. Bayersdorf, *Angew. Makromol. Chem.*, 1980, **85**, 1.
104. D. Braun and F. Bayersdorf, *Angew. Makromol. Chem.*, 1980, **89**, 183.
105. M. Dunky and K. Lederer, *Angew. Makromol. Chem.*, 1982, **102**, 199.
106. D. Braun, *Oesterr. Chem. Z.*, 1985, **86**, 188.
107. D. Braun and W. Pandjojo, *Angew. Makromol. Chem.*, 1979, **80**, 195.
108. D. Braun and V. Legradić, *Angew. Makromol. Chem.*, 1972, **25**, 193.
109. D. Braun and V. Legradić, *Angew. Makromol. Chem.*, 1973, **34**, 35.
110. D. Braun and V. Legradić, *Angew. Makromol. Chem.*, 1974, **35**, 101.
111. D. Braun and V. Legradić, *Angew. Makromol. Chem.*, 1974, **36**, 41.
112. A. J. J. de Breet, W. Dankelmann, W. G. B. Huysmans and J. de Wit, *Angew. Makromol. Chem.*, 1977, **62**, 7.
113. J. R. Ebdon and P. E. Heaton, *Polymer*, 1977, **18**, 971.
114. I. Y. Slonim, S. G. Alekseeva, Y. G. Urman, B. M. Arshava and B. Y. Akselrod, *Vysokomol. Soedin., Ser. A*, 1978, **20**, 1418.
115. B. Tomita and S. Hatono, *J. Polym. Sci., Polym. Chem. Ed.*, 1978, **16**, 2509.
116. B. Meyer and R. Nunlist, *Polym. Prepr., Am. Chem. Soc., Div. Polym. Chem.*, 1981, **22**, 130.
117. H. Schindlbauer and J. Schuster, *Kunststoffe*, 1983, **73**, 325.
118. H. Schindlbauer and J. Anderer, *Angew. Makromol. Chem.*, 1979, **79**, 157.
119. B. Tomita and H. Ono, *J. Polym. Sci., Polym. Chem. Ed.*, 1979, **17**, 3205.
120. W. Y. Lee, *Anal. Chem.*, 1972, **44**, 1284.
121. P. R. Ludlam, *Analyst (London)*, 1973, **98**, 107, 116.
122. D. Braun and W. Pandjojo, *Z. Anal. Chem.*, 1975, **276**, 205.

123. D. Braun and P. Günther, *Kunststoffe*, 1982, **72**, 12.
124. J. F. Walker, 'Formaldehyde', Reinhold, New York, 1964, p. 486.
125. G. Groh and H. Petersen, *Farbe + Lack*, 1981, **87**, 744.
126. A. Berge, S. Gudmundsen and J. Ugelstad, *Eur. Polym. J.*, 1969, **5**, 171.
127. L. Anguiro, *Melliand Textilber.*, 1982, **63**, 522.
128. M. Beroza, *Anal. Chem.*, 1954, **26**, 1970.
129. W. Bitterli and J. M. Sire, *Textilveredlung*, 1976, **11**, 345.
130. C. E. Bricker and W. A. Vail, *Anal. Chem.*, 1950, **22**, 720.
131. R. G. Dickenson and N. W. Jacobson, *J. Chem. Soc., Chem. Commun.*, 1970, 1719.
132. R. R. Miksch, D. W. Anthon, L. Z. Fanning, C. D. Hollowell, K. Revzan and J. Glanville, *Anal. Chem.*, 1981, **53**, 2118.
133. T. Nash, *J. Biotechnol.*, 1953, **55**, 416.
134. F. G. Jahoda, *Collect. Czech. Chem. Commun.*, 1935, **7**, 419.
135. K. Vessely and R. Brdicka, *Collect. Czech. Chem. Commun.*, 1947, **12**, 313.
136. G. A. Crowe, Jr. and C. R. Lynch, *J. Am. Chem. Soc.*, 1950, **72**, 3622.
137. M. Chiavarani, R. Bigatto and W. Conti, *Angew. Makromol. Chem.*, 1978, **70**, 49.
138. M. F. Guiliniano, *Congr. FATIPEC*, 1976, 233.
139. G. Christensen, *Prog. Org. Coat.*, 1977, **5**, 251.
140. R. Nastke, K. Dietrich and W. Teige, *Acta Polym.*, 1980, **31**, 329.
141. D. Braun and P. Günther, *Kunststoffe*, 1982, **72**, 12.
142. D. Braun, P. Günther and W. Pandjojo, *Angew. Makromol. Chem.*, 1982, **102**, 147.
143. R. C. Mackenzie, 'Differential Thermal Analysis', Academic Press, New York, 1970.
144. W. Wendlandt, 'Thermal Methods of Analysis', Wiley, New York, 1964.
145. I. H. Anderson, M. Cawley and W. Steedman, *Br. Polym. J.*, 1971, **3**, 86.
146. D. Braun and P. Günther, *Angew. Makromol. Chem.*, 1984, **128**, 1.
147. G. Camino, L. Operti and L. Trossarelli, *Proc. Int. Conf. Thermal Anal.*, 1982, **7**, 1144.
148. G. Camino, L. Operti and L. Trossarelli, *Polym. Degradation Stab.*, 1983, **5**, 161.
149. H. E. Luben, *Chem. Rundsch.*, 1974, **27**, 3.
150. A. Rothenspieler and R. Heß, *Kunststoffe*, 1972, **62**, 215.
151. W. Sauer and K. Eichler, *Industrieanzeiger*, 1968, **90**, 187.
152. F. Brunnmüller, in 'Ullmanns Encyklopädie der Technischen Chemie', ed. E. Bartholome, E. Biekert and H. Hellmann, Verlag Chemie, Weinheim, 1974, vol. 7, p. 413.
153. Rütgerswerke, *Ger. Pat.* 1 029 150 (1955).
154. BASF, *Ger. Pat.* 2 109 754 (1971).
155. F. J. Bollig and K. H. Decker, *Kunststoffe*, 1980, **70**, 672.
156. *Plastverarbeiter*, 1983, **34**, 53.
157. W. Bauer, *Ger. Pat.* 1 043 628 (1953).
158. D. Braun and P. Günther, *J. Cell. Plast.*, 1985, **21**, 171.
159. D. Braun and P. Günther, *Chem.-Ztg.*, 1985, **109**, 129.

37
Epoxy Resins

KENN HODD

Brunel University, Uxbridge, UK

37.1 INTRODUCTION	667
37.2 SOME COMMERCIALLY IMPORTANT EPOXY RESINS	668
37.2.1 *Glycidyl Resins*	668
37.2.2 *Ene–Oxide Resins*	671
37.3 HOMOPOLYMERIZATION OF EPOXIDES	671
37.3.1 *Introduction*	671
37.3.2 *Boron Trifluoride Catalysis*	672
37.3.3 *Catalysis by Tertiary Amines and Related Compounds*	675
37.3.4 *Photoinitiated Cationic Polymerization of Epoxides*	676
37.4 REAGENTS FOR STEP-GROWTH POLYMERIZATION	678
37.4.1 *Polymerization with Primary and Secondary Amines*	678
37.4.2 *Dicyandiamide Cure of Epoxies*	686
37.4.3 *Diacid Anhydrides*	688
37.4.4 *Isocyanates*	693
37.4.5 *Alcohols and Thiols*	695
37.5 REFERENCES	696

37.1 INTRODUCTION

Epoxy resins comprise a group of crosslinkable materials, which all possess the same type of reactive functional group, the epoxy or oxirane group (**1**); their chemistry and technology have been reported in a number of texts.[1,2]

$$\underset{\text{(1)}}{\overset{\displaystyle O}{\underset{\displaystyle}{\text{C}\!\!-\!\!\text{C}}}}$$

The epoxy group is characterized by its reactivity towards both nucleophilic and electrophilic species and it is thus receptive to a wide range of reagents or curing agents. Such curing agents are of two types: they may be either catalysts or hardeners. Catalysts are usually drawn from tertiary amines or Lewis acids, and they function by initiating the ionic polymerization of the epoxy compound to produce polyether structures. Typically catalysts are used in low concentrations (<1% of the resin's weight). They are often used in association with hardeners in resin formulations. Common examples of hardeners are provided by aliphatic and aromatic amines, and carboxylic anhydrides. These types of compound react directly with the epoxy and other groups of the resin, and so influence the structure of the cured resin. In resin manufacturers' recommended formulations resin:hardener ratios are usually in the band 3:1 to 10:1 by weight.

It is the combination of resin and curing agent which produces the cured thermoset epoxy resin. The process of curing an epoxy resin converts the initially low molecular weight resin into its thermoset form, which is a space network or three-dimensional chemical structure.

The term epoxy resin refers to both the uncured and the cured forms of the resin. Industrial interest and use of these resins resides in both the valuable properties of the cured resin, which include good adhesion to many substrates, relatively high toughness (particularly when rubber modified), good environmental resistance, high electrical resistivity, low shrinkage, *etc.* and the ease with which the curing reaction may be adapted to suit the fabricating process. Thus the curing reaction, which is exothermic, may be induced at room or elevated temperatures or it may be initiated, in the presence of appropriate catalysts, by UV light. The specific curing procedure required to produce a cured resin of optimized performance characteristics is dependent upon the precise combination of resin, curing agent and/or catalyst.

An important technological feature of the curing processes of epoxy resins is that, although they proceed mostly by step-growth reactions, no volatile by-products are generated. This allows the production of coherent void-free structures without recourse to applied pressure during forming or moulding processes.

The combination of good processing characteristics and useful properties has led to the use of epoxy resins in many applications including adhesives,[3] in electronics for encapsulation, potting and printed circuit boards,[4] and in the aerospace industries as matrices for composites.[5]

To meet the differing requirements of this broad sweep of applications, epoxy resins of many types are available, and it is these in combination with the variety of curing agents which contribute the essential versatility of the epoxy system. In the next section some examples of typical epoxy resins are discussed.

37.2 SOME COMMERCIALLY IMPORTANT EPOXY RESINS

37.2.1 Glycidyl Resins

A major type of resin is that derived by the reaction of bis(4-hydroxyphenylene)-2,2-propane (bisphenol A) and 1-chloropropene 2-oxide (epichlorohydrin), in the presence of sodium hydroxide. The structure of the major product, bisphenol A diglycidyl ether (BADGE; **2**), and its condensed forms, is dependent upon the stoichiometry of the reactants. Typically resins are marketed with n in (**2**) in the range 0.2 to 12. The data in Table 1 are indicative of the relationship between n and important properties of the resins, such as their epoxy equivalent weights.[6] Increased attention has been focussed on two aspects of the purity of BADGE-type resins. Firstly to produce resins having n values close to zero, and secondly to produce them with low to vanishingly small concentrations of saponifiable chlorine. This latter improvement reduces the corrosion of electronic devices encapsulated in epoxy resin.[7] Other types of phenol-derived epoxy resins include the diglycidyl ether of bisphenol F (**3**), epoxy novolacs (**4**) and tris(hydroxyphenylene)methane triglycidyl ether (**5**) (see Table 2).[8]

These latter and last resins, because they possess more epoxy groups per resin molecule than the type 2 resins, produce more highly crosslinked resins with higher glass transition temperatures (T_g).

The reactivity of epoxy resins based on polynuclear phenols has been reported.[9,10] Halogenated epoxy resins have also been reported.[4,11–13] Another major group of epoxy resins derived from epichlorohydrin are the resins synthesized by its two-step reaction with an aromatic amine. Of

(**2**) Bisphenol A diglycidyl ether

(**3**) Bisphenol F diglycidyl ether

(**4**) Epoxy novolac

(5) Tris(hydroxyphenylene)methane triglycidyl ether

Table 1 Bisphenol A Diglycidyl Ether Characterization[a,b]

Shell resin	n^c	Number M_w
828	0.2	390
834	0.7	540
1001	2–2.5	980
1004	4–4.5	1550
1007	8.5–9.0	2825
1009	12.0–15.0	4170

[a] Forward with Epoxy Resins, Joint Seminar Shell Chemicals UK Ltd and Anchor Chemical (UK) Ltd., 1986.
[b] Mainly characterized by: (1) epoxide equivalent (g resin/epoxy group), (2) viscosity.
[c] See (2).

Table 2 Typical Liquid Resin Properties[a] of Experimental Epoxy Resins XD-7342.00L and XD-9053.00L[b]

Property	XD-7342.00L	XD-9053.00L
Epoxide equivalent weight	162	220
Specific gravity, 25 °C	1.22	1.22
Viscosity (cSt)[c]		
150 °C	55	550
60 °C	14 000	—
Durran's softening point (°C)	55	85
Volatiles (wt % max.)	1.0	0.5
Hydrolyzable chlorides (wt % max.)	0.1	0.05
Ionic chloride (p.p.m. max.)	10	5
Sodium (p.p.m. max.)	15	5

[a] K. L. Hawthorne and F. C. Henson, *ACS Symp. Ser.*, 1983, **221**, 135.
[b] Based on tris(hydroxyphenyl)methane triglycidyl ether (Dow Chemical Co.).
[c] 1 cSt = 10^{-6} m^2 s^{-1}.

these resins the most important is N,N,N',N'-tetraglycidyl-4,4'-diaminodiphenylenemethane (TGDDM; **6**). King and coworkers have discussed the synthesis of TGDDM and the analysis of the reaction products using HPLC.[14,15] Dobinson and coworkers have reported advances in the synthesis of this type of resin,[16] and the technical consequences of using the resins so produced. Figure 1 compares the HPLC chromatograms of such a resin and a conventional resin; the improvement in the purity of the former is clearly apparent.

Hagnauer et al.[17,18] and Dobas and coworkers[19] have also discussed the implications of purity on the technical aspects of these resins. Table 3 summarizes some details.[19]

(6) N,N,N',N'-Tetraglycidyl-4,4'- diaminodiphenylenemethane

(7) N,N-Diglycidylaniline

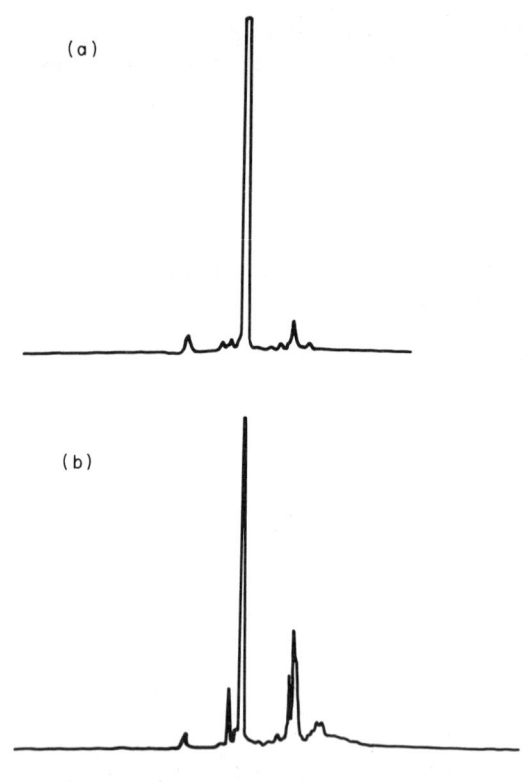

Figure 1 GPC chromatograms of (a) pure and (b) resinous TGDDM, Ciba Geigy EP760 and MY720, respectively (reproduced by permission of the Royal Society of Chemistry from *Br. Polym. J.*, 1986, **18**, 286)

Table 3 Parameters of TGDDM Samples[a]

Sample	Epoxy value (equiv. kg^{-1})	Cl content (%)	Viscosity at 25 °C (Pa s)	M_n (VPO)
A	8.94	0.17	87.4	453
B	8.31	0.29	393	496
C	7.92	0.39	408	494

[a] I. Dobas, S. Lunak, S. Podzimek, M. Mach and V. Spacek, in 'Crosslinked Epoxies', ed. B. Sedlacek and J. Kahovec, Walter de Gruyter, Berlin, 1987, p. 81.

Resins based on the reaction of epichlorohydrin with other amines, such as aniline, to produce *N,N*-diglycidylaniline (DGA; **7**), and 4-hydroxyaniline, to produce *N,N,O*-triglycidyl-4-hydroxyaniline, are also important commercially. Dobinson and coworkers[16] have referred to the synthesis of DGA.

Commercial resins are formulated from more than one resin and other types of resin to be found mixed with amine resins include 1,3-bis(1-glycidyl-5,5-dimethyl-3-hydantoinyl)-2-glycidyloxypropane (**8**) and glycidyl orthophthalate. Table 4 indicates the composition of some typical TGDDM

Table 4 Chemical Constituents in C Fiber–TDGGM–DDS Prepregs[a]

Constituent	Fiberite 934 (wt %)	Narmco 5208 (wt %)
TGDDM	64	71
DDS	25	21
Diglycidyl orthophthalate epoxy (DGOP)	11	0
Glycidyl ether of a bisphenol A novolac (SU-8, Celanese)	0	8
$BF_3 \cdot NH_2Et$	0.4	0

[a] R. J. Morgan and E. T. Mones, *J. Appl. Polym. Sci.*, 1987, **33**, 999.

resin formulations for 'prepregs'. (Prepregs are resin-impregnated fibres for composite production.) Other glycidyl resins reported include epoxy resins based on adamantane[21] and on phosphazene rings.[22]

37.2.2 Ene–Oxide Resins

An alternative approach to the formation of the oxirane group, namely the peroxidation of a carbon–carbon double bond (Scheme 1), generates another major group of epoxy resins, which may be classified as the ene–oxide resins.

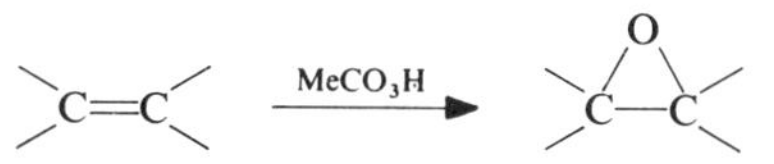

Scheme 1

Various types of resin fall into this category, notably the epoxidized oils[4] and rubbers,[23] and the cycloaliphatic oxides. Of this latter group important examples include vinylcyclohex-3-ene dioxide, dicyclopentadiene dioxide and 3,4-epoxycyclohexylmethyl 3′,4′-epoxycyclohexanecarboxylate (oxycyclohex-3-enylmethyl oxycyclohex-3-enylcarboxylate; **9**).[24]

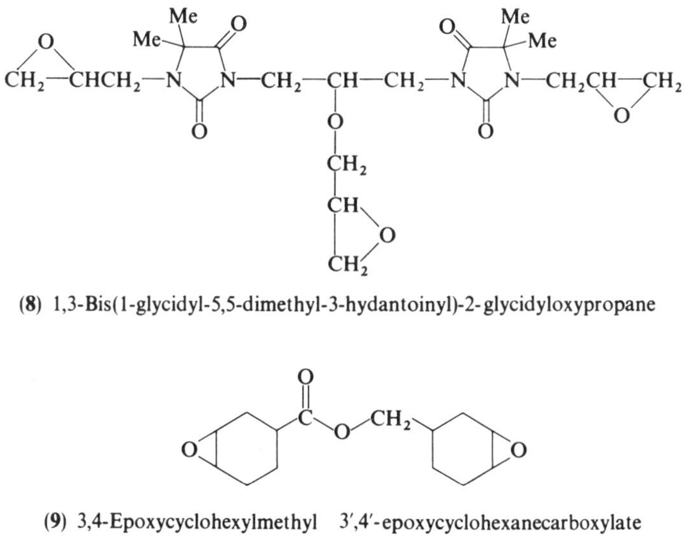

(**8**) 1,3-Bis(1-glycidyl-5,5-dimethyl-3-hydantoinyl)-2-glycidyloxypropane

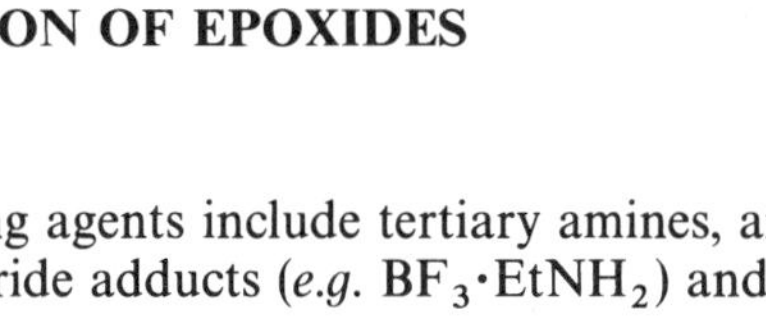

(**9**) 3,4-Epoxycyclohexylmethyl 3′,4′-epoxycyclohexanecarboxylate

37.3 HOMOPOLYMERIZATION OF EPOXIDES

37.3.1 Introduction

Commonly used catalytic curing agents include tertiary amines, and variants such as guanidine and the imidazoles; boron trifluoride adducts (*e.g.* $BF_3 \cdot EtNH_2$) and ammonium complexes of the type $PhNH_3^+ AsF_6^-$.[25-32]

Boron trifluoride–monoethylamine is particularly important for the cure of epoxy resins, and has been identified as a catalyst, an accelerator, and a crosslinking agent.[33-39]

Studies have been reported of the homopolymerization of epoxides and the effects of temperature, catalysts, alcohols and amines.[40-51]

Thus both Lewis acids and Lewis bases are able to catalyze the homopolymerization of epoxy resins. For catalysis by a tertiary amine Narracott (Scheme 2)[52] proposed for the initiating step attack by the amine on the epoxide; propagation then proceeds from the resulting alkoxide anion.

For the homopolymerization of epoxides by Lewis acids Arnold[53] proposed a mechanism involving a cocatalyst (Scheme 3)[54] such as water or an amine to form the initiating species. Propagation then proceeds by a cationic addition process in which the propagating species, the methylene carbonium ion, is stabilized by a gegenion derived from the BF_3 catalyst.[55] Hayase *et*

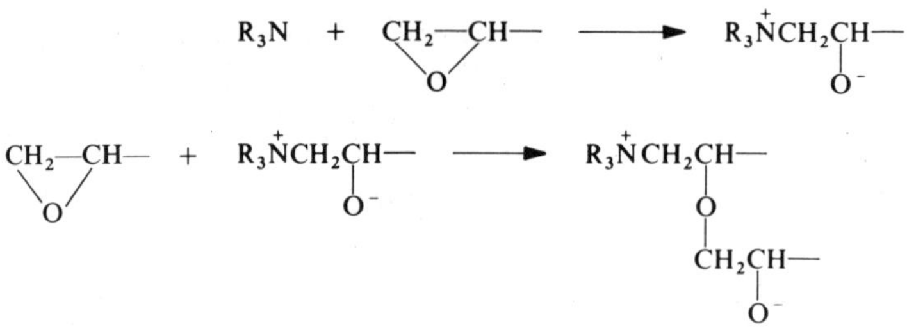

Scheme 2

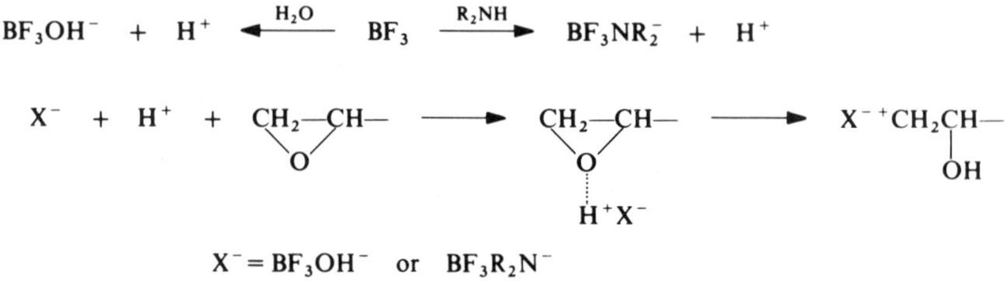

$$R_3N + CH_2\text{—}CH\text{—} \longrightarrow R_3\overset{+}{N}CH_2CH\text{—}$$
$$\underset{O}{\diagup\diagdown}\overset{|}{O^-}$$

$$CH_2\text{—}CH\text{—} + R_3\overset{+}{N}CH_2CH\text{—} \longrightarrow R_3\overset{+}{N}CH_2CH\text{—}$$

with propagation to give $R_3\overset{+}{N}CH_2CH\text{—}$ with pendant $CH_2CH\text{—}O^-$

Scheme 3

$$BF_3OH^- + H^+ \underset{H_2O}{\longleftarrow} BF_3 \overset{R_2NH}{\longrightarrow} BF_3NR_2^- + H^+$$

$$X^- + H^+ + CH_2\text{—}CH\text{—} \longrightarrow CH_2\text{—}CH\text{—} \longrightarrow X^{-\,+}CH_2CH\text{—}$$
$$\overset{|}{H^+X^-}\overset{|}{OH}$$

$$X^- = BF_3OH^- \text{ or } BF_3R_2N^-$$

al.[56] claim to have developed a type of catalyst for the homopolymerization of BADGE resins which initiates a cationic polymerization; the system uses an aluminum complex/silanol catalyst[57] and the silanol is consumed during the polymerization, destroying the ionic species.

Hagnauer and Pearce[18] used DSC to observe the effects of purity on the thermal homopolymerization of TGDDM and their findings are collected in Table 5. They noted that increasing the purity of the resin reduced the T_g of the homopolymerized product and attributed this to a higher level of intramolecular reaction in the purer systems, reducing the crosslink density.

Table 5 DSC Analysis ($5\,°C\,min^{-1}$, N_2)[a]

Sample	Thermal polymerization of TGDDM samples			Curing reaction of TGDDM–DDS (20%) mixtures		
	$-\Delta H$ (cal g^{-1})[b]	T_{exo} (°C)	T_g (°C)	$\Delta H/\Delta H$ (99%)	T_{exo} (°C)	T_g (°C)
H (99%)	314	309	134	1	245	244
G (97%)	316	310	145	0.92	238	240
D (88%)	312	304	156	0.89	233	231
G-120 (78%)	303	291	158	0.83	237	212
MY720 (69%)	279	296	168	0.84	232	202
E (9%)	225	265	181	0.62	228	181

[a] G. L. Hagnauer and P. J. Pearce, *ACS Symp. Ser.*, 1983, **221**, 193.
[b] 1 cal = 4.18 J.

37.3.2 Boron Trifluoride Catalysis

Morgan and Mones[20] have also examined the homopolymerization of TGDDM in the pure and resinous forms (MY720) (see Section 37.2.2), in the presence and absence of boron trifluoride–ethylamine adduct. They used FT–IR to explore this type of polymerization in the temperature range 177–300 °C. In Table 6 the absorption bands are collected, which they used to follow the polymerization process. Figures 2, 3 and 4 summarize the development of the reactions, and, as is apparent from these figures, the rate of epoxide consumption is in the order MY720/BF$_3$–EtNH$_2$ > MY720 > pure TGDDM. Further, they observed that at 177 °C, the cure temperature commonly adopted for TGDDM/DDS/BF$_3$ formulations (see Table 4), the BF$_3$ accelerates the homopolymerization two-fold.

Smith and coworkers[58,59] have also reacted TGDDM in the presence and absence of BF$_3$–EtNH$_2$ and they confirmed the findings of Morgan and Mones regarding the catalytic properties of the boron trifluoride adduct in promoting etherification. Additionally, using FT–IR and ^{19}F NMR they explored the nature of the catalytic species, concluding it to be fluoroboric acid

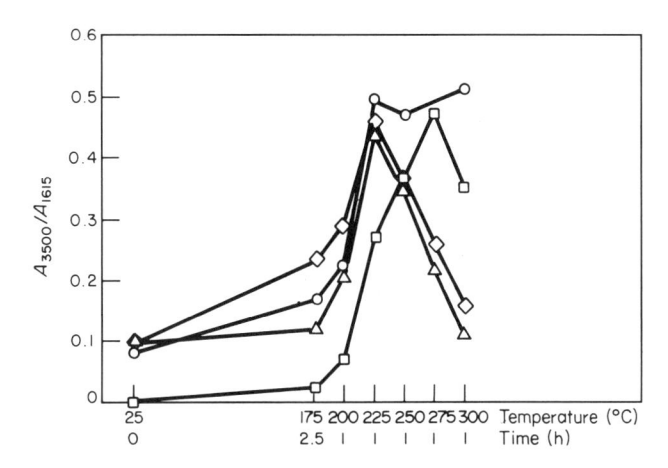

Figure 2 Hydroxyl group IR intensity (A_{3500}/A_{1615}) vs. cure conditions for pure TGDDM (□), MY720 (△), MY720 + 0.4 wt % $BF_3 \cdot NH_2Et$–nonacetone mixed (◇), MY720 + 0.4 wt % $BF_3 \cdot NH_2Et$–acetone mixed (○) (reproduced by permission of John Wiley and Sons from *J. Appl. Polym. Sci.*, 1987, **33**, 999)

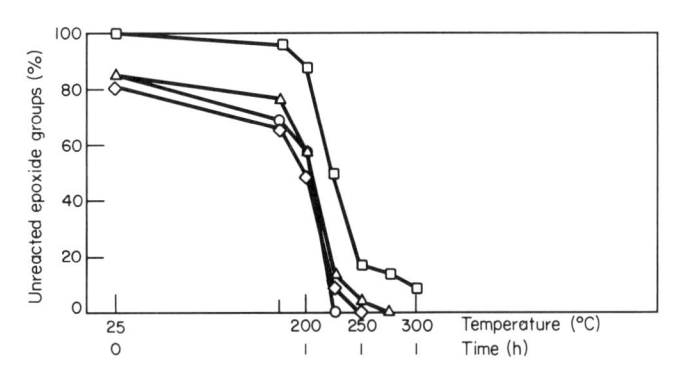

Figure 3 Percent of unreacted epoxide groups vs. cure conditions for pure TGDDM (□), MY720 (△), MY720 + 0.4 wt % $BF_3 \cdot NH_2Et$–nonacetone mixed (◇), MY720 + 0.4 wt% $BF_3 \cdot NH_2Et$–acetone mixed (○) (reproduced by permission of John Wiley and Sons from *J. Appl. Polym. Sci.*, 1987, **33**, 999)

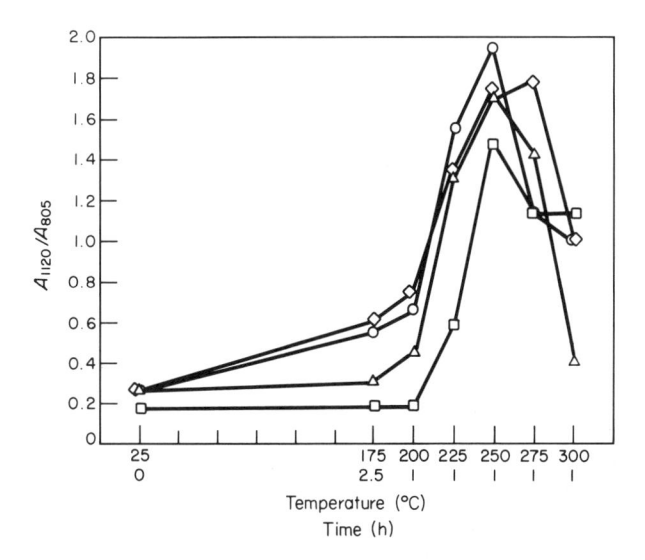

Figure 4 Ether group IR intensity (A_{1120}/A_{805}) vs. cure conditions for pure TGDDM (□), MY720 (△), MY720 + 0.4 wt % $BF_3 \cdot NH_2Et$–nonacetone mixed (◇), MY720 + 0.4 wt% $BF_3 \cdot NH_2Et$–acetone mixed (○) (reproduced by permission of John Wiley and Sons from *J. Appl. Polym. Sci.*, 1987, **33**, 999)

Table 6 FT–IR Absorption Bands Used to Characterize TGDDM Homopolymerization and DDS Cure[a]

Frequency (cm^{-1})	Assignment	Frequency (cm^{-1})	Assignment
906	Epoxide	3410	Secondary amine
1120	Ether	3500	Hydroxyl
1630	Primary amine	815	Phenyl[b]
1720	Carbonyl	1615	Phenyl[b]

[a] R. J. Morgan and E. T. Mones, *J. Appl. Polym. Sci.*, 1987, **33**, 999.
[b] For normalization.

(HBF_4).[60] They further observed fluoroboric acid to be a more effective catalyst than the boron trifluoride adduct.[59] FT–IR has been used to identify the presence of boron compounds including boroxines in cured epoxy resins.[61]

Other information derived from the changes in the IR absorption during the BF_3–MEA cure of TGDDM by Morgan and Mones[20] revealed that the concentrations of carbonyl, ether and hydroxyl groups increased, simultaneously, as the epoxide concentration declined. Further, the increase in carbonyl concentration could be directly correlated to the decrease in epoxide consumption. Above 225 °C the intensities of the ether, hydroxyl and carbonyl bands declined, indicating network degradation.

Four competitive processes (Scheme 4) were proposed by Morgan and Mones[20] to account for the changes observed in the IR absorptions. The first reaction is the basic homopolymerization step, generating chain extension, and which in the impure TGDDM (MY720) is accelerated by the hydroxyl groups. The reaction in impure, as compared to pure, TGDDM is faster (see Figure 2), supporting the contention that the homopolymerization in these systems proceeds, predominantly, through an epoxide–hydroxyl reaction and not through an epoxide–epoxide reaction.

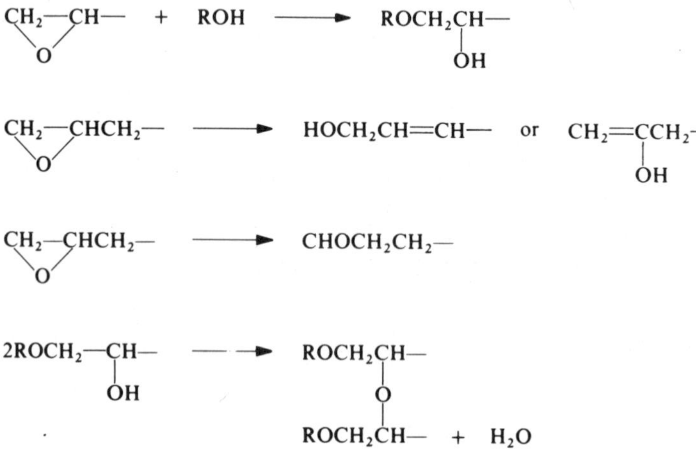

Scheme 4

Morgan and Mones[20] explain the observed increase in hydroxyl concentration by the epoxide isomerization to form allyl alcohols (Scheme 4). This explanation is supported by the observations of other workers.[62,63]

For the isomerization of epoxide to aldehyde, the absence of methyl absorption in the IR led Morgan and Mones[20] to preclude the formation of ketones during the homopolymerization of TGDDM. These workers further explored the chemical structure of the network produced by the homopolymerization of TGDDM by a quantitative analysis of the IR band intensities of the cured resin. They used equations of type (1), where I_{OH_T}, $I_{OH_{23}}$, $I_{E_{23}}$ and $(m.f.)_{TGDDM}$ are the normalized OH band intensities at cure temperature T and at 23 °C, the normalized epoxide band intensity at 23 °C and the mole fraction relative to the pure TGDDM respectively.

$$R_{OH} = \frac{I_{OH_T} - I_{OH_{23}}}{I_{E_{23}} (m.f.)_{TGDDM}} \quad (1)$$

They concluded that only some 25% of the epoxide groups form ether links via the hydroxyl–epoxide chain reaction (Scheme 4), whilst a further 25% isomerize to allylic alcohols, and 50% isomerize to aldehydes.

A yet more detailed analysis produced a description of the spatial distribution in the isomerized TGDDM prior to homopolymerization and this is given in Table 7. In Table 7 the per cent concentration of pair groups is expressed, where (E), (E) = non-isomerized and non-oxidized epoxides; (E), (OH) = epoxy–hydroxyl; (E), (C=O) = epoxy-aldehyde; (OH), (OH) = hydroxyl–hydroxyl; (OH), (C=O) = hydroxyl–aldehyde; and (C=O), (C=O) = aldehyde–aldehyde. Of these combinations of adjacent pairs 36% are (C=O), (C=O) and inactive and do not participate in the homopolymerization or in network-forming processes. Of the 9% (E), (OH) competitive with chain-extending etherification are cyclization reactions forming N-substituted 3-methenyl-4-hydroxymethylmorpholines (10; Scheme 5), and/or 4-substituted 2-hydroxymethyl-1-oxa-4-azacyclohept-3-ene. These types of cyclizations involving epoxide–epoxide[49,64,65] or epoxide–hydroxyl reactions[66] have been reported elsewhere (see also Section 37.4.1, Schemes 12 and 13).

Table 7 Spatial Distribution of Adjacent Pair Groups in Isomerized TGDDM Prior to Homopolymerization[a]

Adjacent pair combinations	Concentration of each pair (%)	Adjacent pair combinations	Concentration of each pair (%)
(E), (E)	9	(OH), (OH)	9
(E), (OH)	9	(OH), (C=O)	18
(E), (C=O)	18	(C=O), (C=O)	36

[a] R. J. Morgan and E. T. Mones, J. Appl. Polym. Sci., 1987, 33, 999.

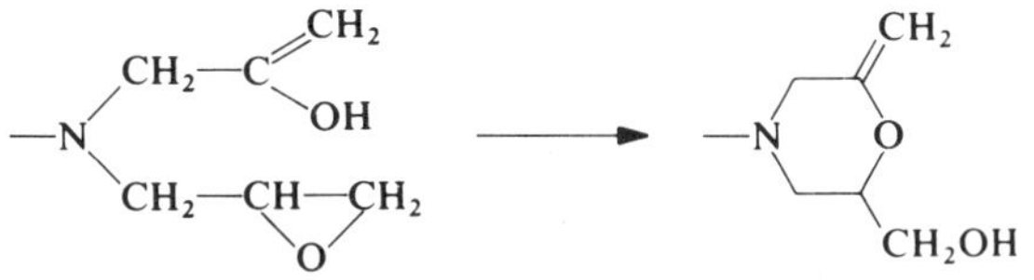

Scheme 5

37.3.3 Catalysis by Tertiary Amines and Related Compounds

A range of tertiary amines and related compounds are available commercially as aids to the control of the cure of epoxy resins. They include 2,4,6-tris(dimethylaminomethyl)phenol, N-benzyl-N,N-dimethylamine and 1,8-diazabicyclo[5.4.0]undec-7-ene (DABCO) and its salts. These compounds are generally used to promote anhydride cures of epoxy resins (see Section 37.4.3), although as strong bases they catalyze the homopolymerization of epoxides (Scheme 2).[67] A group of compounds more commonly used alone to catalyze this homopolymerization, and particularly useful in adhesives technology, are the substituted imidazoles.[68]

Using phenyl glycidyl ether (PGE; 11) as the principal substrate a number of authors have dealt with the polymerization kinetics of epoxides by these catalysts, and the imidazoles studied have included 2-ethyl-4-methylimidazole (EMI),[69-73] 1-methylimidazole (1-MI), 2-methylimidazole (2-MI)[74] and 1-benzyl-2-methylimidazole (BMI) (12).[68]

Farkas and Strom[69] proposed, and the proposal has received support from other researchers,[68,71] that a 1:2 adduct of imidazole:epoxy is formed initially, and it is this adduct which is the catalytic species.

The process of catalysis is probably more complex than this proposal may indicate, for Bressers and Goumans[68] state that they observed multipeaked thermograms in their DSC studies of PGE/BMI reactions and refer to similar observations by Barton et al.,[71] who studied the PGE/EMI system.

However Bressers and Goumans were able to obtain strong evidence for the formation of imidazole/epoxy adducts, for from the early stages of the reaction of PGE and BMI, in a 1:20 molar ratio, they isolated by HPLC two fractions, with retention times of 61 and 69 min, which had molecular weights close to 320 (PGE + BMI = 322). Further the NMR spectra of these two compounds indicated the presence of BMI moieties in each, although the instability of the compounds precluded further analysis. Upon this evidence Bressers and Goumans[68] proposed the polymerization mechanism shown in Scheme 6, in which propagation proceeds anionically from either of the two possible initiating adducts (13) or (14) (cf. Scheme 2).

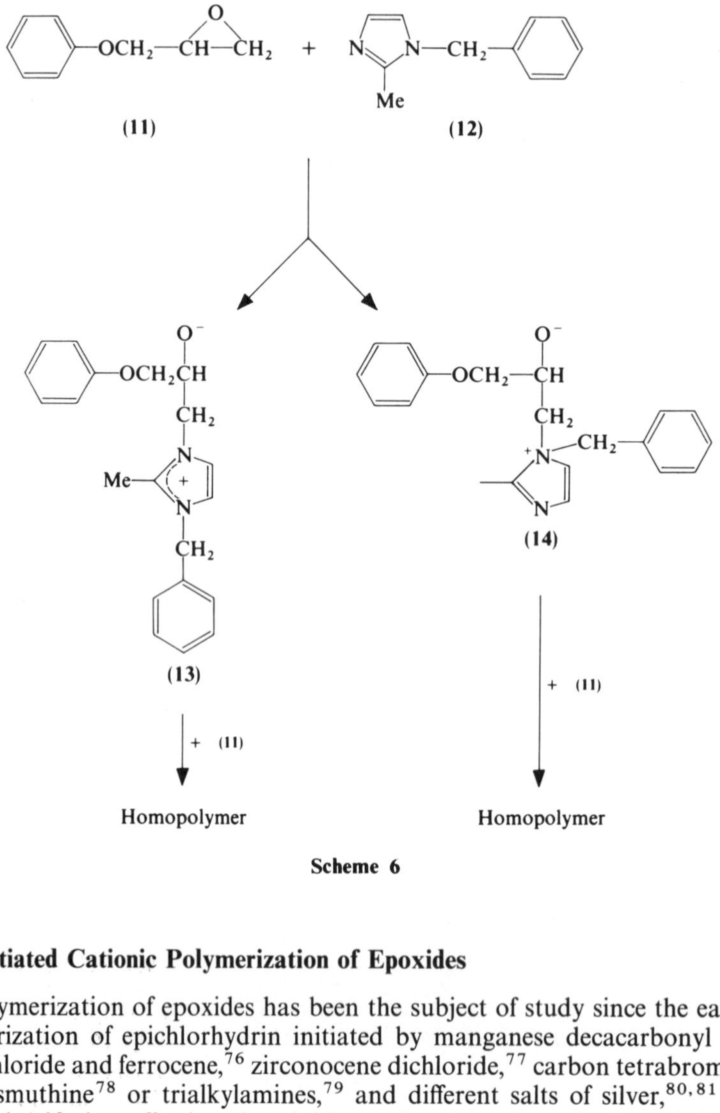

Scheme 6

37.3.4 Photoinitiated Cationic Polymerization of Epoxides

The photopolymerization of epoxides has been the subject of study since the early sixties, when the UV polymerization of epichlorhydrin initiated by manganese decacarbonyl was reported.[75] Titanium tetrachloride and ferrocene,[76] zirconocene dichloride,[77] carbon tetrabromide or iodoform plus triphenylbismuthine[78] or trialkylamines,[79] and different salts of silver,[80,81] and uranium[82] have also been identified as effective photoinitiators for the polymerization of glycidyl ethers and other types of epoxide. Charge transfer complexes formed by, for example, cyclohexene oxide and 1,2,4,5-tetracyanobenzene or pyromellitic dianhydride (1,2,4,5-tetracarboxybenzene dianhydride) polymerize at $-78\,°C$ when irradiated with UV.[83] In another study of a maleic anhydride charge transfer complexed cyclohexene oxide, polymerization was induced by laser irradiation.[84]

The subject of the photoinitiation of epoxides has been reviewed by Crivello,[85-87] and others, with reference to aluminum- and silicon-based catalysts,[88] iron–arene complexes,[89] and cycloaliphatic epoxy polymerization.[90]

The most efficient and effective photoinitiators for the polymerization of epoxides are a range of complex aromatic salts of the Lewis acids of the type discussed in Section 37.3.2. These include arenediazonium salts,[91-98] diaryliodonium salts (15),[99,100] triarylsulfonium salts (16),[87,101-105] dialkylphenacylsulfonium (17),[87,106] dialkyl-4-hydroxyphenylsulfonium salts[107] and trialkylselenonium salts[108] of tetrafluoroboric acid, etc. Their thermal instability has limited the exploitation of the diazonium salts[86] but the other photoinitiators referred to are of utility for the fabrication of, for example, can coatings[86] and photoresists.[85,86]

The virtue of these complex salts is that they do not promote the polymerization of epoxides in the dark, and they are stable even at elevated temperatures (150 °C).[86] Upon irradiation at wave-

$$Ar_2I^+X^- \xrightarrow{h\nu} Ar\cdot + ArI\overset{+}{\cdot}X^-$$

(15)

$$ArI\overset{+}{\cdot}X^- + RH \longrightarrow ArI^+-HX^- + R\cdot$$

$$ArI^+-HX^- \longrightarrow ArI + HX$$

$$X = BF_4, \text{etc.}$$

Scheme 7

$$Ar_3S^+X^- \xrightarrow{h\nu} Ar\cdot + Ar_2S\overset{+}{\cdot}X^-$$

(16)

$$Ar_2S\overset{+}{\cdot}X^- + RH \longrightarrow Ar_2S^+-HX^- + R\cdot$$

$$Ar_2S^+-HX^- \longrightarrow Ar_2S + HX$$

$$X = BF_4, \text{etc.}$$

Scheme 8

lengths of 200–300 nm, they suffer irreversible photolysis involving scission of a carbon–iodine (**15**; Scheme 7), or a carbon–sulfur bond (**16**; Scheme 8).

Reaction of the radical ions so formed with a proton donor, for example a solvent or monomer molecule, releases the protonic acid of the salt (**15**) or (**16**). These acids, HBF_4, HPF_6, $HAsF_6$ and $HSbF_6$, are among the strongest known and are very effective initiators of the curing of epoxy resins by cationic polymerization.

From the observed rates of polymerization of 3,4-epoxycyclohexylmethyl 3′,4′-epoxycyclohexanecarboxylate (**9**) using di-*t*-diphenyliodonium salts as initiators, Crivello and Lam concluded that the strength of these acids was in the order $HBF_4 < HPF_6 < HAsF_6 < HSbF_6$.[109]

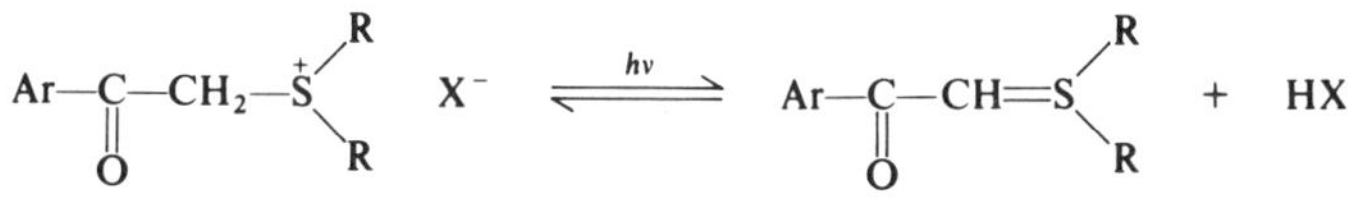

$$Ar-\underset{O}{\overset{}{C}}-CH_2-\overset{+}{S}\underset{R}{\overset{R}{\diagup}} \quad X^- \quad \xrightleftharpoons{h\nu} \quad Ar-\underset{O}{\overset{}{C}}-CH=S\underset{R}{\overset{R}{\diagup}} + HX$$

(17)

$$X = BF_4, \text{etc.}$$

Scheme 9

When exposed to UV light, dialkylphenacylsulfonium salts (**17**) dissociate to form an ylid and a protonic acid (Scheme 9). This process is reversible,[106] and only monomers which are stronger nucleophiles than the ylid will polymerize. The photopolymerization of epoxides is effectively initiated by photolyzed dialkylphenacylsulfonium salts.

Lendwith[110] has proposed an alternative mechanism of photoinitiation to that suggested by Crivello.[87] An important feature of these onium salt photoinitiators is that they are effective catalysts for the polymerization of many types of epoxy compound and resin,[86] and Table 8 lists examples.

Table 8 Photoinitiators for the Cationic Polymerization of Epoxides

Photoinitiator	T_m (°C)	λ_{max} (nm)	Quantum yield (ϕ 313 nm)
Diphenyliodonium BF_4	136	227	—[a]
Di-*t*-butylphenyliodonium AsF_6	169–171	238	0.20[a]
Triphenylsulfonium AsF_6	195–197	230	—[b]
Triphenylselenonium BF_4	183–186	258	0.13[c]

[a] J. V. Crivello and J. H. W. Lam, *Macromolecules*, 1977, **10**, 1307.
[b] J. V. Crivello, *Belg. Pat.* 828 670 (1974).
[c] J. V. Crivello and J. H. W. Lam, *J. Polym. Sci., Polym. Chem. Ed.*, 1980, **18**, 1059.

Using a tack-free cure time to establish the relative reactivities of epoxy compounds towards photoinitiation by diphenyliodonium hexafluoroarsenate, Crivello and Lam[109] identified the following sequence of reactivities: epoxidized dialkenes > epoxidized dialkenes containing acetal and ester linkages > diglycidyl ethers.

Where it is necessary to effect the photopolymerization of epoxides at wavelengths in the longer UV or with visible light, the onium type initiators can be photosensitized to respond to the light of longer wavelengths. For wavelengths longer than 360 nm the following dyes sensitize diaryliodonium salts: acridine yellow, phosphine R, benzoflavin and setoflavin T.[86,111] Polynuclear hydrocarbons, such as perylene, have been found to be effective, in a like manner, with sulfonium salt initiators.[108,112]

DSC techniques have been developed for the rapid and efficient screening of initiators of photopolymerization.[86,113,114] Meijer and Zwiers[115] have compared the use of DSC and phosphorescence spectroscopy to study the photoinitiated polymerization of a number of epoxy resins including 3,4-epoxycyclohexylmethyl 3',4'-epoxycyclohexanecarboylate (9), using di(4-t-butylphenyl)iodonium hexafluoroarsenate as a catalyst.

37.4 REAGENTS FOR STEP-GROWTH POLYMERIZATION

37.4.1 Polymerization with Primary and Secondary Amines

The most commonly used curing agents for epoxides are primary and secondary amines. Scola[116] has reviewed the development of epoxy curing agents, giving particular attention to amines, polyamines and epoxy–amine adducts. The reactions of such amines with epoxides are highly exothermic, which facilitates their study using DSC.[117] Schechter and coworkers[118] have suggested that amine addition to an oxirane ring proceeds with three reactions, as shown in Scheme 10. In such reactions one epoxy ring reacts for each amino proton and so the calculation of the stoichiometry of the reaction of an epoxy resin with an amine would appear simple, provided that the resin's epoxy equivalent weight is known. In practice the 'best cure' formulation may differ from the stoichiometric ratio because of the effect of the epoxy–hydroxyl reaction and relative significance of this reaction *versus* the epoxy–secondary amine reaction. Other factors may also influence the path of the curing reaction, such as the presence of a catalyst. These factors are considered in greater detail in sections below. It is the complexity of the chemistry of the cure of epoxy resins which has necessitated a pragmatic rather than a theoretical approach to the formulation and cure of such systems. Thus formulations and curing procedures are developed by a careful monitoring of the property profiles of the cured resins in relation to their formulations and curing schedules (see Figure 5).[119] There have been many attempts to correlate structure/property relationships in epoxy resins cured with amines.[119–130]

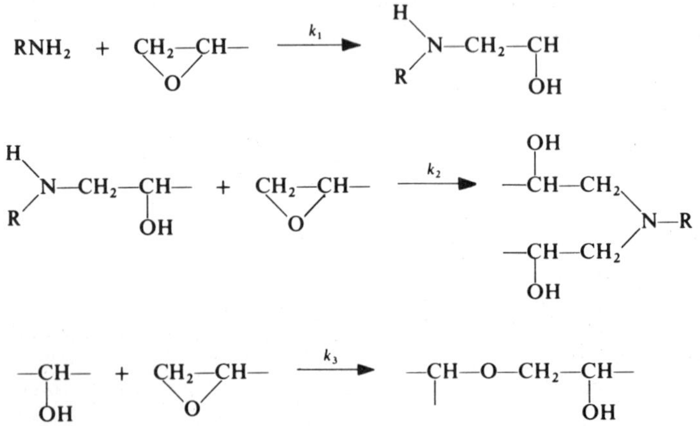

Scheme 10

The aliphatic amines used to cure epoxy resins fall into four main groups:[55] simple aliphatic amines; reactive polyamides; modified aliphatic polyamines; and polyether polyamines.

Aliphatic amines include ethylenediamine (1,2-diaminoethane) and its condensed forms — diethylenetriamine, triethylenetetramine and tetraethylenepentamine. These compounds, which are

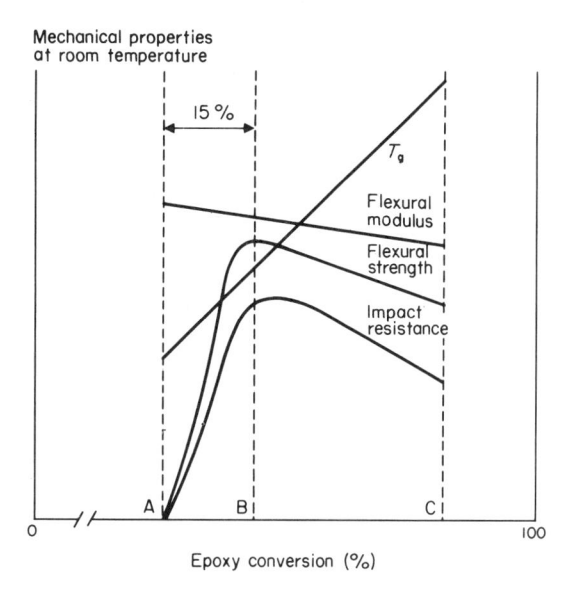

Figure 5 Schematic representation of the development of mechanical properties as a function of the degree of cure (reproduced by permission of Walter de Gruyter, from 'Crosslinked Epoxies', p. 377)

liquids of low viscosity and are freely miscible with, for example, BADGE-type resins, are highly reactive and they are used for ambient temperature cures.

Their high vapour pressures present some handling hazards and modified forms of these and similar amines have been developed. Reactive polyamides, which are formed by the incomplete reaction of polyamines with fatty acids of higher molecular weight, have a much reduced vapour pressure. They retain a high reactivity towards epoxides and are used to impart greater flexibility and toughness to the cured resin.

Modified aliphatic polyamines are also effective for the room temperature cure of epoxy resins. These curing agents are prepared by heating an excess of amine with an epoxy resin. The increase in the molecular weight of the resulting polyamine produces a significant reduction in its vapour pressure.

Polyether polyamines are usually poly(oxypropene) di- and tri-amines. They are used as curing agents for epoxy resins to impart improvements in the toughness and ductility of the cured resin. Towards an amine the epoxy group is monofunctional: however by amine–epoxy reactions (Scheme 10), hydroxyl groups are formed and these may be involved in subsequent network-forming reactions. This involvement will depend upon concentration factors and the rate constants k_1, k_2 and k_3. The relative reactivities of epoxy groups may also be important in determining the route of network formation.

The reactivities of the epoxy groups in BADGE are the same,[131-142] but in other polyepoxy compounds different epoxy groups may have different reactivities.[143]

Since a primary amine has two reactive hydrogens and a secondary amine has one, the ideal ratio for the rate constants k_1 and k_2 is 0.5. Three methods[143] are utilized to determine this ratio: (i) reaction kinetics obtained by monitoring the time change in the concentration of epoxy and/or amino groups; (ii) chromatographic determination of the distribution of free reactants and reaction products; and (iii) critical conversion at the gel point.

The complex time dependence of the concentrations of reactive groups presents formidable obstacles to the kinetic analysis of epoxy–amine systems.[143] Both the amino groups and the hydroxyl groups formed, the proton donors, catalyze the reactions between epoxides and amines, although an amine's catalysis is at least an order of magnitude greater than that of a hydroxyl.[144-148]

For aliphatic amines the etherification reaction does not occur to a significant extent for reactions involving amine and epoxy in stoichiometric ratio (Scheme 10).[118,131,145-152]

HPLC[135] and GPC[152] have been used to obtain quantitative separations of reactants and reaction products. Dusek[143] states that although the data may be somewhat controversial, analysis of the data shows that the ratio of the rate constants k_2/k_1 per mole of amino groups is 0.4–0.6 for aliphatic amines and 0.1–0.3 for aromatic amines. This indicates that in aliphatic amines the primary and secondary amino hydrogens have closely similar reactivities ($k_2/k_1 = 0.5$). Table 9 lists some

Table 9 Ratios of the Primary (k_1) and Secondary (k_2) Rate Constants for the Amine–Epoxy Reactions

Epoxy	Amine	Alcohol	k_1/k_2
BADGE	OMDA	—	0.32[a]
BADGE	DDM	—	0.22[a]
DGA	OMDA	—	0.12[a]
DGA	DDM	—	0.15[a]
TGDDM	DDM	—	0.18–0.26[a]
PGE	BuA	PEO	0.5[b]
PGE	POPPA	PEO	0.2[b]
PGE	POPPA	BuOH	0.3[b]

[a] L. Matejka and K. Dusek, in 'Crosslinked Epoxies', ed. B. Sedlacek and J. Kahovec, Walter de Gruyter, Berlin, 1987, p. 231.
[b] K. Dusek, M. Ilavsky, S. Stokrova, M. Ilavsky and S. Lunak, as above p. 279.
Abbreviations: BADGE, bisphenol A diglycidyl ether; DGA, diglycidylaniline; TGDDM, tetraglycidyldiaminodiphenylmethane; PGE, phenyl glycidyl ether; OMDA, octamethylenediamine; BuA, n-butylamine; BuOH, n-butanol; PEO, poly(ethylene glycol); POPPA, poly(oxypropylene) polyamine.

experimental values for the ratios k_2/k_1,[148,153] determined for an aliphatic amine, a polyether polyamine and an aromatic amine. From these values it is apparent that the substitution effect in aliphatic amines is weak and that it is negative in aromatic amines.[143]

This substitution effect, when highly negative, may be used to form stable epoxy–amine adducts, 2,5-dimethylhexane-2,5-diamine is an example of an amine which will form such adducts.[154] The negative substitution effect delays the onset of gelation on the conversion scale.[143]

In comparison with simple BADGE–aliphatic amine cures, polyether polyamines exhibit retarded cures.[148] This is interpreted as arising from intramolecular hydrogen bonding between secondary amino hydrogens and ether groups and also to a tendency to cyclization.[148]

Kong and coworkers demonstrated that in a polyether polyamine/BADGE cure only amine–epoxy reactions occurred.[155] They further found that the storage modulus, the T_g, and the tensile and yield strengths of the cured resins all pass through a maximum at the point of the stoichiometric ratio of amino hydrogen to epoxy group.[156]

Piperidine (**18**) is a curing agent used widely for epoxy resins and its curing reaction has received attention from a number of research groups.[157–161] A curious feature of the cure is the reported observation[160] that an increase in the cure temperature results in a decline of the T_g of the cured resin resulting. Using DSC Williams et al.[162] have shown that the efficient use of piperidine derives from a cure schedule giving enough time at temperatures below 70–75 °C to ensure that the secondary amine, piperidine, is transformed into a tertiary amine (**19**; Scheme 11), and it is this species which initiates the anionic homopolymerization of the epoxy resin.

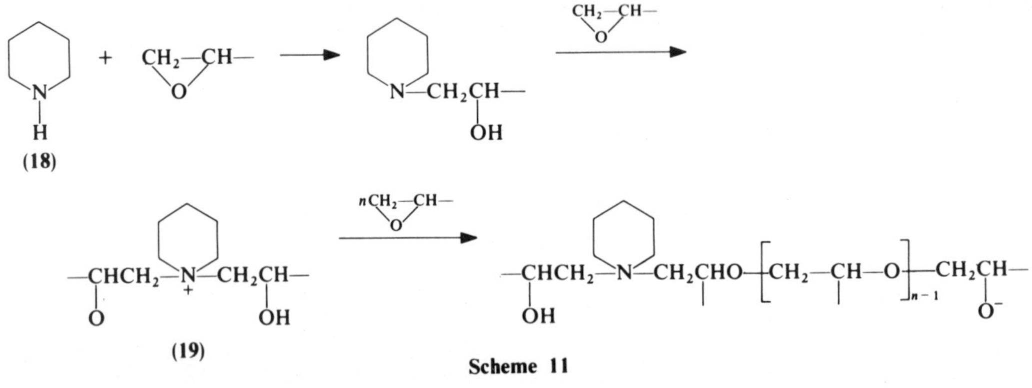

Scheme 11

Conceptually many aromatic amines may be used to cure epoxy resins and typical aromatic amines for this purpose include 4,4'-diaminodiphenylenemethane, 4,4'-diaminodiphenylene sulfone, 3-aminobenzylamine, 3-phenylenediamine and 4,4'-diaminoazodiphenylene.

In practice the combinations studied and reported are restricted and the one most widely used for the matrices of aerospace composites is the combination of TGDDM and 4,4'-diaminodiphenylene sulfone (DDS).[163]

Typically such systems rely not only upon an aromatic diamine to effect the complete cure of the epoxy resins, but they have additions of boron trifluoride adducts as catalysts to complete their cures[163-165] (see also Section 37.3.2). DSC is a useful technique for the study of aromatic amine–epoxy resin cures[163,166-168] and Barton[117] has reviewed the subject. Scheme 10 describes the generalized reactions of primary and secondary amines with epoxy groups and the homopolymerization of the epoxy groups initiated by the hydroxy groups so formed. With aromatic amines it is well established that this last reaction assumes a much greater significance in the epoxy curing process[147,164,169-171] than is the case for aliphatic amines.

This difference arises for a number of reasons, in particular the low basicity of the aromatic amines compared to aliphatic amines reduces their reactivity and this together with their high melting points necessitates high temperature cures.[147] Hagnauer et al.[17] have studied the reaction of TGDDM, in the resinous and the purified forms, with DDS using GPC, HPLC and DSC.

Apicella and coworkers[172] have conducted a related study, giving consideration to the chemorheology of the system.

Using GPC Hagnauer et al.[17] followed the change in composition during cure at 177 °C of a TGDDM/DDS formulation (25% DDS). Figure 6 shows typical GPC chromatograms for reaction times of 0 to 40 min, which are displaced along the ordinate to avoid overlap. In these chromatograms the progressive build-up of the reaction products is evident.

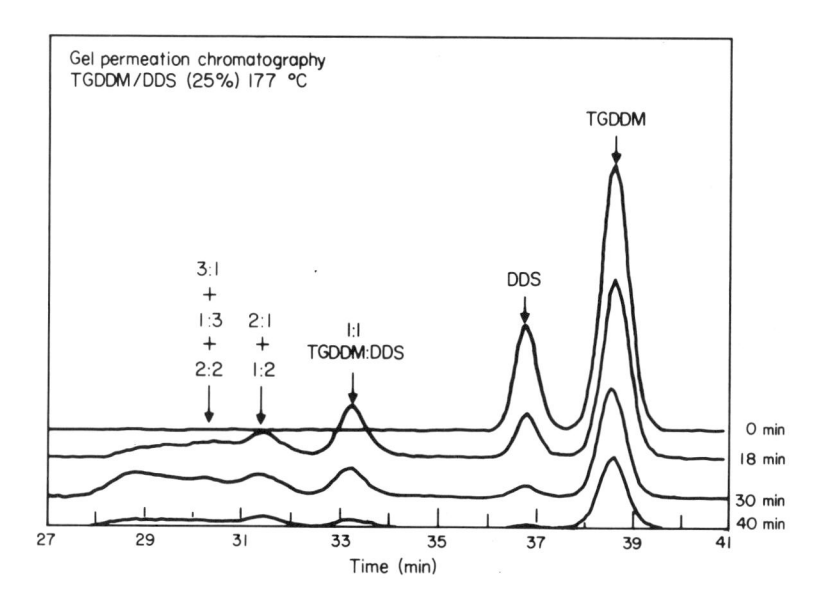

Figure 6 GPC analysis of the TGDDM/DDS (25%) reaction at 177 °C (reproduced by permission of the American Chemical Society, from 'Chemorheology of Thermosetting Polymers', p. 29)

From an analysis of the DSC results[17] it was concluded that for the total heats of reaction of both the commercial and the fractionated TGDDM, for formulations in the range 15 to 37% DDS, the ratio of the total heat of reaction H_t to epoxy concentration was constant, i.e. equation (2) held, where e_{100} is the equivalents of epoxy/100 g. Further it was observed that the heat of reaction for the epoxy–amine reaction was equal to the heat of reaction for the epoxy homopolymerization.

$$\frac{\Delta H_t}{e_{100}} = \text{constant} \qquad (2)$$

Figures 6 and 7(a) and (b) show the progressive development of the reaction during cure. Both figures reveal that after 40 min at 177 °C virtually no DDS remains, although some 20% of TGDDM is unreacted.

This may be explained, for in this formulation the weight of DDS is well below the stoichiometric value (37%). What is more, relatively high concentrations of TGDDM exist well beyond the gel point, which, as is shown in Figure 7, occurs near to 30 min reaction time.

Hagnauer et al.[17] confirmed the assumption that the use of DSC to determine the extent of reaction, a, and the rate of reaction, da, was valid. Figure 7(b) shows plots of the data for these

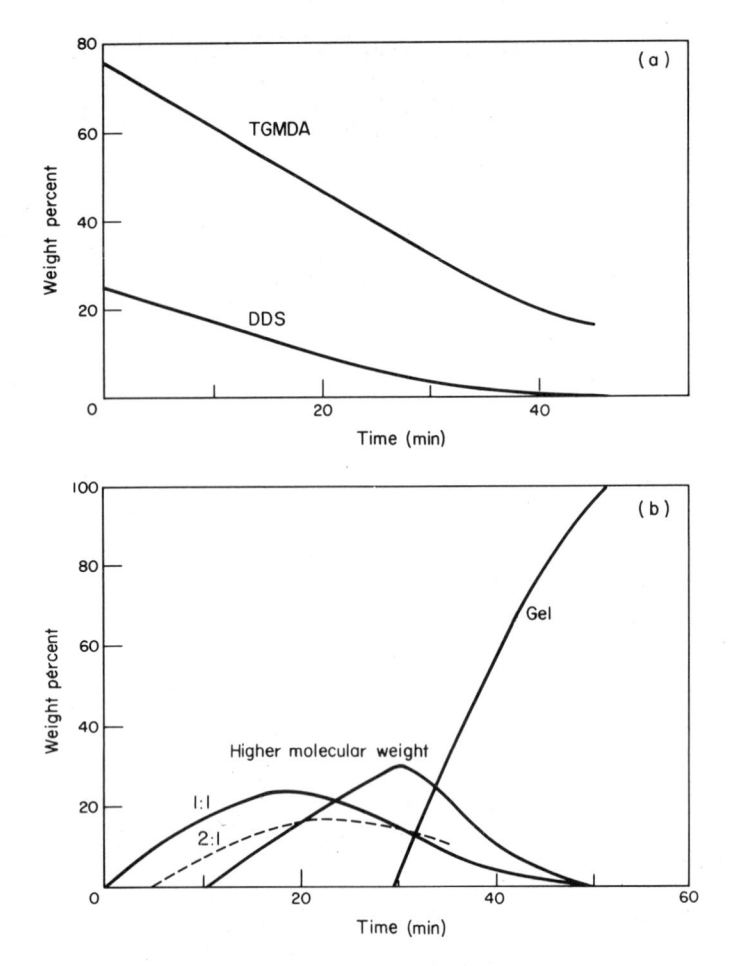

Figure 7 Changes in TGDDM/DDS (25%) with cure time at 177 °C: (a) reactant and (b) reaction product concentrations (reproduced by permission of the American Chemical Society, from 'Chemorheology of Thermosetting Polymers', p. 34)

variables obtained from DSC studies for three formulations (15, 25 and 37% DDS). The effect of the changing stoichiometry upon the rate of reaction is clearly demonstrated.

Kinetic analysis of both GPC and DSC results showed the reaction to be first order with respect to TGDDM and second order with respect to DDS (equation 3).

$$\frac{-d[\text{TGDDM}]}{dt} = k_3[\text{TGDDM}][\text{DDS}]^2 \qquad (3)$$

Hagnauer et al.[17] have also used DSC to study the effects of resin purity on TGDDM/DDS cures. Table 10 lists the rate constants, derived from both the GPC and the DSC results, and the similarity in values is striking. Also noteworthy (see Figure 8) is the increase in the extent of reaction at gel point as the concentration of DDS rises. This may be attributable to a net decrease in the average functionality of this reactant as its concentration increases, which change is due to the

Table 10 GPC Rate Constants for the TGDDM/DDS Reaction at 177 °C[a]

Sample	C_0 (mol kg^{-1})		k_3 (kg^2 mol^{-2} min^{-1})	
	TGDDM	DDS	TGDDM	DDS
TGDDM/DDS (15%)	2.01	0.605	0.0179	0.0179
TGDDM/DDS (25%)	1.78	1.01	0.0184	0.0182
TGDDM/DDS (37%)	1.49	1.49	0.0224	0.0224

[a] G. L. Hagnauer, P. J. Pearce, B. R. LaLiberte and M. E. Roylance, ACS Symp. Ser., 1983, **227**, 33.

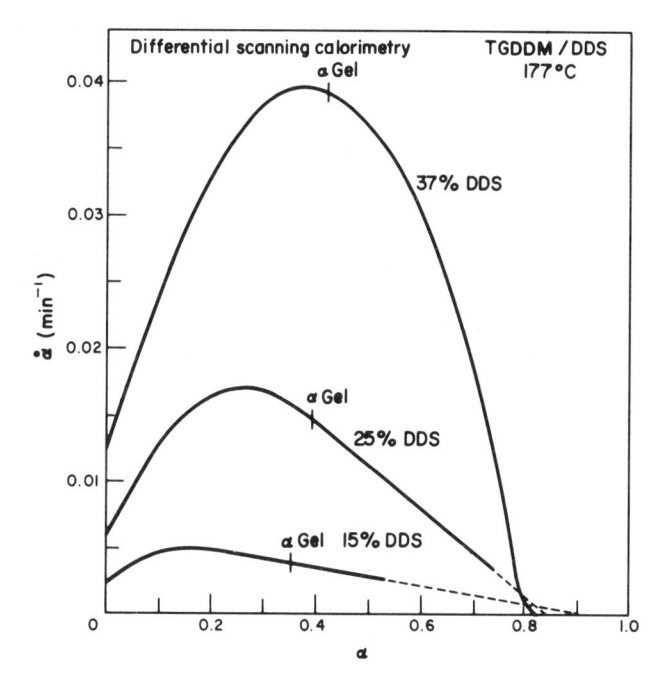

Figure 8 DSC analysis of the rate *vs.* the extent of reaction for different formulations of TGDDM/DDS (reproduced by permission of the American Chemical Society, from 'Chemorheology of Thermosetting Polymers', p. 38)

reducing reactivity of the secondary amine function.[17] The influence which the secondary amino group has upon the cure of TGDDM/DDS formulations has been explored by introducing N,N'-dimethyl-4,4'-diaminodiphenylene sulfone (DMDDS) into the system.[173] The cures were followed at 153 and 177 °C using DSC and FT–IR. In particular FT–IR was used to monitor the rate of decline in the epoxy group absorption at 905 cm^{-1} and the rate of formation of hydroxyls, by following the absorption at 3400 cm^{-1}, during the cure of TGDDM with DDS at 177 °C.

To reduce the complexity of the reaction products in the TGDDM/DDS-type cure, others[174] studied the reaction of N,N-diglycidylaniline (7) and N-methylaniline; N-(3-phenoxy-2-hydroxypropyl)aniline and N-methyl-N-(3-phenoxy-2-hydroxypropylaniline) were synthesized as comparators. Three different reaction paths were considered: addition, etherification, and cyclization. Using a combination of HPLC, MS, NMR and IR, the rate constants for the reactions of (7) with aniline and N-methylaniline were determined.[175]

The ratio of the rate constants k_2/k_1, for reactions for primary and secondary amine–epoxy reactions (Scheme 10), is indicative of the substitution effect of the epoxy groups in diglycidylaniline. For a random reaction this ratio is 0.5 (*cf.* aliphatic primary and secondary amines), Dusek *et al.*[174] report that their results revealed this ratio to be 3 to 16 for DGA, indicating a strongly positive substitution effect. They attribute the increase in reactivity of the epoxy group in the N-methyl-

Scheme 12

aniline–diglycidylaniline reaction product (**20**; Scheme 12), to intramolecular catalysis by its hydroxyl group.

With regard to cyclization, the cyclic compound 1,5-diphenyl-3,7-dihydroxy-1,5-diazacyclooctane (**22**) was isolated from a reaction mixture of DGA and aniline (Scheme 13).[175] Figure 9 shows the kinetic course of reaction between aniline and DGA at 100 °C; evident is the growth in concentration of (**22**), together with the alternating step-growth of DGA and aniline linear oligomers.

Scheme 13

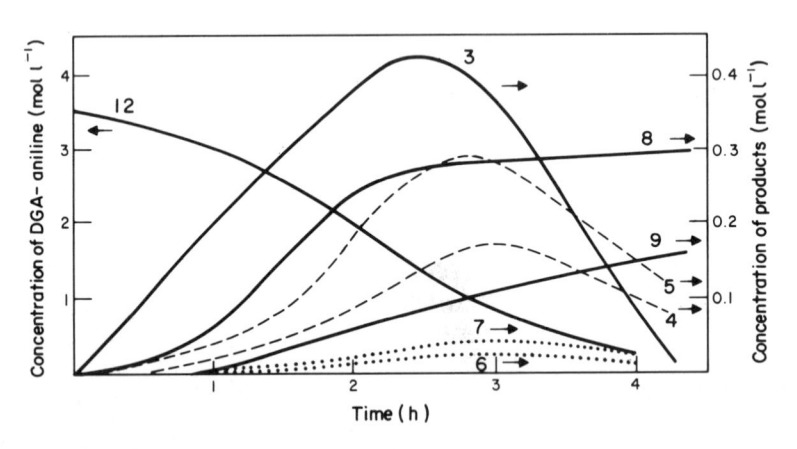

Figure 9 Kinetic course of the equimolar reaction DGA–aniline, $T = 100\,°C$: 1 DGA; 2 aniline; 3 product DGA–A; 4,5 --- isomers A–DGA–A; 6,7 . . . isomers DGA–A–DGA; 8 cycle 22; 9 --- cycle with $t_e = 14.4$ min (reproduced by permission of Walter de Gruyter, from 'Crosslinked Epoxies' p. 249)

Isolated from amongst the reaction products of N-methylaniline and DGA was 4-phenyl-2-(N-methylanilinomethyl)-1-oxa-4-azacycloheptanol (**21**; Scheme 12).[174]

In their studies of DGA with N-ethylaniline, Ancelle and coworkers[176] identified N-phenyl-3-(N-ethylanilinomethyl)-4-hydroxymethylmorpholine. In their model compound studies these workers used the spectroscopic techniques FT–IR and ^{13}C CO/MAS NMR.

The identification of cyclic species, e.g. (**22**), amongst the products of amine–epoxy reactions is supportive of the findings of other authors[17,20,169] that epoxy–hydroxyl group reactions leading to intramolecular cyclizations favouring the formation of morpholine rings may predominate in the later stages of the cure of TGDDM–DDS resin systems (see Scheme 5). Gupta et al.,[177] using ESR spin trapping techniques to study epoxy–amine curing reactions agreed with the conclusion that epoxy–hydroxyl reactions dominate epoxy–epoxy reactions, but found the process to be intermolecular.

As was indicated earlier, it is usual to include a boron trifluoride adduct in resin formulations of the TGDDM–DDS type and such systems have been studied in some detail.[20,163,178-180] FT–IR was used to study the curing reactions of the prepreg reactions of the commercial systems shown in Table 4.[20] Figure 10 shows the observed disappearance of epoxy and primary amino groups and the corresponding formation of secondary amino groups ($I_{NH:SO_2}$, using the SO_2 band of DDS for

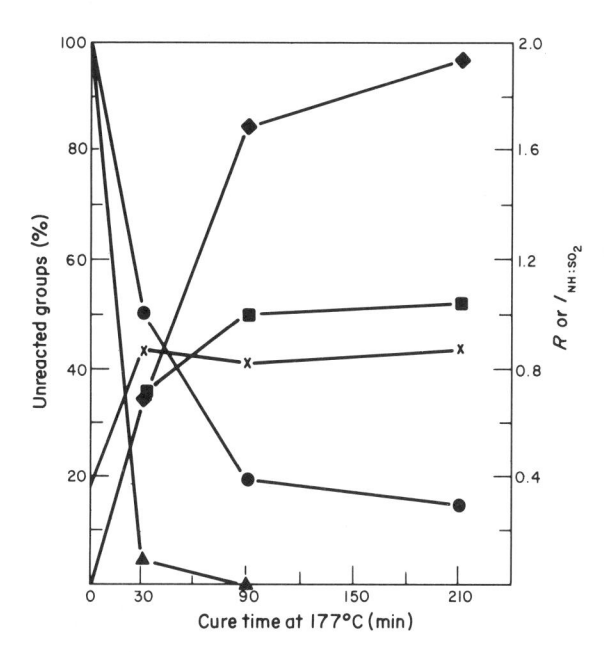

Figure 10 Epoxides (●) and PAs (▲) (% of unreacted groups), R_{OH} (■), R_{ether} (◆), $I_{NH:SO_2}$ (X) vs. cure time at 177 °C for TGDDM–DDS (25 wt % DDS; 0.4% wt % BF$_3$·NH$_2$Et) epoxy (reproduced by permission of John Wiley and Sons from *J. Appl. Polym. Sci.*, 1987, **33**, 999)

normalization purposes), hydroxyl groups (R_{OH}), and ether linkages (R_{ether}) during the cure, at 177 °C, of a TGDDM–DDS formulation (25% DDS and 0.4% BF$_3$·NH$_2$Et). It is apparent that the rate of reaction of both epoxy and primary amino groups is rapid with approximately 45% of the former and 95% of the latter having reacted in 30 min.

To this stage the course of the reaction is similar to the uncatalyzed TGDDM–DDS reactions.[168,176,177] Secondary amine concentration rises during this initial period and then remains constant, indicating that secondary amine–epoxy reactions take no part in network formation in the later stages of the reaction. The secondary amine–epoxy reaction for other aromatic amines, such as 4,4′-diaminodiphenylenemethane (DDM) have been reported to be an order of magnitude slower than the primary amine reaction[181] and other workers[165,177,182] have produced evidence for incomplete secondary amine–epoxy reaction in TGDDM–DDM cures. The incompleteness of the secondary amine–epoxy reaction is attributed in both the BF$_3$-catalyzed[20] and the uncatalyzed[173] TGDDM–DDS cures to constraints imposed by the secondary amino groups being bound into the network.

Morgan and Mones[20] concluded that the epoxy–hydroxyl etherification reaction accounted for some 50% of the total epoxy. Further, they demonstrated, using a range of BF$_3$·EtNH$_2$ concentrations from 0 to 5.0% at 177 °C for 2.5 h, that increasing the BF$_3$ concentration increased the extent of etherification (see Figure 11) through the epoxy–hydroxyl reaction.[164]

At higher temperatures (200–225 °C) it is suggested that the principal catalytic species of boron trifluoride catalyzed cures, BF$_4^-$ EtNH$_3^+$, is deactivated.[183] It is further proposed that the scatter in results for epoxide consumption is attributable to the variability in the chemical composition of BF$_3$·EtNH$_2$ and in its distribution in the resin.[20]

Morgan and Mones[20] generated model network structures for four TGDDM–DDS formulations (0, 12.5, 25 and 35% DDS) using a random number sequence for the reaction of each glycidyl group in a block of 25 TGDDM molecules, for primary amine–epoxy, secondary amine–epoxy and epoxy–hydroxyl reactions. All the networks are complex and possess numerous defects with interconnectivity decreasing at lower DDS percentages.

The T_g values and mechanical response of TGDDM–DDS formulations have been observed to decline rapidly below 20% DDS.[128] BF$_3$·EtNH$_2$ enhances the TGDDM–DDS cure, producing networks with fewer defects and this is reflected in their superior mechanical properties at high temperature, and their improved ductility, which is an order of magnitude greater than the uncatalyzed formulations.[20]

Macosko and coworkers,[184] Dusek *et al.*,[185] and Chu and Seferis[186] have modelled the development of epoxy networks during cure. Gillham *et al.*[187] used the relationships between the extents

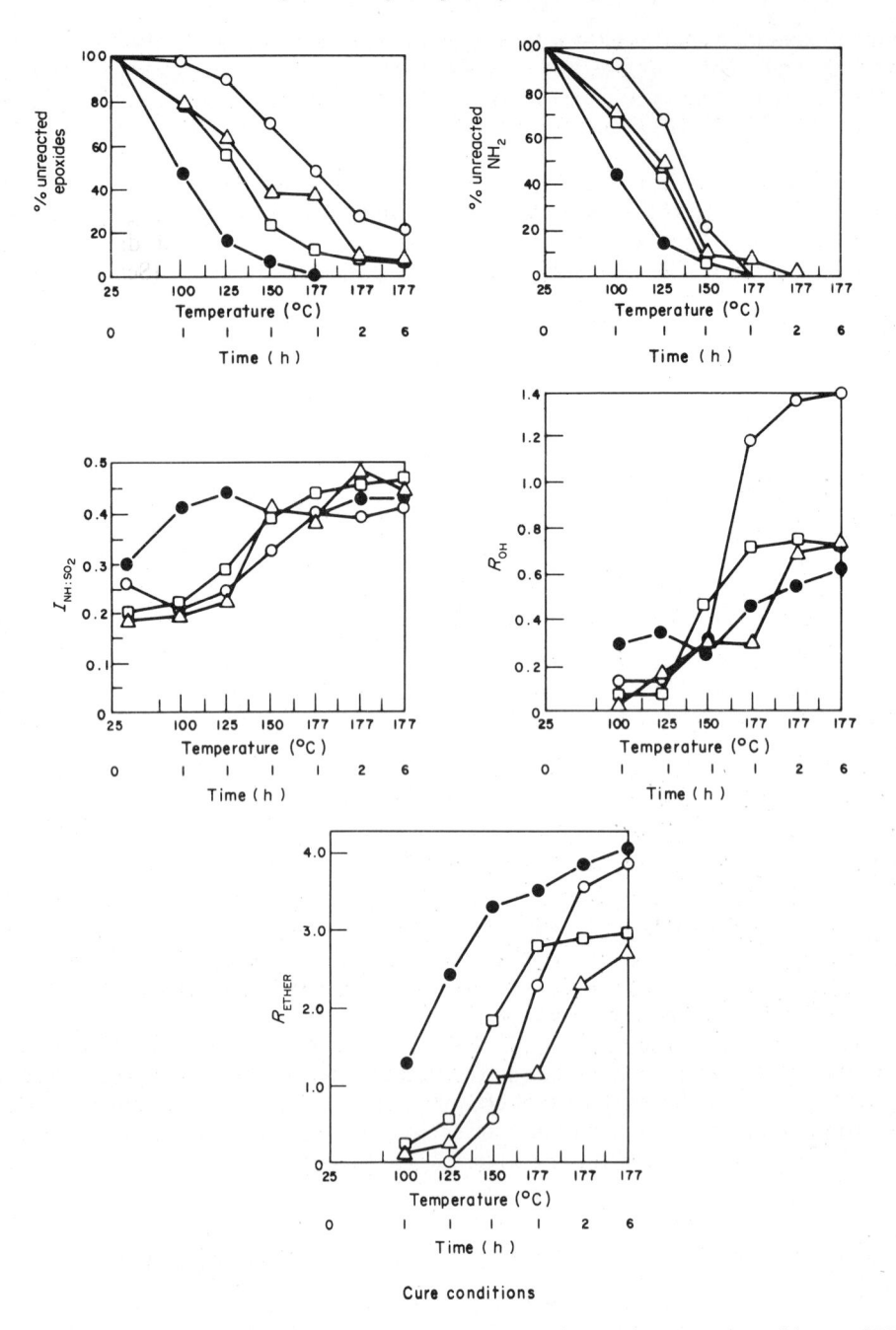

Figure 11 Percent of unreacted epoxide and PA groups, $I_{NH:SO_2}$, R_{ether} and R_{OH} values vs. cure conditions and $BF_3 \cdot NH_2Et$ concentration for TGDDM–DDS (25 wt % DDS) epoxy (reproduced by permission of John Wiley and Sons from *J. Appl. Polym. Sci.*, 1987, **33**, 999)

of conversion at gelation and at vitrification and the isothermal cure temperature to form the basis of a theoretical model of the time–temperature–transformation (TTT) cure diagram, in which the times to gelation and to vitrification during isothermal cure *versus* temperature are predicted.

37.4.2 Dicyandiamide Cure of Epoxides

Cyanoguanidine or dicyandiamide (dicy; **23**) is widely used in one-pack epoxy formulations for prepregs and electrical laminates, for epoxy powder coatings and for adhesives for high temperature cures.[188] Its latency as a curing agent resides in its high melting point (206 °C) and in its low solubility in epoxy resins at ambient temperatures. It is usual to employ a tertiary amine, *e.g.*

benzyldimethylamine, as an accelerator for dicy cures when curing may be effected at temperatures between 100 and 200 °C. Substituted ureas such as 3-(*p*-chlorophenyl)-1,1-dimethylurea (Monuron)[189] may be used as sources of tertiary amines. In the absence of a catalyst curing is only rapid at around 175 °C. HPLC and IR have been used to identify the constituents of prepregs containing dicy.[190,191]

The curing reaction of dicyandiamide and epoxy resins has been extensively studied by a number of workers,[46,192,193] and most recently by Pascault *et al.*[194] In another thorough examination of the curing process Zahir[195] studied the reactions of both cyanamide and dicyandiamide. Both Pascault[194] and Zahir[195] identified dicy as a latent cyanamide (**24**) donor (Scheme 14).

$$H_2NC=NCN \quad \longrightarrow \quad 2NH_2CN$$
$$\underset{NH_2}{|}$$

(**23**)   (**24**)

Scheme 14

The use of cyanamide for prepregs has been recommended.[196] Unlike dicyandiamide, cyanamide is soluble in epoxy resins, but like dicyandiamide it reacts with epoxides to form 2-amino-5-substituted oxazolines (**25**) and the tautomeric 2-imino-5-substituted oxazolidones (**26**; Scheme 15). The aminooxazolines and iminooxazolidones so formed, being basic, will themselves catalyze the cure of epoxy resins.[194]

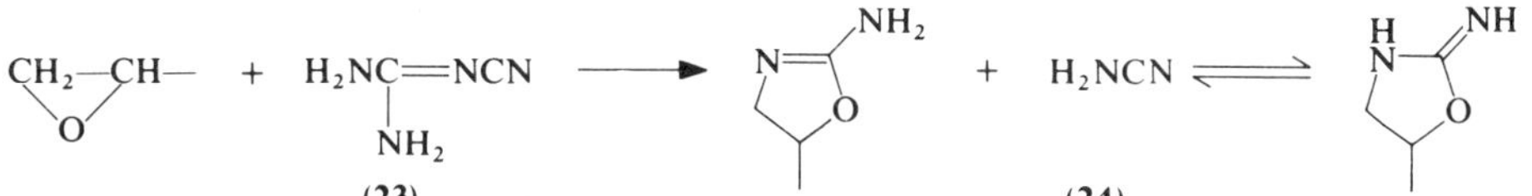

Scheme 15

Zahir[195] summarized the reactions of the dicyandiamide cure and in brief the major network-forming processes are heterocyclization (Scheme 15), and epoxy–amine reactions (Scheme 10). For these latter cyanamide has two and dicy four active hydrogens, although Pascault *et al.*[194] observed dicy to react as if it possessed three reactive hydrogens.

Both Pascault *et al.*,[194] using SEC, HPLC, FT–IR and ^{1}H NMR, and Zahir,[195] using GPC, MS, IR and ^{1}H and ^{13}C NMR, conducted their initial studies of the dicyandiamide/epoxy reaction using phenyl glycidyl ether and this enabled them to separate and identify the reaction products and the range of products which these two groups of workers isolated and identified and these serve to emphasize the significance of reaction Schemes 15 and 16.

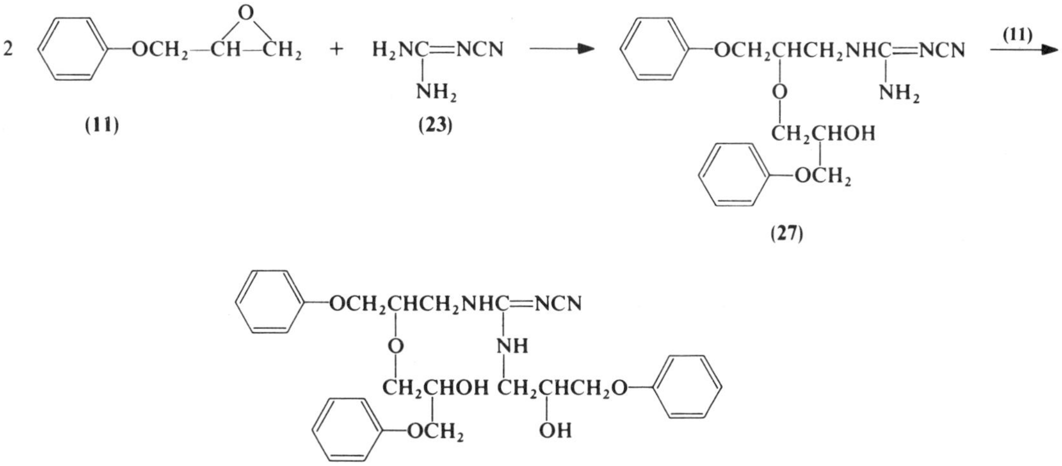

Scheme 16

It is to be noted that whereas the products identified by Pascault et al.[194] were mainly cyanoguanidines (27) and (28), i.e. they were derived from undissociated dicyandiamide, the products reported by Zahir[195] were derived from cyanamide.

Saunders et al.,[197] who followed the change in imino, amido and hydroxyl proton concentrations using IR and NMR, during the reaction of dicyandiamide with phenyl glycidyl ether in the presence of a tertiary amine, also proposed the formation of cyanoguanidines (27) and (28), combined with, as a minor product, polyether formed by a base-catalyzed polymerization.

For the reaction of dicy with BADGE, the reaction path is dependent on the stoichiometry, the catalyst and its concentration and the cure temperature. Plots of the degree of cure as a function of the isothermal cure time at 100, 140 and 160 °C revealed that the position of the curves shifted to shorter times at higher cure temperatures. From DSC thermograms it was also apparent that the heat of reaction of dicy/BADGE mixtures was dependent upon the mix ratio. For 1.0 it was found to be 440 J g^{-1}, and for 0.6 it was 365 J g^{-1}.[194]

For this combination, when cured at high temperature (160 °C) in the presence of 1% BDMA, the epoxy–amine and the epoxy–epoxy reactions compete and in the initial stage the lower dicy ratios favour homopolymerization to a DP of two to three.[194] Amino alcohols formed by the epoxy–amine reaction are a source of protons for the epoxy homopolymerization but fully substituted amino groups prevail, especially at lower dicy ratios and high temperatures, indicating that the reaction of secondary amines (Scheme 10) is not suppressed.[194]

Important changes occur in the IR absorption spectra of the resin formulations during curing, and these have been used to identify the developments in the structure of the network.[194] One such change is the decline in the absorption at 2180 cm^{-1}, indicating that the nitrile group of the dicyandiamide is involved directly in network-forming processes. This decline is accompanied by an increase in absorption at 1740 cm^{-1}, and this has been interpreted[46,197] as being due to the formation of amidic structures. Other absorption bands indicative of structural development in the curing resin referred to include 1680–1560 cm^{-1} (imino), 1120 cm^{-1} (ether) and 1740 cm^{-1} (carbonyl).[194]

This structural development and the rearrangements, involving the formation of linear and cyclic N,N'-substituted ureas, identified[195,197] as occurring in the final stages of cure, may account for the observation[198] that the mechanical strength of an adhesive system based on a commercial formulation of BADGE and dicy increased late in the curing process.

37.4.3 Diacid Anhydrides

After the amines, acid anhydrides constitute the next most commonly used reagent for curing epoxy resins. They are low cost hardeners, which produce polyester-type structures in the cured resin. With glycidyl ether resins they react slowly unless an accelerator is used, and so cure schedules may be long and/or at high temperature. Acid anhydrides exhibit greater reactivity towards cycloaliphatic epoxy resins.[4]

For optimum properties from anhydride cures, which are especially suited to electrical applications,[4] Arnold[56] recommends an anhydride:epoxy ratio of 1:1 for tertiary amine catalyzed systems and 0.85:1 for uncatalyzed formulations.

Commonly used acid anhydrides include phthalic anhydride, tetrahydrophthalic anhydride, hexahydrophthalic anhydride, pyromellitic dianhydride, 3,3,4,4-tetracarboxybenzophenonedianhydride, chlorendic anhydride, methylnadic anhydride, trimellitic anhydride, etc.

In the reaction of an epoxide and an anhydride the anhydride ring-opening step is dependent upon a hydroxyl compound. Tanaka and Kakiuchi[199,200] reported that the rate of reaction was proportional to the concentrations of not only the epoxide and the anhydride, but also of the hydroxyl. Tanaka and Kakiuchi interpreted these dependencies as indicative of the involvement of a termolecular transition state (29) as the rate-determining step (Scheme 17). A particular feature of this mechanism is the simultaneous esterification of both carboxyl groups in the anhydride so that free carboxyl groups are not generated during the resinification.

Fisch and Hofmann[201-203] proposed an alternative reaction process in which esterification of the two carboxyls is sequential (Scheme 18), and etherification is competitive. In this reaction scheme, upon the formation of the monoester, a carboxyl group is liberated. Stevens[204] considered the Tanaka and Kakiuchi mechanism (Scheme 17) to be of low probability due to the large activation entropy required to form the transition state and he proposed an alternative mechanism which was indistinguishable, kinetically, from Scheme 18.

Using IR involving a full spectral peak assignment of the molecular vibrational modes in the

Scheme 17

Scheme 18

range 400 to 4000 cm^{-1}, Stevens identified distinctions between the reaction kinetics of Tanaka and Kakiuchi's and Fisch and Hofmann's schemes.

In his study of the cure of a bisphenol A based resin Stevens explored the influence of both low epoxide/hydroxyl[204] and high epoxide/hydroxyl[205] ratios on the cure mechanism. He concluded that his observations of the cure of a low epoxide/hydroxyl ratio bisphenol A epoxy resin–anhydride system with phthalic anhydride were consistent with Fisch and Hofmann's scheme and also supported free hydroxyl group limited colloidal-type reaction.[204] Stevens' findings of the reaction kinetics, derived from IR studies, are summarized in Figure 12, which he interpreted as the

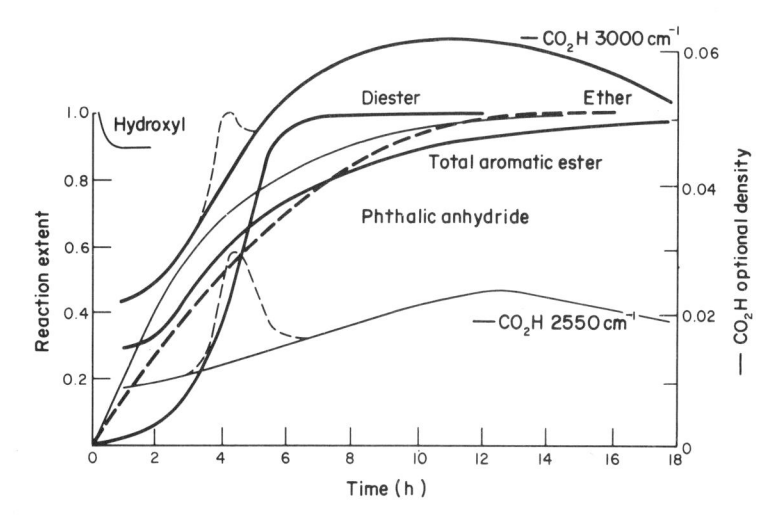

Figure 12 Summary of the reaction kinetics of BADGE (low epoxy/hydroxy)/phthalic anhydride during reaction at 125 °C (reproduced by permission of John Wiley and Sons from *J. Appl. Polym. Sci.*, 1981, **26**, 4259)

resinification proceeding *via* the step-growth reactions of esterification and etherification, which give rise to polyester, polyether and polyester–polyether segments in the resin's structure. A reservoir of carboxyl groups is generated by the reaction of hydroxyl groups with phthalic anhydride. This occurs rapidly, initially, and may peak (dotted line in Figure 12) before settling to a 'steady state' or 'plateau' level. Overall the reactions of phthalic anhydride and the formation of aromatic ester were found to conform to first-order kinetics, confirming Fisch and Hofmann's findings,[201] whilst the etherification reaction continued with half-order kinetics.

Stevens[204] concluded that with the low epoxide/hydroxyl ratio resin about 90% of the anhydride was fully engaged in the resin's structure after cure, the remaining 10% being retained as monoester and was unreacted.

Stevens'[205] findings for the second type of system, which had a high epoxy/hydroxyl ratio resin cured with a phthalic anhydride–tetrahydrophthalic mixture, were broadly similar to those above. The cure programme, however, differed, for it involved a two-step schedule, a gelation stage at 80 °C and a postcure at 120 °C. The study served to emphasize the important role of the hydroxyl group in

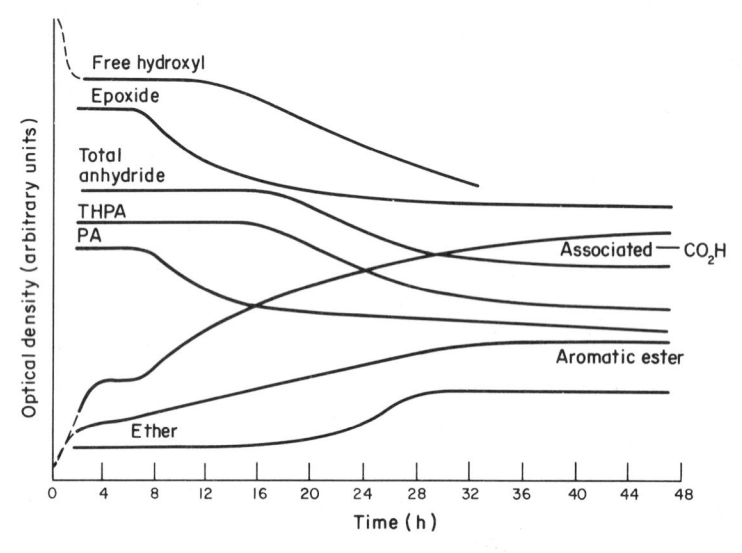

Figure 13 Summary of chemical group optical density behaviour of gelation at 80 °C of BADGE (high epoxy/hydroxyl)/phthalic and tetrahydrophthalic anhydrides (reproduced by permission of John Wiley and Sons from *J. Appl. Polym. Sci.*, 1981, **26**, 4279)

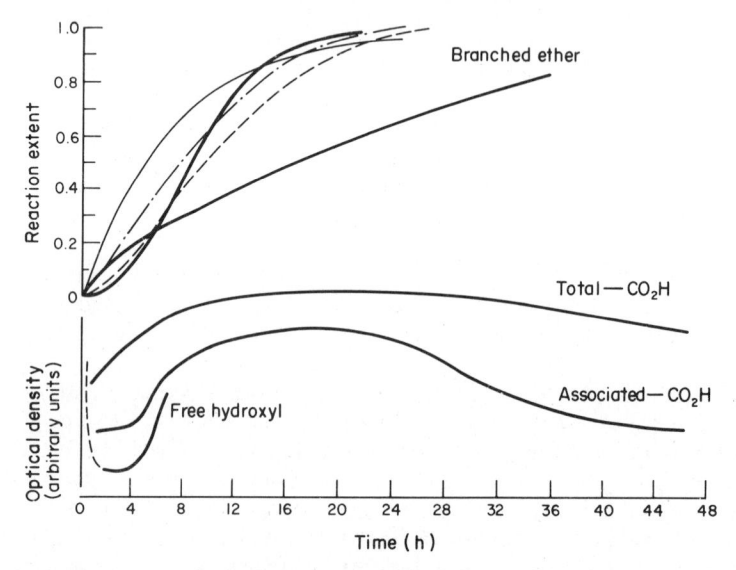

Figure 14 Chemical group reaction extent and optical density behaviour summary of postcure at 120 °C after 18 h gelation at 80 °C: (—) aromatic ester; (—·—) THPA; (---) PA; (—) epoxide (reproduced by permission of John Wiley and Sons from *J. Appl. Polym. Sci.*, 1981, **26**, 4279)

the cure chemistry, for this system had a long induction period with little distinction being observed between the reactivity of the two anhydrides at the gelation stage.

Figure 13 summarizes the chemical group behaviour derived from IR measurements on the reacting resin held at 80 °C, the gelation temperature, and Figure 14 shows similar data for the postcuring stage at 120 °C, which followed 18 h at 80 °C.

The anhydrides' reactions were observed to be first order, as had been previously identified for the phthalic anhydride–low epoxy/hydroxyl resin system.

Stevens refers to the inhomogeneous nature of the reaction[205] and draws attention to the parallel between the cure and the polyesterification reactions of alkyd resins. Solomon et al.[206] have invoked the formation of microgel particles formed from micelles at whose surface rapid polyesterification occurs.

Most anhydride cures of epoxy resins involve a catalyst, especially for low temperature cures. The catalyst facilitates the ring opening of the anhydride, reducing the dependence of the reactivity of the system on the presence of the hydroxyl group, so that long high temperature cures and the polyetherification processes which become dominant at high temperatures are avoided. The catalysts for anhydride–epoxy reactions are usually either Lewis acids or Lewis bases and their use has been reviewed.[207] For Lewis bases such as tertiary amines, Scheme 19 has been proposed.[208,209] In the first step association of the catalyst and anhydride generates a zwitterion (31), which then by nucleophilic attack through the carboxylate anion on the 2-carbon of the oxirane ring of the resin produces the monoester anion (32). The formation of this anion has been identified as the rate-determining step.[209]

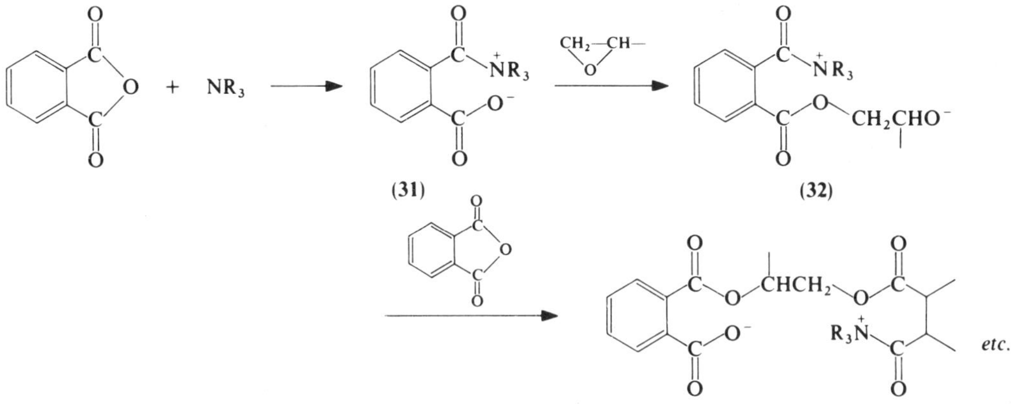

Scheme 19

Matejka and coworkers[210] have deduced the formation of quaternary ammonium ion (see Scheme 20), which is the initiating species, from model compound studies. This ion reacts immediately with the anhydride to form an ester and to liberate the carboxylate ion. Confirmation of this mechanism comes from a statistical treatment of network formation applied to catalyzed anhydride cures,[211] and from DSC and IR studies of the resinification process.[212]

The resinification process is then an alternation of alkoxide anion attack on anhydride and carboxylate anion attack on epoxide, generating a polyester. Lewis bases do not alter this mechanism.[207]

An alternative mechanism[213] involves either solvolysis of the anhydride or reaction of the tertiary amine with a hydroxylated cocatalyst such as water, alcohol or carboxylic acid. Both hydroxyls from the epoxy resin and carboxyls from the hydrolyzed anhydride are likely to be present in resin formulations.

In studying the anhydride cure of a diglycidyl ester, i.e. using hexahydrophthalic anhydride (1,2-dicarboxycyclohexane anhydride) to cure hexahydrophthalic acid diglycidyl ester (bis(1-propenyl) cyclohexane-1,2-dicarboxylate), Steinmann[214] used DSC, FT–IR and the chemical analysis of functional groups to explore the process of network formation. He observed two-stage processes for both the uncatalyzed and the catalyzed anhydride cures of epoxides. The catalysts used in this study were 0.25% of either benzyldimethylamine (BDMA) or 1-methylimidazole (1-MI). From DSC the initial phase of the reaction had a reaction order of 0.5, whereas for the later phase of the reaction it was 1.0, with the transition from one to the other occurring at an extent of reaction close to 60%. The changes which were observed in the major IR absorption bands for the system, viz. anhydride at

1787 cm^{-1}, ester at 1738 cm^{-1} and epoxy at 4545 cm^{-1}, when interpreted kinetically, also exhibited the change of order from 0.5 to 1.0 in the vicinity of 60% reaction.

Ether bands were not observed in the uncatalyzed reaction. Figure 15 shows the changes with time in epoxide, anhydride and carboxyl concentrations for the BDMA-catalyzed cure at 100 °C.[214] Steinman's findings[214] for the uncatalyzed reaction were consistent with Fisch and Hofmann's mechanism.[201-203] The change in reaction order seems to occur close to the gel point. Both Stevens,[204] and Antoon and Koenig[215] have discussed the possibility that this type of reaction becomes diffusion controlled.

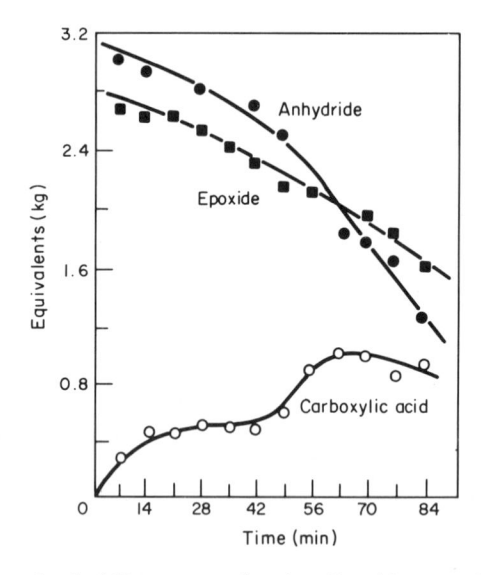

Figure 15 Concentration changes of anhydride, epoxy and carboxylic acid groups for benzyldimethylamine-catalyzed reaction at 100 °C (reproduced by permission of Walter de Gruyter from 'Crosslinked Epoxies', p. 125)

With regard to the catalyzed reaction, if this follows Matejka and coworkers' proposal (Scheme 20),[210] then the amine is irreversibly bound to the epoxide and does not participate in the cure. Alternatively it may be regenerated. Dusek et al.[211] have pointed out that if the critical conversion point is independent of catalyst, then it must be regenerated.

Scheme 20

Steinmann[214] observed in his study of the anhydride cure of an epoxy ester that for the BMDA-catalyzed reaction the critical conversion was concentration dependent, whilst for 1-MI it was independent of concentration. This distinction may be due to the difference in the basicities of these two catalysts, BMDA has a pK_a of 9.53 and 1-MI of 6.95. Other catalysts of value include imidazoles, *e.g.* 2-methylimidazole,[73] 2-phenylimidazole,[216] 2-ethyl-4-methylimidazole,[217] 4-methyl-2-phenylimidazole,[218] 2-heptadylimidazole[219] and 1-(2-hydroxy-3-phenoxypropylimidazole),[220] and other compounds, including quaternary phosphonium salts,[221] phenylphosphonium tetraphenylborate,[222] and aluminum porphyrin coupled quaternary phosphonium bromide.[223]

Triphenylphosphine is used as a catalyst for the rubber modification reactions of epoxy resins.[224] The favoured terminal group of acrylonitrile–butadiene copolymers is carboxyl. Reaction Scheme 21 shows the probable sequence of the formation of rubber-to-epoxy ester bonding.

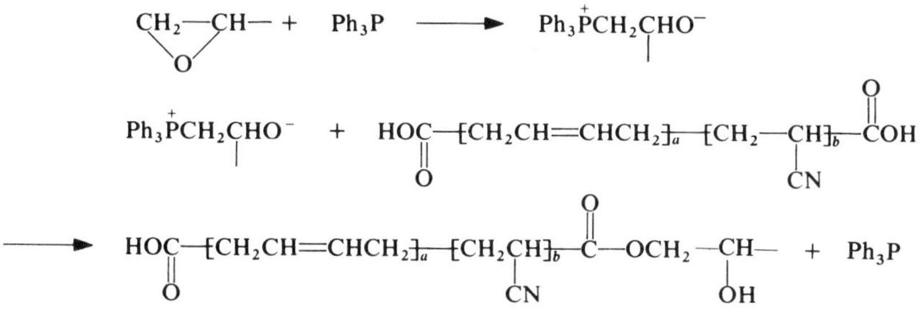

Scheme 21

37.4.4 Isocyanates

Increasing attention is being given to the cure of epoxy resins with isocyanates and the subject has been reviewed.[225] The reaction of an epoxy with an isocyanate forms an oxazolidone (33). In early reports of this reaction attention was directed at the formation of linear polymers, poly(oxazolidone)s (Scheme 22),[226-233] although gels were encountered. Quaternary ammonium halides were found to be effective catalysts for the promotion of oxazolidone formation by Speranza and Peppel,[226] who proposed a mechanism of ring opening for the epoxy involving nucleophilic attack by the halide ion, usually a bromide ion, on the epoxide (Scheme 23). The resulting oxygen anion then adds to the carbonyl carbon of the isocyanate. Ring closure may then produce either an oxazolidone (34), or an N-substituted imino-1,3-dioxolane (35).[230] The facile rearrangement of N-

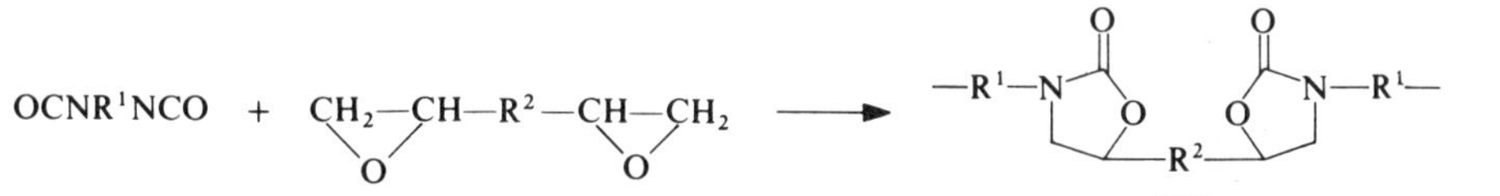

Scheme 22

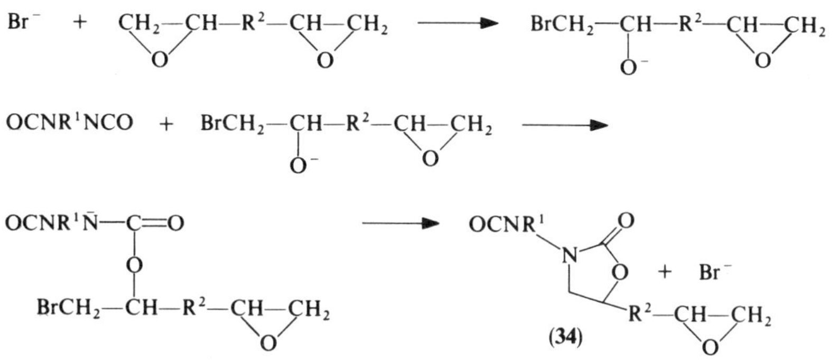

Scheme 23

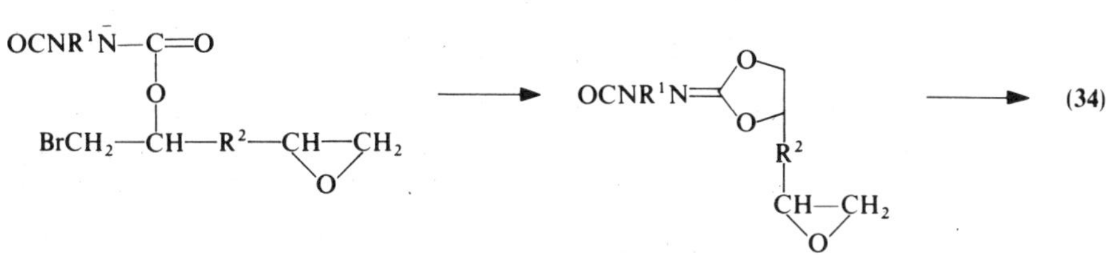

Scheme 24

phenyl-2-imino-1,3-dioxolane to the corresponding 2-oxazolidone under the conditions of reaction has been demonstrated (Scheme 24).[230]

The initial involvement of the catalyst in generating a species for nucleophilic attack on the epoxide ring has received support from Frisch et al.,[234,235] who examined the activity and selectivity of catalysts with respect to the model reaction of phenyl glycidyl ether and phenyl isocyanate to form phenylphenoxymethyloxazolidones. These workers studied the following catalysts: lithium chloride, acetyltrimethylammonium bromide and chloride, aluminum trichloride and Lewis acid/base complex catalysts formed by combining aluminum trichloride with N-methylpyrrolidone, hexamethylphosphoric triamide, tetrahydrofuran or triphenylphosphine oxide, in molar ratio, acid to base, in the range 1:2.54 to 1:10. The Lewis base component of the catalyst increased both the solubility of the Lewis acid in the epoxide and its nucleophilicity, which influenced the first step of the formation of the oxazolidone.

An added feature of the catalysts reported by Frisch[234,235] was that some, particularly the aluminum trichloride–triphenylphosphine oxide (molar ratio 1:2.54), produced a 100% yield of 3-phenyl-5-phenoxymethyloxazolid-2-one (**36**) by reaction of phenyl isocyanate and phenyl glycidyl ether (Scheme 25). From HPLC it appeared that the concentration of the isomeric 3-phenyl-4-phenoxymethyloxazolid-2-one (**37**) was very small. In the synthesis of polyoxazolidones this level of selectivity is not usually achieved, and mixed 4-substituted and 5-substituted oxazolidone structures are observed.[229]

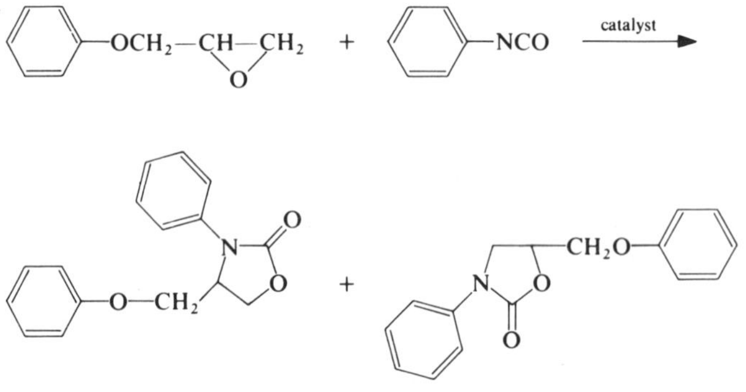

Scheme 25

Crosslinked resins from the cure of epoxides with isocyanates derive from a combination of reactions involving the formation of oxazolidone, urethane, isocyanurate and carbodiimide groups. The absence or presence and proportion of these groupings is dependent upon the stoichiometry and structure of the reactants, the choice of catalyst and the catalyst concentration and the reaction conditions. Kinjo et al.[236] have described the chemistry of thermosetting resins, the ISOX (*is*ocyanurate/*ox*azolidone) resins. They prepared ISOX resins from a range of epoxy resins by reacting them with diphenylenemethane 4,4'-diisocyanate (MDI) in equivalent ratios of epoxy:isocyanate from 1:1 to 1:10, using 1-cyano-2-ethyl-4-methylimidazole as a catalyst, and sequential curing temperatures of 80, 130, 180 and 200 °C.

Clarke[237-239] reports oxazolidone-modified epoxy resins prepared from epoxy novolacs by reaction with either tolylene 2,4-diisocyanate (TDI) (as its dimethyl carbamate)[240] or isocyanuric acid (ICA). Figure 16 shows the idealized structures of these resins.

PIO (*p*oly*i*socyanurate/*o*xazolidone) resins have also been reported.[241,242]

DSC and IR have been used to follow the course of formation of these types of resin.[243,244] The IR absorption frequencies of the more important chemical groupings to be found in these systems during their formation include 2270 cm^{-1} (isocyanate), 1740 cm^{-1} (oxazolidone), 1710 cm^{-1} (isocyanurate) and 910 cm^{-1} (epoxy). The changes in the FT–IR absorption ratios of important bands during the reaction of an equimolar mixture of BADGE and MDI cured at 150 °C exhibit a rapid decline in isocyanate absorption associated with the equally rapid rise in isocyanurate and a slower rise in oxazolidone absorptions. Beyond 2 h reaction time the isocyanurate absorption appeared to decline.[225]

The DTA[245] and DSC[243] thermograms of epoxy/MDI reactions are complex and often exhibit two or more peaks. Figure 17 shows the DSC thermogram of a model compound study of the reaction between MDI and phenyl glycidyl ether, with tetraethylammonium bromide as catalyst, in which the first peak is attributed to trimerization of isocyanate and the second to oxazolidation.[243]

Figure 16 Comparison of bridges joining coupled molecules in modified epoxy novolac: (A) tolylene diisocyanate, (B) isocyanuric acid (reproduced by permission of the American Chemical Society from 'Rubber Modified Thermoset Resins', p. 55)

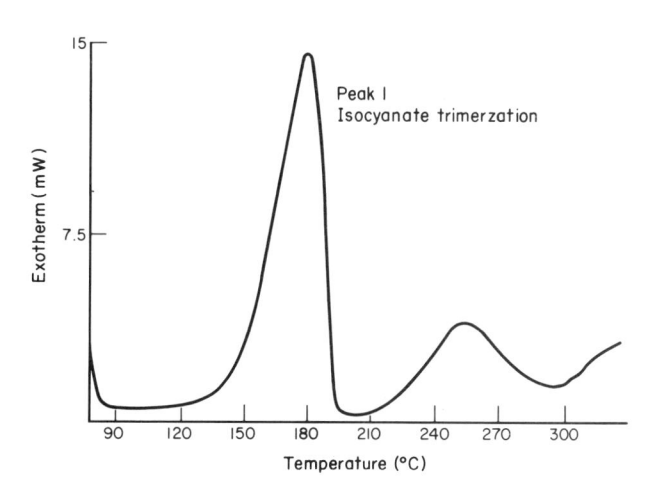

Figure 17 DSC thermogram of the PGE/MDI reaction with tetraethylammonium bromide as catalyst (reproduced by permission of Plenum Press from 'Advances in Polymer Synthesis', p. 251)

37.4.5 Alcohols and Thiols

As has been indicated in previous sections (37.3.2, 37.4.1, *etc.*), epoxy–hydroxyl reactions, *e.g.* Scheme 10, play a role in the resinification of epoxides, but their significance is determined by the precise details of the resin system. Shechter and Wynstra[246] studied the reaction of PGE and isopropanol, in the presence of potassium hydroxide and benzyldimethylamine; they observed the reaction to be slow. The reaction of phenols with epoxides proceeds *via* the phenoxide ion and the process is catalyzed by strong bases,[207] including quaternary derivatives of benzyldimethylamine.[246] The epoxide–alcohol reaction is used to prepare certain glycidyl ethers.[55]

Related to the alcohol–epoxy reactions are those of epoxides with thiols; however, in the latter case the reaction is much more facile and formulations, involving combinations of, for example, BADGE, triethylenediamine, 2,4,6-tri(dimethylaminomethyl)phenol and 2,2-dimercaptodiethyl ether, could be cured at temperatures down to $-25\,°C$.[247]

The reactions of epoxides and thiols have been little reported, but Klemm *et al.*[248] have synthesized linear polymers by reaction of diepoxides, *e.g.* BADGE, with dithiols which is indicative of the latter's greater reactivity towards epoxides, when compared with hydroxyls.

Allied to the thiols are the episulfides and these also afford rapid cures for epoxides when used in combination with, for example, reactive polyamides.[249,250]

37.5 REFERENCES

1. H. L. Lee and K. Neville, 'Handbook of Epoxy Resins', McGraw-Hill, New York, 1982.
2. C. A. May and Y. Tanaka (eds.), 'Epoxy Resins, Chemistry and Technology', Dekker, New York, 1973.
3. H. Dannenberg and C. A. May, in 'Treatise on Adhesion and Adhesives', ed. R. L. Patrick, Dekker, New York, 1973, vol. 3, chap. 2.
4. M. T. Goosey, in 'Plastics for Electronics', ed. M. T. Goosey, Elsevier Applied Science, London, 1985, chap. 4.
5. C. A. May, *ACS Symp. Ser.*, 1985, **285**, 557.
6. R. R. Jay, *Anal. Chem.*, 1964, **36**, 667.
7. F. W. Ainger, J. Brettle, I. Dix and M. T. Goosey, *ACS Symp. Ser.*, 1984, **242**, 313.
8. K. L. Hawthorne and F. C. Henson, *ACS Symp. Ser.*, 1983, **221**, 135.
9. A. L. Cupples, H. Lee and D. G. Stoffey, *Adv. Chem. Ser.*, 1973, **92**, 173.
10. C. S. Chen, B. J. Bulkin and E. M. Pearce, *J. Appl. Polym. Sci.*, 1982, **27**, 1177.
11. F. R. Dannenberg, L. H. Sharpe and H. Schonhorn, *J. Polym. Sci., Part B*, 1965, **3**, 1021.
12. J. R. Griffith, J. G. O'Rear and S. A. Reines, *Chem. Technol.*, 1972, **2**, 311.
13. W. J. Gilwee and Z. Nir, *ACS Symp. Ser.*, 1984, **208**, 321.
14. J. J. King, R. N. Castonguay and J. P. Zizzi, *Natl. SAMPE Tech. Conf.*, 1981, **13**, 53.
15. G. L. Hagnauer and P. J. Pearce, *Org. Coat. Plast. Chem.*, 1982, **46**, 580.
16. M. R. Thoseby, B. Dobinson and C. H. Butt, *Br. Polym. J.* 1986, **18**, 286.
17. G. L. Hagnauer, P. J. Pearce, B. R. Laliberte and M. E. Roylance, *ACS Symp. Ser.*, 1983, **227**, 25.
18. G. L. Hagnauer and P. J. Pearce, *ACS Symp. Ser.*, 1983, **221**, 193.
19. I. Dobas, S. Lunak, S. Podzimek, M. Mach and V. Spacek, in 'Crosslinked Epoxies', ed. B. Sedlacek and J. Kahovec, Walter de Gruyter, Berlin, 1987, p. 81.
20. R. J. Morgan and E. J. Mones, *J. Appl. Polym. Sci.*, 1987, **33**, 999.
21. G. F. Pezdirtz and G. Guidotti, *Appl. Polym. Symp.*, 1973, **22**, 101.
22. M. Sennett, *Polym. Mater. Sci. Eng.*, 1987, **56**, 371.
23. I. R. Gelling, *Rubber Chem. Technol.*, 1985, **58**, 86.
24. A. S. Burhans, *Insulation/Circuits*, Jan. 1979, 73.
25. P. Nowak and M. Saure, *Kunststoffe*, 1964, **54**, 557.
26. F. X. Ventrice, *Mod. Plast.*, 1967, **45**, 201.
27. J. Clark and D. D. Perrin, *Q. Rev., Chem. Soc.*, 1964, **18**, 295.
28. H. K. Hall, *J. Am. Chem. Soc.*, 1957, **79**, 5441.
29. W. A. Henderson and C. A. Streuli, *J. Am. Chem. Soc.*, 1960, **82**, 5791.
30. J. J. Harris and S. C. Temin, *J. Appl. Polym. Sci.*, 1966, **10**, 523.
31. A. S. Sherrand and A. A. Krupnik, *J. Appl. Polym. Sci.*, 1965, **9**, 2707.
32. H. L. Vincent, P. E. Oppliger and C. L. Frye, *Pap. Meet. Am. Chem. Soc., Div. Org. Coat. Plast. Chem.*, 1968, **28**, 504.
33. M. Lidarik and M. Mach, *Plaste Kautsch.*, 1979, **26**, 20.
34. J. D. B. Smith and A. I. Bennett, *Conf. Proc. IEEE Symp. Insulation, Boston*, 1980, 136.
35. I. Soos and J. Biztriczky, *Kautsch. Gummi, Kunstst.*, 1981, **34**, 353.
36. N. C. Paul, P. J. Pearce, D. H. Richards and D. Thompson, *Adhesion*, 1979, **3**, 65.
37. F. Greenspan, C. Johnston and M. Reich, *Mod. Plast.*, 1959, **37**, 142.
38. T. F. Mika, in ref. 2, p. 293.
39. D. M. Simmons and J. H. Verbanc, *J. Polym. Sci.*, 1960, **44**, 303.
40. L. H. Lee, *J. Polym. Sci., Part A*, 1965, **3**, 859.
41. C. C. Price and D. D. Carmelite, *J. Am. Chem. Soc.*, 1966, **88**, 4039.
42. M. F. Sorokin, L. G. Shode and A. B. Shteinpress, *Vysokomol. Soedin., Ser. A*, 1972, **14**, 309.
43. P. V. Sidyakin, *Vysokomol. Soedin., Ser. A*, 1972, **14**, 979.
44. Y. Tanaka and C. A. May, in ref. 2, chap. 2.
45. K. Dusek and M. Bleha, *J. Polym. Sci., Polym. Chem. Ed.*, 1977, **15**, 2393.
46. N. S. Schneider, J. F. Sprouse, G. L. Hagnauer and J. K. Gillham, *Polym. Eng. Sci.*, 1979, **91**, 301.
47. R. J. Morgan, *J. Appl. Polym. Sci.*, 1979, **23**, 2711.
48. U. M. Bokhare and K. S. Gandhi, *J. Polym. Sci., Poly. Chem. Ed.*, 1980, **18**, 857.
49. M. F. Sorokin, L. G. Shode, L. A. Drobrovinskii and G. V. Onosov, *Vysokomol. Soedin., Ser. A*, 1972, **24**, 20.
50. D. A. Whiting and D. E. Kline, *J. Appl. Polym. Sci.*, 1974, **18**, 1043.
51. P. J. Pearce, R. G. Davidson and C. E. M. Morris, *J. Appl. Polym. Sci.*, 1981, **26**, 2363.
52. E. S. Narracott, *Br. Plast.*, 1953, **26**, 120.
53. R. J. Arnold, *Mod. Plast.*, 1964, **41**, 149.
54. P. H. Plesch, 'The Chemistry of Cationic Polymerisation', Pergamon, Oxford, 1963.
55. R. S. Bauer, *ACS Symp. Ser.*, 1985, **285**, 931.
56. S. Hayase, S. Suzuki, M. Wada, Y. Inoue and H. Mitui, *J. Appl. Polym. Sci.*, 1984, **29**, 296.
57. S. Hayase, T. Ito, S. Suzuki and M. Wada, *J. Polym. Sci., Polym. Chem. Ed.*, 1981, **19**, 2185.
58. R. E. Smith, F. N. Larsen and C. L. Long, *J. Appl. Polym. Sci.*, 1984, **29**, 3697.
59. R. E. Smith, F. N. Larsen and C. L. Long, *J. Appl. Polym. Sci.*, 1984, **29**, 3713.
60. R. E. Smith and C. H. Smith, *J. Appl. Polym. Sci.*, 1986, **31**, 929.
61. D. G. Davidson and G. I. Mathys, *J. Appl. Polym. Sci.*, 1983, **28**, 1957.
62. T. L. Jacobs, D. Dankner and H. R. Dankner, *J. Am. Chem. Soc.*, 1958, **80**, 864.
63. D. M. Burness, *J. Org. Chem.*, 1964, **29**, 1862.
64. G. B. Butler, *Polym. Prepr., Am. Chem. Soc., Div. Polym. Chem.*, 1967, **8**, 35.

65. C. A. Byrne, N. S. Schneider and H. Lee, *Coat. Plast. Prepr. Pap. Meet. (Am. Chem. Soc., Div. Org. Coat. Plast. Chem.)*, 1981, **44**, 96.
66. W. Fisch and W. Hofmann, *Plast. Technol.*, 1961, **7**, 28.
67. R. F. Fischer, *J. Polym. Sci.*, 1954, **23**, 497.
68. H. J. L. Bressers and L. Goumans, in ref. 19, p. 223.
69. A. Farkas and P. Strohm, *J. Appl. Polym. Sci.*, 1968, **12**, 159.
70. T. Dearlove, *J. Appl. Polym. Sci.*, 1970, **14**, 1615.
71. J. Barton and P. Shepherd, *Makromol Chem.*, 1975, **176**, 919.
72. F. Riccardi, W. Romanchick and M. M. Jouille, *J. Appl. Polym. Sci.*, 1983, **21**, 1475.
73. F. Riccardi, M. M. Jouille, W. A. Romanchick and A. A. Griscavage, *J. Polym. Sci., Polym. Lett. Ed.*, 1982, **20**, 127.
74. J. Berger and F. Lohse, *J. Appl. Polym. Sci.*, 1985, **30**, 531.
75. V. W. Strohmeier and C. Barbeau, *Macromol. Chem.*, 1965, **81**, 86.
76. K. Kaeriyama, *J. Polym. Sci., Polym. Chem. Ed.*, 1976, **14**, 1547.
77. K. Kaeriyama, *Macromol. Chem.*, 1972, **153**, 229.
78. J. Roteman, *US Pat.* 3 895 954 (1975) (*Chem. Abstr.*, 1975, **83**, 155 786y).
79. O. R. Abolafia, *Ger. Pat.* 2 258 880 (1972). (*Chem. Abstr.*, 1973, **79**, 60 000r).
80. J. E. Kropp, *US Pat.* 3 842 019 (1974).
81. Du Pont de Nemours and Co., *Belg. Pat.* 644 590 (1964).
82. M. Sakamoto, K. Hayashi and S. Okamura, *J. Polym. Sci., Part B*, 1965, **3**, 205.
83. M. Irie, Y. Yamamoto and K. Hayashi, *J. Macromol. Sci., Chem.*, 1975, **9**, 817.
84. R. K. Sadhir, J. D. B. Smith and P. M. Castle, *J. Polym. Sci., Polym. Chem. Ed.*, 1985, **23**, 411.
85. J. V. Crivello, *ACS Symp. Ser.*, 1984, **242**, 3.
86. J. V. Crivello, in 'UV Curing: Science and Technology' ed. P. Pappas, Technology Marketing Corp., Stamford, 1978, p. 23.
87. J. V. Crivello, in 'Developments in Polymer Photochemistry—2' ed. N. S. Allen, Applied Science, London 1982, p. 1.
88. S. Hayase, *Kobunasi*, 1986, **35**, 116.
89. F. Lohse, K. Meier and H. Zweifel, *Proc. Int. Conf. Org. Coat. Sci. Technol. 11th*, 1985, 175.
90. J. V. Koleske, *Pitture Vernici*, 1986, **62**, (4), 40.
91. J. J. Licari, W. Crespeau and P. C. Crespeau *US Pat.* 3 205 157 (1965)
92. S. I. Schlesinger, *US Pat.* 3 708 296 (1973) (*Chem. Abstr.*, 1973, **78**, 85 077p).
93. S. I. Schlesinger, *US Pat.* 3 826 650 (1974) (*Chem. Abstr.*, 1974, **81**, 113 734h).
94. W. R. Watt, *US Pat.* 3 794 576 (1974) (*Chem. Abstr.*, 1974, **81**, 14 799b).
95. J. H. Feinberg, *US Pat.* 3 711 390 (1973) (*Chem. Abstr.*, 1973, **78**, 85 388r).
96. J. H. Feinberg, *US Pat.* 3 816 281 (1974) (*Chem. Abstr.*, 1975, **82**, 4981m).
97. J. H. Feinberg, *US Pat.* 3 817 845 (1974) (*Chem. Abstr.*, 1974, **81**, 71 093r).
98. J. H. Feinberg, *US Pat.* 3 829 369 (1974) (*Chem. Abstr.*, 1975, **82**, 10 025u).
99. J. V. Crivello and H. J. W. Lam, *Macromolecules*, 1977, **10**, 1307.
100. J. V. Crivello and H. J. W. Lam, *J. Polym. Sci., Polym. Symp.*, 1976, **56**, 383.
101. J. V. Crivello and H. J. W. Lam, *J. Polym. Sci., Polym. Chem. Ed.*, 1979, **17**, 977.
102. J. V. Crivello and H. J. W. Lam, *J. Polym. Sci., Polym. Lett. Ed.*, 1979, **17**, 759.
103. J. V. Crivello and H. J. W. Lam, *J. Polym. Sci., Polym. Chem. Ed.*, 1980, **18**, 2677.
104. J. V. Crivello and H. J. W. Lam, *J. Polym. Sci., Polym. Chem. Ed.*, 1980, **18**, 2697.
105. J. V. Crivello and H. J. W. Lam, *J. Polym. Sci., Polym. Chem. Ed.*, 1980, **18**, 1047.
106. J. V. Crivello and H. J. W. Lam, *J. Polym. Sci., Polym. Chem. Ed.*, 1980, **18**, 2877.
107. J. V. Crivello and H. J. W. Lam, *J. Polym. Sci., Polym. Chem. Ed.*, 1980, **18**, 1021.
108. J. V. Crivello and H. J. W. Lam, *J. Polym. Sci., Polym. Chem. Ed.*, 1980, **18**, 1059.
109. J. V. Crivello, H. J. W. Lam and C. N. Volante, *J. Radiat. Curing*, 1977, **4** (3), 2.
110. A. Ledwith, *Polym. Prepr., Am. Chem. Soc., Div. Polym. Chem.*, 1982, **23** (1), 323.
111. J. V. Crivello and H. J. W. Lam, *J. Polym. Sci., Polym. Chem. Ed.*, 1980, **18**, 2441.
112. J. V. Crivello and H. J. W. Lam, *Macromolecules*, 1981, **14**, 441.
113. J. E. Moore, in ref. 86, p. 133.
114. A. R. Shultz and L. D. Stang, in ref. 19, p. 93.
115. E. W. Meijer and R. M. Zwiers, in ref. 19, p. 27.
116. D. A. Scola, *Dev. Reinf. Plast.*, 1984, **4**, 165.
117. J. M. Barton, *Makromol. Chem., Macromol. Symp.*, 1987, **7**, 27.
118. L. Schechter, J. Wynstra and R. P. Kurkjy, *Ind. Eng. Chem.*, 1956, **48**, 94.
119. A. Noordam, J. J. M. H. Wintraecken and G. Walton, in ref. 19, p. 373.
120. T. T. Chiao and R. L. Moore, *Proc. Annu. Conf. Reinf. Plast./Compos. Inst., Soc. Plast. Ind.*, 1974, 16-B, 1.
121. T. T. Chiao, E. S. Jessop and H. A. Newey, *SAMPE Q.*, 1974, **6**, 1.
122. K. Selby and L. E. Millar, *J. Mater. Sci.*, 1975, **10**, 12.
123. G. Pritchard and G. V. Rhoades, *Mater. Sci. Eng.*, 1976, **26**, 1.
124. R. J. Morgan and J. E. O'Neal, *J. Macromol. Sci., Phys.*, 1978, **15**, 139.
125. S. L. Kim, M. D. Skibo, J. A. Manson, R. W. Hertzberg and J. Janiszewski, *Polym. Eng. Sci.*, 1978, **18**, 1093.
126. D. C. Phillips, J. M. Scott and M. Jones, *J. Mater. Sci.*, 1978, **13**, 311.
127. R. J. Morgan, *J. Appl. Polym. Sci.*, 1979, **23**, 2711.
128. R. J. Morgan, J. E. O'Neal and D. B. Millar, *J. Mater. Sci.*, 1979, **14**, 109.
129. S. Yamini and R. J. Young, *J. Mater. Sci.*, 1980, **15**, 1814.
130. J. M. Scott, J. M. Wells and D. C. Phillips, *J. Mater. Sci.*, 1980, **15**, 1436.
131. K. Dusek, M. Ilavsky and S. Lunak, *J. Polym. Sci., Polym. Symp.*, 1975, **53**, 29.
132. S. Lunak and K. Dusek, *J. Polym. Sci., Polym. Symp.*, 1975, **53**, 45.
133. K. Dusek, M. Bleha and S. Lunak, *J. Polym. Sci., Polym. Chem. Ed.*, 1977, **15**, 2393.
134. K. Dusek and M. Ilavsky, *Colloid Polym. Sci.*, 1980, **258**, 605.
135. K. Dusek and M. Ilavsky, *J. Polym. Sci., Polym. Phys. Ed.*, 1983, **21**, 1323.

136. M. Ilavsky, L. M. Bogdanov and K. Dusek, *J. Polym. Sci., Polym. Phys. Ed.*, 1984, **22**, 265.
137. W. Burchard, S. Bantle, M. Muller and A. Reiner, *Pure Appl. Chem.*, 1981, **53**, 1519.
138. W. Burchard, S. Bantle and S. A. Zahir, *Makromol. Chem.*, 1981, **182**, 143.
139. S. Bantle and W. Burchard, *Prepr. IUPAC, Macro 82*, 1982, 531.
140. S. A. Zahir and S. Bantle, *Coat. Plast. Prepr. Pap. Meet. (Am. Chem. Soc., Div. Org. Coat. Plast. Chem.)*, 1982, **46**, 651.
141. J. M. Charlesworth. *J. Polym. Sci., Polym. Phys. Ed.*, 1979, **17**, 1557.
142. J. M. Charlesworth, *J. Polym. Sci., Polym. Phys. Ed.*, 1979, **17**, 1571.
143. K. Dusek, *ACS Symp. Ser.*, 1984, **208**, 3.
144. Y. Tanaka and T. F. Mika, in ref. 2, p. 135.
145. I. T. Smith, *Polymer*, 1961, **3**, 78.
146. Kh. A. Arutunyan, A. D. Tonoyan, S. P. Davtyan, K. A. Rozenberg and N. S. Enikolpyan, *Dokl. Akad. Nauk SSSR*, 1973, **212**, 1128.
147. C: C. Ricardi and R. J. J. Williams, in ref. 19, p. 292.
148. K. Dusek, M. Ilavsky, S. Stokrova, L. Matejka and S. Lunak, in ref. 19, p. 280.
149. L. A. O'Neal and C. P. Cole, *J. Appl. Chem.*, 1956, **6**, 356.
150. H. Dannenberg, *Soc. Plast. Eng. Trans.*, 1963, **3**, 78.
151. K. Horie, H. Hiura, M. Sawada, I. Mita and H. Kambe, *J. Polym. Sci., Part A-1*, 1970, **8**, 1337.
152. C. C. Ricardi, H. E. Adabbo and R. J. J. Williams, *J. Appl. Polym. Sci.*, 1984, **29**, 2481.
153. L. Matejka and K. Dusek, in ref. 19, p. 231.
154. L. Buckley and D. Roylance, *Polym. Eng. Sci.*, 1982, **22**, 166.
155. F. M. Kong, C. M. Walkup and R. J. Morgan, *ACS Symp. Ser.*, 1983, **221**, 211.
156. R. J. Morgan, F. M. Kong and C. M. Walkup, *Polymer*, 1984, **25**, 375.
157. H. Dannenberg and W. R. Harp, *Anal. Chem.*, 1956, **28**, 86.
158. C. H. Klute and W. Viehmann, *J. Appl. Polym. Sci.*, 1961, **5**, 86.
159. J. R. Creedon, *Anal. Calorim.*, 1970, **2**, 185.
160. L. T. Manzione, J. K. Gillham and C. A. McPherson, *J. Appl. Polym. Sci.*, 1987, **26**, 889.
161. T. P. Cuadrado, A. Almarez and R. J. J. Williams, in ref. 19, p. 179.
162. W. D. Bascom, R. L. Cottingham, R. L. Jones and P. Peyser, *J. Appl. Polym. Sci.*, 1975, **19**, 2545.
163. R. J. Morgan, C. M. Walkup and T. H. Hoheisel, *J. Appl. Polym. Sci.*, 1985, **30**, 289.
164. J. Moacanin, M. Cizmecioglu, F. Tsay and R. Gupta, *Coat. Plast. Prepr. Pap. Meet. (Am. Chem. Soc., Div. Org. Coat. Polym. Sci.)*, 1983, **49**, 587.
165. E. T. Mones, C. M. Walkup, J. A. Happe and R. J. Morgan, *Natl. SAMPE Tech. Conf.*, 1982, **14**, 89.
166. J. M. Barton, *Br. Polym. J.*, 1986, **18**, 37.
167. J. M. Barton, *Br. Polym. J.*, 1986, **18**, 44.
168. M. Cizmecioglu and A. Gupta, *SAMPE Q.*, 1982, 16.
169. W. X. Zukas, N. S. Schneider and W. J. McKnight, *Polym. Mater. Sci. Eng.*, 1983, **49**, 588.
170. R. J. Morgan, J. E. Happe and E. T. Mones, *Natl. SAMPE Symp. Exhib.* [*Proc.*], 1983, 596.
171. J. Mijovic, J. Kim and J. Slaby, *J. Appl. Polym. Sci.*, 1984, **29**, 1449.
172. A. Apicella, M. I. Nicolais and P. Passerini, *J. Appl. Polym. Sci.*, 1984, **29**, 2083.
173. J. Moacanin, M. Cizmecioglu, S. D. Hong and A. Gupta, *ACS Symp. Ser.*, 1983, **227**, 84.
174. L. Matejka, S. Pokorny and K. Dusek, in ref. 19, p. 242.
175. L. Matejka, M. Tkuczyk, S. Pokorny and K. Dusek, *Polym. Bull.*, 1986, **15**, 389.
176. J. Ancelle, A. J. Attias, B. Bloch and C. Cavalli, in ref. 19, p. 213.
177. A. Gupta, M. Cismecioglu, D. Coulter, R. H. Liang and A. Yavrouian, *J. Appl. Polym. Sci.*, 1983, **28**, 1011.
178. C. A. May, AFML-TR-76-112, 1976.
179. J. F. Carpenter, McDonnell Aircraft Report, Contract No. N00019-76-C-0138, 1977.
180. R. E. Trujillo and B. P. Engler, Sandia Laboratory Report, SAND 78-1504, 1978.
181. J. P. Bell, *J. Polym. Sci., Part A*, 1970, **8**, 417.
182. C. H. Lau, K. A. Hodd and W. W. Wright, *Br. Polym. J.*, 1986, **18**, 316.
183. J. E. Happe, R. J. Morgan and C. M. Walkup, *Polymer*, 1985, **26**, 827.
184. S. A. Bidstrup and C. W. Macosko, in ref. 19, p. 253.
185. K. Dusek, M. Ilavsky and S. Lunak, Jr., in ref. 19, p. 269.
186. H. S. Chu and J. C. Seferis, in 'The Role of the Polymer Matrix in the Processing and Structural Properties of Composite Materials', ed. J. C. Seferis and L. Nicolais, Plenum Press, New York, 1983, p. 89.
187. J. B. Enns and J. K. Gillham, *J. Appl. Polym. Sci.*, 1983, **28**, 2567.
188. G. P. Speranza and H. G. Waddill, *US Pat.* 4 581 422 (1986) (*Chem. Abstr.*, 1986, **105**, 7487j).
189. G. L. Hagnauer, B. R. LaLiberte and D. A. Dunn, *ACS Symp. Ser.*, 1983, **221**, 229.
190. G. L. Hagnauer and I. Setton, *J. Liq. Chromatogr.*, 1978, **1**, 55.
191. G. L. Hagnauer and D. A. Dunn, *Natl. SAMPE Tech. Conf.*, 1980, **12**, 648.
192. G. L. Hagnauer and D. A. Dunn, *J. Appl. Polym. Sci.*, 1981, **26**, 1846.
193. P. N. Son and C. D. Weber, *J. Appl. Polym. Sci.*, 1974, **17**, 1305.
194. Y. G. Lin, J. Galy, H. Sautereau and J. P. Pascault, in ref. 19, p. 147.
195. S. A. Zahir, *Adv. Org. Coat. Sci. Technol.*, 1982, **4**, 83.
196. Schering AG, *Ger. Pat.* 2 131 929 (1973) (*Chem. Abstr.*, 1973, **78**, 160 483y).
197. T. F. Saunders, M. F. Levy and J. F. Serino, *J. Polym. Sci., Polym. Chem. Ed.*, 1967, **5**, 1609.
198. P. Eyerer, *J. Appl. Polym. Sci.*, 1971, **15**, 3067.
199. Y. Tanaka and H. Kakiuchi, *J. Macromol. Chem.*, 1966, **1**, 307.
200. Y. Tanaka and H. Kakiuchi, *J. Polym. Sci., Part A*, 1964, **2**, 3405.
201. W. Fisch and W. Hofmann, *J. Polym. Sci.*, 1954, **12**, 497.
202. W. Fisch and W. Hofmann, *Makromol. Chem.*, 1961, **54–56**, 8.
203. W. Fisch, W. Hofmann and R. Schmid, *J. Appl. Polym. Sci.*, 1969, **13**, 295.
204. G. C. Stevens, *J. Appl. Polym. Sci.*, 1981, **26**, 4259.
205. G. C. Stevens, *J. Appl. Polym. Sci.*, 1981, **26**, 4279.

206. D. H. Solomon and J. J. Hopwood, *J. Appl. Polym. Sci.*, 1966, **10**, 981.
207. W. C. Mih, *ACS Symp. Ser.*, 1984, **242**, 273.
208. R. F. Fischer, *J. Polym. Sci.*, 1960, **44**, 195.
209. Y. Tanaka and H. Kakiuchi, *J. Appl. Polym. Sci.*, 1963, **7**, 1063.
210. L. Matejka, J. Lovy, S. Pokorny, K. Bouchal and K. Dusek, *J. Polym. Sci., Polym. Chem. Ed.*, 1983, **21**, 2873.
211. K. Dusek, S. Lunak and L. Matejka, *Polym. Bull.*, 1982, **7**, 145.
212. E. W. Crandell and W. C. Mih, *Coat. Plast. Prepr. Pap. Meet. (Am. Chem. Soc., Div. Org. Coat. Plast. Chem.)*, 1982, **47**, 592.
213. M. F. Sorokin, *Lakokras. Mater. Ikh. Primen.*, 1967, **5**, 67.
214. B. Steinmann, in ref. 19, p. 117.
215. M. K. Antoon and J. L. Koenig, *J. Polym. Sci.*, 1981, **19**, 549.
216. Toshiba Corp., *Jpn. Pat.* 82 24 553 (1982) (*Chem. Abstr.*, 1982, **97**, 57 259y).
217. T. Segawa, H. Suzuki, M. Kitamura, M. Numata and K. Nishi, *Ger. Pat.* 3 137 480 (1982) (*Chem. Abstr.*, 1982, **97**, 7395u).
218. Nitto Electrical Industrial Co., *Jpn. Pat.* 82 59 365 (1982) (*Chem. Abstr.*, 1982, **97**, 93 634a).
219. Morton-Norwich Products Inc., *Jpn. Pat.* 82 49 647 (1982) (*Chem. Abstr.*, 1982, **97**, 3987r).
220. K. A. Lyalyushko, M. F. Sorokin and O. V. Sigunova, *Tr. Mosk. Khim.-Tekhnol. Inst. im. D. I. Mendeleeva*, 1980, **110**, 76.
221. J. D. B. Smith, *Coat. Plast. Prepr. Pap. Meet. (Am. Chem. Soc., Div. Org. Coat. Plast. Chem.)*, 1978, **39**, 42.
222. H. Suzuki, M. Sato, T. Muroi and Y. Watanabe, *Prepr. SPE 19th ANTEC*, 1973, 6.
223. T. Aida, K. Sanuki and S. Inoue, *Macromolecules*, 1985, **18**, 1049.
224. W. A. Romanchick, J. E. Sohn and J. F. Geibel, *ACS Symp. Ser.*, 1983, **221**, 86.
225. M. Uribe and K. A. Hodd, in 'Advances in Polymer Synthesis', ed. B. M. Culbertson and J. E. McGrath, Plenum Press, New York, 1986, p. 251.
226. G. Speranza and S. Peppel, *J. Org. Chem.*, 1958, **23**, 1922.
227. S. R. Sandler, F. Berg and G. Kitazawa, *J. Appl. Polym. Sci.*, 1965, **9**, 1994.
228. R. R. Dileone, *J. Polym. Sci., Part A-1*, 1970, **8**, 609.
229. J. E. Herweh and W. Y. Whitmore, *J. Polym. Sci., Part A-1*, 1970, **8**, 2773.
230. K. Gulbins, G. Benzing, R. Maysenholder and K. Herman, *Chem. Ber.*, **93**, 1960, 1975.
231. S. Sandler, *J. Polym. Sci., Part A-1*, 1967, **5**, 1481.
232. Y. Iwakura and S. Iwaza, *J. Digest. Chem.*, 1964, **29**, 379.
233. Y. Iwakura and S. Iwaza, *Bull. Chem. Soc. Jpn.*, 1966, **39**, 2490.
234. A. Sendijarevic, V. Sendijarevic and K. C. Frisch, *Polimeri*, 1986, **7** (6), 159.
235. A. Sendijarevic, V. Sendijarevic and K. C. Frisch, *J. Polym. Sci., Polym. Chem. Ed.*, 1987, **25**, 151.
236. N. Kinjo, S. Numata and T. Koyana, *J. Appl. Polym. Sci.*, 1983, **28**, 1729.
237. J. A. Clarke, *ACS Symp. Ser.*, 1984, **208**, 51.
238. J. A. Clarke, *US Pat.* 3 687 987 (1972).
239. J. A. Clarke, *US Pat.* 3 676 397 (1972).
240. J. L. Bitner, J. L. Rushford, W. S. Rose, D. L. Hunston and C. K. Riew, *J. Adhes.*, 1981, **13**, 3.
241. W. Roth, K. Hanella, V. Steinbach and P. Frosch, *Int. Wiss. Koll. Tech. Hochsch. Ilmenau*, 1983, **28**, 167.
242. V. Steinbach and W. Roth, *Plaste. Kautsch.*, 1983, **30**, 32.
243. M. Uribe and K. A. Hodd, *Thermochim. Acta*, 1984, **77**, 367.
244. J. S. Senger, I. Yilgor and J. E. McGrath, *Coat. Plast. Prepr. Pap. Meet. (Am. Chem. Soc., Div. Org. Coat. Plast. Chem.)*, 1985, **26**, 244.
245. N. S. Gromakov, V. G. Hozin and V. A. Voskvesevski, in 'Thermal Analysis', ed. H. Chihara, Heyden, London, 1977, p. 279.
246. L. Shechter and J. Wynstra, *Ind. Eng. Chem.*, 1956, **48**, 86.
247. R. B. Graver, *J. Paint Technol.*, 1970, **42**, 37.
248. E. Klemm, H.-J. Flammerheim and H.-H. Harold, in ref. 19, p. 55.
249. W. H. Ku and J. P. Bell, *ACS Symp. Ser.*, 1983, **221**, 153.
250. J. P. Bell and W. H. Ku, in ref. 19, p. 3.

38
Polymers with Main-chain Mesogenic Units

ERNO CHIELLINI
University of Pisa, Italy

and

ROBERT W. LENZ
University of Massachusetts, Amherst, MA, USA

38.1	INTRODUCTION	701
38.2	LYOTROPIC MAIN-CHAIN LIQUID CRYSTALLINE POLYMERS	703
	38.2.1 Poly(aromatic amide)s	705
	38.2.2 Polypeptides	707
	38.2.3 Poly(azomethine)s	707
	38.2.4 Poly(alkylisonitrile)s	708
	38.2.5 Polyisocyanates	708
	38.2.6 Poly(organophosphazine)s	708
	38.2.7 Heteroaromatic Mesogen-containing Polymers	708
	38.2.8 Aromatic Polyesters	708
	38.2.9 Cellulose Derivatives	709
38.3	THERMOTROPIC MAIN-CHAIN LIQUID CRYSTALLINE POLYMERS	709
	38.3.1 Fully Aromatic Polyesters	709
	38.3.2 Cellulose Derivatives	714
	38.3.3 Polypeptides	714
	38.3.4 Semiflexible Liquid Crystalline Polymers	714
38.4	REFERENCES	721

38.1 INTRODUCTION

Low molecular weight, liquid crystalline compounds (LCs) have been known for 100 years,[1] but liquid crystalline polymers (LCPs) attained prominence only during the past 30 years. Synthetic and natural polymers with a degree of molecular order, established either in solution (lyotropic) or in the melt (thermotropic), intermediate between that of a solid crystal and that of an isotropic liquid, are termed liquid crystalline, mesomorphic or mesophasic.[2,3]

The structural feature which imparts liquid crystallinity to a low molecular weight compound or a polymer is a sequence of either rigid segments with a high axial ratio[4] or of disc-like structures. For polymers these structures can be present in either the backbone or as pendent groups arranged as a succession of either identical or different repeating units. The location of the mesogen defines two distinct classes of LCP, which are designated as the main-chain and side-chain classes. A schematic representation of these two general classes is given in Figure 1, which also includes a class with the characteristics of both.

The side-chain class of LCPs are described in Volume 5, Chapter 39, so this review is limited to the main-chain of polymers.

Depending upon the specific molecular structure of the LCP, just as for low molecular weight analogs, several different types of supermolecular ordering can occur. These types are characterized by either a one-dimensional orientation, which includes both nematics and twisted nematics or cholesterics, or a two-dimensional layered ordering, which includes the low order smectics (smectic

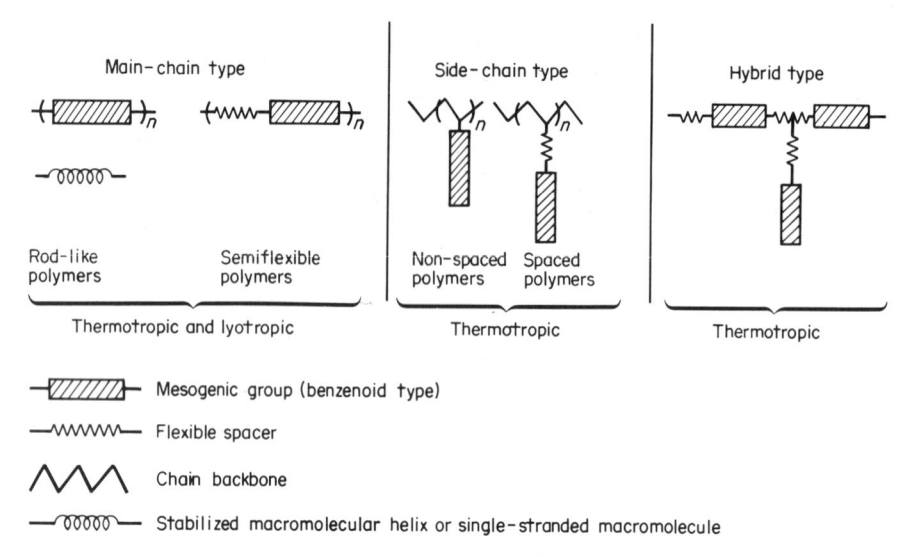

Figure 1 Schematic representation of the assessment of structural component

A and smectic C), or a three-dimensional crystal-like ordering for the high order smectics (smectic B, D, E, F, *etc.*). Each of these types are represented schematically in Figure 2.

From a historical point of view, the initial recognition of the existence of LCPs was made for polypeptides of either biological[5] or synthetic[6] origin, which were found to form LC organizations in different solvents. Subsequently, in the late 1960s, it was observed that synthetic aromatic polyamides with *para*-substituted functional groups were also to be lyotropic,[7] and this discovery led to the development, initially by Du Pont and later by Akzo, of commercial fibers with outstanding high tensile strength and modulus.

The discovery of thermotropic LCPs was made only sometime later, even though a number of polyesters were described in early patents issued to ICI[8] and to Carborundum Co,[9] which were capable of showing thermotropic behavior. Presumably, the fact that no claim to that behavior was inferred in these patents is because the property was so unexpected, and the first assignment of thermotropic character to a polymer appeared in the open literature only in the middle 1970s for main-chain polymers.[10,11] The thermotropic behavior of side-chain polymers[12] was recognized sometime earlier.

At about the same time, theoretical considerations on LCP thermotropism were also put forward[13,14] as a result of the implications contained in the early lattice theory of rigid-rod polymers.[4] According to this theory, an axial ratio of 6.5 or higher is required within a polymer backbone for the spontaneous formation of an LC melt or solution. For such polymers, if the average molecular weight is sufficiently high and accompanied by a reasonably narrow polydispersity, extremely good mechanical properties can be obtained from melt- or solution-processed polymers of this type as a result of the high degree of molecular orientation, which occurs when a polymeric nematic phase is subjected to shear, even at fairly low rates. The preferential alignment occurs parallel to the flow direction and imparts remarkably high strength and stiffness to the processed article. The ease of processing anisotropic melts or solutions, because of their inherently lower viscosities compared to those of equivalent isotropic phases, is an additional major advantage for both lyotropic and thermotropic phases, is an additional major advantage for both lyotropic and thermotropic main-chain LCPs.

As represented in Figure 1, two general types of macromolecules are able to give rise to anisotropic solutions as melts including: (1) rigid-rod polymers and (2) semiflexible polymers. Both can contain mesogens with either rod-like or disc-like (discotic) anisotropy. The rigid-rod polymers can be classified into at least three basic macromolecular structural types, including either (a) a rigid helix, especially the α-helix with a preferential screw-sense as found in polypeptides, or (b) a rigid linear chain containing *trans*-1,4-cyclohexylene units or their analogs as found in cellulose derivatives, or (c) a chain consisting of either linearly substituted aromatic groups or *trans*-1,4-cyclohexylene rings linked by functional groups of relatively short length and containing an even number of atoms in order to retain colinearity. To this last group belong several different types of LCP, whose basic structural elements are represented in Table 1.

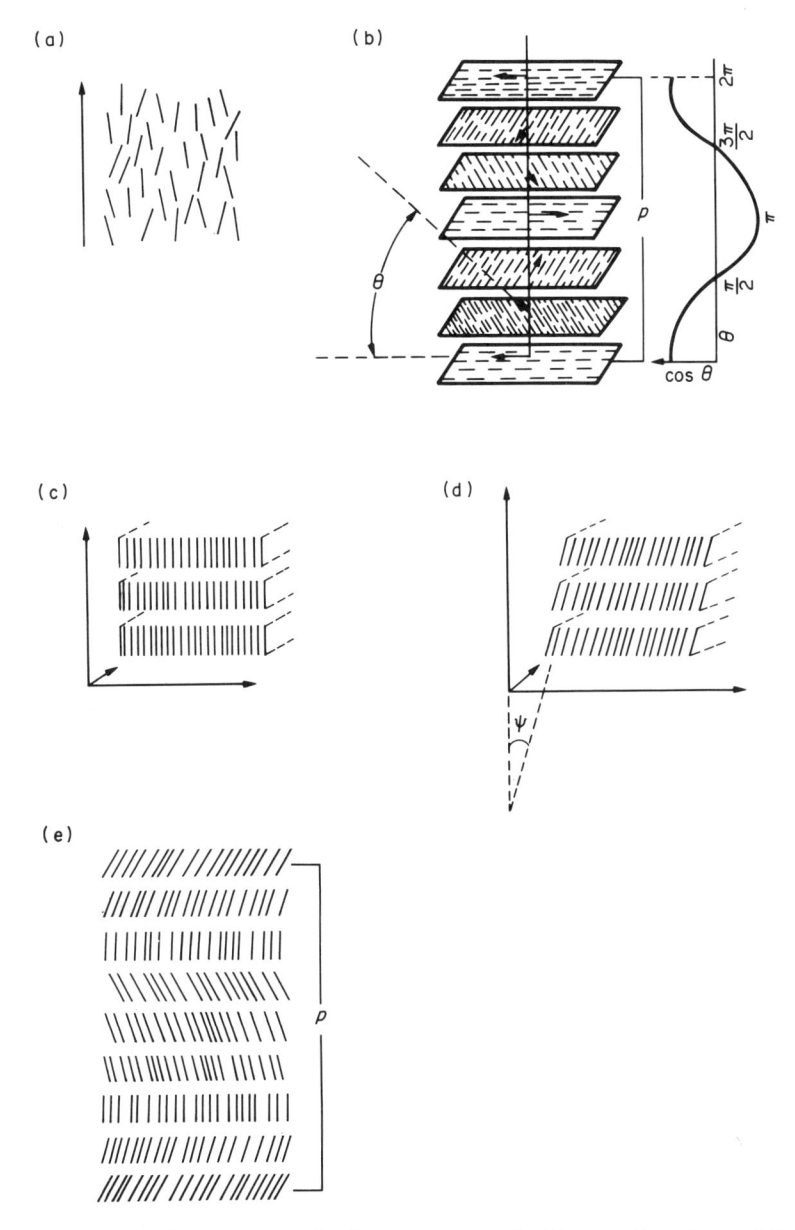

Figure 2 Schematic representation of supermolecular structure in liquid crystalline phases: (a) ordinary nematic, (b) stacking of twisted nematic planes (cholesteric), (c) orthogonal smectic (smectic A), (d) tilted smectic (smectic C) and (e) stacking of twisted smectic C layers (chiral smectic C). θ, twist angle; ψ, tilt angle; and p, helical pitch

38.2 LYOTROPIC MAIN-CHAIN LIQUID CRYSTALLINE POLYMERS

The discovery of supermolecular ordering by polymers in solution was first reported in 1950 for poly(γ-benzyl-L-glutamate) (PBLG).[6] This behavior was described in a more quantitative manner, including a number of observed phenomena associated with LC behavior, a few years later for PBLG in different solvents at different concentrations.[15] In particular, it was concluded that the PBLG macromolecule had an α-helical conformation, and these helices were aligned nearly parallel into nematic domains. However, solution-spun fibers of PBLG in this state did not show exceptional properties,[16] and it was not until such fibers were obtained from aromatic polyamides that the potential of the lyotropic LC state for spinning fibers with extremely good properties was recognized.[7] Since the latter discovery, a large variety of synthetic and semisynthetic condensation polymers have been prepared which contain rigid-rod macromolecules that show this behavior. A collection of the major classes of lyotropic main-chain LCPs is given in Table 2. A few of these have found rewarding industrial applications,[17,18] due especially to their capability for providing fibers

Table 1 Representation of the Basic Structural Elements of Synthetic LCPs Containing Specific Mesogenic Groups within the Main Chain

Benzenoid component	Mesogenic repeating unit	Bridging group
(phenylene) (pyridine) (biphenyl)	—CH=N—	—N=N—
(2,6-naphthylene) (quinoline) (1,5-naphthylene)	—N=N— ↓ O	—CO$_2$—
(benzobisthiazole) (benzobisoxazole)	—CONH—	—CONHNHCO—
(benzobisimidazole)	—OCONH—	—NHCOCONH—

Table 2 Lyotropic Liquid Crystalline Polymers: Classification and Functional Group in the Repeating Unit(s)

Class type	Functional group
Poly(aromatic amide)s	—NHCO—
Poly(α-amino acid)s	—NHCO—
Polyoxamides	—NHCOCONH—
Polyhydrazides	—CONHNHCO—
Polyesters	—CO$_2$—
Poly(azomethine)s	—N=CH—
Polyisonitriles	—C=NR—
Polyisocyanates	—CONR—
Polyphosphazines	—PR$_2$=N—
Heterocycle-containing polymers	
Poly(p-phenylene-2,6-benzobisimidazole)	(structure)
Poly(p-phenylene-2,6-benzobisoxazole)	(structure)
Poly(p-phenylene-2,6-benzobisthiazole)	(structure)
Poly[p-phenylene-2,6-(4-phenylquinoline)]	(structure)

Table 2 (Continued)

Class type	Functional group
Cellulose derivatives[a]	(cellulose structure with CH$_2$OR and OR groups)
Esters (R = —COMe, —NO$_2$, —SO$_2$, —COC$_3$H$_7$)	
Ethers (R = Et, —CH$_2$OH, —C$_2$H$_4$OH, —CH$_2$CHMe(OH), —CH$_2$CO$_2$H, —CH$_2$CH$_2$OEt, —CH$_2$CH$_2$CN)	
Ether–esters (R = —CH$_2$O$_2$CMe, —CH$_2$CH$_2$O$_2$CMe, —CH$_2$CHO$_2$CPh with Me)	
Carbamates (R = PhNHCO—)	

[a] Poly(β-1,4-diglucopyranosylpyranose) derivatives.

with very high tenacity, very high tensile modulus and remarkably low thermal expansion coefficients. Each of these types of polymer are discussed below with regard to their synthesis and structure–property relationships.

38.2.1 Poly(aromatic amide)s

These polymers are usually prepared by the condensation polymerization of an aromatic diacyl dihalide (XCOArCOX) with an aromatic diamine (H$_2$NArNH$_2$) in solution, according to the reaction (1).

$$n\text{XOCArCOX} + n\text{H}_2\text{NAr}'\text{NH}_2 \rightarrow \text{H}{-}[{-}\text{NHAr}'\text{NHCOArCO}{-}]_n{-}\text{OH} + 2n\text{HX} \quad (1)$$

Table 3 Basic Structural Components of Lyotropic Poly(aromatic amide)s

Aromatic diacid	Aromatic diamine	Ref.
HO$_2$C—C$_6$H$_4$—CO$_2$H	H$_2$N—C$_6$H$_4$—NH$_2$	7, 19–21
	1,5-diaminonaphthalene (NH$_2$ groups)	22
	H$_2$N—C$_6$H$_4$—C$_6$H$_4$—NH$_2$ (biphenyl)	22
	H$_2$N—C$_6$H$_3$(Me)—C$_6$H$_3$(Me)—NH$_2$	22
	H$_2$N—C$_6$H$_3$(OMe)—C$_6$H$_3$(OMe)—NH$_2$	22
	H$_2$N—C$_6$H$_4$—N=N—C$_6$H$_4$—NH$_2$	23

Table 3 (*Continued*)

Aromatic diacid	Aromatic diamine	Ref.
HO$_2$C—(pyridine-2,6-diyl)—CO$_2$H	H$_2$N—C$_6$H$_4$—NH$_2$	24, 25
	H$_2$N—C$_6$H$_3$(Cl)—NH$_2$	24
	H$_2$N—C$_6$H$_3$(Me)—C$_6$H$_3$(Me)—NH$_2$	24
HO$_2$C—C$_6$H$_4$—C$_6$H$_4$—CO$_2$H	H$_2$N—C$_6$H$_4$—NH$_2$	26
	H$_2$N—C$_6$H$_4$—C$_6$H$_4$—NH$_2$	23
	H$_2$N—C$_6$H$_3$(OMe)—C$_6$H$_3$(MeO)—NH$_2$	23
HO$_2$C—C$_6$H$_4$—CH=CH—C$_6$H$_4$—CO$_2$H	H$_2$N—C$_6$H$_4$—NH$_2$	27
	H$_2$N—C$_6$H$_3$(Cl)—NH$_2$	27
HO$_2$C—C$_6$H$_4$—N=N—C$_6$H$_4$—CO$_2$H	H$_2$N—C$_6$H$_4$—NH$_2$	23
	H$_2$N—C$_6$H$_4$—N=N—C$_6$H$_4$—NH$_2$	23
HO$_2$C—C$_6$H$_4$—N=N(→O)—C$_6$H$_4$—CO$_2$H	H$_2$N—C$_6$H$_4$—NH$_2$	23
	H$_2$N—C$_6$H$_4$—N=N—C$_6$H$_4$—NH$_2$	23
	HO$_2$C—C$_6$H$_4$—NH$_2$	28

In Table 3 are collected major examples of liquid crystalline aromatic polyamides that give rise to lyotropic solutions in specific, often aggresive, solvents, such as fuming or concentrated sulfuric acid, halosulfonic acids and organic amide solvents, including dimethylacetamide, *N*-methylpyrrolidine and tetramethylurea, each containing either lithium or calcium halides. The concentration at which lyotropic behavior is established depends on the chemical structure and molecular weight of the polymer, the nature of the solvent and temperature. Generally, the effective concentration is in the range of 10–20% by weight of the polymer in the solvent at room temperature. Essentially all of the appropriate polyamides form nematic lyotropic solutions, and the introduction of either a lateral substituent on the diamine,[29] or of 1,3-phenylene units (in very small amounts) or of flexible spacers (especially polymethylene segments) can impart greater solubility to the polymer so that less aggressive solvents can be used.

Table 4 Basic Structural Components of Rigid-rod, not Fully Aromatic Polyamides

Diacid	Diamine	Ref.
HO_2C-CO_2H	3,3'-dimethylbenzidine (H_2N-Ar-Ar-NH_2 with Me substituents)	30
HO_2C-CO_2H	2-chloro-1,4-phenylenediamine	30
HO_2C-Ar-CO_2H	H_2N-NH_2	31
HO_2C-Ar(Cl)-CO_2H	H_2N-NH_2	31, 32

Polyamides based on oxalic acid and aromatic diamines as well as on hydrazine and aromatic diacids also form lyotropic solutions, particularly when aromatic residues with lateral substituents are used (Table 4).[30,31] The substituents play a key role by decreasing the crystallinity of the polymers and thus increasing their solubility so that higher concentrations can be achieved at room temperature. The use of a mixed functionality amide–hydrazide (**1**), based on 4-aminobenzhydrazide and terephthalic acid monomers, has also been extensively studied[33–36] for the formation of lyotropic LCPs.

$$HO + \left(\underset{O}{\underset{\|}{C}} - \text{Ar} - \underset{O}{\underset{\|}{C}} NH - \text{Ar} - \underset{O}{\underset{\|}{C}} NHNH \right)_n H$$

(**1**)

38.2.2 Polypeptides

As discussed above, the ability of polypeptides such as PBLG to form a stable rigid helix in certain solvents is responsible for its lyotropic behavior. Because it contains a chiral center of a specific configuration in each repeating unit, PBLG forms a twisted nematic or cholesteric structure, which is characterized by a helical superstructure whose helicity and pitch length are dependent upon and controlled by external parameters such as temperature, pressure and solvent.[37,38]

38.2.3 Poly(azomethine)s

These polymers contain aromatic rings linked in the *para*-position, such as that derived from *p*-phenylenediamine and terephthalaldehyde, (**2**), which forms a lyotropic nematic solution in sulfuric acid.

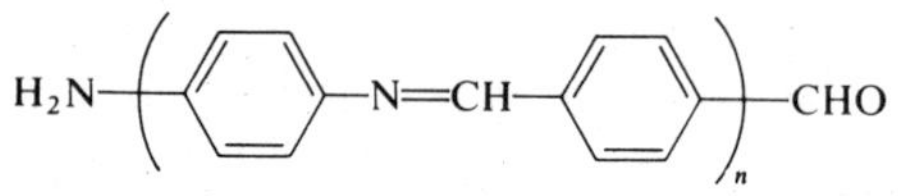

$$H_2N\!\!-\!\!\left(\!\!\left\langle\bigcirc\right\rangle\!\!-\!\!N\!\!=\!\!CH\!\!-\!\!\left\langle\bigcirc\right\rangle\!\!\right)_{\!\!n}\!\!-\!\!CHO$$

(2)

38.2.4 Poly(alkylisonitrile)s

These polymers (3) are prepared by the catalytic polymerization of the corresponding isocyanides[40] and have been found to give lyotropic solutions,[41,42] as expected for a rigid-rod helical structure, which their macromolecular backbones are known to form in certain solvents.

$$-\!\!\left(\!\!\underset{NR}{\overset{C}{\underset{\|}{C}}}\!\!\right)_{\!\!n}\!\!-$$

(3) R = PhCH—, PhCH$_2$CH$_2$—, n-butyl, n-octyl
            |
            Me

38.2.5 Polyisocyanates

These polymers are also characterized by a rigid-rod extended chain structure and have been synthesized with a large variety of pendent R groups as shown in (4). Their lyotropic properties have been found to be dependent upon the structure of the substituents R in homopolymers and on the types of R groups and composition of the copolymers.

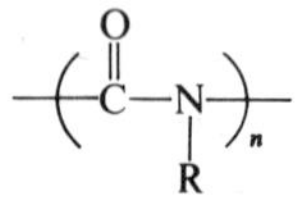

$$-\!\!\left(\!\!\underset{R}{\overset{\overset{O}{\|}}{C}}\!\!-\!\!N\!\!\right)_{\!\!n}\!\!-$$

(4) R = alkyl, arylalkyl

38.2.6 Poly(organophosphazine)s

These polymers (5) have been prepared in a large variety,[46] but only a few cases have been reported to form lyotropic solutions in spite of the rigid-rod nature of their chain backbone. Concentrations of 30–40% are required to achieve that property.[47]

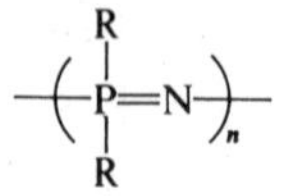

$$-\!\!\left(\!\!\underset{R}{\overset{R}{\underset{|}{\overset{|}{P}}}}\!\!=\!\!N\!\!\right)_{\!\!n}\!\!-$$

(5) R = CF$_3$CH$_2$O—, PhO—

38.2.7 Heteroaromatic Mesogen-containing Polymers

These polymers include a fairly broad class of rod-like LCPs that are capable of forming lyotropic solutions, which can generate fibers with outstanding technological properties. From a structural point of view the majority of the polymers of this class (listed in Table 2) are derived by the reaction of terephthalic acid or its derivatives with a tetrafunctionalized aromatic compound, including poly(p-phenylene-2,6-benzobisthiazole) (PBT)[48,49] and poly(p-phenylene-2,6-benzobisoxazole) (PBO).[48,50,51]

38.2.8 Aromatic Polyesters

When these polymers are constructed of fully aromatic structural units, they are only sparingly soluble even in very aggressive solvents, and, therefore, they are generally unable to form anisotropic solutions. The few cases reported include block copolyesters based on *trans*-1,4-cyclohexyleneoxy, *trans*-1,4-cyclohexylenecarbonyl and terephthaloxyl units of variable compositions, which can form nematic phases in *o*-chlorophenol solutions.[52]

38.2.9 Cellulose Derivatives

A variety of cellulose derivatives, in addition to unsubstituted cellulose itself, constitute a large class of lyotropic main-chain polymers which have rod-like structures with some type of flexibility. The onset of lyotropic behavior is strongly affected by the type of solvent, temperature, molecular weight and degree of substitution of the cellulose derivative. Usually the onset of anisotropic phase formation occurs at concentration levels much higher (20–60%) than those observed for other lyotropic systems. Hydroxypropylcellulose was the first cellulose derivative found to be able to give rise to lyotropic LC behavior,[53,54] but cellulose itself can form lyotropic solutions in mixed organic solvents such as mixtures of chlorinated hydrocarbons with trifluoroacetic acid[55] and mixtures of *N*-methylmorpholine with water.[56,57] High ionic strength solutions containing inorganic salts can also be used for that purpose.[58]

In Table 2 are represented several classes of cellulose derivatives, which are grouped according to the nature of the substituents including esters, ethers, ether–esters and carbamates. The degree of substitution necessary for the level of solubility required for lyotropic behavior may substantially change from one class to the other, and even within the same class, depending on the solvent.

38.3 THERMOTROPIC MAIN-CHAIN LIQUID CRYSTALLINE POLYMERS

As illustrated in Figure 1, thermotropic main-chain LCPs can be obtained from polymers which contain only rigid-rod units or which contain a combination of rigid-rod and flexible units. In Table 5 are listed the major families of rigid-rod and semiflexible type of thermotropic main-chain LCP, including the relevant functional groups present in the repeating units as derived from the condensation polymerization reaction used.

38.3.1 Fully Aromatic Polyesters

By far the major contributions in the area of thermotropic main-chain LCPs have been made in the study of aromatic polyesters with linearly disubstituted arylene groups, especially those in which the substituents are either two hydroxyl groups, or two carboxyl groups or one hydroxyl and one carboxyl group. In Table 6 are listed the most commonly used difunctional aromatic monomers.

The synthesis of these polyesters can be carried out by either melt or low-temperature polycondensation processes using either an ester interchange reaction or a direct esterification with an acyl chloride (Schemes 1 and 2). For the preparation of very high melting polymers, a condensation polymerization can be carried out in a suitable solvent.

By a suitable combination of two or more monomers of a given type in variable proportions, a large variety of high melting copolyesters ($T_m > 350\,°C$) can be obtained by these reactions. The most

$$\text{MeCO}_2\text{ArO}_2\text{CMe} + \text{HO}_2\text{CAr'CO}_2\text{H} \longrightarrow \text{MeC}\overset{O}{\overset{\|}{-}}(-\text{OArO}_2\text{CAr'}\overset{O}{\overset{\|}{C}}-)_n-\text{OH} + (2n-1)\text{MeCO}_2\text{H}$$

$$\text{HOArOH} + \text{PhO}_2\text{CAr'CO}_2\text{Ph} \longrightarrow \text{H}-(-\text{OArO}_2\text{CAr'}\overset{O}{\overset{\|}{C}}-)_n-\text{OPh} + (2n-1)\text{PhOH}$$

Scheme 1 Ester interchange

$$\text{HOArOH} + \text{Cl}\overset{O}{\overset{\|}{C}}\text{Ar'}\overset{O}{\overset{\|}{C}}\text{Cl} \longrightarrow \text{H}-(-\text{OArO}_2\text{CAr'}\overset{O}{\overset{\|}{C}}-)_n-\text{OH} + (2n-1)\text{HCl}$$

Scheme 2 Direct esterification

Table 5 Classification of Thermotropic Main-chain Polymers

	Rigid-rod polymers				Semiflexible polymers	
	Fully aromatic		Helical structure			
Type	Type	Functional group	Type	Functional group	Type	Functional group
Polyesters Poly(ester amide)s		$-CO_2-$ $-CO_2 \sim CONH-$	Poly(α-amino acid) derivatives	$-CONH-$	Polyesters Poly(ester amide)s	$-CO_2-$ $-CO_2 \sim NHCO-$
Polyaldoimines Polyurethanes		$-CH=N-$ $-OCONH-$	Cellulose derivatives	(cellulose ring structure with OR, CH$_2$OR substituents)	Poly(ether ester)s	$-O \sim CO_2-$

Table 6 Basic Components of Rigid-rod Fully Aromatic Polyesters

Dihydroxyarylene	Dicarboxyarylene	Hydroxycarboxyarylene
HO–⬡–OH	HO$_2$C–⬡–CO$_2$H	HO–⬡–CO$_2$H
HO–⬡(X,Y)–OH (X, Y = halogen, alkyl)	HO$_2$C–⬡(X)–CO$_2$H (X = halogen, alkyl)	HO–⬡(X)–CO$_2$H (X = halogen, alkyl)
HO–(naphthalene)–OH	HO$_2$C–⬡–⬡–CO$_2$H	HO–(naphthalene)–CO$_2$H
HO–⬡–⬡–OH	HO$_2$C–(naphthalene)–CO$_2$H	HO–⬡–CH=CHCO$_2$H
HO–⬡(–⬡X)–OH (X = H, halogen, alkyl)	HO$_2$C–⬡–O–⬡–CO$_2$H	HO–(naphthalene)–CO$_2$H
HO–⬡–OH (ortho)	HO$_2$C–⬡–CH$_2$CO$_2$H	
HO–⬡–X–⬡–OH (X = CH$_2$, CMe$_2$, SO$_2$, etc.)	HO$_2$C–⬡–OCH$_2$CH$_2$O–⬡–CO$_2$H	
HO–(anthraquinone)–OH	HO$_2$C–(naphthalene)–CO$_2$H	
	HO$_2$C–⬡–CO$_2$H	

significant examples, of which some have been commercialized as engineering thermoplastics, are reported in Table 7.

The goal in many of these syntheses is to obtain polyesters with the lowest possible melting temperature for better processability in the LC state. The synthetic strategies adopted to that objective[74] are summarized in Scheme 3, which shows four different approaches to achieving this goal, including:

(1) Introduction of suitable lateral substituents on the aromatic rings of the units that are derived from the monomers listed in Table 5. Either the phenyl group, or an alkoxy group, or a halogen atom or an alkyl group is used for that purpose.

Table 7 Basic Structures of Major Examples of Rigid-rod Aromatic Polyesters

Structure[a]	Ref.
—O—⟨⟩—CO₂—⟨⟩—⟨⟩—O₂C—⟨⟩—C(=O)—	59, 60
—O—⟨⟩—CO₂—⟨⟩—O₂C—⟨⟩—CO₂—⟨⟩—O₂C—⟨naph⟩—C(=O)—	61
—O—⟨⟩—CO₂—⟨⟩—O₂C—⟨naph⟩—C(=O)—	62, 63
—O—⟨⟩—CO₂—⟨naph⟩—C(=O)—	64
—O—⟨⟩—CO₂—⟨⟩—O₂C—⟨⟩—CO₂—⟨⟩—⟨⟩—O₂C—⟨⟩—C(=O)—	62
—O—⟨⟩—CO₂—⟨naph⟩—O₂C—⟨⟩—C(=O)—	65
—O—⟨⟩—CONH—⟨⟩—CO₂—⟨naph⟩—C(=O)—	66
—O—⟨⟩—CO₂—⟨⟩—O₂C—⟨naph⟩—CO₂—⟨⟩—C(=O)—	67
—C(=O)—⟨⟩—O₂C—⟨naph⟩—CO₂—⟨⟩—O₂C—⟨⟩—C(=O)—	68
—C(=O)—⟨⟩—CO₂—⟨⟩—CO₂—⟨⟩—⟨⟩—O₂C—⟨⟩—C(=O)—	69
—O—⟨⟩—C(=O)—	70
—O—⟨⟩—CO₂CH₂CH₂O₂C—⟨⟩—C(=O)—	11, 71

Table 7 (Continued)

Structure[a]	Ref.
—O—⟨⟩—O₂C—⟨⟩—CH₂CO₂—⟨⟩—C(=O)—	72
—O—⟨⟩(—C₆H₄X)—O₂C—⟨⟩—C(=O)—	73

[a] The structures are not representative of the chemical composition.

(2) Introduction of flexible spacers of different length and type in a regular or random manner. The latter is illustrated by the copolyester obtained by inserting oxybenzoate units into poly(ethylene terephthalate) by a transesterification process (equation 2).[11,71]

$$HO\text{-}(\text{-}C(=O)\text{-}C_6H_4\text{-}CO_2CH_2CH_2O\text{-})\text{-}H + AcO\text{-}C_6H_4\text{-}CO_2H \longrightarrow$$

$$HO\text{-}(\text{-}C(=O)\text{-}C_6H_4\text{-}CO_2CH_2CH_2O_2C\text{-}C_6H_4\text{-}O\text{-})\text{-}H + AcOH \quad (2)$$

(3) Introduction of different size aromatic units each of which forms a perfectly linear structure, such as 2,6-naphthalene units in combination with *p*-phenylene units.

(a) Introduction of lateral substituents in random fashion

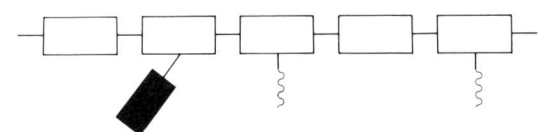

(b) Introduction of flexible spacers within the chain backbone

(c) Introduction of mesogenic counits of different shape and size

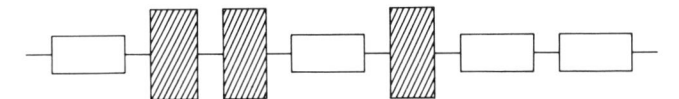

(d) Introduction of colinearity kinks

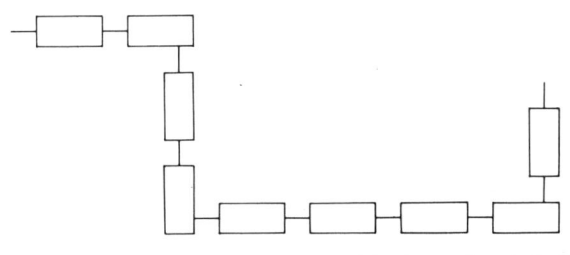

Scheme 3 Representation of the structural designs adopted for improving the bulk tractability of rigid-rod mesomorphic polymers[74]

(4) Introduction of nonlinear aromatic units acting as kinks into the linear rigid-rod structure, but these units also decrease the thermal stability of the LC phase.

38.3.2 Cellulose Derivatives

Cellulose derivatives capable of displaying thermotropic liquid crystalline behavior can be classified as cellulose ether–esters, normally hydroxypropylcellulose, with some additional ester functions localized on the ether branches,[54, 75–80] whose general structure is represented by (6). The molar average degrees of etherification and esterification are 6–8 and 5–6, respectively, for the cellobiose unit. The cellulose backbone appears to be sufficiently stiff to allow the formation of ordered phases with nearly parallel orientation of the macromolecules. Preferential chirality in the repeating units imparts a preferential twist to the parallel assembly of the macromolecules, thus inducing the helical structure typical of cholesteric mesophases.

(6) R = —CH$_2$CHMe,   R' = —CH$_2$CH(Me)OCH$_2$CHMe,   where X = H, MeCO—, CF$_3$CO—, EtCO—, PhCO—
         | |
         OX   OX

Incidence and stability of the mesophases are affected to a significant extent by the average molecular weight of the hydroxypropylcellulose ethers, and for molecular weights lower than 50×10^3 Dalton, no definite mesophase is established.

Selective reflection of the orthogonal incident light is usually caused by the thermotropic cholesteric mesophases. The position of the maximum of reflected light is shifted towards the red with increasing temperature, in contrast with the contraction of cholesteric helix occurring in thermotropic low molar mass LCs.

38.3.3 Polypeptides

Polypeptides analogous to cellulose derivatives (when suitably substituted on the functional groups of the amino acid residues) are able to give rise to anisotropic melts with cholesteric structures. The first reported examples are represented by copolymers of γ-n-alkyl-L-glutamates, as represented by (7).[81] Depending upon the chemical composition, the prepared copolymers form mesophases susceptible of selectively reflecting the visible light. The wavelength of reflected light increases, at any given copolymers composition, with increasing temperature, analogous to the behavior of cellulose derivative mesophases and lyotropic solutions of poly(γ-benzyl-L-glutamate). Such a behavior can be interpreted assuming that the ether–ester branches in cellulose derivatives and the alkyl substituents in copoly(γ-n-alkyl-L-glutamate)s play the role of solvents in essentially concentrated solutions.[82]

—(—NHCHCO—)$_x$—(—NHCHCO—)$_y$—
       | |
  CH$_2$CH$_2$CO$_2$R   CH$_2$CH$_2$CO$_2$R'

(7) a: R = Me,   R' = n-hexyl
     b: R = Me,   R' = n-octyl
     c: R = n-propyl, R' = n-octyl

38.3.4 Semiflexible Liquid Crystalline Polymers

The LCPs belonging to this class are constituted by linear macromolecules with repeating units consisting of a mesogenic group sequenced in a regular fashion by a flexible segment of different length and nature (oligomethylenes, oligoethers and oligosiloxanes). Almost exclusively, they are characterized by the presence of an ester group as a key chemical function recurring in the repeating units. A convenient and summarizing schematic representation of a repeating unit in a semiflexible LCP is reported in Table 8.[83] The first example of this polymer class reported in open literature dates back to the mid-1970s.[10] Since then a great variety of new LCPs have been synthesized and the property–structure correlation thoroughly investigated. In Table 9 the structural features of some major examples of semiflexible LCPs which have been reported in the literature are shown.

Table 8 Schematic Representation of Semiflexible Main-chain LCP[a]

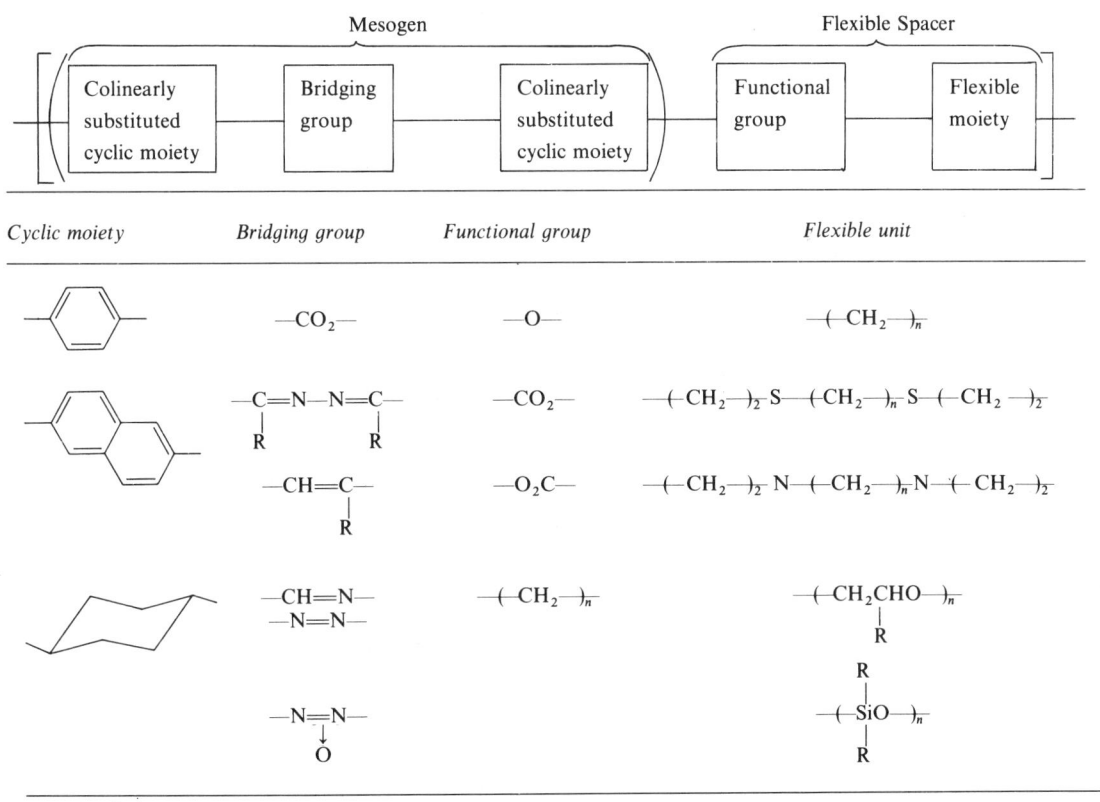

Cyclic moiety	Bridging group	Functional group	Flexible unit
phenyl	$-CO_2-$	$-O-$	$-(CH_2)_n-$
naphthyl	$-C(R)=N-N=C(R)-$	$-CO_2-$	$-(CH_2)_2-S-(CH_2)_n-S-(CH_2)_2-$
	$-CH=C(R)-$	$-O_2C-$	$-(CH_2)_2-N-(CH_2)_n-N-(CH_2)_2-$
cyclohexyl	$-CH=N-$, $-N=N-$	$-(CH_2)_n-$	$-(CH_2CHO)_n-$ with R
	$-N=N(O)-$		$-(SiRO)_n-$ with R

[a] R = H, alkyl.

Analogous to the side-chain LCP, the presence of a spacer in the repeating units plays a fundamental role in decoupling the intramolecular interactions among the mesogens, thus allowing for substantial decreases in the melting and clearing temperatures, accompanied by appreciable solubility in conventional organic solvents. Accordingly most of the polymers belonging to this class have been extensively characterized in bulk by thermal optical analysis, X-ray diffraction and IR at variable temperatures and in solution by spectral methods (NMR, UV) or viscometric and GPC techniques.

It has been possible therefore to determine thermal parameters such as glass transition (T_g) and melting (T_m) temperature and phase transition temperature(s) within the mesomorphic state up to the isotropic melt, and thermodynamic quantities such as ΔH and ΔS relevant to the several thermal transitions. These data, together with phase diagrams and in some cases rheological responses, have allowed semiquantitative relationships to be established with the inherent structural parameters. In this respect, incidence, structure and stability of the polymer mesophases have been related[83] to: (1) extension and nature of the mesogens; (2) nature of the bridging group(s) within the mesogens; (3) nature and relative orientation of the functional groups connecting the mesogens to the flexible spacers; (4) presence of lateral substituents on the benzenoid mesogens; (5) nature, length and parity of the flexible spacers; and (6) enantiomeric purity and relative topology of the chiral center(s), when optically active flexible spacers are used.[82,115,116]

Table 9 Structural Components of Major Main-chain Semiflexible LCPs

Mesogenic moiety	Flexible spacer	n	Interconnecting group	Ref.
–⟨Ph⟩–C(Me)=N–N=C(Me)–⟨Ph⟩–O–	–O–(CH$_2$)$_n$–O–	6–12	Carbonate	84, 85
–⟨Ph⟩–C(Me)=N–N=C(Me)–⟨Ph⟩–O–	–O–(CH$_2$CH$_2$O)$_n$–	2–4	Carbonate	86
–⟨Ph⟩–C(Me)=N–N=C(Me)–⟨Ph⟩–O–	–C(=O)–(CH$_2$)$_n$–C(=O)–	8–12	Ester	10, 87
–⟨Ph⟩–C(Me)=CH–⟨Ph⟩–O–	–C(=O)–(CH$_2$)$_n$–C(=O)–	6–12	Ester	87–89
–⟨Ph⟩–C(Me)=CH–⟨Ph⟩–O–	–O–(CH$_2$CH$_2$O)$_n$–	2–4	Ester	86
–⟨Ph⟩–CH=CH–⟨Ph⟩–C(=O)–	–O–(CH$_2$)$_n$–O–	5–10	Ester	90
–⟨Ph⟩–CH=CH–⟨Ph⟩–O–	–C(=O)–(CH$_2$)$_n$–C(=O)–	10	Ester	91
–⟨Ph⟩–CH=N–⟨Ph⟩–O–	–C(=O)–(CH$_2$)$_n$–C(=O)–	10	Ester	91, 92

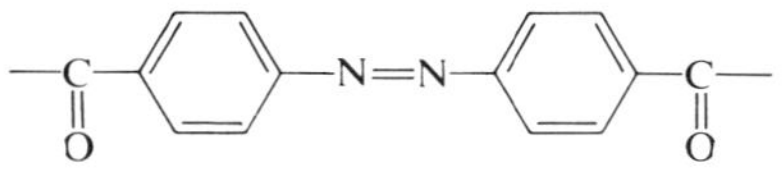	$-O-(CH_2)_n-C(=O)-$	4–16	Ester	93
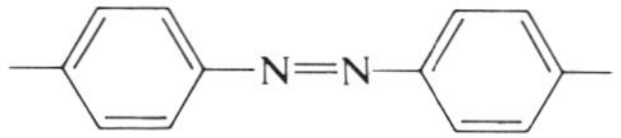	$-O-(CH_2CH_2O)_n-$	2–4	Ether	94
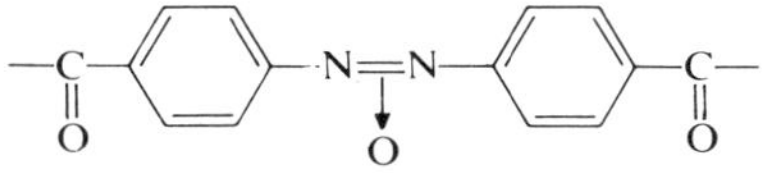	$-O-(CH_2)_n-O-$	4–16	Ester	93
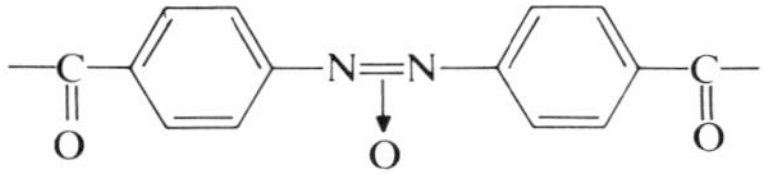	$-O-(CH_2CH_2O)_n-$	2–4	Ester	94
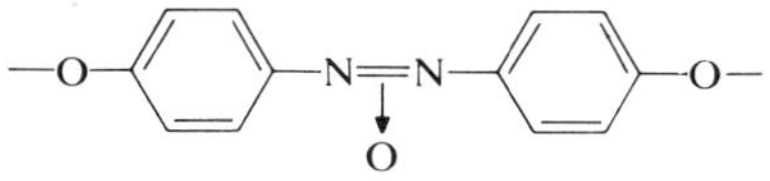	$-C(=O)-(CH_2)_n-C(=O)-$	10	Ester	93
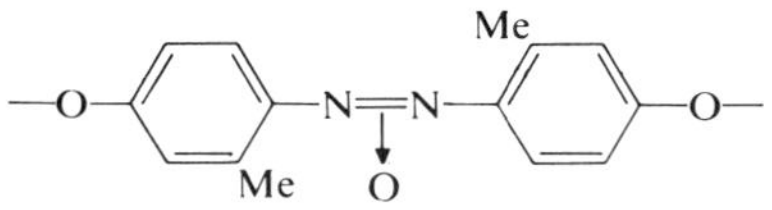	$-C(=O)-(CH_2)_n-C(=O)-$	10	Ester	95
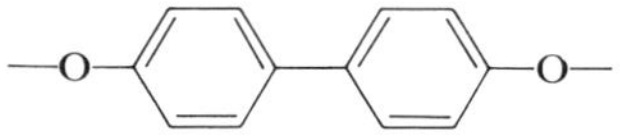	$-C(=O)-(CH_2)_n-C(=O)-$	5–12	Ester	91, 92, 96
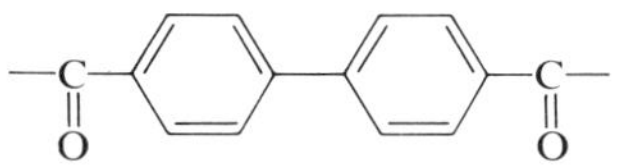	$-O-(CH_2)_n-O-$	2–6	Ester	90
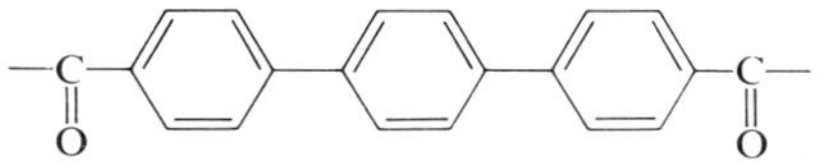	$-O-(CH_2)_n-O-$	2–6	Ester	90

Table 9 (Continued)

Mesogenic moiety	Flexible spacer	n	Interconnecting group	Ref.
4,4'-biphenyl diacyl	$-O-(CH_2CH_2O)_n-$	2–10	Ester	90
phenyl benzoate	$-O-(CH_2)_n-O-$	2–11	Ether	96, 97
phenyl benzoate	$-O_2C-(CH_2)_n-CO_2-$	4–8	Ester	98
phenyl benzoate	$-O-(CH_2CH_2O)_n-$	4	Ether	98
phenyl benzoate	$-O-(CH_2)_3-(SiO)_m-Si(Me)_2-(CH_2)_3-O-$ (Me substituents)	2–5	Siloxane	99
phenyl benzoate (Ph substituted)	$-O-(CH_2)_n-O-$	10	Ether	100
phenyl benzoate (X, Y substituted); X = H, Cl; Y = H, Cl, Me, Br	$-O-(CH_2)_n-O-$	10	Ether	101
phenyl benzoate (X, Y substituted); X = H; Y = Br	$-OCH_2SiO-SiCH_2O-$ (Me substituents)	10	Siloxane	—

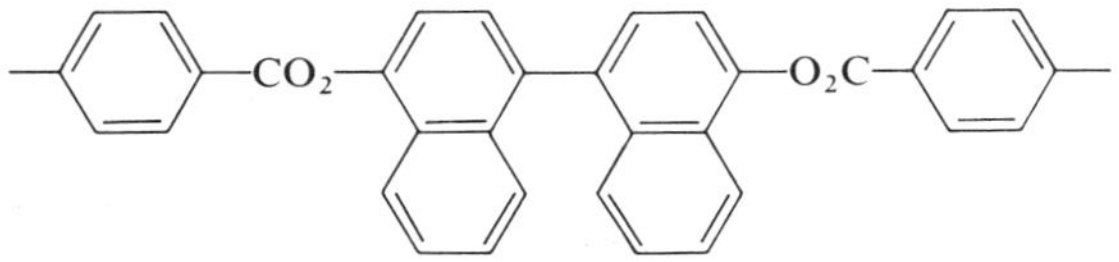	$-O-(CH_2)_n-O-$	10	Ether	100
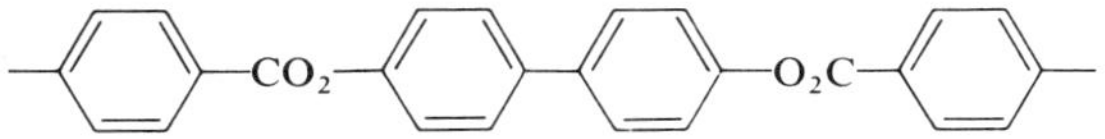	$-O-(CH_2)_n-O-$	9–11	Ether	96–100
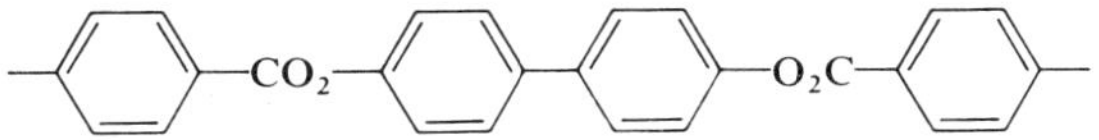	$-O-(CH_2)_3-(SiMe_2O)_n-SiMe_2-(CH_2)_3-O-$	2, 3	Siloxane	103
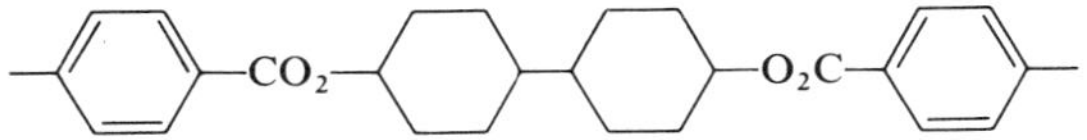	$-O-(CH_2)_3-(SiMe_2O)_n-SiMe_2-(CH_2)_3-O-$	2–5	Siloxane	99
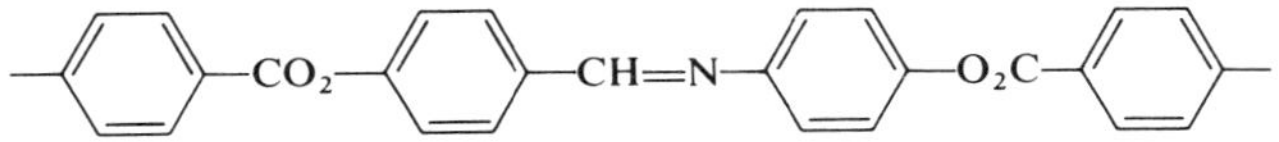	$-O-(CH_2)_3-(SiMe_2O)_n-SiMe_2-(CH_2)_3-O-$	2, 3	Siloxane	103
	$-O-(CH_2)_n-O-$	5–10	Ether	101
	$-OCH_2SiMe_2O-SiMe_2CH_2O-$	—	Siloxane	102
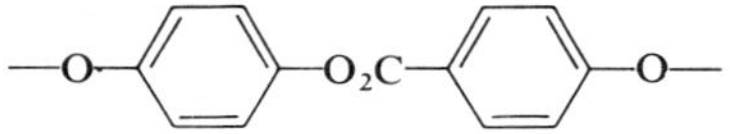	$-C(=O)-(CH_2)_n-C(=O)-$	3–20	Ester	96, 104
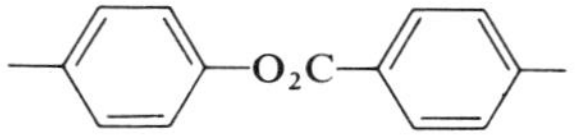	$-O-(CH_2)_n-O-$	4–9	Ether	96

Table 9 (*Continued*)

Mesogenic moiety	Flexible spacer	n	Interconnecting group	Ref.
ph-O₂C-ph-(mesogen 1)	—O—(CH₂)$_n$—O—	copolymer 2–12 / 6–10	Ether	105–107
ph-O₂C-ph-CO₂-ph-C(=O)—	—O—(CH₂)$_n$—O—	2–12	Ester	108
ph-O₂C-ph-CO₂-ph-C(=O)—	—O—(CH₂CH₂O)$_n$—	2–13	Ester	109
ph-O₂C-ph-C(=O)—	—O—(CH₂CH₂O)$_n$—	3	Ester	110
ph-O₂C-ph-C(=O)—	—O—(CH₂)$_n$—O—	2–10	Ester	108, 110
—C(=O)-ph-N=CH-ph-CH=N-ph-C(=O)—	—O—(CH₂)$_n$—O—	2–12	Ester	111
—C(=O)-ph-CH=N-ph-ph-N=CH-ph-C(=O)—	—O—(CH₂)$_n$—O—	12	Ester	112
—O-ph-CO₂-ph-CO₂(CH₂)$_m$O₂C-ph-O—	—C(=O)(CH₂)₂N(R')—R—N(R')(CH₂)₂C(=O)—	6–12	Ester	113
—O-ph-CO₂-ph-CO₂(CH₂)$_n$O₂C-ph-O—	—C(=O)(CH₂)₂S(CH₂)$_m$S(CH₂)₂C(=O)—	6–12	Ester	114

From the above analysis valuable hints can be gained in view of the design and formation of new main-chain LCPs with properties tuned to defined needs.

38.4 REFERENCES

1. F. Reinitzer, *Monatsh. Chem.*, 1888, **9**, 121.
2. O. Lehmanh, *Z. Phys.*, 1889, **4**, 462.
3. A. Fifedei, *Ann. Phys. (Leipzig)*, 1922, **18**, 273.
4. P. J. Flory, *Proc. R. Soc. London, Ser. A*, 1936, **234**, EO.
5. M. C. Rauden and N. W. Pirie, *Proc. R. Soc. London., Ser. B*, 1937, **123**, 274; H. Boedther and N. S. Simmons, *J. Am. Chem. Soc.*, 1958, **80**, 2550.
6. A. Elliot and E. J. Ambrose, *Discuss. Faraday Soc.*, 1950, **9**, 246; C. Robinson, *Trans. Faraday Soc.*, 1956, **52**, 571.
7. S. L. Kwolek, *Fr. Pat.* 1 526 745 (1968); *Br. Pat.* 1 283 064 (1968); *US Pat.* 3 600 350 (1971).
8. I. Goodman, J. E. McIntyre and J. W. Simpson, *Br. Pat.* 989 522 (1962).
9. The Carborundum Co., *Br. Pat.* 1 303 484 (1969).
10. A. Roviello and A. Sirigu, *J. Polym. Sci., Polym. Lett. Ed.*, 1975, **13**, 455.
11. W. J. Jackson and H. F. Kuhfuss, *J. Polym. Sci., Polym. Chem. Ed.*, 1976, **14**, 2093.
12. A. Blumstein and E. C. Hsu, in 'Liquid Crystalline Order in Polymers', ed. A. Blumstein, Academic Press, New York, 1978, p. 105.
13. S. P. Papkov, *Vysokomol. Soedin., Ser. B*, 1973, **15**, 537.
14. A. Ciferri, *Polym. Eng. Sci.*, 1975, **15**, 191.
15. C. Robinson, J. C. Ward and R. B. Beevers, *Discuss. Faraday Soc.*, 1958, **25**, 29.
16. D. G. H. Ballard (Courtaulds Ltd.), *Br. Pat.* 864 962 (1958).
17. Du Pont de Nemours, 'Information Bulletin 6E', 1974.
18. S. R. Allen, A. G. Fillippov, R. J. Farris and E. L. Thomas, in 'The Strength and Stiffness of Polymers', ed. J. R. Schafgen, A. E. Zachariades and R. S. Porter, Dekker, New York, 1988, vol. 9, p. 357.
19. P. W. Morgan, *Macromolecules*, 1977, **10**, 1381.
20. S. L. Kwolek, P. W. Morgan, J. R. Schaijgen and L. W. Gulrich, *Macromolecules*, 1977, **10**, 1390.
21. E. Magat, *Philos. Trans. R. Soc. London, Ser. A*, 1980, **294**, 463.
22. S. L. Kwolek, *US Pat.* 3 671 542 (1972).
23. P. W. Morgan, *US Pat.* 3 804 791 (1972).
24. P. W. Morgan and L. W. Gulrich, *US Pat.* 3 836 498 (1974).
25. N. Ogata, K. Sanui and T. Koyama, *J. Polym. Sci., Polym. Chem. Ed.*, 1981, **19**, 151.
26. Z. B. Li and M. Weda, *J. Polym. Sci., Polym. Chem. Ed.*, 1984, **22**, 3063.
27. P. W. Morgan, *US Pat.* 3 801 528 (1974).
28. S. L. Kwolek, *Br. Pat.* 1 198 081 (1966).
29. S. M. Aharoni, *J. Polym. Sci., Polym. Phys. Ed.*, 1981, **19**, 281.
30. T. C. Pletcher and P. W. Morgan, *J. Polym. Sci., Polym. Chem. Ed.*, 1980 **18**, 643.
31. D. Hartzler and P. W. Morgan, in 'Contemporary Topics in Polymer Science', ed. E. M. Pearce and J. R. Schaefgen, Plenum Press, New York, 1977, vol. 2, p. 19.
32. P. W. Morgan, *J. Polym. Sci., Polym. Symp.*, 1978, **65**, 1.
33. G. Marrucci and A. Ciferri, *J. Polym. Sci., Polym. Lett. Ed.*, 1977, **15**, 643.
34. B. Valents and A. Ciferri, *J. Polym. Sci., Polym. Lett. Ed.*, 1978, **16**, 657.
35. A. N. Cogswell, *Br. Polym. J.*, 1980, **12**, 170.
36. L. L. Chapoy and N. F. La Cour, *Rheol. Acta*, 1980, **19**, 731.
37. I. Weniatsu and Y. Uematsu, *Adv. Polym. Sci.*, 1984, **59**, 38.
38. D. B. Dufore, *J. Appl. Polym. Sci., Appl. Polym. Symp.*, 1985, **41**, 69.
39. B. Milland, A. Thierry and A. Skoulios, *J. Phys., Lett. (Orsay, Fr.)*, 1979, **40**, 607.
40. R. J. M. Nolte and W. Drenth, *Recl. Trav. Chim. Pays-Bas*, 1973, **92**, 788.
41. F. Millich, E. Hellmuth and S. Y. Huang, *J. Polym. Sci., Polym. Chem. Ed.*, 1975, **13**, 2143.
42. S. M. Aharoni, *J. Polym. Sci., Polym. Chem. Ed.*, 1979, **17**, 683.
43. S. M. Aharoni and E. K. Walsh, *Macromolecules*, 1979, **12**, 271.
44. S. M. Aharoni, *Macromolecules*, 1979, **12**, 94; *Macromolecules*, 1981, **14**, 222.
45. S. M. Aharoni, *J. Polym. Sci., Polym. Phys. Ed.*, 1980, **18**, 1439.
46. See for instance: H. R. Allcock, in 'Phosphorus–Nitrogen Compounds', Academic Press, New York, 1972.
47. R. E. Singler, N. S. Schneider and G. L. Hagnauer, *Polym. Eng. Sci.*, 1975, **15**, 321.
48. J. F. Wolfe, B. H. Loo and F. E. Arnold, *Macromolecules*, 1981, **14**, 915.
49. S. G. Chu, S. Venkatraman, G. C. Berry and Y. Einaga, *Macromolecules*, 1981, **14**, 939.
50. G. C. Berry, P. M. Cotts and S. G. Chu, *Polym. J.*, 1981, **13**, 47.
51. E. W. Chow and S. N. Kim, *Macromolecules*, 1981, **14**, 920.
52. M. B. Polk, K. B. Pota, M. Mandu, M. Phingbodhipokkiya and C. Edeogu, *Macromolecules*, 1984, **17**, 129.
53. S. Woebowyz and D. G. Gray, *Mol. Cryst. Liq. Cryst. Lett.*, 1967, **34**, 97.
54. D. G. Gray, *J. Appl. Polym. Sci., Appl. Polym. Symp.*, 1983, **37**, 179.
55. D. L. Patel and R. D. Gilbert, *J. Polym. Sci., Polym. Phys. Ed.*, 1981, **19**, 1231; *J. Polym. Sci., Polym. Phys. Ed.*, 1981, **19**, 1449.
56. P. Nard and J. M. Haudin, *Br. Polym. J.*, 1980, **12**, 174.
57. H. Chauzy, A. Penguy, S. Channis and P. Monzic, *J. Polym. Sci., Polym. Phys. Ed.*, 1980, **18**, 1137.
58. G. Comis, G. Corazza, E. Biouchi, A. Tealdi and A. Ciferri, *J. Polym. Sci., Polym. Lett. Ed.*, 1984, **22**, 273.
59. S. G. Cottis, J. Economy and L. C. Wohrer, *Br. Pat.* 1 499 513 (1975).
60. S. G. Cottis, *Mod. Plast.*, 1975, **7**, 62.

61. W. J. Jackson and J. C. Morris, *Br. Pat.* 2 002 404 (1977).
62. G. W. Calundann, *Br. Pat.* 1 585 511 (1976).
63. W. J. Jackson, *Br. Polym. J.*, 1980, **12**, 154.
64. G. W. Calundann, *Br. Pat.* 2 006 242 (1976).
65. G. W. Calundann, *US Pat.* 4 184 996 (1978).
66. G. W. Calundann, L. F. Charbonneau and A. J. East, *US Pat.* 4 357 917.
67. G. W. Calundann, *US Pat.* 4 130 545 (1978).
68. G. W. Calundann, H. L. Davis, I. J. Gorman and R. M. Mininni, *Ger. Pat.* 2 721 787 (1977).
69. S. G. Cottis, J. Economy and L. C. Wohrer, *Ger. Pat.* 2 507 066 (1976).
70. S. G. Cottis, J. Economy and R. S. Storm, *Ger. Pat.* 2 248 127 (1973).
71. H. F. Kuhfuss and W. J. Jackson, *US Pat.* 3 778 410 (1973).
72. W. J. Jackson and H. F. Kuhfuss, *US Pat.* 4 140 846 (1979).
73. W. J. Jackson, G. G. Gebau and H. F. Kuhfuss, *US Pat.* 4 153 779 (1979).
74. B. P. Griffin and M. K. Cox, *Br. Polym. J.*, 1980, **12**, 147.
75. K. Shimamura, J. L. White and J. F. Fellers, *J. Appl. Polym. Sci.*, 1981, **26**, 2165.
76. S. L. Tseng, A. Valente and D. G. Gray, *Macromolecules*, 1981, **14**, 715.
77. S. M. Aharoni, *J. Polym. Sci., Polym. Lett. Ed.*, 1981, **19**, 495.
78. S. L. Tseng, G. V. Laivins and D. G. Gray, *Macromolecules*, 1982, **15**, 1262.
79. S. N. Bhadani and D. G. Gray, *Makromol. Chem., Rapid Commun.*, 1982, **3**, 449.
80. S. N. Bhadani, S. L. Tseng and D. G. Gray, *Makromol. Chem.*, 1984, **184**, 1727.
81. S. Kasuya, S. Sasaki, J. Watanabe, Y. Fukuda and I. Uematsu, *Polym. Bull. (Berlin)*, 1982, **7**, 241.
82. E. Chiellini and G. Galli, in 'Recent Advances in Liquid Crystalline Polymers', ed. L. L. Chapoy, Elsevier, London, 1985, p. 15.
83. C. K. Ober, J.-I. Jin and R. W. Lenz, *Adv. Polym. Sci.*, 1984, **59**, 103.
84. A. Ronello and A. Sirigu, *Eur. Polym. J.*, 1979, **15**, 61.
85. A. Ronello and A. Sirigu, *Eur. Polym. J.*, 1979, **15**, 423.
86. A. Ronello and A. Sirigu, *Gazz. Chim. Ital.*, 1980, **110**, 403.
87. A. Ronello and A. Sirigu, *Makromol. Chem.*, 1982, **183**, 895.
88. A. Ronello and A. Sirigu, *Makromol. Chem.*, 1979, **180**, 2543.
89. A. Ronello and A. Sirigu, *Makromol. Chem.*, 1980, **181**, 1799.
90. P. Meurisse, C. Noel, L. Monnerie and B. Fayolle, *Br. Polym. J.*, 1981, **13**, 55.
91. A. Blumstein, K. Sivaromakrishnan, S. B. Clough and R. B. Blumstein, *Mol. Cryst. Liq. Cryst. Lett.*, 1979, **49**, 255.
92. A. Blumstein, K. N. Sivaromakrishnan and S. B. Clough, *Polymer*, 1982, **23**, 47.
93. K. Imura, N. Koide and R. Ohta, *Rep. Prog. Polym. Phys. Jpn.*, 1981, **24**, 231.
94. K. Imura, N. Koide, N. Ohta and M. Takeda, *Makromol. Chem.*, 1981, **182**, 2563.
95. A. Blumstein, S. Visalayar, S. Ponrathnan, S. B. Clough and R. B. Blumstein, *J. Polym. Sci., Polym. Phys. Ed.*, 1982, **20**, 877.
96. D. Van Luyen and L. Strzelecki, *Eur. Polym. J.*, 1980, **16**, 303.
97. L. Strzelecki and D. Van Luyen, *Eur. Polym. J.*, 1980, **16**, 299.
98. S. S. Skorokhodov, A. Yu. Buhbin, A. A. Shepelevsky and S. Ya. Frankel, in 'Preprints Macro IUPAC (Florence)', 1980, vol. 2, p. 232.
99. H. Ringsdorf and A. Schneller, *Br. Polym. J.*, 1981, **13**, 43.
100. B.-W. Jo, R. W. Lenz and J.-I. Jin, *Makromol. Chem., Rapid Commun.*, 1982, **3**, 23.
101. S. Antoun, R. W. Lenz and J.-I. Jin, *J. Polym. Sci., Polym. Chem. Ed.*, 1981, **19**, 1901.
102. B.-W. Jo, J.-I. Jin and R. W. Lenz, *Eur. Polym. J.*, 1982, **18**, 233.
103. C. Anguilere, H. Ringsdorf, A. Schneller and R. Zentel, 'Preprints Macro IUPAC (Florence),' 1980, vol. 3, p. 306.
104. L. Strzelecki and L. Liebert, *Eur. Polym. J.*, 1981, **17**, 1271.
105. A. C. Griffin and S. J. Harens, *J. Polym. Sci., Polym. Phys. Ed.*, 1981, **19**, 951.
106. A. C. Griffin and S. J. Harens, *J. Polym. Sci., Polym. Lett. Ed.*, 1980, **18**, 259.
107. A. C. Griffin and S. J. Harens, *Mol. Cryst. Liq. Cryst. Lett.*, 1979, **49**, 239.
108. C. K. Ober, J.-I. Jin and R. W. Lenz, *Polym. J.*, 1982, **14**, 9.
109. G. Galli, E. Chiellini, C. K. Ober and R. W. Lenz, *Makromol. Chem.*, 1982, **183**, 2693.
110. C. K. Ober, R. W. Lenz, G. Galli and E. Chiellini, *Macromolecules*, 1983, **16**, 1034.
111. B. Millaud, T. Thiery, C. Strazielle and A. Skoulios, *Mol. Cryst. Liq. Cryst. Lett.*, 1979, **49**, 299.
112. D. Guillon and A. Skoulios, *Mol. Cryst. Liq. Cryst. Lett.*, 1979, **49**, 119.
113. A. S. Angeloni, M. Laus, C. Castillari, G. Galli, P. Ferruti and E. Chiellini, *Makromol. Chem.*, 1985, **186**, 977.
114. E. Chiellini, G. Galli, A. S. Angeloni, M. Laus and R. Pelligrini, *Liquid Crystals*, 1987, **2**, 529.
115. E. Chiellini and G. Galli, *Faraday Discuss. Chem. Soc.*, 1985, **79**, 241.
116. E. Chiellini and G. Galli, in 'Recent Advances in Mechanistic and Synthetic Aspects of Polymerization', ed. M. Fontanille and A. Guyot, Reidel, Dordrecht, 1987, p. 425.

39
Polymers with Side-chain Mesogenic Units

RUDOLF ZENTEL
Universität Mainz, FRG

39.1 INTRODUCTION	723
39.2 LC POLYMERS WITH ROD-LIKE MESOGENS	724
39.2.1 Polymers with One Mesogen per Repeating Unit	724
39.2.2 Polymers with Two Mesogens per Repeating Unit	726
39.2.3 Polymers with Less Than One Mesogen per Repeating Unit	727
39.2.4 Combined Main-chain/Side-chain Polymers	727
39.2.5 Polymers with Laterally Fixed Mesogens	728
39.2.6 Cross-linked LC Elastomers	730
39.3 LC POLYMERS WITH DISC-LIKE MESOGENS	730
39.4 REFERENCES	730

39.1 INTRODUCTION

Liquid crystalline (LC) polymers are usually prepared by combining formanisotropic structural units (so-called mesogenic groups) and polymer chains (Figure 1). The mesogenic groups used for this purpose are rigid rod-like or disc-like units, which are known to favour LC phases in the case of low molar mass substances. There are two main ways of combining the mesogenic groups and polymer chains. The mesogenic groups can either be incorporated actually into the polymer chains (LC main-chain polymers, see Volume 5, Chapter 38) or they can be attached to flexible polymer chains as side groups (LC side-chain polymers).

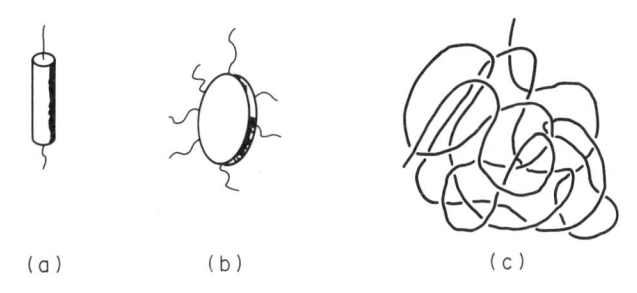

Figure 1 The combination of (a) rod-like, (b) disc-like mesogens and (c) polymer chains results in liquid crystalline polymers

Here we will discuss only LC side-chain polymers, in which the mesogenic groups are linked to the polymer chain as side groups *via* a flexible spacer. In this way a systematic synthesis of LC polymers is possible, by decoupling the polymer chains (random coils) and mesogenic groups (LC phases).[1-3] However, if the mesogenic groups are linked to a polymer chain without a flexible spacer, LC phases are rarely obtained.[4-6]

Thus, the attachment of mesogenic groups to a polymer chain *via* a spacer favours the formation of LC phases. Therefore LC polymers can be obtained by using non-LC monomers. The different types

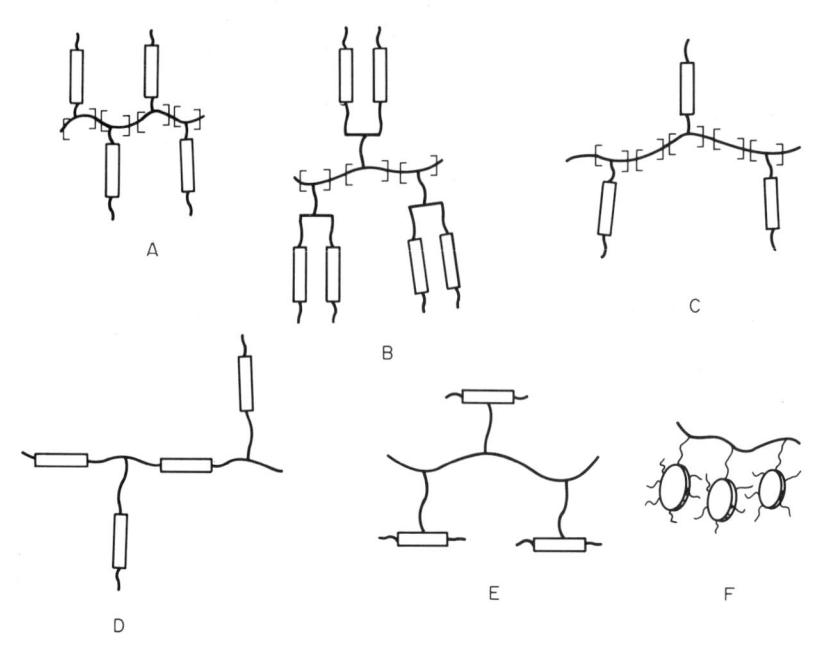

Figure 2 Different possibilities (types A–F) for the attachment of mesogenic groups to polymer chains

of geometrical structure obtained for LC polymers prepared by following this route are illustrated in Figure 2.

In most cases (types A–D), rod-like mesogenic groups are linked at one end to a polymer chain[7] (Figure 2). The polymer chain can be either flexible (types A–C) or rigid (type D). These rigid chains are mesogenic themselves. The ratio of mesogenic groups per repeating unit can be equal to one (type A), greater than one (type B) or less than one (type C). Alternatively, the mesogenic groups can be linked laterally to a polymer chain (type E). Only a few types of LC polymers with disc-like mesogens as side groups (type F) have been prepared.

39.2 LC POLYMERS WITH ROD-LIKE MESOGENS

39.2.1 Polymers with One Mesogen per Repeating Unit

Most of the LC polymers prepared so far belong to type A (Figure 2). The synthesis and properties of these polymers have been summarized in the literature.[8–11] The various chemical structures of the polymer chains, spacer groups and mesogenic groups used are illustrated in Figure 3. The structure and properties of some typical and well-investigated polymers are shown in Table 1.

Table 1 Phase Transitions[13,23,74–78,90,95] of the LC Polymers (**1**)

$$R^1\text{-}\underset{\underset{m}{|}}{\overset{\overset{CH_2}{|}}{C}}\text{-}CO_2(CH_2)_n\text{-}O\text{-}\!\!\left\langle\!\bigcirc\!\right\rangle\!\!\text{-}CO_2\text{-}\!\!\left\langle\!\bigcirc\!\right\rangle\!\!\text{-}R^2$$

(**1**)

R^1	R^2	n	Molecular weight	Phase transitions[a] (°C)
H	OMe	2	39 000	g 62 n 116 i
H	OMe	6	43 000	g 35 s_A 97 n 123 i
Me	OMe	6	>100 000	g 47 n/s 74 n 111 i
Me	OC_4H_9	6	>100 000	g 39 s_A 109 n 114 i

[a] g, Glassy liquid crystal; s_A, smectic A; n, nematic; i, isotropic; n/s, supercooled nematic phase with differing amount of crystalline or smectic phase.

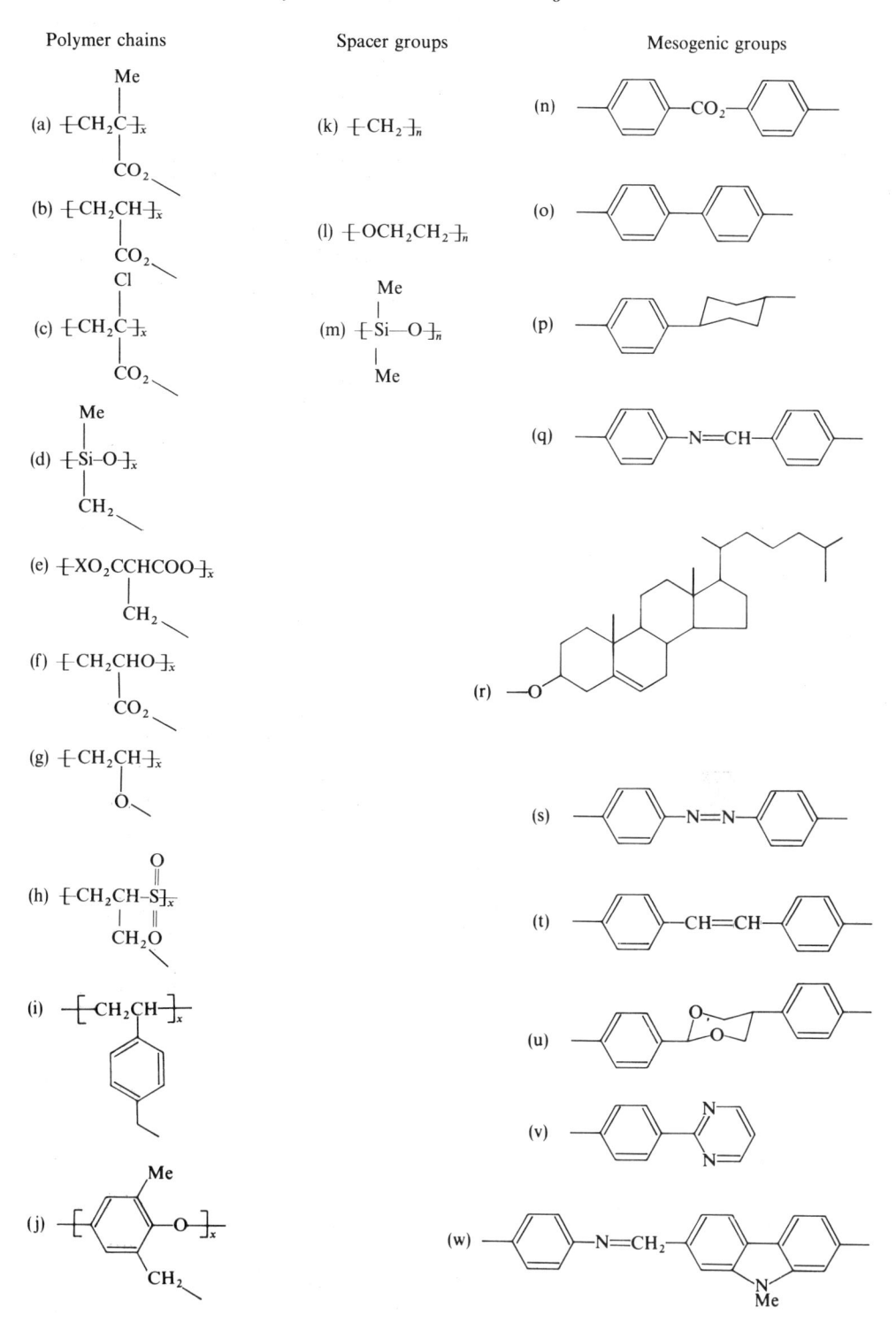

Figure 3 Polymer chains, spacer groups and mesogenic groups that have been used to prepare LC polymers

The polymer chains most commonly used are polymethacrylates (a);[1-3,8,10-22] polyacrylates (b);[8,10,11,18-31] poly(chloroacrylate)s (c);[13,17,32] polysiloxanes (d);[8,33-44] polyesters (e);[7,45] aliphatic polyethers (f);[46,47] poly(vinyl ether)s (g);[48] polysulfones (h);[49] poly(N-acylethyleneimine)s;[115] and more rigid structures, such as polystyrenes (i)[50,51] and aromatic polyethers (j).[52] The flexible spacer groups are usually oligo(methylene) units, but oligo(ethylene oxide)s[7,50,53,72] and oligo(siloxane)s[7] have also been used. The mesogenic groups are usually p-substituted phenyl benzoates (n);[1,8,11,17,26,30,35,44] biphenyls (o);[11,12,18,19,21,35] phenylcyclohexanes (p);[54,55] aromatic azomethines (q);[11,20,23,56] cholesterol derivatives (r);[11,24,57-58] azobenzenes (s);[16,45] stilbenes (t);[16] dioxanes (u);[37,43,59] phenylpyrimidines (v);[22] or substituted carbazoles (w).[60] In copolymers, non-mesogenic dyes have been used as comonomers.[55,61]

Most of the polymethacrylates and polyacrylates are prepared by free-radical polymerization. However, anionic polymerization[62] and group transfer polymerization[15] have also been tried. Polycondensation reactions[45] are used for the polyesters, and a ring-opening polymerization for some of the aliphatic polyethers.[46] In addition, polymer analogous reactions were used for the preparation of the polysiloxanes,[9,33,35,40] some of the polyacrylates and polymethacrylates,[14,25,27,31] and for the aromatic and aliphatic polyethers.[47,52]

The polymers thus prepared exhibit all the types of LC phases known from low molar mass liquid crystals (see Figure 4). Besides nematic phases, cholesteric[11,24,57,58,63-66] and different types of smectic phases[11,20,38,56,67-70] have also been clearly identified. More recently, polymers with chiral smectic C* phases[11,17,26,30,66] were prepared. These LC phases are especially interesting, because they show ferroelectric properties.[71] In contrast to low molar mass liquid crystals, LC polymers do not usually crystallize upon cooling, but form glassy LC phases.

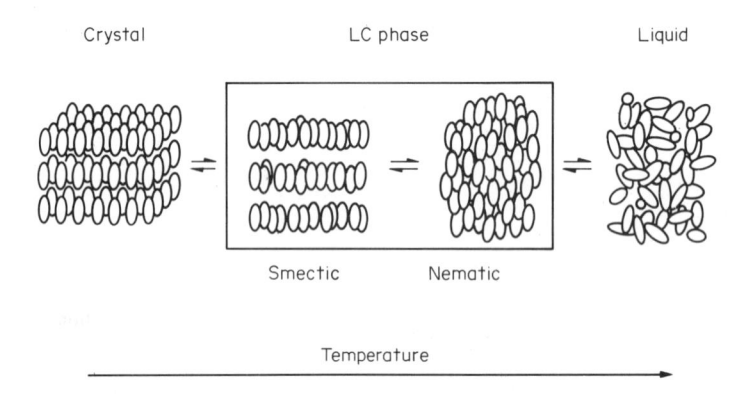

Figure 4 Thermotropic LC phases existing in a temperature range between the crystalline phase and the isotropic liquid

The order parameter of the mesogenic groups has been determined by various methods,[8-10,61] e.g. ^{1}H NMR and ESR spectroscopy,[73-76] and has been found to be in good agreement with the values measured for low molar mass liquid crystals. Structural investigations by X-ray measurements[11,20,32,56,67-70,77-81,113] showed a close analogy to low molar mass liquid crystals, and allowed a clear assignment of the LC phases stated above. Well-oriented LC polymers can be obtained by using electric and magnetic fields, as well as mechanical stretching.[9,10,42,70,82-86] This property of LC polymers can be used, for example, at high temperatures in LC displays.[19,84,87-89] The orientation thus obtained can afterwards be frozen in, below the glass transition temperature.

The differences between the properties of LC side-chain polymers and low molar mass liquid crystals are mainly related to their dynamic aspects. The reduced mobility of the LC side-chain polymers shows up clearly in the measurements of viscosity[10,90,91] and dielectric relaxation.[92-99] The increased orientation times of the mesogenic groups in electric fields,[19,87-89] corresponding to switching times in LC displays, are caused by a coupling of the orientation of the mesogenic groups and the preferred orientation of the polymer chain. The resulting anisotropy of the polymer chain has been established by small angle neutron scattering.[100-103,114]

In order to study the influence of the polymer chains and mesogenic groups on the LC and polymer properties, the ratio of mesogenic groups and repeating units were varied in polymers of types B and C (Figure 2).

39.2.2 Polymers with Two Mesogens per Repeating Unit

In order to enhance the role of the mesogenic groups (LC properties) with respect to the polymer chains, polymers with two mesogens per repeating unit (dimesogenic compounds, type B; Figure 2) were synthesized. They were based on polyitaconate main chains[54,103] (Figure 5) or polysiloxane main chains[7,44,54] (Figure 6; $y=0$). Some of the polyitaconates were prepared by free-radical

Figure 5 Structure and phase transition of a LC polyitaconate[54] (for a definition of the phase transitions, see footnote to Table 1)

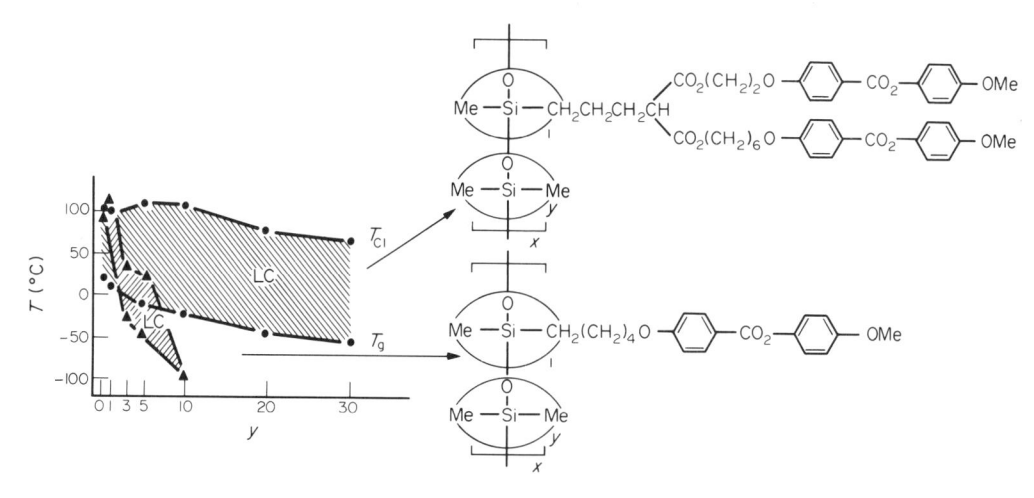

Figure 6 Influence of the content of non-mesogenic dimethylsiloxane units on the phase transitions of monomesogenic and dimesogenic copolysiloxanes[7]

polymerization,[54] but most of the polyitaconates[103] and all polysiloxanes[7,44,54] were prepared by polymer analogous reactions.

39.2.3 Polymers with Less Than One Mesogen per Repeating Unit

Copolymers of type C (Figure 2) were prepared, in order to strengthen the influence of the polymer chains on the properties of the LC polymers. Their synthesis led to the discovery of LC polymers with the low glass transition temperature of poly(dimethylsiloxane) (Figure 6). They were prepared by a polymer analogous reaction.[7,35,54] The 'dilution' of the mesogenic groups along the polymer chain by dimethylsiloxane segments leads to a decrease in the glass transition temperature and the clearing temperature.[7] In the case of monomesogenic compounds, an excess of ten dimethylsiloxane units per mesogen[35] causes destruction of the LC phase. However, the dimesogenic compounds retain their LC phases,[7,44,54] even when larger excesses of the non-mesogenic dimethylsiloxane units are used (Figure 6).

Finally, alternating copolymers[104] were prepared from methyl vinyl ether and maleates (Figure 7).

39.2.4 Combined Main-chain/Side-chain Polymers

An additional possibility for linking mesogenic groups onto one end of a polymer chain is realized in the combined main-chain/side-chain polymers[7,45,105] (type D; Figure 2). In this case, the semiflexible polymer chain is composed of alternating rigid and flexible segments. The structure of these polymers is one of a typical LC main-chain polymer (see Volume 5, Chapter 38). The different ways used to achieve such polymers[105] are illustrated in Figure 8 (for additional structures see also Figure 11).

The polymers[7,45,66,105,110] prepared according to Figure 8 (types D1 and D2) show surprisingly broad LC phases. They are, in most cases, broader than the LC phases of the corresponding pure main-chain or pure side-chain polymers. Polymers of type D1 favour smectic phases,[45,66] whereas

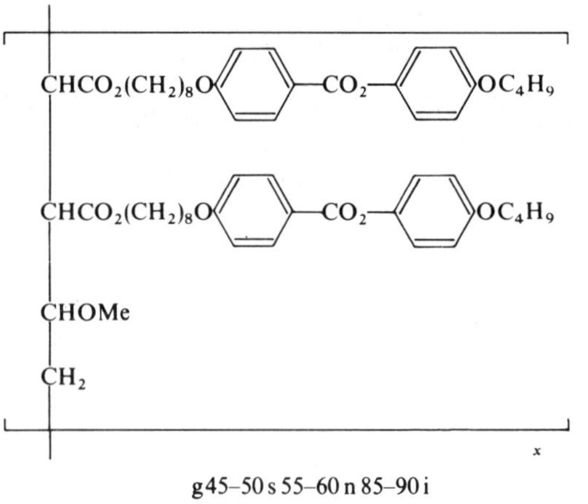

Figure 7 Alternating LC copolymers[104] (for a definition of the phase transitions, see footnote to Table 1)

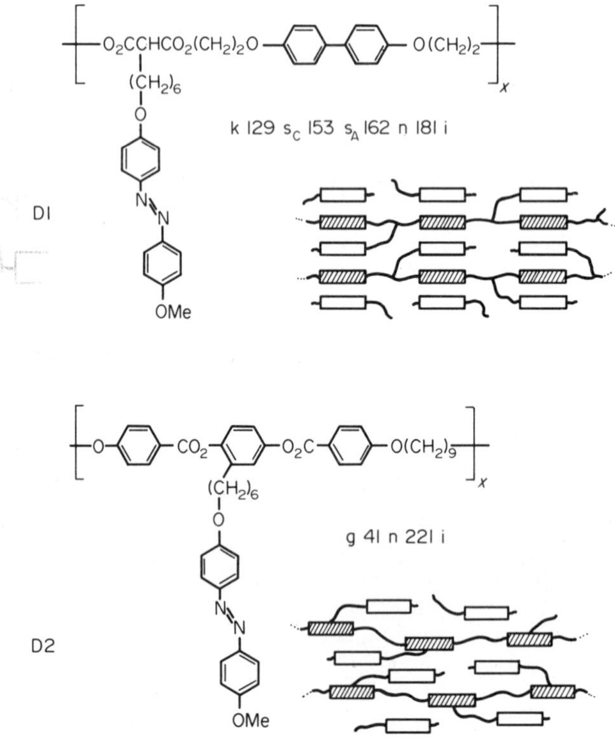

Figure 8 Different ways to realize combined main-chain/side-chain polymers[105] (s_C, smectic C; for a definition of the other phase transitions, see footnote to Table 1)

polymers of type D2 favour nematic phases.[105] Polymers with cholesteric and chiral smectic C* phases[66] have also been realized using this concept.

39.2.5 Polymers with Laterally Fixed Mesogens

In an attempt to vary the way that the mesogenic groups are attached to the polymer chain, polymers with laterally fixed mesogens[106-109] (type E; Figure 2) were prepared. These polymers are especially interesting, because they represent the first thermotropic liquid crystals with biaxial nematic phases.[106,108] One example of these polymers is given in Figure 9.

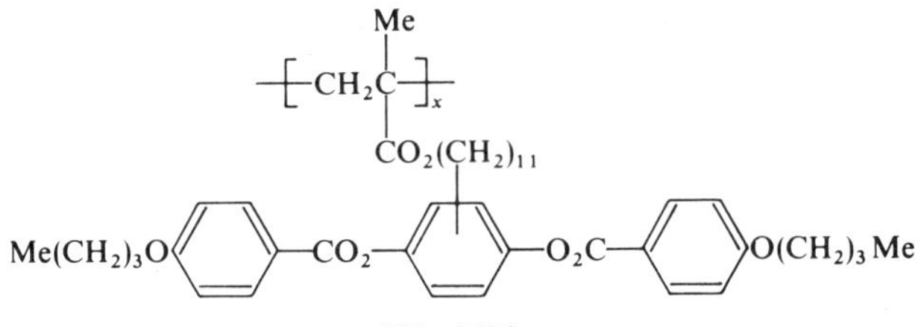

$$\{CH_2\underset{\underset{CO_2(CH_2)_{11}}{|}}{\overset{\overset{Me}{|}}{C}}\}_x$$

Me(CH₂)₃O—⟨⟩—CO₂—⟨⟩—O₂C—⟨⟩—O(CH₂)₃Me

g 292 n 340 i

Figure 9 Structure and phase transition for a LC polymer with laterally fixed mesogens[108] (for a definition of the phase transitions, see footnote to Table 1)

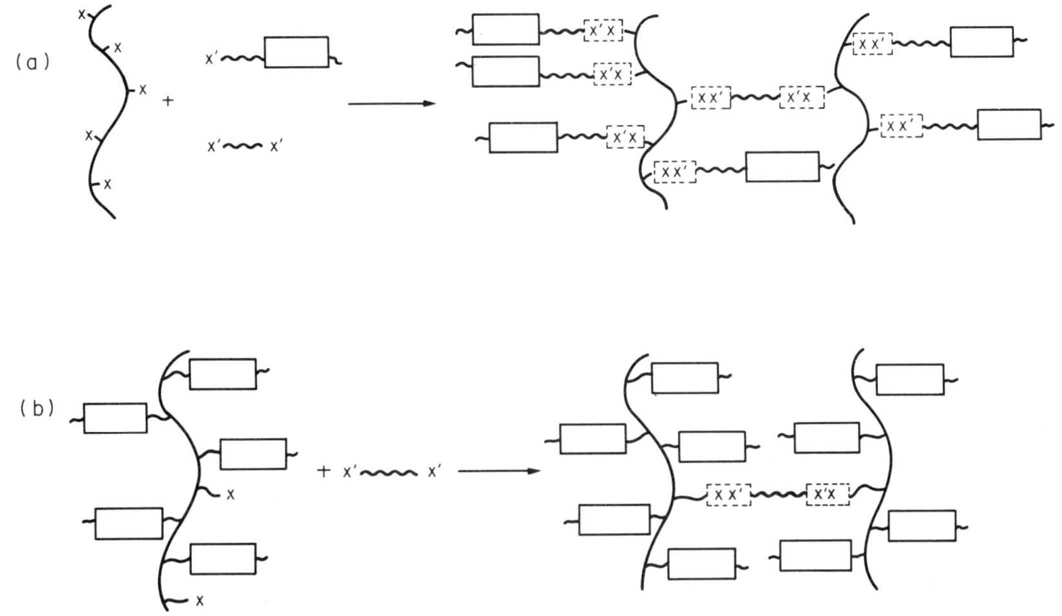

Figure 10 Preparation of lightly cross-linked LC elastomers in one step (a) or two steps (b)

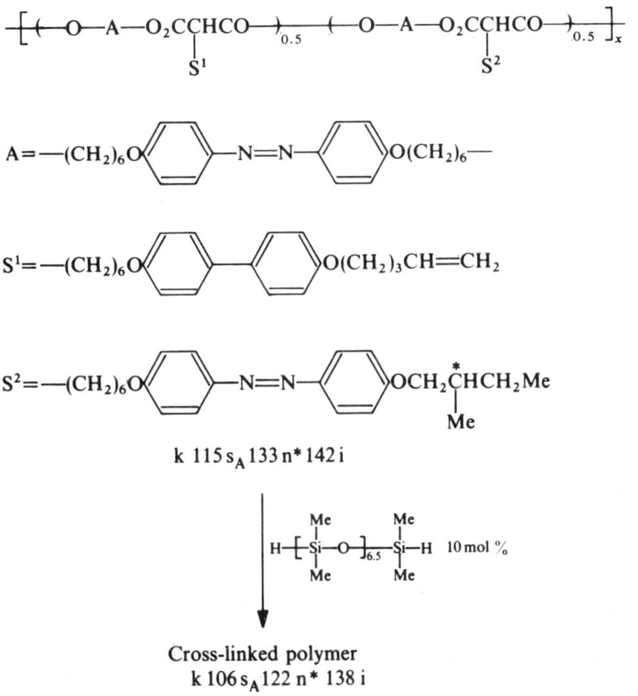

$$\{(-O-A-O_2CCHCO-)_{0.5}(-O-A-O_2CCHCO-)_{0.5}\}_x$$
$$\quad\quad\quad\quad\quad\quad\quad |\quad\quad\quad\quad\quad\quad\quad\quad\quad |$$
$$\quad\quad\quad\quad\quad\quad\quad S^1\quad\quad\quad\quad\quad\quad\quad\quad\; S^2$$

A = —(CH₂)₆O—⟨⟩—N=N—⟨⟩—O(CH₂)₆—

S¹ = —(CH₂)₆O—⟨⟩—⟨⟩—O(CH₂)₃CH=CH₂

S² = —(CH₂)₆O—⟨⟩—N=N—⟨⟩—OCH₂$\overset{*}{C}$HCH₂Me
$\quad\quad\quad\quad\quad\quad\quad\quad\quad\quad\quad\quad\quad\quad\quad\quad\quad\quad\quad\quad |$
$\quad\quad\quad\quad\quad\quad\quad\quad\quad\quad\quad\quad\quad\quad\quad\quad\quad\quad\quad\quad Me$

k 115 s$_A$ 133 n* 142 i

↓ H—[Si(Me)(Me)—O]₆.₅—Si(Me)(Me)—H 10 mol %

Cross-linked polymer
k 106 s$_A$ 122 n* 138 i

Figure 11 Synthesis and phase transitions of a LC elastomer with cholesteric phase[66] (k, crystalline; s$_A$, smectic A; n*, cholesteric; i, isotropic)

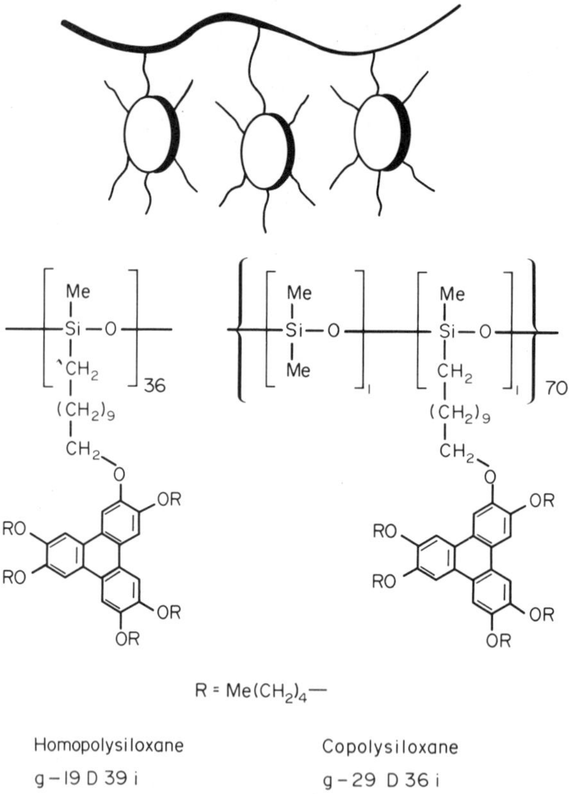

Figure 12 LC side-chain polymers with disc-like mesogenic groups (g, glassy liquid crystal; D, discotic; i, isotropic)

39.2.6 Cross-linked LC Elastomers

By using type A–E polymers (Figure 2), lightly cross-linked LC polymers with elastic properties have been obtained. The first LC elastomers[9,34,42,82,86] were prepared from polysiloxanes in a one-step reaction (Figure 10a). Thus the synthesis of the LC polymers (polymer analogous reaction) and the cross-linking were performed at the same time. LC elastomers based on polymethacrylates,[110] polyacrylates[29,110] and on combined main-chain/side-chain polymers[66,110] have also been prepared. These elastomers were normally prepared in two steps (Figures 10b and 11). The uncross-linked LC polymer is synthesized first and cross-linked afterwards;[110] polymers with laterally fixed mesogens have also been cross-linked in this way.[111] The special interest in these elastomers is due to their very good orientability in the LC phase on applying mechanical stress.[29,42,70,82,85,86]

39.3 LC POLYMERS WITH DISC-LIKE MESOGENS

LC side-chain polymers with discotic phases have been obtained[7,112] by fixing disc-like mesogens to a flexible polymer chain. The first polymers of this type were based on polysiloxane chains and used hexasubstituted triphenylene as the mesogen (Figure 12).

39.4 REFERENCES

1. H. Finkelmann, H. Ringsdorf and J. H. Wendorff, *Makromol. Chem.*, 1978, **179**, 273.
2. V. P. Shibaev and N. A. Platé, *Polym. Sci. USSR (Engl. Transl.)*, 1978, **A19**, 1065.
3. V. P. Shibaev, N. A. Platé and Ya. S. Freidzon, *J. Polym. Sci., Polym. Chem. Ed.*, 1979, **17**, 1655.
4. F. Cser, *J. Phys. (Orsay, Fr.)*, 1979, **40**, C3-459.
5. V. Frosini, G. Levita, D. Lupinacci and P. L. Magagnini, *Mol. Cryst. Liq. Cryst.*, 1981, **66**, 21.
6. J. Horvath, K. Nyitrai, F. Cser and G. Hardy, *Eur. Polym. J.* 1985, **21**, 251.
7. M. Engel, B. Hisgen, R. Keller, W. Kreuder, B. Reck, H. Ringsdorf, H.-W. Schmidt and P. Tschirner, *Pure Appl. Chem.*, 1985, **57**, 1009.

8. H. Ringsdorf and A. Schneller, *Br. Polym. J.*, 1981, **13**, 43.
9. H. Finkelmann and G. Rehage, *Adv. Polym. Sci.*, 1984, **60/61**, 99.
10. V. P. Shibaev and N. V. Platé, *Adv. Polym. Sci.*, 1984, **60/61**, 173.
11. N. A. Platé, R. V. Talroze, Ya. S. Freidzon and V. P. Shibaev, *Polym. J.*, 1987, **19**, 135.
12. H. Finkelmann, M. Happ, M. Portugall and H. Ringsdorf, *Makromol. Chem.*, 1978, **179**, 2541.
13. R. Zentel and H. Ringsdorf, *Makromol. Chem., Rapid Commun.*, 1984, **5**, 393.
14. P. Keller, *Mol. Cryst. Liq. Cryst. Lett.*, 1985, **2**, 101.
15. W. Kreuder, O. W. Webster and H. Ringsdorf, *Makromol. Chem., Rapid Commun.*, 1986, **7**, 5.
16. Y. H. Mariam, K. P. W. Pemawansa and F. Okoh, *Mol. Cryst. Liq. Cryst.*, 1986, **141**, 77.
17. S. Esselin, L. Bosio, C. Noel, G. Decobert and J. C. Dubois, *Liq. Cryst.*, 1987, **2**, 505.
18. V. P. Shibaev, S. G. Kostromin and N. A. Platé, *Eur. Polym. J.*, 1982, **18**, 651.
19. S. G. Kostromin, R. V. Talroze, V. P. Shibaev and N. A. Platé, *Makromol. Chem., Rapid Commun.*, 1982, **3**, 803.
20. R. V. Talroze, V. V. Sinitzyn, V. P. Shibaev and N. A. Platé, *Mol. Cryst. Liq. Cryst.*, 1982, **80**, 211.
21. P. A. Gemmel, G. W. Gray, D. Lacey, A. K. Alimoglu and A. Ledwith, *Polymer*, 1985, **26**, 615.
22. V. Krone and H. Ringsdorf, *Liq. Cryst.*, 1987, **2**, 411.
23. M. Portugall, H. Ringsdorf and R. Zentel, *Makromol. Chem.*, 1982, **183**, 2311.
24. A. M. Mousa, Ya. S. Freidzon, V. P. Shibaev and N. A. Platé, *Polym. Bull. (Berlin)*, 1982, **6**, 485.
25. C. M. Paleos, G. Margomenou–Leonidopoulou, S. E. Filippakis and A. Malliaris, *J. Polym. Sci., Polym. Chem. Ed.*, 1982, **20**, 2267.
26. V. P. Shibaev, M. V. Kozlovsky, L. A. Beresnev, L. M. Blinov and N. A. Platé, *Polym. Bull. (Berlin)*, 1984, **12**, 299.
27. P. Keller, *Macromolecules*, 1984, **17**, 2937.
28. K. P. Naikwadi, A. L. Jadhov, S. Rokushika and H. Hatano, *Makromol. Chem.*, 1986, **187**, 1407.
29. G. R. Mitchell, F. J. Davis and A. Ashman, *Polymer*, 1987, **28**, 639; F. J. Davis, A. Gilbert, J. Mann and G. R. Mitchell, *J. Chem. Soc., Chem. Commun.*, 1986, 1333.
30. G. Decobert, J. C. Dubois, S. Esselin and C. Noel, *Liq. Cryst.*, 1986, **1**, 307.
31. P. Keller, *Macromolecules*, 1987, **20**, 462.
32. G. Decobert, F. Soyer, J. C. Dubois and P. Davidson, *Polym. Bull. (Berlin)*, 1985, **14**, 549.
33. H. Finkelmann and G. Rehage, *Makromol. Chem., Rapid Commun.*, 1980, **1**, 31.
34. H. Finkelmann, H.-J. Kock and G. Rehage, *Makromol. Chem., Rapid Commun.*, 1981, **2**, 317.
35. H. Ringsdorf and A. Schneller, *Makromol. Chem., Rapid Commun.*, 1982, **3**, 557.
36. H. Ringsdorf, H.-W. Schmidt and A. Schneller, *Makromol. Chem., Rapid Commun.*, 1982, **3**, 745.
37. P. A. Gemmel, G. W. Gray and D. Lacey, *Mol. Cryst. Liq. Cryst.*, 1985, **122**, 205.
38. B. Krücke, H. Zaschke, S. G. Kostromin and V. P. Shibaev, *Acta Polym.*, 1985, **36**, 639.
39. M. A. Apfel, H. Finkelmann, G. M. Janini, R. J. Laub, B. H. Lühmann, A. Price, W. L. Roberts, T. J. Shaw, C. A. Smith, *Anal. Chem.*, 1985, **57**, 651.
40. G. W. Gray, D. Lacey, G. Nestor and M. S. White, *Makromol. Chem., Rapid Commun.*, 1986, **7**, 71.
41. M. Mauzac, F. Hardouin, M. Richard, M. F. Achard, G. Sigaud and H. Gasparoux, *Eur. Polym. J.*, 1986, **22**, 137.
42. W. Gleim and H. Finkelmann, *Makromol. Chem.*, 1987, **188**, 1489.
43. Ch. S. Hsu and V. Percec, *Polym. Bull. (Berlin)*, 1987, **17**, 49.
44. S. Diele, S. Oelsner, F. Kuschel, B. Hisgen, H. Ringsdorf and R. Zentel, *Makromol. Chem.*, 1987, **188**, 1993.
45. B. Reck and H. Ringsdorf, *Makromol. Chem., Rapid Commun.*, 1985, **6**, 291.
46. F. Cser, K. Nyitrai, J. Horvath and G. Hardy, *Eur. Polym. J.*, 1985, **21**, 259.
47. C. Pugh and V. Percec, *Polym. Bull. (Berlin)*, 1986, **16**, 521.
48. J. M. Rodriguez-Parada and V. Percec, *J. Polym. Sci., Polym. Chem. Ed.*, 1986, **24**, 1363.
49. D. Braun, R.-P. Herr, N. Arnold, *Makromol. Chem., Rapid Commun.*, 1987, **8**, 359.
50. V. Percec, J. M. Rodriguez-Parada and C. Ericson, *Polym. Bull. (Berlin)*, 1987, **17**, 347.
51. J. V. Crivello, M. Deptolla and H. Ringsdorf, *Liq. Cryst.*, 1988, **3**, 247.
52. C. Pugh and V. Percec, *Polym. Bull. (Berlin)*, 1986, **16**, 513.
53. R. Duran and Ph. Gramain, *Makromol. Chem.*, 1987, **188**, 2001.
54. B. Hisgen, W. Kreuder and H. Ringsdorf, 13. Arbeitstagung Flüssigkristalle (Workshop on Liquid Crystals), 23.03.-25.03.83, Freiburg, FR-Germany.
55. H. Ringsdorf, H.-W. Schmidt, H. Eilingsfeld and K.-H. Etzbach, *Makromol. Chem.*, 1987, **188**, 1355.
56. S. G. Kostromin, V. V. Sinitzyn, R. V. Talroze, V. P. Shibaev and N. A. Platé, *Makromol. Chem., Rapid Commun.*, 1982, **3**, 809.
57. V. P. Shibaev, H. Finkelmann, A. V. Kharitonov, M. Portugall, N. A. Platé and H. Ringsdorf, *Polym. Sci. USSR (Engl. Transl.)*, 1981, **23**, 1029.
58. Ya. S. Freidzon, Y. G. Tropsha, V. P. Shibaev and N. A. Platé, *Makromol. Chem., Rapid Commun.*, 1985, **6**, 625.
59. C. S. Hsu, J. M. Rodriguez-Parada and V. Percec, *Makromol. Chem.*, 1987, **188**, 1017.
60. M. Lux, P. Strohriegl and H. Höcker, *Makromol. Chem.*, 1987, **188**, 811.
61. H. Ringsdorf, H.-W. Schmidt, G. Baur, R. Kiefer and F. Windscheid, *Liq. Cryst.*, 1986, **1**, 319.
62. B. R. Hahn, J. H. Wendorff, M. Portugall and H. Ringsdorf, *Colloid Polym. Sci.*, 1981, **259**, 875.
63. H. Finkelmann, J. Koldehoff and H. Ringsdorf, *Angew. Chem., Int. Ed. Engl.*, 1978, **17**, 935.
64. H. Finkelmann and G. Rehage, *Makromol. Chem., Rapid Commun.*, 1982, **3**, 859.
65. Ya. S. Freidzon, N. J. Boiko, V. P. Shibaev and N. A. Platé, *Eur. Polym. J.*, 1986, **22**, 13.
66. R. Zentel, G. Reckert and B. Reck, *Liq. Cryst.*, 1987, **2**, 83.
67. P. Zugenmaier and J. Mügge, *Makromol. Chem., Rapid Commun.*, 1984, **5**, 11.
68. S. G. Kostromin, V. V. Sinitzyn, R. V. Talroze and V. P. Shibaev, *Polym. Sci. USSR (Engl. Transl.)*, 1984, **26**, 370.
69. Ya. S. Freidzon, N. J. Boiko, V. P. Shibaev, V. V. Tsukruk, V. V. Shilov and Yu. Lipatov, *Polym. Commun.*, 1986, **27**, 190.
70. R. Zentel, G. Schmidt, J. Meyer and M. Benalia, *Liq. Cryst.*, 1987, **2**, 651.
71. L. M. Blinov, V. H. Baikalov, M. J. Barnik, L. A. Beresnev, E. P. Pozhidayev and S. V. Yablonsky, *Liq. Cryst.*, 1987, **2**, 121.
72. S. G. Kostromin, V. P. Shibaev and N. A. Platé, *Liq. Cryst.*, 1987, **2**, 195.
73. Ch. Boeffel, B. Hisgen, U. Pschorn, H. Ringsdorf and H. W. Spiess, *Isr. J. Chem.*, 1983, **23**, 388.

74. K.-H. Wassmer, E. Ohmes, M. Portugall, H. Ringsdorf and G. Kothe, *J. Am. Chem. Soc.*, 1985, **107**, 1511.
75. Ch. Boeffel, H. W. Spiess, B. Hisgen, H. Ringsdorf, H. Ohm and R. G. Kirste, *Makromol. Chem., Rapid Commun.*, 1986, **7**, 777.
76. U. Pschorn, H. W. Spiess, B. Hisgen and H. Ringsdorf, *Makromol. Chem.*, 1986, **187**, 2711.
77. R. Zentel and G. Strobl, *Makromol. Chem.*, 1984, **185**, 2669.
78. P. Davidson, P. Keller and A. M. Levelut, *J. Phys. (Orsay, Fr.)*, 1985, **46**, 939.
79. E. Nachaliel, E. N. Keller, D. Davidov, H. Zimmermann and M. Deutsch, *Phys. Rev. Lett.*, 1987, **58**, 896.
80. R. M. Richardson and N. J. Herring, *Mol. Cryst. Liq. Cryst.*, 1985, **123**, 143.
81. R. Duran, D. Guillon, Ph. Gramain and A. Skoulios, *Makromol. Chem., Rapid Commun.*, 1987, **8**, 181.
82. H. Finkelmann, H.-J. Kock, W. Gleim and G. Rehage, *Makromol. Chem., Rapid Commun.*, 1984, **5**, 287.
83. Z. A. Roganova, A. L. Smolyansky, S. G. Kostromin and V. P. Shibaev, *Eur. Polym. J.* 1985, **21**, 645.
84. H. J. Coles and R. Simon, *Mol. Cryst. Liq. Cryst. Lett.*, 1986, **3**, 37.
85. R. Zentel and M. Benalia, *Makromol. Chem.*, 1987, **188**, 665.
86. J. Schätzle and H. Finkelmann, *Mol. Cryst. Liq. Cryst.*, 1987, **142**, 85.
87. H. Ringsdorf and R. Zentel, *Makromol. Chem.*, 1982, **183**, 1245.
88. H. Finkelmann, U. Kiechle and G. Rehage, *Mol. Cryst. Liq. Cryst.*, 1983, **94**, 343.
89. H. J. Coles and A. I. Hopwood, *Mol. Cryst. Liq. Cryst. Lett.* 1985, **1**, 165.
90. R. Zentel and J. Wu, *Makromol. Chem.*, 1986, **187**, 1727.
91. F. Fabre and M. Veyssie, *Mol. Cryst. Liq. Cryst. Lett.*, 1987, **4**, 99.
92. H. Kresse and R. V. Talroze, *Makromol. Chem., Rapid Commun.*, 1981, **2**, 369; H. Kresse, E. Tennstedt and R. Zentel, *Makromol. Chem., Rapid Commun.*, 1985, **6**, 261.
93. T. I. Borisova, L. L. Burshtein, V. P. Malinovskaya, L. V. Krasner, I. I. Konstantinov and Yu. B. Amerik, *Polym. Sci. USSR (Engl. Transl.)*, 1984, **26**, 2372.
94. T. I. Borisova, L. L. Burshtein, N. A. Nikonorova, Ya. S. Freidzon, V. P. Shibaev and N. A. Platé, *Polym. Sci. USSR (Engl. Transl.)*, 1984, **26**, 1688.
95. R. Zentel, G. Strobl and H. Ringsdorf, *Macromolecules*, 1985, **18**, 960.
96. W. Haase, H. Pranoto and F. J. Bormuth, *Ber. Bunsenges. Phys. Chem.*, 1985, **89**, 1229; H. Pranoto, F. J. Bormuth, W. Haase, U. Kiechle and H. Finkelmann, *Makromol. Chem.* 1986, **187**, 2453.
97. W. Heinrich and B. Stoll, *Colloid Polym. Sci.*, 1985, **263**, 895.
98. G. S. Attard and G. Williams, *J. Mol. Electron.*, 1986, **2**, 107.
99. G. S. Attard, J. J. Moura-Ramos and G. Williams, *J. Polym. Sci., Polym. Phys. Ed.*, 1987, **25**, 1099.
100. R. G. Kirste and H. G. Ohm, *Makromol. Chem., Rapid Commun.*, 1985, **6**, 179.
101. P. Keller, B. Carvalho, J. P. Cotton, M. Lambert, F. Moussa and G. Pépy, *J. Phys., Lett. (Orsay, Fr.)*, 1985, **46**, 1065.
102. F. Moussa, J. P. Cotton, F. Hardouin, P. Keller, M. Lambert, G. Pépy, M. Mauzac and H. Richard, *J. Phys. (Orsay, Fr.)*, 1987, **48**, 1079.
103. P. Keller, *Macromolecules*, 1985, **18**, 2337.
104. P. Keller, *Makromol. Chem., Rapid Commun.*, 1985, **6**, 707.
105. B. Reck and H. Ringsdorf, *Makromol. Chem., Rapid Commun.*, 1986, **7**, 389.
106. F. Hessel and H. Finkelmann, *Polym. Bull. (Berlin)*, 1985, **14**, 375.
107. S. Berg, V. Krone and H. Ringsdorf, *Makromol. Chem., Rapid Commun.*, 1986, **7**, 381.
108. F. Hessel, R. P. Herr and H. Finkelmann, *Makromol. Chem.*, 1987, **188**, 1597.
109. Q. -F. Zhou, H. -M. Li and X. -D. Feng, *Macromolecules*, 1987, **20**, 233.
110. R. Zentel and G. Reckert, *Makromol. Chem.*, 1986, **187**, 1915.
111. R. P. Herr, F. Hessel, D. Pauschinger, J. Schätzle and H. Finkelmann, International Conference 'Liquid Crystal Polymers', Bordeaux, France, July 20–24, 1987, Abstract of Communications.
112. W. Kreuder and H. Ringsdorf, *Makromol. Chem., Rapid Commun.*, 1983, **4**, 807.
113. Yu. S. Lipatov, V. V. Tsukruk and V. V. Shilov, *Rev. Makromol. Chem. Phys.*, 1984, **C24**, 173.
114. A. B. Kunchenko and D. A. Svetogorskii, *J. Phys. (Orsay, Fr.)*, 1986, **47**, 2015.
115. J. M. Rodriguez-Parada and V. Percec, *J. Polym. Sci., Polym. Chem. Ed.*, 1987, **25**, 2269.

Subject Index

AB-PBD, synthesis, 504
AB-PBT, synthesis, 504
Accelerators, curing, polysiloxanes, 603
Acetaldehyde
  novolaks, alkylidene bridges, 625
  phenol resins, production, 614
Acetamide, dimethyl-, polyamides, 376
Acetic anhydride
  polyarylates, 321
  trifluoro-, polyesters, 285
Acetone, 1,3-dichloro-, polycondensation, 4,4'-oxydibenzenesulfinates, 571
Acetylene
  diphenyl-, copolymerization, diacetylene, 469
  phenyl-, copolymerization, diacetylene, 469
  vinyl-, copolymerization, diacetylene, 469
Acidolysis, polyesters, 292, 293
Acrylamide, hydrogen-transfer polymerization, 6, 74
Acrylic acid
  1,4-phenylenedi-
    ethyl ester, crystals, polymerization growth direction, 227
    ethyl ester, polymerization, 225
    methyl ester
      crystals, polymerization growth direction, 227
      photopolymerization, 224
      preparation, 219
Activation energy, thermal polymerization, diynes, 242
Acyl chlorides
  aliphatic, polyamides, 370
  polyarylates, 318
  polyesters, 294
Addition–cyclization polymers, 409
Addition–elimination mechanisms, polyesterification, 277
Addition polymerization, 2
Adhesiveness, polysiloxanes, 601
Adhesives
  epoxy resins, 668
    curing, 686
  polyurethanes, 419
  urea–formaldehyde resins, 662
Adipic acid
  polyesterification, pentaerythrite, 133
  polyesters, 339
Adiponitrile, reactions, with diols, 414
Adipoyl dichloride, polymerization, constitutional isomerism, 98
Aerospace, matrices for composites, epoxy resins, 668
Alcohols, epoxy resins, 695
Alcoholysis, polyesters, 277, 289
Aldehydes
  phenol resins, 614
    production, 614
  polymers, step-growth polymerization, 13
Alkenes, photodimerization, crystalline, 218
Alkenyloxy groups, polysiloxanes, crosslinking, 603
Alkoxy groups
  crosslinking
    elastomers, 600
    polysiloxanes, 602
Alkyl phenol resins, 635
1-Alkynes, copolymerization, diacetylenes, 469
Alkynic resins, 502
Allophanates, 396, 403
  inhibition, 403
  polyurethanes, curing, 418
Alternate copolymers, solid state polycondensation, 203
γ-Alumina, poly(ether ketones), 496
Aluminum trichloride
  curing, epoxy resins, 694
  poly(ether ketones), 486
  production, 485
Amide–hydrazides, aromatic, polyamides, 385
Amides
  activated, polycondensation, 152
  cyclic oligomers, 74, 76
  interchange, aliphatic polyamides, 361
Amido groups, polysiloxanes, crosslinking, 604
Amines
  aliphatic, epoxy resins, 678
  curing, polyurethanes, 416
  primary, epoxy resins, 678
  reaction with isocyanates, kinetics, 437
  secondary, epoxy resins, 678
  siloxanols, polycondensation, 598
  tertiary, epoxy resin catalysts, 675
Aminimides, polyurethanes, 415
Amino acids
  polycondensation, 4
  solid state polycondensation, 204
Amino groups, polysiloxanes, crosslinking, 603
Aminoxy groups, polysiloxanes, crosslinking, 605
Ammonium halides, curing, epoxy resins, 693
Anhydrides
  polyesters, 296
  reaction with oxiranes, 334
Aniline
  o-cyano-, reaction with phenyl isocyanate, kinetics, 439
  N,N-diglycidyl-, epoxy resins, 670
  N,N,O-triglycidyl-4-hydroxy-, epoxy resins, 670
Anthracene
  9,10-bis(chloromethyl)-, copolymerization, 458
  copolymers, synthesis, 458
  dihydro-, benzyl chloride, polymerization, 460
  oil, phenol–formaldehyde polymers, 636
Antifungal properties, polysiloxanes, 601
Antisolvent precipitation, polycarbonates, 353
Aramids, 376
ARDEL D-100
  heat aging, 325
  heat resistance, 324

properties, 325
Arenesulfinates, polycondensation, 571
Arenesulfonic acids, polycondensation, 569
Arenesulfonyl halides
  polycondensation, 564
  self-polycondensation, 563
Arsenic fluoride, polyphenylene synthesis, 466
Aryl bromides, polysulfones, 573
Aryl chlorides, polysulfones, 573
Aryl fluorides, polysulfones, 573
Aryl iodides, polysulfones, 573
ATQ resins, thermal properties, 529
Azelaic acid, polyesters, 339
Azobenzene, mesogenic groups, 726
Azomethines, mesogenic groups, 726

Back-biting process
  cyclic sulfide oligomers, 84
  intramolecular, cyclic oligomers, 64
BADGE resins, homopolymerization, 672
Bakelite, 612
Ballistic trajectories, 124
BATQ resins, thermal properties, 529
Battery separators, phenol–formaldehyde
  polymers, 635
Benzanilides, diamino-, diacid chloride
  copolymerization, constitutional isomerism, 98
Benzene
  1,3-bis(chloromethyl)-2,3,5,6- tetramethyl-,
    copolymerization, anthracene, 458
  1,4-bis(chloromethyl)-2,3,5,6- tetramethyl-,
    copolymerization, anthracene, 458
  1,4-dibromo-, electrochemical reduction, 471
  1,4-dichloro-, polymers, sodium sulfide, 545
  dichloroethyl-, polyarylates, 319
  1,4-dicinnamoyl-, crystal, topochemical
    photopolymerization, 223
  1,2-dimethyl-, copolymerization, xylenes, 457
  1,3-dimethyl-
    copolymerization
      dichloroxylenes, 457
      xylenes, 457
  1,4-dimethyl-
    copolymerization
      dichloroxylenes, 457
      xylenes, 457
  1,4-diphenoxy-, poly(ether ketones), 488
  divinyl-, polyesters, 340
  polymers, sulfur, 544
  1,2,4,5-tetramethyl-
    copolymerization
      anthracenes, 458
      dichloroxylenes, 458
      xylenes, 457
  1,2,4-trimethyl-, copolymerization,
    dichloroxylenes, 457
  1,3,5-trimethyl-, copolymerization,
    dichloroxylenes, 458
Benzenesulfinates, 4,4'-oxydi-,
  polycondensation, 571
Benzenesulfonyl chloride
  4,4'-oxydi-, polycondensation, diphenyl
    ether, 565
  4-phenoxy-, self-polycondensation, 565
Benzenethiol, p-bromo-, copper salt, self-
  condensation, 545
Benzhydrazide
  4-amino-
    polyamides
      constitutional isomerism, 98
      terephthalic acid, 707
Benzoic acid
  4-amino-, polymerization, carbon
    disulfide, 384
  3,5-diisopropyl-4-hydroxy-, polyesters, 285
  hydroxy-
    polyesters, 286
    silylation, bulk polycondensation, 140
  4-hydroxy-
    polycondensation, 4-hydroxy-3,5-dimethoxy-
      benzoic acid, 158
    polyesters, 317
  4-hydroxy-3,5-dimethoxy-, polycondensation,
    4-hydroxybenzoic acid, 158
  4-hydroxy-3-propyl-, polyesters, 285
  3-phenoxy-, poly(ether ketones), 485
  phenyl esters, mesogenic groups, 725
Benzophenone
  4,4'-difluoro-, poly(ether ketones), 491, 492,
    493
  4,4'-dihydroxy-, poly(ether ketones), 494
  2-hydroxy-, polyarylates, UV
    stabilization, 326
  polyesters, 340
Benzoyl chloride
  p-phenoxy-, poly(ether ketones), 485, 487
  rate of polymerization, 37
  m-trimethylsiloxy-, bulk
    polycondensation, 140
Benzyl alcohol
  2,5-dimethyl-, acetate, polymerization, 460
  polymerization, 456
Benzyl bromide, polymerization, 456
Benzyl chloride
  1,1-dimethyl-
    polymerization, 456
    kinetics, 460
  2,5-dimethyl-
    polymerization, 456
    kinetics, 460
  1-ethyl-
    polymerization, 456
    kinetics, 460
  1-methyl-
    polymerization, 456
    kinetics, 460
  polycondensation, kinetics, 458
  polymerization, 455
    termination, 460
  2,3,5,6-tetramethyl-, polymerization, 457
Benzyl compounds
  2,5-dimethyl-, polymerization, 457
  polymerization, 456
  2,3,5,6-tetramethyl-, polymerization, 457
Benzyl fluoride, polymerization, 456
Benzyl halides, polymerization, 456
Benzyl iodide, polymerization, 456
Benzylamine, 3-amino-, epoxy resins,
  curing, 680
4,4'-Biphenol, polyarylates, 317, 320
Biphenyl
  4,4'-dibromo-, electrochemical reduction, 471
  diphenyl ether eutectic mixture,
    polyarylates, 321
  mesogenic groups, 726
Bipolymers, constitutional isomerism, 99
Bisbenzoxazinones, bulk polycondensation,
  chain extension, 133
Bis(carbamoyl chloride), polyureas, 433
Bischloroformates, reactions, with diamines, 414
Biscyclopentadienone, cycloaddition reactions,
  bisacetylenes, 469
2,2-Bis(4-hydroxyphenyl)propane, interfacial
  polycondensation, terephthaloyl
  dichloride, 176

Bismaleimides
  copolymerization, 521
  crosslinking, 520
  synthesis, 518
Bismuth salts, polyurethanes, curing, 420
Bisnadimide, 522
Bisoxazines, bulk polycondensation, chain extension, 133
Bisoxazolines, bulk polycondensation, chain extension, 133
Bisoxazolones, bulk polycondensation, chain extension, 133
Bisphenol
  epoxy resins, 668
  polyarylates, crystallization, 323
  trimethylsilyl-, poly(ether ketones), 494
Bisphenol A
  copolycondensation, terephthaloyl chloride, neopentyl glycol, 263
  diglycidyl ether, 668, 669
  interfacial polymerization, 74
  polyarylates, 317, 319
  polycarbonates, 345
  polyesters, 340
  poly(ether sulfone), bis(4-chlorophenyl) sulfone, 576
  step-growth polymerization, 16
  tetrabromo-, polycarbonates, 346
  tetramethyl-, polycarbonates, 346
3,3'-Bisphenol A
  dichloro-, copolycondensation, phenolphthaleine, terephthaloyl dichloride, 263
  dimethyl-, copolycondensation, phenolphthaleine, terephthaloyl dichloride, 263
Bispyrones, cycloaddition reactions, bisacetylenes, 469
Biurets, 396, 404
  polyurethanes, curing, 418
Block copolymers
  poly(ether ketones), 493
  segmented, polyesters, 276
Blowing agents, polyurethanes, 419
Bond angle, ring–chain equilibria, 93
Bond percolation model, 125
Bonded textile felts, phenol–formaldehyde polymers, 635
Boron trifluoride
  epoxy resins, catalysis, 672
  poly(ether ketones), 485
Branching
  gelation, 119
  poly(phenylene oxide), 474
  step polymerization, 123
Bromination, poly(phenylene oxide), 479
Bulk polymerization, polyureas, reaction injection moulding, 447
Butadienoic acid, 1,4-phenylene-, preparation, 219
1,4-Butanediol, reactions, with 1,6-hexamethylene diisocyanate, 414
Butanol, rate of polymerization, isocyanates, 36
2-Butene, 1,4-dichloro-, polyphenylene synthesis, 467
γ-Butyrolactone, polymerization, 298

Calcium chloride, polyamides, 376, 381
Calixarenes, 624, 625
  t-butyl-, metal ion transport, 629
  conformation, 629
  hydrogen bonds, 627
  linear oligomers, crystal structures, 627
  melting points, 626

t-pentyl-, metal ion transport, 629
  synthesis, 625
Calix(4)arenes, 625
  crystal structure, 627
Can coatings, epoxy resins, 676
Caprolactam
  copolymerization, lactams, 365
  hydrolysis, 361
  polymerization, 298
Carbazoles, mesogenic groups, 726
Carbene radicals
  diynes, solid state polymerization, 243
  photopolymerization, diynes, 242
Carbochain polymers, interfacial polycondensation, 180
Carbodiimides, 398, 407
  isocyanates, catalysis, 399
Carbon dioxide, polyureas, 428
Carbon disulfide, p-aminobenzoic acid, polymerization, 384
Carbon monoxide, polyureas, 435
Carbon oxysulfide, polyureas, 433
Carbonates
  bis(p-nitrophenyl), reaction with 2-(4-aminophenyl)ethylamine, 110
  cyclic oligomers, 75
  diaryl, 348
  methylene-, polycondensation, 4
  poly(ether ketones), 490
  polyureas, 428
Carbonic acid
  dithio-, poly(ether ketones), 488
  thio-, poly(ether ketones), 488
1,1'-Carbonyldiimidazole, polyesters, 285
Carboxy compounds, polyureas, 428
Carboxyl groups, polyesters, 292
Carboxylic acid chlorides, aromatic, polyamides, 376
Carothers equation, 21
Cashew nut shell liquid, phenol–formaldehyde polymers, 636
Casting, polyester resins, 341
Castor oil, phenol–formaldehyde polymers, 635
Catalysts
  curing, polyurethanes, 416
  polyesterification, 279
  polyesters, 280, 290, 293
Catechol
  p-t-butyl-, polyesters, 340
  electrochemical polymerization, 479
Cationic polymerization
  photoinitiated, epoxy resins, 676, 677
Cayley tree approximation, 126
Cellulose, 709, 714
  hydroxy-, 709
  rigid-rod polymers, 702
Chain branching, polysulfones, 584
Chain extension, bulk polycondensation, homopolymers, 133
Chain length
  rate of polymerization, 41
  redistribution, polyesters, 294
Chains
  linear, constitutional regularity, 97–114
  model, ring–chain equilibria, 93
  rotational isomeric state, 93
Chemical resistance, 554
Chlorendic acid, polyesters, 339
Cholesterol, mesogenic groups, 726
Chromium, arenetricarbonyl-, benzyl chloride polymerization, 456
CIBA XU 218, 501
Cinnamic acid
  α-cyano-4-[2-(4-pyridyl)ethenyl]-, crystal,

photopolymerization, 223
photodimerization, 219
  solid state, 217
pyridylethenyl-, ethyl ester,
  photopolymerization, 219, 225
Citraconic acid, polyesters, 339
Clusters
  connected finite, percolation, 125
  finite, percolation, 126
  size, gelation, 122
Coatings
  phenol–formaldehyde polymers, 635
  polyurethanes, 419
  urea–formaldehyde resins, 662
Cobalt, catalysts, oxidative coupling
  polymerization, 474
Colophonium, fiber boards, 634
Comonomers, monofunctional, molecular
  weight, 57
Computer simulations, cyclization, 122
Concentration, polycarbonates, 352
Condensation, polyamides, aliphatic, 359
Condensation polymerization
  growth mechanism, 1
  non-aqueous dispersions, 197–200
Condensation polymers, 1–8
  cyclic oligomers, 63
  structure–property relationships, 98
Conex, 375, 377
Conformation
  cyclic oligomers, 65
  polyesterification, acceptor-catalytic, 146
  ring–chain equilibria, 92
Conjugated polymer chains, single crystals, 233
Connectivity, multifunctional step
  polymerization, 117
Constitutional isomerism
  kinetics, 100, 101, 107
  theory, 99
  thermodynamics, 100
Constitutional regularity
  linear chains, 97–114
  properties, 113
Copolyamides, 265, 266
  amino acids, 74
  aromatic, 384
  dispersion polymerization, 199
  random, solid state polycondensation, 205
Copoly(α-amino acid)s, cyclic oligomers, 74
Copolyarylates, suspension
  polycondensation, 191
Copolycarbonates, 267
Copolycondensates, distributions, 252
Copolycondensation
  applications, 263
  kinetics, 258
  Monte Carlo simulations, 257
  reaction media, 253
  reactions of a different nature, 269
  theoretical studies, 252
  two reactions of the same type, 264
Copoly(ester amide)s, 270, 300
Copoly(ester carbonate)s, 300
Copoly(ester sulfone)s, 265
Copolyesters, 276
  bulk polycondensation, 137, 140
  chiral, 265
  mechanism, 296
Copoly(ether ketone)-*block*-copoly(ether
  ketone sulfone), 268
Copolyetherification, 268
Copoly(β-hydroxyalkanoate)s, 303
Copoly(imide isoindoloquinazolinedione)s, 517
Copolymerization, functional groups,
  reactivity, 120
Copolymers
  bulk polycondensation, 134, 136
  diynes
    crosslinking, 241
    solid state polymerization, 240
  polyesters, 278, 293
  siloxane–urea, 444
  solid state polycondensation, 203
Copolypeptides, alternating, solid state
  polycondensation, 203
Copolythioesters, 265
Copolyureas, 441
  alternating, 434
    bis(carbamoyl chloride)s, 434
  block, bulk syntheses, 448
  bulk polymerization, physical properties, 450
  linear
    aliphatic, heterocyclic aromatic, 442
    aromatic alternating, 442
    molecular weight, 451
  non-linear, 449
Copper, catalysts, oxidative coupling
  polymerization, 474
Copper(II) salts, ion exchanged montmorillonite
  clays, polyphenylene synthesis, 466
Correlation length
  clusters, percolation, 126
  percolation, 126
Cresol
  bishydroxymethyl-, phenol–formaldehyde
    polymers, 621
  phenol–formaldehyde polymers, 636
  rate of polymerization, 36
  resins
    friction materials, 636
    production, 614
o-Cresol, electrochemical polymerization, 479
p-Cresol
  di-*t*-butyl-, polyesters, 340
  melamine–formaldehyde resins, 656
Critical volume fraction, 113
Crosslinking
  elastomers, 600
  multifunctional step polymerization, 117
  phenol–formaldehyde polymers, 621, 635
  polyesters, unsaturated, 338
  polysiloxanes, 601
    alkenyloxy groups, 603
    alkoxy groups, 602
    amido groups, 604
    amino groups, 603
    aminoxy groups, 605
    ketiminoxy groups, 604
  polyurethanes, curing, 416
  urea/formaldehyde systems, 654
Crown ethers
  interfacial polycondensation, 176
  polysulfones, 573
Cryptands
  interfacial polycondensation, 176
  siloxanols, polycondensation, 598
Cryptates, polysulfones, 573
Crystal structure, solid state polymerization,
  diynes, 239
Crystalline melting point, poly(ether
  ketones), 494
Crystalline state polymerization, 217
Crystallization
  bulk polycondensation, 137
  chain-extended crystals, 234
Crystallization-induced reactions, 207, 213
Crystals, diynes, solid state polymerization, 233–246

Subject Index

Cumene, production, 613
Cupric salts, ion exchanged montmorillonite clays, polyphenylene synthesis, 466
Curing
  epoxy resins, 667
    dicyandiamide, 686
  phenol–formaldehyde polymers, 621
  poly(phenylene sulfide), 547, 550
  polyurethanes, 413–425
  urea–formaldehyde resins, 660
Cyclic carbonates, 70
  polyesters, 300
Cyclic compounds, ring–chain equilibria, 91
Cyclic disulfides, 538
  polysulfide synthesis, 541
Cyclic esters, polyesters, 297
Cyclic monosulfides, 538
Cyclic oligoesters, aromatic, 70
Cyclic oligomers, 63–89
  formation, kinetics, 64
  identification, 65
  separation, 65
  solid state polycondensation, 203
Cyclic sulfides, aliphatic–aromatic, 84
Cyclic sulfites, polyesters, 300
Cyclization
  graph theory, 122
  intramolecular, 63
  molecular weight, distribution, 59
Cycloaddition reactions
  isocyanates, 396
  polyphenylene synthesis, 469
Cycloaliphatic oxides, epoxy resins, 671
Cyclobutane polymers, topochemical photopolymerization, 217–231
1,3-Cyclobutanediol, 2,2,4,4-tetramethyl-, polyesters, 303
Cyclododecene, metathesis, ring–chain equilibria, 94
1,3-Cyclohexadiene
  cationic polymerization, 468
  5,6-dihydroxy-, diester, polymerization, 469
Cyclohexanecarboxylic acid
  3,4-epoxycyclohexylmethyl-3′,4′-epoxy- polymerization, 677, 678
  epoxy resins, 671
Cyclohexane, phenyl-, mesogenic groups, 726
1,4-Cyclohexanedimethanol
  polyesters, 303
  polyarylate blends, 327
3-Cyclohexene, vinyl-, dioxide, epoxy resins, 671
trans-1,4-Cyclohexylene, rigid-rod polymers, 702
Cyclolactones, hydroxymethylfurancarboxylic acid, polymerization, 70
Cyclophanes, tricyclic, topochemical photopolymerization, 220
trans-Cyclopropanedicarboxylic acid, polyamides, constitutional isomerism, 98
Cyclosiloxanes, anionic polymerization, 597
Cyclotrimerization, bulk, 139
Cyclotriphosphatriazene, hexachloro-, polyesters, 286

Decanamide, N,N,N',N'-bisdiethyleneimino-, copolycondensation, N,N'-bis(2-aminoethyl)-, 260
Decanediamide
  N,N'-bis(2-aminoethyl)-
    copolycondensation
      N,N,N',N'-bis(diethyleneimino)- decanamide, 260

      bis(2-hydroxyethyl)decanedioate, 260
Decanedioate, bis(2-hydroxyethyl)-, copolycondensation, N,N'-bis(2-aminoethyl)decandiamide, 261
Decarboxylation, polyesters, 301
Degradation reactions, polyesters, 300
Degree of crystallinity, solid state polycondensation, 208
Degree of polymerization, 51
  gelation, 119
  number-average, 51, 121
  weight-average, 51
    percolation, 126
  z-average, 51
  $(z + 1)$-average, 51
Depolymerization, solution, cyclic oligomers, 65
Devolatilization, mechanical, polycarbonates, 353
Diacetates, polyarylates, 320
Diacetonitrile, α,α'-bis(4-acetoxy- 3-methoxybenzylidene)-1,4-phenylene-, four-center photopolymerization, 218
Diacetylenes
  copolymerization, phenylacetylene, 469
  polycyclotrimerization, 469
  polymerization, 218
Diacid anhydrides, curing, epoxy resins, 688
Diacids
  aliphatic, polyesters, 302
  aromatic, hydrazine polyamides, 707
  polyarylates, 321
Diacrylic acid, 1,4-phenylene-, dialkyl ester, photodimerization, 219
Diacyl chlorides
  aromatic, polyamides, 384
  diaminobenzanilide copolymerization, constitutional isomerism, 98
Dialkenes
  crystallographic data, 228
  crystals, photoreactivity, 226
  four-center photopolymerization, topotaxies, 229
  topochemistry, 221
  α type, photopolymerization, 226
Dialkyl halides, polymerization, 299
Diallyl phthalate, polyesters, 340
Diamines
  aliphatic, polyamides, 370
  aromatic, oxalic acid polyamides, 707
  polymerization, dianhydrides, 505
  polyureas, 428
  reactions, with bischloroformates, 414
  N-silylated aromatic, polyamides, 384
Dianhydrides
  polymerization
    diamines, 505, 510
    diisocyanates, 510
    silylated diamines, 511
1,8-Diazabicyclo[5.4.0]undec-7-ene, epoxy resin catalysts, 675
1,5-Diazacyclooctane, 1,5-diphenyl-3,7-dihydroxy-, epoxy resins, 684
Diazonium salts, aryl, photoinitiators, epoxy resins, 676
Dibenzo[18]crown-6, poly(ether ketones), 489
Dibenzofuran, poly(ether ketones), 496
Dibutyltin diacetate, polyurethanes, curing, 420
Dibutyltin dilaurate, polyurethanes, curing, 420
Dicarbene radicals, diynes, solid state polymerization, 243
Dicarboxylic acids
  aliphatic, polyesters, 309
  aromatic, polyesters, 305

diaryl esters, polyarylates, 321
  polyesters, 279
Dicyandiamide, curing, epoxy resins, 686
Dicyclopentadiene
  dioxide, epoxy resins, 671
  polyesters, 340
Diesters, polycondensation, diamines, 512
Diethylene glycol, polyesters, 340
Diffusion, step polymerization, growth models, 124
Diffusion coefficient, interfacial polycondensation, 169
Diglycolide
  polymerization, 298
  tetramethyl-, polymerization, 298
  tetraphenyl, polymerization, 298
Dihalides, reaction with sodium polysulfide, 535
Diisocyanates
  bulk polycondensation
    chain extension, 133
    copolymers, 134
  polyureas, 436
  rate of polymerization, butanol, 36
  reactions, kinetics, 392
Dilactide, polymerization, 298
Dilutability
  urea–formaldehyde resins
    acetone, 660
    water, 660
Dimerization, crystal lattice, 234
Dimethylamine, N-benzyl-, epoxy resin catalysts, 675
Dinitrile carbonates, cyclic, 404
Dinitrile oxalates, cyclic, 404
Dinitrile sulfites, cyclic, 404
Diols
  aliphatic, polyesters, 302, 305
  aromatic, polyesters, 309
  dehydration, 337
  polyesters, 279
Dioxanes, mesogenic groups, 726
1,4,2-Dioxazol-5-one, 3,3'-tetramethylenedi-, reactions, with diols, 414
1,3-Dioxolane, imino-, curing, epoxy resins, 693
Diphenates, polyarylates, 321
Diphenolic acid, phenol–formaldehyde polymers, 635
Diphenols
  diacetate, polyarylates, 320, 321
  polyarylates, 321
4,4'-Diphenoquinone, 3,3',5,5'-tetramethyl-, oxidative coupling polymerization, copper catalysis, 473
Diphenyl carbonate, polyarylates, 321
Diphenyl chlorophosphate, polyesters, 286
Diphenyl oxide, 4, 4'-diamino, bulk polycondensation, pressure, 137
Diphenylene, 4,4'-diaminoazo-, epoxy resins, curing, 680
Dipropylene glycol, polyesters, 340
Dispersion polycondensation, 167–193
Dispersion polymerization
  condensation, kinetics, 200
  particle stability, 199
Dispersion polymers
  liquid reactants, emulsions, 198
  solid reactants, inert liquid, 199
  soluble reactants, 197
Dispersions, non-aqueous, condensation polymerization, 197–200
Displacement polymerization, phenols, 478
Distribution
  functions, molecular weight, 53, 59
  molecular weight, 47–62
    aliphatic polyamides, 361
    constitutional isomerism, 99
    experimental, 58
    gelation, 119
    multifunctional step polymerization, 117
    self-polymerization, 57
    step polymerization, 57, 121
Disulfonyl chloride, polycondensation, aromatic compounds, 566
Dithiols, oxidation, polysulfide synthesis, 540
Diurethanes, polyureas, 435
Diynes
  coloration, 233
  crystals, solid state polymerization, 233–246
  dicarboxylic acid, solid state polymerization, 233
  polymers, crystal structure, 235
  single crystal polymers, ESR, 242
  solid state polymerization
    structure, 235
    substituents, 237
    thermal, 242
10,12-Docosadiyne-1,22-dioic acid, liquid crystals, solid state polymerization, 241
Dodecamethylene diammonium adipate, solid state polycondensation, 207
Drying oils, phenol–formaldehyde polymers, 635
Dyes, polymeric, copolycondensation, 265
Dyson equation, 123

Elastomers, 594
  cold-vulcanizable
    single component, 599
    two-component, 599
  hot-vulcanizable, 599
  polyurethanes, 416
  preparation, 598
Elastoplastics, polyurethanes, 416
Electrical laminates, curing, 686
Electrochemical oxidation
  benzene, 466
  o-terphenyl, 466
  p-terphenyl, 466
Electrochemical polycondensation, 162
Electrochemical polymerization, benzene, 466
Electrochemical syntheses, poly(phenylene oxide), 478
Electronics
  encapsulation, epoxy resins, 668
  epoxy resins, 668
Emulsifiers, interfacial polycondensation, 171
Emulsion polycondensation, 168, 184, 187
End-biting process
  cyclic oligomers, 65
  cyclic sulfide oligomers, 84
Ene–oxide resins, 671
Enzymatic polycondensation, 162
Episulfides, epoxy resins, 696
Epoxides
  applications, 668
  bulk polycondensation, pressure, 138
  polymerization, 667–696
Epoxy compounds, polysulfide synthesis, 539
Epoxy group, polyurethanes, curing, 416, 418
Epoxy novolacs, 668
Epoxy powder coatings, curing, 686
Epoxy resins
  cationic polymerization
    photoinitiation, 676, 677
  curing, dicyandiamide, 686

Subject Index

halogenated, 668
homopolymerization, 671
polynuclear phenols, 668
properties, 669
Equilibrium moisture content, constitutional regularity, 113
Ester groups, polyesters, 292
Esterification
   mild conditions, 285
   polyesters, 277, 279
Esterolysis, 293
Esters
   activated, polycondensation, 147
   cyclic oligomers, 70, 71
   polyamides, aliphatic, 371
Ethane
   1,2-dichloro-, poly(ether ketones), 488
   1,1,2-trichloro-1,2,2-trifluoro-, polyphenylene synthesis, 467
1,2-Ethanediol, polyesters, 340
Ethers
   bis(trimethylsilyl), bulk polycondensation, 585
   cyclic oligomers, 76
   diaryl, 572
   2,5-dimethylbenzyl methyl, polymerization, kinetics, 460
   diphenyl
      polyarylates, 321
      polycondensation, 4,4'-oxydibenzene sulfonyl chloride, 565
      poly(ether ketones), 488
   polyesters, 287
   triglycidyl tris(hydroxyphenylene), epoxy resins, 668
Ethylamine, 2-(4-aminophenyl)-, reaction with bis(p-nitrophenyl) carbonates, 110
Ethylene carbonate, chain extension, bulk polycondensation, 133
Ethylene dichloride
   reactions
      sodium polysulfide, 533
      sodium tetrasulfide, 536
Ethylene glycol
   bulk polycondensation, 134
   cyclic oligomers, 70
   polyesterification with dimethyl terephthalate, 20
Ethylene tetrasulfide, polymers, 533
Ethylenediamine, reaction with bis(p-nitrophenyl) esters, 109
Euler's equation, 125
Exchange reactions
   bulk polycondensation, 134, 137
   ester–ester, 293
   polyesters, 277, 292
Excluded volume forces, multifunctional step polymerization, 127

Fast atom bombardment mass spectrometry, cyclic oligomers, 66
Feynman diagram, 123
Fiber boards, phenol–formaldehyde polymers, 634
Fibers, polyamides, 375
Fillers
   bulk polycondensation, 4,4'-diaminodiphenylmethane, 138
   polysiloxanes, 600
Films, interfacial polycondensation, 172
Filter papers, phenol–formaldehyde polymers, 635
Flash pyrolysis, polymers, mass spectometry, 69

Flexible foams, polyurethanes, 418
Flexible spacers, LC polymers, mesogenic units, 723
Flocculation, condensation polymers, 197
Flory–de Gennes theory, 127
Flory–Stockmayer model
   gelation, 118
   multifunctional step polymerization, 117
Flory–Stockmayer theory, equal reactivity, 121
Fluoroboric acid, catalysis, epoxy resins, 672
Foaming, poly(ether ketones), 491
Foams, urea–formaldehyde resins, 663
Formal
   bis(2-chloroethyl)-, polysulfides, 539
   cyclic oligomers, 76, 79
Formaldehyde
   melamine systems, 655
   phenol–formaldehyde polymers, analysis, 638
   phenol polymers
      history, 612
      synthesis, 612–643
   phenol resins, production, 614
   production, 650
   properties, 615
   rate of polymerization, 36
   reactions
      melamine, 651
      urea, 651
   specification, 615
   toxicology, 637
   urea–formaldehyde resins, analysis, 659
   urea system, 651
Formamide, dimethyl-, polyesters, 287
Formamidinium salts, polyamides, 383
Formox process, 651
Four-center photopolymerization, 217
   dialkenes, topotaxies, 226
   photochemical depolymerization, 229
   quantum yields, 224
   reaction mechanism, 224
   step-growth mechanism, 224
   two-step mechanism, 224
Fractal dimension, 124
Free-radical copolymerization, unsaturated compounds, sulfur dioxide, 562
Friction materials, phenol–formaldehyde polymers, 636
Friedel–Crafts catalysts, benzyl chloride, polymerization, 456, 459
Fumaric acid, polyesters, 339
Functional groups
   reactivity
      multifunctional step polymerization, 120
      rate of polymerization, 35–45
Furancarboxylic acid, hydroxymethyl-, polymerization, cyclic oligomers, 70
Furfural
   phenol–formaldehyde polymers, 636
   phenol resins, production, 614
Furfuryl alcohol, phenol–formaldehyde polymers, 636

Gas chromatography, cyclic oligomers, 65
Gas flow rate, solid state polycondensation, 208
Gel coats, polyester resins, 342
Gel effect, 6
Gel fraction, 119
   order parameter, 126
Gel permeation chromatography, urea–formaldehyde resins, 657
Gel point, 120, 121
   kinetics, 125

multifunctional step polymerization, 121
Gelation
  classical theory, 118
  kinetics, 124
  multifunctional step polymerization, 118
  polyureas, 444, 448
Glass transition temperature
  liquid crystalline polymers, 715
  poly(ether ketones), 494
  solid state, polymerization, 218
Glutamates, γ-n-alkyl-L-, copolymers, 714
Glutamic acid, α-benzyl-, solid state
    polycondensation, 203
Glycidyl resins, 668
Glycolic acid, polyesters, 303
Glycols
  polyesters, 280
  rate of polymerization, 35
Glyoxal, phenol resins, production, 614
Graft copolymers, dispersion polymerization,
    stabilizers, 198
Graph theory, gelation, 118, 119
Grignard reaction, polyphenylene synthesis, 467
Grinding materials, phenol–formaldehyde
    polymers, 636
Group IIA metal oxides, benzyl chloride
    polymerization, 456
Group VIIB metals
  carbonyl complexes, benzyl chloride
    polymerization, 456
  carbonyl halides, benzyl chloride
    polymerization, 456
Growth models, step polymerization, 124
Guanidine, cyano-, curing, epoxy resins, 686, 688

Halides
  aromatic
    halogen displacement, 578
    polycondensation, phenoxides, 574
Halosulfonation, aromatic compounds, 569
Heat of polymerization, diynes, 242
Heat stabilizers, polysiloxanes, 603
α-Helix, rigid-rod polymers, 702
Heteroaromatic polymers, synthesis, 499–529
Heterocumulenes, chemistry, 388
Heterocyclic copolymers, rod-like, 502
Heterocyclic polymers
  rod-like, 502
  synthesis, 500
Hexamethylene diisocyanate
  blocking, 420
  polyureas, 436
  reactions, with 1,4-butanediol, 414
Hexamethylenetetramine
  phenol–formaldehyde polymers, 620
  preparation, 615
Hexamethylphosphoric triamide,
    polyamides, 376
2,5-Hexanediamine, 2,5-dimethyl-, epoxy
    resins, 680
Hexanoic acid
  6-amino-, solid state polycondensation, 204
  salts, polyurethanes, 420
5-Hexen-1-yne, 1,4-diphenyl- N,N-
    dimethylamino-, thermal ring closure, 470
High dilution method, 63
High performance liquid chromatography
  cyclic oligomers, 66
  epoxy resins, 669
Homopolyureas, 441
  bulk polycondensation, 135
  chain extension, bulk polycondensation, 133
  homo-bifunctional monomers, constitutional
      isomerism, 99
  step polymerization, 121
Hydrazides, aromatic, polyamides, 385
Hydrazine, polyamides, aromatic diacids, 707
Hydrogen bonding, solid state polymerization,
    diynes, 239
Hydrogen fluoride, poly(ether ketones), 487
Hydrogen-transfer polymerization,
    acrylamide, 6
Hydroquinone
  diglycidyl ether, bulk polycondensation, 138
  polyarylates, 317, 320
  polyesters, 340
  poly(ether ketones), 491
Hydroxamates, polyurethanes, 414
Hydroxy acids
  acyl azides, polyurethanes, 404
  aliphatic, polyesters, 303
  polycondensation, 4
  polyesters, 279
3-Hydroxy acids, polyesters, 280
Hydroxymethyl compounds
  decomposition, 652
  urea/formaldehyde systems, 651
Hydroxymethylation, phenol–formaldehyde
    polymers, 617
Hyperacidity, calixarenes, 629
Hyperscaling law, 126

Imidazoles
  curing, epoxy resins, 692
  epoxy resin catalysts, 675
Impingement mixing, 447
Inclusion compounds, chain-extended
    polymers, 234
Indane, benzyl chloride, polymerization, 460
Industrial laminates, phenol–formaldehyde
    polymers, 635
Inert gases, solid state polycondensation, 208
Infrared spectroscopy, urea–formaldehyde
    resins, 659
Inorganic polymers, step-polycondensation, 6
Insulation materials, phenol–formaldehyde
    polymers, 634
Interfacial catalysis, 175, 177
  polysulfonates, 179
Interfacial polycondensation, 167–193
  acceptors, 173
  gas–liquid, 168, 180
  kinetics, 171
  liquid–liquid, 168, 170
  molecular weight, 173
  monomers, 173
  non-aqueous binary systems, 174
  non-aqueous systems, 174
  polyurethanes, 414
  side reactions, 173
  solid–liquid, 183
  temperatures, 173
Interfacial polymerization
  aliphatic polyamides, 370
  polyamides, 376
  polyarylates, 318
  polycarbonates, isolation, 352
  polyesters, 295
  polyureas, 446
Interfacial resins
  chemistry, polycarbonates, 347
  polycarbonates, purification, 352
  synthesis, polycarbonates, 349

Subject Index

Interfacial synthesis
  batch, polycarbonates, 349
  batch-continuous, polycarbonates, 349
  continuous, polycarbonates, 350
Interfacial transfer catalysts, 177
Intramolecular rearrangement, oxidative coupling polymerization, phenols, 477
Iodonium salts
  diaryl-, photoinitiators, epoxy resins, 676, 678
  di-t-diphenyl-, photoinitiators, epoxy resins, 677
Isobutyraldehyde, novolaks, alkylidene bridges, 625
Isocyanates
  aromatic, reactive, 389
  blocked, 394
    polyurethanes, 419
  chemistry, 388
  copolymerization, 403
  curing, epoxy resins, 693
  cycloaddition reactions, 396
  dimerization, 399
  homopolymerization, 401
  insertion reactions, 390
  polycondensation, 4
  polymerization, 399
  polymers, 387–411
  polyurethanes, curing, 416
  rate of polymerization, butanol, 36
  reactions, 388
    alcohols, 394
    amines, 396
      kinetics, 437
    catalysis, 394
    ethanol, 390
    kinetics, 389, 394
    mechanism, 394
    phenols, 396
  trimerization, 399
Isocyanurates, 401
Isocyanuric acid, curing, epoxy resins, 694
Isomerization, maleate–fumarate, 336
Isomerizational polycyclization, 140
Isophthalic acid
  diphenyl ester, polyarylates, 322
  polyarylates, 317, 320
    crystallization, 323
  polyesters, 317
Isophthaloyl chloride
  condensation polymerization, 198
  polyarylates, 319
  poly(ether ketones), 487
Isothiocyanates
  cycloaddition reactions, 396
  polymers, 388
  reactions
    kinetics, 390
    protic nucleophiles, 392
Itaconic acid, polyesters, 339

Jacobson–Stockmayer theory, 91
  macrocylization, 64

Ketenes, polymerization, 300
Ketiminoxy groups, polysiloxanes, crosslinking, 604
Kevlar, 376
Kinetics
  copolycondensation, 258
  gelation, 124
  multifunctional step polymerization, 127

Kuhn length, 123
Kureha process, poly(phenylene sulfide), 549

Lactams
  copolymerization, caprolactam, 365
  cyclic
    oligomers, 74, 76
    polycondensation, 4
  polycondensation, 4
Lactic acid
  α-chloro-, polyesters, 304
  polyesterification, 11
Lactones
  cyclic
    oligomers, 70, 71
    polycondensation, 4
  polycondensation, 4
  polymerization, 298
Lamination, polyester resins, 342
Langmuir–Blodgett films, 240
Langmuir–Blodgett technique, polyimide, layers, 517
LARC TPI, 501
Lattice energy, solid state polymerization, diynes, 239
Lead dioxide, catalysts, poly(phenylene oxide), 475
Lead naphthenate, polyurethanes, curing, 420
Lederer–Manasse reaction, phenol–formaldehyde polymers, 612
Leucine, N-carboxy anhydrides, solid state polycondensation, 203
Levy flight trajectories, 124
Lewis acid catalysts
  benzyl chloride, polymerization, 459
  polyarylates, 320
Lewis acids, polyphenylene synthesis, 466
Lewis bases, poly(ether ketones), 486
Light scattering, molecular weight, weight average, 119
Linolenic acid, salts, polyurethanes, 420
Liquid chromatography, cyclic oligomers, 65
Liquid crystalline elastomers, cross-linked, 730
Liquid crystalline phases, diynes, solid state polymerization, 240
Liquid crystalline polymers, 701
  cholesteric phases, 726
  disc-like mesogens, 730
  glassy phases, 726
  lyotropic, 704
  main-chain
    lyotropic, 703
    thermotropic, 709
  mesogenic units, 723
  mesogens, rod-like, 724
  nematic phases, 726
  rigid-rod, 113
  semiflexible, 714
    main-chain, 715, 716
  smectic phases, 726
  thermotropic, 702
Liquid crystallinity, polyamides, 377
Liquid crystals, diynes, solid state polymerization, 241
Liquid–solid transitions, multifunctional step polymerization, 118
Lithium chloride
  polyamides, 376, 381
  poly(ether ketones), 493
Lithium halides, polyesters, 287
Loop formation, gelation, 119
Loops, formation, percolation, 125

Lysine, polymerization, constitutional isomerism, 98

Macallum process, poly(phenylene sulfide), 544
Macrocyclization
  equilibria, 63, 91
  theory, 64
Maleic anhydride, polyesters, 302, 339
Maleimides, thermosetting resins, 502
Malonic dimethyl ester, polyesters, constitutional isomerism, 98
Manganese, catalysts, oxidative coupling polymerization, 474, 475
Manganese dioxide, catalysts, poly(phenylene oxide), 475
Many body statistical mechanics, 123
Mass spectrometry
  cyclic oligomers, 65
  polymers, pyrolysis, 69
Matrix polycondensation, 160
Meisenheimer complexes, polysulfones, 572
Melamine
  formaldehyde systems, 655
  hexa(hydroxymethyl)-, 655
  hydroxymethyl-, 655
  production, 650
  rate of polymerization, 36
  reactions, formaldehyde, 651
Melamine–formaldehyde resins
  analysis, 657
  copolycondensates, 656
  history, 650
  manufacture, 661
  synthesis, 649–663
Melamine resins, 635
Melting point, constitutional regularity, 113
Melting temperature, liquid crystalline polymers, 715
Membranes, interfacial polycondensation, 170
Mesaconic acid, polyesters, 339
Mesitylene, copolymerization, anthracenes, 458
Mesogens
  heteroaromatic polymers, 708
  main-chain, 701–721
  rod-like, LC polymers, 724
  side-chain, 723–730
Metal-containing polymers, interfacial polycondensation, 180
Metal oxides, benzyl chloride, polymerization, 456
Metathesis
  cyclododecene, ring–chain equilibria, 94
  ring–chain equilibria, 94
Methane
  4,4′-bis(chloromethyl)diphenyl-, condensation with diphenylmethane, 456
  copolymerization, anthracene, 458
  bis(2,3,5,6-tetramethylphenyl)-, copolymerization, anthracenes, 458
  4,4′-diaminodiphenyl-
    bulk polycondensation
      N,N'-m-phenylenebismaleimide, 138
      N,N'-tetraglycidyl-4,4′-diaminodiphenylmethane, 138
    poly(ether ether ketones), 492
  4,4′-diaminodiphenylene-
    epoxy resins, 685
    curing, 680
  ditolyl-, polyarylates, 319
  N,N'-tetraglycidyl-4,4′-diaminodiphenyl-, bulk polycondensation, 4,4′-diaminodiphenylmethane, 138

3,3,5,5-tetra(hydroxymethyl)-4,4-dihydroxydiphenyl-, crystallography, 627
Methanesulfonic acid
  poly(ether ketones), 489
  trifluoro-, poly(ether ketones), 488
Methyl methacrylate, polyesters, 340
Methylene, (hydroxyphenylene), oligomers, conformation, 628
4,4′-Methylenediphenylene diisocyanate, polyureas, 436
Mold filling, polyureas, 447
Molding
  polyester resins, 342
  urea–formaldehyde resins, 662
Molding compounds, phenol–formaldehyde polymers, 634
Molecular weight
  aliphatic polyamides, 361
  average, copolycondensation, 254
  distribution, 47–62
    constitutional isomerism, 99
    experimental, 58
    four-center photopolymerization, 224
    gelation, 119
    multifunctional step polymerization, 117
    poly(phenylene oxide), 475
    solid state polycondensation, 209
    step polymerization, 47, 57, 121
    urea–formaldehyde resins, 657
  interfacial polycondensation, 173
  number average, copolycondensation, 254
  poly(ether ketones), 487, 490
  solid state polycondensation, 209
  weight-average, 54, 57
    light scattering, 119
Monte Carlo simulations, polycondensation, 257
Morpholine
  epoxy resins, 684
  3-methenyl-4-hydroxymethyl-, N-substituted, 675
  N-phenyl-3-(N-ethylanilinomethyl)-4-hydroxymethyl-, epoxy resins, 684
Most probable distribution
  molecular weight, 48, 52, 59
  ring–chain equilibria, 92
Multifunctional polymerization, 117
Multifunctional step polymerization, 117–128
Multipolymers, constitutional isomerism, 99

Naphthalene, tetrahydro-, copolymerization, xylenes, 457
Naphthalene diisocyanate, polyureas, 436
Naphthenic acid, salts, polyurethanes, 420
Naturally occurring polymers, 5
Neopentyl glycol
  copolycondensation, terephthaloyl chloride, bisphenol A, 263
  polyesters, 303, 340
Networks
  epoxy resins, 685
  gelation, 125
  multifunctional step polymerization, 117
  tree-like structure, 119
Nickel
  (2,2′-bipyridine-N,N')dichloro-, polyphenylene synthesis, 467
  bis(acetylacetonato)-, polyphenylene synthesis, 467
Nitro groups, poly(ether ketones), 493
Nomex, 375, 376, 377
Noryl, 473
Novolak

continuous production, 631
curing, 622
hexamethylenetetramine cured, 620
high-*ortho*, 634
  synthesis, 633
preparation, 616, 632
reaction mechanism, 623
synthesis, 630
Nuclear magnetic resonance
  urea–formaldehyde resins, 659
  ^{13}C, urea–formaldehyde resins, 658
Nylon-3, hydrogen-transfer polymerization, 6
Nylon 4,6, solid state polycondensation, 215
Nylon 6
  cyclic oligomers, 74
  homopolymer, hydrolytic process, 361
  solid state polycondensation, 204, 207, 214
Nylon 6,6, 12
  dispersion polymerization, 199
  solid state polycondensation, 207, 214
Nylon 6,10, solid state polycondensation, 214
Nylon T4, solid state polycondensation, 215
Nylons, 357
  polyarylate blends, 326

Octanoic acid, salts, polyurethanes, 420
Oligo(alkylene glycol)s, bulk
  polycondensation, 136
Oligo(dimethylsiloxane)s, 139
  bulk polycondensation, 136
Oligoesters, molecular weight, 297
Oligoethers, cyclic, ring-opening
  polymerization, 217
Oligo(ethylene oxide), flexible spacers, 725
Oligo[(2-hydroxy-5-methyl-1,3-phenylene)-
  methylene], melting points, 626
Oligomers, linear, formation, 85
Oligo(methylene) units, flexible spacers, 725
Oligo(siloxane), flexible spacers, 725
Organometallic coupling, polyphenylene
  synthesis, 467
Organotin compounds, catalysts, polyurethane
  curing, 416
Oriented strand board, phenol–formaldehyde
  polymers, 634
Oscillation photographs, photopolymerization,
  crystals, 226
1-Oxa-4-azacycloheptanol, 4-phenyl-2-(*N*-
  methylanilinomethyl)-, epoxy resins, 684
1-Oxa-4-azacyclohept-3-ene, 2-hydroxymethyl-,
  4-substituted, 675
Oxalates, polyurethanes, 414
Oxalic acid
  phenol–formaldehyde polymers,
    synthesis, 630
  polyamides, aromatic diamines, 707
Oxazolidone
  curing, epoxy resins, 693
  phenylphenoxymethyl-, curing, epoxy
    resins, 694
Oxetane
  3,3-bis(bromomethyl)-, solid solution,
    polymerization, 218
  3-bromomethyl-3-ethyl-, solid solution,
    polymerization, 218
Oxidative coupling, polyphenylene, 465
Oxidative coupling polymerization
  phenols
    catalysts, 474
    mechanism, 475
  poly(phenylene oxide), 473
Oxiranes, reaction with anhydrides, 334

Oxychloroformyl groups, polyesterification, 300

Paper impregnation, phenol–formaldehyde
  polymers, 635
[2.2]-Paracyclophane, crystals, 224
Paraformaldehyde
  reaction with metal phenolates, 633
  synthesis, 630
Particle boards, phenol–formaldehyde
  polymers, 634
Pentadecanolide, polymerization, 298
Pentaerythrite, polyesterification, adipic
  acid, 133
1,3-Pentanediol, 2,2,4-trimethyl-,
  polyesters, 303, 340
Percolation, 125
  scaling theory, 120
  theory, 118, 126
Perlon U, 414
Phantom polymers, 124
Phase-transfer catalysts, polyarylates, 319
Phase transitions, percolation, 125
Phenol
  alkyl-, resin production, 614
  allylmethyl-, electrochemical
    polymerization, 479
  2-benzyl-6-methyl-, oxidative coupling
    polymerization, 475
  4-bromo-2,6-dichloro-, displacement
    polymerization, 478
  4-bromo-2,6-dimethyl-, displacement
    polymerization, 478
  4-*t*-butyl-, calixarenes, 625
  cumene process, 613
  3,4-dibromo-2,6-dimethyl-, displacement
    polymerization, 478
  3,4-dibromo-3,5-dimethyl-, displacement
    polymerization, 478
  2,6-dimethyl-
    oxidative coupling polymerization, 475
    copper catalysis, 473
  2,6-diphenyl-, oxidative coupling
    polymerization, 475
  epoxy resins, 695
  extraction, phenol–formaldehyde
    polymers, 638
  formaldehyde polymers
    history, 612
    synthesis, 612–643
  4-halo-, rate of polymerization, 36
  4-halo-2,6-dimethyl-, polymerization, 474
  2-hydroxymethyl-, formaldehyde resins, 617
  4-nitro-
    esters, aminolysis with ethylenediamine, 109
    formaldehyde resins, 629
  4-*t*-octyl-, calixarenes, 625
  phenol–formaldehyde polymers, analysis, 638
  physical properties, 613, 614
  polyarylates, 322
  rate of polymerization, 36
  toxicology, 636
  3,4,5-tribromo-2,6-dimethyl-, displacement
    polymerization, 478
  2,4,6-trichloro-, displacement
    polymerization, 478
  2,4,6-trihalo-, displacement
    polymerization, 478
  2,3,6-trimethyl-, polymerization, 474
  2,4,6-trimethyl-, polymerization, 474
  2,4,6-tris(dimethylaminomethyl)-, epoxy resin
    catalysts, 675
  2,4,6-tris(hydroxymethyl)-,

crystallography, 627
Phenol–formaldehyde polymers, 656
  analysis, 638
  aqueous dispersion, 633
  chromatography, 638
  cyclic oligomers, synthesis, 625
  degradation, 637
  environmental protection, 637
  ESCA, 638
  field desorption mass spectrometry, 638
  gold lacquers, 635
  high-*ortho*-resins, synthesis, 632
  industrial grade, synthesis, 630
  linear oligomers, 624
    synthesis, 624
  microbial degradation, 637
  molecular structure, 615
  oligomers, structurally uniform, 623
  phenol extraction, benzene-caustic process, 638
  preparation
    acidic, 619
    alkaline, 616
    catalysts, 616
  production, water, 632
  reaction mechanism, 615
    computer simulation, 623
    Flory–Stockmeyer model, 623
    statistics, 623
  reactivity, 615
  solid state properties, 626
  spectroscopy, 638
  synthesis, 611–643
  thermal characterization, 642
  thermal incineration, 637
  toxicology, 636
  water solubility, 635
Phenol–furfuryl alcohol–formaldehyde resins, 656
Phenol–melamine–formaldehyde resins, 656
Phenol–urea–formaldehyde resins, 656
Phenol–urea–melamine–formaldehyde resins, 656
Phenolic foam, phenol–formaldehyde polymers, 634
Phenolic resins
  batch process, 631
  market, 612
  production, 613
  toxicology, 636, 637
  uses, 612
Phenolphthalein
  copolycondensation
    dichloro-3,3'- bisphenol A, terephthaloyl dichloride, 263
    dimethyl-3,3'-bisphenol A, terephthaloyl dichloride, 263
  oligoterephthalate, rate of polymerization, 37
Phenolsulfonic acid, phenol–formaldehyde polymers, 635
Phenosolvane process, phenol–formaldehyde polymers, phenol extraction, 638
Phenoxides
  (halophenylsulfonyl)-
    melt polymerization, 581
    self-polycondensation, 580, 582
  polycondensation, aromatic halides, 574
Phenyl esters, aromatic acids, polyamides, 384
Phenyl isocyanates
  insertion reactions, 391
  reactions

catalysts, 395
2-cyanoaniline, 439
2-ethylhexanol, 389
kinetics, 392
3-Phenylenediamine, epoxy resins, curing, 680
4-Phenylenediamine
  2,6-dichloro-, reaction with terephthaloyl dichloride, 112
  poly(azomethine)s, terephthalaldehyde, 707
  reaction with terephthaloyl dichloride, 109, 378
Phillips process, poly(phenylene sulfide), 547
Phosgene
  polycarbonates, 345
  polyesters, synthesis, 319
  poly(ether ketones), 488
  polyureas, 433
Phosphates
  diphenyl-, polyesters, 286
  diphenylchloro-, polyesters, 286
Phosphazenes, cyclic oligomers, 85
Phosphine
  dichloride, triphenyl-, polyesters, 286
  oxide
    carbodiimides, catalysts, 399
    triphenyl-, curing, epoxy resins, 694
    triphenyl-, curing, epoxy resins, 693
Phosphorus oxychloride, polyesters, 286
Phosphorus pentoxide, poly(ether ketones), 489
Photocuring, polyurethanes, 424
Photodimerization, crystalline, alkenes, 218
Photopolymerization
  diynes, solid state, 242
  four-center type, 217, 218
  solid state, diynes, 240
  step-growth mechanism, 218
  topochemical, cyclobutane polymers, 217–231
Photoresists, epoxy resins, 676
2-Phthalate, glycidyl-, epoxy resins, 670
Phthalic acid,
    hexachloroendomethylenetetrahydro, polyesters, 339
Phthalic anhydride
  hexahydro-, curing, epoxy resins, 691
  polyesters, 339
  tetrabromo-, polyesters, 339
  tetrachloro-, polyesters, 339
  tetrahydro-, polyesters, 339
Piperazine, urethane diphenol, 414
Piperidine
  epoxy resins, curing agent, 680
  polyamides, polycondensation, 150
Plasticizers, polysiloxanes, 603
Plywood, phenol–formaldehyde polymers, 634
PMR11, 523
PMR15, 523
Podands, interfacial polycondensation, 176
Polarography, urea–formaldehyde resins, 659
Poly(acetal ketones), 494
Polyacetylenes, solid state synthesis, 201
Polyacrylamides, solid state synthesis, 201
Polyacrylates, mesogenic units, 725
Poly(acyl semicarbazide)s, 405
Poly(*N*-acylethyleneimine), mesogenic units, 725
Poly(β-alanine), 6
Polyalkenes, ring–chain equilibria, 94
Poly(alkylene glycol)s, bulk polycondensation, 134
Poly(alkylene terephthalate)s, 301

Subject Index

solid state polycondensation, 209
Poly(alkylisonitrile)s, 708
Poly(amic acid)s, synthesis, 505
   chemical cyclization, 508
   cyclization, 507
   synthesis
      aprotic polar solvents, 506
      solvents, 507
   thermal cyclization, 507
Polyamidation, 563
   acceptor-catalytic, 146
   interfacial, 174
   kinetics, 359
Polyamide 4, 361
Polyamide 4,6, 366
Polyamide 6, bulk polycondensation, 136
Polyamide 6,6, 366
   batch process, 367
   continuous process, 368
Polyamide 6,10, 368
Polyamide 6,12, 368
Polyamide 11, 366
Polyamide 12, 366
Poly(amide imide)s, 408
Polyamide membranes, 178
Poly(amide sulfone)s, 563, 586
Polyamide-*block*-polyesters, 265
Polyamide-*block*-polypropylene oxide, 266
Polyamide-*block*-polysiloxane, 266
Poly(amide-*co*-urea), 270
Polyamides, 408
   activated, polycondensation, 152
   aliphatic, 357–371
      solid state polycondensation, 204
   aromatic, 375–385, 703
      catalysts, 384
      phenyl esters, 384
      solid state polycondensation, 204
   block copolymers, 369
   bulk polycondensation, 132, 136
   carboxylated, solid state condensation
      polymerization, non-aqueous dispersions, 197
   constitutional isomerism, 98, 109, 110
   constitutional regularity, 112
   copolycondensation, 259
   cyclic aliphatic, 74
   dispersion polymerization, 199
      stabilizers, 198
   dispersions, 198
   emulsion polycondensation, 187
   epoxy resins, 679
   interfacial polycondensation, 178
      gas–liquid, 182
   matrix polycondensation, 160
   partially aromatic, 369
   phosphorylation reaction, 382
   polyamidation, polymer matrices, 160
   polycondensation, 3, 214
      activated, 150
      direct, 154
   rigid-rod, 707
   salt, 367
   solid state polycondensation, 204
   solid state synthesis, 201
   solution polycondensation, acceptor-catalytic, 145
   step-growth polymerization, 13
   suspension polycondensation, 192
   vapour phase, 383
Poly(amido acid)s, polyamidation, direct, 159
Poly(amido ester)s, interfacial transfer catalysts, 179
Poly(amidohydrazide)s, polyamidation, direct, 159
Poly(amidophenylquinoxaline)s, polyamidation, direct, 159
Polyamidrazones, emulsion polycondensation, 187
Polyamines
   aliphatic, epoxy resins, 679
   polyureas, 436
   step-growth polymerization, 16
Poly(α-amino acid)s, cyclic oligomers, 74
Poly(aminoformal), cyclic oligomers, 76
Polyaramide-*block*-polybutadiene, 266
Poly(aromatic amide)s
   liquid crystalline polymers, lyotropic main-chain, 705
   lyotropic, 705
Poly(aryl ether ketone)s, 483
Poly(aryl ether)s, ether cleavage, 493
Polyarylates, 276, 286, 302, 317–327
   AREL D-100, properties, 325
   blends, 326, 327
   clarity, 324
   color stabilization, 324
   conformation, polyesterification, 146
   crystalline, 320
   heat resistance, 324
   hydrolytic stability, 326
   interfacial polycondensation, 178
   manufacture
      comparison, 322
      economics, 322
      process
         complexity, 323
         flexibility, 323
         product quality, 323
         reliability, 323
   mechanical properties, 324
   optical quality, 324
   photo-Fries rearrangement, 325
   properties, 324
   suspension polycondensation, 191
   synthesis, 318
   thermal properties, 324
   thermal stability, 324
   thermal stabilization, 324
   UV stability, 325
Poly(arylene ester sulfone)s, 572, 576, 587
Poly(arylene ether ketone sulfone)s, 562, 587
Poly(arylene ether sulfide sulfone)s, 586
Poly(arylene ether sulfone)s, 581
   bulk polycondensation, 134
   side reactions, 581
Poly(arylene sulfide)s, 564, 570, 586
Poly(arylene sulfide sulfone)s, 586
Poly(arylene sulfone)s, 562
   step-growth polymerization, 17
Poly(azomethine)s, 707
   aromatic, constitutional isomerism, 98
   cyclic oligomers, 85
   liquid crystalline, constitutional isomerism, 98
Poly(*m*-benzamide), solid state polycondensation, 215
Poly(*p*-benzamide), 377, 378
   solid state polycondensation, 204, 215
Polybenzhydrolimides, synthesis, 512
Polybenzimidazoles
   bulk polycondensation, 132
   synthesis, 500, 501, 503
Poly(benzo[1,2-*d*:5,4-*d'*]bisoxazole-2,6-diyl-1,4-phenylene), synthesis, 503
Poly(benzo[1,2-*d*:4,5-*d'*]bisthiazole-2,6-diyl-1,4-

phenylene), synthesis, 503
Polybenzodiimidazoles, synthesis, 505
Polybenzothiazoles, synthesis, 500, 503
Polybenzoxazinediones, 410
Polybenzoxazinones, synthesis, 500
Polybenzoxazoles, synthesis, 500, 503
Poly(γ-benzyl-L-glutamate), 378, 703, 714
Polybenzyls
  branching, 462
  degree of polymerization, 456
  disubstitution, 461
  molecular weight, 455
  NMR, 462
  step-growth polymerization, 17
  structure, 461
    IR spectra, 462
    NMR, 462
  p-substitution, 462
  synthesis, 455–463
  tetrasubstitution, 461
  trisubstitution, 461
Poly(biphenylene sulfide), synthesis, cyclic sulfides, 84
Poly(2,2-bis-4-hydroxyphenyl)propane adipate, 300
Poly(3-bromo-2,6-dimethyl-1,4-phenylene oxide), 479
Poly(2-bromomethyl-6-methyl-1,4-phenylene oxide), 479
Poly(butylene adipate), cyclic oligomers, 70
Poly(butylene isophthalate)
  cyclic oligomers, 68, 70
  separation, 66
Poly(butylene terephthalate), 275, 301, 305, 307
  bulk polycondensation, chain extension, 133
  cyclic oligomers, 70
  polyarylate blends, 327
  solid state polycondensation, 207
    kinetics, 211
Poly(ε-caprolactone), 275
  cyclic oligomers, 70
Poly(carbamoyl sulfonate)s, 420
Polycarbodiimides, 407
Polycarbonate resins
  purification, 352
  synthesis, 349
Polycarbonates, 345–355
  finishing, 353
  granules, 354
  interfacial polycondensation, 178
    gas–liquid, 182
    solid–liquid, 183
  isolation, 352
  linear oligomers, 85
  melt stream, 354
  pellets, 354
  polyarylate blends, 326
  polyesterification, acceptor-catalytic, 145
  polysiloxane block copolymers, interfacial polycondensation, 179
  powder, 354
  properties, 354
  solution polycondensation, 145
  step-growth polymerization, 13
  structure, 346
Poly(chloroacrylates), mesogenic units, 725
Polycondensation
  A–A plus B–B, 59
  activated, 15, 147, 148
  alkali metal arenedisulfinates, dihalides, 563
  arenedisulfonic acids, 563
  arenesulfinates, 571

arenesulfonic acids, 569
arenesulfonyl halides, 564
aromatic halides, phenoxides, 574
bifunctional monomers, models, 37
bulk, 131–141
  apparatus, 132
  catalysts, 132
  copolymers, 134
  high temperatures, 132
  monomer reaction with polymers, 133
  polymer
    participation, 133
    reactions with polymers, 135
  pressure, 137
  silylation, 140
direct, 153
disulfonyl chloride, aromatic compounds, 566
fillers, 138
Friedel–Crafts, 563
history, 2
hydroxybenzoic acids, 158
oligomer formation, mechanism, 63
ordered medium, 161
polyesters, 290
polysiloxanes, 593
polysulfones, 563
solid state, 201–215
  prepolymers, 207
sulfonyl compounds, 572
Polycrystalline aggregates, solid state polycondensation, 203
Poly(1,3-cyclohexadiene), synthesis, 468
Polycyclohexylene
  dehydrogenation, 468
  molecular weight, 469
Polycyclotrimerization, 138
Poly(decamethylene adipate), cyclic oligomers, 70
Poly(3,5-dibromo-2,6-dimethyl-1,4-phenylene oxide), 479
Poly(2,6-dichloro-p-phenylene-terephthalamide), 113
Poly[(1,5-dihydrobenzo[1,2-d:4,5-d']-diimidazole-2,6-diyl)-1,4-phenylene], synthesis, 505
Poly(N,N-dimethylaminohex-5-en-1-yne), synthesis, 470
Poly(2,6-dimethyl-1,4-phenylene oxide), 473
Polydimethylsiloxanes, 593–606
  cyclic oligomers, 85
Poly(4,4'-diphenyl sebacate), 309
Poly(1,4-diphenylene ether sulfone), 565, 569, 570
Polydispersity index, 52, 57, 58, 59
  oligoesters, 297
Poly(2,5-distyrylpyrazine)
  configuration, 218
  crystals, structure, 226
Poly(1,11-dodecadiyne), crosslinking, 241
Polyester resins, 331, 340
  thickening, 337
Poly(ester thioether)s, 300
Polyesteramide-$block$-polyethers, 265
Polyesteramides, copolycondensation, 260
Polyester-$block$-polycarbonates, 269
Polyesterification, 6, 276
  acceptor-catalytic, 145
  rules, 145
  kinetics, 27, 30, 32, 278, 333
  mechanism, 295
  non-equilibrium, 294
Polyesters, 275–309
  acyl chlorides, 294

Subject Index

alcoholysis, 289
aliphatic
  dicarboxylic acids, aromatic diols, 309
  diols, aromatic dicarboxylic acids, 305
  diols and diacids, 302
  hydroxy acids, 303
anhydrides, 296
aromatic, 708, 709
  cyclic oligoesters, 70
  rigid-rod, 711, 712
biosyntheses, 299
bulk polycondensation, 132, 140
classes, 302
condensation polymerization, non-aqueous dispersions, 197
constitutional isomerism, 98, 109
copolyestirifications, 264
cyclic esters, 297
dispersions, 198
dispersion
  polymerization, stabilizers, 198
emulsion polycondensation, 187
esterification, 279
interfacial polycondensation, 178
linear oligomers, 85
liquid crystalline
  polyesterification, 146
  solution polycondensation, 145
mesogenic units, 725
molecular weight, distribution, 59
polyarylate blends, 327
polycondensation, 3
  activated, 150
  direct, 156
polyesterification, acceptor-catalytic, 146
postcondensation, 209
solid state synthesis, 201
solution polycondensation, 145
step-growth polymerization, 13, 16
unsaturated, 331–342
  chemistry, 332
  double bond saturation, 335
  ester interchange, 336
  manufacture, 334
  side reactions, 335
  structure–properties relationships, 339
  synthesis
    catalysis, 333
    kinetics, 332
Poly(α-esters), 278, 300, 303
Poly(ether amide), block copolymers, 369
Poly(ether ether ketones), 491
Poly(ether imides), synthesis, 510
Poly(ether ketones), 483–496, 513
  characterization, 496
  chemistry, 495
  crosslinking, 496
  crystallinity, 484
  double end capping, 487
  inherent viscosity, 484
  isomorphism, 494
  mechanical properties, 487
  miscibility, 494
  molecular weights, 496
  morphology, 496
  production, 484, 489
    electrophilic, 484
  properties, 484, 494
  stabilization, 496
  step-growth polymerization, 17
  sulfonation, 496
  thermal decomposition, 496
Polyether polyamines, epoxy resins, 679
Poly(ether sulfone)s, 563
  aliphatic–aromatic, 585
  monomers, 584
Polyetherification, 563, 572
  kinetics, 579, 597
  mechanism, 597
  methods, 584
  polysulfones, 573
Polyethers
  aliphatic, mesogenic units, 725
  dispersions, 198
  interfacial catalysis, solid–liquid, 183
  interfacial transfer catalysts, 179
  mesogenic units, 725
  polysulfones, rate of polymerization, 36
Polyethersulfone-*block*-polyketone-*block*-polyether, 269
Polyethylene, polycondensation, 3
Poly(ethylene adipate), cyclic oligomers, 70
Poly(ethylene glycol succinate), cyclic oligomers, 70
Poly(ethylene 2-methylsuccinate), ester exchange, 98
Poly(ethylene oxide), matrix polycondensation, 160
Poly(ethylene succinate), pyrolysis-MS, 70
Poly(ethylene terephthalate), 275, 285, 301, 305
  bulk polycondensation, 134, 135, 136, 137
    chain extension, 133
  cyclic oligomers, 70
  dispersion polymerization, 199
  ester-interchange, 12
  polyarylate blends, 327
  polycondensation, bulk, 132
  solid state polycondensation, 205, 207
    kinetics, 210
  transesterification, 137
Poly(ethylene terephthalate-*co*-2-methylsuccinate), crystallization-induced reactions, 214
Poly(ethyleneurea), 431
Polyfunctional monomers, 120
Poly(glycolic acid), cyclic oligomers, 70
Poly(glycolide)s, 304
Polyheteroarylenes
  bulk polycondensation, 131
    pressure, 137
  polycondensation, direct, 159
  solution polycondensation, 145
  suspension polycondensation, 193
  synthesis, 140
Polyheteromacropolycyclic compounds, siloxanols, polycondensation, 598
Poly(hexamethylene sebacate), bulk polycondensation, copolymers, 134
Poly(hexamethylene terephthalate), 285, 301
  bulk polycondensation, 136
  dispersion polymerization, 199
Poly(hexamethyleneadipamide), bulk polycondensation, 136
Poly(hexamethyleneurea), 436
Poly(1,6-hexanamide)
  biaxially oriented, solid state polycondensation, 206
  solid state polycondensation, 204
Polyhydantoins, 409, 501
Polyhydrazides
  activated, polycondensation, 152
  emulsion polycondensation, 187
  interfacial polycondensation, 178
  polyamidation, direct, 159
Poly(β-hydroxyalkanoate), 275, 299, 303
Poly(*m*-hydroxybenzoic acid), pyrolysis, mass

spectrometry, 69
Poly(β-hydroxybutyrate), 275, 299
Poly(hydroxyether)s, interfacial catalysis, 180
Polyimide 2080, 511
Poly(imide ether ketones), 514
Polyimides, 408
   acetylene-terminated, 523
   bulk polycondensation, 132
      pressure, 137
   layers, Langmuir–Blodgett technique, 517
   linear, 514
   low thermal expansion coefficient, 514
   photosensitive, 516
   polycondensation linear, 505
   synthesis, 499–529
      nucleophilic nitro group displacement, 513
   thermoplastic soluble, 514, 515
   thermosetting, 518
      resins, 501
Poly(iminobenzoxazinedione)s, bulk polycondensation, 140
Poly(iminoethylene), polycondensation, 4
Poly(5-iminoimidazolidone)s, 409
Poly(iminoquinazoline)s, bulk polycondensation, 140
Polyisocyanates, 708
Poly(isocyanurateoxazolidone), curing, epoxy resins, 694
Polyisocyanurates, 401, 406
Polyisoimides, synthesis, 508
Poly(isoindoloquinazolinedione)s, synthesis, 500
Poly(4,4′-isopropylidenediphenol carbonate), cyclic oligomers, 70
Poly(4,4′-isopropylidenediphenylene-oxymethylene), cyclic oligomers, 76
Polyketones
   step-growth polymerization, 17
   synthesis, 489
   thermotropic, 489
Poly(lactic acid), cyclic oligomers, 70
Poly(lactide)s, 304
Polymalonates, 302
Polymerization
   A–A plus B–B, 53
   monomeric reactants, 523
Polymethacrylates, mesogenic units, 725
Poly(methoxy-p-phenylene-terephthalamide), 113
Poly(2-methyl-2-ethyl-1,3-propylene sebacate), bulk polycondensation, 134
Poly(methyltrifluoropropylsiloxane), 92
Poly(nitro-p-phenyleneterephthalamide), 113
Poly(1,8-nonadiyne), crosslinking, 241
Poly(nonamethyleneurea), 427, 433
Polynuclear phenols, epoxy resins, 668
Polyols, polyurethanes, curing, 416
Poly(organophosphazine)s, 708
Poly(oxadiazole)s, 385
   bulk polycondensation, pressure, 137
Polyoxalates, 302
Polyoxazolidinediones, 410
Polyoxazolidones, curing, epoxy resins, 693
Poly(2-oxazolidone)s, 405
Poly(oxymethylene), microcrystals, 234
Poly(oxy-1,4-phenylenecarbonyl), 285, 293
Poly(oxy-1,4-phenyleneoxy-1,4-phenylenecarbonyl-1,4-phenylene) — see Poly(oxyalkylene) glycols, 340
Poly(oxypropylenediamine), 451
Poly(parabanic acid)s, 409
Polypeptides
   liquid crystalline polymers, 714
      lyotropic main-chain, 707

polycondensation
   direct, 155
   ordered medium, 161
   rigid-rod polymers, 702
   solid state polycondensation, 204, 205, 206
Polyphenylene
   crosslinked, 471
   phenylated, synthesis, 469
Poly(o-phenylene), synthesis, 466
Poly(m-phenylene), synthesis, 466, 467
Poly(p-phenylene)
   bridging diazo groups, 470
   synthesis, 465, 469
Poly(m-phenylene disulfide), synthesis, 540
Poly(phenylene ether)s, step-growth polymerization, 18
Poly(p-phenylene ether sulfone)s, toughness, 584
Poly(m-phenylene isophthalamide), 377
Poly(phenylene oxide), 473–479
   structure, 479
Poly(phenylene sulfide), 513, 543–559
   bond dissociation energy, 552
   chemical resistance, 554, 555
   chemistry, 545
   copolymers, 558
   cyclic oligomers, 84
   derivatives, 558
   electrical properties, 555, 556
   flame resistance, 554
   flammability, 555
   history, 543
   inherent viscosities, 557
   linear, solid state polycondensation, 205
   mechanical properties, 557
   melt flow values, 551
   melt properties, 556
   processes, 547
   properties, 551, 557
   rate of polymerization, 36
   solid state polycondensation, 202
   solution properties, 556
   step-growth polymerization, 17
   thermal properties, 551
Poly(p-phenylene-2,6-benzobisisothiazole), 708
Poly(p-phenylene-2,6-benzobisisoxazole), 708
Poly(p-phenyleneethylidene), step-growth polymerization, 17
Poly(m-phenyleneisophthalamide), 375–377
Polyphenylenes, synthesis, 465–471
Poly(p-phenyleneterephthalamide), 376, 377, 378
Poly(phenylmethylsiloxanes), cyclic oligomers, 85
Poly(phenylquinoxaline)s, 501, 525
   acetylene-terminated, 529
   bulk polycondensation, pressure, 137
   crosslinked, 527
   glass transition temperatures, 526
   linear, 525
   nadimide-terminated, 528
Poly(phenyltriazine)s, 139
Polyphosphazanes, 6
   cyclic oligomers, 85
Polyphosphonates
   interfacial catalysis, 179
   polycondensation, acceptor-catalytic, 146
Polyphosphoric acid, poly(ether ketone) production, 485
Polyphosphoxanes, 6
Poly(pivalolactone), pyrolysis-MS, 70
Poly(propylene terephthalate)s,

suspensions, 198
Polyquinazolinediones, 409
  synthesis, 500
Polyquinolines, synthesis, 500, 502
Polyquinoxalines, 525
  linear, 525
Polysilathianes, 6
Polysilazanes, 6
Polysiloxanes, 6
  cyclic oligomers, separation, 65
  interfacial polycondensation, gas–liquid, 183
  mesogenic units, 725
  polycarbonate block copolymers, interfacial polycondensation, 179
  step-growth polymerization, 16, 17
Polystilbene, structure, 461
Polystyrene, mesogenic units, 725
Poly(sulfide sulfone)s, 563
Polysulfides
  aliphatic, cyclic oligomers, 83
  aromatic–aliphatic, cyclic oligomers, 67
  bis(2-hydroxyethyl), synthesis, 539
  crosslinking, 537
  curing, 533
  cyclic oligomers, separation, 65
  gas phase chromatography, 583
  inorganic, 534
    synthesis, 534
  interchange reaction, 540
  liquid, preparation, 537
  molecular weight, 537
  oxidation, polysulfones, 562
  rank, 533
  resin modification, 541
  rubbers, 534
  structure, 533
  synthesis, 533–541
  thermal decomposition, 538
  thiol terminated, 537
  toughening, 536
  UV spectra, 535
  utility, 337
  vinyl terminated, 540
Polysulfonamides, emulsion polycondensation, 187
Polysulfonates, interfacial catalysis, 179
Polysulfones, 513, 561–588
  aromatic, 561
  mesogenic units, 725
  monomers, 584
  polyethers, rate of polymerization, 36
  step-growth polymerization, 16, 585
  synthesis, 562
  thermostable, 585
Polysulfonylation, 563, 564
  bulk, 569
  Friedel–Crafts, 564
  high catalyst concentration, 565
  low catalyst concentration, 568
Polysulfoxides, oxidation, 562
Polysulfur nitride, solid state polymerization, 235
Poly[4,4″-(p-terphenylene)amide], solid state polycondensation, 204
Poly(tetramethylene adipate), 300
Poly(1,4-tetramethyleneterephthalamide), solid state polycondensation, 215
Poly(thiocarbonate)s
  interfacial polycondensation, 178
    gas–liquid, 182
Polythiodiglycols, synthesis, 539
Poly(thioester)s
  interfacial polycondensation, 173, 178
    gas–liquid, 182

interfacial transfer catalysts, 179
Polythioetherification, 563
Poly(thioether)s
  interfacial transfer catalysts, 179
  step-growth polymerization, 16
Poly(thioglycolic acid), cyclic dimer, 70
Poly(trimethylene sulfide), cyclic oligomers, 79
Poly(α,ω-undecanoicamide)
  biaxially oriented, 206
  solid state polycondensation, 204
Polyurea, polycondensation, direct, 155
Polyurea-9,1, 427
Polyureas, 387, 404
  aliphatic, 433
  aromatic, 430, 443
  carbonates, 431
  colloids, 447
  constitutional isomerism, 109, 110
  constitutional regularity, 110
  crystallography, 451
  cyclic oligomers, 75
  emulsion polycondensation, 187
  hydrogen-bonding, 451
  interfacial polycondensation, 178
    gas–liquid, 182
  optically–active, 446
  polycondensation, activated, 151
  reaction injection moulding, 447
  sugar residues, 444
  sulfur-containing, 443, 444
  synthesis, 413, 427–452
Poly(urethane amide)s, synthesis, 414
Poly(urethane carbonates)
  cyclic oligomers, 75
  synthesis, 414
Poly(urethane urea), 271
  flexible foams, 419
  reaction injection moulding, 447
  synthesis, 414
Polyurethanes, 387, 403
  anionic dispersions, 424
  bulk polycondensation, 131, 132
  copolymers, diynes, 241
  curing, 413, 415
  cyclic oligomers, 70, 75
  emulsion polycondensation, 187
  interfacial polycondensation, 178
  ionic, 422
  polycondensation, 3
    activated, 151
  reaction injection moulding, 447
  step-growth polymerization, 12, 13
    catalysts, 14, 15
  synthesis, 413
  water-borne, 424
Polyureylenes, 405
Poly(δ-valerolactone), cyclic oligomers, 70
Poly(vinyl ethers), mesogenic units, 725
Poly(vinylpyridine), polyesters, 289
Poly(2-vinylpyridine), matrix polycondensation, 160
Poly(4-vinylpyridine), matrix polycondensation, 160
Polyvinylpyrrolidone, matrix polycondensation, 160
Poly(p-xylylene sulfone), 572
Postcondensation, solid state polycondensation, temperature, 208
Potting, epoxy resins, 668
PPDT — see Poly(p-phenyleneterephthalamide), 376
Precipitation polycondensation, 162
Prepregs

curing, 686
epoxy resins, 671
Pressure, bulk polycondensation, 137
Printed circuit boards, epoxy resins, 668
Printing inks, phenol–formaldehyde polymers, 636
Propane
  1,3-bis(1-glycidyl-5,5-dimethyl-3-hydantoinyl)-2-glycidyloxy-, epoxy resins, 670
  2,2-bis(4-hydroxyphenyl)-, condensation polymerization, terephthalic acid chlorides, 198
1,2-Propanediol, polyesters, 340
1,3-Propanediol, 2,2-dibromomethyl-, polyesters, 340
Propanoic acid
  2-chloro-2-methyl-, polyesters, 304
  3-hydroxy-2,2-dimethyl-, polyesters, 303
Propylene glycol, polyesters, constitutional isomerism, 98
1,2-Propylenediamine, polyamides, constitutional isomerism, 98
Purification, polycarbonates, 352
Pyrazine
  distyryl-, solid state polymerization, 235
  2,5-distyryl-
    photopolymerization, quantum yields, 224
    preparation, 219
    topochemical [2+2] photopolymerization, 218
Pyridine
  2,2'-(1,4-divinylbenzene-β,β'-ylene)di-, photopolymerization, quantum yields, 224
  polyesters, 288
Pyrimidine
  1,3-di(hydroxymethyl)-2-oxohexahydro-, 652
  phenyl-, mesogenic groups, 726
Pyrocatechol, calixarenes, 626
Pyromellitic acid dianhydride
  bulk polycondensation, pressure, 137
  cyclopolycondensation, 12
2-Pyrone, 5,5'-p-phenylenebis-, cycloaddition reactions, bisacetylenes, 469
Pyrrolidone, N-methyl-, polyamides, 376

Quantum field theory, 123
p-Quaterphenyl, 2,2'''-triphenyl-, synthesis, 467
Quinone methides, phenol–formaldehyde polymers, curing, 621
p-Quinquephenyl, 2,2''''-diphenyl-, synthesis, 467

Radel, 561
Radiation curing, polyurethanes, 424
Radius of gyration
  clusters, 123
  excluded volume forces, multifunctional step polymerization, 127
Random fractals, 118
Random polymerization, rate theory, 125
Random walk propagator, 123
Rate of polymerization, functional groups, reactivity, 35–45
Rate theory, random polymerization, 125
Reaction injection moulding
  polyureas, 428
    bulk polymerization, 447
  polyurethanes, 416
Reactivity
  functional groups
    multifunctional step polymerization, 120
    rate of polymerization, 35–45

rate of polymerization, substitution effect, 37
Resols
  ammonia-catalyzed, 620, 635
  curing, acids, 622
  synthesis, 632
    ammonia, 616, 632
    batch process, 632
Resorcinol
  calixarenes, 626
  electrochemical polymerization, 479
  resin production, 614
Reversible topochemical process, 230
Rigid foams, polyurethanes, 419
Rigid-rod polymers, thermotropic main-chain, 710
Rigid-rod type polymers, lattice theory, 702
Ring addition, aliphatic polyamides, 361
Ring–chain equilibria, 91–95
Ring formation, kinetics, 125
Ring forming parameter, cyclization, graph theory, 122
Ring opening, polyamides, aliphatic, 359
Ring opening polymerization
  cyclic oligomers, 63
  polysiloxanes, 593
Ruthenium salts, ion exchanged montmorillonite clays, polyphenylene synthesis, 466

Salicylic acid, phenol–formaldehyde polymers, 635
Scaling hypothesis, 128
Scaling laws, percolation, 126
Scaling theory, percolation, 120
Scavengers, elastomers, 600
Schmidt's rule, 218
Sealants, polyurethanes, 416
Sebacic acid, polyesters, 339
Selenonium salts, trialkyl-, photoinitiators, epoxy resins, 676
Self-condensation, (halophenylsulfonyl)phenoxides, 582
Self-polyetherification, 580
Self-polymerization, molecular weight, distribution, 47
Self-polysulfonylation, 568
Semiflexible polymers, thermotropic main-chain, 710
Sequence length, copolycondensation, 254
Sequence statistics, 107
p-Sexiphenyl, 2,2'''''-diphenyl-, synthesis, 467
Shear modulus, multifunctional step polymerization, 117
Shultz–Flory distribution, molecular weight, 48
Silanediols, cationic polycondensation, 595
Silanes
  methyltriacetoxy-, crosslinking, polysiloxanes, 600
  organo-, condensation, silanols, 598
  polyureas, 435
Silanols
  condensation
    base-catalyzed, 596
    cationic catalysts, 594
    reactions, 594
  homocondensation, thermal, 594
  polycondensation, 594
    catalysts, 598
  reactivity, 594
Silazanes, scavengers, polysiloxanes, 602
Silicon tetrachloride, polyamides, 383

Subject Index

Siloxanes
  bonds, 594
  cyclic oligomers, 84
Siloxanols
  intermolecular reactions, 598
  polycondensation, 594
α,ω-Siloxanediol, oligomers,
    polycondensation, 594
Silver oxide, catalysts, poly(phenylene
    oxide), 475
Silyl ethers, cleavage, 585
Silylation
  gel permeation chromatography, urea–
      formaldehyde resins, 657
  polycondensation, activated, 152
Single crystals, 234
  monomer, 234
  polymerization, 235
  solid state polycondensation, 203
Single electron transfer mechanism, displacement
    polymerization, phenols, 478
Size exclusion chromatography, cyclic
    oligomers, 65
Smoluchowski equation, 128
Sodium bismuthate, catalysts, poly(phenylene
    oxide), 475
Sodium disulfide, 534
Sodium pentasulfide, 534
Sodium polysulfide
  reaction
    dihalides, 535
    ethylene dichloride, 533
  solution, equilibria, 534
Sodium sulfide, polymers, p-
    dichlorobenzene, 545
Sodium tetrasulfide, 534
  reaction with ethylene dichloride, 536
Sol phase, gelation, 125
Solid state polycondensation, 201–215
  by-products, 208
  crystalline order, 206
  kinetics, 206
  low molecular weight by-products, 203
  prepolymers, 207
Solid state polymerization
  diynes
    crystals, 233–246
    kinetics, 241
    substituents, 237
  heterogeneous, diynes, 237
  homogeneous, diynes, 238
Solubility, constitutional regularity, 113
Solution anisotropy, 113
  polyamides, 377
Solution polycarbonate resin
  chemistry, 348
  synthesis, 351
Solution polycondensation, 143–162
  acceptor-catalytic, 144
  characteristics, 144
  high temperature, polyarylates, 319
  low-temperature, polyarylates, 319
  polyarylates, 319
Solution polymerization
  polyamides, 376
  polycarbonates, isolation, 352
  polyureas, 441
Solution process, polycarbonates,
    purification, 352
Solvents
  interfacial polycondensation, 171, 172
  polysulfones, 584

ring–chain equilibria, 93
Spinning
  polyamides, 375
  PPDT, 378
Spray drying, polycarbonates, 353
Stabar, 561
Star-shaped polymers, 122
Steam distillation, polycarbonates, 353
Step copolymerization, 251–272
Step-growth addition polymerization, 7
Step-growth condensation polymerization, 7
Step-growth polymerization, 5, 11–33
  carbonyl reactions, mechanism, 13
  catalysts, 14
  epoxy resins, 678
  constitutional isomerism, 98
  nucleophilic-substitution reactions, 16
  reaction mechanism, 13
Step polyaddition, 6
Step polycondensation, 6
Step polymerization, 92, 118, 121
  bifunctional reactants, 18
  homopolymers, 121
  kinetics, 26
  model, 124
  molecular weight, distribution, 57
  multifunctional, 6
  number-average degree of, 19
  polysulfones, 563, 585
  rate, 27
  reversible, kinetics, 32
Step reactions, vulcanization, 122
Stiffening, polysulfones, 562
Stilan, 483
Stilbazole, photodimerization, 219
Stilbene, mesogenic groups, 726
Stirring, interfacial polycondensation, 172
Stoichiometry, poly(ether ketones), 490
Strain, cyclic oligomers, 65
Styrene
  α-methyl-, polyesters, 340
  polyesters, 340
Substitution effect
  rate of polymerization, 36
  reactivities, 37
Succinic acid, polyesters, 339
Succinic anhydrides, polyesters, 302
Sulfides
  aliphatic, cyclic oligomers, 79
  cyclic oligomers, 78, 81
Sulfites, cyclic, polyurethanes, 414
Sulfonation
  agents, 570
  aromatic compounds, 569
Sulfones
  bis(4-chloro-3-nitrophenyl),
      polycondensation, 571
  bis(4-chlorophenyl), bisphenol A poly(ether
      sulfone), 576
  bis(4-halophenyl), polycondensation, 597
  bis(4-hydroxyphenyl), etherification, 597
  4,4'-diaminodiphenylene, epoxy resins,
      curing, 680
  dibenzyl, 571
  N,N-dimethyl-4,4'-diaminodiphenylene,
      epoxy resins, 683
  diphenyl, poly(ether ketones), 489
  divinyl
    polymerization, 563
    polymers, 588
  unsaturated cyclic, homopolymerization, 562
Sulfonium salts

dialkyl-4-hydroxyphenyl-, photoinitiators, epoxy resins, 676
dialkylphenacyl-, photoinitiators, epoxy resins, 676
triaryl-, photoinitiators, epoxy resins, 676
Sulfonyl chlorides
  aromatic, self-polycondensation, 565
  polycondensation, aromatic compounds, 566
  self-polycondensation, 568
Sulfonyl compounds
  dichloro-, polycondensation, 568
  polycondensation, 572
Sulfonylation, Friedel–Crafts, 564
Sulfoxide, dimethyl, poly(ether ketones), 491
Sulfur
  polymers, benzene, 544
  polysulfide synthesis, 539
Sulfur chloride, polysulfide synthesis, 540
Sulfur dioxide, free-radical copolymerization, unsaturated compounds, 562
Sulfur trioxide, liquid, polyamides, 383
Sumilite FS-1300, 561
Surfaces, solid state polycondensation, 204
Suspension polycondensation, 168, 187, 191
Suspension polyheterocyclization, 193
Swelling, suspension polycondensation, 189
Symmetry, solid state polymerization, diynes, 236

Talpa 1000, 561
Terephthalaldehyde, poly(azomethine), p-phenylenediamine, 707
Terephthalate
  2-hydroxyethyl-, copolycondensation, 2-hydroxy-1-propyl- terephthalate, 259
  2-hydroxy-1-propyl-, copolycondensation, 2-hydroxyethyl- terephthalate, 259
Terephthalic acid
  bis(2-hydroxyethyl) ester, solid state polycondensation, 205
  dimethyl ester
    bulk polycondensation, 134
    polyesterification with ethylene glycol, 20
  diphenyl ester, polyarylates, 322
  polyamides, 4-aminobenzhydrazide, 707
  polyarylates, 320
    crystallization, 323
  polyesters, 317
  polymerization, 299
Terephthalic acid chloride
  condensation polymerization, 198
  copolycondensation
    bisphenol A, neopentyl glycol, 263
    phenolphthaleine, dichloro-3,3′-bisphenol A, 263
    phenolphthaleine, dimethyl-3,3′-bisphenol A, 263
  interfacial polycondensation, 176
  polyamides, constitutional isomerism, 98
  polyarylates, 319
  poly(ether ketones), 487
  reaction
    2,6-dichloro-p-phenylene diamine, 112
    p-phenylenediamines, 109, 378
p-Terphenyl, 2,2″-diphenyl-, synthesis, 467
N,N,N',N'-Tetraglycidyl-4,4′-diamine diphenylenemethane, epoxy resins, 669
Tetramines, bulk polycondensation, 131
Tetramethylenediamine, polyureas, 428
Tetroxane, ring-opening polymerization, 235
TGDDM, 670
  epoxy resins, 680

homopolymerization, 672
  boron trifluoride catalysis, 672
Thermal decomposition, cyclic oligomers, 65
Thermal expansion coefficient, polyimides, 514
Thermal phase transitions, percolation, 126
Thermoanalysis, urea–formaldehyde resins, 660
Thermoelastoplastics, bulk polycondensation, 134
Thermogravimetry, urea–formaldehyde resins, 660
Thermoplastics
  glass-reinforced engineering, high load heat deflection, 553 temperatures
  UL temperature indices, temperatures, 553
Thermotropism, 702
Theta conditions, ring–chain equilibria, 93
Thianthrene, 544
Thin-layer chromatography, urea–formaldehyde resins, 658
Thiokol A, 533
Thiols
  epoxy resins, 695
  reactions with epoxides, polysulfide synthesis, 541
Thionyl chloride, polyamides, 383
Thiophenol, p-halo-, solid state polycondensation, 202
Tin compounds, isocyanate reactions, catalysts, 396
Tin salts, polysiloxanes, polycondensation, 600
Titanates, polyurethanes, curing, 420
Titanium, polysiloxanes, polycondensation, 600, 602
Titanium alkoxides, catalysts, polyesters, 15, 284
Toluene
  diisocyanate
    curing, epoxy resins, 694
    polyureas, 436
    rate of polymerization, 35
  vinyl-, polyesters, 340
p-Toluenesulfonic acid, phenol–formaldehyde polymers, 635
Topochemical photopolymerization
  linear polymers, 219
  products, configuration, 219
  zigzag polymers, 220
Topochemical [2+2] photopolymerization, 218
Topochemical polymerization, 218, 234
  diynes, 235
Topochemical reactions, 234
Topotactic reactions, 234
Toughness, poly(p-diphenylene ether sulfone)s, 584
Toxicity, elastomers, 600
Transesterification
  polycarbonates
    isolation, 353
    resin synthesis, 351
  polyesters, 290
  resin
    chemistry, polycarbonates, 347
    polycarbonates, purification, 352
Transition metals, carbonyl complexes, benzyl chloride polymerization, 456
Triallyl cyanurate, polyesters, 340
Triethylamine, polyesters, 288
Trioxane
  crystals, ring-opening polymerization, 217
  ring-opening polymerization, 235
Tung oil, phenol–formaldehyde polymers, 635
Turbidimetry, urea–formaldehyde resins, 660
Twaron, 376
α-Type crystals

Subject Index

photopolymerization, crystallographic
  interpretation, 226
photodimerization, 217
β-Type crystals, photodimerization, 217
γ-Type crystals, photodimerization, 217

Udel, 561
Ullman ether synthesis, poly(phenylene
  oxide), 478
Ullmann condensation, polysulfones, 573
Ullmann etherification, polysulfones, 573, 576
Ullmann reaction, polyphenylene synthesis, 467
Ultem TR, 501
Ultrason-E, 561
Ultraviolet/visible spectroscopy, urea–
  formaldehyde resins, 659
Undecanoic acid
  11-amino-
    dispersion polymerization,199
    solid state polycondensation, 204
Unsaturated compounds, free-radical
  copolymerization, sulfur dioxide, 562
Upper critical dimension, hyperscaling, 126
Urea
  3-(p-chlorophenyl)-1,1-dimethyl-, curing,
    epoxy resins, 687
  cyclic
    oligomers, 75, 78
    polycondensation, 4
  derivatives, polyureas, 435
  formaldehyde system, 651
  hydroxymethyl-, reactions, 653
  methylene-, cyclic, polycondensation, 4
  phenol–formaldehyde polymers, 634
  polyureas, 435
  production, 650
  rate of polymerization, 36
  reactions
    formaldehyde, 651
    isocyanates, 396
Urea–formaldehyde resins
  analysis, 657
  copolycondensates, 656
  history, 649
  manufacture, 661
  synthesis, 649–663
Urea–furfuryl alcohol–formaldehyde resins, 656
Urea–melamine–formaldehyde resins, 656
Urethane
  alkyd coatings, one-package, 420
  coatings
    aqueous, 422
    one-package moisture-cure, 420
    powder, 424
    radiation-cured, 424
    single-package blocked adduct, 420
    two-package, 421
      amino–formaldehyde cure, 422
      catalyst, 421
      polyol, 421
  cyclic
    oligomers, 75, 78
    polycondensation, 4
  diphenol, piperazine, 414
  gelation, 122
  lacquers, 421
  reactions
    isocyanates, 396
    kinetics, 389
  thermostability, 394
Uretidione, 399
Uretidinedione, 399

Uron, di(hydroxymethyl)-, 653

δ-Valerolactone
  6,6-dimethyl-, polymerization, 298
  polymerization, 298
  3-n-propyl-, polymerization, 298
Victrex, 561
Victrex PEEK, 483
Vinyl compounds, sulfones,
  homopolymerization, 562
Vinyl monomers, polyesters, 340
Viscosity
  multifunctional step polymerization, 117
  polyester resins, 337
Vulcanization
  classical theory, 118
  Flory–Stockmayer theory, 127
  multifunctional step polymerization, 117, 118
  preformed polymers, 121
  step reaction, 122

Wafer boards, phenol–formaldehyde
  polymers, 634
Water, polyamidation, acceptor-catalytic, 147
Weissenberg photographs, photopolymerization,
  crystals, 226
Wirtz–Fittig reaction, polyphenylene
  synthesis, 467

Xanthydrol, 9-phenylene-, poly(ether
  ketones), 487
XD-7342.00L, 669
XD-9053.00L, 669
XYDAR SRT-300
  heat resistance, 324
  mechanical properties, 325
p-Xylene
  copolymerization, anthracenes, 458
  dibromo-, polycondensation, 4,4'-
    oxydibenzene- sulfinates, 571
  α,α'-dichloro-
    copolymerization
      anthracene, 458
      arenes, 457
  2,5-dimethyl-α,α'-dichloro-, copolymerization,
    1,4-dimethylbenzene, 457
  2,3,4,5-tetramethyl-α,α'-dichloro-,
    copolymerization, arenes, 457
Xylenols, resin production, 614

Zinc naphthenate, polyurethanes, curing, 420
Zirconium alkoxides, catalysts, polyesters, 284